What do you need to learn NOW?

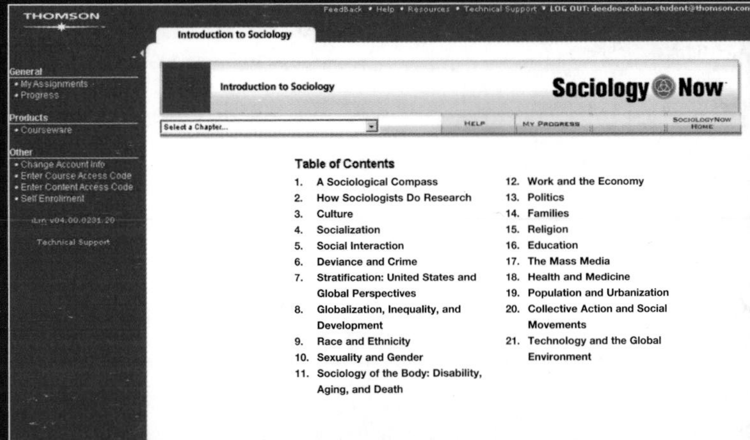

Take charge of your learning with *SociologyNow,* the first assessment-centered student learning tool for sociology!

This new online diagnostic tool could be your key to success in introductory sociology!

SociologyNow will help you:

- create a personalized study plan for each chapter of your introductory text
- understand key concepts in the course
- better prepare for exams—and increase your chances of success

Lift the page for more information.

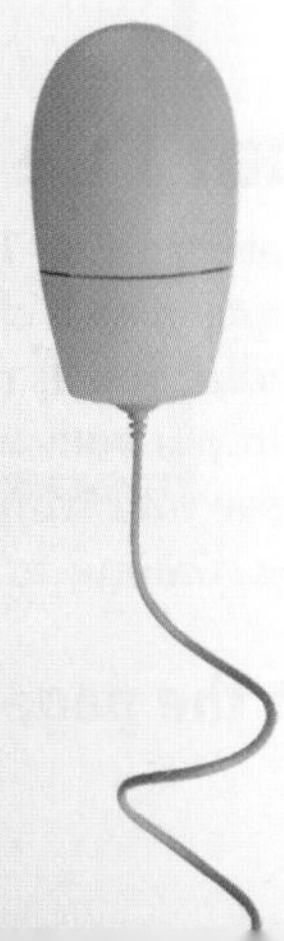

How can you access SociologyNow?

- **Your instructor may have chosen to package the access code card with your new text.** In this case, you'll find the access card within this text, which contains your free four-month pass code, allowing you anytime access to *SociologyNow*.

- **If your instructor did not order the free access code card to be packaged with your text—or if you have a used copy of the text—you can still obtain an access code for a nominal fee.** Just visit the Thomson Wadsworth E-Commerce site at http://sociology.wadsworth.com/brym_lie3e/sociologynow, where easy-to-follow instructions help you purchase your access code.

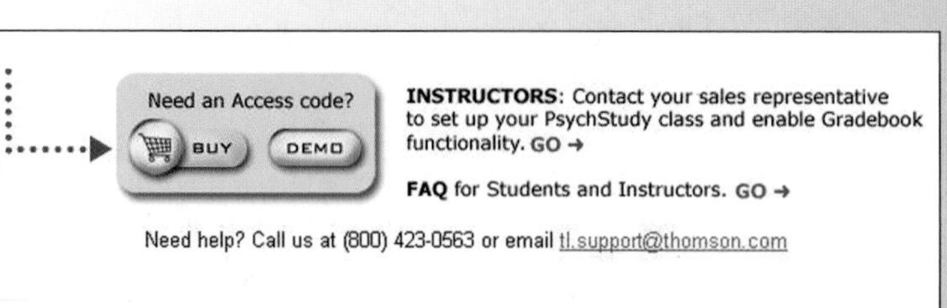

Are you ready to get started? Help is just a click away with SociologyNow™

The exciting program that is firmly grounded in sociology and lets you take diagnostic quizzes, review chapter content, conduct online research, think critically about sociological statistics, watch videos of well-known sociologists as they discuss important concepts, and ultimately improve your success in the course, all on one, easy to use web-based program.

http://sociology.wadsworth.com/brym_lie3e/sociologynow

Log on today!

Thomson Wadsworth ◆ P.O. Box 6904 ◆ Florence, KY 41022
800-423-0563 ◆ Fax 859-647-5020 ◆ Email: review@wadsworth.com

Source code 7TPSOM01

POLITICAL MAP OF THE WORLD

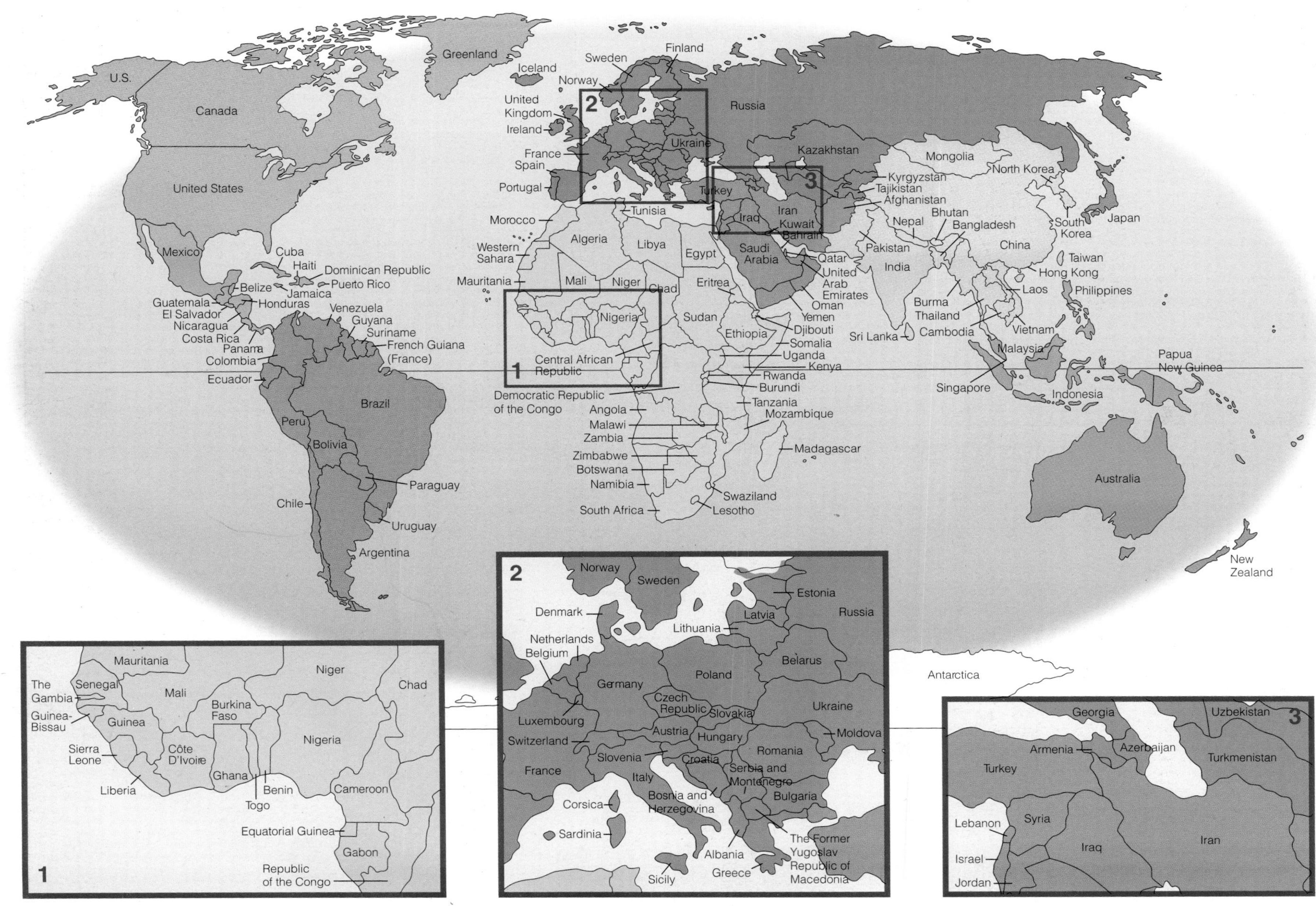

N
NORTH
WEST
EAST
SOUTH
W
E
S

Sociology

Your Compass for a New World

THIRD EDITION

Robert J. Brym
University of Toronto

John Lie
University of California at Berkeley

Australia • Brazil • Canada • Mexico • Singapore
Spain • United Kingdom • United States

Sociology

***Your Compass for a New World,* Third Edition**

Robert J. Brym and John Lie

Senior Sociology Editor: Robert Jucha
Development Editor: Shelley Murphy
Assistant Editor: Elise Smith
Editorial Assistant: Christina Cha
Technology Project Manager: Dee Dee Zobian
Marketing Manager: Wendy Gordon
Marketing Assistant: Gregory Hughes
Marketing Communications Manager: Linda Yip
Project Manager, Editorial Production: Cheri Palmer
Creative Director: Rob Hugel
Print Buyer: Doreen Suruki
Permissions Editor: Kiely Sisk
Production Service: Dan Fitzgerald, Graphic World Publishing Services
Text Designer: Norman Baugher
Photo Researcher: Kathleen Olson
Illustrator: Graphic World Illustration Studio
Cover Designer: Yvo Riezebos
Cover Images: Lawrence Lawry/Getty PictureQuest
Compositor: Graphic World Inc.
Printer: Quebecor World/Dubuque

Printed in the United States of America
1 2 3 4 5 6 7 10 09 08 07 06

Library of Congress Control Number: 2005932101

Student Edition: ISBN 0-495-00684-X
Paper Edition: ISBN 0-495-00848-6

Thomson Higher Education
10 Davis Drive
Belmont, CA 94002-3098
USA

For more information about our products, contact us at:
Thomson Learning Academic Resource Center
1-800-423-0563

For permission to use material from this text or product, submit a request online at **http://www.thomsonrights.com.**
Any additional questions about permissions can be submitted by email to **thomsonrights@thomson.com.**

Dedication

Many authors seem to be afflicted with stoic family members who gladly allow them to spend endless hours buried in their work. I suffer no such misfortune. The members of my family have demanded that I focus on what really matters in life. I think that focus has made this a better book. I am deeply grateful to Rhonda Lenton, Shira Brym, Talia-Lenton-Brym, and Ariella Lenton-Brym. I dedicate this book to them with thanks and love.

ROBERT J. BRYM

For Charis Thompson, Thomas Cussins, Jessica Cussins, and Charlotte Lie, with thanks and love.

JOHN LIE

Robert J. Brym

(pronounced "brim") is an internationally known scholar. He studied in Canada and Israel and received his Ph.D. from the University of Toronto, where he is now on faculty and especially enjoys teaching introductory sociology to more than 1000 students every year. He has won numerous awards for his teaching and scholarly work, which has been translated into half a dozen languages. His main areas of research are in political sociology, race and ethnic relations, and sociology of culture. His major books include *Intellectuals and Politics* (London and Boston: Allen & Unwin, 1980); *From Culture to Power* (Toronto: Oxford University Press, 1989); *The Jews of Moscow, Kiev, and Minsk* (New York: New York University Press, 1994); and *New Society* (Toronto: Nelson, 2004), one of Canada's best-selling introductory sociology textbooks, now in its fourth edition. In 2001, he was co-investigator for the world's first large-scale survey of online dating, sponsored by MSN. From 1992 to 1997, Robert served as editor of *Current Sociology*, the journal of the International Sociological Association, and from 2001 to 2005 he was editor of *East European Jewish Affairs*, published in London. He recently completed a study of the Russian civil service with sociologists at the Institute of Sociology, Russian Academy of Science, and is now conducting a study of suicide bombers.

John Lie

(pronounced "lee") was born in South Korea, grew up in Japan and Hawaii, and attended Harvard University. Currently he is professor of sociology at the University of California at Berkeley, where he also holds the C. K. Cho Professorship. Previously he was professor of sociology at the University of Illinois at Urbana-Champaign and at the University of Michigan. He has also taught at the University of Hawaii at Manoa, the University of Oregon, and Harvard University in the United States, as well as universities in Japan, South Korea, Taiwan, and New Zealand. His primary research interests are comparative macro-sociology and social theory. His major publications include *Blue Dreams: Korean Americans and the Los Angeles Riots* (Cambridge, MA: Harvard University Press, 1995); *Han Unbound: The Political Economy of South Korea* (Stanford, CA: Stanford University Press, 1998); *Multiethnic Japan* (Cambridge, MA: Harvard University Press, 2001); and *Modern Peoplehood* (Cambridge, MA: Harvard University Press, 2004). He has taught introductory sociology classes ranging in size from 3 to more than 700 students in several countries and hopes that this book will stimulate your sociological imagination.

Brief Contents

PART V
Social Change

Contents

PART I
Foundations

PART II
Basic Social Processes

Chapter 3
Culture 62

PART III Inequality

Chapter 7 Deviance and Crime 182

Chapter 8 Stratification: United States and Global Perspectives 212

PART IV Institutions

Chapter 13 Work and the Economy 370

Chapter 14 Politics 402

Chapter 16
Religion 468

Chapter 17
Education 498

PART V
Social Change

Boxes

SOCIOLOGY AT THE MOVIES

SOCIAL POLICY: WHAT DO YOU THINK?

YOU AND THE SOCIAL WORLD

Maps

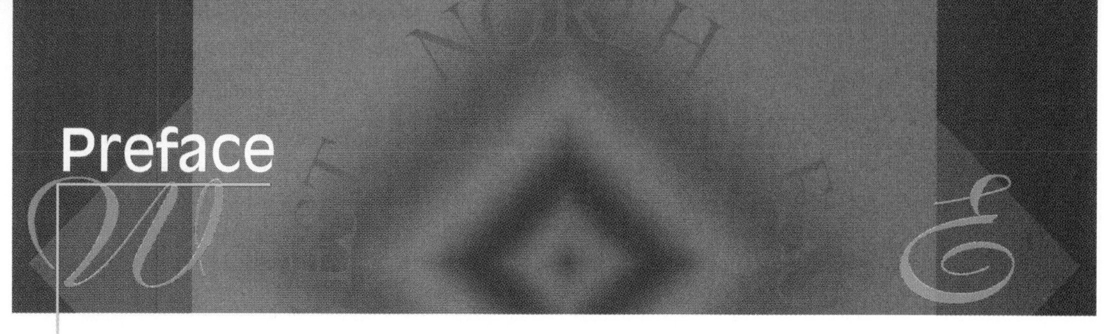

Preface

Why a Compass for a New World?

Soon after European explorers arrived in North and South America, they started calling the twin continents the "New World." Everything was different here. A native population perhaps a hundredth as large as Europe's occupied a territory more than four times larger. The New World was unimaginably rich in resources. European rulers saw that by controlling it they could increase their power and importance. Christians recognized new possibilities for spreading their religion. Explorers discerned fresh opportunities for rewarding adventures. A wave of excitement swelled as word spread of the New World's vast potential and challenges.

Today, it is easy for us to appreciate that wave of excitement—for we, too, have reached the frontiers of a New World. And we are also full of anticipation. Our New World is one of virtually instant long-distance communication, global economies and cultures, weakening nation-states, and technological advances that often make the daily news seem like reports from a distant planet. In a fundamental way, the world is not the same place it was just 50 years ago. Orbiting telescopes that peer to the fringes of the universe, human genetic code laid bare like a road map, fiber-optic cable that carries a trillion bits of information per second, and spacecraft that transport robots to Mars help to make this a New World.

Five hundred years ago, the early European explorers of North and South America set themselves the preliminary task of mapping the contours of the New World. We set ourselves a similar task here. Their frontiers were physical. Ours are social. Their maps were geographical. Ours are sociological. But in terms of functionality, our maps are much like theirs. All maps allow us to find our place in the world and see ourselves in the context of larger forces. Sociological maps, as C. Wright Mills wrote, allow us to "grasp the interplay of [people] and society, of biography and history" (Mills, 1959: 4). This book, then, shows you how to draw sociological maps so you can see your place in the world, figure out how to navigate through it, and perhaps discover how to improve it. It is your sociological compass.

We are not as naive as the early European explorers. Where they saw only hope and bright horizons, minimizing the significance of the violence required to conquer the people of the New World, our anticipation is mixed with dread. Scientific breakthroughs are announced almost daily, but the global environment has never been in worse shape, and AIDS is now the leading cause of death in Africa. Marriages and nations unexpectedly break up and then reconstitute themselves in new and unanticipated forms. We celebrate the advances made by women and racial minorities only to find that some people oppose their progress, sometimes violently. Waves of people suddenly migrate between continents, establishing cooperation but also conflict between previously separated groups. New technologies make work more interesting and creative for some, offering unprecedented opportunities to get rich and become famous; they also make jobs more onerous and routine for others. The standard of living goes up for many people but stagnates for many more.

Is it any wonder that amid all this contradictory news, good and bad, uncertainty about the future prevails? We wrote this book to show undergraduate college students that sociology can help them make sense of their lives, however uncertain they may appear to be. Moreover, we show that sociology can be a liberating practical activity, not just an abstract intellectual exercise. By revealing the opportunities and constraints you face, sociology can help to teach you who you are and what you can become in this particular social and historical context. We cannot know what the future will bring, but we can at least know the choices we confront and the likely consequences of our actions. From this point of view, sociology can help us create the best possible future. That has always been sociology's principal justification, and so it should be today.

Distinctive Features

We have tried to keep sociology's main purpose and relevance front and center in this book. As a result, *Sociology: Your Compass for a New World,* Third Edition, differs from other major introductory sociology textbooks in five ways:

1. **Drawing connections between one's self and the social world.** To varying degrees, all introductory sociology textbooks try to show students how their personal experiences are connected to the larger social world. However, we employ two devices to make these connections clearer than in other textbooks. First, we illustrate key sociological ideas by using examples from popular culture that resonate deeply with student interests and experiences. For example, we conclude our discussion of culture in Chapter 3 by showing how radical subcultures often become commercialized, focusing on the development of rap and heavy metal music. In Chapter 11 we examine the causes and consequences of glamorizing thin bodies in advertising. We analyze Super Bowl XXXVII to highlight key features of Durkheim's theory of religion in Chapter 16. We think these and many other examples speak directly to today's students about important sociological ideas in terms they understand, thus making the connection between self and society clear.

   Second, we developed several unique pedagogical features to draw the connection between students' experiences and the larger social world. **You and the Social World** is a feature that repeatedly challenges students to consider how and why their own lives conform to, or deviate from, various patterns of social relations and actions by collecting and analyzing sociological data. We also enter into a social policy debate in each chapter with a feature entitled **Social Policy: What Do You Think?** Here we set out public policy alternatives on a range of pressing social issues and teach students that sociology can be a matter of the most urgent practical importance. Students also learn they can have a say in the development of public policy. **Sociology at the Movies** takes a universal and popular element of contemporary culture and renders it sociologically relevant. We provide brief reviews of movies, most of them recent releases, and highlight the sociological issues they raise and the sociological insights they contain.
2. **What to think versus how to think.** All textbooks teach students both *what* to think about a subject and *how* to think about it from a particular disciplinary perspective. In our judgment, however, introductory sociology textbooks usually place too much stress on the "what" and not enough on the "how." The result: They sometimes read more like encyclopedias than enticements to look at the world in a new way. We have tipped the balance in the other direction. To be sure, *Sociology: Your Compass for a New World,* Third Edition, contains definitions and literature reviews. It features standard pedagogical aids such as a list of **Chapter Objectives** at the beginning of each chapter, a **Summary,** a list of **Recommended Websites,** a set of **Questions to Consider** at the end of each chapter, and definitions of key terms both in the margins of the text and in a cumulative **Glossary** at the end of the book. However, we devote more space than other authors to showing how sociologists think. We often relate an anecdote to highlight an issue's importance, present contending interpretations of the issue, and then adduce data to judge the merits of the various interpretations. We do not just refer to tables and graphs, we analyze them. When evidence warrants, we reject theories and endorse others. Thus, many sections of the book read more like a simplified journal article than an encyclopedia. If all this sounds just like what sociol-

ogists do professionally, then we have achieved our aim: to present a less antiseptic, more realistic, and therefore intrinsically exciting account of how sociologists practice their craft. Said differently, one of the strengths of this book is that it does not present sociology as a set of immutable truths carved in stone tablets. Instead, it shows how sociologists actually go about the business of solving sociological puzzles.

3. **Objectivity versus subjectivity.** Sociologists since Max Weber have understood that sociologists—indeed, all scientists—are members of society whose thinking and research are influenced by the social and historical context in which they work. Yet most introductory sociology textbooks present a stylized and not very sociological view of the research process. Textbooks tend to emphasize sociology's objectivity and the hypothetico-deductive method of reasoning, for the most part ignoring the more subjective factors that go into the research mix (Lynch and Bogen, 1997). We think this emphasis is a pedagogical error. In our own teaching, we have found that drawing the connection between objectivity and subjectivity in sociological research makes the discipline more appealing to students. It shows how research issues are connected to the lives of real flesh-and-blood women and men and how sociology is related to students' existential concerns. Therefore, in most chapters of *Sociology: Your Compass for a New World,* Third Edition, we feature a **Personal Anecdote** that explains how certain sociological issues first arose in our own minds. We often adopt a narrative style because stories let students understand ideas on an emotional as well as an intellectual level; and when we form an emotional attachment to ideas, they stay with us more effectively than if our attachment is solely intellectual. We place the ideas of important sociological figures in social and historical context. We show how sociological methodologies serve as a reality check, but we also make it clear that socially grounded personal concerns often lead sociologists to decide which aspects of reality are worth checking on in the first place. We believe *Sociology: Your Compass for a New World,* Third Edition, is unique in presenting a realistic and balanced account of the role of objectivity and subjectivity in the research process.
4. **Diversity and a global perspective.** It is gratifying to see how much less parochial American introductory sociology textbooks are today than they were just 20 years ago. Contemporary textbooks highlight gender and race issues. They broaden the student's understanding of the world by comparing the United States with other societies. They show how global processes affect local issues and how local issues affect global processes. *Sociology: Your Compass for a New World,* Third Edition, is no different in this regard. We have made diversity and globalization prominent themes of this book. We make frequent and effective use of crossnational comparisons between the United States and countries as diverse as India and Sweden. We incorporate dozens of original maps that illustrate the distribution of sociological variables globally and regionally, and the relationship among variables across time and space. We remain sensitive to gender and race issues throughout. This has been easy for us because we are members of racial and ethnic minority groups. We are multilingual. We have lived in other countries for extended periods. And we have published widely on countries other than the United States. Robert Brym specializes in the study of Russia, Canada, and, increasingly, Israel, while John Lie's research focuses on South Korea and Japan. As you will see in the following pages, our backgrounds have enabled us to bring greater depth to issues of diversity and globalization than other textbooks.
5. **Currency.** Every book bears the imprint of its time. It is significant, therefore, that the first editions of the leading American introductory sociology textbooks were published in the late 1980s. At that time just over 10 percent of Americans owned personal computers. The World Wide Web did not exist. Genetic engineering was in its infancy. The USSR was a major world power. Nobody could imagine teenage boys committing mass murder at school with semiautomatic weapons. *Sociology: Your Compass for a New World,* Third Edition, is one of the first American introductory sociology textbooks of the 21st century, and it is the most up-to-date. This is reflected

in the currency of our illustrations and references. For instance, we do not just recommend a few websites at the end of each chapter, as is usual in other introductory sociology textbooks. Instead, Web resources form an integral part of this book; fully one-sixth of our citations are of materials on the Web. Throughout the text in the margins you will find small icons indicating a link to either a Web interactive exercise or a Web research project located on the book's companion website, much of which was written by Robert Brym. The icons may also indicate exercises in one of the online resources Wadsworth makes available with its textbooks, such as InfoTrac® College Edition or MicroCase® Online.

The currency of this book is also reflected in the book's theoretical structure. It made sense in the 1980s to simplify the sociological universe for introductory students by claiming that three main theoretical perspectives—functionalism, symbolic interactionism, and conflict theory—pervade all areas of the discipline. However, that approach is no longer adequate. Functionalism is less influential than it once was. Feminism is an important theoretical perspective in its own right. Conflict theory and symbolic interactionism have become internally differentiated. For example, there is no longer a single conflict theory of politics but several important variants. Highly influential new theoretical perspectives, such as postmodernism and social constructionism, have emerged, and not all of them fit neatly into the old categories. *Sociology: Your Compass for a New World,* Third Edition, highlights the contributions of traditional theoretical approaches, but it also notes recent theoretical innovations that are given insufficient attention in other major textbooks.

New in the Third Edition

We have been gratified and moved by the overwhelmingly positive response to previous editions of this book. At the same time, we benefited from the constructive criticisms generously offered by dozens of readers and reviewers. *Sociology: Your Compass for a New World,* Third Edition, is a response to many of their suggestions. Specifically, we have

- expanded the section on careers in sociology in Chapter 1
- developed our discussion of qualitative methods in Chapter 2
- expanded coverage of agents of socialization in Chapter 5
- added a substantial new section on types of societies in Chapter 6
- incorporated a discussion of the 2004 presidential election (including the roles of so-called "527 groups" and felony disenfranchisement in shaping the election outcome) in Chapter 14
- added new theoretical material on spousal abuse in Chapter 15
- added a discussion of the No Child Left Behind Act and a new section on community colleges in Chapter 17
- written eight new reviews of popular movies from a sociological perspective
- created a new data collection and analysis feature (**You and the Social World**)
- highlighted the contribution of the major theoretical perspectives to sociological research even more sharply than in earlier editions
- added subheadings to help guide students through the book
- thoroughly updated the entire book to incorporate the latest research findings

We are delighted with the final product and very much hope our readers will be too.

Supplements

Sociology: Your Compass for a New World, Third Edition, is accompanied by a wide array of supplements prepared to create the best learning environment inside as well as outside the classroom for both the instructor and the student. All the continuing supplements for *Sociology: Your Compass for a New World,* Third Edition, have been thoroughly revised and updated, and several are new to this edition. We invite you to take full advantage of the teaching and learning tools available to you.

For the Instructor

Instructor's Resource Manual. This supplement offers the instructor brief chapter outlines, chapter summaries, chapter-specific summaries, key terms, student learning objectives, extensively detailed chapter lecture outlines, essay/discussion questions, lecture suggestions, student activities, chapter review questions, InfoTrac College Edition discussion exercises, Internet exercises, video suggestions, suggested resources for instructors, and creative lecture and teaching suggestions. Also included is a Resource Integration Guide (RIG), a list of additional print, video, and online resources, and concise user guides for SociologyNow™, InfoTrac College Edition, Turnitin™ and WebTutor™.

Test Bank. This test bank consists of 75–100 multiple-choice questions and 15–20 true-false questions for each chapter of the text, all with answer explanations and page references to the text. Each multiple-choice item indicates the question type (factual, applied, or conceptual). Also included are 10–20 short-answer and 5–10 essay questions for each chapter. All questions are labeled as new, modified, or pickup, so instructors know whether the question is new to this edition of the test bank, modified but picked up from the previous edition of the test bank, or picked up straight from the previous edition of the test bank.

ExamView® Computerized Testing for Macintosh and Windows. Create, deliver, and customize printed and online tests and study guides in minutes with this easy-to-use assessment and tutorial system. ExamView includes a Quick Test Wizard and an Online Test Wizard to guide instructors step by step through the process of creating tests. The test appears on screen exactly as it will print or display online. Using ExamView's complete word processing capabilities, instructors can enter an unlimited number of new questions or edit questions included with ExamView.

Extension: Wadsworth's Sociology Reader Database. Create your own customized reader for your sociology class, drawing from dozens of classic and contemporary articles found on the exclusive Thomson Wadsworth TextChoice database. Using the TextChoice website (http://www.TextChoice.com), you can preview articles, select your content, and add your own original material. TextChoice will then produce your materials as a printed supplementary reader for your class.

Classroom Presentation Tools for the Instructor

JoinIn™ on TurningPoint®. Transform your lecture into an interactive student experience with JoinIn. Combined with your choice of keypad systems, JoinIn turns your Microsoft® PowerPoint® application into audience response software. With a click on a handheld device, students can respond to multiple-choice questions, short polls, interactive exercises, and peer-review questions. You can also take attendance, check student comprehension of concepts, collect student demographics to better assess student needs, and even administer quizzes. In addition, there are interactive text-specific slide sets that you can modify and merge with any your own PowerPoint lecture slides. This tool is available to qualified adopters at http://turningpoint.thomsonlearningconnections.com.

Multimedia Manager Instructor Resource CD: A 2006 Microsoft® PowerPoint® Link Tool. With this one-stop digital library and presentation tool, instructors can assemble, edit, and present custom lectures with ease. The Multimedia Manager contains figures, tables, graphs, and maps from this text, preassembled Microsoft PowerPoint lecture slides, video clips from DALLAS TeleLearning, ShowCase presentational software, tips for teaching, the instructor's manual, and more.

Introduction to Sociology 2006 Transparency Masters. A set of black-and-white transparency masters consisting of tables and figures from Wadsworth's introductory sociology texts is available to help prepare lecture presentations. Free to qualified adopters.

Video. Adopters of *Sociology: Your Compass for a New World,* Third Edition, have several different video options available with the text. Please consult with your Thomson Learning sales representative to determine whether you are a qualified adopter for a particular video.

Wadsworth's Lecture Launchers for Introductory Sociology. An exclusive offering jointly created by Thomson Wadsworth and DALLAS TeleLearning, this video contains a collection of video highlights taken from the *Exploring Society: An Introduction to Sociology Telecourse* (formerly *The Sociological Imagination*). Each 3- to 6-minute video segment has been specially chosen to enhance and enliven class lectures and discussions of 20 key topics covered in the introduction to sociology course. Accompanying the video is a brief written description of each clip, along with suggested discussion questions to help effectively incorporate the material into the classroom. Available on VHS or DVD.

Sociology: Core Concepts Video. Another exclusive offering jointly created by Thomson Wadsworth and DALLAS TeleLearning, this video contains a collection of video highlights taken from *Exploring Society: An Introduction to Sociology Telecourse* (formerly *The Sociological Imagination*). Each 15- to 20-minute video segment will enhance student learning of the essential concepts in the introductory course and can be used to initiate class lectures, discussion, and review. The video covers topics such as the sociological imagination, stratification, race and ethnic relations, social change, and more. Available on VHS or DVD.

CNN® Today Sociology Video Series, Volumes V–VII. Illustrate the relevance of sociology to everyday life with this exclusive series of videos for the introduction to sociology course. Jointly created by Wadsworth and the Cable News Network (CNN), each video consists of approximately 45 minutes of footage originally broadcast on CNN and specifically selected to illustrate important sociological concepts.

Wadsworth Sociology Video Library. Bring sociological concepts to life with videos from Wadsworth's Sociology Video Library, which includes thought-provoking offerings from Films for Humanities, as well as other excellent educational video sources. This extensive collection illustrates important sociological concepts covered in many sociology courses.

Supplements for the Student

SociologyNow™. This online tool provides students with a customized study plan based on a diagnostic "pretest" that they take after reading each chapter. The study plan provides interactive exercises, videos, and other resources to help students master the material. After the study plan has been reviewed, students can then take a "posttest" to monitor their progress in mastering the chapter concepts. Instructors may bundle this product for their students with each new copy of the text for free! If your instructor did not order the free access code card to be packaged with your text—or if you have a used copy of the text—you can still obtain an access code for a nominal fee. Just visit the Thomson Wadsworth E-Commerce site at http://sociology.wadsworth.com/brym_lie3e, where easy-to-follow instructions help you purchase your access code.

Study Guide with Practice Tests. This student study tool contains learning objectives, a list of key terms with page references to the text, detailed chapter outlines, study activities, learning objectives, InfoTrac College Edition discussion exercises, Internet exercises, and practice tests consisting of 25–30 multiple-choice questions, 10–15 true-false questions,

5–10 short-answer questions, and 5 essay questions. All multiple-choice, true-false, short-answer, and essay questions include answer explanations and page references to the text.

Internet-Based Supplements

InfoTrac College Edition with InfoMarks™. Available as a free option with newly purchased texts, InfoTrac College Edition gives instructors and students 4 months of free access to an extensive online database of reliable, full-length articles (not just abstracts) from thousands of scholarly and popular publications going back as much as 22 years. Among the journals available "24/7" are *American Journal of Sociology, Social Forces, Social Research,* and *Sociology.* InfoTrac College Edition now also comes with InfoMarks, a tool that allows you to save your search parameters and your links to specific articles. (Available to North American college and university students only; journals are subject to change.)

WebTutor™ Advantage on WebCT and Blackboard. This Web-based software for students and instructors takes a course beyond the classroom to an anywhere/anytime environment. Students gain access to a full array of study tools, including chapter outlines, chapter-specific quizzing material, interactive games and maps, and videos. With WebTutor Advantage, instructors can provide virtual office hours, post syllabi, track student progress with the quizzing material, and even customize the content to suit their needs.

Wadsworth's Sociology Home Page at http://sociology.wadsworth.com. Combine this text with the exciting range of Web resources on Wadsworth's Sociology Home Page, and you will have truly integrated technology into your learning system. Wadsworth's Sociology Home Page provides instructors and students with a wealth of FREE information and resources, such as Sociology in Action; Census 2000: A Student Guide for Sociology; Research Online; a Sociology Timeline; a Spanish glossary of key sociological terms and concepts; and more.

Turnitin™ Online Originality Checker. This online "originality checker" is a simple solution for professors who want to put a strong deterrent against plagiarism into place and make sure their students are employing proper research techniques. Students upload their papers to their professor's personalized website, and within seconds the paper is checked against three databases—a constantly updated archive of over 4.5 billion webpages; a collection of millions of published works, including a number of Thomson Higher Education texts; and the millions of student papers already submitted to Turnitin.

For each paper submitted, the professor receives a customized report that documents any text matches found in Turnitin's databases. At a glance, the professor can see whether the student has used proper research and citation skills or has simply copied the material from a source and pasted it into the paper without giving credit where credit is due. Our exclusive deal with iParadigms, the producers of Turnitin, gives instructors the ability to package Turnitin with the *Sociology: Your Compass for a New World* Thomson textbook. Please consult with your Thomson Learning sales representative to find out more!

Companion Website for *Sociology: Your Compass for a New World,* Third Edition, at http://sociology.wadsworth.com/brym_lie3e. The book's companion site includes chapter-specific resources for instructors and students. For instructors, the site offers a password-protected instructor's manual, Microsoft PowerPoint presentation slides, and more. For students, there is a multitude of text-specific study aids, including the following:

- Tutorial practice quizzes that can be scored and e-mailed to the instructor
- Web links
- InfoTrac College Edition exercises
- Flash cards
- MicroCase Online data exercises
- Crossword puzzles
- Virtual Explorations

And much more!

Acknowledgments

Anyone who has gone sailing knows that when you embark on a long voyage you need more than a compass. Among other things, you need a helm operator blessed with a strong sense of direction and intimate knowledge of likely dangers. You need crew members who know all the ropes and can use them to keep things intact and in their proper place. And you need sturdy hands to raise and lower the sails. On the voyage to complete this book, our crew demonstrated all these skills. Our acquisitions editor, Bob Jucha, saw this book's promise from the outset, understood clearly the direction we had to take to develop its potential, and on several occasions steered us clear of threatening shoals. We still marvel at how Cheri Palmer, our production project manager; Dan Fitzgerald, our production editor; Dee Dee Zobian, technology product manager; and Elise Smith, assistant editor responsible for the print supplements, were able to keep the many parts of this project in their proper order and prevent the whole thing from flying apart at the seams even in stormy weather. Finally, Shelley Murphy, our developmental editor, and Wendy Gordon, our marketing manager, made this book sail. They knew just when to trim the jib and when to hoist the mainsail. We are deeply grateful to them and to all the members of our crew for a successful voyage.

In preparing this edition we benefited from stimulating discussions with Cynthia Hamlin (Universidade Federal de Pernambuco, Brazil), who heads the team that is preparing the first Brazilian edition of the book, and Steven Rytina (McGill University, Montreal, Canada), who is helping to prepare the second Canadian edition. We thank Murray Straus (University of New Hampshire) for allowing us to reproduce some of his unpublished data from the International Dating Violence Study. And we are grateful to the following colleagues who reviewed the manuscript and revisions for this edition, providing valuable assistance in the development of the Third Edition of *Sociology: Your Compass for a New World:*

Denise Coughlin, *Clinton Community College*
Stan Weeber, *McNeese State University*
Jan Fiola, *Minnesota State University–Moorhead*
Chris Baker, *Walters State Community College*
Becky Ehlts, *Belmont University*
Ande Kidanemariam, *Northeastern State University*
Rex Hargrove, *University of Tennessee*
Thomas Burns, *University of Oklahoma*
Dan Fisher, *University of North Carolina–Greensboro*
Gerardo Marti, *Davidson College*
Donna Abrams, *Kennesaw State University*

We are also grateful to the following colleagues who reviewed the manuscript and provided a wealth of helpful suggestions:

Deborah Abowitz, Bucknell University; Peter Adler, University of Denver; Sarah F. Anderson, Northern Virginia Community College; Carol Bailey, University Center, Rochester, MN; Anne Baird, Morehouse University; Chris Baker, Walters State; Tim Britton, Lenoir Community College; Sherri Ann Butterfield, Rutgers University; William Canack, Middle Tennessee State University; Gregg Carter, Bryant College; Karen Connor, Drake University; Douglas Constance, Sam Houston State University; Stephen Couch, Pennsylvania State University, Schuylkill; Ione DeOllos, Ball State University; Katheryn Dietrich, Texas A&M University; Jan Fiola, Moorhead State University; Juanita Firestone, University of Texas–San Antonio; John Fox, University of Northern Colorado; Ellie Franey, Middle Tennessee State; Phyllis Gorman, Ohio State University; Michael Goslin, Tallahassee Community College; Robert Graham, Lee University; Joanna Grey, Pikes Peak Community College; Ron Hammond, Utah Valley State College; Gary Hampe, University of Wyoming; Eric Hanley, University of Kansas; Emily Ignacio, Loyola University of Chicago; Arthur Jipson, University of Dayton; Robert Kettlitz, Hastings College; Hadley

Klug, University of Wisconsin–Whitewater; Steve Kroll-Smith, University of New Orleans; Jenifer Kunz, West Texas A&M University; Alan Lamb, North Idaho College; Ian Lapp, Monmouth University; Hugh Lena, Providence College; Michael Lovaglia, University of Iowa; Dale Lund, University of Utah Gerontology Center; Steven Lybrand, University of St. Thomas; Duane Matcha, Siena College; Ron Matson, Wichita State University; Christopher Mele, State University of New York–Buffalo; Harry Mersmann, San Joaquin Delta College; Elizabeth Meyer, Pennsylvania College of Technology; Beth Mintz, University of Vermont; Dan Muhwezi, Butler Community College; Virginia Mulle, University of Alaska; Meryl Nason, University of Texas–Dallas; Billye Nipper, Redlands Community College; Nelda Nix-McCray, Community College of Baltimore County; Martin Orr, Boise State University; Hence Parson, Hutchinson Community College; Michael Perez, California State University–Fullerton; Lisa Slattery Rashotte, University of North Carolina–Charlotte; Terry Reuther, Anoka-Ramsey Community College; Luis Salinas, University of Houston; Kent Sandstrom, University of Northern Iowa; Anna Wall Scott, Parkland Community College; Gershon Shafir, University of California–La Jolla; William Smith, Georgia Southern University; Matthew Smith-Lahrman, Dixie College; Joel Snell, Kirkwood Community College; George Stine, Millersville University; Steve Vassar, Minnesota State University, Mankato; Pelgy Vaz, Ft. Hays State University; Peter Venturelli, Valparaiso University; J. Russell Willis, Grambling State University; Ron Wohlstein, Eastern Illinois University.

Robert J. Brym
John Lie

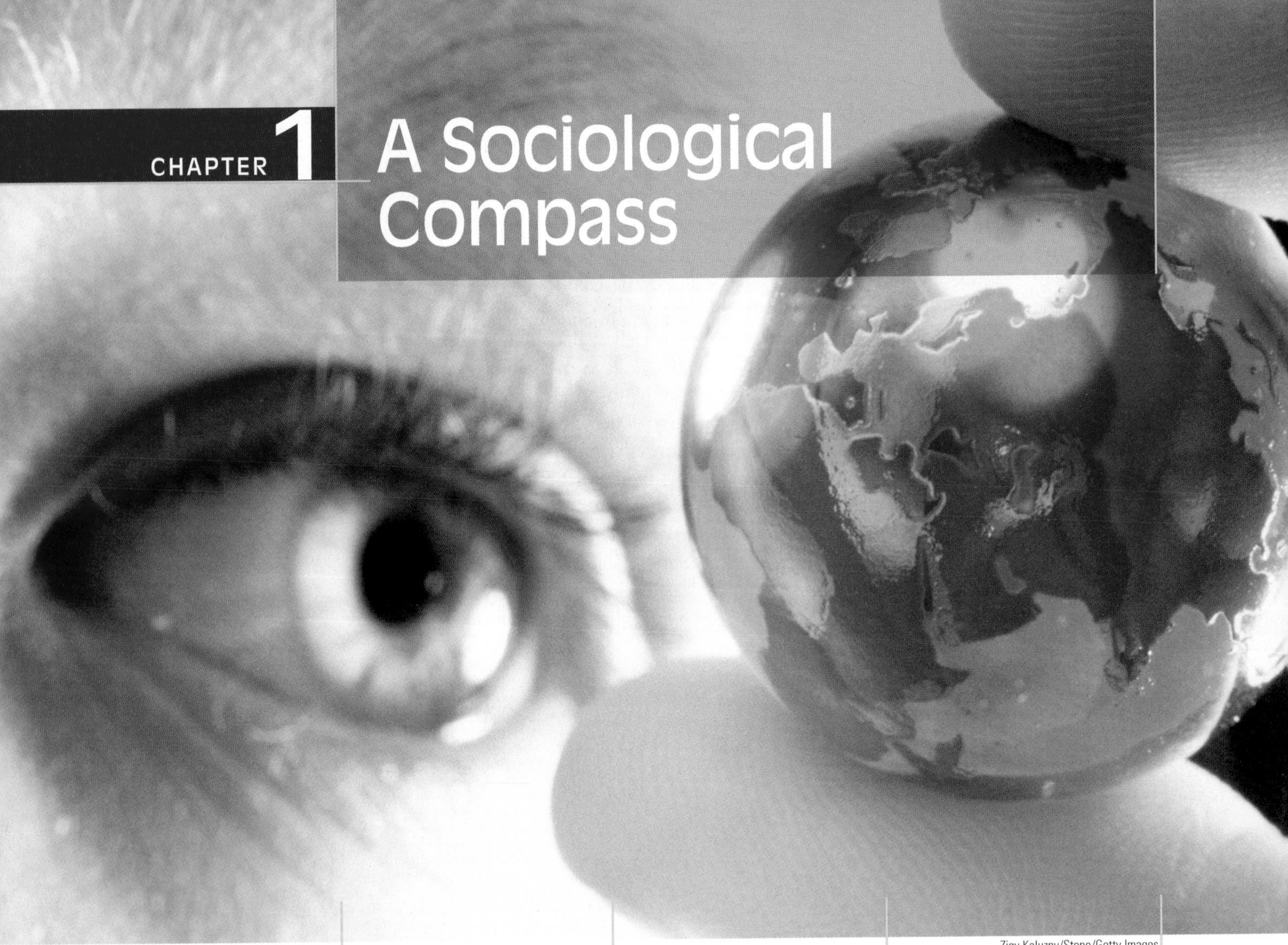

CHAPTER 1

A Sociological Compass

Zigy Kaluzny/Stone/Getty Images

In this chapter, you will learn that:

- The causes of human behavior lie mostly in the patterns of social relations that surround and permeate us.
- Sociology is the systematic study of human behavior in social context.
- Sociologists examine the connection between social relations and personal troubles.
- Sociologists are often motivated to do research by the desire to improve people's lives. At the same time, sociologists adopt scientific methods to test their ideas.
- Sociology originated at the time of the Industrial Revolution. The founders of sociology diagnosed the massive social transformations of their day. They also suggested ways of overcoming social problems created by the Industrial Revolution.
- Today's Postindustrial Revolution similarly challenges us. Sociology clarifies the scope, direction, and significance of social change. It also suggests ways of dealing with the social problems created by the Postindustrial Revolution.
- At the personal level, sociology can help clarify the opportunities and constraints you face. It suggests what you can become in today's social and historical context.

Introduction

Why Robert Brym Decided *Not* to Study Sociology

Personal Anecdote

"When I started college at the age of 18," says Robert Brym, "I was bewildered by the wide variety of courses I could choose from. Having now taught sociology for more than 25 years and met thousands of undergraduates, I am quite sure most students today feel as I did then.

"One source of confusion for me was uncertainty about why I was in college in the first place. Like you, I knew higher education could improve one's chance of finding good work. But, like most students, I also had a sense that higher education is supposed to provide something more than just the training necessary to embark on a career that is interesting and pays well. Several high school teachers and guidance counselors had told me that college was also supposed to 'broaden my horizons' and teach me to 'think critically.' I wasn't sure what they meant, but they made it sound interesting enough to make me want to know more. Thus, I decided in my first year to take mainly 'practical' courses that might prepare me for a law degree (economics, political science, and psychology). I also enrolled in a couple of other courses to indulge my 'intellectual' side (philosophy, drama). One thing I knew for sure. I didn't want to study sociology.

"Sociology, I came to believe, was thin soup with uncertain ingredients. When I asked a few sophomores and juniors in my dorm what sociology is, I received dif-

Sociology Now™

Reviewing is as easy as ❶, ❷, ❸.

Use SociologyNow to help you make the grade on your next exam. When you are finished reading this chapter, go to the Chapter Summary for instructions on how to make SociologyNow work for you.

ferent answers. They variously defined sociology as the science of social inequality, the study of how to create the ideal society, the analysis of how and why people assume different roles in their lives, and the method of figuring out why people don't always do what they are supposed to do. I found all this confusing and decided to forgo sociology for what seemed to be tastier courses.

A Change of Mind

"Despite the opinion I'd formed, I found myself taking no fewer than four sociology courses a year after starting college. That revolution in my life was due in part to the influence of an extraordinary professor I happened to meet just before I began my sophomore year. He set me thinking in an altogether new way about what I could and should do with my life. He exploded some of my deepest beliefs. He started me thinking sociologically.

"Specifically, he first encouraged me to think about the dilemma of all thinking people. Life is finite. If we want to make the most of it we must figure out how best to live. That is no easy task. It requires study, reflection, and the selection of values and goals. Ideally, he said, higher education is supposed to supply students with just that opportunity. Finally, I was beginning to understand what I could expect from college apart from job training.

"The professor also convinced me that sociology in particular could open up a new and superior way of comprehending my world. Specifically, he said, it could clarify my place in society, how I might best maneuver through it, and perhaps even how I might contribute to improving it, however modestly. Before beginning my study of sociology, I had always taken for granted that things happen in the world—and to me—because physical and emotional forces cause them. Famine, I thought, is caused by drought; war, by territorial greed; economic success, by hard work; marriage, by love; suicide, by bottomless depression; rape, by depraved lust. But now this professor repeatedly threw evidence in my face that contradicted my easy formulas. If drought causes famine, why have so many famines occurred in perfectly normal weather conditions or involved some groups hoarding or destroying food so others would starve? If hard work causes prosperity, why are so many hard workers poor? If love causes marriage, why does violence against women and children occur in so many families? And so the questions multiplied.

"As if it were not enough that the professor's sociological evidence upset many of my assumptions about the way the world worked, he also challenged me to understand sociology's unique way of explaining social life. He defined **sociology** as the systematic study of human behavior in social context. He explained that *social* causes are distinct from physical and emotional causes. Understanding social causes can help clarify otherwise inexplicable features of famine, marriage, and so forth. In public school, my teachers taught me that people are free to do what they want with their lives. However, my new professor taught me that the organization of the social world opens some opportunities and closes others, thus constraining our freedom and helping to make us what we are. By examining the operation of these powerful social forces, he said, sociology can help us to know ourselves, our capabilities and limitations. I was hooked. And so, of course, I hope you will be too."

The Power of Sociology

In this chapter we aim to achieve three goals:

Sociology is the systematic study of human behavior in social context.

1. We first illustrate the power of sociology to dispel foggy assumptions and help us see the operation of the social world more clearly. To that end, we examine a phenomenon that at first glance appears to be solely the outcome of breakdowns in individual functioning: suicide. We show that, in fact, social relations powerfully influence

suicide rates. This exercise introduces you to what is unique about the sociological perspective.

2. We show that from its origins, sociological research has been motivated by a desire to improve the social world. Thus, sociology is not just a dry, academic exercise but a means of charting a better course for society. At the same time, however, sociologists adopt scientific methods to test their ideas, thus increasing their validity. We illustrate these points by briefly analyzing the work of the founders of the discipline.
3. We suggest that sociology can help you come to grips with your century, just as it helped the founders of sociology deal with theirs. Today we are witnessing massive and disorienting social changes. Entire countries are becoming unglued. Women are demanding equality with men in all spheres of life. People's wants are increasingly governed by the mass media. Computers are radically altering the way people work and entertain themselves. There are proportionately fewer good jobs to go around. Violence surrounds us. Environmental ruin threatens us. As was the case a century ago, sociologists today try to understand social phenomena and suggest credible ways of improving their societies. By promising to make sociology relevant to you, this chapter should be viewed as an open invitation to participate in sociology's challenge.

"Pacific." Alex Colville. 1967.

But first things first. Before showing how sociology can help you understand and improve your world, we briefly examine the problem of suicide. That will help illustrate how the sociological perspective can clarify and sometimes overturn commonsense beliefs.

The Sociological Perspective

By analyzing suicide sociologically, you can put to a tough test our claim that sociology takes a unique, surprising, and enlightening perspective on social events. After all, suicide appears to be the supremely antisocial and nonsocial act. It is condemned by nearly everyone in society. It is typically committed in private, far from the public's intrusive glare. It is rare. In 2002, there were 11 suicides for every 100,000 Americans (Centers for Disease Control and Prevention, 2005b). When you think about why people commit such acts, you are likely to focus on their individual states of mind rather than on the state of society. In other words, what usually interests us are the aspects of specific individuals' lives that caused them to become depressed or angry enough to commit suicide. We usually do not think about the patterns of social relations that might encourage such actions in general. If sociology can reveal the hidden social causes of such an apparently antisocial and nonsocial phenomenon, there must be something to it!

Sociology Now™

Learn more about **the Sociological Perspective** by going through the Sociological Perspective Learning Module.

The Sociological Explanation of Suicide

At the end of the 19th century, the French sociologist Émile Durkheim, one of the pioneers of the discipline, demonstrated that suicide is more than just an individual act of desperation resulting from psychological disorder, as people commonly believed at the time (Durkheim, 1951 [1897]). Suicide rates, he showed, are strongly influenced by social forces.

Durkheim made his case by examining the association between rates of suicide and rates of psychological disorder for different groups. The idea that psychological disorder

Bettmann/Corbis

Émile Durkheim (1858–1917) was the first professor of sociology in France and is often considered the first modern sociologist. In *The Rules of the Sociological Method* (1895) and *Suicide* (1897), he argued that human behavior is shaped by "social facts," or the social context in which people are embedded. In Durkheim's view, social facts define the constraints and opportunities within which people must act. Durkheim was also keenly interested in the conditions that promote social order in "primitive" and modern societies, and he explored this problem in depth in such works as *The Division of Labor in Society* (1893) and *The Elementary Forms of the Religious Life* (1912).

causes suicide would be supported, he reasoned, only if suicide rates are high where rates of psychological disorder are high, and low where rates of psychological disorder are low. However, his analysis of European government statistics and hospital records revealed nothing of the kind. He discovered that slightly more women than men were in insane asylums, yet four men committed suicide for every woman who did. Jews had the highest rate of psychological disorder among the major religious groups in France. However, they also had the lowest suicide rate. Psychological disorders occurred most frequently when a person reached maturity. Suicide rates, though, increased steadily with advancing age.

So rates of suicide and psychological disorder did not rise and fall together. What then accounts for variations in suicide rates? Durkheim argued that suicide rates vary because of differences in the degree of **social solidarity** in different groups. According to Durkheim, the greater the degree to which a group's members share beliefs and values and the more frequently and intensely they interact, the more social solidarity exists in the group. In turn, the higher the level of social solidarity, the more firmly anchored individuals are to the social world and the less likely they are to commit suicide if adversity strikes. In other words, Durkheim expected groups with a high degree of solidarity to have lower suicide rates than groups with a low degree of solidarity—at least up to a certain point (Figure 1.1).

To support his argument, Durkheim showed that married adults are half as likely as unmarried adults to commit suicide because marriage typically creates social ties and a moral cement that bind the individuals to society. Similarly, women are less likely to commit suicide than men. Why? Women are generally more involved in the intimate social relations of family life. Jews, Durkheim wrote, are less likely to commit suicide than Christians. The reason? Centuries of persecution have turned them into a group that is

Social solidarity refers to (1) the degree to which group members share beliefs and values and (2) the intensity and frequency of their interaction.

Altruistic suicide is Durkheim's term for suicide that occurs in high-solidarity settings, where norms tightly govern behavior. Altruism means devotion to the interests of others. Altruistic suicide is suicide in the group interest.

Egoistic suicide results from a lack of integration of the individual into society because of weak social ties to others.

Anomic suicide is Durkheim's term for suicide that occurs in low-solidarity settings, where norms governing behavior are vaguely defined. Anomie means "without order."

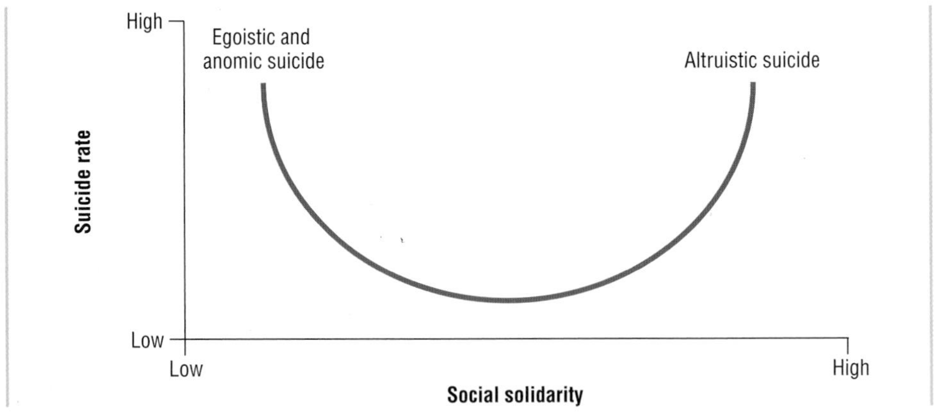

Figure 1.1
Durkheim's Theory of Suicide
Durkheim argued that as the level of social solidarity increases, the suicide rate declines. Then, beyond a certain point, it starts to rise. Hence the U-shaped curve in this graph. Durkheim called suicides that occur in high-solidarity settings *altruistic*. Altruism means devotion to the interests of others. **Altruistic suicide** occurs when norms tightly govern behavior, so individual actions are often in the group interest. For example, when soldiers knowingly give up their lives to protect members of their unit, they commit altruistic suicide out of a deep sense of comradeship. In contrast, suicide that occurs in low-solidarity settings is *egoistic* or *anomic*, said Durkheim. **Egoistic suicide** results from a lack of integration of the individual into society because of weak social ties to others. *Anomie* means "without order." **Anomic suicide** occurs when norms governing behavior are vaguely defined. For example, in Durkheim's view, when people live in a society lacking a widely shared code of morality, the rate of anomic suicide is likely to be high.

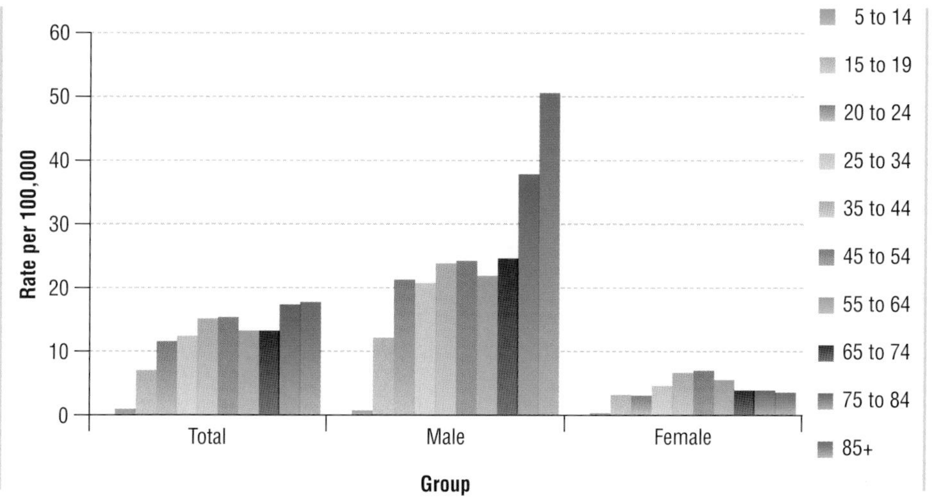

Figure 1.2
Suicide Rate by Sex and Age Cohort, United States, 2002 (per 100,000 people)

Source: U.S. Department of Health and Human Services (2004: 197).

more defensive and socially tightly knit. The elderly are more prone than the young and the middle-aged to take their own lives when faced with misfortune because they are most likely to live alone, to be widowed, and to lack a job and a wide network of friends. In general, Durkheim wrote, "Suicide varies with the degree of integration of the social groups of which the individual forms a part" (Durkheim, 1951 [1897]: 209). Note that his generalization tells us nothing about why any particular individual may take his or her own life. That is a question for psychology. However, it does tell us that a person's likelihood of committing suicide decreases with the degree to which he or she is anchored in society. And it says something surprising and uniquely sociological about how and why suicide rates vary from group to group.

SuperStock

Strong social bonds decrease the probability that a person will commit suicide if adversity strikes.

Durkheim's theory is not just a historical curiosity. It also sheds light on suicide here and now. We noted previously that approximately 11 out of every 100,000 Americans commit suicide each year. However, as the cluster of bars at the far left of Figure 1.2 shows, the suicide rate varies with age, just as it did a century ago in France. The elderly are most likely to commit suicide because they are the least firmly rooted in society. Moreover, among the elderly, suicide is most common among the divorced and widowed (National Center for Injury Prevention and Control, 2000). Figure 1.2 also shows that suicide rates differ between men and women. As in France in the late 1800s, men typically are less involved than women in child care and other duties involving family life and are about four times more likely than women to commit suicide. Research also shows that parts of the United States with high rates of church membership have low suicide rates, whereas areas with high divorce rates have high suicide rates (Breault, 1986). This finding, too, is consistent with Durkheim's theory.

One fact Figure 1.2 does not show is that suicide among young people has become more common over the past half century. For men between the ages of 15 and 24, the suicide rate rose 154 percent (from 6.5 per 100,000 to 16.5 per 100,000) (U.S. Department of Health and Human Services, 2004: 197). Why do you think this happened? Are there ways in which family life, the work world, religion, and other areas of life have changed to weaken young people's ties to society? Can *you* explain increased youth suicide sociologically?

Sociology Now™

Learn more about **Suicide** by going through the Suicide Death Rate per 100,000 Map Exercise.

From Personal Troubles to Social Structures

You have known for a long time that you live in a society. Yet until now, you may not have fully appreciated that society also lives in you: Patterns of social relations affect your innermost thoughts and feelings, influence your actions, and thus help shape who you are.

As we have seen, one such pattern of social relations is the level of social solidarity characteristic of the various groups to which you belong.

Sociologists call stable patterns of social relations **social structures.** One of the sociologist's main tasks is to identify and explain the connection between people's personal troubles and the social structures in which they are embedded. This task is harder work than it may seem at first. In everyday life, we usually see things from our own point of view. Our experiences appear unique to each of us. If we think about them at all, social structures may appear remote and impersonal. To see how social structures operate inside us, we require training in sociology.

An important step in broadening one's sociological awareness involves recognizing that three levels of social structure surround and permeate us. Think of these structures as concentric circles radiating out from you.

Microstructures

Microstructures are patterns of intimate social relations. They are formed during face-to-face interaction. Families, friendship circles, and work associations are all examples of microstructures.

Understanding the operation of microstructures can be useful. Let's say you are looking for a job. You might think you would do best to ask as many close friends and relatives as possible for leads and contacts. However, sociological research shows that people you know well are likely to know many of the same people. After asking a couple of close connections for help landing a job, you would do best to ask more remote acquaintances for leads and contacts. People to whom you are *weakly* connected (and who are weakly connected among themselves) are more likely to know *different* groups of people. Therefore, they will give you more information about job possibilities and ensure that word about your job search spreads farther. You are more likely to find a job faster if you understand "the strength of weak ties" in microstructural settings (Granovetter, 1973).

Macrostructures

Macrostructures are patterns of social relations that lie outside and above your circle of intimates and acquaintances.[1] One important macrostructure is **patriarchy,** the traditional system of economic and political inequality between women and men in most societies. (For exceptions, see Chapter 11, "Sexuality and Gender," and Chapter 15, "Families").

Understanding the operation of macrostructures can also be useful. Consider, for example, one aspect of patriarchy. Most married women who work full time in the paid labor force do more housework, child care, and care for the elderly than their husbands. Governments and businesses support this arrangement insofar as they give little assistance to families in the form of nurseries, after-school programs for children, and nursing homes. Yet the unequal division of work in the household is a major source of dissatisfaction with marriage, especially in families that cannot afford to buy these services privately. Thus, sociological research shows that where spouses share domestic responsibilities equally, they are happier with their marriages and less likely to divorce (Hochschild with Machung, 1989). When a marriage is in danger of dissolving, partners commonly blame themselves and each other for their troubles. However, it should now be clear that forces other than incompatible personalities often put stresses on families. Understanding how the macrostructure of patriarchy crops up in everyday life and doing something to change that structure can help people lead happier lives.

Global Structures

The third level of society that surrounds and permeates us is composed of **global structures.** International organizations, patterns of worldwide travel and communication, and the economic relations between countries are examples of global structures. Global

- **Social structures** are stable patterns of social relations.
- **Microstructures** are the patterns of relatively intimate social relations formed during face-to-face interaction. Families, friendship circles, and work associations are all examples of microstructures.
- **Macrostructures** are overarching patterns of social relations that lie outside and above one's circle of intimates and acquaintances. Macrostructures include classes, bureaucracies, and power systems such as patriarchy.
- **Patriarchy** is the traditional system of economic and political inequality between women and men.
- **Global structures** are patterns of social relations that lie outside and above the national level. They include international organizations, patterns of worldwide travel and communication, and the economic relations between countries.

[1]Some sociologists also distinguish "mesostructures," which are social relations that link microstructures and macrostructures.

structures are increasingly important as inexpensive travel and communication allow all parts of the world to become interconnected culturally, economically, and politically.

Understanding the operation of global structures can be useful, too. For instance, many people are concerned about the world's poor. They donate money to charities to help with famine relief. Some people also approve of the U.S. government giving foreign aid to poor countries. However, many of these same people do not appreciate that charity and foreign aid alone do not seem able to end world poverty. That is because charity and foreign aid have been unable to overcome the structure of social relations between countries that have created and sustain global inequality.

Brown Brothers

C. Wright Mills (1916–62) argued that the sociological imagination is a unique way of thinking. It allows people to see how their actions and potential are affected by the social and historical context in which they exist. Mills employed the sociological imagination effectively in his most important works. For example, *The Power Elite* (1956) is a study of the several hundred men who occupied the "command posts" of major U.S. institutions. It suggests that economic, political, and military power is highly concentrated in U.S. society, which is therefore less of a democracy than we are often led to believe. The implication of Mills's study is that to make our society more democratic, power must be more evenly distributed among the citizenry.

Let us linger on this point for a moment. As you will see in Chapter 9 ("Globalization, Inequality, and Development"), Britain, France, and other imperial powers locked some countries into poverty when they colonized them between the 17th and 19th centuries. Especially in the 1970s and 1980s, the poor (or "developing") countries borrowed money from these same rich countries and Western banks to finance airports, roads, harbors, sanitation systems, and basic health care. Today, the world's developing countries are struggling hopelessly to pay off these loans. In 2002, the world's developing countries paid the developed countries nearly seven times more in interest on loans than they received in official aid (United Nations, 2004: 201). It thus seems that relying exclusively on foreign aid and charity can do little to help solve the problem of world poverty. Understanding how the global structure of international relations created and helps maintain global inequality suggests new policy priorities for helping the world's poor. One such priority might involve campaigning for the cancellation of foreign debt in compensation for past injustices.

As these examples illustrate, personal problems are connected to social structures at the micro, macro, and global levels. Whether the personal problem involves finding a job, keeping a marriage intact, or acting justly to end world poverty, social-structural considerations broaden our understanding of the problem and suggest appropriate courses of action.

The Sociological Imagination

Nearly half a century ago, the great American sociologist C. Wright Mills (1916–62) called the ability to see the connection between personal troubles and social structures the **sociological imagination.** He emphasized the difficulty of developing this quality of mind (Box 1.1). His language is sexist by today's standards, but his argument is as true and inspiring today as it was in the 1950s:

> When a society becomes industrialized, a peasant becomes a worker; a feudal lord is liquidated or becomes a businessman. When classes rise or fall, a man is employed or unemployed; when the rate of investment goes up or down, a man takes new heart or goes broke. When war happens, an insurance salesman becomes a rocket launcher; a store clerk, a radar man; a wife lives alone; a child grows up without a father. Neither the life of an individual nor the history of a society can be understood without understanding both.
>
> Yet men do not usually define the troubles they endure in terms of historical change. . . . The well-being they enjoy, they do not usually impute to the big ups and downs of the society in which they live. Seldom aware of the intricate connection between the patterns of their own lives and the course of world history, ordinary men do not usually know what this connection means for the kind of men they are becoming and for the kind of history-making in which they might take part. They do not possess the quality of mind essential to grasp the interplay of men and society, of biography and history, of self and world. They cannot cope with their personal troubles in such a way as to control the structural transformations that usually lie behind them.

The **sociological imagination** is the quality of mind that enables one to see the connection between personal troubles and social structures.

BOX 1.1
Sociology at the Movies

Minority Report (2002)

The year is 2054 and the place is Washington, D.C. John Anderton (played by Tom Cruise) is a police officer who uses the latest technologies to apprehend murderers *before* they commit their crimes. This remarkable feat is possible because scientists have nearly perfected the use of "Pre-Cogs"—or so, at least, it seems. The Pre-Cog system consists of three psychics whose brains are wired together and who are sedated so they can develop a collective vision about impending murders. Together with powerful computers, the Pre-Cogs are apparently helping to create a crime-free society.

All is well until one of the psychics' visions shows Anderton himself murdering a stranger in less than 36 hours. Suddenly, Anderton is on the run from his own men. Desperate to figure out if the Pre-Cog system is somehow mistaken, he breaks into the system, unwires one of the psychics, and discovers that the three psychics do not always agree about what the future will bring. Sometimes there is a "minority report." Sometimes the minority report is correct. Sometimes people are arrested even though they never would have broken the law. The authorities have concealed this system flaw and allowed the arrest of potentially innocent people in their zeal to create a crime-free society.

And so Anderton comes to realize that not everything is predetermined—that, in his words, "It's not the future if you stop it." And stop it he does. Herein lies an important sociological lesson. Many people believe two contradictory ideas with equal conviction. First, they believe that they are perfectly free to do whatever they want. Second, they believe that the "system" (or "society") is so big and powerful that they are unable to do anything to change it. Neither idea is accurate. As we emphasize throughout this book, various aspects of society exert powerful influences on our behavior; we are not perfectly free. Nonetheless, it is possible to change many aspects of society; we are not wholly determined either. As you will learn, changing various aspects of society is possible under specifiable circumstances, with the aid of specialized knowledge and through great individual and collective effort, so that certain futures can indeed be stopped.

20th Century Fox/Dreamworks/The Kobal Collection

In *Minority Report*, John Anderton (Tom Cruise) works with "Pre-Cogs" to track down criminals before they commit their crimes. When he discovers a flaw in the system, he realizes the future is not entirely fixed and that within limits, we can change it. His insight holds an important sociological lesson for us.

Understanding the social constraints and possibilities for freedom that envelop us requires an active sociological imagination. The sociological imagination urges us to connect our biography with history and social structure—to make sense of our lives against a larger historical and social background and to act in light of our understanding. Have you ever tried to put events in your own life in the context of history and social structure? Did the exercise help you make sense of your life? Did it in any way lead to a life more worth living? Is the sociological imagination a worthy goal?

Although movies are just entertainment to many people, they often achieve by different means what the sociological imagination aims for. Therefore, in each chapter of this book, we review a movie to shed light on topics of sociological importance.

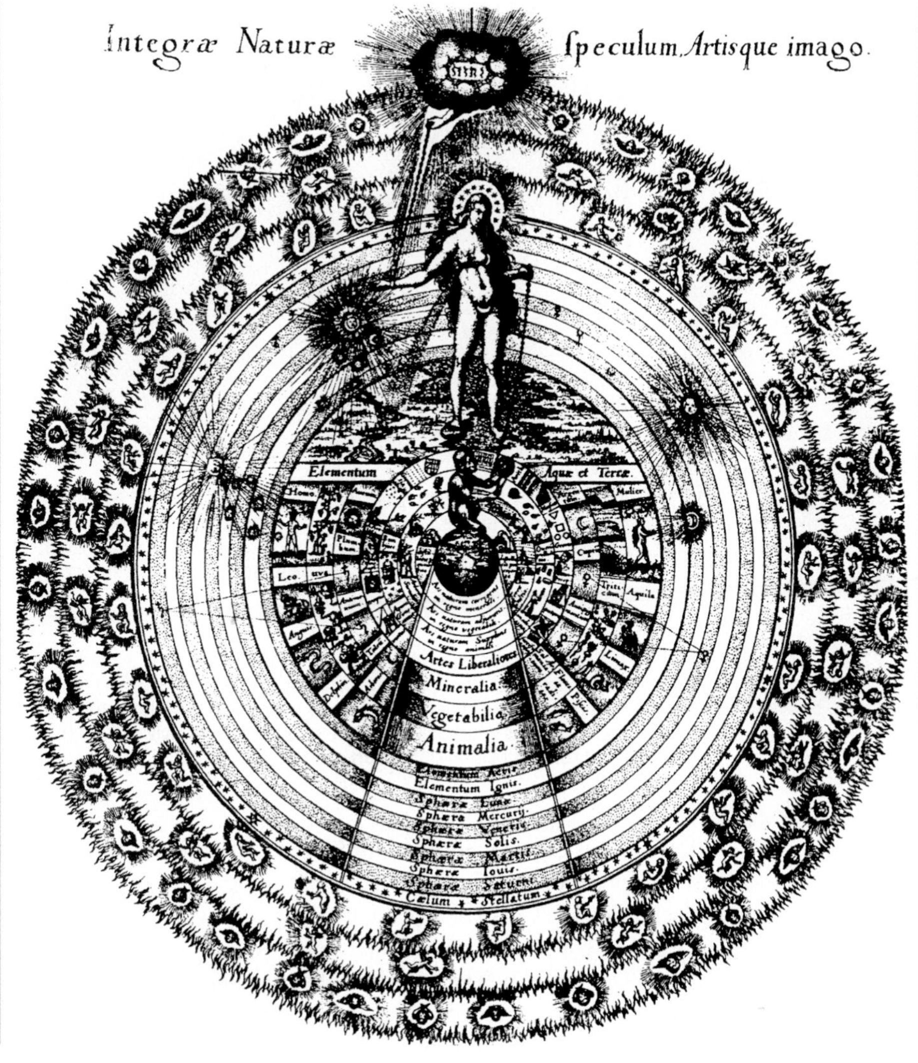

▶Figure 1.3
The European View of the World in 1600

This engraving shows how most educated Europeans pictured the universe in William Shakespeare's time. Note the cloud at the top of the circle. The Hebrew name of God is inscribed on it. God's hand extends from the cloud, holding a chain that is attached to a woman representing Nature. Nature holds a chain in her hand that is connected to "the ape of Nature," representing humankind. The symbolism is clear: God and his intermediary, Nature, control human action. Note also that the engraving arranges everything in a linked hierarchy. The hierarchy includes the mineral, vegetable, and animal kingdoms; the elements; heavenly objects; angels; and so forth. Each level of the hierarchy corresponds to and controls some aspect of the level below it. For example, people believed archangels regulated the movements of the planet Mercury and the movements of Mercury affected human commerce. Similarly, in the medieval view, God ordained a hierarchy of people. The richest people were seen as the closest to God and therefore deserving great privilege. Supposedly, kings and queens ruled because God wanted them to (Tillyard, 1943).

Source: Robert Fludd. *Utriusque Cosmi Maioris Scilicet et Minoris Metaphysica, Physica Atqve Technica Historia.* 1617–19. (Oppenheim, Germany. Johan-Theodori de Bry.) By permission of the Houghton Library, Harvard University.

> What they need . . . is a quality of mind that will help them to [see] . . . what is going on in the world and . . . what may be happening within themselves. It is this quality . . . that . . . may be called the sociological imagination (Mills, 1959: 3–4).

The sociological imagination is a recent addition to the human repertoire. It is only about as old as the United States. True, in ancient and medieval times, some philosophers wrote about society. However, their thinking was not sociological. They believed God and nature controlled society. They spent much of their time sketching blueprints for the ideal society and urging people to follow those blueprints. They relied on speculation rather than evidence to reach conclusions about how society works (▶Figure 1.3).

Origins of the Sociological Imagination

The sociological imagination was born when three modern revolutions pushed people to think about society in an entirely new way:

The Scientific Revolution

The **Scientific Revolution** began about 1550. It encouraged the view that sound conclusions about the workings of society must be based on solid evidence, not just speculation.

The **Scientific Revolution** began in Europe about 1550. It encouraged the view that sound conclusions about the workings of society must be based on solid evidence, not just speculation.

People often link the Scientific Revolution to specific ideas, such as Newton's laws of motion and Copernicus's theory that the Earth revolves around the sun. However,

Paul Almasy/Corbis

Musee du Louvre, Paris/Giraudon, Paris/SuperStock

Eugene Delacroix. *Liberty Leading the People, July 28, 1830.* The democratic forces unleashed by the French Revolution suggested that people are responsible for organizing society and that human intervention can therefore solve social problems. As such, democracy was a foundation stone of sociology.

The Scientific Revolution began in Europe around 1550. Scientists proposed new theories about the structure of the universe and developed new methods to collect evidence so they could test those theories. Shown here is an astrolabe used by Copernicus to solve problems related to the position of the sun, planets, and stars.

science is less a collection of ideas than a method of inquiry. For instance, in 1609 Galileo pointed his newly invented telescope at the heavens, made some careful observations, and showed that his observations fit Copernicus's theory. This is the core of the scientific method: using evidence to make a case for a particular point of view. By the mid-1600s, some philosophers, such as Descartes in France and Hobbes in England, were calling for a science of society. When sociology emerged as a distinct discipline in the 19th century, commitment to the scientific method was one firm pillar of the sociological imagination.

The Democratic Revolution

The **Democratic Revolution** began about 1750. It suggested that people are responsible for organizing society and that human intervention can therefore solve social problems. Four hundred years ago, most Europeans thought otherwise. For them, God ordained the social order. The American Revolution (1775–83) and the French Revolution (1789–99) helped undermine this idea. These democratic political upheavals showed that society could experience massive change in a short period, proved that people could replace unsatisfactory rulers, and suggested that *people* control society. The implications for social thought were profound. For if it were possible to change society by human intervention, then a science of society could play a big role. The new science could help people find ways of overcoming social problems, improving the welfare of citizens, and effectively reaching given goals. Much of the justification for sociology as a science arose out of the democratic revolutions that shook Europe and North America.

The **Democratic Revolution** began about 1750, during which time the citizens of the United States, France, and other countries broadened their participation in government. This revolution also suggested that people organize society and that human intervention can therefore resolve social problems.

The **Industrial Revolution** refers to the rapid economic transformation that began in Britain in the 1780s. It involved the large-scale application of science and technology to industrial processes, the creation of factories, and the formation of a working class. It created a host of new and serious social problems that attracted the attention of many social thinkers.

The Industrial Revolution

The **Industrial Revolution** began about 1780. It created a host of new and serious social problems that attracted the attention of social thinkers. As a result of the growth of industry, masses of people moved from countryside to city, worked agonizingly long hours in crowded and dangerous mines and factories, lost faith in their religions, confronted faceless bureaucracies, and reacted to the filth and poverty of their existence by means of strikes, crime, revolutions, and wars. Scholars had never seen a sociological laboratory like this. The Scientific Revolution suggested that a science of society is possible. The

Diego Rivera. *Detroit Industry,* North Wall. 1932–33. Fresco (detail). Copyright 1997, The Detroit Institute of Arts. The so-called Second Industrial Revolution began in the early 20th century. Wealthy entrepreneurs formed large companies, steel became a basic industrial material, and oil and electricity fueled most industrial production. At the same time, Henry Ford's assembly lines and other mass-production technologies transformed the workplace.

Democratic Revolution suggested that people can intervene to improve society. The Industrial Revolution now presented social thinkers with a host of pressing social problems crying out for a solution. They responded by giving birth to the sociological imagination.

Theory, Research, and Values

Theory without practice cannot survive and dies as quickly as it lives. He who loves practice without theory is like the sailor who boards ship without a rudder and compass and never knows where he may be cast.

LEONARDO DA VINCI (1970)

French social thinker Auguste Comte (1798–1857) coined the term *sociology* in 1838 (Comte, 1975). Comte tried to place the study of society on scientific foundations. He said he wanted to understand the social world as it is, not as he or anyone else imagined it should be. Yet there was a tension in his work, for although Comte was eager to adopt the scientific method in the study of society, he was a conservative thinker, motivated by strong opposition to rapid change in French society. This was evident in his writings. When he moved from his small, conservative hometown to Paris, Comte witnessed the democratic forces unleashed by the French Revolution, the early industrialization of society, and the rapid growth of cities. What he saw shocked and saddened him. Rapid social change was destroying much of what he valued, especially respect for traditional authority. He therefore urged slow change and the preservation of all that was traditional in social life. Thus, scientific methods of research *and* a vision of the ideal society were evident in sociology at its origins.

Although he praised the value of scientific methods, Comte never conducted any research. Neither did the second founder of sociology, British social theorist Herbert Spencer (1820–1903). However, Spencer believed that he had discovered scientific laws governing the operation of society. Strongly influenced by Charles Darwin's theory of evolution, he thought societies were composed of interdependent parts, just like biological organisms. These interdependent parts include families, governments, and the economy. According to Spencer, societies evolve in the same way biological species do. Individuals struggle to survive, and the fittest succeed in this struggle. The least fit die before they can bear offspring. This allows societies to evolve from "barbaric" to "civilized." Deep social inequalities exist in society, but that is just as it should be if societies are to evolve, Spencer suggested (Spencer, 1975 [1897–1906]).

Spencer's ideas, which came to be known as "social Darwinism," were popular for a time in the United States and Great Britain. Wealthy industrialists like the oil baron John D. Rockefeller found much to admire in a doctrine that justified social inequality and trumpeted the superiority of the wealthy and the powerful. Today, few sociologists think that societies are like biological systems. We have a better understanding of the complex economic, political, military, religious, and other forces that cause social change. We know that people can take things into their own hands and change their social environment in ways that no other species can. Spencer remains of interest because he was among the first social thinkers to assert that society operates according to scientific laws—and because his vision of the ideal society nonetheless showed through his writings.

To varying degrees, we see the same tension between belief in the importance of science and a vision of the ideal society in the work of the three giants in the early history of sociology: Karl Marx (1818–83), Émile Durkheim (1858–1917), and Max Weber (pronounced VAY-ber; 1864–1920). During their lifetimes, these three men witnessed various phases of Europe's wrenching transition to industrial capitalism. They wanted to explain the great transformation of Europe and suggest ways of improving people's lives. Like Comte and Spencer, they were committed to the scientific method of research. They actually adopted scientific research methods in their work. However, they also wanted to chart a better course for their societies. The ideas they developed are not just diagnostic tools from which we can still learn, but like many sociological ideas, prescriptions for combating social ills.

The tension between analysis and ideal, diagnosis and prescription, is evident throughout sociology. This becomes clear if we distinguish three important terms: theories, research, and values.

Theory

Sociological ideas are usually expressed in the form of theories. **Theories** are tentative explanations of some aspect of social life. They state how and why certain facts are related. For example, in his theory of suicide, Durkheim related facts about suicide rates to facts about social solidarity. This enabled him to explain suicide as a function of social solidarity. In our broad definition, even a hunch qualifies as a theory if it suggests how and why certain facts are related. As Albert Einstein wrote, "The whole of science is nothing more than a refinement of everyday thinking" (Einstein, 1954: 270).

Research

After sociologists formulate theories, they can conduct research. **Research** is the process of carefully observing social reality, often to "test" a theory or assess its validity. For example, Durkheim collected suicide statistics from various government agencies to see whether the data supported or contradicted his theory. Because research can call the validity of a theory into question, theories are only *tentative* explanations. We discuss the research process in detail in Chapter 2, "How Sociologists Do Research."

A **theory** is a tentative explanation of some aspect of social life that states how and why certain facts are related.

Research is the process of systematically observing reality to assess the validity of a theory.

Values

Before sociologists can formulate a theory, however, they must make certain judgments. For example, they must decide which problems are worth studying. They must make certain assumptions about how the parts of society fit together. If they are going to recommend ways of improving the operation of some aspect of society, they must even have an opinion about what the ideal society should look like. As we will soon see, these issues are shaped largely by sociologists' values. **Values** are ideas about what is right and wrong. Inevitably, values help sociologists formulate and favor certain theories over others (Edel, 1965; Kuhn, 1970 [1962]). Thus, sociological theories may be modified and even rejected due to research, but they are often motivated by sociologists' values.

Durkheim, Marx, and Weber stood close to the origins of the major theoretical traditions in sociology: functionalism, conflict theory, and symbolic interactionism. A fourth theoretical tradition, feminism, has arisen in recent decades to correct some deficiencies in the three long-established traditions. It will become clear as you read this book that many more theories exist in addition to these four. However, because these four traditions have been especially influential in the development of sociology, we present a thumbnail sketch of each one at the beginning.

Sociological Theory and Theorists

Functionalism

Durkheim

Durkheim's theory of suicide is an early example of what sociologists now call functionalism. **Functionalist theories** incorporate four features:

1. They stress that human behavior is governed by stable patterns of social relations, or social structures. For example, Durkheim emphasized how patterns of social solidarity influence suicide rates. The social structures typically analyzed by functionalists are macrostructures.
2. Functionalist theories show how social structures maintain or undermine social stability. That is why functionalists are sometimes called "structural functionalists"; they analyze how the parts of society (structures) fit together and how each part contributes to the stability of the whole (its function). Thus, Durkheim argued that high social solidarity contributes to the maintenance of social order, but the growth of industries and cities in 19th-century Europe lowered the level of social solidarity and contributed to social instability. One aspect of instability, said Durkheim, is a higher suicide rate. Another is frequent strikes by workers.
3. Functionalist theories emphasize that social structures are based mainly on shared values. Thus, when Durkheim wrote about social solidarity, he sometimes meant the frequency and intensity of social interaction, but more often he thought of social solidarity as a kind of moral cement that binds people together.
4. Functionalism suggests that reestablishing equilibrium can best solve most social problems. Durkheim said social stability could be restored in late-19th-century Europe by creating new associations of employers and workers that would lower workers' expectations about what they could expect from life. If more people could agree on wanting less, said Durkheim, social solidarity would rise and there would be fewer strikes and lower suicide rates. Functionalism, then, was a conservative response to widespread social unrest in late-19th-century France. A more liberal or radical response would have been to argue that if people are expressing discontent because they are getting less out of life than they expect, discontent can be lowered by finding ways for them to get more out of life.

Parsons and Merton

Although functionalist thinking influenced American sociology at the end of the 19th century, it was only during the Great Depression of 1929–39 that it took deep root (Russett, 1966). With 30 percent of the labor force unemployed and labor unrest reaching unprecedented levels by 1934, sociologists with a conservative frame of mind were attracted to a

Values are ideas about what is right and wrong.

Functionalist theory stresses that human behavior is governed by relatively stable social structures. It underlines how social structures maintain or undermine social stability. It emphasizes that social structures are based mainly on shared values or preferences and suggests that reestablishing equilibrium can best solve most social problems.

Courtesy of Columbia University

Robert Merton (1910–2003) made functionalism a more flexible theory from the late 1930s to the 1950s. In *Social Theory and Social Structure* (1949), he proposed that social structures are not always functional. They may be dysfunctional for some people. Moreover, not all functions are manifest; some are latent, according to Merton. Merton also made major contributions to the sociology of science, notably in *On the Shoulders of Giants* (1956), a study of creativity, tradition, plagiarism, the transmission of knowledge, and the concept of progress.

Archivo Iconografico, S.A./Corbis

Karl Marx (1818–83) was a revolutionary thinker whose ideas affected not just the growth of sociology but the course of world history. He held that major sociohistorical changes are the result of conflict between society's main social classes. In his major work, *Capital* (1867–94), Marx argued that capitalism would produce such misery and collective strength among workers that they would eventually take state power and create a classless society in which production would be based on human need rather than profit.

theory that focused on how social equilibrium could be restored. Functionalist theory remained popular in the United States for approximately 30 years. It experienced a minor revival in the early 1990s but never regained the dominance it enjoyed from the 1930s to the early 1960s.

Sociologist Talcott Parsons (1902–79) was the foremost American proponent of functionalism. Parsons is best known for identifying how various institutions must work to ensure the smooth operation of society as a whole. He argued that society is well integrated and in equilibrium when the family successfully raises new generations, the military successfully defends society against external threats, schools are able to teach students the skills and values they need to function as productive adults, and religions create a shared moral code among people (Parsons, 1951).

Parsons was criticized for exaggerating the degree to which members of society share common values and to which social institutions contribute to social harmony. This criticism led the other leading functionalist in the United States, Robert Merton (1910–2003), to propose that social structures may have different consequences for different groups of people. Merton noted that some of those consequences may be disruptive or **dysfunctional** (Merton, 1968 [1949]). Moreover, said Merton, while some functions are **manifest** (intended and easily observed) others are **latent** (unintended and less obvious). For instance, a manifest function of schools is to transmit skills from one generation to the next. A latent function of schools is to encourage the development of a separate youth culture that often conflicts with parents' values (Coleman, 1961; Hersch, 1998).

Similarly, to anticipate an argument we make later in this chapter, the manifest function of clothing is to keep people warm in cool weather and cool in warm weather. Yet clothing may be fashionable or unfashionable. The particular style of clothing we adopt may indicate whom we wish to associate with and whom we want to exclude from our social circle. What we wear may thus express the position we occupy in society, how we think of ourselves, and how we want to present ourselves to others. These are all latent functions of clothing.

Dysfunctions are effects of social structures that create social instability.

Manifest functions are visible and intended effects of social structures.

Latent functions are invisible and unintended effects of social structures.

Conflict theory generally focuses on large, macro-level structures, such as the relations between classes. It shows how major patterns of inequality in society produce social stability in some circumstances and social change in others. It stresses how members of privileged groups try to maintain their advantages while subordinate groups struggle to increase theirs. It typically leads to the suggestion that eliminating privilege will lower the level of conflict and increase the sum total of human welfare.

Conflict Theory

The second major theoretical tradition in sociology emphasizes the centrality of conflict in social life. **Conflict theory** incorporates these features:

1. It generally focuses on large, macro-level structures, such as "class relations" or patterns of domination, submission, and struggle between people of high and low standing.
2. It shows how major patterns of inequality in society produce social stability in some circumstances and social change in others.
3. It stresses how members of privileged groups try to maintain their advantages while subordinate groups struggle to increase theirs. From this point of view, social conditions at a given time are the expression of an ongoing power struggle between privileged and subordinate groups.
4. It typically leads to the suggestion that eliminating privilege will lower the level of conflict and increase total human welfare.

Marx

Conflict theory originated in the work of German social thinker Karl Marx. A generation before Durkheim, Marx observed the destitution and discontent produced by the Industrial Revolution and proposed a sweeping argument about the way societies develop (Marx, 1904 [1859]; Marx and Engels, 1972 [1848]). Marx's theory was radically different from Durkheim's. **Class conflict,** the struggle between classes to resist and overcome the opposition of other classes, lies at the center of his ideas.

Marx argued that owners of industry are eager to improve the way work is organized and to adopt new tools, machines, and production methods. These innovations allow them to produce more efficiently, earn higher profits, and drive inefficient competitors out of business. However, the drive for profits also causes capitalists to concentrate workers in larger and larger establishments, keep wages as low as possible, and invest as little as possible in improving working conditions. Thus, said Marx, a large and growing class of poor workers opposes a small and shrinking class of wealthy owners.

Brown Brothers

Max Weber (1864–1920) was Germany's greatest sociologist, and he profoundly influenced the development of the discipline worldwide. Engaged in a lifelong "debate with Marx's ghost," Weber held that economic circumstances alone do not explain the rise of capitalism. As he showed in *The Protestant Ethic and the Spirit of Capitalism* (1904–5), independent developments in the religious realm had unintended, beneficial consequences for capitalist development in some parts of Europe. He also argued that capitalism would not necessarily give way to socialism. Instead, he regarded the growth of bureaucracy and the overall "rationalization" of life as the defining characteristics of the modern age. These themes were developed in *Economy and Society* (1922).

Bettmann/Corbis

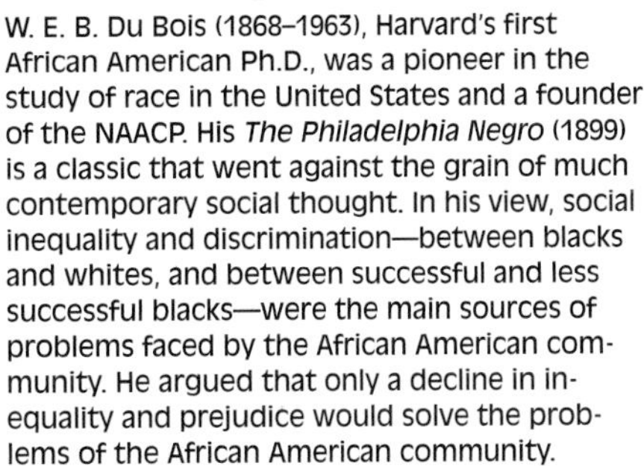
W. E. B. Du Bois (1868–1963), Harvard's first African American Ph.D., was a pioneer in the study of race in the United States and a founder of the NAACP. His *The Philadelphia Negro* (1899) is a classic that went against the grain of much contemporary social thought. In his view, social inequality and discrimination—between blacks and whites, and between successful and less successful blacks—were the main sources of problems faced by the African American community. He argued that only a decline in inequality and prejudice would solve the problems of the African American community.

Marx believed that workers would ultimately become aware of belonging to the same exploited class. He called this awareness "class consciousness." He believed that working-class consciousness would encourage the growth of trade unions and labor parties. According to Marx, these organizations would eventually seek to put an end to private ownership of property, replacing it with a "communist" society, defined as a system in which there is no private property and everyone shares property and wealth.

Weber

Although some of Marx's ideas have been usefully adapted to the study of contemporary society, his predictions about the inevitable collapse of capitalism have been questioned. Max Weber, a German sociologist who wrote his major works a generation after Marx, was among the first to find flaws in Marx's argument (Weber, 1946). Weber noted the rapid growth of the "service" sector of the economy, with its many nonmanual workers and professionals. He argued that many members of these occupational groups stabilize society because they enjoy higher status and income than manual workers employed in the manufacturing sector. In addition, Weber showed that class conflict is not the only driving force of history. In his view, politics and religion are also important sources of historical change (see following discussion). Other writers pointed out that Marx did not understand how investing in technology would make it possible for workers to toil fewer hours under less oppressive conditions. Nor did he foresee that higher wages, better working conditions, and welfare-state benefits would pacify manual workers. Thus, we see that many of the particulars of Marx's theory were called into question by Weber and other sociologists.

Du Bois

Nonetheless, Marx's insights about the fundamental importance of conflict in social life were influential, and still are today. For example, W. E. B. Du Bois (1868–1963) was an early advocate of conflict theory in the United States. For a man writing at the end of the 19th century, Du Bois had a remarkably liberal and even radical frame of

Class conflict is the struggle between classes to resist and overcome the opposition of other classes.

mind. The first African American to receive a Ph.D. from Harvard, Du Bois went to Berlin to hear Max Weber lecture and conducted pioneering studies of race in the United States. He was also a founder of the National Association for the Advancement of Colored People (NAACP) and of the country's second Department of Sociology, at Atlanta University, in 1897.[2]

Du Bois's best-known work is *The Philadelphia Negro.* This book is based on the first major sociological research project conducted in the United States. Du Bois showed that poverty and other social problems faced by African Americans are not due to some "natural" inferiority (which was widely believed at the time) but to white prejudice (Du Bois, 1967 [1899]). He believed that the elimination of white prejudice would reduce racial conflict and create more equality between blacks and whites. Du Bois was also critical of economically successful African Americans. He faulted them for failing to help less fortunate blacks and segregating themselves from the African American community to win acceptance among whites. Du Bois was disappointed with the slow improvement in race relations in the United States. He eventually became a Marxist and near the end of his life moved to Ghana, where he died.

C. Wright Mills

Du Bois was a pioneer in American conflict theory, particularly as it applies to race and ethnic relations. Conflict theory had some advocates in the United States after Du Bois. Most noteworthy is C. Wright Mills, who laid the foundations for modern conflict theory in the United States in the 1950s. Mills conducted pioneering research on American politics and class structure. One of his most important books is *The Power Elite,* a study of the several hundred men who occupy the "command posts" of the American economy, military, and government. He argued that power is highly concentrated in American society, which is therefore less of a democracy than Americans are often led to believe (Mills, 1956).

Exceptions like Mills notwithstanding, conflict theory did not really take hold in the United States until the 1960s, a decade that was rocked by growing labor unrest, antiwar protests, the rise of the Black Power movement, and the first stirrings of feminism. Strikes, demonstrations, and riots were almost daily occurrences in the 1960s and early 1970s, and therefore many sociologists of that era thought conflict between classes, nations, races, and generations was the very essence of society. Many of today's leading sociologists attended graduate school in the 1960s and 1970s and were strongly influenced by the spirit of the times. As you will see throughout this book, they have made important contributions to conflict theory during their professional careers.

Symbolic Interactionism

Weber and the Protestant Ethic

We noted earlier that Weber criticized Marx's interpretation of the development of capitalism. Among other things, Weber argued that early capitalist development was caused not just by favorable economic circumstances. In addition, he said certain *religious* beliefs facilitated robust capitalist growth. In particular, 16th- and 17th-century Protestants believed that their religious doubts could be reduced and a state of grace assured if they worked diligently and lived modestly. Weber called this belief the **Protestant ethic.** He believed it had an unintended effect: People who adhered to the Protestant ethic saved and invested more money than others. Thus, capitalism developed most robustly where the Protestant ethic took hold. He concluded that capitalism did not develop due to the operation of economic forces alone, as Marx argued. Instead, it depended partly on the religious meaning that individuals attached to their work (Weber 1958 [1904–5]). In much of his research, Weber emphasized the importance of empa-

The **Protestant ethic** is the 16th- and 17th-century Protestant belief that religious doubts can be reduced, and a state of grace assured, if people work diligently and live ascetically. According to Weber, the Protestant ethic had the unintended effect of increasing savings and investment and thus stimulating capitalist growth.

---

[2]The country's first Department of Sociology was formed at the University of Kansas in 1892. The country's third and, for decades, most influential Department of Sociology was formed at the University of Chicago in 1899.

thetically understanding people's motives and the meanings they attach to things to gain a clear sense of the significance of their actions. He called this aspect of his approach to sociological research the method of *Verstehen* ("understanding" in German).

The idea that subjective meanings and motives must be analyzed in any complete sociological analysis was only one of Weber's contributions to early sociological theory. Weber was also an important conflict theorist, as you will learn in later chapters of this book. It is enough to note at present that his emphasis on subjective meanings found rich soil in the United States in the late 19th and early 20th centuries because his ideas resonated deeply with the individualism of American culture. A century ago, people widely believed that individual talent and initiative could allow one to achieve just about anything in this land of opportunity. Small wonder then that much of early American sociology focused on the individual or, more precisely, on the connection between the individual and the larger society.

George Herbert Mead (1863–1931) was the driving force behind the study of how individual identity is formed in the course of interaction with other people. The work of Mead and his colleagues gave birth to symbolic interactionism, a distinctively American theoretical tradition that continues to be a major force in sociology today.

The Granger Collection

George Herbert Mead

This was certainly a focus of sociologists at the University of Chicago, the most influential Department of Sociology in the country before World War II. For example, the university's George Herbert Mead (1863–1931) was the driving force behind the study of how the individual's sense of self is formed in the course of interaction with other people. We discuss his contribution in Chapter 4, "Socialization." Here, we note only that the work of Mead and his colleagues gave birth to *symbolic interactionism,* a distinctively American theoretical tradition that continues to be a major force in sociology today.

Functionalist and conflict theories assume that people's group memberships—whether they are rich or poor, male or female, black or white—determine their behavior. This can sometimes make people seem like balls on a pool table: They get knocked around and cannot choose their own destinations. We know from our everyday experience, however, that people are not like that. You often make choices, sometimes difficult ones. You sometimes change your mind. Moreover, two people with similar group memberships may react differently to similar social circumstances. That is because they interpret those circumstances differently.

Erving Goffman

Recognizing these issues, some sociologists focus on the subjective side of social life. They work in the symbolic interactionist tradition, a school of thought that was given its name by sociologist Herbert Blumer (1900–1986), Mead's student at the University of Chicago. **Symbolic interactionism** incorporates these features:

1. A focus on interpersonal communication in micro-level social settings, which distinguishes it from both functionalist and conflict theories.
2. An emphasis on social life as possible only because people attach meanings to things. It follows that an adequate explanation of social behavior requires understanding the subjective meanings people associate with their social circumstances.
3. Stress on the notion that people help to create their social circumstances and do not merely react to them. For example, Canadian-born sociologist Erving Goffman (1922–82), one of the most influential symbolic interactionists, analyzed the many ways people present themselves to others in everyday life so as to appear in the best possible light. Goffman compared social interaction to a carefully staged play, complete with front stage, backstage, defined roles, and a wide range of props. In this play, a person's age, gender, race, and other characteristics may help shape his or her actions, but there is much room for individual creativity as well (Goffman, 1959 [1956]).

Symbolic interactionist theory focuses on interpersonal communication in micro-level social settings. It emphasizes that an adequate explanation of social behavior requires understanding the subjective meanings people attach to their social circumstances. It stresses that people help to create their social circumstances and do not merely react to them. And, by underscoring the subjective meanings people create in small social settings, it validates unpopular and nonofficial viewpoints. This increases our understanding and tolerance of people who may be different from us.

4. Validation of unpopular and nonofficial viewpoints by focusing on the subjective meanings people create in small social settings. This focus increases our understanding and tolerance of people who may be different from us.

To understand symbolic interactionism better, let us briefly return to the problem of suicide. If a police officer discovers a dead person at the wheel of a car that has run into a tree, it may be difficult to establish whether the death was an accident or suicide. Interviewing friends and relatives to discover the driver's state of mind just before the crash may help rule out the possibility of suicide. As this example illustrates, understanding the intention or motive of the actor is critical to understanding the meaning of a social action and explaining it. A state of mind must be interpreted, usually by a coroner, before a dead body becomes a suicide statistic (Douglas, 1967).

For surviving family and friends, suicide is always painful and sometimes embarrassing. Insurance policies often deny payments to beneficiaries in the case of suicide. As a result, coroners are inclined to classify deaths as accidental whenever such an interpretation is plausible. Being human, they want to minimize a family's suffering after such a horrible event. Sociologists therefore believe that suicide rates according to official statistics are about one-third lower than they actually are.

The study of the subjective side of social life reveals many such inconsistencies. It helps us go beyond the official picture, deepening our understanding of how society works and supplementing the insights gained from macro-level analysis. Moreover, by stressing the importance and validity of subjective meanings, symbolic interactionists also increase tolerance for nonofficial, minority, and deviant viewpoints.

Social Constructionism

One variant of symbolic interactionism that has become especially popular in recent years is **social constructionism.** Social constructionists argue that when people interact, they typically assume that things are naturally or innately what they seem to be. However, apparently natural or innate features of life are often sustained by *social* processes that vary historically and culturally. For example, many people assume that differences in the way women and men behave are the result of their respective biological makeups. In contrast, social constructionists show that many of the presumably natural differences between women and men depend on the way power is distributed between them and the degree to which certain ideas about women and men are widely shared (Berger and Luckmann, 1967; see Chapter 11, "Sexuality and Gender"). People usually do such a good job of building natural-seeming social realities in their everyday interactions that they do not notice the materials used in the construction process. Social constructionists identify those materials and analyze how they are pieced together.

In sum, the study of the subjective side of social life helps us get beyond the official picture, deepening our understanding of how society works and supplementing the insights gained from macro-level analysis. By stressing the importance and validity of subjective meanings, symbolic interactionists increase tolerance for minority and deviant viewpoints. By stressing how subjective meanings vary historically and culturally, social constructionists show that many seemingly natural features of social life actually require painstaking acts of social creation.

Feminist Theory

Few women figured prominently in the early history of sociology. The strict demands placed on them by the 19th-century family and the lack of opportunity in the larger society prevented most women from earning a higher-education degree and making major contributions to the discipline. Women who made their mark on sociology in its early years tended to have unusual biographies. Some of these exceptional people introduced gender issues that were largely ignored by Marx, Durkheim, Weber, Mead, and other early sociologists. Appreciation for the sociological contribution of these pioneering women

Social constructionists argue that apparently natural or innate features of life are often sustained by social processes that vary historically and culturally.

has grown in recent years because concern with gender issues has come to form a substantial part of the modern sociological enterprise.

Harriet Martineau

Harriet Martineau (1802–76) is often called the first woman sociologist. Born in England to a prosperous family, she never married. She was able to support herself comfortably from her journalistic writings. Martineau translated Comte into English and wrote one of the first books on research methods. She undertook critical studies of slavery, factory laws, and gender inequality. She was a leading advocate of voting rights and higher education for women and of gender equality in the family. As such, Martineau was one of the first feminists (Martineau, 1985).

Hulton-Deutsch Collection/Corbis

Harriet Martineau (1802–76) was the first woman sociologist. She translated Comte into English and conducted studies on research methods, slavery, factory laws, and gender inequality.

Jane Addams

In the United States in the early 20th century, a few women from wealthy families attended university, received training as sociologists, and wanted to become professors of sociology, but they were denied faculty appointments. Typically, they turned to social activism and social work instead. A case in point is Jane Addams (1860–1935). Addams was cofounder of Hull House, a shelter for the destitute in Chicago's slums, and spent a lifetime fighting for social reform. She also provided a research platform for sociologists from the University of Chicago, who often visited Hull House to interview its clients. In recognition of her efforts, Addams received the ultimate academic award in 1931—the Nobel Prize.

Modern Feminism

Despite its early stirrings, feminist thinking had little impact on sociology until the mid-1960s, when the rise of the modern women's movement drew attention to the many remaining inequalities between women and men. Because of feminist theory's major influence, it may fairly be regarded as sociology's fourth major theoretical tradition. Modern feminism has several variants (see Chapter 11, "Sexuality and Gender"). However, the various strands of **feminist theory** share the following features:

1. Feminist theory focuses on various aspects of patriarchy, the system of male domination in society. Patriarchy, feminists contend, is as important as class inequality, if not more so, in determining a person's opportunities in life.
2. Feminist theory holds that male domination and female subordination are determined not by biological necessity but by structures of power and social convention. From their point of view, women are subordinate to men only because men enjoy more legal, economic, political, and cultural rights.
3. Feminist theory examines the operation of patriarchy in both micro- and macro-level settings.
4. Feminist theory contends that existing patterns of gender inequality can and should be changed for the benefit of all members of society. The main sources of gender inequality include differences in the way boys and girls are reared, barriers to equal opportunity in education, paid work, and politics, and the unequal division of domestic responsibilities between women and men.

Feminist theory claims that patriarchy is at least as important as class inequality in determining a person's opportunities in life. It holds that male domination and female subordination are determined not by biological necessity but by structures of power and social convention. It examines the operation of patriarchy in both micro- and macro-level settings, and it contends that existing patterns of gender inequality can and should be changed for the benefit of all members of society.

The four main theoretical traditions in sociology are summarized in Concept Summary 1.1 (see also Figure 1.4). As you will see in the following pages, sociologists in the United States and elsewhere have applied them to all of the discipline's branches and have elaborated and refined each of them. Some sociologists work exclusively within one tradition. Others conduct research that borrows from more than one tradition. But all sociologists are deeply indebted to the founders of the discipline.

To illustrate how much farther we are able to see using theory as our guide, we now consider how the four traditions we have outlined improve our understanding of an aspect of social life familiar to everyone: the world of fashion.

Concept Summary 1.1
Four Theoretical Traditions in Sociology

Theoretical Tradition	Main Level of Analysis	Main Focus	Main Question
Functionalist	Macro	Values	How do the institutions of society contribute to social stability and instability?
Conflict	Macro	Inequality	How do privileged groups seek to maintain their advantages and subordinate groups seek to increase theirs, often causing social change in the process?
Symbolic interactionist	Micro	Meaning	How do individuals communicate so as to make their social settings meaningful?
Feminist	Macro and Micro	Patriarchy	Which social structures and interaction processes maintain male dominance and female subordination?

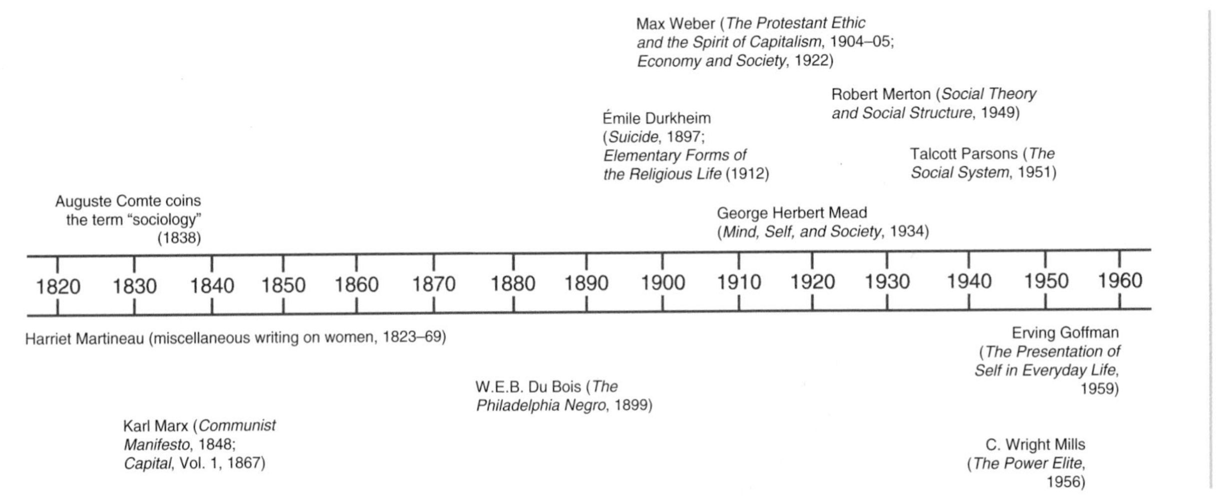

Figure 1.4
A Sociological Time Line of Some Major Figures in the Development of Sociological Theory, 1820–1960

Applying the Four Theoretical Perspectives: The Problem of Fashion

"Oh. Two weeks ago I saw Cameron Diaz at Fred Siegel and I talked her out of buying this truly heinous angora sweater. Whoever said orange is the new pink is seriously disturbed."

ELLE WOODS (ACTOR REESE WITHERSPOON) IN *LEGALLY BLOND* (2001)

In December 2002 the *Wall Street Journal* announced that Grunge might be back (Tkacik, 2002). Since 1998, one of the main fashion trends among white, middle-class, and preteen and young teenage girls was the Britney Spears look: bare midriffs, highlighted hair, wide belts, glitter purses, big wedge shoes, and Skechers "energy" sneakers. But in 2002 a new pop star, Avril Lavigne, was rising in the pop charts. Nominated for a 2003 Grammy Award in the "Best New Artist" category, the 17-year-old skater-punk from the small Canadian town of Napanee (2001 population: 15,132), affects a shaggy, unkempt look (Statistics Canada, 2002). She sports worn-out T-shirts, 1970s-style plaid Western shirts with snaps, low-rise blue jeans, baggy pants, undershirts, ties, backpacks, chain wallets, and, for shoes, Converse Chuck Taylors. The style is similar to the Grunge look of the early 1990s, when Nirvana and Pearl Jam were the big stars on MTV and Kurt Cobain was king of the music world.

Why in late 2002 were the glamorous trends of the pop era giving way in one market segment to "neo-Grunge"? Why, in general, do fashion shifts take place? Sociological theory has interesting things to say on this subject (Davis, 1992).

Until the 1960s, the standard sociological approach to explaining the ebb and flow of fashion trends was *functionalist.* In the functionalist view, fashion trends worked like this: Every season, exclusive fashion houses in Paris and, to a lesser extent, Milan, New York, and London would show new styles. Some of the new styles would catch on among the exclusive clientele of Chanel, Dior, Givenchy, and other big-name designers. The main appeal of wearing expensive, new fashions was that wealthy clients could distinguish themselves from people who were less well off. Thus, fashion performed an important social function. By allowing people of different rank to distinguish themselves from one another, fashion helped preserve the ordered layering of society into classes. ("It is an interesting question," wrote 19th-century American writer Henry David Thoreau in *Walden,* "how far [people] would retain their relative rank if they were divested of their clothes.") By the 20th century, thanks to technological advances in clothes manufacturing, it didn't take long for inexpensive knockoffs to reach the market and trickle down to lower classes. New styles then had to be introduced frequently so that fashion could continue to perform its function of helping to maintain an orderly class system. Hence the ebb and flow of fashion.

The functionalist theory was a fairly accurate account of the way fashion trends worked until the 1960s. Then, fashion became more democratic. Paris, Milan, New York, and London are still hugely important fashion centers today. However, new fashion trends are increasingly initiated by lower classes, minority racial and ethnic groups, and people who spurn "high" fashion altogether. Napanee is, after all, pretty far from Paris, and today big-name designers are more likely to be influenced by the inner-city styles of hip-hop than vice versa. New fashions no longer just trickle down from upper classes and a few high-fashion centers. Upper classes are nearly as likely to adopt lower-class fashion trends that emanate from just about anywhere. As a result, the functionalist theory no longer provides a satisfying explanation of fashion cycles.

Some sociologists have turned to conflict theory as an alternative view of the fashion world. Conflict theorists typically view fashion cycles as a means by which industry owners make big profits. Owners introduce new styles and render old styles unfashionable because they make more money when many people are encouraged to buy new clothes often. At the same time, conflict theorists think fashion keeps people distracted from the many social, economic, and political problems that might otherwise incite them to express dissatisfaction with the existing social order and even rebel against it. Conflict theorists, like functionalists, thus believe that fashion helps maintain social stability. Unlike functionalists, however, they argue that social stability bestows advantages on industrial owners at the expense of nonowners.

Conflict theorists have a point. Fashion is a big and profitable business. Owners *do* introduce new styles to make more money. They have, for example, created The Color Marketing Group (known to insiders as the "Color Mafia"), a committee that meets regularly to help change the national palette of color preferences for consumer products. According to one committee member, the Color Mafia makes sure that "the mass media, . . . fashion magazines and catalogs, home shopping shows, and big clothing chains all present the same options" (Mundell, 1993).

Yet the Color Mafia and other influential elements of the fashion industry are not all-powerful. Remember what Elle Woods said after she convinced Cameron Diaz not to buy that heinous angora sweater: "Whoever said orange is the new pink is seriously disturbed." Like many consumers, Elle Woods *rejected* the advice of the fashion industry. And in fact some of the fashion trends initiated by industry owners flop, one of the biggest being the

Britney Spears vs. Avril Lavigne

Reuters NewMedia Inc./Corbis

David Bergman/Corbis

introduction of the midi-dress (with a hemline midway between knee and ankle) in the mid-1970s. Despite a huge ad campaign, most women simply would not buy it.

This points to one of the main problems with the conflict interpretation: It incorrectly makes it seem like fashion decisions are dictated from above. Reality is more complicated. Fashion decisions are made partly by consumers. This idea can best be understood by thinking of clothing as a form of *symbolic interaction,* a sort of wordless "language" that allows us to tell others who we are and learn who they are.

If clothes speak, sociologist Fred Davis has perhaps done the most in recent years to help us see how we can decipher what they say (Davis, 1992). According to Davis, a person's identity is always a work in progress. True, we develop a sense of self as we mature. We come to think of ourselves as members of one or more families, occupations, communities, classes, ethnic and racial groups, and countries. We develop patterns of behavior and belief associated with each of these social categories. Nonetheless, social categories change over time, and so do we as we move through them and as we age. As a result, our identities are always in flux. We often become anxious or insecure about who we are. Clothes help us express our shifting identities. For example, clothes can convey whether you are "straight," sexually available, athletic, conservative, and much else, thus telling others how you want them to see you and indicating the kinds of people with whom you want to associate. At some point you may become less conservative, sexually available, and so forth. Your clothing style is likely to change accordingly. (Of course, the messages you try to send are subject to interpretation and may be misunderstood.) For its part, the fashion industry feeds on the ambiguities within us, investing much effort in trying to discern which new styles might capture current needs for self-expression.

For example, capitalizing on the need for self-expression among many young girls in the late 1990s, Britney Spears hit a chord. Feminist interpretations of the meaning and significance of Britney Spears are especially interesting in this respect because they focus on the gender aspects of fashion.

Traditionally, feminists have thought of fashion as a form of patriarchy, a means by which male dominance is maintained. They have argued that fashion is mainly a female preoccupation. It takes a lot of time and money to choose, buy, and clean clothes. Fashionable clothing is often impractical and uncomfortable, and some of it is even unhealthy. Modern fashion's focus on youth, slenderness, and eroticism diminishes women by turning them into sexual objects, say some feminists. Britney Spears is of interest to traditional feminists because she supposedly helps lower the age at which girls fall under male domination.

In recent years, this traditional feminist view has given way to a feminist interpretation that is more compatible with symbolic interactionism ("Why Britney Spears Matters," 2001). Some feminists now applaud the "girl power" movement that crystallized in 1996 with the release of the Spice Girls' hit single, "Wannabe." They regard Britney Spears as part of that movement. In their judgment, Spears's music, dance routines, and dress style express a self-assuredness and assertiveness that resonate with the less submissive and more independent role that girls are now carving out for themselves. With her kicks, her shadow boxing, and songs like the 2000 single "Stronger," Spears speaks for the *empowerment* of young women. Quite apart from her musical and dancing talent, then, some feminists think that many young girls are wild about Britney Spears because she helps them express their own social and sexual power. Of course, not all young girls agree. Some, like Avril Lavigne, find Spears "phony" and too much of a "showgirl." They seek "more authentic" ways of asserting their identity through fashion (Pascual, 2002). Still, the symbolic interactionist and feminist interpretations of fashion help us see more clearly the ambiguities of identity that underlie the rise of new fashion trends.

Sociology Now™

Learn more about **Applying the Four Theoretical Perspectives** by going through The Sociological Perspective Video Exercise.

Our analysis of fashion shows that each of the four theoretical perspectives—functionalism, conflict theory, symbolic interactionism, and feminism—can clarify different aspects of a sociological problem. This does not mean that each perspective always has equal validity. Often, the interpretations that derive from different theoretical perspectives are incompatible. They offer *competing* interpretations of the same social reality. It is then necessary to do research to determine which perspective works best for the case at hand. Nonetheless, all four theoretical perspectives usefully illuminate some aspects of the social world. We therefore refer to them often in the following pages.

Web Interactive Exercise: Will Spiritual Robots Replace Humans by 2100?

A Sociological Compass

Our summary of the major theoretical perspectives in sociology suggests that the founders of the discipline developed their ideas in an attempt to solve the great sociological puzzle of their time—the causes and consequences of the Industrial Revolution. This raises two interesting questions. What are the great sociological puzzles of *our* time? How are today's sociologists responding to the challenges presented by the social settings in which *we* live? We devote the rest of this book to answering these questions in depth. In the remainder of this chapter, we outline what you can expect to learn from this book.

Sociology Now™

Learn more about **A Sociological Compass** by going through the Sociological Perspective Animation.

It would be wrong to suggest that the research of tens of thousands of sociologists around the world is animated by just a few key issues. Viewed up close, sociology today is a heterogeneous enterprise enlivened by hundreds of theoretical debates, some focused on small issues relevant to particular fields and geographical areas and others focused on big issues that seek to characterize the entire historical era for humanity as a whole.

Among the big issues, two stand out. Perhaps the greatest sociological puzzles of our time are the causes and consequences of the Postindustrial Revolution and globalization. The **Postindustrial Revolution** is the technology-driven shift from manufacturing to service industries—the shift from employment in factories to employment in offices—and the consequences of that shift for nearly all human activities (Bell, 1976; Toffler, 1990). For example, as a result of the Postindustrial Revolution, nonmanual occupations now outnumber manual occupations, and women have been drawn into the system of higher education and the paid labor force in large numbers. This shift has transformed the way we work and study, our standard of living, the way we form families, and much else. **Globalization** is the process by which formerly separate economies, states, and cultures are becoming tied together and people are becoming increasingly aware of their growing interdependence (Giddens, 1990: 64; Guillén, 2000). Especially in recent decades, rapid in-

The **Postindustrial Revolution** refers to the technology-driven shift from manufacturing to service industries and the consequences of that shift for virtually all human activities.

Globalization is the process by which formerly separate economies, states, and cultures are tied together and people become increasingly aware of their growing interdependence.

HOW TO CLONE A HUMAN

If it works in humans as it has in other mammals, cloning will be technically possible, but also terribly inefficient and risky.

According to experts, producing a single viable clone will require scores of volunteers to donate eggs and carry embryos—most of which will have major abnormalities and never come to term. The clones that do survive could suffer more subtle problems that might show up well after birth. Here's how it might be done.

Egg donors

1 Doctors harvest up to 15 eggs each from up to 40 donors who have been injected with fertility drugs. About 400 eggs are produced

Eggs

Person to be cloned

2 Cells are taken from the cloning candidate

TIME Graphic by Joe Lertola

Many of the scientific and technological advances of the postindustrial era are fraught with dilemmas. For example, most medical scientists hope that the cloning of human cells will soon allow them to repair damaged body parts and grow replacement body parts. A small minority supports the cloning of entire human beings, even though it can create serious medical, ethical, and social problems.

creases in the volume of international trade, travel, and communication have broken down the isolation and independence of most countries and people. Also contributing to globalization is the growth of many institutions that bind corporations, companies, and cultures together. These processes have caused people to depend more than ever on people in other countries for products, services, ideas, and even a sense of identity.

Sociologists agree that globalization and postindustrialism promise many exciting opportunities to enhance the quality of life and increase human freedom. However, they also see many social-structural barriers to the realization of that promise. We can summarize both the promise and the barriers by drawing a compass—a sociological compass (Figure 1.5). Each axis of the compass contrasts a promise with the barriers to its real-

Figure Figure 1.5
A Sociological Compass

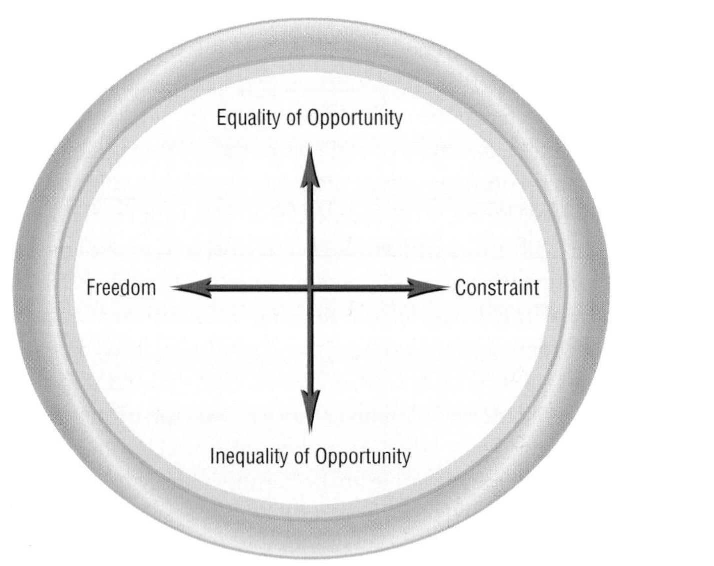

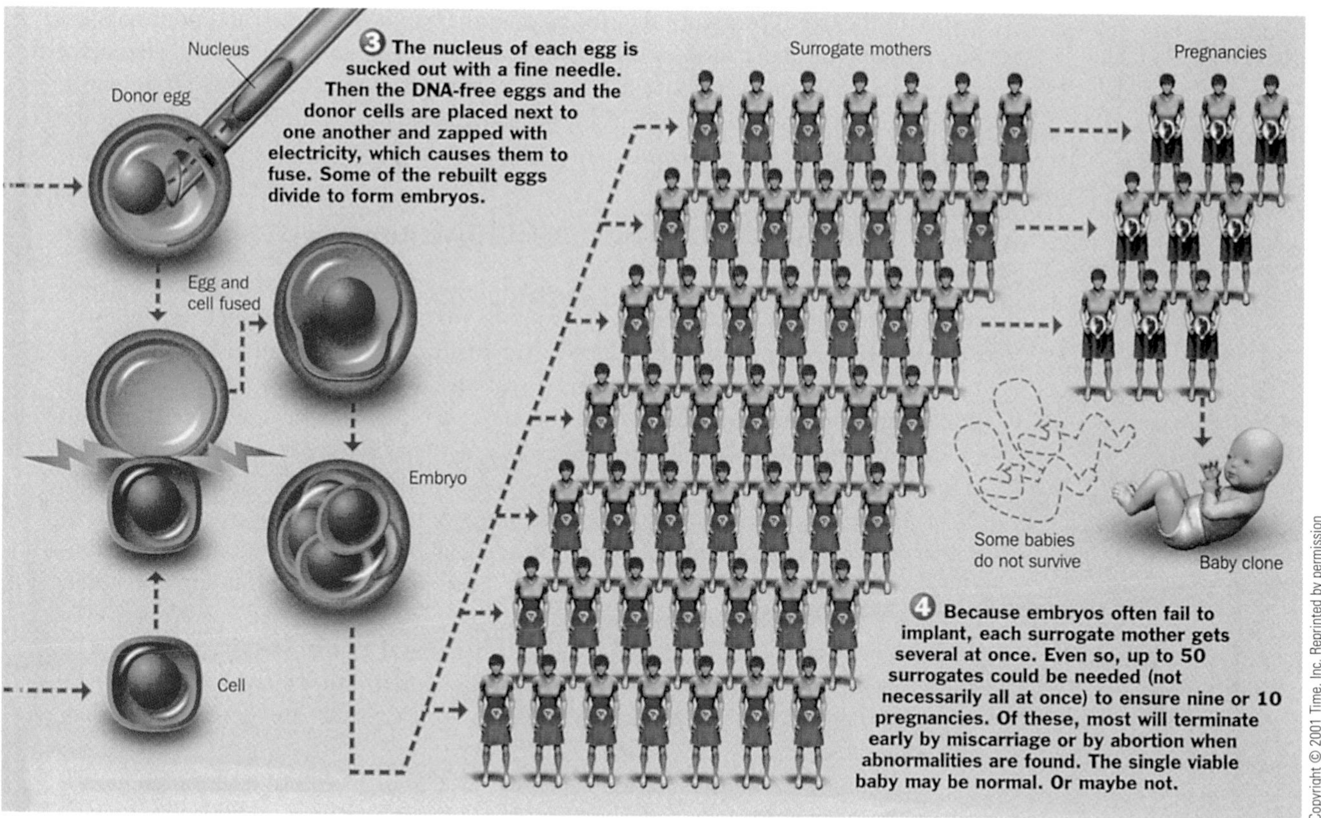

ization. The vertical axis contrasts the promise of equality of opportunity with the barrier of inequality of opportunity. The horizontal axis contrasts the promise of individual freedom with the barrier of constraint on that freedom. Let us consider these axes in more detail, because much of our discussion in the following chapters turns on them.

Equality versus Inequality of Opportunity

Optimists forecast that postindustrialism will provide more opportunities for people to find creative, interesting, challenging, and rewarding work. In addition, the postindustrial era will generate more "equality of opportunity," that is, better chances for *all* people to get an education, influence government policy, and find good jobs.

You will find evidence to support these claims in the following pages. For example, we show that the average standard of living and the number of good jobs are increasing in postindustrial societies such as the United States. Women are making rapid strides in the economy, the education system, and other institutions. Postindustrial societies like the United States are characterized by a decline in discrimination against members of ethnic and racial minorities, while democracy is spreading throughout the world. The desperately poor form a declining percentage of the world's population.

Yet, as you read this book, it will also become clear that all of these seemingly happy stories have a dark underside. For example, it turns out that the number of routine jobs with low pay and few benefits is growing faster than the number of creative, high-paying jobs. Inequality between the wealthiest and poorest Americans has grown in recent decades, as has inequality between the wealthiest and poorest nations. An enormous opportunity gulf still separates women from men. Racism and discrimination are still a part of our world. Our health care system is in crisis just as our population is aging rapidly and most in need of health care. Many of the world's new democracies are only superficially democratic, while Americans and citizens of other postindustrial societies are increasingly cynical about the ability of their political systems to respond to their needs. They are looking for alternative forms of political expression. The absolute number of desperately poor people in the world continues to grow, as does the gap between rich and poor nations. Many people attribute the world's most serious problems to globalization. They have formed organizations and movements—some of them violent—to oppose it. In short, equality of opportunity is an undeniably attractive ideal, but it is unclear whether it is the inevitable outcome of a globalized, postindustrial society.

Individual Freedom versus Individual Constraint

We may say the same about the ideal of freedom. In an earlier era, most people retained their religious, ethnic, racial, and sexual identities for a lifetime, even if they were not particularly comfortable with them. They often remained in social relationships that made them unhappy. One of the major themes of *Sociology: Your Compass for a New World* is that many people are now freer to construct their identities and form social relationships in ways that suit them. To a greater degree than ever before, it is possible to *choose* who you want to be, with whom you want to associate, and how you want to associate with them. The postindustrial and global era frees people from traditional constraints by encouraging virtually instant global communication, international migration, greater acceptance of sexual diversity and a variety of family forms, the growth of ethnically and racially diverse cities, and so forth. For instance, in the past, people often stayed in marriages even if they were dissatisfied with them. Families often involved a father working in the paid labor force and a mother keeping house and raising children without pay. Today, people are freer to end unhappy marriages and create family structures that are more suited to their individual needs.

Again, however, we must face the less rosy aspects of postindustrialism and globalization. In many of the following chapters, we point out how increased freedom is experienced only within certain limits and how social diversity is limited by a strong push to conformity in some spheres of life. For example, we can choose a far wider variety of consumer products than ever before, but consumerism itself increasingly seems a compulsory way of life. Moreover, it is a way of life that threatens the natural environment. Meanwhile, some new technologies, such as surveillance cameras, cause us to modify our behavior and act in more conformist ways. Large, impersonal bureaucracies and standardized products and services dehumanize both staff and customers. The tastes and the profit motive of vast media conglomerates, most of them U.S. owned, govern most of our diverse cultural consumption and arguably threaten the survival of distinctive national cultures. Powerful interests are trying to shore up the traditional nuclear family even though it does not suit some people. As these examples show, the push to uniformity counters the trend toward growing social diversity. Postindustrialism and globalization may make us freer in some ways, but they also place new constraints on us.

Where Do You Fit In?

Our overview of themes in *Sociology: Your Compass for a New World* drives home a point made by Anthony Giddens (1987), renowned British sociologist and adviser to Prime Minister Tony Blair: We live in an era "suspended between extraordinary opportunity . . . and global catastrophe" (166). A whole range of environmental issues, profound inequalities in the wealth of nations and of classes; religious, racial, and ethnic violence; and unsolved problems in the relations between women and men continue to stare us in the face and profoundly affect the quality of our everyday lives.

Despair and apathy are two possible responses to these complex issues, but they are not responses that humans often favor. If it were our nature to give up hope, we would still be sitting around half-naked in the mud outside a cave.

People are more inclined to look for ways of improving their lives, and this period of human history is full of opportunities to do so. We have, for example, advanced to the point where for the first time we have the means to feed and educate everyone in the world. Similarly, it now seems possible to erode some of the inequalities that have always been the major source of human conflict.

Careers in Sociology

Sociology offers useful advice on how to achieve these goals because it is more than just an intellectual exercise. It is also an applied science with practical, everyday uses, especially in the realms of research, teaching, and **public policy,** the creation of laws and regulations by organizations and governments (Box 1.2). That is because sociologists are trained not just to see what is, but to see what is possible.

Perhaps 20,000 people have sociology Ph.D.s in the United States. Hundreds of thousands have sociology M.A.s and B.A.s. To see what you can do with a sociology degree, let's divide people with sociological training into two groups: those with a B.A. in sociology and those with an M.A. or Ph.D. in sociology.

A sociology B.A. improves one's understanding of the diverse social conditions affecting men and women, people with different sexual orientations, and people from different countries, regions, classes, races, and ethnic groups. Therefore, people with a B.A. in sociology tend to be attracted to jobs that require good "people skills" and that are involved in managing and promoting social change (American Sociological Association, 2005; Stephens, 1999). Americans with sociology B.A.s work in:

- Government services: federal, state, and local government planning and management in such areas as housing, labor, transportation, and agriculture
- Publishing, journalism, and public relations: writing, research, editing, and sales
- College administration: admissions, alumni relations, and placement
- Health services: family planning, substance abuse, rehabilitation counseling, health planning, hospital admissions, and insurance
- Nongovernmental organizations (NGOs): union organizing, environmental groups, international development agencies, and political parties
- Teaching: elementary and secondary education (following teacher certification)
- Social services: rehabilitation, case management, recreation, administration, and counseling youth and the elderly
- Community work: social service and nonprofit organizations, child-care and community development agencies
- Corrections: probation and parole
- Business: advertising, marketing and consumer research, insurance, real estate, personnel, training, and sales

Most people with a graduate degree in sociology teach and conduct research in colleges and universities, with research being a more important component of the job in larger and more prestigious institutions. They earn good salaries. In July 2003, average earnings in the United States stood at about $38,000 for all occupations, $49,000 for all white-collar occupations, $67,000 for all college and university professors, and $68,000 for sociology professors alone (U.S. Department of Labor, 2004: 73–4).

Web Research Project: The Uses of Sociology

Many sociologists do not teach. Instead, they conduct research and give policy advice in a wide range of settings outside the system of higher education. On the whole, opportunities for research and policy work are growing faster than teaching opportunities. In the federal government, for example, sociologists are employed as researchers and policy consultants in the Department of Health and Human Services, the National Institute of Aging, the National Institute of Mental Health, the National Institute of Drug Abuse, the Bureau of the Census, the Department of Agriculture, the General Accounting Office, the National Science Foundation, Housing and Urban Development, the Peace Corps, and the Centers for Disease Control and Prevention, among other agencies. Sociologists also conduct research and policy analysis in NGOs including the World Bank, the National Academy of Sciences, the Social Science Research Council, the Children's Defense Fund, Common Cause, and many professional and public-interest associations. In the private sector, you can find sociologists practicing their craft in firms specializing in public opinion polling, management con-

Public policy involves the creation of laws and regulations by organizations and governments.

BOX 1.2

SOCIAL POLICY: WHAT DO YOU THINK?

Are Corporate Scandals a Problem of Individual Ethics or Social Policy?

In 2002 the Sloan School of Management at the Massachusetts Institute of Technology conducted a survey of 600 graduates as part of its 50th anniversary observance. Sixty percent of survey respondents said that honesty, integrity, and ethics are the main factors making a good corporate leader. Most alumni believe that living a moral professional life is more important than pulling in large paychecks and generous perks. "Demonstrate daily that your word is your bond and always try to give something back to your community and those less fortunate than yourself," said Michael Campbell, a 1976 Sloan graduate and currently president of Nova Technology Corporation in Portsmouth, New Hampshire (Goll, 2002).

Unfortunately, the behavior of American executives sometimes fails to reflect Mr. Campbell's high ethical standards. In 2002, for example, investigators uncovered the biggest corporate scandals ever to rock America. Things got so bad that Andy Grove, a founder of Intel, said he was "embarrassed and ashamed" to be a corporate executive in America today (quoted in Hochberg, 2002), and the Wall Street investment firm of Charles Schwab ran a highly defensive television ad claiming to be "almost the opposite of a Wall Street firm." What brought about such astonishing statements was this: Corporate giants, including Enron, WorldCom, Tyco, Global Crossings, and Adelphia Communications, were shown to have engaged in accounting fraud to make their earnings appear higher than they actually were. This practice kept their stock prices artificially high—until investigators made public what was going on, at which time their stock prices took a nosedive. Ordinary stockholders lost hundreds of billions of dollars. Many company employees lost their pensions (because they had been encouraged or compelled to place their retirement funds in company stock) and their jobs (because their companies soon filed for bankruptcy). In contrast, accounting fraud greatly benefited senior executives. They had received stock options as part of their compensation packages. If you own stock options, you can buy company stock whenever you want at a fixed low price, even if the market price for the stock is much higher. Then you can sell your stock at a high price. Senior executives typically exercised their stock options *before* the stocks crashed, netting them billions of dollars in profit.

Can we rely on individual morality or ethics to show senior executives how to behave responsibly, that is, in the long-term interest of their companies and society as a whole? Ethics courses have been taught at nearly all of the country's business schools for years, but as the acting dean of the Haas School of Business (University of California-Berkeley) recently noted, these courses can't "turn sinners into saints. . . . If a company does a lot of crazy stuff but its share price continues to rise, a lot of people will look the other way and not really care whether senior management is behaving ethically or not" (quoted in Goll, 2002).

Because individual ethics often seem weak in the face of greed, some observers have suggested that new public policies, that is, laws and regulations passed by organizations and governments, are required to regulate executive compensation. For example, some people think that the practice of granting stock options to senior executives should be outlawed, stiff jail terms should be imposed on anyone who engages in accounting fraud, and strong legal protection should be offered to anyone who "blows the whistle" on executive wrongdoing.

Sociology helps us see what may appear to be personal issues in the larger context of public policy. Even our tendency to act ethically or unethically is shaped in part by public policy—or the lack of it. Therefore, we review a public policy debate in each chapter of this book. It is good exercise for the sociological imagination, and it will help you gain more control over the forces that shape your life.

sulting, market research, standardized testing, and "evaluation research." Evaluation researchers assess the impact of particular policies and programs before or after they go into effect. They conduct trials, surveys, focus groups, and statistical analyses. They are therefore experts in managing social change. Note too that with a graduate degree and some work experience, opportunities greatly expand for sociologists to work at the managerial and administrative levels.

Sociology also has benefits for people who do not work as sociologists. A sociology degree is excellent preparation for post-B.A. studies in a variety of fields, including law, urban planning, industrial relations, social work, politics, educational administration, and community organizing. We can see the benefits of a sociological education by compiling a list of some of the famous practical idealists who studied sociology in college. That list includes several former heads of state, among them President Fernando Cardoso of Brazil, President Tomas Masaryk of Czechoslovakia, Prime Minister Edward Seaga of Jamaica, and President Ronald Reagan of the United States. Famous Americans with sociology degrees include Senator Barbara Mikulski (Maryland); members of congress Shirley Chisholm (New York), Maxine Waters (Los Angeles), and Tim Holden (Pennsylvania);

Mayors Wellington Webb (Denver), Brett Schundler (Jersey City) and Annette Strauss (Dallas); civil rights leaders Rev. Martin Luther King, Rev. Jesse Jackson, Rev. Ralph Abernathy, and Roy Wilkins; Cardinal Theodore McCarrick (Washington, D.C.); community organizer Saul Alinsky; Nobel Prize winners Jane Addams, Saul Bellow, and Emily Balch; Secretary of Labor Francis Perkins; and Chief Justice Richard Barajas (Texas Supreme Court). Many people regard point guard Steve Nash of the Phoenix Suns as the best team player in professional basketball today. His agent claims he is "the most colorblind person I've ever known" (Robbins, 2005). Arguably, Nash's sociology degree contributed to his performance on the court by helping him better understand the importance of groups and diverse social conditions in shaping human behavior.

Although sociology offers no easy solutions as to how the goal of improving society may be accomplished, it does promise a useful way of understanding our current predicament and seeing possible ways of dealing with it, of leading us a little farther away from the mud outside the cave. You sampled sociology's ability to tie personal troubles to social-structural issues when we discussed suicide. You reviewed the major theoretical perspectives that enable sociologists to connect personal factors with social structures. When we outlined the half-fulfilled promises of postindustrialism and globalization, you saw sociology's ability to provide an understanding of where we are and where we can head.

We frankly admit that the questions we raise in this book are tough to answer. Sharp controversy surrounds them all. However, we are sure that if you try to grapple with them, you will enhance your understanding of your society's, and your own, possibilities. In brief, sociology can help you figure out where you fit into society and how you can make society fit you.

Summary

Sociology Now™

Reviewing is as easy as ➊, ➋, ➌.

➊ Before you do your final review, take the SociologyNow diagnostic quiz to help you identify the areas on which you should concentrate. You will find information on how to access SociologyNow on the foldout at the front of the textbook.

➋ As you review, take advantage of SociologyNow's study aids to help you master the topics in this chapter.

➌ When you are finished with your review, take SociologyNow's post-test to confirm you are ready to move on to the next chapter.

1. What does the sociological study of suicide tell us about society and about sociology?

   Durkheim noted that suicide is an apparently nonsocial and antisocial action that people often but unsuccessfully try to explain psychologically. In contrast, he showed that suicide rates are influenced by the level of social solidarity of the groups to which people belong. This theory suggests that a distinctively *social* realm influences all human behavior.

2. What is the sociological perspective?

   The sociological perspective analyzes the connection between personal troubles and three levels of social structure: microstructures, macrostructures, and global structures.

3. How are values, theories, and research related?

   Values are ideas about what is right and wrong. Values often motivate sociologists to define which problems are worth studying and to make initial assumptions about how to explain sociological phenomena. A theory is a tentative explanation of some aspect of social life. It states how and why specific facts are connected. Research is the process of carefully observing social reality to test the validity of a theory. Sociological theories may be modified and even rejected due to research, but they are often motivated by sociologists' values.

4. What are the major theoretical traditions in sociology?

   Sociology has four major theoretical traditions. Functionalism analyzes how social order is supported by macrostructures. The conflict approach analyzes how social inequality is maintained and challenged. Symbolic interactionism analyzes how meaning is created when people communicate in micro-level settings. Feminism focuses on the social sources of patriarchy in both macro- and micro-level settings.

5. What were the main influences on the rise of sociology?

   The rise of sociology was stimulated by the Scientific, Industrial, and Democratic Revolutions. The Scientific Revolution encouraged the view that sound conclusions about the workings of society must be based on solid evidence, not just speculation. The Democratic Revolution suggested that people are responsible for organizing society and that human intervention can therefore solve social problems. The Industrial Revolution created a host of new and serious social problems that attracted the attention of many social thinkers.

6. **What are the main influences on sociology today and what are the main interests of sociology?**

   The Postindustrial Revolution is the technology-driven shift from manufacturing to service industries. Globalization is the process by which formerly separate economies, states, and cultures are becoming tied together and people are becoming increasingly aware of their growing interdependence. The causes and consequences of postindustrialism and globalization form the great sociological puzzles of our time. The tensions between equality and inequality of opportunity, and between freedom and constraint, are among the chief interests of sociology today.

Questions to Consider

1. Do you think the promise of freedom and equality will be realized in the 21st century? Why or why not?
2. In this chapter, you learned how variations in the level of social solidarity affect the suicide rate. How do you think variations in social solidarity might affect other areas of social life, such as criminal behavior and political protest?
3. Is a science of society possible? If you agree that such a science is possible, what are its advantages over common sense? What are its limitations?

Web Resources

Companion Website for This Book

http://sociology.wadsworth.com

Begin by clicking on the Student Resources section of the website. Choose "Introduction to Sociology" and the Brym and Lie book cover. Next, select the chapter you are currently studying from the pull-down menu. From the Student Resources page you will have easy access to InfoTrac® College Edition, MicroCase Online exercises, additional web links, and many resources to aid you in your study of sociology, including practice tests for each chapter.

Recommended Websites

For an inspiring essay on the practice of the sociological craft by one of America's leading sociologists, see Gary T. Marx's "Of Methods and Manners for Aspiring Sociologists: 36 Moral Imperatives," on the World Wide Web at http://web.mit.edu/gtmarx/www/37moral.html. This article was originally published in The *American Sociologist* 28 (1997): 102–125.

SocioWeb is a comprehensive guide to sociological resources on the World Wide Web at http://www.socioweb.com.

For descriptions of departments of sociology at universities throughout the world, visit http://www.socioweb.com/directory/university-departments.

The American Sociological Association (ASA) is the main professional organization of sociologists in the United States. The ASA website is at http://www.asanet.org. See particularly "Careers in Sociology" at http://www.asanet.org/student/career/homepage.html and "Do You Want to Enhance Your Workforce? Employ the Sociological Advantage" at http://www.asanet.org/pubs/brochures/employhome.html.

CHAPTER 2 How Sociologists Do Research

Richard Lord/PhotoEdit

In this chapter, you will learn that:

- Scientific ideas differ from common sense and other forms of knowledge. Scientific ideas are assessed in the clear light of systematically collected evidence and public scrutiny.
- Sociological research depends not just on the rigorous testing of ideas but also on creative insight. Thus, the objective and subjective phases of inquiry are both important in good research.
- The main methods of collecting sociological data include systematic observations of natural social settings, experiments, surveys, and the analysis of existing documents and official statistics.
- Each data collection method has characteristic strengths and weaknesses. Each method is appropriate for different kinds of research problems.

Science and Experience

OTTFFSSENT

Personal Anecdote

"Okay, Mr. Smarty Pants, see if you can figure this one out." That's how Robert Brym's 11-year-old daughter, Talia, greeted him one day when she came home from school. "I wrote some letters of the alphabet on this sheet of paper. They form a pattern. Take a look at the letters and tell me the pattern."

Robert took the sheet of paper from Talia and smiled confidently. "Like most North Americans, I'd had a lot of experience with this sort of puzzle," says Robert. "For example, most IQ and SAT tests ask you to find patterns in sequences of letters, and you learn certain ways of solving these problems. One of the commonest methods is to see if the 'distance' between adjoining letters stays the same or varies predictably. For example, in the sequence ADGJ, two letters are missing within each adjoining pair. Insert the missing letters and you get the first 10 letters of the alphabet: A(BC)D(EF)G(HI)J.

"This time, however, I was stumped. On the sheet of paper, Talia had written the letters OTTFFSSENT. I tried to use the distance method to solve the problem. Nothing worked. After 10 minutes of head scratching, I gave up."

"The answer's easy," Talia said, clearly pleased at my failure. "Spell out the numbers 1 to 10. The first letter of each word—one, two, three, and so forth—spells OTTFFSSENT. Looks like you're not as smart as you thought. See ya." And with that she bounced off to her room.

"Later that day, it dawned on me that Talia had taught me more than just a puzzle. She had shown me that experience sometimes prevents people from seeing things. My experience in solving letter puzzles by using certain set methods obviously kept me from solving the unusual problem of OTTFFSSENT. Said differently,

Sociology Now™

Reviewing is as easy as ❶, ❷, ❸.

Use SociologyNow to help you make the grade on your next exam. When you are finished reading this chapter, go to the Chapter Summary for instructions on how to make SociologyNow work for you.

Figure 2.1
How Research Filters Perception

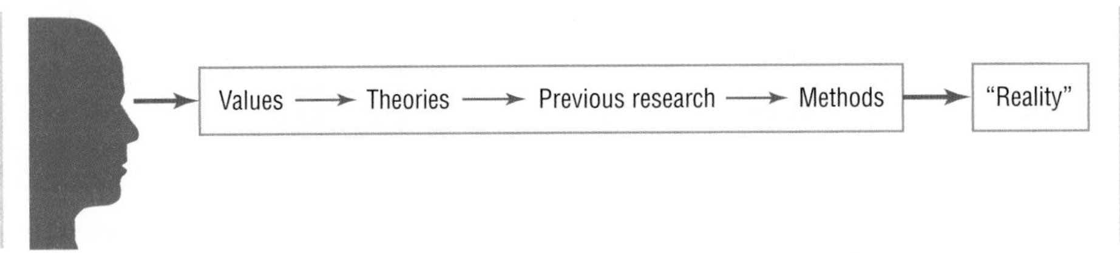

> reality (in this case, a pattern of letters) is not just a thing 'out there' we can learn to perceive 'objectively.' As social scientists have appreciated for more than a century, experience helps determine how we perceive reality, including what patterns we see and whether we are able to see patterns at all" (Hughes, 1967: 16).

The fact that experience filters perceptions of reality is the single biggest problem for sociological research. In sociological research, the filtering occurs in four stages (Figure 2.1). First, as noted in Chapter 1, the real-life experiences and passions of sociologists motivate much research. That is, our *values* often help us decide which problems are worth investigating. These values may reflect the typical outlook of our class, race, gender, region, and historical period. Second, our values lead us to formulate and adopt favored *theories* for interpreting and explaining those problems. Third, sociologists' interpretations are influenced by *previous research,* which we consult to find out what we already know about a subject. And fourth, the *methods* we use to gather data mold our perceptions. The shape of our tools often helps to determine which bits of reality we dig up.

Given that values, theories, previous research, and research methods filter our perceptions, you are right to conclude we can never perceive society in a pure or objective form. What we can do is use techniques of data collection that minimize bias. We can also clearly and publicly describe the filters that influence our perceptions. Doing so enables us to eliminate obvious sources of bias. It also helps others see biases we miss and try to correct for them. The end result is a more accurate perception of reality than is possible by relying exclusively on blind prejudice or common sense.[1]

It is thus clear that a healthy tension pervades all sociological scholarship. On one hand, researchers generally try to be objective to perceive reality as clearly as possible. They follow the rules of scientific method and design data collection techniques to minimize bias. On the other hand, the values and passions that grow out of personal experience are important sources of creativity. As Max Weber said, we choose to study "only those segments of reality which have become significant to us because of their value-relevance" (Weber, 1964 [1949]: 76). So objectivity and subjectivity each play an important role in science, including sociology. Oversimplifying a little, we can say that while objectivity is a reality check, subjectivity leads us to define which aspects of reality are worth checking on in the first place.

Most of this chapter is about the reality check. It explores how sociologists try to adhere to the rules of scientific method. We first contrast scientific and nonscientific thinking. Next, we discuss the steps involved in the sociological research process. We then describe the main methods of gathering sociological data and the decisions that have to be made during the research process. In the final section, we return to the role of subjectivity in research.

Scientific versus Nonscientific Thinking

In science, seeing is believing. In everyday life, believing is seeing. In other words, in everyday life our biases easily influence our observations. This often leads us to draw incorrect conclusions about what we see. In contrast, scientists, including sociologists, develop ways of collecting, observing, and thinking about evidence that minimize their chance of drawing biased conclusions.

---

[1]Some scholars think it is possible to examine data without any preconceived notions and then formulate theories on the basis of this examination. However, they seem to form a small minority (Medawar, 1996: 12–32).

On what basis do you decide statements are true in everyday life? In the following list we describe 10 types of nonscientific thinking (Babbie, 2000 [1973]). As you read about each one, ask yourself how frequently you think unscientifically. If you often think unscientifically, this chapter is for you.

Archivo Iconografico, S.A./Corbis

Perhaps the first major advance in modern medicine took place when doctors stopped using unproven interventions in their treatment of patients. One such intervention involved using leeches to bleed patients, shown here in a medieval drawing.

Sociology Now™

Learn more about **Observation** by going through The Role of the Observer Animation.

1. "Chicken soup helps get rid of a cold. *It worked for my grandparents, and it works for me.*" This statement represents knowledge based on tradition. Although some traditional knowledge is valid (sugar will rot your teeth), some is not (masturbation will not blind you). Science is required to separate valid from invalid knowledge.
2. "Weak magnets can be used to heal many illnesses. *I read all about it in the newspaper.*" This statement represents knowledge based on *authority.* We often think something is true because we read it in an authoritative source or hear it from an expert. But authoritative sources and experts can be wrong. For example, 19th-century Western physicians commonly "bled" their patients with leeches to draw "poisons" from their bodies. This often did more harm than good. As this example suggests, scientists should always question authority to arrive at more valid knowledge.
3. "The car that hit the cyclist was dark brown. I was going for a walk last night when *I saw the accident.*" This statement represents knowledge based on *casual observation.* Unfortunately, we are usually pretty careless observers. That is why good lawyers can often trip up eyewitnesses in courtrooms. Eyewitnesses are rarely certain about what they saw. In general, uncertainty can be reduced by observing in a conscious and deliberate manner and by recording observations. That is just what scientists do.
4. "If you work hard, you can get ahead. *I know because several of my parents' friends started off poor but are now comfortably middle class.*" This statement represents knowledge based on *overgeneralization.* For instance, if you know a few people who started off poor, worked hard, and became rich, you may think that any poor person may become rich if he or she works hard enough. You may not know about the more numerous poor people who work hard and remain poor. Scientists, however, sample cases that are representative of entire populations. This practice enables them to avoid overgeneralization. They also avoid overgeneralization by repeating research, which ensures that they don't draw conclusions from an unusual set of research findings.
5. "I'm right because *I can't think of any contrary cases.*" This statement represents knowledge based on *selective observation.* Sometimes we unconsciously ignore evidence that challenges our firmly held beliefs. Thus, you may actually know some people who work hard but remain poor. However, to maintain your belief that hard work results in wealth, you may keep them out of mind. The scientific requirement that evidence be drawn from representative samples of the population minimizes bias arising from selective observation.
6. "Mr. Smith is poor even though he works hard, but that's because he's disabled. Disabled people are the only *exception to the rule* that if you work hard you can get ahead." This statement represents knowledge based on *qualification.* Qualifications or "exceptions to the rule" are often made in everyday life, and they are made in science, too. The difference is that in everyday life, qualifications are easily accepted as valid, whereas in scientific inquiry they are treated as statements that must be carefully examined in the light of evidence.
7. "If the winner of the Super Bowl comes from the old National Football League, the stock market will go up over the following year. If the Super Bowl winner is from the old American Football League, the stock market will go down over the following year." Remarkably, the tendency described by this statement is factually quite accurate. The stock market has followed the Super Bowl winner 81 percent of the time (Bloom, 2003). However, the view that the Super Bowl winner has an *effect* on the stock mar-

Even Albert Einstein, often hailed as the most intelligent person of the 20th century, sometimes ignored evidence in favor of pet theories. However, the social institution of science, which makes ideas public and subjects them to careful scrutiny, often overcomes such bias.

ket is based on *illogical reasoning.* That is because rare sequences of events sometimes happen just by chance, not because one event causes another. For example, it is possible to flip a coin 10 times and have it come up heads each time. On average, this pattern will occur once every 1,024 times you flip a coin 10 times. It is illogical to believe this phenomenon is a result of anything other than chance. Scientists refrain from such illogical reasoning. They also use statistical techniques to distinguish between events that are probably a result of chance and those that are not. (By the way, the connection between the Super Bowl winner and the direction of the stock market is not just an improbable but chance recurrence of an event. The stock market posts a positive year 70 percent of the time and the traditionally stronger former NFL teams have won 75 percent of Super Bowls, so the two trends substantially overlap.)

8. *"I just can't be wrong."* This statement represents knowledge based on *ego defense.* Even scientists may be passionately committed to the conclusions they reach in their research because they have invested much time, energy, and money in them. It is other scientists—more accurately, the whole institution of science, with its commitment to publishing research results and critically scrutinizing findings—that puts strict limits on ego defense in scientific understanding.
9. *"The matter is settled once and for all."* This statement represents knowledge based on the *premature closure of inquiry,* which involves deciding all the relevant evidence has been gathered on a particular subject. Science, however, is committed to the idea that all theories are only temporarily true. Matters are never settled.
10. *"There must be supernatural forces at work here."* This statement represents knowledge based on *mystification.* When we can find no rational explanation for a phenomenon, we may attribute it to forces that cannot be observed or fully understood. Although such forces may exist, scientists remain skeptical. They are committed to discovering observable causes of observable effects.

Conducting Research

The Research Cycle

Sociological research seeks to overcome the kind of unscientific thinking described in the previous section. It is a cyclical process that involves six steps (Figure 2.2).

First, the sociologist must *formulate a research question.* A research question must be stated so it can be answered by systematically collecting and analyzing sociological data.

Figure 2.2
The Research Cycle

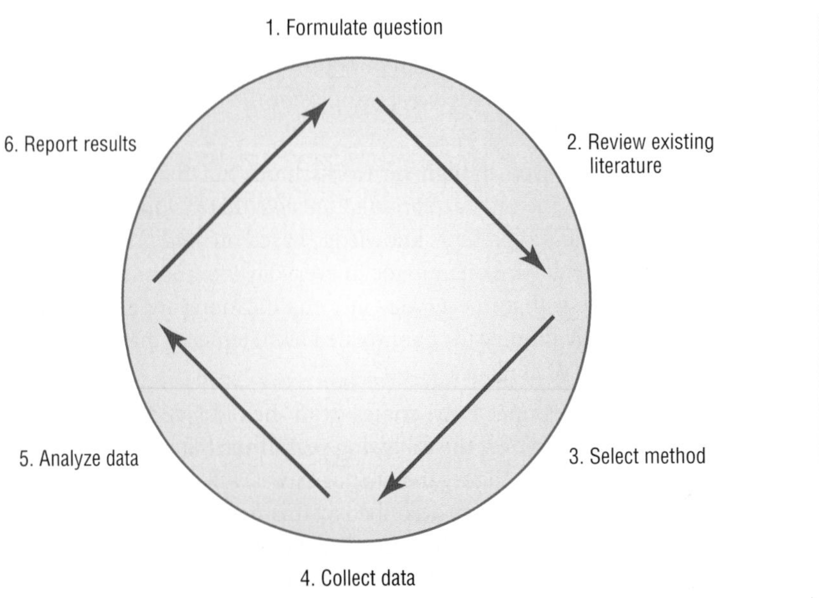

Sociological research cannot determine whether God exists or what is the best political system. Answers to such questions require faith more than evidence. Sociological research can determine why some people are more religious than others and which political systems create more opportunities for higher education. Answers to such questions require evidence more than faith.

Second, the sociologist must *review the existing research literature.* Researchers must elaborate their research questions in the clear light of what other sociologists have already debated and discovered. Why? Because reading the relevant sociological literature stimulates researchers' sociological imaginations, allows them to refine their initial questions, and prevents duplication of effort.

Selecting a research method is the third step in the research cycle. As you will see detailed later in the chapter, each data collection method has strengths and weaknesses. Each method is therefore best suited to studying a different kind of problem. When choosing a method, one must keep these strengths and weaknesses in mind. (In the ideal but, unfortunately, infrequent case, several methods are used simultaneously to study the same problem. This approach can overcome the drawbacks of any single method and increase confidence in one's findings.)

The fourth stage of the research cycle involves *collecting the data* by observing subjects, interviewing them, and reading documents produced by or about them. Many researchers think this stage is the most exciting of the research cycle because it brings them face-to-face with the puzzling sociological reality that so fascinates them.

Other researchers find the fifth step of the research cycle, *analyzing the data,* the most challenging. During data analysis, you can learn things that nobody ever knew before. During this step, data confirm some of your expectations and confound others, requiring you to think creatively about familiar issues, reconsider the relevant theoretical and research literature, and abandon pet ideas.

Data may be qualitative or quantitative—they may take the form of words or numbers, respectively—and different analytical techniques are required for the two types of data. *Qualitative analysis* is usually aimed at understanding patterns of social relationships in small-scale social settings and the meanings people attach to those relationships. Sociologists who engage in qualitative analysis often systematically observe people in their natural settings over extended periods and record their observations in "field notes" that form the basis for their generalizations. *Quantitative analysis* associates specific social qualities (types and degrees of knowledge, attitudes, and behaviors) with discrete quantities (numbers). This procedure allows sociologists to analyze relationships among various features of social life with more precision than is possible using qualitative techniques. For example, they can use statistical methods to determine whether certain features of social life are associated with one another and, if so, then exactly how they are associated.

Sociology Now™

Learn more about **Qualitative Research** by going through the Qualitative Field Research Learning Module.

Research is not much use to the sociological community, the subjects of the research, or the wider society if researchers do not *publicize the results* in a report, a scientific journal, or a book. This is the research cycle's sixth step. Publication serves another important function, too. It allows other sociologists to scrutinize and criticize the research. On this basis, they can formulate new and more sophisticated questions for the next round of research.

Ethical Considerations

Throughout the research cycle, researchers must be mindful of the need to *respect their subjects' rights.* This means, in the first instance, that researchers must do their subjects no harm. This is the subject's right to safety. Second, research subjects must have the right to decide whether their attitudes and behaviors may be revealed to the public and, if so, in what way. This is the subject's right to privacy. Third, researchers cannot use data in a way that allows them to be traced to a particular subject. This is the subject's right to confidentiality. Fourth, subjects must be told how the information they supply will be used. They must also be allowed to judge the degree of personal risk involved in answering questions. This is the subject's right to informed consent.

Carol Wainio. *We Can Be Certain.* 1982. Research involves taking the plunge from speculation to testing ideas against evidence.

Courtesy of Carol Wainio, London, Ontario, Canada

Sociology Now™

Learn more about **Ethical Issues** by going through the Ethical Issues in Social Research Learning Module.

Bearing in mind this thumbnail sketch of the research cycle, let us now examine sociology's major research methods. These methods include the examination of existing documents and official statistics, experiments, surveys, and participant observation. We begin by describing participant-observation and related research methods.

The Main Methods of Sociological Research

Field Research

Eikoku News Digest published a column in the late 1990s by a supposed expert observer of English social life. It advised Japanese residents of England how to act English. One column informed readers that once they hear the words "You must come round for dinner!" they will have been accepted into English social life and must therefore know how to behave appropriately. The column advised readers to obey the following dinner party rules:

1. Arrive 20 minutes late because in England dinner always takes 20 minutes longer to prepare than expected.
2. Don't compliment the décor because in England everyone else's taste in décor is considered dreadful.
3. Bring a cheap wine because the English can't taste the difference.
4. Praise the food by saying "mmmm." The larger the number of "m"s, the greater the compliment.
5. After the meal, don't say it was good. "If someone asks how your dinner was, do not praise the food in detail. They will deduce that the conversation was boring. . . . And do not talk about the interesting conversation you had. They will assume that the food was particularly unpleasant" ("Going Native: Dinner Parties," 2003).

This advice was almost certainly meant as a joke. However, its absurdity serves to caution all students of social life about the dangers of drawing conclusions based on casual observation. By observing others casually, we can easily get things terribly wrong.

Some sociologists undertake **field research,** or research based on the observation of people in their natural settings. The field researcher goes wherever people meet: from the Italian American slum to the intensive care unit of a major hospital; from the white teenage heavy-metal gang in small-town New Jersey and the alternative hard rock scene in Chicago to the audience of a daytime TV talk show (Chambliss, 1996; Gaines, 1990; Grindstaff, 1997; Schippers, 2002; Whyte, 1981 [1943]). However, when they go into the field, researchers come prepared with strategies to avoid getting things terribly wrong.

Field research is research based on the observation of people in their natural settings.

One such strategy is *detached observation.* This approach involves classifying and counting the behavior of interest according to a predetermined scheme. For example, a sociologist wanted to know more about how sex segregation originates. He observed children of different ages playing in summer camps and day-care centers in Canada and Poland. He recorded the sex composition of play groups, the ages at which children started playing in sex-segregated groups, and the kinds of play activities that became sex segregated (Richer, 1990). Similarly, two sociologists wanted to know more about why some American college students don't participate in class discussions. They sat in on classes and recorded the number of students who participated, the number of times they spoke, and the sex of the instructor and the students who spoke (Karp and Yoels, 1976). In both research projects, the sociologists were interested in knowing how young people express gender in everyday behavior, and they knew what to look for before they entered the field.

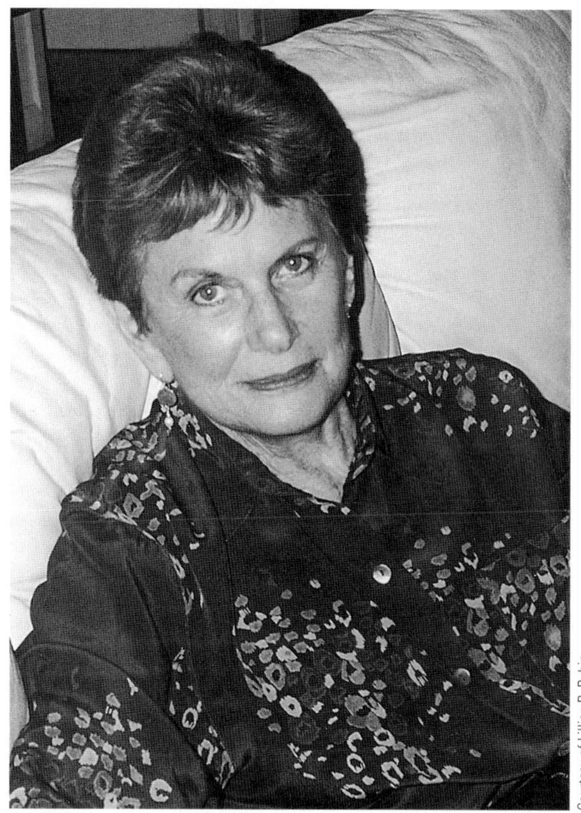

Courtesy of Lillian B. Rubin

Lillian Rubin (University of California at Berkeley) is one of the most talented participant-observation researchers in the United States. One of her most widely acclaimed works is *Families on the Fault Line* (1994), which gives voice to the voiceless by investigating how race, ethnicity, and gender divide the working class. For example, she sensitively captures the ambivalence that working-class men feel about their wives working in the paid labor force. She also weaves her interviews into a revealing story about white ethnic pride as a reaction to the economic upheavals of the 1970s and 1980s and the demands of minorities.

Two main problems confound direct observation. First, the presence of the researcher may itself affect the behavior of the people being observed. Sociologists sometimes call this the *Hawthorne effect,* because researchers at the Western Electric Company's Hawthorne factory in the 1930s claimed to find that workers' productivity increased no matter how they changed their work environment. Productivity increased, they said, just because the researchers were paying attention to the workers.[2] Similarly, some students may participate less in classroom discussion if researchers come in and start taking notes; the presence of the researchers may be intimidating. The second problem with direction observation is that the meaning of the observed behavior may remain obscure to the researcher. The twitch of an eye may be an involuntary muscle contraction, an indication of a secret being kept, a sexual come-on, a parody of someone else twitching an eye, and so forth. We cannot know what it means simply by observing the eye twitch.

To understand what an eye twitch (or any other behavior) means, we must be able to see it in its social context and from the point of view of the people we are observing. Anthropologists and a growing number of sociologists spend months or even years living with people so they can learn their language, values, mannerisms—their entire culture—and develop an intimate understanding of their behavior. This sort of research is called **ethnographic** when it describes the entire way of life of a people (*ethnos* means "nation" or "people" in Greek) (Burawoy et al., 2000; Gille and Riain, 2002; Geertz, 1973).

In rare cases, ethnographic researchers have "gone native," actually giving up their research role and becoming members of the group they are studying. Going native is of no value to the sociological community because it does not result in the publication of new findings. However, going native is worth mentioning because it is the opposite of detached observation. Usually, field researchers develop techniques for collecting data between the two extremes of detached observation and going native. The field method they employ most often is participant observation.

An **ethnographic** researcher spends months or even years living with people to learn their language, values, and mannerisms—their entire culture—and develop an intimate understanding of their behavior.

[2]Subsequent analysis questioned the existence of a productivity effect in the Hawthorne study (Franke and Kaul, 1978). However, the general principle derived from the Hawthorne study—that social science researchers can influence their subjects—is now widely accepted (Webb, Campbell, Schwartz, and Sechrest, 1966).

Participant Observation

Sociologists engage in **participant observation** when they attempt to observe a social milieu objectively *and* take part in the activities of the people they are studying (Lofland and Lofland, 1995 [1971]). By participating in the lives of their subjects, researchers are able to see the world from their subjects' point of view. This method allows them to achieve a deep and sympathetic understanding of people's beliefs, values, and motives. In addition, participant observation requires that sociologists step back and observe their subjects' milieu from an outsider's point of view. This helps them see their subjects more objectively. In participant-observation research, then, tension exists between the goals of subjectivity and of objectivity. As you will see, however, this is a healthy tension that enhances our understanding of many social settings.

The Professional Fence

A well-known example of participant-observation research is Carl B. Klockars's analysis of the professional "fence," a person who buys and sells stolen goods (Klockars, 1974). Among other things, Klockars wanted to understand how criminals can knowingly hurt people and live with the guilt. Are criminals capable of this because they are "sick" or unfeeling? Klockars came to a different conclusion by examining the case of Vincent Swaggi (a pseudonym).

Swaggi buys cheap stolen goods from thieves and then sells them in his store for a handsome profit. His buying is private and patently criminal. His selling is public and, to his customers, appears to be legal. Consequently, Swaggi faces the moral dilemma shared by all criminals to varying degrees. He has to reconcile the different moral codes of the two worlds he straddles, canceling out any feelings of guilt he derives from conventional morality.

"The way I look at it, I'm a businessman," says Swaggi. "Sure I buy hot stuff, but I never stole nothing in my life. Some driver brings me a couple of cartons, though, I ain't gonna turn him away. If I don't buy it, somebody else will. So what's the difference? I might as well make money with him instead of somebody else." Swaggi thus denies responsibility for his actions. He also claims his actions never hurt anyone:

> Did you see the paper yesterday? You figure it out. Last year I musta had $25,000 wortha merchandise from Sears. In this city last year they could'a called it Sears, Roebuck, and Swaggi. Just yesterday I read where Sears just had the biggest year in history, made more money than ever before. Now if I had that much of Sears's stuff, can you imagine how much they musta lost all told? Millions, must be millions. And they still had their biggest year ever. . . . You think they end up losing when they get clipped? Don't you believe it. They're no different from anybody else. If they don't get it back by takin' it off their taxes, they get it back from insurance. Who knows, maybe they do both.

And if he has done a few bad things in his life, then, says Swaggi, so has everyone else. Besides, he's also done a lot of good. In fact, he believes his virtuous acts more than compensate for the skeletons in his closet. Consider, for example, how he managed to protect one of his suppliers and get him a promotion at the same time:

> I had this guy bringin' me radios. Nice little clock radios, sold for $34.95. He worked in the warehouse. Two a day he'd bring me, an' I'd give him fifteen for the both of 'em. Well, after a while he told me his boss was gettin' suspicious 'cause inventory showed a big shortage. . . . So I ask him if anybody else is takin' much stuff. He says a couple of guys do. I tell him to lay off for a while an' the next time he sees one of the other guys take somethin' to tip off the boss. They'll fire the guy an' clear up the shortage. Well he did an' you know what happened? They made my man assistant shipper. Now once a month I get a carton delivered right to my store with my name on it. Clock radios, percolators, waffle irons, anything I want fifty off wholesale (quoted in Klockars, 1974: 135–61).

Participant observation involves carefully observing people's face-to-face interactions and actually participating in their lives over a long period, thus achieving a deep and sympathetic understanding of what motivates them to act in the way they do.

Lessons in Method

Without Klockars's research, we might think that all criminals are able to live with their guilt only because they are pathological or lack empathy for their fellow human beings. But thanks partly to Klockars's research, we know better. We understand that criminals are able to avoid feeling guilty about their actions and get on with their work because they weave a blanket of rationalizations for their criminal activities. These justifications make their illegal activities appear morally acceptable and normal, at least to the criminals themselves. We understand this aspect of criminal activity better because Klockars spent 15 months befriending Swaggi and closely observing him on the job. He interviewed Swaggi for about 400 hours, taking detailed "field notes" most of the time. He then included his descriptions, quotations, and insights in a book that is now considered a minor classic in the sociology of crime and deviance (Klockars, 1974).

Why are observation *and* participation necessary in participant-observation research? Because sociological insight is sharpest when researchers stand both inside and outside the lives of their subjects. Said differently, we see more clearly when we move back and forth between inside and outside.

By immersing themselves in their subjects' world and by learning their language and their culture in depth, insiders are able to experience the world just as their subjects do. Subjectivity can, however, go too far. After all, "natives" are rarely able to see their cultures with much objectivity, and inmates of prisons and mental institutions do not have access to official information about themselves. It is only by regularly standing apart and observing their subjects from the point of view of outsiders that researchers can raise analytical issues and see things to which their subjects are blind or are forbidden from seeing.

Objectivity can also go too far. Observers who try to attain complete objectivity will often not be able to make correct inferences about their subjects' behavior. That is because they cannot fully understand the way their subjects experience the world and cannot ask them about their experiences. Instead, observers who seek complete objectivity must rely only on their own experiences to impute meaning to a social setting. Yet the meaning a situation holds for observers may differ from the meaning it holds for their subjects.

In short, opting for pure observation or pure participation compromises the researcher's ability to see the world sociologically. Instead, participant observation requires the researcher to keep walking a tightrope between the two extremes of objectivity and subjectivity.

It is often difficult for participant-observers to gain access to the groups they wish to study. They must first win the confidence of their subjects, who must feel at ease in the presence of the researcher before they behave naturally. *Reactivity* occurs when the researcher's presence influences the subjects' behavior (Webb, Campbell, Schwartz, and Sechrest, 1966). Reaching a state of nonreactivity requires patience and delicacy on the researcher's part. It took Klockars several months to meet and interview 60 imprisoned thieves before one of them felt comfortable enough to recommend that he contact Swaggi. Klockars had to demonstrate genuine interest in the thieves' activities and convince them he was no threat to them before they opened up to him. Often, sociologists can minimize reactivity by gaining access to a group in stages. At first, researchers may simply attend a group meeting. After a time, they may start to attend more regularly. Then, when their faces are more familiar, they may strike up a conversation with some of the friendlier group members. Only later will they begin to explain their true motivation for attending.

Klockars and Swaggi are both white men. Their similarity made communication between them easier. In contrast, race, gender, class, and age differences sometimes make it difficult, and occasionally even impossible, for some researchers to study some groups. There are many participant-observation studies in which social differences between sociologists and their subjects were overcome and resulted in excellent research (e.g., Liebow, 1967; Stack, 1974). On the other hand, one can scarcely imagine a sociologist nearing retirement conducting participant-observation research on youth gangs or an African American sociologist using this research method to study the Ku Klux Klan.

Most participant-observation studies begin as **exploratory research,** during which researchers have at first only a vague sense of what they are looking for, and perhaps no sense at all of what they will discover in the course of their study. They are equipped only with some hunches based on their own experience and their reading of the relevant research literature. They try, however, to treat these hunches as hypotheses. **Hypotheses** are unverified but testable statements about the phenomena that interest researchers. As they

Exploratory research is an attempt to describe, understand, and develop theory about a social phenomenon in the absence of much previous research on the subject.

A **hypothesis** is an unverified but testable statement about the relationship between two or more variables.

immerse themselves in the life of their subjects, researchers' observations constitute sociological data that allow them to reject, accept, or modify their initial hypotheses. Indeed, researchers often purposely seek out observations that enable them to determine the validity and scope of their hypotheses ("From previous research I know elderly people are generally more religious than young people and that seems to be true in this community too. But does religiosity vary among people of the same age who are rich, middle class, working class, and poor? If so, why? If not, why not?").

Purposively choosing observations results in the creation of a *grounded theory,* which is an explanation of a phenomenon based not on mere speculation but on the controlled scrutiny of one's subjects (Glaser and Straus, 1967).

Methodological Problems

Measurement

The great advantage of participant observation is that it lets researchers get "inside the minds" of their subjects and discover their view of the world in its full complexity. The technique is especially valuable when little is known about the group or phenomenon under investigation and the sociologist is interested in constructing a theory about it. But participant observation has drawbacks, too. To understand them, we must say a few words about measurement in sociology.

When researchers think about the social world, they use mental constructs or concepts such as "race," "class," and "gender." Concepts that can have more than one value are called **variables.** Height and wealth are variables. Perhaps less obviously, affection and perceived beauty are variables, too. Just as one can be 5′ 7″ or 6′ 2″, rich or poor, one can be passionately in love with, or indifferent to, the girl next door on the grounds that she is beautiful or plain. In each case, we know we are dealing with a variable because height, wealth, affection, and perceived beauty can take different values.

Sociology Now™

Learn more about **Variables** by going through the Understanding Variables Learning Module.

Once researchers identify the variables that interest them, they must decide which real-world observations correspond to each variable. Should "class," for example, be measured by determining people's annual income? Or should it be measured by determining their accumulated wealth, years of formal education, or some combination of these or other indicators of rank? **Operationalization** is the act of deciding which observations link to which variables.

Sociological variables can sometimes be measured by casual observation. It is usually pretty easy to tell whether someone is a man or a woman, and participant-observers can learn a great deal more about their subjects through extended discussion and careful observation. When researchers find out how much money their subjects earn, how satisfied they are with their marriages, whether they have ever been the victims of a criminal act, and so forth, they are measuring the values of the sociological variables embedded in their hypotheses.

Typically, researchers must establish criteria for assigning values to variables. At exactly what level of annual income can someone be considered "upper class"? What are the precise characteristics of settlements that allow them to be characterized as "urban"? What features of a person permit us to say she is a "leader"? Answers to such questions all involve measurement decisions.

Here we confront a big problem. In any given research project, participant-observers typically work alone and usually investigate only one group or one type of group. Thus, when we read their research results we must be convinced of three things if we are to accept their findings: (1) The findings extend beyond the single case examined; (2) the researcher's interpretations are accurate; and (3) another researcher would interpret things in the same way. Let us examine each of these points in turn (Figure 2.3).

- A **variable** is a concept that can take on more than one value.
- **Operationalization** is the procedure by which researchers establish criteria for assigning observations to variables.
- **Reliability** is the degree to which a measurement procedure yields consistent results.

Reliability

Would another researcher interpret or measure things in the same way? This is the problem of **reliability.** If a measurement procedure repeatedly yields consistent results, we consider it reliable. However, in the case of participant observation, usually only one per-

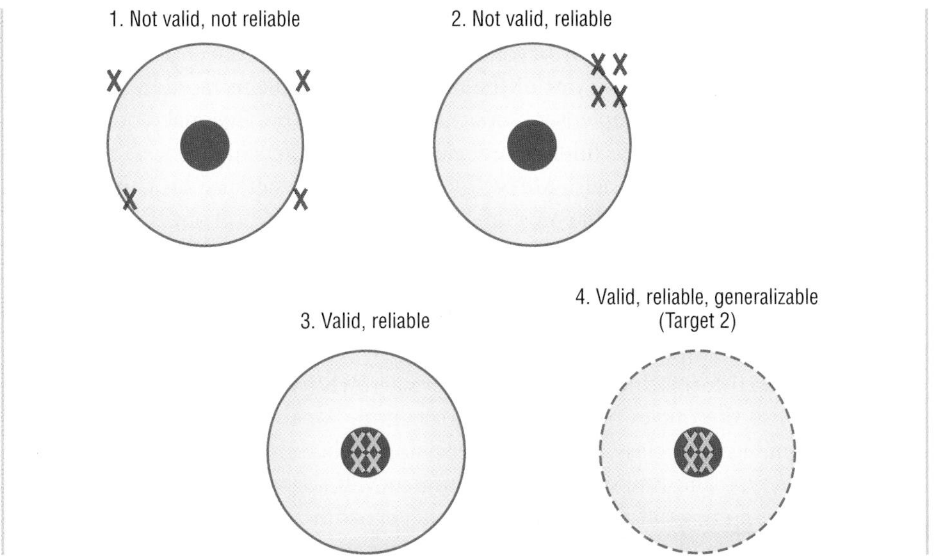

Figure 2.3
Measurement as Target Practice: Validity, Reliability, and Generalizability Compared
Validity, reliability, and generalizability may be explained by drawing an analogy between measuring a variable and firing at a bull's-eye. In case 1, above, shots (measures) are far apart (not reliable) and far from the bull's-eye (not valid). In case 2, shots are close to each other (reliable) but far from the bull's-eye (not valid). In case 3, shots are close to the bull's-eye (valid) and close to each other (reliable). In case 4, we use a second target. Our shots are again close to each other (reliable) and close to the bull's-eye (valid). Because our measures are valid and reliable in cases 3 and 4, we conclude that our results are generalizable.

son does the measuring in only one setting. Therefore, we have no way of knowing whether repeating the procedures would yield consistent results.

Validity

Are the researcher's interpretations accurate? This is the problem of **validity.** If a measurement procedure measures exactly what it is supposed to, then it is valid. All valid measures are reliable. However, not all reliable measures are valid. Measuring a person's shoe with a ruler may give us a reliable indicator of that individual's shoe size because the ruler repeatedly yields the same results. However, regardless of consistency, shoe size as measured by a ruler is an invalid measure of a person's annual income. Similarly, you may think you are measuring annual income by asking people how much they earn. Another interviewer at another time may get exactly the same result when posing the same question. Despite this reliability, however, respondents may understate their true income. (A *respondent* is a person who answers the researcher's questions.) Our measure of annual income may therefore lack validity. Perfectly consistent measures may, in other words, have little truth-value.

Participant-observers have every right to think they are on solid ground when it comes to the question of validity. If anyone can tell whether respondents are understating their true income, surely it is someone who has spent months or even years getting to know everything about their lifestyle. Still, doubts may arise if the criteria used by the participant-observers to assess the validity of their measures are all *internal* to the settings they are investigating. Our confidence in the validity of researchers' measures increases if we are able to use *external* validation criteria. Consider age. Asking people their age is one way of determining how old they are. The problem with this measure is that people tend to exaggerate their age when they are young and minimize it when they are old. A more valid way of determining people's age is to ask them about their year of birth. Year of birth is generally reported more accurately than responses to the question "How old are you?" It is therefore a more valid measure of age. A still more valid measure of age can be found in the "year of birth" entry on people's birth certificates. The point is that validity increases if we have some external check on our measure of age.

Sociology Now™

Learn more about **Measurement** by going through the Levels of Measurement Learning Module.

Web Interactive Exercise: Are IQ and SAT Tests Valid?

Validity is the degree to which a measure actually measures what it is intended to measure.

Generalizability and Causality

Do the research findings apply beyond the specific case examined? This is the problem of **generalizability,** and it is one of the most serious problems faced by participant-observation studies. For example, Klockars studied just one professional fence in depth. Can we safely conclude that his findings are relevant to all professional fences? Do we dare apply his insights to all criminals? Are we foolhardy if we generalize his conclusions to nearly all of us on the grounds that most of us commit deviant acts at one time or another and must deal with feelings of guilt? None of this is clear from Klockars's research, nor are questions of generalizability clearly answered by many participant-observation studies, because they usually are studies of single cases.

Related to the issue of generalizability is that of **causality,** the analysis of causes and their effects. Information on how widely or narrowly a research finding applies can help us establish the causes of a social phenomenon. For instance, we might want to know how gender, race, class, parental supervision, police surveillance, and other factors shape the type and rate of juvenile delinquency. If so, we require information on types and rates of criminal activity among teenagers with a variety of social characteristics and in a variety of social settings. A participant-observation study of crime is unlikely to provide that sort of information; it is more likely to clarify the process by which a specific group of people in a single setting learn to become criminals. Indeed, researchers who conduct participant-observation studies tend not to think in somewhat mechanical, cause-and-effect terms at all.[3] They prefer instead to view their subjects as engaged in a fluid process of social interaction. As a result, participant observation is not the preferred method for discovering the general causes of social phenomena.

In sum, participant-observation research has both strengths and weaknesses. The technique is especially useful in exploratory research, constructing grounded theory, creating internally valid measures, and developing a sympathetic understanding of the way people see the world. It is often deficient when it comes to establishing reliability, generalizability, and causality. As you will soon learn, these are the strengths of surveys and (with the exception of generalizability) experiments.

Only a small percentage of sociologists conduct experiments. Nonetheless, experiments are important because they set certain standards that other more popular methods try to match. We can show this by discussing experiments concerning the effects of TV on real-world violence.

Experiments

In the mid-1960s, about 15 years after the introduction of commercial TV in the United States, rates of violent crime began to increase dramatically. Some people were not surprised. The first generation of American children exposed to high levels of TV violence virtually from birth had reached their midteens. TV violence, some commentators said, legitimized violence in the real world, making it seem increasingly normal and acceptable. As a result, they concluded, American teenagers in the 1960s and subsequent decades were much more likely than pre-1960s teens to commit violent acts.

Social scientists soon started investigating the connection between TV and real-world violence using experimental methods. An **experiment** is a carefully controlled artificial situation that allows researchers to isolate hypothesized causes and measure their effects precisely (Campbell and Stanley, 1963). Experiments use a special procedure called **randomization** to create two similar groups. Randomization involves assigning individuals to the groups by chance processes; the hypothesized cause is then introduced to only one of

- **Generalizability** exists when research findings apply beyond the specific case examined.
- **Causality** refers to the analysis of causes and their effects.
- An **experiment** is a carefully controlled artificial situation that allows researchers to isolate hypothesized causes and measure their effects precisely.
- **Random** means "by chance"—for example, having an equal and nonzero probability of being sampled. **Randomization** involves assigning individuals to groups by chance processes.

[3] On philosophical grounds, some researchers avoid the terms "cause" and "effect." We use these terms because they are widely accepted and easy to understand. Moreover, we do not want to introduce philosophical complications in an elementary treatment of the subject.

the groups. By comparing the state of the two groups before and after only one of the groups has been exposed to the hypothesized cause, an experiment can determine whether the presumed cause has the predicted effect.

Here is how an experiment on the effects of TV violence on aggressive behavior might work:

1. *Selection of subjects.* Researchers advertise in local newspapers for parents willing to allow their children to act as research subjects. Researchers then select 50 children for the experiment.
2. *Random assignment of subjects to experimental and control groups.* At random, each child draws a number from 1 to 50 from a box. The researchers assign children who draw odd numbers to the **experimental group.** This group will be exposed to a violent TV program during the experiment. They assign children who draw even numbers to the **control group.** This group will not be exposed to a violent TV program during the experiment.
3. Note that randomization and repetition make the experimental and control groups similar. That is, by assigning subjects to the two groups using a chance process and repeating the experiment many times, researchers ensure that the experimental and control groups are likely to have the same proportion of boys and girls, members of different races, children highly motivated to participate in the study, and so forth. Random assignment eliminates bias by allowing a chance process and only a chance process to decide which group each child is assigned to.
4. *Measurement of dependent variable in experimental and control groups.* Researchers put small groups of children in a room and give them toys to play with. They observe the children through a one-way mirror, rating each child in terms of the aggressiveness of his or her play. This rating is the child's pretest score on the **dependent variable,** aggressive behavior. The dependent variable is the effect in any cause-and-effect relationship.
5. *Introduction of independent variable to experimental group.* The researchers show children in the experimental group an hour-long TV show in which many violent and aggressive acts take place. They do not show the film to children in the control group. In this experiment, the violent TV show is the **independent variable.** The independent variable is the presumed cause in any cause-and-effect relationship.
6. *Remeasurement of dependent variable in experimental and control groups.* Immediately after the children see the TV show, the researchers again observe the children in both groups at play. Each child's play is given a second aggressiveness rating-the posttest score.
7. *Assessment of experimental effect.* Posttest minus pretest scores are calculated for both the experimental and control groups. If the posttest minus pretest score for the experimental group is significantly greater than the posttest minus pretest score for the control group, the researchers conclude that the independent variable (watching violent TV) has a significant effect on the dependent variable (aggressive behavior). This conclusion is warranted because the introduction of the independent variable is the only difference between the experimental and control groups.

As this example shows, an experiment is a precision instrument for isolating the single cause of theoretical interest and measuring its effect in an exact and repeatable way. But high reliability and the ability to establish causality come at a steep price. Cynics sometimes say experimental sociology allows researchers to know more and more about less and less. Many sociologists argue that experiments are highly artificial situations. They believe that removing people from their natural social settings usually lowers the validity of one's findings. These misgivings are evident in experimental studies of the effects of TV violence.

Sociology Now™

Learn more about **Experiments** by going through the Independent and Dependent Variables Animation.

An **experimental group** in an experiment is the group that is exposed to the independent variable.

A **control group** in an experiment is the group that is not exposed to the independent variable.

A **dependent variable** is the presumed effect in a cause-and-effect relationship.

An **independent variable** is the presumed cause in a cause-and-effect relationship.

David Tumely/Corbis

When children fight at home, an adult is often present to intervene. By repeatedly separating the children and not sanctioning their aggressive behavior, the adult can teach them that fighting is unacceptable. In contrast, experiments on the effect of TV on aggressive behavior lack validity, in part because they may sanction violence and may even encourage it.

Experiments on TV Violence

The author of a recent review of the English-language literature found that 55 percent of the laboratory experiments conducted to date offer no support for the view that TV violence causes aggressive behavior. Some 16 percent of the experiments offer mixed support and just 28 percent offer support. The most supportive studies tend to use questionable measures of aggression. Supportive studies that focused on children had only about one-fifth as many subjects as those that did not support the argument (Freedman, 2002: 62, 67).

Only a small minority of laboratory experiments show that watching violent TV increases violent behavior in the short term. Moreover, in the real world, violent behavior usually means attempting to harm another person physically. Shouting, hitting a doll, or kicking a toy is just not the same thing. In fact, such acts may enable children to relieve frustrations in a fantasy world, thus lowering their chance of acting violently in the real world. What is more, in a laboratory situation, aggressive behavior may be encouraged because it is legitimized. Simply showing a violent TV program may suggest to subjects how the experimenter expects them to behave during the experiment. Subjects who are influenced by the prestige of the researcher and the scientific nature of the experiment compound this problem. They will try to do what is expected of them in order not to appear poorly adjusted. Finally, aggressive behavior is not punished or controlled in the laboratory setting as it is in the real world. If a boy watching *Power Rangers* stands up and delivers a karate kick to his younger brother, a parent is likely to take action to prevent a recurrence. Such action teaches the boy not to engage in aggressive behavior. This deterrence does not happen in the lab, where the lack of disciplinary control may facilitate unrealistically high levels of aggression (Felson, 1996; Freedman, 2002).

Field and Natural Experiments

In an effort to overcome the validity problem yet retain some of the benefits of experimental design, researchers have conducted experiments in natural settings. In such experiments, researchers forgo strict randomization of subjects. They compare groups that are already quite similar. They either introduce the independent variable themselves (this is called a *field experiment*) or observe what happens when the independent variable is introduced to one of the groups in the normal course of social life (this is called a *natural experiment*).

Some field experiments on media effects compare boys in institutionalized settings. The researchers expose half the boys to violent TV programming. Measures of aggressiveness taken before and after the introduction of violent programming allow researchers to calculate its effect on behavior. Other natural experiments compare rates of aggressive behavior in towns with and without TV service. A recent reanalysis of the 10 most rigorous field experiments found that only three of them yielded results even slightly supporting the view that TV violence causes aggression (Freedman, 2002: 106).

Jeff Greenberg/PhotoEdit

Researchers collect information using surveys by asking people in a representative sample a set of identical questions. People interviewed on a downtown street corner do *not* constitute a representative sample of American adults. That is because the sample does not include people who live outside the urban core, it underestimates the number of elderly and disabled people, it does not take into account regional diversity, and so forth.

In part because of the validity problems noted in the preceding section, researchers have not demonstrated that TV violence generally encourages violent behavior. Some researchers conclude that TV violence may have a small effect on a small percentage of viewers; others conclude that the effects are too weak to be detected (Felson, 1996: 123; Freedman, 2002: 200–1).

The extent of the effect is unclear partly because the experimental method makes it difficult to generalize from the specific groups studied to the entire population. The subjects of an experiment on media effects may be white, middle-class people from a college town in the Midwest who read newspaper ads and are in a position to take a day off to participate in the experiment. Such a group is hardly representative of the American population as a whole. But experimentalists are rarely concerned that their subjects are representative of an entire population. As we will see in the following section, one of the strong points of surveys is that they allow us to make safer generalizations.

Surveys

Sampling

Surveys are part of the fabric of everyday life in America. You see surveys in action when a major TV network conducts a poll to discover the percentage of Americans who approve of the President's performance, when someone phones to ask about your taste in breakfast cereal, and when an advice columnist asks her readers, "If you had to do it over again, would you have children?" In every survey, people are asked questions about their knowledge, attitudes, or behavior, in either a face-to-face interview, telephone interview, or paper-and-pencil format.

Remarkably, advice columnist Ann Landers found that fully 70 percent of parents would not have children if they could make the choice again. She ran a shocking headline saying so. Should we have confidence in her finding? Hardly. As the letters from her readers indicated, many of the people who answered her question were angry with their children. All 10,000 respondents felt at least strongly enough about the issue to take the trouble to mail in their replies at their own expense. Like all survey researchers, Ann Landers aimed to study part of a group—a **sample**—to learn about the whole group—the **population** (in this case, all American parents). The problem is that she received replies from a *voluntary response sample,* a group of people who chose *themselves* in response to a general appeal. People who choose themselves are unlikely to be representative of the population of interest. In contrast, a *representative sample* is a group of people chosen so their characteristics closely match those of the population of interest. The difference in the quality of knowledge we can derive from the two types of samples cannot be overstated. Thus, a few months after Ann Landers conducted her poll, a scientific survey based on a representative sample found that 91 percent of American parents *would* have children again (Moore, 1995: 178).

How can survey researchers draw a representative sample? You might think that setting yourself up in a public place like a shopping mall and asking willing passersby to answer some questions would work. However, this sort of *convenience sample,* which chooses the people who are easiest to reach, is also highly unlikely to be representative. Most peo-

A **survey** asks people questions about their knowledge, attitudes, or behavior, in either a face-to-face interview, telephone interview, or paper-and-pencil format.

A **sample** is the part of the population of research interest that is selected for analysis.

A **population** is the entire group about which a researcher wishes to generalize.

AP/Wide World Photos

Homeless people may be interested in public policy, but public policy will ignore them if they are not counted in the census.

ple who go to malls earn above-average income. Moreover, a larger proportion of homemakers, retired people, and teenagers visit malls than can be found in the American population as a whole. Convenience samples are almost always unrepresentative.

To draw a representative sample, respondents cannot select themselves, as in the Ann Landers case. Nor can the researcher choose respondents, as in the mall example. Instead, respondents must be chosen at random, and an individual's chance of being chosen must be known and greater than zero. A sample with these characteristics is known as a **probability sample.** To draw a probability sample, you first need a *sampling frame,* which is a list of all the people in the population of interest. You also need a randomizing method, which is a way of ensuring that every person in the sampling frame has a known and nonzero chance of being selected.

Up-to-date membership lists of organizations are useful sampling frames if you want to survey members of organizations. But if you want to investigate, say, the religious beliefs of Americans, then the membership lists of places of worship are inadequate because many Americans do not belong to such institutions. In such cases, you might turn to another frequently used sampling frame, the telephone directory. The telephone directory is now available for the entire country on CD-ROM. However, even the telephone directory lacks the names and addresses of some poor and homeless people (who do not have phones) and some rich people (who have unlisted phone numbers). Computer programs are available that dial residential phone numbers at random, including unlisted numbers. However, that still leaves some households that will be excluded from any survey relying on the telephone directory as a sampling frame.

As the example of the telephone directory shows, few sampling frames are perfect. Even one of the largest and most expensive surveys in the world, the U.S. census, missed an estimated 2.1 percent of the population in 1990. Although minority groups composed only about a quarter of the U.S population, members of minority groups composed half of the undercounted population because most of the undercount was in poor sections of big cities. For African Americans, the undercount was 4.8 percent. The figure for Native Americans was 5.0 percent. Some 5.2 percent of Hispanic Americans were not counted in the 1990 census (Anderson and Feinberg, 2000: 90) (Box 2.1). Nevertheless, researchers maximize the accuracy of their generalizations by using the least biased sampling frames available and adjusting their analyses and conclusions to take account of known sampling bias.

Once a sampling frame has been chosen or created, individuals must be selected by a chance process. One way to do this is by picking, say, the 10th person on your list and then every 20th (or 30th, or 100th) person after that, depending on how many people you need in your sample. A second method is to assign the number 1 to the first person in the sampling frame, the number 2 to the next person, and so on. Then, you create a separate list of random numbers by using a computer or consulting a table of random numbers, which you can find at the back of almost any elementary statistics book. Your list of random numbers should have as many entries as the number of people you want in your sample. The individuals whose assigned numbers correspond to the list of random numbers are the people in your sample.

Sociology Now™

Learn more about **Sampling** by going through the Types of Sample Designs Learning Module.

Sample Size and Statistical Significance

How many respondents do you need in a sample? That depends on how much inaccuracy you are willing to tolerate. Large samples give more precise results than small samples. For most sociological purposes, however, a random sample of 1,500 people will give acceptably accurate results, even if the population of interest is the entire adult population of the United States. More precisely, if you draw 20 random samples of 1,500

In a **probability sample**, the units have a known and nonzero chance of being selected.

BOX 2.1

SOCIAL POLICY: WHAT DO YOU THINK?

The Politics of the U.S. Census

When John Lie was teaching at the University of Oregon in the early 1990s, several Spanish-speaking sociologists he knew were busy counting the number of Hispanic (or Latino) migrant agricultural workers in the state's rural areas. Why is it important for the government to pay sociologists to count Spanish-speaking migrant workers or, for that matter, other residents of the United States?

Counting the number of Americans may seem a simple matter. Most people do not have trouble counting the number of people in a classroom, so what is the big deal about conducting a national census? Well, imagine counting the number of people at a rock concert or a major sports event. Not only would it take a long time to count them one by one, but the crowd is constantly in motion, making it still harder to count accurately. If it is difficult to count thousands of people in one place, you can appreciate how hard it is to count nearly 300 million Americans. At a given time, many Americans are on the move. They may be traveling or living abroad. They may be driving or flying within the United States, camping in the Sierras, or stuck in an elevator in New York. It is little short of a wonder, then, that the U.S. Census Bureau succeeded in counting about 98.5 percent of the nation's population in 2000.

Why do we need to know how many Americans there are? The census is important because many important decisions are based on population figures. For example, the number of congressional seats is decided on the basis of population figures. So is the amount of money each state government receives from the federal government. Decisions about everything from school budgets to highway construction rely on census counts. Thus, if census figures are lower than the actual population in a particular area, it can be a serious liability for the people living there.

Census undercounting is especially problematic in the case of racial and ethnic minorities. For example, some recent Hispanic American immigrants may not get counted because they have difficulty communicating with census takers who do not speak Spanish. Other Hispanic Americans may worry that census takers are undercover police agents looking for undocumented migrants. Because of these problems, the Spanish-speaking sociologists working in Oregon in the early 1990s were trying to arrive at a more accurate count of the Hispanic American population in the state.

Because so many political decisions are based on the census count, the U.S. Census Bureau faces much political controversy. For example, some people think it does not matter much if approximately 1.5 percent of the United States population (more than 4 million people in 2000) are not counted. They believe it is a waste of money to create a more accurate census. Other people say that a more accurate census is important because the groups that are undercounted are, in effect, the victims of discrimination. Thus, a 2001 U.S. Census Monitoring Board study found that 31 states, and the District of Columbia, would lose $4.1 billion in federal funding as a result of the census undercount. One of the biggest losers was New York City (Scott, 2001).

Among people who think a more accurate census is necessary, controversy exists about how best to deal with undercounting. Democrats favor using sampling techniques to estimate the number and characteristics of the undercounted. They cite the findings of a blue-ribbon panel appointed by the National Academy of Sciences in the 1990s. The panel determined that sampling could produce more accurate results than the current census. However, on January 25, 1999, the Supreme Court rejected a federal plan to supplement the census with a sample. Republicans oppose sampling, fearing it could be manipulated to give desired results. Instead, some Republicans have suggested that census information be collected in as many as 33 languages, including English braille. They have also suggested that more money be spent on marketing and outreach to increase the response rate (Anderson, 1999; Anderson and Feinberg, 2000; Choldin 1994; Democratic National Committee, 1998; "Supreme Court," 1999).

Aside from Hispanic Americans, what other groups of people are particularly susceptible to census undercounting in your opinion? If you were in charge of the U.S. Census Bureau, what steps would you take to ensure a more accurate count of the population? Do you think the characteristics of the census takers in the field, particularly their race and ethnicity, affect the census count? If so, how? Do you think that supplementing the census with a sample survey could increase the accuracy of the census? Does sampling introduce more risk of political manipulation than a straight count?

individuals each, 19 of them will be accurate within 2.5 percent. Imagine, for example, that 50 percent of a random sample of 1,500 respondents say they think the President of the United States is doing a good job. We can be reasonably confident that only 1 in 20 random samples of that size will yield results less than 47.5 percent or more than 52.5 percent. This finding leads us to conclude that the actual percentage of people in the *population* who think the President is doing a good job is probably between 47.5 percent and 52.5 percent. When we read that a finding is **statistically significant,** it usually means we can expect similar findings in 19 out of 20 samples of the same size. Said differently, researchers in the social sciences are conventionally prepared to tolerate a 5 per-

Statistical significance exists when a finding is unlikely to occur by chance, usually in 19 out of every 20 samples of the same size.

cent chance that the characteristics of a population are actually different from the characteristics of their sample ($\frac{1}{20}$ = 5%).

In sum, probability sampling enables us to conduct surveys that permit us to generalize from a part (the sample) to the whole (the population) within known margins of error. Now let us consider the validity of survey data.

Types of Surveys and Interviews

We can conduct a survey in three main ways. Sometimes, a *self-administered questionnaire* is used. For example, a form containing questions and permitted responses may be mailed to the respondent and returned to the researcher through the mail system. The main advantage of this method is its low cost. One drawback of this method is an unacceptably low *response rate.* (The response rate is the number of people who answer the questionnaire divided by the number of people asked to do so, expressed as a percentage.) Another is that if you mail questionnaires, an interviewer is not present to explain problematic questions and response options to the respondent. *Face-to-face interviews* are therefore generally preferred over mailed questionnaires. In this type of survey, questions and allowable responses are presented to the respondent by the interviewer during a meeting. However, training interviewers and sending them to conduct interviews is expensive. That is why *telephone interviews* have become increasingly popular over the past two or three decades. They can elicit relatively high response rates and are relatively inexpensive to administer.

Survey Questions and Validity

Sociology Now™

Learn more about **Questionnaire Construction** by going through the Questionnaire Construction Coached Problem.

Questionnaires can contain two types of questions. A *closed-ended question* provides the respondent with a list of permitted answers. Each answer is given a numerical code so that the data can later be easily input into a computer for statistical analysis. *Open-ended questions* allow respondents to answer questions in their own words. They are particularly useful in exploratory research, where the researcher does not have enough knowledge to create a meaningful and complete list of possible answers. Open-ended questions are more time-consuming to analyze than closed-ended questions, although computer programs for analyzing text make the task much easier.

Researchers want the answers elicited by surveys to be valid, to actually measure what they are supposed to. To maximize validity, researchers must guard against several dangers. We have already considered one threat to validity in survey research: *undercounting* some categories of the population because of an imperfect sampling frame. Even if an individual is contacted, however, he or she may refuse to participate in the survey. This is the second threat to validity in survey research: *nonresponse.* If nonrespondents differ from respondents in ways that are relevant to the research topic, then the conclusions one draws from the survey may be in jeopardy. For instance, some alcoholics may not want to participate in a survey on alcohol consumption because they regard the topic as sensitive. If so, a measure of the rate of alcohol consumption taken from the sample would not be an accurate reflection of the rate of alcohol consumption in the population. Actual alcohol consumption in the population would be higher than the rate reflected in the sample.

Survey researchers pay careful attention to nonresponse. They try to discover whether nonrespondents differ systematically from respondents so that they can take this into account before drawing conclusions from their sample. They must also take special measures to ensure that the response rate remains acceptably high—generally, 70 percent or more of people contacted. Proven tactics ensure a high response rate. Researchers can notify potential respondents about the survey in advance. They can remind them to complete and mail in survey forms. They can get universities and other prestigious institutions to sponsor the survey. They can stress the practical and scientific value of the research. And they can give people small rewards, such as a dollar or two, for participating.

If respondents do not answer questions completely and accurately, then a third threat to validity is present: *response bias.* The survey may focus on sensitive, unpopular, or illegal behavior. As a result, some respondents may not be willing to answer questions honestly. The interviewer's attitude, gender, or race may suggest that some responses are preferred rather than others. This can elicit biased responses. Some of these problems can be overcome by carefully selecting and training interviewers and closely supervising their work. Response bias on questions about sensitive, unpopular, or illegal behavior can be minimized by having such questions answered in private. For example, the General Social Survey (GSS) is a nationwide survey that has been conducted by the National Opinion Research Center at the University of Chicago nearly every year since 1972. It is one of the most important ongoing surveys in the United States, and we will refer to GSS data many times in the following chapters. Nearly every year the GSS measures the opinions, social characteristics, and behaviors of a representative sample of 1,500 or more American adults. Interviews are conducted face-to-face in people's households. Since 1988, the GSS has asked questions about how many sexual partners the respondent has had in the past year, the relation of those sex partners to the respondent, and the gender of the sex partners. But rather than having the interviewer ask these questions, almost certainly causing response bias, the respondent is given a card containing the questions. He or she completes the card in private, places it in an envelope provided by the interviewer, seals the envelope, and is assured that the interviewer will not read the card. Researchers believe this procedure minimizes response bias (Smith, 1992). However, some response bias remains insofar as men apparently tend to exaggerate how many sex partners they have when they are asked about this subject in surveys (McConaghy, 1999: 311–14).

Fourth, validity may be compromised because of *wording effects.* That is, the way questions are phrased or ordered can influence and invalidate responses. Experienced survey researchers have turned questionnaire construction into a respected craft. Increasingly, they refine the lessons learned from experience with evidence from field experiments. These experiments divide samples into two or more randomly chosen subsamples. Different question wording or ordering is then administered to the people in each subsample so that wording effects can be measured. Detected problems can then be resolved in future research.

Both experience and field experiments suggest that survey questions must be specific and simple. They should be expressed in plain, everyday language. They should be phrased neutrally, never leading the respondent to a particular answer and never using inflammatory terms. Because people's memories are often faulty, questions are more likely to elicit valid responses if they focus on important, singular, current events rather than less salient, multiple, past events. Breaking these rules lowers the validity of survey findings (Converse and Presser, 1986; Ornstein, 1998).

Causality

A survey is not an ideal instrument for conducting exploratory research. It cannot provide the kind of deep and sympathetic understanding one gains from participant observation. On the other hand, surveys produce results from which we can confidently generalize. If properly crafted, they provide valid measures of many sociologically important variables. Because they allow the same questions to be asked repeatedly, surveys enable researchers to establish the reliability of measures with relative ease. And finally, as we will now see, survey data are useful for discovering relationships among variables, including cause-and-effect relationships.

Recall how causality is established in experiments. Randomly assigning subjects to experimental and control groups makes the two groups similar. Exposing only the experimental group to an independent variable lets the researcher say the independent variable alone is probably responsible for any measured effect. That conclusion is warranted because the effects of irrelevant variables have been removed by randomization. In surveys, too, the effects of independent variables can be measured. However, the effects of irrelevant variables are removed not by randomization but by manipulating the survey data.

Reading Tables

One of the most useful tools for manipulating survey data is the **contingency table.** A contingency table is a cross-classification of cases by at least two variables that allows you to see how, if at all, the variables are associated. This might sound complex, but it's really not. It's as simple as the corners of your classroom.

A contingency table is a cross-classification of cases by at least two variables that allows you to see how, if at all, the variables are associated.

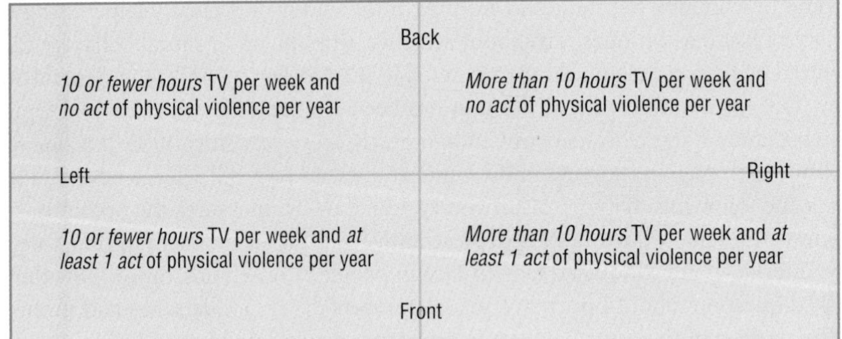

Figure 2.4
Turning a Classroom into a Contingency Table

To interpret a table, you must pay careful attention to the way it is percentaged, that is, exactly what it is that adds up to 100 percent. This table shows that 52 percent *of all students who watch TV fewer than 10 hours a week* commit zero violent acts per year, whereas 46 percent *of all students who watch TV 10 or more hours a week* commit zero violent acts per year. You know this because each category of the variable "TV Viewing" equals 100 percent. (The actual number of students in each category of the variable "TV Viewing" is given in the row labeled "Total frequency.")

Table 2.1
TV Viewing by Aggressiveness (in percent)

		TV VIEWING		
		<10 Hours per Week	10+ Hours per Week	Percentage Difference
Aggressiveness	*0 Violent acts per year*	52	46	6
	1+ Violent acts per year	48	54	6
	Total frequency (n)	130	70	
	Total percent	100	100	

Let's say we want to test the hypothesis that the number of hours one spends watching TV per week (the independent variable) increases the frequency of one's violent behavior (the dependent variable). We can test this hypothesis by first asking the students in your class who watch TV more than 10 hours a week to stand by the right wall and the other students to stand by the left wall. We can then ask the students who committed at least one act of physical violence against another person in the past year to move to the front of the room and the others to move to the back. This procedure would, in effect, create a contingency table in the four corners of your classroom. The students would be simultaneously classified (or "cross-classified") by how much TV they watch and their physical aggressiveness (Figure 2.4).

An **association** exists between two variables if the value of one variable changes with the value of the other. For example, if the percentage of students who committed an act of physical violence in the past year is higher among frequent TV viewers than among infrequent TV viewers, an association exists between the two variables. The greater the percentage difference between frequent and infrequent TV viewers, the stronger the association. In Table 2.1, for instance, the difference is 6 percent.

The existence of such an association does not by itself prove that watching TV causes physical aggression. The association may exist for other reasons. For example, it may be that men are more aggressive than women because of the way they are brought up. They may just happen to watch more TV than women do, too.

We can test the hypothesis that watching TV increases physical aggression by creating a second contingency table. Continuing with our classroom example, we can ask all the women to leave the room. In effect, this breaks our original contingency in two, allowing us to examine the association between TV viewing and aggressiveness within a category of a third variable, gender. The third variable, gender, acts as a **control variable,** meaning that we have manipulated the data to remove the effect of gender from the original association.

An **association** exists between two variables if the value of one variable changes with the value of the other.

A **control variable** is a variable whose influence is removed from the association between an independent and a dependent variable.

Table 2.2 shows TV viewing by aggressiveness for men only. It says that 40 percent of men who watch TV infrequently and 40 percent of men who watch TV frequently committed no acts of physical violence in the past year. The percentage difference between these two groups of men is zero. Said differently, once we remove the effect of gender by means

Table 2.2
TV Viewing by Aggressiveness, Men Only (in percent)

		TV VIEWING		
		<10 Hours per Week	10+ Hours per Week	Percentage Difference
Aggressiveness	*0 Violent acts per year*	40	40	0
	1+ Violent acts per year	60	60	0
	Total frequency (n)	50	50	
	Total percent	100	100	

of statistical control, an association no longer exists between the independent and dependent variables (TV viewing and physical violence, respectively). This finding obliges us to conclude that the original association in Table 2.1 is **spurious,** or accidental. In our example, then, watching TV in and of itself does not seem to cause physical violence. Instead, the association between watching TV and committing acts of physical violence is due to the fact that men happen to watch more TV and are more physically aggressive than women.

In general, to conclude that the association between an independent and a dependent variable is nonspurious, or causal, three conditions must hold:

- An association must exist between the two variables.
- The presumed cause must occur before the presumed effect.
- When a control variable is introduced, the original association must not disappear.

If an initial association disappears once a control variable is introduced, the association is spurious. If an initial association stays the same after we introduce a control variable, then we tentatively conclude that the association is causal. We say "tentatively" because other variables may be responsible for the association. If we control for these other variables and find that the association persists, then we will have greater confidence that the association is causal. In an experiment, all extraneous variables are eliminated by randomization. In the analysis of survey data, the best we can hope for is the elimination, by means of statistical control, of those variables that might plausibly explain the original association. This leaves us with a genuine causal association.

We illustrate our argument in Figure 2.5. The top half of the figure shows the original association between watching TV and physical aggression. The arrow indicates the ex-

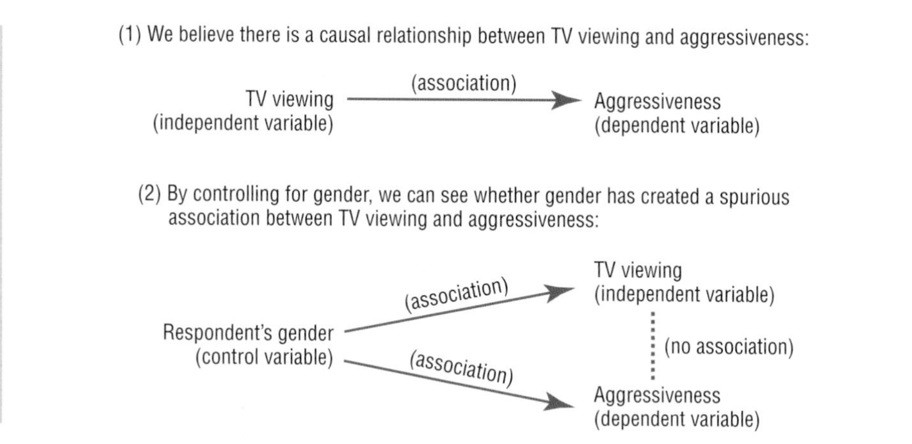

Figure 2.5
Testing an Association for Spuriousness

A **spurious association** exists between an independent and a dependent variable when the introduction of a causally prior control variable makes the initial association disappear.

Box 2.2

YOU AND THE SOCIAL WORLD

Thinking Causally

Each of the following paragraphs describes associations among three variables. For each paragraph, draw a diagram similar to Figure 2.6 identifying the independent variable, the dependent variable, and the control variable. Then, in less than 100 words, offer a plausible explanation of how the three variables are causally related and suggest the next step a researcher might want to take to further explore the relationship between the independent and dependent variables.

- A sociologist discovers that the intensity of civil war over the past 50 years is higher in poor countries than in rich countries. She then divides rich countries into democracies and nondemocracies and does the same with poor countries. In the case of both rich and poor countries, she finds no difference between democracies and nondemocracies in the intensity of civil war.
- A sociologist compares the earnings of women and men working full-time in the paid labor force and discovers that women earn 60 percent of what men earn. He then compares unmarried women and men working full-time in the paid labor force and discovers that unmarried women earn 75 percent of what unmarried men earn.
- In a survey of adults in the United States, a sociologist finds that 10 percent of the population express prejudice against black Americans. The figure falls to 5 percent among Americans with a college degree and rises to 20 percent among Americans without a high school diploma.

istence of a presumed causal relationship. The bottom half of the figure shows what happens when we control for gender: The original association disappears, as suggested by the broken line. The arrows in the bottom half of Figure 2.5 show the actual relationships among the variables.

The analysis of survey data involves more than just searching for nonspurious associations. Many interesting and unexpected things can happen when a two-variable association is elaborated by controlling for a third variable (Hirschi and Selvin, 1972). The original association may remain unchanged. It may strengthen. It may weaken. It may disappear or weaken in only some categories of the control variable. It may even change direction entirely. Data analysis is therefore full of surprises, and accounting for the outcomes of statistical control requires a lot of creative theoretical thinking (Box 2.2).

Web Research: Online Data Analysis

Analysis of Existing Documents and Official Statistics

Apart from participant observation, experiments, and surveys, there is a fourth important sociological research method: the analysis of existing documents and official statistics. What do existing documents and official statistics have in common? They are created by people other than the researcher for purposes other than sociological research.

The three types of existing documents that sociologists have mined most widely and deeply are diaries, newspapers, and published historical works. For example, one of the early classics of American sociology, a study of Polish immigrants, is based on a close reading of immigrants' diaries and letters (Thomas and Znaniecki, 1958 [1918–20]). More recently, sociologists have made outstanding contributions to the study of political protest by systematically classifying 19th- and early-20th-century French, Italian, and British newspaper accounts of strikes and demonstrations (Tilly, Tilly, and Tilly, 1975).

In recent decades, sociologists have tried to discover the conditions that led some countries to dictatorship and others to democracy, some to economic development and others to underdevelopment, some to become thoroughly globalized and others to remain less tied to global social processes. In trying to answer such broad questions, they have had to rely on published histories as their main source of data. No other method would allow the breadth of coverage and depth of analysis required for such comparative and historical work. For example, Barrington Moore spent a decade reading the histories of Britain,

France, Russia, Germany, China, India, and other countries to figure out the social origins of dictatorship and democracy in the modern world (Moore, 1967). Immanuel Wallerstein canvassed the history of virtually the entire world to make sense of why some countries become industrialized while others remain undeveloped (Wallerstein, 1974–89). What distinguishes this type of research from purely historical work is the kinds of questions posed by the researchers. Moore and Wallerstein asked the same kinds of big, theoretical questions (and used the same kinds of research methods) as Karl Marx and Max Weber. They have inspired a generation of younger sociologists to adopt a similar approach. Comparative-historical research is therefore one of the growth areas of the discipline.

Charles Tilly by John Sheretz. ©1997 CASAS

Charles Tilly (Columbia University) is one of the most prolific and respected sociologists in the world. He specializes in the study of large-scale social change and its relation to contentious politics in western Europe, using existing documents such as newspapers, administrative reports, and secondary historical works as sources of evidence. His pathbreaking work has helped to reorient the study of state formation and social movements. Major works include *The Rebellious Century, 1830–1930*, with Louise Tilly and Richard Tilly (1975); *From Mobilization to Revolution* (1978); *Big Structures, Large Processes, Huge Comparisons* (1985); *The Contentious French* (1986); and *Roads from Past to Future* (1997).

Census data, police crime reports, and records of key life events are perhaps the most frequently used sources of official statistics. The first U.S. census was taken in 1790, and censuses have been conducted at regular intervals since then. The modern census tallies the number of U.S. residents and classifies them by place of residence, race, ethnic origin, occupation, age, and hundreds of other variables. The Federal Bureau of Investigation (FBI) publishes an annual *Uniform Crime Report* that tallies the number of crimes in the United States and classifies them by location and type of crime, by age, sex, and race of offenders and victims, and by other variables. And the Centers for Disease Control and Prevention (CDC) regularly publish "vital statistics" reports on births, deaths, marriages, and divorces by sex, race, age, and so forth.

Existing documents and official statistics have four main advantages over other types of data. First, they can save the researcher time and money because they are usually available at no cost in libraries or on the World Wide Web. (See the "Recommended Websites" at the end of the chapter for useful Web resources containing official statistics.) Second, official statistics usually cover entire populations and are collected using rigorous and uniform methods, thus yielding high-quality data. Third, existing documents and official statistics are especially useful for historical analysis. The analysis of data from these sources is the only sociological method that does not require live subjects. Fourth, because the method does not require live subjects, reactivity is not a problem; the researcher's presence does not influence the subjects' behavior.[4]

Existing documents and official statistics, however, share one big disadvantage. These data sources are not created with the researchers' needs in mind. They often contain biases that reflect the interests of the individuals and organizations that created them. Therefore, they may be less than ideal for research purposes and must always be treated cautiously. For instance, if law enforcement officials decide to patrol minority-group neighborhoods more than majority-group neighborhoods, their action may result in an increase in the number of apprehended minority-group criminals over a given period. However, the increase may not be the result of a rise in the underlying crime rate. It may be due to the administrative decision of the officials to increase patrols in certain neighborhoods. It follows that official crime statistics are not ideal measures of crime rates, especially for certain types of crime (see Chapter 7, "Deviance and Crime").

To illustrate further the potential bias of official statistics, consider how researchers used to compare the well-being of Americans with that of people living in other countries.

---

[4]When researchers finish analyzing survey data, they typically deposit computer-readable files of the data in an archive, which allows other researchers to conduct secondary analyses of survey data years later. Such data are widely used. Government departments do not collect them, but they have all of the advantages of official statistics listed previously, although they are based on samples rather than populations. The largest social science data archive is at the University of Michigan's Inter-University Consortium for Political and Social Research (ICPSR). The ICPSR website, at http://www.icpsr.umich.edu, allows visitors to conduct elementary data analysis online.

BOX 2.3
Sociology at the Movies

Kinsey (2004)

In early-20th-century New Jersey, Alfred Kinsey's father sermonized that the telephone and the automobile were the devil's work. In his opinion, these conveniences increased interaction between young men and women, thereby promoting impure thoughts, petting, and all manner of sexual perversion.

Not surprisingly, the adolescent Alfred rejected his father's Puritanism and petty tyranny. He escaped to Harvard to study biology and zoology. He devoted 20 years to collecting and analyzing 100,000 specimens of the gall wasp, but underneath his mania for counting, classifying, and marveling at natural diversity, his rebellion against sexual repression and imposed sexual uniformity never ended.

In the 1930s, now a full professor of zoology at Indiana University, Kinsey began to investigate human sexual behavior with the same fervor he had formerly invested in the gall wasp. Between 1938 and 1963, he and his associates conducted 18,216 in-depth interviews that formed the basis of two best-selling volumes on human sexual behavior that astounded the American public and put Kinsey on the cover of *Time* magazine (Kinsey, Pomeroy, and Martin, 1948; Kinsey, Pomeroy, Martin, and Gebhard, 1953). In an era when masturbation, contraception, and premarital sex were widely considered sins, Kinsey's work sparked a revolution in attitudes toward sex by showing that even far more scandalous practices—extramarital affairs, homosexuality, and so forth—were commonplace. For many Americans, his findings were liberating. For others, they were filthy lies that threatened to undermine the moral fiber of the nation. Both reactions, and the life of the man who caused them, are brilliantly captured in *Kinsey* (2004), starring Liam Neeson in the title role.

Equipped with the information in this chapter, you can appreciate that Kinsey's methods were primitive and biased by modern sociological standards (Ericksen, 1998). We single out four main problems:

- *Sampling.* Kinsey relied on what we today call a "convenience sample" of respondents. He and his colleagues interviewed accessible volunteers rather than a randomized and representative sample of the American population. About one-third of Kinsey's respondents had a known "sexual bias." They were prostitutes, members of secretive homosexual communities, patients in mental hospitals, residents of homes for unwed mothers, and the like. Two-thirds of these people were convicted felons. Five percent were male prostitutes. But even if we eliminate respondents with a sexual bias, we do not have a representative sample. For example, 84 percent of the men without a sexual bias went to college. Most of them were from the Midwest, especially Indiana (Gebhard and Johnson, 1979). In Kinsey's defense, scientific sampling was in its infancy when he did his research. Still, we are obliged to conclude that it is difficult to generalize from Kinsey's work because his sample is unrepresentative.

- *Questionnaire design.* Kinsey required that his interviewers memorize long questionnaires including 350 or more questions. He encouraged them to adapt the wording and ordering of the questions to suit the "level" of the respondent and the natural flow of conversation that emerged during the interview. Yet much research now shows that even subtle changes in question wording and ordering can produce sharply different results. A question about frequency of masturbation per month yields means between 4 and 15 depending on how the question is phrased (Bradburn and Sudman, 1979). To avoid such problems, modern researchers prefer standardized questions. They also prefer questionnaires considerably briefer than Kinsey's because asking 350 questions can take hours and often results in "respondent fatigue," a desire on the part of respondents to offer quick and easy answers (as opposed to considered,

For years, researchers used a measure called Gross Domestic Product Per Capita (GDPpc). GDPpc is the total dollar value of goods and services produced in a country in a year divided by the number of people in the country. It was a convenient measure because all governments regularly published GDPpc figures.

Researchers were aware of a flaw in GDPpc. The cost of living varies from one country to the next. A dollar can buy you a cup of coffee in many American restaurants, but the same cup of coffee will cost you $6 in a Tokyo restaurant. GDPpc looks at how many dollars you have, not at what the dollars can buy. Therefore, researchers were happy when governments started publishing an official statistic called Purchasing Power Parity (PPP), which takes into account the cost of goods and services in each country.

Significantly, however, both PPP and GDPpc ignore two serious problems. First, GDPpc and PPP can increase while most people in a society are worse off. This situation occurred in the United States in the 1980s. The richest people in the country earned all of truthful responses) so they can end the interview as quickly as possible.

- *Interviewing.* People are generally reluctant to discuss sex with strangers, and Kinsey and his associates have often been praised for making their respondents feel at ease talking about the most intimate details of their personal life. Yet to establish rapport with respondents, Kinsey and his colleagues did not remain neutral. They expressed empathy with the pains and frustrations many respondents expressed, often reassuring them that their sexual histories were normal and decent. Today, researchers frown upon any departure from neutrality in the interview situation because it may influence respondents to answer questions in a less than truthful way. The reassurance and empathy expressed by Kinsey and his associates may have led some respondents to offer exaggerated reports of their behavior.
- *Data analysis.* It is unclear how Kinsey decided whether the effect of one variable on another was significant. He rarely used statistical tests for this purpose. He never introduced control variables to see if observed associations between variables were spurious. Moreover, he saw no problem in lumping together data collected over decades. Yet between 1938, when Kinsey started collecting data, and 1953, when his second book was published, the United States experienced unprecedented social change fueled by depression and boom, war and peace. Sexual attitudes and behavior undoubtedly changed, and one may wonder whether it is meaningful to analyze respondents from the late 30s and the early 50s together.

Liam Neeson in *Kinsey*.

20th Century Fox/American Zoetrope/The Kobal Collection

Since Kinsey, researchers have conducted more than 750 scientific surveys of the sexual behavior of Americans. Today, using modern research methods, we are able to describe and explain sexual behavior more accurately and insightfully than did Kinsey and his pioneering colleagues. We know that many of the details of Kinsey's writings are suspect. But we also know that despite the serious methodological problems summarized previously, his basic finding is accurate. The sexual behavior of Americans is highly diverse. As Kinsey says in the movie, "Variation is the only reality."

Herein, too, lies an important lesson about the relationship between subjectivity and objectivity in research. Clearly, Kinsey's biography and his passions helped to shape his innovative scientific agenda. There is nothing unusual in that; all good scientists are passionate about their work, and their agendas are often rooted in their biographies. Like Kinsey, they try to be objective; but even if they fail, they can rely on the scientific community to uncover biases and discover the imaginative and valid core of every good theory. Without human emotions grounded in our subjectivity, there could never be a quest for truth, and without research methods that improve our objectivity, there could never be a science.

the newly created wealth, whereas the incomes of most Americans decreased. Any measure of well-being that ignores the *distribution* of well-being in society is biased toward measuring the well-being of the well-to-do. Second, in some countries the gap in well-being between women and men is greater than in others. A country like Kuwait ranks quite high on GDPpc and PPP. However, women benefit far less than do men from that country's prosperity. A measure of well-being that ignores the gender gap is biased toward measuring the well-being of men.

This story has a happy ending. Realizing the biases in official statistics like GDPpc and PPP, social scientists at the United Nations (UN) created two new measures of well-being in the mid-1990s. First, the Human Development Index (HDI) combines PPP with a measure of average life expectancy and average level of education. The reasoning of the UN social scientists is that people living in countries that distribute well-being more equitably will live longer and be better educated. Second, the Gender Empowerment Measure

Table 2.3
Rank of Countries by Four Measures of Well-Being, 2002

Gross Domestic Product	Purchasing Power Parity	Human Development Index	Gender Empowerment Measure
1. Luxembourg	1. Luxembourg	1. Norway	1. Norway
2. Switzerland	2. Norway	2. Sweden	2. Sweden
3. Japan	3. Ireland	3. Australia	3. Denmark
4. Norway	4. United States	4. Canada	4. Finland
5. Denmark	5. Denmark	5. Netherlands	5. Netherlands

Adapted from National Energy Information Centre (2005); United Nations (2004: 139, 221).

(GEM) combines the percent of parliamentary seats, good jobs, and earned income controlled by women.

Table 2.3 lists the countries ranked first through fifth on all four measures of well-being we have mentioned. As you can see, the list of the top five countries differs for each measure. There is no "best" measure. Each measure has its own bias, to which researchers must be sensitive, as they must whenever they use official statistics.

The Importance of Being Subjective

In the following chapters, we show how participant observation, experiments, surveys, and the analysis of existing documents and official statistics are used in sociological research. You are well equipped for the journey. By now, you should have a pretty good idea of the basic methodological issues that confront any sociological research project. You should also know the strengths and weaknesses of some of the most widely used data collection techniques (Concept Summary 2.1).

Our synopsis of sociology's "reality check" should not obscure the fact that sociological research questions often spring from real-life experiences and the pressing concerns of the day. But before sociological analysis, we rarely see things as they are. We see them as we are. Then, a sort of waltz begins. Subjectivity leads, objectivity follows. When the dance is finished, we see things more accurately (Box 2.3).

Feminism provides a prime example of this process. Here is a *political* movement of people and ideas that has helped to shape the sociological *research* agenda over the past 35 or 40 years. The division of labor in the household, violence against women, the effects of child-rearing responsibilities on women's careers, the social barriers to women's participation in politics and the armed forces, and many other related concerns were sociological "nonissues" before the rise of the modern feminist movement. Sociologists did not study these problems. Effectively, they did not exist for the sociological community (although they did of course exist for women). But subjectivity led. Feminism as a political movement brought these and many other concerns to the attention of the American public. Objectivity followed. Large parts of the sociological community began doing rigorous research on feminist-inspired issues and greatly refined our knowledge about them.

The entire sociological perspective began to shift as a growing number of scholars abandoned gender-biased research (Eichler, 1988; Tavris, 1992). Thus, *male centeredness,* or approaching sociological problems from an exclusively male perspective, is now less common than it used to be. For instance, it is less likely in 2006 than in 1976 for a sociologist to study work but ignore unpaid housework as one type of labor. Similarly, *overgeneralization,* or using data on one sex to draw conclusions about all people, is now generally frowned upon. Today, for example, few researchers would be inclined to make claims about the social factors influencing health based on a sample of men only. In addition, *gender-blindness,* or excluding gender as an independent variable, is becoming less com-

Concept Summary 2.1
Strengths and Weaknesses of Four Research Methods

Method	Strengths	Weaknesses
Participant observation	Allows researchers to get "inside" the minds of their subjects and discover their worldview; useful for exploratory research and the discovery of grounded theory; high internal validity	Low reliability; low external validity; low generalizability; not very useful for establishing cause-and-effect relationships
Experiment	High reliability; excellent for establishing cause-and-effect relationships	Low validity for many sociological problems (field and natural experiments somewhat better); problems with generalizability
Survey	Good reliability; useful for establishing cause-and-effect relationships; good generalizability	Some problems with validity (but techniques exist for boosting validity)
Analysis of existing documents and official statistics	Often inexpensive and easy to obtain, provides good coverage; useful for historical analysis; nonreactive	Often contains biases reflecting the interests of their creators and not the interests of the researcher

mon. Thus, 30 years ago, many researchers failed to notice that the experiences of elderly men and women often differ radically because women tend to live longer and are poorer than men on average. That sort of mistake is less common today. Finally, applying a *double standard,* or assuming that women and men should necessarily be assessed on the basis of different criteria, is now viewed as problematic by many sociologists. Increasingly, for example, we understand that there is nothing inevitable about husbands being the only breadwinners and wives being the only nurturers.

As these advances in sociological thinking show, and as has often been the case in the history of the discipline, objective sociological knowledge has been enhanced as a result of subjective experiences. And so the waltz continues. As in *Alice in Wonderland,* the question now is, "Will you, won't you, will you, won't you, join the dance?"

Summary

SociologyNow™

Reviewing is as easy as ❶, ❷, ❸.

❶ Before you do your final review, take the SociologyNow diagnostic quiz to help you identify the areas on which you should concentrate. You will find information on how to access SociologyNow on the foldout at the front of the textbook.

❷ As you review, take advantage of SociologyNow's study aids to help you master the topics in this chapter.

❸ When you are finished with your review, take SociologyNow's post-test to confirm you are ready to move on to the next chapter.

1. **What is the aim of science and how is it achieved?**

   The aim of science is to arrive at knowledge that is less subjective than other ways of knowing. A degree of objectivity is achieved by testing ideas against systematically collected data and leaving research open to public scrutiny.

2. **Does science have a subjective side?**

   It does. The subjective side of the research enterprise is no less important than the objective side. Creativity and the motivation to study new problems from new perspectives arise from individual passions and interests.

3. **What methodological issues must be addressed in any research project?**

   To maximize the scientific value of a research project, one must address issues of reliability (consistency in measurement), validity (precision in measurement), generalizability (assessing the applicability of findings beyond the case studied), and causality (assessing cause-and-effect relations among variables).

4. **What is participant observation?**

   Participant observation is one of the main sociological methods. It involves carefully observing people's face-to-face interactions and actually participating in their lives over a long period of time. Participant observation is particularly useful for exploratory research, constructing grounded theory, and validating measures on the basis of internal criteria. Issues of external validity, reliability, generalizability, and causality make participant observation less useful for other research purposes.

5. **What is an experiment?**

   An experiment is a carefully controlled artificial situation that allows researchers to isolate hypothesized causes and measure their effects by randomizing the allocation of subjects to experimental and control groups and exposing only the experimental group to an independent variable. Experiments get high marks for reliability and analysis of causality, but issues of validity and generalizability make them less than ideal for many research purposes.

6. **What is a survey?**

   In a survey, people are asked questions about their knowledge, attitudes, or behavior, in either a face-to-face interview, telephone interview, or paper-and-pencil format. Surveys rank high on reliability and validity as long as researchers train interviewers well, phrase questions carefully, and take special measures to ensure high response rates. Generalizability is achieved through probability sampling and the analysis of causality by means of data manipulation.

7. **What are the advantages and disadvantages of using official documents and official statistics as sources of sociological data?**

   Existing documents and official statistics are inexpensive and convenient sources of high-quality data. However, they must be used cautiously because they often reflect the biases of the individuals and organizations that create them rather than the interests of the researcher.

Questions to Consider

1. **What is the connection between objectivity and subjectivity in sociological research?**
2. **What criteria do sociologists apply to select one method of data collection over another?**
3. **What are the methodological strengths and weaknesses of various methods of data collection?**

Web Resources
Companion Website for This Book

http://sociology.wadsworth.com

Begin by clicking on the Student Resources section of the website. Choose "Introduction to Sociology" and then the Brym and Lie book cover. Next, select the chapter you currently are studying from the pull-down menu. From the Student Resources page you will have easy access to InfoTrac® College Edition, MicroCase Online exercises, additional web links, and many resources to aid you in your study of sociology, including practice tests for each chapter.

Recommended Websites

Bill Trochim at Cornell University has put together a comprehensive and impressive sociological research methods course on the World Wide Web. Visit it at http://trochim.human.cornell.edu.

For a comprehensive listing of websites devoted to qualitative research, go to http://www.nova.edu/ssss/QR/web.html.

"Statistics Every Writer Should Know" is an exceptionally clear presentation of basic statistics on the World Wide Web at http://www.robertniles.com/stats.

The World Wide Web contains many rich sources of official statistics. In preparing this book, we relied heavily on data from the websites of the U.S. Census Bureau (http://www.census.gov), the U.S. Department of Labor's Bureau of Labor Statistics (http://stats.bls.gov), the National Center for Health Statistics (http://www.cdc.gov/nchs/fastats/default.htm), the FBI's Uniform Crime Reports (http://www.fbi.gov /ucr/ucr.htm), and the UN (http://www.un.org).

Appendix | Four Statistics You Should Know

In this book we sometimes report the results of sociological research in statistical form. You need to know four basic statistics to understand this material:

1. *Mean* (or arithmetic average). Imagine we know the height and annual income of the first nine people who entered your sociology classroom today. The height and income data are arranged in ▶Table 2.4. From this table, you can calculate the mean by summing the values for each student, or *case,* and dividing by the number of cases. For example, the nine students are a total of 609 inches tall. Dividing 609 by 9, we get the mean height: 67.7 inches.
2. *Median.* The mean can be deceiving when some cases have exceptionally high or low values. For example, in Table 2.4, the mean income is $37,667, but because one lucky fellow has an income of $200,000, the mean is higher than the income of seven of the nine students. It is therefore a poor measure of the center of the income distribution. The median is a better measure. If you order the data from the lowest to the highest income, the median is the value of the case at the

Table 2.4
The Height and Annual Income of Nine Students

Student	Height (inches)	Income ($ thousands)
1	67	5
2	65	8
3	60	9
4	64	12
6	68	15
7	70	20
8	69	30
5	72	40
9	74	200

midpoint. The median income in our example is $15,000. Four students earn more than that, four earn less. (Note: If you have an even number of cases, the midpoint is the *average* of the middle two values.)

3. *Correlation.* We have seen how valuable contingency tables are for analyzing relationships among variables. However, for variables that can assume many values, such as height and income, contingency tables become impracticably large. In such cases, sociologists prefer to analyze relationships among variables using *scatterplots.* Markers in the body of the graph indicate the score of each case on both the independent and dependent variables. The pattern formed by the markers is inspected visually and through the use of statistics. The strength of the association between the two variables is measured by a statistic called the *correlation coefficient* (signified as *r*). The value of *r* can vary from −1.0 to 1.0. If the markers are scattered around a straight, upward sloping trend line, *r* takes a positive value. A positive *r* suggests that as the value of one variable increases, so does the value of the other (Figure 2.6, scatterplot 1). If the markers are scattered around a straight, downward sloping trend line, *r* takes a negative value. A negative *r* suggests that as the value of one variable increases, the value of the other decreases (Figure 2.6, scatterplot 2). Whether positive or negative, the magnitude (or absolute value) of *r* decreases the more widely scattered the markers are from the line. If the degree of scatter is very high, *r* = 0. That is, no association exists between the variables (Figure 2.6, scatterplot 3). However, a low *r* or an *r* of zero may derive from a relationship between the two variables that does not look like a straight line. It may look like a curve. As a result, it is always necessary to inspect scatterplots visually and not just rely on statistics such as *r* to interpret the data.
4. A *rate* lets you compare the values of a variable among groups of different size. For example, let's say 1,000 women got married last year in a city of 100,000 people and 2,000 women got married in a city of 300,000 people. If you want to compare the likelihood of women getting married in the two cities, you have to divide the number of women who got married in each city by the total number of women in each city. Because 1,000/100,000 = 0.01 or 1 percent, and 2,000/300,000 = 0.00666 or 0.67 percent, we can say that the *rate* of women marrying is higher in the first city even though fewer women got married there last year. Note that rates are often expressed in percentage terms. In general, dividing the number of times an event occurs (e.g., a woman getting married) by the total number of people to whom the event could occur in principle (e.g., the number of women in a city) will give you the rate at which an event occurs.

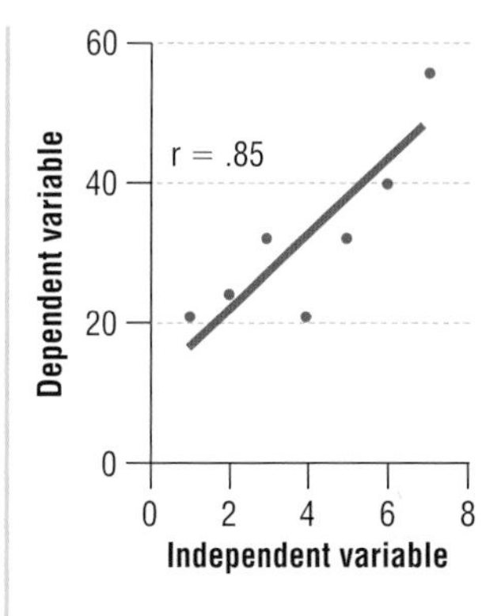

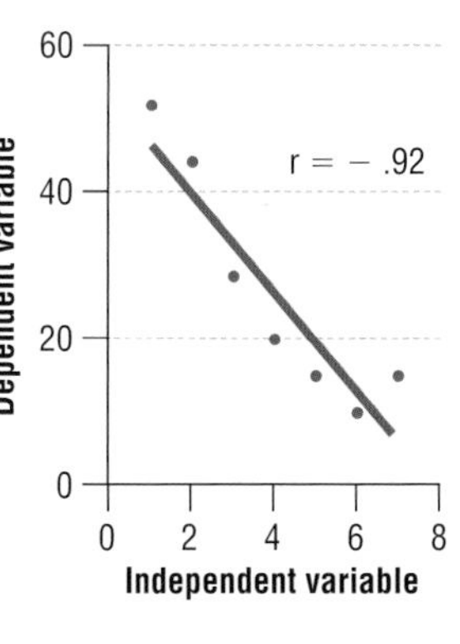

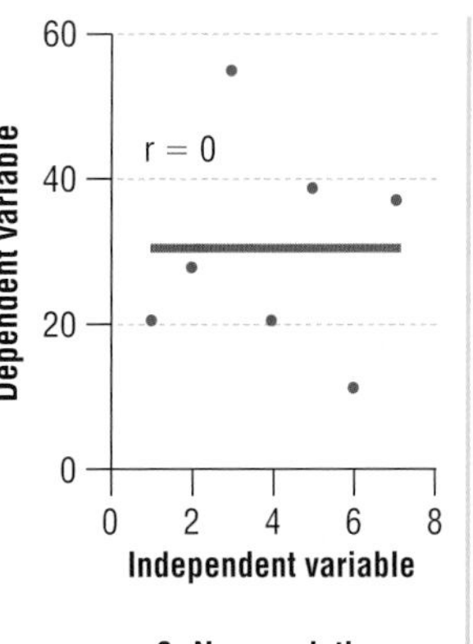

Figure 2.6
Correlation

CHAPTER 3 Culture

Canadian Press Picture Archives/Kevin Frayer

In this chapter, you will learn that:

- Culture is the sum of shared ideas, practices, and material objects that people create to adapt to, and thrive in, their environments.
- Humans have thrived in their environments because of their unique ability to think abstractly, cooperate with one another, and make tools.
- Although sociologists recognize that biology sets broad human limits and potentials, most sociologists do not believe that specific human behaviors and social arrangements are biologically determined.
- In some respects, the development of culture makes people freer. For example, culture has become more diversified and consensus has declined in many areas of life, allowing people more choice in how they live.
- In other respects, the development of culture puts limits on who we can become. For example, the culture of buying consumer goods has become a virtually compulsory national pastime. Increasingly, therefore, people define themselves by the goods they purchase.

Culture as Problem Solving

If you follow professional baseball, you probably know that star shortstop Nomar Garciaparra can take 10 seconds to repeatedly pull up his batting gloves and kick the dirt with the toes of his cleats before he swings the bat. He believes that this routine brings him luck. Garciaparra has other superstitious practices as well. For example, he never changes his cap. And although his name is really Anthony, he adopted "Nomar," his father's name spelled backward, for good luck. Garciaparra's nervous prebatting dance, as well as his other superstitious practices, make some people chuckle. But they put Garciaparra at ease. They certainly didn't hurt his league-leading .372 batting average in 2000. As Garciaparra says: "I have some superstitions, definitely, and they're always going to be there. I think a lot of people have them in baseball. . . . [It] definitely helps because it gets you in the mind set" ("Garciaparra Explains his Superstitions," 2000).

Like soldiers going off to battle, college students about to write final exams, and other people in high-stress situations, athletes invent practices to help them stop worrying and focus on the job at hand. Some wear a lucky piece of jewelry or item of clothing. Others say special words or a quick prayer. Still others cross themselves.

And then there are those who engage in more elaborate rituals. For example, two sociologists interviewed 300 college students about their superstitious practices before final exams. One student felt that she would do well only if she ate a sausage and two eggs sunny-side up on the morning of each exam. She had to place the sausage vertically on the left side of her plate and the eggs to the right of the sausage so they formed the "100" percent she was aiming for (Albas and Albas, 1989). Of course, the ritual had more direct influence on her cholesterol level than on her grades. Yet indirectly it may have had the de-

Culture Can Solve Practical Problems
Nomar Garciaparra creates a little culture . . . and then knocks one out of Fenway Park.

Andrew Woolley

sired effect. To the degree it helped to relieve her anxiety and relax her, she may have done better on her exams.

When some people say "culture," they refer to opera, ballet, art, and fine literature. For sociologists, however, this definition is too narrow. Sociologists call opera, and so forth, **high culture** to distinguish it from **popular** or **mass culture.** Although high culture is consumed mainly by upper classes, popular or mass culture is consumed by all classes. Sociologists define **culture** in the general sense very broadly as all the ideas, practices, and material objects that people create to deal with real-life problems. For example, when Nomar Garciaparra developed the practice of pulling at his gloves and the college student invented the ritual of preparing for exams by eating sausage and eggs arranged just so, they were creating culture in the sociological sense. These practices helped Garciaparra and the student deal with the real-life problem of high anxiety.

Similarly, tools help people solve the problem of how to plant crops and build houses. Religion helps people face the problem of death and how to give meaning to life. Tools and religion are also elements of culture because they, too, help people solve real-life problems. Note, however, that religion, technology, and many other elements of culture differ from the superstitions of Garciaparra and the college student in one important respect. Superstitions are often unique to the individuals who create them. In contrast, religion and technology are widely shared. They are even passed on from one generation to the next. How does cultural sharing take place? By means of communication and learning. Thus, shared culture is *socially* transmitted. It requires a society to persist. (In turn, a **society** is a number of people who interact, usually in a defined territory, and share a culture.) We conclude that culture is composed of the socially transmitted ideas, practices, and material objects that enable people to adapt to, and thrive in, their environments.

- **High culture** is culture consumed mainly by upper classes.
- **Popular culture** (or **mass culture**) is culture consumed by all classes.
- **Mass culture** (*see* Popular culture).
- **Culture** is the sum of socially transmitted practices, languages, symbols, beliefs, values, ideologies, and material objects that people create to deal with real-life problems. Cultures enable people to adapt to, and thrive in, their environments.
- A **society** is composed of people who interact, usually in a defined territory, and share a culture.

The Origins and Components of Culture

You can appreciate the importance of culture for human survival by considering the predicament of early humans about 100,000 years ago. They lived in harsh natural environments. They had poor physical endowments, being slower runners and weaker fighters than many other animals. Yet, despite these disadvantages, they survived. More than that: They prospered and came to dominate nature. This was possible largely because they were the smartest creatures around. Their sophisticated brains enabled them to create cultural survival kits of enormous complexity and flexibility. These cultural survival kits contained three main tools. Each tool was a uniquely human talent. Each gave rise to a different element of culture.

Symbols

The first tool in the human cultural survival kit was **abstraction,** the capacity to create general ideas, or ways of thinking not linked to particular instances. **Symbols,** for example, are one important type of idea. They are things that carry particular meanings. Languages, mathematical notations, and signs are all sets of symbols. Symbols allow us to classify experience and generalize from it. For example, we recognize that we can sit on many objects but that only some of those objects have four legs, a back, and space for one person. We distinguish the latter from other objects by giving them a name: "chairs." By the time a baby reaches the end of her first year, she has heard that word repeatedly and understands that it refers to a certain class of objects. True, chimpanzees have been taught how to make some signs with their hands. In this way, they have learned a few dozen words and how to string together some simple phrases. Yet even these extraordinarily intelligent animals cannot learn rules of grammar, teach other chimps much of what they know, or advance beyond the vocabulary of a very young human (Pinker, 1994a). Abstraction beyond the most rudimentary level is a uniquely human capacity. The ability to abstract enables humans to learn and transmit knowledge in a way no other animal can.

Sociology Now™

Learn more about **Norms** by going through the Norms Video Exercise.

Norms and Values

Cooperation is the second main tool in the human cultural survival kit. It is the capacity to create a complex social life. This is accomplished by establishing **norms,** or generally accepted ways of doing things. When we raise children and build schools, we are cooperating to reproduce and advance the human race. When we create communities and industries, we are cooperating by pooling resources and encouraging people to acquire specialized skills. This enables them to accomplish things that no person could possibly do on his or her own. An enormous variety of social arrangements and institutions, ranging from health-care systems to forms of religious worship to political parties, demonstrates the advanced human capacity to cooperate and follow norms. Animals, including insects, cooperate to varying degrees, but this occurs more due to instinct rather than the learning of norms. And although there is plenty of apparently noncooperative behavior in the world, people who engage in war, crime, and revolution must cooperate and respect norms or fail to achieve their survival aims. The bank robber who is left stranded by his getaway driver will be caught; the navy captain whose sailors mutiny in time of war will lose the battle.

Material and Nonmaterial Culture

Production is the third main tool in the human cultural survival kit. It involves making and using tools and techniques that improve our ability to take what we want from nature. Such tools and techniques are known as **material culture** because they are tangible, while the symbols, norms, and other elements of **nonmaterial culture** are not. All animals take from nature to subsist, and an ape may sometimes use a rock to break another object. But only humans are sufficiently intelligent and dexterous to *make* tools and use them to produce everything from food to computers. Understood in this sense, production is a uniquely human activity.

Concept Summary 3.1 illustrates each of the basic human capacities and their cultural offshoots with respect to three types of human activity: medicine, law, and religion. It shows, for all three types of activity, how abstraction, cooperation, and production give rise to specific kinds of ideas, norms, and elements of material culture. In medicine, theoretical

- **Abstraction** is the human capacity to create general ideas, or ways of thinking that are not linked to particular instances. For example, languages, mathematical notations, and signs allow us to classify experience and generalize from it.
- A **symbol** is anything that carries a particular meaning, including the components of language, mathematical notations, and signs. Symbols allow us to classify experience and generalize from it.
- **Cooperation** is the human capacity to create a complex social life.
- **Norms** are generally accepted ways of doing things.
- **Production** is the human capacity to make and use tools. It improves our ability to take what we want from nature.
- **Material culture** is composed of the tools and techniques that enable people to get tasks accomplished.
- **Nonmaterial culture** is composed of symbols, norms, and other nontangible elements of culture.

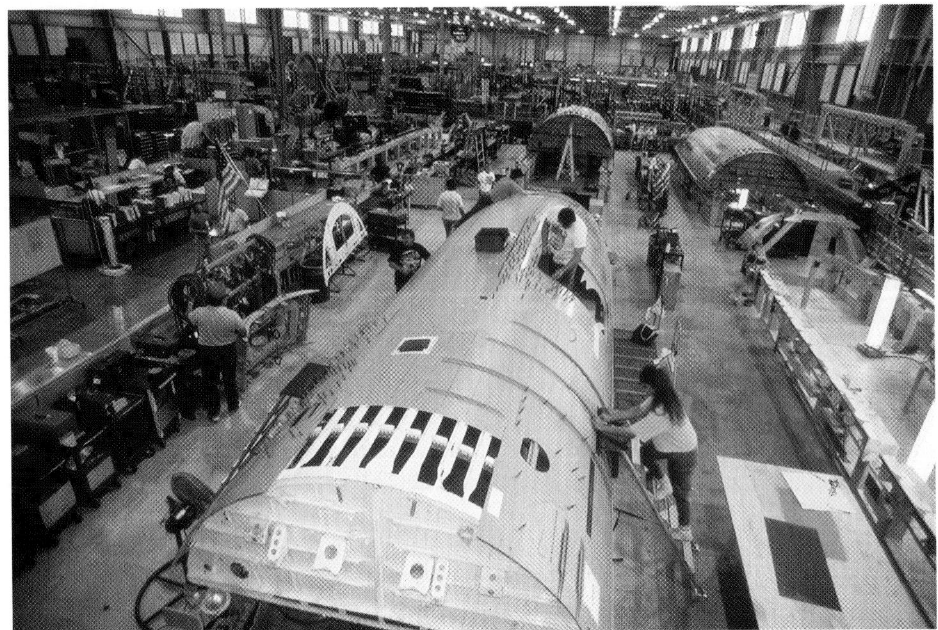

By acquiring specialized skills, people are able to accomplish things that no person could possibly do on his or her own.

Mark Richards/PhotoEdit

ideas about the way our bodies work are evaluated using norms about how to test theories experimentally. Experimentation, in turn, results in the production of new medicines and therapies. These are part of material culture. In law, values (shared ideas about what is right and wrong) are embodied in a legal code, consisting of norms defining illegal behavior and punishments for breaking the law. The application of the law requires the creation of courts and jails, which are also part of material culture. Religious folklore—traditional ideas about how the universe was created, the meaning of life, and so forth—is expressed in religious customs regarding how to worship and how to treat fellow human beings. Religious folklore and customs can give rise to material culture that includes churches, their associated art and architecture, and so forth. As these examples suggest, the capacity for abstraction, cooperation, and production are evident in all spheres of culture.

Sociology Now™

Learn more about **the Culture** by going through the Definition of Culture Video Exercise.

Sanctions, Taboos, Mores, and Folkways

In concluding this discussion of the origins of culture, we must note that people are usually rewarded when they follow cultural guidelines and punished when they do not. These rewards and punishments aimed at ensuring conformity are known as **sanctions.** Taken

CONCEPT SUMMARY 3.1
The Building Blocks of Culture

	HUMAN CAPACITIES		
	Abstraction	**Cooperation**	**Production**
	↓	↓	↓
	Ideas	**Norms**	**Material Culture**
	↓	↓	↓
	↓	↓	↓
		CULTURAL ACTIVITIES	
Medicine	Theories	Experiments	Treatments
Law	Values	Laws	Courts, jails
Religion	Religious folklore	Religious customs	Church art

Source: Adapted from Bierstedt (1963).

Sanctions are rewards and punishments intended to ensure conformity to cultural guidelines.

together they are called the system of **social control.** Rewards (or positive sanctions) include everything from praise and encouragement to money and power. Punishments (or negative sanctions) range from avoidance and contempt to arrest, physical violence, and banishment. **Taboos** are among the strongest norms. When someone violates a taboo, it causes revulsion in the community, and punishment is severe. Incest is one of the most widespread taboos. Breaking other core norms does not cause revulsion, but most people still feel that such norms are essential for the survival of their group or their society. Sociologist William Graham Sumner (1940 [1907]) called such core norms **mores** (the Latin word for "customs," pronounced MORE-ays). Sumner called the least important norms **folkways.** They evoke the least severe punishment. If a man walks down the street wearing nothing on the lower half of his body, he is violating a more. If he walks down the street wearing nothing on the top half his body, he is violating a folkway.

Despite efforts to control them, people often reject elements of existing culture and create new elements of culture. Reasons for this are discussed later in this chapter and in Chapter 7 ("Deviance and Crime"), Chapter 18 ("The Mass Media"), and Chapter 21 ("Collective Action and Social Movements"). Here it is enough to say that just as social control is needed to ensure stable patterns of interaction, so resistance to social control is needed to ensure cultural innovation and social renewal. Stable but vibrant societies are able to find a balance between social control and cultural innovation.

Culture and Biology

"Nature, Mr. Allnut, is what we are put in the world to rise above."

ROSE SAYER (ACTOR KATHARINE HEPBURN) IN *THE AFRICAN QUEEN* (1951)

The Evolution of Human Behavior

We have seen how the human capacity for abstraction, cooperation, and production enables us to create culture and makes us distinctively human. This capacity is built on a solid biological foundation. Biology, as every sociologist recognizes, sets broad human limits and potentials, including the potential to create culture.

However, some students of human behavior go a step further. Practitioners of sociobiology and evolutionary psychology claim that human brain structure and genes—chemical units that carry traits from parents to children—account not just for physical characteristics but also for specific behaviors and social practices (Wilson, 1975; Pinker, 2002; Tooby and Cosmides, 1992). From their point of view, genes, for example, determine not just whether our eyes are blue or brown but also many aspects of our social behavior. This kind of argument has become increasingly popular since the early 1970s. Most sociologists disagree with it. We therefore devote a few paragraphs to the argument's misconceptions (see also Chapter 11, "Sexuality and Gender").

Evolutionary psychology's starting point is Charles Darwin's theory of evolution. Darwin (1859) observed wide variations in the physical characteristics of members of each species. For example, some deer can run quickly. Others run slower. Because of such variations, the "fittest" members of each species—the quicker deer, for example—are more likely to survive long enough to have offspring. Therefore, concluded Darwin, the species characteristics that endure are those that increase the survival chances of the species.

Male Promiscuity, Female Fidelity, and Other Myths

Contemporary evolutionary psychologists make similar arguments about human behavior and social arrangements. Typically, *they first identify a supposedly universal human behavioral trait.* For example, they claim that men are more likely than women to want many sexual partners.

They next offer an explanation as to why this behavior increases survival chances. Thus, to continue with our example, they account for supposedly universal male promiscuity

The system of **social control** is the sum of sanctions in society by means of which conformity to cultural guidelines is ensured.

Taboos are among the strongest norms. When someone violates a taboo, it causes revulsion in the community, and punishment is severe.

Mores are core norms that most people believe are essential for the survival of their group or society.

Folkways are the least important norms and violating them evokes the least severe punishment.

Table 3.1
Number of Sexual Partners by Respondent's Sex, United States, 2002 (in percent)

	RESPONDENT'S SEX	
Number of Sexual Partners	Male	Female
0 or 1	79	90
More than 1	21	10
Total	100	100
N	1004	1233

Source: National Opinion Research Center (2004).

Table 3.2
Number of Sexual Partners by Respondent's Sex, United States, 2002, Married Respondents Only (in percent)

	RESPONDENT'S SEX	
Number of Sexual Partners	Male	Female
0 or 1	95	99
More than 1	5	1
Total	100	100
N	499	534

Source: National Opinion Research Center (2004).

and female fidelity as follows. Every time a man ejaculates, he produces hundreds of millions of sperm. In contrast, a woman typically releases only one egg per month between puberty and menopause in periods when she is not pregnant. Evolutionary psychologists claim that because of these sex differences, men and women develop different strategies to increase the chance they will reproduce their genes. Because a woman produces few eggs, she improves her chance of reproducing her genes if she has a mate who stays around to help and protect her during those few occasions when she is pregnant, gives birth, and nurses a small infant. Because a man's sperm is so plentiful, he improves his chance of reproducing his genes if he tries to impregnate as many women as possible. In short, women's desire for a single mate and men's desire for many sexual partners is simply the way men and women play out the game of survival of the fittest. Even male rapists, writes one evolutionary psychologist, may just be "doing the best they can to maximize their [reproductive] fitness" (Barash, 1981: 55).

The final part of the evolutionary psychologists' argument is that the behavior in question cannot easily be changed. The characteristics that maximize the survival chances of a species supposedly get encoded or "hardwired" in our genes. It follows that what exists is necessary.

Most sociologists and many biologists and psychologists are critical of the reasoning of evolutionary psychologists (Gould and Lewontin, 1979; Lewontin, 1991; Schwartz, 1999). In the first place, *some behaviors discussed by evolutionary psychologists are not universal and some are not even that common.* Consider male promiscuity. Is it true that men are promiscuous and women are not? The data tell a different story. According to the 2002 General Social Survey, only a minority of adult American men (21 percent) claimed they had more than one sexual partner in the previous year (Table 3.1). The figure for adult American women was significantly lower (10 percent). However, if we consider married adults only, the figures fall to 5 percent for men and 1 percent for women, a much smaller difference (Table 3.2). Apparently, certain *social* arrangements such as the institution of marriage account for variations in male promiscuity. There is no *universal* propensity to male promiscuity.

Still, 11 percent more men than women said they had more than one sexual partner in the year preceding the survey. Among unmarried people, the male-female difference was 14 percent. Researchers have identified two main reasons for the difference (McConaghy, 1999: 311–14). First, men seem to be somewhat more likely than women to have sexual relations with members of their own sex, and men who do so are more likely to have many sexual partners than are women who do so.[1] This contradicts the evolutionary psychologists' argument, which ties male promiscuity to male reproductive strategies. Second, men

[1]There were too few respondents in the 2000 GSS to be able to draw this conclusion with confidence, so we base it on cumulative data from 1988–2002 (n = 13,981). In this period, 9.1 percent of men with more than one sexual partner had sexual relations with other men, while 7.7 percent of women with more than one sexual partner had sexual relations with other women.

apparently tend to exaggerate how many sexual partners they have when asked about this subject in surveys because our culture puts a premium on male sexual performance. This, too, is bad news for the evolutionary psychologists, who would like us to believe that men are actually and naturally promiscuous, not influenced by cultural standards to simply say they are.

In general, then, the evolutionary psychologists' claim about male promiscuity and female fidelity is false. So are many of their other claims about so-called behavioral constants or universals.

The second big problem with evolutionary psychology is that nobody has ever verified that specific behaviors and social arrangements are associated with specific genes (for an exception, see the discussion of language later). Therefore, when it comes to supporting their key argument, evolutionary psychologists have little to stand on apart from a fragile string of maybes and possibilities: "[W]e *may* have to open our minds and admit the *possibility* that our need to maximize our [reproductive] fitness *may* be whispering somewhere deep within us and that, *know it or not,* most of the time we are heeding these whisperings" (Barash, 1981: 31; our emphasis). Maybe. Then again, maybe not.

The National Post, Toronto, Canada, 2000

When scientists announced they had finished sequencing the human genome on June 26, 2000, some people thought all human characteristics could be read from the human genetic "map." They cannot. The functions of most genes are still unknown. Moreover, because genes mutate randomly and interact with environmental (including social) conditions, the correspondence between genetic function and behavioral outcome is highly uncertain.

Finally, even if researchers eventually discover an association between particular genes and particular behaviors, it would be wrong to conclude that variations among people are due just to their genes. As one of the world's leading biologists writes, "Variations among individuals within species are a unique consequence of both genes *and environment* in a constant interaction . . . [and] random variation in growth and division of cells during development" (Lewontin, 1991: 26–7; our emphasis). Genes *never* develop without environmental influence. The genes of a human embryo, for example, are profoundly affected by whether the mother consumes the recommended daily dose of calcium or nearly overdoses daily on crack cocaine. And what the mother consumes is, in turn, determined by many social factors. Even if one inherits a mutant cancer gene, the chance of developing cancer is strongly influenced by diet, exercise, tobacco consumption, and factors associated with occupational and environmental pollution.[2] It follows that the pattern of your life is not entirely hardwired by your genes. Changes in social environment do produce physical and, to an even greater degree, behavioral change. However, to figure out the effects of the social environment on human behavior we have to abandon the premises of evolutionary psychology and develop specifically sociological skills for analyzing the effects of social structure and culture.

Language and the Sapir-Whorf Thesis

One important field in which biological thinking has been influential in recent years is the study of language. A **language** is a system of symbols strung together to communicate thought. Equipped with language, we can share understandings, pass experience and knowledge from one generation to the next, and make plans for the future. In short, language allows culture to develop. Consequently, sociologists commonly think of language as a cultural invention that distinguishes humans from other animals.

Is Language Innate or Learned?

Yet MIT cognitive scientist Steven Pinker, a leading figure in the biological onslaught, says culture has little to do with our acquisition of language. In his view, "people know

[2]Some cancers are more heritable than others, but even the most heritable cancers seem to be much more strongly influenced by environmental than genetic factors (Fearon, 1997; Hoover, 2000; Kevles, 1999; Lichtenstein et al., 2000; Remennick, 1998).

A **language** is a system of symbols strung together to communicate thought.

how to talk in more or less the sense that spiders know how to spin webs" (Pinker, 1994: 18). Language, says Pinker, is an "instinct."

Pinker bases his radical claim on the observation that most people can easily create and understand sentences that have never been uttered before. We even invent countless new words (including the word "countless," which was invented by Shakespeare). We normally develop this facility quickly and without formal instruction at an early age. This suggests that people have a sort of innate recipe or grammar for combining words in patterned ways. In support, he discusses cases of young children with different language backgrounds who were brought together in settings as diverse as Hawaiian sugar plantations in the 1890s and Nicaraguan schools for the deaf in the 1970s and who spontaneously created their own language system and grammatical rules.

If children are inclined to create grammars spontaneously at a young age, we can also point to seats of language in the brain; damage to certain parts of the brain impairs one's ability to speak, although intelligence is unaffected. Moreover, scientists have identified a gene that may help wire these seats of language into place. A few otherwise healthy children fail to develop language skills. They find it hard to articulate words and they make a variety of grammatical errors when they speak. If these language disorders cannot be attributed to other causes, they are diagnosed as Specific Language Impairment (SLI). Recently it was discovered that a mutation of a gene known as FOXP2 is associated with SLI (Pinker, 2001). Only when the gene is normal do children acquire complex language skills at an early age. From these and similar observations, Pinker concludes that language is not so much learned as it is grown. Should we believe him?

The Social Roots of Language

From a sociological point of view, there is nothing problematic about the argument that we are biologically prewired to acquire language and create grammatical speech patterns. What is sociologically interesting, however, is how the social environment gives these predispositions form. We know, for example, that young children go through periods of rapid development, and if they do not interact symbolically with others during these critical periods, their language skills remain permanently impaired (Sternberg, 1998: 312). This suggests that our biological potential must be unlocked by the social environment to be fully realized. Language must be learned. The environment is in fact such a powerful influence on language acquisition that even a mutated FOXP2 gene doesn't seal one's linguistic fate. Up to half of children with SLI recover fully with intensive language therapy (Shanker, 2002).

In an obvious sense, all language is learned, even though our potential for learning and the structure of what we can learn is rooted in biology. Our use of language depends on which language communities we are part of. You say tomāto and I say tomăto, but Luigi says *pomodoro* and Shoshanna says *agvaniya.* But what exactly is the relationship between our use of language, the way we think, and our social environment? We now turn to just that question.

The Sapir-Whorf Thesis and Its Critics

In the 1930s, linguists Edward Sapir and Benjamin Lee Whorf first proposed that experience, thought, and language interact in what came to be known as the **Sapir-Whorf thesis.** The Sapir-Whorf thesis holds that we experience certain things in our environment and form concepts about those things (path 1⟶2 in Figure 3.1). We then develop language to express our concepts (path 2⟶3). Finally, language itself influences how we see the world (path 3⟶1).

The **Sapir-Whorf thesis** holds that we experience certain things in our environment and form concepts about them. We then develop language to express our concepts. Finally, language itself influences how we see the world.

Whorf saw speech patterns as "interpretations of experience" (path 1⟶2⟶3; Whorf, 1956: 137). This seems uncontroversial. The Garo of Burma, a rice-growing people, distinguish many types of rice. Nomadic Arabs have more than 20 different words for camel (Sternberg, 1998 [1995]: 305). Verbal distinctions among types of rice and camels are necessary for different groups of people because these objects are important in their environment.[3] As a matter of necessity, they distinguish among many different types of what we may regard as "the same" object. Similarly, terms that apparently refer to the same things or people may change to reflect a changing reality. For example, a committee used to be headed by a "chairman." Then, when women started entering the paid labor force in large numbers in the 1960s and some of them became committee heads, the term changed to "chairperson" or simply "chair." In such cases, we see clearly how the environment or experience influences language.

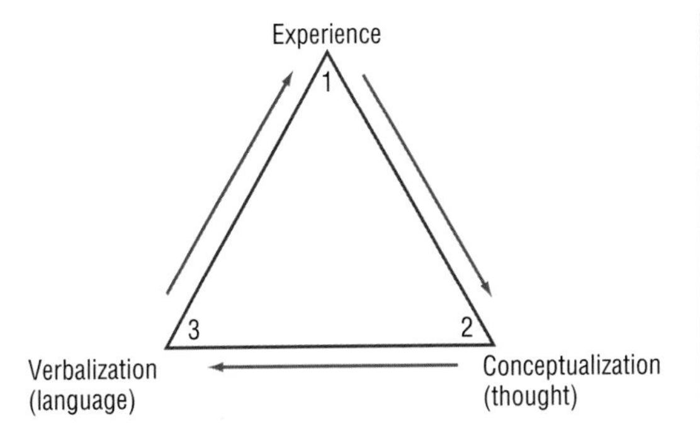

Figure 3.1
The Sapir-Whorf Thesis

It is equally uncontroversial to say that people must think before they can speak (path 2⟶3). Anyone who has struggled for just the right word or rewritten a sentence to phrase a thought more precisely knows this.

The controversial part of the Sapir-Whorf thesis is path 3⟶1. In what sense does language *in and of itself* influence the way we experience the world? In the first wave of studies based on the Sapir-Whorf thesis, researchers focused on whether speakers of different languages perceived color in different ways. By the 1970s, they concluded that they did not. People who speak different languages may have a different number of basic color terms, but everyone with normal vision is able to see the full visible spectrum. There are two words for blue in Russian and only one in English, but this does not mean that English speakers are somehow handicapped in their ability to distinguish shades of blue.

In the 1980s and 1990s, researchers found some effects of language on perception. For example, the German word for key is masculine, whereas the Spanish word for key is feminine. When German and Spanish speakers are asked to describe keys, German speakers tend to use words like "hard," "heavy," and "jagged," whereas Spanish speakers use words such as "lovely," "shiny," and "shaped." Apparently, the gender of the noun in and of itself influences how people see the thing to which the noun refers (Minkel, 2002). Still, the degree to which language itself influences thought is a matter of controversy. Some men use terms like "fox," "babe," "bitch," "ho," and "doll" to refer to women. These terms are deeply offensive to many people. They certainly reflect and reinforce underlying inequalities between women and men. Some people assert that these terms *in and of themselves* influence people to think of women simply as sexual objects, but social scientists have yet to demonstrate the degree to which they do so.

We conclude that biological thinking about culture has both benefits and dangers. On the one hand, biology helps us see more clearly the broad limits and potentials of human creativity. On the other hand, some scholars have managed to get themselves trapped in a biological straitjacket. They fail to appreciate how the social environment unlocks biological potentials and creates enormous variation in their cultural expression. Analyzing culture in all its variety and showing how cultural variations are related to variations in social structure are jobs for the sociologist. In the rest of this chapter, we show how sociologists go about these jobs. We begin by first considering how it is possible to observe culture in a relatively unbiased fashion.

---

[3]Whorf wrote that the Inuit ("Eskimos") have seven words for different types of snow, but the claim is misleading. Inuktitut (the Inuit language) has no single word for snow—or bear, or people, or fish—because the language combines adjectives and nouns into new terms. It is this grammatical feature of Inuktitut that allows separate words for "snow on the ground," "water-soaked snow," "snowbank around the house," and even "snow that has been peed on." Meanwhile, in English, too, we have many words for types of snow: sleet, hail, powder, slush, hardpack, flurry, and so forth (Crucefix, 2003; Minkel, 2002).

Many Westerners find the Indian practice of cow worship bizarre. However, cow worship performs a number of useful economic functions and is in that sense entirely rational. By viewing cow worship exclusively as an outsider (or, for that matter, exclusively as an insider), we fail to see its rational core.

Culture and Ethnocentrism: A Functionalist Analysis of Culture

Culture, despite its central importance in human life, is often invisible. That is, people tend to take their own culture for granted; it usually seems so sensible and natural that they rarely think about it (Box 3.1). In contrast, people are often startled when confronted by cultures other than their own. That is, the ideas, norms, and techniques of other cultures frequently seem odd, irrational, and even inferior.

Judging another culture exclusively by the standards of one's own is known as **ethnocentrism.** Ethnocentrism impairs sociological analysis. This can be illustrated by a practice that seems bizarre to many Westerners: cow worship among Hindu peasants in India.

Hindu peasants refuse to slaughter cattle and eat beef because, for them, the cow is a religious symbol of life. Pinup calendars throughout rural India portray beautiful women with the bodies of fat, white cows, milk jetting out of each teat. Cows are permitted to wander the streets, defecate on the sidewalks, and stop to chew their cud in busy intersections or on railroad tracks, causing traffic to come to a complete halt. In Madras, police stations maintain fields where stray cows that have fallen ill can graze and be nursed back to health. The government even runs old-age homes for cows, where dry and decrepit cattle are kept free of charge. All this seems utterly inscrutable to most Westerners, for it takes place amid poverty and hunger that could presumably be alleviated if only the peasants would slaughter their "useless" cattle for food instead of squandering scarce resources feeding and protecting them.

According to anthropologist Marvin Harris, however, ethnocentrism misleads many Western observers (Harris, 1974: 3–32). Cow worship, it turns out, is an economically rational practice in rural India. For one thing, Indian peasants can't afford tractors, so cows are needed to give birth to oxen, which are in high demand for plowing. For another, the cows produce hundreds of millions of pounds of recoverable manure, about half of which is used as fertilizer and half as a cooking fuel. With oil, coal, and wood in short supply, and with the peasants unable to afford chemical fertilizers, cow dung is, well, a godsend. What is more, cows in India don't cost much to maintain because they eat mostly food that isn't fit for human consumption. And they represent an important source of protein and a livelihood for members of low-ranking castes, who have the right to dispose of the bodies of dead cattle. These "untouchables" eat beef and form the workforce of India's large leathercraft industry. The protection of cows by means of cow worship is thus a perfectly sensible and highly efficient economic practice. It seems irrational only when judged by the standards of Western agribusiness.

Harris's analysis of cow worship in rural India is interesting for two reasons. First, it illustrates how functionalist theory can illuminate otherwise mysterious social practices. Harris uncovers a range of latent functions performed by cow worship, thus showing how a particular social practice has unintended and nonobvious consequences that make social order possible. Second, we can draw an important lesson about ethnocentrism from Harris's analysis. If you refrain from judging other societies by the standards of your own, you will have taken an important first step toward developing a sociological understanding of culture.

Ethnocentrism is the tendency to judge other cultures exclusively by the standards of one's own.

BOX 3.1
Sociology at the Movies

Austin Powers: International Man of Mystery (1997), *The Spy Who Shagged Me* (1999), and *Goldmember* (2002)

All Austin Powers movies are about super-spy Austin Powers battling his arch-enemy Dr. Evil and his plan to destroy the world. For instance, in *The Spy Who Shagged Me* (1999), Austin Powers and Dr. Evil—both played by Mike Myers—happen to have been frozen in the 1960s and unthawed in the 1990s. Dr. Evil figures he can defeat Austin Powers if he returns to 1969 in a time machine and steals the legendary sexual energy, or "mojo," from Powers' still-frozen body. However, the mojo-less Powers remains determined to save the world and, along the way, recapture the mojo that makes him "deadly to his enemies" and "irresistible to women." The film, like others in this series, is half satire and half tribute to the hugely popular spy movies and TV series of the 1960s and 1970s, especially, the Sean Connery–era James Bond films.

Austin Powers films are funny because we are well aware of the enormous cultural changes that have taken place over the past three or four decades, whereas Austin Powers is not. We laugh at his velvet jumpsuit, frilly shirt, heavy necklace, and big dark-framed glasses. We roar at his use of 1960s slang ("Yeah, baby!"). And what about his efforts to prove himself an expert at "shagging"? We might regard his attempts at seduction as blatant sexual harassment. If he could read our minds, Austin Powers would no doubt call us "uptight."

Apart from being funny, the Austin Powers movies have sociological significance. They remind us that no culture is static. Cultural changes that occur within even a few short decades can be profound. Many of the fashions, expressions and behaviors we take for granted and think of as "cool" today will likely seem ridiculous to us tomorrow. You might even consider putting together a scrapbook of today's fads and fashions, to be opened in just a few years. Inevitably, when the time comes to open your "time capsule," you will experience a mixture of nostalgia and amusement.

By making the contemporary world a foreign world to Austin Powers, these movies invite us to turn a critical eye on our own culture. This is no easy task. As the anthropologist Ralph Linton (1936) observed many years ago, "The last thing a fish would ever notice would be water." Much of the sociological value of the Austin Powers movies is that they make us notice the water.

The Everett Collection

Mike Myers as Austin Powers in *Goldmember* (2002)

Web Interactive Exercise: How Is Social Inequality Justified in Our Culture?

The Two Faces of Culture: Freedom and Constraint

Culture has two faces. First, it provides us with an opportunity to exercise our *freedom.* We create elements of culture in our everyday life to solve practical problems and express our needs, hopes, joys, and fears.

However, creating culture is just like any other act of construction in that we need raw materials to get the job done. The raw materials for the culture we create consist of cultural elements that either existed before we were born or were created by other people since our birth. We may put these elements together in ways that produce something genuinely new. But there is no other well to drink from, so existing culture puts limits on what we can think and do. In that sense, culture *constrains* us. This is culture's second face.

Culture as Freedom

Because culture can be seen both as an opportunity for freedom and as a source of constraint, we will examine both faces of culture. We begin with the view that culture is an opportunity for freedom. We first establish that people are not just passive recipients but active producers and interpreters of culture. Next, we show that the range of cultural choices available to us has never been greater, because we live in a society that is characterized by unparalleled cultural diversity. We then show how globalization processes contribute to the diversification of culture and broaden the range of cultural choices open to us. We argue that this has led to the emergence of a new, "postmodern" era of culture. After developing the idea that culture is a source of freedom, we turn to culture's flipside as a source of social constraint.

Cultural Production and Symbolic Interactionism

Until the 1960s, many sociologists argued that culture is simply a "reflection" of society. Using the language introduced in Chapter 2, we can say that they regarded culture as a dependent variable. Harris's analysis of rural Indians certainly fits that mold. In Harris's view, the social necessity of protecting cows caused the *cultural* belief that cows are holy.

In recent decades, the symbolic interactionist tradition we discussed in Chapter 1 has influenced many sociologists of culture. Symbolic interactionists are inclined to regard culture as an *independent* variable. In their view, people do not accept culture passively; we are not empty vessels into which society pours a defined assortment of beliefs, symbols, and values. Instead, we actively produce and interpret culture, creatively fashioning it and attaching meaning to it in accordance with our diverse needs.

British literary critic Richard Hoggart (1958) and social historian E. P. Thompson (1968) wrote pioneering works emphasizing how people produce and interpret culture. Hoggart and Thompson showed how working-class people shape the cultural *environments* in which they live. For instance, religious ideas and secular reading materials may be created for members of the working class by people in higher-class positions—"from the outside," as it were. What then happens, according to Hoggart and Thompson, is that members of the working class make sense of these elements of culture on their own terms. In general, audiences always change ideas to make them meaningful to themselves. This line of thought was developed by sociologist Stuart Hall (1980) and his colleagues, who showed how people mold culture to fit their sense of self. It gave rise to the field of "cultural studies," which overlaps the sociology of culture (Griswold, 1994; Long, 1997; Wolff, 1999). Later in this chapter and again in Chapter 18 ("The Mass Media"), we take up some of the themes introduced by Hoggart, Thompson, and Hall.

Cultural Diversity

The idea that people actively produce and interpret culture implies that, to a degree, we are at liberty to choose how culture influences us Let us linger a moment on the question of why we enjoy that freedom today more than ever before.

Part of the reason we are increasingly able to choose how culture influences us is that a greater diversity of culture is available from which to choose. Thus, the proportion of the population that consists of immigrants is higher in the United States than in all but three other countries (Israel, Australia, and Canada). Moreover, ethnically and racially, the United States is a more heterogeneous society now than at any point in its history. According to the United States Census Bureau, more than 28 percent of the United States population was nonwhite or Hispanic in 2000. Over the next 50 years, Hispanic and Asian American groups are projected to grow more than 200 percent. African and Native American groups are projected to grow 60–70 percent. Meanwhile, the white non-Hispanic group will grow less than 6 percent. It will begin to shrink after 2030. Sometime around 2060, white non-Hispanics will form a *minority* of the United States population (U.S. Census Bureau, 2000g).

The United States continues to diversify culturally.

The cultural diversification of American society is evident in all aspects of life, from the growing popularity of Latino music to the increasing influence of Asian design in clothing and architecture to the ever-broadening international assortment of foods consumed by most Americans. Marriage between people of different ethnic groups is widespread, and marriage between people of different races is increasingly common. For example, about half of Asian Americans and a tenth of African Americans now marry outside their racial group (Stanfield, 1997). At the political level, however, cultural diversity has become a source of conflict. This is nowhere more evident than in the debates that have surfaced in recent years concerning curricula in the American educational system.

Until recent decades, the American educational system stressed the common elements of American culture, history, and society. Students learned the story of how European settlers overcame great odds, prospered, and forged a unified nation out of diverse ethnic and racial elements. School curricula typically neglected the contributions of nonwhites and non-Europeans to America's historical, literary, artistic, and scientific development. Moreover, students learned little about the less savory aspects of American history, many of which involved the use of force to create a strict racial hierarchy that persists to this day, albeit in modified form (see Chapter 10, "Race and Ethnicity"). History books did not deny that African Americans were enslaved and that force was used to wrest territory from Native Americans and Mexicans. They did, however, make it seem as if these unfortunate events were part of the American past, with few implications for the present. The history of the United States was presented as a history of progress involving the *elimination* of racial privilege.

Multiculturalism

Supporters of **multiculturalism** argue that the curricula of America's public schools and colleges should reflect the country's ethnic and racial diversity and recognize the equality of all cultures.

For the past several decades, advocates of **multiculturalism** have argued that school and college curricula should present a more balanced picture of American history, culture, and society—one that better reflects the country's ethnic and racial diversity in the past and its growing ethnic and racial diversity today (Ball, Berkowitz, and Mzamane, 1998). A multicultural approach to education highlights the achievements of nonwhites and non-Europeans in American society. It gives more recognition to the way European settlers came to dominate non-white and non-European communities, stresses how racial domination resulted in persistent social inequalities, and encourages elementary-level instruction in Spanish in California, Texas, New Mexico, Arizona, and Florida, where a substantial minority of people speak Spanish at home. (About one in seven Americans over the age of 5 speaks a language other than English at home. Of these people, more than half speak Spanish. Most Spanish speakers live in the states just listed.)

Sociology Now™

Learn more about **Multiculturalism** by going through the Multiculturalism Animation.

Most critics of multiculturalism do not argue against teaching cultural diversity. What they fear is that multiculturalism is being taken too far (Glazer, 1997; Schlesinger, 1991; Stotsky, 1999). Specifically, they say multiculturalism has three negative consequences:

1. Critics believe that multicultural education hurts minority students by forcing them to spend too much time on noncore subjects. To get ahead in the world, they say, one needs to be skilled in English and math. By taking time away from these subjects, multicultural education impedes the success of minority-group members in the work world. (Multiculturalists counter that minority students develop pride and self-esteem from a curriculum that stresses cultural diversity. They argue that this helps minority students get ahead in the work world.)
2. Critics also believe that multicultural education causes political disunity and results in more interethnic and interracial conflict. Therefore, they want schools and colleges to stress the common elements of the national experience and highlight Europe's contribution to American culture. (Multiculturalists reply that political unity and interethnic and interracial harmony simply maintain inequality in American society. Conflict, they say, while unfortunate, is often necessary to achieve equality between majority and minority groups.)
3. Finally, critics of multiculturalism complain that it encourages the growth of **cultural relativism.** Cultural relativism is the opposite of ethnocentrism. It is the belief that all cultures and all cultural practices have equal value. The trouble with this view is that some cultures oppose the most deeply held values of most Americans (Box 3.2). Other cultures promote practices that most Americans consider inhumane. Should we respect racist and antidemocratic cultures, such as the apartheid regime that existed in South Africa from 1948 until 1992? How about female circumcision, which is still widely practiced in Somalia, Sudan, and Egypt? Or the Australian aboriginal practice of driving spears through the limbs of criminals (Garkawe, 1995)? Critics argue that by promoting cultural relativism, multiculturalism encourages respect for practices that are abhorrent to most Americans. (Multiculturalists reply that cultural relativism need not be taken to such an extreme. *Moderate* cultural relativism encourages tolerance, and it should be promoted.)

Sociology Now™

Learn more about **Cultural Relativism** by going through the Ethnocentrism versus Cultural Relativism Learning Module.

Clearly, multiculturalism is a complex and emotional issue requiring much additional research and debate. It is worth pondering here, however, because it says something important about the state of American culture today and, more generally, about how world culture has developed since our remote ancestors lived in tribes. Even 50 years ago, the American ideal was to create one new culture out of many—*E pluribus unum.* Today, multiculturalism stands for the opposite—creating many cultures out of one. Nor is this shift unique to the United States. In general, as we will soon see, cultures tend to become more heterogeneous over time, with important consequences for everyday life.

The Rights Revolution: A Conflict Analysis of Culture

Cultural relativism is the belief that all cultures have equal value.

What are the social roots of cultural diversity and multiculturalism? Conflict theory suggests where to look for an answer. Recall from Chapter 1 the central argument of conflict theory: Social life is an ongoing struggle between more and less advantaged groups. Privileged groups try to maintain their advantages while subordinate groups struggle to increase theirs. And sure enough, if we probe beneath cultural diversification and multi-

BOX 3.2

SOCIAL POLICY: WHAT DO YOU THINK?

Female Genital Mutilation: Cultural Relativism or Ethnocentrism?

The World Health Organization defines female genital mutilation as "all procedures involving partial or total removal of the external female genitalia or other injury to the female genital organs whether for cultural or other nontherapeutic reasons" (World Health Organization [WHO], 2001). Elderly women who lack medical training usually perform these procedures.

Female genital mutilation results in pain, humiliation, psychological trauma, and loss of sexual pleasure. In the short term it is associated with infection, shock, injury to neighboring organs, and severe bleeding. In the long term it is associated with infertility, chronic infections in the urinary tract and reproductive system, increased susceptibility to hepatitis B and HIV/AIDS, and so forth.

Although frequently associated with Islam, female genital mutilation is rare in many predominantly Muslim countries. It is a social custom, not a religious practice. It is nearly universal in Somalia, Djibouti, and Egypt, and common in other parts of Africa. Over 132 million women and girls worldwide have undergone female genital mutilation. About two million girls are at risk of undergoing it every year (Ahmad, 2000; WHO, 2001).

Female genital mutilation is typically performed as a rite of passage on girls between the ages of 4 and 14. In some cultures, people think it enhances female fertility. Furthermore, they commonly assume that women are naturally "unclean" and "masculine" inasmuch as they possess a vestige of a "male" sex organ, the clitoris. From this point of view, women who have not experienced genital mutilation are more likely to demonstrate "masculine" levels of sexual interest and activity. They are less likely to remain virgins before marriage and faithful within it. Accordingly, some people think that female genital mutilation lessens or eradicates sexual arousal in women.

One reaction to female genital mutilation is the "human rights perspective." In this view, the practice is simply one manifestation of gender-based oppression and the violence that women experience in societies worldwide. Adopting this perspective, the United Nations has defined female genital mutilation as a form of violence against women. This perspective is also reflected in a growing number of international, regional, and national agreements that commit governments to preventing female genital mutilation, assisting women at risk of undergoing it, and punishing people who commit it. In the United States, the penalty for conducting female genital mutilation is up to five years in prison. The law stresses that "belief . . . that the operation is required as a matter of custom or ritual" is irrelevant in determining its illegality (United States Code, 1998).

Proponents of a second perspective on female genital mutilation are cultural relativists. They regard the human rights perspective as ethnocentric. The cultural relativists view interventions that interfere with the practice as little more than neo-imperialist attacks on African cultures. From their point of view, all talk of "universal human rights" denies cultural sovereignty to less powerful peoples. Moreover, opposition to female genital mutilation undermines tolerance and multiculturalism while reinforcing racist attitudes. Accordingly, cultural relativists argue that we should affirm the right of other cultures to practice female genital mutilation even if we regard it as destructive, senseless, oppressive, and abhorrent. We should respect the fact that other cultures regard female genital mutilation as meaningful and as serving useful functions.

Which of these perspectives do you find more compelling? Do you believe that certain principles of human decency transcend the particulars of any particular culture? If so, what are those principles? If you do not believe in the existence of any universal principles of human decency, then does anything go? Would you agree that, say, genocide is acceptable if most people in a society favor it? Or are there limits to your cultural relativism? In a world where supposedly universal principles often clash with the principles of particular cultures, where do you draw the line?

culturalism, we find what has been called the **rights revolution,** the process by which socially excluded groups have struggled to win equal rights under the law and in practice.

After the outburst of nationalism, racism, and genocidal behavior among the combatants in World War II, the United Nations proclaimed the Universal Declaration of Human Rights in 1948. Its preamble reads in part:

> Whereas recognition of the inherent dignity and of the equal and inalienable rights of all members of the human family is the foundation of freedom, justice and peace in the world. . . . Now, therefore The General Assembly proclaims this Universal Declaration of Human Rights as a common standard of achievement for all peoples and all nations, to the end that every individual and every organ of society, keeping this Declaration constantly in mind, shall strive by teaching and education to promote respect for these rights and freedoms and by progressive measures, national and international, to secure their universal and effective recognition and observance. (United Nations, 1998)

The **rights revolution** is the process by which socially excluded groups have struggled to win equal rights under the law and in practice since the 1960s.

Fanned by such sentiment, the rights revolution was in full swing by the 1960s. Today, women's rights, minority rights, gay and lesbian rights, the rights of people with special needs, constitutional rights, and language rights are key parts of our political discourse. Due to the rights revolution, democracy has been widened and deepened (see Chapter 14, "Politics"). The rights revolution is by no means finished. Many categories of people are still discriminated against socially, politically, and economically. However, in much of the world, all categories of people now participate more fully than ever before in the life of their societies (Ignatieff, 2000).

The rights revolution raises some difficult issues. For example, some members of groups that have suffered extraordinarily high levels of discrimination historically, such as Native and African Americans, have demanded reparations in the form of money, symbolic gestures, land, and political autonomy (see Chapter 10, "Race and Ethnicity"). Much controversy surrounds the extent of the obligation of current citizens to compensate past injustices.

Such problems notwithstanding, the rights revolution is here to stay and it affects our culture profoundly. Specifically, the rights revolution fragments American culture by (1) legitimizing the grievances of groups that were formerly excluded from full social participation and (2) renewing their pride in their identity and heritage. Our history books, our literature, our music, our use of languages, our very sense of what it means to be American have diversified culturally. White, male, heterosexual property owners of northern European origin are still disproportionately influential in the United States, but our culture is no longer dominated by them in the way it was just four decades ago.

From Diversity to Globalization

In preliterate or tribal societies, cultural beliefs and practices are virtually the same for all group members. For example, many tribal societies organize **rites of passage.** These are cultural ceremonies that mark the transition from one stage of life to another (e.g., baptisms, confirmations, weddings) or from life to death (e.g., funerals). These religious rituals involve elaborate body painting, carefully orchestrated chants and movements, and so forth. They are conducted in public, and no variation from prescribed practice is allowed. Culture is homogeneous (Durkheim, 1976 [1915]).

In contrast, preindustrial western Europe and North America were rocked by artistic, religious, scientific, and political forces that fragmented culture. The Renaissance, the Protestant Reformation, the Scientific Revolution, the French and American Revolutions—between the 14th and 18th centuries, all of these movements involved people questioning old ways of seeing and doing things. Science placed skepticism about established authority at the very heart of its method. Political revolution proved there was nothing ordained about who should rule and how they should do so. Religious dissent ensured that the Catholic Church would no longer be the supreme interpreter of God's will in the eyes of all Christians. Authority and truth became divided as never before.

Cultural fragmentation picked up steam during industrialization, as the variety of occupational roles grew and new political and intellectual movements crystallized. Its pace is quickening again today in the postindustrial era. This is due to globalization, the process by which formerly separate economies, states, and cultures are being tied together and people are becoming increasingly aware of their growing interdependence.

The roots of globalization are many. International trade and investment are expanding. Even a business as "American" as McDonald's now reaps 60 percent of its profits from outside the United States, and its international operations are expected to grow at four times the rate of its U.S. outlets (Commins, 1997). At the same time, members of different ethnic and racial groups are migrating and coming into sustained contact with one another. A growing number of people date, court, and marry across religious, ethnic, and racial lines. Influential "transnational" organizations such as the International Monetary Fund, the World Bank, the European Union, Greenpeace, and Amnesty International are multiplying. Relatively inexpensive international travel and communication make contacts between people from diverse cultures routine. The mass media make Vin Diesel and

Rites of passage are cultural ceremonies that mark the transition from one stage of life to another (e.g., baptisms, confirmations, weddings) or from life to death (e.g., funerals).

Figure 3.2
The Effect of Globalization on Corn Flakes

The idea of globalization first gained prominence in marketing strategies in the 1970s. In the 1980s, companies like Coca-Cola and McDonald's expanded into non-Western countries to find new markets. Today, Kellogg's markets products in more than 160 countries. Basmati Flakes cereal was first produced by the Kellogg's plant in Tajola, India, in 1992.

Courtesy of Kelloggs

Survivor nearly as well known in Warsaw as in Wichita. MTV brings rock music to the world via MTV Latino, MTV Brazil, MTV Europe, MTV Asia, MTV Japan, MTV Mandarin, and MTV India (Hanke, 1998). Globalization, in short, destroys political, economic, and cultural isolation, bringing people together in what Canadian media analyst Marshall McLuhan (1964) called a "global village." As a result of globalization, people are less obliged to accept the culture into which they are born and freer to combine elements of culture from a wide variety of historical periods and geographical settings. Globalization is a schoolboy in Bombay, India, listening to Bob Marley on his MP3 player as he rushes to slip into his Levis, wolf down a bowl of Kellogg's Basmati Flakes, and say good-bye to his parents in Hindi because he's late for his English-language school (Figure 3.2).

The Rise of English and the Decline of Indigenous Languages

A good indicator of the influence and extent of globalization is the spread of English since 1600. In 1600, English was the mother tongue of 4–7 million people. Not even all people in England spoke it. Today, 750 million to 1 billion people speak English worldwide, over half as a second language. With the exception of the many varieties of Chinese, English is the most widespread language on earth. Over half the world's technical and scientific periodicals are written in English. English is the official language of the Olympics, of the Miss Universe contest, of navigation in the air and on the seas, and of the World Council of Churches (Figure 3.3).

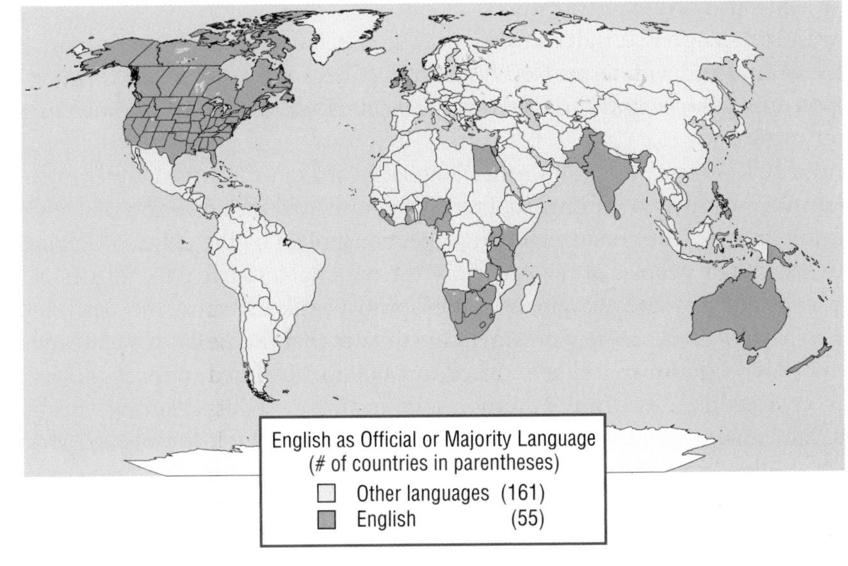

Figure 3.3
English as Official or Majority Language

Sources: United Nations Educational, Scientific and Cultural Organization (2001); Central Intelligence Agency (2002).

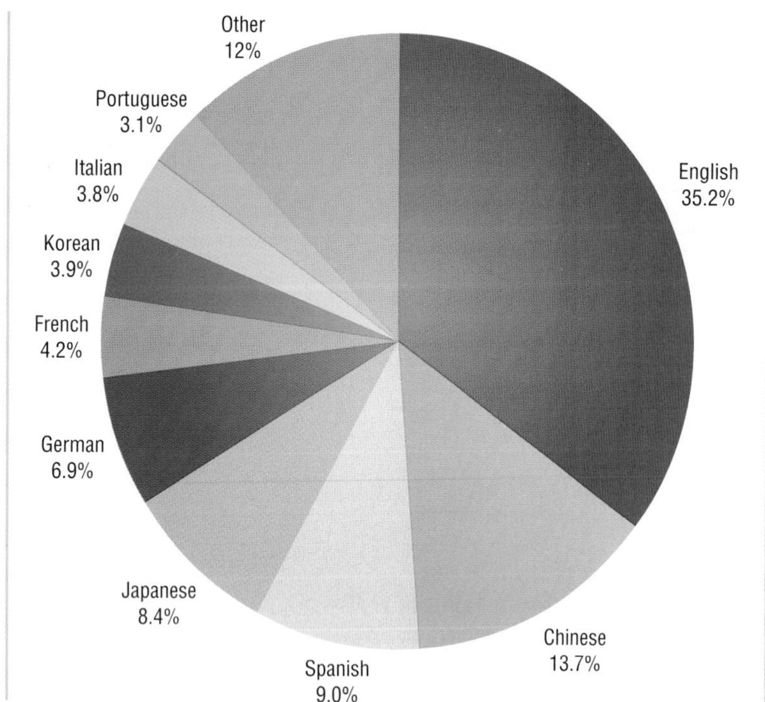

▸Figure 3.4
Internet Use by Language Group, September 2004

Source: Global Reach (2004).

English is dominant because Britain and the United States have been the world's most powerful and influential countries—economically, militarily, and culturally—for 200 years. (Someone once defined "language" as a dialect backed up by an army.) In recent decades, the global spread of capitalism, the popularity of Hollywood movies and American TV shows, and widespread access to instant communication via telephone and the Internet have increased the reach of the English language (▸Figure 3.4). There are now more speakers of excellent English in India than in Britain, and when a construction company jointly owned by German, French, and Italian interests undertakes a building project in Spain, the language of business is English (McCrum, Cran, and MacNeil, 1992).

Even in Japan, where relatively few people speak the language, English words are commonly used and Japanese words that are badly translated into English often become popular. The result is what is commonly known as "Japlish." Sometimes the results are unintelligible to a native English speaker. "Push to my nose! I might be changing to you?" says the catchy sign in a T-shirt store in Tokyo's Ueno district. Certain computer terms are more comprehensible to a native English speaker. For example, when you learn to open a computer file's *ai-kon* (icon) you are told to *daburu-kurikku* (double-click) the *mausu* (mouse) (Delmos, 2002; Kristof, 1997).

In view of the extensive use of English in Japan, *The Japanese Times,* one of Tokyo's four English-language daily newspapers, ran a story a few years ago noting the pressures of globalization and suggesting that it might be time for Japan to switch to English. However, there is an official backlash. To limit the anglicization of Japanese, the Ministry of Health and Welfare banned excessive use of English in its documents a couple of years ago. The Ministry of Education is now replacing many English words in official documents—words such as *sukeemu* (scheme), *eensenchibu* (incentive), *deribatibu* (derivative), and *identyityi* (identity). Whether official pronouncements will have much effect on the way English and Japlish are used in advertising and on the streets is, however, another question entirely. As one Japanese newspaper pointed out, given the popularity of English words, it's doubtful there will be much *foro-uppu* (follow-up).

Japanese teenagers consider English and Japlish the very height of fashion. Their conversation is laced with terms like *chekaraccho* (Check it out, Joe), *disu* (diss, or show disrespect toward), *denjarasu* (dangerous), *wonchu* (I want you), and *hi mentay* (high maintenance).

Meanwhile, several thousand other languages around the world are being eliminated due to the influence of English, French, Spanish, and the languages of a few other colonizing nations. The endangered languages are spoken by the tribes of Papua New Guinea, the Native peoples of the Americas, the national and tribal minorities of Asia, Africa, and Oceania, and marginalized European peoples such as the Irish and the Basques. The Linguistic Society of America estimates that the 5,000 to 6,000 languages spoken in the world today will be reduced to 1,000 to 3,000 in a century. Much of the culture of a people—its prayers, humor, conversational styles, technical vocabulary, myths, and so on—is expressed through language. Therefore, the loss of language amounts to the disappearance of tradition and perhaps even identity. These are often replaced by the traditions and identity of the colonial power, with television playing an important role in the transformation (Woodbury, 2003).

Owen Franklin/Corbis

A hallmark of postmodernism is the combining of cultural elements from different times and places. Architect I. M. Pei unleashed a storm of protest when his 72-foot glass pyramid became an entrance to the Louvre in Paris. It created a postmodern nightmare in the eyes of some critics.

Aspects of Postmodernism

Some sociologists think so much cultural fragmentation and reconfiguration has taken place in the last few decades that a new term is needed to characterize the culture of our times: **postmodernism.** Scholars often characterize the last half of the 19th century and the first half of the 20th century as the era of modernity. During this hundred-year period, belief in the inevitability of progress, respect for authority, and consensus around core values characterized much of Western culture. In contrast, postmodern culture involves an eclectic mixing of elements from different times and places, the erosion of authority, and the decline of consensus around some core values. Let us consider each of these aspects of postmodernism in turn.

Blending Cultures

An eclectic mixing of elements from different times and places is the first aspect of postmodernism. In the postmodern era, it is easier to create individualized belief systems and practices by blending facets of different cultures and historical periods. Consider religion. In the United States today, there are many more ways of worshiping than there used to be. The latest edition of the *Encyclopedia of American Religions* lists more than 2,100 religious groups, and one can easily construct a personalized religion involving, say, belief in the divinity of Jesus and yoga (Melton, 1996 [1978]). In the words of one journalist: "In an age when we trust ourselves to assemble our own investment portfolios and cancer therapies, why not our religious beliefs?" (Creedon, 1998). Nor are religious beliefs and practices drawn just from conventional sources. Even fundamentalist Christians who believe that the Bible is the literal word of God often supplement Judeo-Christian beliefs and practices with less conventional ideas about astrology, psychic powers, communication with the dead, and so forth. This is clear from a series of questions that were asked in the 1989 General Social Survey (▶Figure 3.5). Individuals thus draw on religions much like consumers shop in a mall. They practice religion *à la carte.* Meanwhile, churches, synagogues, and other religious institutions have diversified their menus in order to appeal to the spiritual, leisure, and social needs of religious consumers and retain their loyalties in the competitive market for congregants and parishioners (Finke and Stark, 1992).

Postmodernism is characterized by an eclectic mixing of cultural elements and the erosion of consensus.

The mix-and-match approach we see when it comes to religion is evident in virtually all spheres of culture. Purists may scoff at this sort of cultural blending. However, it probably has an important positive social consequence. It seems likely that people who engage

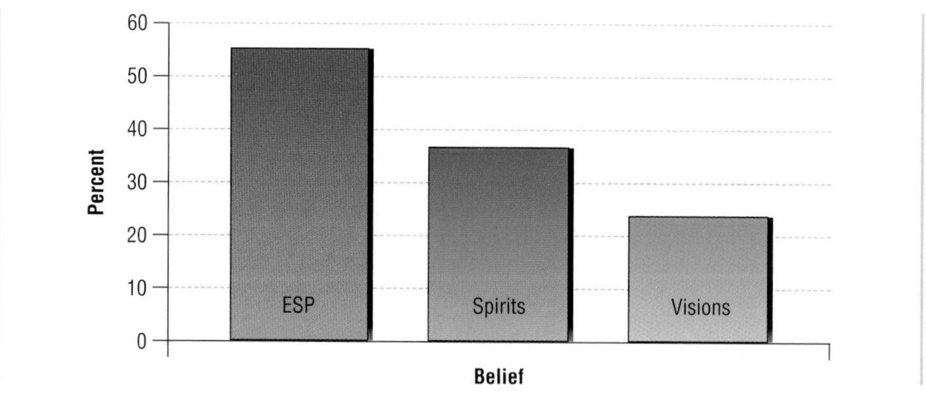

Figure 3.5
Unconventional Beliefs Among Christian Fundamentalists, United States (in percent; $n = 312$).

"How often have you had any of the following experiences: Felt in touch with someone when they were far away from you ('ESP')? Felt as though you were really in touch with someone who had died ('spirits')? Seen events that happened at a great distance as they were happening ('visions')?" Responses are shown for respondents who said they were Protestant or Catholic and believed that the Bible was the literal word of God.

National Opinion Research Center, 1999. General Social Survey, 1972–98.

in cultural blending are usually more tolerant and appreciative of ethnic, racial, and religious groups other than their own.

Erosion of Authority

The second aspect of postmodernism is the erosion of authority. Half a century ago, Americans were more likely than they are today to defer to authority in the family, schools, politics, medicine, and so forth. As the social bases of authority and truth have multiplied, however, we are more likely to challenge authority. Authorities once widely respected, including parents, physicians, and politicians, have come to be held in lower regard by many people. In the 1950s, Robert Young played the firm, wise, and always-present father in the TV hit *Father Knows Best.* Fifty years later, Homer Simpson plays a fool in *The Simpsons.* In the 1950s, three quarters of Americans expressed confidence in the federal government's ability to do what is right "just about always" or "most of the time." Fifty years later, the figure stood at just one third (Figure 3.6). The rise of Homer Simpson and the decline of confidence in government both reflect the society-wide erosion of traditional authority (Nevitte, 1996).

Instability of Core American Values

The decline of consensus around core values is the third aspect of postmodernism. More than half a century ago, sociologist Robin M. Williams, Jr., identified a dozen core American values (Williams, 1951). Americans, he wrote, value:

1. achievement and success
2. individualism
3. activity and work
4. efficiency and practicality
5. science and technology
6. progress
7. material comfort
8. humanitarianism
9. freedom
10. democracy
11. equality
12. the groups to which they belong above other groups.

Many Americans still believe in these values. Today, however, the values of Americans and all other people are less likely to remain stable over the course of their adult lives, and consensus has broken down on some issues.

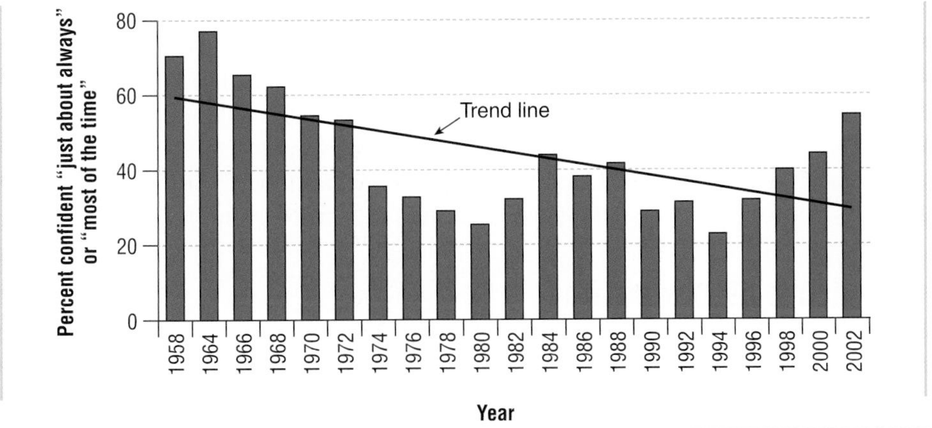

Figure 3.6
Confidence in Washington, 1958–2002 (in percent)

The U.S. National Election Study periodically asks: "How much of the time do you think you can trust the government in Washington to do what is right—just about always, most of the time, only some of the time or almost never?" Shown here is the percent who answered "nearly always" or "most of the time." Note that although the trend in confidence is downward, countertrends were evident during the administrations of Presidents Reagan, Clinton, and George W. Bush. Presidents Reagan and Clinton were very popular, whereas President Bush's administration spanned the terrorist attacks of September 11, 2001, and the wars in Afghanistan and Iraq.

Source: Sapiro, Rosenstone, and the National Election Studies (2004).

Value instability is evident in voting patterns, for example. In the middle of the 20th century, the great majority of adults remained loyal to one political party from one election to the next. By the third quarter of the 20th century, however, specific issues and personalities had eclipsed party loyalty as the driving forces of American politics (Nie, Verba, and Petrocik, 1976). Today, people are more likely to vote for different parties in succeeding elections than they were in 1950.

We can also illustrate the decline of consensus by considering the fate of "Big Historical Projects." For most of the past 200 years, consensus throughout the world was built around Big Historical Projects. Various political and social movements convinced people they could take history into their own hands and create a glorious future just by signing up. German Nazism was a Big Historical Project. Its followers expected the Reich to enjoy 1,000 years of power. Communism was an even bigger Big Historical Project, mobilizing hundreds of millions of people for a future that promised to end inequality and injustice for all time. However, the biggest and most successful Big Historical Project was not so much a social movement as a powerful idea—the belief that progress is inevitable, that life will always improve, due mainly to the spread of democracy and scientific innovation.

The 20th century was unkind to Big Historical Projects. Russian communism lasted 74 years. German Nazism endured a mere 12. And the idea of progress fell on hard times as a hundred million soldiers and civilians died in wars, the forward march of democracy took wrong turns into fascism, communism, and regimes based on religious fanaticism, and pollution due to urbanization and industrialization threatened the planet. In the postmodern era, more and more people recognize that apparent progress, including scientific advances, often have negative consequences (Scott, 1998). As the poet e e cummings once wrote, "nothing recedes like progress."

Postmodernism has many parents, teachers, politicians, religious leaders, and not a few university professors worried. Given the eclectic mixing of cultural elements from different times and places, the erosion of authority, and the decline of consensus around some core values, how can we make binding decisions? How can we govern? How can we teach children and adolescents the difference between right and wrong? How can we transmit accepted literary tastes and artistic standards from one generation to the next? These are the kinds of issues that plague people in positions of authority today.

Although their concerns are legitimate, many authorities seem not to have considered the other side of the coin. The postmodern condition, as we have described it, empowers ordinary people and makes them more responsible for their own fate. It frees people to adopt religious, ethnic, and other identities they are comfortable with, as opposed to identities imposed on them by others. It makes them more tolerant of difference. That is no small matter in a world torn by group conflict. And the postmodern attitude encourages healthy skepticism about rosy and naive scientific and political promises.

Culture as Constraint

We noted previously that culture has two faces. One we labeled "freedom," the other "constraint." Diversity, globalization, the rights revolution, and postmodernism are all aspects of the new freedoms that culture allows us today. We now examine several aspects of culture that act as constraining forces on our lives.

Values

Value Change in the United States and Globally

Although people in much of the world are freer than ever to choose their values, powerful social forces still constrain their choices. These constraints result in the formation of distinct value clusters that change over time, but only gradually.

Since 1981, researchers involved in the World Values Survey (WVS) have interviewed people around the world to identify value patterns and measure persistence and change in values. The WVS is the premier source of information on this subject, and it will prove useful to begin our discussion of constraints on culture by considering some of its findings. The complete WVS data set is based on surveys conducted between 1981 and 1997 of nearly 166,000 people from 65 countries comprising 75 percent of the world's population. However, Ronald Inglehart and Wayne Baker (2000) found they could usefully array the respondents' values along just two value dimensions. We call these value dimensions: *traditional/modern* and *materialist/postmaterialist.*

In the traditional/modern value dimension, respondents who hold *traditional* values tend to say that God is important in their life, abortion is never justifiable, and it is more important for a child to learn obedience and religious faith than independence and determination. They are also inclined to have a strong sense of national pride and to favor more respect for authority. At the other extreme, respondents who hold *modern* values tend to say that God is not important in their lives, abortion is justifiable, and it is more important for children to learn independence and determination than obedience and religious faith. They are also inclined to have a weak sense of national pride and to oppose more respect for authority.

In the materialist/postmaterialist value dimension, respondents who hold *materialist* values tend to give priority to economic and physical security over self-expression and quality of life. They tend to describe themselves as not very happy. They tend to be people who say they never have signed, and never will sign, a petition. They are also inclined not to trust people and to think that homosexuality is never justifiable. At the other extreme, respondents who hold *postmaterialist* values tend to give priority to self-expression and quality of life. They tend to describe themselves as very happy. They tend to be people who have signed, and would again sign, a petition. They are also inclined to trust people and to think that homosexuality is justifiable.

Inglehart and Baker assigned numerical scores to every possible response to the values listed previously. They then calculated each respondent's total score on each of the two value dimensions. They did this for two points in time—the year of the first WVS survey and the year of the most recent WVS survey. Next, they calculated average scores on each value dimension for all of the respondents in each country. Finally, they drew a graph that displayed each country's average score on both value dimensions at the two time points. ▶Figure 3.7 shows the results for the United States, Canada, Mexico, Japan, Russia, Sweden, and Nigeria. We focus on these countries because they typify important trends in the complete data set, which is too complex to discuss here.

How can we explain the position and movement over time of countries along the two value dimensions? In brief, two sets of social forces—one economic, the other religious—influence where countries are located on the two value dimensions. We can see this by examining Figure 3.7 closely.

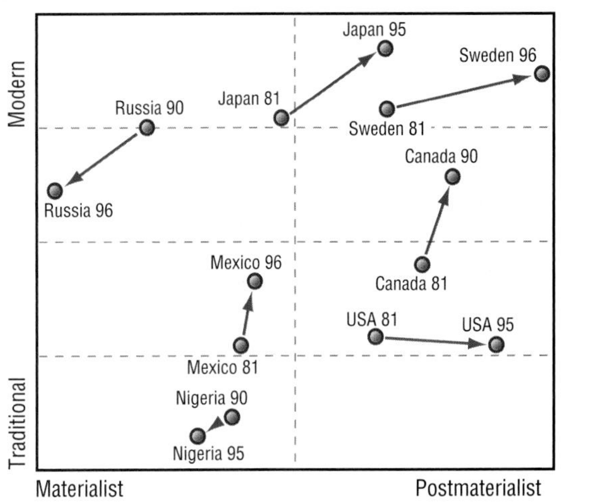

Figure 3.7
Value Change in Seven Countries, 1981–1997

Ronald Inglehart and Wayne Baker (2000), "Modernization, Cultural Change and the Persistence of Traditional Values," American Sociological Review 65:19–51. Copyright © 2000 American Sociological Association. Reprinted with permission of the publisher and author.

The first thing we notice about Figure 3.7 is that countries in the postmaterialist right half are highly developed economically. The countries in the materialist left half are not. Thus, there is a positive correlation between a country's level of economic development and its level of postmaterialism. A low level of economic development pushes people's values in a materialist direction, forcing them to pay more attention to physical and economic security, and so forth. A high level of economic development pushes people's values in a postmaterialist direction, allowing them to pay more attention to self-expression, the quality of life, and so forth.

Pay attention next to the direction of the arrows in Figure 3.7. The five countries that moved in a postmaterialist direction between the first and second survey dates (Japan, Sweden, Canada, Mexico, and the United States) all experienced considerable economic growth during the period. In contrast, the two countries that moved in a materialist direction between the first and second survey dates (Russia and Nigeria) experienced economic decline. We conclude that it is not just the *level* of economic development but the *direction* of economic change that influences the level of postmaterialism in a society.

If level of economic development and direction of economic change are the main factors that distinguish countries along the materialist/postmaterialist value dimension, religion is the main factor that distinguishes them along the traditional/modern value dimension. Predominantly Protestant countries (such as Sweden) and countries with a predominantly Confucian religious history (such as Japan) tend to be in the top quarter of Figure 3.7. They are the most modern countries in the sense that they are the most secular, the most in favor of children learning independence and determination, and so forth. Countries with a large Muslim population (e.g., Nigeria) tend to be in the bottom quarter. They are the most traditional countries in the sense that they are the most religious, the most in favor of children learning obedience and religious faith, and so forth. (For example, in 1995, 87 percent of Nigerian respondents said God was extremely important in their lives, compared with just 5 percent of Japanese respondents.) Predominantly Catholic countries (e.g., Mexico) and countries that are composed mainly of Catholics and Protestants (e.g., Canada and the United States) fall between these two extremes.

Finally, if we look again at the direction of the arrows in Figure 3.7, we see that three countries (Russia, Nigeria, and the United States) became more traditional between the first and most recent WVS surveys. Four countries (Japan, Sweden, Canada, and Mexico) became more modern. The countries that became more traditional during this period experienced the biggest revival of interest in religion, particularly fundamentalist religion. We conclude that it is not just the *predominance* of one religion or another that influences the level of modernism in a society, but also the *direction* of religious change.

These findings are interesting for two main reasons. First, they show that values are not randomly distributed across populations and that people are not free to choose whatever values they want. Instead, values cluster along identifiable dimensions, one traditional/modern, the other materialist/postmaterialist. Second, the findings are interesting insofar as they show that values cluster in the way they do because they are influenced by powerful social forces, one economic, the other religious. Americans, for example, score high on the postmaterialist dimension because we live in a country with a highly developed economy. Yet the United States is also the most traditional of the world's highly developed countries because religion remains a stronger force here than in any other highly developed nation (see Chapter 16, "Religion").

Cultural Lag

The WVS findings also say something interesting about **cultural lag,** the tendency of symbolic culture to change more slowly than material culture (Ogburn, 1966 [1922]). Many sociologists used to group values along just the traditional/modern value dimension and argue that values generally become more modern as societies develop economically (see Chapter 9, "Globalization, Inequality, and Development"). If the values of a particular country did not modernize as expected, sociologists often attributed it to cultural lag. However, it may be that cultural lag is a way of describing something that cannot otherwise be explained. As we have seen, once we allow for the existence of two value dimensions, it becomes possible to explain the value position of apparent anomalies, such as the United States, which ranks high on postmaterialism but low on modernism because of its level of economic development and religious history. From this point of view, cultural lag is less an explanation than an invitation to search for additional variables that might account for the lag.

Value Persistence

Previously we listed a dozen core American values identified by Robin M. Williams, Jr., in the 1950s (p. 82). We noted the subsequent erosion of consensus on some of those values. Erosion is not, however, disappearance. Most of the core American values identified by Williams persist, albeit to a degree and in forms that have been altered by more than half a century of social change. Insofar as they persist, these values act as constraints on our lives.

Consider in this connection the fact that most Americans still unquestioningly value science and technology, efficiency and practicality, and material progress that makes life easier. What form do these values take today? How do they affect our lives? Although we cannot answer these questions exhaustively here, we can gain some insight by taking a look at how we use time and how we consume goods and services (see also Chapter 22, "Technology and the Global Environment").

The Regulation of Time

The positive valuation of science, technology, efficiency, and practicality has led to what Max Weber called **rationalization.** Rationalization, in Weber's usage, means: (a) the application of the most efficient means to achieve given goals and (b) the unintended, negative consequences of doing so. Rationalization is evident in the way our use of time has evolved since the 14th century.

In the 14th century, an upsurge in demand for textiles in Europe caused loom owners to look for ways of increasing productivity. To that end, they imposed longer hours on loom workers. They also turned to a new technology for assistance: the mechanical clock. They installed public clocks in town squares. The clocks, known as *Werkglocken* ("work clocks") in German, signaled the beginning of the workday, the timing of meals, and quitting time.

The mechanical clock had earlier been used to impose a rigid schedule in Benedictine monasteries (Zerubavel, 1981: 31–40). The monks embraced the precise daily rhythm of prayers. In contrast to the monks, however, German workers resisted the regimentation of their lives. They were accustomed to enjoying many holidays and a fairly flexible and vague work schedule regulated only approximately by the seasons and the rising and setting of the sun. The strict regime imposed by the work clocks made life harder. The workers staged uprisings to silence the clocks—but to no avail. City officials sided with the employers and imposed fines for ignoring the *Werkglocken*. Harsher penalties, including death, were imposed on anyone trying to use the clocks' bells to signal a revolt (Thompson, 1967).

Now, nearly 700 years later, many people are, in effect, slaves of the *Werkglock*. This is especially true of big-city North American couples who are employed full-time in the

Cultural lag is the tendency of symbolic culture to change more slowly than material culture.

Rationalization is the application of the most efficient means to achieve given goals and the unintended, negative consequences of doing so.

paid labor force and have preteen children. For them, life often seems an endless round of waking up at 6:30 a.m., getting everyone washed and dressed, preparing the kids' lunches, getting them out the door in time for the school bus or the car pool, driving to work through rush-hour traffic, facing the speedup at work that resulted from the recent downsizing, driving back home through rush-hour traffic, preparing dinner, taking the kids to their soccer game, returning home to clean up the dishes and help with homework, getting the kids washed, brushed, and into bed, and (if you haven't brought some office work home) grabbing an hour of TV before collapsing, exhausted, for 6½ hours before the story repeats itself. In 1998, married couples with children under the age of 6 worked for pay 16 hours a week more than they did in 1969 (U.S. Department of Labor, 1999: 100). Managers are more likely than any other category of workers to be working for pay 49 hours a week or more. Next come sales personnel who earn commissions, transportation workers (especially truck drivers), and professionals (Rones, Ilg, and Gardner, 1997: 9). Life is less hectic for residents of small towns, unmarried people, couples without small children, retirees, and the unemployed. But the lives of most people are so packed with activities that time must be carefully regulated, each moment precisely parceled out so that we may tick off item after item from an ever-growing list of tasks that need to be completed on schedule (Schor, 1992).

The Everett Collection

Have we come to depend too heavily on the *Werkglock*?

After more than 600 years of conditioning, it is unusual for people to rebel against the clock in the town square anymore. In fact, we now wear a watch on our wrist without giving it a second thought, as it were. This signifies that we have accepted and internalized the regime of the *Werkglock*. Allowing clocks to precisely regulate our activities seems the most natural thing in the world—which is a pretty good sign that the internalized *Werkglock* is, in fact, a product of culture.

Is the precise regulation of time rational? It certainly is rational as a means of ensuring the goal of efficiency. Minding the clock maximizes how much work you get done in a day. The regulation of time makes it possible for trains to run on schedule, university classes to begin punctually, and business meetings to start on time. Yet one restaurant in Japan has installed a punch-clock for its customers. The restaurant offers all you can eat for 35 yen per minute. As a result, "the diners rush in, punch the clock, load their trays from the buffet table, and concentrate intensely on efficient chewing and swallowing, trying not to waste time talking to their companions before rushing back to punch out. This version of fast food is so popular that as the restaurant prepares to open at lunchtime, Tokyo residents *wait in line*" (Gleick, 2000 [1999]: 244; emphasis in the original). Meanwhile, in New York and Los Angeles, some upscale restaurants have gotten in on the act. An increasingly large number of business clients are so pressed for time that they feel the need to pack in two half-hour lunches with successive guests. The restaurants oblige, making the resetting of tables "resemble the pit-stop activity at the Indianapolis 500" (Gleick, 2000 [1999]: 155).

Minding the clock is rational in one sense. It ensures the goal of applying technology to ensure efficiency and practicality. But is it rational as an end in itself? For many people, it is not. They complain that the precise regulation of time has gotten out of hand. Life has simply become too hectic for many people to enjoy. In this sense, a *rational means* (the *Werkglock*) has been applied to a given goal (maximizing work) but has led to an *irrational end* (a hectic life).

Rationalization

This, in a nutshell, is Max Weber's thesis about the rationalization process. Weber claimed that rationality of means has crept into all spheres of life, even language, leading to unintended consequences that dehumanize and constrain us (Figure 3.8). As our use of time shows, rationalization enables us to do just about everything more efficiently, but

Max Weber likened the modern era to an "iron cage." Sociology promises to teach us both the dimensions of the cage and the possibilities for release.

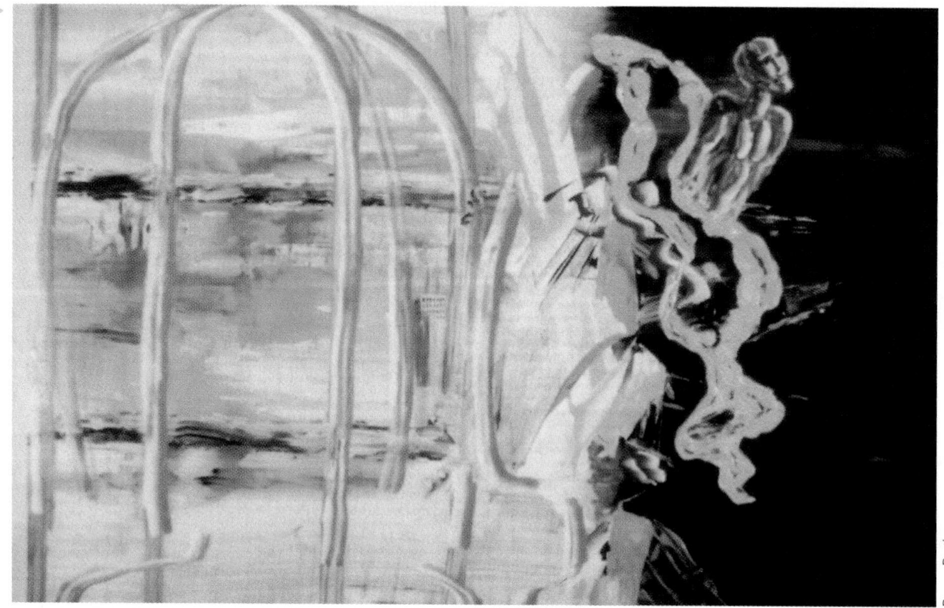

Brown Brothers

at a steep cost. In Weber's view, rationalization is one of the most constraining aspects of culture. In his view, it makes life in the modern world akin to living inside an "iron cage."

Web Research Project: Consumerism

Consumerism

Another constraining aspect of culture is consumerism. **Consumerism** is the tendency to define ourselves in terms of the goods and services we purchase. It is the contemporary form of valuing "material progress that makes life easier," which Robin M. Williams, Jr., identified as a core American value in the 1950s.

Sociology Now™

Learn more about **Consumerism** by going through the Number of Shopping Centers in the U.S. Map Exercise.

In 1998, apparel sales in North America were lagging. As a result, the GAP launched a new ad campaign to help revitalize sales. The company hired Hollywood talent to create a slick and highly effective series of TV spots for khaki pants. According to the promotional material for the ad campaign, the purpose of the ads was to "reinvent khakis," that is, to stimulate demand for the pants. In *Khakis rock,* "skateboarders and in-line skaters dance, glide, and fly to music by the Crystal Method." In *Khakis groove,* "hip-hop dancers

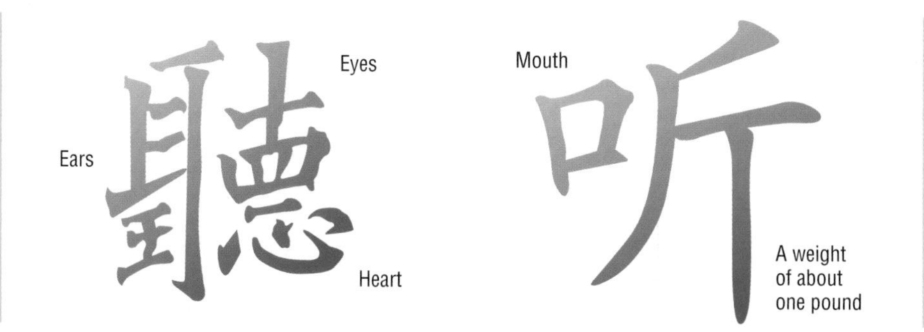

Figure 3.8
The Rationalization of Chinese Script

Reprinted here are the Chinese characters for "listening" (*t'ing*) in traditional Chinese script (left) and simplified, modern script (right). Each character is composed of several word-symbols. In classical script, listening is depicted as a process involving the eyes, the ears, and the heart. It implies that listening demands the utmost empathy and involves the whole person. In contrast, modern script depicts listening as something that merely involves one person speaking and the other "weighing" speech. Modern Chinese script has been rationalized. Has empathy been lost in the process?

Consumerism is the tendency to define ourselves in terms of the goods we purchase.

What is being sold here, the pants or the attitude?

Bill Aron/PhotoEdit

throw radical moves to the funky beat of Bill Mason." In *Khakis swing,* "two couples break away from a crowd to demonstrate swing techniques to the vintage sounds of Louis Prima" (Gap.com, 1999).

About 55 seconds of each ad featured the dancers. During the last 5 seconds, the words "GAP khakis" appeared on the screen. The GAP followed a similar approach in its 2000 ad campaign, inspired by the musical *West Side Story.* The 30-second spots replaced the play's warring street gangs, the Jets and the Sharks, with fashion factions of their own, the Khakis and the Jeans. Again, most of the ad was devoted to the riveting dance number. The pants were mentioned for only a few seconds at the end.

As the imbalance between stylish come-on and mere information suggests, the people who created the ads understood well that it was really the appeal of the dancers that would sell the pants. They knew that to stimulate demand for their product, they had to associate the khakis with desirable properties such as youth, good health, coolness, popularity, beauty, and sex. As an advertising executive said in the 1940s: "It's not the steak we sell. It's the sizzle."

Because advertising stimulates sales, there is a tendency for business to spend more on advertising over time (▶Figure 3.9). Because advertising is widespread, most people unquestioningly accept it as part of their lives. In fact, many people have *become* ads. Thus, when your father was a child and quickly threw on a shirt, allowing a label to hang out, your grandmother might admonish him to "tuck in that label." In contrast, many people today proudly display consumer labels as marks of status and identity. Advertisers teach us to associate the words "Gucci" and "Nike" with different kinds of people, and when

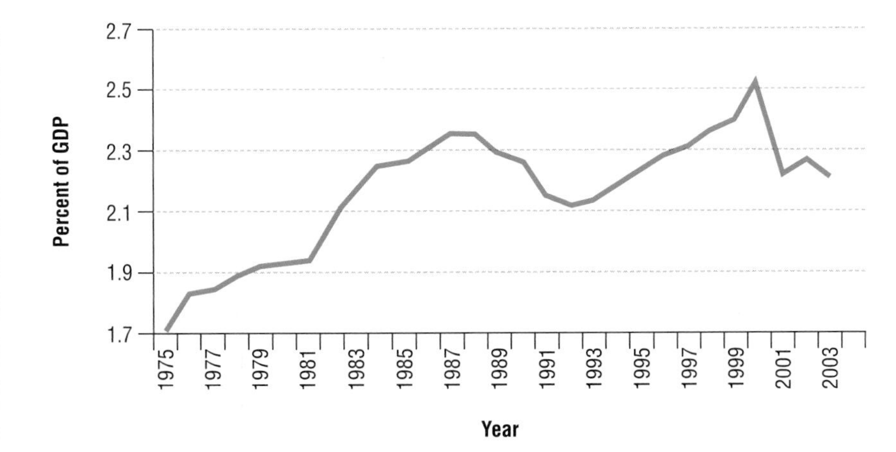

▶Figure 3.9
Advertising as Percent of Gross Domestic Product (GDP), United States, 1975–2003

Source: Television Bureau of Advertising (2005).

Box 3.3

YOU AND THE SOCIAL WORLD

Labeling Yourself and Others

Carefully observe the way you and your sociology classmates dress. Write a 500-word essay answering the following questions: What are the main distinctive styles of dress in your classroom? To what degree do conspicuous labels help to distinguish styles? What are the characteristics of the people who adopt different styles of dress (class, race, values, lifestyle, etc.)? Do your friends wear styles and display labels similar to those you wear? How about people you dislike? To what degree are styles of dress and clothing labels cultural artifacts that increase the solidarity of social groups and segregate them from other groups?

people display these labels on their clothes they are telling us something about the kind of people they are. Advertising becomes us (Box 3.3).

The rationalization process, when applied to the production of goods and services, enables us to produce more efficiently, to have more of just about everything than our parents did. But it is consumerism, the tendency to define our selves in terms of the goods we purchase, that ensures that all the goods we produce will be bought. Of course, we have lots of choice. We can select from dozens of styles of running shoes, cars, toothpaste, and all the rest. We can also choose to buy items that help to define us as members of a particular **subculture,** adherents of a set of distinctive values, norms, and practices within a larger culture. But regardless of individual tastes and inclinations, nearly all of us have one thing in common: We tend to be good consumers. We are motivated by advertising, which is based on the accurate insight that people will likely be considered cultural outcasts if they fail to conform to stylish trends. By creating those trends, advertisers push us to buy even if we must incur large debts to do so (Schor, 1999). That is why North Americans' "shop-till-you-drop" lifestyle prompted French sociologist Jean Baudrillard to remark pointedly that even what is best in America is compulsory (Baudrillard, 1988 [1986]). And that is why we say that consumerism, like rationalization, acts as a powerful constraint on our lives.

From Counterculture to Subculture

In concluding our discussion of culture as a constraining force, we note that consumerism is remarkably effective at taming countercultures. **Countercultures** are subversive subcultures. They oppose dominant values and seek to replace them. The hippies of the 1960s formed a counterculture and so do environmentalists today.

Countercultures rarely pose a serious threat to social stability. Most often, the system of social control, of rewards and punishments, keeps countercultures at bay. In our society, consumerism acts as a social control mechanism that normally prevents countercultures from disrupting the social order. It does that by transforming deviations from mainstream culture into means of making money and by enticing rebels to become entrepreneurs (Frank and Weiland, 1997).Two examples from popular music subcultures will help illustrate the point.

Ozzy Osbourne is the godfather of heavy metal. Beginning in the late 1960s, he and his band, Black Sabbath, inspired Metallica, Kiss, Judas Priest, Marilyn Manson, and others to play loud, nihilistic music, reject conventional morality, embrace death and violence, and spark youthful rebellion and parental panic. In 1982, he bit the head off a bat during a performance and urinated on the Alamo. He was given rabies shots for the former and arrested for the latter. Around the same time, Tipper Gore, wife of the future presidential candidate, formed the Parental Music Resource Committee to fight against violence and sex in the lyrics of popular music. Osbourne was one of the committee's prin-

- A **subculture** is a set of distinctive values, norms, and practices within a larger culture.
- A **counterculture** is a subversive subculture that opposes dominant values and seeks to replace them.

cipal targets. The "Prince of Darkness," as he was often called, was about as rebellious a figure as one could imagine in 1982.

Reuters NewMedia Inc./Corbis

The Osbournes

Flash-forward 20 years. Osbourne, now 53, has the sixth most popular show on American television among 18- to 34-year-olds, just behind *Survivor* in the ratings. MTV placed a dozen cameras throughout his Beverly Hills mansion, and every Tuesday night viewers get to see everything going on in the Osbourne household for half an hour. According to *USA Today,* it turns out that Osbourne is "a lot like anyone's adorable dad. Shuffles a bit. Forgets things. Worries about the garbage. Snores on the couch while the TV blares. Walks the dog" (Gundersen, Keveney, and Oldenburg: 1A). There is a lot of swearing in the Osbourne family. His 17-year-old daughter sports pink hair, and his 16-year-old son wears dark nail polish. But CNN's Greta Van Susteren said she found the Osbournes "charming," and Rosie O'Donnell said to Ozzy's wife, Sharon: "What I love most about [your show] is not only the relationship you have with Ozzy—and you obviously adore each other—but the honesty with which you relate to your children. The love is so evident between all of you. It's heartwarming" (Gundersen, Keveney, and Oldenburg: 2A). Sharon and Ozzy Osbourne were invited to dinner at the White House in 2002, and in the same year, former Republican Vice-President Dan Quayle (who in 1992 criticized the TV character Murphy Brown for deciding to have a baby out of wedlock) admitted, "There are some very good lessons being transmitted" by the Osbournes (Beck, 2002). *The Osbournes,* it turns out, is a comfort to many people. It proves that the frightening rejection of mainstream culture in the 70s and 80s was just a passing phase and that things of eternal value—especially the nuclear family and commercialism—remain intact. Ozzy Osbourne has thus been transformed from the epitome of rebellion against society to a family man, a small industry, and a media icon.

The development of hip-hop also illustrates the commercialization of rebellion (Brym, 2001). Originating in the squalor of inner-city American ghettos in the 70s, hip-hop was at first a highly politicized rebellion. Early hip-hop artists glorified the mean streets of the inner city and held the police, the mass media, and other pillars of white society in utter contempt, blaming them for arbitrary arrests, the political suppression of black activists, and the malicious spreading of lies about African Americans. However, by the time *Public Enemy* became a hit in the late 1980s, MTV had aired its first regular program devoted to the genre and much of hip-hop's audience was composed of white middle-class youth. Hip-hop artists were quick to see the potential of commercialization. Soon, Wu-Tang Clan had its own line of clothes, Versace was marketing clothing influenced by ghetto styles, and Puff Daddy (who later had a makeover as P. Diddy) was reminding his audience: "N_____ get money, that's simply the plan" (from his 1999 CD, *Forever*). No less than heavy metal and punk, hip-hop's radicalism gave way to the lures of commercialism.

In sum, the fates of heavy metal and hip-hop are testimony to the capacity of postmodern culture to constrain expressions of freedom, individualism, dissent, and rebellion (Frank and Weiland, 1997). They are compelling illustrations of postmodern culture's second face.

Summary

Sociology Now™

Reviewing is as easy as ❶, ❷, ❸.

❶ Before you do your final review, take the SociologyNow diagnostic quiz to help you identify the areas on which you should concentrate. You will find information on how to access SociologyNow on the foldout at the front of the textbook.
❷ As you review, take advantage of SociologyNow's study aids to help you master the topics in this chapter.
❸ When you are finished with your review, take SociologyNow's post-test to confirm you are ready to move on to the next chapter.

1. **What are the main components of culture and what is culture's main function?**

   Culture is composed of various types of ideas (such as symbols, language, values, and beliefs), norms of behavior, and human-made material objects. The ability to create symbols, cooperate, and make tools has enabled humans to thrive in their environments.

2. **What role does biology, as opposed to culture, play in shaping specific human behaviors and social arrangements?**

   Biology sets broad limits and potentials on human behavior and social arrangements. However, most of the variation in human behavior and social arrangements is due to social forces. The roles of biology and culture are illustrated by the development of language skills. We are biologically prewired to acquire language and create grammatical speech patterns. However, the social environment gives form to these predispositions. We know, for example, that people can never acquire language proficiency unless they are exposed to language in the first years of life. Moreover, our use of language is shaped by the language community or communities to which we belong.

3 **What is the ideal vantage point for analyzing a culture?**

   We can see the contours of culture most sharply if we are neither too deeply immersed in it nor too much removed from it. Understanding culture requires refraining from taking your own culture for granted and from judging other cultures by the standards of your own.

4. **What does it mean to say that culture has "two faces"?**

   First, culture provides us with increasing opportunities to exercise our freedom in some respects. The rights revolution, multiculturalism, globalization, and postmodernism reflect this tendency. Second, culture constrains us in other respects, putting limits on what we can become. The shift of values along traditional/modern and materialist/postmaterialist dimensions, the growth of rationalization, and the spread of consumerism reflect this tendency.

5. **What is the multiculturalism debate?**

   Advocates of multiculturalism want school and college curricula to reflect the country's growing ethnic and racial diversity. They also want school and college curricula to stress that all cultures have equal value. They believe that multicultural education will promote self-esteem and economic success among members of racial minorities. Critics fear that multiculturalism results in declining educational standards. They believe that multicultural education causes political disunity and interethnic and interracial conflict. They argue that it promotes an extreme form of cultural relativism.

6. **What is the "rights revolution"?**

   The rights revolution is the process by which socially excluded groups have struggled to win equal rights under the law and in practice. In full swing by the 1960s, the rights revolution has come to involve the promotion of women's rights, minority rights, gay and lesbian rights, the rights of people with special needs, constitutional rights, and language rights. The rights revolution fragments American culture by legitimizing the grievances of groups that were formerly excluded from full social participation and renewing their pride in their identity and heritage.

7. **What has caused the "globalization" of culture?**

   The globalization of culture has resulted from the growth of international trade and investment, ethnic and racial migration, influential "transnational" organizations, and inexpensive travel and communication.

8. **What is postmodernism?**

   Postmodernism involves an eclectic mixing of elements from different times and places, the decline of authority, and the erosion of consensus around core values.

9. **What accounts for a country's predominant value pattern and what accounts for how it changes?**

   A country's value pattern is shaped by its level of economic development and direction of economic change, and by its predominant religion and direction of religious change. Specifically, a high level of economic development and economic growth are associated with postmaterialist values, whereas a low level of economic development and negative growth are associated with materialist values. Moreover, the predominance of Christian and Confucian religions and the absence of strong fundamentalist movements are associated with modern values, whereas the predominance of Islam and the presence of strong fundamentalist movements are associated with traditional values.

10. **What is rationalization?**

    Rationalization involves the application of the most efficient means to achieve given goals and the unintended, negative consequences of doing so. Rationalization is evident in the increasingly regulated use of time and in many other areas of social life.

11. What is consumerism?

    Consumerism is the tendency to define ourselves in terms of the goods we purchase. Excessive consumption puts limits on who we can become and constrains our capacity to dissent from mainstream culture.

Questions to Consider

1. We imbibe culture but we also create it. What elements of culture have you created? Under what conditions were you prompted to do so? Was your cultural contribution strictly personal or was it shared with others? Why?
2. Select a subcultural practice that seems odd, inexplicable, or irrational to you. By interviewing members of the subcultural group and reading about them, explain how the subcultural practice you chose makes sense to members of the subcultural group.
3. Do you think the freedoms afforded by postmodern culture outweigh the constraints it places on us? Why or why not?

Web Resources

Companion Website for This Book

http://sociology.wadsworth.com

Begin by clicking on the Student Resources section of the website. Choose "Introduction to Sociology" and finally the Brym and Lie book cover. Next, select the chapter you are currently studying from the pull-down menu. From the Student Resources page you will have easy access to InfoTrac® College Edition, MicroCase Online exercises, additional web links, and many resources to aid you in your study of sociology, including practice tests for each chapter.

Recommended Websites

Benjamin Barber "Jihad vs. McWorld," on the World Wide Web at http://www.theatlantic.com/politics/foreign/barberf.htm, is a brief, masterful analysis of the forces that are simultaneously making world culture more homogeneous and more heterogeneous. The article was originally published in *The Atlantic Monthly* (March 1992). For the full story, see Benjamin Barber, *Jihad vs. McWorld: How Globalism and Tribalism are Reshaping the World* (New York: Ballantine Books, 1996).

Adbusters is an organization devoted to analyzing and criticizing consumer culture. Its provocative website is at http://adbusters.org.

Sharon Zupko's Cultural Studies Center is our favorite site on the sociology of popular culture. Visit it at http://www.popcultures.com.

The Resource Center for Cyberculture Studies is an organization devoted to studying emerging cultures on the World Wide Web. Its website is at http://www.com.washington.edu/rccs.

CHAPTER 4 Socialization

Zena Holloway/Getty Images

In this chapter, you will learn that:

- The view that social interaction unleashes human potential is supported by studies showing that children raised in isolation do not develop normal language and other social skills.
- While the socializing influence of the family decreased in the 20th century, the influence of schools, peer groups, and the mass media increased.
- People's identities change faster, more often, and more completely than they did just a couple of decades ago; the self has become more plastic.
- The main socializing institutions often teach children and adolescents contradictory lessons, making socialization a more confusing and stressful process than it used to be.
- Declining parental supervision and guidance, increasing assumption of adult responsibilities by youth, and declining participation in extracurricular activities are transforming the character of childhood and adolescence today.

Social Isolation and Socialization

One day in 1800, a 10- or 11-year-old boy walked out of the woods in southern France. He was filthy, naked, and unable to speak and had not been toilet trained. After the police took him to a local orphanage, he repeatedly tried to escape and refused to wear clothes. No parent ever claimed him. He became known as "the wild boy of Aveyron." A thorough medical examination found no major physical or mental abnormalities. Why, then, did the boy seem more animal than human? Because, until he walked out of the woods, he apparently had been raised in isolation from other human beings (Shattuck, 1980).

Similar horrifying reports lead to the same conclusion. Occasionally a child is found locked in an attic or a cellar, where he or she saw another person for only short periods each day to receive food. Like the wild boy of Aveyron, such children rarely develop normally. Typically, they remain disinterested in games. They cannot form intimate social relationships with other people. They develop only the most basic language skills.

Some of these children may suffer from congenitally subnormal intelligence. It is uncertain how much and what type of social contact they had before they were discovered. Some may have been abused. Therefore, their condition may not be a result of just social isolation. However, these examples do at least suggest that the ability to learn culture and become human is only a potential. To be actualized, **socialization** must unleash this human potential. Socialization is the process by which people learn their culture. They do so by (1) entering and disengaging from a succession of roles and (2) becoming aware of

Socialization is the process by which people learn their culture. They do so by entering and disengaging from a succession of roles and becoming aware of themselves as they interact with others.

© Martin Rogers/Stock Boston

In the 1960s, researchers Harry and Margaret Harlow placed baby rhesus monkeys in various conditions of isolation to witness and study the animals' reactions. Among other things, they discovered that baby monkeys raised with an artificial mother made of wire mesh, a wooden head, and the nipple of a feeding tube for a breast were later unable to interact normally with other monkeys. However, when the artificial mother was covered with a soft terry cloth, the infant monkeys clung to it in comfort and later revealed less emotional distress. Infant monkeys preferred the cloth mother even when it gave less milk than the wire mother. The Harlows concluded that emotional development requires affectionate cradling.

themselves as they interact with others. A **role** is the behavior expected of a person occupying a particular position in society.

Convincing evidence of the importance of socialization in unleashing human potential comes from a study conducted by René Spitz (Spitz, 1945; 1962). Spitz compared children who were being raised in an orphanage with children who, for medical reasons, were being raised in a nursing home. Both institutions were hygienic and provided good food and medical care. However, the children's mothers cared for them in the nursing home, whereas just 6 nurses cared for the 45 orphans in the orphanage. The orphans therefore had much less contact with other people. Moreover, from their cribs, the nursing home infants could taste a slice of society. They saw other babies playing and receiving care. They saw mothers, doctors, and nurses talking, cleaning, serving food, and giving medical treatment. In contrast, the nurses in the orphanage would hang sheets from the cribs to prevent the infants from seeing the activities of the institution. Depriving the infants of social stimuli for most of the day apparently made them less demanding.

Social deprivation had other effects too. Because of the different patterns of child care described previously, by the age of 9 to 12 months the orphans were more susceptible to infections and had a higher death rate than the children in the nursing home. By the time they were 2 to 3 years old, all the children from the nursing home were walking and talking, compared with fewer than 8 percent of the orphans. Normal children begin to play with their own genitals by the end of their first year. Spitz found that the orphans began this sort of play only in their fourth year. He took this behavior as a sign that they might have an impaired sexual life when they reached maturity. This outcome had occurred in rhesus monkeys raised in isolation. Spitz's natural experiment thus amounts to quite compelling evidence for the importance of childhood socialization in making us fully human. Without childhood socialization, most of our human potential remains undeveloped.

Sociology Now™

Learn more about **Socialization** by going through The Definition of Socialization Video Exercise.

The Crystallization of Self-Identity

The formation of a sense of self continues in adolescence, which is a particularly turbulent period of rapid self-development. Consequently, many people can remember experiences from their youth that helped to crystallize their self-identity. Do you? Robert Brym clearly recalls one such defining moment.

Personal Anecdote

"I can date precisely the pivot of my adolescence," says Robert. "I was in grade 10. It was December 16. At 4 p.m. I was a nobody and knew it. Half an hour later, I was walking home from school, delighting in the slight sting of snowflakes melting on my upturned face, knowing I had been swept up in a sea change.

"About 200 students sat impatiently in the auditorium that last day of school before the winter vacation. We were waiting for Mr. Garrod, the English teacher who headed the school's drama program, to announce the cast of *West Side Story.* I was hoping for a small speaking part and was not surprised when Mr. Garrod failed to read my name as a chorus member. However, as the list of remaining characters grew shorter, I became despondent. Soon only the leads remained. I knew an unknown kid in grade 10 couldn't possibly be asked to play Tony, the male lead. Leads were almost always reserved for more experienced Grade 12 students.

A **role** is the behavior expected of a person occupying a particular position in society.

"Then the thunderclap. 'Tony,' said Mr. Garrod, 'will be played by Robert Brym.'

"'Who's Robert Brym?' whispered a girl two rows ahead of me. Her friend merely shrugged in reply. If she had asked *me* that question, I might have responded similarly. Like nearly all 15-year-olds, I was deeply involved in the process of figuring out exactly who I was. I had little idea of what I was good at. I was insecure about my social status. I wasn't sure what I believed in. In short, I was a typical teenager. I had only a vaguely defined sense of self.

"A sociologist once wrote that 'the central growth process in adolescence is to define the self through the clarification of experience and to establish self-esteem'" (Friedenberg, 1959: 190). From this point of view, playing Tony in *West Side Story* turned out to be the first section of a bridge that led me from adolescence to adulthood. Playing Tony raised my social status in the eyes of my classmates, made me more self-confident, taught me I could be good at something, helped me to begin discovering parts of myself I hadn't known before, and showed me that I could act rather than merely be acted upon. In short, it was through my involvement in the play (and, subsequently, in many other plays throughout high school) that I began to develop a clear sense of who I am."

The crystallization of self-identity during adolescence is just one episode in a lifelong process of socialization. To paint a picture of the socialization process in its entirety, we must first review the main theories of how one's sense of self develops during early childhood. We then discuss the operation and relative influence of society's main socializing institutions, or "agents of socialization": families, schools, peer groups, and the mass media. In these settings, we learn, among other things, how to control our impulses, think of ourselves as members of different groups, value certain ideals, and perform various roles. You will see that these institutions do not always work hand in hand to produce happy, well-adjusted adults. Often, they give mixed messages and are at odds with each other. That is, they teach children and adolescents different and even contradictory lessons. You will also see that although recent developments give us more freedom to decide who we are, they can make socialization more disorienting than ever before. Finally, in the concluding section of this chapter, we examine how decreasing supervision and guidance by adult family members, increasing assumption of adult responsibilities by youth, and declining participation in extracurricular activities are changing the nature of childhood and adolescence today. Some analysts even say that childhood and adolescence are vanishing before our eyes. Thus, the main theme of this chapter is that the development of one's self-identity is often a difficult and stressful process–and it is becoming more so.

It is during childhood that the contours of one's self are first formed. We therefore begin by discussing the most important social-scientific theories of how the self originates in the first years of life.

Theories of Childhood Socialization

Socialization begins soon after birth. Infants cry out, driven by elemental needs, and are gratified by food, comfort, or affection. Because their needs are usually satisfied immediately, at first they do not seem able to distinguish themselves from their main caregivers, usually their mothers. However, social interaction soon enables infants to begin developing a self-image or sense of **self**—a set of ideas and attitudes about who they are as independent beings.

Freud

The **self** consists of one's ideas and attitudes about who one is.

Sigmund Freud proposed the first social-scientific interpretation of the process by which the self emerges (Freud, 1962 [1930]; 1973 [1915–17]). Freud, an Austrian, was the founder of psychoanalysis. He referred to the part of the self that demands immediate

Corbis

Sigmund Freud (1856–1939) was the founder of psychoanalysis. Many issues have been raised about the specifics of his theories. Nevertheless, his main sociological contribution was his insistence that the self emerges during early social interaction and that early childhood experience exerts a lasting impact on personality development.

gratification as the **id.** According to Freud, a self-image begins to emerge as soon as the id's demands are denied. For example, at a certain point, parents usually decide not to feed and comfort a baby every time she wakes up in the middle of the night. The parents' refusal at first incites howls of protest. Eventually, however, the baby learns certain practical lessons from the experience: to eat more before going to bed, to sleep for longer periods, and to go back to sleep if she wakes up. Equally important, the baby begins to sense that her needs differ from those of her parents, that she has an existence independent of others, and that she must somehow balance her needs with the realities of life.

Because of many such lessons in self-control, including toilet training, the child eventually develops a sense of what constitutes appropriate behavior and a moral sense of right and wrong. Soon a personal conscience or, to use Freud's term, a **superego,** crystallizes. The superego is a repository of cultural standards. In addition, the child develops a third component of the self, the **ego.** According to Freud, the ego is a psychological mechanism that, in well-adjusted individuals, balances the conflicting needs of the pleasure-seeking id and the restraining superego.

In Freud's view, the emergence of the superego is a painful and frustrating process. In fact, said Freud, to get on with our daily lives we have to repress memories of denying the id immediate gratification. Repression involves storing traumatic memories in a part of the self we are not normally aware of: the **unconscious.** Repressed memories influence emotions and actions even after they are stored away. Particularly painful instances of childhood repression may cause various types of psychological problems later in life that require therapy to correct. However, some repression is the cost of civilization. As Freud said, we cannot live in an orderly society unless we deny the id (Freud, 1962 [1930]).

Criticisms of Freud's Analysis

Researchers have called into question many of the specifics of Freud's argument. Three criticisms stand out:

1. *The connections between early childhood development and adult personality are more complex than Freud assumed.* Freud wrote that when the ego fails to balance the needs of the id and the superego, individuals develop personality disorders. Typically, he said, this situation occurs if a young child is raised in an overly repressive atmosphere. To avoid later psychiatric problems, Freud and his followers recommended raising young children in a relaxed and permissive environment. Such an environment is characterized by prolonged breast-feeding, nursing on demand, gradual weaning, lenient and late bladder and bowel training, frequent mothering, and freedom from restraint and punishment. However, sociological research reveals no connection between these aspects of early childhood training and the development of well-adjusted adults (Sewell, 1958). One group of researchers who were influenced by Freud's theories tracked people from infancy to the age of 32 and made *incorrect* predictions about personality development in two-thirds of the cases. They "had failed to anticipate that depth, complexity, problem-solving abilities, and maturity might derive from painful [childhood] experiences" (Coontz, 1992: 228).
2. *Many sociologists criticize Freud for gender bias in his analysis of male and female sexuality.* According to Freud (1977 [1905]), women who are psychologically normal are immature and dependent on men because they envy the male sexual organ. Women who are mature and independent he classified as abnormal. However, Freudians have not collected any experimental or survey data showing that boys are more independent than girls because of the latter's envy of the male sexual organ. Nor is it clear why young girls must define themselves in relation to young boys by focusing on lack of a

- The **id**, according to Freud, is the part of the self that demands immediate gratification.
- The **superego**, according to Freud, is a part of the self that acts as a repository of cultural standards.
- The **ego**, according to Freud, is a psychological mechanism that balances the conflicting needs of the pleasure-seeking id and the restraining superego.
- The **unconscious**, according to Freud, is the part of the self that contains repressed memories that we are not normally aware of.

penis. There is no reason that young girls' sexual self-definitions cannot focus positively on their own reproductive organs, including their unique ability to bear children. Freud simply assumed that men are superior to women and then invented a speculative theory that justified gender differences.

3. *Sociologists often criticize Freud for neglecting socialization after childhood.* Freud believed that the human personality is fixed by about the age of 5. However, sociologists have shown that socialization continues throughout the life course (Box 4.1). We devote much of this chapter to exploring socialization after early childhood.

Despite the shortcomings listed previously, the sociological implications of Freud's theory are profound. His main sociological contribution was his insistence that the self emerges during early social interaction and that early childhood experience exerts a lasting impact on personality development. As we will now see, American sociologists and social psychologists took these ideas in a still more sociological direction.

Cooley's Symbolic Interactionism

More than a century ago, American sociologist Charles Horton Cooley introduced the idea of the **looking-glass self,** making him a founding father of the symbolic interactionist tradition and an early contributor to the sociological study of socialization. Cooley observed that when we interact with others, they gesture and react to us. This allows us to imagine how we appear to them. We then judge how others evaluate us. Finally, from these judgments we develop a self-concept or a set of feelings and ideas about who we are. In other words, our feelings about who we are depend largely on how we see ourselves evaluated by others. Just as we see our physical body reflected in a mirror, so we see our social selves reflected in people's gestures and reactions to us (Cooley, 1902).

The implications of Cooley's argument are intriguing. Consider, for example, that the way other people judge us helps determine whether we develop a positive or negative self-concept. Among other things, having a negative self-concept is associated with low achievement in school and college (Hamachek, 1995). Some students' poor performance in school may partly be a result of teachers evaluating them negatively, perhaps because of the students' class or race. Similarly, the way others evaluate us helps determine the size of the discrepancy between our self-concept and the person we would like to be. To compensate for a large discrepancy between actual and ideal self-concept, some people may engage in compulsive buying sprees or out-of-control collecting or binge gift giving. Young women seem to be more prone to such behavior than other categories of the population. This behavior may reflect the difficulty many young women have of achieving the ideals of body weight and body shape that are promoted by the mass media (Benson, 2000). Here we have examples of what came to be known as symbolic interactionism—the idea that in the course of face-to-face communication, people engage in a process of attaching meaning to things.

Mead

George Herbert Mead (1934) took up and developed the idea of the looking-glass self. Like Freud, Mead noted that a subjective and impulsive aspect of the self is present from birth. Mead called it simply the **I.** Again like Freud, Mead argued that a repository of culturally approved standards emerges as part of the self during social interaction. Mead called this objective, social component of the self the **me.** However, whereas Freud focused on the denial of the id's impulses as the mechanism that generates the self's objective side, Mead drew attention to the unique human capacity to "take the role of the other" as the source of the me.

Mead understood that human communication involves seeing yourself from other people's point(s) of view. How, for example, do you interpret your mother's smile? Does

The **looking-glass self** is Cooley's description of the way our feelings about who we are depend largely on how we see ourselves evaluated by others.

The **I**, according to Mead, is the subjective and impulsive aspect of the self that is present from birth.

The **me**, according to Mead, is the objective component of the self that emerges as people communicate symbolically and learn to take the role of the other.

BOX 4.1
Sociology at the Movies

Monster (2003)

Socialization is above all a developmental process. Especially in childhood, it involves the learning, mentoring, sharing, and mutual affirmation that take place when children interact with parents, other family members, and friends. It is a fragile process, the outcome of which is vulnerable to derailment by abusive treatment or serious lapses in care. Resilient children can compensate for harsh influences to varying degrees, depending on temperament, intelligence, and sheer luck (Luthar, 2003).

Some children are not resilient, or the circumstances they face deprive them of anything resembling normal socialization. *Monster* explores what can happen when a tragic history of neglect and abuse completely overwhelms a child's capacity for resilience and there is no one to intervene until it is far too late. It is the real-life story of Aileen Wuornos (played by Charlize Theron), who was executed in 2002 for the serial killings of seven men who had each picked her up as a roadside prostitute. When she was a child, Wuornos was beaten and abandoned by the adults responsible for her care. By the time she was 9 years old, she had learned to sell sex. She had been socialized into a life of prostitution by the repeated molestations she suffered as a child at the hands of strangers, family members, and neighborhood "friends."

Monster begins in Florida, where, as an adult incapable of imagining a better life, Wuornos contemplates suicide. Her last john gave her $5, so she decides to spend it on a farewell drink at the closest bar. At the bar, she meets an 18-year-old woman, Selby Wall (played by Christina Ricci), who was sent to Florida by her family so that an aunt could try to "cure" her of her lesbianism. The two begin a relationship that might have saved Wuornos if she had had the emotional resources to rise to it. In her love for Selby and, more important, because she is loved by Selby, Wuornos demonstrates a spark of resilience, a struggle to master her smoldering rage, and a sincere attempt to connect with another human being despite the chronic fear and distrust of others that were beaten into her as a child. She fails. What happens instead is both tragic and twisted. Wuornos attempts to provide for the limited needs of herself and Selby by robbing the men who pick her up, stealing their cars, and killing them.

Monster does not try to justify Wuornos's crimes. Nor does it relieve her of responsibility for them. Instead, it explains her crimes sociologically. It widens responsibility for the murders Wuornos committed to the society and the social relationships that so completely failed her as a child.

Charlize Theron and Christina Ricci as Aileen Wuornos and Selby Wall in *Monster.*

it mean "I love you," "I find you humorous," or something else entirely? According to Mead, you can find the answer by using your imagination to take your mother's point of view for a moment and see yourself as she sees you. In other words, you must see yourself objectively as a "me" to understand your mother's communicative act. All human communication depends on being able to take the role of the other, wrote Mead. The self thus emerges from people using symbols such as words and gestures to communicate. It follows that the "me" is not present from birth. It emerges only gradually during social interaction.

Mead's Four Stages of Development: Role-Taking.

Unlike Freud, Mead did not think that the emergence of the self was traumatic. On the contrary, he thought it was fun. Mead saw the self as developing in four stages of role-taking. At first, children learn to use language and other symbols by *imitating* important people in their lives, such as their mother and father. Mead called such people **significant others.** Second, children pretend to *be* other people. That is, they use their imaginations to role-play in games such as "house," "school," and "doctor." Third, about the time they reach the age of 7, children learn to play complex games that require them to simultaneously take the role of *several* other people. In baseball, for example, the infielders have to be aware of the expectations of everyone in the infield. A shortstop may catch a line drive. If she wants to make a double play, she must almost instantly be aware that a runner is trying to reach second base and that the person playing second base expects her to throw there. If she hesitates, she probably cannot execute the double play. Once a child can think in this complex way, she can begin the fourth stage in the development of the self, which involves taking the role of what Mead called the **generalized other.** Years of experience may teach an individual that other people, employing the cultural standards of their society, usually regard her as funny or temperamental or intelligent. A person's image of these cultural standards and how they are applied to her is what Mead meant by the generalized other.

Since Mead, psychologists have continued to study childhood socialization. For example, they have identified the stages in which thinking and moral skills develop from infancy to the late teenage years. Let us briefly consider some of their most important contributions.

Piaget

The Swiss psychologist Jean Piaget divided the development of thinking (or "cognitive") skills during childhood into four stages (Piaget and Inhelder, 1969). In the first two years of life, he wrote, children explore the world only through their five senses. Piaget called this the "sensorimotor" stage of cognitive development. At this point in their lives, children's knowledge of the world is limited to what their senses tell them. They cannot think using symbols.

According to Piaget, children begin to think symbolically between the ages of 2 and 7, which he called the "preoperational" stage of cognitive development. Language and imagination blossom during these years. However, children at this age are still unable to think abstractly. Piaget illustrated this by asking a series of 5- and 6-year-olds to inspect two identical glasses of colored water. He then asked them whether the glasses contained the same amount of colored water. All of the children said the glasses contained the same amount. Next, the children watched Piaget pour the water from one glass into a wide, low beaker and the water from the other glass into a narrow, tall beaker. Obviously, the water level was higher in the second beaker although the volume of water was the same in both containers. Piaget then asked each child whether the two beakers contained the same amount of water. Nearly all of the children said that the second beaker contained more water. Clearly, the abstract concept of volume had no meaning for them.

In contrast, most 7- or 8-year-old children understood that the volume of water was the same in both beakers, despite the different water levels. This suggests that abstract thinking begins at about the age of 7. Moreover, between the ages of 7 and 11, children are able to see the connections between causes and effects in their environment. Piaget called this the "concrete operational" stage of cognitive development. Finally, by about the age of 12, children develop the ability to think more abstractly and critically. This behavior marks the beginning of what Piaget called the "formal operational" stage of cognitive development.

Significant others are people who play important roles in the early socialization experiences of children.

The **generalized other,** according to Mead, is a person's image of cultural standards and how they apply to him or her.

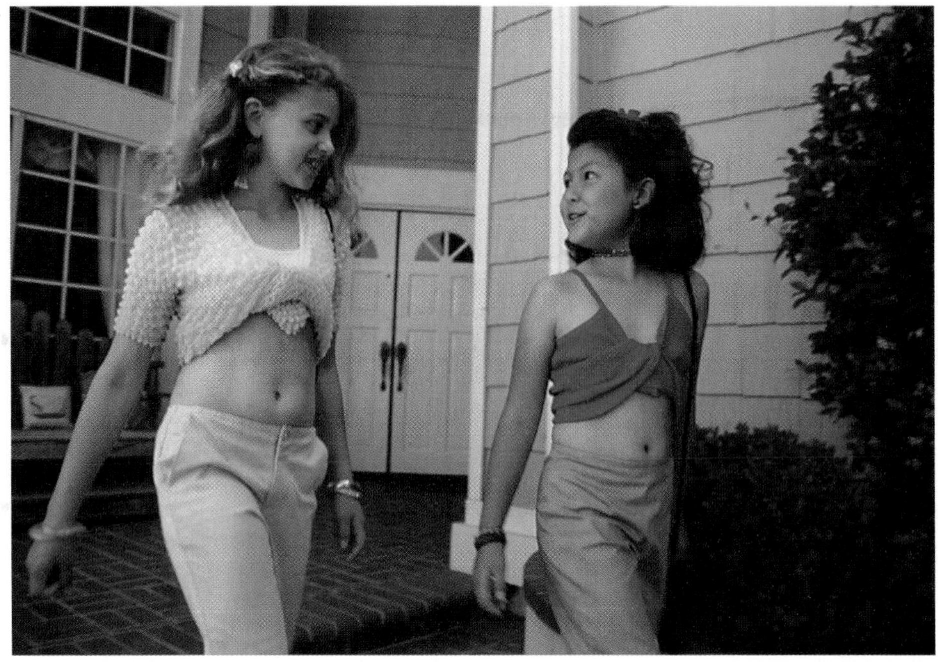

Myrleen F. Cate/PhotoEdit

Much socialization takes place informally, with the participants being unaware that they are being socialized. These girls are learning gender roles as they go to the mall dressed like Britney Spears.

Kohlberg

Lawrence Kohlberg, an American social psychologist, took Piaget's ideas in a somewhat different direction. He showed how children's *moral* reasoning—their ability to judge right from wrong—also passes through developmental stages (Kohlberg, 1981). Kohlberg argued that young children distinguish right from wrong based only on whether something gratifies their immediate needs. At this stage of moral growth, which Kohlberg labeled the "preconventional" stage, what is "right" is simply what satisfies the young child. For example, from the point of view of a 2-year-old, it is entirely appropriate to grab a cookie from a playmate and eat it. An abstract moral concept like theft has no meaning for the very young child.

Teenagers, in contrast, begin to think about right and wrong in terms of whether specific actions please their parents and teachers and are consistent with cultural norms. This phase is the "conventional" stage of moral growth in Kohlberg's terminology. At this stage, a child understands that theft is a proscribed act and that getting caught stealing will result in punishment.

Some people never advance beyond conventional morality. Others, however, develop the capacity to think abstractly and critically about moral principles. This is Kohlberg's "postconventional" stage of moral development. At this stage, one may ponder the meaning of such abstract terms as freedom, justice, and equality. One may question whether the laws of one's society or the actions of one's parents, teachers, or other authorities conform to lofty moral principles. For instance, a 19-year-old who believes the settlement of Europeans in North America involved the theft of land from native peoples is thinking in postconventional moral terms. Such an adolescent is applying abstract moral principles independently and is not merely accepting them as interpreted by authorities.

Vygotsky

Modern psychology has done much to reveal the cognitive and moral dimensions of childhood development. However, from a sociological point of view, the main problem with this body of research is that it minimizes the extent to which society shapes the way we think. Thus, most psychologists assume that people pass through the same stages of mental development and think in similar ways, regardless of the structure of their society and their position in it. Many sociologists disagree with these assumptions.

A few psychologists do, too. The Belarusian psychologist Lev Vygotsky and the American educational psychologist Carol Gilligan offer the most sociological approaches to thinking about cognitive and moral development, respectively. For Vygotsky, ways of thinking are determined not so much by innate factors as they are by the nature of the social institutions in which individuals grow up. Consider, for example, the contrast between ancient China and ancient Greece. In part because of complex irrigation needs, the rice agriculture of ancient southern China required substantial cooperation among neighbors. This form of agriculture had to be centrally organized by an elaborate hierarchy within a large state. Harmony and social order were therefore central to ancient Chinese life. Ancient Chinese thinking, in turn, tended to stress the importance of mutual social obligation and consensus rather than debate. Ancient Chinese philosophies focused on the way in which wholes, not analytical categories, caused processes and events. In contrast, the hills and seashores of ancient Greece were suited more to small-scale herding and fishing than large-scale, centrally organized agriculture. Ancient Greek society was less socially complex than that of ancient China. It was more politically decentralized and gave its citizens more personal freedom. As a result, ancient Greek thinking stressed personal agency. Debate was an integral part of politics. Philosophies tended to be analytical, which means, among other things, that processes and events were viewed as the result of discrete categories rather than whole systems. Markedly different civilizations grew up on these different cognitive foundations; ways of thinking depended less on innate characteristics than on the structure of society (Cole, 1995; Nisbett, Peng, Choi, and Norenzayan, 2001; Vygotsky, 1987).

Courtesy of Carol Gilligan. Photo by Jerry Bauer

In her research, Carol Gilligan attributes differences in the moral development of boys and girls to the different cultural standards parents and teachers pass on to them. By emphasizing that moral development is socially differentiated and does not follow universal rules, Gilligan has made a major sociological contribution to our understanding of childhood development.

Gilligan

In a like manner, Gilligan emphasized the sociological foundations of moral development in her studies of American boys and girls. She attributed differences in the moral development of boys and girls to the different cultural standards parents and teachers pass on to them (Gilligan, 1982; Gilligan, Lyons, and Hanmer, 1990; Brown and Gilligan, 1992). For example, Gilligan found that unlike boys, girls suffer a decline in self-esteem between the ages of 5 and 18. She attributed this decline to their learning our society's cultural standards over time. Specifically, our society tends to define the ideal woman as eager to please and therefore nonassertive. Most girls learn this lesson as they mature, and their self-esteem suffers. The fact that girls encounter more male teachers and fewer female teachers and other authority figures as they mature reinforces this lesson, according to Gilligan. Subsequent research failed to find a decline in the self-esteem of teenage girls, although it did find that boys tend to score somewhat higher than girls on self-esteem (Kling et al., 1999).

Influenced more by the approaches of Vygotsky and Gilligan than of Piaget and Kohlberg, we now assess the contribution of various agents of socialization to the development of the self. These agents of socialization include families, schools, peer groups, and the mass media. We emphasize differences in socialization among societies, social groups, and historical periods. Our approach to socialization, therefore, is rigorously sociological.

Agents of Socialization

Families

Freud and Mead understood well that the family is the most important agent of **primary socialization,** the process of mastering the basic skills required to function in society during childhood. They argued that for most babies, the family is the world. This is as true to-

Primary socialization is the process of acquiring the basic skills needed to function in society during childhood. Primary socialization usually takes place in a family.

Spencer Grant/PhotoEdit

The family is still an important agent of socialization, although its importance has declined since the 19th century.

day as it was 100 years ago. The family is well suited to providing the kind of careful, intimate attention required for primary socialization. The family is a small group. Its members are in frequent face-to-face contact. Child abuse and neglect exist, but most parents love their children and are therefore highly motivated to care for them. These characteristics make most families ideal even today for teaching small children everything from language to their place in the world.

The family into which one is born also exerts an *enduring* influence over the course of one's entire life. Consider the long-term effect of the family's religious atmosphere, for instance. We used data from the General Social Survey (GSS) to examine respondents who, in their youth, had mothers who attended church once a month or more. As adults, 62 percent of these respondents attended church once a month or more themselves. In contrast, only 38 percent of them attended church once a month or less (Table 16.2, p. 494). Clearly, the religious atmosphere of the family into which one is born exerts a strong influence on one's religious practice as an adult.

Despite the continuing importance of the family in socialization, things have changed since Freud and Mead wrote their important works in the early 1900s. They did not foresee how the relative influence of various socialization agents would alter during the next century. The influence of some socialization agents increased, whereas the influence of others—especially the family—declined.

The socialization function of the family was more pronounced a century ago, partly because adult family members were more readily available for child care than they are today. As industry grew across America, families left farming for city work in factories and offices. Especially after the 1950s, many women had to work outside the home for a wage to maintain an adequate standard of living for their families. Fathers, for the most part, did not compensate by spending more time with their children. In fact, because divorce rates have increased and many fathers have less contact with their children after divorce, children probably see less of their fathers on average now than they did a century ago. As a result of these developments, child care—and therefore child socialization—became a big social problem in the 20th century.

Schools

For children older than the age of 5, the child-care problem was resolved partly by the growth of the public school system, which was increasingly responsible for **secondary socialization,** or socialization outside the family after childhood. American industry needed better trained and educated employees. Therefore, by 1918, every state required children to attend school until the age of 16 or the completion of grade 8. By the beginning of the 21st century, more than four-fifths of Americans older than the age of 25 had graduated from high school and about a quarter had graduated from college. By these standards, Americans are the most highly educated people in the world.

Although schools help to prepare students for the job market, they do not necessarily give them an accurate picture of what the job market requires. In 1992, for example, a nationwide survey highlighted the mismatch between the ambitions of American high school students and the projected needs of the American economy in 2005 (Schneider and Stevenson, 1999: 77–8). The number of high school students wanting to become lawyers and judges was five times the projected number needed. The number who wanted to become writers, artists, entertainers, and athletes was 14 times higher than expected openings in 2005. At the other extreme, in 2005 there were five times more administrative and clerical jobs than students interested in such work in 1992. Seven times more service jobs were available than teenagers who wanted them. American high school students, it seems safe to say, often have unrealistically high ex-

Secondary socialization is socialization outside the family after childhood.

pectations about the kinds of jobs they are likely to get when they finish their education (Figure 4.1).

Class, Race, and Conflict Theory

Instructing students in academic and vocational subjects is just one part of the school's job. In addition, a **hidden curriculum** teaches students what will be expected of them in the larger society once they graduate. The hidden curriculum teaches them how to be conventionally "good citizens." Most parents approve of this instruction. According to a survey conducted in the United States and the highly industrialized countries of Europe in 1998, the capacity of schools to socialize students is more important to the public than all academic subjects except mathematics (Galper, 1998).

What is the content of the hidden curriculum? In the family, children tend to be evaluated on the basis of personal and emotional criteria. As students, however, they are led to believe that they are evaluated solely on the basis of their performance on impersonal, standardized tests. They are told that similar criteria will be used to evaluate them in the world of work. The lesson is, of course, only partly true. As you will see in Chapter 10 ("Race and Ethnicity"), Chapter 11 ("Sexuality and Gender"), and Chapter 17 ("Education"), not just performance, but also class, gender, and racial criteria help to determine success in school and in the work world. But the accuracy of the lesson is not the issue here. The important point is that the hidden curriculum has done its job if it convinces students that they are judged on the basis of performance alone. Similarly, a successful hidden curriculum teaches students punctuality, respect for authority, the importance of competition in leading to excellent performance, and other conformist behaviors and beliefs that are expected of good citizens, conventionally defined.

The idea of the hidden curriculum was first proposed by conflict theorists, who, you will recall, see an ongoing struggle between privileged and disadvantaged groups whenever

Sociology Now™

Learn more about **Agents of Socialization** by going through the Agents of Socialization Learning Module.

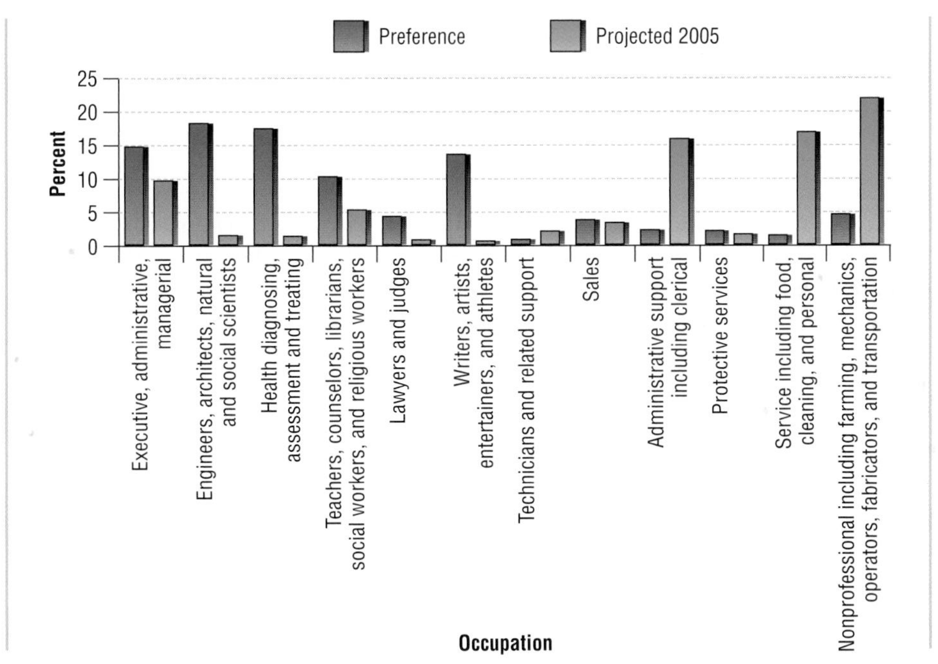

Figure 4.1

Adolescent Job Preferences and Projected Jobs in Paid Labor Force, United States, 2005 (in percent)

Source: From *The Ambitious Generation* by B. Schneider and D. Stevenson, © 1999 Yale University Press. Reprinted by permission.

A **hidden curriculum** teaches students what will be expected of them as conventionally good citizens once they leave school.

Learning disciplined work habits is an important part of the socialization that takes place in schools.

SuperStock

they probe beneath the surface of social life (Willis, 1984 [1977]). From the point of view of conflict theory, some poor and racial-minority students accept the hidden curriculum, thereby learning to act like conventionally good citizens. Other such students reject the hidden curriculum, consequently doing poorly in school and eventually entering the work world near the bottom of the socioeconomic hierarchy. In either case, the hidden curriculum helps sustain the overall structure of society, with all its privileges and disadvantages.

The Self-Fulfilling Prophecy

Why do some poor and racial-minority students reject the hidden curriculum? Because their experience and the experience of their friends, peers, and family members may make them skeptical about the ability of school to open job opportunities for them. As a result, they rebel against the authority of the school. Expected to be polite and studious, they openly violate rules and neglect their work.

Believing that school does not lead to economic success can act as a **self-fulfilling prophecy,** an expectation that helps cause what it predicts. W. I. Thomas and Dorothy Swaine Thomas had a similar idea in stating what became known as the **Thomas theorem:** "Situations we define as real become real in their consequences" (Thomas, 1966 [1931]: 301). For example, believing that school will not help you get ahead may cause you to do poorly in school, and you are more likely to end up near the bottom of the class structure if you perform poorly in school.

Teachers, for their part, can also develop expectations that turn into self-fulfilling prophecies. In one famous study, two researchers informed teachers in a primary school that they were going to administer a special test to the pupils to predict intellectual "blooming." In fact, the test was just a standard intelligence quotient (IQ) test. After the test, they told teachers which students they could expect to become high achievers and which they could expect to become low achievers. In fact, the researchers assigned pupils to the two groups at random. At the end of the year, the researchers repeated the IQ test. They found that the students singled out as high achievers scored significantly higher than those singled out as low achievers. Because the only difference between the two groups of students was that teachers expected one group to do well and the other to do poorly, the researchers concluded that teachers' expectations alone influenced students' performance (Rosenthal and Jacobson, 1968). The clear implication of this research is

A **self-fulfilling prophecy** is an expectation that helps bring about what it predicts.

The **Thomas theorem** states: "Situations we define as real become real in their consequences."

Gender segregation during schoolyard play.

Paul Conklin/PhotoEdit

that if a teacher believes that poor or minority children are likely to do poorly in school, chances are they will.

The Functions of Peer Groups

A second socialization agent whose importance increased in the 20th century is the **peer group.** Peer groups consist of individuals who are not necessarily friends but are about the same age and of similar status. (**Status** refers to a recognized social position that an individual can occupy.) Peer groups help children and adolescents separate from their families and develop independent sources of identity. They are especially influential over lifestyle issues such as appearance, social activities, and dating. In fact, from middle childhood through adolescence, the peer group is often the dominant socializing agent.

As you probably learned from your own experience, conflict often exists between the values promoted by the family and those promoted by the adolescent peer group. Adolescent peer groups are controlled by youth. Through these groups young people begin to develop their own identities by rejecting some parental values, experimenting with new elements of culture, and engaging in various forms of rebellious behavior, including the consumption of alcohol, drugs, and tobacco (▶Figure 4.2). In contrast, parents control families. They represent the values of childhood. Under these circumstances, such issues as tobacco, drug, and alcohol use, hair and dress styles, political views, music, and curfew time are likely to become points of conflict between the generations.

We should not, however, overstate the significance of adolescent–parent conflict. For one thing, the conflict is usually temporary. Once adolescents mature, the family exerts a more enduring influence on many important issues. Research shows that families have more influence than peer groups over the educational aspirations and the political, social, and religious preferences of adolescents and college students (Davies and Kandel, 1981; Milem, 1998; Sherkat, 1998).

A second reason why we should not exaggerate the extent of adolescent–parent discord is that peer groups are not just sources of conflict. They also help to *integrate* young people into the larger society. A recent study of preadolescent children in a small city in the Northwest illustrates this point. Over a period of 8 years, sociologists Patricia and Peter Adler conducted in-depth interviews with schoolchildren between the ages of 8 and 11.

One's **peer group** is composed of people who are about the same age and of similar status as the individual. The peer group acts as an agent of socialization.

Status refers to a recognized social position that an individual can occupy.

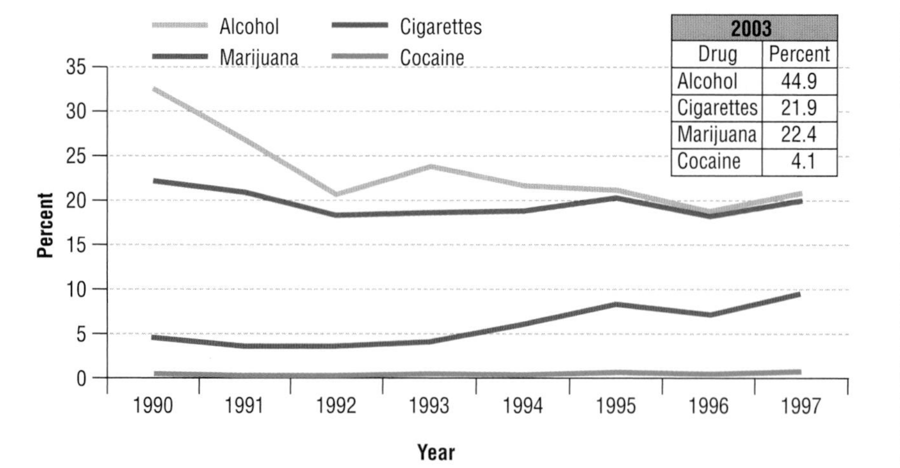

2003	
Drug	Percent
Alcohol	44.9
Cigarettes	21.9
Marijuana	22.4
Cocaine	4.1

Figure 4.2
Percentage of Young Americans Who Used Cigarettes, Alcohol, Marijuana, or Cocaine in Month Prior to Survey, 1990–97 (Ages 12–17) and 2003 (Grades 9–12)

Source: U.S. Department of Health and Human Services (1999: 22; 2004b: 49, 55, 57, 59).

They lived in a well-to-do community in the Northwest composed of about 80,000 whites and 10,000 Hispanics and other racial minority group members (Adler and Adler, 1998). In each school they visited, they found a system of cliques arranged in a strict hierarchy, much like the arrangement of classes and racial groups in adult society. In schools with a substantial number of Hispanics and other nonwhites, cliques were divided by race. Nonwhite cliques were usually less popular than white cliques. In all schools, the most popular boys were highly successful in competitive and aggressive achievement-oriented activities, especially athletics. The most popular girls came from well-to-do and permissive families. One of the main bases of their popularity was that they had the means and the opportunity to participate in the most interesting social activities, ranging from skiing to late-night parties. Physical attractiveness was also an important basis of girls' popularity. Thus, elementary school peer groups prepared these youngsters for the class and racial inequalities of the adult world and the gender-specific criteria that would often be used to evaluate them as adults, such as competitiveness in the case of boys and attractiveness in the case of girls. (For more on gender socialization, see the discussion of the mass media in the following section and Chapter 11, "Sexuality and Gender.") What we learn from this research is that the function of peer groups is not just to help adolescents form an independent identity by separating them from their families. In addition, peer groups teach young people how to adapt to the ways of the larger society.

Web Research Project: Male Socialization, Pornography, and Women

The Mass Media

Like the school and the peer group, the mass media have also become increasingly important socializing agents since the 20th century. The mass media include TV, radio, movies, videos, CDs, audio tapes, the Internet, newspapers, magazines, and books.

The fastest growing mass medium is the Internet. Worldwide, the number of Internet users jumped from 40 million in 1995 to 280 million in 2000 and more than 800 million in 2005 (Figure 4.3). However, TV viewing consumes more of the average American's time than any other mass medium. In 1992 the A. C. Nielsen Company, which measures audience size, estimated that more than 98 percent of American households owned a TV. On average, each TV was turned on for seven hours a day. The University of Maryland's 1993–95 "Americans' Use of Time" project collected national survey data showing that watching TV was the most time-consuming waking activity for women between the ages of 18 and 24 and the second most time-consuming waking activity for men in the same age group (Figure 4.4). Survey research shows that American adults watched more TV in the 1970s than in the 1960s, more in the 1980s than in the 1970s, and more in the early 1990s than in the 1980s. Since the mid-1990s, however, Internet use has been eating into

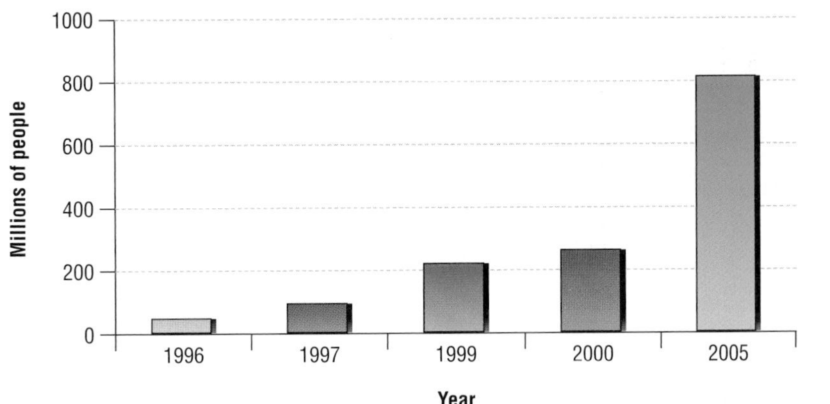

▸Figure 4.3
Number of Internet Users, 1996–2005

Source: "Face of the Web. . . ." (2000); "Internet Growth" (1999); "Internet Usage Statistics. . . ." (2005).

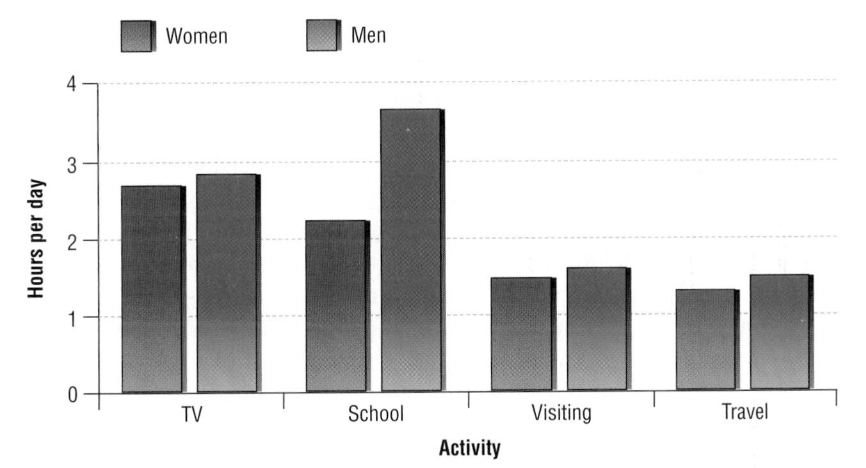

▸Figure 4.4
Top Four Waking Activities of American Women and Men, Ages 18–24, 1993–95 (hours per day)

Source: "The Children's Hours" by John P. Robinson and Suzanne Bianchi, American Demographics, Vol. 20, No. 4, December 1997. Used with permission.

TV viewing hours, especially among more highly educated Americans. Heavy users of TV are concentrated among socially disadvantaged groups, and that trend is intensifying over time (Hao, 1994; Robinson and Bianchi, 1997) (▸Table 4.1).

Children and adolescents use the mass media for entertainment and stimulation. The mass media also help young people cope with anger, anxiety, and unhappiness. Finally, the cultural materials provided by the mass media help young people construct their identities—for example, by emulating the appearance and behavior of appealing movie stars, rock idols, and sports heroes. In performing these functions, the mass media offer youth much choice. Many Americans have access to scores of radio stations and TV channels, hundreds of magazines, thousands of CD titles, hundreds of thousands of books, and millions of websites. Most of us can gain access to hip-hop, heavy metal, or Haydn with equal ease. Thus, whereas adolescents have little choice over how they are socialized by their family and their school, the very proliferation of the mass media gives them more say over which media messages will influence them. To a degree, the mass media allow adolescents to engage in what sociologist Jeffrey Jensen Arnett (1995) calls **self-socialization,** or choosing socialization influences from the wide variety of mass media offerings.

Gender Roles, the Mass Media, and the Feminist Approach to Socialization

Although people are to some extent free to choose socialization influences from the mass media, they choose some influences more often than others. Specifically, they tend to choose influences that are more pervasive, fit existing cultural standards, and are made es-

Self-socialization involves choosing socialization influences from the wide variety of mass media offerings.

Table 4.1
Hours of TV Viewing per Week by Highest Year of School Completed, United States, 2002 ($N = 903$)

Hours per Day Watching TV	HIGHEST YEAR OF SCHOOL COMPLETED				
	0–11	12	13–14	15–16	17–20
0–1	15.3	17.9	27.6	32.7	46.6
2	16.0	24.3	31.3	30.2	24.3
3–4	34.0	36.8	23.8	30.9	21.4
5+	34.7	21.1	17.3	6.2	7.8
Total	100.0	100.0	100.0	100.0	100.0
n	144	280	214	162	103

Source: National Opinion Research Center (2004).

pecially appealing by those who control the mass media. We can illustrate this point by considering how feminist sociologists analyze gender roles. **Gender roles** are widely shared expectations about how males and females are supposed to act. They are of special interest to feminist sociologists, who claim that people are not born knowing how to express masculinity and femininity in conventional ways. Instead, say feminist sociologists, people *learn* gender roles, in part through the mass media.

The learning of gender roles through the mass media begins when small children learn that only a kiss from Snow White's Prince Charming will save her from eternal sleep. It continues in magazines, romance novels, TV, advertisements, music, and the Internet. It is big business. For example, Harlequin Enterprises of Toronto dominates the production and sale of romance novels worldwide. The company sells more than 160 million books a year in 23 languages and more than 100 national markets. About 1 in every 6 mass-market paperbacks sold in North America is a Harlequin romance. The average romance reader spends $800 a year on the genre. Most readers of Harlequin romances consume between 3 and 20 books a month. A central theme in these romances is the transformation of women's bodies into objects for men's pleasure. In the typical Harlequin romance, men are expected to be the sexual aggressors. They are typically more experienced and promiscuous than women. Women are expected to desire love before intimacy. They are assumed to be sexually passive, giving only subtle cues to indicate their interest in male overtures. Supposedly lacking the urgent sex drive that preoccupies males, women are often held accountable for moral standards and contraception (e.Harlequin.com, 2000; Jensen, 1984; Grescoe, 1996) (Figure 4.5).

Boys and girls do not passively accept such messages about appropriate gender roles. They often interpret them in unique ways and sometimes resist them. For the most part, however, they try to develop skills that will help them perform gender roles in a conventional way (Eagley and Wood, 1999: 412–13). Of course, conventions change. What children learn about femininity and masculinity today is less sexist than what they learned just a few generations ago. For example, comparing *Cinderella* and *Snow White* with *Mulan,* we immediately see that children who watch Disney movies today are sometimes presented with more assertive and heroic female role models than the passive heroines of the 1930s and 1940s. However, we must not exaggerate the amount of change in gender socialization. *Cinderella* and *Snow White* are still popular movies. Moreover, for every *Mulan* there is a *Little Mermaid,* a movie that simply modernizes old themes about female passivity and male conquest. In the end, the Little Mermaid's salvation comes through her marriage. The heroic, gutsy, smart, and enterprising protagonists in nearly all children's movies are still boys (Douglas, 1994: 296–7).

Sociology Now™

Learn more about **Gender Roles** by going through the Gender Roles and Videos Animation.

As the learning of gender roles through the mass media suggests, not all media influences are created equal. We may be free to choose which media messages influence us, but most people are inclined to choose the messages that are most widespread, most closely aligned with existing cultural standards, and made most enticing by the mass media. As feminist sociologists remind us, in the case of gender roles, these messages support conventional expectations about how males and females are supposed to act.

Professional Socialization

When a person enters the full-time paid labor force, and especially when he or she learns a professional role, secondary socialization enters a new phase that may be more or less stressful. Stresses are most evident during training, particularly if professional demands are at odds with early socialization.

Gender roles are widely shared expectations about how males and females are supposed to act.

ized, thinking only of survival, escape, and their growing hatred of the guards. If they were thinking as college students, they could have walked out of the experiment at any time. Some of the prisoners did in fact beg for parole. However, by the fifth day of the experiment, they were so programmed to think of themselves as prisoners that they returned docilely to their cells when their request for parole was denied.

The Palo Alto experiment suggests that your sense of self and the roles you play are not as fixed as you may think. Radically alter your social setting and, like the college students in the experiment, your self-conception and patterned behavior are likely to change too. Such change is most evident among people undergoing resocialization in total institutions. However, the sociological eye is able to observe the flexibility of the self in all social settings, including those that greet the individual in adult life.

Jonathan Blair/Corbis

Not all initiation rites, or "rites of passage," involve resocialization, in which powerful socializing agents deliberately cause rapid change in people's values, roles, and self-conception, sometimes against their will. Some rites of passage are a normal part of primary and secondary socialization and signify merely the transition from one status to another. Here, an Italian family celebrates the first communion of a young boy.

Socialization Across the Life Course

Adult Socialization

Although we form our basic personality and sense of identity early in life, socialization continues in adulthood. Adult socialization is necessary for four main reasons (Mortimer and Simmons, 1978: 425–7):

Oleg Popov/Reuters/Corbis

Private Lyndie England was born in a small West Virginia town and joined the army after finishing high school. She became infamous when photographs were made public showing her and other American soldiers abusing Iraqi prisoners at Abu Ghraib prison in obvious contravention of international law. "She's never been in trouble. She's not the person that the photographs point her out to be," said her childhood friend, Destiny Gloin (quoted in "Woman soldier. . . . ," 2004). Ms. Gloin was undoubtedly right. Private England at Abu Ghraib was not Lyndie England back in high school. As in the Palo Alto prison experiment, she was transformed by a structure of power and a culture of intimidation that made the prisoners seem subhuman. Here, a former prisoner of Abu Ghraib prison shows a photo of Private England abusing prisoners.

Sociology Now™

Learn more about **Roles** by going through the Roles and Status Learning Module.

1. *Adult roles are often discontinuous.* That is, contradictory expectations are associated with earlier and later roles. For instance, we socialize children to be nonresponsible, submissive, and asexual, whereas we expect adults to be responsible, assertive, and sexually active. Similarly, we expect adults to be independent and productive, whereas we expect elderly retired people to be more dependent and less productive. We need adult socialization to overcome these role discontinuities. People must learn what others expect of them in their new adult roles.
2. *Some adult roles are largely invisible.* Adult roles are often hidden to people too young to perform them. For example, children have a very limited knowledge of what it means to be married and only a vague sense of what specific occupational roles involve. Adult socialization is required to overcome such role invisibility.
3. *Some adult roles are unpredictable.* People know about the role changes mentioned previously before they occur. To help us learn a predictable new role, we typically engage in **anticipatory socialization,** which involves beginning to take on the norms and behaviors of the role to which we aspire. For instance, 10-year-olds who become fans of the TV program *Friends* engage in anticipatory socialization insofar as the program teaches them what it can be like to be single and in their 20s. However, in adult life many events cause *unpredictable* role change. They include falling in love and marrying someone from a different ethnic, racial, or religious group; separation and divorce; the sudden death of a spouse; job loss and long-term unemployment; forced international migration; and the transition from peace to war. Unpredictable role changes caused by such events require adult socialization.
4. *Adult roles change as we mature.* Forces outside the individual shape the three types of role change just listed. The fourth and final role change that demands adult socialization is mainly the result of inner developmental processes. For example, as children and parents mature, family roles change. When people reach middle age, they may grow more aware of their eventual demise and begin to question their way of life. Hence the so-called "midlife crisis." These kinds of role changes also demand adult socialization.

In the process of learning life's final role—that of the terminally ill person—one can see all four reasons for adult socialization operating. The role of the terminally ill person is characterized by *discontinuity* in the sense that we expect elderly retired people to find new ways of enjoying life, yet, at some point, we expect them to prepare for imminent death. Learning to accept death requires socialization. *Invisibility* characterizes the role of the terminally ill person because our culture makes a great effort to deny death and keep us at a distance from it. We celebrate youth in movies and in advertising. We typically use technology to prolong life even when we know death is imminent. We have no holiday like the Mexican Day of the Dead, which helps young people learn to accept death. We normally consign terminally ill people to die apart from us, in a hospital and not at home, as is common in less industrialized societies. The fact that the role of the terminally ill person is thus rendered largely invisible makes socialization for dying necessary. Finally, although one may think of death as an inevitable part of the course of life, discovering the imminence of one's death is always a surprise. Learning the role of the terminally ill person thus requires socialization because it combines *maturation* and *unpredictability.*

The simultaneous operation of role discontinuity, invisibility, maturation, and unpredictability makes the role of the terminally ill person very hard to learn. Research shows, however, that people are more likely to accept the role if they receive an unambiguous prognosis. In addition, acceptance is greater when primary caregivers (family, friends, physicians) encourage terminally ill people to accept that treatment will no longer help and that they should focus on the alleviation of pain and suffering. Finally, people are more likely to learn the role of the terminally ill person if their physicians lack affiliation with a teaching hospital. Physicians associated with a teaching hospital are more likely to be involved in research on how to prolong life and are therefore reluctant to encourage death (Prigerson, 1992). In short, role clarity, encouragement of role acceptance by signif-

Anticipatory socialization involves beginning to take on the norms and behaviors of a role to which one aspires but does not yet occupy.

icant others, and social distance from people who thwart role learning are associated with acceptance of the role of the terminally ill person. Indeed, these factors are associated with the successful learning of all adult roles. They help overcome all four of the difficulties of adult association discussed previously.

The Flexible Self

Web Interactive Exercise: Identity and Community in Cyberspace

Older sociology textbooks acknowledge that the development of the self is a lifelong process. They note that when young adults enter a profession and get married, they must learn new occupational and family roles. If they marry someone from an ethnic, racial, or religious group other than their own, they are likely to adopt new cultural values or at least modify old ones. Retirement and old age present an entirely new set of challenges. Giving up a job, seeing children leave home and start their own families, and losing a spouse and close friends—all these changes later in life require people to think of themselves in new ways and to redefine who they are.

In our judgment, however, older treatments of adult socialization underestimate the plasticity or flexibility of the self. We believe that today, people's identities change faster, more often, and more completely than they did just a couple of decades ago. One important factor contributing to the growing flexibility of the self is globalization. As we saw in Chapter 3, people are now less obliged to accept the culture into which they were born. Because of globalization, they are freer to combine elements of culture from a wide variety of historical periods and geographical settings.

A second factor increasing our freedom to design our selves is our growing ability to fashion new bodies from old. People have always defined themselves partly in terms of their bodies; your self-conception is influenced by whether you are a man or a woman, tall or short, healthy or ill, conventionally good-looking or plain. But our bodies used to be fixed by nature. People could do nothing to change the fact that they were born with certain features and grew older at a certain rate.

Now, however, you can change your body, and therefore your self-conception, radically and virtually at will—if, that is, you can afford it. Some examples:

- Bodybuilding, aerobic exercise, and weight reduction regimes are more popular than ever.
- Sex-change operations, while infrequent, are no longer a rarity.
- Plastic surgery allows people to buy new breasts, noses, lips, eyelids, and hair and to remove unwanted fat, skin, and hair from various parts of their body. Nearly 3.4 million plastic surgeries took place in the United States in 2001, 80 percent on women. Although the annual number of reconstructive operations increased 55 percent between 1992 and 2003, the number of cosmetic enhancements soared by more than 1000 percent over the same period according to the American Society of Plastic Surgeons (▸Figure 4.6). In 2001, the number of cosmetic enhancements exceeded the number of reconstructive surgeries for the first time. The top four cosmetic procedures in 2001 were nose reshaping (370,698 patients), liposuction (275,463 patients), eyelid surgery (238,213 patients), and breast augmentation (219,883 patients) (American Society of Plastic Surgeons, 2003; MacCarthy, 1999).
- The use of collagen, Botox, and other nonsurgical procedures to remove facial wrinkles is increasingly common. In fact, a growing number of plastic surgeons are organizing Botox parties. They are sort of like Tupperware parties with needles. At a typical Botox party, a plastic surgeon invites about 10 clients to an evening of champagne, chocolate truffles, brie, and botulinum toxin type A. When injected in the face at a cost of \$250 to \$300 per injection, the toxin relaxes muscles and erases signs of aging for a few months. Men composed 14 percent of all Botox users in 2001 (American Society of Plastic Surgeons, 2003). The *Wall Street Journal* reports that some male lawyers "get treatments a week before a big trial to appear less angry and more sympathetic to jurors" (Zimmerman, 2002: B3).

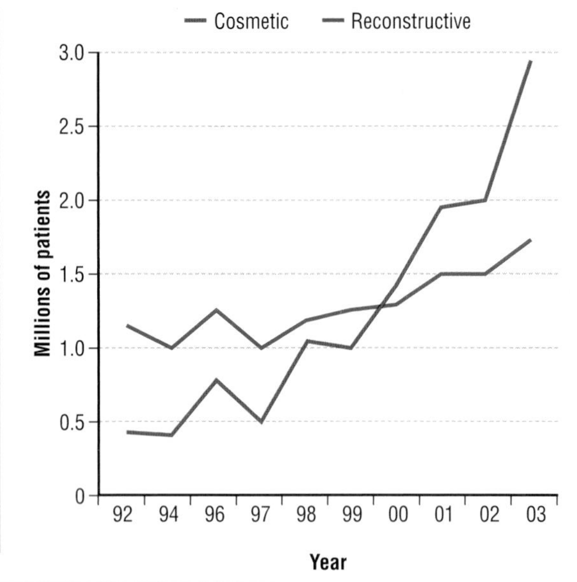

Figure 4.6
Cosmetic and Reconstructive Surgeries, United States, 1992–2003

Note: Does not include nonsurgical procedures.
Source: American Society of Plastic Surgeons (2004).

- Organ transplants are routine. At any given time, about 50,000 Americans are waiting for a replacement organ. Brisk illegal international trade in human hearts, lungs, kidneys, livers, and eyes enables well-to-do people to enhance and extend their lives (Rothman, 1998).
- In 2002 Kevin Warwick, a professor of cybernetics at Reading University in the United Kingdom, became the first cyborg (part man, part computer) when he had a computer chip implanted in his wrist and connected to about 100 neurons. A wire runs under the skin up his arm to a point just below his elbow, where a junction box allows data from his neurons to be transmitted wirelessly to a computer. He plans on implanting a similar device in his wife. Through an Internet connection, they will each be able to feel what the other feels in his or her arm. Warwick predicts that before 2020, more sophisticated implants will allow a primitive form of telepathy (Akin, 2002).
- As if all this is not enough to change how people think of themselves, Dr. Robert J. White of Case Western Reserve University School of Medicine in Cleveland has started to do whole-body transplants. In 1998 he removed the head of a rhesus monkey and connected it by tubes and sutures to the trunk of another monkey. The new entity lived and gained consciousness, although it was paralyzed below the neck because there is no way yet to connect the millions of neurons bridging the brain and the spinal column. Could the same operation be done on a human? No problem, says White. Because the human body is larger and we know more about human than monkey anatomy, the operation would in fact be simpler than it is on monkeys. And, notes White, because research on spinal regeneration is advancing rapidly, it is only a matter of time before creating a fully functional human out of one person's head and another's body will be possible (Browne, 1998; ABC Evening News, April 30, 1998). We used to think of our selves as congruent with our bodies, but the two have now become disjointed. As a result, the formerly simple question "Who are you?" has grown complex.

Identity and the Internet

Further complicating the process of identity formation today is the growth of the Internet and its audiovisual component, the World Wide Web. In the 1980s and early 1990s most observers believed that social interaction by means of computer would involve only the exchange of information between individuals. It turns out they were wrong. Computer-assisted social interaction can profoundly affect how people think of themselves (Brym and Lenton, 2001; Dibbell, 1993; Wellman et al., 1996; Haythornwaite and Wellman, 2002).

Internet users interact socially by exchanging text, images, and sound via e-mail, messaging services (such as ICQ and MSN Messenger), Internet phone, video conferencing, computer-assisted work groups, and online dating services. In the process, they form **virtual communities.** Virtual communities are associations of people, scattered across the country or the planet, who communicate via computer and modem about subjects of common interest.

A **virtual community** is an association of people, scattered across the country, continent, or planet, who communicate via computer and modem about a subject of common interest.

Some virtual communities are short-lived and loosely structured in the sense that they have few formal rules and people quickly drift in and out of them. Chat groups are typical of this genre. Other virtual communities are more enduring and structured. For example, discussion groups cater to people's interest in specialized subjects such as Latino culture, BMWs, dating, or white-water canoeing. Still other virtual communities are highly structured, with many formal rules and relatively stable membership. For example, MUDs (multiple user dimensions) are computer programs that allow people to role-play and engage in a sort of collective fantasy. These programs define the aims and rules of the virtual community and the objects and spaces it contains. Users around the world log on to the MUD from their computers and define their character—their identity—any way they wish. They interact with other users by exchanging text messages or by having their "avatars" (graphical representations) act and speak for them.

Regardless of the degree to which virtual communities are structured, their members form social relationships. They exchange confidences, give advice, share resources, get emotionally involved, and talk sex. Although their true identities are usually concealed, some people who meet online decide to meet and interact in real life.

A large and growing number of people are finding that virtual communities affect their identities in profound ways. Specifically, because virtual communities allow interaction using concealed identities, people are free to assume new identities and are encouraged to discover parts of themselves they were formerly unaware of. In virtual communities, shy people can become bold, normally assertive people can become voyeurs, old people can become young, straight people can become gay, and women can become men.

Take Doug, a Midwestern college junior interviewed by sociologist Sherry Turkle. Doug plays four characters distributed across three different MUDs: a seductive woman, a macho cowboy type, a rabbit who wanders its MUD introducing people to each other, and a fourth character "I'd rather not even talk about because my anonymity there is very important to me. Let's just say that I feel like a sexual tourist." Doug often divides his computer screen into separate windows, devoting a couple of windows to MUDs and a couple to other applications. This setup allows him, in his own words, to

> [s]plit my mind. . . . I can see myself as being two or three or more. And I just turn on one part of my mind and then another when I go from window to window. I'm in some kind of argument in one window and trying to come on to a girl in a MUD in another, and another window might be running a spreadsheet program or some other technical thing for school. . . . And then I'll get a real-time message . . . that's RL [real life]. . . . RL is just one more window . . . and it's not usually my best one (quoted in Turkle, 1995: 13).

Turkle (1995: 14) comments:

> [I]n the daily practice of many computer users, windows have become a powerful metaphor for thinking about the self as a multiple, distributed system. The self is no longer simply playing different roles in different settings at different times, something that a person experiences when, for example, she wakes up as a lover, makes breakfast as a mother, and drives to work as a lawyer. The life practice of windows is that of a decentered self that exists in many worlds and plays many roles at the same time. . . . MUDs . . . offer parallel identities, parallel lives.

Experience on the Internet thus reinforces our main point. In recent decades, the self has become increasingly flexible, and people are freer than ever to shape their selves as they choose.

However, this freedom comes at a cost, particularly for young people. In concluding this chapter, we consider some of the socialization challenges American youth faces today. To set the stage for this discussion, we first examine the emergence of "childhood" and "adolescence" as categories of social thought and experience some 400 years ago.

Dilemmas of Childhood and Adolescent Socialization

In preindustrial societies, children were thought of as small adults. From a young age, they were expected to conform as much as possible to the norms of the adult world largely because children were put to work as soon as they could contribute to the welfare of their families. Often, this meant doing chores by the age of 5 and working full time by the age of 10 or 12. Marriage, and thus the achievement of full adulthood, was common by the age of 15 or 16.

Until the late 1600s, children in Europe and North America fit this pattern. Not until the late 1600s did the idea of childhood as a distinct stage of life emerge. At that time, the feeling grew among well-to-do Europeans and North Americans that boys should be allowed to play games and receive an education that would allow them to develop the emotional, physical, and intellectual skills they would need as adults. Girls continued to be treated as "little women" (the title of Louisa May Alcott's 1869 novel) until the 19th century. Most working-class boys did not enjoy much of a childhood until the 20th century. Only in the last century did the idea of childhood as a distinct and prolonged period of life become universal in the West (Ariès, 1962).

The Emergence of Childhood and Adolescence

The idea of childhood emerged when and where it did because of social necessity and social possibility. Prolonged childhood was *necessary* in societies that required better-educated adults to do increasingly complex work because it gave young people a chance to prepare for adult life. Prolonged childhood was *possible* in societies where improved hygiene and nutrition allowed most people to live more than 35 years, the average life span in Europe in the early 1600s. In other words, before the late 1600s, most people did not live long enough to permit the luxury of childhood. Moreover, there was no social need for a period of extended training and development before the comparatively simple demands of adulthood were thrust upon young people.

In general, wealthier and more complex societies whose populations enjoy a long average life expectancy stretch out the pre-adult period of life. For example, in Europe in 1600 most people reached mature adulthood by the age of about 16. In contrast, in the United States today, most people are considered to reach mature adulthood only around the age of 30, by which time they have completed their formal education, married, and "settled down." Once teenagers were relieved of adult responsibilities, a new term had to be coined to describe the teenage years: "adolescence." Subsequently, the term "young adulthood" entered popular usage as an increasingly large number of people in their late teens and 20s delayed marriage to attend college.

Although these new terms describing the stages of life were firmly entrenched in North America by the middle of the 20th century, some of the categories of the population they were meant to describe soon began to change dramatically. Somewhat excitedly, a number of analysts began to write about the "disappearance" of childhood and adolescence altogether (Friedenberg, 1959; Postman, 1982). While undoubtedly overstating their case, these social scientists identified some of the social forces responsible for the changing character of childhood and adolescence in recent decades. We examine these social forces in the concluding section of this chapter.

Problems of Childhood and Adolescent Socialization Today

Three social forces have done much to change the socialization patterns of American youth over the past 40 years or so: (1) declining adult supervision and guidance; (2) increasing mass media and peer group influence; and (3) the increasing assumption of substantial adult responsibilities to the neglect of extracurricular activities (Box 4.2). Let us consider each of these developments in turn.

Declining Adult Supervision and Guidance

In a six-year, in-depth study of American adolescence, Patricia Hersch wrote that "in all societies since the beginning of time, adolescents have learned to become adults by observing, imitating and interacting with grown-ups around them" (Hersch, 1998: 20). However, in contemporary America, notes Hersch, adults are increasingly absent from the lives of adolescents. Why? According to Hersch, it is because "American society has left its children behind as the cost of progress in the workplace" (Hersch, 1998: 19). What she means is that more American adults are working longer hours than ever before. Consequently, they have less time to spend with their children than they used to. We ex-

Box 4.2

YOU AND THE SOCIAL WORLD

Changing Patterns of Adolescent Socialization

Ask yourself and a parent the following questions: When you were between the ages of 10 and 17, how often were you at home or with friends but without adult supervision? How often did you have to prepare your own meals or take care of a younger sibling while your parent or parents were at work? How many hours a week did you spend cleaning house? How many hours a week did you have to work at a part-time job to earn spending money and save for college? How many hours a week did you spend on extracurricular activities associated with your school? What about on TV viewing and other mass media use? In about 500 words, write a comparison of your parents' and your own socialization experiences during adolescence. Chances are, many more of your waking hours outside of school were spent without adult supervision and/or assuming substantial adult responsibilities such as those listed previously. Compared with your parents, you are unlikely to have spent much time on extracurricular activities associated with your school but quite a lot of time viewing TV and using other mass media. What consequences have these different patterns of socialization had for your life and that of your parent?

amine some reasons for the increasing demands of paid work in Chapter 13 ("Work and the Economy"). Here, we stress a major consequence for American youth: Young people are increasingly left alone to socialize themselves and build their own community. This community sometimes revolves around high-risk behavior. To be sure, more is involved in high-risk behavior than socialization patterns (Box 4.3). However, not coincidentally, the peak hours for juvenile crime are between 3 p.m. and 6 p.m. on weekdays—that is, after school and before most parents return home from work (Hersch, 1998: 362). Also of significance in this connection is that girls are less likely to engage in juvenile crime than boys, partly because parents tend to supervise and socialize their sons and daughters differently (Hagan, Simpson, and Gillis, 1987). Parents typically exert more control over girls, supervising them more closely and socializing them to avoid risk. These research findings suggest that many of the teenage behaviors commonly regarded as problematic result from declining adult guidance and supervision.

Increasing Media Influence

Declining adult supervision and guidance also leaves American youth more susceptible to the influence of the mass media and peer groups. As one parent put it, "When they hit the teen years it is as if they can't be children anymore. The outside world has invaded the school environment" (quoted in Hersch, 1998: 111). In an earlier era, family, school, church, and community usually taught young people more or less consistent beliefs and values. Now, however, the mass media offer a wide variety of cultural messages, many of which differ from each other and from those taught in school and at home. The result for many adolescents is confusion (Arnett, 1995). Should the 10-year-old girl dress modestly or in a sexually provocative fashion? Should the 14-year-old boy devote more time to attending church, synagogue, temple, or mosque—or to playing electric guitar in the garage? Should you just say no to drugs? The mass media and peer groups often pull young people in different directions from the school and the family, leaving them uncertain about what constitutes appropriate behavior and making the job of growing up more stressful than it used to be.

Declining Extracurricular Activities and Increasing Adult Responsibilities

As the opening anecdote about Robert Brym's involvement in high school drama illustrates, extracurricular activities are important for adolescent personality development. They provide opportunities for students to develop concrete skills and thereby make sense of the world and their place in it. In schools today, academic subjects are too

BOX 4.3
SOCIAL POLICY: WHAT DO YOU THINK?

Socialization versus Gun Control

On April 20, 1999, Columbine High School in Littleton, Colorado, was the scene of a mass killing by two students. The shooters, Dylan Klebold and Eric Harris, murdered 13 of their fellow students and then turned their guns on themselves.

After the massacre at Columbine High School, newspapers, magazines, Internet chat rooms, and radio and TV talk shows were abuzz with the problem of teenage violence. What is to be done? people asked. One solution that seems obvious to many people outside of the United States—and to an increasing number of Americans—is to limit the availability of firearms. Their reasoning is simple. All advanced industrial societies except the United States restrict gun ownership. Only the United States has a serious problem with teenagers shooting one another. Other countries have problems with teenage violence. However, because guns are not readily available, teenage violence does not lead to mass killings in, say, Canada, Australia, Britain, or Japan. According to a 1995 Canadian government report, the rate of homicide using firearms per 100,000 people was 2.2 in Canada, 1.8 in Australia, 1.2 in Japan, 1.3 in Britain, and 9.3 in the United States (Department of Justice, Canada, 1995).

In the United States, however, most political discussions about teenage violence focus on the problem of socialization, not on gun control. Soon after the Littleton tragedy, for example, the House of Representatives passed a "juvenile crime bill." It cast blame on the entertainment industry, especially Hollywood movies, and the decline of "family values." Henry Hyde, an Illinois Republican, complained, "People were misled and disinclined to oppose the powerful entertainment industry" (quoted in Lazare, 1999: 57). Tom DeLay, Republican congressman from Texas, worried: "We place our children in daycare centers where they learn their socialization skills . . . under the law of the jungle" (quoted in Lazare, 1999: 58). In other words, according to these politicians, teenage massacres result from poor childhood socialization: the corrupting influence of Hollywood movies, and declining family values.

Some politicians, including Hyde and DeLay, want to reintroduce Christianity into public schools to help overcome this presumed decay. DeLay thus reported an e-mail message he received. It read: "'Dear God, why didn't you stop the shootings at Columbine?' And God writes, 'Dear student, I would have, but I wasn't allowed in school'" (quoted in Lazare, 1999: 57–8). One consequence of the Littleton massacre was not a gun control bill, but a bill to display the Ten Commandments in public schools.

Spencer Grant/PhotoEdit

Citizens of most developed countries must purchase a license before they can possess firearms and buy ammunition. Licensing allows officials to require that applicants pass a safety course and a background check. This practice lowers the risk that firearms will be used for illegal purposes.

What do you think? Is the problem of students shooting each other a problem of socialization, lack of gun control, or a combination of both? In answering this question, think about the situation in other countries and refer back to the discussion of media influence in Chapter 2.

often presented as disconnected bits of knowledge that lack relevance to the student's life. Drama, music, and athletics programs are often better at giving students a framework within which they can develop a strong sense of self because they are concrete activities with clearly defined rules. By training and playing hard on a football team, mastering electric guitar, or acting in plays, you can learn something about your physical, emotional, and social capabilities and limitations, about what you are made of, and what you can and cannot do. Adolescents require these types of activities for healthy self-development.

However, if you are like most young Americans today, you spend fewer hours per week on extracurricular activities associated with school than your parents did when they went to school. Educators estimate that only about a quarter of today's high school students take part in extracurricular activities such as sports, drama, and music (Hersch, 1998). Many of them are simply too busy with household chores, child-care responsibilities, and part-time jobs to enjoy the benefits of school activities outside the classroom. The

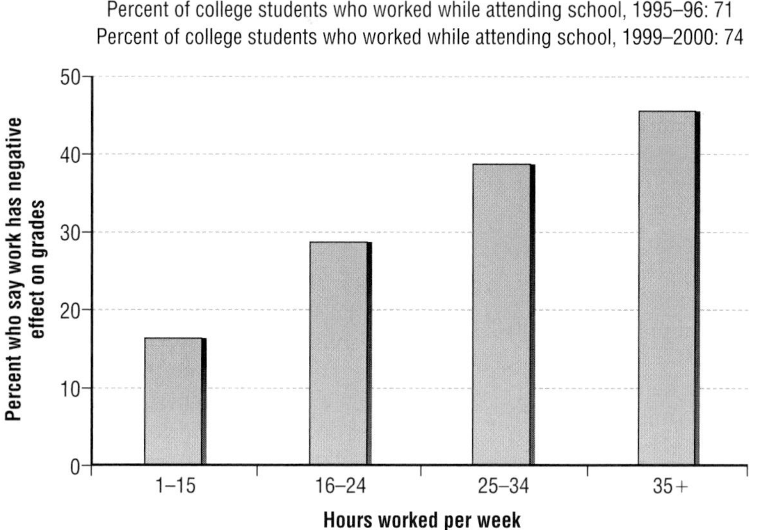

Figure 4.7
Negative Effect of Employment on Grades among Full-Time College Students Who Work, United States, 1999–2000

Source: Marklein (2002)

need for part-time or even full-time work increases when adolescents enter college, often with negative consequences for their grades (Figure 4.7).

"The Vanishing Adolescent"

Some analysts wonder whether the assumption of so many adult responsibilities, the lack of extracurricular activities, declining adult supervision and guidance, and increasing mass media and peer group influence are causing childhood and adolescence to disappear. As early as 1959, one sociologist spoke of "the vanishing adolescent" in American society (Friedenberg, 1959). More recently, another commentator remarked: "I think that we who were small in the early sixties were perhaps the last generation of Americans who actually had a childhood, in the . . . sense of . . . a space distinct in roles and customs from the world of adults, oriented around children's own needs and culture rather than around the needs and culture of adults" (Wolf, 1997: 13; also Postman, 1982). Childhood and adolescence became universal categories of social thought and experience in the 20th century. Under the impact of the social forces discussed previously, the experience and meaning of childhood and adolescence now seem to be changing radically.

Summary

Sociology Now™

Reviewing is as easy as ❶, ❷, ❸.

❶ Before you do your final review, take the SociologyNow diagnostic quiz to help you identify the areas on which you should concentrate. You will find information on how to access SociologyNow on the foldout at the front of the textbook.

❷ As you review, take advantage of SociologyNow's study aids to help you master the topics in this chapter.

❸ When you are finished with your review, take SociologyNow's post-test to confirm you are ready to move on to the next chapter.

1. Why is social interaction necessary?

   Studies show that children raised in isolation do not develop normally. This finding corroborates the view that social interaction unleashes human potential.

2. What are the major theories of childhood socialization?

   Freud called the part of the self that demands immediate gratification the id. He argued that a self-image begins to emerge when the id's demands are denied. Because of many lessons in self-control, a child eventually develops a sense of what constitutes appropriate behavior, a moral sense of right and wrong, and a personal conscience or superego. The superego is a repository of cultural standards. A third component of the self, the ego, develops to balance the demands of the id and superego.

Like Freud, Mead noted that an impulsive aspect of the self is present from birth. He called it the "I." Developing Cooley's idea of the "looking-glass self," Mead also argued that a repository of culturally approved standards emerges as part of the self during social interaction. Mead called it the "me." However, he drew attention to the unique human capacity to "take the role of the other" as the source of the me. People develop, he wrote, by first imitating and pretending to be their significant others, then learning to play complex games that require understanding several roles simultaneously, and finally developing a sense of cultural standards and how they apply.

Since the early 20th century, Piaget and Kohlberg have contributed to our understanding of cognitive and moral socialization, but Vygotsky and Gilligan have done more to underline the social conditions that account for variations in cognitive and moral development. Specifically, their work suggests that gender and economic and political structures shape socialization patterns.

3. **How has the influence of various social agents changed over the past century?**

   The increasing socializing influence of schools, peer groups, and the mass media has been matched by the decreasing socializing influence of the family.

4. **Why is adult socialization necessary?**

   Adult socialization is necessary because roles change. Specifically, roles mature and are often discontinuous, invisible, and unpredictable.

5. **In what sense is the self more flexible than it used to be?**

   People's self-conceptions are subject to more flux now than they were even a few decades ago. Cultural globalization, medical advances, and computer-assisted communication are among the factors that have made the self more plastic.

6. **What social forces have caused and are causing change in the character and experience of childhood and adolescence?**

   Childhood as a distinct stage of life emerged for well-to-do boys in the late 1600s when life expectancy started to increase and boys had to be trained for more complex work tasks. Girls were treated as "little women" until the 19th century, and most working-class boys first experienced childhood only in the 20th century. Once teenagers were relieved of adult responsibilities, the term "adolescence" was coined to describe the teenage years. Subsequently, the term "young adulthood" entered popular usage as an increasingly large number of people in their late teens and 20s delayed marriage to attend college.

   Today, decreasing parental supervision and guidance, the increasing assumption of substantial adult responsibilities by children and adolescents, declining participation in extracurricular activities, and increased mass media and peer group influence are causing changes in the character and experience of childhood and adolescence. According to some analysts, childhood and adolescence as they were known in the first half of the 20th century are disappearing.

Questions to Consider

1. Do you think of yourself in a fundamentally different way from the way your parents (or other close relatives or friends at least 20 years older than you) thought of themselves when they were your age? Interview your parents, relatives, or friends to find out. Pay particular attention to the way in which the forces of globalization may have altered self-conceptions over time.
2. Have you ever participated in an initiation rite in college, the military, or in a religious organization? If so, describe the ritual rejection, ritual death, and ritual rebirth that made up the rite. Do you think that the rite increased your identification with the group you were joining? Did it increase the sense of solidarity—the "we" feeling—of group members?
3. List the contradictory lessons that different agents of socialization taught you as an adolescent. How have you resolved these contradictory lessons? If you have not, how do you intend to do so?

Web Resources

Companion Website for This Book

http://sociology.wadsworth.com

Begin by clicking on the Student Resources section of the website. Choose "Introduction to Sociology" and finally the Brym and Lie book cover. Next, select the chapter you are currently studying from the pull-down menu. From the Student Resources page you will have easy access to InfoTrac® College Edition, MicroCase Online exercises, additional web links, and many resources to aid you in your study of sociology, including practice tests for each chapter.

Recommended Websites

For a slide show and discussion of questions regarding Zimbardo's Stanford Prison Experiment, visit http://www.prisonexp.org.

For the socialization experiences that characterize different generations, go to a major search engine on the Web, such as Google or Yahoo, and search for "teenagers," "generation X," "baby boomers," "the elderly," and so forth.

Initiation rites (or rites of passage) are conveniently summarized in the online version of the *Encarta* encyclopedia. Go to the Encarta search engine at http://encarta.msn.com and search for "rites of passage."

CHAPTER 5 Social Interaction

Henry Diltz/Corbis

In this chapter, you will learn that:

- Social interaction involves people communicating face-to-face, acting and reacting in relation to each other. The character of every social interaction depends on people's distinct positions in the interaction (statuses), their standards of conduct (norms), and their sets of expected behaviors (roles).
- Humor, fear, anger, grief, disgust, love, jealousy, and other emotions color social interactions. However, emotions are not as natural, spontaneous, authentic, and uncontrollable as we commonly believe. Various aspects of social structure influence the texture of our emotional life.
- Nonverbal means of communication, including facial expressions, gestures, body language, and "status cues," are as important as language in social interaction.
- People interact mainly out of fear, envy, or trust. Domination, competition, and cooperation give rise to these three emotions.
- Sociological theories focus on six aspects of social interaction: (1) the way people exchange valued resources; (2) the way they maximize gains and minimize losses; (3) the way they interpret, negotiate, and modify norms, roles, and statuses; (4) the way they manage the impressions they give to others; (5) the way preexisting norms influence social interaction; and (6) the way status hierarchies influence social interaction.

What Is Social Interaction?

In the early decades of the 20th century, the service personnel on ocean liners and trains were mostly men. When some commercial airlines first started operating in Germany in the 1910s and the United States in the 1920s, hiring cabin boys and stewards therefore seemed the natural thing to do.

Things began to change a little in the 1930s, and even more by the early 1950s. The government tightly regulated the airline industry at the time. It decided where and when planes could fly and how much they could charge. On transatlantic flights, the government even decided the allowable amount of passenger legroom and the number and types of courses that constituted a meal. This regulation made all the airlines pretty much identical. How then could one airline stand out from the others and thereby win more business? The airlines came up with a creative answer. In the 1950s, they started hiring large numbers of women as stewardesses (known today by the gender-neutral term "flight attendants"). They outdid one another training and marketing stewardesses as glamorous sex objects, using them to lure the still largely male clientele to fly with them as opposed to their competitors.

The plan required the establishment of a new form of **social interaction,** the creation of a novel way of people to communicate face-to-face and to act and react in relation to each other. As is generally the case, this social interaction was structured around specific statuses, roles, and norms.

Sociology Now™

Learn more about **Social Interaction** by going through the Social Interaction: The Ropes Course, Video Exercise.

The Structure of Social Interaction

Status

In everyday speech, "status" means prestige, but when sociologists say "status" they mean recognized positions occupied by interacting people. Flight attendants and passengers occupy distinct statuses. Note that each person occupies many statuses. Thus, an in-

Sociology Now™

Reviewing is as easy as ❶, ❷, ❸.

Use SociologyNow to help you make the grade on your next exam. When you are finished reading this chapter, go to the Chapter Summary for instructions on how to make SociologyNow work for you.

Social interaction involves people communicating face-to-face and acting and reacting in relation to other people. It is structured around norms, role, and statuses.

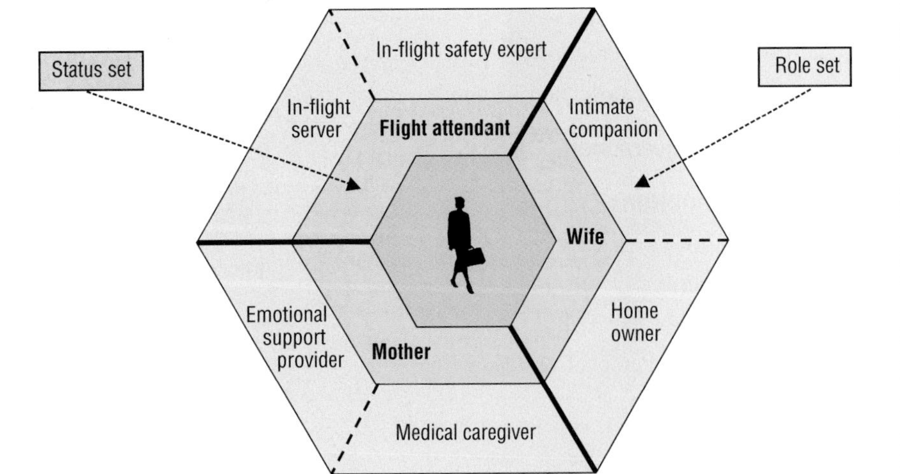

▸Figure 5.1
Role Set and Status Set

dividual may be a flight attendant, a wife, and a mother at the same time. Sociologists call the entire ensemble of statuses occupied by an individual a **status set.** If a status depends on the capabilities and efforts of the individual, it is an **achieved status.** If it does not depend on the capabilities and efforts of the individual, it is an **ascribed status.** "Daughter" is an ascribed status, "flight attendant" an achieved status. Some statuses matter more than others for a person's identity. One's **master status** is the status that is most influential in shaping one's life at a given time.

Sociology Now™

Learn more about **Roles and Status** by going through the Roles and Status Learning Module.

Roles

Social interaction also requires *roles,* or sets of expected behaviors. While people *occupy* statuses, they *perform* roles. A **role set** is a cluster of roles attached to a single status. For example, someone occupying the status of flight attendant may play the roles of in-flight safety expert and server (▸Figure 5.1).

Sociology Now™

Learn more about **Roles** by going through the Roles in Education Video Exercise.

Norms

Finally, social interaction requires *norms,* or generally accepted ways of doing things. Some norms are *prescriptive.* They suggest what a person is expected to do to while performing a particular role. Other norms are *proscriptive.* They suggest what a person is expected *not* to do while performing a particular role. Norms often change over time. At one point in time, some norms are universal, whereas others differ from situation to situation and from role to role.

Case Study: Stewardesses and Their Clientele

- A **status set** is the entire ensemble of statuses occupied by an individual.
- An **achieved status** is a status that depends on the capabilities and efforts of the individual.
- An **ascribed status** is a status that does not depend on the capabilities and efforts of the individual.
- One's **master status** is the status that is most influential in shaping one's life at a given time.
- A **role set** is a cluster of roles attached to a single status.

Let us consider these three elements of social interaction in the context of the evolution of the stewardess's job.

The Changing Role of Stewardess

In 1930, Boeing Air Transport hired Ellen Church, the world's first stewardess (United Airlines, 2003). Trained as a nurse, she wore her white uniform on all flights, which says something about the nature of her role. Flying was much more dangerous than it is today. Although Ellen Church served coffee and sandwiches, her main role was to reassure apprehensive flyers that they were in safe hands in the event of an emergency requiring medical attention. (Although one can imagine that seeing a uniformed nurse walking around might have had the opposite effect on some passengers.) With the introduction of pressurized cabins and other safety features, nurse's uniforms gave way to tailored suits. However, the real revolution in the role of stewardess was signaled by the first of a series of radical uniform changes in 1965. An advertising executive persuaded now defunct

Braniff International Airways to hire a leading fashion designer to redesign its stewardesses' uniforms. The new op-art pastels and hemlines six inches above the knee were a sensation. Everyone wanted to fly Braniff. Braniff stock rose from $24 to $120 per share. Soon all the airlines were in on the act. Advertising reflected the new expectations surrounding the stewardess's role. "Does your wife know you're flying with us?" one Braniff ad teased. A National Airlines ad used this blunt come on: "I'm Linda. Fly me." Pan Am's radio commercials asked: "How do you like your stewardesses?" Continental, which painted its planes with splashes of gold and whose stewardesses wore golden uniforms, advertised itself as "The Proud Bird with the Golden Tail" and later emphasized, "We really move our tail for you." Movies and books solidified the stewardess's new role as sex object. The 1965 movie *Boeing, Boeing* featured Tony Curtis juggling three stewardess girlfriends on different flight schedules. *Coffee, Tea or Me,* a novel published in 1967, advertised itself as an exposé of the stewardess's life behind the scenes, "the uninhibited memoirs of two airline stewardesses" according to the book jacket. It was translated into 12 languages and sold 3 million copies (Handy, 2003).

Bettmann/Corbis

Braniff International Airlines ad featuring stewardesses in designer uniforms.

The Everett Collection

Dany Saval, Tony Curtis, and Thelma Ritter in *Boeing, Boeing* (1965).

The Enforcement of Norms

The airlines specified and enforced many norms pertaining to the stewardess's role. The expectations of passengers helped to reinforce those norms. For example, until 1968 stewardesses had to be single. Until 1970 they could not be pregnant. They had to be attractive, have a good smile, and achieve certain standards on IQ and other psychological tests. In 1954, American Airlines imposed a mandatory retirement age of 32 that became the industry standard. Stewardesses had to have a certain "look"—slim, wholesome, and not too buxom. They were assigned an ideal weight based on their height and figure. Preflight weigh-ins ensured they did not deviate from the ideal. All stewardesses had to wear girdles, and supervisors did a routine "girdle check" by flicking an index finger against a buttock. Weight and height standards were not abolished until 1982 (Lehoczky, 2003). Then there was the question of "personality." Stewardesses were expected to be charming and solicitous at all times. They also had to at least appear "available" to the clientele.

Role Conflict and Role Strain

Role conflict occurs when two or more statuses held at the same time place contradictory role demands on a person. Today's female flight attendants experience role conflict to the degree that working in the airline industry requires frequent absences from

Role conflict occurs when two or more statuses held at the same time place contradictory role demands on a person.

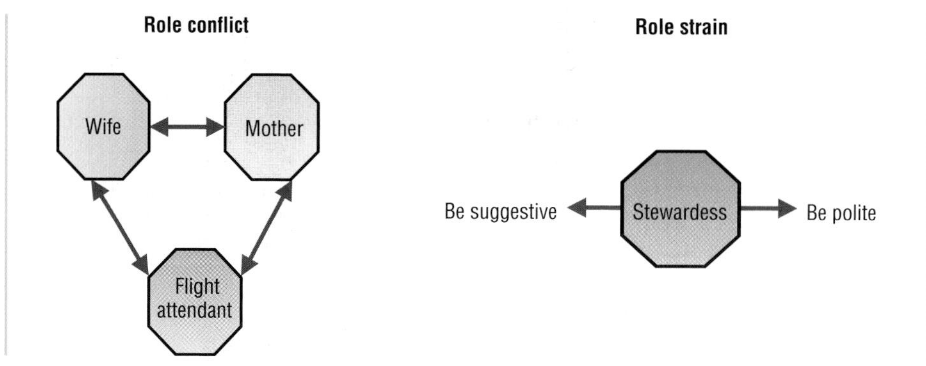

Figure 5.2
Role Conflict and Role Strain

Role conflict takes place when different role demands are placed on a person by two or more statuses held at the same time. How might a flight attendant experience role conflict due to the contradictory demands of the statuses diagrammed in the figure?

Role strain occurs when incompatible role demands are placed on a person in a single status. Why was the status of stewardess in the 1960s and 1970s high in role strain?

home, whereas being a mother and a wife require spending considerable time at home. In contrast, in the 1950s and 1960s role conflict was minimal. Back then, sick children and demanding husbands could hardly interfere with the performance of the stewardess role because the airlines did not allow stewardesses to be mothers or wives. On the other hand, the demands and expectations placed on the stewardess in the 1960s maximized role strain. **Role strain** occurs when incompatible role demands are placed on a person in a single status. For instance, constantly having to be suggestive while also politely warding off unwanted, impolite, and even crude overtures made the stewardess role a lot more stressful than it appeared in the ads (Figure 5.2).

Change in Status

From the 1950s to the late 1970s or early 1980s, the role of stewardess was certainly glamorous. Dating a stewardess was not that different from dating an actress, and people often stared enviously at stewardesses marching proudly through an airport terminal on their way to a presumably exotic location, sporting the latest fashions and hairstyles. On the other hand, the pay was anything but glamorous. Top pay for a starting American Airlines stewardess in 1960 was $24,400 in 2002 dollars. Stewardesses typically had to live four or five to an apartment and, as one stewardess said, "If you wanted to eat you had to find a boyfriend real quick" (quoted in Handy, 2003: 220). Moreover, working conditions were far from ideal. One stewardess described her job to Newsweek in 1968 as "food under your fingernails, sore feet, complaints and insults" (quoted in Handy, 2003: 220).

In the past couple of decades, the status of the stewardess (the position of the stewardess in relation to others) has changed. In the era of shoe searches, deep discount no-frills service, and packaged peanut snacks, little of the glamour remains. However, in the 1960s and 1970s what is now called the Association of Flight Attendants won changes in rules regarding marriage, pregnancy, retirement, and the hiring of men. Stewardesses, once considered sex objects, became flight attendants, and what often used to be a two-year stint leading to marriage became a career.

What Shapes Social Interaction?

Norms, roles, and statuses such as those just illustrated are the building blocks of all face-to-face communication. Whenever people communicate face-to-face, these building blocks structure their interaction. This argument may be hard to swallow because it runs counter to common sense. After all, we typically think of our interactions as outcomes of

Role strain occurs when incompatible role demands are placed on a person in a single status.

our emotional states. For example, we interact differently with people depending on whether they love us, make us angry, or make us laugh. More precisely, we usually think of emotions as deeply personal states of mind that are evoked involuntarily and result in uncontrollable action. We all know that nobody loves, gets angry, or laughs in quite the same way as anyone else; and literature and the movies are full of stories about how love, anger, and laughter often seem to happen by chance and easily spill outside the boundaries of our control. So can we truthfully say that norms, roles, and statuses shape our interactions? We answer this question in the affirmative in the next section. As you will see, our emotions are not as unique, involuntary, and uncontrollable as we are often led to believe. Underlying the turbulence of emotional life is a measure of order and predictability governed by sociological principles.

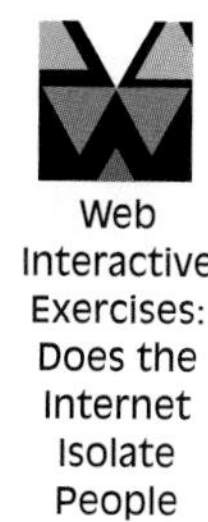

Web Interactive Exercises: Does the Internet Isolate People Socially?

Just as building blocks need cement to hold them together, norms, roles, and statuses require a sort of "social cement" to prevent them from falling apart and to turn them into a durable social structure. What is the nature of the cement that holds the building blocks of social life together? Asked differently, exactly how is social interaction maintained? This is the most fundamental sociological question one can ask, for it is really a question about how social structures and society as a whole are at all possible.

There are three main ways of maintaining social interaction and thereby cementing social structures and society as a whole: by means of domination, competition, and cooperation. In this chapter's second section, we investigate each of these modes of interaction in detail. First, however, we turn to the problem of emotions, beginning with laughter and humor.

The Sociology of Emotions

Laughter and Humor

Every joke is a tiny revolution.

GEORGE ORWELL

Robert Provine (2000) and his research assistants eavesdropped on 1200 conversations of people laughing in public places such as shopping malls. When they heard someone laughing, they recorded who laughed (the speaker, the listener, or both) and the gender of the speaker and the listener. To simplify things, they eavesdropped only on two-person groups, or "dyads." Table 5.1 summarizes some of Provine's findings.

Provine found that in general, speakers laugh more often than listeners do (79.8 percent vs. 54.7 percent of incidents; percentages do not total 100 percent because the speaker, the audience, or both may laugh during an incident). Moreover, interesting patterns emerged when he considered the gender of the speaker and the listener. Women, it turns out, laugh more than men do in everyday conversations. The biggest discrepancy in laugh-

Table 5.1
Laughter and Gender in 1200 Dyads

Dyad	Episodes	PERCENT LAUGHING		Percent Difference
		Speaker	Listener	
Male speaker, male listener	275	75.6	60.0	15.6
Female speaker, female listener	502	86.0	49.8	36.2
Male speaker, female listener	238	66.0	71.0	−5.0
Female speaker, male listener	185	88.1	38.9	49.2
Total	1200	79.8	54.7	25.1

Adapted from *Laughter: A Scientific Investigation* by Robert R. Provine, copyright © 2000 by Robert R. Provine. Used by permission of Viking Penguin, a division of Penguin Group (USA), Inc.

Sipkin Corey/Sygma/Corbis

Chris Rock onstage.

ing occurs when the speaker is a woman and the listener is a man. In that case, women laugh more than twice as often as men do (88.1 percent vs. 38.9 percent). However, even when a man speaks and a woman listens, the woman is more likely to laugh than the man is (71 percent vs. 66 percent). In contrast, men get more laughs than women.

Some people might think Provine's findings confirm the stereotype of the giggling female. Others might interpret his data as confirming the view that when dealing with men, women have more to laugh at. A sociologist, however, would notice that the gender distribution of laughter fits a more general pattern. In social situations where people of different statuses interact, laughter is unevenly distributed across the status hierarchy. People with higher status get more laughs and people with lower status laugh more, which is perhaps why class clowns are nearly always boys. It is also why a classic sociological study of laughter among staff members in a psychiatric hospital discovered "downward humor" (Coser, 1960). At a series of staff meetings, the psychiatrists averaged 7.5 witticisms, the residents averaged 5.5, and the paramedics averaged a mere 0.7. Moreover, the psychiatrists most often made the residents the target of their humor, whereas the residents and the paramedics targeted the patients or themselves. Laughter in everyday life, it turns out, is not as spontaneous as we may think. It is often a signal of dominance or subservience.

Humor and Social Status

Much social interaction takes place among status equals—among members of the same national or racial group, for example. If status equals enjoy a privileged position in the larger society, they often direct their humor at perceived social inferiors. White Americans of northern European origin make jokes about "Polacks" and blacks. The English laugh about the Irish and, more recently, the Welsh. The French howl at the Belgians. The Canadians tell "Newfie" jokes (about Newfoundlanders). And the Russians make jokes about the impoverished and oppressed Chukchi people of northern Siberia ("When a Chukchi man comes back from hunting, he first wants his supper on the table. Then he wants to make love to his wife. Then he wants to take off his skis"). Similarly, when people point out that the only good thing about having Alzheimer's disease is that you can hide your own Easter eggs, they are making a joke about a socially marginal and powerless group. This sort of joke is appropriately called a "'put-down.'" It has the effect of excluding outsiders, making you feel superior, and reinforcing group norms and the status hierarchy itself.

"Ain't nothing more horrifying than a bunch of poor white people," Chris Rock once quipped. "They blame n______ for everything. . . . 'Space shuttle blew up! Them damn n______, that's what it was!'" Chris Rock is, of course, an African American, a member of a group that is disadvantaged in American society. Disadvantaged people often laugh at the privileged majority, but not always. Nation of Islam leader Louis Farrakhan organized the Million Man March in Washington, D.C., in 1996 as a celebration of black solidarity and pride. One of the speakers at the march was Marion Barry, the black mayor of Washington, who had been arrested on cocaine charges six years earlier. Here is what Chris Rock had to say about the incident in front of a black Washington audience: "Marion Barry at the Million Man March. You know what that means? That means even at our finest hour, we had a crackhead onstage!" (Farley, 1998). When members of disadvantaged groups are not laughing at the privileged majority, they are typically laughing at themselves.

Humor and the Structure of Society

People sometimes direct humor against government. Some scholars argue that the more repressive a government and the less free its mass media, the more widespread antigovernment jokes are (Davies, 1998). Sometimes, humor merely has a political edge, "political" here being understood broadly as having to do with the distribution of power and privilege in society (discussed previously). And sometimes, humor seems to have no political content at all; one would be hard pressed to discern the political significance of a chicken crossing a road to get to the other side. Yet all jokes in a sense, even those about chickens crossing the road, are "little revolutions," as George Orwell once remarked, for all jokes invert or pervert reality. They suddenly and momentarily let us see beyond the serious, taken-for-granted world. Analyzed sociologically, jokes even enable us to see the structure of society that lies just beneath our laughter (Zijderveld, 1983).

Emotion Management

My marriage ceremony was chaos, unreal, completely different than I imagined it would be. . . . My sister didn't help me get dressed or flatter me, and no one in the dressing room helped until I asked. I was depressed. I wanted to be so happy on our wedding day. I never ever dreamed how anyone could cry at their wedding . . . [Then] from down the long aisle we looked at each other's eyes. His love for me changed my whole being from that point. When we joined arms I was relieved. The tension was gone. From then on, it was beautiful.

I was in the sixth grade at the time my grandfather died. I remember being called to the office of the school where my mother was on the phone from New York (I was in California). She told me what had happened and all I said was, "Oh." I went back to class and a friend asked me what had happened and I said, "Nothing." I remember wanting very much just to cry and tell everyone what had happened. But a boy doesn't cry in the sixth grade for fear of being called a sissy. So I just went along as if nothing happened while deep down inside I was very sad and full of tears.

IN HOCHSCHILD (1983: 59–60, 67)

Some scholars think that emotions are like the common cold. In both cases, an external disturbance causes a reaction that we experience involuntarily. The external disturbance may involve exposure to a particular virus that causes us to catch cold, or exposure to a grizzly bear attack that causes us to experience fear. In either case, we cannot control our body's patterned response. Emotions, like colds, just happen to us (Thoits, 1989: 319).

The trouble with this argument is that we can and often do *control* our emotions. Emotions do not just happen to us; we manage them. If a grizzly bear attacks you in the woods, you can run as fast as your legs will carry you or you can calm yourself, lie down, play dead, and silently pray for the best. You are more likely to survive the grizzly bear attack if you control your emotions and follow the second strategy.[1] You will also temper your fear with a new emotion: hope (◗Figure 5.3).

When we manage our emotions, we tend to follow certain cultural "scripts," like the culturally transmitted knowledge that lying down and playing dead gives you a better chance of surviving a grizzly bear attack. That is, we usually know the culturally designated emotional response to a particular external stimulus and we try to respond appropriately. If we don't succeed in achieving the culturally appropriate emotional response, we are likely to feel guilty, disappointed, or (as in the case of the grizzly bear attack) something much worse.

The reminiscences quoted at the beginning of this section illustrate typical emotional response processes. In the first case, the bride knew she was not experiencing her wedding day the way she was supposed to, that is, the way her culture defined as appropriate. So by an act of will she locked eyes with the groom and pulled herself out of her depression. In the second case, the schoolboy entirely repressed his grief over his grandfather's death so as not to appear a "sissy" in front of his classmates. As these examples suggest, emotions pervade all social interaction, but they are not, as we commonly believe, spontaneous and

[1]Standard advice for polar and black bear attacks is to yell and fight back.

Figure 5.3
How We Get Emotional

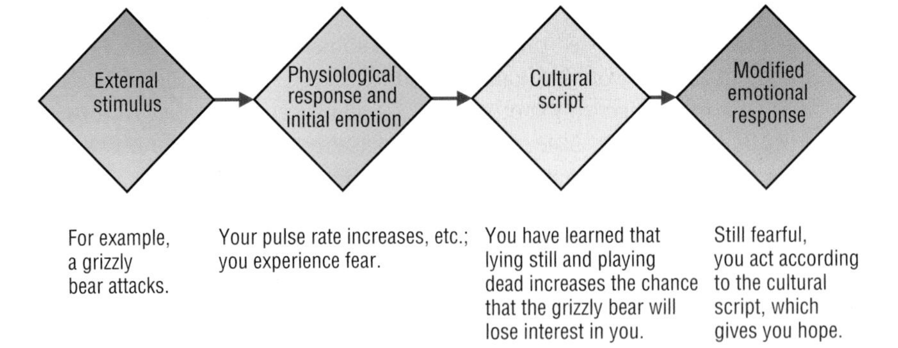

uncontrollable reactions to external stimuli. Rather, the norms of our culture and the expectations of the people around us pattern our emotions.

Sociologist Arlie Russell Hochschild is one of the leading figures in the study of **emotion management.** In fact, she coined the term. She argues that emotion management involves people obeying "feeling rules" and responding appropriately to the situations in which they find themselves (Hochschild, 1979; 1983). So, for example, people talk about the "right" to feel angry, and they acknowledge that they "should" have mourned a relative's death more deeply. We have conventional expectations not only about what we should feel but also about how much we should feel, how long we should feel it, and with whom we should share our feelings. Moreover, feeling rules vary from one category of the population to the next. For example, Hochschild claims that "women, Protestants, and middle-class people cultivate the habit of suppressing their own feelings more than men, Catholics, and lower-class people do" (Hochschild, 1983: 57). This variation exists because our culture invites women to focus more on feeling than action and men to focus more on action than feeling. Similarly, Protestantism invites people to participate in an inner dialogue with God, whereas the Catholic Church offers sacrament and confession, which allow and even encourage the expression of feeling. Finally, more middle-class than lower-class people are employed in service occupations in which the management of emotions is an important part of the job. Hence they are more adept at suppressing their own feelings.

D. Le Strat/Sygma/Corbis

A Catholic taking confession.

Emotion Labor

Hochschild distinguishes emotion management (which everyone does in their personal life) from **emotion labor** (which many people do as part of their job and for which they are paid). For example, teachers, sales clerks, nurses, and flight attendants must be experts in emotion labor. They spend a considerable part of their work time dealing with other people's misbehavior, anger, rudeness, and unreasonable demands. They spend another part of their work time in what is essentially promotional and public relations work on behalf of the organizations that employ them. In all these tasks, they carefully manage their own emotions while trying to render their clientele happy and orderly. Hochschild estimates that in the United States, nearly half the jobs women do and one-fifth of the jobs men do involve substantial amounts of emotion labor.

Emotion management involves people obeying "feeling rules" and responding appropriately to the situations in which they find themselves.

Emotion labor is emotion management that many people do as part of their job and for which they are paid.

Across occupations, different types of emotion labor are required. For instance, according to Hochschild, the flight attendant and the bill collector represent two extremes. If an appropriate motto for the flight attendant is "please, placate, and promote," the bill collector's motto might be "control, coerce, and collect." Flight attendants seek to smooth ruffled feathers, ensure comfort and safety, and encourage passengers to fly the same airline on their next trip. But as Hochschild found when she studied a debt collection office, bill collectors seek to get debtors riled up, ensure their discomfort, and intimidate them to the point where they pay what they owe. As the head of the office once shouted to his employees: "I don't care if it's Christmas or what goddamn holiday! You tell those people to get that money in!" Or on another occasion: "Can't you get madder than *that?* Create alarm!" (quoted in Hochschild, 1983: 141, 146).

Notwithstanding this variation, all jobs requiring emotion labor have in common the fact that "they allow the employer, through training and supervision, to exercise a degree of control over the emotional activities of employees" (Hochschild, 1983: 147). Moreover, as the focus of the economy shifts from the production of goods to the production of services, the market for emotion labor grows. More and more people are selected, trained, and paid for their skill in emotion labor. Emotion labor becomes a commodity that employers buy in much the same way a furniture manufacturer buys fabric to upholster chairs. The emotional life of workers—or at least the way they openly express their feelings—is increasingly governed by the organizations for which they work and is therefore less and less spontaneous and authentic.

Emotions in Historical Perspective

Social structure impinges on emotional experiences in many ways. As we have seen, status hierarchies influence patterns of laughter. Cultural scripts and the expectations of others influence the way we manage our emotions in personal life. The growth of the economy's service sector requires more emotion labor, turns it into a commodity, and decreases the ability of people to experience emotions spontaneously and authentically. In these and other ways, the commonsense view of emotions as unique, spontaneous, uncontrollable, authentic, natural, and perhaps even rooted in biology proves to be misguided.

We can glean additional evidence of the impact of society on our emotional life from historical studies. It turns out that feeling rules take different forms under different social conditions, which vary historically. Three examples from the social history of emotions help illustrate the point:

- *Grief.* Among other factors, the "crude death rate" (the annual number of deaths per 1000 people in a population) helps determine our experience of grief (Lofland, 1985). In Europe as late as 1600, life expectancy was only about 35 years. Many infants died at birth or in their first year of life. Infectious diseases decimated entire populations. The medical profession was in its infancy. The risk of losing family members, especially babies, was thus much greater than today. One result of this situation was that people invested less emotionally in their children than we typically do. Their grief response to child deaths was shorter and less intense than ours; the mourning period was briefer and people became less distraught. As health conditions improved and the infant mortality rate fell over the years, emotional investment in children increased. It intensified especially in the 19th century, when women starting having fewer babies on average as a result of industrialization (see Chapter 20, "Population and Urbanization"). As emotional investment in children increased, grief response to child deaths intensified and lasted longer. (Incidentally, the greater emotional intensity of family life due to smaller average family size also seems to have increased sibling rivalry and jealousy [Thoits, 1989: 335].)
- *Anger.* Industrialization and the growth of competitive markets in 19th-century North America and Europe turned the family into an emotional haven from a world increasingly perceived as heartless (see Chapter 15, "Families"). In keeping with the enhanced emotional function of the family, anger control, particularly by women, became increasingly important for the establishment of a harmonious household. The early 20th century witnessed mounting labor unrest and the growth of the service sector. Avoiding anger thus became an important labor relations goal. This trend influenced

Painting of medieval feast showing rude manners by today's standards.

family life too. Child-rearing advice manuals increasingly stressed the importance of teaching children how to control their anger (Stearns and Stearns, 1985; 1986).

- *Disgust.* Manners in Europe in the Middle Ages were utterly disgusting by our standards. Even the most refined aristocrats spat in public and belched shamelessly during banquets (with the king in attendance, no less). Members of high society did not flinch at scratching themselves and passing gas at the dinner table, where they ate with their hands and speared food with knives. What was acceptable then causes revulsion now because feeling rules have changed. Specifically, manners began to change with the emergence of the modern political state, especially after 1700. The modern political state raised armies and collected taxes, imposed languages, and required loyalty. All this coordination of effort necessitated more self-control on the part of the citizenry. Changes in standards of public conduct—signaled by the introduction of the fork, the nightdress, the handkerchief, the spittoon, and the chamber pot—accompanied the rise of the modern state. Good manners also served to define who had power and who lacked it. For example, there is nothing inherently well mannered about a father sitting at the head of the table carving the turkey and children waiting to speak until they are spoken to. These rules about the difference between good manners and improper or disgusting behavior were created to signify the distribution of power in the family by age and gender (Elias, 1994 [1939]; Scott, 1998).

In Chapter 15 ("Families"), we tell a similar story about romantic love, which became a major criterion in marriage decisions only in industrialized societies (Swidler, 1980). That story leads us to much the same conclusion we arrive at in this chapter: Although emotions form an important part of all social interactions, they are not universal, nor are they constant. They have histories and deep sociological underpinnings in statuses, roles, and norms. Bearing these important lessons in mind, we may now turn to the second main task of this chapter, analyzing the "social cement" that binds together the building blocks of social life and turns them into durable social structures.

Modes of Social Interaction

Web Research Project: Conversation Analysis

Interaction as Competition and Exchange

Have you ever been in a conversation where you can't get a word in edgewise? If you are like most people, this situation is bound to happen from time to time. The longer this kind of one-sided conversation persists, the more neglected you feel. You may make increasingly less subtle attempts to turn the conversation your way. But if you fail, you may decide to end the interaction altogether. If this experience repeats itself—if the person you are talking to consistently monopolizes conversations—you are likely to want to avoid getting into conversations with him or her in the future. Maintaining interaction (and maintaining a relationship) requires that both parties' need for attention is met.

Most people do not consistently try to monopolize conversations. If they did, there wouldn't be much talk in the world. In fact, "turn-taking" is one of the basic norms that govern conversations; people literally take turns talking to make conversation possible. Nonetheless, a remarkably large part of all conversations involves a subtle competition for attention. Consider the following snippet of dinner conversation:

> John: "I'm feeling really starved."
> Mary: "Oh, I just ate."
> John: "Well, I'm feeling really starved."
> Mary: "When was the last time you ate?"

Charles Derber recorded this conversation (Derber, 1979: 24). John starts by saying how hungry he is. The attention is on him. Mary replies that she is not hungry, and the attention shifts to her. John insists he is hungry, shifting attention back to him. Mary finally allows the conversation to focus on John by asking him when he last ate. John thus "wins" the competition for attention.

Derber recorded 1500 conversations in family homes, workplaces, restaurants, classrooms, dormitories, and therapy groups. He concluded that Americans usually try to turn conversations toward themselves. They usually do so in ways that go unnoticed. Nonetheless, says Derber, the typical conversation is a covert competition for attention. In Derber's words, there exists

> a set of extremely common conversational practices which show an unresponsiveness to other's topics and involve turning them into one's own. Because of norms prohibiting blatantly egocentric behavior, these practices are often exquisitely subtle. . . . Although conversationalists are free to introduce topics about themselves, they are expected to maintain an appearance of genuine interest in those about others in a conversation. A delicate face-saving system requires that people refrain from openly disregarding others' concerns and keep expressions of disinterest from becoming visible (Derber, 1979: 23).

Derber is careful to point out that conversations are not winner-take-all competitions. Unless both people in a two-person conversation receive some attention, the interaction is likely to cease. As such, conversation typically involves the *exchange* of attention (Box 5.1).

Exchange and Rational Choice Theories

The idea that social interaction involves trade in attention and other valued resources is the central insight of **exchange theory** (Blau, 1964; Homans, 1961). Exchange theorists argue that *all* social relationships involve a literal give and take. From this point of view, when people interact they exchange valued resources, including attention, pleasure, approval, prestige, information, and money. If you give a lot to others, you expect a lot in return, and people who receive a lot from others are under social pressure to return a lot. The more often an action is rewarded, the more often it is repeated; and stable social structures derive from persistent interactions. In other words, with payoffs, relationships endure and can give rise to various organizational forms. Without payoffs, relationships end.

A variant of this approach is **rational choice theory** (Coleman, 1990; Elster, 1996; Hechter, 1987). Rational choice theory focuses less on the resources being exchanged than the way interacting people weigh the benefits and costs of interaction. According to rational choice theory, interacting people always try to maximize benefits and minimize costs. Businesspeople want to keep their expenses to a minimum so they can keep their profits as high as possible. Similarly, everyone wants to gain the most from their interactions—socially, emotionally, and economically—while paying the least. This holds true even for intimate relations, including marriage: "When men and women decide to marry or have

Exchange theory holds that social interaction involves trade in valued resources.

Rational choice theory focuses on the way interacting people weigh the benefits and costs of interaction. According to rational choice theory, interacting people always try to maximize benefits and minimize costs.

Box 5.1

YOU AND THE SOCIAL WORLD

Competing for Attention

You can observe the competition for attention yourself. Record a couple of minutes of conversation in your dorm, home, or workplace. Then play it back. Write a 500-word essay evaluating each statement in the conversation. Does the statement try to change who is the subject of the conversation? Or does it say something about the *other* conversationalist(s) or ask them about what they said? How does not responding or merely saying "uh-huh" in response operate to shift attention? Are other conversational techniques especially effective in shifting attention? Who "wins" the conversation? What is the winner's gender, race, and class position? Is the winner popular or unpopular? Do you think a connection exists between the person's status in the group and his or her ability to win?

children or divorce, they attempt to maximize their utility by comparing benefits and costs. So they marry when they expect to be better off than if they remained single, and they divorce if that is expected to increase their welfare" (Becker, 1992: 46).

Undoubtedly, one can explain many types of social interaction in terms of exchange and rational choice theories. However, some types of interaction cannot be explained in these terms. For example, people get little or nothing of value out of some relationships, yet they persist. Slaves remain slaves not because they are well paid or because they enjoy the work but because they are forced to do it. Some people remain in abusive relationships because their abusive partner keeps them socially isolated and psychologically dependent. They lack the resources needed to get out of the abusive relationship.

At the other extreme, people often act in ways they consider fair or just even if it does not maximize their personal gain (Frank, 1988; Gamson, Fireman, and Rytina, 1982). Some people even engage in altruistic or heroic acts from which they gain nothing. They do so even though altruism and heroism can sometimes place them at considerable risk. Consider a woman who hears a drowning man cry for help and decides to risk her life to save him (Lewontin, 1991: 73–4). Some analysts assert that such a hero is willing to save the drowning man because she thinks the favor may be returned in the future. To us, this seems far-fetched. In the first place, the probability that today's rescuer will be drowning someday and that the man who is drowning today will be present to save her is close to zero. Second, a man who cannot swim well enough to save himself is just about the last person you would want to rescue you if the need arose. Third, heroes typically report that they decided to act in an instant, before they had a chance to weigh any costs and benefits. Heroes respond to cries for help based on emotion (which, physiologists tell us, takes $\frac{1}{125}$ of a second to register in the brain), not calculation (which takes seconds or even minutes).

When people behave fairly or altruistically, they are interacting with others based on *norms* they have learned—norms that say they should act justly and help people in need, even if substantial costs are attached. Such norms are for the most part ignored by exchange and rational choice theorists. Exchange and rational choice theorists assume that most of the norms governing social interaction are like the "norm of reciprocity," This norm states that you should try to do for others what they try to do for you, because if you do not, then others will stop doing things for you (Homans, 1950). But social life is richer than this narrow view suggests. Interaction is not all selfishness.

Moreover, as you will now see, we cannot assume what people want, because norms (as well as roles and statuses) are not presented to us fully formed. Nor do we mechanically accept norms when they *are* presented to us. Instead, we constantly negotiate and modify norms—as well as roles and statuses—as we interact with others. We will now explore this theme by considering the ingenious ways in which people manage the impressions they give to others during social interaction.

Interaction as Symbolic

The best way of impressing [advisors] with your competence is asking questions you know the answer to. Because if they ever put it back on you, "Well what do you think?" then you can tell them what you think and you'd give a very intelligent answer because you knew it. You didn't ask it to find out information. You ask it to impress people.

A THIRD-YEAR MEDICAL STUDENT

Bettmann/Corbis

Impression management involves manipulating the way you present yourself so that others will view you in the best possible light. Politicians should be adept at impression management because their success depends heavily on voters' opinions of them.

Soon after they enter medical school, students become adept at managing the impression they make on other people. As Jack Haas and William Shaffir (1987) show in their study of professional socialization, students adopt a new, medical vocabulary and wear a white lab coat to set themselves apart from patients. They try to model their behavior after that of doctors who have authority over them. They may ask questions to which they know the answers so that they can impress their teachers. When dealing with patients, they may hide their ignorance under medical jargon to maintain their authority. By engaging in these and related practices, medical students reduce the distance between their premedical-school selves and the role of doctor. By the time they finish medical school, they have reduced the distance so much that they no longer see any difference between who they are and the role of doctor. They come to take for granted a fact they once had to socially construct—the fact that they are doctors (Haas and Shaffir, 1987: 53–83).

Haas and Shaffir's study is an application of symbolic interactionism, a theoretical approach introduced in Chapter 1. Symbolic interactionists regard people as active, creative, and self-reflective. Whereas exchange theorists assume what people want, symbolic interactionists argue that people create meanings and desires in the course of social interaction. According to Herbert Blumer (1969), symbolic interactionism is based on three principles. First, "human beings act toward things on the basis of the meaning which these things have for them." Second, "the meaning of a thing" emerges from the process of social interaction. Third, "the use of meanings by the actors occurs through a process of interpretation" (Blumer, 1969: 2; see also Berger and Luckmann, 1966; Strauss, 1993).

Sociology Now™

Learn more about **Impression Management** by going through the Impression Management Animation.

Dramaturgical Analysis: Role-Playing

While there are several distinct approaches to symbolic interactionism (Denzin, 1992), probably the most widely applied approach is **dramaturgical analysis.** As first developed by sociologist Erving Goffman (1959 [1956]), dramaturgical analysis takes literally Shakespeare's line from *As You Like It:* "All the world's a stage and all the men and women merely players."

From Goffman's point of view, we are constantly engaged in role-playing. This is most evident when we are "front stage," that is, in public settings. Just as being front stage in a drama requires the use of props, set gestures, and memorized lines, so does acting in public space. A server in a restaurant, for example, must dress in a uniform, smile, and recite fixed lines ("How are you? My name is Sam and I'm your server today. May I get you a drink before you order your meal?"). When the server goes "backstage," he or she can relax from the front stage performance and discuss it with fellow actors ("Those kids at table six are driving me nuts!"). Thus, we often distinguish between our public roles and our "true" selves. Note, however, that even backstage we engage in role-playing and impression management. It's just that we are less likely to be aware of it. For instance, in the kitchen, a server may try to present herself in the best possible light to impress another server so that she can eventually ask him out for a date. Thus, the implication of dramaturgical

Dramaturgical analysis views social interaction as a sort of play in which people present themselves so that they appear in the best possible light.

BOX 5.2
Sociology at the Movies

Miss Congeniality (2000)

Scene: A New Jersey schoolyard in 1982. An 8-year-old schoolyard bully is picking a fight with another, smaller boy. Unexpectedly, a girl comes to the rescue, telling the bully to back off. The following dialogue ensues:

> Bully: "If you weren't a girl, I'd beat your face off."
> Girl: "If *you* weren't a girl, I'd beat *your* face off.'
>
> Bully: "You calling me a girl?"
> Girl: "You called *me* one."

Whereupon the bully takes a swing at the girl, which she neatly evades, and she proceeds to deck him. She then approaches the other boy, and says sweetly:

> Girl: "Forget those guys. They're just jealous. You're funny. You're smart. Girls like that."
> Other Boy: "Well I don't like you. Now everyone thinks I need a girl to fight for me. You are a dork brain."

Girl punches other boy in nose. End of scene.

Almost predictably, the girl grows up to become a tough-talking and tomboyish undercover agent (played by Sandra Bullock) without a boyfriend. The plot thickens when Bullock is forced to take an undercover assignment as a contestant in a beauty contest. Someone is plotting a terrorist act during the pageant and she has to find out who it is. First, however, she has to undergo a role change, something far deeper than a mere makeover. A beauty consultant (played by Michael Caine) teaches her how to walk like a stereotypical women, wear makeup, and dress to kill. In her interaction with the other contestants, she begins to learn how to behave in a conventionally feminine way. She even becomes a finalist in the beauty pageant. In the end, she gets the bad guy, captures the heart of the handsome FBI agent (played by Benjamin Bratt), and wins the pageant's "Miss Congeniality" award. Her true self emerges and everyone goes home happy.

Castle Rock/Fortis/The Kobal Collection

Sandra Bullock in *Miss Congeniality* (2000).

Sandra Bullock does not have a monopoly on this theme, which is at least as old as Cinderella. In Hollywood as in fairy tales, the emergence of one's "true self" is often the resolution of the conflict that animates the story. Yet life rarely comes in such neat packages. The sociological study of social interaction shows how we balance different selves in front stage and backstage performances, play many roles simultaneously, get pulled in different directions by role strain and role conflict, and distance ourselves from some of our roles. To make matters even more complex and dynamic, sociology underlines how we continuously enter new stages, roles, conflicts, strains, and distancing maneuvers as we mature. This social complexity makes our "true self" not a thing we discover once and for all time but a work in progress. The resolution of every conflict that animates our lives is temporary. *Miss Congeniality* is an entertaining escape from reality's messiness, but a poor guide to life as we actually live it. For that we need sociology.

analysis is that there is no single self, just the ensemble of roles we play in various social contexts (Box 5.2). Servers in restaurants have many roles off the job. They play on basketball teams, sing in church choirs, and hang out with friends at shopping malls. Each role is governed by norms about what kinds of clothes to wear, what kind of conversation to engage in, and so on. We play on many front stages in everyday life.

Role distancing involves giving the impression that we are just "going through the motions" and actually lack serious commitment to a role.

We do not always do so enthusiastically. If a role is stressful, we may engage in role distancing. **Role distancing** involves giving the impression that we are just "going through the motions" but actually lack serious commitment to a role. Thus, when people think a role they are playing is embarrassing or beneath them, they typically want to give their peers the impression that the role is not their "true" self. My parents force me to sing in the church choir; I'm working at McDonald's just to earn a few extra dollars, but I'll be going back to college next semester; this old car I'm driving is just a loaner—these are the kinds of rationalizations one offers when distancing oneself from a role.

Ethnomethodology

Goffman's view of social interaction seems cynical. He portrays people as inauthentic or constantly playing roles, but never really being themselves. His mindset is not much different from that of Holden Caufield, the antihero of J. D. Salinger's *The Catcher in the Rye.* To Holden Caulfield, all adults seem "phony," "hypocritical," and "fake," constantly pretending to be people they are not.

However discomforting Goffman's cynicism may be, we should not overlook his valuable sociological point: The stability of social life depends on our adherence to norms, roles, and statuses. If that adherence broke down, social life would become chaotic. Take something as simple as walking down a busy street. Hordes of pedestrians rush toward you, yet rarely collide with you. Collision is avoided because of a norm that nobody actually teaches and few people are aware of but almost everyone follows. If someone blocks your way, you move to the right. When you move to the right and the person walking toward you moves to the right (which is your left), you avoid bumping into each other. Similarly, consider the norm of "civil inattention." When we pass people in public, we may establish momentary eye contact out of friendliness but we usually look away quickly. According to Goffman, such a gesture is a "ritual" of respect, a patterned and expected action that affirms our respect for strangers. Just imagine what would happen if you fixed your stare at a stranger for a few seconds longer than the norm. Rather than being seen as respectful, your intention might be viewed as rude, intrusive, or hostile. One could witness an extreme case of this behavior in the American South in the 1950s, where racists routinely engaged in long, ritualized "hate stares" at African Americans (Griffin, 1961). If an African American were bold enough to stare back, conflict would inevitably erupt. Thus, we would not even be able to walk down a street in peace were it not for the existence of certain unstated norms. These and many other norms, some explicit and some not, make an orderly social life possible (Goffman, 1963; 1971).

By emphasizing how we construct social reality in the course of interaction, symbolic interactionists downplay the importance of norms and understandings that *precede* any given interaction. **Ethnomethodology** tries to correct this shortcoming. Ethnomethodology is the study of the methods ordinary people use, often unconsciously, to make sense of what others do and say. Ethnomethodologists stress that everyday interactions could not take place without *preexisting* shared norms and understandings. The norm of moving to the right to avoid bumping into an oncoming pedestrian and the norm of civil inattention are both examples of preexisting shared norms and understandings.

To further illustrate the importance of preexisting shared norms and understandings, Harold Garfinkel conducted a series of experiments. In one such experiment he asked one of his students to interpret a casual greeting in an unexpected way (Garfinkel, 1967: 44):

> Acquaintance: [waving cheerily] How are you?
> Student: How am I in regard to what? My health, my finances, my schoolwork, my peace of mind, my . . . ?
> Acquaintance: [red in the face and suddenly out of control] Look! I was just trying to be polite. Frankly, I don't give a damn how you are.

As this example shows, social interaction requires tacit agreement between the actors about what is normal and expected. Without shared norms and understandings, no sus-

Ethnomethodology is the study of how people make sense of what others do and say by adhering to preexisting norms.

tained interaction can occur. People are likely to get upset and end an interaction when one violates the assumptions underlying the stability and meaning of daily life.

Assuming the existence of shared norms and understandings, let us now inquire briefly into the way people communicate in face-to-face interaction. This issue may seem trivial. However, as you will soon see, having a conversation is actually a wonder of intricate complexity. Even today's most advanced supercomputer cannot conduct a natural-sounding conversation with a person (Kurzweil, 1999: 61, 91).

Verbal and Nonverbal Communication

Fifty years ago an article appeared in the British newspaper *News Chronicle*, trumpeting the invention of an electronic translating device at the University of London. According to the article, "As fast as [a user] could type the words in, say, French, the equivalent in Hungarian or Russian would issue forth on the tape" (quoted in Silberman, 2000: 225). The report was an exaggeration, to put it mildly. It soon became a standing joke that if you ask a computer to translate "The spirit is willing, but the flesh is weak" into Russian, the output would read "The vodka is good, but the steak is lousy." Today, we are closer to high-quality machine translation than we were in the 1950s. However, a practical Universal Translator exists only on *Star Trek*.

The Social Context of Language

The main problem with computerized translation systems is that computers find it difficult to make sense of the *social and cultural context* in which language is used. The same words may mean different things in different settings, so computers, lacking contextual cues, routinely botch translations. For this reason metaphors are notoriously problematic for computers. The following machine translation, which contains both literal and metaphorical text, illustrates this point:

English original:

> Babel Fish is a computerized translation system (at http://world.altavista.com) that is available on the World Wide Web. You can type a passage in a window and receive a nearly instant translation in one of four languages. Simple, literal language is translated fairly accurately. But when understanding requires an appreciation of social context, as most of our everyday speech does, the computer can quickly get you into a pickle. What a drag!

Machine translation from English to Spanish:

> El pescado de Babel es un sistema automatizado de la traducción que está disponible en el World Wide Web (en http://world.altavista.com). Usted puede pulsar un paso en un Window y recibir una traducción casi inmediata en uno de cuatro lenguajes. El lenguaje simple, literal se traduce bastante exactamente. Pero cuando la comprensión requiere un aprecio del contexto social, como la mayoría de nuestro discurso diario, el ordenador puede conseguirle rápidamente en una salmuera. Una qué fricción!

Machine translation from Spanish back to English:

> The fish of Babel is an automated system of the translation that is available in the World Wide Web (at http://world.altavista.com). You can press a passage in a Window and receive an almost immediate translation in one of four languages. The simple, literal language is translated rather exactly. But when the understanding requires an esteem of the social context, like most of our daily speech, the computer can obtain to him in a brine quickly. One what friction!

Despite the complexity involved in accurate translation, human beings are much better at it than computers. Why is this so? A hint comes from computers themselves. Machine translation works best when applications are restricted to a single social context—say, weather forecasting or oil exploration. In such cases, specialized vocabularies and meanings specific to the context of interest can be built into the program. Ambiguity is thus reduced and computers can "understand" the meaning of words well enough to translate them with reasonable accuracy. Similarly, humans must be able to reduce ambiguity and make sense of words to become good translators. They do so by learning the nuances of meaning in different cultural and social contexts over an extended period of time.

Mastery of one's own language happens the same way. People are able to understand one another not just because they are able to learn words—computers can do that well enough—but because they can learn the social and cultural contexts that give words meaning. They are greatly assisted in that task by *nonverbal* cues.

Let us linger for a moment on the question of how nonverbal cues enhance meaning. Sociologists, anthropologists, and psychologists have identified numerous nonverbal means of communication that establish context and meaning. The most important types of nonverbal communication involve the use of facial expressions, gestures, body language, and status cues.

Facial Expressions, Gestures, and Body Language

The April 2000 issue of *Cosmopolitan* magazine featured an article advising female readers on "how to reduce otherwise evolved men to drooling, panting fools." Basing his analysis on the work of several psychologists, the author of the article first urges readers to "[d]elete the old-school seductress image (smoky eyes, red lips, brazen stare) from your consciousness." Then, he writes, you must "[u]pload a new inner temptress who's equal parts good girl and wild child." This involves several steps, including the following: (1) Establish eye contact by playing sexual peek-a-boo. Gaze at him, look away, peek again, and so forth. By interrupting the intensity of your gaze, you heighten his anticipation of the next glance. The trick is to hold his gaze long enough to rouse his interest yet briefly enough to make him want more. Three seconds of gazing followed by five seconds of looking away seems to be the ideal. (2) Sit down with your legs crossed to emphasize their shapeliness. Your toes should be pointed toward the man who interests you and should reach inside the 3-foot "territorial bubble" that defines his personal space. (3) Speak quietly. The more softly you speak, the more intently he must listen. Speaking just above a whisper will grab his full attention and force him to remain fixed on you. (4) Invade his personal space and enter his "intimate zone" by finding an excuse to touch him. Picking a piece of lint off his jacket and then leaning in to tell him in a whisper what you've done ought to do the trick. Then you can tell him how much you like his cologne. (5) Raise your arm to flip your hair. The gesture subliminally beckons him forward. (6) Finally, smile—and when you do, tilt your head to reveal your neck because he'll find it exciting (Willardt, 2000). If things progress, another article in the same issue of *Cosmopolitan* explains how you can read his body language to tell whether he is lying (Dutton, 2000).

Whatever we may think of the soundness of *Cosmopolitan's* advice or the image of women and men it tries to reinforce, this example drives home the point that social interaction typically involves a complex mix of verbal and nonverbal messages. The face alone is capable of more than 1000 distinct expressions reflecting the whole range of human emotion. Arm movements, hand gestures, posture, and other aspects of body language send many more messages to one's audience (Wood, 1999 [1996]) (▸Figure 5.4).

Despite the wide variety of facial expressions in the human repertoire, most researchers believed until recently that the facial expressions of six emotions are similar across cultures. These six emotions are happiness, sadness, anger, disgust, fear, and surprise (Ekman, 1978). A smile, it was believed, looks and means the same to advertising executives in Manhattan and members of an isolated tribe in Papua New Guinea. Researchers concluded that the facial expressions that express these basic emotions are reflexes rather than learned responses.

Especially since the mid-1990s, however, some researchers have questioned whether a universally recognized set of facial expressions reflects basic human emotions. Among other things, critics have argued that "facial expressions are not the readout of emotions but displays that serve social motives and are mostly determined by the presence of an audience" (Fernandez-Dols, Sanchez, Carrera, and Ruiz-Belda, 1997: 163). From this point of view, a smile will reflect pleasure if it serves a person's interest to present a smiling face

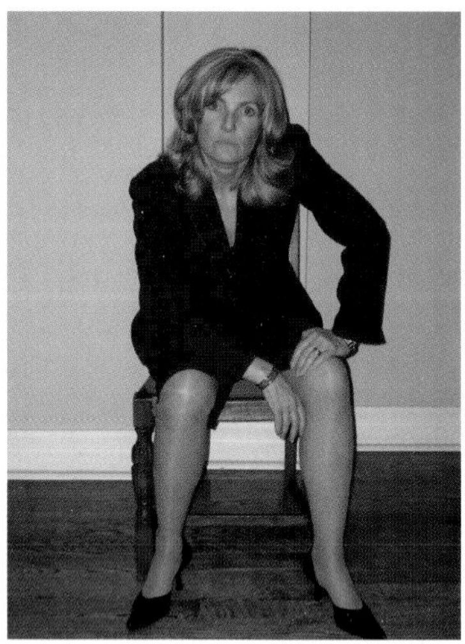

Robert J. Brym

Figure 5.4
Body Language
Among other things, body language communicates the degree to which people conform to gender roles, or widely shared expectations about how males or females are supposed to act. In these photos, which postures suggest power and aggressiveness? Which suggest pleasant compliance? Which are "appropriate" to the sex of the person?

to his or her audience. On the other hand, a person may be motivated to conceal anxiety by smiling or to conceal pleasure by suppressing a smile.

Some people are better at deception than others. Most people find it hard to deceive others because facial expressions are hard to control. If you have ever tried to stop yourself from blushing, you will know what we mean. Sensitive analysts of human affairs—not just sociologists trained in the fine points of symbolic interaction, but police detectives, lawyers, and other specialists in deception—can often see through phony performances. They know that a crooked smile, a smile that lasts too long, or a smile that fades too quickly may suggest that something fishy is going on beneath the superficial level of impression management. Still, smooth operators can fool experts. Moreover, experts can be mistaken. Crooked smiles and the like may be the result of innocent nervousness, not deception.

No gestures or body postures mean the same thing in all societies and all cultures. In our society, people point with an outstretched hand and an extended finger. However, people raised in other cultures tip their head or use their chin or eyes to point out something. We nod our heads "yes" and shake "no," but others nod "no" and shake "yes."

Finally, we must note that in all societies, people communicate by manipulating the space that separates them from others (Hall, 1959; 1966). This point is well illustrated in our *Cosmopolitan* example, where women are urged to invade a man's "personal space" and "intimate zone" to arouse his interest. Sociologists commonly distinguish four zones that surround us. The size of these zones varies from one society to the next. In North America, an intimate zone extends about 18 inches from the body. It is restricted to people with whom we want sustained, intimate physical contact. A personal zone extends from about 18 inches to 4 feet away. It is reserved for friends and acquaintances. We tolerate only a little physical intimacy from such people. The social zone is situated in the area roughly 4 to 12 feet away from us. Apart from a handshake, no physical contact is permitted from people we restrict to this zone. The public zone starts around 12 feet from our bodies. It is used to distinguish a performer or a speaker from an audience.

Status Cues

Aside from facial expressions, gestures, and body language, a second type of nonverbal communication takes place by means of **status cues,** or visual indicators of other people's social position. Goffman (1959 [1956]) observed that when individuals come into contact, they typically try to acquire information that will help them define the situation and make interaction easier. This goal is accomplished in part by attending to status cues. Elijah Anderson (1990) developed this idea by studying the way African Americans and European Americans interact on the street in two adjacent urban neighborhoods. Members of both groups visually inspect strangers before concluding that they are not dangerous. They make assumptions about others on the basis of skin color, age, gender, companions, clothing, jewelry, and the objects they carry with them. They evaluate the movements of strangers, the time of day, and other factors to establish how dangerous they might be. In general, children pass inspection easily. White women and white men are treated with greater caution, but not as much caution as African American women and African American men. Urban dwellers are most suspicious of African American male teenagers. People are most likely to interact verbally with individuals who are perceived as the safest.

Although status cues may be useful in helping people define the situation and thus greasing the wheels of social interaction, they also pose a social danger, for status cues can quickly degenerate into **stereotypes,** or rigid views of how members of various groups act regardless of whether individual group members really behave that way. Stereotypes create social barriers that impair interaction or prevent it altogether. For instance, police officers in some states routinely stop young African American male drivers without cause to check for proper licensing, possession of illegal goods, and so forth. In this case, a social cue has become a stereotype that guides police policy. Young black males, the great majority of whom never commit an illegal act, view this police practice as harassment. Racial stereotyping therefore helps to perpetuate the sometimes poor relations between the African American community and law enforcement officials.

As these examples show, face-to-face interaction may at first glance appear to be straightforward and unproblematic. Most of the time it is. However, underlying the taken-for-granted surface of human communication is a wide range of cultural assumptions, unconscious understandings, and nonverbal cues that make interaction possible.

Status cues are visual indicators of other people's social position.

Stereotypes are rigid views of how members of various groups act, regardless of whether individual group members really behave that way.

Power and Conflict Theories of Social Interaction

In our discussion of the social cement that binds statuses, roles, and norms together, we have made four main points:

1. One of the most important forces that cements social interaction is the competitive exchange of valued resources. People communicate to the degree they get something

Stereotypes are rigid views of how members of various groups act, regardless of whether individual group members really behave that way. Stereotypes create social barriers that impair interaction or prevent it altogether. Which stereotypes about Japanese people are reinforced by this American World War II poster?

Louseous Japanicas

The first serious outbreak of this lice epidemic was officially noted on December 7, 1941, at Honolulu, T. H. To the Marine Corps, especially trained in combating this type of pestilence, was assigned the gigantic task of extermination. Extensive experiments on Guadalcanal, Tarawa, and Saipan have shown that this louse inhabits coral atolls in the South Pacific, particularly pill boxes, palm trees, caves, swamps and jungles.

Flame throwers, mortars, grenades and bayonets have proven to be an effective remedy. But before a complete cure may be effected the origin of the plague, the breeding grounds around the Tokyo area, must be completely annihilated.

Appeared in *Leatherneck*, March, 1945

valuable out of the interaction. Simultaneously, however, they must engage in a careful balancing act. If they compete too avidly and prevent others from getting much out of the social interaction, communication will break down. This is exchange and rational choice theory in a nutshell.

2. Nobody hands values, norms, roles, and statuses to us fully formed, nor do we accept them mechanically. We mold them to suit us as we interact with others. For example, we constantly engage in impression management so that others will see the roles we perform in the best possible light. This is a major argument of symbolic interactionism and its most popular variant, dramaturgical analysis.
3. Norms do not emerge entirely spontaneously during social interaction, either. In general form, they exist before any given interaction takes place. Indeed, sustained interaction would be impossible without preexisting shared understandings. This is the core argument of ethnomethodology.
4. Nonverbal mechanisms of communication greatly facilitate social interaction. These mechanisms include facial expressions, hand gestures, body language, and status cues.

We now want to highlight a fifth point that has been lurking in the background of our discussion up until now. **Conflict theories of social interaction** emphasize that when people interact, their statuses are often arranged in a hierarchy. People on top enjoy more power than those on the bottom. The degree of inequality strongly affects the character of social interaction between the interacting parties (Bourdieu, 1977; Collins, 1982; Kemper, 1978; 1987; Molm, 1997) (Concept Summary 5.1).

Conflict theories of social interaction emphasize that when people interact, their statuses are often arranged in a hierarchy. Those on top enjoy more power than those on the bottom. The degree of inequality strongly affects the character of social interaction between the interacting parties.

Power is the probability that one actor within a social relationship will be in a position to carry out his or her own will despite resistance.

Max Weber (1947: 152) defined **power** as "the probability that one actor within a social relationship will be in a position to carry out his [or her] own will despite resistance." We can clearly see how the distribution of power affects interaction by examining male-female interaction. Women are typically socialized to assume subordinate positions in life, whereas men assume superordinate positions. As we saw in Chapter 4 ("Socialization"), this distribution of power is evident in the way men usually learn to be aggressive and competitive and women learn to be cooperative and supportive (see also Chapter 11, "Sexuality and Gender"). Because of this learning, men often dominate conversations. Thus, conversation analyses conducted by

BOX 5.3

SOCIAL POLICY: WHAT DO YOU THINK?

Allocating Time Fairly in Class Discussions

When John Lie was Chair of the Department of Sociology at the University of Illinois (Urbana-Champaign), he often heard student complaints. Sometimes they were reasonable. Sometimes they were not. A particularly puzzling complaint came from a self-proclaimed feminist taking a women's studies class. She said: "The professor lets the male students talk in class. They don't seem to have done much of the reading, but the professor insists on letting them say something even when they don't really have anything to say." John later talked to the professor, who claimed she was only trying to let different opinions come out in class.

Policy debates often deal with important issues at the state, national, and international levels. However, they may revolve around everyday social interaction. For instance, as this chapter's discussion of Deborah Tannen's work suggests, gender differences in conversational styles have a big impact on gender inequality. Thus, many professors use class participation to evaluate students. Your grade may depend in part on how often you speak up and whether you have something interesting to say. But Tannen's study suggests that men tend to speak up more often and more forcefully than women. Men are more likely to dominate classroom discussions. Therefore, does the evaluation of class participation in assigning grades unfairly penalize female students? If so, what policies can you recommend that might overcome the problem?

One possibility is to eliminate class participation as a criterion for student evaluation. However, most professors would object to this approach on the grounds that good discussions can demonstrate students' familiarity with course material, sharpen their ability to reason logically, and enrich everyone's educational experience. A college lacking energetic discussion and debate would not be much of an educational institution.

A second option is to systematically encourage women to participate in classroom discussion. A third option is to allot equal time for women and men or to allot each student equal time. Criticisms of such an approach come readily to mind. Shouldn't time be allocated only to people who have done the reading and have something interesting to say? The woman who complained to John Lie made that same point. Encouraging everyone to speak or forcing each student to speak for a certain number of minutes, even if the student has nothing interesting to contribute, would probably be boring or frustrating for better-prepared students.

As you can see, the question of how time should be allocated in class discussions has no obvious solution. In general, the realm of interpersonal interaction and conversation is an extremely difficult area in which to impose rules and policies. So what should your professor do to ensure that class discussion time is allocated fairly?

Concept Summary 5.1
Theories of Social Interaction

Theory	Focus of Attention	Principal Theorist(s)
Exchange theory	Exchange of valued resources	Homans, Blau
Rational choice theory	Maximization of gains and minimization of losses	Coleman, Hechter
Symbolic interactionism	Interpretation, negotiation, and modification of norms, rules, and statuses	Blumer, Denzin
Dramaturgical analysis	Impression management	Goffman
Ethnomethodology	Influence of preexisting norms	Garfinkel
Conflict theory	Influence of status hierarchies	Bourdieu, Collins

Deborah Tannen show that men are more likely than women to engage in long monologues and interrupt when others are talking (Tannen, 1994a; 1994b) (Box 5.3). They are also less likely to ask for help or directions because doing so would imply a reduction in their authority. Much male-female conflict results from these differences. A stereotypical case is the lost male driver and the helpful female passenger. The female passenger, seeing that the male driver is lost, suggests that they stop and ask for directions. The male driver does not want to ask for directions because he thinks that would make him look incompetent. If both parties remain firm in their positions, an argument is bound to result.

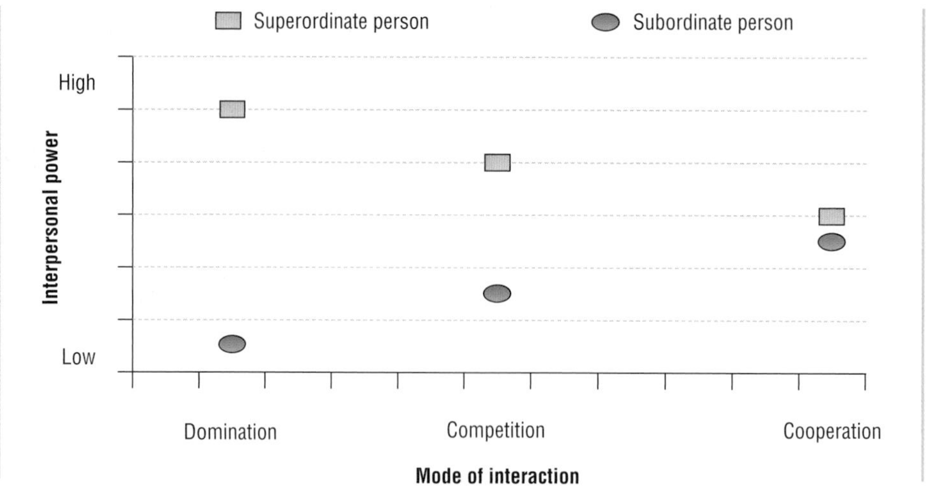

Figure 5.5
Interpersonal Power by Mode of Interaction

Types of Interaction

To get a better grasp of the role of power in social interaction, let us consider two extreme cases and a case that lies at the midpoint between the extremes (Figure 5.5 and Concept Summary 5.2). **Domination** represents one extreme type of interaction. In social interaction based on domination, nearly all power is concentrated in the hands of people of similar status, whereas people of different status enjoy almost no power. Guards versus inmates in a concentration camp and landowners versus slaves on plantations in the antebellum South were engaged in social interaction based on domination. In extreme cases of domination, subordinates live in a state of near-constant fear.

Concept Summary 5.2
Main Modes of Interaction

Mode of Interaction	Domination	Competition	Cooperation
Level of inequality	High	Medium	Low
Characteristic emotion	Fear	Envy	Trust
Efficiency	Low	Medium	High

The other extreme involves interaction based on **cooperation.** Here, power is more or less equally distributed among people of different status. Cooperative interaction is based on feelings of trust. As we will see in Chapter 15 ("Families"), marriages are happier when spouses share housework and child care equitably. Perceived inequity breeds resentment and dissatisfaction. It harms intimacy. It increases the chance that people will have extramarital affairs and divorce. In contrast, a high level of trust between spouses is associated with marital stability and enduring love (Wood, 1999 [1996]).

Between the two extremes of (a) interaction based on domination and (b) interaction based on cooperation is interaction based on **competition.** In this mode of interaction, power is unequally distributed, but the degree of inequality is less than in systems of domination. Most of the social interactions analyzed by exchange and rational choice theorists are of this type. If trust is the prototypical emotion of relationships based on cooperation, and fear is the characteristic emotion of subordinates involved in relationships based on domination, envy is an important emotion in competitive interaction.

Significantly, the mode of interaction in an organization strongly influences its efficiency or productivity, that is, its ability to achieve its goals at the least possible cost. Thus, African American slaves on plantations in the antebellum South and Jews in Nazi concentration camps were usually regarded as slow and inept workers by their masters (Collins, 1982: 66–9). This characterization was not just a matter of prejudice. Slavery is inefficient because, in the final analysis, fear of coercion is the only motivation for slaves to work. Yet, as psychologists have known for more than half a century, punishment is a far less effective motivator than reward (Skinner, 1953). Slaves hate the tedious and often backbreaking labor, they get little in exchange for it, and therefore they typically work with less than maximum effort.

Domination is a mode of interaction in which nearly all power is concentrated in the hands of people of similar status. Fear is the dominant emotion in systems of interaction based on domination.

Cooperation is a basis for social interaction in which power is more or less equally distributed between people of different status. The dominant emotion in cooperative interaction is trust.

Competition is a mode of interaction in which power is unequally distributed, but the degree of inequality is less than in systems of domination. Envy is an important emotion in competitive interactions.

Erich Lessing/Art Resource, NY

Gustav Klimt, "The Kiss," 1907–8. Social interaction is not based entirely on selfishness. Love, for example, is based on trust.

Charles & Josette Lenars/Corbis

Society fits together like a set of nested Russian dolls, with face-to-face interaction constituting the smallest doll in the set.

In a competitive mode of interaction, subordinates receive more benefits, including prestige and money. Prestige and money are stronger motivators than the threat of coercion. Thus, if bosses pay workers reasonably well and treat them with respect, they will work more efficiently than slaves, even if they do not particularly enjoy their work or identify with the goals of the company. Knowing that they can make more money by working harder and that their efforts are appreciated, workers will often put in extra effort (Collins, 1982: 63–5).

As Randall Collins and others have shown, however, the most efficient workers are those who enjoy their work and identify with their employer (Collins, 1982: 60–85; Lowe, 2000). Giving workers a bigger say in decision making, encouraging worker creativity, and ensuring that salaries and perks are not too highly skewed in favor of those on top all help to create high worker morale and foster a more cooperative work environment. Company picnics, baseball games, and, in Japan, the singing of company songs before the workday begins all help workers feel they are in harmony with their employer and are playing on the same team. Similarly, although sales meetings and other conferences have an instrumental purpose (the discussion of sales strategies, new products, etc.), they also offer opportunities for friendly social interaction that increase workers' identification with their employer. When workers identify strongly with their employers, they will be willing to undergo self-sacrifice, take the initiative, and give their best creative effort, even without the prospect of increased material gain.

Micro, Meso, Macro, and Global Structures

At several points in this chapter we referred to norms, roles, and statuses as the "building blocks" of social life. These building blocks form the microstructures within which face-to-face interaction takes place. In concluding, we add that sustained micro-level interaction often gives rise to higher-level, "meso" structures, such as networks, groups, and organizations. In the next chapter, we examine these intermediate-level structures. Then, in Part IV of this book, we show how these intermediate-level structures can form macro-level structures known as "institutions."

Society, it will emerge, fits together like a set of nested Russian dolls, with face-to-face interaction constituting the smallest doll in the set. Big structures set limits to the behavior of small structures. However, it is within small structures that people interpret, negotiate, and modify their immediate social settings, thus giving big structures their dynamism and their life.

Summary

Sociology Now™

Reviewing is as easy as ❶, ❷, ❸.

❶ Before you do your final review, take the SociologyNow diagnostic quiz to help you identify the areas on which you should concentrate. You will find information on how to access SociologyNow on the foldout at the front of the textbook.

❷ As you review, take advantage of SociologyNow's study aids to help you master the topics in this chapter.

❸ When you are finished with your review, take SociologyNow's post-test to confirm you are ready to move on to the next chapter.

1. What is social interaction?

   Social interaction involves verbal and nonverbal communication between people acting and reacting to each other. It is ordered by norms, roles, and statuses.

2. Don't emotions govern all social interaction? Aren't emotions natural, spontaneous, and largely uncontrollable?

   Emotions do form an important part of all social interactions. However, they are less spontaneous and uncontrollable than we commonly believe. For example, your status in an interaction and in the larger society affects how much you laugh and what you laugh at. Similarly, people manage their emotions in personal life and at work according to "feeling rules" that reflect historically changing cultural standards and the demands of organizations.

3. In what sense is social interaction based on competition?

   When we interact socially, we exchange valued resources—everything from attention and pleasure to prestige and money. However, because people typically try to maximize their rewards and minimize their losses, social interaction may be seen as a competition for scarce resources.

4. Is competition the only basis of social interaction?

   No, it is not. People may interact cooperatively and altruistically because they have been socialized to do so. They may also maintain interaction based on domination. Thus, the three major modes of interaction—domination, competition, and cooperation—are based respectively on fear, envy, and trust.

5. How do symbolic interactionists analyze social interaction?

   Symbolic interactionists focus on how people create meaning in the course of social interaction and on how they negotiate and modify roles, statuses, and norms. Symbolic interactionism has several variants. For example, dramaturgical analysis is based on the idea that people play roles in their daily lives in much the same way as actors on stage. When we are front stage, we act publicly, sometimes from ready-made scripts. Backstage, we relax from our public performances and allow what we regard as our "true" selves to emerge (even though we engage in role performances backstage, too). We may distance ourselves from our roles when they embarrass us, but role-playing nonetheless pervades social interaction. Together with various norms of interaction, role-playing enables society to function. Ethnomethodology is another symbolic interactionist approach to social interaction. It analyzes the methods people use to make sense of what others do and say. It insists on the importance of preexisting shared norms and understandings in making everyday interaction possible.

8. Is all social interaction based on language?

   No, it is not. Nonverbal communication, including socially defined facial expressions, gestures, body language, and status cues, are as important as verbal communication in conveying meaning.

9. Is domination the most efficient basis of social interaction?

   Not usually. One might expect slaves to be highly efficient because they must do what their masters dictate. However, slaves typically expend minimal effort because they are rewarded poorly. Efficiency increases in competitive environments where rewards are linked to effort. It increases further in cooperative settings where status differences are low and people enjoy their work and identify with their organization.

Questions to Consider

1. Draw up a list of your current and former girlfriends or boyfriends. Indicate the race, religion, age, and height of each person on the list. How similar or different are you from the people with whom you have chosen to be intimate? What does this list tell you about the social distribution of intimacy? Is love blind? What criteria other than race, religion, age, and height might affect the social distribution of intimacy?
2. Is it accurate to say that people always act selfishly to maximize their rewards and minimize their losses? Why or why not?
3. In what sense (if any) is it reasonable to claim that all of social life consists of role acting and that we have no "true self," just an ensemble of roles?

Web Resources
Companion Website for This Book

http://sociology.wadsworth.com

Begin by clicking on the Student Resources section of the website. Choose "Introduction to Sociology" and finally the Brym and Lie book cover. Next, select the chapter you are currently studying from the pull-down menu. From the Student Resources page you will have easy access to InfoTrac® College Edition, MicroCase Online exercises, additional web links, and many resources to aid you in your study of sociology, including practice tests for each chapter.

Recommended Websites

If you need convincing that social interaction on the Internet can have deep emotional and sociological implications, read Julian Dibbell's "A Rape in Cyberspace," on the World Wide Web at http://www.levity.com/julian/bungle.html. This compelling article is especially valuable for showing how social structure emerges in virtual communities. Originally published in *The Village Voice* (21 December 1993, 36–42).

For social interaction on the World Wide Web, visit "The MUD Connector" at http://www.mudconnect.com.

The Society for the Study of Social Interaction is a professional organization of sociologists who study social interaction. Visit their website at http://sun.soci.niu.edu/~sssi.

CHAPTER 6 Social Collectivities: From Groups to Societies

Yellow Dog Productions/Getty Images

In this chapter, you will learn that:

- We commonly explain the way people act in terms of their interests and emotions. However, sometimes people act against their interests and suppress their emotions because various social collectivities (groups, networks, bureaucracies, and societies) exert a powerful influence on what people do.
- We live in a surprisingly small world. Only a few social ties separate us from complete strangers.
- The patterns of social ties through which emotional and material resources flow form social networks. Information, communicable diseases, social support, and many other resources typically spread through social networks.
- People who are bound together by interaction and a common identity form social groups. Groups impose conformity on members and draw boundary lines between those who belong and those who do not.
- Bureaucracies are large, impersonal organizations that operate with varying degrees of efficiency. Efficient bureaucracies keep hierarchy to a minimum, distribute decision making to all levels of the bureaucracy, and keep lines of communication open between different units of the bureaucracy.
- Societies are collectivities of interacting people who share a culture and a territory. As societies evolve, the relationship of humans to nature changes, with consequences for population size, the permanence of settlements, the specialization of work tasks, labor productivity, and social inequality.
- Although various social collectivities constrain our freedom, we can also use them to increase our freedom. Networks, groups, organizations, and entire societies can be mobilized for good or evil.

Beyond Individual Motives

The Holocaust

Personal Anecdote

In 1941, the large stone and glass train station was one of the proudest structures in Smolensk, a provincial capital of about 100,000 people on Russia's western border. Always bustling, it was especially busy on the morning of June 28. For besides the usual passengers and well-wishers, hundreds of Soviet Red Army soldiers were nervously talking, smoking, writing hurried letters to their loved ones, and sleeping fitfully on the station floor waiting for their train. Nazi troops had invaded the nearby city of Minsk in Belarus a couple of days before. The Soviet soldiers were being positioned to defend Russia against the inevitable German onslaught.

Robert Brym's father, then in his 20s, had been standing in line for nearly 2 hours to buy food when he noticed flares arching over the station. Within seconds, Stuka bombers, the pride of the German air force, swept down, releasing their bombs just before pulling out of their dive. Inside the station, shards of glass, blocks of stone, and mounds of earth fell indiscriminately on sleeping soldiers and nursing mothers alike. Everyone panicked. People trampled over one another to get out. In minutes, the train station was rubble.

Sociology Now™

Reviewing is as easy as ❶, ❷, ❸.

Use SociologyNow to help you make the grade on your next exam. When you are finished reading this chapter, go to the Chapter Summary for instructions on how to make SociologyNow work for you.

Nearly two years earlier, Robert's father had managed to escape Poland when the Nazis invaded his hometown near Warsaw. Now, he was on the run again. By the time the Nazis occupied Smolensk a few weeks after their dive-bombers destroyed its train station, Robert's father was deep in the Russian interior serving in a workers' battalion attached to the Soviet Red Army.

"My father was one of 300,000 Polish Jews who fled eastward into Russia before the Nazi genocide machine could reach them," says Robert. "The remaining 3 million Polish Jews were killed in various ways. Some died in battle. Many more, like my father's mother and younger siblings, were rounded up like diseased cattle and shot. However, most of Poland's Jews wound up in the concentration camps. Those deemed unfit were shipped to the gas chambers. Those declared able to work were turned into slaves until they could work no more. Then they, too, met their fate. A mere 9 percent of Poland's 3.3 million Jews survived World War II. The Nazi regime was responsible for the death of 6 million Jews in Europe (Burleigh, 2000).

"One question that always perplexed my father about the war was this: How was it possible for many thousands of ordinary Germans—products of what he regarded as the most advanced civilization on earth—to systematically murder millions of defenseless and innocent Jews, Roma ('Gypsies'), homosexuals, and mentally disabled people in the death camps?" To answer this question adequately, we must borrow ideas from the sociological study of networks, groups, and bureaucracies.

How Social Groups Shape Our Actions

How could ordinary German citizens commit the crime of the century? The conventional, nonsociological answer is that many Nazis were evil, sadistic, or deluded enough to think that Jews and other undesirables threatened the existence of the German people. Therefore, in the Nazi mind, the innocents had to be killed. This answer is given in the 1993 movie *Schindler's List,* and in many other accounts. Yet, it is far from the whole story. Sociologists emphasize three other factors.

Norms of Solidarity

When we form relationships with friends, lovers, spouses, teammates, and comrades-in-arms, we develop shared ideas, or "norms of solidarity," about how we should behave toward them to sustain the relationships. Because these relationships are emo-

German industrialist Oskar Schindler (Liam Neeson, center), searches for his plant manager Itzhak Stern among a trainload of Polish Jews about to be deported to Auschwitz-Birkenau in *Schindler's List.* The movie turns the history of Nazism into a morality play, a struggle between good and evil forces. It does not probe into the sociological roots of good and evil.

The Everett Collection

tionally important to us, we sometimes pay more attention to norms of solidarity than to the morality of our actions. For example, a study of the Nazis who roamed the Polish countryside to shoot and kill Jews and other "enemies" of Nazi Germany found that the soldiers often did not hate the people they systematically slaughtered, nor did they have many qualms about their actions (Browning, 1992). They simply developed deep loyalty to each other. They felt they had to get their assigned job done or face letting down their comrades. Thus, they committed atrocities partly because they just wanted to maintain group morale, solidarity, and loyalty. They committed evil deeds not because they were extraordinarily bad but because they were quite ordinary—ordinary in the sense that they acted to sustain their friendship ties and to serve their group, just like most people. It is the power of norms of solidarity that helps us understand how soldiers are able to undertake many unpalatable actions. As one soldier says in the 2001 movie *Blackhawk Down:* "When I go home people will ask me: 'Hey, Hoot, why do you do it, man? Why? Are you some kinda war junkie?' I won't say a goddamn word. Why? They won't understand. They won't understand why we do it. They won't understand it's about the men next to you. And that's it. That's all it is."

The case of the Nazi regime may seem extreme, but other instances of going along with criminal behavior uncover a similar dynamic at work. Why do people rarely report crimes committed by corporations? Employees may worry about getting reprimanded or fired if they become "whistleblowers," but they also worry about letting down their coworkers. Why do gang members engage in criminal acts? They may seek financial gain, but they also regard crime as a way of maintaining a close social bond with their fellow gang members (Box 6.1).

A study of the small number of Polish Christians who helped save Jews during World War II helps clarify why some people violate group norms (Tec, 1996). The heroism of these Polish Christians was not correlated with their educational attainment, political orientation, religious background, or even attitudes toward Jews. In fact, some Polish Christians who helped save Jews were quite anti-Semitic. Instead, these Christian heroes were for one reason or another estranged or cut off from mainstream norms. Because they were poorly socialized into the norms of their society, they were freer not to conform and instead act in ways they believed were right. We could tell a roughly similar story about corporate whistleblowers or people who turn in their fellow gang members. They are disloyal from an insider's point of view but heroic from an outsider's point of view, often because they have been poorly socialized into the group's norms.

Obedience to Structures of Authority

Structures of authority tend to render people obedient. Most people find it difficult to disobey authorities because they fear ridicule, ostracism, and punishment. This was strikingly demonstrated in an experiment conducted by social psychologist Stanley Milgram (1974). Milgram informed his experimental subjects that they were taking part in a study on punishment and learning. He brought each subject to a room where a man was strapped to a chair. An electrode was attached to the man's wrist. The experimental subject sat in front of a console. It contained 30 switches with labels ranging from "15 volts" to "450 volts" in 15-volt increments. Labels ranging from "slight shock" to "danger: severe shock" were pasted below the switches. The experimental subjects were told to administer a 15-volt shock for the man's first wrong answer and then increase the voltage each time he made an error. The man strapped in the chair was in fact an actor. He did not actually receive a shock. As the experimental subject increased the current, however, the actor began to writhe, shouting for mercy and begging to be released. If the experimental subjects grew reluctant to administer more current, Milgram assured them the man strapped in the chair would be fine and insisted that the success of the experiment depended on the subject's obedience. The subjects were, however, free to abort the experiment at any time. Remarkably, 71 percent of experimental subjects were prepared to administer shocks of 285 volts or more, even though the switches at that level were labeled "intense shock," "extreme intensity shock," and "danger: severe shock"

BOX 6.1

SOCIAL POLICY: WHAT DO YOU THINK?

Group Loyalty or Betrayal?

Group cohesion led Nazi soldiers to commit genocide. Group loyalty led many ordinary German citizens to support them. Although the Nazis are an extreme case, ordinary people often face a stark choice between group loyalty and group betrayal.

Glen Ridge is an affluent suburb of 7800 people in northern New Jersey: white, orderly, and leafy. It is the hometown of *the* all-American boy, Tom Cruise. It was also the site of a terrible rape case in 1989. A group of 13 teenage boys lured a sweet-natured young woman with an IQ of 49 and the mental age of a second-grader into a basement. There, four of them raped her while three others looked on; six left when they realized what was going to happen. The rapists used a baseball bat and a broomstick. The boys were the most popular students in the local high school. They had everything going for them. They were every mother's dream, every father's pride. Nor was the young woman a stranger to them. Some of them had known her since she was 5 years old, when they convinced her to lick the point of a ballpoint pen that had been coated in dog feces.

What possessed these boys to gang-rape a helpless young woman? And how can we explain the subsequent actions of many of the leading citizens of Glen Ridge? It was weeks before anyone reported the rape to the police and years before the boys went to trial. At trial, many members of the community rallied behind the boys, blaming and ostracizing the rape victim. The courts eventually found three of the four young men guilty of first-degree rape, but they were allowed to go free for eight years while their cases were appealed and received only light sentences in 1997. With good behavior, two were released from jail in 1999 and one in 1998. Why did members of the community refuse to believe the clear-cut evidence? What made them defend the rapists? Why did the boys get off so easily?

Bernard Lefkowitz (1997b) interviewed 250 key players and observers in the Glen Ridge Rape case. Ultimately, he indicted the *community* for the rape. He concluded that "[the rapists] adhered to a code of behavior that mimicked, distorted, and exaggerated the values of the adult world around them," while "the citizens supported the boys because they didn't want to taint the town they treasured" (Lefkowitz, 1997b: 493). What were some of the community values the elders upheld and the boys aped?

The subordination of women. All of the boys grew up in families where men were the dominant personalities. Only one of them had a sister. Not a single woman occupied a position of authority in Glen Ridge High School. The boys classified their female classmates either as "little mothers" who fawned over them or "bad girls" who were simply sexual objects.

Lack of compassion for the weak. According to the minister of Glen Ridge Congregational Church, "Achievement was honored and respected almost to the point of pathology, whether it was the achievements of high school athletes or the achievements of corporate world conquerors." Adds Lefkowitz: "Compassion for the weak wasn't part of the curriculum" (Lefkowitz, 1997b: 130).

Tolerance of male misconduct. The boys routinely engaged in delinquent acts, including one spectacular trashing of a house. However, their parents always paid damages, covered up the misdeeds, and rationalized them with phrases like "Boys will be boys." Especially because they were town football heroes, many people felt they could do no wrong.

Intense group loyalty. "The guys prized their intimacy with each other far above what could be achieved with a girl," writes Lefkowitz (1997b: 146). The boys formed a tight clique, and team sports reinforced group solidarity. Under such circumstances, the probability of someone "ratting" on his friends was very low. In the end, of course, there was a "rat." His name was Charles Figueroa. He did not participate in the rape but he was an athlete, part of the jock clique, and therefore aware of what had happened.

Significantly, he was one of the few black boys in the school, tolerated because of his athletic ability, but never trusted because of his race and often called a n_____ by his teammates behind his back. This young man's family was highly intelligent and morally sensitive. He was the only one to have the courage to betray the group (Lefkowitz, 1997a; 1997b).

Lefkowitz presents a strong indictment of the community as a whole and the values it upheld. Beyond that, however, he raises the important question of where we ought to draw the line between group loyalty and group betrayal. Considering your own group loyalties, are there times when you regret not having spoken up? Are there times when you regret not having been more loyal? What is the difference between these two types of situations? Can you specify criteria for deciding when loyalty is required and when betrayal is the right thing to do? You may have to choose between group loyalty and betrayal on more than one occasion, so thinking about these criteria—and clearly understanding the values for which your group stands—will help you make a more informed choice.

and despite the fact that the actor appeared to be in great distress at this level of current (Figure 6.1).

Milgram's experiment teaches us that as soon as we are introduced to a structure of authority, we are inclined to obey those in power. This is the case even if the authority structure is new and highly artificial, even if we are free to walk away from it with no penalty, and even if we think that by remaining in its grip we are inflicting terrible pain on

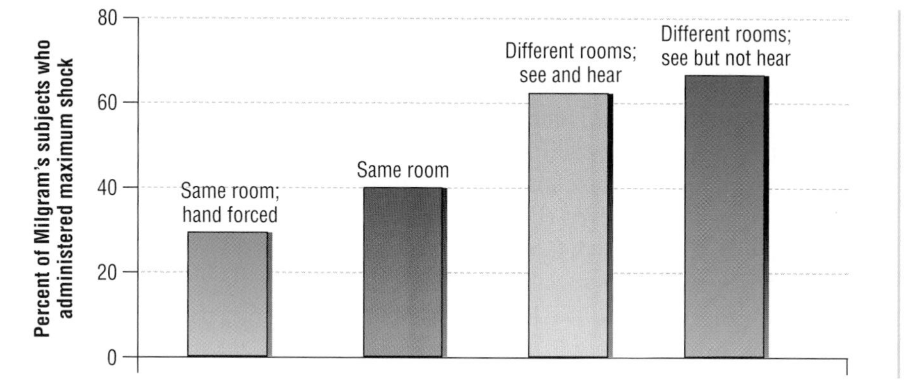

Figure 6.1
Obedience to Authority Increases with Separation from the Negative Effects of One's Actions

Milgram's experiment supports the view that separating people from the negative effects of their actions increases the likelihood of compliance. When subject and actor were in the same room and the subject was told to force the actor's hand onto the electrode, 30 percent of subjects administered the maximum, 450-volt shock. When subject and actor were merely in the same room, 40 percent of subjects administered the maximum shock. When subject and actor were in different rooms but the subject could see and hear the actor, 62.5 percent of subjects administered the maximum shock. When subject and actor were in different rooms and the actor could be seen but not heard, 65 percent of subjects administered the maximum shock.

Source: Bar graph based on information in chapter 4, "Closeness of the Victim," from *Obedience to Authority* by Stanley Milgram. Copyright © 1994 by Stanley Milgram. Reprinted by permission of HarperCollins Publishers, Inc.

another human being. In this context, the actions and inactions of German citizens in World War II become more understandable if no more forgivable.

Time Inc./Time Life Pictures/Getty Images

Corporate whistleblowers were widely celebrated in 2002 as news of corporate and bureaucratic scandals proliferated in the mass media.

Bureaucratic Organization

Bureaucracies are highly effective structures of authority. The Nazi genocide machine was so effective partly because it was bureaucratically organized. As Max Weber (1978) defined the term, a **bureaucracy** is a large, impersonal organization composed of many clearly defined positions arranged in a hierarchy. A bureaucracy has a permanent, salaried staff of qualified experts and written goals, rules, and procedures. Staff members always try to find ways of running their organization more efficiently. "Efficiency" means achieving the bureaucracy's goals at the least cost. The goal of the Nazi genocide machine was to kill Jews and other undesirables. To achieve that goal with maximum efficiency, the job was broken into many small tasks. Most officials performed only one function, such as checking train schedules, organizing entertainment for camp guards, maintaining supplies of Zyklon B gas, and removing ashes from the crematoria. The full horror of what was happening eluded many officials or at least could be conveniently ignored as they concentrated on their jobs, most of them far removed from the gas chambers and death camps in occupied Poland. Many factors account for variations in Jewish victimization rates across Europe during World War II. One factor was bureaucratic organization. Not coincidentally, the proportion of Jews killed was highest not in the Nazi-controlled countries where the hatred of Jews was most intense (e.g., Romania), but in countries where the Nazi bureaucracy was best organized (e.g., Holland) (Bauman, 1991 [1989]; Sofsky, 1997 [1993]).

A **bureaucracy** is a large, impersonal organization composed of many clearly defined positions arranged in a hierarchy. A bureaucracy has a permanent, salaried staff of qualified experts and written goals, rules, and procedures. Staff members always try to find ways of running the bureaucracy more efficiently.

In short, the sociological reply to the question posed by Robert's father is that it was not just blind hatred but the nature of groups and bureaucracies that made it possible for the Nazis to kill innocent people so ruthlessly.

We commonly think *individual* motives prompt our actions, and for good reason. As we saw in Chapter 5 ("Social Interaction"), we often make rational calculations to maximize gains and minimize losses. In addition, our deeply held emotions partly govern our behavior. However, this chapter asks you to make a conceptual leap beyond the individual motives that prompt us to act in certain ways. We ask you to consider the way four kinds of social *collectivities* shape our actions: networks, groups, bureaucracies, and societies. The limitations of an analysis based exclusively on individual motives should be clear from our discussion of the social roots of evil. The advantages of considering how social collectivities affect us should become clear as you read this chapter. We begin by considering the nature and effects of social networks.

Networks

It's a Small World

The *Internet Movie Database* (2003) contains information on the half-million actors who have ever performed in a commercially released movie. While this number is large, you might be surprised to learn that socially they form a small world. We can demonstrate this fact by first selecting an actor who is not an especially big star—someone like Kevin Bacon. We can then use the *Internet Movie Database* to find out which other actors have ever been in a movie with him (University of Virginia, 2003). Acting in a movie with another actor constitutes a link. A single link ties 1,506 actors to Kevin Bacon. Among them is Julia Roberts, who starred with Bacon in *Flatliners* (1990). Two links tie another 119,754 actors to Bacon. They have never been in a movie with Bacon but they have been in a movie with *another* actor who has been in a movie with him. Brittany Murphy, for example, was in *Monsters* (2003) with Charlize Theron, and Theron was in *Trapped* (2002) with Bacon. Remarkably, more than 85 percent of the half-million actors in the database have one, two, or three links to Bacon. Nearly 99 percent of the half-million actors have four or fewer links to him. The greater the number of links, the further back in time and the farther afield one must go. We conclude that although film acting stretches back more than a century and has involved people in many countries, the half-million people who have ever acted in films form a pretty small world.

What is true for the world of film actors turns out to be true for the rest of us, too. Jeffrey Travers and Stanley Milgram (1969) conducted a famous study in which they asked 300 randomly selected people in Nebraska and Kansas to mail a document to a complete stranger, a stockbroker in Boston. However, the people could not mail the document directly to the stockbroker. They had to mail it to a person they knew on a first-name basis, who in turn could send it only to a person *he or she* knew on a first-name basis, and so forth. Travers and Milgram defined this passing of a letter from one person to another as a link, or a "degree of separation." Most people thought it would take many degrees of separation, perhaps hundreds, to get the letter to the Boston stockbroker. Remarkably, however, the average number was about six. Following publication of the study, the idea became widespread that there are no more than six degrees of separation between any two people in the United States. A 1990 play by John Guare called *Six Degrees of Separation*, a 1993 movie by the same name (starring Stockard Channing, Will Smith, and Donald Sutherland), and the 1994 game *Six Degrees of Kevin Bacon* invented by three Pennsylvania college students helped to popularize the idea. An attempt to apply the idea to the entire world via the Internet is under way at Columbia University's Department of Sociology. You can participate in the study by visiting http://smallworld.columbia.edu.

You may not have played *Six Degrees of Kevin Bacon,* but you have probably said "It's a small world" more than once. And you were right. Meeting a complete stranger who turns out to be separated from us by just a couple of links is not uncommon. Most people are surprised about how small their world is. Think of how many people you know. You have family members, friends, acquaintances, and work colleagues. They probably total more than a few hundred people. Given the more than 298 million Americans in 2006, how is it possible that only about six links separate any of us from complete strangers?

Reuters NewMedia Inc./Corbis

Among all actors who have ever performed in a movie, Kevin Bacon is the 1161st most central. That is, 1160 other actors have a smaller average number of ties to all other actors. Still, nearly a quarter of all people who have ever acted in a movie are separated from Bacon by just one or two links, whereas more than 85 percent are separated by fewer than four links.

Network Analysis

The short sociological answer is that we are enmeshed in overlapping sets of social relations, or "social networks." Although any particular individual may know a small number of people, his or her family members, friends, coworkers, and others know many more people who extend far beyond that individual's "personal network." So, for example, the authors of this textbook are likely to be complete strangers to you. Yet your professor may know one of us or at least know someone who knows one of us. Probably no more than three links separate us from you. Put differently, although our personal networks are small, they lead quickly to much larger networks. We live in a small world because our social networks connect us to the larger world.

What is a social network? A **social network** is a bounded set of individuals linked by the exchange of material or emotional resources, everything from money to friendship. The patterns of exchange determine the boundaries of the network. Members exchange resources more frequently with each other than with nonmembers. They also think of themselves as network members. Social networks may be formal (defined in writing) or informal (defined only in practice). The people you know personally form the boundaries of your personal network. However, each of your network members is linked to other people. This is what connects you to people you have never met, creating a "small world" that extends far beyond your personal network.

The study of social networks is not restricted to ties among individuals (Berkowitz, 1982; Wasserman and Faust, 1994; Wellman and Berkowitz, 1997 [1988]). The units of analysis, or "nodes," in a network can be individuals, groups, organizations, and even countries. Thus, social network analysts have examined everything from intimate relationships among lovers to diplomatic relations among nations. For example, in Chapter 9 ("Globalization, Inequality, and Development") we show how patterns in the flow of international trade divide the world into three major trading blocs, with the United States, Germany, and Japan at the center of each bloc. In Chapter 13 ("Work and the Economy"), we show how American corporate networks create alliances that are useful for exchanging information and influencing government. In both cases, our analysis of social networks teaches us something new and unexpected about the social bases of economic and political affairs.

A **social network** is a bounded set of individuals who are linked by the exchange of material or emotional resources. The patterns of exchange determine the boundaries of the network. Members exchange resources more frequently with each other than with nonmembers. They also think of themselves as network members. Social networks may be formal (defined in writing), but they are more often informal (defined only in practice).

Unlike organizations, most networks lack names and offices. There is a Boy Scouts of America but no American Trading Bloc. In a sense, networks lie beneath the more visible collectivities of social life, but that makes them no less real or important. Some analysts claim that we can gain only a partial sense of why certain things happen in the social world by focusing on highly visible collectivities. From their point of view, the whole story requires probing below the surface and examining the network level. The study of social networks clarifies a wide range of social phenomena, including how people find jobs, how information, innovations, and communicable diseases spread, and how some people exert influence over others. To illustrate further the value of network analysis, we now focus on each of these issues in turn.

Finding a Job

Former president Bill Clinton was a master networker. He and his colleagues referred to his wide circle of friends and acquaintances as "Friends of Bill." His personal network included people stretching from his hometown of Hope, Arkansas, all the way to college friends from Georgetown, Oxford, and Yale Law School (including his wife). Friends of Bill were critical in providing financial, political, and moral support during his political career. They in turn benefited not only from the reflected glory of his fame and power but also by receiving favors, including cabinet and ambassadorial appointments.

Bill Clinton is not alone in having cultivated his social network. The old saying "It's not what you know but who you know" contains much sociological truth. For example, many bright and inquisitive students have passing thoughts about becoming professors. What might transform that momentary interest into a lifelong career is often a mentor who encourages the student and provides him or her with useful advice, including introductions to other ambitious students, stimulating professors, great books, and graduate programs.

Many people learn about important events, ideas, and opportunities from their social networks. Friends and acquaintances often introduce you to everything from an interesting college course or a great restaurant to a satisfying occupation or a future spouse. Of course, social networks are not the only source of information, but they are highly significant.

Consider how people find jobs. Do you look in the "Help Wanted" section of your local newspaper, scan the Internet, or walk around certain areas of town looking for "Employee Wanted" signs? Although these strategies are common, people often learn about employment opportunities from other people. But what kind of people? According to Mark Granovetter (1973), you may have strong or weak ties to another person. You have strong ties to people who are close to you, such as family members and friends. You have weak ties to mere acquaintances, such as people you meet at parties and friends of friends. In his research, Granovetter found that weak ties are more important than strong ties in finding a job, which is contrary to common sense. One might reasonably assume that a mere acquaintance would not do much to help you find a job, whereas a close friend or relative would make a lot more effort in this regard. However, by focusing on the flow of

Parents can help their graduating children find jobs by getting them "plugged into" the right social networks. Here, in the 1968 movie *The Graduate,* a friend of the family advises Dustin Hoffman that the future lies in the plastics industry.

The Everett Collection

information in personal networks, Granovetter found something different. Mere acquaintances are more likely to provide useful information about employment opportunities than friends or family members because people who are close to you typically share overlapping networks. Therefore, the information they can provide about job opportunities is often redundant. In contrast, mere acquaintances are likely to be connected to *diverse* networks. They can therefore provide information about many different job openings and make introductions to many different potential employers. Moreover, because people typically have more weak ties than strong ties, the sum of weak ties holds more information about job opportunities than the sum of strong ties. These features of personal networks allow Granovetter to conclude that the "strength of weak ties" lies in their diversity and abundance.

Urban Networks

We rely on social networks for a lot more than job information. Consider everyday life in the big city. We often think of big cities as cold and alienating places where few people know one another. In this view, urban acquaintanceships tend to be few and functionally specific; we know someone fleetingly as a bank teller or a server in a restaurant but not as a whole person. Even dating often involves a long series of brief encounters. In contrast, people often think of small towns as friendly, comfortable places where everyone knows everyone else (and everyone else's business). Indeed, some of the founders of sociology emphasized just this distinction. Notably, German sociologist Ferdinand Tönnies (1988 [1887]) contrasted "community" with "society." According to Tönnies, a community is marked by intimate and emotionally intense social ties, whereas a society is marked by impersonal relationships held together largely by self-interest. A big city is a prime example of a society in Tönnies's judgment.

Tönnies's view prevailed until network analysts started studying big city life in the 1970s. Where Tönnies saw only sparse, functionally specific ties, network analysts found elaborate social networks, some functionally specific and some not. For example, Barry Wellman and his colleagues studied personal networks in Toronto, Canada (Wellman, Carrington, and Hall, 1997 [1988]). They found that each Torontonian had an average of about 400 social ties, including immediate and extended kin, neighbors, friends, and coworkers. These ties provided everything from emotional aid (e.g., visits after a personal tragedy) and financial support (e.g., small loans) to minor services (e.g., fixing a car) and information of the kind Granovetter studied. Strong ties that last a long time are typically restricted to immediate family members, a few close relatives and friends, and a close coworker or two. Beyond that, however, people relied on a wide array of ties for different purposes at different times. Downtown residents sitting on their front stoop on a summer evening, sipping soda and chatting with neighbors as the kids play road hockey, may be less common than it was 50 years ago. However, the automobile, public transportation, the telephone, and the Internet help people stay in close touch with a wide range of contacts for a variety of purposes (Haythornwaite and Wellman, 2002). Far from living in an impersonal and alienating world, these Torontonians' lives are network rich. Research conducted elsewhere in North America reveals much the same pattern of urban life.

Scientific Innovation

One of the advantages of network analysis is its focus on people's actual social relationships rather than their abstract attributes, such as their age, gender, or occupation. This focus is useful for helping us understand many aspects of social life. For example, how does information spread? You might think that people with a particular set of occupational attributes, such as leading scientists in a particular field of study, first gain and then distribute information to other scientists in their field. At a crude level, this fact is true. However, new information does not diffuse evenly throughout a scientific community. Instead, it flows through networks of friends, coresearchers, and people who have studied together, only later spreading to the broader community (Rogers, 1995). Networks

Box 6.2

YOU AND THE SOCIAL WORLD

Networks and Health

Social networks exert a powerful influence on health in general. People who are well integrated into cohesive social networks of family, extended kin, and friends are less likely to suffer heart attacks, complications during pregnancy, and so forth. Surprisingly, although more social contacts may expose you to more germs, research shows that you are *less* likely to come down with an infectious disease the more social contacts you have (Jones, Gallagher, and McFalls, 1995: 109–11).

If you are skeptical about the claim that social contacts decrease the likelihood of infectious disease, try this exercise to see how social networks affect your health and that of your classmates. Ask everyone in your sociology class to answer two questions: (1) How many times have you caught cold or had the flu during the past three months? (2) How many relatives and family members did you see face-to-face at least three times in the last three months? After you have collected the answers to these questions, tally the responses. (If your class is large, you will want to do this exercise with a group of your classmates.) Create a 2 × 2 table showing how, if at all, the frequency of face-to-face contact with relatives and family members influences the likelihood of catching cold or the flu.

We hypothesize that the greater the contact, the less the chance of catching cold or the flu. In 250–500 words, explain whether your results support or refute our hypothesis or whether they are inconclusive. Outline how researchers might pursue this line of inquiry to determine whether your results are credible or idiosyncratic.

also shape scientific influence because scientists in a social network tend to share similar scientific beliefs and are thus more open to some influences than others (Friedkin, 1998).

From HIV/AIDS to the Common Cold

Another example of the usefulness of focusing on concrete social ties rather than abstract attributes comes from the study of how communicable diseases spread. HIV/AIDS was widely considered a "gay disease" in the 1980s. HIV/AIDS spread rapidly in the gay community during that decade. However, network analysis helped to show that the characterization of HIV/AIDS as a gay disease was an oversimplification (Watts, 2003). The disease did not spread uniformly throughout the community. Rather, it spread along the friendship and acquaintanceship networks of people first exposed to it. Meanwhile, in India and parts of Africa, HIV/AIDS did not initially spread among gay men at all. Instead, it spread through a network of long-distance truck drivers and the prostitutes who catered to them. Again, concrete social networks—not abstract categories like "gay men" or "truck drivers"—track the spread of the disease (Box 6.2).

Sociology Now™

Learn more about **HIV/AIDS** by going through the # of AIDS Cases Map Exercise.

The Building Blocks of Social Networks: Dyads and Triads

Researchers often use mathematical models and computer programs to analyze social networks. However, no matter how sophisticated the mathematics or the software, network analysts begin from an understanding of the basic building blocks of social networks.

The most elementary network form is the **dyad,** a social relationship between two nodes or social units (e.g., people, firms, organizations, countries, etc.). A **triad** is a social relationship among three nodes. The difference between a dyad and a triad may seem small. However, the social dynamics of these two elementary network forms are fundamentally different, as sociologist Georg Simmel showed early in the 20th century (Simmel, 1950) (Figure 6.2).

- A **dyad** is a social relationship between two "nodes," or social units (e.g., people, firms, organizations, countries, etc.).
- A **triad** is a social relationship among three "nodes," or social units (e.g., people, firms, organizations, countries, etc.).

Dyads

In a dyadic relationship such as a marriage, both partners tend to be intensely and intimately involved. Moreover, the dyad needs both partners to live but only one to die. A marriage, for example, can endure only if both partners are intensely involved; if one partner ceases active participation, the marriage is over in practice if not in law. This need for

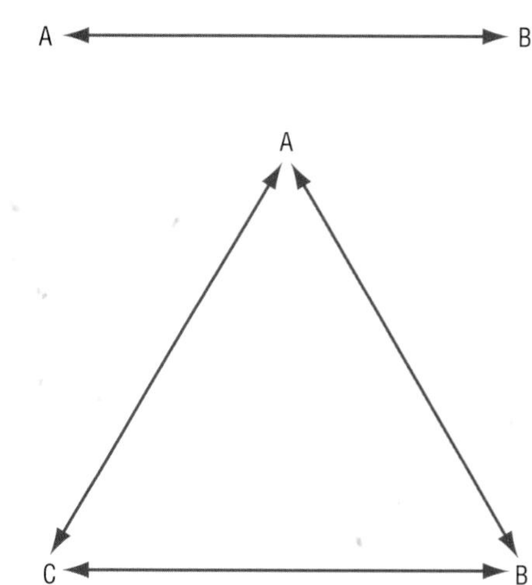

Characteristics of the dyad:
1. Both partners are intensely absorbed in the relationship.
2. The dyad needs both partners to live but only one to die.
3. No "free riders" are possible.
4. Neither partner can deny responsibility by shifting it to a larger collectivity.

Characteristics of the triad:
1. Intensity and intimacy are reduced.
2. The triad restricts individuality by allowing a partner to be constrained for the collective good. A partner may be outvoted by a majority, for example.
3. Coalitions are possible.
4. Third-party mediation of conflict between two partners is possible.
5. Third-party exploitation of rivalry between two partners is possible.
6. A third-party divide-and-conquer strategy is possible.
7. "Free riders" are possible.
8. It is possible to shift responsibility to the larger collectivity.

Figure 6.2
Dyad and Triad

intense involvement on the part of both partners is also why a dyad can have no "free riders," or partners who benefit from the relationship without contributing to it. Finally, in a dyadic relationship, the partners must assume full responsibility for all that transpires. Neither partner can shift responsibility to some larger collectivity because no larger collectivity exists beyond the relationship between the two partners.

Sociology Now™

Learn more about **Dyads and Triads** by going through the Group Size Effects Animation.

Triads

In contrast, when a third person (or other social unit) enters the picture, thereby creating a triad, relationships tend to be less intimate and intense. Equally significantly, the triad restricts individuality by allowing one partner to be constrained for the collective good. This situation occurs when a majority outvotes one partner. The existence of a triad also allows coalitions or factions to form. Furthermore, it allows one partner to mediate conflict or exploit rivalry between the other two to achieve dominance. Thus, the introduction of a third partner makes possible a whole new set of social dynamics that are structurally impossible in a dyadic relationship.

Web Interactive Exercise: Does the Internet Isolate People Socially?

Groups

Love and Group Loyalty

Although intensity and intimacy characterize dyadic relationships, outside forces often destroy them. For instance, the star-crossed lovers in *Romeo and Juliet* are torn between their love for one another and their loyalty to the feuding Montague and Capulet families. In the end, Romeo and Juliet lie dead, victims of the feud.

Love thwarted by conflicting group loyalty is the stuff of many tragic plays, novels, and movies. Most audiences have no problem grasping the fact that group loyalty is often more powerful than romantic love. However, why group loyalty holds such power over us is unclear. The sociological study of groups provides some useful answers.

Sociology Now™

Learn more about **Social Groups** by going through the How Social Groups Shape Our Actions Data Experiment.

Varieties of Group Experience

Social groups are composed of one or more networks of people who identify with one another and adhere to defined norms, roles, and statuses. We usually distinguish social groups from **social categories,** in which people share similar status but do not identify with one another. Coffee drinkers form a social category. They do not, however, normally share norms and identify with each other. In contrast, members of a group, such as a family, sports team, or college, are aware of shared membership. They identify with the collectivity and think of themselves as members.

A **social group** is composed of one or more networks of people who identify with one another and adhere to defined norms, roles, and statuses.

A **social category** is composed of people who share a similar status but do not identify with one another.

Brian Leng/Corbis

Primary groups such as families agree upon norms, roles, and statuses that are not set down in writing. Social interaction creates strong emotional ties, extends over a long period, involves a wide range of activities, and results in group members knowing one another well.

Primary Groups and Secondary Groups

Many kinds of social groups exist. However, sociologists make a basic distinction between primary and secondary groups. In **primary groups,** norms, roles, and statuses are agreed upon but are not put in writing. Social interaction creates strong emotional ties. It extends over a long period, involves a wide range of activities, and results in group members knowing one another well. The family is the most important primary group.

Secondary groups are larger and more impersonal than primary groups. Compared with primary groups, social interaction in secondary groups creates weaker emotional ties, extends over a shorter period, involves a narrow range of activities, and results in most group members having at most a passing acquaintance with one another. Your sociology class is an example of a secondary group.

Bearing these distinctions in mind, we can begin to explore the power of groups to ensure conformity.

Sociology Now™

Learn more about **Primary Groups and Secondary Groups** by going through the Primary Groups and Secondary Groups Data Experiment.

Primary groups are groups whose members agree upon norms, roles, and statuses without putting them in writing. Social interaction leads to strong emotional ties, extends over a long period, involves a wide range of activities, and results in group members knowing one another well.

Secondary groups are larger and more impersonal than primary groups. Compared with primary groups, social interaction in secondary groups creates weaker emotional ties, extends over a shorter period, involves a narrow range of activities, and results in most group members having at most a passing acquaintance with one another.

Group Conformity

Television's first reality TV show was *Candid Camera.* In an early episode, an unsuspecting man waited for an elevator. When the elevator door opened, he found four people, all confederates of the show, facing the elevator's back wall. Seeing the four people with their backs to him, the man at first hesitated. He then tentatively entered the elevator. However, rather than turning around so he would face the door, he remained facing the back wall, just like the others. The scene was repeated several times. Men and women, black and white, all behaved the same. Confronting unanimously bizarre behavior, they chose conformity over common sense.

Conformity is an integral part of group life, and primary groups generate more pressure to conform than secondary groups. Strong social ties create emotional intimacy. They also ensure that primary group members share similar attitudes, beliefs, and information. Beyond the family, friendship groups (or cliques) and gangs demonstrate these features. Group members tend to dress and act alike, speak the same "lingo," share the same likes and dislikes, and demand loyalty, especially in the face of external threat. Conformity ensures group cohesion.

A classic study of U.S. soldiers in World War II demonstrates the power of conformity to get people to face extreme danger. Samuel Stouffer and his colleagues (1949) showed that primary group cohesion was the main factor motivating soldiers to engage in combat. Rather than belief in a cause, such as upholding liberty or fighting the evils of Nazism, the feeling of camaraderie, loyalty, and solidarity with fellow soldiers supplied the principal motivation to face danger. As Brigadier General S. L. A. Marshall (1947: 160–1) famously wrote: "A man fights to help the man next to him. . . . Men do not fight for a cause but because they do not want to let their comrades down." As such, if you want to create a great military force, you need to promote group solidarity and identity. Hence the importance of uniforms, anthems, insignia, flags, drills, training under duress, and instilling hatred of the enemy.

Asch's Experiment

A famous experiment conducted by social psychologist Solomon Asch half a century ago also demonstrates how group pressure creates conformity (Asch, 1955). Asch assembled seven men. One of them was the experimental subject. The other six were Asch's confederates. Asch showed the seven men a card with a line drawn on it. He then showed them a second card with three lines of varying length drawn on it (▶Figure 6.3). One by one, he asked the confederates to judge which line on card 2 was the same length as the line on card 1. The answer was obvious. One line on card 2 was much shorter than the line on card 1. One line was much longer. One was exactly the same length. Yet, as instructed by Asch, all six confederates said that either the shorter or the longer line was the same length as the line on card 1. When it came time for the experimental subject to make his judgment, he typically overruled his own perception and agreed with the majority. Only 25 percent of Asch's experimental subjects consistently gave the right answer. Asch thus demonstrated how easily group pressure can overturn individual conviction and result in conformity.

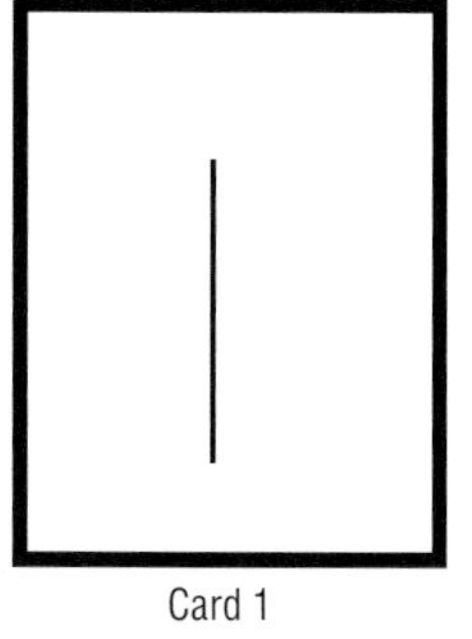

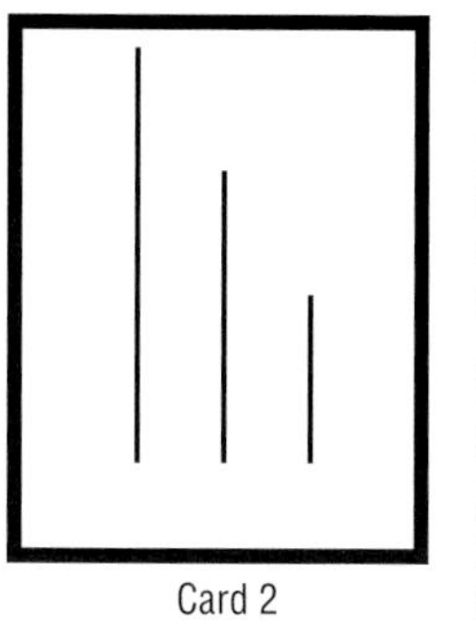

▶Figure 6.3
The Asch Experiment

What Affects Group Conformity?

Asch's work and subsequent research show that several factors affect the likelihood of conformity (Sternberg, 1998: 499–500). First, the likelihood increases as *group size* increases to three or four members. For groups larger than four, the likelihood of conformity generally does not increase. Second, as *group cohesiveness* increases, so does the likelihood of conformity. Where greater intimacy and sharing of values occur, group members are less likely to express dissent. Third, *social status* affects the likelihood of conformity. People with low status in a group (e.g., because of their gender or race) are less likely to dissent than people with high status. Fourth, *culture* matters. People in individualistic societies like the United States tend to conform less than people in collectivist societies like China. Fifth, the *appearance of unanimity* affects the likelihood of conformity. Even one dissenting voice greatly increases the chance that others will dissent.

Groupthink

The power of groups to ensure conformity is often a valuable asset. Armies could not function without willingness to undergo personal sacrifice for the good of the group, nor could sports teams excel. However, being a "good team player" can have a downside because the consensus of a group can sometimes be misguided or dangerous. Dissent might save the group from making mistakes, but the pressure to conform despite individual misgivings—sometimes called **groupthink** (Janis, 1972)—can lead to disaster.

Great managers are able to encourage frank and open discussion, assess ideas based on their merit, develop a strategy that incorporates the best ideas voiced, and *then* create consensus on how to implement the ideas. Inadequate managers feel they know it all. They rationalize their plan of action, squelch dissent, and fail to examine alternatives. They create a group culture that inhibits people from expressing their misgivings and use the fact that people do not want to appear disloyal to *impose* consensus on the group. High-stress situations—a war room, an operating room—often do not allow a more democratic managerial style. Therefore, it is precisely in high-stress situations that the dangers of groupthink are greatest.

Examples of Groupthink

Arguably, groupthink was operating when President Roosevelt and his advisers refused to believe the Japanese would bomb Pearl Harbor, President Kennedy decided to embark on his ill-fated invasion of Cuba, President Johnson escalated the bombing of North Vietnam, and President Nixon decided to cover up the Watergate break-in. Groupthink was also at work in high-level meetings preceding the Space Shuttle *Columbia* disaster in 2003. Transcripts of those meetings show that the official from the National Aeronautics and Space Administration (NASA) who ran shuttle management meetings, a nonengineer, believed from the outset that foam insulation debris could not damage the

Groupthink is group pressure to conform despite individual misgivings.

spacecraft. She dismissed the issue and cut off discussion when an engineer expressed his concerns. The others present quickly fell into line with the nonengineer running the meeting (Wald and Schwartz, 2003). A few days later, damage caused by foam insulation debris caused Columbia to break apart on reentry into the Earth's atmosphere.

Other famous examples of how the lack of a single dissenting voice can result in tragedy come from two homicide cases that grabbed the world's attention. In 1964 in Queens, New York, 28-year-old Kitty Genovese parked her car after returning home from work. When she got out, a man grabbed and stabbed her. She screamed for help. For 35 minutes, at least 38 middle-class, law-abiding neighbors watched from darkened windows as the man repeatedly attacked Genovese and stabbed her 17 times. Finally, one neighbor called the police, but only after he had called a friend and asked what to do. Some of the neighbors later pled ignorance. Others said they thought it was just a lovers' quarrel or "some kids having fun." Still others admitted they did not want to get involved (Gado, 2003). A similar thing happened in 1993 near Liverpool, England. Two 10-year-old boys abducted 2-year-old James Bulger from a shopping mall. They took him on a long, aimless walk, torturing him along the way—dropping him on his head and kicking him in the ribs. Motorists and pedestrians saw the toddler crying, noticed his wounds, and even witnessed some of the violence. "A persuading kick" was the way one motorist later described the blow to the ribs (Scott, 2003). Nobody called the police. These cases illustrate "bystander apathy." As the number of bystanders increases, the likelihood of any one bystander helping another decreases because the greater the number of bystanders, the less responsibility any one individual feels. This behavior shows that people usually take their cues for action from others and again demonstrates the power of groups over individuals. If no one else in a large collectivity responds, most people figure nothing is wrong.

Sociology Now™

Learn more about **Bystander Apathy** by going through the Bystander Apathy Video Exercise.

Inclusion and Exclusion: In-groups and Out-groups

If a group exists, some people must not belong to it. Accordingly, sociologists distinguish **in-group** members (those who belong) from **out-group** members (those who do not). Members of an in-group typically draw a boundary separating themselves from members of the out-group, and they try to keep out-group members from crossing the line. Anyone who has gone to high school knows all about in-groups and out-groups. They have seen firsthand how race, class, athletic ability, academic talent, and physical attractiveness act as boundaries separating groups. Sadly, only in the movies can someone in a high school out-group get a chance to return to school as a young adult and use her savvy to become a member of the in-group (*Never Been Kissed* [1999]).

Group Boundaries: Competition and Self-Esteem

Why do group boundaries crystallize? One theory is that group boundaries emerge when people compete for scarce resources. For example, old immigrants may greet new immigrants with hostility if the latter are seen as competitors for scarce jobs (Levine and Campbell, 1972). Another theory is that group boundaries emerge when people are motivated to protect their self-esteem. From this point of view, drawing group boundaries allows people to increase their self-esteem by believing that out-groups have low status (Tajfel, 1981).

Both theories are supported by the classic experiment on prejudice, *The Robber's Cave Study* (Sherif et al., 1988 [1961]). Researchers brought two groups of 11-year-old boys to a summer camp at Robber's Cave State Park in Oklahoma in 1954. The boys were strangers to one another and for about a week the two groups were kept apart. They swam, camped, and hiked. Each group chose a name for itself and the boys printed their group's name on their caps and T-shirts. Then the two groups met. A series of athletic competitions were set up between them. Soon, each group became highly antagonistic toward the other. Each group came to hold the other in low esteem. The boys ransacked cabins,

In-group members are people who belong to a group.

Out-group members are people who are excluded from the in-group.

started food fights, and stole various items from members of the other group. Thus, under competitive conditions, the boys drew group boundaries starkly and quickly.

The investigators next stopped the athletic competitions and created several apparent emergencies whose solution required cooperation between the two groups. One such emergency involved a leak in the pipe supplying water to the camp. The researchers assigned the boys to teams composed of members of *both* groups. Their job was to inspect the pipe and fix the leak. After engaging in several such cooperative ventures, the boys started playing together without fighting. Once cooperation replaced competition and the groups ceased to hold each other in low esteem, group boundaries melted away as quickly as they had formed. Significantly, the two groups were of equal status—the boys were all white, middle-class, and 11 years old—and their contact involved face-to-face interaction in a setting where norms established by the investigators promoted a reduction of group prejudice. Social scientists today recognize that all these conditions must be in place before the boundaries between an in-group and an out-group fade away (Sternberg, 1998: 512).

Hulton-Deutsch Collection/Corbis

Natural or artificial boundaries—rivers, mountains, highways, railway tracks—typically separate groups or communities.

Dominant Groups

The boundaries separating groups often seem unchangeable and even "natural." In general, however, dominant groups construct group boundaries in particular circumstances to further their goals (Barth, 1969; Tajfel, 1981). Consider Germans and Jews. By the early 20th century, Jews were well integrated into Germany society. They were economically successful, culturally innovative, and politically influential, and many of them considered themselves more German than Jewish. In 1933, the year Hitler seized power, 44 percent of marriages involving at least one German Jew were to a non-Jew. In addition, some German Jews converted before marrying non-Jewish Germans (Gordon, 1984). Yet, although the boundary separating Germans from Jews was quite weak, the Nazis chose to redraw and reinforce it. Defining a Jew as anyone who had at least one Jewish grandparent, they passed a whole series of anti-Jewish laws and, in the end, systematically slaughtered the Jews of Europe. The division between Germans and Jews was not "natural." It came into existence because of its perceived usefulness to a dominant group.

Groups and Social Imagination

So far, we have focused almost exclusively on face-to-face interaction in groups. However, people also interact with other group members in their imagination. Take reference groups, for example. A **reference group** is composed of people against whom an individual evaluates his or her situation or conduct. Put differently, members of a reference group function as "role models." The classic case of a reference group is Theodore Newcomb's (1943) study of students at Bennington College in Vermont, an institution well known for its liberalism. Although nearly all the students came from politically conservative families, they tended to become more liberal with every passing year. Twenty years later, they remained liberals. Newcomb argued that their liberalism grew because over time they came to identify less with their conservative parents (members of their primary group) and more with their liberal professors (their reference group).

The Influence of Reference Groups

A reference group is composed of people against whom an individual evaluates his or her situation or conduct.

Interestingly, reference groups may influence us even though they represent a largely imaginary ideal. The advertising industry promotes certain body ideals that many people try to emulate, although we all know that hardly anyone looks like a runway model or a

JPL/NASA

Is it possible to imagine everyone in the world as a community? Why or why not? Under what conditions might it be possible?

Barbie doll (see Chapter 11, "Sexuality and Gender," and Chapter 12, "Sociology of the Body: Disability, Aging, and Death").

We should not exaggerate the influence of reference groups, however; for despite their influence, most people continue to highly value the opinions of in-group members. Thus, Bennington College professors did not influence individual students one by one. Rather, groups of students came to admire and emulate the faculty reference group. Similarly, individual girls do not dream of looking like Barbie. Rather, groups of girls come to share the body ideal represented by the Barbie role model. Reference groups are important influences, but evidence gathered by social psychologists points to the preponderant power of in-groups in determining which reference groups matter to us (Wilder, 1990).

Imagined Communities

We have to exercise our imagination vigorously to participate in the group life of a society like ours because much social life in a complex society involves belonging to secondary groups without knowing or interacting with most group members. For an individual to interact with any more than a small fraction of the nearly 300 million people living in this country is impossible. Only 0.35 percent of Americans received a Christmas card from President Bush in 2002 (Gladwell, 2002). Nonetheless, most Americans feel a strong emotional bond to their fellow citizens and their president. Similarly, think about the employees and students at your school. They know they belong to the same secondary group and many of them are probably fiercely loyal to it. Yet how many people at your school have you met? You have probably met no more than a small fraction of the total. One way to make sense of the paradox of intimacy despite distance is to think of your college or the United States as an "imagined community." They are imagined because you cannot possibly meet most members of the group and can only speculate about what they must be like. They are nonetheless communities because people believe strongly in their existence and importance (Anderson, 1991).

Many secondary groups are **formal organizations,** secondary groups designed to achieve explicit objectives. In complex societies like ours, the most common and influential formal organizations are bureaucracies. We now turn to an examination of these often frustrating but necessary organizational forms.

Bureaucracy

Bureaucratic Inefficiency

At the beginning of this chapter, we noted that Weber regarded bureaucracies as the most efficient type of secondary group. This runs against the grain of common knowledge. In everyday speech, when someone says "bureaucracy," people commonly think of bored clerks sitting in small cubicles spinning out endless trails of "red tape" that create needless waste and frustrate the goals of clients (Figure 6.4 and Box 6.3). The idea that bureaucracies are efficient may seem very odd.

Real events often reinforce the common view. Consider, for instance, the case of the *Challenger* space shuttle, which exploded shortly after takeoff on January 28, 1986, killing all seven crewmembers. The weather was cold, and the flexible "O-rings" that were supposed to seal the sections of the booster rockets had become rigid, which allowed burning gas to leak. The burning gas triggered the explosion. Some engineers at NASA and at the company that manufactured the O-rings knew they would not function properly in cold weather. However, this information did not reach NASA's top bureaucrats:

Formal organizations are secondary groups designed to achieve explicit objectives.

> [The] rigid hierarchy that had arisen at NASA . . . made communication between departments formal and not particularly effective. [In the huge bureaucracy,] most communication was done through memos and reports. Everything was meticulously documented, but critical details tended to get lost in the paperwork blizzard. The result was that the upper-level managers were kept informed about possible problems with the O-rings . . . but they never truly understood the seriousness of the issue (Pool, 1997: 257).

As this tragedy shows, then, bureaucratic inefficiencies can sometimes have tragic consequences. Indeed, some of the lessons of the 1986 disaster appear to have gone unheeded. In 2003 the shuttle *Columbia* was destroyed on reentry into the Earth's atmosphere, again killing all seven astronauts on board. NASA administrator Sean O'Keefe was roundly criticized for bureaucratic mismanagement. The critics demanded to know why O'Keefe had not received internal NASA e-mails that expressed safety concerns about damage caused by debris during the takeoff. As U.S. Representative Anthony Weiner told O'Keefe: "I read this stuff before you did. That's crazy" (quoted in Stenger, 2003).

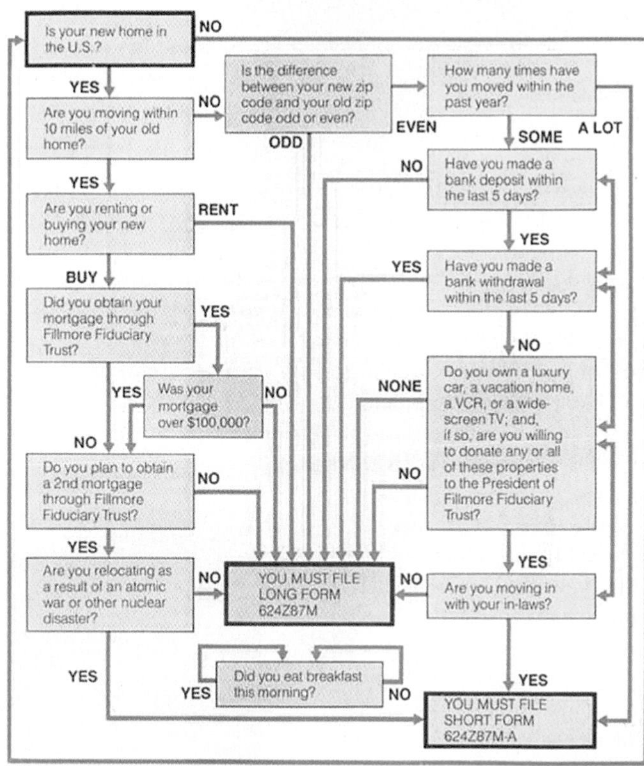

Figure 6.4
Red Tape

The 1980s Infocom game "Bureaucracy" satirized the conventional view of bureaucratic red tape.

Source: Infocom (2003).

How can we square the reality of bureaucratic inefficiencies—even tragedies—with Weber's view that bureaucracies are the most efficient type of secondary group? The answer is twofold. First, we must recognize that when Weber wrote about the efficiency of bureaucracy, he was comparing it to older organizational forms. These operated on the basis of either traditional practice ("We do it this way because we've always done it this way") or the charisma of their leaders ("We do it this way because our chief inspires us to do it this way"). Compared with such "traditional" and "charismatic" organizations, bureaucracies *are* generally more efficient. Second, we must recognize that Weber thought bureaucracies could operate efficiently only in the ideal case. He wrote extensively about some of bureaucracy's less admirable aspects in the real world. In other words, he understood that reality is often messier than the ideal case. So should we. In reality, bureaucracies vary in efficiency. Therefore, rather than proclaiming bureaucracy efficient or inefficient, we should find out what makes bureaucracies work well or poorly. We can then apply this knowledge to improving the operation of bureaucracies.

Dehumanization, Ritualism, Oligarchy, and Inertia

Traditionally, sociologists have lodged four main criticisms against bureaucracies. First is the problem of **dehumanization.** Rather than treating clients and personnel as people with unique needs, bureaucracies sometimes treat clients as standard cases and personnel as cogs in a giant machine. This treatment frustrates clients and lowers worker morale. Second is the problem of **bureaucratic ritualism** (Merton, 1968 [1949]). Bureaucrats sometimes get so preoccupied with rules and regulations they make it difficult for the organization to fulfill its goals. Third is the problem of **oligarchy,** or "rule of the few" (Michels, 1949 [1911]). Some sociologists have argued that in all bureaucracies power tends to become increasingly concentrated in the hands of a few people at the top of the organizational pyramid. This tendency is particularly problematic in political orga-

Dehumanization occurs when bureaucracies treat clients as standard cases and personnel as cogs in a giant machine. This treatment frustrates clients and lowers worker morale.

Bureaucratic ritualism involves bureaucrats getting so preoccupied with rules and regulations that they make it difficult for the organization to fulfill its goals.

Oligarchy means "rule of the few." All bureaucracies have a supposed tendency for power to become increasingly concentrated in the hands of a few people at the top of the organizational pyramid.

BOX 6.3
Sociology at the Movies

Ikiru (1952)

Mounds of paperwork, cluttered office desks, long lines of complaining citizens, and indifferent clerks who quietly shuffle papers—this image of bureaucracy can be found in all modern and postmodern societies. Few people are without a story or two of frustrating struggles against one bureaucracy or another. Most movies depict bureaucracy as perpetually mired in red tape and as an impersonal, soulless machine.

Ikiru, directed by the Japanese filmmaker Akira Kurosawa, is a profound portrait of the individual versus bureaucracy. Many film critics consider it one of the top 10 films of the 20th century.

At the beginning of the film, the main character, Kanji Watanabe, seems little more than a living corpse. As a minor clerk in a large city bureaucracy, he spends much of the day plodding through documents. He is a true bureaucratic ritualist. He does not see and does not seem to care about the people he is supposed to be serving. He simply shuffles paper.

One day, however, Watanabe learns he is suffering from stomach cancer and has only a year left to live. Without a word, he leaves his job of 30 years. He decides to devote his remaining time to finding meaning in life. But he is alone in the world. His wife is dead. His son is indifferent. His coworkers are strangers. So he decides to go to a bar for the first time in his life and drink himself into oblivion. He finds the experience meaningless.

Then Watanabe spots a pretty young woman from his office. Perhaps she can divert his attention from his looming mortality? In the end, she does, though not in the way Watanabe expected. She inspires him to do something small that will make the world a better place. He hears about a struggle to create a small park for children in his neighborhood. Soon, we find him devoting all of his energy to turning the idea of the park into a reality. Ironically, he spends much of his time battling an uncooperative bureaucracy staffed by uncaring and indifferent officials.

Brandom Films/The Everett Collection

In *Ikiru,* Kanji Watanabe, played by Takashi Shimura, finds solace from bureaucracy and the prospect of death by helping to create a small park for children in his neighborhood.

In the end, Watanabe dies. Initially, those who had fought with him vow to go on, to realize the dead man's dream. Soon, however, the rhythm of bureaucratic life resumes. Nearly everyone returns to his or her role as a functionary. As a charismatic leader of a small social movement, the hero briefly made an impact on his society. In the end, however, the wheels of bureaucracy grind on.

Can you think of a situation in which you or someone you know attempted to challenge and reform a bureaucracy? Can change come from within the bureaucracy or does it need to come from outside? Can individuals overcome bureaucratic inertia? Or are we doomed to have the wheels of bureaucracy roll over us?

nizations because it hinders democracy and renders leaders unaccountable to the public. Fourth is the problem of **bureaucratic inertia.** Bureaucracies are sometimes so large and rigid that they lose touch with reality and continue their policies even when their clients' needs change. Like the *Titanic,* they are so big they find it difficult to shift course and steer clear of dangerous obstacles. Two main factors underlie bureaucratic inefficiency: size and social structure.

Bureaucratic inertia refers to the tendency of large, rigid bureaucracies to continue their policies even when their clients' needs change.

Size

Consider size first. Something can be said for the view that bigger is almost inevitably more problematic. Some of the problems caused by size are evident even when you remember some of the differences between dyads and triads. When only two people are involved in a relationship, they may form a strong social bond. If they do, communication is direct and sometimes unproblematic. Once a third person is introduced,

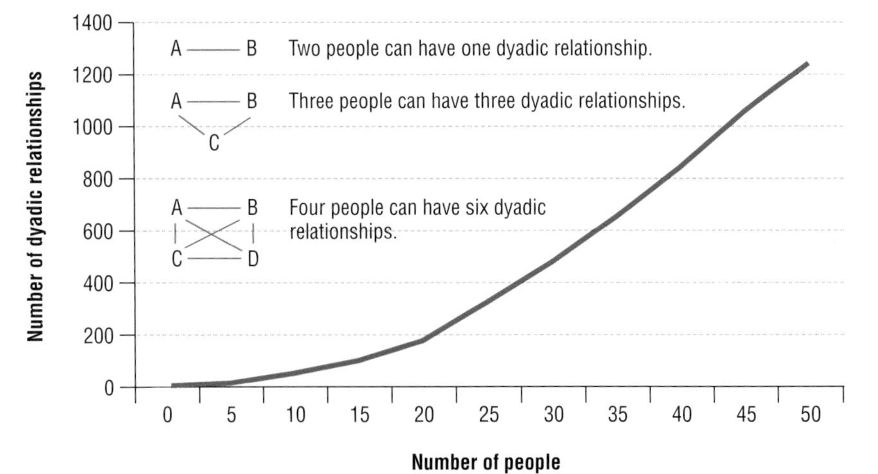

Figure 6.5
Number of Possible Dyadic Relationships by Number of People in Group

Note: The number of dyadic relationships = $(n^2 - n)/2$, where n is the number of people.

however, a secret may be kept, a coalition of two against one may crystallize, and jealousy may result.

Problems can multiply in groups of more than three people. For example, as Figure 6.5 shows, only one dyadic relationship can exist between two people, whereas three dyadic relationships can exist among three people and six dyadic relationships among four people. The number of *potential* dyadic relationships increases exponentially with the number of people. Hence, 300 dyadic relationships are possible among 25 people and 1225 dyadic relationships are possible among 50 people. The possibility of clique formation, rivalries, conflict, and miscommunication rises as quickly as the number of possible dyadic social relationships in an organization.

Social Structure

The second factor underlying bureaucratic inefficiency is social structure. Figure 6.6 shows a typical bureaucratic structure. Note that it is a hierarchy. The bureaucracy has a head. Below the head are three divisions. Below the divisions are six departments. As you move up the hierarchy, the power of the staff increases. Note also the lines of communication that join the various bureaucratic units. Departments report only to their divisions. Divisions report only to the head.

Usually, the more levels in a bureaucratic structure, the more difficult communication becomes. That is because people have to communicate indirectly, through department and division heads, rather than directly with each other. Information may be lost, blocked, reinterpreted, or distorted as it moves up the hierarchy, or an excess of information may cause top levels to become engulfed in a "paperwork blizzard" that prevents them from clearly seeing the needs of the organization and its clients. Bureaucratic heads may have only a vague and imprecise idea of what is happening "on the ground" (Wilensky, 1967).

Consider also what happens when the lines of communication directly joining departments or divisions are weak or nonexistent. As the lines joining units in Figure 6.6 sug-

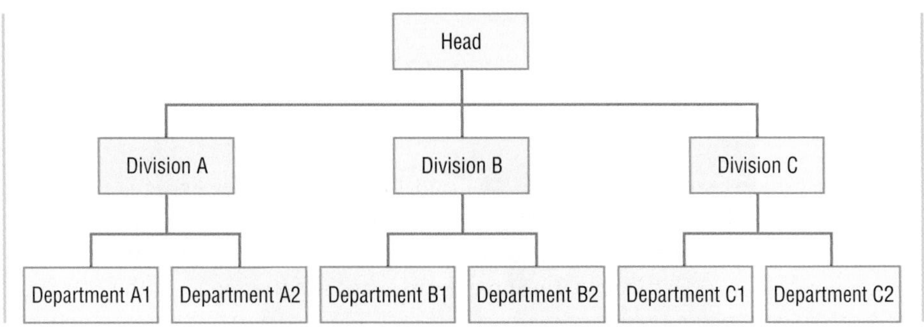

Figure 6.6
Bureaucratic Structure

gest, department A1 may have information that could help department B1 do its job better but may have to communicate that information indirectly through the division level. At the division level, the information may be lost, blocked, reinterpreted, or distorted. Thus, just as people who have authority may lack information, people who have information may lack the authority to act on it directly (Crozier, 1964 [1963]).

Later we consider some ways of overcoming bureaucratic inefficiency. As you will see, these typically involve establishing patterns of social relations that flatten the bureaucratic hierarchy and cut across the sort of bureaucratic rigidities illustrated in Figure 6.6. As a useful prelude to this discussion, we first note some shortcomings of Weber's analysis of bureaucracy. Weber tended to ignore both bureaucracy's "informal" side and the role of leadership in influencing bureaucratic performance. Yet, as you will learn, it is precisely by paying attention to such issues that we can make bureaucracies more efficient.

Bureaucracy's Informal Side

Social Relations Within Bureaucracies

Weber was concerned mainly with the formal structure, or "chain of command," in a bureaucracy. He paid little attention to the social networks that underlie the chain of command.

Evidence for the existence of social networks and their importance in the operation of bureaucracies goes back to the 1930s. Officials at the Hawthorne plant of the Western Electric Company near Chicago wanted to see how various aspects of the work environment affected productivity. They sent social scientists in to investigate. Among other things, researchers found that workers in one section of the plant had established a norm for daily output. Workers who failed to meet the norm were helped by coworkers until their output increased. Workers who exceeded the norm were chided by coworkers until their productivity fell. Company officials and researchers previously regarded employees merely as individuals who worked as hard or as little as they could in response to wage levels and work conditions. However, the Hawthorne study showed that employees are members of social networks that regulate output (Roethlisberger and Dickson, 1939).

In the 1970s, Rosabeth Moss Kanter conducted another landmark study of informal social relations in bureaucracies (Kanter, 1977). Kanter studied a corporation in which most women were sales agents. They were locked out of managerial positions. However, she did not find that the corporation discriminated against women as a matter of policy. She did find a male-only social network whose members shared gossip, went drinking, and told sexist jokes. The cost of being excluded from the network was high: To get good

Informal interaction is common even in highly bureaucratic organizations. A water cooler, for example, can be a place for exchanging information and gossip, and even a place for decision making.

Rob Lewine/Corbis

raises and promotions, one had to be accepted as "one of the boys" and be sponsored by a male executive, which was impossible for women. Thus, despite a company policy that did not discriminate against women, an informal network of social relations ensured that the company discriminated against women in practice.

Despite their overt commitment to impersonality and written rules, bureaucracies rely profoundly on informal interaction to get the job done (Barnard, 1938; Blau, 1963 [1955]). This fact is true even at the highest levels. For example, executives usually decide important matters in face-to-face meetings, not in writing or via phone. That is because people feel more comfortable in intimate settings, where they can get to know "the whole person." Meeting face-to-face, people can use their verbal and nonverbal interaction skills to gauge other people's trustworthiness. Socializing—talking over dinner, for example—is an important part of any business because the establishment of trust lies at the heart of all social interactions that require cooperation (Gambetta, 1988).

Leadership

Apart from overlooking the role of informal relations in the operation of bureaucracies, Weber also paid insufficient attention to the issue of leadership. Weber thought the formal structure of a bureaucracy largely determines how it operates. However, sociologists now realize that leadership style also has a big bearing on bureaucratic performance (Barnard, 1938; Ridgeway, 1983).

Research shows that the least effective leader is the one who allows subordinates to work things out largely on their own, with almost no direction from above. This is known as ***laissez-faire* leadership,** from the French expression "let them do." Note, however, that *laissez-faire* leadership can be effective under some circumstances. It works best when group members are highly experienced, trained, motivated, and educated and when trust and confidence in group members are high. In such conditions, a strong leader is not really needed for the group to accomplish its goals.

At the other extreme is **authoritarian leadership.** Authoritarian leaders demand strict compliance from subordinates. They are most effective in a crisis such as a war or the emergency room of a hospital. They may earn grudging respect from subordinates for achieving the group's goals in the face of difficult circumstances, but they rarely win popularity contests.

Democratic leadership offers more guidance than the *laissez-faire* variety but less control than the authoritarian type. Democratic leaders try to include all group members in the decision-making process, taking the best ideas from the group and molding them into a strategy that all can identify with. Except for crisis situations, democratic leadership is usually the most effective leadership style.

In sum, contemporary researchers have modified Weber's characterization of bureaucracy in two main ways. First, they have stressed the importance of informal social networks in shaping bureaucratic operations. Second, they have shown that democratic leaders are most effective in noncrisis situations because they tend to widely distribute decision-making authority and rewards. As you will now see, these lessons are important when it comes to thinking about how to make bureaucracies more efficient.

Overcoming Bureaucratic Inefficiency

In the business world, large bureaucratic organizations sometimes find themselves unable to compete against smaller, innovative firms, particularly in industries that are changing quickly (Burns and Stalker, 1961). This situation occurs partly because innovative firms tend to have flatter and more democratic organizational structures, such as the network illustrated in Figure 6.7. Compare the flat network structure in Figure 6.7 with the traditional bureaucratic structure in Figure 6.6. Note that the network structure has fewer levels than the traditional bureaucratic structure. Moreover, in the network structure, lines of communication link all units. In the traditional bureaucratic structure, information flows only upward.

***Laissez-faire* leadership** allows subordinates to work things out largely on their own, with almost no direction from above. It is the least effective type of leadership.

Authoritarian leadership demands strict compliance from subordinates. Authoritarian leaders are most effective in a crisis such as a war or the emergency room of a hospital.

Democratic leadership offers more guidance than the *laissez-faire* variety but less control than the authoritarian type. Democratic leaders try to include all group members in the decision-making process, taking the best ideas from the group and molding them into a strategy with which all can identify. Outside of crisis situations, democratic leadership is usually the most effective leadership style.

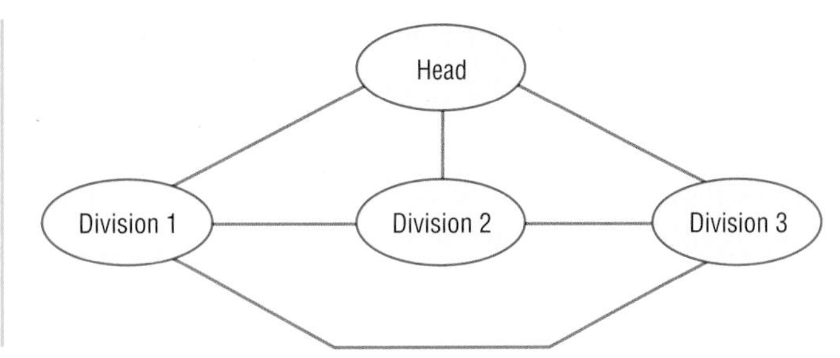

Figure 6.7
Network Structure

Much evidence suggests that flatter bureaucracies with decentralized decision making and multiple lines of communication produce more satisfied workers, happier clients, and bigger profits (Kanter, 1989). Some of this evidence comes from Sweden and Japan. Beginning in the early 1970s, corporations such as Volvo and Toyota were at the forefront of bureaucratic innovation in those countries. They began eliminating middle-management positions. They allowed worker participation in a variety of tasks related to their main functions. They delegated authority to autonomous teams of a dozen or so workers that were allowed to make many decisions themselves. They formed "quality circles" of workers to monitor and correct defects in products and services. As a result, product quality, worker morale, and profitability improved. Today, these ideas have spread well beyond the Swedish and Japanese automobile industry and are evident in such American corporate giants as General Motors, Ford, Boeing, and Caterpillar.

In the 1980s and 1990s, companies outside the manufacturing sector introduced similar bureaucratic reforms, again with positive effects. Consider the case of Bob R., who works as a field technician for Bell South. His company has created small teams of field technicians that are each responsible for keeping a group of customers happy. The technicians set their own work schedule. They figure out when they need to do preventive maintenance, when they need to conduct repairs, and when it is time to try to sell customers new services. Bob describes his new job as follows:

> I've been at the company for twenty-three years, and I always thought we were overmanaged, overcontrolled, and oversupervised. They treated us like children. We're having a very good time under the new system. They've given us the freedom to work on our own. This is the most intelligent thing this company has done in years. It's fun. (quoted in Hammer, 1999: 87)

Managers and workers in many industries have offered similar testimonies to the benefits of more democratic, network-like structures.

Organizational Environments

If flatter organizations are more efficient, why aren't all bureaucracies flatter? Mainly, say sociologists, because of the environment in which they operate. An **organizational environment** is composed of a host of economic, political, and cultural factors that lie outside an organization and affect the way it works (Aldrich, 1979; Meyer and Scott, 1983). Some organizational environments are conducive to the formation of flatter, network-like bureaucracies. Others are not. We can illustrate the effects of organizational environments by discussing two cases that have attracted much attention in recent years: the United States and Japan.

The Japanese Organizational Environment

In the 1970s, American business bureaucracies tended to be more hierarchical than their Japanese counterparts. This was one reason why worker dissatisfaction was high and labor productivity low in the United States. In Japan, where corporate decision making was more decentralized, worker morale and productivity were high (Dore, 1983). Several aspects of the organizational environment help to explain Japanese/American differences in the 1970s. Specifically:

An **organizational environment** is composed of a host of economic, political, cultural, and other factors that lie outside an organization and affect the way it works.

- *Japanese workers were in a position to demand and achieve more decision-making authority than U.S. workers.* After World War II, the proportion of Japanese workers in unions increased, whereas the proportion of American workers in unions declined (see Chapter 21, "Collective Action and Social Movements"). Unions gave Japanese workers more clout than their American counterparts enjoyed.

- *International competition encouraged bureaucratic efficiency in Japan.* Many big Japanese corporations matured in the highly competitive post–World War II international environment. Many big American corporations had originated earlier, in an international environment with few competitors. Thus, Japanese corporations had a bigger incentive to develop more efficient organizational structures (Harrison, 1994).
- *The availability of external suppliers allowed Japanese firms to remain lean.* Many large American companies matured when external sources of supply were scarce. For example, when IBM entered the computer market in the 1950s, it had to produce all components internally because nobody else was making them. This situation led IBM to develop a large, hierarchical bureaucracy. In contrast, Japanese computer manufacturers could rely on many external suppliers in the 1970s. Therefore, they could develop flatter organizational structures (Podolny and Page, 1998).

The U.S. Organizational Environment Today

Today, Japanese/American differences have substantially decreased because most big businesses in America have introduced Japanese-style bureaucratic reforms (Tsutsui, 1998). For instance, Silicon Valley, the center of the American computer industry today, is full of companies that fit the "Japanese" organizational pattern. These companies originated in the 1980s and 1990s, when external suppliers were abundant and international competitiveness was intense. In addition, American companies started to copy Japanese business structures because they saw them as successful (DiMaggio and Powell, 1983). We thus see how changes in the organizational environment help account for convergence between Japanese and American bureaucratic forms.

The experience of the United States over the past few decades holds out hope for increasing bureaucratic efficiency and the continued growth of employee autonomy and creativity at work. It does not mean, however, that bureaucracies in Japan and the United States will be alike in all respects in 20 or 50 or 100 years. The organizational environment is unpredictable, and sociologists are just beginning to understand its operation. It is therefore anyone's guess how far convergence will continue.

Societies

Networks, groups, and bureaucracies are embedded in **societies,** collectivities of interacting people who share a culture and a territory.[1] Like smaller collectivities, societies help shape human action. They influence the kind of work we do and how productively we work. They mold patterns of class, gender, racial, and ethnic inequality. They impinge on the way religious, family, and other institutions operate. They affect the way we govern and how we think of ourselves.

Despite the pervasiveness of these influences, most people are blind to them. We tend to believe that we are free to do what we want. Yet the plain fact is that societies affect even our most personal and intimate choices. For example, deciding how many children to have is one of the most intensely private and emotional issues a woman must face. So why is it that tens of millions of women have decided in the space of just a few decades to have an average of two babies instead of six, or eight babies instead of four? Why do so many individuals make almost exactly the same private decision at almost precisely the same historical moment? The answer is that certain identifiable social conditions prompt them to reach the same conclusion—in this case, to have fewer or more babies. And so it is with most decisions. Identifiable social conditions increase the chance that we will choose one course of action over another.

The relationship between people and nature is the most basic determinant of how societies are structured and therefore how people's choices are constrained. Accordingly, researchers have identified six stages of human evolution, each characterized by a shift in the

Societies are collectivities of interacting people who share a culture and a territory.

[1]For "virtual societies" on the Internet, however, a shared territory is unnecessary.

Table 6.1
The Transformation of Human Societies over the Past 100,000 Years

Type of Society	Foraging	Horticultural, Pastoral	Intensive Agricultural	Industrial	Postindustrial	Postnatural
Approximate years since origin	100,000	10,000	5,000	225	60	30
Revolutionary technology	Simple hand tools	Domestication	Plow	Steam engine	Computer	Recombinant DNA
Productivity, division of labor, population size, permanence of settlements (rank)	6	5	4	3	2	1
Gender equality (rank)	1	2	4	3	2	2
Class equality (rank)	1	2	3	3, then 2	2, but increasing	?

relationship between people and nature. As we review each of these stages, note what happens to the human/nature relationship: With each successive stage, people are less at the mercy of nature and transform it more radically. The changing relationship between people and nature has huge implications for all aspects of social life (Table 6.1). Let us identify these implications as we sketch the evolution of human society in bold strokes.

Foraging Societies

Until about 10,000 years ago, all people lived in **foraging societies.** They lived by searching for wild plants and hunting wild animals (Lenski, Nolan, and Lenski, 1995; O'Neil, 2004; Sahlins, 1972). They were passively dependent on nature, taking whatever it made available and transforming it only slightly to meet their needs. They built simple tools such as baskets, bows and arrows, spears, and digging sticks. They might burn grasslands to encourage the growth of new vegetation and attract game. But they neither planted crops nor domesticated many animals.

Most foragers lived in temporary encampments, and when food was scarce they migrated to more bountiful regions. Harsh environments could support one person per 10–50 square miles. Rich environments could support 10–30 people per square mile. Foraging communities or bands averaged about 25–30 people but could be as large as 100. "Aquatic foragers," such as those on the western coast of North America, concentrated on fishing and hunting marine mammals. "Equestrian foragers," such as the Great Plains Indians of North America, hunted large mammals from horseback. "Pedestrian foragers" engaged in diversified hunting and gathering on foot and could be found on all continents.

Until the middle of the 20th century, most social scientists thought that foragers lived brief, grim lives. In their view, foragers were engaged in a desperate struggle for existence that was typically cut short by disease, starvation, pestilence, or some other force of nature. We now know that this characterization says more about the biases of early anthropologists than the lives that foragers actually lived. Consider the !Kung of the Kalahari desert in southern Africa, who maintained their traditional way of life until the 1960s (Lee, 1979). Young !Kung did not fully join the workforce until they reached the age of 20. Adults worked only about 15 hours a week. Due mainly to disease, children faced a much smaller chance of surviving childhood than is the case in contemporary society, but about 10 percent of the !Kung were older than 60, the same percentage as Americans in the early 1970s. It thus seems that the !Kung who survived childhood lived relatively long, secure, leisurely, healthy, and happy lives. Nor were they unique. The tall totem poles, ornate wood carvings, colorful masks, and elaborate clothing of the Kwakiutl and the Haida on Canada's west coast serve as beautiful reminders that many foragers had the leisure time to invest considerable energy in ornamentation.

Foraging societies are those in which people live by searching for wild plants and hunting wild animals. Such societies predominated until about 10,000 years ago. Inequality, the division of labor, productivity, and settlement size are very low in such societies.

Equestrian foragers were hierarchical, male dominated, and warlike, especially after they acquired rifles in the 19th century. However, the social structure of pedestrian foragers—the great majority of all foragers—was remarkably nonhierarchical. They shared what little wealth they had, and women and men enjoyed approximately equal status.

The elaborate and colorful design of this Haida totem pole from British Columbia, Canada, suggests that the Haida, like many foragers, had the leisure time to invest considerable energy in ornamentation.

Dave Jacobs/Index Stock Imagery

Pastoral and Horticultural Societies

Substantial social inequality became widespread about 10,000 years ago, when some bands began to domesticate various wild plants and animals, especially cattle, camels, pigs, goats, sheep, horses, and reindeer (Lenski, Nolan, and Lenski, 1995; O'Neil, 2004). By using hand tools to garden in highly fertile areas (**horticultural societies**) and herding animals in more arid areas (**pastoral societies**), people increased the food supply and made it more dependable. Nature could now support more people. Moreover, pastoral and horticultural societies enabled fewer people to specialize in producing food and more people to specialize in constructing tools and weapons, making clothing and jewelry, and trading valuable objects with other bands. Some families and bands accumulated more domesticated animals, cropland, and valued objects than others. As a result, pastoral and horticultural societies developed a higher level of social inequality than was evident in most foraging societies.

As wealth accumulated, feuding and warfare grew, particularly among pastoralists. Men who controlled large herds of animals and conducted successful predatory raids acquired much prestige and power and came to be recognized as chiefs. Some chiefs formed large, fierce, mobile armies. The Mongols and the Zulus were horse pastoralists who conquered large parts of Asia and Africa, respectively.

Most pastoralists were nomadic, with migration patterns dictated by the needs of their animals for food and water. Some pastoralists migrated regularly from the same cool highlands in the summer to the same warm lowland valleys in the winter and were able to establish villages in both locations. Horticulturalists often established permanent settlements beside their croplands. These settlements might include several hundred people. However, the founding of large permanent settlements, including the first cities, took place only with the development of intensive agriculture.

Horticultural societies are those in which people domesticate plants and use simple hand tools to garden. **Pastoral societies** are those in which people domesticate cattle, camels, pigs, goats, sheep, horses, and reindeer. Such societies first emerged about 10,000 years ago. Horticulture and pastoralism increase the food supply and make it more dependable. This increases average settlement size and permanence, the division of labor, productivity, and inequality above the levels typical of foraging societies.

Agricultural Societies

Especially in the fertile river valleys of the Middle East, India, China, and South America, human populations flourished much so that about 5000 years ago, they could no longer be sustained by pastoral and horticultural techniques. It was then that **agricultural societies** originated. The plow was invented to harness animal power for more intensive and efficient agricultural production. The plow allowed farmers to plant crops over much larger areas and dig below the topsoil, bringing nutrients to the surface and thus increasing yield (Lenski, Nolan, and Lenski, 1995; O'Neil, 2004).

Agricultural societies are those in which plows and animal power are used to substantially increase food supply and dependability as compared with horticultural and pastoral societies. Agricultural societies first emerged about 5000 years ago. Average settlement size and permanence, the division of labor, productivity, and inequality are higher in agricultural societies than in horticultural and pastoral societies.

Because the source of food was immobile, many people now built permanent settlements, and because people were now able to produce considerably more food than was necessary for their own subsistence, surpluses were sold in village markets. Some of these centers became towns and then cities, home to rulers, religious figures, soldiers, craft workers, and government officials. The population of some agricultural societies numbered in the millions.

The crystallization of the idea of private property was one of the most significant developments of the era. Among pedestrian foragers, there was no private ownership of land

A medieval engraving of peasants harvesting and their lord and master supervising.

Archivo Iconografico, S. A./Corbis

or water. Among horticulturalists, particular families might be recognized as having rights to some property, but only while they were using it. If the property was not in use, they were obliged to share it or give it to a family that needed it. In contrast, in societies that practiced intensive agriculture, powerful individuals succeeded in having the idea of individual property rights legally recognized. It was now possible to buy land and water, to call them one's own, and to transmit ownership to one's offspring. One could now become rich and, through inheritance, make one's children rich.

Ancient civilizations thus became rigidly divided into classes. Royalty surrounded itself with loyal landowners, protected itself with professional soldiers, and justified its rule with the help of priests, part of whose job was to convince ordinary peasants that the existing social order was God's will. Government officials collected taxes, and religious officials collected tithes, thus enriching the upper classes with the peasantry's surplus production. In this era, inequality between women and men also reached its historical high point (Boulding, 1976).

Industrial Societies

Stimulated by international exploration, trade, and commerce, the Industrial Revolution began in Britain in the 1780s. A century later, it had spread to all of western Europe, much of North America, Japan, and Russia. It involved the use of fuel—at first, water power and steam—to drive machines and thereby greatly increase productivity, the quantity of things that could be produced with a given amount of effort.

Industrial societies use machines and fuel to greatly increase the supply and dependability of food and finished goods. The first such societies emerged about 225 years ago in Great Britain. Productivity, the division of labor, and average settlement size increased substantially in industrial societies compared with agricultural societies. While social inequality was substantial during early industrialism, it declined as the industrial system matured.

If you have ever read a Charles Dickens novel such as *Oliver Twist,* you know that hellish working conditions and deep social inequalities characterized early **industrial societies.** Work in factories and mines became so productive that owners amassed previously unimaginable fortunes, but ordinary laborers worked 16-hour days in dangerous conditions and earned barely enough to survive. They struggled for the right to form and join unions and expand the vote to all adult citizens, hoping to use union power and political influence to win improvements in the conditions of their existence. At the same time, new technologies and ways of organizing work made it possible to produce ever more goods at a lower cost per unit. This made it possible to meet many of the workers' demands and raise living standards for the entire population.

Increasingly, businesses required a literate, numerate, and highly trained work force. To raise profits, they were eager to identify and hire the most talented people. They en-

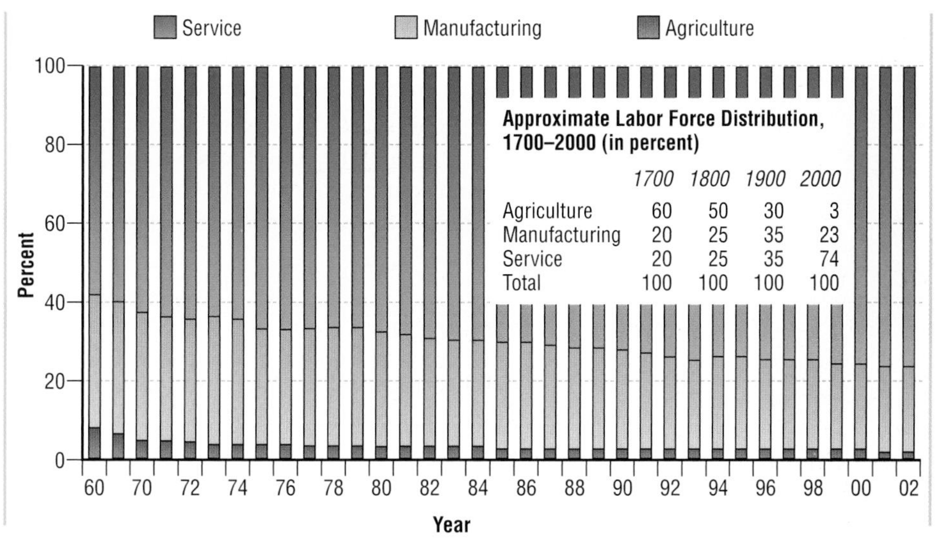

	1700	1800	1900	2000
Agriculture	60	50	30	3
Manufacturing	20	25	35	23
Service	20	25	35	74
Total	100	100	100	100

Figure 6.8
Civilian Employment by Economic Sector, United States, 1960–2002 (in percent)

Source: De Long (1998: 23); U.S. Department of Labor (2003a).

couraged everyone to develop their talents and rewarded them for doing so by paying higher salaries. Even inequality between women and men began to decrease because of the demand for talent and women's struggles to enter the paid workforce on an equal footing with men. Why hire an incompetent man over a competent woman when you can profit more from the services of a capable employee? Put in this way, women's demands for equality made good business sense. For all these reasons, class and gender inequality declined as industrial societies matured.

Postindustrial Societies

In the early 1970s, sociologist Daniel Bell (1973) argued that industrial society was rapidly becoming a thing of the past. According to Bell, just as agriculture had given way to manufacturing as the driving force of the economy in the 19th century, so did manufacturing give way to service industries by the mid-20th century, resulting in the birth of **postindustrial societies.**

Even in preagricultural societies, a few individuals specialized in providing services rather than producing goods. For example, a person considered adept at tending to the ill, forecasting the weather, or predicting the movement of animals might be relieved of hunting responsibilities to focus on these services. However, such jobs were rare because productivity was low. Nearly everyone had to do physical work for the tribe to survive. Even in early agricultural societies, it took 80 to 100 farmers to support one nonfarmer (Hodson and Sullivan, 1995 [1990]: 10). Only at the beginning of the 19th century in industrialized countries did productivity increase to the point where a quarter of the labor force could be employed in services. Shortly after World War II, the United States became the first country in the world in which more than half of civilian employees worked in the service sector. By 1960, all of the highly industrialized countries had reached that threshold. In 2002, 76.7 percent of civilian employees in the United States worked in the service sector (Figure 6.8).

Postindustrial societies are those in which most workers are employed in the service sector and computers spur substantial increases in the division of labor and productivity. Shortly after World War II, the United States became the first postindustrial society. Gender inequality is reduced in postindustrial societies, partly because so many women are brought into the system of higher education and into the paid labor force. Class inequality increases in some postindustrial societies.

In postindustrial societies, women have been recruited to the service sector in disproportionately large numbers, and that has helped to ensure a gradual increase in equality between women and men in terms of education, income, and other indicators of rank (see Chapter 11, "Sexuality and Gender"). The picture with respect to inequality between classes is more complex. Nearly all postindustrial societies have experienced increases in class inequality. The United States, however, has experienced a bigger increase in inequality than any other postindustrial country, whereas France, with less inequality in 2000 than in 1977, has bucked the trend (Smeeding, 2004). We discuss the reasons for these different patterns in Chapter 8 ("Stratification: United States and Global Perspectives").

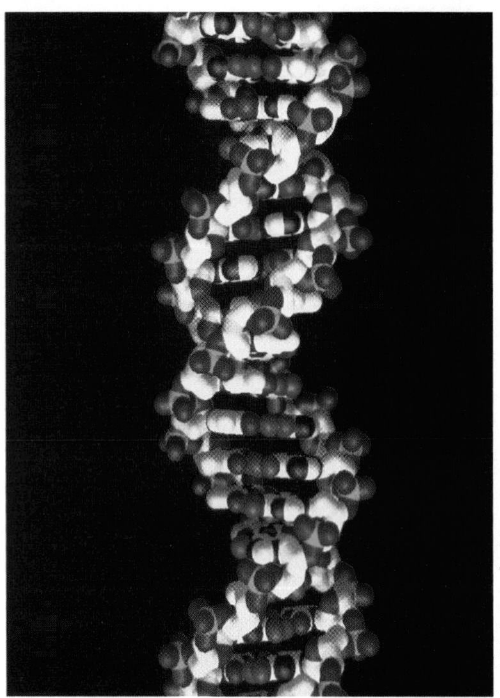

▲ The double helix of DNA.

Rapid change in the composition of the labor force during the final decades of the 20th century was made possible by the computer. The computer automated many manufacturing and office procedures. It created jobs in the service sector as quickly as it eliminated them in manufacturing. The computer is to the service sector as the steam engine was to manufacturing, the plow was to intensive agriculture, domestication was to horticulture and pastoralism, and simple hand tools were to foraging.

Postnatural Societies

On February 28, 1953, two men walked into a pub in Cambridge, England, and offered drinks all around. "We have discovered the secret of life!" proclaimed one of the men. He was James Watson. With his colleague, Francis Crick, he had found the structure of deoxyribonucleic acid, or DNA, the chemical that makes up genes. During cell division, a single DNA molecule uncoils into two strands. New, identical molecules are formed from each strand. In this way, growth takes place and traits are passed from one generation to the next. It was one of the most important scientific discoveries ever (Watson, 1968).

By the early 1970s, scientists were beginning to develop techniques for manipulating DNA (so-called **recombinant DNA**). Soon they could cut a segment out of a DNA strand and join the remaining sections together, or they could take a DNA strand and connect it to segments of DNA from another living thing. This meant that scientists could now create new life forms, a capability that had until then been restricted in the popular imagination to God alone. Enthusiasts proclaimed a "second genesis" as they began to speculate about the potential of the new technology to rid the world of hereditary disease, feed the hungry with higher-yield, disease-resistant farm products, and even create more intelligent, beautiful, and athletic children. For many millions of years, nature had selected the "fittest" living things for survival. Now it seemed possible for humans to speed up natural selection, thus escaping the whims of nature and creating a more perfect society under their control. The invention of recombinant DNA marked the onset of a new social era—what we prefer to call the era of **postnatural society** (Dyson, 1999; Watson, 2000).

We consider some of the perils of postnatural society in detail in Chapter 22 ("Technology and the Global Environment"). Here we want only to emphasize that genetic engineering could easily result in increased social inequality. For example, the technology for creating more perfect babies will undoubtedly be expensive, so rich countries and rich people are most likely to benefit from it. Princeton University biologist Lee Silver and Nobel Prize–winning physicist Freeman Dyson go so far as to speculate that the ultimate result of genetic engineering will be several distinct human species. People who are in a position to take full advantage of genetic engineering will be better looking, more intelligent, less likely to suffer from disease, and more athletic. People who are not so fortunate will have to face nature's caprice in handing out talents and disadvantages, just like our foraging ancestors did (Brave, 2003). The main and perhaps only effective safeguard against such an outcome is true democracy, which would allow ordinary people to decide which risks are worth taking and how the benefits of genetic engineering should be distributed within and across populations (Häyry and Lehto, 1998).

Recombinant DNA involves removing a segment of DNA from a gene or splicing together segments of DNA from different living things, thus effectively creating a new life form.

Postnatural societies are those in which genetic engineering enables people to create new life forms. While genetic engineering holds out much promise for improving productivity, feeding the poor, ridding the world of disease, and so forth, social inequality could increase in postnatural societies unless people democratically decide on the acceptable risks of genetic engineering and the distribution of its benefits.

Freedom and Constraint in Social Life

Throughout this chapter, we have emphasized the capacity of networks, groups, bureaucracies, and societies to constrain human behavior. As we have seen, such social collectivities can even encourage dangerously high levels of conformity, compel people to act against their better judgment, dominate people in a vise of organizational rigidities, and affect the level of social inequality in society.

We stressed the constraining aspect of social collectivities because we wanted to counter the commonsense view that motives alone determine the way people act. Now, however, in conclusion, it would serve us well to remind you that people often have two options other than bowing to the will of their social collectivities: "exit" and "voice" (Hirschman, 1970). In some circumstances, they can leave the social collectivities to which they belong (exit). In other circumstances, they can struggle against the constraints their social collectivities seek to impose on them (voice). As French philosopher Jean-Paul Sartre once remarked, it is always possible to say no, even to the worst tyrant. Less dramatically but no less importantly, knowledge, including sociological knowledge, can increase the ability of people to resist the constraints imposed on them. Recall the Milgram experiment we discussed at the beginning of this chapter, in which subjects administered what they thought were painful shocks to a person just because the experimenters told them to. When the experiment was replicated years later, many of the subjects refused to go along with the demands of the experimenters. Some invoked the example of the Nazis to justify their refusal to comply. Others mentioned Milgram's original experiment. Their knowledge, some of it perhaps gained in sociology courses, enabled them to resist unreasonable demands (Gamson, Fireman, and Rytina, 1982).

Paradoxically, to succeed in challenging social collectivities, people must sometimes form a new social collectivity themselves. Half a century ago, Seymour Martin Lipset, Martin A. Trow, and James S. Coleman (1956) conducted a classic sociological study that made just this point. They investigated the remarkable case of the International Typographical Union—remarkable because in the 1950s it was the outstanding exception to the tendency of trade union bureaucracies to turn into oligarchies, or organizations run by the few. The International Typographical Union remained democratic because the nature of printing as an occupation and an industry made the resources for democratic politics more widely available than is typical in trade unions. Strong local unions that valued their autonomy had founded the international union. The local and regional markets typical of the printing industry at the time strengthened their autonomy. At the same time, strong factions in the union prevented any one faction from becoming dominant. Finally, robust social networks on the shop floor enabled ordinary printers to fight for their rights and resist the slide into oligarchy and dull obedience. What this case illustrates is the way people can form social collectivities to counteract other social collectivities. Embedded in social relations, we can use them for good or evil.

Sociology Now™

Learn more about **Groups** by going through the Group Dynamics Learning Module.

Summary

Sociology Now™

Reviewing is as easy as ➊, ➋, ➌.

1. Before you do your final review, take the SociologyNow diagnostic quiz to help you identify the areas on which you should concentrate. You will find information on how to access SociologyNow on the foldout at the front of the textbook.
2. As you review, take advantage of SociologyNow's study aids to help you master the topics in this chapter.
3. When you are finished with your review, take SociologyNow's post-test to confirm you are ready to move on to the next chapter.

1. Do people act the way they do only because of their interests and emotions?

   People's motives are important determinants of their actions, but social collectivities also influence the way we behave. Because of the power of social collectivities, people sometimes act against their interests, values, and emotions.

2. Is it a small world?

   It is a small world. Most people interact repeatedly with a small circle of family members, friends, coworkers, and other strong ties. However, our personal networks overlap with other social networks, which is why only a few links separate us from complete strangers.

3. What is network analysis?

   Network analysis is the study of the concrete social relations linking people. By focusing on concrete ties, network analysts often come up with surprising results. For example, network analysis has demonstrated the strength of weak ties in job searches, explained patterns in the flow of information and communicable disease, and demonstrated that a rich web of social affiliations underlies urban life.

4. **What are groups?**

   Groups are clusters of people who identity with each other. Primary groups involve intense, intimate, enduring relations; secondary groups involve less personal and intense ties; and reference groups are groups against which people measure their situation or conduct. Groups impose conformity on members and seek to exclude nonmembers.

5. **Is bureaucracy just "red tape"? Is it possible to overcome bureaucratic inefficiency?**

   Although bureaucracies often suffer from various forms of inefficiency, they are generally efficient compared with other organizational forms. Bureaucratic inefficiency increases with size and degree of hierarchy. By flattening bureaucratic structures, decentralizing decision-making authority, and opening lines of communication among bureaucratic units, efficiency can often be improved.

6. **How accurate is Weber's analysis of bureaucracy?**

   Social networks underlie the chain of command in all bureaucracies and affect their operation. Weber ignored this aspect of bureaucracy. He also downplayed the importance of leadership in the functioning of bureaucracy. However, research shows that democratic leadership improves the efficiency of bureaucratic operations in noncrisis situations, authoritarian leadership works best in crises, and *laissez-faire* leadership is the least effective form of leadership in nearly all situations.

7. **What impact does the organizational environment have on bureaucracy?**

   The organizational environment influences the degree to which bureaucratic efficiency can be achieved. For example, bureaucracies are less hierarchical where workers are more powerful, competition with other bureaucracies is high, and external sources of supply are available.

8. **How have societies evolved over the past 100,000 years?**

   Over the past 100,000 years, growing human domination of nature has increased the supply and dependability of food and finished goods, productivity, the division of labor, and the size and permanence of human settlements. Class and gender inequality increased until the 19th century and then began to decline. Class inequality began to increase in some societies in the last decades of the 20th century and may continue to increase in the future.

   In foraging societies, people lived by searching for wild plants and hunting wild animals. Horticultural and pastoral societies emerged about 10,000 years ago. In horticultural societies, people domesticated plants and used simple hand tools to garden. In pastoral societies, people domesticated cattle, camels, pigs, goats, sheep, horses, and reindeer. Agricultural societies first emerged about 5,000 years ago. In such societies, people used plows and animal power to produce food. Great Britain was the first society to industrialize, beginning about 225 years ago. Industrial societies used machines and fuel to greatly increase the supply and dependability of food and finished goods. Shortly after World War II, the United States became the first postindustrial society. In postindustrial societies, most workers are employed in the service sector and computers spur substantial increases in the division of labor and productivity. Some societies may be said to have entered a "postnatural" phase in the early 1970s, when genetic engineering became possible. Genetic engineering enables people to create new life forms, holding out much promise for improving productivity, feeding the poor, and ridding the world of disease, and so forth, and much uncertainty as to whether these benefits will be equitably distributed.

9. **What does the sociological analysis of networks, groups, and bureaucracies tell us about the possibility of human freedom?**

   Networks, groups, bureaucracies, and societies influence and constrain everyone. However, people can also use these social collectivities to increase their freedom. In this sense, social collectivities are a source of both constraint and freedom.

Questions to Consider

1. **Would you have acted any differently from ordinary Germans if you were living in Nazi Germany? Why or why not? What if you were a member of a Nazi police battalion? Would you have been a traitor to your group? Why or why not?**
2. **If you were starting your own business, how would you organize it? Why? Base your answer on theories and research discussed in this chapter.**

Web Resources

Companion Website for This Book

http://sociology.wadsworth.com

Begin by clicking on the Student Resources section of the website. Choose "Introduction to Sociology" and finally the Brym and Lie book cover. Next, select the chapter you are currently studying from the pull-down menu. From the Student Resources page you will have easy access to InfoTrac® College Edition, MicroCase Online exercises, additional web links, and many resources to aid you in your study of sociology, including practice tests for each chapter.

Recommended Websites

Sociologists at Columbia University have organized "The Small World Project." They are trying to extend Stanley Milgram's ideas to the entire wired world. To participate in this study, visit http://smallworld.columbia.edu.

A key excerpt from Max Weber's classic essay on bureaucracy is available at http://www2.pfeiffer.edu/lridener/DSS/Weber/bureau.html.

Yahoo! sponsors many sociology discussion groups. To join, visit http://dir.groups.yahoo.com/dir/Science/Social_Sciences/Sociology.

CHAPTER 7

Deviance and Crime

The Everett Collection

In this chapter, you will learn that:

- Deviance and crime vary among cultures, across history, and from one social context to another.
- Rather than being inherent in the characteristics of individuals or actions, deviance and crime are socially defined and constructed. The distribution of power is especially important in the social construction of deviance and crime.
- Following dramatic increases in the 1960s and 1970s, crime rates eased in the 1980s and fell in the 1990s, mainly because of more effective policing, a declining proportion of young men in the population, and a booming economy.
- Statistics show that a disproportionately large number of African Americans are arrested, convicted, and imprisoned, mainly because of bias in the way crime statistics are collected, the low social standing of the African American community, and racial discrimination in the criminal justice system.
- Many theories of deviance and crime exist. Each theory illuminates a different aspect of the process by which people break rules and are defined as deviants and criminals.
- As in deviance and crime, conceptions of appropriate punishment vary culturally and historically.
- In some respects, modern societies are characterized by less conformity than premodern societies, but in other respects they tolerate less deviance.
- Imprisonment is one of the main forms of punishment in industrial societies; in the United States the prison system has grown quickly in the past 30 years, and punishment has become harsher.
- Fear of crime is increasing, but it is based less on rising crime rates than on manipulation by commercial and political groups that benefit from it.
- There are cost-effective and workable alternatives to the punishment regime currently in place in the United States.

The Social Definition of Deviance and Crime

The 2004–05 TV season was much like any other. Crime was the subject of 10 of the top 20 prime-time network TV programs. They pulled in about 117 million American viewers a week (Nielsen Media Research, 2005). Because millions of additional viewers watched fictional crime shows on cable, local, daytime, and late-night TV and because the news is full of crime stories, one might reasonably conclude that the United States is a society obsessed with crime.

As one might expect in such a society, punishment is also a big issue. This fact is evident from the more than 2.1 million people in state and federal prisons and local jails—a number that is increasing by 50,000 to 80,000 per year. The United States has more people behind bars than any other country on earth. In fact, over 10 percent more people are behind bars in the United States (2006 population: about 298 million) than in China and India combined (2006 population: about 2.4 billion). State prisons in California alone hold more criminals in their grip than do Japan, Germany, France, Great Britain, the Netherlands, and Singapore combined. As of 2000, the United States had the highest incarceration rate (the number of people imprisoned per 100,000 population) of any country in the world (Schlosser, 1998; The Sentencing Project, 1997, 2001; U.S. Census Bureau, 2005).

Why is there so much concern with crime in the United States and why is there so much imprisonment? Because we commonly think of criminals as "the bad guys," we might be excused for thinking that the United States simply contains a disproportionately large number of bad people who have broken the law. For the sociologist, however, this statement is an oversimplification.

Consider the fact that in 1872, Susan B. Anthony—whose image graced the original dollar coin—was arrested and fined. What was her crime? The criminal indictment charged that she "knowingly, wrongfully and unlawfully voted for a representative to the Congress of the United States." Justice Ward Hunt advised the jury, "There is no question

---

Sociology Now™

Reviewing is as easy as ❶, ❷, ❸.

Use SociologyNow to help you make the grade on your next exam. When you are finished reading this chapter, go to the Chapter Summary for instructions on how to make SociologyNow work for you.

Police arrest Martin Luther King, Jr., on September 4, 1958, in Montgomery, Alabama. Considered by many a deviant in his time, King is today hailed by most Americans as a hero. This shift in opinion suggests how the social definition of deviance may change over time.

for the jury, and the jury should be directed to find a verdict of guilty" (quoted in Flexner, 1975: 170). In the late 1950s and early 1960s, Martin Luther King, Jr., whose birthday we now celebrate as a federal holiday, was repeatedly arrested. What was his crime? He marched in the streets of Birmingham, Alabama, and other Southern cities for African Americans' civil rights, including their right to vote.

Susan B. Anthony and Martin Luther King, Jr., were considered deviant and criminal in their lifetimes. Few Americans in the 1870s thought women should be allowed to vote. Anthony disagreed. In acting on her deviant belief, she committed a crime. Similarly, in the 1950s most people in the American South believed in white superiority. They expressed that belief in many ways, including so-called Jim Crow laws that prevented many African Americans from voting. Martin Luther King, Jr., and other participants in the Civil Rights movement challenged the existing law and were therefore arrested.

Most of you would consider the sexist and racist society of the past, rather than Anthony and King, deviant or criminal. That is because norms and laws have changed dramatically. The 19th Amendment guaranteed women's suffrage in 1920. The Voting Rights Act of 1965 guaranteed voting rights for African Americans. Today, anyone arguing that women or African Americans should not be allowed to vote is considered deviant. *Preventing* them from voting would result in arrest.

As the examples of Anthony and King suggest, definitions of deviance and crime change over time. Thus, homosexuality used to be considered a crime and then a sickness, but an increasing proportion of people in the United States and elsewhere now recognize homosexuality as a legitimate sexual orientation (Greenberg, 1988). Similarly, acts that are right and heroic for some people are wrong and treacherous for others. Ask any of the 13,000 law enforcement officials who were brought in to restore order in Los Angeles after the eruption of the 1992 race riot, and they will almost certainly tell you that the people who engaged in mass looting and violence were all common criminals. Ask a politically sophisticated radical African American, such as Sanyika Shakur (a.k.a. "Monster" Kody Scott), a former leader of the notorious Los Angeles gang the Crips, and he will tell you the riot was largely a political reaction to the oppression and powerlessness of inner-city blacks (Shakur, 1993: 381). Sociologists, however, will not jump to hasty conclusions. Instead, they will try to understand how social definitions, social relationships, and social conditions led to the rioting and the labeling of the rioters as criminals by the police.

We take such an approach to deviance and crime in this chapter. We first discuss how deviance and crime are socially defined. We then analyze crime patterns in the United States—who commits crimes and what accounts for changing crime rates over time. Next, we assess the major theories of deviance and crime. Finally, we examine the social determinants of different types of punishment.

The Difference Between Deviance and Crime

- **Deviance** occurs when someone departs from a norm.
- **Informal punishment** involves a mild sanction that is imposed during face-to-face interaction, not by the judicial system.
- People who are **stigmatized** are negatively evaluated because of a marker that distinguishes them from others.
- **Formal punishment** takes place when the judicial system penalizes someone for breaking a law.

Deviance involves breaking a norm. If you were the only man in a college classroom full of women, you probably would not be considered deviant. However, if a man were to use a women's restroom, we would regard him as deviant. That is because deviance is not merely departure from the statistical average. It implies violating an accepted rule of behavior.

Many deviant acts go unnoticed or are considered too trivial to warrant punishment. However, people who are observed committing more serious acts of deviance are typically punished, either informally or formally. **Informal punishment** is mild. It may involve raised eyebrows, gossip, ostracism, "shaming," or **stigmatization** (Braithwaite, 1989). When people are stigmatized, they are negatively evaluated because of a marker that distinguishes them from others (Goffman, 1963). For example, until recently, people with physical or mental disabilities were often treated with scorn or as a source of amusement. **Formal punishment** results from people breaking laws, which are norms stipulated and enforced by government bodies. For example, criminals may be formally punished by having to serve time in prison or perform community service.

Types of Deviance and Crime

Sociologist John Hagan (1994) usefully classifies various types of deviance and crime along three dimensions (Figure 7.1). The first dimension is the *severity of the social response.* At one extreme, homicide and other serious forms of crime result in the most severe negative reactions, such as life imprisonment or capital punishment. At the other end of the spectrum, slight deviations from a norm, such as wearing a nose ring, will cause some people to do little more than express mild disapproval.

High agreement
Very harmful
Severe
Confusion, apathy
Relatively harmless
Mild
Agreement about the norm
Evaluation of social harm
Severity of social response

Figure 7.1
Types of Deviance and Crime

Source: *From Crime and Disrepute* by John Hagan. Copyright © 1994 Pine Forge Press. Reprinted by permission of Sage Publications.

The second dimension of deviance and crime is the *perceived harmfulness* of the deviant or criminal act. Some deviant acts, such as rape, are generally seen as very harmful, whereas others, such as tattooing, are commonly regarded as being of little consequence. Note that actual harmfulness is not the only issue here. *Perceived* harmfulness is. Coca-Cola got its name because in the early part of the 20th century, it contained a derivative of cocaine. Now cocaine is an illegal drug because people's perceptions of its harmfulness changed.

The third characteristic of deviance is the *degree of public agreement* about whether an act should be considered deviant. For example, people disagree about whether smoking marijuana should be considered a crime, especially because it may have therapeutic value in treating pain associated with cancer. In contrast, virtually everyone agrees that murder is seriously deviant. Note, however, that even the social definition of murder varies over time and across cultures and societies. At the beginning of the 20th century, Inuit ("Eskimo") communities sometimes allowed newborns to freeze to death. Life in the far north was precarious. Killing newborns was not considered a punishable offense if community members agreed that investing scarce resources in keep-

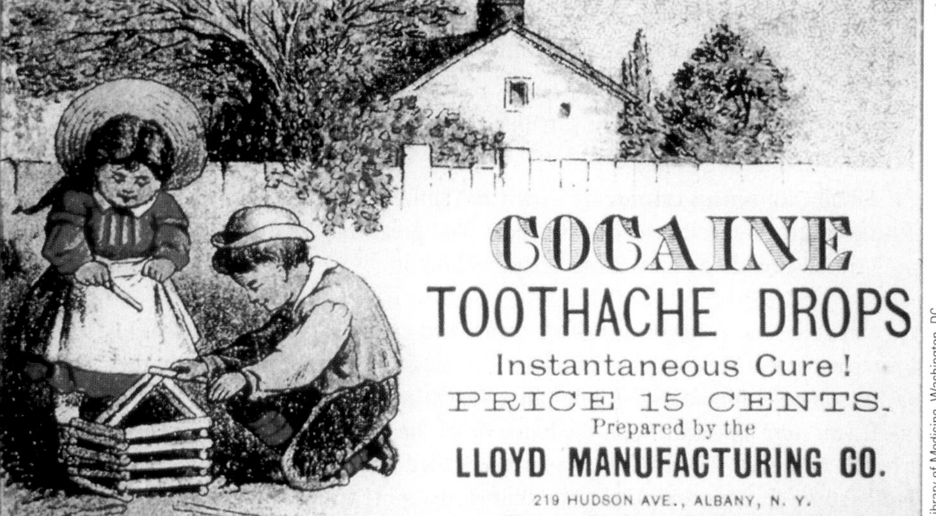

National Library of Medicine, Washington, DC

One of the determinants of the seriousness of a deviant act is its *perceived* harmfulness. Perceptions vary historically, however. For instance, until the early part of the 20th century, people considered cocaine a medicine. It was an ingredient of Coca-Cola and toothache drops and was commonly given to children in these forms.

Transvestites dress in clothing generally considered appropriate to members of the opposite sex. Is transvestitism a social diversion, a social deviation, a conflict crime, or a consensus crime? Why?

WeeGEE/ICP/Getty Images

Sociology Now™

Learn more about **the Difference between Deviance and Crime** by going through The Difference between Deviance and Crime Animation.

A **social diversion** is a minor act of deviance that is generally perceived as relatively harmless and that evokes, at most, a mild societal reaction such as amusement or disdain.

Social deviations are noncriminal departures from norms that are nonetheless subject to official control. Some members of the public regard them as somewhat harmful, whereas other members of the public do not.

Conflict crimes are illegal acts that many people consider harmful to society. However, many people think they are not very harmful. They are punishable by the state.

Consensus crimes are illegal acts that nearly all people agree are bad and harm society greatly. The state inflicts severe punishment for consensus crimes.

ing the newborn alive could endanger everyone's well-being. Similarly, whether we classify the death of a miner as an accident or murder depends on the kind of worker-safety legislation in existence. Some societies have more stringent worker-safety rules than others, and deaths considered accidental in some societies are classified as criminal offenses in others. So we see that even when it comes to consensus crimes, social definitions are variable.

As Figure 7.1 shows, Hagan's analysis allows us to classify four types of deviance and crime:

1. **Social diversions** are minor acts of deviance such as participating in fads and fashions. People usually perceive such acts as harmless. At most they evoke a mild societal reaction such as amusement or disdain, because many people are apathetic or unclear about whether social diversions are in fact deviant.
2. **Social deviations** are more serious acts. Large proportions of people agree that they are deviant and somewhat harmful, and they usually are subject to institutional sanction.
3. **Conflict crimes** are deviant acts that the state defines as illegal, but the definition is controversial in the wider society.
4. Finally, **consensus crimes** are widely recognized to be bad in themselves. There is little controversy over their seriousness. The great majority of people agree that such crimes should be met with severe punishment.

We conclude that conceptions of deviance and crime vary substantially over time and among societies. Most of us regard it as harmless when some people dye their hair purple. In contrast, in medieval Japan hairstyle was an important expression of people's status. If you were a peasant with the hairstyle of the *samurai* (warrior caste), you could be arrested and even killed because you had called the entire social order into question. Under some circumstances, an issue that seems quite trivial to us can be a matter of life and death to others.

Power and the Social Construction of Crime and Deviance

To truly understand deviance and crime, you have to study how people socially construct norms and laws. The school of sociological thought known as *social constructionism* emphasizes that various social problems, including crime, are *not* inherent in certain actions themselves. Instead, some people are in a position to create norms and pass laws that stigmatize other people. Therefore, one must study how norms and laws are created (or "constructed") to understand why particular actions get defined as deviant or criminal in the first place (Chapter 1, "A Sociological Compass").

Power is a crucial element in the social construction of deviance and crime. Power, you will recall from Chapter 5 ("Social Interaction"), is "the probability that one actor within a social relationship will be in a position to carry out his [or her] own will despite resistance" (Weber, 1947: 152). An "actor" may be an entire social group. Relatively powerful groups are generally able to create norms and laws that suit their interests. Relatively powerless social groups usually are unable to do so.

The powerless, however, often struggle against stigmatization. If their power increases, they may succeed in their struggle. We can illustrate the importance of power in the social construction of crime and deviance by considering crimes against women and white-collar crime.

New York Public Library

▲ The 1873 Comstock Law was meant to stop trade in "obscene literature" and "immoral articles." It was targeted against "dirty books," birth control devices, abortion, and information on sexuality and sexually transmitted diseases—all perfectly legal today. In this 1915 cartoon, Robert Minor satirizes the morality underlying the Comstock Law. The caption reads: "Your honor, this woman gave birth to a naked child!"

Crimes Against Women

In the previous section we argued that definitions of crime are usually constructed to bestow advantages on the more powerful members of society and disadvantages on the less powerful. As you will learn in detail in Chapter 11 ("Sexuality and Gender"), women are generally less powerful than men in all social institutions. Has the law therefore been biased against women? We believe it has.

Until recently, many types of crimes against women—including rape—were largely ignored in the United States and most other parts of the world. Admittedly, "aggravated rape" involving strangers was sometimes severely punished. But "simple rape," which involved a friend or an acquaintance, was rarely prosecuted. And "marital rape" was viewed as a contradiction in terms, as if it were logically impossible for a married woman to be raped by her spouse. In her research, Susan Estrich (1987) found that rape law was not taught at American law schools in the 1970s. Law professors, judges, police officers, rapists, and even victims did not think simple rape was "real rape." Similarly, judges, lawyers, and social scientists rarely discussed physical violence against women and sexual harassment until the 1970s. Governments did not collect data on the topic, and few social scientists showed any interest in what has now become a large and important area of study.

Today, the situation has improved. To be sure, as Diana Scully's (1990) study of convicted rapists shows, rape is still associated with a low rate of prosecution. Rapists often hold women in contempt and do not regard rape as a real crime. Yet efforts by Estrich and others to have all forced sex defined as rape have raised people's awareness of date, acquaintance, and marital rape. Rape is prosecuted more often now than it used to be. The same is true for violence against women and sexual harassment.

Why the change? In part because women's position in the economy, the family, and other social institutions has improved over the past 30 years. Women now have more autonomy in the family, earn more, and enjoy more political influence. They also created a movement for women's rights that heightened concern about crimes disproportionately affecting them. For instance, until recently male sexual harassment of female workers was considered normal. Following Catharine MacKinnon's pathbreaking work on the subject, however, feminists succeeded in having the social definition of sexual harassment transformed (MacKinnon, 1979). Sexual harassment is now considered a social deviation and, in some circumstances, a crime. Increased public awareness of the extent of sexual harassment has probably made it less common. We thus see how social definitions of crimes against women have changed with a shift in the distribution of power.[1]

White-Collar Crime

White-collar crime refers to illegal acts "committed by a person of respectability and high social status in the course of his [or her] occupation" (Sutherland, 1949: 9). Such crimes include embezzlement, false advertising, tax evasion, insider stock trading, fraud, unfair labor practices, copyright infringement, and conspiracy to fix prices and restrain trade. Sociologists often contrast white-collar crimes with **street crimes.** The latter include arson, burglary, robbery, assault, and other illegal acts. Street crimes are committed disproportionately by people from lower classes, whereas white-collar crime is committed disproportionately by people from middle and upper classes.

White-collar crime is underreported. A recent Federal Bureau of Investigation (FBI) study notes that local law enforcement agencies are responsible for reporting white-collar crime but only on a voluntary basis (Barnett, n.d.). Because they receive no funding for compiling the data, few law enforcement agencies do. The FBI rarely bothers to analyze the data and publish results. Thus, for 1997–99, the most recent period for which data seem to be available, local agencies covering a mere 12 percent of the U.S. population reported white-collar crime data. A disproportionately large number of these agencies were from small- and medium-sized jurisdictions, that is, outside the big cities where corporate headquarters are most often found. Only a narrow range of white-collar crimes are reported. Many of the most serious types of white-collar crime fall under federal jurisdiction and, therefore, are not reported (e.g., environmental crimes). Many of the reported crimes are petty (e.g., passing bad checks). Some analysts would not consider some of the reported infractions as white-collar crime (e.g., welfare fraud). One may justifiably ask whether so haphazard and biased a reporting scheme has any sociological value at all.

Despite underreporting, many sociologists think white-collar crime is costlier to society than street crime. Consider that armed robbers netted perhaps $400 million in the 1980s, but the savings and loan scandal, in which bankers mismanaged funds and committed fraud, cost the American public $500–$600 *billion* during that decade (Brouwer, 1998). Nonetheless, white-collar criminals, including corporations, are prosecuted relatively infrequently, and they are convicted even less often. This is true even in extreme cases, where white-collar crimes result in environmental degradation or death due, for example, to the illegal relaxation of safety standards. The police and the FBI routinely pursue burglars, but, typically, many of the guilty parties in the savings and loan scandal of the 1980s were not even charged with a misdemeanor.

White-collar crime results in few prosecutions and still fewer convictions for two main reasons. First, much white-collar crime takes place in private and is therefore difficult to detect. For example, corporations may illegally decide to fix prices and divide markets, but executives make these decisions in boardrooms and private clubs that are not generally subject to police surveillance. Second, corporations can afford legal experts, public relations firms, and advertising agencies that advise their clients on how to bend laws, build up their corporate image in the public mind, and influence lawmakers to pass laws "without teeth" (Blumberg, 1989; Clinard and Yeager, 1980; Hagan, 1989; Sherrill, 1997; Sutherland, 1949).

Governments also commit serious crimes. However, punishing political leaders is difficult (Chambliss, 1989). Authoritarian governments often call their critics "terrorists" and even torture people who are fighting for democracy, but such governments rarely have to account for their deeds (Herman and O'Sullivan, 1989). Some analysts argue that even the

White-collar crime refers to an illegal act committed by a respectable, high-status person in the course of work.

Street crimes include arson, burglary, assault, and other illegal acts disproportionately committed by people from lower classes.

[1]Significantly, black rapists of white women receive much more severe punishments than white rapists of white women (LaFree, 1980). This pattern suggests that race is still an important power factor in the treatment of crime, a subject we have much to say about in the following pages.

U.S. government, in spite of its democratic ideals, sometimes behaves in a manner that may be regarded as criminal. For example, while the United States was engaged in a "war on drugs" in the late 1980s, the CIA participated in the drug trade to help arm the right-wing Contra military forces in Nicaragua (Scott and Marshall, 1991).

In sum, white-collar crime is underreported, underdetected, underprosecuted, and underconvicted because it is the crime of the powerful and the well-to-do. The social construction of crimes against women has changed over the past 30 years, partly because women have become more powerful. In contrast, the social construction of white-collar crime has changed little since 1970 because upper classes are no less powerful now than they were then.

Measuring Crime

Some crimes are more common than others, and rates of crime vary over place, over time, and among different social groups. We will now describe some of these variations. Then we will review the main sociological explanations of crime and deviance.

First, a word about crime statistics. Much crime is not reported to the police. For example, many common assaults go unreported because the assailant is a friend or a relative of the victim. Similarly, many rape victims are reluctant to report the crime because they are afraid they will be humiliated and stigmatized by making it public. Moreover, authorities and the wider public decide which criminal acts to report and which to ignore. For instance, if the authorities decide to crack down on drugs, more drug-related crimes will be counted, not because more drug-related crimes occur but because more drug criminals are apprehended. Third, many crimes are not incorporated in major crime indexes published by the FBI. Excluded are many so-called **victimless crimes,** such as prostitution and illegal drug use, which involve violations of the law in which no victim steps forward and is identified. Also excluded from the indexes are most white-collar crimes.

Recognizing these difficulties, students of crime often supplement official crime statistics with other sources of information. **Self-report surveys** are especially useful. In such surveys, respondents are asked to report their involvement in criminal activities, either as perpetrators or victims. In the United States, the main source of data on victimization is the National Crime Victimization Survey, conducted by the U.S. Department of Justice twice annually since 1973 and involving a nationwide sample of about 80,000 people in 43,000 households (Rennin, 2002). Among other things, such surveys show about the same rate of serious crime (e.g., murder and nonnegligent manslaughter) as do official statistics but two to three times the rate of less serious crime, such as assault.

A definitive international self-report survey was conducted in 2000 in 17 countries, including the United States (van Kesteren, Mayhew, and Nieuwbeerta, 2001). It found that 38 percent of the approximately 34,000 respondents were victims of crime in the year preceding the survey. The victimization rate ranged from a high of 58 percent in Australia to a low of 22 percent in Japan, with the United States somewhat above average at 42 percent. Examining the percentage distribution of victims within countries, the researchers found that the United States was just above average with respect to burglary and theft, just below average with respect to contact crime (robberies, sexual incidents, and assaults and threats), and considerably below average with respect to vehicular crime (Figure 7.2).

Survey data are influenced by people's willingness and ability to discuss criminal experiences frankly. Therefore, indirect measures of crime are sometimes used as well. For instance, sales of syringes are a good index of the use of illegal intravenous drugs. Indirect measures are unavailable for many types of crime, however.

Victimless crimes involve violations of the law in which no victim has stepped forward and been identified.

In **self-report surveys**, respondents are asked to report their involvement in criminal activities, either as perpetrators or victims.

Crime Rates

Bearing these caveats in mind, what does the official record show? *Every hour* during 2003, law enforcement agencies in the United States received verifiable reports on an average of about 2 murders or nonnegligent manslaughters, 11 rapes, 47 robberies, 98 aggravated as-

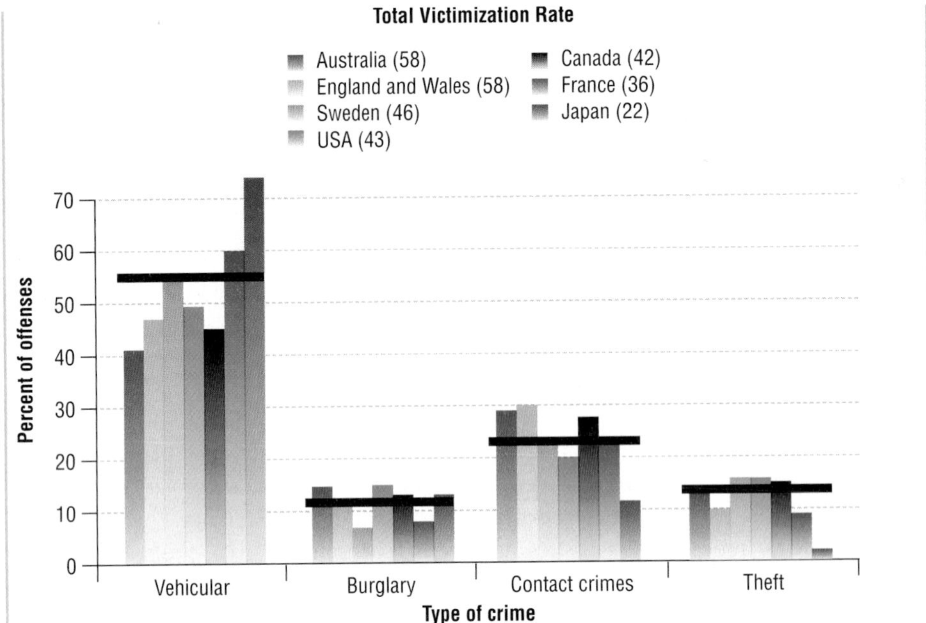

▶Figure 7.2
Victimization: Percent of Offenses by Type of Crime, Seven Countries, 2000 (percent of population victimized by all crimes)

Note: Contact crimes include robberies, sexual incidents, and assaults and threats. Horizontal lines indicate international average for each type of crime for all 17 countries in the survey. Thirty-eight percent of the population of all 17 countries were victimized in the year preceding the survey.
Source: van Kesteren, Mayhew, and Nieuwbeerta (2001: 38, 40).

Sociology Now™

Learn more about **Crime Rates** by going through the Measuring Crime Rates Data Experiment.

saults, 144 motor vehicle thefts, 246 burglaries, and 802 larceny-thefts (U.S. Federal Bureau of Investigation [FBI], 2003). Between 1960 and 1992, the United States experienced a roughly 500 percent increase in the rate of violent crime, including murder and nonnegligent manslaughter, rape, robbery, and aggravated assault. (Remember, the *rate* refers to the number of cases per 100,000 people.) Over the same period, the rate of major property crimes—motor vehicle theft, burglary, and larceny-theft—increased about 150 percent.

Although these statistics are alarming, we can take comfort from the fact that the long crime wave that began in the early 1960s and continued to surge in the 1970s eased in the 1980s and decreased in the 1990s. The good news is evident in ▶Figure 7.3 and ▶Figure 7.4, which show trends in violent and property crime between 1978 and 2003. Except for aggravated assault, the major crime rates for 1990 were about the same as or lower than the major crime rates for 1980. After about 1990, the rates for all forms of major crime began to fall significantly. The rate of murder and nonnegligent manslaughter, for instance, fell 33 percent between 1991 and 2003, and the burglary rate fell 32 percent. The results of the ongoing National Victimization Survey mirror these

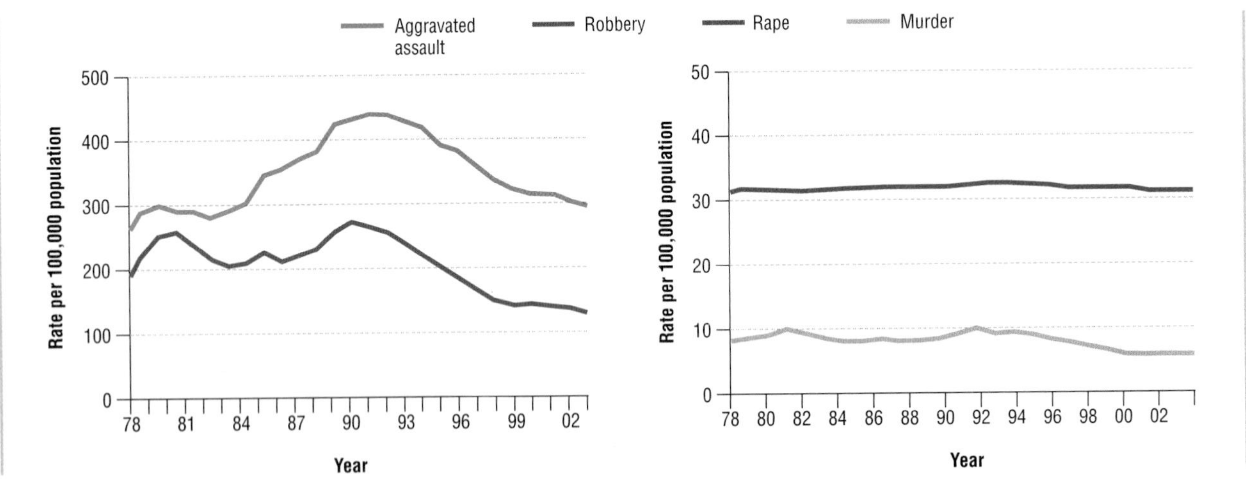

▶Figure 7.3
Violent Crime, United States, 1978–2003, Rate per 100,000 Population

Source: U.S. Federal Bureau of Investigation (1999, 2002, 2003).

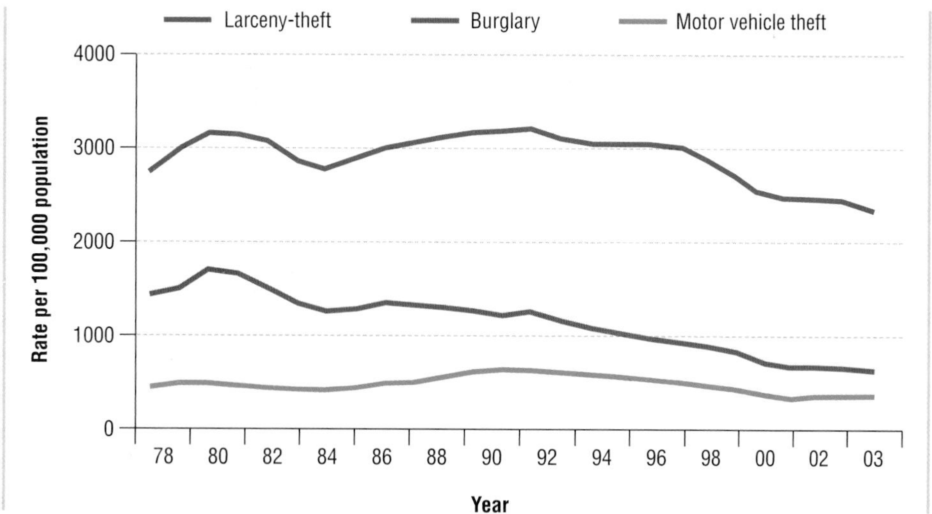

▶Figure 7.4
Property Crime, United States, 1978–2003, Rate per 100,000 Population

Source: U.S. Federal Bureau of Investigation (1999, 2002, 2003).

trends (Rennison, 2002). The 2001 criminal victimization rate was the lowest since the survey began in 1973. The rate fell about 50 percent between 1993 and 2001. This decrease means that there were only about half the number of crime victims per 1000 people in the United States in 2001 as in 1993.

Sociology Now™

Learn more about **Crime Rates** by going through the Measuring Crime Animation.

Why the Decline?

Sociologists usually mention four factors in explaining the decline. First, in the 1990s, governments put more police on the streets, and many communities established their own systems of surveillance and patrol. This trend inhibited street crime. Second, young men are most prone to street crime, but America is aging and the proportion of young men in the population has declined. Third, the economy boomed in the 1990s. Usually, crime rates fluctuate with unemployment rates. When fewer people have jobs, more crime occurs. With an unemployment rate below 5 percent for much of the decade, economic conditions in the United States favored less crime. Finally, and more controversially, some researchers have recently noted that the decline in crime started 19 years after abortion was legalized in the United States. Proportionately fewer unwanted children were in the population beginning in 1992, and unwanted children are more crime prone than wanted children because they tend to receive less parental supervision and guidance (Donahue and Levitt, 2001; Hochstetler and Shover, 1997; Holloway, 1999; LaFree, 1998; Skolnick, 1997) (▶Figure 7.5).

Sociology Now™

Learn more about **Crime Rates** by going through the Violent Crimes Map Exercise.

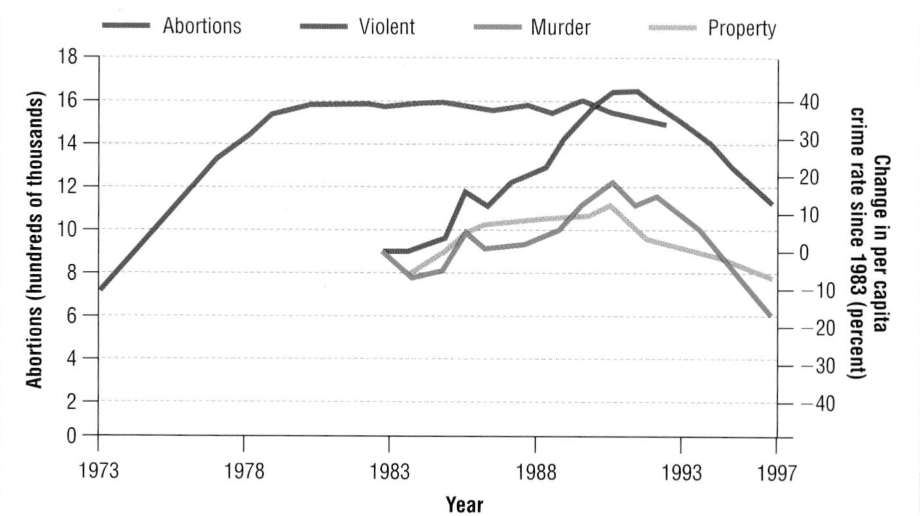

▶Figure 7.5
Abortions and Crime, United States, 1973–1997

Source: Margueritte Holloway from "The Aborted Crime Wave?" *Scientific American*, Vol. 281, No. 6, pp. 23–24. Copyright © 1999 Sarah Donelson.

Table 7.1
Arrests by Age Cohort and Race, United States, 2001

	Percent of Population*	Percent of Arrests
Sex		
Male	49.1	76.8
Female	50.9	23.2
Total	100.0	100.0
Age Cohort		
Under 10	14.0	0.2
10–14	7.3	5.0
15–19	7.1	21.0
20–24	6.8	20.0
25–29	6.8	12.5
30–34	7.2	10.7
35–39	8.2	10.0
40–44	8.1	9.0
45–49	7.2	5.8
50–54	6.1	3.0
55–59	4.8	1.4
60+	16.4	1.3
Total	100.0	100.1*
Racial Group		
White	75.1	70.6
Black	12.3	27.0
American Indian and Alaskan Native	0.9	1.3
Asian and Pacific Islander	3.7	1.2
*Other****	8.0	—
Total	100.0	100.1*

*According to the 2000 census.
**Does not equal 100.0 because of rounding.
*** "Other" includes people who declare two or more races. The race classification used by the U.S. Census Bureau (left column) differs from that used by the FBI in its *Uniform Crime Report* (right column). Therefore, the two columns are only approximately comparable.
Sources: Calculated from U.S. Census Bureau (2003); FBI (2003).

Please note: We have not claimed that putting more people in prison and imposing tougher penalties for crime help to account for lower crime rates. We explain why these actions do not generally result in lower crime rates when we discuss **social control** (methods of ensuring conformity) and punishment. We also probe one of the most fascinating questions raised by the FBI statistics: If crime rates steadied in the 1980s and decreased in the 1990s, what accounts for the exploding prison population and our widespread and growing fear of crime over the past 20 years?

Criminal Profiles

Gender and Age

According to FBI statistics, 77 percent of all persons arrested in the United States in 2003 were men. In the violent crime category, men accounted for 82 percent of arrests (FBI, 2003). As in all things, women, and especially teenage women, are catching up, albeit slowly. Men are still 3.3 times more likely than women to be arrested. However, with every passing year women compose a slightly bigger percentage of arrests. This change is partly because, in the course of socialization, traditional social controls and definitions of femininity are less often being imposed on women (Chapter 11, "Sexuality and Gender").

Most crime is committed by people who have not reached middle age. As Table 7.1 shows, in 2003 Americans between the ages of 15 and 39 accounted for 74 percent of arrests. The 15- to 19-year-old age cohort is the most crime prone.

Race

Table 7.1 also shows that crime has a distinct racial distribution. Although the U.S. Census Bureau classified 75.1 percent of the U.S. population as white in 2000, whites accounted for only 70.6 percent of arrests in 2001. For African Americans, the story is reversed. They accounted for 27.0 percent of arrests but composed only 12.3 percent of the population.

Most sociologists agree that the disproportionately high arrest, conviction, and incarceration rates of African Americans are as a result of three main factors: bias in the way crime statistics are collected, the low class position of blacks in American society, and racial discrimination in the criminal justice system (Hagan, 1994). The statistical bias is largely because of the absence of data on white-collar crimes in the official crime indexes. Because white-collar crimes are committed disproportionately by whites, official crime indexes make it seem as if blacks commit a higher proportion of all crimes than they actually do.

The low class standing of African Americans means that they experience twice the unemployment rate of whites, three times the rate of child poverty, and more than three times the rate of single motherhood. All these factors are associated with higher crime rates; the great majority of poor people are law abiding, but poverty and its associated disabilities are associated with elevated crime rates. The effect of poverty on crime rates is much the same for blacks and whites, but the problem worsened for the African American community in the last quarter of the 20th century. During this period, the U.S. economy was massively restructured and budgets for welfare and inner-city schools were massively cut. Many manufacturing plants in or near U.S. inner cities were shut down in the 1970s and 1980s, causing high unemployment among local residents, a disproportionately large

Social control refers to the social sanctions by means of which conformity to cultural guidelines is ensured.

The Everett Collection

In the 1950s, a sort of "racial profiling" was being commonly applied to rising ethnic minorities such as Puerto Ricans. In the 1957 classic *12 Angry Men*, the character played by Henry Fonda convinces the members of a jury to overcome their prejudices, examine the facts dispassionately, and allow a disadvantaged minority youth accused of murdering his father to go free.

number of whom were African Americans. Many young African Americans, with little prospect of getting a decent education and finding meaningful work, turned to crime as a livelihood and a source of prestige and self-esteem (Sampson and Wilson, 1995).

Finally, as Jerome Miller has convincingly shown, the criminal justice system efficiently searches out African American males for arrest and conviction (Miller, 1996: 48–88). Many white citizens are more zealous in reporting African American than white offenders. Many police officers are more eager to arrest African Americans than whites. Court officials are less likely to allow African Americans than whites to engage in plea bargaining. Fewer African Americans than whites can afford to pay fines that would prevent them from being jailed. Especially since the onset of the war on drugs in the 1980s, African Americans have been targeted, arrested, sentenced, and imprisoned in disproportionate numbers (Box 7.1). The fact that some 40 percent of the U.S. prison population consists of African American men is not just the result of their criminal activity. In the mid-1990s the crime rate of African American men was not much different from their crime rate in 1980, but their imprisonment rate rose more than 300 percent during that period (Tony, 1995).

Sociology Now™

Learn more about **Plea Bargaining** by going through the Criminal Justice System: Plea Bargaining Video Exercise.

Explaining Deviance and Crime

Lep: "I remember your li'l ass used to ride dirt bikes and skateboards, actin' crazy an' shit. Now you want to be a gangster, huh? You wanna hang with real muthaf_____ and tear shit up, huh? . . . Stand up, get your l'il ass up. How old is you now anyway?"

Kody: "Eleven, but I'll be twelve in November."

SANYIKA SHAKUR (1993: 8)

"Monster" Scott Kody eagerly joined the notorious gang the Crips in South Central Los Angeles in 1975 when he was in grade 6. He was released from Folsom Prison on parole in 1988, at the age of 24. Until about three years before his release, he was one of the most ruthless gang leaders in Los Angeles and the California prison system. In 1985, however,

BOX 7.1

SOCIAL POLICY: WHAT DO YOU THINK?

The War on Drugs

Did your high school conduct random drug searches? Did you have to take a Breathalyzer test at your prom? Increasingly, companies are demanding that employees take urine and other tests for drug use. The war on drugs, symbolized by Nancy Reagan's plea to "Just Say No" during the 1980s, thus continues in the United States.

One consequence of the continuing war on drugs is the stiff penalties imposed on drug offenders. For example, if you are caught selling one vial of crack or one bag of heroin, your sentence is 5 to 25 years, depending on what state you are in. New York drug laws are toughest. If you are caught selling two or more ounces of heroin or owning four ounces of cocaine in New York, you will receive the maximum prison sentence of life in prison even as a first-time offender. The authorities make more than 1.5 million arrests every year for drug-related offenses, including 700,000 for the sale or possession of marijuana (Massing et al., 1999: 11–20).

Despite all of these arrests, most people think our drug control policy is ineffective. The U.S. government spends 18 times more on drug control now than it did in 1980 ($18 billion versus $1 billion). Eight times as many Americans are in jail today for drug-related offenses (400,000 versus 50,000 in 1980). Yet an estimated 4 million hardcore drug users are living in the United States (Massing et al., 1999: 32). What should we do?

Rather than continuing the war on drugs, some sociologists suggest it is time to think of alternative policies. We can, for example, estimate the effectiveness of four major policies on drug control: controlling the drug trade abroad, stopping drugs at the border, arresting drug traders and users, and preventing and treating drug use. In one major government-funded study, "[t]reatment was found to be seven times more cost-effective than law enforcement, ten times more effective than interdiction [stopping drugs at the border], and twenty-three times more effective than attacking drugs at their source" (quoted in Massing et al., 1999: 14). Yet the U.S. government spends less than 10 percent of its $18 billion drug-control budget on prevention and treatment. Over two thirds of the money is spent on reducing the supply of drugs (Massing et al., 1999:14).

Another more radical option is to seek limited legalization of drugs. Two arguments support this proposal. First, the United States' major foray into the control of substance abuse—the prohibition of alcohol during the 1920s and early 1930s—turned into a fiasco. It led to an increase in the illegal trade in alcohol and the growth of the Mafia. Second, the Netherlands, for example, has succeeded in decriminalizing marijuana use. Even after it became legal, no major increase in the use of marijuana or more serious drugs such as heroin took place (Massing et al., 1999: 28–9).

Clearly, the citizens of the United States need to discuss drug policy in a serious way. Just saying "no" and spending most of our drug-control budget on trying to curb the supply of illegal drugs are ineffective policies (Reinarman and Levine, 1999).

he decided to reform. He adopted the name of Sanyika Shakur, became a black nationalist, and began a crusade against gangs. Few people in his position have chosen that path. In Kody's heyday, about 30,000 gang members roamed Los Angeles County. Today there are more than 150,000. It is estimated that in 2002 there were 21,500 youth gangs in the United States with 731,500 members ("Gangs," 2005).

What makes the criminal life so attractive to so many young men and women? In general, why do deviance and crime occur at all? Sociologists have proposed dozens of explanations. However, we can group them into two basic types. **Motivational theories** identify the social factors that *drive* people to commit deviance and crime. **Constraint theories** identify the social factors that *impose* deviance and crime (or conventional behavior) on people. Later we examine three examples of each type of theory. Before doing so, however, we want to stress that becoming a *habitual* deviant or criminal is a learning process that occurs in a social context. Motive and lack of constraint may ignite a single deviant or criminal act, but repeatedly engaging in that act requires the learning of a deviant or criminal role.

Sociology Now™

Learn more about **Explaining Deviance and Crime** by going through the Perspectives on Deviance Learning Module.

Motivational theories identify the social factors that drive people to commit deviant and criminal acts.

Constraint theories identify the social factors that impose deviance and crime (or conventional behavior) on people.

Learning the Deviant Role: The Case of Marijuana Users

Howard S. Becker, a giant in the sociological study of deviance, analyzed this learning process in a classic study of marijuana users (Becker, 1962: 41–58). In 1948 and 1949, Becker financed his Ph.D. studies at the University of Chicago by playing piano in local jazz bands. He used the opportunity to do participant-observation research, carefully observing his fellow musicians, informally interviewing them in depth, and writing up detailed field notes after performances. All told, Becker observed and interviewed 50 jazz musicians who smoked marijuana.

Becker found that his fellow musicians had to pass through a three-stage learning process before becoming regular marijuana users. Failure to pass a stage meant failure to learn the deviant role and become a regular user. These are the three stages:

1. *Learning to smoke the drug in a way that produces real effects.* First-time marijuana smokers do not ordinarily get high. To do so, they must learn how to smoke the drug in a way that ensures a sufficient dose to produce intoxicating effects (taking deep drags and holding one's breath for a long time). This process takes practice, and some first-time users give up, typically claiming that marijuana has no effect on them and that people who claim otherwise are just fooling themselves. Others are more strongly encouraged by their peers to keep trying. If they persist, they are ready to go to stage 2.
2. *Learning to recognize the effects and connect them with drug use.* Those who learn the proper smoking technique may not recognize that they are high, or they may not connect the symptoms of being high with smoking the drug. They may get very hungry, laugh uncontrollably, play the same song for an hour on end, and yet still fail to realize that these are symptoms of intoxication. If so, they will stop using the drug. Becker found, however, that his fellow musicians typically asked experienced users how they knew whether they were high. Experienced users identified the symptoms of marijuana use and helped novices make the connection between what they were experiencing and smoking the drug. Once they made that connection, novices were ready to advance to stage 3.
3. *Learning to enjoy the perceived sensations.* Smoking marijuana is not inherently pleasurable. Some users experience a frightening loss of self-control ("paranoia"). Others feel dizzy, uncomfortably thirsty, itchy, forgetful, or dangerously impaired in their ability to judge time and distance. If these negative sensations persist, marijuana use will cease. However, Becker found that experienced users typically helped novices redefine negative sensations as pleasurable. They taught novices to laugh at their impaired judgment, take special pleasure in quenching their deep thirst, and find deeper meaning in familiar music. If and only if novices learned to define the effects of smoking as pleasurable did they become habitual marijuana smokers.

So we see that becoming a regular marijuana user involves more than just motive and opportunity. In fact, learning any deviant or criminal role requires a social context like the one Becker describes. Experienced deviants or criminals must teach novices the "tricks of the trade." Bearing this fact in mind, we may now examine the two main types of theories that seek to explain deviance and crime—those that ask what motivates people to break rules and those that ask how social constraints sometimes fail to prevent rules from getting broken.

Motivational Theories

Durkheim's Functional Approach

In one of the first sociological works on deviance, Émile Durkheim (1964 [1895]) wrote that deviance is normal. What did he mean by this apparently contradictory statement? He meant that deviance is necessary or functional, and, therefore, it exists in all societies. What functions does deviance perform? According to Durkheim, deviance gives people the opportunity to define what is moral and what is not. Our reactions to deviance range from scorn to outrage and our punishments from raised eyebrows to the death penalty. But all of our reactions have one thing in common: They clarify moral boundaries, allowing us to draw the line between right and wrong. This clarification is useful in two ways. First, it promotes the unity of society or its "social solidarity." Second, by pushing against the limits of our tolerance, some deviance encourages healthy social change. Today's deviance may be tomorrow's morality, so some acts that violate norms suggest

Concept Summary 7.1
Merton's Strain Theory of Deviance

		INSTITUTIONALIZED MEANS		
		Accept	Reject	Create New
	Accept	Conformity	Innovation	—
Cultural goals	Reject	Ritualism	Retreatism	—
	Create New	—	—	Rebellion

Source: Adapted from Merton (1938).

new paths for moral development. The functional necessity of deviance derives from these benefits, wrote Durkheim.

Strain Theory

Durhkeim also argued that the absence of clear norms—"anomie"—can result in elevated rates of suicide and other forms of deviant behavior (Chapter 1, "A Sociological Compass"). Robert Merton's **strain theory,** summarized in Concept Summary 7.1, extends Durkheim's insight (Merton, 1938).

Merton argued that cultures often teach people to value material success. Just as often, however, societies do not provide enough legitimate opportunities for everyone to succeed. Therefore, some people experience strain. Most of them will force themselves to adhere to social norms despite the strain (Merton called this "conformity"). The rest adapt in one of four ways. They may drop out of conventional society ("retreatism"). They may reject the goals of conventional society but continue to follow its rules ("ritualism"). They may protest against convention and support alternative values ("rebellion"). Or they may find alternative and illegitimate means of achieving their society's goals ("innovation"); that is, they may become criminals. The American Dream of material success starkly contradicts the lack of opportunity available to poor youths, said Merton. Therefore, poor youths sometimes engage in illegal means of attaining legitimate ends. Merton would say that "Monster" Scott Kody became an innovator at the age of 11 and a rebel at the age of 21.

Subcultural Theory

A second type of motivational theory, known as **subcultural theory,** emphasizes that adolescents like Kody are not alone in deciding to join gangs. Many similarly situated adolescents make the same kind of decision, rendering the formation and growth of the Crips and other gangs a collective adaptation to social conditions. Moreover, this collective adaptation involves the formation of a subculture with distinct norms and values. Members of this subculture reject the legitimate world that they feel has rejected them (Cohen, 1955).

The literature emphasizes three features of criminal subcultures. First, depending on the availability of different subcultures in their neighborhoods, delinquent youths may turn to different types of crime. In some areas, delinquent youths are recruited by organized crime, such as the Mafia. In areas that lack organized crime networks, delinquent youths are more likely to create violent gangs. Thus, the relative availability of different subcultures influences the type of criminal activity to which one turns (Cloward and Ohlin, 1960).

A second important feature of criminal subcultures is that their members typically spin out a whole series of rationalizations for their criminal activities. These justifications make their illegal activities appear morally acceptable and normal, at least to the members of the subculture. Typically, criminals deny personal responsibility for their actions ("What I did harmed nobody"). They condemn those who pass judgment on them ("I'm no worse than anyone else"). They claim their victims get what they deserve ("She had it coming to her"). And they appeal to higher loyalties, particularly to friends and family ("I had to do it because he dissed my gang"). The creation of such justifications and rationalizations enables criminals to clear their consciences and get on with the job. Sociologists call such rationalizations **techniques of neutralization** (Sykes and Matza, 1957) (see the case of the professional fence in Chapter 2, "How Sociologists Do Research").

Finally, although deviants depart from mainstream culture, they are strict conformists when it comes to the norms of their own subculture. They tend to share the same beliefs, dress alike, eat similar food, and adopt the same mannerisms and speech patterns. Whether among professional thieves (Conwell, 1937) or young gang members (Short and

Strain theory holds that people may turn to deviance when they experience strain. Strain results when a culture teaches people the value of material success and society fails to provide enough legitimate opportunities for everyone to succeed.

Subcultural theory argues that gangs are a collective adaptation to social conditions. Distinct norms and values that reject the legitimate world crystallize in gangs.

Techniques of neutralization are the rationalizations that deviants and criminals use to justify their activities. Techniques of neutralization make deviance and crime seem normal, at least to the deviants and criminals themselves.

Strodtbeck, 1965), deviance is strongly discouraged *within* the subculture. Paradoxically, deviant subcultures depend on internal conformity.

The main problem with strain and subcultural theories is that they exaggerate the connection between class and crime. Many self-report surveys find at most a weak tendency for criminals to come disproportionately from lower classes. Some self-report surveys report no such tendency at all, especially among young people and for less serious types of crime (Weis, 1987). A stronger correlation exists between *serious street crimes* and class. Armed robbery and assault, for instance, are more common among people from lower classes. A stronger correlation also exists between *white-collar* crime and class. Middle- and upper-class people are most likely to commit white-collar crimes. Thus, generalizations about the relationship between class and crime must be qualified by considering the severity and type of crime (Braithwaite, 1981). Note also that because official statistics are concerned only with street crime, they usually exaggerate class differences. Moreover, lower-class neighborhoods generally have more police surveillance.

Learning Theory

Apart from exaggerating the association between class and crime, strain and subcultural theories are problematic because they tell us nothing about which adaptation someone experiencing strain will choose. Even when criminal subcultures beckon ambitious adolescents who lack opportunities to succeed in life, only a minority join up. Most adolescents who experience strain and have the opportunity to join a gang reject the life of crime and become conformists and ritualists, to use Merton's terms. Why?

Edwin Sutherland (1939) addressed both the class and choice problems more than 60 years ago by proposing a third motivational factor in what he called the theory of **differential association.** The theory of differential association is still one of the most influential ideas in the sociology of deviance and crime. In Sutherland's view, a person learns to favor one adaptation over another as a result of his or her life experiences or socialization. Specifically, everyone is exposed to both deviant and nondeviant values and behaviors as they grow up. If you happen to be exposed to more deviant than nondeviant experiences, chances are you will learn to become a deviant yourself. You will come to value a particular deviant lifestyle and consider it normal. Everything depends, then, on the exact mix of deviant and conformist influences a person faces. For example, a substantial body of participant observation and survey research has failed to discover widespread cultural values prescribing crime and violence in the inner city (Sampson, 1997: 39). Most inner-city residents follow *conventional* norms, which is one reason why most inner-city adolescents do not learn to become gang members. Those who do become gang members tend to grow up in very specific situations and contexts that teach them the value of crime.

The Everett Collection

According to Edwin Sutherland's theory of differential association, people who are exposed to more deviant than nondeviant experiences as they grow up are likely to become deviants. As in *The Sopranos,* having family members and friends in the Mafia predisposes one to Mafia involvement.

Significantly, the theory of differential association holds for people in all class positions. For instance, Sutherland applied the theory of differential association in his path-breaking research on white-collar crime. He noted that white-collar criminals, like their counterparts on the street, learn their skills from associates and share a culture that rewards rule breaking and expresses contempt for the law (Sutherland, 1949).

Constraint Theories

Motivational theories ask how some people are driven to break norms and laws. Constraint theories, in contrast, pay less attention to people's motivations. The kinds of questions they pose are: How are deviant and criminal labels imposed on some people? How do various forms of social control fail to impose conformity on them? How does the distribution of power in society shape deviance and crime?

Differential association theory holds that people learn to value deviant or nondeviant lifestyles depending on whether their social environment leads them to associate more with deviants or nondeviants.

Labeling Theory: A Symbolic Interactionist Approach

Symbolic interactionism focuses on the meanings people attach to objects, actions, and other people in the course of their everyday lives. As we establish meanings, we put labels on things; you call the object in your hand a book and the streaker at a football game a deviant. While labels are often convenient, the trouble with applying them to people is that they may stick irrespective of the actual behavior involved. We may persist in our belief that a person is deviant even when the person ceases to act in a deviant way. Our labeling itself may then cause more deviance. This is the chief insight of **labeling theory**—that deviance results not just from the actions of the deviant but also from the responses of others, who define some actions as deviant and other actions as normal (Becker, 1962).

If an adolescent misbehaves in high school a few times, teachers and the principal may punish him. However, his troubles really begin if the school authorities and the police label him a "delinquent." Surveillance of his actions will increase. Actions that authorities would normally not notice or would define as of little consequence are more likely to be interpreted as proof of his delinquency. He may be ostracized from nondeviant cliques in the school and eventually be socialized into a deviant subculture. Over time, immersion in the deviant subculture may lead the adolescent to adopt "delinquent" as his **master status,** or overriding public identity. More easily than we may care to believe, what starts out as a few incidents of misbehavior can get amplified into a criminal career because of labeling (Matsueda, 1988, 1992).

The important part that labeling plays in who gets caught and who gets charged with crime was demonstrated more than 30 years ago by Aaron Cicourel (1968). Cicourel examined the tendency to label rule-breaking adolescents "juvenile delinquents" if they came from families in which the parents were divorced. He found that police officers tended to use their discretionary powers to arrest adolescents from divorced families more often than adolescents from intact families who committed similar delinquent acts. Judges, in turn, tended to give more severe sentences to adolescents from divorced families than to adolescents from intact families who were charged with similar delinquent acts. Sociologists and criminologists then collected data on the social characteristics of adolescents who were charged as juvenile delinquents, "proving" that children from divorced families were more likely to become juvenile delinquents. Their finding reinforced the beliefs of police officers and judges. Thus, the labeling process acted as a self-fulfilling prophecy.

Control Theory

All motivational theories assume that people are good and that special circumstances are required to make them bad. In contrast, a popular type of constraint theory assumes that people are bad and that special circumstances are required to make them good. That is because, according to **control theory,** the rewards of deviance and crime are many. Proponents of this approach argue that nearly everyone wants fun, pleasure, excitement, and profit. Moreover, they say that if we could get away with it, most of us would commit deviant and criminal acts to get more of these valued things. For control theorists, the reason most of us do not engage in deviance and crime is that we are prevented from doing so. The reason deviants and criminals break norms and laws is that social controls are insufficient to ensure their conformity.

Travis Hirschi and Michael Gottfredson first developed the control theory of crime (Hirschi, 1969; Gottfredson and Hirschi, 1990). They argued that adolescents are more prone to deviance and crime than adults because they are incompletely socialized and therefore lack self-control. Adults and adolescents may both experience the impulse to break norms and laws, but adolescents are less likely to control that impulse. Gottfredson and Hirschi went on to show that adolescents who are most prone to delinquency are likely to lack four types of social control. They tend to have few social *attachments* to parents, teachers, and other respectable role models; few legitimate *opportunities* for educa-

- **Labeling theory** holds that deviance results not so much from the actions of the deviant as from the response of others, who label the rule breaker a deviant.
- One's **master status** is one's overriding public identity.
- **Control theory** holds that the rewards of deviance and crime are ample. Therefore, nearly everyone would engage in deviance and crime if they could get away with it. The degree to which people are prevented from violating norms and laws accounts for variations in the level of deviance and crime.

tion and a good job; few *involvements* in conventional institutions; and weak *beliefs* in traditional values and morality. Because of the lack of control stemming from these sources, they are relatively free to act on their deviant impulses.

Other sociologists have applied control theory to gender differences in crime. They have shown that girls are less likely to engage in delinquency than boys because families typically exert more control over girls, supervising them more closely and socializing them to avoid risk (Hagan, Simpson, and Gillis, 1987; Peters, 1994). Sociologists have also applied control theory to different stages of life. Just as weak controls exercised by family and school are important in explaining why some adolescents engage in deviant or criminal acts, job and marital instability make some adults more likely to be unable to resist the temptations of deviance and crime (Sampson and Laub, 1993).

Labeling and control theories have little to say about why people regard certain kinds of activities as deviant or criminal in the first place. For the answer to that question, we must turn to conflict theory, a third type of constraint theory.

The Conflict Theory of Deviance and Crime

The day after Christmas, 1996, JonBenét Ramsey was found strangled to death in the basement of her parents' $800,000 home in Boulder, Colorado. The police found no footprints in the snow surrounding the house and no sign of forced entry. The FBI concluded that nobody had entered the house during the night when, according to the coroner, the murder took place. The police found a ransom note saying that the child had been kidnapped. A linguistics expert from Vassar College later compared the note with writing samples of the child's mother. In a 100-page report, the expert concluded that the child's mother was the author of the ransom note. Authorities also determined that all of the materials used in the crime had been purchased by the mother. Finally, investigators discovered that JonBenét had been sexually abused. Although by no means an open-and-shut case, enough evidence was available to cast a veil of suspicion over the parents. Yet, apparently because of the lofty position of the Ramsey family in their community, the police treated them in an extraordinary way. On the first day of the investigation, the commander of the Boulder Police Detective Division designated the Ramseys an "influential family" and ordered that they be treated as victims, not suspects (Oates, 1999: 32). The father was allowed to participate in the search for the child. In the process, he may have contaminated crucial evidence. The police also let him leave the house unescorted for about an hour, which led to speculation that he might have disposed of incriminating evidence. Because the Ramseys were millionaires, they were able to hire accomplished lawyers, who prevented the Boulder police from interviewing them for four months, and a public relations team that reinforced the idea that the Ramseys were victims. A grand jury decided on October 13, 1999, that nobody would be charged with the murder of JonBenét Ramsey.

Regardless of the innocence or guilt of the Ramseys, the way their case was treated adds to the view that the law applies differently to rich and poor. Such differentiation is the perspective of **conflict theory.** In brief, conflict theorists maintain that the rich and the powerful impose deviant and criminal labels on the less powerful members of society, particularly those who challenge the existing social order. Meanwhile, they are usually able to use their money and influence to escape punishment for their own misdeeds.

Steven Spitzer (1980) conveniently summarizes this school of thought. He notes that capitalist societies are based on private ownership of property. Moreover, their smooth functioning depends on the availability of productive labor and respect for authority. When thieves steal, they challenge private property. Theft is therefore a crime. When so-called "bag ladies" and drug addicts drop out of conventional society, they are defined as deviant because their refusal to engage in productive labor undermines a pillar of capitalism. When young, politically volatile students or militant trade unionists strike or otherwise protest against authority, they also represent a threat to the social order and are defined as deviant or criminal.

Conflict theories of deviance and crime hold that deviance and crime arise out of the conflict between the powerful and the powerless.

Concept Summary 7.2
The Main Theories of Deviance and Crime

Theory	Sociologists	Summary
Motivational theories	*Identify the social factors that drive people to deviance and crime*	
Strain theory	Merton	Societies do not provide enough legitimate opportunities for everyone to succeed, resulting in strain, one reaction to which is to find alternative and illegitimate means of achieving society's goals.
Subcultural theory	Cohen, Cloward, Ohlin	Emphasizes the *collective* adaptations to strain, such as the formation of gangs and organized crime, and the degree to which these collective adaptations have distinct norms and values that reject the nondeviant or noncriminal world.
Learning theory	Sutherland	People become deviants or criminals—or fail to do so—because of "differential association" (i.e., they are exposed to, and therefore learn, deviant and criminal values to varying degrees).
Constraint theories	*Identify the social factors that impose deviance and crime (or conventional behavior) on people*	
Labeling theory	Becker, Matsueda, Cicourel	Deviance and crime result not just from the actions of the deviant or criminal but also from the responses of others, who define some actions as deviant and other actions as normal.
Control theory	Hirschi and Gottfredson	Deviants and criminals tend to be people with few social attachments to parents, teachers, and other respectable role models, few legitimate opportunities for education and a good job, few involvements in conventional institutions, and weak beliefs in traditional values and morality. The lack of control stemming from these sources leaves them relatively free to act on their deviant impulses.
Conflict theory	Spitzer	The rich and the powerful impose deviant and criminal labels on the less powerful members of society, particularly those who challenge the existing social order. Meanwhile, they are usually able to use their money and influence to escape punishment for their own misdeeds.

Of course, says Spitzer, the rich and the powerful engage in deviant and criminal acts, too. But he adds that they tend to be dealt with more leniently. Industries can grievously harm people by damaging the environment, yet serious charges are rarely brought against the owners of industry. White-collar crimes are less severely punished than street crimes, regardless of the relative harm they cause. Compare burglary and fraud, for example. Fraud almost certainly costs society more than burglary. But burglary is a street crime committed mainly by lower-class people, whereas fraud is a white-collar crime committed mainly by middle- and upper-class people. Not surprisingly, in 1992, 82 percent of people tried for burglary in the United States were sentenced to prison, and served an average of approximately 26 months, whereas only about 46 percent of people tried for fraud went to prison, and served an average of 14 months (Reiman, 1995: 125). Laws and norms may change along with shifts in the distribution of power in society. However, according to conflict theorists, definitions of deviance and crime and punishments for misdeeds are always influenced by who is on top.

And so we see that many theories contribute to our understanding of the social causes of deviance and crime. Some forms of deviance and crime are better explained by one theory than another. Different theories illuminate different aspects of the process by which people are motivated to break rules and become defined as rule breakers. Our overview should make it clear that no one theory is best. Instead, taking many theories into account allows us to develop a fully rounded appreciation of the complex processes surrounding the social construction of deviance and crime (Concept Summary 7.2).

Trends in Criminal Justice

Social Control

Web Interactive Exercise: The Mass Media and Gun Control

No discussion of crime and deviance would be complete without considering in some depth the important issues of social control and punishment. All societies seek to ensure that their members obey norms and laws. All societies impose sanctions on rule breakers. However, the *degree* of social control varies over time and from one society to the next. *Forms* of punishment also vary. We now focus on trends in criminal justice, that is, on how social control and punishment have changed historically.

Consider first the difference between preindustrial and industrial societies. Beginning in the late 19th century, many sociologists argued that preindustrial societies are characterized by strict social control and high conformity, whereas industrial societies are characterized by less stringent social control and low conformity (Tönnies, 1957 [1887]). Similar differences were said to characterize small communities versus cities. As an old German proverb says, "City air makes you free."

This point of view holds much truth. Whether they are fans of opera or reggae or connoisseurs of fine wine or marijuana, city dwellers in industrial societies find belonging to a group or subculture of their choice easier than do people in small preindustrial communities (see, for example, the discussion of homosexual communities in Chapter 11, "Sexuality and Gender"). In general, the more complex a society, the less likely it is that many norms will be widely shared. In fact, in a highly complex society such as the United States today, finding an area of social life in which everyone is alike or where one group can impose its norms on the rest of society without resistance is difficult. The existence of more than 2100 different religious groups in the United States today speaks volumes about the extent of social diversity in our society (Melton, 1996 [1978]).

Nonetheless, some sociologists believe that social control has intensified over time, at least in some ways. They recognize that individuality and deviance have increased but only within quite strict limits, beyond which it is now *more* difficult to move. In their view, many crucial aspects of life have become more regimented, not less.

Much of the regimentation of modern life is tied to the growth of capitalism and the state. Factories require strict labor regimes, with workers arriving and leaving at a fixed time and, in the interim, performing fixed tasks at a fixed pace. Workers initially rebelled against this regimentation because they were used to enjoying many holidays and a flexible and vague work schedule regulated only approximately by the seasons and the rising and setting of the sun. But they had little alternative as wage labor in industry overtook feudal arrangements in agriculture (Thompson, 1967). Meanwhile, institutions linked to the growth of the modern state or regulated by it—armies, police forces, public schools, health-care systems, and various other bureaucracies—also demanded strict work regimes, curricula, and procedures. These institutions existed on a much smaller scale in preindustrial times or did not exist at all. Today they penetrate our lives and sustain strong norms of belief and conduct (Foucault, 1977 [1975]).

Electronic technology makes it possible for authorities to exercise more effective social control than ever before. With millions of cameras mounted in public places and workplaces, some sociologists say we now live in a "surveillance society" (Lyon and Zureik, 1996). Spy cameras enable observers to see deviance and crime that would otherwise go undetected and take quick action to apprehend rule breakers. Moreover, people tend to alter their behavior when they are aware of the presence of spy cameras. For example, attentive shoplifters migrate to stores lacking electronic surveillance. On factory floors and in offices, workers display more conformity to management-imposed work norms. On college campuses, students are inhibited from engaging in organized protests (Boal, 1998).

Thanks to computers and satellites, intelligence services in the United States, Britain, Canada, Australia, and New Zealand now monitor all international telecommunications traffic, always on the lookout for threats. As easily as you can find the word "anomie" in your sociology essay using the search function of your word processor, the National

Security Agency (NSA) can scan digitized telephone and e-mail traffic in many languages for key words and word patterns that suggest unfriendly activity (Omega Foundation, 1998). However, the system, known as Echelon, is also used to target sensitive business and economic secrets from western Europe, and some people, including Senator Frank Church, have expressed the fear that it could be used on the American people, robbing them of their privacy. Meanwhile, credit information on 95 percent of American consumers is available for purchase, the better to tempt you with credit cards, marketing ploys, and junk mail. When you browse the Web, information about your browsing patterns is collected in the background by many of the sites you visit, again largely for marketing purposes. Most large companies monitor and record their employees' phone conversations and e-mail messages. These instances are all efforts to regulate behavior, enforce conformity, and prevent deviance and crime more effectively using the latest technologies available (Garfinkel, 2000).

Sociology Now™

Learn more about **Criminal Justice System** by going through the Criminal Justice System Learning Module.

A major development in social control that accompanied industrialization was the rise of the prison. Today, prisons figure prominently in the control of criminals the world over. Americans, however, have a particular affinity for the institution, as we will now see.

The Prison

When he was 22, Robert Scully was sent to San Quentin Prison for robbery and dealing heroin. Already highly disturbed, he became more violent in prison and attacked another inmate with a makeshift knife. As a result, Scully was shipped off to Corcoran Prison, one of the new maximum-security facilities that the state of California began opening in the early 1980s. He was thrown into solitary confinement. In 1990, Scully was transferred to the new "supermax" prison at Pelican Bay. There, he occupied a cell the size of a bathroom. It had a perforated sheet metal door. He received food through a hatch. Even exercise was solitary. When he was released on parole in 1994, he had spent nine years in isolation.

Web Research Project: Does Prison Deter Criminals?

One night in 1995 Scully was loitering around a restaurant with a friend. The owner, fearing a robbery, called the police. Deputy Sheriff Frank Rejo, a middle-aged grandfather looking forward to retirement, soon arrived at the scene. He asked to see a driver's license. As Scully's friend searched for it, Scully pulled out a sawed-off shotgun and shot Rejo in the forehead. Scully and his friend were apprehended by police the next day.

Robert Scully was already involved in serious crime before he got to prison, but he became a murderer in San Quentin, Corcoran, and Pelican Bay—a pattern known to sociologists for a long time. Prisons are agents of socialization, and new inmates often become more serious offenders as they adapt to the culture of the most hardened, long-term prisoners (Wheeler, 1961). In Scully's case, psychologists and psychiatrists called in by the defense team said that things had gone even further. Years of sensory deprivation and social isolation had so enraged and incapacitated Scully that thinking through the consequences of his actions became impossible. He had regressed to the point where his mental state was that of an animal able to act only on immediate impulse (Abramsky, 1999).

In preindustrial societies criminals who committed serious crimes were put to death, often in ways that seem cruel by today's standards. One method involved hanging the criminal with starving dogs.

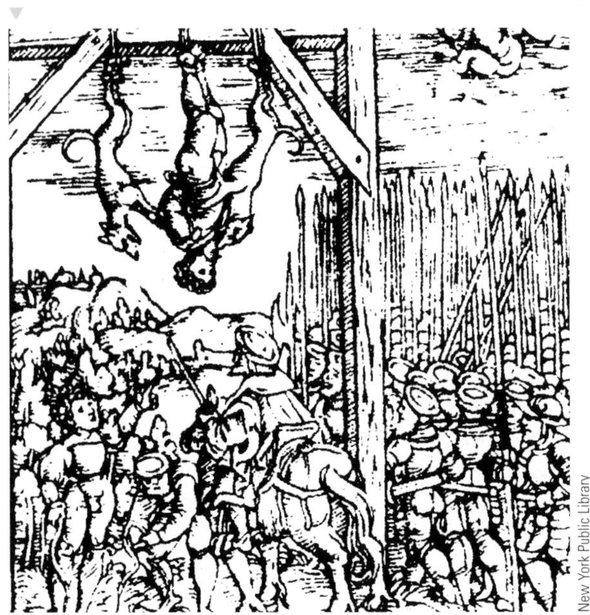

New York Public Library

Origins of Imprisonment

Because prison often turns criminals into worse criminals, pondering the institution's origins, development, and current dilemmas is worthwhile. As societies industrialized, imprisonment became one of the most important forms of punishment for criminal behavior (Garland, 1990; Morris and Rothman, 1995). In preindustrial societies, criminals were publicly humiliated, tortured, or put to death, depending on the severity of their transgression. In the industrial era, depriving criminals of their freedom by putting them in prison seemed less harsh and more "civilized" (Durkheim, 1973 [1899–1900]).

Goals of Incarceration

Some people still take a benign view of prisons, even seeing them as opportunities for *rehabilitation*. They believe that prisoners, while serving time, can be taught how to be productive citizens upon release. In the United States, this view predominated in the 1960s and early 1970s, when many prisons sought to reform criminals by offering them psychological counseling, drug therapy, skills training, college education, and other programs that would help at least the less violent offenders get reintegrated into society.

In 1966, 77 percent of Americans believed that the main goal of prison was to rehabilitate prisoners; by 1994 only 16 percent held that opinion (Bardes and Oldendick, 2003: 183). Today, the great majority of Americans scoff at the idea that prisons can rehabilitate criminals. We have adopted a much tougher line, as the case of Robert Scully shows. Some people see prison as a means of *deterrence*. In this view, people will be less inclined to commit crimes if they know they are likely to get caught and serve long and unpleasant prison terms. Others think of prisons as institutions of *revenge*. They think that depriving criminals of their freedom and forcing them to live in poor conditions is fair retribution for their illegal acts. Still others see prisons as institutions of *incapacitation*. From this viewpoint, the chief function of the prison is simply to keep criminals out of society as long as possible to ensure they can do no more harm (Feeley and Simon, 1992; Simon, 1993; Zimring and Hawkins, 1995).

No matter which of these views predominates, one thing is clear: The American public has demanded that more criminals be arrested and imprisoned. And it has gotten what it wants (Gaubatz, 1995; Savelsberg, 1994). The nation's incarceration rate rose substantially in the 1970s, doubled in the 1980s, and doubled again in the 1990s.

Moral Panic

What happened between the early 1970s and the present to so radically change the U.S. prison system? In a phrase, the United States was gripped by **moral panic.** The fear that crime posed a grave threat to society's well-being motivated wide sections of the American public, including lawmakers and officials in the criminal justice system (Cohen, 1972; Goode and Ben-Yehuda, 1994). The government declared a war on drugs, which resulted in the imprisonment of hundreds of thousands of nonviolent offenders. Sentencing got tougher, and many states passed "three strikes and you're out" laws. This law put three-time felons in prison for life. The death penalty became increasingly popular. As Figure 7.6 shows, support for capital punishment more than doubled between 1965 and 1994, from 38 percent to 80 percent of the population, although it fell to 71 percent by 2005 (also Figure 7.7). In addition, in 2003 the number of inmates on death row dropped for the first time in a generation.

The fall in the percentage of Americans supporting the death penalty is due to two main factors. First, an investigative series in the *Chicago Tribune* in 1999 examined all 285 Illinois capital trials since 1997 and found an astonishing number of disbarred defense counsels, lying prosecutors, pseudoscientific evidence, and corrupt informants. This series led the Republican governor of Illinois, George Ryan, a longtime death-penalty advocate, to declare a state moratorium on capital executions pending an investigation of the Illinois judicial system. In the first half of 2000 many other important Republicans began to question the wisdom of the death penalty publicly. In 2003 Governor Ryan pardoned four men and commuted the death sentence of the remaining 163 men and 4 women on death row in Illinois. Second, the Vatican expressed opposition to the death penalty in the 1997 edition of its catechism, which led many American Catholics to speak openly against the death penalty for the first time. In 2005, the Supreme Court narrowed the class of people eligible for execution by excluding juvenile offenders; it had earlier excluded the mentally retarded (Liptak, 2003; 2005; Seeman, 2000; Wilgoren, 2003). (See pages 207–208 for a discussion of capital punishment.)

Despite the recent decline in support for the death penalty, evidence of the moral panic has been evident in crime *prevention*, too. For example, many well-to-do Americans

A **moral panic** occurs when many people fervently believe that some form of deviance or crime poses a profound threat to society's well-being.

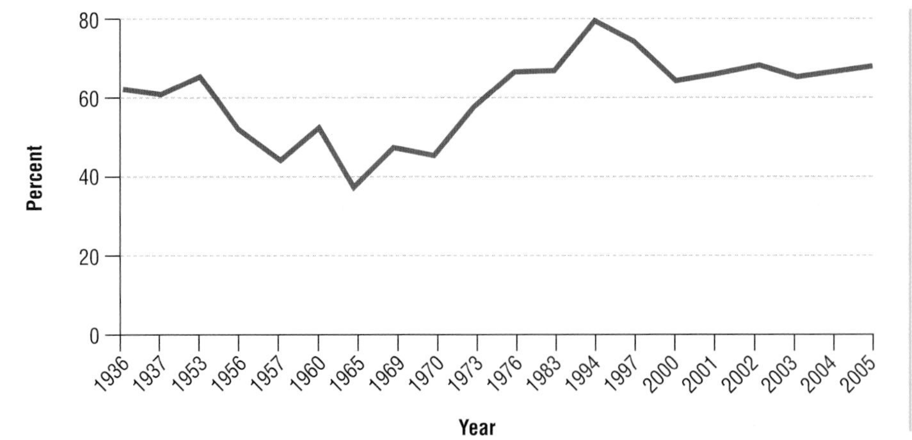

Figure 7.6
Belief in Capital Punishment, United States, 1965–2005 (percent "for"): "Do you believe in capital punishment, that is, the death penalty, or are you opposed to it?"

Note: Where more than one poll is available in a given year, the yearly average is shown.
Sources: Bardes and Oldendick (2003); Bureau of Justice Statistics (2002); Maguire and Pastore (1988: 138); Newport (2000); PollingReport.com (2005).

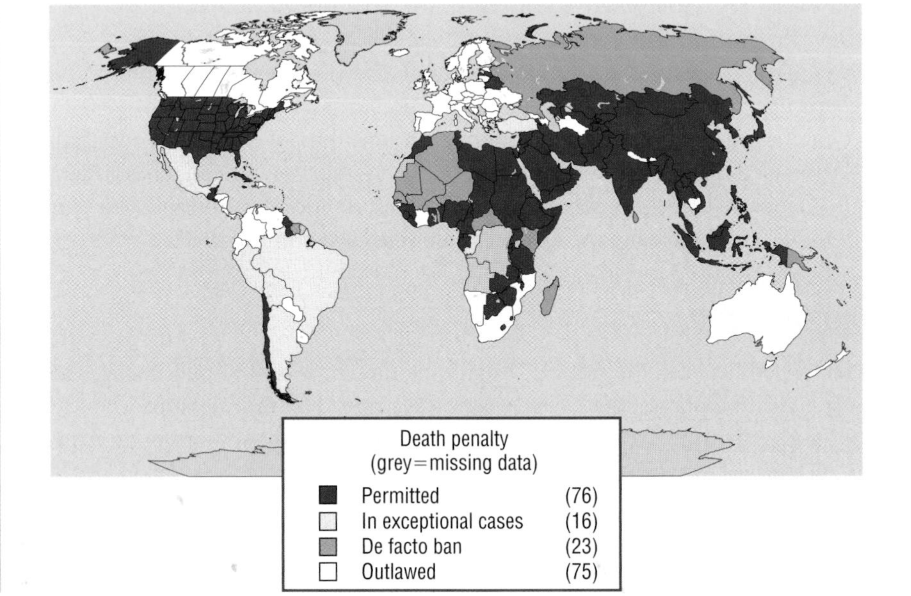

Figure 7.7
The Death Penalty Worldwide, 2005

The most controversial punishment is the death penalty, or capital punishment. Organizations such as Amnesty International believe that the death penalty is a serious human rights violation. Much research shows that it is not an effective deterrent, and concern continues to rise regarding racial bias and wrongful conviction in its use. Despite the controversy, 48 percent of the world's countries allow capital punishment for at least some types of crimes. How can you explain the global distribution of these laws? The data suggest a positive relationship between economic development and abolishing the death penalty, but there are obvious exceptions. Notably, the United States, one of the most developed countries in the world, still allows the death penalty. Several African countries, although poor, outlaw capital punishment. What factors other than economic development might account for these exceptions? How, if at all, do you think the death penalty is related to other types of human rights violations?

Source: Infoplease (2005).

had walls built around their neighborhoods, restricting access to residents and their guests. They hired private security police to patrol perimeters and keep potential intruders at bay. Middle- and upper-class Americans installed security systems in their homes and steel bars in their basement windows. Many people purchased handguns in the belief that they would enhance their personal security. The number of handguns in the United States in 2006 is estimated at about 200 million. Some states even passed laws allowing people to conceal handguns on their person. In short, over the past 20 years or so, Americans have prepared themselves for an armed invasion and have decided to treat criminals much more toughly than in the past (Box 7.2).

BOX 7.2
Sociology at the Movies

Bowling for Columbine (2002)

The homicide rate in the United States is 3 to 10 times higher than the homicide rate in the other 20 or so wealthy, highly industrialized countries. In 2001, 15,980 Americans were murdered, which is five times more than the number of people killed in the terrorist attacks on September 11 of that year. Firearms were used in nearly two-thirds of the murders committed in the United States in 2001 (U.S. Federal Bureau of Investigation, 2002).

In *Bowling for Columbine,* documentary filmmaker Michael Moore describes the magnitude of the problem and helps us figure out why Americans kill one another with such extraordinary frequency. Moore succeeds in his first, descriptive task by combining hilarity with horror. He takes us to a bank that gives away rifles instead of toasters to people who open an account. ("Don't you think it's a little dangerous to have all these guns in a bank?" he asks a teller.) We see security camera footage from the 1999 massacre at Columbine High School in Littleton, Colorado, where two students shot and killed 13 fellow students and then committed suicide. We listen while Moore interviews Charlton Heston, then president of the National Rifle Association (NRA). Heston cannot answer when Moore asks why a man who has never been personally threatened and who now lives behind a perimeter wall and a locked security gate in a protected neighborhood with security patrols feels the need to keep a loaded gun in the house. These and many other scenes, some tragic and others absurd, provide compelling evidence that guns are as American as apple pie. Not for nothing did the International Documentary Film Association name this film the best documentary ever made.

When it comes to *explaining* the unusually high American homicide rate, however, the film is less successful. Moore examines three main explanations and rejects them all. First, Americans may be violent because the mass media—TV, video games, movies, and popular music—are full of violence and influence young people in particular to engage in violent acts in the real world. Moore dismisses that argument on the grounds that Canadians are exposed to almost exactly the same mass media influences as Americans yet have a homicide rate only one-third as high. Second, Americans may have a high homicide rate because they have a violent history that has bred a violent culture. He also rejects that argument, not because the conquest and settlement of the United States was not violent, but because other countries, including Germany and the United Kingdom, also have violent pasts yet boast low homicide rates today. Third is the argument that Americans have a high homicide rate because guns are so readily available in this country. Moore finds fault with that argument too. He says that Canadians, for example, have the same rate of firearm ownership as Americans but only one-third the homicide rate. Ultimately, we are left with a brilliant description of a problem but no clear explanation of its origins or solution.

The trouble is that Moore got some of his figures wrong. The rate of firearm ownership is actually more than twice as high in the United States as in Canada. About 17 percent of Canadian households versus 38 percent of American households have at least one firearm owner (Smith, 1999; Government of Canada, 2002). In general, a strong correlation exists between firearm ownership and homicide, not just crossnationally but within the United States. Thus, jurisdictions that have restricted firearm ownership in the United States have experienced an almost immediate decline in the homicide rate. For example, in 1976 the District of Columbia enacted a new gun control law that gave residents 60 days to register their firearms. Thereafter, newly acquired handguns became illegal if unregistered. Surrounding areas of Maryland and Virginia in the same metropolitan area did not enact the new gun control law. In the District of Columbia, gun-related homicides fell 25 percent between 1976 and 1985. In the surrounding areas of Maryland and Virginia, the number of gun-related homicides did not change significantly (Bogus, 1992). Available data on homicide point unmistakably to a smoking gun—and it is a smoking gun.

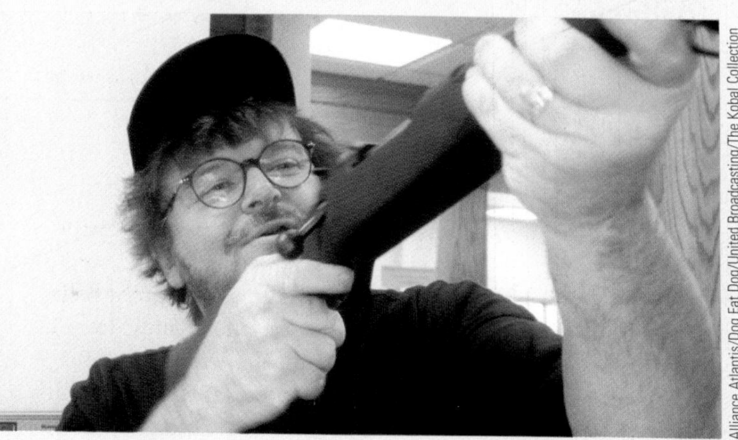

Michael Moore in *Bowling for Columbine.*

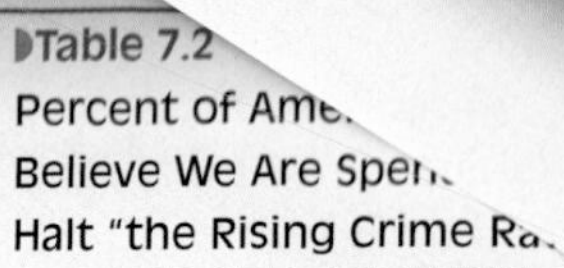

Box 7.3

YOU AND THE SOCIAL WO

Moral Panic

The 2002 General Social Survey presented American adults with the following question: "We are faced with many problems in this country, none of which can be solved easily or inexpensively. I'm going to name some of these problems, and for each one I'd like you to tell me whether you think we're spending too much money on it, too little money, or about the right amount. First, are we spending too much, too little, or about the right amount on halting the rising crime rate?"

▶Table 7.2 shows how respondents who believed that we were spending "too little" to halt the "rising" crime rate were distributed across several variables. Which categories of the population were most inclined to say that we were spending too little to halt crime? Why were these categories of the population so inclined? Can you make a case for the view that the people most in favor of spending more to halt crime were most tightly gripped by moral panic? Or is it more reasonable to conclude that it was the people who were most exposed to violent crime who were most inclined to want to spend more to halt it? You will find ▶Figure 7.8 helpful in thinking about these questions. How do your social characteristics and your exposure to violent crime affect *your* attitude toward spending more money to fight crime?

▶**Table 7.2**
Percent of Ame… Believe We Are Spen… Halt "the Rising Crime Ra…

	%	
Gender		
Male	52	589
Female	62	732
Highest year of schooling completed		
0–11	63	189
12	66	404
13+	51	726
Age		
18–29	50	259
30–39	62	279
40–49	57	271
50–59	62	193
60–69	60	141
701	57	154
Region		
New England	61	69
Middle Atlantic	58	204
South Atlantic	59	236
East North Central	58	218
West North Central	50	113
East South Central	62	99
West South Central	64	125
Mountain	48	88
Pacific	54	169
Total annual family income		
$0–49,999	58	722
$50,0001	56	550
Race/ethnicity		
White	55	1053
Black	69	182
Hispanic	60	99
Other	53	47

Source: National Opinion Research Center (2004).

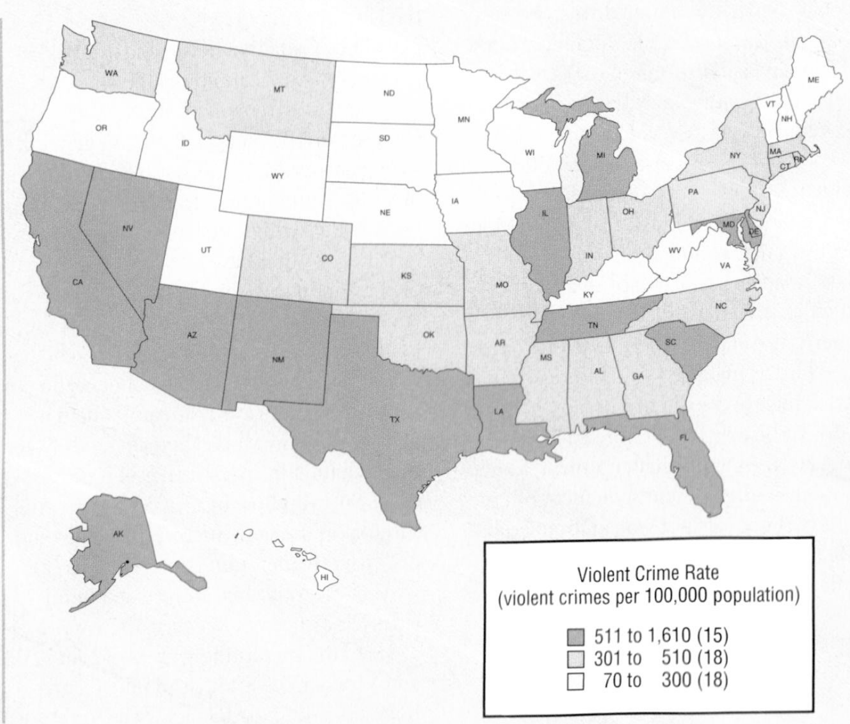

▶**Figure 7.8**
Violent Crime, United States, 2003

Source: U.S. Federal Bureau of Investigation (2003).

Are you part of the moral panic? Have you and your family taken special precautions to protect yourself from the "growing wave" of criminality in the United States? Even if you are not part of the moral panic, chances are you know someone who is (Box 7.3). Therefore, to put things in perspective, you will need to recall an important fact from our discussion of recent trends in crime rates: According to FBI statistics, the moral panic of

BOX 7.2
Sociology at the Movies

Bowling for Columbine (2002)

The homicide rate in the United States is 3 to 10 times higher than the homicide rate in the other 20 or so wealthy, highly industrialized countries. In 2001, 15,980 Americans were murdered, which is five times more than the number of people killed in the terrorist attacks on September 11 of that year. Firearms were used in nearly two-thirds of the murders committed in the United States in 2001 (U.S. Federal Bureau of Investigation, 2002).

In *Bowling for Columbine,* documentary filmmaker Michael Moore describes the magnitude of the problem and helps us figure out why Americans kill one another with such extraordinary frequency. Moore succeeds in his first, descriptive task by combining hilarity with horror. He takes us to a bank that gives away rifles instead of toasters to people who open an account. ("Don't you think it's a little dangerous to have all these guns in a bank?" he asks a teller.) We see security camera footage from the 1999 massacre at Columbine High School in Littleton, Colorado, where two students shot and killed 13 fellow students and then committed suicide. We listen while Moore interviews Charlton Heston, then president of the National Rifle Association (NRA). Heston cannot answer when Moore asks why a man who has never been personally threatened and who now lives behind a perimeter wall and a locked security gate in a protected neighborhood with security patrols feels the need to keep a loaded gun in the house. These and many other scenes, some tragic and others absurd, provide compelling evidence that guns are as American as apple pie. Not for nothing did the International Documentary Film Association name this film the best documentary ever made.

When it comes to *explaining* the unusually high American homicide rate, however, the film is less successful. Moore examines three main explanations and rejects them all. First, Americans may be violent because the mass media—TV, video games, movies, and popular music—are full of violence and influence young people in particular to engage in violent acts in the real world. Moore dismisses that argument on the grounds that Canadians are exposed to almost exactly the same mass media influences as Americans yet have a homicide rate only one-third as high. Second, Americans may have a high homicide rate because they have a violent history that has bred a violent culture. He also rejects that argument, not because the conquest and settlement of the United States was not violent, but because other countries, including Germany and the United Kingdom, also have violent pasts yet boast low homicide rates today. Third is the argument that Americans have a high homicide rate because guns are so readily available in this country. Moore finds fault with that argument too. He says that Canadians, for example, have the same rate of firearm ownership as Americans but only one-third the homicide rate. Ultimately, we are left with a brilliant description of a problem but no clear explanation of its origins or solution.

The trouble is that Moore got some of his figures wrong. The rate of firearm ownership is actually more than twice as high in the United States as in Canada. About 17 percent of Canadian households versus 38 percent of American households have at least one firearm owner (Smith, 1999; Government of Canada, 2002). In general, a strong correlation exists between firearm ownership and homicide, not just crossnationally but within the United States. Thus, jurisdictions that have restricted firearm ownership in the United States have experienced an almost immediate decline in the homicide rate. For example, in 1976 the District of Columbia enacted a new gun control law that gave residents 60 days to register their firearms. Thereafter, newly acquired handguns became illegal if unregistered. Surrounding areas of Maryland and Virginia in the same metropolitan area did not enact the new gun control law. In the District of Columbia, gun-related homicides fell 25 percent between 1976 and 1985. In the surrounding areas of Maryland and Virginia, the number of gun-related homicides did not change significantly (Bogus, 1992). Available data on homicide point unmistakably to a smoking gun—and it is a smoking gun.

Michael Moore in *Bowling for Columbine.*

Box 7.3

YOU AND THE SOCIAL WORLD

Moral Panic

The 2002 General Social Survey presented American adults with the following question: "We are faced with many problems in this country, none of which can be solved easily or inexpensively. I'm going to name some of these problems, and for each one I'd like you to tell me whether you think we're spending too much money on it, too little money, or about the right amount. First, are we spending too much, too little, or about the right amount on halting the rising crime rate?"

Table 7.2 shows how respondents who believed that we were spending "too little" to halt the "rising" crime rate were distributed across several variables. Which categories of the population were most inclined to say that we were spending too little to halt crime? Why were these categories of the population so inclined? Can you make a case for the view that the people most in favor of spending more to halt crime were most tightly gripped by moral panic? Or is it more reasonable to conclude that it was the people who were most exposed to violent crime who were most inclined to want to spend more to halt it? You will find Figure 7.8 helpful in thinking about these questions. How do your social characteristics and your exposure to violent crime affect *your* attitude toward spending more money to fight crime?

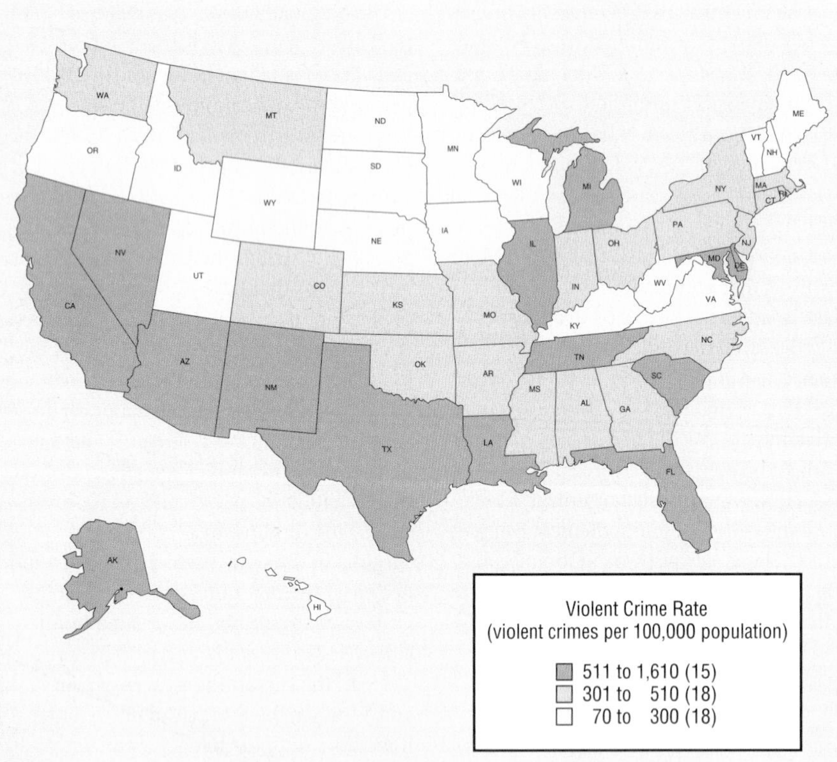

Figure 7.8
Violent Crime, United States, 2003

Source: U.S. Federal Bureau of Investigation (2003).

Table 7.2
Percent of American Adults Who Believe We Are Spending Too Little to Halt "the Rising Crime Rate," 2002

	%	*N*
Gender		
Male	52	589
Female	62	732
Highest year of schooling completed		
0–11	63	189
12	66	404
13+	51	726
Age		
18–29	50	259
30–39	62	279
40–49	57	271
50–59	62	193
60–69	60	141
701	57	154
Region		
New England	61	69
Middle Atlantic	58	204
South Atlantic	59	236
East North Central	58	218
West North Central	50	113
East South Central	62	99
West South Central	64	125
Mountain	48	88
Pacific	54	169
Total annual family income		
$0–49,999	58	722
$50,0001	56	550
Race/ethnicity		
White	55	1053
Black	69	182
Hispanic	60	99
Other	53	47

Source: National Opinion Research Center (2004).

Are you part of the moral panic? Have you and your family taken special precautions to protect yourself from the "growing wave" of criminality in the United States? Even if you are not part of the moral panic, chances are you know someone who is (Box 7.3). Therefore, to put things in perspective, you will need to recall an important fact from our discussion of recent trends in crime rates: According to FBI statistics, the moral panic of recent decades occurred during a period when all major crime indexes stabilized and then *decreased* quite dramatically.

Why then the panic? Who benefits from it? We may mention several interested parties. First, the mass media benefit from moral panic because it allows them to rake in hefty profits. They publicize every major crime because crime draws big audiences, and big audiences mean more revenue from advertisers. Fictional crime programs draw tens of millions of additional viewers to their TVs, as the statistics cited at the beginning of this chapter show. Second, the crime prevention and punishment industry benefits from moral panic for much the same reason. Prison construction and maintenance firms, firearms manufacturers, and so forth are all big businesses that flourish in a climate of moral panic. Such industries want Americans to own more guns and imprison more people, so they lobby hard in Washington and elsewhere for relaxed gun laws and invigorated prison construction programs. Third, some formerly depressed rural regions of the United States have become highly dependent on prison construction and maintenance for the economic well-being of their citizens. The Adirondack region of northern New York State is a case in point. Nearly 30 percent of the people who moved to upstate New York in the 1990s were prison inmates (Staples, 2004). Fourth, the criminal justice system is a huge bureaucracy with millions of employees. They benefit from moral panic because increased spending on crime prevention, control, and punishment secures their jobs and expands their turf. Finally, and perhaps most important, moral panic is useful politically. Since the early 1970s, many politicians have based entire careers on get-tough policies. Party allegiance and ideological orientation matter less than you might think here; plenty of liberal Republicans (such as former governor Nelson Rockefeller of New York) and Democrats (such as former governor Mario Cuomo of New York) have done as much to build up the prison system as have conservative Republicans (Schlosser, 1998).

Alternative Forms of Punishment

The two most contentious issues concerning the punishment of criminals are: (1) Should the death penalty be used to punish the most violent criminals? (2) Should less serious offenders be incarcerated in the kinds of prisons we now have? In concluding this chapter, let us briefly consider each of these issues.

Capital Punishment

Although the United States has often been at the forefront of the struggle for human rights, it is one of the few highly industrialized societies to retain capital punishment for the most serious criminal offenders.

Although the death penalty ranks high as a form of revenge, whether it serves as a deterrent is questionable for two reasons. First, murder is often committed in a rage, when the perpetrator is not thinking entirely rationally. In such circumstances the murderer is unlikely to coolly consider the costs and consequences of his or her actions. Second, if rational calculation of consequences does enter into the picture, the perpetrator is likely to know that very few murders result in the death sentence. More than 15,000 murders take place in the United States every year. Only about 175 death sentences are handed out. Thus, a murderer has about a 1 percent chance of being sentenced to death. The chance that he or she actually will be executed is even smaller.

Because the death penalty is not likely to deter many people unless the probability of its use is high, some people take these figures as justification for sentencing more violent offenders to death. However, one must remember that capital punishment as it is actually practiced is hardly a matter of blind justice. This fact is particularly evident if we consider the racial distribution of people who are sentenced to death and executed. Murdering a white person is much more likely to result in a death sentence than murdering a black person. For example, in Florida in the 1970s, an African American who killed a white person was 40 times more likely to receive the death penalty than an African American who killed another African American. Moreover, a white person who murders a black person very

Sister Helen Prejean is a Catholic nun from Louisiana. Since 1981 she has been one of the country's most outspoken critics of the death penalty. Her book *Dead Man Walking* reflects on her experience with inmates on death row and raises important questions about the death penalty. *Dead Man Walking* was made into a critically acclaimed film starring Sean Penn and Susan Sarandon in 1995. See also Prejean (2005).

The Everett Collection

rarely gets sentenced to death, but a black person who murders a white person is one of the types of people most likely to get the death penalty. Thus, of the 80 white people who murdered African Americans in Florida in the 1970s, not one was charged with a capital crime. In Texas, 1 out of 143 was charged with a capital crime (Haines, 1996; Tonry, 1995; Black, 1989). Given this patent racial bias, we cannot view the death penalty as a justly administered punishment.

Sometimes people favor capital punishment because it saves money. They argue that killing someone outright costs less than keeping the person alive in prison for the rest of his or her life. However, after trials and appeals, a typical execution costs the taxpayer up to six times *more* than a 40-year stay in a maximum-security prison (Haines, 1996).

Sociology Now™

Learn more about **Capital Punishment** by going through the Death Penalty Map Exercise.

Finally, in assessing capital punishment, one must remember that mistakes are common. Nearly 40 percent of death sentences since 1977 have been overturned because of new evidence or mistrial (Haines, 1996).

Incarcerating Less Serious Offenders in Violent, "No Frills" Prisons

Most of the increase in the prison population over the past 20 years is because of the conviction of nonviolent criminals. Many of them were involved in drug trafficking, and many of them are first-time offenders. The main rationale for imprisoning such offenders is that incarceration presumably deters them from repeating their offense. Supposedly, it also deters others from engaging in crime. Arguably then, the streets become safer by isolating criminals from society.

Unfortunately for the hypothesis that imprisoning more people lowers the crime rate, available data show a weak relationship between the two variables. True, between 1980 and 1986 the number of inmates in U.S. prisons increased 65 percent and the number of victims of violent crime decreased 16 percent, which is what one would expect to find if incarceration deterred crime. However, between 1986 and 1991, the prison population increased 51 percent and the number of victims of violent crime *increased* 15 percent—just the opposite of what one would expect to find if incarceration deterred crime. The same sort of inconsistency is evident if we examine the relationship between incarceration and crime across states. For example, in 1992 Oklahoma had a high incarceration rate and a low crime rate, whereas Mississippi had a low incarceration rate and a high crime rate. These cases fit the hypothesis that imprisonment lowers the crime rate. However,

Louisiana had a high incarceration rate and a high crime rate, whereas North Dakota had a low incarceration rate and a low crime rate, which is the opposite of what one would expect to find if incarceration deterred crime (Mauer, 1994). We can only conclude that, contrary to popular opinion, prison does not consistently deter criminals or lower the crime rate by keeping criminals off the streets.

However, prison often teaches inmates to behave more violently. The case of Robert Scully, who graduated from robbery to killing a police officer thanks to his experiences in the California prison system, is one example. Budgets for general education, job training, physical exercise, psychological counseling, and entertainment have been cut. Brutality in the form of solitary confinement, hard labor, and physical violence is increasing. The result is a prison population that is increasingly enraged, incapacitated, lacking in job skills, and more dangerous upon release than upon entry into the system. Massachusetts governor William F. Weld captured the spirit of the times when he said that prisons ought to be "a tour through the circles of hell" where inmates should learn only "the joys of busting rocks" (quoted in Abramsky, 1999). However, the new regime of U.S. prisons may have an effect just the opposite of that intended by Governor Weld. Between 1999 and 2010, an estimated 3.5 million first-time releases are expected from U.S. prisons (Abramsky, 1999). We may therefore be on the verge of a real crime wave, one that will have been created by the very get-tough policies that were intended to deter crime. Ominously, the homicide rate spiked upward in 2001. This increase resulted from the downturn in the economy and the rising number of inmates being released from state and federal prisons, which increased from 474,300 in 1995 to 635,000 in 2001. According to Sgt. John Pasquarello of the Los Angeles Police Department, "Prison is basically a place to learn crime, so when these guys come out, we see many of them getting back into drug operations, and this leads to fights and killings" (quoted in Butterfield, 2001).

Rehabilitation and Reintegration

Is there a reasonable alternative to the kinds of prisons we now have? Although saying so may be unpopular, anecdotal evidence suggests that institutions designed to rehabilitate criminals and reintegrate them into society can work, especially for less serious offenders. They also cost less than the kind of prison system we have created.

Those are the conclusions some people have drawn from experience at McKean, a medium-security correctional facility opened in Bradford, Pennsylvania, in 1989. Dennis Luther, the warden at McKean, is a maverick who has bucked the trend in American corrections. Nearly half the inmates at McKean are enrolled in classes, many of them earning licenses in masonry, carpentry, horticulture, barbering, cooking, and catering that will help them get jobs when they leave. Recreation facilities are abundant, and annual surveys conducted in the prison show that inmates get into less trouble the greater their involvement in athletics. The inmates run self-help groups and teach adult continuing education. Good behavior is rewarded. If a cellblock receives high scores for cleanliness and orderliness during weekly inspection, the inmates in the cellblock get special privileges, such as the use of TV and telephones in the evening. Inmates who consistently behave well are allowed to attend supervised picnics on Family Days, which helps them adjust to life on the outside. Inmates are treated with respect and are expected to take responsibility for their actions. For example, after a few minor incidents in 1992 Luther restricted inmates' evening activities. The restriction was meant to be permanent, but some inmates asked Luther if he would do away with the restriction provided the prison was incident free for 90 days. Luther agreed, and he has never had to reimpose the restrictions.

The effects of these policies are evident throughout McKean. The facility is clean and orderly. Inmates do not carry "shanks" (homemade knives). The per-inmate cost to taxpayers is below average for medium-security facilities and 28 percent lower than the average for all state prisons, partly because few guards are needed to maintain order. In McKean's first six years of operation, no escapes, homicides, sexual assaults, or suicides occurred. Inmates and staff members were victims of a few serious assaults, but the *annual* rate of assault at McKean is equal to the *weekly* rate of assault at other state prisons of about the same size. Senior staff members and a local parole officer claim that McKean inmates return to prison far less often than inmates of other institutions (Worth, 1995).

Thus, a cost-effective and workable alternative to the current prison regime may exist, at least for less serious offenders. Furthermore, some aspects of the McKean approach possibly could have beneficial effects in overcrowded, maximum-security prisons, where violent offenders are housed and gangs proliferate. Dennis Luther thinks so, but we do not really know because it has not been tried. Nor is it likely to be tried anytime soon given the current climate of public opinion.

Summary

Sociology Now™

Reviewing is as easy as ❶, ❷, ❸.

❶ Before you do your final review, take the SociologyNow diagnostic quiz to help you identify the areas on which you should concentrate. You will find information on how to access SociologyNow on the foldout at the front of the textbook.

❷ As you review, take advantage of SociologyNow's study aids to help you master the topics in this chapter.

❸ When you are finished with your review, take SociologyNow's post-test to confirm you are ready to move on to the next chapter.

1. **What are deviance and crime? What determines how serious a deviant or criminal act is?**

   Deviance involves breaking a norm. Crime involves breaking a law. Both crime and deviance evoke societal reactions that help define the seriousness of the rule-breaking incident. The seriousness of deviant and criminal acts depends on the severity of the societal response to them, their perceived harmfulness, and the degree of public agreement about whether they should be considered deviant or criminal. Acts that rank lowest on these three dimensions are called social diversions. Next come social deviations and then conflict crimes. Consensus crimes rank highest.

2. **Are definitions of deviance and crime the same everywhere and at all times?**

   No. Definitions of deviance and crime vary historically and culturally. These definitions are socially defined and constructed. They are not inherent in actions or the characteristics of individuals.

3. **In what sense is power a key element in defining deviance and crime?**

   Powerful groups are generally able to create norms and laws that suit their interests. Less powerful groups are usually unable to do so. For example, the increasing power of women has led to greater recognition of crimes committed against them. However, no similar increase has occurred in the prosecution of white-collar criminals because the distribution of power between classes has not changed much in recent decades.

4. **Where do crime statistics come from?**

   Crime statistics come from official sources, self-report surveys, and indirect measures. Each source has its strengths and weaknesses.

5. **How has the rate of crime changed in the United States over the past four decades?**

   A crime wave occurred in the 1960s and 1970s. The crime rate began to taper off in the 1980s and decreased substantially in the 1990s because of more policing, a smaller proportion of young men in the population, a booming economy, and perhaps a decline in the number of unwanted children resulting from the availability of abortion.

6. **Why do African Americans experience disproportionately high arrest, conviction, and incarceration rates?**

   African Americans experience these disproportionately high rates because of bias in the way crime statistics are collected, the low social standing of the African American community, and racial discrimination in the criminal justice system.

7. **What are the main types of theories of deviance and crime?**

   Theories of deviance and crime include motivational theories (i.e., strain theory, subcultural theory, and the theory of differential association) and constraint theories (i.e., labeling theory, control theory, and conflict theory). Different theories illuminate different aspects of the process by which people are motivated to break rules and become defined as rule breakers.

8. **Do all societies control their members in the same way?**

   All societies seek to ensure that their members obey norms and laws by imposing sanctions on rule breakers. However, the degree and form of social control vary historically and culturally. For example, although some sociologists say that social control is weaker and deviance is greater in industrial societies than in preindustrial societies, other sociologists note that in some respects social control is greater.

9. **How important is imprisonment as a form of punishment in modern industrial societies? What do prisons accomplish?**

   The prison is one of the most important forms of punishment in modern industrial societies. Since the 1980s the incarceration rate has increased in the United States. Prisons now focus less on rehabilitation than on isolating and incapacitating inmates.

10. **What is a "moral panic"?**

    A moral panic occurs when many people fervently believe that some form of deviance or crime poses a profound threat to society's well-being. For example, a moral panic about crime has engulfed the United States, although crime rates have been moderating in recent decades. In all aspects of crime prevention and punishment, most Americans have taken a "get-tough" stance. Some commercial and political groups benefit from the moral panic over crime and therefore encourage it.

11. **What are some of the problems with the death penalty as a form of punishment?**

    Although the death penalty ranks high as a form of revenge, its effectiveness as a deterrent is questionable. Moreover, the death penalty is administered in a racially biased manner, does not save money, and sometimes results in tragic mistakes.

12. **Does the rehabilitation of criminals ever work?**

    Rehabilitative correctional facilities are cost-effective. They do work, especially for less serious offenders. However, they are unlikely to become widespread given the current political climate.

Questions to Consider

1. Has this chapter changed your view of criminals and the criminal justice system? If so, how? If not, why?
2. Do you think different theories are useful in explaining different types of deviance and crime? Or do you think that one or two theories explain all types of deviance and crime, whereas other theories are not very illuminating? Justify your answer using logic and evidence.
3. Do TV crime shows and crime movies give a different picture of crime in the United States than this chapter gives? What are the major differences? Which picture do you think is more accurate? Why?

Web Resources

Companion Website for This Book

http://sociology.wadsworth.com

Begin by clicking on the Student Resources section of the website. Choose "Introduction to Sociology" and finally the Brym and Lie book cover. Next, select the chapter you are currently studying from the pull-down menu. From the Student Resources page you will have easy access to InfoTrac® College Edition, MicroCase Online exercises, additional web links, and many resources to aid you in your study of sociology, including practice tests for each chapter.

Recommended Websites

The FBI's website at http://www.fbi.gov is a rich resource on crime in the United States. For official statistics, click on "Uniform Crime Reports."

For a measure of how widespread corruption is in the governments of every country in the world, visit http://www.nationmaster.com/graph-T/gov_cor.

On the relationship between crime and the mass media, visit http://www.criminology.fsu.edu/cjlinks/media2.html.

You can find useful data on incarceration in the United States and globally at http://www.sentencingproject.org/pubs_02.cfm.

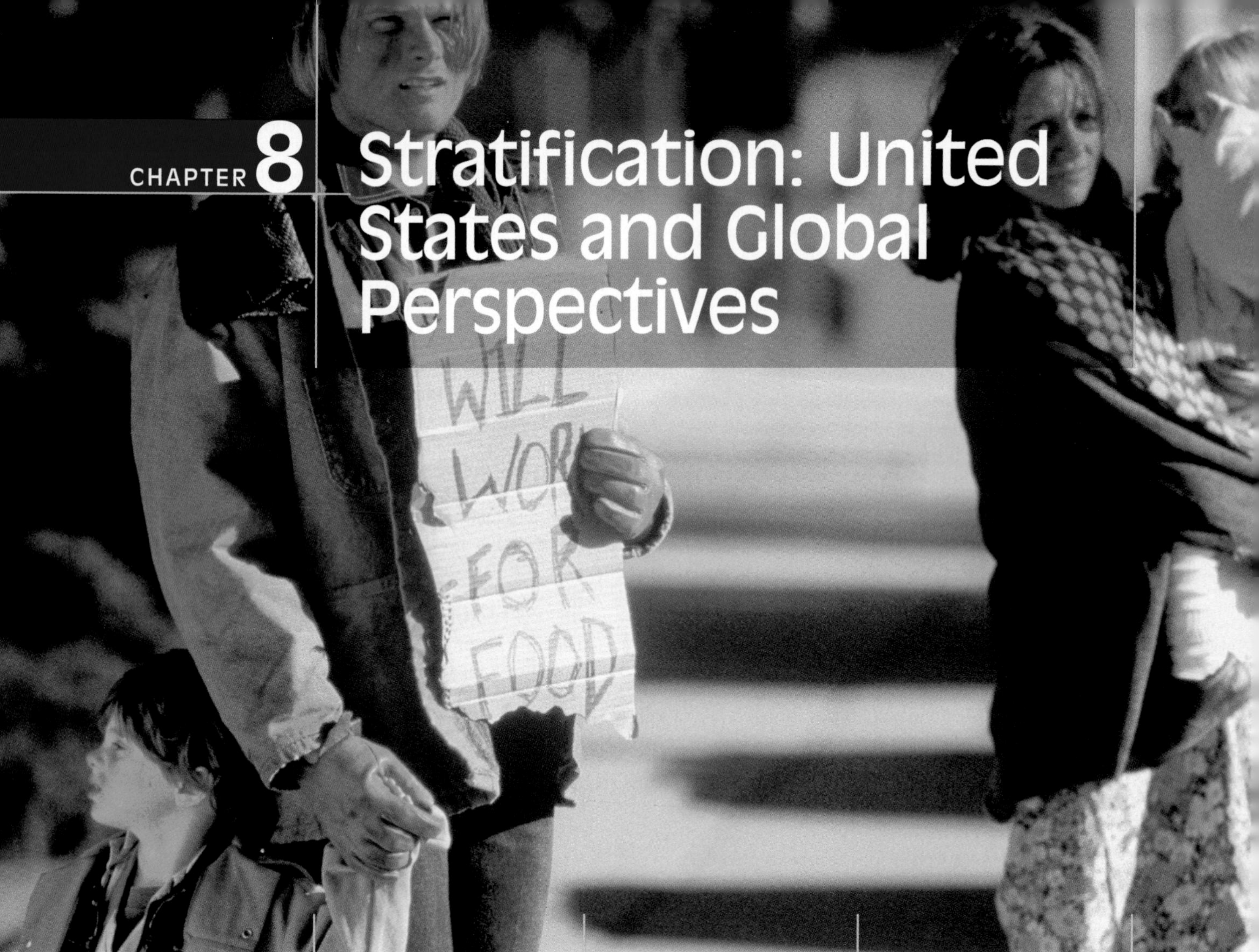

CHAPTER 8 Stratification: United States and Global Perspectives

Bruce Ayers/Getty Images

In this chapter, you will learn that:

- Income inequality in the United States has been increasing since the mid-1970s.
- Income inequality is higher in the United States than in any other postindustrial society.
- As societies develop, inequality increases at first. Then, after passing the early stage of industrialization, inequality in society declines. In the postindustrial stage of development, inequality appears to increase again.
- Most theories of social inequality focus on its economic roots.
- Prestige and power are important noneconomic sources of inequality.
- Some sociologists used to think that talent and hard work alone determine one's position in the socioeconomic hierarchy. Now, sociologists think that being a member of certain social categories limits one's opportunities for success. In this sense, social structure shapes the distribution of inequality.

Social Stratification: Shipwrecks and Inequality

Writers and filmmakers sometimes tell stories about shipwrecks and their survivors to make a point about social inequality. They use the shipwreck as a literary device. It allows them to sweep away all trace of privilege and social convention. What remains are human beings stripped to their essentials, guinea pigs in an imaginary laboratory for the study of wealth and poverty, power and powerlessness, esteem and disrespect.

The tradition began with Daniel Defoe's *Robinson Crusoe,* first published in 1719. Defoe tells the story of an Englishman marooned on a desert island. His strong will, hard work, and inventiveness turn the poor island into a thriving colony. Defoe was one of the first writers to portray capitalism favorably. He believed that people get rich if they possess the virtues of good businessmen—and stay poor if they do not.

The 1975 Italian movie *Swept Away* tells almost exactly the opposite story. (An inferior 2002 remake of the film starred Madonna.) In the movie, a beautiful woman, one of the idle rich, boards her yacht for a cruise in the Mediterranean. She treats the hardworking deckhands in a condescending and abrupt way. The deckhands do their jobs but seethe with resentment. Then comes the storm. The yacht is shipwrecked. Only the beautiful woman and one handsome deckhand remain alive, swept up on a desert island. Now equals, the two survivors soon have passionate sex and fall in love.

Sociology Now™

Reviewing is as easy as ❶, ❷, ❸.

Use SociologyNow to help you make the grade on your next exam. When you are finished reading this chapter, go to the Chapter Summary for instructions on how to make SociologyNow work for you.

The Granger Collection

Horatio Alger, Jr., wrote more than 100 books from the post–Civil War era until the end of the 19th century. He inspired Americans with tales of how courage, faith, honesty, hard work, and a little luck could help young people rise from rags to riches. Alger's novels lack appreciation of the social determinants of social stratification. Nonetheless, his ideas live on. Since 1947, they have been perpetuated by the Horatio Alger Association of Distinguished Americans. The Association honors the achievements of "outstanding individuals in our society who have succeeded in the face of adversity" according to its website (http://www.horatioalger.com). On April 6, 2001, President George W. Bush welcomed Horatio Alger National Scholars and Association Members to the White House.

All is well until the day of their rescue. As soon as they return to the mainland, the woman resumes her haughty ways. She turns her back on the deckhand, who is reduced again to the role of a common laborer. Thus, the movie sends the audience three harsh messages. First, it is possible to be rich without working hard because one can inherit wealth. Second, one can work hard without becoming rich. Third, something about the structure of society causes inequality, for inequality disappears only on the desert island, without society as we know it.

The most recent movie on the shipwreck-and-inequality theme is *Titanic.* At one level, the movie shows that class differences are important. For example, in first class, living conditions are luxurious, whereas in third class they are cramped. Indeed, on the *Titanic*, class differences spell the difference between life and death. After the *Titanic* strikes the iceberg off the coast of Newfoundland, the ship's crew prevents second- and third-class passengers from entering the few available lifeboats. They give priority to rescuing first-class passengers.

As the tragedy of the *Titanic* unfolds, however, another contradictory theme emerges. Under some circumstances, we learn, class differences can be insignificant. In the movie, the sinking of the *Titanic* is the backdrop to a fictional love story about a wealthy young woman in first class and a working-class youth in the decks below. The sinking of the *Titanic* and the collapse of its elaborate class structure give the young lovers an opportunity to cross class lines and profess their devotion to one another. At one level, then, *Titanic* is an optimistic tale that holds out hope for a society where class differences matter little, a society much like that of the American Dream.

Web Interactive Exercise: Are the Rich Getting Richer and the Poor Getting Poorer?

Robinson Crusoe, Swept Away, and *Titanic* raise many of the issues we address in this chapter. What are the sources of social inequality? Do determination, industry, and ingenuity shape the distribution of advantages and disadvantages in society, as *Robinson Crusoe* suggests? Or is *Swept Away* more accurate? Do certain patterns of social relations underlie and shape that distribution? Is *Titanic*'s first message still valid? Does social inequality still have big consequences for the way we live? What about *Titanic*'s second message? Can people act to decrease the level of inequality in society? If so, how?

Sociology Now™

Learn more about **Social Stratification** by going through the Social Stratification in the United States Data Experiment.

To answer these questions, we first sketch the pattern of social inequality in the United States and globally. We pay special attention to change over time. We then critically review the major theories of **social stratification,** the way society is organized in layers, or strata. We assess these theories in the light of logic and evidence. From time to time, we take a step back and identify issues that need to be resolved before we can achieve a more adequate understanding of social stratification, one of the fundamentally important aspects of social life.

Patterns of Social Inequality

Wealth

Social stratification refers to the way society is organized in layers, or strata.

How long would it take you to spend a million dollars? If you spent $1000 a day, it would take you nearly three years. How long would it take you to spend a *billion* dollars? If you spent $1000 a day, you couldn't spend the entire sum in a lifetime. It would take nearly 3000 years to spend a billion dollars at the rate of $1000 a day. Thus, a billion dollars is an almost unimaginably large sum of money. Yet in 1995 George Soros, an American hedge

Table 8.1
The 25 Richest Americans, 2004

Name	Net Worth ($ billion)	Source
1. Bill Gates	$48	Microsoft Corp.
2. Warren Buffet	$41	Berkshire Hathaway
3. Paul Allen	$20	Microsoft Corp.
4. Helen Walton	$18	Wal-Mart stores (inheritance)
4. John Walton	$18	Wal-Mart stores (inheritance)
4. Alice Walton	$18	Wal-Mart stores (inheritance)
4. S. Robson Walton	$18	Wal-Mart stores (inheritance)
4. Jim Walton	$18	Wal-Mart stores (inheritance)
9. Michael Dell	$14.2	Dell Computer Corp.
10. Lawrence Ellison	$13.7	Oracle Corp.
11. Steven Ballmer	$12.6	Microsoft Corp.
12. Abigail Johnson	$12	mutual funds (inheritance)
13. Barbara Cox Anthony	$11.3	Cox Enterprises (inheritance)
14. Anne Cox Chambers	$11.3	Cox Enterprises (inheritance)
15. John Kluge	$11	Metromedia Co.
16. Pierre Omidyar	$10.4	eBay
17. Jacqueline Mars	$10	Mars Inc. (inheritance)
17. John Franklin Mars	$10	Mars Inc. (inheritance)
17. Forrest Edward Mars Jr.	$10	Mars Inc. (inheritance)
20. Sumner Redstone	$8.1	Viacom Inc.
21. Carl Icahn	$7.6	leveraged buyouts
22. Philip H. Knight	$7.4	Nike
23. Charles Ergen	$7.3	EchoStar
24. George Soros	$7.2	hedge funds
25. Donald Edward Newhouse	$7.0	publishing (inheritance)

Note: Inherited family fortune: 48%. Women: 20%. Women who inherited family fortune: 100%.
Source: Forbes.com (2004).

fund manager and currency speculator, earned $1.5 billion. In contrast, the annual income of a full-time, minimum-wage worker was $8840. The earnings of George Soros in 1995 were enough to hire nearly 170,000 minimum-wage workers for a year.

George Soros is not the richest person in America. In 2004, he was in 24th place, with $7.2 billion in accumulated wealth. We list the 25 richest Americans in Table 8.1. Their net worth ranges from $7 billion to $48 billion.

Your wealth is what you own. For most adults, it includes a house (minus the mortgage), a car (minus the car loan), and some appliances, furniture, and savings (minus the credit card balance). Owning a nice house and a good car and having a substantial sum of money invested securely enhances your sense of well-being. You know you have a "cushion" to fall back on in difficult times; if you have children, you know you do not have to worry about paying for their college education; you will be able to make ends meet during retirement. Wealth can also give you more political influence. Campaign contributions to political parties and donations to favorite political causes increase the chance that policies you favor will become law. Wealth even improves your health. Because you can afford to engage in leisure pursuits, turn off stress, consume high-quality food, and employ superior medical services, you are likely to live a healthier and longer life than someone who lacks these advantages.

Unfortunately, sociologists and other social scientists have neglected the study of wealth, partly because reliable data on the subject are hard to come by. Americans are not required to report their wealth. Therefore, wealth figures are sparse and based mainly on a few sample surveys and analyses of people who pay estate tax. The best available estimates are, however, startling. In the mid-1990s, the richest 1 percent of American households owned nearly 39 percent of all national wealth, whereas the richest 10 percent owned almost 72 percent. In contrast, the poorest 40 percent of American households owned a meager 0.2 percent of all national wealth. The bottom 20 percent had a negative net worth, which means they owed more than they owned.

Patterns of Wealth Inequality

Wealth inequality has been increasing since the early 1980s. Some 62 percent of the increase in national wealth in the 1990s went to the richest 1 percent of Americans, and fully 99 percent of the increase went to the richest 20 percent. The United States has now surpassed all other highly industrialized countries in wealth inequality. Between 50 percent and 80 percent of the net worth of American families now derives from transfers and bequests, usually from parents (Hacker, 1997; Keister, 2000; Keister and Moller, 2000; Levy, 1998; Spilerman, 2000; Wolff, 1996 [1995]).

Wealth inequality is also significant because only a modest correlation exists between income and wealth. Some wealthy people have low annual income and some people with high annual income have little accumulated wealth. As such, annual income may not be the best measure of a person's well-being, and policies that seek to redistribute income from the wealthy to the poor, such as income-tax laws, may not get at the root of economic inequality because income redistribution has little effect on the distribution of wealth. Black-white inequality in wealth is especially stark, so income-based policies have the least effect on it (Oliver and Shapiro, 1995; Conley, 1999).

Income

Sociology Now™

Learn more about **Income** by going through the Median Household Money Income Map Exercise.

Your income is what you earn in a given period. In the United States and other societies, there is less inequality in income than in the distribution of wealth. Nonetheless, income inequality is steep. Thankfully, precise and detailed information on income inequality is readily available because individuals must report their income to the government, and sociologists have mined income figures well.

Students of social stratification often divide populations into categories of unequal size that differ in their lifestyle. These are often called "income classes." Table 8.2 shows how American households were divided into income classes in 2001.

Alternatively, sociologists divide populations into a number of equal-sized statistical categories, usually called "income strata." Figure 8.1 adopts this approach. It divides the country's households into five 20-percent income strata, from the top 20 percent of income earners down to the bottom 20 percent. It shows how total national income was divided among each of these fifths in 1974 and 2003. It also shows the share of national income that went to the top 5 percent of income earners in these two years.

Table 8.2
Income Classes, Households, United States, 2001

Income class	Percent of Households	Annual Household Income
Upper upper	1.0	$1 million+
Lower upper	12.4	$100,000–$999,999
Upper middle	22.5	$57,500–$99,999
Average middle	18.8	$37,500–$57,499
Lower middle or working	22.7	$20,000–$37,499
Lower	22.6	$0–$19,999
Total	100.0	

Note: The U.S. Census Bureau does not provide breakdowns of incomes that are more than $100,000. Therefore, we estimated the breakpoint between the top two classes.
Source: U. S. Census Bureau (2002b).

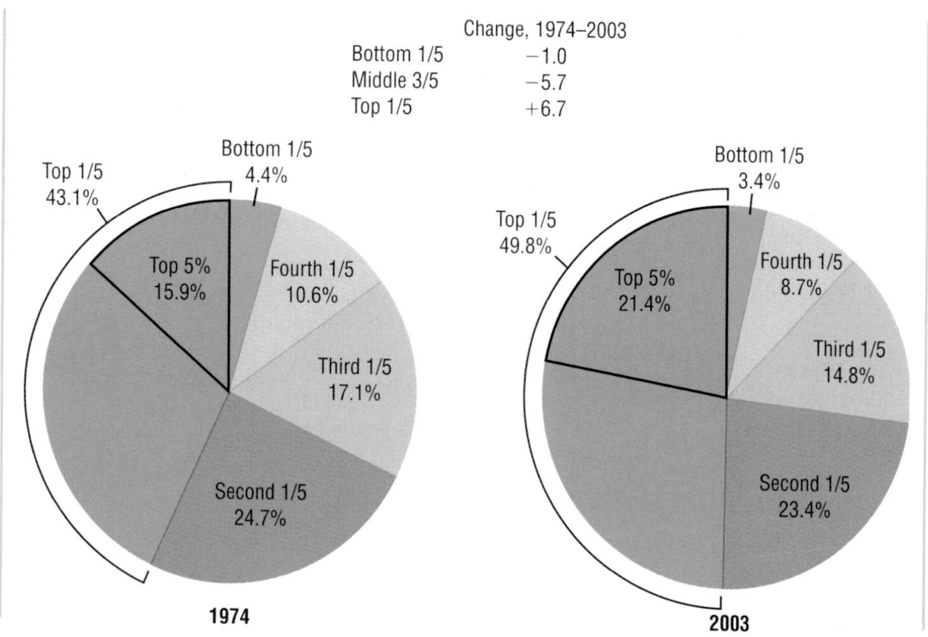

Figure 8.1
The Distribution of National Income Among Households, United States, 1974 and 2003

Source: U.S. Census Bureau (2004a).

Figure 8.1 illustrates three important facts. First, income inequality has been increasing in the United States for three decades. In 1974 the top fifth of households earned 9.8 times more than the bottom fifth. By 2003, the top fifth of households earned 14.6 times more than the bottom fifth. Second, in 2003 the top 20 percent of households earned nearly as much as the remaining 80 percent—about half of all national income. Third, the middle 60 percent of income earners have been "squeezed" during the past three decades, with their share of national income falling from 52.5 percent to 46.9 percent of the total. We conclude that for more than a quarter of a century, the rich have been getting relatively richer, whereas middle-income earners and the poor have been getting relatively poorer in the United States. (Note also that average earnings adjusted for inflation, or "average purchasing power," rose from the 1820s to the early 1970s and started declining after that [Henwood, 1999]).

What do these statistics mean? How do they reflect the everyday lives of real men and women? To flesh out the numbers, we now present brief sketches of ordinary people, most of them living in Silicon Valley (Santa Clara County), California. We do not focus on Silicon Valley because it is typical. It is not. It is, however, the center of the U.S. computer industry, the heart of America's most recent economic boom, and an extreme version of what has been happening to patterns of social stratification in the United States as a whole. According to some analysts, it indicates how social stratification may develop in this country in coming decades. In particular, Silicon Valley illustrates well the growing gap between the "haves" and the "have-nots" in American society.

Income Classes

Sociologists often divide society's upper class into two categories, the "upper-upper class" and the "lower-upper class" (see Table 8.2). The upper-upper class, comprising less than 1 percent of the U.S. population, used to be described as "old money" because people in that class inherited most of their wealth. Moreover, most of it was originally earned in older industries such as banking, insurance, oil, real estate, and automobiles. Old money inhabited elite neighborhoods on the East Coast—places like Manhattan and Westchester Counties (New York); Fairfield, Somerset, and Bergen Counties (Connecticut); and Arlington (Virginia). It still does. Members of this class send their children to expensive private schools and high-prestige colleges. They belong to exclusive private clubs. They are overwhelmingly white and non-Hispanic. They live in a different world from most Americans (Baltzell, 1964).

In the past couple of decades, however, and especially in the 1990s, a substantial amount of "new money" entered the upper-upper class. Booming high-tech industries created new opportunities for entry. Among those with new money, wealth is based less on inheritance than talent. New money is concentrated in high-tech meccas in the West—in places like San Francisco, San Mateo County, Santa Clara County (all in California), and King County (Washington) (Whitman, 2000). Larry Ellison, the flamboyant CEO of Oracle Corporation, the world's leading supplier of information management software, is perhaps the outstanding example of the new breed. The adopted son of a Chicago couple of modest means, Ellison started Oracle in 1977 with $1200 after dropping out of college. In 2004, his personal net worth was $13.7 billion and *Forbes* magazine ranked him the 10th richest person in the United States. Such success stories notwithstanding, one thing remains constant: New members of the upper-upper class are still overwhelmingly white and non-Hispanic (Rothman and Black, 1998).

Web Research Project: The Super Rich

How Green Is the Valley?

Let us take a moment to consider the history of new money in Santa Clara County, California, popularly known as Silicon Valley. John Doerr, an appropriately named venture capitalist in Menlo Park, California, has called Silicon Valley "the largest legal creation of wealth in the history of the planet" (quoted in Goodell, 1999: 65). He may be right. Fertilized by defense spending, Stanford University in Palo Alto spawned the region's electronics industry in the 1950s. By 1980, electronics had replaced prunes and apricots as the area's main product. Economic growth skyrocketed with the spread of the personal computer in the 1980s and the growing popularity of the Internet in the 1990s. High-tech industry in the Valley diversified and soon included not just microchip, computer, and software manufacturers, but telecommunications and genetic engineering firms, too.

By 1999, Santa Clara County was the home of 13 billionaires, several hundred people worth $25 million or more each, and 17,000 people worth more than $1 million each. These figures exclude the value of people's homes, which is no trifle in Santa Clara County, where the median single-family detached house price was $555,000 in 2000. Home prices eased in the wake of the economic downturn that began in 2000, but they reached $582,000 in 2004, more than triple the national average (Bernstein, 2000; Federal Reserve Bank of San Francisco, 2000; McCallister, 2003; 2005; Richtel, 2002; "Santa Clara County," 2002; Stacey, 1991: 20–6).

Fueled by high-tech industries, the 1990s witnessed the longest economic boom in American history. The mass media trumpeted the many success stories of that remarkable decade. Often submerged beneath the good news, however, was a more sobering reality: Although high-tech industries helped to change patterns of social stratification in the United States, the changes were not always positive.

Reuters NewMedia, Inc./Corbis

Reuters NewMedia, Inc./Corbis

Bill Gates, the world's richest man, lives in a house with more than 66,000 square feet of floor space. It is valued at more than $53 million.

The "Poor Rich"

The high-tech industry of the 1990s helped to create a new division at the bottom of the lower-upper class—the "poor rich." Comprising more than 12 percent of the U.S. population, members of the lower-upper class earn between $100,000 and $999,999 a year (see Table 8.2), yet are "struggling" to get by. They rely mainly on earnings, not inheritance, for their wealth and income. They are typically highly educated. They work as entrepreneurs, physicians, corporate lawyers, computer engineers, and so forth. In the 1990s many of these people grew even richer from their high-tech stock market investments. Their ranks swelled because of the growth of high-tech industries.

If you find it difficult to understand the predicament of the poor rich, consider Liliana and Peter Townshend (Goodell, 1999). They aspire to be among the millionaires of Silicon Valley, yet they are far from their goal. Peter, 28, is a lawyer in one of the top law firms in the Valley. He earned $120,000 in 1999. Liliana, 27, was starting her own e-commerce business. They bought their new 3500-square-foot home in 1998 for $720,000. Yet, when you step inside the house, the first thing you notice is that it is bare. A few scraps of furniture—hand-me-downs and items picked up at yard sales—dot the house. Their only luxuries are big new TVs on each floor. The Townshends regularly work 12 to 15 hours a day. Peter has been hospitalized twice for exhaustion. Liliana often wakes up at 3 a.m. in a panic that her business will fail. Peter is out of shape, he feels guilty about not making time to visit his parents, and Liliana complains about the marriage because she and Peter hardly ever see each other. They have two big mortgages on their house, and Liliana still owes $60,000 in student loans. Liliana buys food at Costco discount stores. When she visits her family in Los Angeles, she loads her car with groceries because food is so expensive in the Valley. "I feel like the poorest of the poor in Silicon Valley," says Liliana.

Like many other poor rich, Liliana and Peter Townshend struggle despite their high family income because they have assumed a mountain of debt and live in an area where the cost of living is very high. The Silicon Valley economic boom flooded the area with new residents who drove up demand for housing and just about everything else. Rent for a studio apartment in a bad neighborhood is $1000 a month. Gasoline costs almost 40 percent more than in the rest of the United States. Things got worse during the economic downturn that began in 2000. Some 127,000 jobs were lost between the first quarter of 2001 and the second quarter of 2002, most of them in the computer and communications sector (Fisher, 2003; Markoff and Richtel, 2002).

The Middle Class and Downward Mobility

The high-tech industry boom squeezed the middle class and encouraged downward mobility. The "middle class" consists of the nearly 65 percent of American households that earn more than $20,000 but less than $100,000 a year. Conventionally, the middle class is divided into roughly equal thirds: the "upper middle class," the "average middle class," and the "lower middle class," or "working class" (see Table 8.2).

Over a lifetime, an individual may experience considerable movement up or down the stratification system. Sociologists call this movement **vertical social mobility.** Movement up the stratification system (upward mobility) is a constant theme in American literature and lore. Much sociological research has been conducted on the subject (Box 8.1). Sociologists have studied movement down the stratification system (downward mobility) less often. Since the early 1960s, when sociologists started measuring social mobility reliably in the United States as a whole, more upward than downward mobility has occurred. However, in the early 1980s the gap between upward and downward mobility started to shrink as about a quarter of Americans reported deterioration in their economic situation (Hauser et al., 2000; Hout, 1988; Newman, 1988: 7, 21). In general, downward mobility increases during periods of economic recession and especially during periods of economic restructuring, such as the United States experienced in the 1980s and early 1990s. In those years, layoffs and plant closings were common as computerized production became widespread and well-paying manufacturing jobs in steel, autos, and other industries were lost to low-wage countries such as Mexico and China. At

Vertical social mobility refers to movement up or down the stratification system.

BOX 8.1
Sociology at the Movies

Sweet Home Alabama (2002)

Sweet Home Alabama is a Cinderella story with a twist: The successful heroine from humble beginnings gets the handsome prince but is not sure he is truly what she wants.

In the seven years since Melanie Carmichael (Reese Witherspoon) left her small-town Alabama home, she has achieved impressive upward social mobility. Beginning as a daughter of the working class, she has become a world-famous fashion designer in New York City. As the film begins, the mayor's son is courting Melanie. Andrew (Patrick Dempsey) proposes to her in Tiffany's, the upscale jewelry store that epitomizes upper-class consumption in the popular imagination. She says yes, but before she can marry him she has to clear up a not-so-minor detail: She needs a divorce from Jake (Josh Lucas), the childhood sweetheart she left behind.

Most of the story unfolds back in rural Alabama, in a town where friends climb the local water tower to drink beer and watch the folks pass by below, where major social events include Civil War reenactments and catfish festivals, and where special hospitality is shown by offering guests hot pickles "right out of the grease." Melanie finds herself caught between two classes and two subcultures, and the film follows her struggle to reconcile her conflicting identities. Her dilemma will require her to acknowledge and reconnect with her mother (Mary Kay Place), who lives in a trailer park, while standing up to her future mother-in-law, the mayor of New York City (Candice Bergen).

In the end, Melanie returns to Jake, while Andrew, briefly heartbroken, pleases his mother by marrying a woman of his own class. Melanie's homecoming does not, however, require that she return to life in a trailer park. She discovers that while she was in New York, Jake transformed himself. The working-class "loser" built a successful business as a glassblower. This change allows Melanie to imagine an upwardly mobile future by Jake's side.

Touchstone/The Kobal Collection/Ovino, Peter

Patrick Dempsey and Reese Witherspoon in *Sweet Home Alabama* (2002).

Sweet Home Alabama sends the message that people are happiest when they marry within their own class and subculture. That message is comforting because it helps the audience reconcile itself to two realities. First, although many people may want to "marry up," most Americans do not in fact succeed in doing so. We tend to marry within our own class (Kalmijn, 1998: 406–8). Second, marrying outside your class and subculture is likely to be unsettling insofar as it involves abandoning old norms, roles, and values and learning new ones. It is therefore in some sense a relief to learn you're better off marrying within your own class and subculture, especially since you will probably wind up doing just that.

There is an ideological problem with this message, however. Staying put in your own class and subculture denies the American Dream of upward mobility. *Sweet Home Alabama* resolves the problem by holding out the promise of upward mobility without having to leave home, as it were. Melanie and the transformed Jake can enjoy the best of both worlds, moving up the social hierarchy together without forsaking the community and the subculture they cherish. *Sweet Home Alabama* achieves a happy ending by denying the often difficult process of adapting to a new subculture as one experiences social mobility.

the same time, computer technology and office reorganization allowed companies to fire many of their middle managers (see Chapter 13, "Work and the Economy"). Not even the computer industry was immune.

Take the case of David Patterson (Newman, 1988: 1–7). After growing up in the slums of Philadelphia, he managed to earn a business degree and land a good managerial job in California's thriving computer industry in the 1970s. His company transferred him to New York to take a more important executive job in the early 1980s. Two years later, he was fired in the midst of an industrywide shakedown. After nine months of failing to find work, he and his wife were forced to sell their house and move to a modest apartment in a nearby town. Their two teenage children grew furious with David. His wife began to express subtle doubts about his desire to find a new job. The family's upper-middle-class friends gradually stopped calling. When he listened to the news, David heard about all the plant closings and business restructuring taking place across the United States. He knew there were good economic reasons for his plight. Nevertheless, as the months wore on, he grew depressed and started to ask: What's wrong with me? What have I done wrong? Like many people who lose their jobs, he forgot about the ups and downs of the nation's economy and blamed himself for his fate.

In Silicon Valley, even people who are employed in solid, middle-class jobs feel squeezed. In 1999, Dan Hingle, 35, was a quality-assurance engineer at a start-up company. He earned $50,000 a year but could afford to live only in a trailer park. "With $200,000, you can begin to approach a middle-class life," says Hingle. "How many people have jobs that pay $200,000? Not many. So people move out of the area, to where they can afford to buy a house, and commute an hour or two to work every day. That's fine, but then you're spending three or four hours on the road, and it's real easy to start hating life" (quoted in Goodell, 1999). In Los Altos, starting pay for police officers is around $40,000 a year. Only one of the town's 33 police officers can afford to live in Los Altos. In neighboring Los Gatos, 42 of 45 officers live out of town. In Palo Alto, the comparable figure is 0 of 95. Officer Thomas Joy drives into Los Altos once a week, sleeps on his mother's couch for four nights, then drives home to his family. For the most part, teachers, nurses—all the people needed to run essential services in Silicon Valley—feel the same sort of squeeze.

The New Poor

High-tech industry lowered the value of unskilled work, swelling the ranks of the poor. Thad Wingate earned almost $29,000 in 1999 as a driver for a Silicon Valley courier service. By national standards, that made him a member of the working, or lower middle, class. By Silicon Valley standards, however, that pushed him into the lower class. (Nationally, lower-class households earn less than $20,000 a year.) In fact, Wingate lived in a shelter for homeless people in San Jose. In addition to the ex-convicts, recovering alcoholics, and mentally ill people, other residents of the shelter included a male nurse, a middle-aged trucker, a Puerto Rican woman in a McDonald's uniform, and other people who work full time but are poor. San Jose's shelters and soup kitchens are booming.

Unskilled workers who are employed by big Silicon Valley corporations are victims of two increasingly popular corporate strategies: subcontracting and outsourcing (Bernstein, 2000). These strategies involve big corporations hiring smaller companies that in turn hire and supervise staff to do many of the manufacturing, janitorial, secretarial, and other routine jobs. Big corporations favor this approach because it allows them to easily hire and fire workers as circumstances dictate. Some corporations like this approach because it allows them to rely on subcontractors to keep wages down without getting involved in messy labor disputes. The 5500 unionized janitors in Silicon Valley are mainly Mexican immigrants who earned at most about $18,000 in 2000, only about $1000 above the official poverty line for a family of four (Reed, 2000). To afford food and shelter, many of them work second jobs and bunch into tiny apartments with friends and relatives. Twenty-two-year-old Alfredo Morales is a janitor who sleeps in an iron-casting shop in San Jose. He sends most of his wages back to his family in Mexico. He plans on moving back to Mexico as soon as he acquires some computer skills after hours. "I wouldn't bring my family here," he says. "It would be bringing them to greater misery" (Avery, 2000).

In sum, high-tech industry in Silicon Valley has encouraged much upward mobility. It has given a big boost to median income. However, it has also pushed up the cost of living and widened the gulf between the very rich and just about everyone else. In Silicon Valley, the middle class finds its standard of living deteriorating, and the lower class is down and out. Silicon Valley is admittedly an extreme case. However, as our statistics on the distribution of national income show, growing income inequality is a countrywide

Saturday Night, June 3, 2000. "The War Next Door," p. 48 story and photographs by Ami Vitale

Saturday Night, June 3, 2000. "The War Next Door," p. 48 story and photographs by Ami Vitale

Inequality on a global scale. Left: An amputated Angolan boy, victim of a land mine, hops home in Luanda. Right: U.S. and Canadian children relax at home in Luanda's North American enclave, built by Exxon for its supervisors and executives and their families.

story that has been developing for more than 30 years. The stories sketched in this section put a human face on the statistics. Moreover, they suggest what may lie ahead as high-tech industries come to dominate the American economy.

Global Inequality

International Differences

Just as income and wealth vary widely *within* countries, so do they vary *between* countries. For example, the United States is one of the richest countries in the world. Angola is a world apart. Angola is an African country of 11 million people. About 650,000 of its citizens have been killed in a civil war that has been raging since 1975, when the country gained independence from Portugal. Angola is one of the poorest nations on earth. Most Angolans live in houses made of cardboard, tin, and cement blocks. Most of these houses lack running water. The average income is about $1000 a year. There is one telephone for every 140 people and one TV for every 220 people. Inflation runs at about 90 percent per year. Adding to the misery of Angola's citizens are the millions of land mines that lay scattered throughout the countryside, regularly killing and maiming innocent passersby. Approximately 85 percent of the population survive on subsistence agriculture.

Sociology Now™

Learn more about **International Differences** by going through the Theories of Global Stratification Learning Module.

Angola itself is a highly stratified society, however. In fact, the gap between rich and poor is much wider than in the United States because multinational companies such as Exxon and Chevron drill for oil in Angola. Oil exports account for nearly half the country's wealth. In the coastal capital of Luanda, an enclave of North Americans who work for Exxon and Chevron live in gated and heavily guarded communities containing luxury homes, tennis courts, swimming pools, maids, and sport utility vehicles (SUVs). Here, side by side in the city of Luanda, people form a gulf that is as large as that separating Angola from the United States.

Some countries, like the United States, are rich. Others, like Angola, are poor. When sociologists study such differences *between* countries, they are studying **global inequality.** However, it is possible for country A and country B to be equally rich, whereas inside country A, the gap between rich and poor is greater than inside country B. When sociologists study such differences *within* countries, they are studying **crossnational variations in internal stratification.**

Global inequality refers to differences in the economic ranking of countries.

Crossnational variations in internal stratification are differences between countries in their stratification systems.

Consider global inequality for a moment. The United States, Canada, Japan, Australia, and a dozen or so other European countries including Germany, France, and the United Kingdom are the world's richest postindustrial societies. The world's poorest countries cover much of Africa, Latin America and the Caribbean, and Asia. Inequality between rich and poor countries is staggering. Nearly one-fifth of the world's population

Table 8.3
United Nations Indicators of Human Development, Top 10 and Bottom 10 Countries, 2002

Country and Overall Rank	Life Expectancy (years)	Adult Literacy (percent)	Gross Domestic Product per Capita (PPP $ US)
1. Norway	78.9	99.0	36,600
2. Sweden	80.0	99.0	26,050
3. Australia	79.1	99.0	28,260
4. Canada	79.3	99.0	29,480
5. Netherlands	78.3	99.0	29,100
6. Belgium	78.7	99.0	27,570
7. Iceland	79.7	99.0	29,750
8. United States	77.0	99.0	35,750
9. Japan	81.5	99.0	26,940
10. Ireland	76.9	99.0	36,360
168. Democratic Republic of the Congo	41.4	62.7	650
169. Central African Republic	39.8	48.6	1,170
170. Ethiopia	45.5	41.5	780
171. Mozambique	38.5	46.5	1,050
172. Guinea-Bissau	45.2	39.4	710
173. Burundi	40.8	50.4	630
174. Mali	48.5	19.0	930
175. Burkina Faso	45.8	12.8	1,100
176. Niger	46.0	17.1	800
177. Sierra Leone	34.3	36.0	520
World	66.9	n.a.	7,804

Note: PPP = purchasing power parity; n.a. = not available.
Source: United Nations (2004: 139–42).

lacks adequate shelter, and more than one-fifth lacks safe water. About one-third of the world's people are without electricity and more than two-fifths lack adequate sanitation. In the United States in 2002, there are 646 wired phone lines and 488 cell phones for every 1000 people, but in Angola there are 6 wired phone lines and 9 cell phones for every 1000 people. In 2002, annual health expenditure was $4887 in the United States, $70 in Angola (United Nations, 2004). People living in poor countries are also more likely than people in rich countries to experience extreme suffering on a mass scale. For example, because of political turmoil in many poor countries, an estimated 20 to 22 million people have been driven from their homes by force in recent years (Hampton, 1998). There are still about 27 million slaves in Mozambique, Sudan, and other African countries (Bales, 1999; 2002) (Table 8.3).

Sociology Now™

Learn more about **Global Inequality** by going through the Global Stratification Data Experiment.

We devote much of Chapter 9 ("Globalization, Inequality, and Development") to analyzing the causes, dimensions, and consequences of global inequality. You will learn that much of the wealth of the rich countries has been gained at the expense of the poor countries. Specifically, beginning centuries ago, the rich European countries turned large parts of the world into colonies that were used both as captive markets and as sources of cheap labor and raw materials. Even after the colonies gained political independence in the 20th century, most of them remained economically dependent on the rich countries. This helped to keep them poor. In Chapter 9, we also analyze the unique features of the few former colonies that have been able to escape poverty. Our analysis leads us to suggest possible ways of dealing with world poverty, one of the most vexing social issues. For the moment, having merely described the problem of global inequality, we focus on a second type of international difference: crossnational variations in internal stratification.

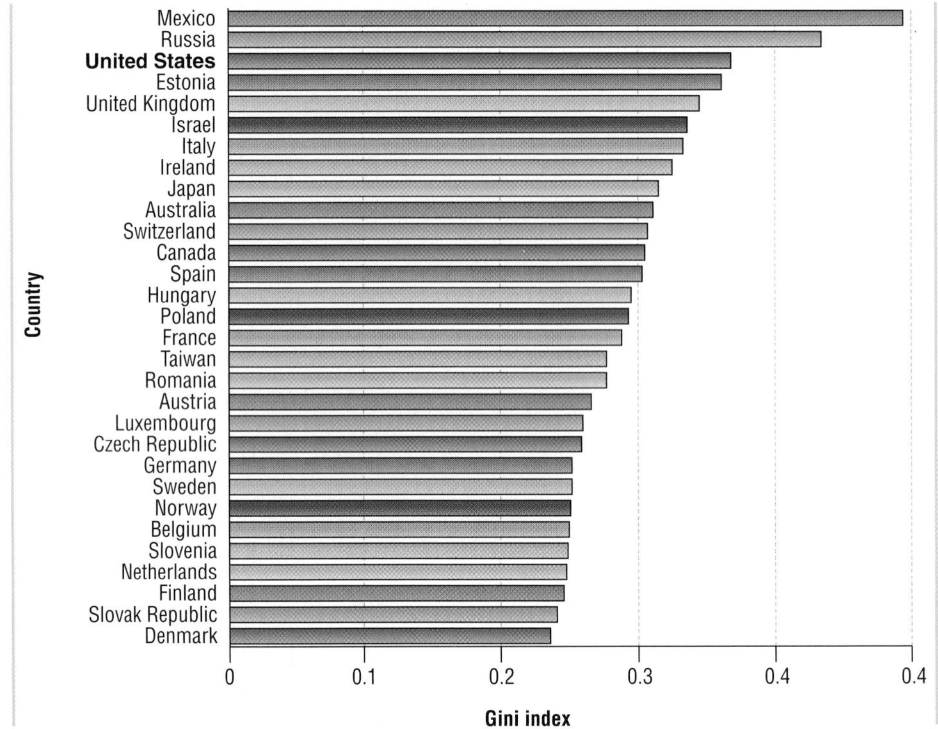

Figure 8.2
Household Income Inequality, 30 Countries, circa 2000

Source: Smeeding (2004).

Measuring Internal Stratification

Levels of global inequality aside, how does internal stratification differ from one country to the next? We can answer this question by first examining the **Gini index,** named after the Italian economist who invented it. The Gini index is a measure of income inequality. Its value ranges from 0 to 1. A Gini index of 0 indicates that every household in the country earns exactly the same amount of money. At the opposite pole, a Gini index of 1 indicates that a single household earns the entire national income. These are theoretical extremes. In the real world, most countries have Gini indexes between 0.2 and 0.4.

Figure 8.2 shows the Gini index using the most recent crossnational income data available for 30 countries. Of the 30 countries, Denmark has the lowest Gini index (0.236), whereas the United States has the highest Gini index among highly developed, rich countries (0.368). Among the 30 countries, only Russia and Mexico have a level of income inequality higher than that of the United States.

Economic Development

What accounts for crossnational differences in internal stratification, such as those described in the previous section? Later in this chapter, you will learn that *political factors* explain some of the differences. For the moment, however, we focus on how *socioeconomic development* affects internal stratification.

You will recall from Chapter 6 ("Social Collectivities: From Groups to Societies") that over the course of human history, as societies became richer and more complex, the level of social inequality changed. Let us recap these changes now (Lenski, 1966; Lenski, Nolan, and Lenski, 1995; Figure 8.3).

The **Gini index** is a measure of income inequality. Its value ranges from 0 (which means that every household earns exactly the same amount of money) to 1 (which means that all income is earned by a single household).

Foraging Societies

For the first 90,000 years of human existence, people lived in nomadic bands of fewer than 100 people. To survive, they hunted wild animals and foraged for wild edible plants. Some foragers and hunters were undoubtedly more skilled than others, but they did not hoard food. Instead, they shared food to ensure the survival of all band members. They

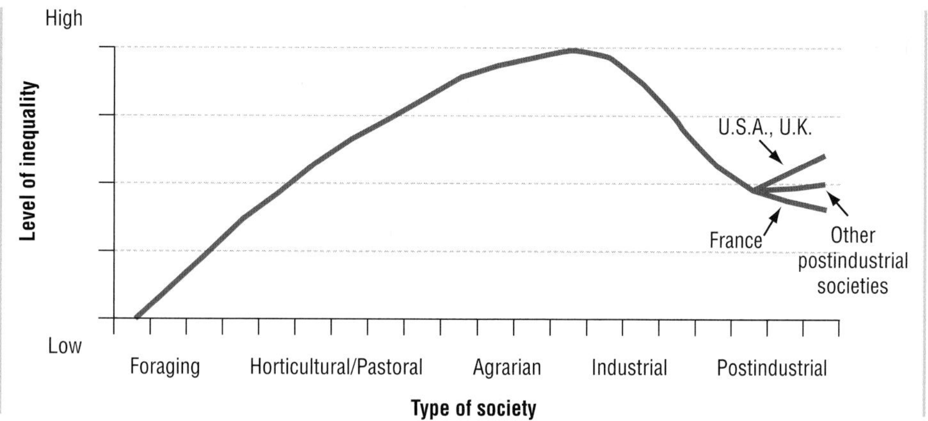

Figure 8.3
Inequality and Development

produced little or nothing above what they required for subsistence. There were no rich and no poor.

Horticultural and Pastoral Societies

About 10,000 years ago, people established the first agricultural settlements. These settlements were based on horticulture (the use of small hand tools to cultivate plants) and pastoralism (the domestication of animals). These technological innovations enabled people to produce wealth, a surplus above what they needed for subsistence. A small number of villagers controlled the surplus. Thus, significant social stratification emerged.

Agrarian Societies

About 5000 years ago, people developed plow agriculture. By attaching oxen and other large animals to plows, farmers could increase the amount they produced. Again thanks to technological innovation, surpluses grew. With more wealth came still sharper social stratification.

Agrarian societies developed religious beliefs justifying steeper inequality. People came to believe that kings and queens ruled by "divine right." They viewed large landowners as "lords." Moreover, if you were born a peasant, you and your children were likely to remain peasants. If you were born a lord, you and your children were likely to remain lords. In the vocabulary of modern sociology, we say that stratification in agrarian societies was based more on **ascription** than **achievement.** That is, a person's position in the stratification system was determined more by the features he or she was born with ("ascribed characteristics") than his or her accomplishments ("achieved characteristics"). Another way of saying this is that little social mobility took place.

A nearly purely ascriptive society existed in agrarian India. Society was divided into **castes,** four main groups and many subgroups arranged in a rigid hierarchy. Being born into a particular caste meant you had to work in the distinctive occupations reserved for that caste and marry someone from the same or an adjoining caste. The Hindu religion strictly reinforced the system (Srinivas, 1952). For example, Hinduism explained people's place in the caste system by their deeds in a previous life. If you were good, you were presumably rewarded by being born into a higher caste in your next life. If you were bad, you were presumably punished by being born into a lower caste. Belief in the sanctity of caste regulated even the most mundane aspects of life. Thus, someone from the lowest caste could dig a well for a member of the highest caste, but once the well was dug, the well digger could not so much as cast his shadow on the well. If he did, the well was considered polluted and upper-caste people were forbidden to drink from it.

Caste systems have existed in industrial times. For example, the system of **apartheid** existed in South Africa from 1948 until 1992. The white minority enjoyed the best jobs and other privileges, and they consigned the large black majority to menial jobs. It also prevented marriage between blacks and whites and erected separate public facilities for

An **ascription-based** stratification system is one in which the allocation of rank depends on the characteristics a person is born with.

An **achievement-based** stratification system is one in which the allocation of rank depends on a person's accomplishments.

A **caste** system is an almost pure ascription-based stratification system in which occupation and marriage partners are assigned on the basis of caste membership.

Apartheid was a caste system based on race that existed in South Africa from 1948 until 1992. It consigned the large black majority to menial jobs, prevented marriage between blacks and whites, and erected separate public facilities for members of the two races. Asians and people of "mixed race" enjoyed privileges between these two extremes.

Vincent Van Gogh. *The Potato Eaters* (1889). Most people in agrarian societies were desperately poor. In the early 1800s in Ireland, for example, potatoes supplied about 80 percent of the peasant's diet. On average, each peasant consumed about 10 potatoes a day. The economic surplus was much larger than in horticultural and pastoral societies, but much of the surplus wound up in the hands of royalty, the aristocracy, and religious authorities.

Art Resource, NY

members of the two races. Asians and people of "mixed race" enjoyed privileges between these two extremes. However, apartheid was an exception. For the most part, industrialism causes a *decline* in inequality.

Industrial Societies

At first, the Industrial Revolution that began in the late 1700s did little to lower the level of social stratification. But improvements in the technology and social organization of manufacturing allowed people to produce more goods at a lower cost per unit, making a rise in living standards possible. Business leaders realized they could profit most by identifying, training, and hiring the most talented people, and offered higher wages to recruit them. Workers used union power and growing political influence to win improvements in the conditions of their existence. As a result, social mobility became more widespread than ever before. Stratification declined as industrial societies developed.

Postindustrial Societies

Since the 1970s, social inequality has increased in nearly all postindustrial societies. Among rich countries, the United States exhibited the most social inequality in the 1970s, and since then inequality has grown faster in the United States than in any other rich country. The concentration of wealth in the hands of the wealthiest 1 percent of Americans is higher today than at any time in the past 100 years. The gap between rich and poor is bigger today than it has been for at least 50 years.

Technological factors are partly responsible for the trend toward growing inequality in the United States and elsewhere. Many high-tech jobs have been created at the top of the stratification system over the past few decades. These jobs pay well. At the same time new technologies have made many jobs routine. Routine jobs require little training, and they pay poorly. Because the number of routine jobs is growing more quickly than the number of jobs at the top of the stratification system, the overall effect of technology today is to increase the level of inequality in society. One reason for the high level of inequality in the United States is that we have proportionately more low-wage jobs than any other rich country.

A second factor responsible for the growing trend toward social inequality is government policy. Through tax and social welfare policies, governments are able to prevent big income transfers to the rich. Among the rich postindustrial societies, however, only the French government has intervened to *lower* the level of inequality since the 1970s. The government that has done least to moderate the growth of inequality is that of the United States (Centre for Economic Policy Research, 2002; Smeeding, 2004).

These, then, are the basic patterns and trends in the history of social stratification in the United States and globally. Bearing these descriptions in mind, we must now probe more deeply into the ways sociologists have explained social stratification.

Theories of Stratification

Marx

We begin with the theory of Karl Marx, who formulated a sweeping theory of social and historical development 150 years ago. The "engine" of Marx's theory—the driving force of history, in his view—is the interaction of society's class structure with its technological base. Marx's work represents the first major sociological theory of social stratification, so it is worth considering in detail.

In medieval western Europe, peasants worked small plots of land owned by landlords. Peasants were legally obliged to give their landlords a set part of the harvest and to continue working for them under any circumstance. In turn, landlords were required to protect peasants from marauders. They were also obliged to open their storehouses and feed the peasants if crops failed. This arrangement was known as **feudalism,** or serfdom. The peasants were called serfs.

According to Marx, by the late 1400s several forces were beginning to undermine feudalism. Most important was the growth of exploration and trade, which increased the demand for many goods and services in commerce, navigation, and industry. By the 1600s and 1700s some urban craftsmen and merchants had opened small manufacturing enterprises and saved enough capital to expand production. However, they faced a big problem. To increase profits, they needed more workers whom they could hire in periods of high demand and fire during slack times. Yet the biggest potential source of workers—the peasantry—was legally bound to the land. Thus, feudalism had to be destroyed so peasants could be turned into workers. In Scotland, for example, enterprising landowners recognized that they could make more money raising sheep and selling wool than by having their peasants till the soil. So they turned their cropland into pastures, forcing peasants off the land and into the cities. The former peasants had no choice but to take jobs as urban workers.

In Marx's view, relations between workers and capitalists at first encouraged rapid technological change and economic growth. After all, capitalists wanted to adopt new tools, machines, and production methods so they could produce more efficiently and earn higher profits. But this had unforeseen consequences. In the first place, some capitalists were driven out of business by more efficient competitors and forced to become members of the working class. Together with former peasants pouring into the cities from the countryside, this caused the working class to grow. Second, the drive for profits motivated capitalists to concentrate workers in larger and larger factories, keep wages as low as possible, and invest as little as possible in improving working conditions. Thus, as the capitalist class grew richer and smaller, the working class grew larger and more impoverished, wrote Marx.

Marx thought that workers would ultimately become aware of belonging to an exploited class. Their sense of **class consciousness** would, he wrote, encourage the growth of unions and workers' political parties. These organizations would eventually try to create a communist system in which there would be no private wealth. Instead, under communism everyone would share wealth, said Marx (Marx, 1904 [1859]; Marx and Engels, 1972 [1848]).

We must note several points about Marx's theory. First, a person's class is determined by the *source* of his or her income, or to use Marx's phrase, by one's "relationship to the

Feudalism was a legal arrangement in preindustrial Europe that bound peasants to the land and obliged them to give their landlords a set part of the harvest. In exchange, landlords were required to protect peasants from marauders and open their storehouses and feed the peasants if crops failed.

Class consciousness refers to being aware of membership in a class.

means of production." For example, members of the capitalist class (or **bourgeoisie**) own means of production, including factories, tools, and land. However, they do not do any physical labor. They are thus in a position to earn profits. In contrast, members of the working class (or **proletariat**) do physical labor. However, they do not own means of production. They are thus in a position to earn wages. The *source* of income, not the amount, distinguishes classes in Marx's view.

A second noteworthy point about Marx's theory is that it recognizes more than two classes in any society. For example, Marx discussed the **petty bourgeoisie,** a class of small-scale capitalists who own means of production but employ only a few workers or none at all. This situation forces them to do physical work themselves. In Marx's view, however, members of the petty bourgeoisie are bound to disappear as capitalism develops because they are economically inefficient. Just two great classes characterize every economic era, said Marx—landlords and serfs during feudalism, bourgeoisie and proletariat during capitalism.

Finally, some of Marx's predictions about the development of capitalism turned out to be wrong. Nevertheless, Marx's ideas about classes have stimulated thinking and research on social stratification to this day, as you will see in the next section.

A Critique of Marx

Marx's ideas strongly influenced the development of sociological conflict theory (see Chapter 1). Today, however, more than 120 years after Marx's death, sociologists generally agree that Marx did not accurately foresee some specific aspects of capitalist development:

- Industrial societies did not polarize into two opposed classes engaged in bitter conflict. Instead, a large and heterogeneous middle class of "white-collar" workers has emerged. Some of them are nonmanual employees. Others are professionals. Many of them enjoy higher income, wealth, and status than manual workers. With a bigger stake in capitalism than manual workers, nonmanual employees and professionals have generally acted as a stabilizing force in society.
- Marx correctly argued that investment in technology makes it possible for capitalists to earn high profits. However, he did not expect investment in technology also to make it possible for workers to earn higher wages and toil fewer hours under less oppressive conditions. Yet that is just what happened. Their improved living standard tended to pacify workers, as did the availability of various welfare-state benefits such as unemployment insurance.
- Communism took root not where industry was most highly developed, as Marx predicted, but in semi-industrialized countries such as Russia in 1917 and China in 1948. Moreover, instead of evolving into classless societies, new forms of privilege emerged under communism. For example, in communist Russia, although income was more equal than in the West, membership in the Communist Party, and particularly in the so-called *nomenklatura,* a select group of professional state managers, brought special privileges. These included luxurious country homes, free trips abroad, exclusive access to stores where they could purchase scarce Western goods at nominal prices, and so forth. According to a Russian quip from the 1970s, "Under capitalism, one class exploits the other, but under communism it's the other way around."

Weber

Writing in the early 1900s, Max Weber foretold most of these developments. For example, he did not think communism would create classlessness. He also understood the profound significance of the growth of the middle class. As a result, Weber developed an approach to social stratification much different from Marx's.

Weber, like Marx, saw classes as economic categories (Weber, 1946 [1922]: 180–95). However, he did not think a single criterion—ownership versus nonownership of property—determines class position. Class position, wrote Weber, is determined by one's

- The **bourgeoisie** are owners of the means of production, including factories, tools, and land. They do not do any physical labor. Their income derives from profits.
- The **proletariat,** in Marx's usage, is the working class. Members of the proletariat perform physical labor but do not own means of production. They are thus in a position to earn wages.
- The **petty bourgeoisie,** in Marx's usage, is the class of small-scale capitalists who own means of production but employ only a few workers or none at all, forcing them to do physical work themselves.

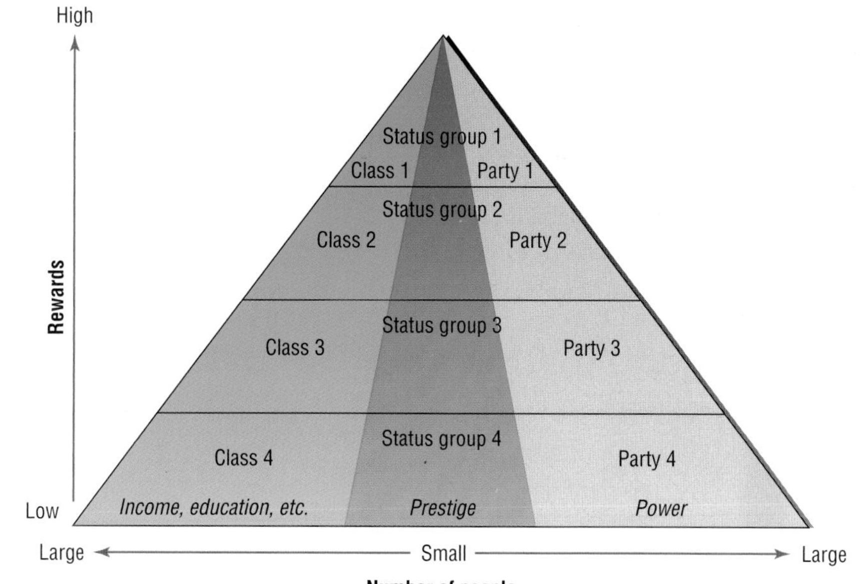

▶Figure 8.4
Weber's Stratification Scheme

"market situation," including the possession of goods, opportunities for income, level of education, and degree of technical skill. From this point of view, Weber defined four main classes: large property owners, small property owners, propertyless but relatively highly educated and well-paid employees, and propertyless manual workers. Thus, white-collar employees and professionals emerge as a large class in Weber's scheme of things.

If Weber broadened Marx's idea of class, he also recognized that two types of groups other than classes have a bearing on the way a society is stratified: status groups and parties. **Status groups** differ from one another in the prestige or social honor they enjoy and in their lifestyle. Consider members of a particular minority ethnic community who have recently immigrated. They may earn relatively high income but endure relatively low prestige. The longer-established members of the majority ethnic community may look down on them as vulgar "new rich." If their cultural practices differ from that of the majority ethnic group, their lifestyle may also become a subject of scorn. Thus, the position of the minority ethnic group in the social hierarchy does not derive just from its economic position but also from the esteem in which it is held.

In Weber's usage, **parties** are not just political groups but, more generally, organizations that seek to impose their will on others. Control over parties, especially large bureaucratic organizations, does not depend just on wealth or another class criterion. One can head a military, scientific, or other bureaucracy without being rich, just as one can be rich and still have to endure low prestige.

So we see why Weber argued that in order to draw an accurate picture of a society's stratification system, one must analyze classes, status groups, *and* parties as somewhat independent bases of social inequality (▶Figure 8.4 and ▶Table 8.4). But to what degree are they independent of one another? Weber said that the importance of status groups as a basis of stratification is greatest in precapitalist societies. Under capitalism, classes and parties (especially bureaucracies) become the main bases of stratification.

An American Perspective: Functionalism

Marx and Weber were Germans who wrote their major works between the 1840s and the 1910s. Inevitably, their theories bear the stamp of the age in which they wrote. The next major developments in the field occurred in mid-20th-century America. Just as inevitably, these innovations were colored by the optimism, dynamism, and prejudices of that time and place.

Status groups differ from one another in terms of the prestige or social honor they enjoy and in terms of their lifestyle.

Parties, in Weber's usage, are organizations that seek to impose their will on others.

Table 8.4
Mean Annual Earnings, Full-Time Workers, and Prestige Scores, Selected Occupations, United States

Occupation	Median Hourly Income, 2003	Prestige Score, 1989
Airplane pilots	$120,589	61
Physicians	117,664	86
Lawyers	99,798	75
Aerospace engineers	82,113	72
College and university professors	66,945	74
Computer programmers	59,940	61
Police officers and detectives	49,032	60
Plumbers, pipefitters, and steamfitters	46,047	45
High school teachers	45,254	66
Electrical and electronic technicians	43,856	51
Preschool teachers	31,589	55
Hairdressers and cosmetologists	26,704	36
Janitors	22,691	22
Secretaries	21,035	46
Security guards	20,804	42
Taxi drivers and chauffeurs	20,253	28
Restaurant cooks	19,107	31
Sewing machine operators	17,643	28
Waiters, waitresses, and bartenders	14,209	28

Note: Median annual income for the occupations listed in the table is taken from a labor force survey conducted by the federal government. Occupational prestige scores are based on the GSS. Respondents were asked to rank occupations in terms of the prestige attached to them. Prestige scores for all occupations range from 17 (miscellaneous food preparation occupations) to 86 (physicians). The correlation between median income and prestige scores for the occupations listed here is strong ($r = .83$).
Source: Inter-University Consortium for Political and Social Research (1992); U.S. Department of Labor (2004).

Consider first in this connection the **functional theory of stratification,** proposed by Kingsley Davis and Wilbert Moore at the end of World War II (Davis and Moore, 1945). Davis and Moore observed that some jobs are more important than others. A judge's work, for example, contributes more to society than the work of a janitor. This presents society with a big problem: How can people be motivated to undergo the long training they need to serve as judges, physicians, and engineers and in other important jobs? After all, higher education is expensive. One cannot earn much money during training. One must study long and hard rather than seek pleasure. Clearly, an incentive is needed to motivate the most talented people to train for the most important jobs. That "something," said Davis and Moore, is money. More precisely, social stratification is necessary (or "functional") because the prospect of high material rewards motivates people to undergo the sacrifices needed to get a higher education. Without substantial inequality, they conclude, the most talented people would have no incentive to become judges and so forth.

Although the functional theory of stratification may at first seem plausible, we can conduct what Max Weber called a "thought experiment" to uncover one of its chief flaws. Imagine a society with just two classes of people—physicians and small family farmers. The farmers grow food. The physicians tend the ill. Then, one day, a rare and deadly virus strikes. The virus has the odd property of attacking only physicians. Within weeks, our imaginary society has no more physicians. As a result, the farmers are much worse off. Cures and treatments for their ailments are no longer available. Soon the average farmer lives fewer years than his or her predecessors. The society is less well off, although it survives.

The **functional theory of stratification** argues that (a) some jobs are more important than others, (b) people have to make sacrifices to train for important jobs, and (c) inequality is required to motivate people to undergo these sacrifices.

Now imagine the reverse. Again we have a society composed of physicians and farmers. Again a rare and lethal virus strikes. This time, however, the virus has the odd prop-

According to the functional theory of stratification, "important" jobs require more training than "less important" jobs. The promise of big salaries motivates people to undergo that training. Therefore, the functionalists conclude, social stratification is necessary. As the text makes clear, however, one of the problems with the functional theory of stratification is that it is difficult to establish which jobs are important, especially when one takes a historical perspective.

erty of attacking only farmers. Within weeks, the physicians' stores of food are depleted. A few more weeks, and the physicians start dying of starvation. The physicians who try to become farmers catch the virus and die. Within months, society no longer exists. Who then does the more important work: physicians or farmers? Our thought experiment suggests that farmers do, for without them society cannot exist.

From a historical point of view, we can say that *none* of the jobs regarded by Davis and Moore as "important" would exist without the physical labor done by people in "unimportant" jobs throughout the ages. To sustain the witch doctor in a tribal society, hunters and gatherers had to produce enough for their own subsistence plus a surplus to feed, clothe, and house the witch doctor. To sustain the royal court in an agrarian society, serfs had to produce enough for their own subsistence plus a surplus to support the lifestyle of members of the royal family. Government and religious authorities have taken surpluses from ordinary working people for thousands of years by means of taxes, tithes, and force. Among other things, these surpluses were used to establish the first institutions of higher learning in the 13th century. From these, modern universities developed.

So we see that the question of which occupations are most important is not as clear-cut as Davis and Moore make it seem. To be sure, physicians today earn a lot more money than small family farmers, and they also enjoy a lot more prestige. But that is not because their work is more important in any objective sense of the word. (More discussion about why physicians and other professionals earn more than nonprofessionals can be found in Chapter 17, "Education").

Other problems with the functional theory of stratification were noted soon after Davis and Moore published their article (Tumin, 1953). We mention two of the most important criticisms here. First, the functional theory of stratification stresses how inequality helps society discover talent. However, it ignores the pool of talent lying undiscovered *because of* inequality. Bright and energetic adolescents may be forced to drop out of high school to help support themselves and their families. Capable and industrious high school graduates may be forced to forgo a college education because they cannot afford it. Inequality may encourage the discovery of talent, but only among those who can afford to take advantage of the opportunities available to them. For the rest, inequality prevents talent from being discovered.

A final problem with the functional theory of stratification is its failure to examine how people pass advantages from generation to generation. Like *Robinson Crusoe,* the functional theory correctly emphasizes that talent, hard work, and sacrifice often result in high occupational attainment and high material rewards. However, once people attain high class standing, they can use their power to maintain their position and promote the interests of their families regardless of their children's talent. For example, inheritance allows parents to transfer wealth to children irrespective of their talent. As far back as the late 18th century, 90 percent of the 100 wealthiest Virginians had inherited the bulk of their estates from their parents and grandparents (Pessen, 1984: 171). Glancing back at Table 8.1, we see that 48 percent of the 25 largest personal fortunes in the United States are based on inheritance. The Walton children are tied as the fourth richest people in America not because of their talent but because their father gave them his Wal-Mart empire. Even rich people who do not inherit large fortunes often start near the top of the stratification system. Bill Gates, for example, is the richest person in the world. He did not inherit his fortune. However, his father was a partner in one of the most successful law firms in Seattle. Gates himself went to the most exclusive and expensive private schools in the city, followed by a stint at Harvard. In the late 1960s, his high school was one of the first in the nation to boast a computer terminal connected to a nearby university mainframe. Gates's early fascination with computers dates from this period. Gates is without doubt a highly talented man, but surely the advantages with which he was born, and not just his talents, helped to elevate him to his present lofty status (Wallace and Erickson, 1992). Recent research shows that the importance of inheritance in determining class position has increased in the United States in recent decades, is considerably higher than previously thought, and is higher in the United States than in other highly industrialized countries (Björklund and Jäntti, 2000; Bowles and Gintis, 2002; Levine and Mazumder, 2002). Therefore, an adequate theory of stratification clearly must take inheritance into account, just as it must recognize how inequality prevents the discovery of talent and avoid making untenable assumptions about which jobs are important and which are not.

Social Mobility: Theory and Research

They built the cabin from rough logs cut from the wilderness. It is small and drafty. The flame in the solitary lamp gutters when the wind blows. Outside, a cougar howls. Inside, the frontier family tends a small child lying on a simple quilt. Someday, the child will become president of the United States.

Is he Abraham Lincoln? Or Andrew Jackson perhaps? It hardly matters. What matters is the moral of the story. As every schoolchild learns, in the United States anyone can become president, no matter how humble his—who knows? perhaps someday her—origins. The opportunity to rise to the top is the crux of the American Dream (Pessen, 1984). To many observers, it is the main difference between the United States and Europe.

Two hundred years ago, many observers viewed Britain and other European societies as based on class privilege. In this view, if you were born into a certain class, you were destined to go to certain schools, speak with a certain accent, and take a certain type of job. Some people managed to rise above their class origins, but not many. In contrast, the United States was widely viewed as the land of golden opportunity. With few people and abundant natural resources, it already enjoyed the highest standard of living in the world. Its western frontier beckoned adventurous migrants with gold rushes and vast tracts of fertile land. The United States came to be viewed as a land where hard work and talent could easily overcome humble origins. Almost anyone could strike it rich, it seemed. If Europe was based on class privilege, the United States was widely regarded as classless.

The rate of upward social mobility may have been higher in the United States than in Europe in the 19th century. As we will see later, however, there was little difference between American and European mobility rates by the second half of the 20th century. It is revealing in this connection to contrast the biographies of Abraham Lincoln and Andrew Jackson with those of the front-running presidential hopefuls from both parties in 2000 and 2004. All of the latter (George W. Bush, John McCain, Steve Forbes, Al Gore, Bill

Bradley, John Kerry, Howard Dean) were born into millionaire families and attended elite colleges. Nonetheless, the *idea* that America is classless has persisted.[1]

Blau and Duncan: The Status Attainment Model

In 1967, that idea was incorporated in Peter Blau and Otis Dudley Duncan's *The American Occupational Structure.* This book became one of the most influential works in American sociology. Blau and Duncan set themselves the task of figuring out the relative importance of inheritance versus individual merit in determining one's place in the stratification system. To what degree is one's position based on ascription—that is, inheriting wealth and other advantages from one's family? To what degree is one's position based on achievement—that is, applying one's own talents to life's tasks? Blau and Duncan's answer was plain: Stratification in America is based mainly on individual achievement.

Blau and Duncan abandoned the European tradition of viewing the stratification system as a set of distinct groups. Marx, you will recall, distinguished two main classes by the source of their income. He was sure the bourgeoisie and the proletariat would become class conscious and take action to assert their class interests. Similarly, Weber distinguished four main classes by their market situation. He saw class consciousness and action as potentials that each of these classes might realize in some circumstances. In contrast, Blau and Duncan saw little if any potential for class consciousness and action in the United States, which is why they abandoned the entire vocabulary of class. For them, the stratification system is not a system of distinct classes at all, but a continuous hierarchy or ladder of occupations with hundreds of rungs. Each occupation—each rung on the ladder—requires different levels of education and generates different amounts of income.

To reflect these variations in education and earnings, Blau and Duncan created a **socioeconomic index of occupational status (SEI).** Using survey data, they found the average earnings and years of education of men employed full time in various occupations. They combined these two averages to arrive at an SEI score for each occupation. (Similarly, other researchers combined income, education, and occupational prestige data to construct an index of **socioeconomic status [SES].**)

Next, Blau and Duncan used survey data to find the SEI of each respondent's current job, first job, and father's job, as well as the years of formal education completed by the respondent and the respondent's father. They showed how all five of these variables were related (Figure 8.5). Their main finding was that the respondents' own achievements (years of education and SEI of first job) had much more influence on their current occupational status than did ascribed characteristics (father's occupation and years of education). Blau and Duncan concluded that the United States is a relatively open society in which individual merit counts for more than family background. This finding partly vindicated the functionalists and confirmed the core idea of the American Dream.

Blau and Duncan's study influenced a whole generation of stratification researchers in the United States and, to a lesser degree, other countries (Boyd et al., 1985; Erikson and Goldthorpe, 1992; Featherman and Hauser, 1978; Featherman, Jones, and Hauser, 1975; Grusky and Hauser, 1984). Their approach to studying social stratification, which focuses on the effects of family background and educational level on occupational achievement, became known as the "status attainment model."

Subsequent research confirmed that the rate of social mobility for men in the United States is high and that most mobility has been upward. Since the early 1970s, however, substantial downward mobility has occurred. In 1978, 23 percent of adult men born into the bottom fifth of the stratification system made it into the top fifth. Today, the figure is just 10 percent. A person born into the top fifth today is more than five times more likely than a person born in the bottom fifth to end up in the top fifth ("Meritocracy in America," 2005).

Blau and Duncan's **socioeconomic index of occupational status (SEI)** combines, for each occupation, average earnings and years of education of men employed full time in the occupation.

Socioeconomic status (SES) combines income, education, and occupational prestige data in a single index of one's position in the socioeconomic hierarchy.

[1]Actually, even the rise of Lincoln and Jackson from poverty to the most powerful position in the land was not unaided. Lincoln's wife was the daughter of a banker. Jackson's wife was the daughter of a relatively well-to-do owner of a boarding house.

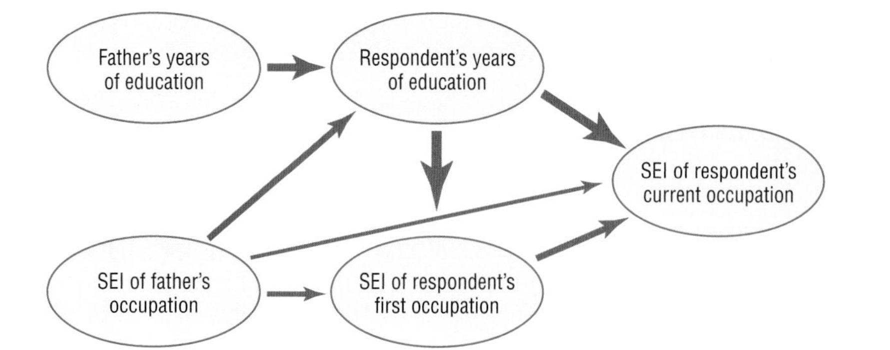

Figure 8.5
Blau and Duncan's Model of Occupational Achievement
Note: Arrows indicate cause-and-effect relationships between variables, with the arrowheads pointing to effects. The association between father's education and occupation is not illustrated here for simplicity's sake. The thicker the line, the stronger the relationship between variables. The diagram shows that the respondents' own achievements (years of education and socioeconomic index of occupational status [SEI] of first job) had much more influence on their current occupational status than did ascribed characteristics (father's occupation and years of education).

Researchers also found that mobility for men within a single generation (**intragenerational mobility**) is generally modest. Few people move "from rags to riches" or fall from the top to the bottom of the stratification system in a lifetime. On the other hand, mobility for men over more than one generation (**intergenerational mobility**) can be substantial.

In addition, research showed that most social mobility is the result of change in the occupational structure. One of the most dramatic changes during the late 19th and early 20th centuries was the decline of agriculture and the rise of manufacturing. This decline caused a big decrease in the number of farmers and a corresponding surge in the number of factory workers. A second dramatic change, especially apparent during the last third of the 20th century, was the decline of manufacturing and the rise of the service sector. This change caused a big drop in the number of manual workers and a corresponding surge in the number of white-collar service workers. Mobility due to such changes in the occupational structure is known as **structural mobility.** Just as high tide raises all ships and low tide causes all ships to fall, structural mobility is a powerful force drawing individuals away from old occupations and into new ones.

Finally, research shows that there are only small differences in rates of social mobility among the highly industrialized countries. The United States does not have an exceptionally high rate of upward social mobility (Lipset and Bendix, 1963; Erikson and Goldthorpe, 1992; Grusky and Hauser, 1984). In fact, some countries, such as Australia and Canada, enjoy even higher upward mobility rates than the United States (Tyree, Semyonov, and Hodge, 1979).

A Critique of Blau and Duncan

As we have seen, Blau and Duncan's approach to studying social stratification stimulated much research and led to important sociological insights. However, beginning in the 1970s their stratification model was criticized from many angles. Most of the criticisms converged on a single point: Blau and Duncan's theory ignored much that was interesting and important in the study of social stratification.

To fully appreciate the significance of this criticism, one must know that Blau and Duncan sampled only men who were employed full time. Women were excluded on the grounds that most of them were homemakers and not in the paid labor force. Whatever the merits of this exclusion in 1962, the year of Blau and Duncan's survey, it made little sense even 10 years later. It makes still less sense today, when the great majority of adult women are in the paid labor force (England, 1992).[2]

Intragenerational mobility is social mobility that occurs within a single generation.

Intergenerational mobility is social mobility that occurs between generations.

Structural mobility refers to the social mobility that results from changes in the distribution of occupations.

[2]The inclusion of married women in stratification studies raises the question of whether they should be located in terms of their own labor force characteristics or in terms of their own *and their husbands'* labor force characteristics. We share Breen and Rottman's (1995: 62–7) opinion that the proper unit of analysis should be the household, not the individual. Each spouse is affected by the other's labor force characteristics, and decision making is usually made at the household, not the individual, level. See also Clement and Myles (1994) and Crompton and Mann (1986).

Similarly, Blau and Duncan did not sample part-time and unemployed workers. This omission may not have been serious given the low unemployment of the early 1960s. However, it became serious in the 1970s and 1980s, when the unemployment rate rose. Compounding the problem is that the proportion of jobs that are part time have increased steadily from the 1970s onward.

Finally, a disproportionately large number of unemployed and part-time workers are African American or Hispanic American and think of themselves as members of the working class. Therefore, these groups were underrepresented in Blau and Duncan's study. To say that Blau and Duncan sampled only white middle-class men would be an exaggeration. But the percentage of women, working-class people, African Americans, and Hispanic Americans in their sample was smaller than the percentage in the U.S. population (Miller, 1998; Sørenson, 1992).[3]

This fact raises a basic question about Blau and Duncan's study: How can one make valid claims about the relative importance of ascription versus achievement in American society when many of the most disadvantaged people in the society are excluded from one's analysis? The short answer is that one cannot. As Blau and Duncan claimed, merit may be more important than inheritance in determining the social position of white middle-class men with full-time jobs. However, their study has little to say about the determinants of stratification among most adult Americans, who are not white middle-class men employed full time.

Group Barriers: Race and Gender

Subsequent research yielded both good news and bad news for the Blau and Duncan status attainment model. The good news: Research confirmed that the process of status attainment is much the same for women and minorities as it is for white men. Years of schooling influence status attainment more than does father's occupation whether one examines white men, women, African Americans, or Hispanic Americans. The bad news for the Blau and Duncan status attainment model is that if you compare people *with the same level of education and similar family backgrounds,* women and members of minority groups tend to attain lower status than white men (Featherman and Hauser, 1976; Hout, 1988; Hout and Morgan, 1975; McClendon, 1976; Stolzenberg, 1990; Tienda and Lii, 1987). These findings suggest that one cannot adequately explain status attainment by examining only the characteristics of *individuals,* such as their years of education and father's occupation. One must also examine the characteristics of *groups,* such as whether some groups face barriers to mobility, regardless of the individual characteristics of their members (Horan, 1978). Such group barriers include racial and gender discrimination and being born in neighborhoods that make upward mobility unusually difficult because of poor living conditions. For instance, women and African Americans who are employed full time earn less on average than white men with the same level of education (see Chapter 10, "Race and Ethnicity," and Chapter 11, "Sexuality and Gender"). The existence of such group disadvantages suggests that American society is not as open or "meritocratic" as Blau and Duncan made it out to be. Group barriers to mobility, such as gender and race, do exist.

Could class in the Marxist or Weberian sense act like race and gender, bestowing advantages and disadvantages on entire groups of people and perhaps even helping to shape their political views? Some sociologists think so. Their research shows that parents' wealth, education, and occupation are more important determinants of a person's occupation than Blau and Duncan's research suggest (Jencks et al., 1972; Rytina, 1992). Other sociologists, dissatisfied with the Blau and Duncan model, have revisited and updated the Marxist and Weberian concepts of class to make them more relevant to the late 20th century. We now briefly review their work.

---

[3]In fact, they devoted special attention to showing that nonwhites had less incentive to seek higher education than did whites because the nonwhite higher-education graduate suffered more occupational discrimination than did less-educated nonwhites.

Table 8.5
Wright's Typology of Classes, United States, 1980

Owners of Means of Production	Nonowners of Production				
	Skill Assets				
	(+)	(>0)	(−)		
Bourgeoisie (hire, don't work), 2%	Expert manager, 4%	Semicredentialed manager, 6%	Uncredentialed manager, 2%	(+)	Organizational assets
Small employers (hire, work), 6%	Expert supervisor, 4%	Semicredentialed supervisor, 7%	Uncredentialed supervisor, 7%	(>0)	
Petty bourgeoisie (work, do not hire), 7%	Expert nonmanager, 3%	Semicredentialed worker, 12%	Proletarian, 40%	(−)	

Source: Wright (1985: 88).

The Revival of Class Analysis

An adequate theory of class stratification must do two things. First, it must specify criteria that distinguish a *small number* of distinct classes. Why small? Because the larger the number of classes specified by the theory, the more its picture of the stratification system will resemble Blau and Duncan's occupational ladder with its hundreds of rungs. Therefore, the less it will capture *class* differences in economic opportunities, political outlooks, and cultural styles. Second, an adequate theory of class stratification must spark research that demonstrates *substantial gaps* between classes in economic opportunities, political outlooks, and cultural styles. In other words, research must show that such differences are larger *between* classes than *within* them.

In the 1980s and early 1990s sociologists on both sides of the Atlantic developed theories that meet the first requirement. In the United States, Erik Olin Wright updated Marx. In Britain, John Goldthorpe updated Weber. Both scholars created new "class maps" that specified criteria for distinguishing a small number of classes.

On the second requirement, the work of Goldthorpe and especially Wright has been less successful. Not enough research has been conducted to show that the classes distinguished by Wright and Goldthorpe differ substantially in terms of economic opportunities, political outlooks, and cultural styles. Especially sparse is research that would allow us to judge which of the two theories is superior in this regard. Nonetheless, as we will see, the research literature offers some tantalizing hints. Wright and Goldthorpe appear to be on the right track, and Goldthorpe's Weberian approach may be the more promising of the two.

Wright

Wright's update of Marx's class scheme is illustrated in Table 8.5 (Wright, 1985; 1997). Like Marx, Wright's basic distinction is between property owners and nonowners. The former earn profits, the latter wages and salaries. Wright also distinguishes large, medium, and small owners. They differ from one another in terms of how much property they own and whether they have many employees, a few employees, or none at all.

If there are three propertied classes in Wright's theory, nine classes lack property. These wage and salary earners differ from one another in two ways. First, they have different "skill and credential levels": Some wage and salary earners have more training and education than others. Second, wage and salary earners differ from one another in terms of their "organization assets": Some of them enjoy more decision-making authority than others. The two extremes among the nonpropertied are "expert managers," who have high skill and credential levels combined with high organizational assets, and "proletarians," who have low skill and credential levels combined with no organizational assets. The percentages in Table 8.5 show the proportion of the American labor force Wright found in each class in a survey he conducted in 1980.

Goldthorpe

For Goldthorpe, different classes are characterized by different "employment relations" (Goldthorpe, Llewellyn, and Payne, 1987 [1980]; Erikson and Goldthorpe, 1992). Goldthorpe's basic division in employment relations is among employers, self-employed

Table 8.6
Goldthorpe's Typology of Classes

Service classes	
I.	Higher-grade professionals, administrators, and officials; managers in large industrial enterprises; large proprietors
II.	Lower-grade professionals, administrators, and officials; higher-grade technicians; managers in small industrial establishments; supervisors of nonmanual employees
Intermediate classes	
IIIa.	Routine nonmanual employees, higher grade (administration and commerce)
IIIb.	Routine nonmanual employees, lower grade (sales and service)
IVa.	Small proprietors, artisans, and so forth, with employees
IVb.	Small proprietors, artisans, and so forth, without employees
IVc.	Farmers and smallholders; other self-employed workers in primary production
V.	Lower-grade technicians; supervisors of manual workers
Working Classes	
VI.	Skilled manual workers.
VIIa.	Semiskilled and unskilled manual workers not in primary production
VIIb.	Agricultural and other workers in primary production

Source: Goldthorpe (1987 [1980]).

people, and employees. He then makes finer distinctions within each of these broad groupings. For instance, he distinguishes between large and small employers and between self-employed people in agriculture and those outside of agriculture. Employees involved in service relationships, such as professionals, and those who have labor contracts, such as factory workers, are also viewed by Goldthorpe as different **classes.** He makes still finer distinctions based on educational and supervisory criteria. The result is an 11-class model of the stratification system (Table 8.6).

Goldthorpe's class schema differs from Wright's in several important respects, two of which we mention here. First, the proletariat is a large and undifferentiated mass in Wright's model, amounting to 40 percent of the U.S. labor force. In contrast, Goldthorpe defines three classes of workers that are distinguished by skill and sector. Second, consistent with Marx's theory, Wright says that large employers form a separate class. Goldthorpe, however, groups large employers with senior managers, professionals, administrators, and officials. He believes that the common features of these occupational groups—level of income and authority, political interests, and lifestyle—transcend the Marxist divide between owners and nonowners.

Research should eventually help decide the merits of Wright's versus Goldthorpe's theories, but the jury is still out on this question (Crompton, 1993). To date, only a few attempts have been made to determine empirically how well the theories perform in comparison with each other. British sociologists have published the most comprehensive analysis of this issue to date (Marshall, Newby, Rose, and Vogler, 1988). They determined statistically whether Wright's or Goldthorpe's class models do a better job of explaining people's social mobility, voting intentions, class consciousness, and so forth. For all these variables, they found Goldthorpe's model superior.

Class in Marx's sense of the term is determined by one's relationship to the means of production. In Weber's usage, class is determined by one's "market situation." Wright distinguishes classes on the basis of relationship to the means of production, amount of property owned, organizational assets, and skill. For Goldthorpe, classes are determined mainly by one's "employment relations."

Noneconomic Dimensions of Class

Prestige and Power

The theories we have just reviewed focus on occupations, production relations, and employment relations. In short, they all emphasize the *economic* sources of inequality. However, as Weber correctly pointed out, inequality is not based on money alone. It is also based on prestige and power. These important dimensions of inequality have been some-

Sociology Now™

Learn more about **Prestige and Power** by going through the Power and Authority Learning Module.

what neglected in recent writings on stratification. Therefore, in the following section we discuss the political side of stratification. We focus on prestige or honor in this section.

Weber, you will recall, said that status groups differ from one another in terms of their lifestyles and the honor in which they are held. Here we may add that members of status groups signal their rank by means of material and symbolic culture. They seek to distinguish themselves from others by displays of "taste" in fashion, food, music, literature, manners, and travel.

The difference between "good taste," "common taste," and "bad taste" is not inherent in cultural objects themselves. Rather, cultural objects that are considered to be in the best taste are generally those that are least accessible.

To explain the connection between taste and accessibility, let us compare Bach's *The Well-Tempered Clavier* with Gershwin's *Rhapsody in Blue.* A survey by French sociologist Pierre Bourdieu showed different social groups prefer these two musical works (Bourdieu, 1984: 17). Well-educated professionals, high school teachers, professors, and artists prefer *The Well-Tempered Clavier.* Less-well-educated clerks, secretaries, and junior commercial and administrative executives favor *Rhapsody in Blue.* Why? The two works are certainly very different types of music. Gershwin evokes the jazzy dynamism of big-city America early in the 20th century, whereas Bach evokes the almost mathematically ordered courtly life of early-18th-century Germany. But one would be hard pressed to argue that *The Well-Tempered Clavier* is intrinsically *superior* music. Both are great art. Why then do more highly educated people prefer *The Well-Tempered Clavier* to *Rhapsody in Blue*? According to Bourdieu, during their education they acquire specific cultural tastes associated with their social position. These tastes help to distinguish them from people in other social positions. Many of them come to regard lovers of Gershwin condescendingly, just as many lovers of Gershwin come to think of Bach enthusiasts as snobs. These distancing attitudes help the two status groups remain separate.

Bach's music is complex, and to really appreciate it one may require some formal instruction. Many other elements of "high culture," such as opera and abstract art, are similarly inaccessible to most people because fully understanding them requires special education. However, education is not the only factor that makes some cultural objects less accessible than others. Purely financial considerations also enter the picture. A Mercedes costs four times more than a Ford, and a winter ski trip to Aspen can cost four times more than a week in a modest motel near the beach in Fort Lauderdale. Of course, one can get

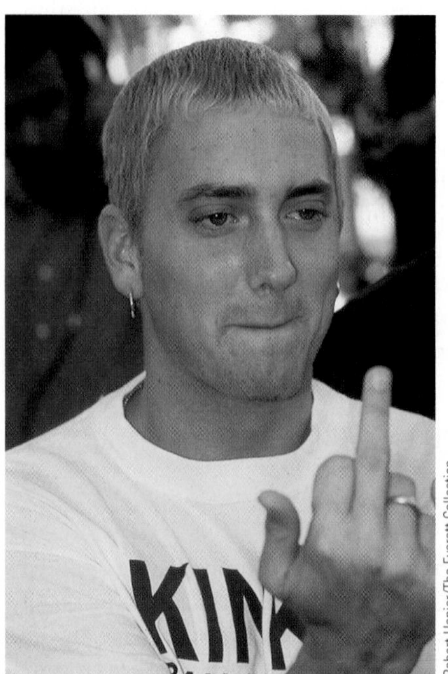

Robert Hepier/The Everett Collection

Gerrit Greve/Corbis

The differences among "good taste," "common taste," and "bad taste" are not inherent in cultural objects themselves. Rather, cultural objects that are considered to be in the best taste are generally the least accessible. Left: Eminem. Right: Johann Sebastian Bach.

from point A to point B quite comfortably in a Ford and have a perfectly enjoyable winter vacation in south Florida. Still, most people would prefer the Mercedes and the trip to Aspen at least partly because they signal higher status. Access to tasteful cultural objects, then, is as much a matter of cost as of education.

Status and Clothing Styles

Often, rich people engage in conspicuous displays of consumption, waste, and leisure not because they are necessary, useful, or pleasurable but simply to impress their peers and inferiors (Veblen, 1899). This is evident if we consider how clothing acts as a sort of language that signals one's status to others (Lurie, 1981).

For thousands of years, certain clothing styles have indicated rank. In ancient Egypt, only people in high positions were allowed to wear sandals. The ancient Greeks and Romans passed laws controlling the type, number, and color of garments one could wear and the type of embroidery with which they could be trimmed. In medieval Europe various aspects of dress were also regulated to ensure that certain styles were specific to certain groups.

European laws governing the dress styles of different groups fell into disuse after about 1700 because a new method of control emerged as Europe became wealthier. From the 18th century onward, the *cost* of clothing came to designate a person's rank. Expensive materials, styles that were difficult to care for, heavy jewelry, and superfluous trimmings became all the rage. Rich people did not wear elaborate powdered wigs, heavy damasked satins, the furs of rare animals, diamond tiaras, and patterned brocades and velvets for comfort or utility. Such raiment was often hot, stiff, heavy, and itchy. One could scarcely move in many of these getups. And that was just their point—to prove not only that the wearer could afford enormous sums for handmade finery, but also that he or she did not have to work to pay for them.

Today, we have different ways of using clothes to signal status. For instance, designer labels loudly proclaim the dollar value of garments. Furthermore, a great variety and quantity of clothing are required to maintain appearances. Thus, the well-to-do athletic type may have many different and expensive outfits that are "required" for jogging, hiking, cycling, aerobics, golf, and tennis. In fact, many people who really cannot afford to obey the rules of conspicuous consumption, waste, and leisure feel compelled to do so anyway. As a result, they go into debt to maintain their wardrobes. Doing so helps them maintain prestige in the eyes of associates and strangers alike, even if their economic standing secretly falters.

Politics and the Plight of the Poor

Politics is a second noneconomic dimension of stratification that has received insufficient attention in recent work on inequality. Yet political life has a profound impact on the distribution of opportunities and rewards in society. Politics can reshape the class structure by changing laws governing people's right to own property. Less radically, politics can change the stratification system by entitling people to various welfare benefits and by redistributing income through tax policies. We discuss the social bases of politics in detail in Chapter 14 ("Politics"). Here we illustrate the effect of politics on inequality by considering how government policy has affected the plight of America's poor (the standard definition of poverty is discussed later).

The two main currents of American opinion on the subject of poverty correspond roughly to the Democratic and Republican positions. Broadly speaking, most Democrats want the government to play an important role in helping to solve the problem of poverty. Most Republicans want to reduce government involvement with the poor so people can solve the problem themselves. At various times, each of these approaches to poverty has dominated public policy.

We can see the effect of government policy on poverty by examining fluctuations in the poverty rate over time. The **poverty rate** is the percentage of Americans who fall

The **poverty rate** is the percentage of people living below the poverty threshold, which is three times the minimum food budget established by the U.S. Department of Agriculture.

▸Figure 8.6
Poverty Rate, Individuals, United States, 1961–2003 (percent)

Between 1961 and 2001, the poverty rate fell in 17 of the 20 years in which the president was a Democrat (85%) but in only 11 of the 24 years in which the president was a Republican (46%).

Source: U.S. Census Bureau (2004d).

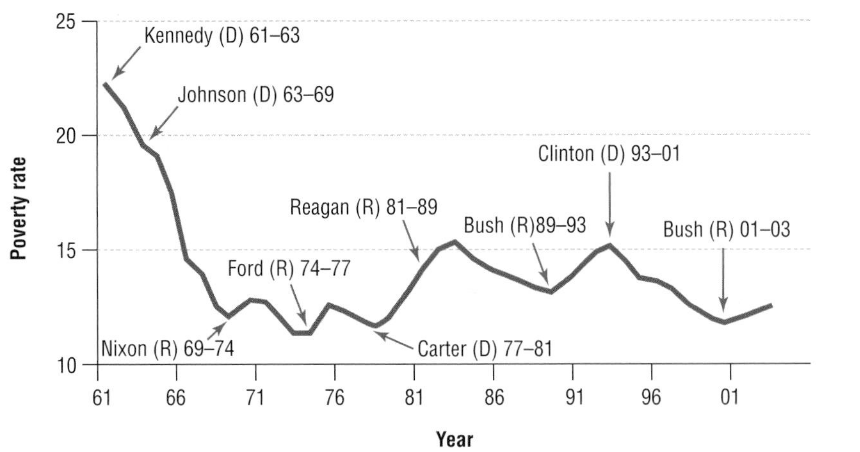

Sociology Now™

Learn more about **the Poverty Rate** by going through the % of persons below the Poverty Level Map Exercise.

below the "poverty threshold." To establish the poverty threshold, the U.S. Department of Agriculture first determines the cost of an economy food budget. The poverty threshold is then set at three times that budget. It is adjusted for the number of people in the household, the annual inflation rate, and whether individual adult householders are younger than 65 years of age. In 2003 the poverty threshold for individual adult householders under the age of 65 was $9573 per year. For a family of four with two children under the age of 18, the poverty threshold was $18,660 (U.S. Census Bureau, 2004c). ▸Figure 8.6 shows the percentage of Americans who lived below the poverty threshold from 1961 to 2001. Between the early 1960s and the late 1970s, the poverty rate dropped dramatically from about 22 percent to around 11 percent. Then between 1980 and 1984, it jumped to about 15 percent, fluctuating in the 11 percent to 15 percent range after that. In 2003, 35.9 million Americans were living in poverty, 12.5 percent of the population.

Dorothea Lange. "Migrant Mother, Nipomo California, 1936." During the Great Depression (1929–39), about one-third of Americans were unemployed and many of the people lucky enough to have jobs were barely able to make ends meet.

Government Policy and the Poverty Rate in the United States

Fluctuations in the poverty rate are related to political events. Thus, between 1961 and 2003 the poverty rate fell in 17 of the 20 years in which the president was a Democrat (85 percent) but in only 11 of the 24 years in which the president was a Republican (46 percent). The poverty rate fluctuates as a result of economic conditions, but these figures show that the policies of the party in power also have an important bearing on the poverty rate.

The 1930s: The Great Depression

More broadly, we can identify three policy initiatives that have been directed at the problem of poverty. The first dates from the mid-1930s. During the Great Depression (1929–39), nearly one-third of Americans were unemployed and many of the people lucky enough to have jobs were barely able to make ends meet. Remarkably, an estimated 68 percent of Americans were poor in 1940 (O'Hare, 1996: 13). In the middle of the Depression, Franklin Roosevelt was elected president. In response to the suffering of the American people and the large, violent strikes of the era, he introduced such programs as Social Security, Unemployment Insurance, and Aid to

Families with Dependent Children (AFDC) (see Chapter 21, "Collective Action and Social Movements"). For the first time, the federal government took responsibility for providing basic sustenance to citizens who were unable to do so themselves. Because of Roosevelt's "New Deal" policies and rapidly increasing prosperity in the decades after World War II, the poverty rate fell dramatically, reaching 19 percent in 1964.

The 1960s: The War on Poverty

The second antipoverty initiative dates from the mid-1960s. In 1964, President Lyndon Johnson declared a "war on poverty." This initiative was in part a response to a new wave of social protest. Millions of southern blacks who had migrated to northern and western cities in the 1940s and 1950s were unable to find jobs. In some census tracts in Detroit, Chicago, Baltimore, and Los Angeles, black unemployment ranged from 26 to 41 percent in 1960. Suffering extreme hardship, many African Americans demanded at least enough money from the government to allow them to subsist. Some helped to organize the National Welfare Rights Organization to pressure the government to give relief to more poor individuals. Others took to the streets as race riots rocked the nation in the mid-1960s. President Johnson soon broadened access to AFDC and other welfare programs (Piven and Cloward, 1993 [1971]; 1977: 264–361). As a result, in 1973 the poverty rate dropped to 11.1 percent, the lowest it has ever been in this country.

The 1980s: "War Against the Poor"

Finally, the third initiative aimed at the poverty problem dates from 1980. The mood of the country had shifted by the time President Ronald Reagan took power that year. Reagan assumed office after the social activism and rioting of the 1960s and 1970s had died down. He was elected in part by voters born in the 1950s and 1960s. These so-called "baby boomers" expected their standard of living to increase as quickly as that of their parents. Many of them were deeply disappointed when things did not work out that way. Real household income (earnings minus inflation) remained flat in the 1970s and 1980s. Even that discouraging performance was achieved thanks only to the mass entry of women into the paid labor force. Reagan explained this state of affairs as the result of too much government. In his view, big government inhibits growth. In contrast, cutting government services and the taxes that fund them stimulates economic growth. For example, Reagan argued that welfare causes long-term dependency on government handouts, which, he claimed, worsens the problem of poverty rather than solving it. As Reagan was fond of saying, "We fought the War on Poverty and poverty won" (quoted in Rank, 1994: 7). The appropriate solution, in Reagan's view, was to reduce relief to the poor. Cut welfare programs, he said, and welfare recipients will be forced to work. Cut taxes and taxpayers will spend more, thus creating jobs.

Reagan's message fell on receptive ears. The War on Poverty was turned into what one sociologist called a "war against the poor" (Gans, 1995). Because a large proportion of welfare recipients were African Americans and Hispanic Americans, some analysts have argued that the war against the poor was fed by racist sentiment (Quadagno, 1994). The invidious stereotype of a young unmarried black woman having a baby in order to collect a bigger welfare check became common. The AFDC budget, expenditures for employee training, and many other government programs were cut sharply. Poverty rates rose. The number of homeless people jumped from 125,000 in 1980 to 402,000 in 1987–88, and after declining during the economic boom of the 1990s started to surge to record levels in 2001 (Belluck, 2002; Jencks, 1994). One of the main reasons for this increase was the erosion of government support for public housing (Rossi, 1989; Liebow, 1993).[4]

---

[4]Just as the Reagan administration was cutting welfare, it lowered the top personal tax bracket. This change substantially increased the amount of disposable income in the hands of the wealthiest Americans (Phillips, 1990). In this way, politics helped to increase the level of inequality in American society.

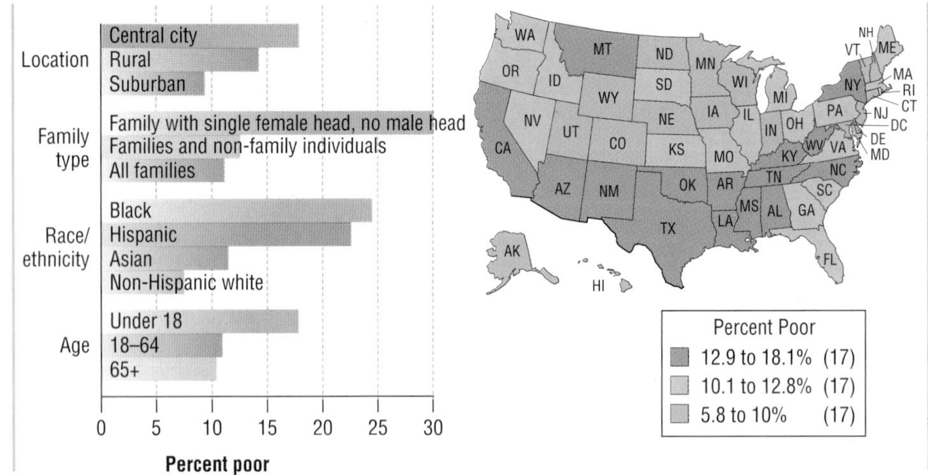

Figure 8.7
Poverty in the United States, 2003

Source: U.S. Census Bureau (2004b).

Poverty Myths

Research conducted in the 1990s showed that many of the beliefs underlying the war against the poor are inaccurate:

- *Myth 1: The overwhelming majority of poor people are African American or Hispanic American single mothers with children.* The fact is that while close to a quarter of African Americans and Hispanic Americans were poor in 2002, fully 68 percent of the poor were non-Hispanic whites (Figure 8.7). Moreover, 38 percent of the poor lived in married-couple families. Twenty-eight percent lived alone, with nonrelatives, or in male-headed families with no wife present. Just 34 percent lived in families with a female head and no husband (Proctor and Dalaker, 2003).
- *Myth 2: People are poor because they do not want to work.* Among poor people older than the age of 15, 38 percent *did* work in 2002, 11 percent of them full-time and year-round. Moreover, many poor people are simply too young or too old to work, as suggested by the fact that 45 percent of poor people are younger than 18 or older than 65 (Proctor and Dalaker, 2003). Finally, many poor people are unable to work for reasons of health or disability—or because they are single parents who have to stay at home to care for their children due to the unavailability of affordable child care. The following comment of a 32-year-old never-married mother of two is typical of welfare recipients' attitude to work: "I feel better about myself when I'm working than when I'm not. Even if I had a job and every penny went to living from payday to payday, it doesn't bother me because I feel like a better person because I am going to work" (quoted in Rank, 1994: 111).
- *Myth 3: Poor people are trapped in poverty.* In fact, the poverty population is dynamic. People are always struggling to move out of poverty. They often succeed, at least for a time. Only about 12 percent of the poor remain poor five or more years in a row (O'Hare, 1996: 11).
- *Myth 4: Welfare encourages married women with children to divorce so they can collect welfare, and it encourages single women on welfare to have more children.* Because some welfare payments increase with the number of children in the family, some people believe that welfare mothers give birth to more children to get more money from the government. In fact, women on welfare have a lower birthrate than women in the general population (Rank, 1994). Moreover, welfare payments are very low and recipients therefore suffer severe economic hardship. This is hardly an incentive to go on welfare. In the words of a 51-year-old divorced welfare mother: "I can't see anybody that would ever settle for something like this just for the mere fact of getting a free ride, because it's not worth it" (quoted in Rank, 1994: 168). As for the argument that people with children often divorce to collect welfare, the most rigorous study of wel-

BOX 8.2

SOCIAL POLICY: WHAT DO YOU THINK?

Redesigning Welfare

One of the most important experiments in American history is taking place today. Wisconsin is at the forefront of the experiment. In 1986 the state began testing pilot programs aimed at reducing dependency and increasing the economic self-sufficiency of families. Ten years later, President Clinton signed a federal welfare law replacing Aid to Families with Dependent Children, or AFDC, with a program of time limits and work requirements. In 1997 Wisconsin introduced a tough new plan called W-2. It required nearly all welfare recipients to work. It also imposed a strict five-year limit on how long one could receive welfare assistance. So it came about that the age of welfare passed and the age of "workfare" dawned first in Wisconsin. Nationally, the welfare caseload dropped 47 percent between 1994 and 1999. In Wisconsin it dropped 91 percent.

Pointing to the drop in the welfare caseload, some observers consider the Wisconsin experiment a brilliant success. Closer inspection reveals mixed results (Newman, 1999; Massing, 1999). Roughly one-third of former AFDC recipients in Wisconsin are working and earning $9 or $10 an hour, well above minimum wage. Getting by on this kind of wage is extraordinarily hard, even if it is an improvement over welfare (Ehrenreich, 2001). Another third of former AFDC recipients are working at menial jobs close to minimum wage. The bottom one-third have few if any skills. Many of them suffer psychological, drug, or alcohol problems. For the most part, they have not been able to find work and many of them cannot pay rent. They live in Wisconsin's shelters and rely on handouts for food. The city's shelters have become so crowded that the Red Cross runs overflow sites in church basements during the winter. In 1999 the state of Wisconsin published the results of a survey of mothers who had left the welfare system in the first quarter of 1998. Sixty-nine percent said they were just barely making do. Thirty-three percent said they could not afford child care. Seventy-five percent of those who found jobs lost them within nine months. Only 16 percent enjoyed earnings above the poverty threshold. The proportion of poor children living with two married parents increased in Wisconsin and nationally between 1995 and 2000, partly because of welfare-to-work laws that encourage poor women to marry for the sake of greater financial stability. However, most of these arrangements appear to be high-conflict, unstable families that are no better for the children than single-parent families (Harden, 2001).

The mixed results of the Wisconsin experiment force us to reconsider welfare policy. The direction of welfare policy is clear: The day before Valentine's Day 2003, the U.S. House of Representatives passed a bill imposing still stricter work requirements on poor people who receive cash assistance from the federal government. However, the results of this get-tough policy are mixed at best. Most poor Americans without a job cannot work because they are too young, too old, too disabled, or have at least one young child at home and no male adult in the household. Many single mothers with children lack the skills and child-care services that would help them get a job. Thus, without more child care and job training, workfare's ability to eliminate poverty in America is questionable. Meanwhile, former AFDC recipients who do find work barely manage to scrape by. Nationally in 2002, they were working a 35-hour week at just under $8 an hour on average, giving them an annual income of $14,560 (Edelman, 2002).

Another problem is that politicians designed and implemented the current welfare policy in the mid-1990s, during an economic boom. The number of people on welfare fell nationally by 60 percent between 1996 and 2002, but then the welfare rolls started growing again because the country was in recession and low-wage jobs became scarce. As early as 2001, many states had to scrap the five-year limit on receiving welfare benefits because so many people who could not find work were in desperate need (Associated Press, 2002; Kilborn, 2001; Pear, 2003). Whether a program designed for good times can survive the bad is unclear.

We conclude that the challenge Americans face in the 21st century is to develop innovative programs that can prevent welfare dependency without punishing the powerless and the destitute.

fare recipients in the United States to date found that welfare programs "have little effect on the likelihood of marriage and divorce" (Rank, 1994: 169).

- *Myth 5: Welfare is a strain on the federal budget.* "Means-tested" welfare programs require that recipients meet an income test to qualify. Such programs accounted for a mere 6 percent of the federal budget in 2001 (Executive Office, 2000). In percentage terms, this amount is substantially less than what is spent by western European governments for similar programs.[5]

[5]Incidentally, a study published by the Federal Reserve Bank of Cleveland found that eliminating the new-birth increment would save only 3 percent of the AFDC budget (Powers, 1994).

Box 8.3

YOU AND THE SOCIAL WORLD

Perceptions of Class

We expect that you have had some strong reactions to our review of sociological theories and research on social stratification. You may therefore find it worthwhile to reflect more systematically on your own attitudes to social inequality and the attitudes of people unlike you. One way of doing that is to ask an acquaintance who is least like you in terms of class position the following questions, and then answer the following questions yourself. You can then compare the two sets of answers with the American averages given in the text:

- Do you consider the family in which you grew up to have been lower class, working class, middle class, or upper class?
- Do you think the gaps between classes in American society are big, moderate, or small?
- How strongly do you agree or disagree with the view that big gaps between classes are needed to motivate people to work hard and maintain national prosperity? (strongly agree, agree, neither, disagree, strongly disagree)
- How strongly do you agree or disagree with the view that inequality persists because it benefits the rich and the powerful? (strongly agree, agree, neither, disagree, strongly disagree)
- How strongly do you agree or disagree with the view that inequality persists because ordinary people do not join together to eliminate it? (strongly agree, agree, neither, disagree, strongly disagree)

In 500 words, compare your perceptions and evaluations of the American class structure with those of your acquaintance and the American public in general. How do you account for your views, the views of your acquaintance, and the views of the American public?

We conclude that sociological evidence does not support many of the arguments people use to justify current welfare reforms, nor are current welfare reforms enjoying unqualified success in reducing poverty (Box 8.2). Western Europe, with a poverty rate roughly half that of ours, seems to be having more success in this regard. That is because the governments of western Europe have established job-training and child-care programs that allow poor people to take jobs with livable wages and benefits. Most of the 35.9 million Americans living in poverty are eager to work. Providing them with the required skills and the opportunity to work could possibly bring the poverty rate down to western European levels. Simply denying them welfare seems to do little to help alleviate poverty.[6]

Perception of Class Inequality in the United States

Surveys show that few Americans have trouble placing themselves in the class structure when asked to do so (Box 8.3). The General Social Survey (GSS) has been asking Americans almost annually since 1972 whether they consider themselves lower class, working class, middle class, or upper class. Of the nearly 44,000 respondents over the years, a mere 279 (0.6 percent) said either that they did not know which class they are in or that they are not members of any class. Just over 3 percent said they are upper class, and just over 5 percent said they are lower class. About 46 percent said they are working class, and about the same percentage said they are middle class. These percentages changed little from 1972 to 2002 (National Opinion Research Center, 2004; see also Jackman and Jackman, 1983; Vanneman and Cannon, 1987).

Sociology Now™

Learn more about **Class Structure** by going through the American Class Structure Learning Module.

If Americans see the stratification system as divided into classes, they also know that the gaps between classes are relatively large. For instance, one study compared respondents in New Haven, Connecticut, and London, England. The study revealed that Americans correctly perceive more inequality in their society than the British do in theirs (Bell and Robinson, 1980; Robinson and Bell, 1978).

[6]Whether denying welfare even helps the taxpayer is unclear. To our knowledge, nobody has compared the direct tax saving gained from cutting welfare with the indirect cost of dealing with the consequences of widespread poverty, such as bigger Medicaid bills, larger budgets for police and prison services, and so forth.

Do we think these big gaps between classes are needed to motivate people to work hard, thus increasing their own wealth and the wealth of the nation? Some Americans think so, but most do not. A survey conducted in 18 countries asked more than 22,000 respondents (including nearly 1200 Americans) whether large differences in income are necessary for national prosperity. Americans were among the most likely to *disagree* with that view (Pammett, 1997: 77).

So we know we live in a class-divided society. We also tend to think that deep class divisions are not necessary for national prosperity. Why then do we think inequality continues to exist? The 18-nation survey cited previously sheds light on this issue. One of the survey questions asked respondents how strongly they agreed or disagreed with the view that "inequality continues because it benefits the rich and powerful." Most Americans agreed with that statement. Only 23 percent disagreed with it in any way. Another question asked respondents how strongly they agreed or disagreed with the view that "inequality continues because ordinary people don't join together to get rid of it." Again, most Americans agreed. Only 30 percent disagreed in any way (Pammett, 1997: 77–8).

Government's Role in Reducing Inequality

Despite widespread awareness of inequality and considerable dissatisfaction with it, most Americans are opposed to the government playing an active role in reducing inequality. Most of us do not want government to provide citizens with a basic income. We tend to oppose government job-creation programs. We even resist the idea that government should reduce income differences through taxation (Pammett, 1997: 81). Most Americans remain individualistic and self-reliant. On the whole, we persist in the belief that opportunities for mobility are abundant and that it is up to the individual to make something of those opportunities by means of talent and effort (Kluegel and Smith, 1986).

Significantly, however, all the attitudes summarized previously vary by class position. For example, discontent with the level of inequality in American society is stronger at the bottom of the stratification system than at the top. The belief that American society is full of opportunities for upward mobility is stronger at the top of the class hierarchy than at the bottom. One finds considerably less opposition to the idea that government should reduce inequality as one moves down the stratification system. These findings permit us to conclude that if Americans allow inequality to persist, it is because the *balance* of attitudes—and of power—favors continuity over change. We take up this important theme again in Chapter 14, where we discuss the social roots of politics.

Summary

Sociology Now™

Reviewing is as easy as ❶, ❷, ❸.

❶ Before you do your final review, take the SociologyNow diagnostic quiz to help you identify the areas on which you should concentrate. You will find information on how to access SociologyNow on the foldout at the front of the textbook.

❷ As you review, take advantage of SociologyNow's study aids to help you master the topics in this chapter.

❸ When you are finished with your review, take SociologyNow's post-test to confirm you are ready to move on to the next chapter.

1. What is the difference between wealth and income? How are they distributed in the United States?

Wealth is assets minus liabilities. Income is the amount of money earned in a given period. Substantial inequality of both wealth and income exists in the United States, but inequality of wealth is greater. Both types of inequality have increased over the past quarter of a century. The United States leads the other highly industrialized countries in both measures of inequality.

2. How does inequality change as societies develop?

Inequality increases as societies develop from the foraging to the early industrial stage. With increased industrialization, inequality declines. In the early stages of postindustrialism, inequality then increases in some countries (e.g., the United States) but not in others where governments take a more active role in redistributing income (e.g., France).

3. **What are the main differences between Marx's and Weber's theories of stratification?**

   Marx's theory of stratification distinguishes between classes on the basis of their role in the productive process. It predicts inevitable conflict between bourgeoisie and proletariat and the birth of a communist system. Weber distinguished between classes on the basis of their "market relations." His model of stratification included four main classes. He argued that class consciousness may develop under some circumstances but is by no means inevitable. Weber also emphasized prestige and power as important noneconomic sources of inequality.

4. **What is the functional theory of stratification?**

   Davis and Moore's functional theory of stratification argues that (1) some jobs are more important than others, (2) people have to make sacrifices to train for important jobs, and (3) inequality is required to motivate people to undergo these sacrifices. In this sense, stratification is "functional."

5. **What is Blau and Duncan's theory of stratification?**

   Blau and Duncan viewed the stratification system as a ladder with hundreds of occupational rungs. Rank is determined by the income and prestige associated with each occupation. On the basis of their studies, they concluded that the United States enjoys an achievement-based stratification system. However, many sociologists subsequently concluded that being a member of certain social categories limits one's opportunities for success. In this sense, social structure shapes the distribution of inequality.

6. **How did Wright and Goldthorpe update the ideas of Marx and Weber, respectively?**

   Wright updated Marx's class schema by distinguishing three classes of property owners (based on capitalization and number of employees) and nine classes of nonowners of property (based on skill levels and organizational assets). Goldthorpe revised Weber's class schema by distinguishing classes on the basis of their "employment relations" and then making finer distinctions on the basis of economic sector, skill, and so forth.

7. **Is stratification based only on economic criteria?**

   No. People often engage in conspicuous consumption, waste, and leisure to signal their position in the social hierarchy. Moreover, politics often influences the shape of stratification systems by changing the distribution of income, welfare entitlements, and property rights.

8. **How do Americans view the class system?**

   Most Americans are aware of the existence of the class system and their place in it. They believe that large inequalities are not necessary to achieve national prosperity. Most Americans also believe that inequality persists because it serves the interests of the most advantaged members of society and because the disadvantaged do not join together to change things. However, most Americans disapprove of government intervention to lower the level of inequality.

Questions to Consider

1. How do you think the American and global stratification systems will change over the next 10 years? Over the next 25 years? Why do you think these changes will occur?
2. Why do you think most Americans oppose more government intervention to reduce the level of inequality in society? In answering this question, think about the advantages that inequality brings to many people and the resources at their disposal for maintaining inequality.
3. Compare the number and quality of public facilities such as playgrounds and libraries in various parts of your community. How is the distribution of public facilities related to the socioeconomic status of neighborhoods? Why does this relationship exist?

Web Resources

Companion Website for This Book

http://sociology.wadsworth.com

Begin by clicking on the Student Resources section of the website. Choose "Introduction to Sociology" and finally the Brym and Lie book cover. Next, select the chapter you are currently studying from the pull-down menu. From the Student Resources page you will have easy access to InfoTrac® College Edition, MicroCase Online exercises, additional web links, and many other resources to aid you in your study of sociology, including practice tests for each chapter.

Recommended Websites

For information on the 400 richest people in America, visit the annual compilation of *Forbes* business magazine at http://www.forbes.com/lists/forbes400/2004/09/22/rl04land.html.

An excellent analysis of Americans' real earnings from the 1870s to the 1990s can be found at http://www.panix.com/~dhenwood/Stats_earns.html.

How rich or poor are you compared with everyone else in the world? To find out, visit the Global Rich List at http://www.globalrichlist.com.

For recent government data and analysis of poverty in the United States, see http://www.census.gov/prod/2003pubs/p60-222.pdf.

For United Nations data on global inequality, see the *United Nations Development Report 2004* at http://hdr.undp.org/reports/global/2004.

CHAPTER 9

Globalization, Inequality, and Development

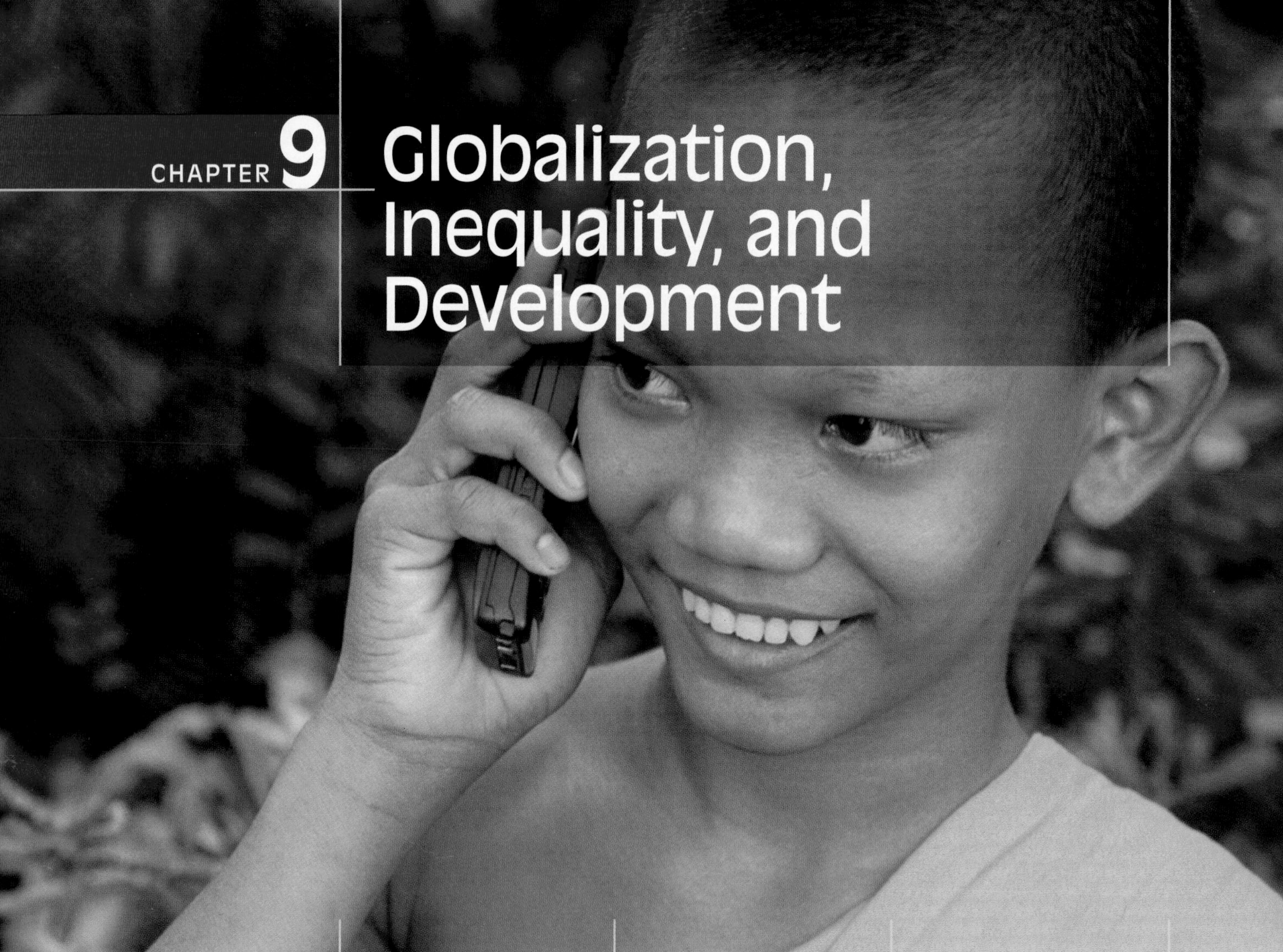

Jeremy Horne/Getty Images

In this chapter, you will learn that:

- People and institutions across the planet are becoming increasingly aware of, and dependent on, one another. Sociologists call this tendency "globalization."
- Globalization creates a world that is more homogeneous in some ways and more localized in others. Globalization also generates its own opposition.
- Globalization is a process that became significant along with the development of capitalism and world exploration about 500 years ago.
- Global inequality has increased tremendously since industrialization and is still increasing in some respects today.
- Global inequality has two competing explanations. One stresses how the deficiencies of some societies contribute to their own lack of economic growth. The other stresses how the history of social relations among countries enriched some nations at the expense of others.
- For identifiable reasons, some non-Western countries have successfully industrialized.
- Globalization has both benefits and disadvantages. Various reforms—especially more democratic participation in economic decision making—can increase the benefits.

The Creation of a Global Village

Suppose you decide to travel to Europe. You might check the Internet to buy an inexpensive ticket. You would then get your passport and perhaps buy a guidebook. Depending on the kind of person you are, you might spend a lot of time planning and preparing for the trip or you might just pack the basics—the passport, the ticket, a knapsack full of clothes, your credit card—and embark on an adventure.

How different things were just 25 years ago. Then, you probably would have gone to see a travel agent first, as most people did when they wanted airline tickets. Next, you would have had to make sure you had not only a valid passport, but also visas for quite a few countries. Obtaining a visa was a tedious process. You had to drive to an embassy or a consulate or mail in your passport. Then you had to wait days or weeks to receive the visa. Today, visas are required for fewer countries.

The next step in organizing the European trip would involve withdrawing money from your bank account. ATMs were still rare. You had to stand in line at the bank before getting to a teller. Next, you had to take the withdrawn money to the office of a company that sold traveler's checks because banks did not sell them. When you arrived in Europe, you needed to have local currency, which you could buy only in large banks and from moneychangers. Few college students had credit cards 25 years ago. Even if you were one of the lucky few, you could use it only in large stores and restaurants in large cities. If you ran out of cash and traveler's checks, you were in big trouble. You could look for another American tourist and try to convince him or her to take your personal check. Alternatively, you could go to a special telephone for international calls, phone home, and have money wired to a major bank for you. That was time consuming and expensive. Today, most people have credit cards and ATM cards. Even in small European towns, you can charge most of your shopping and restaurant bills on your credit card without using local currency. If you need lo-

Chuck Savage/Corbis

Ariel Skelley/Corbis

Currency conversion, then and now.

cal currency, you go to an ATM and withdraw money from your home bank account or charge it to your credit card; the ATM automatically converts your dollars to local currency.

Most European cities and towns today have many American-style supermarkets. Speakers of English are also numerous, so social interaction with Europeans is easier. You would have to try hard to get very far from a McDonald's. It is also easy to receive news and entertainment from back home on CNN and MTV. In contrast, each country you visited 25 years ago would have featured a distinct shopping experience. English speakers were rarer. American fast-food outlets were practically nonexistent. Apart from the *International Herald-Tribune,* which was available only in larger towns and cities, American news was hard to come by. TV featured mostly local programming. You might see an American show now and then, but it would not be in English.

Clearly, the world seems a much smaller place today than it did 25 years ago. Some people go so far as to say that we have created a "global village." But what exactly does that mean? Is the creation of a global village uniformly beneficial? Or does it have a downside too? We now explore these questions in depth.

Sociology Now™

Learn more about **Globalization** by going through the Globalization Video Exercise.

The Triumphs and Tragedies of Globalization

As suggested by our two imaginary trips to Europe, separated by a mere 25 years, people throughout the world are now linked together as never before (Table 9.1 and Figure 9.1). Consider these facts:

- The growth in tourism is one indicator of our denser international ties. In 1980 just 3.5 percent of the world's population traveled internationally as tourists. By 2001 that figure had more than tripled to 11.3 percent.
- Many more international organizations and agreements now span the globe. In 1980 about 14,000 international organizations existed in the world. By 1999, there were three and a half times as many. Individual nation-states give up some of their independence when they join international organizations or sign international agreements. For example, when the United States, by far the world's most powerful country, entered the North American Free Trade Agreement (NAFTA) with Canada and Mexico in 1992, it agreed that trade disputes would be adjudicated by a three-country tribunal. The autonomy of nation-states has eroded somewhat with the creation of many "transnational" bodies and treaties.
- International telecommunication has become easy and inexpensive. In 1930 a three-minute telephone call from New York to London cost nearly $250 and only a minority of Americans had telephones in their homes. Today the same call costs as little as 15¢, and telephones, including cell phones, seem to be everywhere. The Internet did not exist in 1980, but by 2002 it was composed of 165 million servers connecting

Table 9.1
Indicators of Globalization, early 1980s–circa 2003

	1980–81	1998–2003	Percentage Change
Foreign direct investment as percentage of gross domestic product	4.6[1]	8.8[5]	91.3
International tourist arrivals as percentage of world population	3.5[1]	11.3[6]	222.9
Air passengers (millions)	748[1]	1,656[5]	121.4
Air freight and mail (billions of tons/km)	331	1245	275.8
Internet hosts (millions)	0[1]	165[7]	Undefined
Number of international organizations	14,273[2]	50,373[4]	252.9
Annual entries on globalization in Sociological Abstracts	89[1]	1,009[3]	1,033.7

Note: [1]1980, [2]1981, [3]1998, [4]1999, [5]2000, [6]2001, [7]2003.
Sources: Guillén (2001); International Civil Aviation Organization (2002); Internet Software Consortium (2002); U.S. Census Bureau (2002); World Bank (2002); World Tourism Organization (2002); "International Organizations by Year and Type," Table 2 (2001).

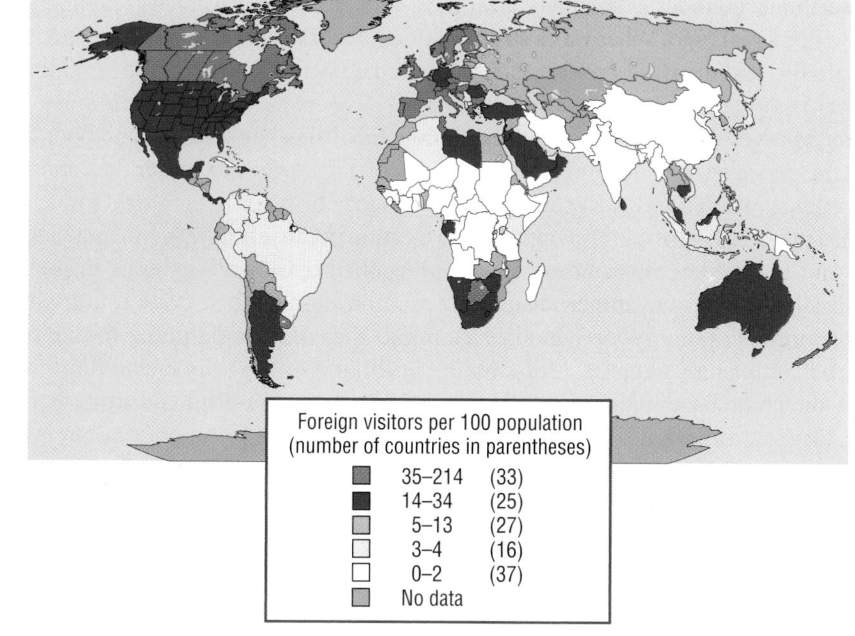

Figure 9.1
Foreign Visitors per 100 Population

International tourism leads to the sense that the world forms one society. However, exposure to international tourism is higher in some countries than in others. This map shows the number of foreign visitors who traveled to each country per 100 members of the local population. Which countries are most exposed to international tourism? Which countries are least exposed? What consequences might differential exposure have for people's self-identity?

Note: Data are for 1998.

United Nations Educational, Scientific, and Cultural Organization (2001).

nearly 600 million people from around the world through e-mail, file transfers, websites, and videoconferencing.

- International trade and investment has increased rapidly. For example, from 1980 to 2000, foreign direct investment as a percent of gross domestic product (GDP) nearly doubled.[1]

Sociology Now™

Learn more about **Telecommunications** by going through the Mobil Telephones Map Exercise.

These and other indicators point unmistakably to the "globalization" of the planet, which we defined in Chapter 1 as the process by which formerly separate economies, states, and cultures are being tied together and people are becoming increasingly aware of their growing interdependence (Giddens, 1990: 64; Guillén, 2000). Sociologists have taken

[1]Gross national product (GNP) is the total dollar value of goods and services produced in a country in a year. It allocates goods and services based on the nationality of the owners. GDP is a similar measure but it allocates goods and services based on the location of the owners. So, for example, goods and services produced overseas by foreign subsidiaries would be included in GNP but not in GDP.

Technology speeds globalization.

Brian A. Vikander/Corbis

Roy McMahon/Corbis

note; the number of published sociology articles on globalization increased 11-fold between 1980 and 1998.

The benefits of the rapid movement of capital, commodities, culture, and people across national boundaries should be clear from comparing our imaginary trips to Europe today and 25 years ago. What was a struggle for an American traveler 25 years ago is easy today. In this and many other ways, globalization has transformed and improved the way we live.

Yet, not everyone is happy with globalization. Inequality between rich and poor countries remains staggering. In some respects, it is increasing. Arguably, rather than spreading the wealth, globalized industries and technologies may be turning the world into a more unequal place. Many people also oppose globalization because it may be hurting local cultures and the natural environment. Some antiglobalization activists even suggest that globalization is a form of **imperialism,** the economic domination of one country by another. From their point of view, globalization puts the entire world under the control of powerful commercial interests. Moreover, it contributes to the "homogenization" of the world, the cultural domination of less powerful by more powerful countries. It is one thing, they say, for Indonesians and Italians to have closer ties to Americans, but is it desirable that they become *like* Americans?

In the next section, we first explore what globalization is and how it affects our everyday lives. To make the impact of globalization concrete, we trace the global movement of two commodities familiar to everyone: money and athletic shoes. This exercise illuminates the many ways in which far-flung individuals are bound together. Next, we consider the causes of globalization, emphasizing the importance of political, economic, and technological factors. We then analyze whether globalization is forcing different parts of the world to become alike. We also explore how the very process of globalization generates its own opposition. Finally, we take a longer view and examine the roots of globalization historically.

The second task we set ourselves is to examine the nature and causes of global inequality. We note that the gap between rich and poor countries is wide and that by some measures it is getting wider. We then discuss the major theories that seek to explain global inequality and conclude that poor countries are not doomed to remain poor. We close by considering what people can do to alleviate global inequality and poverty.

Globalization

Globalization in Everyday Life

Imperialism is the economic domination of one country by another.

Has anyone ever asked you, "What's that got to do with the price of tea in China?" The question is of course rhetorical. The person who asked it was really saying that whatever you were talking about is as *irrelevant* as the price of tea in China.

Yet globalization implies that the price of tea in China—and the condition of the rain forest in Brazil, and the outcome of the war in the Russian province of Chechnya, and the revival of traditional Celtic dance in Ireland—actually do influence your life. Everything influences everything else in a globalized world.

To explore this idea, Barbara Garson (2001) traced a small sum of money she invested in the Chase Manhattan Bank. Garson discovered that the bank almost immediately lent some of her money (along with much additional money) to Caltex, a large U.S. corporation, for the construction of an oil refinery in Thailand. Initially, Thai officials opposed the refinery, mainly because Thai fishers and farmers believed the refinery would pollute and eventually destroy their livelihood and their villages. However, intensive lobbying by American politicians on behalf of Caltex eventually led Thai officials to relent and allow construction. Caltex secured the assistance of the politicians by using some of Garson's money for political party contributions. Caltex helped change the minds of Thai officials by using some of Garson's money for bribes. Some of Garson's money was used to help construct the oil refinery, which did contribute to the destruction of the natural environment and the displacement of farmers and fishers. Eventually, because of an economic downturn in Thailand, the refinery was shut down. When it closed, many Thai oil workers became unemployed. In short, Garson learned that her small investment, together with many other small investments, had huge implications for people's everyday lives on the other side of the planet: "Wherever my money went, it changed the landscape and it changed lives," wrote Garson (2001: 319). Garson's money got caught up in the globalization of the world.

Sandy Felsenthal/Corbis

Michael Jordan helps globalize Nike.

Nike, Michael Jordan, and Global Commodity Chains

The difference between Garson and most of us is that we do not often appreciate that our actions have implications for people far away. Yet, when we buy a commodity, we often tap into a **global commodity chain,** "a [worldwide] network of labor and production processes, whose end result is a finished commodity" (Hopkins and Wallerstein, 1986: 159).

We can better understand the web of global social relations by tracing the way one commodity—athletic shoes—binds consumers and producers in a global commodity chain. Until the 1970s, most sports shoes were manufactured in the United States. Today, corporations such as Nike produce all their shoes abroad. Manufacturing plants moved from the United States because foreign governments eliminated many of the laws, regulations, and taxes that had acted as barriers to foreign investment and trade. Consequently, Nike and other manufacturers started setting up overseas plants, where they could take advantage of low labor costs. The result was a new international division of labor. High-wage management, finance, design, and marketing services were concentrated in the United States and other advanced industrial countries, and low-wage manufacturing in the less developed, industrializing countries (Fröbel, Heinrichs, and Kreye, 1980) (see Chapter 13, "Work and the Economy"). Not just Nike, but General Motors, General Electric, and many other large corporations closed some or all of their plants in the United States (a high-wage country) and established factories in Mexico, Indonesia, and other developing economies.

A **global commodity chain** is a worldwide network of labor and production processes whose end result is a finished commodity.

The new international division of labor yielded high profits. For example, in the 1990s, "Nike spent $5.95 to make a pair of shoes in Indonesia, where workers were paid 14 cents an hour, then sold the shoes in the United States for between $49 and $125" (LaFeber, 1999: 126). Nike's Indonesian workers were not just poorly paid. They had to work as much as 6 hours a day overtime. Reports of beatings and sexual harassment by managers were common. When the workers tried to form a union to protect themselves, union organizers were fired and the military was brought in to restore order (LaFeber, 1999: 142). In 1976 Nike manufactured 70 million shoes in Indonesia but paid its 25,000 workers an average of just over $2 a day. At the same time, Michael Jordan became a spokesman for Nike and was paid a $20 million endorsement fee, more than the combined yearly wages of all the Indonesian workers who made the shoes (LaFeber, 1999: 107).

When someone buys a pair of Nike athletic shoes, they insert themselves into a global commodity chain. Of course, the buyer does not create the social relations that exploit Indonesian labor and enrich Michael Jordan. Still, it would be difficult to deny the buyer's part, however small, in helping those social relations persist.

Sociology makes us aware of the complex web of social relations and interactions in which we are embedded. The sociological imagination allows us to link our biography with history and social structure. Globalization extends the range of that linkage, connecting our biography with global history and global social structure. We live in a world where the price of tea in China is increasingly relevant to our lives.

The Sources of Globalization

Few people doubt the significance of globalization. Although social scientists disagree on its exact causes, most of them stress the importance of technology, politics, and economics.

Technology

Technological progress has made it possible to move things and information over long distances quickly and inexpensively. The introduction of commercial jets radically shortened the time necessary for international travel, and its cost dropped dramatically after the 1950s. Similarly, various means of communication, such as telephone, fax, and e-mail, allow us to reach people around the globe inexpensively and almost instantly. Whether we think of international trade or international travel, technological progress is an important part of the story of globalization. Without modern technology, it is hard to imagine how globalization would be possible.

Sociology Now™

Learn more about **Technology and Globalization** by going through the Technology and Change Learning Module.

Politics

Globalization could not occur without advanced technology, but advanced technology by itself could never bring globalization about. Think of the contrast between North Korea and South Korea. Both countries are about the same distance from the United States. You have probably heard of major South Korean companies like Hyundai and Samsung and may have met people from South Korea or their descendants, Korean Americans. Yet, unless you are an expert on North Korea, you will have had no contact with North Korea and its people. We have the same technological means to reach the two Koreas. Yet, although we enjoy strong relations and intense interaction with South Korea, we lack ties to North Korea. The reason is political. The South Korean government has been a close ally of the United States since the Korean War in the early 1950s and has sought greater political, economic, and cultural integration with the outside world. North Korea, in an effort to preserve its authoritarian political system and socialist economic system, has remained isolated from the rest of the world. As this example shows, politics is important in determining the level of globalization.

Economics

Finally, economics is an important source of globalization. As we saw in our discussion of global commodity chains and the new international division of labor, industrial capitalism is always seeking new markets, higher profits, and lower labor costs. Put differently, capitalist competition has been a major spur to international integration (Stopford and Strange, 1991; Gilpin, 2001).

Transnational corporations—also called multinational or international corporations—are the most important agents of globalization in the world today. They are different from traditional corporations in five ways (LaFeber, 1999; Gilpin, 2001):

1. Traditional corporations rely on domestic labor and domestic production. Transnational corporations depend increasingly on foreign labor and foreign production.
2. Traditional corporations extract natural resources or manufacture industrial goods. Transnational corporations increasingly emphasize skills and advances in design, technology, and management.
3. Traditional corporations sell to domestic markets. Transnational corporations depend increasingly on world markets.
4. Traditional corporations rely on established marketing and sales outlets. Transnational corporations depend increasingly on massive advertising campaigns.
5. Traditional corporations work with or under national governments. Transnational corporations are increasingly autonomous of national governments.

John Van Hasselt/Corbis Sygma

"I want to emphasize that the embassy and the various U.S. government agencies in Washington will keep the interests of Philip Morris and the other American cigarette manufacturers in the forefront of our daily concerns." (Commercial counselor, U.S. embassy, Seoul, South Korea, in a 1986 memo to the public affairs manager of Philip Morris Asia)

Technological, political, and economic factors do not work independently in leading to globalization. For example, governments often promote economic competition to help transnational corporations win global markets. Consider Philip Morris, the company that makes Marlboro cigarettes (Barnet and Cavanagh, 1994). Philip Morris introduced the Marlboro brand in 1954. It soon became the country's best-selling cigarette, partly because of the success of an advertising campaign featuring the Marlboro Man. The Marlboro Man symbolized the rugged individualism of the American frontier, and he became one of the most widely recognized icons in American advertising. Philip Morris was the smallest of the country's six largest tobacco companies in 1954, but it rode on the popularity of the Marlboro Man to become the country's biggest tobacco company by the 1970s.

In the 1970s, the antismoking campaign began to have an impact, leading to slumping domestic sales. Philip Morris and other tobacco companies decided to pursue globalization as a way out of the doldrums. Economic competition and slick advertising alone did not win global markets for American cigarette makers, however. The tobacco companies needed political influence to make cigarettes one of the country's biggest and most profitable exports. To that end, the U.S. trade representative in the Reagan administration, Clayton Yeutter, worked energetically to dismantle trade barriers in Japan, Taiwan, South Korea, and other countries. He threatened legal action for breaking international trade law and said the United States would restrict Asian exports unless these countries allowed the sale of American cigarettes. Such actions were critically important in globalizing world trade in cigarettes. As the commercial counselor at the U.S. embassy in Seoul, South Korea, said to the public affairs manager of Philip Morris Asia in a 1986 memo: "I want to emphasize that the embassy and the various U.S. government agencies in Washington will keep the interests of Philip Morris and the other American cigarette manufacturers in the forefront of our daily concerns" (quoted in Frankel, 1996). As the case of Philip Morris illustrates, economics and politics typically work hand in hand to globalize the world.

Transnational corporations are large businesses that rely increasingly on foreign labor and foreign production; skills and advances in design, technology, and management; world markets; and massive advertising campaigns. They are increasingly autonomous of national governments.

A World Like the United States?

We have seen that globalization links people around the world, often in ways that are not obvious. We have also seen that the sources of globalization lie in closely connected technological, economic, and political forces. Now let us consider one of the consequences of

Figure 9.2
The Size and Influence of the U.S. Economy, 2000

This map will help you gauge the enormous importance of the United States in globalization because it emphasizes just how large the U.S. economy is. The economy of each U.S. state is as big as that of a whole country. Specifically, this map shows how the gross domestic product (GDP) of various countries compares with that of each state. For example, the GDP of California is equal to that of France, the GDP of New Jersey is equal to that of Russia, and the GDP of Texas is equal to that of Canada.

Source: *Globe and Mail*, March 8, 2003, p. F1. Reprinted with permission of The Globe and Mail.

globalization, the degree to which globalization is "homogenizing" the world and, in particular, making the whole world look like the United States (Figure 9.2).

Much impressionistic evidence supports the view that globalization homogenizes societies. Many economic and financial institutions around the world now operate in roughly the same way. For instance, transnational organizations such as the World Bank and the International Monetary Fund (IMF) have imposed economic guidelines for developing countries that are similar to those governing advanced industrial countries. In the realm of politics, the United Nations (UN) engages in global governance, whereas Western ideas of democracy, representative government, and human rights have become international ideals (Box 9.1). In the domain of culture, American icons circle the planet: supermarkets, basketball, Hollywood movies, Disney characters, Coca-Cola, CNN, McDonald's, and so forth.

McDonaldization

Indeed, one common shorthand expression for the homogenizing effects of globalization is **McDonaldization.** George Ritzer (1996: 1) defines McDonaldization as "the process by which the principles of the fast-food restaurant are coming to dominate more and more sectors of American society as well as of the rest of the world." The idea of McDonaldization extends Weber's concept of rationalization, the application of the most efficient means to achieve given ends (see Chapter 3, "Culture"). Because of McDonaldization, says Ritzer, the values of efficiency, calculability, and predictability have spread from the United States to the entire planet and from fast-food restaurants to virtually all spheres of life.

McDonaldization is a form of rationalization. Specifically, it refers to the spread of the principles of fast-food restaurants, such as efficiency, predictability, and calculability, to all spheres of life.

As Ritzer shows, McDonald's has lunch down to a science. The ingredients used to prepare your meal must meet minimum standards of quality and freshness. Each food item contains identical ingredients. Each portion weighs the same and is prepared according to a uniform and precisely timed process. McDonald's expects customers to spend as little time as possible eating the food—hence the drive-through window, chairs designed to be comfortable for only about 20 minutes, and small express outlets in subways and department stores where customers eat standing up or on the run. McDonald's is even field-testing self-service kiosks in which an automated machine cooks and bags French fries while a vertical grill takes patties from the freezer and grills them to your liking (Carpenter, 2003). In short, McDonald's executives have carefully thought through every aspect of your lunch. They have turned its preparation into a model of rationality. With the goal of making profits, they have optimized food preparation to make it as fast and as inexpensive as possible. Significantly, McDonald's now does most of its business outside the United States. You can find McDonald's restaurants in nearly every country in the

BOX 9.1

SOCIAL POLICY: WHAT DO YOU THINK?

Should the United States Promote World Democracy?

"In starting and waging a war, it is not right that matters, but victory," said Adolf Hitler (quoted in "A Survey," 1998: 10). The same mindset rationalizes state brutality today, from the Serbian attack on Bosnia and Kosovo to the Indonesian war against East Timor. Cherished ideals, such as political democracy and human rights, are trampled on daily by dictatorships and military governments.

Opinions differ as to what the U.S. government should do about this situation. One influential argument is that of Samuel Huntington. He argues that the United States and other Western nations should not be ethnocentric and impose Western values on people in other countries. If there are violations of democratic principles and human rights in these countries, they express in some way the indigenous values of those people. We should not intervene to stop nondemocratic forces and human rights abuse abroad.

Critics of Huntington argue that the ideals of democracy and human rights can be found in non-Western cultures, too. If we explore Asian or African traditions, for instance, we find "respect for the sacredness of life and for human dignity, tolerance of differences, and a desire for liberty, order, fairness and stability" (quoted in "A Survey," 1998: 10). Although Asian and African despots champion supposedly traditional values, people in Asia, Africa, and elsewhere struggle for democracy and human rights. For example, beginning in the mid-1970s, the Indonesian government tried to smother democracy in East Timor, but the East Timorese fought valiantly for democracy.

What do you think the role of the United States should be? Should the United States government promote democracy and human rights? Or should we avoid what Huntington regards as ethnocentrism? Because foreign aid to authoritarian regimes may help nondemocratic forces and thereby stifle human rights, should we give foreign aid to such regimes?

world. McDonaldization has come to stand for the global spread of values associated with the United States and its business culture.

Sociology Now™

Learn more about **McDonaldization** by going through the McDonaldization of Society Video Exercise.

Glocalization and Symbolic Interactionism

Despite the appeal of the concept of McDonaldization, anyone familiar with symbolic interactionism should be immediately suspicious of sweeping claims about the homogenizing effects of globalization. After all, it is a central principle of symbolic interactionism that people create their social circumstances and do not merely react to them, that they negotiate their identities and do not easily settle for identities imposed on them by others. Accordingly, some analysts find fault with the view that globalization is making the world a more homogeneous place based on American values. They argue that people always interpret globalizing forces in terms of local conditions and traditions. Globalization, they say, may in fact sharpen some local differences. They have invented the term **glocalization** to describe the simultaneous homogenization of some aspects of life and the strengthening of some local differences under the impact of globalization (Shaw, 2000). They note, for example, that McDonald's serves different foods in different countries (Watson, 1997). A typical burger at a Taiwanese McDonald's comes with betel nuts. Vegetarian burgers are the norm at an Indian McDonald's and kosher burgers are the norm at an Israeli McDonald's. A Dutch McDonald's serves the popular McKrocket, made of 100 percent beef ragout fried in batter. In Hawaii, McDonald's routinely serves Japanese ramen noodles with their burgers. Although the Golden Arches may suggest that the world is becoming the same everywhere, once we go through them and sample the fare, we find much that is unique.

Glocalization is the simultaneous homogenization of some aspects of life and the strengthening of some local differences under the impact of globalization.

Regionalization

Those who see globalization merely as homogenization also ignore the **regionalization** of the world, the division of the world into different and often competing economic, political, and cultural areas. The argument here is that the institutional and cultural integration of countries often falls far short of covering the whole world. Figure 9.3 illustrates one aspect of regionalization. It shows that world trade is not evenly distributed around the planet or dominated by just one country. Instead, three main trading blocs

Regionalization is the division of the world into different and often competing economic, political, and cultural areas.

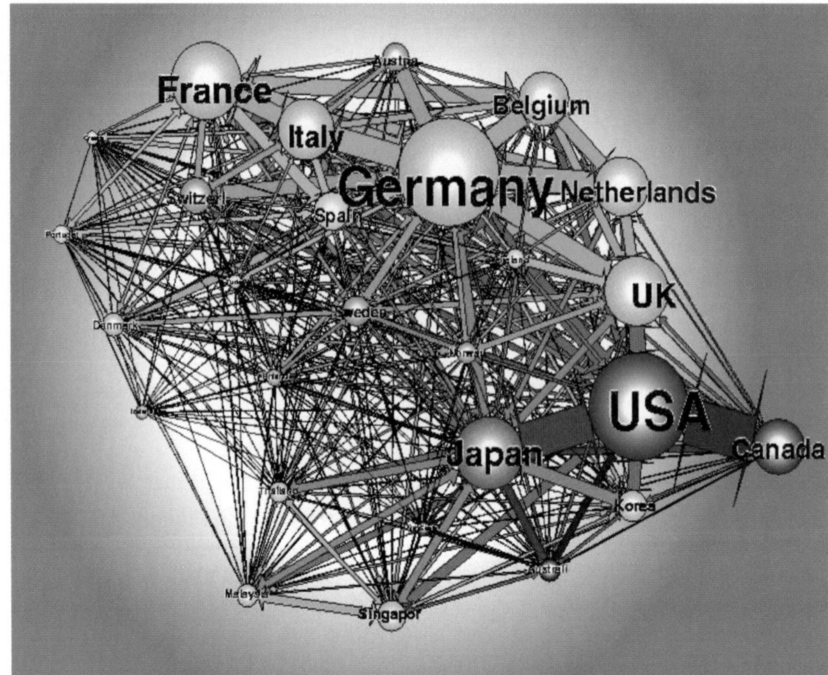

Figure 9.3
The Regionalization of World Trade

This graphic illustrates the network of world trade in 1992. The thickness of lines shows the volume of trade between countries. Colors distinguish regional trading blocs. Note that most world trade took place *within* regional trading blocs, with the United States, Germany, and Japan at the center of each of the three main blocs.

Source: Reprinted with permission of Max Planck Institute for the Study of Society.

exist—an Asian bloc dominated by Japan, a North American bloc dominated by the United States, and a western European bloc dominated by Germany. Most world trade takes place *within* each of these blocs. Each bloc competes against the others for a larger share of world trade. Politically, we can see regionalization in the growth of the European Union. Most western European bloc countries now share the same currency, the euro, and they coordinate economic, political, military, social, and cultural policies. Similarly, Islam has become a unifying religious, political, and cultural force over vast stretches of Asia and Africa. The expansion of Islam, especially in its politically active form, is a form of regional integration, too.

We conclude that globalization does not have a simple, one-way, and inevitable consequence. In fact, as you will now learn, its impact is often messy. Globalization has generated much criticism and opposition, unleashing a growing antiglobalization movement.

Globalization and Its Discontents: Antiglobalization and Anti-Americanism

In 1992 political scientist Benjamin Barber published an important book titled *Jihad vs. McWorld.* Barber argued that globalization (the making of what he called "McWorld") was generating an antiglobalization reaction, which he called *jihad. Jihad* means "striving" or "struggle" in Arabic. Traditionally, Muslims use the term to mean perseverance in achieving a high moral standard. The term can also suggest the idea of a holy war against those who harm Muslims. In the latter sense, it represents an Islamic fundamentalist reaction to globalization. The most spectacular and devastating manifestations of fundamentalist Islamic *jihad* were the September 11, 2001, jet hijackings that led to the crash of an airliner in Pennsylvania and the destruction of the World Trade Center in New York City and part of the Pentagon. About 3000 people, the great majority Americans, lost their lives as a result of these attacks. The operatives of the al Qaeda network sought to roll back the forces of globalization by attacking what they thought symbolized the global reach of godless, American capitalism.

Islamic fundamentalism is the most far-reaching and violent of many reactions against globalization throughout the world. Another well-known example involves Mexican peasants in the southern state of Chiapas, who staged an armed rebellion against the Mexican government in the 1990s. The government provoked the peasants by turning over land, which was long regarded as communally owned, to commercial farmers so that they could increase their exports to the United States and Canada under the terms of NAFTA. The uprising involved violence and lasted several years.

Finally, we should mention the antiglobalization movement in the advanced industrial countries (Klein, 2000). In 1994 the governments of 134 countries set up the World Trade Organization (WTO) to encourage and referee global commerce. When the WTO met in Seattle in December 1999, 40,000 union activists, environmentalists, supporters of worker and peasant movements in developing countries, and other opponents of transnational corporations staged protests that caused property damage and threatened to disrupt the proceedings. The police and the National Guard replied with concussion grenades, tear gas, rubber bullets, and mass arrests.

Subsequent meetings of the WTO and allied organizations in other countries met with the same sort of protest on the part of antiglobalization forces. These protests were for the most part nonviolent, often using street drama to make their point. For example, in spring 2001, 34 heads of government from North, Central and South America, and the Caribbean gathered in Québec City to discuss the economic integration of the Americas and related matters. Among other tactics, protesters built large wooden catapults that sent volleys of miniature teddy bears at the riot police.

These examples illustrate that although globalization is far from universally welcome, the antiglobalization movement has many currents. Some are extremely violent, some nonviolent. Some reject only what they regard as the excesses of globalization, others reject globalization in its entirety. This complexity supports our view that globalization is not a simple process with predictable consequences. It is a multifaceted phenomenon, the outcome of which is unclear.

The History of Globalization

The extent of globalization since about 1980 is unprecedented in world history. Sociologist Martin Albrow therefore argues that the "global age" is only a few decades old (Albrow, 1997). He dates it from the spread of global awareness and skepticism about the benefits of modernization. However, Albrow and others are inclined to exaggerate the extent of globalization (Hirst and Thompson, 1999; Gilpin, 2001). The nation-state is still a major center of power in the world. National borders remain important. Truly transnational corporations are rare; most concentrate their business in a single country. Most foreign trade occurs between advanced industrial countries and, as we have seen, within distinct regional groups of advanced industrial countries. Many developing countries are poorly integrated into the global economy. Most people in the world have little or no access to the advanced technologies that exemplify globalization, such as e-mail. Cultural differences remain substantial across the planet.

Furthermore, globalization is not as recent as Albrow would have us believe. Anthony Giddens (1990) argues that globalization is the result of industrialization and modernization, which picked up pace in the late 19th century. And in fact a strong case can be made that the world was highly globalized 100 or more years ago. In the late 19th century, people could move across national borders without passports. The extent of international trade and capital flow in the late 20th century only restored the level achieved before World War I (1914–18) (Hirst and Thompson, 1999).

World War I and the Great Depression (1929–39) undermined the globalization of the late 19th and early 20th centuries. They incited racism, protectionism, and military buildup and led to Nazi and Communist dictatorships that culminated in World War II (1939–45) (Hobsbawm, 1994; James, 2001). International trade and investment plummeted between 1914 and 1945, and governments erected many new barriers to the free movement of people and ideas. A longer view of globalization suggests that it is not an inevitable and linear process. Periods of accelerated globalization—including our own—can end.

An even longer historical view leads to additional insight. Some sociologists, such as Roland Robertson (1992), note that globalization is as old as civilization itself and is in fact the *cause* of modernization rather than the other way around. Archaeological remains show that long-distance trade began 5000 years ago. People have been migrating across continents and even oceans for thousands of years. Alexander the Great conquered vast stretches of Europe, western Asia, and northern Africa, while Christianity and Islam spread far beyond their birthplace in the Middle East. All these forces contributed to globalization in Robertson's view.

So is globalization 25, 125, or 5000 years old, as Albrow, Giddens, and Robertson respectively suggest? That question has no "correct" answer. The answer depends on what we want globalization to mean. A short historical view—whether 125 or 25 years—misses the long-term developments that have brought the world's people closer to one another. Yet if we think of globalization as a 5000-year-old phenomenon, the definition and value of the term is diluted. After all, the price of tea in China *was* irrelevant 5000 years ago. Few people knew what tea was.

We prefer to take an intermediate position and think of globalization as a roughly 500-year-old phenomenon. We regard the establishment of colonies and the growth of capitalism as the main forces underlying globalization, and both **colonialism** and capitalism began about 500 years ago, symbolized by Columbus's voyage to the Americas. As you will now see, the main advantages of thinking of globalization in this way is that it tells us much about the causes of development or industrialization and the growing gap between rich and poor countries and people. These topics will now be the focus of our attention.

Colonialism refers to the control of developing societies by more developed, powerful societies.

Global Inequality

Personal Anecdote

John Lie decided to study sociology because he was concerned about the poverty and dictatorships he had read about in books and observed during his travels in Asia and Latin America. "Initially," says John, "I thought I would major in economics. In my economics classes, I learned about the importance of birth-control programs to cap population growth, efforts to prevent the runaway growth of cities, and measures to spread Western knowledge, technology, and markets to people in less economically developed countries. My textbooks and professors assumed that if only the less developed countries would become more like the West, their populations, cities, economies, and societies would experience stable growth. Otherwise, the developing countries were doomed to suffer the triple catastrophe of overpopulation, rapid urbanization, and economic underdevelopment.

"Equipped with this knowledge, I spent a summer in the Philippines working for an organization that offered farmers advice on how to promote economic growth. I assumed that, as in North America, farmers who owned large plots of land and used high technology would be more efficient and better off. Yet I found the most productive villages were those in which most farmers owned *small* plots of land. In such villages, there was little economic inequality. The women in these villages enjoyed low birthrates and the inhabitants were usually happier than the inhabitants of villages in which there was more inequality.

"As I talked with the villagers, I came to realize that farmers who owned at least some of their own land had an incentive to work hard. The harder they worked, the more they earned. With a higher standard of living, they didn't need as many children to help them on the farm. In contrast, in villages with greater inequality, many farmers owned no land but leased it or worked as farmhands for wealthy landlords. They didn't earn more for working harder, so their productivity and their standard of living were low. They wanted to have more children to increase household income.

"Few Filipino farms could match the productivity of high-tech American farms because even large plots were small by American standards. Much high-tech agricultural equipment would have been useless there. Imagine trying to use a harvesting machine in a plot not much larger than some suburban backyards.

"Thus, my Western assumptions turned out to be wrong. The Filipino farmers I met were knowledgeable and thoughtful about their needs and desires. When I started listening to them I started understanding the real world of economic development. It was one of the most important sociological lessons I ever learned."

Let us begin our sociological discussion of economic development by examining trends in levels of global inequality. We then discuss two theories of development and underdevelopment, both of which seek to uncover the sources of inequality among nations. Next, we analyze some cases of successful development. Finally, we consider what we can do to alleviate global inequality in light of what we learn from these successful cases.

Levels of Global Inequality

We learned in Chapter 8 that the United States is a highly stratified society. If we shift our attention from the national to the global level, we find an even more dramatic gap between rich and poor. In a Manhattan restaurant, pet owners can treat their cats to $100-a-plate birthday parties. In Cairo (Egypt) and Manila (the Philippines), garbage dumps are home to entire families who sustain themselves by picking through the refuse. People who travel outside the 20 or so highly industrialized countries of North America, western Europe, Japan, and Australia often encounter scenes of unforgettable poverty and misery. The UN calls the level of inequality worldwide "grotesque" (United Nations, 2002: 19).

Sociology Now™

Learn more about **Global Inequality** by going through the Global Stratification Data Experiment.

The average income of citizens in the highly industrialized countries far outstrips that of citizens in the developing societies. Yet because poor people live in rich countries and rich people live in poor countries, these averages fail to capture the extent of inequality between the richest of the rich and the poorest of the poor. Noting that the total worth of the world's 358 billionaires equals that of the world's 2.3 billion poorest people (45 percent of the world's population) is perhaps more revealing. The three richest people in the world own more than the combined GDP of the 48 least developed countries. The richest 1 per-

Photo by Miro Cernetig, *Globe & Mail*, Toronto, Canada

A half-hour's drive from the center of Manila, the capital of the Philippines, an estimated 70,000 Filipinos live on a 55-acre mountain of rotting garbage, 150 feet high. It is infested with flies, rats, dogs, and disease. On a lucky day, residents can earn up to $5 retrieving scraps of metal and other valuables. On a rainy day, the mountain of garbage is especially treacherous. In July 2000 an avalanche buried 300 people alive. People who live on the mountain of garbage call it "The Promised Land."

Table 9.2
Global Priorities: Annual Cost of Various Goods and Services

Good/Service	Annual Cost (in billion $)
Basic education for everyone in the world	6
Cosmetics in the United States	8
Water and sanitation for everyone in the world	9
Ice cream in Europe	11
Reproductive health for all women in the world	12
Perfumes in Europe and the United States	12
Basic health and nutrition for everyone in the world	13
Pet foods in Europe and the United States	17
Business entertainment in Japan	35
Cigarettes in Europe	50
Alcoholic drinks in Europe	105
Narcotic drugs in the world	400
Military spending in the world	780

Note: Items in italics represent estimates of what they would cost to achieve. Other items represent estimated actual cost.

Source: United Nations (1998b: 37)

cent of the world's population earns as much income as the bottom 57 percent. The top 10 percent of U.S. income earners earn as much as the poorest 2 billion people in the world (United Nations, 2002).

According to the UN, 800 million people in the world are malnourished and 4 billion people—two-thirds of the world's population—are poor, which is defined as lacking the ability to obtain adequate food, clothing, shelter, and other basic needs. According to the UN (1994: 2): "A fifth of the developing world's population goes hungry every night, a quarter lack access to even a basic necessity like safe drinking water, and a third live in a state of abject poverty—at such a margin of human existence that words simply fail to describe it." The citizens of the 20 or so rich, highly industrialized countries spend more on cosmetics, alcohol, ice cream, or pet food than it would take to provide basic education, water and sanitation, or basic health and nutrition for everyone in the world (Table 9.2).

Of the 1.3 billion people around the world living on $1 a day or less, 1 billion of them are women. Of the estimated 854 million illiterate adults in the world, 64 percent of them are women (United Nations, 2002). Moreover, racial and other minority groups often fare worse than their majority-group counterparts in the developing countries. For example, almost all South African whites are literate, compared with a literacy rate of 50 percent among blacks. The average life expectancy for whites in South Africa is 70 years, whereas the corresponding figure for blacks is 59 (United Nations, 1998).

Trends in Global Inequality

Has global inequality increased or decreased over time? Figure 9.4 helps answer that question. Between 1975 and 2000, the annual income gap between the 20 or so richest countries and the rest of the world grew enormously. Average income in two regions—sub-Saharan Africa and central and eastern Europe and the former Soviet Union—actually fell. Table 9.3 focuses on individual rather than regional income. It shows that the

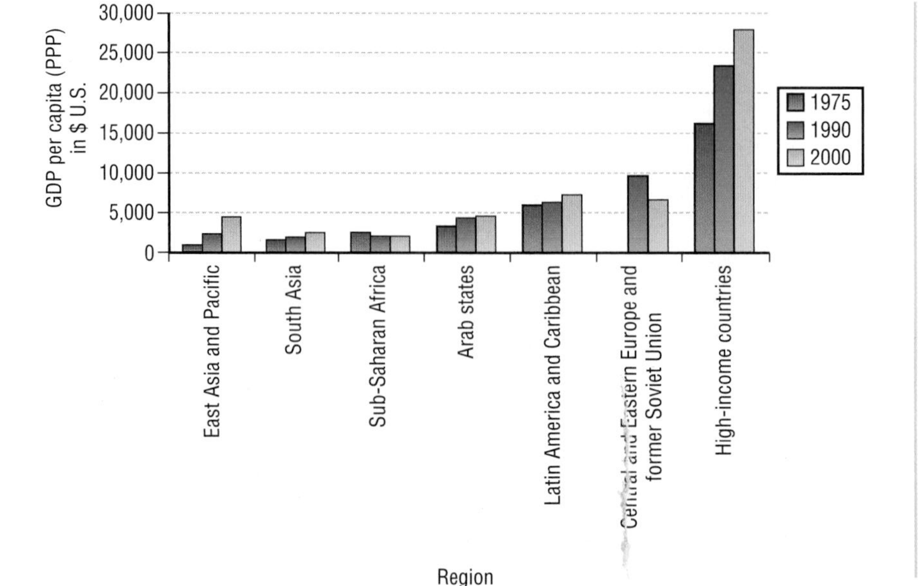

Figure 9.4
Gross Domestic Product per Capita, World Regions, 1975–2000 (2000 purchasing power parity [PPP], $ US)

Source: United Nations (2002: 19)

Table 9.3
Trends in Global Inequality, 1820–1992

	YEAR						
	1820	1870	1910	1929	1950	1970	1992
Share of world income going to top 10%	42.8	47.6	50.9	49.8	51.3	50.8	53.4
Share of world income going to bottom 20%	4.7	3.8	3.0	2.9	2.4	2.2	2.2
Number of people living on $1 a day or less, adjusted for inflation (%)	3.9	75.4	65.6	56.3	54.8	35.6	23.7
Number of people living on $1 a day or less, adjusted for inflation (number)	887	954	1128	1150	1376	1305	1294

Source: Centre for Economic Policy Research (2002: 57).

Table 9.4
Number of People Living on Less Than $1 a Day, 1990–1999

	PERCENT		NUMBER (MILLIONS)	
Region	1990	1999	1990	1999
Sub-Saharan Africa	47.7	46.7	242	300
East Asia (excluding China)	27.6	14.2	452	260
South Asia	44.0	36.9	495	490
Latin America	16.8	15.1	74	77
Eastern Europe and Central Asia	1.6	3.6	7	17
Middle East and North Africa	2.4	2.3	6	7
Total	28.1	24.5	916	936

Source: From Human Development Report 2002 by United Nations Development Program, copyright © 2002 by the United Nations Development Program. Used by permission of Oxford University Press.

share of world income going to the top 10 percent of individuals increased from 42.8 percent in 1820 to 53.4 percent in 1992. Over that same period, the share of world income going to the bottom 20 percent of individuals fell from 4.7 percent to just 2.2 percent. That trend has continued since 1992. On the slightly brighter side, the number of people in the world living in extreme poverty ($1 a day or less) peaked in 1950 and then started to decline gradually. In percentage terms, the number of people living in extreme poverty fell from 83.9 percent in 1820 to 23.7 percent in 1992. However, as Table 9.4 shows, if we consider only the less developed countries, the number of people living on $1 a day or less *increased* by 20 million in the 1990s. Even by the most optimistic interpretation, these figures are little cause for joy. In 1999, 47 percent of the world's population lived on $2 a day or less.

Statistics never speak for themselves. We need theories to explain them. Let us now outline and critically assess the two main theories that seek to explain the origins and persistence of global inequality.

Theories of Development and Underdevelopment

Modernization Theory: A Functionalist Approach

Two main sociological theories claim to explain global inequality. The first, **modernization theory,** is a variant of functionalism. According to modernization theory, global inequality results from various dysfunctional characteristics of poor societies themselves. Specifically, modernization theorists say the citizens of poor societies lack sufficient *capi-*

Modernization theory holds that economic underdevelopment results from poor countries lacking Western attributes. These attributes include Western values, business practices, levels of investment capital, and stable governments.

From Eric Hobsbawm's *The Age of Empire* (Vintage Books). Copyright © 1987 by E. J. Hobsbawm

In 1893 leaders of the British mission pose before taking over what became Rhodesia and is now Zimbabwe. To raise a volunteer army, every British trooper was offered about 9 square miles of native land and 20 gold claims. The Matabele and Mashona peoples were subdued in a three-month war. Nine hundred farms and 10,000 gold claims were granted to the troopers and about 100,000 cattle were looted, leaving the native survivors without a livelihood. Forced labor was subsequently introduced by the British so that the natives could pay a £2 a year tax.

tal to invest in Western-style agriculture and industry. They lack rational, Western-style *business techniques* of marketing, accounting, sales, and finance. As a result, their productivity and profitability remain low. They lack stable, Western-style *governments* that could provide a secure framework for investment. Finally, they lack a Western *mentality,* which stresses the need for savings, investment, innovation, education, high achievement, and self-control in having children (Inkeles and Smith, 1976; Rostow, 1960). Societies characterized by these dysfunctions are poor. It follows that people living in rich countries can best help their poor cousins by transferring Western culture and capital to them and eliminating the dysfunctions. Only then will the poor countries be able to cap population growth, stimulate democracy, and invigorate agricultural and industrial production. Government-to-government foreign aid can accomplish some of this. Much work also needs to be done to encourage Western businesses to invest directly in poor countries and to increase trade between rich and poor countries.

Dependency Theory: A Conflict Approach

Proponents of **dependency theory,** a variant of conflict theory, have been quick to point out the chief flaw in modernization theory (Baran, 1957; Cardoso and Faletto, 1979; Wallerstein, 1974–89). For the last 500 years, the most powerful countries in the world deliberately impoverished the less powerful countries. Focusing on internal characteristics blames the victim rather than the perpetrator of the crime. It follows that an adequate theory of global inequality should not focus on the internal characteristics of poor countries themselves. Instead, it ought to follow the principles of conflict theory and focus on patterns of domination and submission—specifically in this case, on the relationship between rich and poor countries. That is just what dependency theory does.

According to dependency theorists, less global inequality existed in 1500 and even in 1750 than today. However, beginning around 1500, the armed forces of the world's most powerful countries subdued and then annexed or colonized most of the rest of the world. Around 1780 the industrial revolution began. It enabled the western European countries, Russia, Japan, and the United States to amass enormous wealth, which they used to extend their global reach. They forced their colonies to become a source of raw materials, cheap labor, investment opportunities, and markets for the conquering nations. The colonizers thereby prevented industrialization and locked the colonies into poverty.

In the decades following World War II, nearly all of the colonies in the world became politically independent. However, dependency theorists say that exploitation by direct political control was soon replaced by new means of achieving the same end: substantial foreign investment, support for authoritarian governments, and mounting debt.

Substantial Foreign Investment

Multinational corporations invested heavily in the poor countries to siphon off wealth in the form of raw materials and profits. True, they created some low-paying jobs in the process. They created many more high-paying jobs in the rich countries where the raw materials were used to produce manufactured goods. They also sold part of the manufactured goods back to the poor, unindustrialized countries for additional profit.

Dependency theory views economic underdevelopment as the result of exploitative relations between rich and poor countries.

Support for Authoritarian Governments

According to dependency theorists, multinational corporations and rich countries continued their exploitation of the poor countries in the postcolonial period by giving economic and military support to local authoritarian governments. These governments managed to keep their populations subdued most of the time. When that was not possible, Western governments sent in troops and military advisers, engaging in what became known as "gunboat diplomacy." The term itself was coined in colonial times. In 1839 the Chinese rebelled against the British importation of opium into China, and the British responded by sending a gunboat up the Yangtze River, starting the Opium War. The war resulted in Britain winning control of Hong Kong and access to five Chinese ports; what started as gunboat diplomacy ended as a rich feast for British traders. In the postcolonial period, the United States has been particularly active in using gunboat diplomacy in Central America. A classic case is Guatemala in the 1950s (LaFeber, 1993). In 1952 the democratic government of Guatemala began to redistribute land to impoverished peasants. Some of the land was owned by the United Fruit Company, a United States multinational corporation and the biggest landowner in Guatemala. Two years later, the U.S. Central Intelligence Agency (CIA) backed a right-wing coup in Guatemala, preventing land reform and allowing the United Fruit Company to continue its highly profitable business as usual.

Mounting Debt

The governments of the poor countries struggled to create transportation infrastructures (airports, roads, harbors, etc.), build up their education systems, and deliver safe water and at least the most basic health care to their people. To accomplish these tasks, they had to borrow money from Western banks and governments. So it came about that debt—and the interest payments that inevitably accompany debt—grew every year. Crushing interest payments leave governments of poor countries with too little money for development tasks. Foreign aid helps, but not much. In 2002, foreign aid to the world's developing countries was only one-seventh the amount that the developing countries paid to Western banks in loan interest (United Nations, 2004: 201).

Effects of Foreign Investment

Almost all sociologists agree that dependency theorists are correct on one score. Since about 1500, Spain, Portugal, Holland, Britain, France, Italy, the United States, Japan, and Russia treated the world's poor with brutality to enrich themselves. They rationalized their actions by claiming they were bringing "civilization" to the "savages" and inventing other such stories, such as the notion of the Dutch colonizers in early-17th-century Brazil that "there is no sin south of the equator." Colonialism did have a devastating economic and human impact on the poor countries of the world. In the postcolonial era, the debt burden has crippled the development efforts of many poor countries.

That said, a big question remains that research has not yet fully answered: Do foreign investment and liberalized trade policies have positive or negative effects today? Much hinges on the answer to this question. Modernization theorists want more foreign investment in poor countries and freer trade because they think that will promote economic growth and general well-being. They want trade and investment barriers to be dropped so free markets can bring prosperity to everyone. Dependency theorists diametrically oppose this strategy. They think foreign investment drains wealth out of poor countries. Therefore, they want the poor countries to revolt against the rich countries, throw up barriers to free trade and investment, and find their own paths to economic well-being.

Over the past couple of decades, researchers have carefully examined the effects of free trade and foreign investment. The results of their analyses depend partly on which variables they include in their analyses, which time periods they analyze, which countries they consider, and which statistical techniques they use (Bornschier and Chase-Dunn, 1985;

Centre for Economic Policy Research, 2002; DeSoya and Oneal, 1999; Firebaugh and Beck, 1994; Weisbrot et al., 2001; Weisbrot and Baker, 2002). Not surprisingly, if you throw different countries, variables, time periods, and statistical techniques into the mix, you bake entirely different cakes. Yet, a recent summary of research in this area cautiously reaches two conclusions (Centre for Economic Policy Research, 2002). First, in the 1980s and 1990s but not in the 1960s and 1970s, openness to international trade and foreign investment generally seems to have stimulated economic growth. Second, in most well-documented cases, openness to international trade and foreign investment also increased inequality. Few benefits of economic growth went to the most needy. This fact was true for China, Malaysia, Thailand, Chile, Colombia, and Mexico. On the other hand, openness to international trade and foreign investment did not always increase inequality. No such increase took place in Taiwan or (with qualifications) in Brazil or Indonesia.[2]

These findings suggest that lumping all periods of history and all countries together when considering the effects of openness on economic growth and inequality is a mistake. After all, countries have different histories and different social structures. They may adopt a variety of economic policies that influence the effects of international trade and foreign direct investment in different ways. Consequently, international trade and foreign direct investment may have different effects in different times and places. Historical, social-structural, and policy factors matter greatly in determining how a particular country responds to international trade and foreign direct investment.

Core, Periphery, and Semiperiphery

Immanuel Wallerstein proposes a variation on this theme. He argues that capitalist development has resulted in the creation of an integrated "world system" composed of three tiers. First are the **core** capitalist countries (such as the United States, Japan, and Germany), which are major sources of capital and technology. Second are the **peripheral** countries (the former colonies), which are major sources of raw materials and cheap labor. Third are the **semiperipheral** countries (such as South Korea, Taiwan, and Israel), consisting of former colonies that are making considerable headway in their attempts to become prosperous. To give just one dramatic illustration of this progress, South Korea and the African country of Ghana were among the poorest nations in the world in 1960. Ghana still is. But South Korea, the recipient of enormous foreign investment and aid, was nine times wealthier than Ghana in 1997 (as measured by gross national product [GNP] per capita; calculated from World Bank, 1999c: 193). Comparing the unsuccessful peripheral countries with the more successful semiperipheral countries presents us with a useful natural experiment. The comparison suggests circumstances that help some poor countries overcome the worst effects of colonialism.

Sociology Now™

Learn more about **Core, Periphery, and Semiperiphery** by going through the World Systems Theory Animation.

The semiperipheral countries differ from the peripheral countries in four main ways (Kennedy, 1993: 193–227; Lie, 1998).

Type of Colonialism

Around the turn of the 20th century, Taiwan and Korea became colonies of Japan. They remained so until 1945. However, in contrast to the European colonizers of Africa, Latin America, and other parts of Asia, the Japanese built up the economies of their colonies. They established transportation networks and communication systems. They built steel, chemical, and hydroelectric power plants. After Japanese colonialism ended, Taiwan and South Korea were thus at an advantage compared with Ghana, for example, at the time Britain gave up control of that country. South Korea and Taiwan could use the Japanese-built infrastructure and Japanese-trained personnel as springboards to development.

- The **core** capitalist countries are rich countries, such as the United States, Japan, and Germany, that are the major sources of capital and technology in the world.
- The **peripheral** countries are former colonies that are poor and are major sources of raw materials and cheap labor.
- The **semiperipheral** countries, such as South Korea, Taiwan, and Israel, consist of former colonies that are making considerable headway in their attempts to industrialize.

[2]The effect of free trade and foreign investment on highly developed countries is similarly diverse. In the 1980s and 1990s, the United States and the United Kingdom experienced increasing inequality as a result of openness, whereas Canada, France, and Germany did not. In the latter countries, government policies prevent inequality from growing.

Seoul, capital of South Korea. South Korea is one of the semiperipheral countries that are making considerable headway in their attempts to become prosperous.

Brent Patterson/Corbis

Geopolitical Position

Although the United States was the leading economic and military power in the world by the end of World War II, it began to feel its supremacy threatened in the late 1940s by the Soviet Union and China. Fearing that South Korea and Taiwan might fall to the communists, the United States poured unprecedented aid into both countries in the 1960s. It also gave them large, low-interest loans and opened its domestic market to Taiwanese and South Korean products. Because the United States saw Israel as a crucially important ally in the Middle East, it also received special economic assistance. Other countries with less strategic importance to the United States received less help in their drive to industrialize.

State Policy

A third factor that accounts for the relative success of some countries in their efforts to industrialize and become prosperous concerns state policies. As a legacy of colonialism, the Taiwanese and South Korean states were developed on the Japanese model. They kept workers' wages low, restricted trade union growth, and maintained quasi-military discipline in factories. Moreover, by placing high taxes on consumer goods, limiting the import of foreign goods, and preventing their citizens from investing abroad, they encouraged their citizens to put much of their money in the bank. This situation created a large pool of capital for industrial expansion. The South Korean and Taiwanese states also gave subsidies, training grants, and tariff protection to export-based industries from the 1960s onward. (Tariffs are taxes on foreign goods.) These policies did much to stimulate industrial growth. Finally, the Taiwanese and South Korean states invested heavily in basic education, health care, roads, and other public goods. A healthy and well-educated labor force combined with good transportation and communication systems laid solid foundations for economic growth.

Social Structure

Taiwan and South Korea are socially cohesive countries, which makes it easy for them to generate consensus around development policies. It also allows them to get their citizens to work hard, save a lot of money, and devote their energies to scientific education.

Social solidarity in Taiwan and South Korea is based partly on the sweeping land reform they conducted in the late 1940s and early 1950s. By redistributing land to small farmers, both countries eliminated the class of large landowners, who usually oppose industrialization. Land redistribution got rid of a major potential source of social conflict. In contrast, many countries in Latin America and Africa have not undergone land reform. The United States often intervened militarily in Latin America to prevent land reform because U.S. commercial interests profited handsomely from the existence of large plantations (LaFeber, 1993).

Another factor underlying social solidarity in Taiwan and South Korea is that neither country suffered from internal conflicts like those that wrack Africa south of the Sahara desert. British, French, and other western European colonizers often drew the borders of African countries to keep antagonistic tribes living side-by-side in the same jurisdiction so as to foment tribal conflict. Keeping tribal tensions alive made it possible to play one tribe off against another. That made it easier for imperial powers to rule. This policy led to much social and political conflict in postcolonial Africa. Today, the region suffers from frequent civil wars, coups, and uprisings. It is the most conflict-ridden area of the world. For example, the civil war in the Democratic Republic of the Congo has been raging since 1998 and has so far resulted in 3.5 million deaths. This high level of internal conflict acts as a barrier to economic development in sub-Saharan Africa.

In sum, certain conditions seem to permit foreign investment to have positive economic effects. Postcolonial countries that enjoy a solid industrial infrastructure, strategic geopolitical importance, strong states with strong development policies, and socially cohesive populations are in the best position to join the ranks of the rich countries in the coming decades. We may expect countries that have *some* of these characteristics to experience some economic growth and increase in the well-being of their populations in the near future. Such countries include Chile, Thailand, Indonesia, Mexico, and Brazil. In contrast, African countries south of the Sahara are in the worst position of all. They have inherited the most damaging consequences of colonialism and enjoy few of the conditions that could help them escape the history that has been imposed on them.

Neoliberal versus Democratic Globalization

Globalization and Neoliberalism

For some political and economic leaders, the road sign that marks the path to prosperity reads "neoliberal globalization." **Neoliberal globalization** is a policy that promotes private control of industry, minimal government interference in the running of the economy, the removal of taxes, tariffs, and restrictive regulations that discourage the international buying and selling of goods and services, and the encouragement of foreign investment. Advocates of neoliberal globalization resemble the modernization theorists of a generation ago. They believe that if only the poor countries would emulate the successful habits of the rich countries, they would prosper as well.

Many social scientists are skeptical of this prescription. For example, Nobel Prize–winning economist and former chief economist of the World Bank Joseph E. Stiglitz (2002) argues that the World Bank and other international economic organizations often impose outdated policies on developing countries, putting them at a disadvantage vis-à-vis developed countries. In the African country of Mozambique, for example, foreign debt was four and a half times the GNP in the mid-1990s. This means that the amount of money Mozambique owed to foreigners was four and a half times more than the value of goods and services produced by all the people of Mozambique in a year. Facing an economic crisis, Mozambique sought relief from the World Bank and the IMF. The response of the IMF was to argue that Mozambique's government-imposed minimum wage of less than $1 a day was "excessive." The IMF also recommended that Mozambique spend twice as much as its education budget and four times as much as its health budget on interest payments to service its foreign debt (Mittelman, 2000: 104). How such crippling policies might help the people of Mozambique is unclear.

Neoliberal globalization is a policy that promotes private control of industry, minimal government interference in the running of the economy, the removal of taxes, tariffs, and restrictive regulations that discourage the international buying and selling of goods and services, and the encouragement of foreign investment.

Does Neoliberalism Work?

Does historical precedent lead us to believe that neoliberalism works? To the contrary, with the exception of Great Britain, neoliberalism was *never* a successful development strategy in the early stages of industrialization. Germany, the United States, Japan, Sweden, and other rich countries became highly developed economically between the second half of the 19th century and the early 20th century. South Korea, Taiwan, Singapore, and other semiperipheral countries became highly developed economically in the second half of the 20th century. Today, China and, to a lesser degree, India are industrializing quickly. These countries did not pursue privatization, minimal government intervention in the economy, free trade, and foreign investment in the early stages of industrialization. Rather, the governments of these countries typically intervened to encourage industrialization. They protected infant industries behind tariff walls, invested public money heavily to promote national industries, and so forth. Today, China and India maintain among the highest barriers to international trade in the world (Chang, 2002; Gerschenkron, 1962; Laxer, 1989).

Typically, it is only after industrial development is well under way and national industries can compete on the global market that countries begin to advocate neoliberal globalization to varying degrees. As one political economist writes, "The sensible ones liberalise in line with the growth of domestic capacities—they try to expose domestic producers to enough competition to make them more efficient, but not enough to kill them" (Wade in Wade and Wolf, 2002: 19). The exception that proves the rule is Great Britain, which needed less government involvement to industrialize because it was the first industrializer and had no international competitors. Even the United States, one of the most vocal advocates of neoliberal globalization today, invested a great deal of public money subsidizing industries and building infrastructure (roads, schools, ports, airports, electricity grids, etc.) in the late 19th and early 20th centuries. It was an extremely protectionist country until the end of World War II. As late as 1930 the Smoot-Hawley Tariff Act raised tariffs on foreign goods 60 percent. Today, the United States still subsidizes large corporations in a variety of ways and maintains substantial tariffs on a whole range of foreign products including agricultural goods and softwood lumber. In this, the United States is little different from Japan, France, and other rich countries.

In sum, there is good reason to be skeptical about the benefits of neoliberal globalization for poor countries (Bourdieu, 1998; Brennan, 2003). Yet, as you will now see, globalization can be reformed so that its economic and technological benefits are distributed more uniformly throughout the world.

Globalization Reform

In the film *About Schmidt* (2002), Jack Nicholson plays Warren Schmidt, a former insurance executive. Retirement leaves Schmidt with little purpose in life. His wife dies. His adult daughter has little time or respect for him. He feels his existence lacks meaning. Then, while watching TV one night, Schmidt is moved to support a poor child in a developing country. He decides to send a monthly $27 check to sponsor a Tanzanian boy named Ndugu. He writes Ndugu long letters about his life. While Schmidt's world falls apart before our eyes, his sole meaningful human bond is with Ndugu. At the end of the movie, Schmidt cries as he looks at a picture Ndugu drew for him: an adult holding a child's hand.

It does not take a Warren Schmidt to find meaning in helping the desperately poor. Many people contribute to charities that help developing countries in a variety of ways. Many more people contribute development aid indirectly through the taxes they pay to the federal government. On the whole, we think our contributions are extraordinarily generous. On average, Americans believe we are spending 24 percent of our budget on aid to poor countries ("Are We Stingy?" 2004). As a result of this perception, few Americans believe we are doing too little to help the world's poor. The General Social Survey (GSS) periodically asks respondents whether the United States is spending too much, too little, or about the right amount on 17 items, including foreign aid. ▶Table 9.5 shows the results

Table 9.5
National Priorities, United States, 2002 (in percent)

"We are faced with many problems in this country, none of which can be solved easily or inexpensively. I'm going to name some of these problems, and for each one I'd like you to tell me whether you think we're spending too much money on it, too little money, or about the right amount. First, are we spending too much, too little, or about the right amount on . . . "

Priority	Percent Answering "Too Little"
1. Improving and protecting the nation's health	74.9
2. Improving the nation's education system	73.9
3. Social Security	60.8
4. Improving and protecting the environment	60.0
5. Assistance for child care	59.1
6. Dealing with drug addiction	59.1
7. Halting the rising crime rate	57.4
8. Solving problems of the big cities	45.4
9. Mass transportation	37.0
10. Scientific research	36.4
11. Highways and bridges	35.7
12. Parks and recreation	35.0
13. Improving the conditions of blacks	32.7
14. The military, armaments, and defense	31.3
15. Welfare	21.2
16. Space exploration program	11.8
17. Foreign aid	6.7

Source: National Opinion Research Center (2004).

for 2002. Foreign aid ranks a distant last on Americans' list of priorities. Fewer than 7 percent of Americans think we are doing too little when it comes to foreign aid.

Perceptions are, however, at odds with reality. In 2002 a World Bank official compared the subsidies rich countries gave to farms and businesses within their borders with the amount of development aid they gave developing countries. He concluded that "[t]he average cow [in a rich country] is supported by three times the level of income of a poor person in Africa" (quoted in Schuettler, 2002). At a conference in Rio de Janeiro in 1992, the world's governments decided on an official development assistance target of 0.7% of GNP. By 2003, only five countries had reached that goal: Norway, Denmark, the Netherlands, Luxembourg, and Sweden. The United States ranked last among the 22 rich nations at 0.14 percent of GNP, about one-seventh of Norway's percentage ("The U.S. and Foreign Aid Assistance," 2005). More recently, the tsunami disaster of December 26, 2004, took the lives of more than 200,000 people in Indonesia, Sri Lanka, and other countries in the Indian Ocean. Our response was hailed by many people as a clear example of American generosity. Yet we ranked 27th in the world—behind Greece and the Czech Republic—in our relative contribution to tsunami relief (Table 9.6).

Some people claim that the United States makes up for its shortfall through generous *private* donations to the less developed countries. That, too, is an exaggeration. In 2003, the United States gave about 14¢ for every $100 of national income to poor countries in government aid and about another 6¢ in private aid for every $100 of national income. That still places the United States at or near the bottom of the list of rich countries in terms of overall aid contributions (Clemens, Radelet, and Roodman, 2004; Kristof, 2005).

Foreign Aid

Even if the United States quintupled its foreign aid budget to meet UN guidelines, some foreign aid as presently delivered is not an effective way of helping the least developed countries. For one thing, most U.S. foreign aid goes to strategically important friends, not to the most needy countries. Egypt, Russia, and Iraq top the list of U.S. aid recipients. In addition, some charities waste a lot of money on administration and overhead expenses. Finally, food aid often has detrimental effects on poor countries. One expert writes, "As well as creating expensive addictions to nonindigenous cereals, and discouraging export-production of items like corn and rice in which the U.S.A. is expanding its own

Table 9.6
Foreign Aid by Country, 2003, and Tsunami Relief to February 10, 2005

FOREIGN AID, 2003		TSUNAMI RELIEF TO FEBRUARY 2005	
Country	Aid as % of GNP	Country	Aid per $1 billion GDP
1. Norway	.92	1. Kuwait	$2.71 million
2. Denmark	.84	2. Australia	1.83
3. Netherlands	.81	3. Norway	1.77
4. Luxembourg	.80	4. Qatar	1.57
5. Sweden	.70	5. New Zealand	1.04
6. Belgium	.61	6. Netherlands	1.01
7. France	.41	7. Denmark	0.71
7. Ireland	.41	8. Ireland	0.69
9. Switzerland	.38	9. Switzerland	0.68
10. Finland	.34	10. Finland	0.66
10. United Kingdom	.34	11. Sweden	0.63
12. Germany	.28	12. Canada	0.54
13. Canada	.25	13. Germany	0.50
13. Australia	.25	14. Austria	0.40
13. Spain	.25	15. United Arab Emirates	0.37
16. New Zealand	.23	16. Saudi Arabia	0.36
17. Greece	.21	17. Luxembourg	0.31
17. Portugal	.21	18. Hong Kong	0.29
19. Austria	.20	19. Taiwan	0.27
19. Japan	.20	**20. United States**	**0.26**
21. Italy	.16		
22. United States	**.14**		

Source: Becker (2005); "Category: Disasters" (2005); "The US and Foreign Aid Assistance" (2005).

trade, [U.S. food aid] has frequently served as a major disincentive to the efforts of local farmers to grow food even for domestic consumption" (Hancock, 1989: 169).

We conclude that foreign aid is often beneficial. The poorest countries need more of it. But strict government oversight is required to ensure that foreign aid is not wasted and that it is directed to truly helpful projects, such as improving irrigation and sanitation systems, helping people acquire better farming techniques, and expanding educational and health services (Payer, 1982; Colson, 1982). Increasing the amount of foreign aid and redesigning its delivery can thus help mitigate some of the excesses of neoliberal globalization.

Debt Cancellation

Many analysts argue that the world's rich countries and banks should simply write off the debt owed to them by the developing countries in recognition of historical injustices. They reason that the debt burden of the developing countries is so onerous that it prevents them from focusing on building up economic infrastructure, improving the health and education of their populations, and developing economic policies that can help them emerge from poverty. This proposal for blunting the worst effects of neoliberal globalization may be growing in popularity among politicians in the developed countries. For example, former president Bill Clinton, British Prime Minister Tony Blair, and Canadian Prime Minister Paul Martin, among others, support the idea.

Tariff Reduction

A third reform proposed in recent years involves the elimination of tariffs by the *rich* countries. Many of these tariffs prevent developing countries from exporting goods that could earn them money for investment in agriculture, industry, and infrastructure. The

BOX 9.2
Sociology at the Movies

Three Kings (1999)

It is 1991. A few months earlier, the army of Iraq invaded Kuwait, a small country in the enviable position of possessing 10 percent of the world's oil reserves. Now the United States and its allies have declared victory over Iraq and its brutal leader, Saddam Hussein. They have pushed Iraqi forces out of Kuwait. American soldiers celebrate the restoration of the Kuwaiti people's freedom and the destruction of the Iraqi dictator's military might. The United States glories in its role as the friend of democracy and the scourge of oppressors everywhere.

Then the picture gets complicated. Some Americans discover a map on a captured Iraqi soldier. It leads to a bunker containing gold bars that Saddam Hussein looted from Kuwait. The Americans decide to do something for their families. As Major Archie Gates (played by George Clooney) says, "Saddam stole it from the [Kuwaiti] sheiks. I have no problem stealing it from Saddam." He assembles three trusted comrades (the "three kings" of the movie's title and, ironically, the Christmas carol), a Humvee, and some arms and other supplies. They head out to the secret bunker, located in a village.

What awaits the American soldiers is wholly unexpected. Remnants of Saddam's elite Revolutionary Guard protect the bunker. They are of no great danger to the Americans because a cease-fire has been declared. However, the villagers despise the Guard and Saddam. They have organized an armed resistance against the Iraqi leader and his troops. They cheer the arrival of the American soldiers. In response, the Revolutionary Guard shoots, imprisons, and tortures the rebels.

At first, the Americans ignore the plight of the freedom fighters because U.S. forces are under strict orders not to get involved. Their job as defined by the U.S. government—securing the region's oil—is done. Suddenly, the friend of democracy and the scourge of oppressors everywhere is seen to be more self-interested than idealistic. The United States has ended the Iraqi military threat to the West's oil supply. (Saudi Arabia, with more than one-quarter of the world's oil reserves, was next on Saddam's hit list.) Democracy, the movie says, is a nice ideal when it suits American interests, but the United States is prepared to scatter that ideal to the winds when it is not in American interests.

Warner Brothers/The Kobal Collection/Close, Murray

Ice Cube, Mark Wahlberg, and George Clooney in *Three Kings* (1999).

In the movie, Archie Gates and his comrades develop sympathies for the village rebels. They help them overcome the Revolutionary Guard and lead them to sanctuary in neighboring Iran. In a tense standoff with a U.S. general and his troops at the Iraq-Iran border, Gates trades his knowledge of where the gold bars are stored for the safe passage of the villagers across the border. We leave the movie feeling that although the American government plays the game of oil politics, some American citizens are idealistic and sympathetic to democracy and the oppressed everywhere.

In the real world, the United States abandoned the freedom fighters of Iraq and the fight for democracy. Saddam Hussein ruthlessly suppressed the rebels in the south. The United States also protected the undemocratic regimes of Kuwait and Saudi Arabia. Kuwait allows just 10 percent of its population—adult men whose families lived in the country before 1920 or who have been Kuwaiti citizens for more than 30 years—to vote in elections. That figure is at least 10 percent better than Saudi Arabia, where *no* elections take place above the municipal level (Central Intelligence Agency, 2002). The West's oil supply is secure, although oil wealth in the region remains concentrated in the hands of the sheiks, their families, and their close supporters. Most of the Arab world remains impoverished as resentment seethes against regimes like those of Kuwait and Saudi Arabia, fueling Islamic fundamentalist militancy.

Bush administration recently proposed lifting all tariffs on textiles and apparel produced in the Western hemisphere by 2008 (Becker, 2003). That sort of move, if broadened to include agricultural goods and the entire world, could help stimulate economic growth in the developing countries. To date, however, there is little room for optimism in this regard. In 2002 a less developed country like Mexico gave its farmers an average subsidy of $1000 a year, whereas the United States gave its farmers an average of subsidy of $16,000 a year. The comparable figures for western Europe and Japan were $17,000 and $27,000, respectively. Talks to lower government subsidies to Western farmers broke down in 2003.

Democratic Globalization

The final reform we wish to consider involves efforts to help spread democracy throughout the developing world. A large body of research shows that democracy lowers inequality and promotes economic growth (Pettersson, 2003; Sylwester, 2002; United Nations, 2002). Democracies have these effects for several reasons. They make it more difficult for elite groups to misuse their power and enhance their wealth and income at the expense of the less well-to-do. They increase political stability, thereby providing a better investment climate. Finally, because they encourage broad political participation, democracies tend to enact policies that are more responsive to people's needs and benefit a wide range of people from all social classes. For example, democratic governments are more inclined to take steps to avoid famines, protect the environment, build infrastructure, and ensure basic needs like education and health. These measures help create a population better suited to pursue economic growth.

Although democracy has spread in recent years, by 2002 only 80 countries with 55 percent of the world's population were considered fully democratic by one measure (United Nations, 2002: 2; see Chapter 14, "Politics"). For its part, the United States has supported as many antidemocratic as democratic regimes in the developing world. Especially between the end of the World War II in 1945 and the collapse of the Soviet Union in 1991, the U.S. government gave military and financial aid to many antidemocratic regimes, often in the name of halting the spread of Soviet influence. These actions often generated unexpected and undesirable consequences, or what the CIA came to call "blowback" (Johnson, 2000). For example, in the 1980s the U.S. government supported Saddam Hussein when it considered Iraq's enemy, Iran, the greater threat to U.S. security interests. The United States also funded Osama bin Laden when he was fighting the Soviet Union in Afghanistan. Only a decade later, these so-called allies turned into our worst enemies (Chomsky, 1991; Johnson, 2000; Kolko, 2002) (see Box 9.2 for an example of half-hearted American support for democracy and its consequences).

In sum, we have outlined four reforms that could change the nature of neoliberal globalization and turn it into what we would like to call "democratic globalization." These reforms include offering stronger support for democracy in the developing world, contributing more and better foreign aid, forgiving the debt owed by developing countries to the rich countries, and eliminating tariffs that restrict exports from developing countries. These kinds of policies could plausibly help the developing world overcome the legacy of colonialism and join the ranks of the well-to-do. They would do much to ensure that the complex process we call globalization would benefit humanity as a whole.

Summary

Sociology Now™

Reviewing is as easy as ❶, ❷, ❸.

❶ Before you do your final review, take the SociologyNow diagnostic quiz to help you identify the areas on which you should concentrate. You will find information on how to access SociologyNow on the foldout at the front of the textbook.

❷ As you review, take advantage of SociologyNow's study aids to help you master the topics in this chapter.

❸ When you are finished with your review, take SociologyNow's post-test to confirm you are ready to move on to the next chapter.

1. **What is globalization and why is it taking place?**

   Globalization is the growing interdependence and mutual awareness of individuals and economic, political, and social institutions. It is a response to many forces, some technological (e.g., the development of inexpensive means of rapid international communication), others economic (e.g., burgeoning international trade and investment), and still others political (e.g., the creation of transnational organizations that limit the sovereign powers of nation states).

2. **What are the consequences of globalization?**

   Globalization has complex consequences, some of which are captured by the idea of "glocalization," which denotes the homogenization of some aspects of life and the simultaneous sharpening of some local differences. In addition, globalization evokes an antiglobalization reaction.

3. **How long is the history of globalization?**

   Globalization has a long history. Its origins can be traced to the beginning of long-distance migration and trade. However, it was the beginning of European exploration and capitalism about 500 years ago that really put the spurs to globalization.

4. **What are the main trends in global inequality and poverty?**

   Global inequality and poverty are staggering and in some respects getting worse. The income gap between rich and poor countries and between rich and poor individuals has grown worldwide since the 19th century. In recent decades the number of desperately poor people has declined absolutely and in percentage terms worldwide, but it has increased in the less developed countries.

5. **What are the main sociological theories of economic development?**

   Modernization theory argues that global inequality occurs as a result of some countries lacking sufficient capital, Western values, rational business practices, and stable governments. Dependency theory counters that global inequality results from the exploitative relationship between rich and poor countries. An important test of the two theories concerns the effect of foreign investment on economic growth, but research on this subject is equivocal. Apparently, historical, social-structural, and policy factors matter greatly in determining how a particular country responds to international trade and foreign direct investment.

6. **What are the characteristics of formerly colonized countries that emerged from poverty?**

   The poor countries best able to emerge from poverty have a colonial past that left them with industrial infrastructures. They also enjoy a favorable geopolitical position. They implement strong, growth-oriented economic policies, and they have socially cohesive populations.

7. **Can neoliberal globalization be reformed?**

   Neoliberal globalization can be reformed so that the benefits of globalization are more evenly distributed throughout the world. Possible reforms include offering stronger support for democracy in the developing world, contributing more and better foreign aid, forgiving the debt owed by developing countries to the rich countries, and eliminating tariffs that restrict exports from developing countries.

Questions to Consider

1. How has globalization affected your life, family, and town? What would life be like in a place that has not been affected by globalization?
2. Think of a commodity you consume or use everyday—bananas, coffee, shoes, cell phones, and so forth—and find out where and how it was manufactured, transported, and marketed. How does your consumption of the commodity tie you into a global commodity chain? How does your consumption of the commodity affect other people in other parts of the world in significant ways?
3. Should Americans do anything to alleviate global poverty? Why or why not? If you think Americans should be doing something to help end global poverty, then what should we do?

Web Resources

Companion Website for This Book

http://sociology.wadsworth.com

Begin by clicking on the Student Resources section of the website. Choose "Introduction to Sociology" and finally the Brym and Lie book cover. Next, select the chapter you are currently studying from the pull-down menu. From the Student Resources page you will have easy access to InfoTrac® College Edition, MicroCase Online exercises, additional web links, and many other resources to aid you in your study of sociology, including practice tests for each chapter.

Recommended Websites

UN statisticians have created indicators of human development for every country. The indicators for 2004 are available at http://hdr.undp.org/reports/global/2004/pdf/hdr04_HDI.pdf.

Naomi Klein's *No Logo: Taking Aim at the Brand Bullies* (New York: HarperCollins, 2000) has been called the manifesto of the antiglobalization movement. For updates, visit http://www.nologo.org.

The World Bank has numerous reports and data on the state of the global economy as well as on different nations at http://www.worldbank.org.

CHAPTER 10 Race and Ethnicity

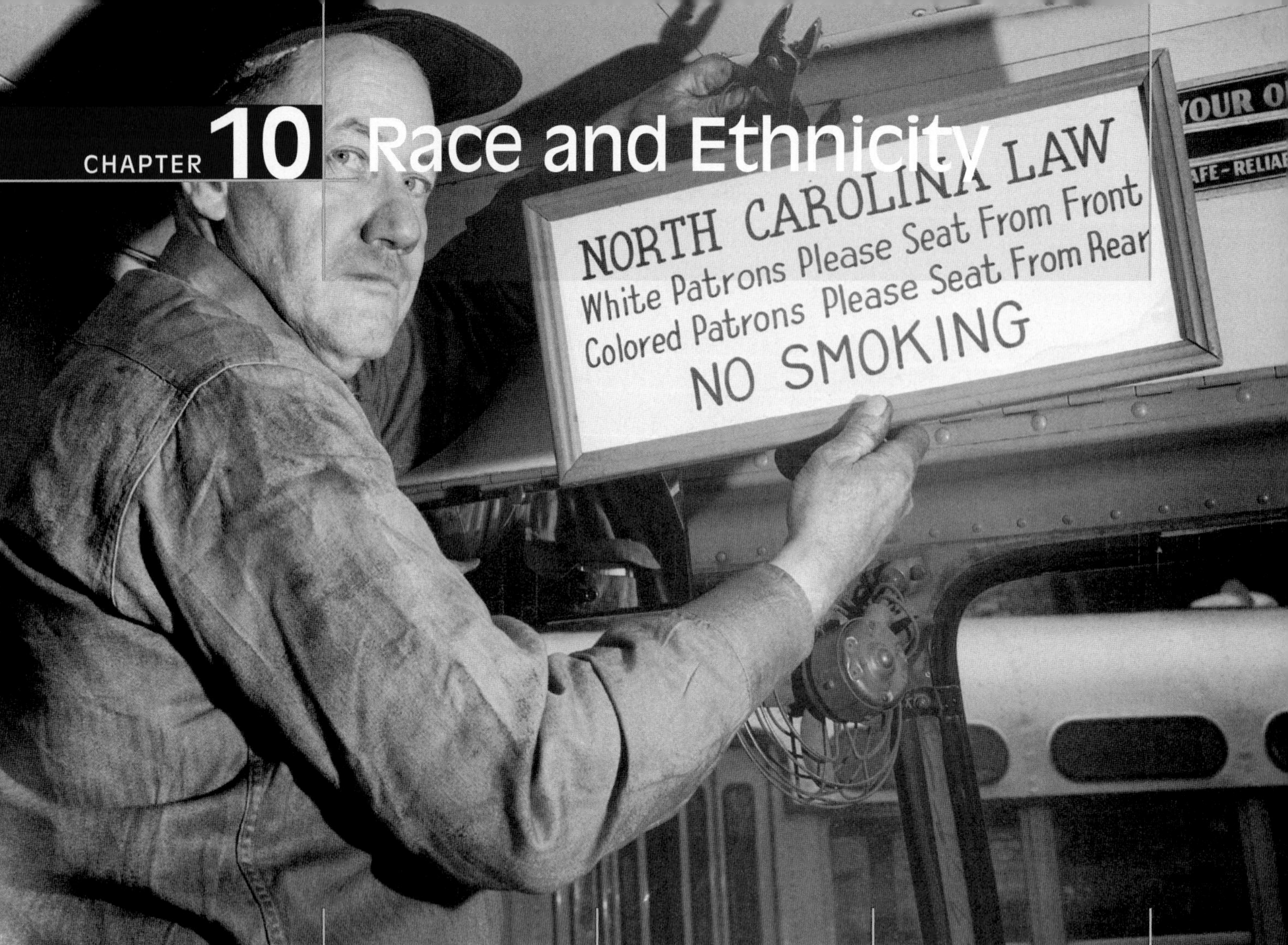

Jack Moebes/Corbis

In this chapter, you will learn that:

- Race and ethnicity are socially constructed ideas. We use them to distinguish people based on perceived physical or cultural differences, with profound consequences for their lives.
- Racial and ethnic labels and identities change over time and place. Relations between racial and ethnic groups help to shape these labels and identities.
- In the United States, members of some racial and ethnic groups are blending over time. However, this tendency is weaker among members of highly disadvantaged groups, such as African Americans and Native Americans.
- Identifying with a racial or ethnic group can be economically, politically, and emotionally advantageous for some people.
- A high level of immigration gives new life to older racial and ethnic communities. High levels of racial and ethnic inequality are likely to persist in the United States for the foreseeable future.

Defining Race and Ethnicity

The Great Brain Robbery

Dr. Samuel George Morton of Philadelphia was the most distinguished scientist in the United States 150 years ago. When he died in 1851, the *New York Tribune* wrote that "probably no scientific man in America enjoyed a higher reputation among scholars throughout the world, than Dr. Morton" (quoted in Gould, 1996 [1981]: 81).

Among other things, Morton collected and measured human skulls. The skulls came from various times and places. Their original occupants were members of different races. Morton believed he could show that the bigger your brain, the smarter you were. He packed BB-sized shot into a skull until it was full. Next, he poured the shot from the skull into a graduated cylinder. He then recorded the volume of shot in the cylinder. Finally, he noted the race of the person from whom each skull came, which, he thought, allowed him to draw conclusions about the average brain size of different races.

As he expected, Morton found that the races ranking highest in the social hierarchy had the biggest brains, whereas those ranking lowest had the smallest brains. He claimed that the people with the biggest brains were whites of European origin. Next were Asians. Then came Native Americans. The people at the bottom of the social hierarchy—and those with the smallest brains—were African Americans.

Morton's research had profound sociological implications because he claimed to show that the system of social inequality in the United States and throughout the world had natural, biological roots. If, on average, members of some racial groups are rich and

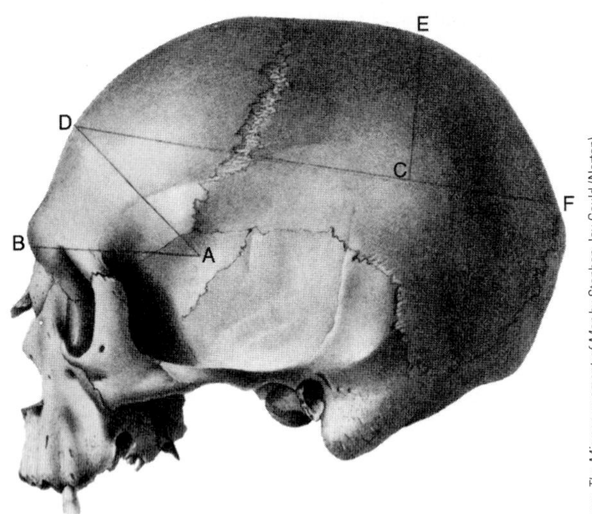

From *The Mismeasurement of Man* by Stephen Jay Gould (Norton)

In the 19th century, brain size was falsely held to be one of the main indicators of intellectual capacity. Average brain size was incorrectly said to vary by race. Researchers who were eager to prove the existence of such correlations are now widely regarded as practitioners of a racist quasi-science.

others poor, some highly educated and others illiterate, some powerful and others powerless, it was, said Morton, because of differences in brain size and mental capacity. Moreover, because he used science to show that Native Americans and African Americans *naturally* rest at the bottom of the social hierarchy, his ideas were used to justify two of the most oppressive forms of domination and injustice: colonization and slavery.

Despite claims to scientific objectivity, not a shred of evidence supported Morton's ideas. For example, in one of his three main studies, Morton measured the capacity of skulls robbed from Egyptian tombs. He found the average volume of black people's skulls to be 4 cubic inches smaller than the average volume of white people's skulls. This finding seemed to prove his case. Today, however, we know that three main issues compromise his findings:

- Morton claimed to be able to distinguish the skulls of white and black people by their shapes. However, even today archaeologists cannot precisely determine race by skull shape. Therefore, whether the skulls Morton identified as "Caucasian" belonged to white people and those identified as "Negroid" and "Negro" belonged to black people is unclear.
- Morton's skulls formed a small, unrepresentative sample. Morton based his conclusions on only 72 specimens—a very small number on which to base any generalization. Moreover, those 72 skulls are not representative of the skulls of white and black people in ancient Egypt or any other time and place. They are just the 72 skulls to which Morton happened to have access. For all we know, they may be highly unusual.
- Even if we ignore these first two problems, 71 percent of the skulls Morton identified as Negroid or Negro were women's, compared with only 48 percent of the skulls he identified as Caucasian. Yet, women's bodies are on average smaller than men's bodies. To make a fair comparison, Morton would have had to make sure that the sex composition of the white and black skulls was identical. He did not. Instead, he biased his findings in favor of finding larger white skulls. When we compare Morton's black and white female skulls, the white skulls are only two cubic inches bigger. When we compare his black and white male skulls, the white skulls are one cubic inch *smaller* (Gould, 1996 [1981]: 84, 91, 92).

Scientifically speaking, Morton's findings are meaningless. Yet, they were influential for a long time. Some people still believe them. For example, less than 40 years ago, the author of an article about race in the *Encyclopedia Britannica,* the world's most authoritative general reference source, wrote that blacks have "a rather small brain in relation to their size" (Buxton, 1963: 864A). That claim was repeated in a controversial book published by a leading American press in the mid-1990s (Rushton, 1995). Yet, no more evidence exists today than in 1850 that whites have bigger brains than blacks.

Race, Biology, and Society

Biological arguments about racial differences have grown more sophisticated over time. However, their scientific basis is just as shaky now as it always was.

In medieval Europe, some aristocrats saw blue veins underneath their pale skin. However, they could not see blue veins underneath the peasants' suntanned skin. They concluded that the two groups must be racially distinct. The aristocrats called themselves "blue bloods." They ignored the fact that the color of blood from an aristocrat's wound was just as red as the blood from a peasant's wound.

About 80 years ago, some Americans expressed the belief that racial differences in average IQ scores are based in biology. On average, Jews scored below non-Jews on IQ tests in the 1920s. This was used as an argument against Jewish immigration. "America must be kept American," proclaimed President Calvin Coolidge as he signed the 1924 Immigration Restriction Act (quoted in Gould, 1996 [1981]: 262). More recently, African Americans have on average scored below European Americans on IQ tests. Some people say this justifies cutting budgets for schools in the inner city, where many African Americans live. Why invest good money in inner-city schooling, such people ask, if low IQ scores are rooted in biology and therefore fixed (Herrnstein and Murray, 1994)? However, the people who argued against Jewish immigration and better education for inner-city African Americans ignored two facts. First, Jewish IQ scores rose as Jews moved up the class hierarchy and could afford better education. Second, enriched educational facilities have routinely boosted the intellectual development and academic achievement of inner-city African American children (Campbell and Ramey, 1994; Frank Porter Graham Child Development Center, 1999; Gould, 1996 [1981]; Hancock, 1994; Steinberg, 1989 [1981]). Much evidence shows that the social environment in which one is raised and educated has a big impact on IQ and other standardized test scores. The evidence that racial differences in IQ scores are biologically based is about as strong as evidence showing that aristocrats have blue blood (Fischer et al., 1996; Maume, Cancio, and Evans, 1996; Chapter 17, "Education").[1]

If one cannot reasonably maintain that racial differences in average IQ scores are based in biology, what about differences in singing ability or athletic prowess or crime rates? For example, some people insist that genetically, African Americans are better than whites at singing and sports, and more prone to crime. Does any evidence exist to support this belief?

At first glance, the supporting evidence might seem strong. Consider sports. Aren't about 85 percent of National Basketball Association (NBA) players and 75 percent of National Football League (NFL) players black? Don't West African–descended blacks hold the 200 fastest 100-meter times, all under 10 seconds? Don't North and East Africans regularly win 40 percent of the top international distance-running honors, yet represent only a fraction of 1 percent of the world's population (Entine, 2000)? Although these facts are undeniable, the argument for the genetic basis of black athletic superiority begins to falter once we consider two additional points. First, no gene linked to general athletic superiority has been identified. Second, athletes of African descent do not perform unusually well in many sports, such as swimming, hockey, cycling, tennis, gymnastics, and soccer. The idea that people of African descent are generally superior athletes is simply untrue.[2]

Prejudice, Discrimination, and Sports

Sociologists have identified certain *social* conditions leading to high levels of participation in sports (as well as entertainment and crime). These conditions operate on all groups of people, whatever their race. Specifically, people who face widespread prejudice and discrimination often enter sports, entertainment, and crime in disproportionately large numbers for lack of other ways to improve their social and economic position. For such people, other avenues of upward mobility tend to be blocked. (**Prejudice** is an attitude that judges a person on his or her group's real or imagined characteristics. **Discrimination** is unfair treatment of people because of their group membership.) Thus, in the United States, Irish, Jews, Italians, Puerto Ricans, and African Americans formed successive waves of high-crime groups in the late 19th and 20th centuries. Their crime rates declined only

Prejudice is an attitude that judges a person according to his or her group's real or imagined characteristics.

Discrimination is unfair treatment of people because of their group membership.

[1]Although sociologists commonly dispute a genetic basis of mean intelligence for *races*, evidence suggests that *individual* differences in intelligence are partly genetically transmitted (Bouchard et al., 1990; Lewontin, 1991: 19–37; Scarr and Weinberg, 1978; Schiff and Lewontin, 1986).

[2]Genetic differences may lead some groups to have physical characteristics that lend themselves to excellence in *particular* sports, but that argument is different from the general argument we are criticizing here (Entine, 2000).

Duomo/Corbis

Athlete-heroes like Shaquille O'Neal are often held up as role models for African American youth, even though the chance of "making it" as a professional athlete is much less than the chance of getting a college education and succeeding as a professional. The cultural emphasis on African American sports heroes also has the effect of reinforcing harmful and incorrect racial stereotypes about black athletic prowess and, conversely, intellectual inferiority.

Annie Sachs/Archive Photo/Getty Images

Colin Powell is of African, English, Irish, Scottish, Arawak Indian, and Jewish ancestry. Like an increasingly large number of Americans, it is difficult to say that he is a member of any particular race or ethnic group.

to the degree that their economic and social standing improved (Bell, 1960). Similarly, prejudice and discrimination against American Jews did not began to decline appreciably until the 1950s. Until then, Jews played a prominent role in professional sports. For instance, when the New York Knicks played their first game on November 1, 1946, beating the Toronto Huskies 68–66, the starting lineup consisted of Ossie Schechtman, Stan Stutz, Jake Weber, Ralph Kaplowitz, and Leo "Ace" Gottlieb—an all-Jewish squad (National Basketball Association, 2000). Koreans in Japan today are subject to much prejudice and discrimination. They often pursue careers in sports and entertainment. In contrast, Koreans in the United States face less prejudice and discrimination. Few of them become athletes and entertainers. Instead, they are often said to excel in engineering and science. As these examples show, social circumstances have a big impact on criminal, athletic, and other forms of behavior.

The idea that people of African descent are genetically superior to whites in athletic ability complements the idea that they are genetically inferior to whites in intellectual ability. Both ideas have the effect of reinforcing black/white inequality. Although the United States has fewer than 10,000 elite, professional athletes, many millions of pharmacists, graphic designers, lawyers, systems analysts, police officers, nurses, and people in other interesting occupations that offer steady employment and good pay live in the United States. By promoting only the Shaquille O'Neals of the world as suitable role models for African American youth, the idea of "natural" black athletic superiority and intellectual inferiority in effect asks black Americans to bet on a high-risk proposition—that they will make it in professional sports. At the same time, it deflects attention from a much safer bet—that they can achieve upward mobility through academic excellence (Doberman, 1997).[3]

Sociology Now™

Learn more about **Prejudice and Discrimination** by going through the Prejudice, Discrimination, and Racism Animation.

An additional problem with the argument that genes determine the specific behaviors of different racial groups is that one cannot neatly distinguish races based on genetic differences. A high level of genetic mixing has taken place among people of various races throughout the world. In the United States, for instance, white male slave owners often raped female African American slaves. Similarly, many African Americans had children with Native Americans in the 18th and 19th centuries. Today, racial intermarriage is no longer a rarity, and it is becoming more common. About 50 percent of Asian Americans and 10 percent of African Americans now marry outside their racial group (Stanfield, 1997). A growing number of Americans are similar to Tiger Woods and Colin Powell. Woods claims he is of "Cablinasian" ancestry—part Caucasian, part black, part Indian (Native American), and part Asian. Colin Powell's ancestry includes African, English, Irish, Scottish, Arawak Indian, and Jewish. As these examples illustrate, the difference between "black," "white," "Asian," and so forth is often anything but clear-cut.

[3]The genetic argument also belittles athletic activity itself by denying the role of training in developing athletic skill.

Chinese *Japanese*

HOW TO TELL YOUR FRIENDS FROM THE JAPS

Of these four faces of young men (*above*) and middle-aged men (*below*) the two on the left are Chinese, the two on the right Japanese. There is no infallible way of telling them apart, because the same racial strains are mixed in both. Even an anthropologist, with calipers and plenty of time to measure heads, noses, shoulders, hips, is sometimes stumped. A few rules of thumb—not always reliable:

▶ Some Chinese are tall (average: 5 ft. 5 in.). Virtually all Japanese are short (average: 5 ft. 2 1/3 in.).

▶ Japanese are likely to be stockier and broader-hipped than short Chinese.

▶ Japanese—except for wrestlers—are seldom fat; they often dry up and grow lean as they age. The Chinese often put on weight, particularly if they are prosperous (in China, with its frequent famines, being fat is esteemed as a sign of being a solid citizen).

▶ Chinese, not as hairy as Japanese, seldom grow an impressive mustache.

▶ Most Chinese avoid horn-rimmed spectacles.

▶ Although both have the typical epicanthic fold of the upper eyelid (which makes them look almond-eyed), Japanese eyes are usually set closer together.

▶ Those who know them best often rely on facial expression to tell them apart: the Chinese expression is likely to be more placid, kindly, open: the Japanese more positive, dogmatic, arrogant.

In Washington, last week, Correspondent Joseph Chiang made things much easier by pinning on his lapel a large badge reading "Chinese Reporter–NOT *Japanese*—Please."

▶ Some aristocratic Japanese have thin, aquiline noses, narrow faces and, except for their eyes, look like Caucasians.

▶ Japanese are hesitant, nervous in conversation, laugh loudly at the wrong time.

▶ Japanese walk stiffly erect, hard-heeled. Chinese, more relaxed, have an easy gait, sometimes shuffle.

Chinese *Japanese*

Time magazine explains how to make arbitrary racial distinctions, 1941.

The Social Construction of Race

Many scholars believe we all belong to one human race, which originated in Africa (Cavalli-Sforza, Menozzi, and Piazza, 1994). They argue that subsequent migration, geographical separation, and inbreeding led to the formation of more or less distinct races. However, particularly in modern times, humanity has experienced so much intermixing that race as a biological category has lost nearly all meaning. Some biologists and social scientists therefore suggest we drop the term "race" from the vocabulary of science (Angier, 2000).

Most sociologists, however, continue to use the term "race" because *perceptions* of race affect the lives of most people profoundly. Everything from your wealth to your health is influenced by whether others see you as African American, white, Asian American, Native American, or something else. Race as a *sociological* concept is thus an invaluable analyti-

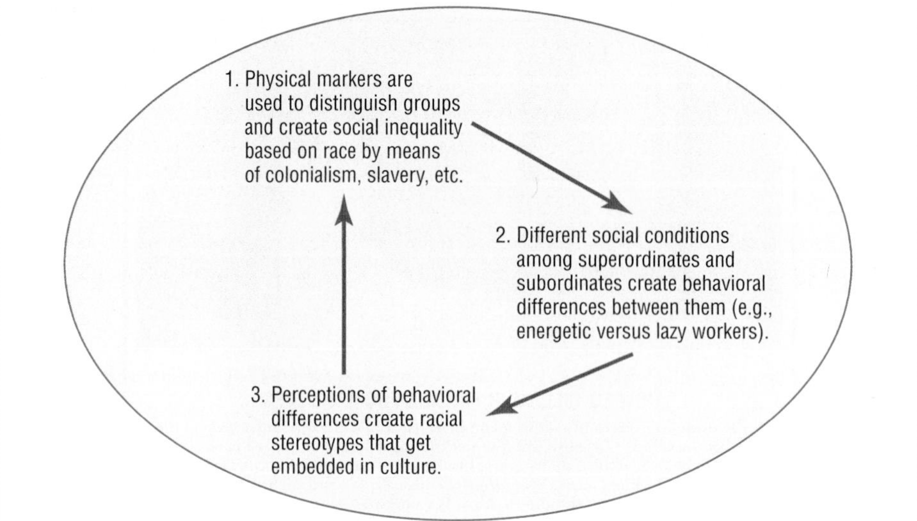

Figure 10.1
The Vicious Circle of Racism

cal tool—if the user remembers that it refers to socially significant physical differences, such as skin color, rather than biological differences that determine behavioral traits.

Said differently, perceptions of racial difference are socially constructed and often arbitrary (Ferrante and Brown, 2001 [1998]). The Irish and the Jews were regarded as "blacks" by many people 100 years ago, and today many northern Italians still think of southern Italians from Sicily and Calabria as "blacks" (Gilman, 1991; Ignatiev,1995; Roediger, 1991). In the United States during World War II, some people made arbitrary physical distinctions between Chinese allies and Japanese enemies that helped to justify the American policy of placing Japanese Americans in internment camps. These examples show that racial distinctions are social constructs, not biological "givens."

Why Race Matters

Finally, then, we can define **race** as a social construct used to distinguish people in terms of one or more physical markers, usually with profound effects on their lives. However, this definition raises an interesting question. If race is merely a social construct and not a useful biological term, why are perceptions of physical difference used to distinguish groups of people in the first place? Why, in other words, does race matter? Most sociologists believe race matters because it allows social inequality to be created and maintained. The English who colonized Ireland, the Americans who went to Africa looking for slaves, and the Germans who used the Jews as a scapegoat to explain their deep economic and political troubles after World War I all set up systems of racial domination. (A **scapegoat** is a disadvantaged person or category of people whom others blame for their own problems.) Once colonialism, slavery, and concentration camps were established, behavioral differences developed between subordinates and superordinates. For example, African American slaves and Jewish concentration camp inmates, with little motivating them to work hard except the ultimate threat of the master's whip, tended to do only the minimum work necessary to survive. Their masters noticed this and characterized their subordinates as inherently slow and unreliable workers (Collins, 1982: 66–9). In this way, racial stereotypes are born. The stereotypes then embed themselves in literature, popular lore, journalism, and political debate. This reinforces racial inequalities (Figure 10.1). We thus see that race matters to the degree that it helps to create and maintain systems of social inequality.

- **Race** is a social construct used to distinguish people in terms of one or more physical markers, usually with profound effects on their lives.
- A **scapegoat** is a disadvantaged person or category of people whom others blame for their own problems.
- An **ethnic group** is composed of people whose perceived cultural markers are deemed socially significant. Ethnic groups differ from one another in terms of language, religion, customs, values, ancestors, and the like.

Ethnicity, Culture, and Social Structure

Race is to biology as ethnicity is to culture. A race is a category of people whose perceived *physical* markers are deemed socially significant. An **ethnic group** is composed of people whose perceived *cultural* markers are deemed socially significant. Ethnic groups differ from one another in terms of language, religion, customs, values, and ancestors. However, just as physical distinctions do not *cause* differences in the behavior of various races, cultural distinctions are often not by themselves the major source of differences in the behavior of various ethnic groups. In other words, ethnic values and other elements of ethnic culture have less of an effect on the way people behave than we commonly believe because *social-structural* differences typically underlie cultural differences.

An example will help drive home the point. Many ethnic groups may be found in the African American community (Waters, 2000). By far the largest is composed of descendants of former Southern slaves. Another is composed of Jamaican and other Caribbean or West Indian immigrants and their descendants. People often say that West Indians emphasize hard work, saving, investment, and education (Glazer and Moynihan, 1963: 35). Descendants of Southern black slaves, they claim, lack a culture emphasizing these values. As a result (the argument continues), descendants of Southern black slaves suffer from higher unemployment, lower income, and higher crime rates.

To our knowledge, nobody ever conducted a scientific survey comparing the values of these two groups. The contrast may therefore be false or exaggerated. However, for the sake of argument, let us accept the accuracy of the value contrast. Is it a sufficient explanation for behavioral differences between the two ethnic groups? It is not, according to sociologist Stephen Steinberg (1989 [1981]: 275–80; also Kalmijn, 1996). Immigration records show that the first wave of West Indian immigrants who came to the United States in the 1920s was 89 percent literate. That figure is higher than the literacy rate for the U.S. population as a whole at the time. More than 40 percent of the immigrants were skilled workers. Many of them were highly educated professionals. Thus, the first West Indian immigrants to the United States occupied a much higher class position than descendants of Southern black slaves. The latter were nearly all unskilled workers in the 1920s. If the West Indians exhibited greater interest in education and saving, it was undoubtedly related to their class position. Skilled and literate people usually exhibit these traits more than unskilled and illiterate people. The latter often understand only too well that they have few opportunities to pursue higher education and a professional career.

That class was more important than ethnic values in determining economic success is also evident if we compare West Indian immigrants in New York and London in the 1950s and 1960s. The two groups came from the same place at the same time and presumably shared many values. However, the New York West Indians enjoyed more economic success than the London West Indians. Why? Largely because the percentage of white-collar and professional workers was nearly three times higher (27 percent) among the New Yorkers when they arrived in the United States. The New Yorkers started with a big advantage and capitalized on it.

Most sociologists stress how social-structural conditions rather than values determine the economic success or failure of racial and ethnic groups (Abelmann and Lie, 1995; Brym with Fox, 1989: 103–19; Lieberson, 1980; Portes and Rumbaut, 1990; Steinberg, 1989 [1981]: 82–105, 270–5). For example, people often praise Jews, Koreans, and other economically successful groups for emphasizing education, family, and hard work. People less commonly notice, however, that American immigration policy is highly selective. For the most part, the Jews and Koreans who arrived in the United States were literate, urbanized, and skilled. Some even came with financial assets. They confronted much prejudice and discrimination but far less than that reserved for descendants of Southern blacks. These *social-structural* conditions facilitated Jewish and Korean success. They gave members of these groups a firm basis on which to build and maintain a culture emphasizing education, family, and other middle-class virtues.

We conclude that it is misleading to claim that "[r]ace and ethnicity . . . are quite different, since one is biological and the other is cultural" (Macionis, 1997 [1987]: 321). As we have seen, both race and ethnicity are rooted in social structure, not in biology and culture. The biological and cultural aspects of race and ethnicity are secondary to their sociological character. Moreover, the distinction between race and ethnicity is not as simple as the difference between biology and culture. As noted previously in relation to the Irish and the Jews, groups once socially defined as races may be later redefined as ethnic, even though they do not change biologically. Biology and culture do not determine whether a group is viewed as a race or an ethnic group, but social definitions do. The interesting question from a sociological point of view is why social definitions of race and ethnicity change. We now consider that issue.

Race and Ethnic Relations

Labels and Identity

Personal Anecdote

John Lie moved with his family from South Korea to Japan when he was a baby. He moved from Japan to Hawaii when he was 10 years old, and again from Hawaii to the American mainland when he started college. The move to Hawaii and the move to the mainland changed the way John thought of himself in ethnic terms.

In Japan, the Koreans form a minority group. (A **minority group** is a group of people who are socially disadvantaged, though they may be in the numerical majority, like the blacks in South Africa.) Before 1945, when Korea was a colony of Japan, some Koreans were brought to Japan to work as miners and unskilled laborers. The Japanese thought the Koreans who lived there were beneath and outside Japanese society (Lie, 2001). Not surprisingly, then, Korean children in Japan, including John, were often teased and occasionally beaten by their Japanese schoolmates. "The beatings hurt," says John, "but the psychological trauma resulting from being socially excluded by my classmates hurt more. In fact, although I initially thought I was Japanese like my classmates, my Korean identity was literally beaten into me.

"When my family immigrated to Hawaii, I was sure things would get worse. I expected Americans to be even meaner than the Japanese. (By Americans, I thought only of white European Americans.) Was I surprised when I discovered that most of my schoolmates were not white European Americans, but people of Asian and mixed ancestry! Suddenly I was a member of a numerical majority. I was no longer teased or bullied. In fact, I found that students of Asian and non-European origin often singled out white European Americans (called *haole* in Hawaiian) for abuse. We even had a "beat up *haole* day" in school. Given my own experiences in Japan, I empathized somewhat with the white Americans. But I have to admit that I also felt a great sense of relief and an easing of the psychological trauma associated with being Korean in Japan.

"As the years passed, I finished public school in Hawaii. I then went to college in Massachusetts and got a job as a professor in Illinois. I associated with, and befriended, people from various racial and ethnic groups. My Korean origin became a less and less important factor in the way people treated me. There was simply less prejudice and discrimination against Koreans during my adulthood in the United States than in my early years in Japan. I now think of myself less as Japanese or Korean than as American. When I lived in Illinois, I sometimes thought of myself as a Midwesterner. Now that I have changed jobs and moved to California, my self-conception may shift again; my identity as an Asian American may strengthen given the large number of Asians who live in California. Clearly, my ethnic identity has changed over time in response to the significance others have attached to my Korean origin. I now understand what the French philosopher Jean-Paul Sartre meant when he wrote that 'the anti-Semite creates the Jew'" [Sartre, 1965 [1948]: 43].

Sociology Now™

Learn more about **Race and Ethnic Relations** by going through the Race and Ethnic Relations Data Experiment.

A **minority group** is a group of people who are socially disadvantaged, even if they are in the numerical majority (e.g., the blacks of South Africa).

Heather Titus/Photo Resource Hawaii

The United States is becoming a more ethnically and racially diverse society as a result of a comparatively high immigration rate and a comparatively low birthrate among non-Hispanic whites. The U.S. Census Bureau estimates that around 2060 a minority of Americans will be non-Hispanic whites.

The Formation of Racial and Ethnic Identities

The details of John Lie's life are unique. But experiencing a shift in racial or ethnic identity is common. Social contexts, and in particular the nature of one's relations with members of other racial and ethnic groups, shape and continuously reshape one's racial and ethnic identity. Change your social context, and your racial and ethnic self-conception eventually changes too (Miles, 1989; Omi and Winant, 1986).

Consider Italian Americans. Around 1900 Italian immigrants thought of themselves as people who came from a particular town or perhaps a particular province, such as Sicily or Calabria. They did not usually think of themselves as Italians. Italy became a unified country only in 1861. A mere 40 years later, far from all of its citizens identified with their new nationality. In the United States, however, officials and other residents identified the newcomers as "Italians." The designation at first seemed odd to many of the new immigrants. However, over time it stuck. Immigrants from Italy started thinking of themselves as Italian Americans because others defined them that way. A new ethnic identity was born (Yancey, Ericksen, and Leon, 1976).

As symbolic interactionists emphasize, the development of racial and ethnic labels and ethnic and racial identities is typically a process of negotiation. For example, members of a group may have a racial or ethnic identity, but outsiders may impose a new label on them. Group members then reject, accept, or modify the label. The negotiation between outsiders and insiders eventually results in the crystallization of a new, more or less stable ethnic identity. If the social context changes again, the negotiation process begins anew.

Case Study: The Diversity of the "Hispanic American" Community

You can witness the formation of new racial and ethnic labels and identities in the United States today. Consider, for instance, the terms "Hispanic American" and "Latino" (Darder and Torres, 1998). "Latino" (male) and "Latina" (female) seem to be the preferred usage in the West and "Hispanic American" in the East. "Latino/a" is the more inclusive term because it encompasses Brazilians, who speak Portuguese, but we use "Hispanic American" here because that is the term used by the U.S. Census Bureau. In any case, people scarcely

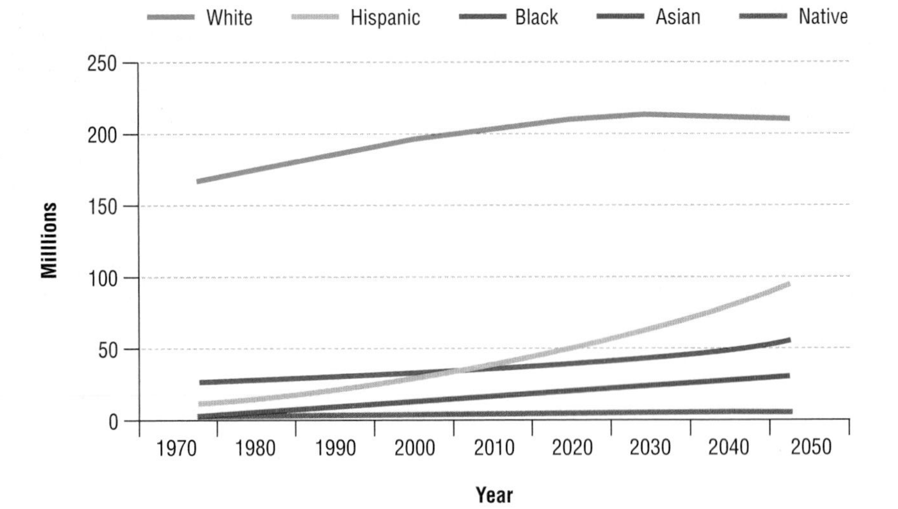

Figure 10.2
Racial and Ethnic Composition, United States, 1970–2050 (projected)

Source: U.S. Census Bureau (1999e; 2002e).

Web Research Project: The Social Characteristics of Hispanic Americans

used these terms 30 or 40 years ago. Now they are common. According to the U.S. Census Bureau, nearly 39 million Hispanic Americans lived in the United States in 2003. Because of continuing robust immigration and relatively high fertility, the Bureau predicts they will number more than 96 million in 2050. They are the second fastest growing ethnic category in the country (next to Asian Americans) and as of 2003 they were the second biggest, next to non-Hispanic whites (Figure 10.2).

But who is a Hispanic American? Table 10.1 makes it clear that Hispanic Americans form a heterogeneous population. In 2002, 67 percent of Hispanic Americans were of Mexican origin, nearly 9 percent of Puerto Rican origin, almost 4 percent of Cuban origin, and 14 percent were from El Salvador, the Dominican Republic, and other countries in Latin America. Mexican Americans have been concentrated in the South and the West, especially California and Texas, where they are known as "Chicanos." Puerto Ricans are found mainly in the Northeast, especially in New York. Cubans reside principally in the South, especially Miami. Other Hispanic Americans are divided fairly evenly among the Northeast, South, and West. Relatively few Hispanic Americans live in the Midwest.

Besides varying degrees of knowledge of the Spanish language, what do members of these groups have in common? One survey shows that most of them do *not* want to be called "Hispanic American." Instead, they prefer being referred to by their national origin—as Cuban Americans, Puerto Rican Americans, Mexican Americans, and so forth. Many Hispanic Americans born in the United States want to be called simply "Americans" (de la Garza et al., 1992).

Table 10.1
Population by Hispanic Origin and Region, United States, 2002 (in percent)

	Mexican	Puerto Rican	Cuban	Other	Total
Northeast	2.4	58.0	13.3	30.2	13.3
Midwest	8.7	8.5	3.0	4.8	7.7
South	34.3	27.0	75.1	32.6	34.8
West	54.6	6.4	8.5	32.4	44.2
Total	100.0	99.9*	99.9*	100.0	100.0
Percent of all Hispanic Americans	67.0	8.6	3.7	20.7	100.0
N *(in 000s)*	25,074	3,222	1,376	7,765	37,438

*Due to rounding
Source: U.S. Census Bureau (2002a).

The Formation of Ethnic Enclaves

Members of these groups also enjoy different cultural traditions, occupy different positions in the class hierarchy, and vote differently. For example, there are many more middle-class and professional people among Americans of Cuban origin than among members of other major Hispanic groups in the United States. That is because a large wave of middle-class Cubans fled Castro's revolution and arrived in the Miami area in the late 1950s and early 1960s. There they formed an **ethnic enclave.** An ethnic enclave is a geographical concentration of ethnic group members who establish businesses that serve and employ mainly members of the ethnic group and reinvest profits in community businesses and organizations (Portes and Manning, 1991). Many Cubans who arrived in subsequent waves of immigration were poor, but because they could rely on a well-established and prosperous Cuban American ethnic enclave for jobs and other forms of support, most of them soon achieved middle-class status. The fact that Cuban immigrants shared strong procapitalist and anticommunist values with the American public also helped their integration into the larger society. Not surprisingly given their background, Cuban Americans are more likely to vote Republican than Americans of Mexican and Puerto Rican origin.

Immigrants from Mexico and Puerto Rico tend to be members of the working class who have not completed high school. Their children typically achieve educational levels similar to that of non-Hispanic whites (Bean and Tienda, 1987). However, they have not reached the average level of prosperity enjoyed by Cuban Americans because of the lower-class origins of the communities and resulting weaker ethnic enclave formations. Chicanos and Puerto Rican Americans are more likely than Cuban Americans to vote Democrat.

What Unifies the Hispanic American Community?

So we see that the Hispanic American community is highly diverse. In fact, not even knowledge of the Spanish language unifies Hispanic Americans. Some people who are commonly viewed as Hispanic American do not speak Spanish. Haitians, for example, speak French or a regional dialect of French. Brazilians speak Portuguese. Still other people who are commonly viewed as Hispanic American in the United States reject the label. Chief among them are Mayans, indigenous Central Americans whom the Spanish colonized.[4]

Despite this internal diversity, however, the term "Hispanic American" is more widely used than it was 30 or 40 years ago for three main reasons:

- Many Hispanic Americans find it politically useful. Recognizing that power flows from group size and unity, Hispanic Americans have created national organizations to promote the welfare of their entire community. One of these is the National Council of La Raza, a nonprofit organization established in 1968 to "reduce poverty and discrimination, and improve life opportunities for Hispanic Americans" ("The National Council of La Raza," 2000). Thus, for certain purposes, Hispanic Americans from Cuba, Mexico, Puerto Rico, and other places find it convenient to play down what separates them and accentuate what they have in common.
- A second reason the term "Hispanic American" is becoming more common is that the government finds it useful for data collection and public policy purposes. Like "Asian American," "African American," and "Native American," "Hispanic American" is a convenient administrative term. For example, by collecting census data on the number of people who identify themselves as Hispanic American, the government is better able to allocate funding for Spanish-language instruction in schools, ensure diversity in the workplace, and take other public policy actions that reflect the changing racial and ethnic composition of American society (Box 10.1).

An **ethnic enclave** is a spatial concentration of ethnic group members who establish businesses that serve and employ mainly members of the ethnic group and reinvest profits in community businesses and organizations.

[4]Similarly, particularly in the West, many indigenous Americans reject the label "Native American" as official, "white" terminology and proudly call themselves "Indians."

BOX10.1

SOCIAL POLICY: WHAT DO YOU THINK?

Bilingual Education

Ron Unz is an opponent of bilingual education. He argues that a "quarter of all the children in California public schools are classified as not knowing English. . . . Of the ones who don't know English in any given year, only 5 or 6 percent learn English. Since the goal of the system, obviously, should be to make sure that these children learn English, we're talking about a system with an annual failure rate of 95 percent. . . . Many of my friends are foreign immigrants. They came here when they were a variety of different ages. All of them agree that little children or even young teenagers can learn another language quickly, though only 5 percent of these children in California are learning English each year. And that's what I define as failure" (quoted in Public Broadcasting System, 1997).

In response, James Lyons of the National Association for Bilingual Education says, "It is not the case that bilingual education is failing children. There are poor bilingual education programs, just as there are poor programs of every type in our schools today. But bilingual education has made it possible for children to have continuous development in their native language, while they're in the process of learning English, something that doesn't happen overnight, and it's made it possible for children to learn math and science at a rate equal to English-speaking children while they're in the process of acquiring English" (quoted in Public Broadcasting System, 1997).

About 8 percent of all public school students are enrolled in bilingual education programs in the United States. That amounts to more than 3.7 million children. Bilingual education programs cost taxpayers hundreds of millions of dollars. Is the expenditure worth it? Whatever the expense, is the ideal of bilingual education worth pursuing? The debate between Unz and Lyons touches on some important points concerning these issues. On the one hand, many bilingual education programs are not very effective. Furthermore, English is the main language spoken in the United States and should be taught to non–English speakers to ensure their economic progress; a less than fluent speaker of English is bound to do poorly not only in school but also in most workplaces. On the other hand, advocates argue that bilingual education helps nonnative speakers of English adjust to schools and keep up in other subjects. Bilingual education also recognizes the importance of learning a native language, such as Spanish, rather than simply assimilating to the dominant English-speaking culture. Some people argue that when students value their native language skills, their self-esteem increases. Later, this improves their economic success.

What do you think? Should we make every public student learn only in English? What would we gain and lose by such a policy? Is it worthwhile keeping bilingual programs for students who wish to preserve their language skills? Should we encourage native language preservation just for large groups, such as Spanish-speaking students, or for small groups as well, such as students of Chinese or Russian origin? Perhaps in this age of globalization, everyone should participate in a bilingual education program. Would that make Americans less ethnocentric and better able to participate in global affairs?

- Finally, "Hispanic American" is an increasingly popular label because non-Hispanic Americans find it convenient. Central and South America are composed of about 25 countries, and many ethnic divisions exist within those countries. It would be a tough job for anyone to keep all of those countries and ethnic groups straight in everyday speech. Lumping all Hispanic Americans together makes life easier for the majority group, although lack of sensitivity to a person's specific origins is not infrequently offensive to minority-group members.

So we see that "Hispanic American" is a new ethnic label and identity. It did not spring fully formed one day from the culture of the group to which it refers. It was created out of social necessity and is still being socially constructed (Portes and Truelove, 1991). We can say the same about *all* ethnic labels and identities, even those that may seem most fixed and natural, such as "white" (Lieberson, 1991; Lieberson and Waters, 1986; Waters, 1990).

Ethnic and Racial Labels: Choice versus Imposition

The idea that race and ethnicity are socially constructed does not mean that everyone can always choose their racial or ethnic identity freely. The degree to which people can exercise such freedom of choice varies widely from one society to the next. Moreover, in a given society at a given time, different categories of people are more or less free to choose. To illustrate these variations, we next discuss the way the government of the former Soviet

Union imposed ethnicity on the citizens of that country. We then contrast imposed ethnicity in the former Soviet Union with the relative freedom of ethnic choice in the United States. In the United States' case, we also underline the social forces that make it easier to choose one's ethnicity than one's race.

State Imposition of Ethnicity in the Soviet Union

Until it formally dissolved in 1991, the Soviet Union was the biggest and one of the most powerful countries in the world (Chapter 13, "Work and the Economy," and Chapter 14, "Politics"). Stretching over two continents and 11 time zones, it was composed of 15 republics—Russia, Ukraine, Kazakhstan, and so forth—with a combined population larger than the United States. In each republic, the largest ethnic group was the so-called "titular" ethnic group of the republic: Russians in Russia, Ukrainians in Ukraine, Kazakhs in Kazakhstan, and so forth. Over 100 minority ethnic groups, or "nationalities," lived in the republics. As ▶Figure 10.3 shows, in some republics the combined number of minority ethnic group members was greater than the titular ethnic group population.

The vast size and ethnic heterogeneity of the Soviet Union required that its leaders develop strategies for preventing the country from falling apart at the seams. One such strategy involved weakening the boundaries between the republics so "a new historical community, the Soviet people" could come into existence (Bromley, 1982 [1977]: 270). The creation of a countrywide educational system and curriculum, the spread of the Russian language, and the establishment of propaganda campaigns trumpeting remarkable national achievements helped to create a sense of unity among many Soviet citizens.

A second strategy promoting national unity involved the creation of a system allowing power and privilege to be shared among ethnic groups. This was accomplished administratively through the "internal passport" system. Beginning in the 1930s, Soviet governments issued identity papers or internal passports to all citizens at the age of 16. The fifth entry in each passport noted the bearer's ethnicity. Adolescents were obliged to adopt the ethnicity of their parents. Only if the parents were of different ethnic backgrounds could a 16-year-old choose the ethnicity of the mother or the father.

The internal passport system enabled officials to apply strict ethnic quotas in recruiting people to institutions of higher education, professional and administrative positions, and political posts. Ethnic quotas were even used to determine where people could reside. Thus, ethnicity became critically important in determining some of the most fundamental aspects of one's life. To ensure the loyalty of the disparate and far-flung republics to the central government, officials granted advantages to members of titular ethnic groups living in their own republics. You enjoyed the best opportunities for educational, occupational, and political advancement if you were a Russian living in Russia, a Ukrainian living in Ukraine, and so forth. If you happened to be a member of a titular ethnic group living outside your republic, you were at a disadvantage in this regard. And if you happened to be a member of a nontitular ethnic group, you were most disadvantaged (Brym with Ryvkina, 1994: 6–16; Karklins, 1986; Zaslavsky and Brym, 1983).

By thus organizing many basic social processes along ethnic lines, the government imposed ethnic labels on its citizens. Despite efforts to create a new "Soviet people," traditional ethnic identities remained strong. In 1991, when the Soviet Union ceased to exist, Russians knew they were Russians and Jews knew they were Jews mainly because the ethnicity entry in their internal passports had circumscribed their opportunities in life for more than 60 years. Only in 1997 did the Russian government finally introduce new internal passports without the notorious fifth entry, thus bringing an end to the era of administratively imposed ethnicity in that country.

Ethnic and Racial Choice in the United States

The situation in the United States is vastly different from that in the Soviet Union before 1991. Americans are freer to choose their ethnic identity than citizens of the Soviet Union were. The people with the most freedom to choose their ethnic or racial identity

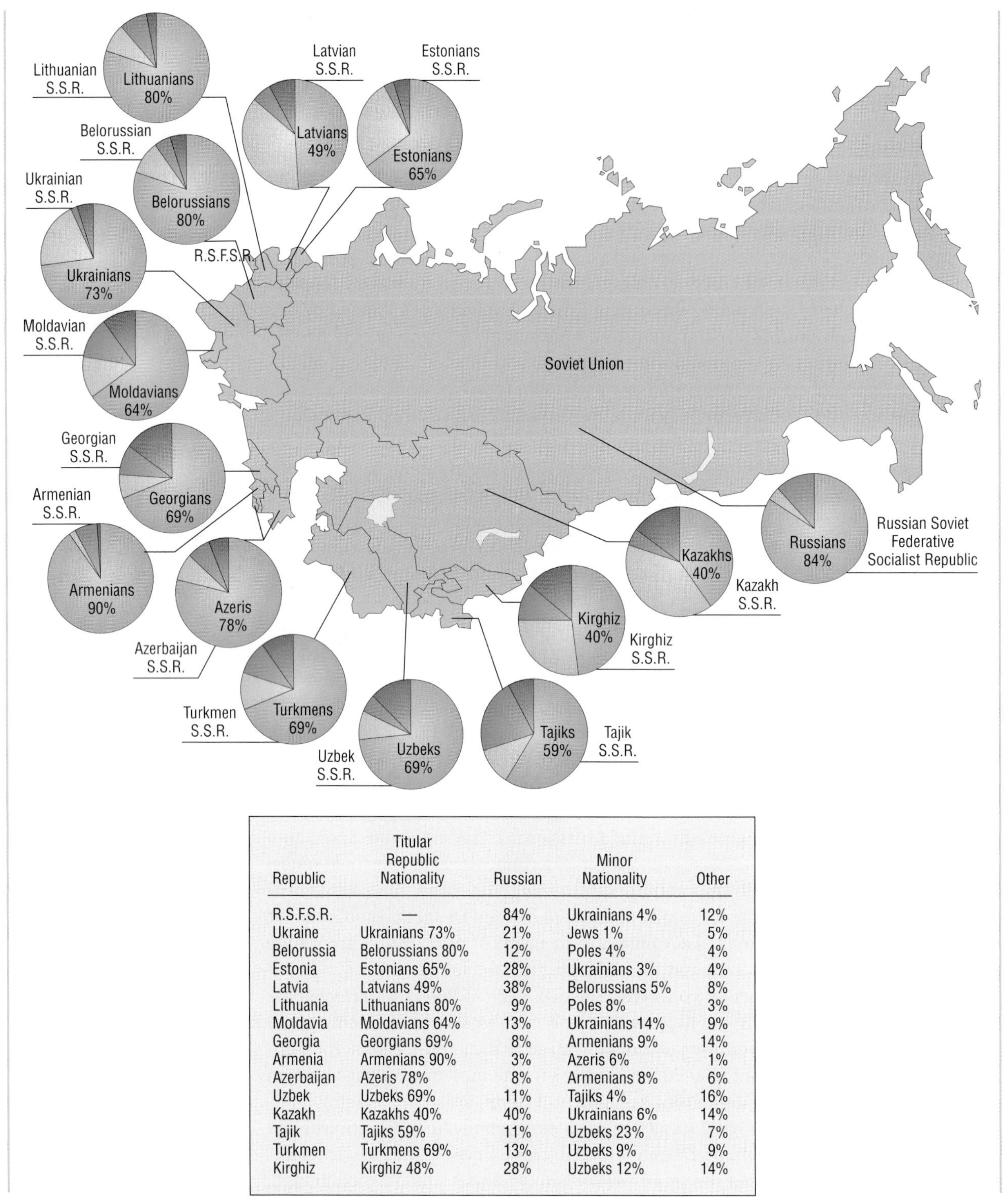

Republic	Titular Republic Nationality	Russian	Minor Nationality	Other
R.S.F.S.R.	—	84%	Ukrainians 4%	12%
Ukraine	Ukrainians 73%	21%	Jews 1%	5%
Belorussia	Belorussians 80%	12%	Poles 4%	4%
Estonia	Estonians 65%	28%	Ukrainians 3%	4%
Latvia	Latvians 49%	38%	Belorussians 5%	8%
Lithuania	Lithuanians 80%	9%	Poles 8%	3%
Moldavia	Moldavians 64%	13%	Ukrainians 14%	9%
Georgia	Georgians 69%	8%	Armenians 9%	14%
Armenia	Armenians 90%	3%	Azeris 6%	1%
Azerbaijan	Azeris 78%	8%	Armenians 8%	6%
Uzbek	Uzbeks 69%	11%	Tajiks 4%	16%
Kazakh	Kazakhs 40%	40%	Ukrainians 6%	14%
Tajik	Tajiks 59%	11%	Uzbeks 23%	7%
Turkmen	Turkmens 69%	13%	Uzbeks 9%	9%
Kirghiz	Kirghiz 48%	28%	Uzbeks 12%	14%

Figure 10.3
Ethnic Groups in the Soviet Union by Republic, 1979

Source: Perry-Castañeda Library Map Collection (2000).

are white Americans whose ancestors came from Europe more than two generations ago (Waters, 1990). For example, identifying oneself as an Irish American no longer has negative implications, as it did in, say, 1900. Then, in a city like Boston, where a substantial number of Irish immigrants were concentrated, the Anglo-Protestant majority typically regarded working-class Irish Catholics as often drunk, inherently lazy, and born superstitious. This strong anti-Irish sentiment, which often erupted into conflict, meant that the Irish found it difficult to escape their ethnic identity even if they wanted to. Since then, however, Irish Americans have followed the path taken by many other white European groups. They have achieved upward mobility and blended into the majority. As a result, Irish Americans no longer find their identity imposed on them. Instead, they may *choose* whether to march in the St. Patrick's Day parade, enjoy the remarkable contributions of Irish authors to English-language literature and drama, and take pride in the athleticism and precision of Riverdance. For them, ethnicity is largely a *symbolic* matter, as it is for the other white European groups that have undergone similar social processes. Herbert Gans defines **symbolic ethnicity** as "a nostalgic allegiance to the culture of the immigrant generation, or that of the old country; a love for and a pride in a tradition that can be felt without having to be incorporated in everyday behavior" (Gans, 1979b: 436).

Bettmann/Corbis

As Malcolm X noted, it does not matter to a racist whether an African American is a professor or a panhandler, a genius or a fool, a saint or a criminal. Where racism is common, racial identities are compulsory and at the forefront of one's self-identity.

Racism and Identity

At the other extreme, most African Americans lack the freedom to enjoy symbolic ethnicity. They may well take pride in their cultural heritage. However, their identity as African Americans is not an option, because racism imposes it on them daily. **Racism** is the belief that a visible characteristic of a group, such as skin color, indicates group inferiority and justifies discrimination. **Institutional racism** is bias that is inherent in social institutions and is often not noticed by members of the majority group. We see institutional racism in practice when police single out African Americans for car searches, department stores tell their floorwalkers to keep a sharp eye on African Americans in particular, and banks reject African American mortgage applications more often than those from white Americans of the same economic standing. In his autobiography, the black militant Malcolm X poignantly noted how both individual and institutional racism can impose racial identity on people. He described one of his black Ph.D. professors as "one of these ultra-proper-talking Negroes" who spoke and acted snobbishly. "Do you know what white racists call black Ph.D.s?" asked Malcolm X. "He said something like, 'I believe that I happen not to be aware of that . . .' And I laid the word down on him, loud: 'N_____!'" (Malcolm X, 1965: 284). Malcolm X's point is that it does not matter to a racist whether an African American is a professor or a panhandler, a genius or a fool, a saint or a criminal. Where racism is common, racial identities are compulsory and at the forefront of one's self-identity.

In sum, political and social processes structure the degree to which people are able to choose their ethnic and racial identities. Americans today are freer to choose their ethnic identity than citizens of the Soviet Union used to be. Members of *ethnic* minority groups in the United States today are freer to choose their identity than members of *racial* minority groups.

The contrast between Irish Americans and African Americans also suggests that relations between racial and ethnic groups can take different forms. For example, racial and ethnic groups can blend as a result of residential integration, friendship, and intermar-

Symbolic ethnicity is a nostalgic allegiance to the culture of the immigrant generation, or that of the old country, that is not usually incorporated into everyday behavior.

Racism is the belief that a visible characteristic of a group, such as skin color, indicates group inferiority and justifies discrimination.

Institutional racism is bias that is inherent in social institutions and is often not noticed by members of the majority group.

riage, or they can remain separate because of hostility. We now discuss several theories that explain why forms of racial and ethnic relations vary over time and from place to place.

Theories of Race and Ethnic Relations

Ecological Theory

Nearly a century ago, Robert Park proposed an influential theory of how race and ethnic relations change over time (Park, 1914; 1950). His **ecological theory** focuses on the struggle for territory. It distinguishes five stages in the process by which conflict between ethnic and racial groups emerges and is resolved:

1. *Invasion.* One racial or ethnic group tries to move into the territory of another. The territory may be as large as a country or as small as a neighborhood in a city.
2. *Resistance.* The established group tries to defend its territory and institutions against the intruding group. It may use legal means, violence, or both.
3. *Competition.* If the established group does not drive out the newcomers, the two groups begin to compete for scarce resources. These resources include housing, jobs, public park space, and political positions.
4. *Accommodation and Cooperation.* Over time, the two groups work out an understanding of what they should segregate, divide, and share. **Segregation** involves the spatial and institutional separation of racial or ethnic groups. For example, the two groups may segregate churches, divide political positions in proportion to the size of the groups, and share public parks equally.
5. *Assimilation.* **Assimilation** is the process by which a minority group blends into the majority population and eventually disappears as a distinct group. Park argued that assimilation is bound to occur as accommodation and cooperation allow trust and understanding to develop. Eventually, goodwill allows ethnic groups to fuse socially and culturally. Where two or more groups formerly existed, only one remains. Park agreed with the memorable image of America as "God's crucible, the great Melting Pot where all the races of Europe are melting and re-forming" (Zangwill, 1909: 37).

Park's theory stimulated important and insightful research (e.g., Suttles, 1968). However, it is more relevant to some ethnic groups than others. It applies best to whites of European origin. As Park predicted, many whites of European origin stopped thinking of themselves as Italian American, Irish American, or German American after their families were in the United States for three or four generations. Today, they think of themselves just as "whites" (Lieberson, 1991). Over time, they achieved rough equality with members of the majority group and, in the process, began to blend in with them. The story of the Irish is fairly typical. During the first half of the 20th century, Irish Americans experienced much upward mobility. By the middle of the century, they earned about as much as Anglo-Protestants. The tapering off of working-class Irish immigration prevented the average status of the group from falling. As their status rose, Irish Americans increasingly intermarried with members of other ethnic groups. Use of the Irish language, Gaelic, virtually disappeared. The Irish became less concentrated in particular cities and less segregated in certain neighborhoods. Finally, conflict with majority-group Americans declined (Greeley, 1974). With variations, a similar story may be told about Italian Americans and German Americans.

However, the story does not apply to all Americans. Park's theory gives a too optimistic account of the prospects for assimilation of African Americans, Native Americans, Mexican Americans, and Chinese Americans, among others. It also fails to take into account the persistence of ethnicity among some middle-class whites of European origin into the third and fourth generation after immigration. For reasons we will now explore, many members of racial minorities and some white people of European origin seem fixed in Park's third and fourth stages.

The **ecological theory** of ethnic succession argues that ethnic groups pass through five stages in their struggle for territory: invasion, resistance, competition, accommodation and cooperation, and assimilation.

Segregation involves the spatial and institutional separation of racial or ethnic groups.

Assimilation is the process by which a minority group blends into the majority population and eventually disappears as a distinct group.

Internal Colonialism and the Split Labor Market

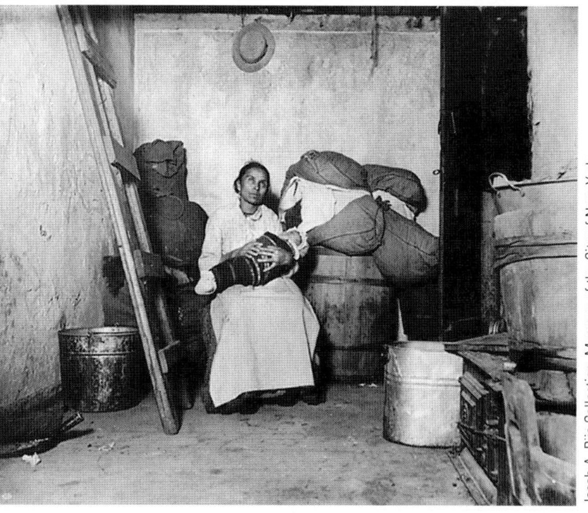

Jacob A. Riis Collection, Museum of the City of New York

Many white Americans lived in poor urban ghettos in the 19th and early 20th centuries. However, a larger proportion of them experienced upward mobility than was the case for African Americans in the second half of the 20th century. This photo shows the home of an Italian rag picker in a New York tenement in the late 19th century.

The main weakness of Park's theory is that it pays insufficient attention to the *social-structural* conditions that prevent some groups from assimilating. Robert Blauner examined one such condition, which he called **internal colonialism** (Blauner, 1972; Hechter, 1974). *Colonialism* involves people from one country invading another. In the process, the invaders change or destroy the native culture. They gain virtually complete control over the native population. They develop the racist belief that the natives are inherently inferior. They confine the natives to work they consider demeaning. Internal colonialism involves the same processes but within the boundaries of a single country. It prevents assimilation by segregating the colonized in terms of jobs, housing, and social contacts ranging from friendship to marriage. To varying degrees, Russia, China, France, Great Britain, Canada, Australia, the United States, and other countries have engaged in internal colonialism.

Edna Bonacich developed a second important theory that focuses on social-structural conditions hindering the assimilation of some groups. According to Bonacich, racial identities are reinforced in **split labor markets.** Where low-wage workers of one race and high-wage workers of another race compete for the same jobs, high-wage workers are likely to resent the presence of low-wage competitors. Conflict is bound to result. Consequently, racist attitudes develop or get reinforced. The effects of split labor markets and internal colonialism on racial and ethnic identity persist for some time even after these social-structural conditions cease to exist.

Let us examine how these theories apply to the United States by considering four racial and ethnic groups: Native Americans, Mexican Americans, African Americans, and Chinese Americans.

Native Americans

The treatment of Native Americans by European settlers in the 19th century involved expulsion and genocide. **Expulsion** is the forcible removal of a population from a territory claimed by another population. **Genocide** is the intentional extermination of an entire population defined as a race or a people.

Once the British effectively won the battle for control of North America in the late 18th century, Indian/white relations focused on land issues (Cornell, 1988; Marger, 2003: 175–85). Native Americans owned their land communally, not privately. The British (and their successors, the Americans) gave them two options to change this state of affairs. First, the Indians could model themselves after the European settlers, become private landowners, and farm the land. Second, they could sell their land to European settlers; and if they refused, as they often did, they faced the might of a technologically superior army. In different guises, forcing the Indians to become like the European settlers and forcing them off their land became the two dominant strategies that informed American policy for centuries.

Early American policy treated the various Native American tribes like independent political entities. Accordingly, Indians and settlers signed treaties defining the land belonging to each group. Indians typically entered into these treaties because they were forced to do so or did not understand their true meaning. Treaty violations on the part of the land-hungry settlers were common, and clashes inevitably ensued. The American solution was to renegotiate the treaties and push the Indians farther west.

In 1830 the United States government passed the Indian Removal Act. It called for the relocation of all Native Americans to land set aside for them west of the Mississippi. For the next decade, white European Americans fought a series of wars against various Native

Internal colonialism involves one race or ethnic group subjugating another in the same country. It prevents assimilation by segregating the subordinate group in terms of jobs, housing, and social contacts.

In **split labor markets**, low-wage workers of one race and high-wage workers of another race compete for the same jobs. High-wage workers are likely to resent the presence of low-wage competitors, and conflict is bound to result. Consequently, racist attitudes develop or get reinforced.

Expulsion is the forcible removal of a population from a territory claimed by another population.

Genocide is the intentional extermination of an entire population defined as a "race" or a "people."

In the late 19th and early 20th centuries, North American governments, with the help of various religious groups, tried to eradicate native religions, languages, and cultures by taking Indian children from their parents and placing them in boarding schools where they could be "civilized." Here we see one Thomas Moore before and after such schooling in the early 20th century.

American tribes. Relying on superior military technology and troop strength, the U.S. Army easily won. In one notorious incident, the "Trail of Tears," the U.S. Army rounded up all 16,000 Cherokees, held them for months in camps infested with disease, and then marched them to Oklahoma. Four thousand Cherokees died on the trek. Thanks to such tactics, General Thomas Jesup could report in 1838 that

> [t]he villages of the Indians have all been destroyed; and their cattle, horses, and other stock, with nearly all their other property, taken or destroyed. The swamps and hammocks have been every where penetrated, and the whole country traversed from the Georgia line to the southern extremity of Florida; and the small bands who remain dispersed over the extensive region, have nothing of value left but their rifles (quoted in Wallace, 1993: 99).

The effective end of the war against the Indians came in 1890 with the slaughter of hundreds of Sioux at Wounded Knee in South Dakota (Brown, 1970). Gradually, the remaining "small bands" were placed on reservations under the rule of the Bureau of Indian Affairs. On the reservations, they were segregated from the majority population. Good jobs, health facilities, and opportunities for educational advancement were scarce.

What war could not accomplish, disease and the extermination of the buffalo did. European settlers brought measles, influenza, cholera, typhoid, and malaria to North America. Native Americans had no immunity against those diseases. The settlers also killed some 15 million buffalo for meat and hides, thus destroying the Indians' most important source of food, clothing, and shelter. Between 1800 and 1900, the Native American population was cut in half.

Sociology Now™

Learn more about **Assimilation** by going through the Native American Assimilation Video Exercise.

In the late 19th century, the government adopted a policy of forced assimilation. It sold some Indian land to non-Indians and assigned some of it to individual Indians willing to farm like the settlers. This policy partly destroyed tribal life. With the help of various religious groups, the government also tried to eradicate native religions, languages, and cultures by taking Native American children from their parents and placing them in boarding schools where they could be "civilized."

The administration of Franklin D. Roosevelt adopted a more liberal policy in the 1930s and 1940s. It prohibited the further breakup of Indian lands and encouraged Native self-rule and cultural preservation. However, this was only a brief deviation from traditional policy. In the 1950s the government reverted to form. It proposed to end the reservation system, deny the sovereign status of the tribes, cut off all government services, and stop protecting Indian lands held in trust for the tribes.

The proposal backfired. It was never implemented because of strong resistance on the part of the Native American community. By the 1960s a full-fledged Red Power movement

Table 10.2
Casino Revenue per Tribal Member, Selected Native American Tribes, 2001

Tribe	Population	Annual Casino Revenue per Member	Federal Aid per Member
Mashantucket Pequot	677	$1,624,815	$2,304
Santa Ynez	159	$1,257,862	$8,360
Miccosukee Tribe	400	$250,000	$20,560
Seminole Tribe	2,817	$87,862	$8,540
Mississippi Choctaw	8,823	$25,047	$5,717
Hopi Tribe	11,267	$0	$2,006
Navajo Nation	260,010	$0	$912

Source: Barlett and Steele (2002: 40).

had emerged. The black Civil Rights movement helped to inspire it. Bitter resentment against centuries of oppression fueled it. The reservation system, which segregated Native Americans and helped preserve their unique identities, gave it an organizational base. The Red Power movement transcended tribal differences and spoke in the name of all Native Americans because it faced what it saw as a common enemy that had treated all tribes with equal disdain and similar intent over many years. Indian migration to cities invigorated this supertribal consciousness, because the ethnic groups with which Indians came into contact did not recognize "fine" distinctions between Cherokee and Navajo, Choctaw and Sioux. Galvanized by these conditions, the Red Power movement organized a series of occupations, sieges, sit-ins, marches, and demonstrations between 1969 and 1972. These efforts finally pushed the government to address the needs and rights of Native Americans (Cornell, 1988; Nagel, 1996).

In recent decades Native Americans have used the legal system to fight for political self-determination and the protection of their remaining lands. Their ability to do so has been enhanced by the discovery of valuable resources on reservations, including oil, natural gas, coal, and uranium. In addition, Indians have established many enterprises on reservations in recent years. Casinos are the most important of these economically. They have generated much wealth for a few tribes, but they have also created glaring inequalities between rich and poor. Small, rich tribes have used part of their casino revenue to gain political influence in Washington. Consequently, they receive most federal aid per capita, whereas large, poor tribes receive relatively little (Table 10.2).

Despite new sources of wealth, many Native Americans still suffer the consequences of internal colonialism. According to the 2000 U.S. Census, Native Americans form a growing population numbering 4.1 million people. (Two and a half million reported only "American Indian or Alaska Native" identity, and another 1.6 million reported some other identity *in addition* to that.) Nearly three-quarters of Native Americans live in the West and the South. Five states—California, Oklahoma, Arizona, Texas, and New Mexico—account for nearly 42 percent of the Native American population. Over half of all Native Americans live in urban areas, and that proportion is growing (Ogunwole, 2002).

Urban Indians are less impoverished than those who live on reservations, but even in the cities they fall below the national average in terms of income, education, occupation, employment, health care, and housing (Marger, 2003: 188–93). The median family income of Native Americans is just over half the national average, and the poverty rate is nearly three times the national average. On reservations, the unemployment rate is nearly 50 percent, and in urban areas Indians often lack the skills and qualifications to secure steady jobs. Much discrimination and stereotyping continue to hamper their progress.

Chicanos

We can tell a similar story about Mexican Americans, or Chicanos. Motivated by the desire for land, the United States went to war with Mexico in 1848. The United States won. Arizona, California, New Mexico, Utah, and parts of Colorado and Texas became part of the United States. Americans justified the war in much the same way as they did the conquest of Native Americans. As an editorial in the *New York Evening Post* of the time put it:

> The Mexicans are Indians—Aboriginal Indians. . . . They do not possess the elements of an independent national existence. The Aborigines of this country have not attempted and cannot attempt to exist independently along side of us. Providence has so ordained it, and it is folly not to recognize the fact. The Mexicans are Aboriginal Indians, and they must share the destiny of their race (quoted in Steinberg, 1989 [1981]: 22).

For the past 150 years, millions of Chicanos have lived in the U.S. Southwest (Camarillo, 1979). Specifically, more than two-thirds of Hispanic Americans are of Mexican origin and nearly 80 percent of Hispanic Americans live in the South and the West (Ramirez and de la Cruz, 2003: 2). Mainly because of discrimination, Chicanos are socially, occupationally, and residentially segregated from white European Americans. They were not forced onto reservations like Native Americans. However, until the 1970s, most Chicanos lived in ghettos, or *barrios,* far from white European American neighborhoods. Many of them still do. Many white European Americans continue to regard Chicanos as social inferiors. As a result, Chicanos interact mainly with other Chicanos. They still experience much job discrimination and work mainly as agricultural and unskilled laborers with few prospects for upward mobility. (Ironically, some Mexicans who now work in California, Texas, and other states from which their ancestors were expelled are often called "illegal migrants.") About half of Chicanos 25 years or older have not graduated from high school. The comparable figure for non-Hispanic whites is 11 percent. Among Hispanic Americans, Chicanos are least likely to work in managerial or professional occupations, least likely to earn $50,000 a year or more, and second most likely (next to Puerto Rican Americans) to live in poverty (Ramirez and de la Cruz, 2003: 5–6). Thus, as with Native Americans, high levels of occupational, social, and residential segregation prevent Chicanos from assimilating. Instead, especially since the 1960s, many Chicanos have taken part in a movement to renew their culture and protect and advance their rights (Gutiérrez, 1995).

African Americans

Most features of the internal colonialism model can explain the obstacles to African American assimilation too. For although the lands of Africa were not invaded and incorporated into the United States, many millions of Africans were brought here by force and enslaved. **Slavery** is the ownership and control of people. By about 1800, 24 million Africans had been captured and transported on slave ships to North, Central, and South America. As a result of violence, disease, and shipwreck, only 11 million survived the passage. Fewer than 10 percent of those 11 million arrived in the United States. However, because the birthrate of African slaves in the United States was higher than elsewhere in the Americas, nearly 30 percent of the black population in the New World was living in the United States by 1825. By the outbreak of the Civil War, 4.4 million black slaves lived in the United States. The cotton and tobacco economy of the American South depended completely on their labor (Patterson, 1982).

Slavery kept African Americans segregated from white society. Even after slavery was legally banned in 1863, they remained a race apart. So-called Jim Crow laws kept blacks from voting, attending white schools, and in general participating equally in many social institutions. In 1896 the U.S. Supreme Court approved segregation when it ruled that separate facilities for blacks and whites were legal as long as they were of nominally equal quality (*Plessy v. Ferguson*). Most African Americans remained unskilled workers throughout this period.

Slavery is the ownership and control of people.

In the late 19th and early 20th centuries, a historic opportunity to integrate the black population into the American mainstream presented itself. This was a period of rapid industrialization. The government could have encouraged African Americans to migrate northward and westward and get jobs in the new factories, where they could have enjoyed job training, steady employment, and better wages. But United States policymakers chose instead to encourage white European immigration. Between 1880 and 1930, 23 million Europeans came to the United States to work in the expanding industries. While white European immigrants made their first strides on the path to upward mobility, the opportunity to integrate the black population quickly and completely into the American mainstream was squandered (Steinberg, 1989 [1981]: 173–200).

Some jobs in northern and western industries did go to African Americans, who migrated from the South in substantial numbers from the 1910s onward. They were able to compete against European immigrants in the labor market by accepting low wages. Here we see the operation of a classic split labor market of the type that Edna Bonacich analyzed. The split labor market fueled deep resentment, animosity, and even antiblack riots on the part of working-class whites. These feelings solidified racial identities, both black and white (Bonacich, 1972).

Despite this conflict, black migration northward and westward continued because social and economic conditions in the South were even worse. Already by the 1920s, the world center of jazz had shifted from New Orleans to Chicago. This shift as much as anything signaled the permanence and vitality of the new black communities. By the mid-1960s about 4 million African Americans were living in the urban centers of the North and the West.

Jacob Lawrence. *The Migration of the Negro, Panel No. 57.* 1940–1941. Jacob Lawrence's "The Great Migration" series of paintings illustrates the mass exodus of African Americans from the South to the North in search of a better life. Lawrence's parents were among those who migrated in the first wave of the great migration (1916–1919).

The migrants from the South tended to congregate in low-income neighborhoods, where they sought inexpensive housing and low-skill jobs. Slowly—more slowly than was the case for white European immigrants—their situation improved. Many children of migrant blacks finished high school. Some finished college. Others established ethnic enterprises. Still others got jobs in the civil service. Residential segregation in poor neighborhoods decreased, and there was even some intermarriage with members of the white community (Lieberson, 1980).

In the 1960s, some sociologists observed these developments and expected African Americans to continue moving steadily up the social class hierarchy. As we will soon see, this optimism was only partly justified. Social-structural impediments, some new and some old, prevented many African Americans from achieving the level of prosperity and assimilation enjoyed by white Europeans. Thus, in the mid-1960s about one-third of African Americans lived in poverty and the proportion is virtually unchanged today. In concluding this chapter, we carry the story of African Americans and other racial minorities forward to the present.

Chinese Americans

In 1882 Congress passed an act prohibiting the immigration of three classes of people into the United States for 10 years: lunatics, idiots, and Chinese. The act was extended for another decade in 1892, made permanent in 1907, and repealed only in 1943, when Congress established a quota of a grand total of 105 Chinese immigrants per year. Earlier, the California gold rush of the 1840s and the construction of the transcontinental railroads in the 1860s had drawn tens of thousands of Chinese immigrants into the United States. The great majority of them were young men who worked as unskilled laborers. However, until 1965, less than 100,000 Chinese were living in the United States.

Museum of History and Industry, Seattle

Anti-Chinese riot in Seattle, February 8, 1886, in front of the New England Hotel on Main and First Avenue.

They were the objects of one of the most hostile anti-ethnic movements in American history.

Extraordinary prejudice and ignorance greeted the Chinese in the United States, and the film and tourist industries did much to reinforce fear of the "yellow peril" in the first half of the 20th century. For example, in the 1920s, guides would take tourists to San Francisco's Chinatown and warn them to stick together and keep their eyes peeled for Chinese hatchet men, always eager to chop the heads off unsuspecting white folk. To prove the "danger," the guides paid Chinese men to look sinister and dart in and out of the shadows of dimly lit alleys, knives and hatchets at the ready. Tourists were shown false opium dens, told phony stories about brothels populated by white women who had been enslaved by the Chinese, and misinformed that certain cuts of meat in Chinese butcher shops were rat carcasses (Takaki, 1989).

It was, however, a split labor market that caused anti-Chinese prejudice to boil over into periodic race riots and laws aimed at keeping Chinese immigrants out of the United States. The American government allowed Chinese immigrants into California to do backbreaking work such as railway construction and mining with hand tools. The immigrants received wages well below those of white workers. As split labor market theory predicts, where competition for jobs between Chinese and white workers was intense, trouble brewed and ethnic and racial identities were reinforced. Recent research shows that anti-Chinese activity was especially high where white workers were geographically concentrated and successful in creating anti-Chinese organizations. San Francisco was the epicenter of anti-Chinese sentiment and activity (Fong and Markham, 2002).

Chinese Americans have experienced considerable upward mobility in the past half century. More than 30 percent of Chinese Americans now marry whites (Marger, 2003: 385). However, a social-structural factor—split labor markets—did much to prevent such mobility and assimilation until the middle of the 20th century. Similarly, our sketch of Native Americans, Chicanos, and African Americans shows that a second social circumstance—internal colonialism—helps to explain why upward mobility and assimilation are often hampered, contrary to the prediction of Park's ecological theory. The groups that have had most trouble achieving upward mobility in the United States are those that were subjected to slavery and expulsion from their native lands. Expulsion and slavery left a legacy of racism that created social-structural impediments to assimilation, such as forced segregation in low-status jobs and low-income neighborhoods. By focusing on factors like these, we arrive at a more realistic picture of the state of race and ethnic relations in the United States than is afforded by ecological theory alone.

Some Advantages of Ethnicity

The theories of internal colonialism and split labor markets emphasize how social forces outside a racial or ethnic group force its members together, preventing their assimilation into the larger society. It focuses on the disadvantages of race and ethnicity. Moreover, it deals only with the most disadvantaged minorities. The theory has less to say about the *internal* conditions that promote group cohesion and in particular about the value of group membership. Nor does it help us to understand why some white European Americans continue to participate in the life of their ethnic communities, even if their families have been in the country more than two or three generations.

A review of the sociological literature suggests that three main factors enhance the value of ethnic group membership for some Americans—even white European Americans—who have lived in the country for many generations. These factors are economic, political, and emotional.

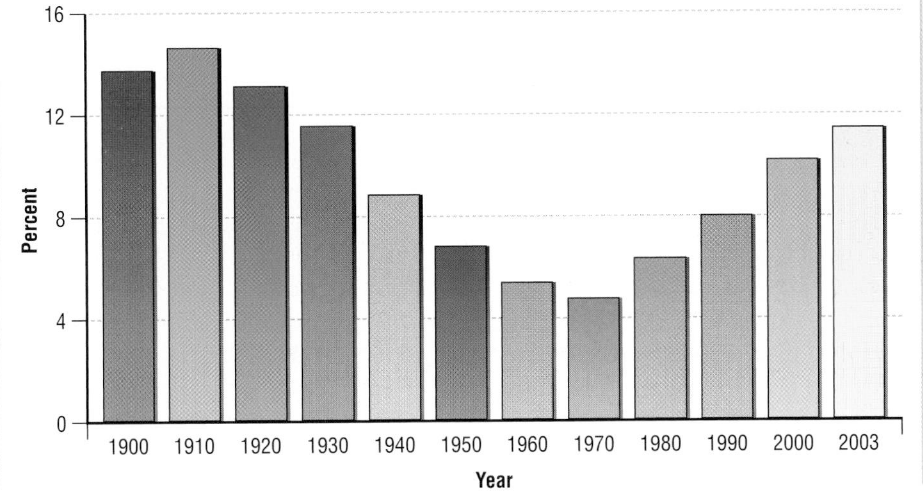

Figure 10.4
Percent Foreign Born, United States, 1900–2003

Source: United States Bureau of the Census (1993: 2; 1999e; 2001a; 2002c; 2004: 7, 44)

Economic Advantages of Ethnic Group Membership

The economic advantages of ethnicity are most apparent for immigrants, who comprised nearly 12 percent of the American population in 2003 (Figure 10.4). However, some economic advantages extend into the third generation and beyond. Immigrants often lack extensive social contacts and fluency in English. Therefore, they commonly rely on members of their ethnic group to help them find jobs and housing. In this way, immigrant communities become tightly knit. Community solidarity is also an important resource for "ethnic entrepreneurs." These are businesspeople who operate largely within their ethnic community. They draw on their community for customers, suppliers, employees, and credit. They can pass on their businesses to their children, who in turn can pass the businesses on to the next generation. In this way, strong economic incentives encourage some people to remain ethnic group members, even beyond the immigrant generation (Bonacich, 1973; Light, 1991; Portes and Manning, 1991).

David Turnley/Corbis

Until the passage of the 1965 Hart-Cellar Act, immigration from Asia, Africa, and Latin America was sharply restricted. Today, most American immigrants come from Latin America and Asia.

Political Advantages of Ethnic Group Membership

Ethnic group membership can be politically useful. Consider, for instance, the way some European Americans reacted to the Civil Rights movement of the 1960s. Civil rights legislation opened new educational, housing, and job opportunities for African Americans. It also led to the liberalization of immigration laws. Until the passage of the 1965 Hart-Cellar Act, immigration from Asia and Latin America was sharply restricted. Afterward, most immigrants came from these regions (Table 10.3). Some white European Americans felt threatened by the improved social standing of African Americans and non-European immigrants. As a result, "racial minorities and white ethnics became polarized on a series of issues relating to schools, housing, local government, and control over federal programs" (Steinberg, 1989 [1981]: 50). Not coincidentally, many European Americans experienced renewed interest in their ethnic roots just at this time. Many sociologists believe that the white ethnic revival of the 1960s and 1970s was a reaction to political conflicts with African, Asian, Native, and Hispanic Americans. Such conflicts helped to strengthen ethnic group solidarity.

Table 10.3
Top 10 Countries of Origin of Foreign-Born Americans, 1900, 1960, and 2000 (in thousands; percent of total foreign-born in parentheses)

1900		1960		2000	
Germany	2,663 (25.8)	Italy	1,257 (12.9)	Mexico	7,418 (26.1)
Ireland	1,615 (15.6)	Germany	990 (10.2)	China	1,391 (4.9)
Canada	1,179 (11.4)	Canada	953 (9.8)	Philippines	1,222 (4.3)
Great Britain	1,167 (11.3)	Great Britain	765 (7.9)	India	1,007 (3.5)
Sweden	582 (5.6)	Poland	748 (7.7)	Cuba	952 (3.4)
Italy	484 (4.7)	Soviet Union	691 (7.1)	Vietnam	863 (3.0)
Soviet Union	423 (4.1)	Mexico	576 (5.9)	El Salvador	765 (2.7)
Poland	383 (3.7)	Ireland	338 (3.5)	Korea	701 (2.5)
Norway	336 (3.2)	Hungary	245 (2.5)	Dominican Republic	692 (2.4)
Austria	275 (2.7)	Czechoslovakia	228 (2.3)	Canada	678 (2.4)
Other	1,234 (11.9)	Other	2,947 (30.2)	Other	12,690 (44.7)
Total	10,341 (100.0)	Total	9,738 (100.0)	Total	28,379 (100.0)

Sources: Calculated from U.S. Census Bureau (1997; 2001b: 9).

Emotional Advantages of Ethnic Group Membership

Like economic benefits, the emotional advantages of ethnicity are most apparent in immigrant communities but can endure for generations. Speaking the ethnic language and sharing other elements of one's native culture are valuable sources of comfort in an alien environment. Even beyond the second or third generation, however, ethnic group membership can perform significant emotional functions. For example, some ethnic groups, such as the Jews, have experienced unusually high levels of prejudice and discrimination involving expulsion and attempted genocide. For people who belong to such groups, the resulting trauma is so severe that it can be transmitted for several generations. In such cases, ethnic group membership offers security in a world still seen as hostile long after the threat of territorial loss or annihilation has disappeared (Bar-On, 1999). Another way in which ethnic group membership offers emotional support beyond the second generation is by providing a sense of rootedness. Especially in a highly mobile, urbanized, technological, and bureaucratic society such as ours, ties to an ethnic community can be an important source of stability and security (Isajiw, 1978).

Transnational Communities

The three factors just listed make ethnic group membership useful to some European Americans whose families have been in the country for many generations. It is also important to note in this connection that retaining ethnic ties beyond the second generation has never been easier. Inexpensive international communication and travel allow ethnic group members to maintain strong ties to their motherland in a way that was never possible in earlier times.

Immigration used to involve cutting all or most ties to one's country of origin. Travel by sea and air was expensive, long-distance telephone rates were prohibitive, and the occasional letter was about the only communication most immigrants had with their relatives in the old country. Lack of communication encouraged assimilation. Today, however, ties to the motherland are often maintained in ways that sustain ethnic culture. For example, nearly 400,000 Jews have emigrated from the former Soviet Union to the United States since the early 1970s. They frequently visit relatives in the former Soviet Union and Israel, speak with them on the phone, and use the Internet to exchange e-mail with them. They also receive Russian-language radio and TV broadcasts, act as conduits for foreign investment, and send money to relatives abroad (Brym with Ryvkina, 1994; Markowitz, 1993; Remennick, 2002). This sort of intimate and ongoing connection with the motherland is typical of most recent immigrant communities in the United States. Thanks to inexpensive international travel and communication, some ethnic groups have become **transnational communities** whose boundaries extend between countries. The Ticuani Potable Water Committee in Brooklyn, New York, has been raising money for the farming community of Ticuani in the Mixteca region of Mexico for decades. Its seal reads "Por el Progreso de Ticuani: Los Ausentes Siempre Presentes. Ticuani y New York" ["For the Progress of Ticuani: The Absent Ones Always Present. Ticuani and New York"]. The phrase "the absent ones always present" nicely captures the essence of transnational communities, whose growing number and vitality facilitate the retention of ethnic group membership beyond the second generation (Portes, 1996).

Lewis H. Hines/Hulton Archive/Getty Images

In the late 19th and early 20th centuries, Ellis Island in New York harbor was the point of entry of more than 12 million Europeans to the United States. Today, more than one-third of all Americans can trace their origins to a person who passed through Ellis Island. Here, a customs official attaches labels to the coats of a German immigrant family at the Registry Hall on Ellis Island in 1905.

In sum, ethnicity remains a vibrant force in American society for a variety of reasons. Even some white Americans whose families arrived in this country generations ago have reason to identify with their ethnic group. Bearing this in mind, what is the likely future of race and ethnic relations in the United States? We conclude by offering some tentative answers to that question.

The Future of Race and Ethnicity

At 2:30 a.m. on June 7, 1998, James Byrd, Jr., was walking home along a country road near Jasper, Texas. Three men stopped and offered Byrd a ride. But instead of taking him home, they forced Byrd out of their pickup truck. They beat him until he was unconscious. They then chained his ankles to the back of the truck and dragged him along the jagged road for nearly 3 miles. The ride tore Byrd's body into more than 75 pieces.

What was the reason for the murder? Byrd was a black man. His murderers are "white supremacists." They want the United States to be a white-only society, and they are prepared to use violence to reach their goal. The murderers expressed no remorse during their trial. Two of them wore tattoos suggesting membership in the racist Aryan Nation or the Ku Klux Klan. As such, they are part of a growing problem in the United States. More than 750 racist and neo-Nazi organizations have sprung up in the country (Southern Poverty Law Center, 2005). In 2003 the Federal Bureau of Investigation (FBI) recorded 7489 **hate crimes,** or criminal incidents motivated by a person's race, religion, or ethnicity. Each incident may involve multiple offenses, such as assault and property damage. A total of 8715 offenses were recorded. By far the most frequent victims of hate crimes are African Americans, who were the object of nearly 35 percent of all offenses (U.S. Federal Bureau of Investigation, 2004).

White supremacists form only a tiny fraction of the American population. Many more Americans engage in subtle forms of racism. Thus, sociologists Joe Feagin and Melvin Sikes (1994) interviewed a sample of middle-class African Americans in 16 cities. They found that their respondents often had trouble hailing a cab. If they arrived in a store before a white customer, a clerk commonly served them afterward. When they shopped, store security often followed them around to make sure they did not shoplift. Police officers often stop middle-class African American men in their cars without apparent reason. Blacks are less likely than whites of similar means to receive mortgages and other loans (Oliver and Shapiro, 1995). In short, African Americans continue to suffer high levels of racial prejudice and discrimination, whether overt or covert (Hacker, 1992; Shipler, 1997).

Transnational communities are communities whose boundaries extend between countries.

Hate crimes are criminal acts motivated by a person's race, religion, or ethnicity.

The Declining Significance of Race?

Still, sociologist William Julius Wilson (1980 [1978]) and others believe race is declining in significance as a force shaping the lives of African Americans. Wilson argues that the Civil Rights movement helped to establish legal equality between blacks and whites. In 1954 the Supreme Court ruled against earlier decisions permitting school segregation (*Brown v. Board of Education*). The 1964 Civil Rights Act outlawed discrimination in public housing, employment, and the distribution of federal funds. It also supported school integration. The 1965 Voting Rights Act prohibited the systematic exclusion of blacks from the political process. The 1968 Civil Rights Act banned racial discrimination in housing. These reforms allowed a large black middle class to emerge, says Wilson. Today, one-third of the African American population is middle class.

Many facts support Wilson's view that the social standing of blacks has improved. For example, in 1947 median family income among blacks was only 51 percent that of whites. In 2002 it stood at 64 percent (Figure 10.5). The proportion of whites with 4 or more years of college tripled from 1960 to 2000. In the same period, the proportion of blacks with 4 or more years of college increased more than fivefold (see Chapter 17, Figure 17.3). In 2002, 8.4 percent fewer blacks lived below the poverty line than in 1980, but the poverty rate among whites remained unchanged (U.S. Census Bureau, 2004: 452). Public opinion polls suggest that whites are becoming more tolerant of blacks and that Americans now feel "warmer" toward African Americans than toward Hispanic and Asian Americans (Thernstrom and Thernstrom, 1997) (Figure 10.6, Figure 10.7). Wilson, citing similar data, admits that the gap between blacks and whites remains substantial. Still, he stresses, it is shrinking.

Web Interactive Exercise: Are African Americans Making Progress?

For Wilson, the one-third of African Americans who live below the poverty line are little different from other Americans in similar economic circumstances. He therefore calls for "color-blind" public policies that aim to improve the class position of the poor, such as job training and health care. He opposes race-specific policies, such as **affirmative action,** that give preference to minority group members if equally qualified people are available for a position. He feels that such policies disproportionately help middle-class African Americans while keeping poor African Americans poor (Wilson, 1996).

Despite the points noted previously, some sociologists think Wilson exaggerates the declining significance of race. For Wilson's critics, racism remains a big barrier to black progress. Their case for the continuing impact of race is strengthened by statistical analyses of data on racial differences in wages and housing patterns. For example, they showed

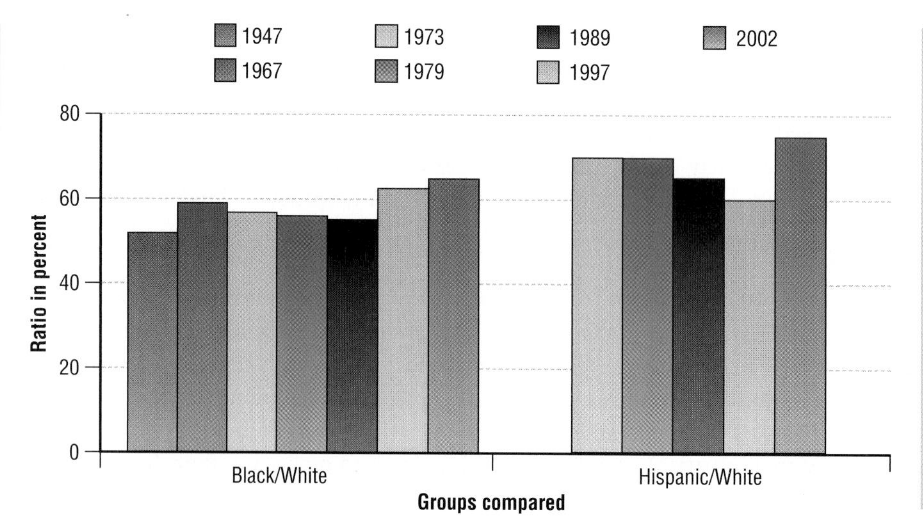

Figure 10.5
Median Family Income Ratios, Black/White and Hispanic/White, United States, 1947–2002

Source: Mishel, Bernstein, and Schmitt (1999: 45); U.S. Census Bureau (2004: 443).

Affirmative action is a policy that gives preference to minority group members if equally qualified people are available for a position.

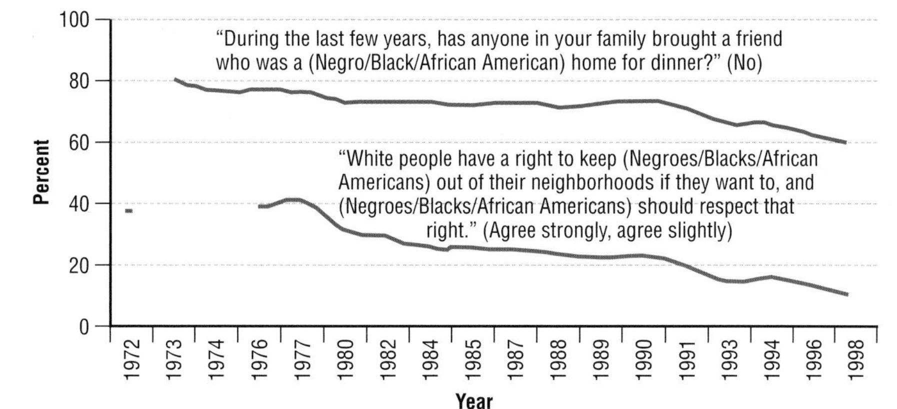

▶Figure 10.6
White Prejudice and Discrimination against Blacks, United States, 1972–1998

Source: National Opinion Research Center, 2004. General Social Survey, 1972–2002. Copyright © 2004 NORC. Used with permission.

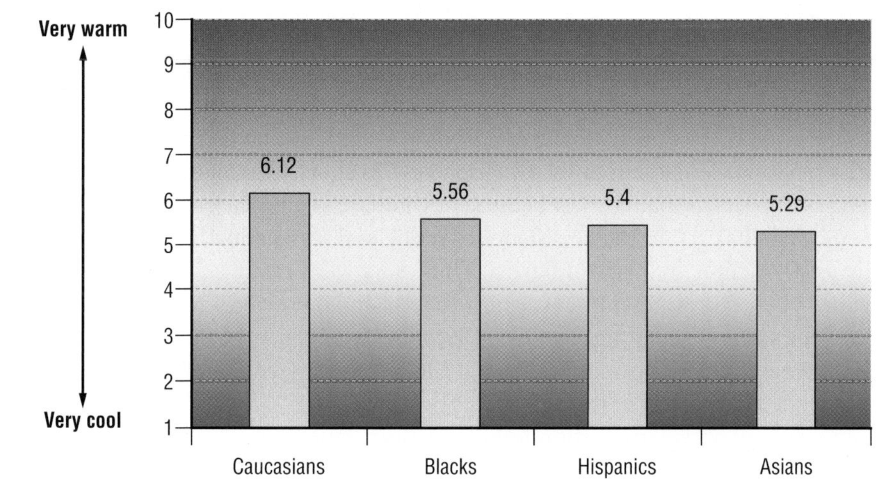

▶Figure 10.7
"How warm or cool do you feel towards Caucasian, African, Hispanic, and Asian Americans?" United States, 2002 ($N = 2765$)

Source: National Opinion Research Center, 2004. General Social Survey, 1972–2002. Copyright © 2004 NORC. Used with permission.

that race had a *stronger* impact on wages in 1985 than in 1976, probably because the government retreated from antidiscrimination initiatives in the 1980s (Cancio, Evans, and Maume, 1996). Moreover, many African Americans continue to live in inner-city ghettos. There they experience high rates of poverty, crime, divorce, teenage pregnancy, and unemployment.

Wilson says ghettos persist not so much because of racism but for three economic and class reasons. First, since the 1970s older manufacturing industries have closed down in cities where the black working class was concentrated. This increased unemployment and poverty. Second, many middle- and working-class African Americans with good jobs moved out of the inner city. This deprived young people of successful role models they could emulate. Third, the exodus of successful blacks eroded the inner-city's tax base at precisely the same time that conservative federal and state governments were cutting budgets for public services. This added to the destitution of inner-city residents (Wilson, 1996).

Although the economic and class factors discussed by Wilson undoubtedly explain much, they do not explain the persistently high level of residential segregation among the many *middle-class* blacks who left the ghettos since the 1960s. They moved to the suburbs, yet their neighborhoods are nearly as segregated as those in the inner city. That is why people sometimes call them "ghettos with grass."

Sociologists calculate a "segregation index" that measures the extent of the problem. The index has a value of zero if the percentage of nonwhites living on each city block is the same throughout the city. This is an unsegregated distribution. The index has a value of 100 if all nonwhites would have to move to produce an unsegregated distribution. In 1970, the segregation index in the 11 northern cities with the biggest black populations stood at 84.5 percent. By 1990, it had fallen to 77.8 percent. This means that in 1990 the

Table 10.4
The 10 Most Racially Segregated Major Metropolitan Areas in the United States

City	Segregation Index
Detroit	87.6
Chicago	85.8
Cleveland	85.1
Milwaukee	82.8
New York	82.2
Philadelphia	77.2
St. Louis	77.0
Los Angeles–Long Beach	73.1
Birmingham	71.7
Baltimore	71.4

Note: These data are for 1990.
Source: Massey and Denton (1993).

black population was only slightly less segregated than it was 20 years earlier. In the 9 southern cities with the biggest black populations, the index of segregation fell from 75.5 percent to 66.5 percent over the same period (Table 10.4). These figures suggest only modest improvement in housing segregation in recent decades.

As sociologists Douglas Massey and Nancy Denton (1993) show, black segregation in housing persists because of racism. Specifically, white homeowners are likely to move elsewhere if "too many" blacks move into a neighborhood. Meanwhile, real estate agents and mortgage lenders sometimes withhold information, refuse loans, and otherwise discourage blacks from moving into certain areas in order to protect real estate values.

The experience of Hispanic and especially Asian Americans is different. They are less segregated in housing than African Africans. Moreover, they are becoming desegregated faster. There is only one major exception to this pattern, and it supports the general argument about the continuing significance of race. Puerto Ricans are nearly as segregated in housing as blacks, and their situation is improving just as slowly. Massey and Denton (1987) say that this is because the white majority typically views them as "black Hispanics." In short, race matters in housing, but being black appears to matter most because of widespread antiblack sentiment.

Some analysts explain the persistence of residential segregation and other aspects of racism as a reaction to black progress. Thus, survey research shows that European Americans often express resentment against the real and perceived advantages enjoyed by African Americans (Kinder and Sanders, 1996). In particular, arguments for affirmative action generate strong opposition (Bobo and Kluegel, 1993) (see Chapter 17, "Education," for an extended discussion of this subject).

Supporters of affirmative action feel that African Americans should get preference if equally qualified people apply for a job or college entrance. In this way, the historical injustices of slavery, segregation, and discrimination can be corrected. Many whites object. They say that affirmative action is a form of "reverse discrimination," or bias against European Americans. For them, America stands for equal opportunity, not racial preference. Most blacks favor affirmative action, but some oppose it. Opponents argue that affirmative action demeans their accomplishments. It brands them as the "best black" rather than the best person for a job. Affirmative action, they say, contributes to belief in black inferiority (Carter, 1991).

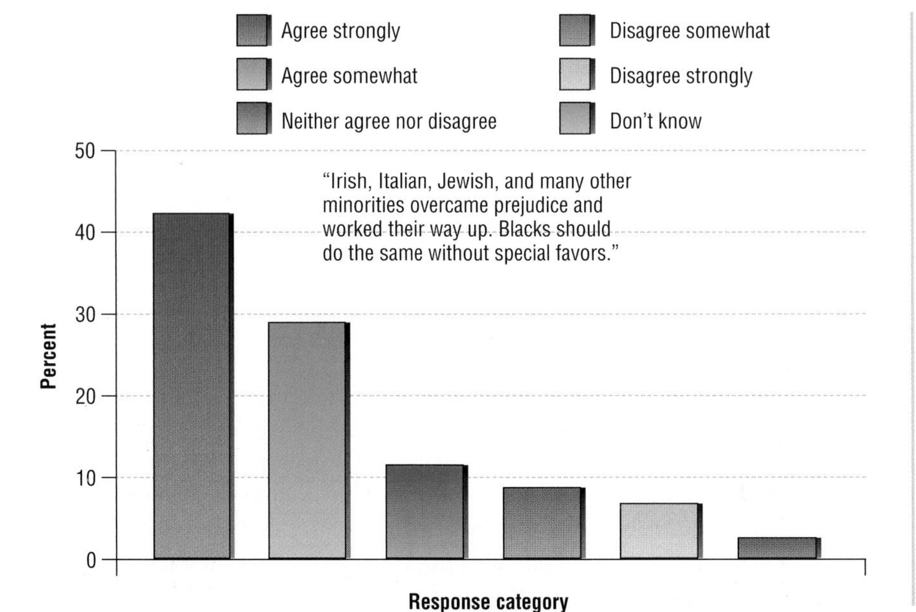

Figure 10.8
Opposition to Affirmative Action, United States

Note: Data are for 1998.
Source: National Opinion Research Center, 2004. General Social Survey, 1972–2002. Copyright © 2004 NORC. Used with permission.

Today's political climate is unfavorable to affirmative action. According to the 1998 General Social Survey (GSS), less than 15 percent of Americans think blacks need "special favors" to work their way up (▸Figure 10.8). Congress has ruled against giving preference to minority-owned firms in awarding small government contracts. California has eliminated affirmative action policies in state colleges.

Immigration and the Renewal of Racial and Ethnic Communities

If high levels of racism and inequality ensure the persistence of racial and ethnic identity, so does immigration. A steady flow of new immigrants gives new life to racial and ethnic groups. Immigrants bring with them knowledge of languages, appreciation of group culture, and a sense of community that might disappear if an ethnic or racial group were cut off from its origins. Seen in this light, we can expect many vibrant racial and ethnic communities to invigorate American life for a long time. Not since early in the 20th century has the immigration rate been so high, and not since the 1930s has such a large percentage of Americans consisted of people born in other countries (see Figure 10.4 and ▸Figure 10.9).

Of all the broad racial and ethnic categories used by the U.S. Census Bureau, the fastest growing is "Asian American," numbering nearly 12 million in the 2000 census. Most people in this category arrived after the mid-1960s (▸Table 10.5). That was when legislators eliminated racist selection criteria and instead designed an immigration law that emphasizes the importance of choosing newcomers who can make a big economic contribution to the country. Many Asian immigrants are middle-class professionals and businesspeople. For example, nearly 60 percent of Asian-Indian American adults are col-

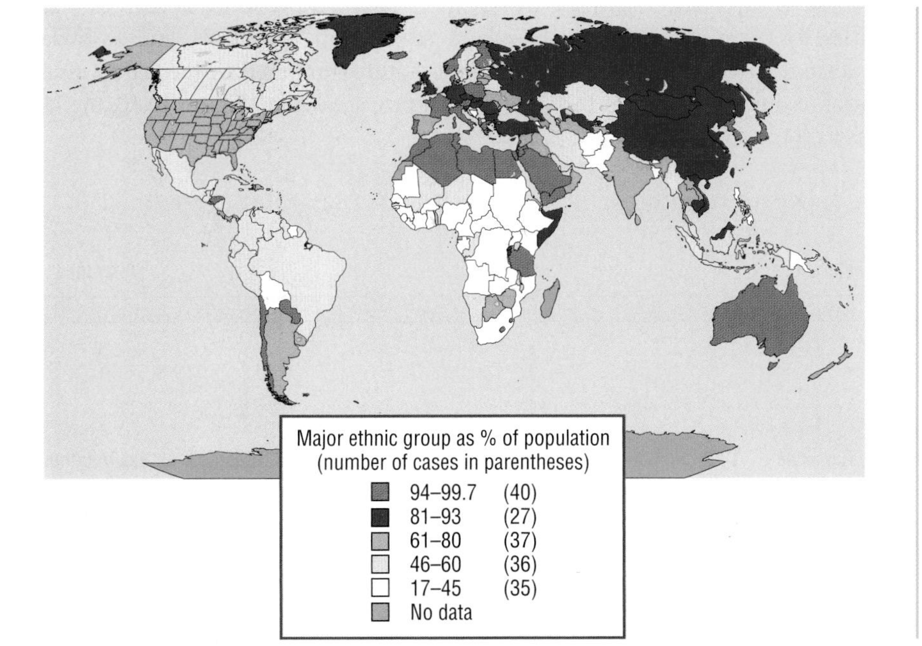

▸Figure 10.9

Percent of Population Accounted for by Largest Ethnic Group

Today the world has more than 5000 ethnic groups. No country is ethnically homogeneous, and in many countries the largest ethnic group forms less than half the population. Ethnic diversity has increased in recent years because of international migration. This map illustrates the world's ethnic diversity by showing the percent of each country's population that is accounted for by the country's largest ethnic group. Which countries are most diverse? Which are least diverse? How might a country's level of ethnic diversity affect its population's self-identity? How ethnically diverse is the United States compared with other countries?

Sources: "Ethnic Groups in the World," *Scientific American*. On the World Wide Web at http://www.sciam.com/1998/0998issue/0998numbers.html (December 4, 2001); *CIA World Factbook 2001*. On the World Wide Web at http://www.cia.gov/cia/publications/factbook/ (January 10, 2002).

Table 10.5
Asian Americans, 2000

Group	Number (millions)	Percent of total
Chinese	2.7	22.7
Filipino	2.4	20.1
Indian	1.8	16.0
Vietnamese	1.1	9.4
Korean	1.1	9.0
Japanese	0.8	6.7
Other	1.9	16.1
Total	11.9 *	100.0

*As a result of rounding
Sources: Barnes and Bennett (2002); U.S. Census Bureau (n.d.).

lege graduates, and a remarkable one-third hold graduate or professional degrees. More generally, Chinese Americans, Filipino Americans, Asian-Indian Americans, and Japanese Americans[5] earn above-median income, have a below-average poverty rate, and are more likely than non-Hispanic whites to hold a college degree. Considering only the American born, one may add Korean Americans to this list (Marger, 2003: 362, 367).

Some people hold up Asian Americans as models for other disadvantaged groups. Their formula is disarmingly simple: Emulate the Asians—work hard, keep your family intact, make sure your kids go to college—and you will surely succeed economically. This argument, however, has two main problems. First, some substantial Asian American groups, including several million Vietnamese, Cambodians, Hmong, and Laotians, do not fit the "Asian model." Members of these groups came to the United States as political refugees after the Vietnam War and they are disproportionately poor and unskilled, with a poverty rate higher than that of African Americans. Thus, no universal "Asian model" exists. Second, most economically successful Asian Americans were selected as immigrants precisely because they possessed educational credentials, skills, or capital that could benefit the American economy. They arrived on American shores with advantages, not liabilities. Saying that unskilled Chicanos or the black descendants of slaves should follow their example ignores the very social-structural brakes on mobility and assimilation that sociologists have discovered and emphasized in their research.

A Vertical Mosaic

In 1800 the United States was a society based on slavery, expulsion, and segregation. Two centuries later, we are a society based on segregation, pluralism, and assimilation. (**Pluralism** is the retention of racial and ethnic culture combined with equal access to basic social resources.) Thus, on the scale of tolerance shown in Figure 10.10, the United States has come a long way in the past 200 years.

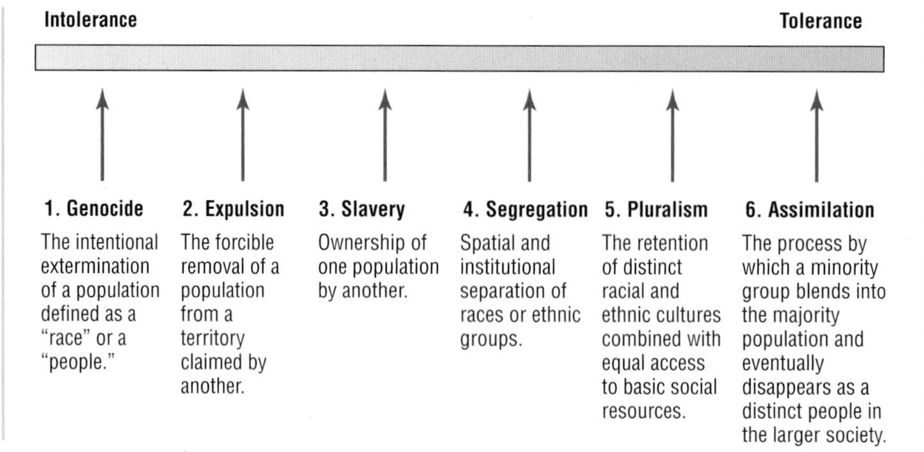

Figure 10.10
Six Degrees of Separation: Types of Ethnic and Racial Group Relations

Source: Adapted from Kornblum (1997 [1998]: 385).

Pluralism is the retention of racial and ethnic culture combined with equal access to basic social resources.

[5]Little Japanese immigration occurred after the 1920s, however.

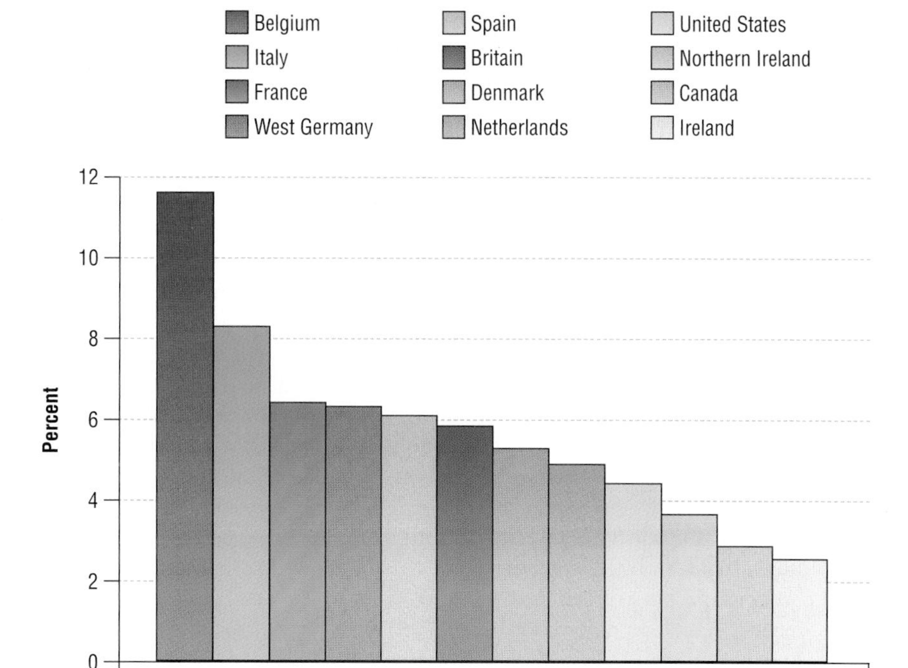

▶Figure 10.11
Percent Opposed to Person of Another Race, Immigrants, or Foreign Workers Living Next Door, 12 Postindustrial Countries

Note: Data are for 1990.

Source: Data from *The Decline of Deference* by Neil Nevitte, © 1996 Broadview Press.

In comparison with most other countries, the United States also is a relatively tolerant land. In the 1990s, racial and ethnic tensions in some parts of the world erupted into wars of secession and attempted genocide. Conflict between Croats, Serbs, and other ethnic groups broke Yugoslavia apart. Russia is fighting a bloody war against its Chechen ethnic minority. In Rwanda in 1991, Hutu militia and soldiers massacred many thousands of Tutsi civilians. Three years later, Tutsi soldiers massacred about 800,000 Hutu civilians (Box 10.2). Comparing the United States with such poor countries may seem to stack the deck in favor of concluding that the United States is a relatively tolerant society. However, even when we compare the United States with other rich, stable, postindustrial countries, ours seems relatively tolerant by some measures (▶Figure 10.11).

Sociology Now™

Learn more about **Genocide** by going through the Genocide: Mike Jacobs' Story Video Exercise.

Because of such factors as intermarriage and immigration, the growth of tolerance in the United States is taking place in the context of increasing ethnic and racial diversity. By the time today's first-year college student is 75 or 80 years old, non-Hispanic whites will form a minority of the U.S. population for the first time in 350 years. The United States will be even more of a racial and ethnic mosaic than it is now (▶Figure 10.12 and ▶Figure 10.13).

If present trends continue, however, the racial and ethnic mosaic will be arrayed vertically. Some groups, such as African Americans, Native Americans, Puerto Rican Americans, Chicanos, and some Asian American groups will be disproportionately clustered at the bottom. They will remain among the most disadvantaged groups in the country, enjoying less wealth, income, education, good housing, health care, and other social rewards than other ethnic and racial groups.

Policy initiatives could decrease the verticality of the American ethnic mosaic, thus speeding up the movement from segregation to pluralism and assimilation for the country's most disadvantaged groups. Apart from affirmative action programs, equality would be furthered by more job training, improvements in public education, and subsidized health and child care. These programs would be of greatest benefit to the most disadvantaged Americans. However, as we noted previously, and as we elaborate in subsequent chapters, the country does not seem much in the mood for such expensive reforms at this time (see especially Chapter 15, "Families," and Chapter 17, "Education"). The United States is likely to remain a vertical mosaic for some time to come.

BOX 10.2
Sociology at the Movies

Hotel Rwanda (2004)

In just a few days in 1994, the Hutus of Rwanda massacred 800,000 Tutsis—more than a tenth of Rwanda's population—with guns, machetes, hammers, and spears. Most of the world watched the attempted genocide with horror but did nothing. After all, Rwanda is in Africa, and as a United Nations peacekeeper explains to the manager of the Hotel Des Milles Collines in *Hotel Rwanda,* most of the world thinks of Africans as dung.

Hotel Rwanda is based on the true story of how the hotel manager, Paul Rusesabagina (Don Cheadle), used cunning and bribery to save more than 1200 Tutsis and sympathetic Hutus by protecting them in his hotel. Rusesabagina, a Hutu married to a Tutsi (Sophie Okonedo) is not the only hero of the piece. The UN peacekeeper (Nick Nolte) stands in for Romeo Dallaire, the Canadian general who led a tiny 500-man force credited with saving the lives of 20,000 Rwandans. Contemplating the horror outside the hotel gates, one of Rusesabagina's trusted employees asks: "Why are people so cruel?" Rusesabagina replies: "Hatred. Insanity. I don't know." Rusesabagina and Dallaire remind us that in an insane world, not everyone must succumb to madness.

Hotel Rwanda is a beautifully acted and heartbreaking movie. Because it unblinkingly shows the world its responsibility for failing to respond to genocide, it is worthwhile propaganda. But it is poor sociology because it leaves the viewer with the impression that "hatred" and "insanity" explain what the Hutus did to the Tutsis—or that one simply cannot know why people periodically kill one another in the name of ethnicity.

But we can know. Hutus and Tutsis existed as somewhat distinct ethnic groups for centuries before 1994. The Hutus were mainly farmers and the Tutsis mainly cattle herders. The Tutsis were the ruling minority, yet they spoke the same language as the Hutus, shared the same religious beliefs, lived side by side, often intermarried with them, and never came into serious conflict with them.

When the Belgians took over Rwanda in 1916, they made ethnic divisions far more inflexible than they had been. Now one *had*

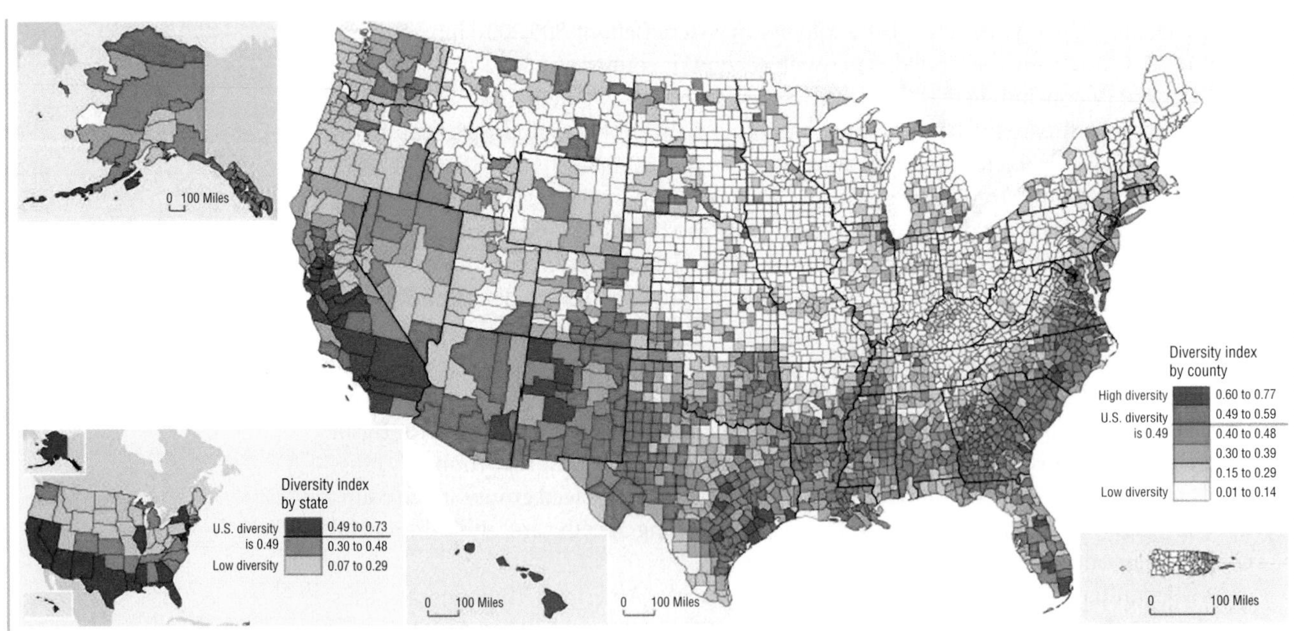

Figure 10.12
Ethnic and Racial Diversity in the United States, 2000

The diversity index reports the percentage of times two randomly selected people would differ by race/ethnicity. Working with percents expressed as ratios (e.g., 63 percent = 0.631), the index is calculated in three steps: (1) Square the percent for each group, (2) sum the squares, and (3) subtract the sum from 1.00. Eight groups were used for the index: 1. White, not Hispanic; 2. Black or African American; 3. American Indian or Alaska Native (AIAN); 4. Asian; 5. Native Hawaiian and Other Pacific Islander (NHOPI); 6. Two or more races, not Hispanic; 7. Some other race, not Hispanic; and 8. Hispanic or Latino. People indicating Hispanic origin who also indicated Black, AIAN, Asian, or NHOPI were counted only in their race group (0.5 percent of the population). They were not included in the Hispanic group.

Source: United States Bureau of the Census (2001b).

Paul Rusesabagina (played Don Cheadle) and his wife (played by Sophie Okonedo) protect their children in *Hotel Rwanda.*

to be a Tutsi to serve in an official capacity, and the Belgians started distinguishing Tutsis from Hutus by measuring the width of their noses; Tutsi noses, they arbitrarily proclaimed, are thinner. It was a preposterous policy (not least because half the population of Rwanda is of mixed, Hutu-Tutsi ancestry) and it served to sharply increase animosity between the two ethnic groups (Organization of African States, 2000: 10).

Before the Belgians left Rwanda in 1962, they encouraged power sharing between the Tutsis and the Hutus, but by this time the damage had been done. The Tutsis objected to any loss of power, and civil war broke out. Tutsi rebels fled to Uganda, and when Rwanda proclaimed independence, the Hutu majority took power. Then, in the early 1990s, descendants of the Tutsi rebels, backed by the United States and Britain, tried to overthrow the Hutu government, backed by France and Belgium. (Western interest in the region is high because it is rich in minerals [Rose, 2001]). The 1994 genocide erupted when the plane of the Hutu president was shot down in mysterious circumstances, killing him.

Belgium, the United States, Britain, and France must, then, bear responsibility for stoking the flames of ethnic conflict in Rwanda and not just for standing by when the conflict degenerated into genocide. While Belgium and the United States have at least apologized to Rwanda for looking the other way, most people continue to believe that the Rwandans alone are responsible for what transpired in their country in 1994. Unfortunately, *Hotel Rwanda* helps to reinforce that misconception.

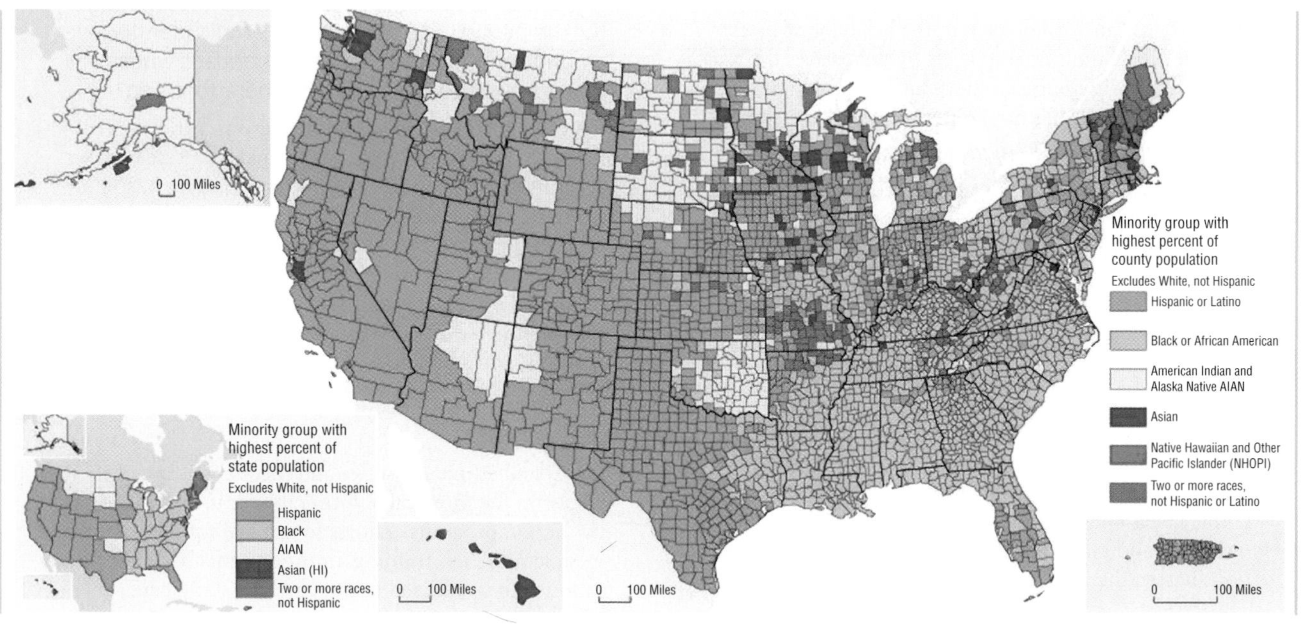

Figure 10.13

Prevalence of Hispanics and Nonwhite Minorities, United States, 2000

Percent for the "Some other race, not Hispanic" group was not highest in any state or country. People of Hispanic origin who were not white were counted in the Hispanic group and were also counted in the black, AIAN, Asian, and NHOPI group they indicated. Each of these people was counted twice in the comparison of percentages (+0.5 percent of the population).

Source: U.S Census Bureau (2001b).

Summary

Sociology Now™

Reviewing is as easy as ❶, ❷, ❸.

❶ Before you do your final review, take the SociologyNow diagnostic quiz to help you identify the areas on which you should concentrate. You will find information on how to access SociologyNow on the foldout at the front of the textbook.

❷ As you review, take advantage of SociologyNow's study aids to help you master the topics in this chapter.

❸ When you are finished with your review, take SociologyNow's post-test to confirm you are ready to move on to the next chapter.

1. **Is "race" a meaningful term?**

   Some biologists suggest that "race" is not a meaningful or useful term because the biological differences that distinguish races do not predict differences in social behavior. However, sociologists retain the term because *perceptions* of racial difference have important consequences for people's lives. Whether one is seen as belonging to one race or another affects one's health, wealth, and many others aspects of one's life.

2. **What is the difference between race and ethnicity?**

   A race is a category of people whose perceived *physical* markers are deemed socially significant. An ethnic group is a category of people whose perceived *cultural* markers are deemed socially significant. Just as physical distinctions do not cause differences in the behavior of races, cultural distinctions are often not by themselves the major source of differences in the behavior of various ethnic groups. *Social-structural* differences are typically the most important sources of differences in social behavior.

3. **What is meant by "the social construction of race and ethnicity"?**

   When sociologists speak of the social construction of race and ethnicity they mean that race and ethnicity are not fixed and are not inherent in people's biological makeup or cultural heritage. Rather, the way race and ethnicity are perceived and expressed depends on the history and character of race and ethnic relations in particular social contexts. These social contexts shape the way people formulate (or "construct") their perceptions and expressions of race and ethnicity. Thus, racial and ethnic labels and identities are variables. They change over time and place. For example, cordial group relations hasten the blending of labels and identities.

4. **What is Robert Park's theory of race and ethnic relations?**

   Robert Park's ecological theory of race and ethnic relations focuses on the way racial and ethnic groups struggle for territory and eventually blend into one another. He divides this struggle into five stages: invasion, resistance, competition, accommodation and cooperation, and assimilation. The main problem with his theory is that some groups get "stuck" between stages 3 and 5 and Park offers no explanation for why this happens.

5. **What are the main social-structural theories that supplement Park's theory?**

   The theory of internal colonialism and the theory of split labor markets usefully supplement Park's theory. The theory of internal colonialism highlights the way invaders change or destroy native culture, gaining virtually complete control over the native population, developing the racist belief that natives are inherently inferior, and confining natives to work that the invaders consider demeaning. Internal colonialism prevents assimilation by segregating the colonized in terms of jobs, housing, and social contacts ranging from friendship to marriage. The theory of split labor markets highlights the way low-wage workers of one race and high-wage workers of another race may compete for the same jobs. In such circumstances, high-wage workers are likely to resent the presence of low-wage competitors. Conflict is bound to result, and racist attitudes are likely to develop or get reinforced. The effects of split labor markets and internal colonialism on racial and ethnic identity persist for some time even after these social-structural conditions cease to exist. Thus, racial and ethnic groups may blend over time as members of society become more tolerant. However, this tendency is weak among members of highly disadvantaged groups because they remain highly segregated in jobs, housing, and social contacts. This is a historical legacy of slavery, expulsion, legalized segregation, and split labor markets.

6. **Aside from the historical legacy of internal colonialism and split labor markets, do other reasons exist for the persistence of racial and ethnic identity, even among some white European Americans whose ancestors came to this country generations ago?**

   Identifying with a racial or ethnic group can have economic, political, and emotional benefits. These benefits account for the persistence of ethnic identity in some white European American families, even after they have been in the United States more than two or three generations. In addition, high levels of immigration renew racial and ethnic communities by providing them with new members who are familiar with ancestral languages, customs, and so forth.

7. **What is the future of race and ethnicity in the United States?**

   Racial and ethnic identities and inequalities are likely to persist in the foreseeable future. In addition to affirmative action programs, efforts to achieve equality could include more job training, improvements in public education, and subsidized health care and child care. However, the country does not seem to favor these expensive reforms at this time.

Questions to Consider

1. **How do you identify yourself in terms of your race or ethnicity? Do conventional ethnic and racial cate-**

gories, such as black, white, Hispanic, and Asian, "fit" your sense of who you are? If so, why? If not, why not?

2. Do you think racism is becoming more serious in the United States and worldwide? Why or why not? How do trends in racism compare with trends in other forms of prejudice, such as sexism? What accounts for similarities and differences in these trends?
3. What are the costs and benefits of ethnic diversity in your college? Do you think it would be useful to adopt a policy of affirmative action to make the student body and the faculty more ethnically and racially diverse? Why or why not?

Web Resources

Companion Website for This Book

http://sociology.wadsworth.com

Begin by clicking on the Student Resources section of the website. Choose "Introduction to Sociology" and finally the Brym and Lie book cover. Next, select the chapter you are currently studying from the pull-down menu. From the Student Resources page you will have easy access to InfoTrac® College Edition, MicroCase Online exercises, additional web links, and many other resources to aid you in your study of sociology, including practice tests for each chapter.

Recommended Websites

For a stimulating analysis of transnational ethnic communities by Alejandro Portes, former president of the American Sociological Association, go to http://www.prospect.org/archives/25/25port.html.

For FBI statistics and analysis of hate crimes in the United States, see the annual reports at http://www.fbi.gov/ucr/01hate.pdf.

Basic information from the 2000 U.S. Census and the 2002 Current Population Survey on major ethnic categories in the United States is available on the Web. For American Indian and Alaska Natives, see http://www.census.gov/prod/2002pubs/c2kbr01-15.pdf. For Asian and Pacific Islanders, see http://www.census.gov/prod/2002pubs/c2kbr01-16.pdf and http://www.census.gov/prod/2003pubs/p20-540.pdf. For Hispanic Americans, go to http://www.census.gov/population/documentation/twps0056/tab01.pdf. See also the projections of state populations by race and Hispanic origin from 1995 to 2005 at http://www.census.gov/population/projections/state/stpjrace.txt. A gorgeous and extremely informative website devoted to the history of African American migrations can be found at http://www.inmotionaame.org.

CHAPTER 11 Sexuality and Gender

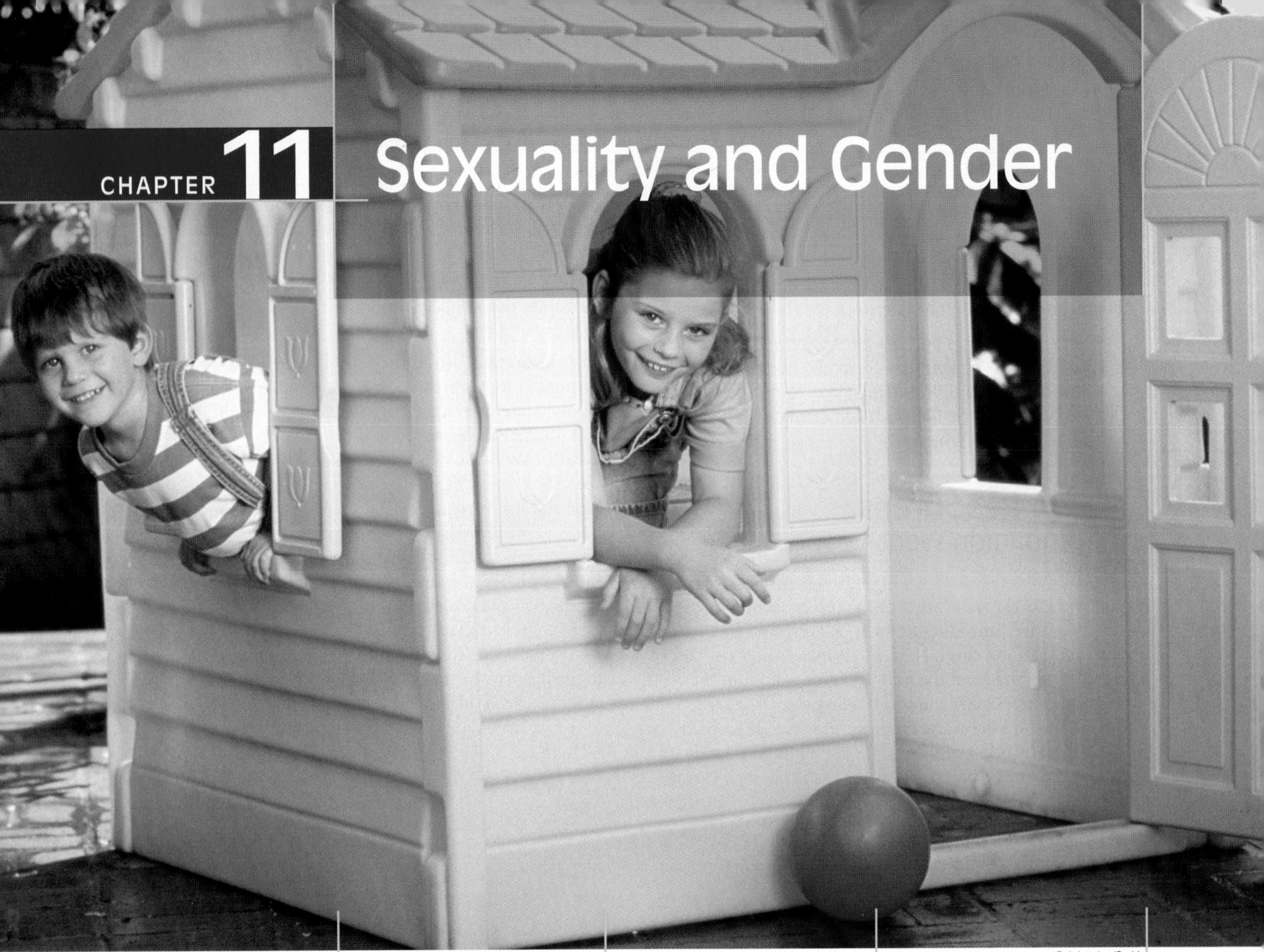

Rob Lewine/Corbis

In this chapter, you will learn that:

- While biology determines sex, social structure and culture largely determine gender, or the expression of culturally appropriate masculine and feminine roles.
- The social construction of gender is evident in the way parents treat babies, teachers treat pupils, and the mass media portray ideal body images.
- The social forces pushing people to assume conventionally masculine or feminine roles are compelling.
- The social forces pushing people toward heterosexuality operate with even greater force.
- The social distinction between men and women serves as an important basis of inequality in the family and the workplace.
- Male aggression against women is rooted in gender inequality.

Sex versus Gender

Is It a Boy or a Girl?

On April 27, 1966, identical 8-month-old twin boys were brought to a hospital in Winnipeg, Canada, to be circumcised. An electrical cauterizing needle—a device used to seal blood vessels as it cuts—was used for the procedure. However, because of equipment malfunction or doctor error, the needle entirely burned off one baby's penis. The parents desperately sought medical advice. No matter whom they consulted, they were given the same prognosis. As one psychiatrist summed up the baby's future, "He will be unable to consummate marriage or have normal heterosexual relations; he will have to recognize that he is incomplete, physically defective, and that he must live apart" (quoted in Colapinto, 1997: 58).

One evening, more than half a year after the accident, the parents, now deeply depressed, were watching TV. They heard Dr. John Money, a psychologist from Johns Hopkins Medical School in Baltimore, say that he could *assign* babies a male or female identity. Money had been the driving force behind the creation of the world's first "sex change" clinic at Johns Hopkins. He was well known for his research on **intersexed** infants, babies born with ambiguous genitals resulting from a hormone imbalance in the womb or some other cause. About 18 out of every 100,000 babies are born with this condition (Sax, 2002). It was Money's opinion that infants with "unfinished genitals" should be assigned a sex by surgery and hormone treatments, and reared in accordance with their newly assigned sex. According to Money, these strategies would lead to the child developing a self-identity consistent with its assigned sex.

The Winnipeg couple wrote to Dr. Money, who urged them to bring their child to Baltimore. After consulting with various physicians and with Money, the parents agreed

Sociology Now™

Reviewing is as easy as ❶, ❷, ❸.

Use SociologyNow to help you make the grade on your next exam. When you are finished reading this chapter, go to the Chapter Summary for instructions on how to make SociologyNow work for you.

Intersexed people are born with ambiguous genitals because of a hormone imbalance in their mother's womb or some other cause.

to have their son's sex reassigned. In anticipation of what would follow, they stopped cutting his hair, dressed him in feminine clothes, changed his name to Brenda, and in all other respects started treating him like a girl. Surgical castration was performed when the child was 22 months old. Early reports of Brenda's development indicated success. In contrast to her brother, she was said to disdain cars, gas pumps, and tools. She was supposedly fascinated by dolls, a dollhouse, and a doll carriage. Her mother reported that at the age of 4½, she took pleasure in her feminine clothing. She received dolls and skipping ropes for presents. She became a Girl Scout. At puberty, she began taking regular doses of the female hormone, estrogen.

The "twins case" generated worldwide attention after Dr. Money made the case public at the 1972 meeting of the American Association for the Advancement of Science in Washington, D.C. The experiment, he said, was an unqualified success. Brenda was feminine in manner and appearance. According to *Time* magazine, this proved that "conventional patterns of masculine and feminine behavior can be altered." Furthermore, Money's work cast "doubt on the theory that major sexual differences, psychological as well as anatomical, are immutably set by the genes at conception" (quoted in Colapinto, 1997: 66). Dr. Money subsequently reported good progress in 1978 and 1985. Textbooks in medicine and the social sciences were rewritten to incorporate Money's reports.

Then, in March 1997, two researchers dropped a bombshell when they published an article showing that Brenda had in fact struggled against her imposed girlhood from the start. Basing their judgments on a review of medical records and extensive interviews with the child's parents and the twins themselves, they documented that Brenda tried to tear off her first dress, wanted a toy razor like her brother's, refused to play with makeup and her toy sewing machine, insisted on playing with her brother's dump trucks and Tinker Toys, and strongly preferred to urinate standing up, even though it made a mess. She refused to play with girls. By the age of 7, she said she wanted to be a boy. She experienced academic failure and ridicule from her classmates, who called her "Cavewoman." At age 9, she had a nervous breakdown. At age 14, she attempted suicide (Colapinto, 2001: 96, 262).

In 1980, Brenda learned the details of her sex reassignment from her father. Unable to suffer her imposed sexual identity any longer, she stopped taking estrogen. In 1983, at age 16, she decided to have her sex reassigned once again. She had her breasts surgically removed, a penis surgically constructed, and took the name David. At age 25, David married a woman and adopted her three children. That did not, however, end his ordeal. In May 2004, at the age of 38, David Reimer committed suicide.

Sociology Now™

Learn more about **Sex versus Gender** by going through the Sex versus Gender Animation.

Sociology Now™

Learn more about **Gender Identity and Gender Role** by going through Sex: The Biological Dimension Learning Module.

Gender Identity and Gender Role

The twins case introduces the first big question of this chapter. What makes us male or female? Of course, part of the answer is biological. Your **sex** depends on whether you were born with distinct male or female genitals and a genetic program that released either male or female hormones to stimulate the development of your reproductive system.

However, the twins case also shows that more is involved in becoming male or female than biological sex differences. Recalling his life as a girl, David Reimer said, "[E]veryone is telling you that you're a girl. But you say to yourself, 'I don't *feel* like a girl.' You think girls are supposed to be delicate and *like* girl things—tea parties, things like that. But I like to *do* guy stuff. It doesn't match" (quoted in Colapinto, 1997: 66; our emphasis). As this quotation suggests, being male or female involves not just biology but also certain "masculine" and "feminine" feelings, attitudes, and behaviors. Accordingly, sociologists distinguish biological sex from sociological **gender.** Your gender is composed of the feelings, attitudes, and behaviors typically associated with being male or female. **Gender identity** is your identification with, or sense of belonging to, a particular sex—biologically, psychologically, and socially. When you behave according to widely shared expectations about how males or females are supposed to act, you adopt a **gender role.**

- Your **sex** depends on whether you were born with distinct male or female genitals and a genetic program that released either male or female hormones to stimulate the development of your reproductive system.
- Your **gender** is your sense of being male or female and your playing masculine and feminine roles in ways defined as appropriate by your culture and society.
- **Gender identity** is one's identification with, or sense of belonging to, a particular sex—biologically, psychologically, and socially.
- A **gender role** is the set of behaviors associated with widely shared expectations about how males or females are supposed to act.

The Social Learning of Gender

Contrary to first impressions, the twins case suggests that, unlike sex, gender is not determined just by biology. Research shows that babies first develop a vague sense of being a boy or a girl at about the age of 1. They develop a full-blown sense of gender identity between the ages of 2 and 3 (Blum, 1997). We can therefore be confident that baby Brenda already knew he was a boy when he was assigned a female gender identity at the age of 22 months. He had, after all, been raised as a boy by his parents and treated as a boy by his brother for almost two years. He had seen boys behaving differently from girls on TV and in storybooks. He had played only with stereotypical boys' toys. After his gender reassignment, the constant presence of his twin brother reinforced those early lessons on how boys ought to behave. In short, baby Brenda's *social* learning of his gender identity was already far advanced by the time he had his sex-change operation. Many researchers believe that if gender reassignment occurs before the age of 18 months, it will usually be successful (Creighton and Mihto, 2001; Lightfoot-Klein et al., 2000).[1] However, once the social learning of gender takes hold, as with baby Brenda, it is apparently difficult to undo, even by means of reconstructive surgery, hormones, and parental and professional pressure. The main lesson we draw from this story is not that biology is destiny but that the social learning of gender begins very early in life.

The first half of this chapter helps you better understand what makes us male or female. We first outline two competing theories of gender differences. The first theory argues that gender is inherent in our biological makeup and is merely reinforced by society. The second argues that gender is constructed mainly by social influences. For reasons outlined later, we side with the second viewpoint.

After establishing our theoretical approach, we examine how people learn gender roles during socialization in the family and at school. Then we show how everyday social interactions and advertising reinforce gender roles.

We next discuss how members of society enforce **heterosexuality**—the preference for members of the opposite sex as sexual partners. For reasons that are still poorly understood, some people resist and even reject the gender roles that are assigned to them based on their biological sex. When this occurs, negative sanctions are often applied to get them to conform or to punish them for their deviance. Members of society are often eager to use emotional and physical violence to enforce conventional gender roles.

The second half of the chapter examines one of the chief consequences of people learning conventional gender roles. Gender, as currently constructed, creates and maintains social inequality. We illustrate this in two ways. We first investigate why gender is associated with an earnings gap between women and men in the paid labor force. We then show how gender inequality encourages sexual harassment and rape. In concluding our discussion of sexuality and gender, we discuss some social policies that sociologists have recommended to decrease gender inequality and improve women's safety.

Theories of Gender

Most arguments about the origins of gender differences in human behavior adopt one of two perspectives. Some analysts see gender differences as a reflection of naturally evolved tendencies and argue that society must reinforce those tendencies if it is to function smoothly. Sociologists call this perspective **essentialism** (Weeks, 1986). That is because it views gender as part of the nature, or "essence," of one's biological and social makeup. Other analysts see gender differences mainly as a reflection of the different social positions occupied by women and men. Sociologists call this perspective **social constructionism.**

Heterosexuals are people who prefer members of the opposite sex as sexual partners.

Essentialism is a school of thought that sees gender differences as a reflection of biological differences between women and men.

Social constructionism is a school of thought that sees gender differences as a reflection of the different social positions occupied by women and men.

[1] Still, a movement has emerged to allow intersexed people to either choose their sex when they reach puberty or continue living with ambiguous genitals if they wish to do so.

That is because it views gender as "constructed" by social structure and culture. Conflict, feminist, and symbolic interactionist theories focus on various aspects of the social construction of gender.

Essentialism

Sociobiologists and evolutionary psychologists have proposed one popular essentialist theory. They argue that humans instinctively try to ensure that their genes are passed on to future generations. Men and women supposedly develop different strategies to achieve that goal. Presumably, women have a bigger investment than men in ensuring the survival of their offspring because women produce only a small number of eggs during their reproductive years and, at most, can give birth to about 20 children each. It is therefore in a woman's best interest to maintain primary responsibility for her genetic children and to seek out the single mate who can best help support and protect them. In contrast, men can produce as many as a billion sperm per ejaculation, and this feat can be replicated every day or two (Saxton, 1990: 94–5). To maximize their chance of passing on their genes to future generations, men must have many sexual partners.

According to sociobiologists and evolutionary psychologists, as men compete with other men for sexual access to many women, competitiveness and aggression emerge. Women, says one evolutionary psychologist, are greedy for money, whereas men want casual sex with women, treat women's bodies as their property, and react violently to women who incite male sexual jealousy. These are supposedly "universal features of our evolved selves" that contribute to the survival of the human species (Buss, 1998). Thus, from the point of view of sociobiology and evolutionary psychology, gender differences in behavior are based in biological differences between women and men.

Functionalism and Essentialism

Functionalists reinforce the essentialist viewpoint when they claim that traditional gender roles help to integrate society (Parsons, 1942). In the family, wrote Talcott Parsons, women traditionally specialize in raising children and managing the household. Men traditionally work in the paid labor force. Each generation learns to perform these complementary roles by means of *gender role socialization.*

For boys, noted Parsons, the essence of masculinity is a series of "instrumental" traits such as rationality, self-assuredness, and competitiveness. For girls, the essence of femininity is a series of "expressive" traits such as nurturance and sensitivity to others. Boys and girls first learn their respective gender traits in the family as they see their parents going about their daily routines. The larger society also promotes *gender role conformity.* It instills in men the fear that they won't be attractive to women if they are too feminine, and it instills in women the fear that they won't be attractive to men if they are too masculine. In the functionalist view, then, learning the essential features of femininity and masculinity integrates society and allows it to function properly.

A Critique of Essentialism from the Conflict and Feminist Perspectives

Conflict and feminist theorists disagree sharply with the essentialist account. They have lodged four main criticisms against it.

1. *Essentialists ignore the historical and cultural variability of gender and sexuality.* Wide variations exist in what constitutes masculinity and femininity. Moreover, the level of gender inequality, the rate of male violence against women, the criteria used for mate selection, and other gender differences that appear universal to the essentialists vary widely too. This variation deflates the idea that there are essential and universal behav-

ioral differences between women and men. Three examples help illustrate the point:

- In societies with low levels of gender inequality, the tendency decreases for women to stress the good provider role in selecting male partners, as does the tendency for men to stress women's domestic skills (Eagley and Wood, 1999).
- When women become corporate lawyers or police officers or take other jobs that involve competition or threat, their production of the hormone testosterone is stimulated, causing them to act more aggressively. Aggressiveness is partly role related (Blum, 1997: 158–88).
- Literally hundreds of studies conducted mainly in North America show that women are developing traits that were traditionally considered masculine. Women have become considerably more assertive, competitive, independent, and analytical in the last 35 years or so (Biegler, 1999; Duffy, Gunther, and Walters, 1997; Nowell and Hedges, 1998; Twenge, 1997).

Frans Lemmens/Getty Images

Definitions of "male" and "female" traits vary across societies. The ceremonial dress of male Wodaabe nomads in Niger may appear "feminine" by conventional North American standards.

As these examples show, gender differences are not constants and they are not inherent in men and women. They vary with social conditions.

2. *Essentialism tends to generalize from the average, ignoring variations within gender groups.* On average, women and men do differ in some respects. For example, one of the best-documented gender differences is that men are on average more verbally and physically aggressive than women. However, when essentialists say that men are *inherently* more aggressive than women, they make it seem as if that is true of all men and all women. As Figure 11.1 shows, it is not. When trained researchers measure verbal or physical aggressiveness, scores vary widely within gender groups. There is considerable overlap in aggressiveness between women and men. Thus, many women are more aggressive than the average man and many men are less aggressive than the average woman.
3. *Little or no evidence directly supports the essentialists' major claims.* For example, sociobiologists and evolutionary psychologists have not identified any of the genes that,

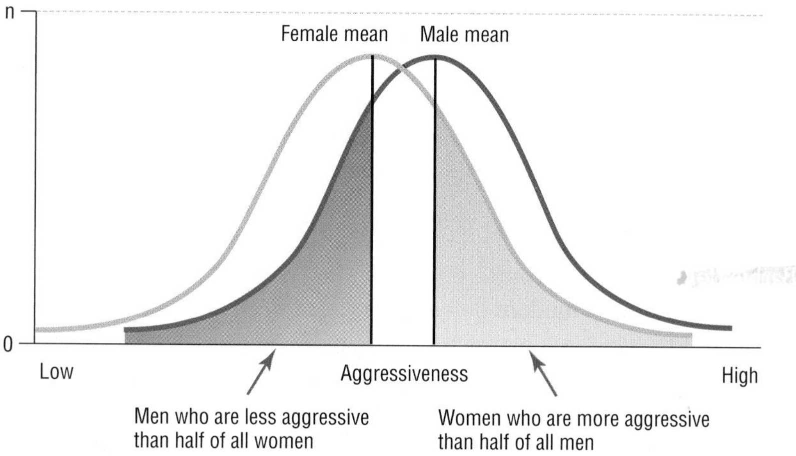

Figure 11.1
The Distribution of Aggressiveness among Men and Women

they claim, cause male jealousy, female nurturance, the unequal division of labor between men and women, and so forth.

4. *Essentialists' explanations for gender differences ignore the role of power.* Essentialists assume that existing behavior patterns help ensure the survival of the species and the smooth functioning of society. However, as conflict and feminist theorists argue, their assumption overlooks the fact that men are usually in a position of greater power and authority than women.

Conflict theorists dating back to Marx's collaborator, Friedrich Engels, have located the root of male domination in class inequality (Engels, 1970 [1884]). According to Engels, men gained substantial power over women when preliterate societies were first able to produce more than their members needed for their own subsistence. At that point, some men gained control over the economic surplus. They soon devised two means of ensuring that their offspring would inherit the surplus. First, they imposed the rule that only men could own property. Second, by means of socialization and force, they ensured that women remained sexually faithful to their husbands. As industrial capitalism developed, Engels wrote, male domination increased because industrial capitalism made men still wealthier and more powerful while it relegated women to subordinate, domestic roles.

Feminist theorists doubt that male domination is so closely linked to the development of industrial capitalism. For one thing, they note that gender inequality is greater in agrarian than in industrial capitalist societies. For another, male domination is evident in societies that call themselves socialist or communist. These observations lead many feminists to conclude that male domination is rooted less in industrial capitalism than in the patriarchal authority relations, family structures, and patterns of socialization and culture that exist in most societies (Lapidus, 1978: 7).

Despite this disagreement, conflict and feminist theorists concur that behavioral differences between women and men result less from any essential differences between them than from men being in a position to forward their interests over the interests of women. From the conflict and feminist viewpoints, functionalism, sociobiology, and evolutionary psychology may themselves be seen as examples of the exercise of male power, that is, as rationalizations for male domination and sexual aggression.

Social Constructionism and Symbolic Interactionism

Essentialism is the view that masculinity and femininity are inherent and universal traits of men and women, whether because of biological or social necessity or some combination of the two. In contrast, social constructionism is the view that *apparently* natural or innate features of life, such as gender, are actually sustained by *social* processes that vary historically and culturally. As such, conflict and feminist theories may be regarded as types of social constructionism. So may symbolic interactionism. Symbolic interactionists, you will recall, focus on the way people attach meaning to things in the course of their everyday communication. One of the things to which people attach meaning is what it means to be a man or a woman. We illustrate the symbolic interactionist approach by first considering how boys and girls learn masculine and feminine roles in the family and at school. We then show how gender roles are maintained in the course of everyday social interaction and through advertising in the mass media.

Web Interactive Exercise: Does Liberalism Cause Sex?

Gender Socialization

Barbie dolls have been around since 1959. Based on the creation of a German cartoonist, Barbie is the first modern doll modeled after an adult. (Lili, the German original, became a pornographic doll for men.) Some industry experts predicted that mothers would never buy dolls with breasts for their little girls. Were *they* wrong! Mattel now sells about 10 million Barbies and 20 million accompanying outfits annually. The Barbie trademark is worth a billion dollars.

What do American girls learn when they play with Barbie? The author of a website devoted to Barbie undoubtedly speaks for millions when she writes, "Barbie was more than a doll to me. She was a way of living: the Ideal Woman. When I played with her, I could make her do and be *anything* I wanted. Never before or since have I found such an ideal method of living vicariously through anyone or anything. And I don't believe I am alone. I am certain that most people have, in fact, lived their dreams with Barbie as the role player" (Elliott, 1995).

One dream that Barbie stimulates among many girls concerns body image. After all, Barbie is a scale model of a woman with a 40-18-32 figure. The scales that come with Workout Barbie always register a lithe 110 pounds. She can choose from many hundreds of outfits. And, judging from the Barbie sets available, she divides her days mainly between personal hygiene and physical fitness.[2] All this attention to appearance and physical perfection seems to be largely for the benefit of Ken. So when American girls play with Barbie, they learn to want to be slim, blond, and shapely and to exist mainly to please a pleasant man. The Scandinavian rock group Aqua put it well in their 1997 top-10 hit, "Barbie Girl":

Pascal Le Segretain/Corbis Sygma

A Barbie doll

Make me walk
Make me talk
I can act like a star
I can beg on my knees . . .
You can touch
You can play
If you say
"I'm always yours."

(Mattel tried to sue Aqua for its social commentary. A Los Angeles judge tossed the case out of court in May 1998.[3])

A comparable story, with competition and aggression its theme, could be told about how boys' toys, such as GI Joe, teach stereotypical male roles. True, a movement to market more gender-neutral toys arose in the 1960s and 1970s. However, it has now been overtaken by the resumption of a strong tendency to market toys based on gender. As *The Wall Street Journal* pointed out, "gender-neutral is out, as more kids' marketers push single-sex products" (Bannon, 2000: B1). For example, in 2000, Toys 'R' Us took the wraps off a new store design that included a store directory featuring "Boy's World" and "Girl's World." The Boy's World section listed action figures, sports collectibles, radio remote-controlled cars, Tonka trucks, boys' role-playing games, and walkie-talkies. The Girl's World section listed Barbie dolls, baby dolls, collectible horses, play kitchens, housekeeping toys, girls' dress-up, jewelry, cosmetics, and bath and body products.

Yet toys are only part of the story of gender socialization, and hardly its first or final chapter. Research conducted in the early 1970s showed that from birth, infant boys and girls who are matched in length, weight, and general health are treated differently by parents, and fathers in particular. Girls tend to be identified as delicate, weak, beautiful, and cute, boys as strong, alert, and well coordinated (Rubin, Provenzano, and Lurra, 1974). Recent attempts to update and extend this research show that although parents' gender-stereotyped perceptions of newborns have declined, especially among fathers, they have not disappeared (Fagot, Rodgers, and Leinbach, 2000; Gauvain et al., 2002; Karraker, Vogel, and Lake, 1995). When viewing videotape of a 9-month-old infant, experimental subjects tend to label its startled reaction to a stimulus as "anger" if the child has earlier

[2]We say "mainly" because in recent years Mattel has released a few Barbie sets that portray Barbie as a professional.
[3]Ironically, Lene Nystrom, the Norwegian lead singer of Aqua, had breast implants in 2000. She upset many women when she told Norway's leading daily newspaper, "I just want to be more feminine" (quoted in "Aqua Singer," 2000).

Reuters News Media, Inc./Corbis

A movement to market more gender-neutral toys emerged in the 1960s and 1970s. However, it has now been overtaken by the resumption of a strong tendency to market toys based on gender.

been identified by the experimenters as a boy, and as "fear" if it has earlier been identified as a girl, *whatever the infant's actual sex* (Condry and Condry, 1976). Parents, and especially fathers, are more likely to encourage their sons to engage in boisterous and competitive play and discourage their daughters from doing likewise. Parents tend to encourage girls to engage in cooperative, role-playing games (MacDonald and Parke, 1986). These different play patterns lead to the heightened development of verbal and emotional skills among girls. They lead to more concern with winning and the establishment of hierarchy among boys (Tannen, 1990). Boys are more likely than girls to be praised for assertiveness, and girls are more likely than boys to be rewarded for compliance (Kerig, Cowan, and Cowan, 1993). Given this early socialization, it seems perfectly "natural" that boys' toys stress aggression, competition, spatial manipulation, and outdoor activities, while girls' toys stress nurturing, physical attractiveness, and indoor activities (Hughes, 1995 [1991]). Still, what seems natural must be continuously socially reinforced. Presented with a choice between playing with a tool set and a dish set, preschool boys are about as likely to choose one as the other—unless the dish set is presented as a girl's toy and they think their fathers would view playing with it as "bad." Then, they tend to pick the tool set (Raag and Rackliff, 1998).

It would take someone who has spent very little time in the company of children to think they are passive objects of socialization. They are not. Parents, teachers, and other authority figures typically try to impose their ideas of appropriate gender behavior on children. But children creatively interpret, negotiate, resist, and self-impose these ideas all the time. Gender, we might say, is something that is done, not just given (West and Zimmerman, 1987). This is nowhere more evident than in the way children play.

Gender Segregation and Interaction

Consider the fourth- and fifth-grade American classroom that sociologist Barrie Thorne (1993) observed. The teacher periodically asked the children to choose their own desks. With the exception of one girl, they always segregated *themselves* by gender. The teacher then drew upon this self-segregation in pitting the boys against the girls in spelling and math contests. These contests were marked by cross-gender antagonism and expression of within-gender solidarity. Similarly, when children played chasing games in the schoolyard, groups often *spontaneously* crystallized along gender lines. These games had special names, some of which, like "chase and kiss," had clear sexual meanings. Provocation, physical contact, and avoidance were all sexually charged parts of the game.

Although Thorne found that contests, chasing games, and other activities often involved self-segregation of boys and girls, she saw many cases of boys and girls playing together. She also noticed quite a lot of "boundary crossing." Boundary crossing involves boys playing stereotypically "girls'" games and girls playing stereotypically "boys'" games. The most common form of boundary crossing involved girls who were skilled at sports that were central to the boys' world—like soccer, baseball, and basketball. If girls demonstrated skill at these activities, boys often accepted them as participants. Finally, Thorne noticed occasions where boys and girls interacted without strain and without strong gender identities coming to the fore. For instance, activities requiring cooperation, such as a group radio show or art project, lessened attention to gender. Another situation that lessened strain between boys and girls, causing gender to recede in importance, was when adults organized mixed-gender encounters in the classroom and in physical education periods. On such occasions, adults legitimized cross-gender contact. Mixed-gender interaction was also more common in less public and crowded settings. Thus, boys and girls were more likely to play together and in a relaxed way in the relative privacy of their neighborhoods. By contrast, in the school yard, where they were under the scrutiny of their peers, gender segregation and antagonism were more evident.

David Young-Wolff/PhotoEdit

In her research on schoolchildren, sociologist Barrie Thorne noticed quite a lot of "boundary crossing" between boys and girls. Most commonly, boys accepted girls as participants in soccer, baseball, and basketball games if girls demonstrated skill at these sports.

In sum, Thorne's research makes two important contributions to our understanding of gender socialization. First, children are actively engaged in the process of constructing gender roles. They are not merely passive recipients of adult demands. Second, although schoolchildren tend to segregate themselves by gender, boundaries between boys and girls are sometimes fluid and sometimes rigid, depending on social circumstances. In other words, the content of children's gendered activities is by no means fixed.

This is not to suggest that adults have no gender demands and expectations. They do, and their demands and expectations make important contributions to gender socialization. For instance, in most schools, teachers and guidance counselors still expect boys to do better in the sciences and math. They expect girls to achieve higher marks in English. Parents, for their part, tend to reinforce these stereotypes in their evaluation of different activities (Eccles, Jacobs, and Harold, 1990). Significantly, research comparing mixed- and single-sex schools shows that girls do much better in the latter. Sharlene Hesse-Biber and Gregg Lee Carter (2000: 99–100) summarize this research and are worth quoting at length:

> [In girls-only schools], female cognitive development is greater; female occupational aspirations and their ultimate attainment are increased; female self-confidence and self-esteem are magnified. Moreover, . . . females receive better treatment in the classroom; they are more likely to be encouraged to explore—and to have access to—wider curriculum opportunities; and teachers have greater respect for their work. Finally, females attending single-sex schools have . . . more egalitarian attitudes toward the role of women in society than do their counterparts in mixed-sex schools. . . . Single-sex schools accrue these benefits for girls for a variety of reasons . . . : (1) a diminished emphasis on "youth culture," which centers on athletics, social life, physical attractiveness, heterosexual popularity, and negative attitudes toward academics; (2) the provision of more successful same-sex role models (the top students in all subjects and all extracurricular activities [are] girls); (3) a reduction in sex bias in teacher-student interaction (there are [no] boys around [who] can be "favored"); and (4) elimination of sex stereotypes in peer interaction (generally, cross-sex peer interaction in school involves male dominance, male leadership, and, often, sexual harassment).

Adolescents must usually start choosing courses in school by the age of 14 or 15. By then, their **gender ideologies** are well formed. Gender ideologies are sets of interrelated ideas about what constitutes appropriate masculine and feminine roles and behavior. One aspect of gender ideology becomes especially important around grades 9 and 10: adoles-

A **gender ideology** is a set of ideas about what constitutes appropriate masculine and feminine roles and behavior.

Table 11.1
Ten Leading Occupational Categories of Employed Women 16 Years and Older, United States, 2004

Occupational Category	Total Employed Women (in thousands)	Percent Women	Women's Median Weekly Earnings ($)	Ratio of Women's to Men's Earnings
Office and administrative support (e.g., secretaries)	14,781	75.9	636	88.9
Sales and related (e.g., sales clerks)	7,878	49.3	464	62.1
Education, training, and library (e.g., teachers)	5,796	72.5	729	76.3
Management	5,344	26.4	871	87.9
Health-care practitioner and technical (e.g., nurses)	4,922	72.9	808	71.7
Food preparation and serving	4,084	56.4	339	88.3
Personal care and service (e.g., hair dressers)	3,484	73.9	380	76.0
Business and financial	3,172	53.2	746	74.1
Production (e.g., factory workers)	2,875	30.4	405	67.8
Health-care support (e.g., caregivers for the elderly)	2,609	89.3	402	88.7
Total for all occupations	64,728	46.5	573	80.4

Note: There is a negative correlation (−.16) between percent women in an occupational category and women's median weekly earnings. This means that there is a tendency for women to earn less in occupations where they form a higher percentage of the workforce. The correlation would be much stronger if we considered detailed occupations rather than broad occupational categories, because broad occupational categories obscure gender differences (e.g., "health-care practitioner and technical" includes physicians, most of whom are men, and nurses, most of whom are women).
Source: Department of Labor (2004).

cents' ideas about whether, as adults, they will focus mainly on the home, paid work, or a combination of the two. Adolescents usually make course choices with gender ideologies in mind. Boys are strongly inclined to consider only their careers in making course choices. Most girls are inclined to consider both home responsibilities and careers, although a minority considers only home responsibilities and another minority considers only careers. Consequently, boys tend to choose career-oriented courses, particularly in math and science, more often than girls. In college, the pattern is accentuated. Young women tend to choose easier courses that lead to lower-paying jobs because they expect to devote a large part of their lives to childrearing and housework (Hochschild with Machung, 1989: 15–18; Machung, 1989). The effect of these choices is to sharply restrict women's career opportunities and earnings in science and business (Reskin and Padavic, 2002 [1994]) (Table 11.1). We examine this problem in depth in the second half this chapter.

The Mass Media and Body Image

The social construction of gender does not stop at the school steps. Outside school, children, adolescents, and adults continue to negotiate gender roles as they interact with the mass media. If you systematically observe the roles played by women and men in TV programs and ads one evening, you will probably discover a pattern noted by sociologists since the 1970s. Women will more frequently be seen cleaning house, taking care of children, modeling clothes, and acting as objects of male desire. Men will more frequently be seen in aggressive, action-oriented, and authoritative roles. The effect of these messages on viewers is much the same as that of the Disney movies and Harlequin romances we discussed in Chapter 4 ("Socialization"). They reinforce the normality of traditional gender roles. As you will now see, many people even try to shape their bodies after the body images portrayed in the mass media.

Courtesy of the White Rock Beverage Company

Courtesy of the White Rock Beverage Company

The "White Rock Girl," featured on the logo of the White Rock Beverage Company, dropped 15 pounds between 1894 *(left)* and 1947 *(right)*.

The human body has always served as a sort of personal billboard that advertises gender. However, historian Joan Jacobs Brumberg (1997) makes a good case for the view that the importance of body image to our self-definition has grown over the past century. Just listen to the difference in emphasis on the body in the diary resolutions of two typical white, middle-class American girls, separated by a mere 90 years. From 1892: "Resolved, not to talk about myself or feelings. To think before speaking. To work seriously. To be self restrained in conversation and actions. Not to let my thoughts wander. To be dignified. Interest myself more in others." From 1982: "I will try to make myself better in any way I possibly can with the help of my budget and baby-sitting money. I will lose weight, get new lenses, already got new haircut, good makeup, new clothes and accessories" (quoted in Brumberg, 1997: xxi).

As body image became more important for one's self-definition in the course of the 20th century, the ideal body image became thinner, especially for women. Thus, the first American "glamour girl" was Mrs. Charles Dana Gibson, who was famous in advertising and society cartoons in the 1890s and 1900s as the "Gibson Girl." According to the Metropolitan Museum of Art's Costume Institute, "[e]very man in America wanted to win her" and "every woman in America wanted to be her. Women stood straight as poplars and tightened their corset strings to show off tiny waists" (Metropolitan Museum of Art, 2000). As featured in the *Ladies Home Journal* in 1905, the Gibson Girl measured 38-27-45—certainly not slim by today's standards. During the 20th century, however, the ideal female body type thinned out. The "White Rock Girl," featured on the logo of the White Rock beverage company, was 5′4″ and weighed 140 pounds in 1894. In 1947 she had slimmed down to 125 pounds. By 1970 she was 5′8″ and 118 pounds (Peacock, 2000).

The low-calorie and diet food industry promotes an ideal of slimness that is often impossible to attain and that generates widespread body dissatisfaction.

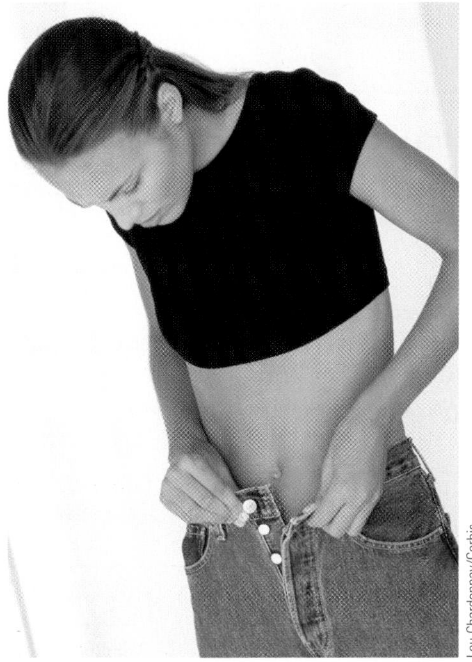
Lou Chardonnay/Corbis

Why did body image become more important to people's self-definition during the 20th century? Why was slimness stressed? Part of the answer to both questions is that more Americans grew overweight as their lifestyles became more sedentary. As they became better educated, they also grew increasingly aware of the health problems associated with being overweight. The desire to slim down was, then, partly a reaction to bulking up. But that is not the whole story. The rake-thin models who populate modern ads are not promoting good health. They are promoting an extreme body shape that is virtually unattainable for

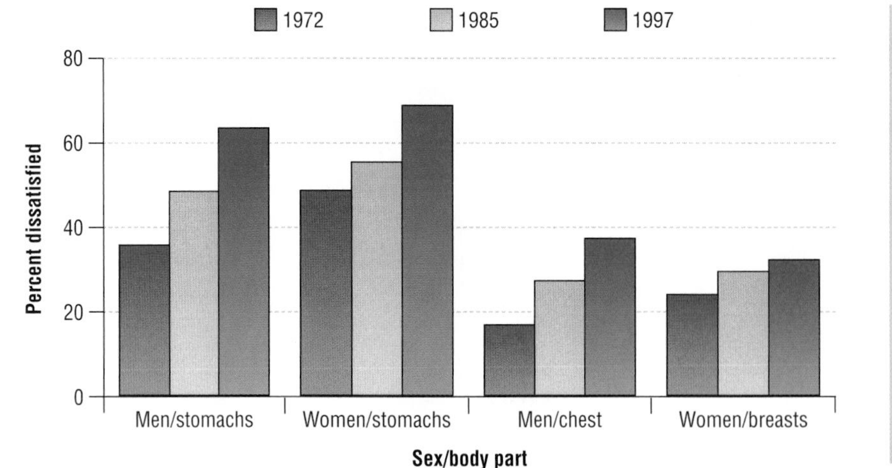

▶Figure 11.2
Body Dissatisfaction, United States, 1972–1997 (in percentage, *N* = 4000)

Note: The *N* of 4000 refers only to the 1997 survey. The number of respondents in the earlier surveys was not given.

Source: "The 1997 Body Image Survey Results" by David M. Garner. *Psychology Today*, Vol. 30, No. 1, pp. 30–44. Reprinted with permission from *Psychology Today* magazine. Copyright © 1997 Sussex Publishers, Inc.

most people. They do so because it is good business. In 1990 the United States diet and low-calorie frozen entrée industry alone enjoyed revenues of nearly $700 million. Some 65 million Americans spent upward of $30 billion in the diet and self-help industry in the pursuit of losing weight. The fitness industry generated $43 billion in revenue, and the cosmetic surgery industry another $5 billion (Hesse-Biber, 1996: 35, 39, 51, 53). Bankrolled by these industries, advertising in the mass media blankets us with images of slim bodies and makes these body types appealing. Once people become convinced that they need to develop bodies like the ones they see in ads, many of them are really in trouble because these body images are very difficult for most people to attain.

Survey data show just how widespread dissatisfaction with our bodies is and how important a role the mass media play in generating our discomfort. For example, a 1997 survey of North American college graduates showed that 56 percent of women and 43 percent of men were dissatisfied with their overall appearance (Garner, 1997). Only 3 percent of the dissatisfied women, but 22 percent of the dissatisfied men, wanted to gain weight. This reflects the greater desire of men for muscular, stereotypically male physiques. Most of the dissatisfied men, and even more of the dissatisfied women (89 percent), wanted to lose weight. This reflects the general societal push toward slimness and its greater effect on women.

▶Figure 11.2 reveals gender differences in body ideals in a different way. It compares women's and men's attitudes toward the appearance of their stomachs. It also compares women's attitudes toward their breasts with men's attitudes toward their chests. It shows, first, that women are more concerned about their stomachs than men are. Second, it shows that by 1997 men were more concerned about their chests than women were about their breasts. Clearly, then, people's body ideals are influenced by their gender. Note also that Figure 11.2 shows trends over time. North Americans' anxiety about their bodies increased substantially between 1972 and 1997.[4]

▶Table 11.2 suggests that advertising is highly influential in creating anxiety and insecurity about appearance, particularly about body weight. Here we see that in 1997 nearly 30

▶Table 11.2
The Influence of Fashion Models on Feelings about Appearance, North America (in percent; *N* = 4000)

	Men	Women	Extremely Dissatisfied Women
I always or often:			
Compare myself to models in magazines	12	27	43
Carefully study the shape of models	19	28	47
Very thin or muscular models make me:			
Feel insecure about my weight	15	29	67
Want to lose weight	18	30	67

Note: Data are for 1997.
Source: Adapted from Garner (1997).

[4]Slimness is somewhat less important for African American women (Molloy and Herzberger, 1998.)

percent of North American women compared themselves with the fashion models they saw in advertisements, felt insecure about their own appearance, and wanted to lose weight as a result. Among women who were dissatisfied with their appearance, the percentages were much larger, with about 45 percent making comparisons with fashion models and two-thirds feeling insecure and wanting to lose weight. It seems safe to conclude that fashion models stimulate body dissatisfaction among many North American women.

Body dissatisfaction, in turn, motivates many women to diet. Because of anxiety about their weight, 84 percent of North American women said they had dieted in the 1997 survey. The comparable figure for men was 54 percent. Just how important is it for people to achieve their weight goals? According to the survey, it's a life or weight issue: 24 percent of women and 17 percent of men said they would trade more than three years of their lives to achieve their weight goals.

Body dissatisfaction prompts some people to take dangerous and even life-threatening measures to reduce. In the 1997 survey, 50 percent of female smokers and 30 percent of male smokers said they smoked to control their weight. Other surveys suggest that between 1 percent and 5 percent of American women suffer from anorexia, or refusal to eat enough to remain healthy. About the same percentage of American female college students suffer from bulimia, or regular self-induced vomiting. For college men, the prevalence of bulimia is between 0.2 percent and 1.5 percent (Averett and Korenman, 1996: 305–6). In the United Kingdom, eating disorders are just as common, and the British Medical Association warned that celebrities such as *Ally McBeal* star Calista Flockhart were contributing to a rise in anorexia and bulimia. However, U.K. magazine editors are taking some responsibility for the problem. They recognize that waif-thin models are likely influencing young women to feel anxious about their weight and shape. As a result, the editors recently drew up a voluntary code of conduct that urges them to monitor the body images they portray, impose a minimum size for models, and use models of varying shapes and sizes ("British Magazines," 2000). Whether similar measures are adopted in the United States remains to be seen.

Sociology Now™

Learn more about **Eating Disorders** by going through the Eating Disorders Animation.

Male/Female Interaction

The gender roles that children learn in their families, at school, and through the mass media form the basis for their social interaction as adults. For instance, by playing team sports, boys tend to learn that social interaction is most often about competition, conflict, self-sufficiency, and hierarchical relationships (leaders versus led). They understand the importance of taking center stage and boasting about their talents (Messner, 1995 [1989]). Because many of the most popular video games for boys exclude female characters, use women as sex objects, or involve violence against women, they reinforce some of the most unsavory lessons of traditional gender socialization (Dietz, 1998). On the other hand, by playing with dolls and baking sets, girls tend to learn that social interaction is most often about maintaining cordial relationships, avoiding conflict, and resolving differences of opinion through negotiation. They understand the importance of giving advice and not promoting themselves or being bossy.

Because of these early socialization patterns, misunderstandings between men and women are common. A stereotypical example: Harold is driving around lost. However, he refuses to ask for directions because doing so would amount to an admission of inadequacy and therefore a loss of status. Meanwhile, it seems perfectly "natural" to Sybil to want to share information, so she urges Harold to ask for directions. The result: conflict between Harold and Sybil (Tannen, 1990: 62).

Gender-specific interaction styles also have serious implications for who is heard and who gets credit at work. Here are some examples uncovered by Deborah Tannen's research (1994a: 132–59):

- A female office manager doesn't want to seem bossy or arrogant. She is eager to preserve consensus among her coworkers. So she spends a good deal of time soliciting their opinions before making an important decision. She asks questions, listens attentively, and offers suggestions. She then decides. But her boss perceives her approach as indecisive and incompetent. He wants to recruit leaders for upper-management positions, so he overlooks the woman and selects an assertive man for a senior job that just opened up.

- A female technical director at a radio station wants to help a nervous new male soundboard operator do a good job. However, she is sensitive to the possibility that giving him direct orders may make him feel incompetent and cause him to do worse. So instead of instructing him, she starts a conversation about Macintosh computers, something he knows a lot about. This makes him feel capable and relaxed, and he sits back and puts his feet up. She then talks about some technical issues. She is careful to put everything in the context of an upcoming show (something the new soundboard operator couldn't possibly know about) rather than general technical knowledge (something he should have). Because of her sensitive management style, the show goes off without a hitch. Thankfully, the technical director's male supervisor did not come into the studio when she was making queries and the soundboard operator had his feet up. If the supervisor had arrived then, he could easily have concluded that the technical director was so incompetent that she had to get information from a subordinate who had just been hired.
- Male managers are inclined to say "I" in many situations, where female managers are inclined to say "we"—as in "I'm hiring a new manager and I'm going to put him in charge of my marketing division" or "This is what I've come up with on the Lakehill deal." This sort of phrasing draws attention to one's personal accomplishments. In contrast, Tannen heard a female manager talking about what "we" had done when in fact she had done all the work alone. This sort of phrasing camouflages women's accomplishments.

Sociology Now™

Learn more about **the Glass Ceiling** by going through the Glass Ceiling Animation.

The contrasting interaction styles illustrated previously often result in female managers not getting credit for competent performance. That is why they sometimes complain about a **glass ceiling,** a social barrier that makes it difficult for them to rise to the top level of management. As we will soon see, factors other than interaction styles, such as outright discrimination and women's generally greater commitment to family responsibilities, also support the glass ceiling. Yet gender differences in interaction styles play an independent role in constraining women's career progress.

- The **glass ceiling** is a social barrier that makes it difficult for women to rise to the top level of management.
- **Transgendered** people are individuals who break society's gender norms by defying the rigid distinction between male and female.
- **Transsexuals** believe they were born with the "wrong" body. They identify with, and want to live fully as, members of the "opposite" sex, and to do so they often change their appearance or resort to medical intervention.
- **Homosexuals** are people who prefer sexual partners of the same sex. People usually call homosexual men gay and homosexual women lesbians.
- **Bisexuals** are people who prefer sexual partners of both sexes.

Homosexuality

The preceding discussion outlines some powerful social forces pushing us to define ourselves as conventionally masculine or feminine in behavior and appearance. For most people, gender socialization by the family, the school, and the mass media is compelling and is sustained by daily interactions. A minority of people, however, resist conventional gender roles. For example, **transgendered** people are individuals who break society's gender norms by defying the rigid distinction between male and female. About 1 in every 5,000 to 10,000 people in North America is transgendered. Some transgendered people are **transsexuals.** Transsexuals believe that they were born with the "wrong" body. They identify with, and want to live fully as, members of the "opposite" sex. They often take the lengthy and painful path to a sex change operation. About 1 in every 30,000 people in North America is a transsexual (Nolen, 1999). **Homosexuals** are people who prefer sexual partners of the same sex, and **bisexuals** are people who prefer sexual partners of both sexes. People usually call homosexual men gay and homosexual women lesbians. The most comprehensive survey of sexuality in the United States shows that 2.8 percent of American men and 1.4 percent of American women think of themselves as homosexual or bisexual. However, 10.1 percent of men and 8.6 percent of women (a) think of themselves as homosexual or bisexual, (b) have had some same-sex experience, or (c) have had some same-sex desire (Laumann, Gagnon, Michael, and Michaels, 1994: 299) (Table 11.3).

Table 11.3
Homosexuality in the United States (in percent; N = 3432)

	Men	Women
Identified themselves as homosexual or bisexual	2.8	1.4
Had sex with person of same sex in past 12 months	3.4	0.6
Had sex with person of same sex at least once since puberty	5.3	3.5
Felt desire for sex with person of same sex	7.7	7.5
Had some same-sex desire or experience or identified themselves as homosexual or bisexual	10.1	8.6

Note: Data are for 1992.
Source: Michael, Gagnon, Laumann, and Kolata (1994: 40).

Homosexuality has existed in every society. Some societies, such as ancient Greece, have encouraged it. More frequently, however, homosexual acts have been forbidden. Until the late 18th and early 19th centuries in western Europe and the United States, "unnatural acts" such as sodomy were punishable by death. However, homosexuals were not identified as a distinct category of people until the 1860s. That was when the term "homosexuality" was coined. The term "lesbian" is of even more recent vintage.

We do not yet understand well why some individuals develop homosexual orientations. Some scientists think the reasons are mainly genetic, others think they are chiefly hormonal, while still others point to life experiences during early childhood as the most important factor. We do know that sexual orientation does not appear to be a choice. According to the American Psychological Association, it "emerges for most people in early adolescence without any prior sexual experience. . . . It is not changeable" ("American Psychological Association," 1998).

In any case, sociologists are less interested in the origins of homosexuality than in the way it is socially constructed, that is, in the wide variety of ways it is expressed and repressed (Foucault, 1990 [1978]; Weeks, 1986). It is important to note in this connection that homosexuality has become less of a stigma over the past century. Two factors are chiefly responsible for this, one scientific, the other political. In the 20th century, sexologists—psychologists and physicians who study sexual practices scientifically—first recognized and stressed the wide diversity of existing sexual practices. The American sexologist Alfred Kinsey was among the pioneers in this field. He and his colleagues interviewed thousands of men and women. In the 1940s they concluded that homosexual practices were so widespread that homosexuality could hardly be considered an illness affecting a tiny minority (Kinsey, Pomeroy, and Martin, 1948).

Peter Dejong/AP/Wide World Photos

On April 1, 2001, the Netherlands recognized full and equal marriage rights for homosexual couples. Within hours, Dutch citizens were taking advantage of the new law. The Dutch law is part of a worldwide trend to legally recognize long-term same-sex unions.

Sexologists, then, provided a scientific rationale for belief in the normality of sexual diversity. However, it was sexual minorities themselves who provided the social and political energy needed to legitimize sexual diversity among an increasingly large section of the public. Especially since the middle of the 20th century, gays and lesbians have built large communities and subcultures, particularly in major urban areas like New York and San Francisco. They have gone public with their lifestyles. They have organized demonstrations, parades, and political pressure groups to express their self-confidence and demand equal rights with the heterosexual majority. This has done much to legitimize homosexuality and sexual diversity in general.

Yet antipathy to homosexuals is widespread, and it is so strong among some people that they are prepared to back up their beliefs with force. A study of about 500 young adults in the San Francisco Bay area (probably the most sexually tolerant area in the United States) found that 1 in 10 admitted physically attacking or threatening people they believed were homosexuals. Twenty-four percent reported engaging in antigay name-calling. Among male respondents, 18 percent reported acting in a violent or threatening way and 32 percent reported name-calling. In addition, a third of those who had *not* engaged in antigay aggression said they would do so if a homosexual flirted with, or propositioned, them (Franklin, 1998).

Opposition to people who don't conform to conventional gender roles is strong at all stages of the life cycle. For example, when you were a child, did you ever laugh at a girl who, say, liked to climb trees and play with toy trucks? Did you ever call such a girl a "tomboy"? When you were a child, did you ever tease a boy who, say, liked to bake muffins while listening to Mozart? Did you ever call such a boy a "sissy"? If so, your behavior was not unusual. Children are typically strict about enforcing conventional gender roles. They often apply sanctions against playmates who deviate from convention.

Among adults, opposition is just as strong. What is your attitude today toward transgendered people, transsexuals, and homosexuals? Do you, for example, think that sexual relations between adults of the same sex are always, or almost always, wrong? If so, you are again not unusual. According to the 2002 General Social Survey, 60 percent of Americans believe that sexual relations between adults of the same sex are always, or almost always, wrong (National Opinion Research Center, 2004).

Due to widespread animosity toward homosexuals, many people who have wanted sex with members of the same sex, or who have had sex with someone of the same sex, do not identify themselves as gay, lesbian, or bisexual. The operation of norms against homosexuality is evident in figures on the geographical distribution of people who identify themselves as homosexuals or bisexuals. More than 9 percent of the residents of America's 12 largest cities (excluding suburbs) identify themselves as homosexual or bisexual. That figure falls to just over 4 percent in the cities ranked 13–100 by size (again excluding suburbs), and it drops to a little over 1 percent in rural areas. Why? Because small population centers tend to be less tolerant of homosexuality and bisexuality. People with same-sex desires are therefore less likely to express and develop homosexual and bisexual identities in smaller communities. They are inclined to migrate to larger and more liberal cities where supportive, established gay communities exist (Michael, Gagnon, Laumann, and Kolata, 1994: 178, 182).

Research suggests that some antigay crimes may result from repressed homosexual urges on the part of the aggressor (Adams, Wright, and Lohr, 1998). From this point of view, aggressors are **homophobic,** or afraid of homosexuals, because they cannot cope with their own, possibly subconscious, homosexual impulses. Their aggression is a way of acting out a denial of these impulses. However, although this psychological explanation may account for some antigay violence, it seems inadequate when set alongside the finding that fully half of all young male adults admitted to some form of antigay aggression in the San Francisco Bay area study previously cited. An analysis of the motivations of these San Franciscans showed that some of them did commit assaults to prove their toughness and heterosexuality. Others committed assaults just to alleviate boredom and have fun. Still others believed they were defending themselves from aggressive sexual propositions.

Homophobic people are afraid of homosexuals.

BOX 11.1

SOCIAL POLICY: WHAT DO YOU THINK?

Hate Crime Law and Homophobia

On October 7, 1998, Matthew Shepard, an undergraduate at the University of Wyoming, went to a campus bar in Laramie. From there, he was lured by Aaron James McKinney and Russell Henderson, both 21, to an area just outside town. McKinney and Henderson apparently wanted to rob Shepard. They wound up murdering him. They used the butt of a gun to beat Shepard's head repeatedly. According to the prosecutor at the trial of the two men, "As [Shepard] lay there bleeding and begging for his life, he was then bound to the buck fence" (quoted in CNN, 1998). The murderers left Shepard there in near-freezing temperatures. When a passerby saw Shepard several hours later, he thought the nearly dead man was a "scarecrow or a dummy set there for Halloween jokes" (quoted in CNN, 1998). Shepard died five days later.

One issue raised by Shepard's death concerns the definition of hate crime. Hate crimes are criminal acts motivated by a person's race, religion, or ethnicity. If hate motivates a crime, the law requires that the perpetrator be punished more severely than otherwise. For example, assaulting a person during an argument generally carries a lighter punishment than assaulting a person because he is an African American. Furthermore, the law (18 United States Code 245) permits federal prosecution of a hate crime only "if the crime was motivated by bias based on race, religion, national origin, or color, and the assailant intended to prevent the victim from exercising a 'federally protected right' [e.g., voting, attending school, etc.]" (Human Rights Campaign, 1999). This definition excludes crimes motivated by the sexual orientation of the victim. According to the FBI, if crimes against gays, lesbians, and bisexuals were defined as hate crimes, they would have composed 14 percent of the total (Federal Bureau of Investigation, 2002; Human Rights Campaign, 1999). Matthew Shepard was 1 of 33 antigay murders in the United States in 1998, up from 14 the year before. Do you think that crimes motivated by the victim's sexual orientation are the same as crimes motivated by the victim's race, religion, or ethnicity? If so, why? If not, why not? Do you think crimes motivated by the sexual orientation of the victim should be included in the legal definition of hate crime? If so, why? If not, why not?

A fourth group acted violently because they wanted to punish homosexuals for what they perceived as moral transgressions (Franklin, 1998). It seems clear, then, that antigay violence is not just a question of abnormal psychology but a broad, cultural problem with several sources.

On the other hand, anecdotal evidence suggests that opposition to antigay violence is also growing in America. The 1998 murder of Matthew Shepard in Wyoming led to a public outcry. In the wake of his murder, some people called for a broadening of the definition of hate crime to include antigay violence (Box 11.1). The 1999 movie *Boys Don't Cry* also raised awareness of the problem of antigay violence (Box 11.2).

In sum, strong social and cultural forces lead us to distinguish men from women and heterosexuals from homosexuals. We learn these distinctions throughout the socialization process, and we continually construct them anew in our daily interactions. Most people use positive and negative sanctions to ensure that others conform to conventional heterosexual gender roles. Some people resort to violence to enforce conformity and punish deviance.

Our presentation also suggests that the social construction of conventional gender roles helps to create and maintain social inequality between women and men. In the remainder of this chapter, we examine the historical origins and some of the present-day consequences of gender inequality.

Joseph Sohm, ChromoSohm, Inc./Corbis

Especially since the middle of the 20th century, gays and lesbians have built large communities and subcultures, particularly in major urban areas such as New York and San Francisco. They have gone public with their lifestyles. They have organized demonstrations, parades, and political pressure groups to express their self-confidence and demand equal rights with the heterosexual majority. This has done much to legitimize homosexuality and sexual diversity in general.

BOX 11.2
Sociology at the Movies

Boys Don't Cry (1999)

Boys Don't Cry is the true story of Teena Brandon, played by Hilary Swank, who won the Best Actress Oscar for her portrayal. Teena is a young woman in Lincoln, Nebraska, with short hair and a sexual-identity problem. She wants a sex-change operation but can't afford one. So she takes the inexpensive route. She stuffs a sock down the front of her jeans and goes to a bar to socialize with women. The women respond warmly to the soft-spoken and kindly man she appears to be. Soon, she decides to move to Falls City, change her name to Brandon Teena, and pass as a man.

In Falls City, Brandon develops a romantic relationship with a woman by the name of Lana Tisdel (Chloe Sevigny). At first, Lana thinks Brandon is a man. She falls in love with Brandon. The men she has known until then were violent and demeaning toward women. In contrast, Brandon is gentle, romantic, and thoughtful. At some point, Lana discovers that Brandon is a woman. But by then it doesn't matter. Love conquers all.

However, when two male members of Lana's family discover the truth about Teena, they are outraged. The very idea of two women in love sickens them and violates everything they have been taught to believe about the natural order of things and about their own sexuality. A few days later, the two men murder Teena. The local police may be sympathetic to their view of the natural order, because they fail to jail the two men, suggesting a certain sympathy toward them. Teena committed no crime. Her only transgression was that she wanted to be a man.

Fox Searchlight/The Kobal Collection/Matlock, Bill

Hilary Swank in *Boys Don't Cry* (1999).

Gender Inequality

The Origins of Gender Inequality

Sociology Now™

Learn more about **the Origins of Gender Inequality** by going through the Gender Inequality Video Exercise.

Contrary to what essentialists say, men have not always enjoyed much more power and authority than women. Substantial inequality between women and men has existed for only about 6000 years. It was socially constructed. Three major sociohistorical processes account for the growth of gender inequality. Let us briefly consider each of them.

Long-Distance Warfare and Conquest

The anthropological record suggests that women and men were about equal in status in foraging societies, the dominant form of society for 90 percent of human history. Rough gender equality was based on the fact that women produced a substantial amount of the band's food, up to 80 percent in some cases (Chapter 15, "Families"). The archaeological record from Old Europe tells a similar story. Old Europe is a region stretching roughly from Poland in the north to the Mediterranean island of Crete in the south, and from Switzerland in the west to Bulgaria in the east (◗Figure 11.3). Between 7000 and 3500 B.C.E., men and women enjoyed approximately equal status throughout the region. In fact, the religions of the region gave primacy to fertility and creator goddesses. Kinship was traced through the mother's side of the

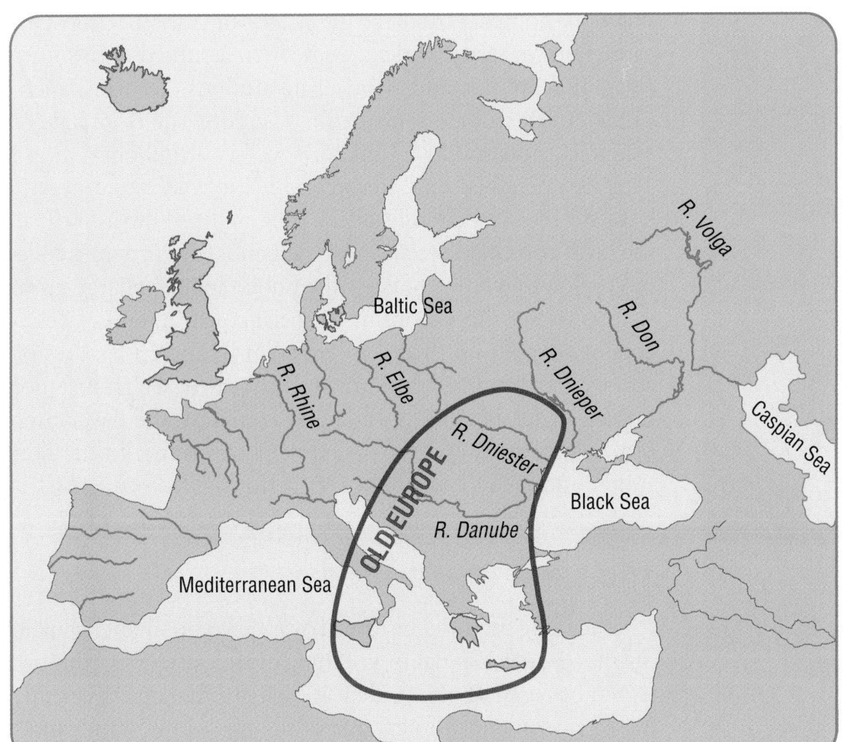

Figure 11.3
Old Europe

Source: Gimbutas (1982: 16).

family. Then, sometime between 4300 and 4200 B.C.E., all this began to change. Old Europe was invaded by successive waves of warring peoples from the Asiatic and European northeast (the Kurgans) and the deserts to the south (the Semites). Both the Kurgan and Semitic civilizations were based on a steeply hierarchical social structure in which men were dominant. Their religions gave primacy to male warrior gods. They acquired property and slaves by conquering other peoples and imposed their religions on the vanquished. They eliminated, or at least downgraded, goddesses as divine powers. God became a male who willed that men should rule women. Laws reinforced women's sexual, economic, and political subjugation to men. Traditional Judaism, Christianity, and Islam all embody ideas of male dominance, and they all derive from the tribes who conquered Old Europe in the fifth millennium B.C.E. (Eisler, 1995 [1987]; see also Lerner, 1986).

Plow Agriculture

Long-distance warfare and conquest catered to men's strengths and so greatly enhanced male power and authority. Large-scale farming using plows harnessed to animals had much the same effect. Plow agriculture originated in the Middle East around 5000 years ago. It required that strong adults remain in the fields all day for much of the year. It also reinforced the principle of private ownership of land. Because men were on average stronger than women, and because women were restricted in their activities by pregnancy, nursing, and childbirth, plow agriculture made men more powerful socially. Thus, land was owned by men and ownership was typically passed from father to eldest son (Coontz and Henderson, 1986).

The Separation of Public and Private Spheres

In the agricultural era, economic production was organized around the household. Men may have worked apart from women in the fields, but the fields were still part of the *family* farm. In contrast, during the early phase of industrialization, men's work moved

Women's domestic role was idealized in the 19th century.

out of the household and into the factory and the office. Most men became wage or salary workers. Some men assumed decision-making roles in economic and political institutions. Yet while men went public, most women remained in the domestic or private sphere. The idea soon developed that this was a natural division of labor. This idea persisted until the second half of the 20th century, when a variety of social circumstances, ranging from the introduction of the birth control pill to women's demands for entry into college, finally allowed women to enter the public sphere in large numbers.

So we see that according to social constructionists, gender inequality derives not from any inherent biological features of men and women but from three main sociohistorical circumstances: the impact of far-ranging warfare and conquest, the development of plow agriculture, and the assignment of women to the domestic sphere and men to the public sphere during the early industrial era.

The Earnings Gap Today

After reading this brief historical overview, you might be inclined to dismiss gender inequality as a thing of the past. If so, your decision would be hasty. That is evident if we focus first on the earnings gap between men and women, one of the most important expressions of gender inequality today. In 2001, women over the age of 15 working full-time in the paid labor force earned only 76 percent of what men earned (U.S. Department of Labor, 2002b). Four main factors contribute to the gender gap in earnings (Bianchi and Spain, 1996; England, 1992):

Sociology Now™

Learn more about **Gender Inequality** by going through the Gender Inequality Data Experiment.

Sociology Now™

Learn more about **Gender Inequality and Work** by going through the Gender Inequality and Work Learning Module.

Gender discrimination involves rewarding men and women differently for the same work.

The **female–male earnings ratio** is women's earnings expressed as a percentage of men's earnings.

1. *Gender discrimination.* In February 1985, when Microsoft, the software giant, employed about 1000 people, it hired its first two female executives. According to a well-placed source who was involved in the hiring, both women got their jobs because Microsoft was trying to win a United States Air Force contract. Under the government's guidelines, Microsoft didn't have enough women in top management positions to qualify. The source quotes then-29-year-old Bill Gates, president of Microsoft, as saying: "Well, let's hire two women because we can pay them half as much as we will have to pay a man, and we can give them all this other 'crap' work to do because they are women" (quoted in Wallace and Erickson, 1992: 291).

   This incident is a clear illustration of **gender discrimination,** rewarding women and men differently for the same work. Gender discrimination has been illegal in the United States since 1964. It has not disappeared, however. For example, between 1992 and 2003, when the nation's birthrate dropped 9 percent, the number of women claiming discrimination on the job because they were pregnant jumped 39 percent. A growing number of women claim they have been unfairly denied promotion, fired, and in some cases urged to terminate pregnancies to keep their jobs (Armour, 2005).

   Despite gender discrimination, antidiscrimination laws have helped to increase the **female–male earnings ratio,** that is, women's earnings as a percentage of men's earnings. The female–male earnings ratio increased 17.3 percent between 1960 and 2000. At that rate of improvement, women will be earning as much as men by 2050, around the time most first-year college students today retire (calculated from Feminist.com, 1999; U.S. Department of Labor, 2000b).

2. *Impact of heavy domestic responsibilities.* Raising children can be one of the most emotionally satisfying experiences in life. However, it is so exhausting and time-consuming and requires so many interruptions due to pregnancy and illness that it substantially decreases the time one can spend getting training and doing paid work. Because women are disproportionately involved in childrearing, they suffer the brunt of this economic reality. Women also do more housework and elderly care than men.

Specifically, in most countries, including the United States, women do between two-thirds and three-quarters of all unpaid child care, housework, and care for the elderly (Boyd, 1997: 55). As a result, they devote fewer hours to paid work than do men, experience more labor-force interruptions, and are more likely than men to take part-time jobs. Part-time jobs pay less per hour and offer fewer benefits than full-time work. Even when they work full-time in the paid labor force, women continue to shoulder a disproportionate share of domestic responsibilities, working, in effect, a double shift (Hochschild with Machung, 1989; see Chapter 15, "Families"). This affects how much time they can devote to their jobs and careers, with negative consequences for their earnings (Mahony, 1995; Waldfogel, 1997).

3. *Concentration of women in low-wage occupations and industries.* Often, the courses that women select in high school and college tend to limit them to jobs in low-wage occupations and industries. Thus, although women have made big strides since the 1970s, especially in managerial employment, they are still concentrated in lower-paying clerical and service occupations and underrepresented in higher-paying business, scientific, and manual occupations. Lower earnings are associated with occupations where women are concentrated (Hesse-Biber and Carter, 2000: 114–73) (see Table 11.1).
4. *Undervaluation of women's labor.* Work done by women is commonly considered less valuable than work done by men because it is viewed as involving fewer skills (Figart and Lapidus, 1996; Sorenson, 1994). Compare telephone installers and repairers, 98.2 percent of whom were men in 2004, with prekindergarten and kindergarten teachers, 97.7 percent of whom were women. The man who installed phones earned an average of $813 a week, while the woman who taught and played with 5-year-olds earned an average of $515 (U.S. Department of Labor, 2004). It is, however, questionable whether it takes less training and skill to teach a young child the basics of counting, reading, cooperation, and sharing than it does to install a phone. As this example suggests, we apply somewhat arbitrary standards to reward different occupational roles. In our society, these standards systematically undervalue the kind of skills needed for jobs where women are concentrated.

Janette Beckman/Corbis

Although women have entered many traditionally "male" occupations since the 1970s, they are still concentrated in lower-paying clerical and service occupations and underrepresented in higher-paying manual occupations.

Sociology Now™

Learn more about **Gender Inequality and Work** by going through the Women-Owned Firms Map Exercise.

We thus see that the gender gap in earnings is based on several *social* circumstances rather than any inherent difference between women and men. This means that people can reduce the gender gap if they want to. Later, we discuss social policies that could create more equality between women and men. But first, to stress the urgency of such policies, we explain how the persistence of gender inequality encourages sexual harassment and rape.

Male Aggression Against Women

Serious acts of aggression between men and women are common. The great majority are committed by men against women. For example, in 2003, 93,433 rapes of women were reported to the police in the United States (U.S. Federal Bureau of Investigation, 2004: 70). The rate of rape is especially high among young singles. Thus, in a survey of acquaintance and date rape in American colleges, 7 percent of men admitted that they had attempted or committed rape in the past year. Eleven percent of women said they were victims of attempted or successful rape (Koss, Gidycz, and Wisniewski, 1987) (▶Figure 11.4).

Why do men commit more frequent (and more harmful) acts of serious aggression against women than women commit against men? It is not because men on average are *physically* more powerful than women. Greater physical power is more likely to be used to

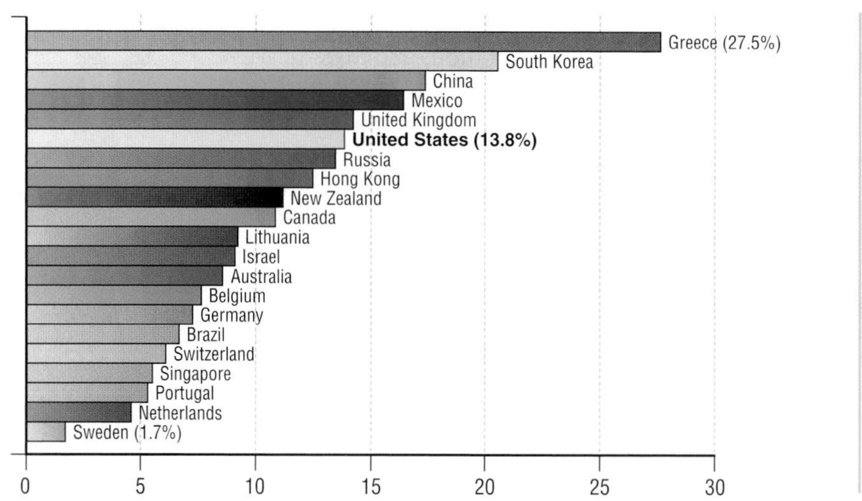

Figure 11.4
Percent of University Students Who Severely Assaulted a Dating Partner in the Past Year, by Country, 2001–05 (N = 6700)

Percent of students who used a knife or gun on a partner, punched or hit partner with something that could hurt, choked partner, slammed partner against wall, beat up partner, burned or scalded partner, or kicked partner in year preceding the survey.

Note: Some of the American data were collected in 1998.
Source: International Dating Violence Study. Custom tabulation courtesy of Murray A. Straus.

Web Research Project: Martial Rape

commit acts of aggression when norms justify male domination and men have much more *social* power than women. When women and men are more equal socially, and norms justify gender equality, the rate of male aggression against women is lower. This is evident if we consider various types of aggressive interaction, including rape and sexual harassment (see also the discussion of wife abuse in Chapter 15, "Families").

Rape

Some people think that rapists suffer a psychological disorder that compels them to achieve immediate sexual gratification even if violence is required. Others think rape occurs because of flawed communication. They believe some rape victims give mixed signals to their assailants by, for example, drinking too much and flirting with them.

Such explanations are not completely invalid. Interviews with victims and perpetrators show that some rapists do suffer from psychological disorders. Other offenders do misinterpret signals in what they regard as sexually ambiguous situations (Hannon, Hall, Kuntz, Van Laar, and Williams, 1995). But such cases account for only a small proportion of the total. Men who rape women are rarely mentally disturbed, and it is abundantly clear to most assailants that they are doing something their victims strongly oppose.

What then accounts for rape being as common as it is? A sociological answer is suggested by the fact that rape is sometimes not about sexual gratification at all. Some rapists cannot ejaculate. Some cannot even achieve an erection. Significantly, however, all rape involves domination and humiliation as principal motives. It is not surprising, therefore, that some rapists are men who were physically or sexually abused in their youth. They develop a deep need to feel powerful as psychological compensation for their early powerlessness. Other rapists are men who, as children, saw their mothers as potentially hostile figures who needed to be controlled or as mere objects available for male gratification. They saw their fathers as emotionally cold and distant. Raised in such an atmosphere, rapists learn not to empathize with women. Instead, they learn to want to dominate them (Lisak, 1992).

Psychological factors aside, certain *social* situations also increase the rate of rape. One such situation is war. In war, conquering male soldiers often feel justified in wanting to humiliate the vanquished, who are powerless to stop them. Rape is often used for this purpose, as was especially well documented in the ethnic wars that accompanied the breakup of Yugoslavia in the 1990s (Human Rights Watch, 1995).

Aggressiveness is also a necessary and important part of police work. Spousal abuse is therefore common among police officers. One U.S. study found that 37 percent of anonymously interviewed police wives reported spousal abuse. Several other surveys of police officers put the figure in the 40 percent range (Roslin, 2000). "It's a horrible, horrible problem," says Penny Harrington, former chief of police in Portland, Oregon, and now head of

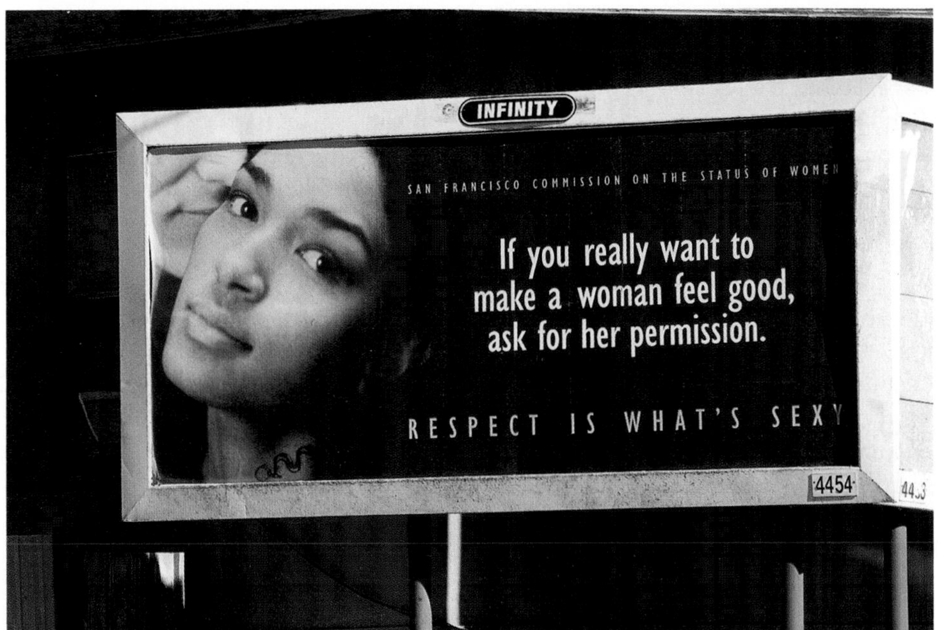

Rachel Epstein/PhotoEdit

A San Francisco billboard suggests that men still need reminding that no means no.

the Los Angeles–based National Center for Women and Policing. "Close to half of all 911 calls are due to family violence," says Harrington. "If the statistics are true, you've got a two-in-five chance of getting a batterer coming to answer your call" (quoted in Roslin, 2000: 46).

The relationship between male dominance and rape is also evident in research on college fraternities. Many college fraternities tend to emphasize male dominance and aggression as a central part of their culture. Thus, sociologists who have interviewed fraternity members have shown that most fraternities try to recruit members who can reinforce a macho image and avoid any suggestion of effeminacy and homosexuality. Research also shows that fraternity houses that are especially prone to rape tend to sponsor parties that treat women in a particularly degrading way. Thus, by emphasizing a very narrow and aggressive form of masculinity, some fraternities tend to facilitate rape on college campuses (Boswell and Spade, 1996; Martin and Hummer, 1989; Sanday, 1990).

Another social circumstance that increases the likelihood of rape is participation in athletics. Of course, few athletes are rapists. However, there are proportionately more rapists among men who participate in athletics than among nonathletes (Welch, 1997). That is because many sports embody a particular vision of masculinity in North American culture: competitive, aggressive, and domineering. By recruiting men who display these characteristics and by encouraging the development of these characteristics in athletes, sports can contribute to off-field aggression, including sexual aggression. Furthermore, among male athletes, there is a distinct hierarchy of sexual aggression. Male athletes who engage in contact sports are more prone to be rapists than other athletes. There are proportionately even more rapists among athletes involved in collision and combative sports, notably football (Welch, 1997).

Rape, we conclude, involves using sex to establish dominance. The incidence of rape is highest in situations where early socialization experiences predispose men to want to control women, where norms justify the domination of women, and where a big power imbalance between men and women exists.

Sexual Harassment

There are two types of sexual harassment. **Quid pro quo harassment** takes place when sexual threats or bribery are made a condition of employment decisions. (The Latin phrase *quid pro quo* means "something for something.") **Hostile-environment sexual harassment** involves sexual jokes, comments, and touching that interfere with work or

Quid pro quo sexual harassment takes place when sexual threats or bribery are made a condition of employment decisions.

Hostile-environment sexual harassment involves sexual jokes, comments, and touching that interfere with work or create an unfriendly work setting.

create an unfriendly work setting. Research suggests that relatively powerless women are the most likely to be sexually harassed. Moreover, sexual harassment is most common in work settings that exhibit high levels of gender inequality and a culture justifying male domination of women. Specifically, women who are young, unmarried, and employed in nonprofessional jobs are most likely to become objects of sexual harassment, particularly if they are temporary workers, the ratio of women to men in the workplace is low, and the organizational culture of the workplace tolerates sexual harassment (Rogers and Henson, 1997; Welsh, 1999).

Ultimately, then, male aggression against women, including sexual harassment and rape, is encouraged by a lesson most of us still learn at home, in school, at work, through much of organized religion, and in the mass media—that it is natural and right for men to dominate women. To be sure, recent decades have witnessed important changes in the way women's and men's roles are defined. Nevertheless, in the world of paid work, in the household, in government, and in all other spheres of life, men still tend to command substantially more power and authority than women. Daily patterns of gender domination, viewed as legitimate by most people, get built into our courtship, sexual, family, and work norms. From this point of view, male aggression against women is simply an expression of male authority by other means.

This does not mean that all men endorse the principle of male dominance, much less that all men are inclined to rape or engage in other acts of aggression against women. Many men favor gender equality, and most men never rape or abuse a woman. However, the fact remains that many aspects of our culture legitimize male dominance, making it seem valid or proper. For example, pornography, jokes at the expense of women, and whistling and leering at women might seem mere examples of harmless play. At a subtler, sociological level, however, they are assertions of the appropriateness of women's submission to men. Such frequent and routine reinforcements of male authority increase the likelihood that some men will consider it their right to assault women physically or sexually if the opportunity to do so exists or can be created. "Just kidding" has a cost. For instance, researchers have found that college men who enjoy sexist jokes are most likely to report engaging in acts of sexual aggression against women (Ryan and Kanjorski, 1998).

We thus see that male aggression against women and gender inequality are not separate issues. Gender inequality is the foundation of aggression against women. In concluding this chapter, we consider how gender inequality can be decreased in the coming decades. As we proceed, you should bear in mind that gender equality is not just a matter of justice. It is also a question of safety.

Toward 2050

The 20th century witnessed growing equality between women and men in many countries. In the United States, the decline of the family farm made children less economically useful and more costly to raise. As a result, women started having fewer children. The industrialization of America, and then the growth of the economy's service sector, increased demand for women in the paid labor force (Figure 11.5). This gave them substantially more economic power and also encouraged them to have fewer children. The legalization and availability of contraception made it possible for women to exercise unprecedented control over their own bodies. The women's movement fought for, and won, increased rights for women on a number of economic, political, and legal fronts. All these forces brought about a massive cultural shift, a fundamental reorientation of thinking on the part of many Americans about what women could and should do in society.

One indicator of the progress of women is the Gender Empowerment Measure (GEM). The GEM is computed by the United Nations. It takes into account women's share of seats in parliament (the House of Representatives in the United States), women's share of administrative, managerial, professional, and technical jobs, and women's earning power. A score of 1.0 indicates equality with men on these three dimensions.

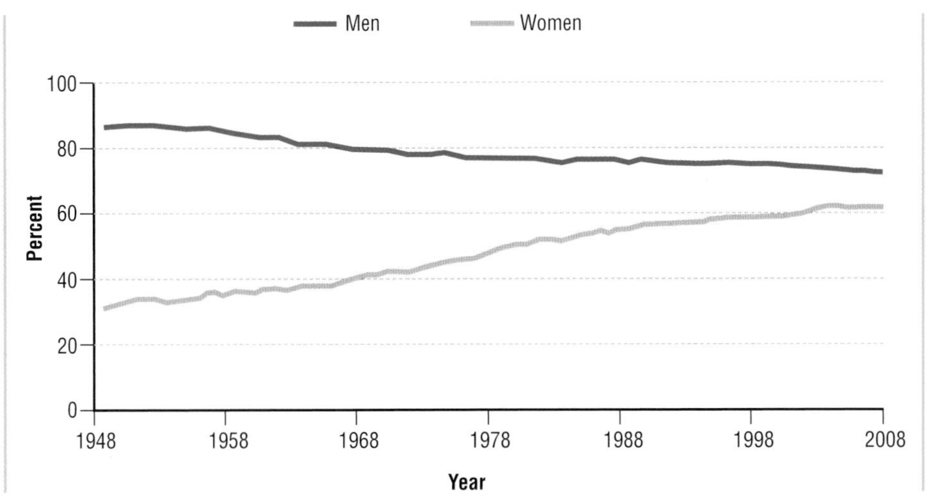

Figure 11.5
Labor Force Participation Rate by Sex, United States, 1948–2000 and 2001–2010 projected (in percentage)

Note: Figures are expressed as percentage of men and women 16 years of age and older. From 2010 to 2050, male and female labor force participation rates are expected to decline, and the gap between the two is expected to grow slightly for the first time since statistics were first collected.

Source: U.S. Department of Labor (1998b, 1999b, 2003a).

As Figure 11.6 shows, the Scandinavian countries (Norway, Sweden, Denmark, and Finland) were the most gender-egalitarian countries in the world in 1998. They had GEM scores ranging from 0.908 to 0.801. This means that Scandinavian women are roughly between 80 and 90 percent of the way to equality with men on the three dimensions tapped by the GEM. The United States ranked 14th in the world, with a GEM score of 0.769. This means that American women are about three-quarters of the way to equality with men. In general, there is more gender equality in rich than in poor countries. Thus, the top five countries are all rich. In contrast, the lowest GEM scores are found mainly in poor Asian and African countries. This suggests that gender equality is a function of economic development.

However, our analysis of the GEM data suggests that there are some exceptions to the general pattern. They show that gender equality is also a function of government policy (Brym et al., 2005). Thus, in some of the former communist countries of eastern Europe (such as Poland, Hungary, Slovakia, the Czech Republic, and Latvia), gender equality is *higher* than one would expect given their level of economic development. Meanwhile, in some of the Islamic countries (such as the United Arab Emirates, Bahrain, Saudi Arabia, and Algeria), gender equality is *lower* than one would expect given their level of economic

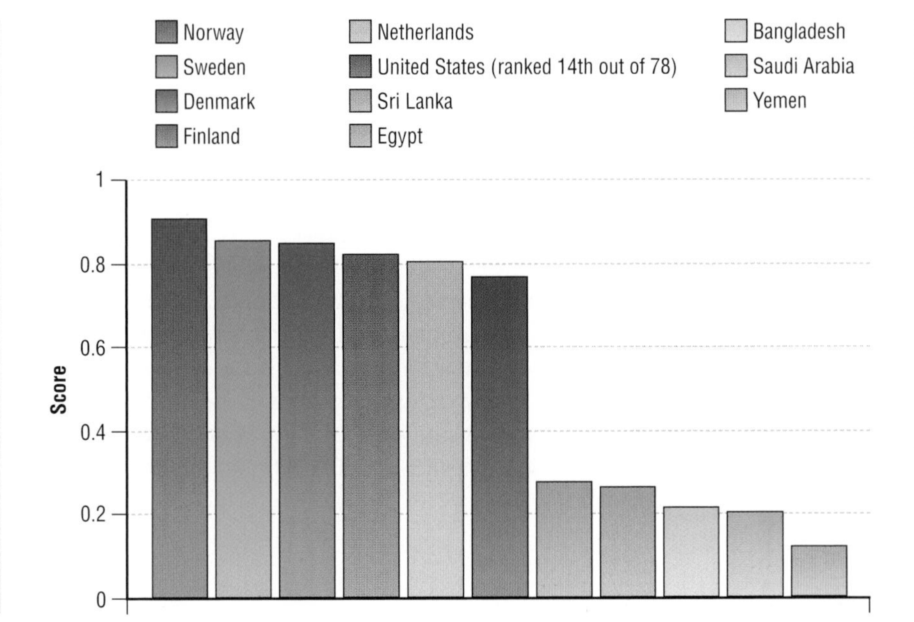

Figure 11.6
Countries with Highest and Lowest Scores on Gender Empowerment Measure (GEM), 2004

Source: United Nations (2004).

development. These anomalies exist because the former communist countries made gender equality a matter of public policy, whereas many Islamic countries do just the opposite. To cite an extreme case, in 1996 authorities in Afghanistan made it illegal for girls to attend school and women to work in the paid labor force.

The GEM figures suggest that American women still have a long way to go before they achieve equality with men. We have seen, for example, that the gender gap in earnings is shrinking but will not disappear until 2050—and then only if it continues to diminish at the 1960–2000 rate. That is a big if, because progress is never automatic.

In 1963, Congress passed the Equal Pay Act. It requires equal pay for the same work. Soon after, Congress passed Title VII of the Civil Rights Act. It prohibits employers from discriminating against women. These laws were important first steps in diminishing the gender gap in earnings. Since the mid-1960s, people in favor of closing the gender gap have recognized that we need additional laws and social programs to create gender equality.

Socializing children at home and in school to understand that women and men are equally adept at all jobs is important in motivating girls to excel in nontraditional fields. **Affirmative action,** which involves hiring more qualified women to diversify organizations, is important in helping to compensate for past discrimination in hiring.[5] However, without in any way belittling the need for such initiatives, we should recognize that their impact will be muted if women continue to shoulder disproportionate domestic responsibilities and if occupations containing a high concentration of women continue to be undervalued in monetary terms.

Two main policy initiatives will probably be required in coming decades to bridge the gender gap in earnings. One is the development of a better child-care system. The other is the development of a policy of comparable worth. Let us consider both these issues in turn.

Child Care

High-quality, government-subsidized, affordable child care is widely available in most western European countries, but not in the United States (see Chapter 15, "Families"). As a result, many American women with small children are either unable to work outside the home or are able to do so on only a part-time basis.

For women who have small children and work outside the home, child-care options and the quality of child care vary by social class (Annie E. Casey Foundation, 1998; Gormley, 1995; Murdoch, 1995). Affluent Americans can afford to hire nannies and to send their young children to expensive day-care facilities that enjoy stable, relatively well-paid, and well-trained staffs and a high ratio of caregivers to children. These features yield high-quality child care. In contrast, the day-care centers, nursery schools, and preschools to which middle-class Americans typically send their children have higher staff turnover, relatively poorly paid, poorly trained staff, and a lower ratio of caregivers to children. Fewer than a third of American children in child care attend such facilities, however. More than two-thirds—mainly from lower-middle-class and poor families—use family child-care homes or rely on the generosity of extended-family members or neighbors. Overall, the quality of child care is lowest in family child-care homes.

A third of all day-care facilities in the United States do not meet children's basic health and safety needs. This is true of about 12 percent of day-care centers, 13 percent of family child-care providers regulated by government, and 50 percent of family child-care providers unregulated by government (calculated from Annie E. Casey Foundation, 1998; Gormley, 1995; Murdoch, 1995). In 1997 veterinary assistants earned a median wage of \$7.34 an hour, parking lot attendants \$6.38, and child-care workers \$6.12. One interpretation of these figures is that our society considers tending pets and cars more important than looking after young children.

Affirmative action involves hiring a woman if equally qualified men and women are available for a job, thus compensating for past discrimination.

[5]Affirmative action has also been applied to other groups that experience high levels of discrimination, such as African, Native, and Hispanic Americans.

The welfare/day-care initiative taken by 29 states in the late 1990s adds weight to this interpretation. Governments throughout the country recognized the crisis in child care. They also wanted to get people off welfare and into the workforce. So, in 1997, they began recruiting thousands of welfare mothers to start at-home, for-profit day-care facilities, thus hoping to kill two birds—welfare and child care—with one stone. However, these welfare mothers were given few resources and little training. Most of them had little formal education and lived in substandard housing. Many of them could have undoubtedly made good child-care providers. That, however, would have required years of training, substantial subsidies, and ongoing support from outside sources (Dickerson, 1998).

Many companies, schools, and religious organizations in the United States provide high-quality day care. However, until the situation previously described changes, women, particularly those in the middle and lower classes, will continue to suffer economically from the lack of accessible, affordable day care.

Comparable Worth

In the 1980s, researchers found that women earn less than men partly because jobs in which women are concentrated are valued less than jobs in which men are concentrated. They therefore tried to establish gender-neutral standards by which they could judge the dollar value of work. These standards include such factors as the education and experience required to do a particular job and the level of responsibility, amount of stress, and working conditions associated with it. Researchers felt that by using these criteria to compare jobs in which women and men were concentrated, they could identify pay inequities. The underpaid could then be compensated accordingly. In other words, women and men would receive equal pay for jobs of **comparable worth,** even if they did different jobs.

A number of states have adopted laws requiring equal pay for work of comparable worth. Minnesota leads the country in this regard. However, the laws do not apply to most employers ("Comparable Worth," 1990). Moreover, some comparable-worth assessments have been challenged in the courts. The courts have been reluctant to agree that the devaluation of jobs in which women are concentrated is a form of discrimination (England, 1992: 250). Only broad, new federal legislation is likely to change this state of affairs. However, no federal legislation on comparable worth is on the drawing boards. Most business leaders seem opposed to such laws because their implementation would cost many billions of dollars.

The Women's Movement

Improvements in the social standing of women do not depend just on the sympathy of government and business leaders. Progress on this front has always depended in part on the strength of the organized women's movement. This is likely to be true in the future too. In concluding this chapter, it is therefore fitting to consider the state of the women's movement and its prospects.

The first wave of the women's movement emerged in the 1840s. Drawing a parallel between the oppression of black slaves and the oppression of women, first-wave feminists made a number of demands, chief among them the right to vote. They finally achieved that goal in 1920, as the result of much demonstrating, lobbying, organizing, and persistent educational work.

In the mid-1960s, the second wave of the women's movement started to grow. Second-wave feminists were inspired in part by the successes of the Civil Rights movement. They felt that women's concerns were largely ignored in American society despite persistent and pervasive gender inequality. Like their counterparts more than a century earlier, they held demonstrations, lobbied politicians, and formed women's organizations to further their cause. They advocated equal rights with men in education and employment, the elimination of sexual violence, and women's control over reproduction. One

Comparable worth refers to the equal dollar value of different jobs. It is established in gender-neutral terms by comparing jobs in terms of the education and experience needed to do them and the stress, responsibility, and working conditions associated with them.

The first wave of the women's movement emerged in the 1840s. The movement achieved its main goal—the right to vote for women—in 1920 as a result of much demonstrating, lobbying, organizing, and persistent educational work.

SuperStock

focus of their activities was mobilizing support for the Equal Rights Amendment (ERA) to the Constitution. The ERA stipulates equal rights for men and women under the law. The ERA was approved by the House of Representatives in 1971 and the Senate in 1972. However, it fell 3 states short of the 38 needed for ratification in 1982. Since then, no further attempt has been made to ratify the ERA.

Beyond the basic points of agreement noted earlier, there is considerable intellectual diversity in the modern feminist movement concerning ultimate goals. Three main streams may be distinguished (Tong, 1989):

1. *Liberal feminism* is the most popular current in the women's movement today. Its advocates believe that the main sources of women's subordination are learned gender roles and the denial of opportunities to women. Liberal feminists advocate nonsexist methods of socialization and education, more sharing of domestic tasks between women and men, and extending to women all of the educational, employment, and political rights and privileges that men enjoy.
2. *Socialist feminists* regard women's relationship to the economy as the main source of women's disadvantages. They believe that the traditional nuclear family emerged along with inequalities of wealth. In their opinion, once men possessed wealth, they wanted to ensure that their property would be transmitted to their children, particularly their sons. They accomplished this in two ways. First, men exercised complete economic control over their property, thus ensuring that it would not be squandered and would remain theirs and theirs alone. Second, they enforced female monogamy, thus ensuring that their property would be transmitted only to *their* offspring. Thus, according to socialist feminists, the economic and sexual oppression of women has its roots in capitalism. Socialist feminists also assert that the reforms proposed by liberal feminists are inadequate. That is because they can do little to help working-class women, who are too poor to take advantage of equal educational and work opportunities. Socialist feminists conclude that only the elimination of private property and the creation of economic equality can bring about an end to the oppression of all women.
3. *Radical feminists,* in turn, find the reforms proposed by liberals and the revolution proposed by socialists inadequate. Patriarchy—male domination and norms justifying that domination—is more deeply rooted than capitalism, say the radical femi-

Joseph Sohm, ChromoSohm, Inc./Corbis

The second wave of the women's movement started to grow in the mid-1960s. Members of the movement advocated equal rights with men in education and employment, the elimination of sexual violence, and women's control over reproduction.

nists. After all, patriarchy predates capitalism. Moreover, it is just as evident in self-proclaimed communist societies as it is in capitalist societies. Radical feminists conclude that the very idea of gender must be changed to bring an end to male domination. Some radical feminists argued that new reproductive technologies, such as *in vitro* fertilization, are bound to be helpful in this regard because they can break the link between women's bodies and childbearing (see Chapter 15, "Families"). But the revolution envisaged by radical feminists goes beyond the realm of reproduction to include all aspects of male sexual dominance. From their point of view, pornography, sexual harassment, restrictive contraception, rape, incest, sterilization, and physical assault must be eliminated for women to reconstruct their sexuality on their own terms.

This thumbnail sketch by no means exhausts the variety of streams of contemporary feminist thought. For example, since the mid-1980s, *antiracist* and *postmodernist* feminists have criticized liberal, socialist, and radical feminists for generalizing from the experience of white women and failing to see how women's lives are rooted in particular historical and racial experiences (hooks, 1984). These new currents have done much to extend the relevance of feminism to previously marginalized groups.

Partly due to the political and intellectual vigor of the women's movement, some feminist ideas have gained widespread acceptance in American society over the past three decades (▶Table 11.4). For example, the General Social Survey showed that 82 percent of Americans approved of married women working in the paid labor force in 1998. In 2002, 78 percent thought women were as well suited to politics as men. In 1996, 64 percent of Americans regarded women's rights issues as important or very important. Support for these ideas has grown in recent decades. Still, many people, especially men, oppose the women's movement. In fact, in recent years several antifeminist men's groups have sprung up to defend traditional male privileges.[6] It is apparently difficult for some men to accept feminism because they feel that the social changes advocated by feminists threaten their traditional way of life and perhaps even their sexual identity.

[6]Profeminist men's groups, such as the National Organization for Men Against Sexism, also exist but seem to have a smaller membership (National Organization, 2000).

Table 11.4
Attitudes to Women's Issues, United States, 1972–2002 (in percent)

	1972–1982	1983–1987	1996	1998	2002
Approve of married women working in paid labor force	70	80	83	82	—
Women suited for politics	54	63	78	77	78
Women's rights issue important/one of the most important	—	58	64	—	—
Favor preferential hiring of women	—	—	27	—	—
Think of himself/herself as a feminist	—	—	22	—	—
Women can best improve their position through women's rights groups	—	15	—	—	—

Source: National Opinion Research Center (2004).

Personal Anecdote

Our own experience suggests that traditional patterns of gender socialization weigh heavily on many men. For example, John Lie grew up in a patriarchal household. His father worked outside the home, and his mother stayed home to do nearly all the housework and child care. "I remember my grandfather telling me that a man should never be seen in the kitchen," recalls John, "and it is a lesson I learned well. In fact, everything about my upbringing—the division of labor in my family, the games I played, the TV programs I watched—prepared me for the life of a patriarch. I vaguely remember seeing members of the 'women's liberation movement' staging demonstrations on the TV news in the early 1970s. Although I was only about 11 or 12 years old, I recall dismissing them as slightly crazed, bra-burning man haters. Because of the way I grew up and what I read, heard, and saw, I assumed that the existing gender division of labor was natural. Doctors, pilots, and professors should be men, I thought, and people in the 'caring' professions, such as nurses and teachers, should be women.

"But socialization is not destiny," John insists. "Entirely by chance, when I got to college I took some courses taught by female professors. It is embarrassing to say so now, but I was surprised that they were so much brighter, more animated, and more enlightening than my male high school teachers had been. In fact, I soon realized that many of my best professors were women. I think this is one reason why I decided to take the first general course in women's studies offered at my university. It was an eye opener. I soon became convinced that gender inequalities are about as natural and inevitable as racial inequalities. I also came to believe that gender equality could be as enriching for men as for women. Sociological reflection overturned what my socialization had taught me. Sociology promised—and delivered. I think many college-educated men have similar experiences today, and I hope I now contribute to their enlightenment."

Summary

Sociology Now™

Reviewing is as easy as ❶, ❷, ❸.

❶ Before you do your final review, take the SociologyNow diagnostic quiz to help you identify the areas on which you should concentrate. You will find information on how to access SociologyNow on the foldout at the front of the textbook.
❷ As you review, take advantage of SociologyNow's study aids to help you master the topics in this chapter.
❸ When you are finished with your review, take SociologyNow's post-test to confirm you are ready to move on to the next chapter.

1. Are sex and gender rooted in nature?

   While sex refers to certain anatomical and hormonal features of a person, gender refers to the culturally appropriate expression of masculinity and femininity. Sex is largely rooted in nature, although people can change their sex by undergoing a sex-change operation and hormone therapy. In contrast, social as well as biological forces strongly influence gender. Sociologists study the way social conditions affect the expression of masculinity and femininity.

2. What are some of the major social forces that channel people into performing culturally appropriate gender roles?

   Various agents of socialization channel people into performing culturally approved gender roles. The family, the school, and the mass media are among the most important of these agents of socialization. Once the sex of children is known (or assumed), parents and teachers tend to treat boys and girls differently in terms of the kind of play, dress, and learning they encourage. The mass media reinforce the learning of masculine and feminine roles by making different characteristics seem desirable in boys and girls, men and women.

3. Aside from agents of socialization, are there other social forces that influence the expression of masculinity and femininity?

   Yes. One of the most important nonsocialization forces that influences the expression of masculinity and femininity is the level of social inequality between men and women. High levels of gender inequality encourage more traditional or conventional gender roles. There are fewer differences in gender roles where low levels of gender inequality prevail. Historically, high levels of gender inequality have been encouraged by far-ranging warfare and conquest, plow agriculture, and the separation of public and private spheres. Each of these changes enhanced male power and added a layer of what we now consider tradition to gender roles. Today, we can see the influence of gender inequality on gender roles by examining male aggression against women, which tends to be high where men are much more socially powerful than women and low where there is greater gender equality.

4. What is homosexuality and why does it exist?

   Homosexuals are people who prefer sexual partners of the same sex. We do not yet well understand the causes of homosexuality—whether it is genetic, hormonal, psychological, or some combination of the three. We do know that homosexuality does not appear to be a choice and that it emerges for most people in early adolescence without prior sexual experience. Sociologists are in any case more interested in the way homosexuality is expressed and repressed. For example, they have studied how, in the 20th century, scientific research and political movements have made the open expression of homosexuality more acceptable. Sociologists have also studied the ways in which various aspects of society reinforce heterosexuality and treat homosexuality as a form of deviance subject to tight social control.

5. How does the existence of sharply defined gender roles influence men's and women's income?

   One important consequence of strict gender differentiation is the existence of a big earnings gap between women and men. The gender gap in earnings derives from outright discrimination against women, women's disproportionate domestic responsibilities, women's concentration in low-wage occupations and industries, and the undervaluation of work typically done by women.

6. How might the gender gap in earnings be reduced or eliminated?

   Among the major reforms that could help eliminate the gender gap in earnings and reduce the overall level and expression of gender inequality are (a) the development of an affordable, accessible system of high-quality day care and (b) the remuneration of men and women on the basis of their work's actual worth.

Questions to Consider

1. By interviewing your family members and using your own memory, compare the gender division of labor in (a) the households in which your parents grew up and (b) the households in which you grew up. Then, imagine the gender division of labor you would like to see in the household you hope to live in 10 years from now. What accounts for change over time in the gender division of labor in these households? Do you think your hopes are realistic? Why or why not?
2. In your own case, rank the relative importance of your family, your schools, and the mass media in your gender socialization. What criteria do you use to judge the importance of each socialization agent?
3. Systematically note the roles played by women and men on TV programs and ads one evening. Is there a gender division of labor on TV? If so, describe it.
4. Are you a feminist? If so, which of the types of feminism discussed in this chapter do you find most appealing? Why? If not, what do you find objectionable

about feminism? In either case, what is the ideal form of gender relations in your opinion? Why do you think this form is ideal?

Web Resources
Companion Website for This Book

http://sociology.wadsworth.com

Begin by clicking on the Student Resources section of the website. Choose "Introduction to Sociology" and then the Brym and Lie book cover. Next, select the chapter you are currently studying from the pull-down menu. From the Student Resources page you will have easy access to InfoTrac® College Edition, MicroCase Online exercises, additional web links, and many other resources to aid you in your study of sociology, including practice tests for each chapter.

Recommended Websites

For a useful list of resources on gender and sexuality on the World Wide Web, go to http://www.georgetown.edu/crossroads/asw/gender.html.

The National Committee on Pay Equity is a coalition of more than 180 organizations working to eliminate sex- and race-based wage discrimination and to achieve pay equity. Visit their website at http://www.feminist.com/fairpay.

On the United Nations GEM, discussed in the text, see http://www.undp.org/hdro/98gem.htm.

In 1997, a conference was held in San Diego to analyze what the participants called the "National Sex Panic." For the proceedings of the conference on RealAudio, go to http://www.managingdesire.org/sexpanic/sexpanicindex.html.

CHAPTER 12 Sociology of the Body: Disability, Aging, and Death

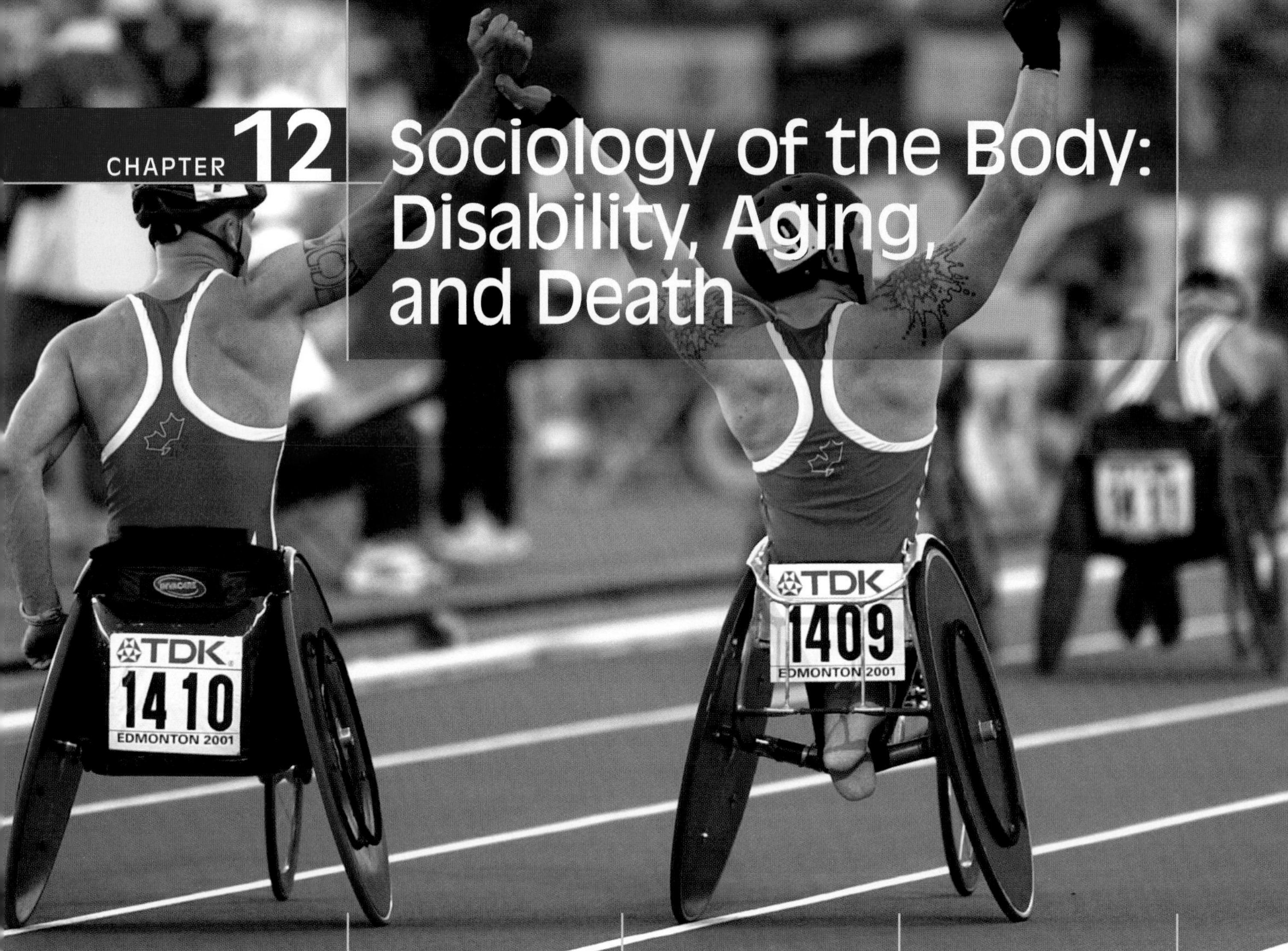

Tim Waele/Isosport/Corbis

In this chapter, you will learn that:

- Seemingly "natural" features of the human body, such as height and weight, have social causes and consequences of far-reaching importance.
- Enhancing one's body image to conform to prevailing norms became especially important in urban, industrial societies.
- People have defined and dealt with disability in different ways in different times and places.
- Age is an important basis of social stratification. However, the correlation between age and the command of resources is far from perfect, and political conflicts shape the degree to which any given age cohort exercises resource control.
- Although prejudice and discrimination against older people are common in the United States, the elderly have wielded increasing political power in recent decades.
- The way people die reflects the nature of the society in which they live. This is evident in our attitudes toward death, euthanasia, and funerals.

Bob Dole's Body

Two categories of people can identify Bob Dole. First are those who remember him chiefly as an influential U.S. senator from Kansas, vice presidential candidate in 1976, and presidential candidate in the race against Bill Clinton in 1996. Second are those who remember him mainly as a figure in popular TV ads for Viagra and Pepsi between 1999 and 2001. Age divides these two categories. People who think of Bob Dole mainly as a political figure tend to be 30 years old or more. People who think of Bob Dole chiefly as an advertising icon tend to be under 30.

For the two categories of people, Bob Dole's body has different meanings. If you are 30 or older, you probably know that Dole has a disabled right arm, the result of a World War II combat wound. Dole was a leading advocate for the rights of disabled people. Not surprisingly, he became a sort of poster boy for disabled Americans.

If you are younger than 30, you probably remember Dole as a poster boy of a different sort. In a 1999 ad, Dole said Viagra helped him overcome erectile dysfunction. He promised it could do the same for all men suffering that disability. Having thus established himself in the public consciousness as a man of considerable vitality despite his 75 years, Dole was next used to promote the Pepsi generation. In the Pepsi commercial, he sits in an armchair in a darkened room, his dog by his side, watching Britney Spears dance on TV. He smiles, entranced by Spears's gyrations. The dog watches too. He gets excited and barks whereupon Dole, still fixated on Spears, mutters, "Easy, boy," leaving it unclear whether he is talking to the dog or his Viagra-enhanced self. The ad ends with the slogan, "The Joy of Pepsi."

For everyone, regardless of age, Bob Dole is a hero who has overcome a physical disability. However, people of different ages focus on different body parts when they think about his body. He is therefore a different kind of hero to different age cohorts. His body means different things to different people.

Sociology Now™

Reviewing is as easy as ❶, ❷, ❸.

Use SociologyNow to help you make the grade on your next exam. When you are finished reading this chapter, go to the Chapter Summary for instructions on how to make SociologyNow work for you.

Matthew Mendelsohn/Corbis

Bob Dole's body has come to stand for distinct disabilities. Long known for his impaired right arm, he has also come to be associated with erectile dysfunction.

The example of Bob Dole may be amusing but it is not trivial, for in it lies embedded a sociological principle that is the main lesson of this chapter. The human body is not just a wonder of biology. It is also a sociological wonder and a subject of growing interest in the discipline (Turner, 1996 [1984]). Its parts, its disabilities, its aging, and its death mean different things and have different consequences for different cultures, historical periods, and categories of people. For example, a person's height, weight, and attractiveness may seem to be facts of nature. Closer inspection reveals, however, that the standards by which we define a "normal" or "desirable" body vary historically. Moreover, a person's height, weight, and perceived attractiveness influence his or her annual income, health, likelihood of getting married, and much else. We explore the relationship between body characteristics and social status in the next section.

The relationship between the body and society is especially clear in the case of disability. The very definition of what constitutes a disability has varied over time and place. So have strategies for dealing with disabilities. In the following pages, we illustrate these variations. Among other things, we show that the dominant treatment tendency since the 19th century has involved the rehabilitation and integration of the disabled into "normal" society. Then, in the early 20th century, some governments tried to eliminate disabled people altogether. Finally, in the late 20th century, disabled people began to assert their dignity and normality. One consequence of this new attitude is a vigorous move toward self-help and the establishment of independent communities of the disabled.

In this chapter's final section, we turn to the problems of aging and death. We show that aging is not just a natural process of growth and decline. Age is one basis of social inequality, so in a certain sense it is a trip through the stratification system. Moreover, it is a trip fraught with dangers. Elderly people face prejudice and discrimination. The systems for providing personal care and adequate pensions are in trouble. Poverty is a looming possibility for some and a bitter reality for others. We survey each of these issues and conclude by training our sociological eye on the ultimate social problem: death and dying.

Society and the Human Body

The Body and Social Status

Height

In an experiment, four people of the same height and roughly similar appearance were introduced to a group of students. The first person was introduced as a fellow undergraduate, the second as a graduate student, the third as an assistant professor, and the fourth as a professor. Members of the group were asked to rank the four people in terms of their height. Despite the fact that all four were of equal stature, the students estimated that the professor was the tallest, the assistant professor next tallest, then the graduate student, and finally the undergraduate. Apparently believing that physical stature reflects social stature, the students correlated social status with height ("Short Guys Finish Last," 1995–1996).

Is this perception accurate? Do tall people really tend to enjoy high social status? And why are some people tall in the first place? We can begin to answer these questions by first acknowledging that genes are an important determinant of any particular individual's height. However, different *populations* are approximately the same genetically. A complex series of *social* factors determines the average height of any population, whether the population consists of members of a country, a class, a racial or ethnic group, and so forth. Moreover, a complex series of social consequences flow from differences in height.

Figure 12.1 shows some of the main social causes and consequences of height. For purposes of illustration, consider the impact of family income on stature. Average family

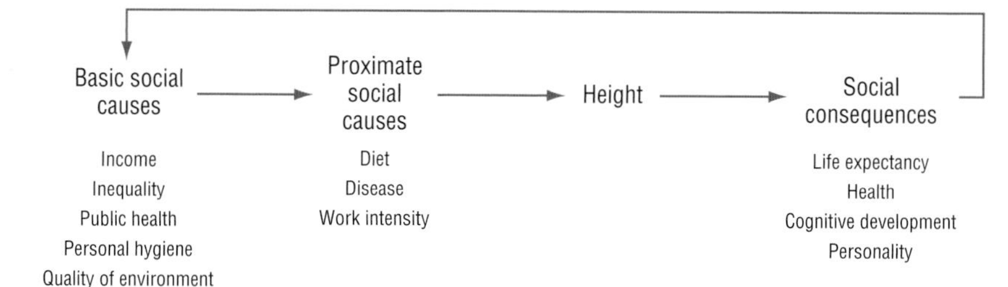

Figure 12.1
Selected Social Causes and Consequences of Height in Human Populations

Source: Adapted from Steckel (1995: 1908)

income is the single most important determinant of the quality of one's diet (especially protein consumption). Higher family income translates into a higher-quality diet. In turn, the quality of one's diet during childhood strongly influences one's stature. Thus, Americans born in 1975 are on average three centimeters taller than Americans who were born in 1880 (Steckel, 1995: 1920). More dramatically, Japanese men were on average five centimeters taller at the end of the 20th century than they were at midcentury (French, 2001). Norwegian men were on average nearly eight centimeters taller in 1984 than in 1761 (Floud, Wachter, and Gregory, 1990). Whether we look at Mexican, Japanese, or Swedish immigrants to the United States, American-born children are taller than their parents born abroad (Roberts, 1995). In all these cases, the main cause of growing stature is the same: A higher standard of living led to an improved diet, which allowed the human body to come closer to realizing its full growth potential. The correlation between per capita family income and average height across many countries is very strong ($r = 0.82$ or higher) (Steckel, 1995: 1912).

As one might expect, within countries there is also a correlation between stature and class position. Class differences in height are smaller than they were centuries ago. Even today, however, members of upper classes are on average taller than members of middle classes, who are in turn taller than members of working classes. The Scandinavian countries are the exceptions that prove the rule. There are no differences in stature between classes in Scandinavia. That is because class inequality is less pronounced in Sweden, Norway, and the rest of Scandinavia than anywhere else in the world (Floud, Wachter, and Gregory, 1990; Kingston, 1997).

The *consequences* of stature are important too. Scrutiny of many sources, ranging from U.S. Army records since the Civil War to all Norwegian x-ray records from the 1950s, reveals that, on average, tall people live longer than others. Tall people also earn more than others and tend to reach the top of their profession more quickly (Kingston, 1997; McCulloch, 2003). Significantly, in only 5 of the 22 United States presidential elections since 1900 did the shorter candidate win ("Short Guys Finish Last," 1995–1996). At least part of the reason short people tend to be less successful in some ways than tall people is that they experience subtle discrimination based on height (Box 12.1).

Weight

What is true for height is also true for body weight. Body weight influences status because of the cultural expectations we associate with it. Thus, one study of more than 10,000 young American adults found that overweight women tend to complete four fewer months of school than women who are not overweight. They are also 20 percent less likely to be married. An overweight woman's household is likely to earn nearly $7000 less per year than the household of a woman who is not overweight. Overweight women are 10 percent more likely to live in poverty. The consequences of being overweight are less serious for men. Still, overweight men are 11 percent less likely to be married than men who are not overweight (Gortmaker et al., 1993; see also Averett and Korenman, 1996).

Interestingly, the negative effects of being overweight are evident even for women matched in terms of their social and economic backgrounds. This suggests the need to revise the simple, conventional view that poverty encourages obesity. It is certainly true that poor women have fewer opportunities and resources that would allow them to eat health-

Box 12.1

YOU AND THE SOCIAL WORLD

Height Discrimination

The idea that short people experience discrimination based on height may seem far-fetched. However, your own attitudes and experience may help drive the point home. Answer the following questions in about 500 words:

- Is it important that you choose a spouse who is taller than you? How do you think your response is related to your sex?
- Why do you think the overwhelming majority of people believe that boyfriends should be taller than their girlfriends and husbands should be taller than their wives?
- Observe the leaders of your sports teams, friendship circles, college tutorials, and other groups to which you belong. Record the approximate height of the leaders in each group and the approximate height of nonleaders. Is there a systematic difference in height between leaders and nonleaders? How do you account for any observed difference or lack of observed difference?
- Where do you place in the status hierarchy based on height? Have you ever felt advantaged or disadvantaged because of your height?

Bridgeman Art Library, London/SuperStock

Venus and Adonis, Rubens, 1615. Venus, the Greek goddess of beauty, as depicted by Peter Paul Rubens four centuries ago. Would Venus need a tummy tuck and a membership in a diet club to be considered beautiful today?

ier diets, get more exercise, and bring down their weight. For example, if you live in a poor, high-crime neighborhood, it is dangerous to go out for a speed walk, and you may not be able to afford anything more nutritious than high-calorie fast food when you go out for a meal. However, the reverse is also true: Obesity in and of itself tends to make women poorer. This conclusion seems reasonable because, for women matched in terms of their social and economic backgrounds, being overweight still has negative effects on income. There is apparently a "reciprocal relationship" between obesity and social class, with each variable affecting the other (Gortmaker et al., 1993).[1]

In preindustrial societies, people generally favored well-rounded physiques because they signified wealth and prestige. Not surprisingly, beautiful women as depicted by the great artists of the past tend to be on the heavy side from our contemporary perspective.

In contrast, in our society, being well rounded usually signifies undesirability. Being overweight in the contemporary United States has become a source of negative stereotyping and even outright discrimination. Many of us think of overweight people as less attractive, industrious, and disciplined than thin people.

The percentage of overweight people is big and growing quickly in the United States. In the period 1998–2000, 26 percent of the American population was obese, compared with just 3 percent in Japan. The American figure reached 31 percent in 2000, a 35 percent increase since 1994. Americans are the most overweight people in the world (Critser, 2003; Tarmann, 2003) (Figure 12.2). This means that a large and expanding percentage of the population lives with a stigma that affects their life chances, due partly to the availability of inexpensive high-calorie fast food anywhere/anytime and the food companies' annual advertising budget of $33 billion in 2000 (Buckley, 2003; Nestle, 2002; Goffman, 1963).

[1]Research on more than 13,000 American adolescents in 2003 showed that obesity is also associated with race. Being overweight is less common as you move up the class structure for girls of all races, but the decline is less pronounced for black girls than it is for white, Hispanic, and Asian girls. The pattern was less extreme among boys, with more obesity among Hispanic and black boys than white and Asian boys even at high household income and education levels (Gorden-Larsen, 2003).

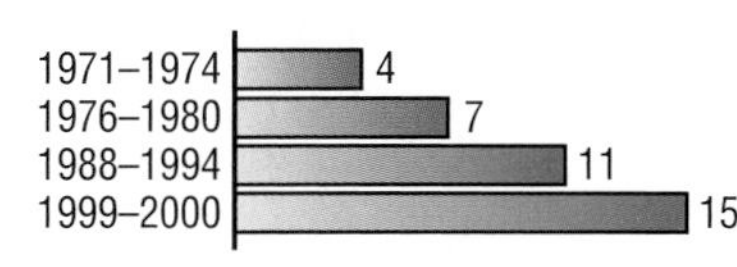

Percent of American children between the ages of 6 and 11 who were obese, selected years

Country	Percent
Japan	3
Italy	9
France	10
Canada	15
Australia	21
United Kingdom	21
United States	26

Percent of adults who were obese, selected countries, 1998–2000

Figure 12.2
Obesity among Children and Adults

Obese children have a body mass index (BMI) at or above the 95th percentile of the sex-specific BMI growth charts. Obese adults have a BMI of 30 or above. (BMI = weight in kilograms divided by the square of height in meters.)

Source: Allison Tarman, "Obesity in the U.S.: Reaching a Critical Mass," in Mary M. Kent and Mark Mather, "What Derives U.S. Population Growth," *Population Bulletin 57*, No. 4, December 2002. Used with permission.

Sociology of the Body

The Body and Society

Our discussion of height and weight should make it clear that our bodies are not just biologically but also socially defined. Let us now develop this point by considering how social forces influence the way we manipulate our body image. We then turn to an analysis of disability—how we define it and how disabled and nondisabled people deal with it.

In the United States and other highly developed countries, people tend to think they have rights over their own bodies. Typically, the feminist movement asserts the right of every woman to control her own reproductive functions through birth control, for example (Gordon, 1990 [1976]). Yet people have not always endorsed this view (and some Americans still contest it). For instance, slaves' bodies were the property of their owners. In most slave societies, masters could use their slaves' bodies for anything they wanted, including hard labor and sex. Similarly, in deeply patriarchal societies, a husband effectively owned his wife's body. Because his wife's body was at his disposal, a husband could rape his wife with impunity.

Despite the widespread view that people have rights over their own bodies, most people do not treat their bodies in wildly idiosyncratic ways. Instead, norms of body practice influence us. Catholic priests are defined partly by sexual abstinence. Male Jews and Muslims are typically defined partly by circumcision. Many of the most important social distinctions—gender, race, age, tribe, and so forth—are "written" on the body by different styles of dress, jewelry, tattoos, cosmetics, and so forth. People have always attempted to affect their body shape and appearance, but they do so according to principles laid out by society.

For social, economic, and technological reasons, enhancing one's body image to conform to prevailing norms became especially important in urban, industrial societies. Specifically:

- *Socially,* urbanized societies present people with many more opportunities to meet and interact with strangers. This increases the need for the kinds of status cues and impression management techniques that can make social interaction easier (see Chapter 5, "Social Interaction"). Manipulating body image helps grease the wheels of social interaction in complex societies by making it clear to strangers exactly who you are.
- *Economically,* industrialized societies enable people to afford body enhancement. For example, peasants in preindustrial societies had no access to relatively inexpensive, mass-produced clothing, something we take for granted.
- *Technologically,* we have created many new techniques for transfiguring the body. Consider something as basic as your teeth and gums. Until recently, the daily practice of brushing, much less of flossing, was uncommon. Two hundred years ago, a 40-year-old might boast a nice set of wooden dentures; poor people were of-

Many of the most important social distinctions—gender, race, age, tribe, and so forth—are "written" on the body by different styles of dress, jewelry, tattoos, cosmetics, and so forth. People have always attempted to affect their body shape and appearance, but they do so according to principles laid out by society.

Sinibaldi/Corbis

Rob Lewine/Corbis

Until recently, the daily practice of brushing, much less of flossing, was uncommon. Two hundred years ago, a 40-year-old might boast a nice set of implanted animal teeth or wooden dentures; poor people were often toothless. Today, dental hygiene, dentistry, and orthodontics allow people to enjoy a set of white, straight teeth for a lifetime, with bridges, crowns, and implants indistinguishable from the real thing.

ten toothless and without false teeth. Today, dental hygiene, dentistry, and orthodontics allow many people to enjoy a set of straight, white teeth for a lifetime, with bridges, crowns, and implants indistinguishable from the real thing. The rise of plastic surgery is another popular technological innovation that allows us to transfigure our bodies (see Chapter 3, "Culture," and Haiken, 1997).

Thus, social, economic, and technological forces have transformed the way we manipulate our body image. As you will now see, they have also radically altered the way we deal with disability.

Disability

The Social Construction of Disability

Pity the poor lefty, for centuries considered inferior. About 400 years ago, the Catholic Church declared left-handed people servants of the Devil. It burned some of them at the stake. Then it forced lefties to become right-handed in school. In Japan as recently as the early 20th century, left-handedness in a wife was grounds for divorce. Natives in Papua New Guinea don't let their left thumbs touch their beer mugs because they believe that would poison the beer. Maori women in New Zealand weave ceremonial cloth with the right hand because they believe that using the left desecrates the cloth. Some African tribes along the Niger River do not allow women to prepare food with the left hand for fear of being poisoned. Almost universally, people have considered left-handedness a handicap, so much so that the sentiment has been embedded in many languages. In Russian, to do something *na levo* means to do it under the table or illegally, but literally it means "on the left." In English, the word "left" derives from an Old English word that means "weak" or "worthless." Also in English, "gauche" means "ill-mannered"—but the French from which it is derived means "left." (In contrast, "adroit" means "proper" in English—but the French from which it is derived means "to the right.") In Latin, "right" is *dexter* (as in the English "dexterous," a desirable attribute), whereas "left" is *sinister*, which of course means "evil" in English. *Linkisch* is German for "leftish"—and "awkward."

To us, negative attitudes toward left-handedness seem like so much nonsense. We don't think of left-handed people—roughly 10 percent of the population—as **impaired** or deficient in physical or mental capacity. Nor do we think of them as **disabled** or incapable of performing within the range of "normal" human activity. The fact that so many people once thought otherwise suggests that definitions of disability are not based on self-evident biological realities. Instead, they vary socially and historically. Note also that some people, but not others, consider a 4-foot-tall person disabled, and that most people must be convinced by advertising that erectile dysfunction in a 75-year-old man is a disability. These examples suggest not only that definitions of disability differ across societies and historical periods but that in any one time and place people may disagree over these definitions.

- **Impaired** people are considered deficient in physical or mental capacity.
- **Disabled** people are incapable of performing within the range of "normal" human activity.
- **Rehabilitation** involves curing disabilities to the extent possible through medical and technological intervention; trying to improve the lives of the disabled by means of care, training, and education; and integrating the disabled into "normal" society.

Rehabilitation and Elimination

Modern Western approaches to disability emerged in the 19th century. All scientists and reformers of the time viewed disability as a self-evident biological reality. Some scientists and reformers sought the **rehabilitation** of the disabled. Rehabilitation involves curing disabilities to the extent possible through medical and technological intervention. It also involves trying to improve the lives of the disabled by means of care, training, and education. Finally, it involves integrating the disabled into "normal" society (Stiker, 1999 [1982]; Terry and Urla, 1995). Their desire for rehabilitation motivated the establishment of schools for the blind, the widespread use of prosthetics, the construction of wheelchair-accessible buildings, and so forth. It also prompted the passage of laws that benefit the disabled. In the United States, the Architectural Barriers Act (1968) ensures that all federally funded buildings are accessible to disabled people. The Americans with Disabilities Act (1990) prohibits discrimination against people with disabilities and ensures equal access to employment, state and local government services, public accommodations, commercial facilities, and transportation. These laws have done much to help integrate disabled people into "normal" society.

Other scientists and reformers took a different tack. They sought to eliminate disability altogether by killing the disabled or sterilizing them and preventing them from having offspring. The Nazis adopted this approach in Germany beginning in 1933. They engineered the sterilization and killing of the mentally "deficient" and the physically "deviant," including the blind and the deaf (Proctor, 1988).

One of the ugliest chapters in our history involves the federally funded, forced sterilization of native North American women from the 1920s to the 1970s. The "disability" these women were alleged to suffer from was that they were native North Americans who were deemed by physicians to be having too many babies. Tubal ligations and hysterectomies were performed as birth control on many thousands of native North Americans, some of them minors, without their informed consent. In two cases, doctors told 15-year-old girls that they were having their tonsils out and then proceeded to remove their ovaries. It was only in 1975 that Congress passed laws prohibiting the use of federal funds to force women to undergo abortion or sterilization. By then, however, tremendous damage had been inflicted on the Native American population. By 1982, when 15 percent of white American women of childbearing age had been sterilized, the figure for Native American women was about 40 percent (DeFine, 1997; England, n.d.; Johansen, 1998).[2]

Ablism

Perhaps a tenth of the world's people identify themselves as disabled or are characterized as such by others (Priestly, 2001). Because the human environment is structured largely around the norms of the able-bodied, disabled people face many disadvantages. Their deprivations are still greater if they are elderly, women, or members of a lower class or a disadvantaged racial or ethnic group (Barnes, Mercer, and Shakespeare, 1999).

Specifically, people routinely stigmatize the disabled, negatively evaluating them because of a visible characteristic that supposedly sets them apart from others. People also routinely employ stereotypes when dealing with the disabled, expecting them to behave according to a rigid and often inaccurate view of how everyone with their disability acts. The resulting prejudice and discrimination against the disabled is called **ablism.** A historical example of ablism is the widespread belief among 19th-century Western educators that blind people were incapable of high-level or abstract thought. Due to this prejudice, the blind were systematically discouraged from pursuing intellectually challenging tasks and occupations. Similarly, an 1858 article in the *American Annals of the Deaf and Dumb* held that "the deaf and dumb are guided almost wholly by instinct and their animal passions. They have no more opportunity of cultivating the intellect and reasoning facilities than the savages of Patagonia or the North American Indians" (quoted in Groce, 1985: 102). Racists think of racial minorities as naturally and incurably inferior. Ablists think of disabled people in the same way. As the preceding quotation suggests, racists and ablists are often the same people.

Ablism involves more than active prejudice and discrimination. It also involves the largely unintended *neglect* of the conditions of disabled people. This point should be clear to anyone who has to get around in a wheelchair. Many buildings were constructed without the intention of discriminating against people in wheelchairs, yet they are extremely inhospitable to them. Impairment becomes disability when the human environ-

Ablism is prejudice and discrimination against disabled people.

---

[2]The figure for African American women was 24 percent and for Puerto Rican American women 35 percent. These percentages suggest a strong class and racial bias in the sterilization of American women (DeFine, 1997).

Lightscapes Photography, Inc./Corbis

The social environment turns an impairment into a disability. Architecture and urban planning that neglect nonstandard modes of mobility make life difficult for people who depend on wheelchairs.

ment is constructed largely on the basis of ablism. Ablism exists through both intention and neglect.

The Normality of Disability

In 1927 science fiction writer H. G. Wells published a short story called "The Country of the Blind" (Wells, 1927). It provocatively reversed the old saying that "in the country of the blind, the one-eyed man is king." In the story, the protagonist, Nuñez, survives an avalanche high in the Andes. When he revives in a mountain valley, he discovers that he is on the outskirts of an isolated village whose members are all blind due to a disease that struck 14 generations earlier. For them, words like "see," "look," and "blind" have no meaning.

Because he can see, Nuñez feels vastly superior to the villagers; he thinks he is their "Heaven-sent King and master." Over time, however, he realizes that his sight places him at a disadvantage vis-à-vis the villagers. Their sense of hearing and touch are more highly developed than his, and they have designed their entire community for the benefit of people who cannot see. Nuñez stumbles where his hosts move gracefully, and he constantly rants about seeing—which only proves to his hosts that he is out of touch with reality. The head of the village concludes that Nuñez is "an idiot. He has delusions; he can't do anything right." In this way, Nuñez's vision becomes a disability. He visits a doctor who concludes there is only one thing to do. Nuñez must be cured of his ailment. As the doctor says:

> Those queer things that are called eyes . . . are diseased . . . in such a way as to affect his brain. They are greatly distended, he has eyelashes, and his eyelids move, and consequently his brain is in a state of constant irritation and distraction. . . . I think I may say with reasonable certainty that, in order to cure him complete, all that we need to do is a simple and easy surgical operation—namely, to remove these irritant bodies.

Thus, Wells suggests that in the country of the blind, the man who sees must lose his vision or be regarded as a raving idiot.

Wells's tale is noteworthy because it makes blindness seem utterly normal. Its depiction of the normality of blindness comes close to the way many disabled people today think of their disabilities—not as a form of deviance but as a different form of normality. As one blind woman wrote: "If I were to list adjectives to describe myself, blind would be only one of many, and not necessarily the first in significance. My blindness is as intrinsically a part of me as the shape of my hands or my predilection for salty snacks. . . . The most valuable insight I can offer is this: blindness is normal to me" (Kleege, 1999: 4) (Box 12.2).

The idea of the normality of disability has partly supplanted the rehabilitation ideal that, as we saw, originated in the 19th century. Able-bodied reformers led the rehabilitation movement. They represented and assisted disabled people. Disabled people themselves participated little in efforts to improve the conditions of their existence. This situation began to change in the 1960s. Inspired by other social movements of the era, notably the Civil Rights movement, disabled people began to organize themselves (Shapiro, 1993; Campbell and Oliver, 1996). The founding of the Disabled Peoples' International in 1981 and inclusion of the rights of the disabled in the United Nations Universal Declaration of Human Rights in 1985 signified the growth—and growing legitimacy—of the new movement not just in the United States but globally. Since the 1980s, disabled people have begun to assert their autonomy and the "dignity of difference" (Oliver, 1996; Charlton, 1998). Rather than requesting help from others, they insist on self-help. Rather than seeing disability as a personal tragedy, they see it as a social problem. Rather than regarding themselves as deviant, they think of themselves as inhabiting a different but quite normal world.

BOX 12.2
Sociology at the Movies

Shallow Hal (2001)

Hal is so shallow that when someone asks him if he would prefer a woman with one breast or half a brain, he replies, "How big is the breast?" His balding sidekick, Mauricio (who sprays on most of his hair), isn't any deeper. They are obnoxious men with below-average looks and sexist values, always on the make but with nothing to offer any woman with even half a brain. They are oblivious to their character flaws, and that is what makes them so funny.

Then one day, Hal (played by Jack Black) gets trapped in an elevator with self-help guru Tony Robbins and everything changes. After a brief conversation, Robbins hypnotizes Hal and tells him to see only the inner beauty of women. Suddenly, Hal's "luck" with women improves. He meets Rosemary, an ex–Peace Corps volunteer who now works in a children's hospital. When Hal looks at her, she is an intelligent and generous soul from a wealthy family and she just happens to look like Gwyneth Paltrow. When *we* look at her, she is an intelligent and generous soul from a wealthy family and she just happens to weigh 300 pounds. (Rosemary is played by Gwyneth Paltrow alternatively wearing and not wearing a "fat suit.")

All is well until Mauricio (*Seinfeld's* Jason Alexander) breaks the hypnotic spell. Hal then sees all of Rosemary's 300 pounds and must make the decision of his life. Should he continue with his shallow ways or go for inner beauty? He makes the right choice, and in the end we see Rosemary carrying Hal to his car as they leave for a Peace Corps assignment.

The Farrelly brothers, Peter and Robert, made this movie. They also made *Dumb and Dumber* (1994), a buddy movie about two guys with low IQs and *Me, Myself, and Irene* (2000), about a schizophrenic police officer. Are the lives of the Farrelly brothers just one big sick joke? Have they devoted themselves to poking fun at people with disabilities? Quite the contrary. What is striking about their movies, and particularly *Shallow Hal,* is that they make us see the *normality* of disability. *Shallow Hal* is densely populated by people with disabilities and oddities who turn out to be just like everyone else. There is the little girl in Rosemary's hospital with horrific facial scars from third-degree burns; the cross-dressing receptionist at the fancy restaurant where Hal first sees Rosemary's true girth; Rosemary's former boyfriend, who suffers from chronic and highly visible psoriasis; Hal's best friend, Mauricio, who has a vestigial tail at the base of his spine; and, most importantly, Walt (Rene Kirby). Walt has a curvature of the spine so severe he must walk on all fours. However, Walt doesn't use a wheelchair. He skis, rides horses, cycles, and does acrobatics. He is a successful businessman who sells his software company to Microsoft for a fortune. He loves life and he loves women and he can laugh at himself. Consider this great pickup line, delivered at a nightclub: "Hey, Sally, I've got a leash. Would you like to take me for a walk?" To which Sally, greatly amused, replies, "Come on, boy." The male viewer may be excused for wondering why *he* could never come up with such a clever opener.

The point of all this is not just that people with disabilities are ordinary folk with all the problems, quirks, deficiencies, and virtues of everyone else—and a good deal more to contend with besides—but that many of us are odd for not being able to see their normality. The Farrelly brothers may be trying to tell us that the person with the biggest disability in *Shallow Hal* is Hal as we first meet him, and that the people with the biggest disability in real life are all the "normal" people who think like him, equating difference with inferiority. Some audience members may think Rosemary is lucky for winning Hal, but in fact Hal is the one who really lucks out.

Gwyneth Paltrow and Jack Black in *Shallow Hal* (2001).

The deaf community typifies the new challenge to ablism. Increasingly, deaf people share a "collective identity" with all other deaf people (Becker, 1980: 107). Members of the deaf community have a common language and culture, and they tend to marry other deaf people (Davis, 1995: 38). Rather than feeling humiliated by the seeming disadvantage of deafness, they take pride in their condition. Indeed, many people in the deaf community are eager to remain deaf even if medical treatment can "cure" them (Lane, 1992). As Roslyn Rosen, former president of the National Association of the Deaf, put it: "I'm happy with who I am. . . . I don't want to be 'fixed.' . . . In our society everyone agrees that whites have an easier time than blacks. But do you think a black person would undergo operations to become white?" (quoted in Donick, 1993: 38).

Aging

Sociology of Aging

Disability affects some people. Aging affects us all. Many people think of aging as a natural process that inevitably thwarts our best attempts to delay death. Sociologists, however, see aging in a more complex light. For them, aging is also a process of socialization, or learning new roles appropriate to different stages of life (see Chapter 4, "Socialization"). The sociological nature of aging is also evident in the fact that its significance varies from one society to the next. That is, different societies attach different meanings to the progression of life through its various stages. Menopause, for example, occurs in all mature women. In the United States, we often see it as a major life event. Thus, the old euphemism for menopause was the rather dramatic expression "change of life." In contrast, menopause is a relatively minor matter in Japan. Moreover, while menopausal American women frequently suffer hot flashes, menopausal Japanese women tend to complain mainly about stiff shoulders (Lock, 1993). In many Western countries, complaining about stiff shoulders is a classic symptom of having just given birth. As this example shows, the stages of life are not just natural processes but events deeply rooted in society and culture. As we will see, the same holds for death.

Sociology Now™

Learn more about **the Sociology of Aging** by going through the Then and Now: Aging in the Movies.

Aging and the Life Course

All individuals pass through distinct stages of life, which, taken together, sociologists call the **life course.** These stages are often marked by **rites of passage,** or rituals signifying the transition from one life stage to another (Fried and Fried, 1980). Baptism, confirmation, the bar mitzvah and bat mitzvah, high school graduation, college convocation, the wedding ceremony, and the funeral are among the best-known rites of passage in the United States. Rituals do not mark some transitions in the life course, however. For example, in North America, people often complain about the "terrible twos," when toddlers defy parental demands in their attempt to gain autonomy. Similarly, when some American men reach the age of about 40, they experience a "midlife crisis," in which they attempt to defy the passage of time and regain their youth. (It never works.)

Some of the stages of the life course are established not just by norms but by law. For example, most societies have laws that stipulate the minimum age for smoking tobacco, drinking alcohol, driving a vehicle, and voting. Although the United States has outlawed mandatory retirement, most societies continue to have a legal retirement age. Moreover, the duration of each stage of life differs from one society and historical period to the next. For example, there are no universal rules about when one becomes an adult. In preindustrial societies, adulthood arrived soon after puberty. In Japan, one becomes an adult at 20. In the United States, adulthood arrives at 18 (the legal voting age) or 21 (the legal drinking age). Not so long ago, the legal voting age was 21 and the legal drinking age was 18 in the United States.

The **life course** refers to the distinct phases of life through which people pass. These stages vary from one society and historical period to another.

A **rite of passage** is a ritual that marks the transition from one stage of life to another.

Even the number of life stages varies historically and across societies. For instance, childhood was a brief stage of development in medieval Europe (Ariès, 1962). In contrast, childhood is a prolonged stage of development in rich societies today, and adolescence is a new phase of development that was virtually unknown just a few hundred years ago (Gillis, 1981). Increased life expectancy and the need for a highly educated labor force made childhood and adolescence possible and necessary. (**Life expectancy** is the average age at death of the members of a population.)

Finally, although some life-course events are universal—birth, puberty, marriage, and death—not all cultures attach the same significance to them. Thus, ritual practices marking these events vary. For example, formal puberty rituals in many preindustrial societies are extremely important because they mark the transition to adult responsibilities. However, adult responsibilities do not immediately follow puberty in industrial and postindustrial societies because of the introduction of a prolonged period of childhood and adolescence. Therefore, formal puberty rituals are less important in such societies.

Schwartzwald Lawrence/Corbis Sygma

Many people find romance and marriage between people widely separated by age problematic and even repulsive. Woody Allen's fourth wife, Soon-Yi Previn, is 35 years younger than he and is the adopted daughter of his third wife, Mia Farrow. His romance with Previn caused an uproar.

Age Cohort

As you pass through the life course, you learn new patterns of behavior that are common to people about the same age as you. Sociologically speaking, a category of people born in the same range of years is called an **age cohort.** For example, all Americans born between 1980 and 1989 form an age cohort. **Age roles** are patterns of behavior that we expect of people in different age cohorts. Age roles form an important part of our sense of self and others (Riley, Foner, and Waring, 1988). As we pass through the stages of the life course, we assume different age roles. To put it simply, a child is supposed to act like a child, an elderly person like an elderly person. We may find a 5-year-old dressed in a suit cute but look askance at a lone 50-year-old riding on a merry-go-round. "Act your age" can be an admonishment to people of all ages who do not conform to their age roles. Many age roles are informally known by character types, such as the "rebellious teenager" and "wise old woman." We formalize some age roles by law. For instance, the establishment of minimum ages for smoking, drinking, driving, and voting formalizes certain aspects of the adolescent and adult roles.

Sociology Now™

Learn more about **Life Expectancy** by going through the Average Life Expectancy Map Exercise.

We find it natural that children in the same age cohort, such as preschoolers in a park, should play together or that people of similar age cluster together at parties. Conversely, many people find romance and marriage between people widely separated by age problematic and even repulsive.

Differences between age cohorts are sufficiently large in the United States that some sociologists regard youth culture as a distinct subculture. Adolescents and teenagers—divided though they may be by gender, class, race, and ethnicity—frequently share common interests in music, movies, and so forth.

Generation

A **generation** is a special type of age cohort. Many people think of a generation as people born within a 15- to 30-year span. Sociologists, however, usually define generation more narrowly. From a sociological point of view, a generation is composed of members of an age cohort who have unique and formative experiences during youth. Age cohorts are statistically convenient categories, but most members of a generation are conscious of belonging to a distinct age group. For example, "baby boomers" are North Americans who were born in the prosperous years from 1946 to 1964. Most of them came of age between the mid-1960s and the early 1970s. Common experiences that bind them include major historical events (the war in Vietnam, the Watergate break-in, the subsequent resignation of President Nixon) and popular music (the songs of Bob Dylan,

- **Life expectancy** is the average age at death of the members of a population.
- An **age cohort** is a category of people born in the same range of years.
- **Age roles** are norms and expectations about the behavior of people of different age cohorts.
- A **generation** is an age group that has unique and formative historical experiences.

Woodstock, 1969. A generation is composed of members of an age cohort who have unique and formative experiences during their youth.

the Beatles, and the Rolling Stones). "Generation X"—identified with the 1980s—followed the baby boomers. Members of Generation X faced a period of slower economic growth and a job market glutted by the baby boomers. Consequently, many of them resented having to take so-called "McJobs" when they entered the labor force. Thus, Douglas Coupland, the novelist who coined the term "Generation X," cuttingly defined a McJob as a "low-pay, low-prestige, low-dignity, low-benefit, no-future job in the service sector. Frequently considered a satisfying career choice by people who have never held one" (Coupland, 1991: 5).

Tragedies may help to crystallize the feeling of being a member of a particular generation. For instance, when you are elderly you will probably remember where you were when you heard about the terrorist attacks of September 11, 2001. Such memories may someday distinguish you from those who are too young to remember this tragic series of events. The assassination of President Kennedy may play a similar role in the memory of your parents.

The crystallization of a generational "we-feeling" among youth is a quite recent phenomenon. The very ideas of youth and adolescence gained currency only in the 19th century. This was due to growth in life expectancy, extended schooling, and other factors. Middle-class youth culture often challenged tradition and convention (Gillis, 1981). It was marked by a sense of adventure and rebellion, especially against parents, that manifested itself in political liberalism and cultural radicalism.

Finally, we note that generations sometimes play a major role in history. Revolutionary movements, whether in politics or the arts, are sometimes led by members of a young generation who aggressively displace members of an older generation (Eisenstadt, 1956; Mannheim, 1952).

Aging and Inequality

Age Stratification

Age stratification refers to social inequality between age cohorts. It exists in all societies, and we can observe it in everyday social interaction. For example, there is a clear status hierarchy in most high schools. On average, seniors enjoy higher status than sophomores, and sophomores enjoy higher status than freshmen.

The very young are often at the bottom of the stratification system. In preindustrial societies, people sometimes killed infants so populations would not grow beyond the ability of the environment to support them. Facing poverty and famine, parents sometimes abandoned children. Many developing countries today are overflowing with orphans and street children. During the early stages of Western industrialization, adults brutally exploited children. For instance, the young chimney sweeps in *Mary Poppins* may look cute, but during the Industrial Revolution skinny "climbing boys" as young as 4 were valued in Britain because they could squeeze up crooked chimney flues no more than a foot or two in diameter 12 hours a day. Space was tight, so they worked naked, and they rubbed their elbows, knees, noses, and other protrusions raw against the soot, which was often hot. The first description of job-related cancer appeared in an article published in 1775. "Soot-wart," as it was then known, killed chimney sweeps as young as 8 (Nuland, 1993: 202–5).

Even in rich countries, poverty is more widespread among children than adults. According to the 2000 U.S. census, for example, childhood poverty exceeds poverty among adults by 71 percent (U.S. Department of Health and Human Services, 2002) (Figure 12.3). The United States is also distinguished by having the highest child poverty rate among the world's two dozen richest countries (Bradbury and Jäntti, 2001).

Age stratification is social inequality between age cohorts.

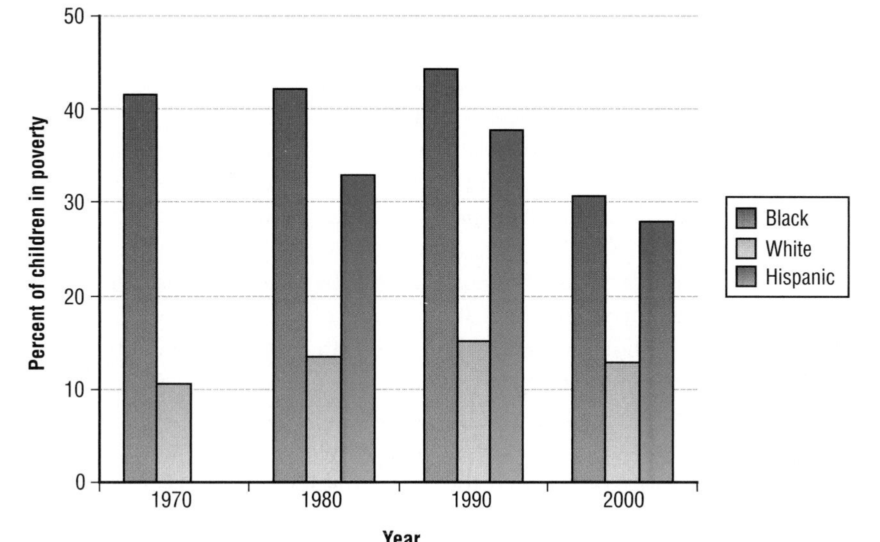

Figure 12.3
Child Poverty by Race and Ethnicity, United States, 1970–2000

Source: U.S. Department of Health and Human Services (2002).

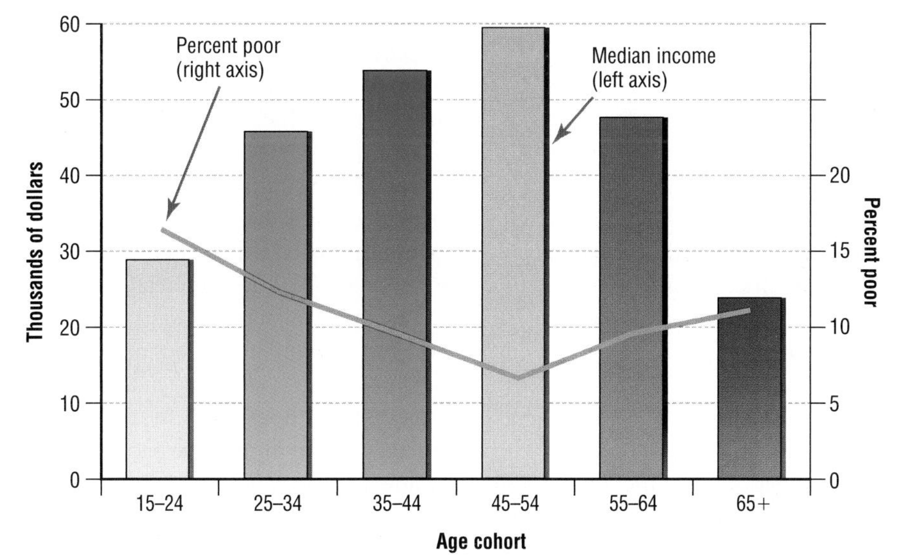

Figure 12.4
Median Income and Percent Poor by Age Cohort, United States, 2002

Source: U.S. Census Bureau (2005: 444, 453).

Gerontocracy

Some people believe that ancient China and other preindustrial societies were **gerontocracies,** or societies in which elderly men ruled, earned the highest income, and enjoyed the most prestige. Even today, people in some industrialized countries pay more attention to age than Americans do. In South Korean corporations, for instance, when a new manager starts work, everyone in the department who is older than the new manager may resign or be reassigned. Given the importance of age seniority in South Korea, it is considered difficult for a manager to hold authority over older employees. Older employees in turn find it demeaning to be managed by a younger boss (Lie, 1998).

However, although some societies may approximate the gerontocratic model, its extent has been exaggerated. Powerful, wealthy, and prestigious leaders are often mature, but not the oldest, people in a society. The United States today is typical of most societies, past and present, in this regard. For example, in the United States, median income gradually rises with age, reaching its peak in the 45–54 age cohort. Median income then declines for the oldest age cohorts (Figure 12.4). Prestige and power follow the same course. A similar pattern is evident in South Korea.

A **gerontocracy** is a society ruled by elderly people.

In general, true gerontocracy is rare. Like King Lear, the elderly often give up power and become marginalized. Even in traditional societies that held the elderly in high esteem, aging was not usually seen as an unambiguous good. After all, aging denotes physical and mental decline and the nearness of death. Ambivalence about aging—especially as people reach the oldest age cohorts—is a cultural universal. As one historian writes: "Youth has always and everywhere been preferred to old age. Since the dawn of history, old people have regretted [loss of] their youth and young people have feared the onset of old age" (Minois, 1989 [1987]).

Just as true gerontocracy is uncommon, so is rule by youth. True, as noted previously, relatively young age cohorts sometimes supply most of a country's political leadership. This happened in revolutionary France in the late 18th century, revolutionary Russia in the early 20th century, and revolutionary China in the mid-20th century. However, youthful ruling cadres may become gerontocracies in their own right, especially in nondemocratic societies. This was the case with the Russian Communist leadership in the 1980s and the Chinese Communist leadership in the 1990s; the then-young generation that had grabbed power half a century earlier still clung on as senility approached.

Theories of Age Stratification

The Functionalist View

How can we explain age stratification? *Functionalists* observe that in preindustrial societies, family, work, and community were tightly integrated (Parsons, 1942). People worked in and with their family, and the family was the lifeblood of the community. However, industrialization separated work from family. It also created distinct functions for different age cohorts. Thus, whereas traditional farming families lived and worked together on the farm, the heads of urban families work outside the home. Children worked for their parents in traditional farming families, but in urban settings they attend schools. At the same time, industrialization raised the standard of living and created other conditions that led to increased life expectancy. The cohort of retired elderly people thus grew. And so it came about that various age cohorts were differentiated in the course of industrialization.

At least in principle, social differentiation may exist without social stratification. But, according to the functionalists, age stratification did develop in this case because different age cohorts performed functions of differing value to society. For example, in preindustrial societies, the elderly were important as a storehouse of knowledge and wisdom. With industrialization, their function became less important and so their status declined. Age stratification, in the functionalist view, reflects the importance of each age cohort's current contribution to society, with children and the elderly distinctly less important than adults employed in the paid labor force. Moreover, all societies follow much the same pattern. Their systems of age stratification "converge" under the force of industrialization.

Conflict Theory

Conflict theorists agree with the functionalists that the needs of industrialization generated distinct categories of youth and the elderly. They disagree, however, on two points. First, they dispute that age stratification reflects the functional importance of different age cohorts (Gillis, 1981). Instead, they say, age stratification stems from competition and conflict. Young people may participate in a revolutionary overthrow and seize power. The elderly may organize politically to decrease their disadvantages and increase their advantages in life. In other words, power and wealth do not necessarily correlate with the roles the functionalists regard as more or less important; competition and conflict may redistribute power and wealth between age cohorts. The second criticism lodged by conflict theorists concerns the problem of convergence. Conflict theorists suggest that political struggles can make a big difference in how much age stratification exists in a society. We saw, for example, that child poverty is higher in the United States than in other rich countries. That is because in other rich countries, particularly in continental western Europe, successful working-class political parties have struggled to implement generous child welfare measures and employment policies that lower the poverty level. This suggests that the fortunes of age cohorts are shaped by other forms of inequality, such as class stratification (Gillis, 1981; Graff, 1995).

Symbolic Interactionist Theory

Symbolic interactionists focus on the meanings people attach to age-based groups and age stratification. They stress that the way in which people understand aging is nearly always a matter of interpretation. Symbolic interactionists have done especially important research in community studies of the elderly. They have also helped us to understand better the degree and nature of prejudice and discrimination against the elderly. For example, one study examined how movies from the 1940s to the 1980s contributed to the negative stereotyping of the elderly, particularly women. Among other things, it found that young people were overrepresented numerically in the movies (as compared with their representation in the general population) and tended to be portrayed as leading active, vital lives. Elderly women were underrepresented numerically and tended to be portrayed as unattractive, unfriendly, and unintelligent (Brazzini et al., 1997).

Social Problems of the Elderly

If you've been to south Florida lately, you have a good idea of what the age composition of the United States will look like in 50 years. ▶Figure 12.5 shows how the elderly have grown as a percentage of the United States population since 1900 and how this age cohort is expected to grow until 2050. In 1900, only about 4 percent of the U.S. population were 65 and older. Today, the figure is around 13 percent. By 2040, nearly 21 percent of Americans will be elderly. Thereafter, their weight in the U.S. population will start to decline. ▶Figure 12.6 shows the geographical distribution of Americans 65 years and older.

Sociology Now™

Learn more about **the Aging Population** by going through the Aging Population Data Experiment.

Many sociologists of aging refer to elderly people who enjoy relatively good health—usually people between the ages of 65 and 74—as the "young old" (Neugarten, 1974; Laslett, 1991 [1989]). They refer to people 85 and older as the "old old." Figure 12.5 shows that the young old are expected to decline as a percentage of the American population after 2030. In contrast, the proportion of the old old is expected to continue increasing.

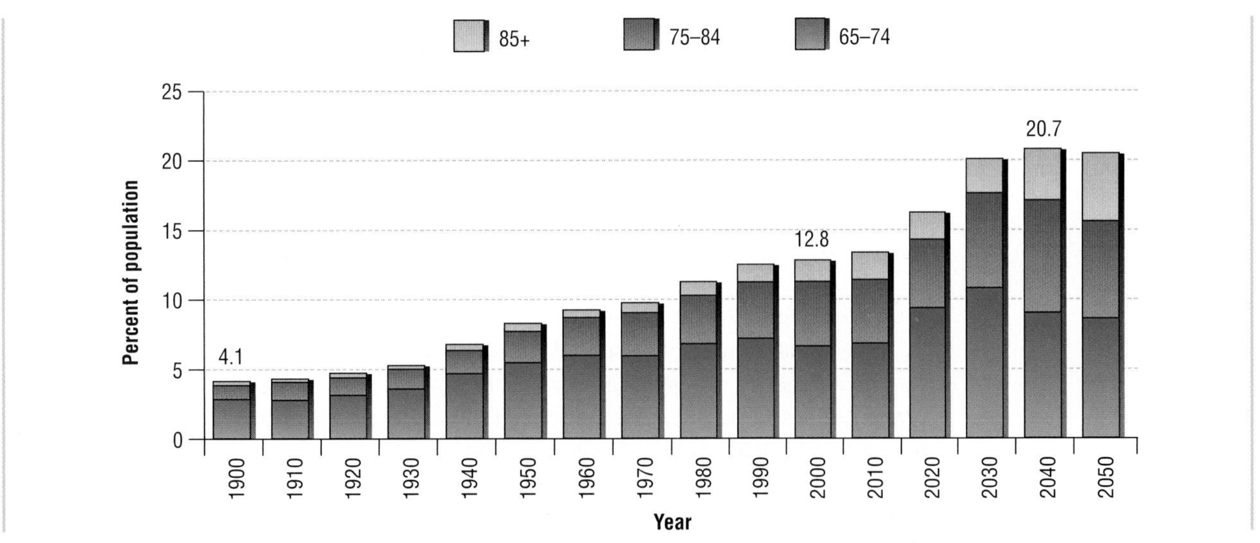

▶Figure 12.5
Elderly as Percent of U.S. Population, 1900–2050 (projected)

Source: U.S. Administration on Aging (1999).

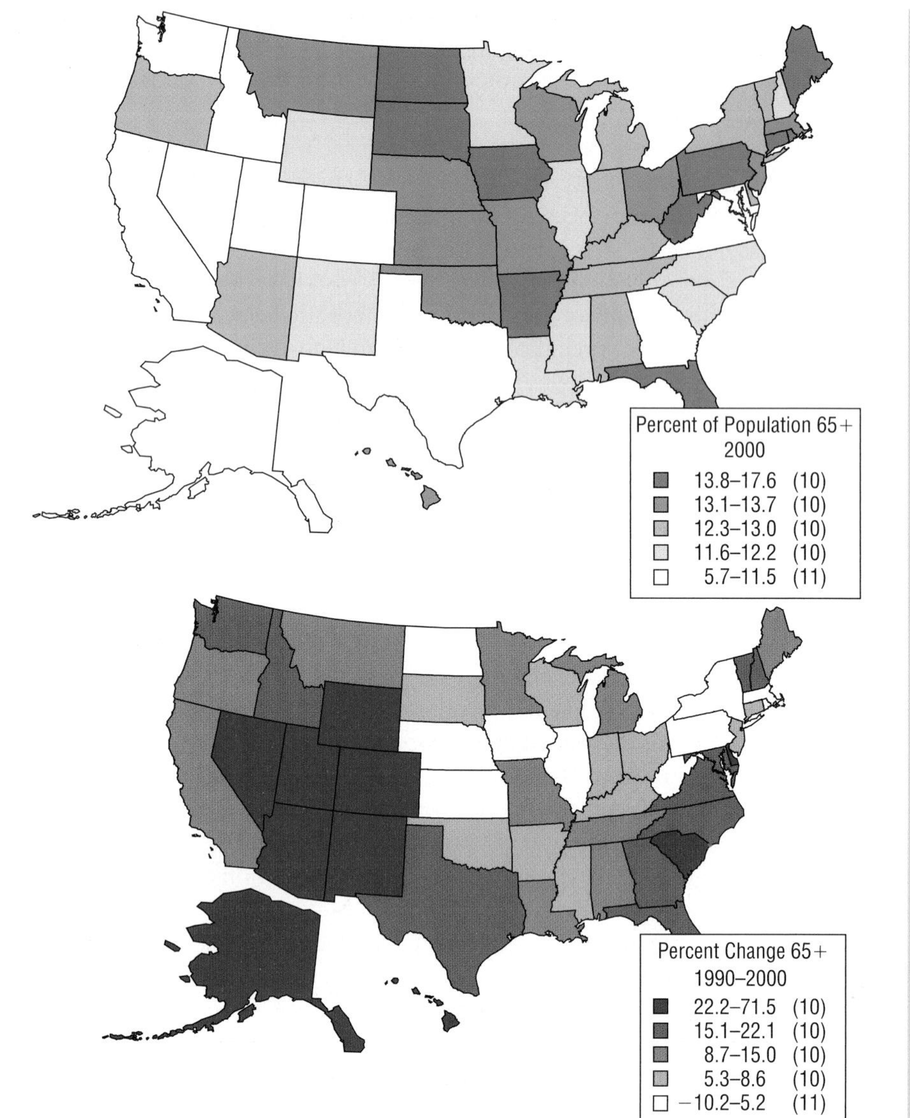

Figure 12.6
Percent of Population Age 65+, 2000, and Percent Change in Population Age 65+, 1990–2000, United States

The top map shows the percent of the population of each state that was 65 or older in 2000. The bottom map shows the percent change in the 65+ population in each state between 1990 and 2000. Do you see a pattern in the data? What kinds of states have a small percentage of elderly residents? What kinds of states have a large percentage of elderly residents? In what kinds of states is the percentage of elderly people growing most quickly? In what kinds of states is the percentage of elderly people growing least quickly or shrinking?

Source: AARP (2001: 11).

The reasons for the rising proportion of elderly people are clear. On the one hand, fertility rates have been declining in virtually all industrial societies. That is, women are having fewer babies and population growth has slowed, if not actually declined, in rich societies (see Chapter 20, "Population and Urbanization"). On the other hand, life expectancy has been increasing because of medical advances, better welfare provisions, and other factors.

Aging and Poverty

The rising number of the old old concerns many people because they are most likely to suffer general physiological decline, life-threatening diseases, social isolation, and poverty. It is also significant that the sex ratio (the number of men divided by the number of women) falls with age. That is, because women live longer than men on average, there are more women than men among the elderly. This imbalance is most marked in the oldest age cohorts. In large part, therefore, poverty and related problems among the oldest Americans are partly gender issues.

Economic inequality between elderly women and men is largely the result of women's lower earning power when they are young. Women are entering the paid workforce in increasing numbers, but there are still more women than men who are homemakers and do not work for a wage. Therefore, fewer women than men enjoy employer pensions when they retire. Moreover, as we saw in Chapter 11 ("Sexuality and Gender"), women who are in the paid labor force tend to earn less than men. When they retire, their employer pensions are generally inferior. As a result, the people most in need—the most elderly women—receive the fewest retirement benefits.

In addition to the old old and women, the categories of elderly people most likely to be poor include African Americans, people living alone, and people living in rural areas (Siegel, 1996). But declining income and poverty are not the only social problems faced by the elderly. In addition, the elderly are sometimes socially segregated in nursing homes, seniors' apartment buildings, and subdivisions with a high proportion of retired people.

A Shortage of Caregivers

Another problem facing the elderly is the looming shortage of caregivers. In 2001, home-care agencies and institutional care settings such as nursing homes employed 2.1 million caregivers in the United States. The U.S. Bureau of Labor Statistics expects a 58 percent rise in demand for such workers between 1998 and 2008. However, such workers are increasingly hard to find and increasingly hard to keep on the job because the work is difficult and pays little. You might think new workers could easily be recruited from the ranks of former welfare recipients. The problem here is that the government requires two weeks of pre-employment training for direct-care aides, and this runs afoul of Congress's 1996 welfare reform, which discourages such training for former welfare recipients (DeFrancis, 2002). Thus, along with the increasing strain on Social Security, Medicare, and Medicaid, the scarcity of caregivers has become a major concern in our aging society.

Among rich countries, the United States is a relatively young society. The proportion of the old to the total population is comparatively low. This is due to the relatively high fertility rate and the constant influx of young immigrants. In contrast, Japan faces an acute aging crisis. It has one of the longest life expectancies in the world (85 for women, 78 for men). It has one of the world's lowest fertility rates (on average, each woman has 1.33 children). It has limited the number of immigrant workers. Consequently, there is a large and growing imbalance between the increasing number of elderly who must be supported through pensions and the declining number of young and middle-aged people in the paid labor force who must support them. In 2002, 3.6 people between the ages of 20 and 64 supported each person older than 65. By 2025, the ratio is expected to fall to 1.9 (Moffett, 2003). Consequently, many Japanese people have become anxious about the future. Older Japanese people will likely have to save more, spend less, and delay their retirement.

Ageism

Especially in a society that puts a premium on vitality and youth, such as the United States, being elderly is a social stigma. **Ageism** is prejudice about, and discrimination against, elderly people. Ageism is evident, for example, when elderly men are stereotyped as "grumpy." Ageism affects women more than men. Thus, the same person who considers some elderly men "distinguished looking" may disparage elderly women as "haggard" (Banner, 1992).

Often, however, elderly people do not conform to the negative stereotypes applied to them. In the United States, 65 is usually taken to be the age when people become elderly. (Sixty-five used to be the mandatory retirement age.) But just because someone is 65 or older does not mean he or she is decrepit and dependent. On the contrary, most people who retire from an active working life are far from being a tangle of health problems and a burden on society. This is due to the medical advances of recent decades, the healthier lifestyles followed by many elderly people, and the improved financial status of the elderly.

Specifically, the housing arrangements of elderly people are not usually desolate and depressing (Hochschild, 1973; Myerhoff, 1978). Fewer than 5 percent of Americans 65 years and older lived in nursing homes in 2000, and many of these were of high quality. For people 85 years and older, the figure was more than 18 percent (AARP, 2001: 7). In 2000, fully 70 percent of Americans 75 years and older lived in single-family detached

Ageism is prejudice and discrimination against elderly people.

Stone/Getty Images

Most people who retire from an active working life are far from being a tangle of health problems and a burden on society. This is due to the medical advances of recent decades, the healthier lifestyles followed by many elderly people, and the improved financial status of the elderly.

homes. Moreover, surveys show that most Americans want to stay in their own home as long as possible. This desire is strongest among the oldest Americans and is increasing over time. More than 80 percent of Americans 45 years and older say they would prefer to modify their homes and have help at home should such assistance become necessary (Bayer and Harper, 2000).

French social critic Simone de Beauvoir argued, "There is only one solution if old age is not to be an absurd parody of our former life, and that is to go on pursuing ends that give our existence meaning—devotion to individuals, to groups or to causes, social, political, intellectual, or creative work" (Beauvoir, 1972 [1970]: 540). Indeed, this is precisely what many elderly people are doing now, as they work, pursue education and training, enjoy travel and leisure, and maintain ties with their family members and fellow senior citizens while living on their own.

The Power and Wealth of the Elderly

Many elderly people are able to enjoy their retirement because they own assets aside from their home, such as investments. Most elderly people receive private and public pensions. Nearly 13 percent of Americans older than 65 work in the paid labor force, nearly half of them full-time (AARP, 2001: 11). As a result, although the poverty rate does increase somewhat for people older than 54, the poverty rate among people 65 and older is only about two-thirds the poverty rate of people in the 15–24 age cohort (see Figure 12.4). Poverty among the elderly has fallen sharply since the 1960s. Finally, in comparison with other age cohorts, the elderly are better served by the social welfare system. Social Security and other programs geared to the elderly are relatively generous (Box 12.3). The elderly are well covered by Medicare or health plans tied to the pension plan of their former union or employer.

One reason for the relative economic security of the elderly is that they are well organized politically. Their voter participation rate is above average, and they are overrepresented among those who hold positions of political, economic, and religious power. Many groups seek to improve the status of the elderly, the Gray Panthers being perhaps the best known among them. The AARP (formerly the American Association of Retired Persons) is an effective lobby in Washington, D.C. (Morris, 1996). In part because of the activism of elderly people, discrimination based on age has become illegal in the United States. In fact, activism on the part of the elderly may have led to a redistribution of resources away from young people. For example, educational funding has declined, but funding has increased for medical research related to diseases disproportionately affecting the elderly.

Sociology Now™

Learn more about **the Aging Population** by going through the Aging Population Learning Module.

Death and Dying

It may seem odd to say so, but the ultimate social problem the elderly must face is their own demise. Why are death and dying *social* problems and not just religious, philosophical, and medical issues?

For one thing, attitudes toward death vary widely across time and place. So do the settings within which death typically takes place. Although individuals have always dreaded death, in most traditional societies, such as Europe until early modern times, most people accepted it (Ariès, 1982). That is partly because most people apparently believed in life after death, whether in the form of a continuation of life in heaven or in cyclical rebirth. What also made death easier to accept was that the dying were not isolated from other people. They continued to interact with household members and neighbors, who offered them continuous emotional support. Finally, because the dying had previous experience giving emotional support to other dying people, they could more easily accept death as part of everyday life.

BOX 12.3

SOCIAL POLICY: WHAT DO YOU THINK?

The Social Security Crisis

One of the greatest triumphs of public policy since the 19th century has been the development of the public health system. It led to a substantial increase in life expectancy. Another major public policy achievement is social welfare, especially as it applies to the elderly. The combination of Social Security, Medicare, and other government programs goes a long way toward ensuring that the elderly are not doomed to poverty and illness. It explains why, in spite of the abolition of a mandatory retirement age, the average age at retirement for American men fell from 69 in 1950 to less than 62 in 2000. Because more Americans feel financially secure in their old age, they can retire earlier.

A longer and more secure life span is a wonderful thing. However, some scholars and policymakers worry that a major crisis is looming. We may not be able to afford government programs for the elderly. That is because of the expected retirement of the "baby boom" generation, people born between 1946 and 1964. Many Americans born in this period contributed to Social Security and other measures to support the aged. As they begin to retire from the active labor force, however, fewer Americans will be contributing to government coffers. Currently, there are roughly three workers for every retiree. Analysts expect the ratio to reach 2:1 by 2030 (Urban Institute, 1998).

Some scholars argue that economic growth and higher immigration could offset the expected decline in the active labor force. However, others argue that we also need more concrete measures to deal with the expected crisis in government support for the elderly. Some of the possible policy proposals include the following:

- Encourage people to work longer, thereby having more workers support retirees.
- Increase national savings.
- Lower health-care costs, especially the disproportionately higher burden of medical costs for the elderly.
- Provide government support only for the truly needy, thereby eliminating or lowering Social Security and other federal benefits for the well-off.
- Allow a portion of Social Security payments to be invested in the stock market, which has had a relatively high rate of return historically but also poses more risk because an individual may retire when the stock market is in a slump.

What do you think? Should we worry about the expected crisis in Social Security and other government programs that support the elderly? If you expect the potential crisis to arise in your lifetime, what should you be doing now to avert it? What are the advantages and disadvantages of each of the listed policy proposals? What kind of lifestyle do you expect to lead when you are in your 60s and 70s? Do you think you will be working full-time? Or will you be fully or partly retired?

In contrast, in the United States today, dying and death tend to be separated from everyday life. Most terminally ill patients want to die peacefully and with dignity at home, surrounded by their loved ones. Yet about 80 percent of Americans die in hospitals. Often, hospital deaths are sterile, noiseless, and lonely (Nuland, 1993). Dying used to be public. It is now private. The frequent lack of social support makes dying a more frightening experience for many people (Elias, 1985 [1982]). In addition, our culture celebrates youth and denies death (Becker, 1973). We use diet, fashion, exercise, makeup, and surgery to prolong youth or at least the appearance of youth. This makes us less prepared for death than our ancestors were.

Our reluctance to accept death is evident from the many euphemisms we use as a means of distancing ourselves from it. People used to say that the dead had "entered the Pearly Gates" or had gone to "sing in God's heavenly choir." We are now more likely to say that the dead have "passed away" or "gone to meet their maker" in "a better place" or that they lie in "their final resting place." Sometimes we use humorous expressions as a distancing mechanism. We say that people have "croaked" or "kicked the bucket" or "cashed in their chips" or that they are now "pushing up daisies." These and other similar expressions allow us to separate ourselves symbolically from the horror of death.

Sociology Now™

Learn more about **Death and Dying** by going through the Death and Dying Animation.

Psychiatrist Elisabeth Kübler-Ross's analysis of the stages of dying in contemporary America also suggests how reluctant we are to accept death (Kübler-Ross, 1969). She based her analysis on interviews with patients who were told that they had an incurable disease. At first, the patients went into *denial*, refusing to believe that their death was imminent. Then they expressed *anger*, seeing their demise as unjust. *Negotiation* followed; they pleaded with God or with fate to delay their death. Then came *depression*, when they re-

John Hillary/Reuters/Corbis

Dr. Jack Kevorkian is the most prominent advocate of euthanasia in the United States. See Table 12.1 for research on Kevorkian and an analysis of euthanasia statistics.

signed themselves to their fate but became deeply despondent. Only then did the patients reach the stage of *acceptance*, when they put their affairs in order, expressed regret over not having done certain things when they had the chance, and perhaps spoke about going to heaven.

Euthanasia

The reluctance of many Americans to accept death is clearly evident in the debate over euthanasia, also known as mercy killing or assisted suicide (Rothman, 1991). The very definitions of life and death are no longer clear-cut (Lock, 2002). Various medical technologies, including machines able to replace the functions of the heart and lungs, can prolong life beyond the point that was possible in the past. This raises the question of how to deal with people who are near death. In brief, is it humane or immoral to hasten the death of terminally ill patients?

The American Medical Association's Council on Ethical and Judicial Affairs (AMA-CEJA) says it is the duty of doctors to withhold life-sustaining treatment if that is the wish of a mentally competent patient. The AMA-CEJA also endorses the use of effective pain treatment even if it hastens death. As a result, doctors and nurses make decisions every day about who will live and who will die (Zussman, 1992; 1997). Public opinion polls show that about three quarters of Americans favor this practice (Benson, 1999).

Euthanasia involves a doctor prescribing or administering medication or treatment that is *intended* to end a terminally ill patient's life. It is therefore a more active form of intervention than those noted in the preceding paragraph. Public opinion polls show that about two-thirds of Americans favor physician-assisted euthanasia (Benson, 1999). Between 33 percent and 60 percent of American doctors (depending on the survey) say they would be willing to perform euthanasia if it were legal. Nearly 30 percent of American doctors have received a euthanasia request but only 6 percent have ever complied with such a request (Meier et al., 1998; Finsterbusch, 2001). The AMA-CEJA, the Catholic Church, some disabled people, and other groups oppose euthanasia.

Sociology Now™

Learn more about **Euthanasia** by going through the % who believe euthanasia is ok Map Exercise.

Euthanasia is legal in the Netherlands and may become legal in some other countries in the next few years. In Oregon, a physician-assisted suicide law, the Death with Dignity Act, took effect in October 1997. It allows doctors to prescribe a lethal dose of drugs to terminally ill patients who choose not to prolong their suffering. The law stipulates that before a doctor can give a patient barbiturates to end the patient's life, two physicians must concur that the patient is terminally ill and has less than six months to live. Patients must request euthanasia three times, both orally and in writing. Finally, patients must swallow the barbiturates themselves. In a little more than six years of operation (to the end of 2003), only 171 people have taken advantage of Oregon's physician-assisted suicide law (Table 12.1). As of this writing, the U.S. Department of Justice has filed preliminary paperwork on an appeal that, if successful, would make physician-assisted suicide illegal in Oregon. The outcome of this case is as yet unknown.

What is clear is that, in the words of a 1997 ruling of the U.S. Supreme Court, "[t]hroughout the Nation, Americans are engaged in an earnest and profound debate about the morality, legality and practicality of physician assisted suicide" (Longwood College Library, 2000). Euthanasia is bound to become a major political issue in coming decades as medical technologies for prolonging life improve, the number of elderly people increases, and the cost of medical care skyrockets. Extending the lives of terminally ill patients by all means possible will be upheld as an ethical imperative by some people. Others will regard it as immoral because it increases suffering and siphons scarce resources away from other pressing medical needs.

Euthanasia (also known as mercy killing and assisted suicide) involves a doctor prescribing or administering medication or treatment that is intended to end a terminally ill patient's life.

The Business of Dying

In every society, a rite of passage surrounds death. We pay our last respects to the departed, soothe the pain of our loss, and affirm our resolve to carry on. People have a deep emotional need to bury or cremate the dead, which is why we go to extraordinary lengths

Table 12.1
Characteristics of People Who Hastened Their Death under Oregon's Death with Dignity Act (N = 171) Compared with People in Oregon Who Died of the Same Diseases (N = 53,544), 1998–2003

Variable	Physician-Assisted Suicide	Other Deaths
Median Age	67	76
Race		
White	97	97
Asian	3	1
Other	0	2
Sex		
Male	54	50
Female	47	50
Marital Status		
Married	44	49
Widowed	24	33
Divorced	25	14
Never married	8	4
Unknown	0	0
Education		
Less than high school	9	25
High school graduation	30	43
Some college	21	18
Baccalaureate or higher	40	14
Unknown	0	0
Insurance		
Private	52	63
Medicare or Medicaid	48	35
None	0	2
Percentage diagnosed with terminal illness leading to death in six months	100	n.a.
Median duration of physician/patient relationship (weeks)	13	n.a.
Median time between first request and death (weeks)	39	n.a.
Median time between ingestion and unconsciousness (minutes)	5	n.a.

Note: n.a. = not available or not applicable. A study of 69 people who died with Dr. Jack Kevorkian's assistance found that only 25 percent were terminally ill, and a disproportionately large number were socially isolated women (divorced or never married) (Roscoe, Gragovic, and Cohen, 2000). These findings suggest that *in the absence of clinical safeguards,* some groups are especially vulnerable to euthanasia and its misuse. Data from the Oregon case support the view that safeguards help to alleviate much of the problem. Every one of the physician-assisted suicides was diagnosed with a terminal illness that would have resulted in death within 6 months. Overall, those who elected suicide had relationships of considerable duration (on average, more than 3 months) with the physicians who helped them die. A long period (nearly 10 months on average) elapsed after their initial request, giving them an opportunity to reconsider their choice. Their level of education was relatively high. Those who elected physician-assisted suicide were not disproportionately women. Nor were they (as some critics suggest) desperate to die because of lack of medical insurance; all of them had insurance. Note, however, that the people who elected suicide appear to have been somewhat more socially isolated than those who died in Oregon of the same underlying diseases. Specifically, they were somewhat less likely to be married, somewhat more likely to be never-married, and considerably more likely to be widowed or divorced.

Source: Oregon Department of Human Services (2004).

to recover bodies even under difficult and dangerous circumstances, including war and natural disaster.

The way we die reflects the nature of our society and culture. The United States is a capitalist, business-oriented society. Not surprisingly, therefore, funerals are big business—a $20 billion a year industry in 1999 (Wiegand and Gibson, 1999). The average undertaker's bill in the late 1990s was $4700. Adding other expenses such as flowers and cemetery charges, the average funeral and burial bill grew to $7800 (Mitford, 1998 [1963]).

There are two main reasons why funerals are so expensive. First, big corporations are supplanting small family operations in the funeral industry. The undisputed giant in this field, Services Corp. International (SCI), now controls about 10 percent of the U.S. funeral industry (along with 15 percent in Britain and 25 percent in Australia). Concentration of ownership lowers competition and results in higher prices, as a New York City report recently documented (New York City Department of Consumer Affairs, 2001). The second main reason why funerals cost so much is that people are vulnerable when their loved ones die, and the funeral industry takes full advantage of their vulnerability. In journals such as *Mortuary Management,* funeral directors can learn how to make the bereaved feel they can make up for any real or imagined neglect of the deceased by spending lavishly on the funeral. In industry workshops they can learn how to bamboozle the bereaved in other ways. A tape from a Florida workshop, "How to Add $1,400 to Your Cremation Calls," explains the importance of requiring "identification viewing":

> Yes, it is self-serving. But if the family hasn't selected one of your more expensive cremation caskets, make sure you show Mom's body in a cardboard box, because someone in the family is bound to say, "Maybe we should get something a little bit nicer." (Quoted in Wiegand and Gibson, 1999)

In Sacramento, "a little bit nicer" can run as high as $24,000. We conclude that funerals, no less than other social processes involving the body, bear the imprint of the society in which they take place.

Summary

Sociology Now™

Reviewing is as easy as ➊, ➋, ➌.

➊ Before you do your final review, take the SociologyNow diagnostic quiz to help you identify the areas on which you should concentrate. You will find information on how to access SociologyNow on the foldout at the front of the textbook.

➋ As you review, take advantage of SociologyNow's study aids to help you master the topics in this chapter.

➌ When you are finished with your review, take SociologyNow's post-test to confirm you are ready to move on to the next chapter.

1. **The human body is a biological wonder. In what sense is it a sociological wonder too?**

   The body's parts, its disabilities, its aging, and its death mean different things and have different consequences for different cultures, historical periods, and categories of people. The human body cannot be fully understood without appreciating its sociological dimension.

2. **What is the connection between body type and social status?**

   Because low status influences diet and other factors, it is associated with people of short stature and people who are overweight. In turn, people of short stature and people who are overweight tend to receive fewer social rewards because of their body type.

3. **In what sense do people have rights over their own bodies?**

   Most people believe that they should and do have rights over their own bodies. Advances in medical technology and changing social norms encourage us to transform our bodies through surgery, prosthesis, and other means. However, we do not treat our bodies in idiosyncratic ways. Norms of body practice influence the way we treat our bodies.

4. **Aren't disabilities defined similarly everywhere and at all times? Haven't disabilities always been handled in the same way?**

   No. The definition of disability varies over time and place. For example, some people used to consider left-handedness a disability, but we do not share that view. As far as treatment is concerned, we also see much variation. Disabled people have traditionally suffered much prejudice and discrimination, but attempts were made beginning in the 19th century to integrate and rehabilitate them. Some governments sought to eliminate disabled people from society in the 20th century. Recently, disabled people have begun to organize themselves, assert the normality of disability, and form communities of disabled people.

5. **What is sociological about the aging process?**

   People attach different meanings to aging in different societies and historical periods. Thus, the stages of life vary in number and significance across societies.

6. **Is there a positive correlation between age and status?**

   Although it is true that the young have been, and still are, disadvantaged in many ways, it is rarely true that the eldest people in society are the best off. In most societies, including the United States, people of middle age have the most power and economic clout.

7. **What are the main approaches to age stratification?**

   Functionalist theory emphasizes that industrialization led to the differentiation of age cohorts and the receipt of varying levels of reward by each age cohort based on its functional importance to society. This supposedly results in the convergence of age stratification systems in all industrialized societies. Conflict theory stresses the way competition and conflict can result in the redistribution of rewards between age cohorts and the divergence of age stratification systems. Symbolic interactionists focus not on these macrosociological issues but on the meanings that people attach to different age cohorts.

8. **How is the United States aging?**

   The population of the United States is aging rapidly. By 2040 more than a fifth of Americans will be 65 or older. The fastest-growing age cohort among the elderly is composed of people 85 years and older. The ratio of men to women falls with age.

9. **How are elderly people faring economically in the United States?**

   Economically speaking, elderly Americans are faring reasonably well, partly because they have considerable political power. For example, poverty is less widespread among people older than 65 than among people younger than 45. Among the elderly, poverty is most widespread for those 85 and older, women, African Americans, people living alone, and people living in rural areas.

10. **If elderly people in the United States are doing reasonably well economically and politically, then does this mean that they don't face significant problems in our society?**

    Elderly Americans still face much prejudice and discrimination. Moreover, there exist crises in the provision of adequate care to the elderly and in the pension system.

11. **What is sociological about death and dying?**

    Different cultures attach different meanings to death and dying. Norms and commercial interests affect how we deal with these processes.

Questions to Consider

1. If you do not already use a wheelchair, borrow one and try to get around campus for a few hours. If you cannot borrow a wheelchair, pretend you are in one. Draw up an inventory of difficulties you face. How would the campus have to be redesigned to make access easier? Alternatively, if you are not already overweight, borrow some big clothes and add padding to make you look overweight. Go shopping for a few hours and see if sales clerks and other shoppers treat you differently from the way you are normally treated. Record your observations.
2. What generation do you belong to? What is the age range of people in your generation? What are some of the defining historical events that have taken place in your generation? How strongly do you identify with your generation? Ask the same questions of someone of a different race, gender, or class. How do these variables influence the experience of a generation?
3. Phone three funeral homes and request the least expensive pricing for cremation, a casket, and interment. Chances are you will be told that you must come in to discuss particulars. Insist that you are interested only in finding out the least expensive prices for these three items so that no additional conversation or viewing of options is necessary. It is still unlikely that you will be given prices over the phone. If so, ask why prices can't be given over the phone. Why do *you* think funeral directors are reluctant to discuss prices without a face-to-face meeting?

Web Resources

Companion Website for This Book

http://sociology.wadsworth.com

Begin by clicking on the Student Resources section of the website. Choose "Introduction to Sociology" and then the Brym and Lie book cover. Next, select the chapter you are currently studying from the pull-down menu. From the Student Resources page you will have easy access to InfoTrac® College Edition, MicroCase Online exercises, additional web links, and many other resources to aid you in your study of sociology, including practice tests for each chapter.

Recommended Websites

How fast are you aging? Find out by answering the questionnaire at http://www.msnbc.com/modules/quizzes/lifex.asp. This website also gives tips on how you can live longer.

Visit the AARP website at http://www.aarp.org to find out about the many issues faced by retirees and the individual and collective actions they are taking to deal with these issues. Also of interest on this site are the list of research and reference resources at http://www.aarp.org/indexes/reference.html#center and the essay on "How to Write a Research Paper in Gerontology" by Harry R. Moody at http://research.aarp.org/ageline/modhome.html.

You can find much valuable material on the Americans with Disabilities Act (1990) at http://www.usdoj.gov/crt/ada/adahom1.htm.

Michael Kearl maintains one of the most frequently visited webpages on the sociology of death and dying at http://www.trinity.edu/mkearl/death.html.

CHAPTER 13

Work and the Economy

Luis Castaneda/Getty Images

In this chapter, you will learn that:

- Three work-related revolutions—one in agriculture, one in industry, and one in the provision of services—have profoundly altered the way people sustain themselves and the way people live.
- In the past few decades, the number of "good" jobs has grown, but "bad" jobs have become even more numerous.
- With varying degrees of success, people seek to control work through unions, professional organizations, corporations, and markets.
- The growth of large corporations and global markets has shaped the transformation of work in recent decades and will shape the choices you face as a member of the labor force and a citizen.

The Promise and History of Work

Salvation or Curse?

The computerization of the office began in earnest about 20 years ago. Soon, the image of the new office was as familiar as a Dilbert cartoon. It was a checkerboard of 8′ × 8′ cubicles. Three padded walls, head high, framed each cubicle. Inside, a computer terminal sat on a desk. A worker quietly tapped away at a keyboard, seemingly entranced by the glow of a video screen.

Sociologist Shoshana Zuboff visited many such offices soon after they were computerized. She sometimes asked the office workers to draw pictures capturing their job experience before and after computerization. The pictures were strikingly similar. Smiles changed to frowns, mobility became immobility, sociability was transformed into isolation, freedom turned to regimentation. We reprint two of the workers' pictures on page 372. Work automation and standardization emerge from these drawings as profoundly degrading and inhuman processes (Zuboff, 1988).

The image conveyed by these drawings is only one view of the transformation of work in the information age. There is another, and it is vastly different. Bill Gates argues that computers reduce our work hours. They make goods and services less expensive by removing many distribution costs of capitalism. (Think of Amazon.com, which reduces the need for bookstores.) They allow us to enjoy our leisure time more (Gates with Myhrvold and Rinearson, 1996). This vision is well captured by the arresting December 1999 cover of *Wired* magazine, reprinted here on page 372. According to *Wired,* computers liberate us. They allow us to become more mobile and more creative. Computerized work allows our imaginations to leap and our spirits to soar.

These strikingly different images form the core questions of the sociology of work, and they will be our focus in this chapter. Is work a salvation or a curse? Or is it perhaps both at once? Is it more accurate to say that work has become more of a salvation or a

One view of the effects of computers on work. Shoshana Zuboff asked office workers to draw pictures representing how they felt about their jobs before and after a new computer system was introduced. Here are "before" and "after" pictures drawn by two office workers. Notice how even the flower on one worker's desk wilted after the new computer system was introduced.

"Before I was able to get up and hand things to people without having someone say, what are you doing? Now, I feel like I am with my head down, doing my work."

"My supervisor is frowning because we shouldn't be talking. I have on the stripes of a convict. It's all true. It feels like a prison in here."

From *In the Age of the Smart Machine: The Future of Work and Power* by Shoshana Zuboff, © 1988 by Basic Books, Inc. Reprinted by permission of Basic Books, a Member of Persens Books, L.L.C.

Another View of the Effect of Computers on Work. *Wired* magazine is always high on the benefits of computer technology.

Courtesy of *Wired* Magazine, Condé Nast Publications, Dec. 1999, James Porto photo.

curse over time? Or is work a salvation for some and a curse for others?

To answer these questions, we first sketch the evolution of work from preagricultural to postindustrial times (for a more detailed account, see Chapter 6, "Social Collectivities: From Groups to Societies"). As you will see, we have experienced three work-related revolutions in the past 10,000 years. Each revolution has profoundly altered the way we sustain ourselves and the way we live. Next, we examine how job skills have changed over the past century. We also trace changes in the number and distribution of "good" and "bad" jobs over time and project these changes nearly a decade into the future. We then analyze how people have sought to control work through unions, professional organizations, corporations, and markets. Finally, we place our discussion in a broader context. The growth of large corporations and markets on a global scale has shaped the transformation of work over the past quarter century. Understanding these transformations will help you understand the work-related choices you face as both a member of the labor force and a citizen.

Three Revolutions

One of the great engineering and construction feats of the late 20th century was the "Chunnel" linking France and Britain under the English Channel. Here, construction workers carry an I-beam.

The **economy** is the institution that organizes the production, distribution, and exchange of goods and services. Conventionally, analysts divide the economy into three sectors. The *primary* sector includes farming, fishing, logging, and mining. In the *secondary* sector, raw materials are turned into finished goods; manufacturing takes place. Finally, in the *tertiary* sector, services are bought and sold. These services include the work of nurses, teachers, lawyers, hairdressers, computer programmers, and so forth. Often, the three sectors of the economy are called the "agricultural," "manufacturing," and "service" sectors. We follow that practice here.

Three truly revolutionary events have taken place in the history of human labor. In each revolution, a different sector of the economy rose to dominance (Gellner, 1988; Lenski, 1966). These three revolutions were:

1. *The agricultural revolution.* Nearly all humans lived in nomadic tribes until about 10,000 years ago. Then, people in the fertile valleys of the Middle East, Southeast Asia, and South America began to herd cattle and grow plants using simple hand tools. Stable human settlements spread in these areas. About 5000 years ago, farmers invented the plow. By attaching plows to large animals, they substantially increased the land under cultivation. **Productivity**—the amount produced for every hour worked—soared.
2. *The Industrial Revolution.* International exploration, trade, and commerce stimulated the growth of markets from the 15th century on. **Markets** are social relations that regulate the exchange of goods and services. In a market, prices are established by how plentiful goods and services are (supply) and how much they are wanted (demand). About 225 years ago, the steam engine, railroads, and other technological innovations greatly increased the ability of producers to supply markets. This was the era of the Industrial Revolution. Beginning in England, the Industrial Revolution spread to western Europe, North America, Russia, and Japan within a century, making manufacturing the dominant economic sector.
3. *The revolution in services.* Service jobs were rare in preagricultural societies and few in agricultural societies because nearly everyone had to do physical work for human societies to survive. As productivity increased, however, service-sector jobs proliferated. By automating much factory and office work, the computer accelerated this shift in the last third of the 20th century. In the United States today, more than three-quarters of the labor force is employed in the service sector.

Sociology Now™

Learn more about **the Division of Labor** by going through The Division of Labor Animation.

The Social Organization of Work

Besides increasing productivity and causing shifts between sectors in employment, the agricultural, industrial, and service revolutions altered the way work was socially organized. For one thing, the **division of labor** increased. That is, work tasks became more specialized with each successive revolution. In preagricultural societies, there were four main jobs: hunting wild animals, gathering wild edible plants, raising children, and tending to the tribe's spiritual needs. In contrast, a postindustrial society like the United States boasts tens of thousands of different kinds of jobs.

In some cases, increasing the division of labor involves creating new skills. Some new jobs even require long periods of study. Foremost among these are the professions. In other cases, increasing the division of labor involves breaking a complex range of skills into a series of simple routines. For example, a hundred years ago, a butcher's job involved knowing how to dissect an entire cow. In today's meat-packing plants there are large-stock scalpers, belly shavers, crotch busters, gut snatchers, gut sorters, snout pullers, ear cutters, eyelid removers, stomach washers (also known as belly bumpers), hind-leg pullers, front-leg toenail pullers, and oxtail washers. A different person performs each routine. The tasks are repetitive and require a narrow range of skills.

- The **economy** is the institution that organizes the production, distribution, and exchange of goods and services.
- **Productivity** refers to the amount of goods or services produced for every hour worked.
- **Markets** are social relations that regulate the exchange of goods and services. In a market, the prices of goods and services are established by how plentiful they are (supply) and how much they are wanted (demand).
- The **division of labor** refers to the specialization of work tasks. The more specialized the work tasks in a society, the greater the division of labor.

If the division of labor increased as one work revolution gave way to the next, then social relations among workers also changed. In particular, work relations became more hierarchical. Although work used to be based on cooperation among equals, it now involves superordinates exercising authority and subordinates learning obedience. Owners oversee executives. Executives oversee middle managers. Middle managers oversee ordinary workers. Increasingly, work hierarchies are organized bureaucratically. That is, clearly defined positions and written goals, rules, and procedures govern the organization of work.

Clearly, the increasing division of labor changed the nature of work in fundamental ways. But did the *quality* of work improve or worsen as jobs became more specialized? That is the question we now address.

The Quality of Work: "Good" versus "Bad" Jobs

Personal Anecdote

John Lie once got a job as a factory worker in Honolulu. "The summer after my sophomore year in high school," John recalls, "I decided it was time to earn some money. I had expenses, after all, but only an occasionally successful means of earning money: begging my parents. Scouring the Help Wanted ads in the local newspaper, I soon realized I wasn't really qualified to do anything in particular. Some friends at school suggested I apply for work at a pineapple-canning factory. So I did.

"At the factory, an elderly man asked me a few questions and hired me. I was elated—but only for a moment. A tour of the factory floor ruined my mood. Row upon row of conveyor belts carried pineapples in various states of disintegration. Supervisors pushed the employees to work faster yet make fewer mistakes. The sickly sweet smell, the noise, and the heat were unbearable. After the tour, the interviewer announced I would get the graveyard shift (11 p.m. to 7 a.m.) at minimum wage.

"The tour and the prospect of working all night finished me off. Now dreading the prospect of working in the factory, I wandered over to a mall. I bumped into a friend there. He told me a bookstore was looking for an employee (nine to five, no pineapple smell, and air-conditioned, although still minimum wage). I jumped at the chance. Thus, my career as a factory worker ended before it ever began.

"A dozen years later, just after I got my Ph.D., I landed one of my best jobs ever. I was teaching for a year in South Korea. However, my salary hardly covered my rent. I needed more work desperately. Through a friend of a friend, I found a second job as a business consultant in a major corporation. I was given a big office with a panoramic view of Seoul and a personal secretary who was both charming and efficient. I wrote a handful of sociological reports that year on how bureaucracies work, how state policies affect workers, how the world economy had changed in the past two decades, and so on. I got to accompany the president of the company on trips to the United States. I spent most of my days reading books. I also went for long lunches with colleagues and took off several afternoons a week to teach."

What is the difference between a "good" job and a "bad" job, as these terms are usually understood? As this anecdote illustrates, bad jobs don't pay much and require the performance of routine tasks under close supervision. Working conditions are unpleasant and sometimes dangerous. Bad jobs require little formal education. In contrast, good jobs often require higher education. They pay well. They are not closely supervised, and they encourage the worker to be creative in pleasant surroundings. Other distinguishing features of good and bad jobs are not apparent from the anecdote. Good jobs offer secure employment, opportunities for promotion, health insurance, and other fringe benefits. In a bad job, you can easily be fired, you receive few if any fringe benefits, and the prospects for promotion are few. That is why bad jobs are often called "dead-end" jobs.

Most jobs fall between the two extremes sketched here. They have some mix of good and bad features. But what can we say about the overall mix of jobs in the United States? Are there more good than bad jobs? And what does the future hold? Are good or bad jobs likely to become more plentiful? What are your job prospects? These are tough questions, not least because some conditions that influence the mix of good and bad jobs are unpredictable. Nonetheless, sociological research sheds some light on these issues.

The Everett Collection

Charlie Chaplin's 1929 movie *Modern Times* was a humorous critique of the factory of his day. In the movie, Chaplin gets a tick and moves like a machine on the assembly line. He then gets stuck on a conveyor belt and run through a machine. Finally, he is used as a test dummy for a feeding machine. The film thus suggests that workers were being used for the benefit of the machines rather than the machines being used for the benefit of the workers.

The Deskilling Thesis

One view of how jobs are likely to develop was proposed more than 30 years ago by Harry Braverman (1974). Braverman argued that capitalists are always eager to organize work to maximize their profits. Therefore, they break complex tasks into simple routines. They replace labor with machines wherever possible. They exert increasing control over workers to make sure they do their jobs more efficiently. As a result, work tends to become **deskilled** over time. In the 1910s, for example, Henry Ford introduced the assembly line with just this aim in mind. The assembly line enabled Ford to produce affordable cars for a mass market. It also forced workers to do highly specialized, repetitive tasks requiring little skill at a pace set by their supervisors. Around the same time, Frederick W. Taylor developed the principles of **scientific management.** After analyzing the movements of workers as they did their jobs, Taylor trained them to eliminate unnecessary actions and greatly improve their efficiency. Workers became cogs in a giant machine known as the modern factory.

Many criticisms were lodged against Braverman's deskilling thesis in the 1970s and 1980s. Perhaps the most serious criticism was that he was not so much wrong as irrelevant. That is, even if his characterization of factory work was accurate (and we will see later that in some respects it was not), factory workers represent only a small proportion of the labor force. They represent a smaller proportion with every passing year. In 1974, the year Braverman's book was published, less than a third of the U.S. labor force was employed in manufacturing. By 1998, the figure had fallen to a little more than a fifth. Science fiction writer Isaac Asimov's claim that the factory of the future will employ only a man and a dog is clearly an exaggeration. (The man will be there to feed the dog, said Asimov. The dog will be there to keep the man away from the machines.) However, the manufacturing sector is shrinking and the service sector is expanding. The vital question, according to some of Braverman's critics, is not whether jobs are becoming worse in manufacturing but whether good jobs or bad jobs are growing in services, the sector that accounts for three-quarters of U.S. jobs today.

Shoshana Zuboff's analysis of office workers, mentioned at the beginning of this chapter, made it appear that Braverman's insights apply beyond the factory walls (Zuboff, 1988). She argued that the computerization of the office in the 1980s involved increased supervision of deskilled work. And she was right, at least in part. The computer did eliminate many jobs and routinize others. It allowed supervisors to monitor every keystroke, thus taking worker control to a new level. Today, employees who consistently fall behind a prescribed work pace or use their computers for personal e-mail, surfing the Web, and other pastimes can easily be identified and then retrained, disciplined, or fired. In the 1980s and 1990s, some analysts feared that good jobs in manufacturing were being replaced by bad jobs in services. From this point of view, the entire labor force was experiencing a downward slide (Bluestone and Harrison, 1982; Rifkin, 1995).

Sociology Now™

Learn more about **Employment** by going through the % of civilian labor force that is unemployed Map Exercise.

Deskilling refers to the process by which work tasks are broken into simple routines requiring little training to perform. Deskilling is usually accompanied by the use of machinery to replace labor wherever possible and increased management control over workers.

Scientific management is a system of improving productivity developed in the 1910s by Frederick W. Taylor. After analyzing the movements of workers as they did their jobs, Taylor trained them to eliminate unnecessary actions and greatly improve their efficiency.

Part-Time Work

Adding to the fear of a downward slide was the growth of part-time work. The proportion of part-time workers in the U.S. labor force increased 46 percent between 1957 and 1996 and another 25 percent between 1996 and 2004. In 2004, a quarter of all people in the U.S. paid labor force were part-timers, working less than 35 hours a week (Tilly, 1996: 1–4; U.S. Department of Labor, 2005b).

For two reasons, the expansion of part-time work is not a serious problem in itself. First, some part-time jobs are good jobs in the sense we defined earlier. Second, some people want to work part-time and can afford to do so. For example, some people who want a job also want to devote a large part of their time to family responsibilities. Part-time work affords them that flexibility. Similarly, many high school and college students work part-time and are happy to do so. Perhaps you have joined the ranks of part-time retail clerks and fast-food servers to help pay for your college education or to earn money for a car or a vacation during March break.

Although the growth of part-time jobs is not problematic for voluntary part-time workers or people who have good part-time jobs, an increasingly large number of people depend on part-time work for the necessities of full-time living. And the plain fact is that most part-time jobs are bad jobs. Thus, part-time workers make up two-thirds of the people working at or below minimum wage. Moreover, the fastest-growing category of part-time workers is composed of *involuntary* part-timers. Today, according to official statistics, a quarter of part-time workers want to be working more hours. And official statistics underestimate the scope of the problem. Surveys show that about a third of women officially classified as voluntary part-time workers would work more hours if good child care or elder care were available (Henson, 1996; Tilly, 1996: 1–4). Among unemployed workers, 83 percent want full-time work (U.S. Department of Labor, 2005b).

The downside of part-time work is not only economic, however. Nor is it just a matter of coping with the dull routine, the routinization and standardization of procedures analyzed in our discussion of "McDonaldization" in Chapter 9 ("Globalization, Inequality, and Development"). If you've ever had a part-time job, you know that one of its most difficult aspects involves maintaining your self-respect in the face of low pay, few if any benefits, little security, low status, and no chance for creativity. In the words of Dennis, a McDonald's employee interviewed by sociologist Robin Leidner: "This isn't really a job. . . . It's about as low as you can get. Everybody knows it" (quoted in Leidner, 1993: 182). And, Dennis might have added, nearly everybody lets you know they know it.

Sociology Now™

Learn more about **McDonaldization** by going through the McDonaldization of Society Video Exercise.

Katherine Newman (1999: 89) studied fast-food workers in Harlem, many of whom are teenagers working part-time. These workers are trained, Newman noted, to keep smiling no matter how demanding or rude their customers may be. The trouble is you can only count backward from a hundred so many times before feeling utterly humiliated. Anger almost inevitably boils over. According to Natasha, a young fast-food worker whom Newman interviewed:

> It's hard dealing with the public. There are good things, like old people. They sweet. But the younger people around my age are always snotty. They think they better than you because they not working [here]. . . . They told us that we just suppose to walk to the back and ignore it, but when they in your face like that, you get so upset you have to say something. . . . I got threatened with a gun one time. 'Cause this customer had threw a piece of straw paper in the back and told me to pick it up like I'm a dog. I said, "No." And he cursed at me. I cursed at him back and he was like, "Yeah, next time you won't have nothing to say when I come back with my gun and shoot your ass." Oh, excuse me. (Quoted in Newman, 1999, 90–91)

In inner-city ghettos like Harlem, the difficulty of maintaining one's dignity as a fast-food worker is compounded by the high premium most young people place on independence, autonomy, and respect. Young people enjoy few opportunities for high-quality education and upward mobility in the inner city. Unable to derive dignity from such conventional achievements, they tend to define their self-worth by "macho" behavior codes (Anderson, 1990). The problem this creates for teenagers who take jobs in fast-food restaurants is that their constant deference to customers violates the norms of inner-city youth culture. Therefore, fast-food workers are typically stigmatized by their peers. They are frequently the brunt of insults and ridicule. Newman recounts the case of one fast-food worker in Harlem who tried valiantly to keep his job a secret from his friends and acquaintances. He wouldn't tell his friends where he was going when he left for work. He took a long, circuitous route to work so his friends wouldn't know where he was headed. He kept his uniform in a bag and put it on only at work. He lied to his friends when they asked him where he got his spending money. He even hid behind a large freezer when his friends came in for a burger (Newman, 1999: 97).

Fast-food workers in Harlem undoubtedly represent an extreme case of the indignity endured by part-timers. However, the problem exists in various guises in most part-time jobs. For instance, if you work as a "temp" in an office you are more likely than other office workers to be the victim of sexual harassment. You are especially vulnerable to unwanted advances because you lack power in the office and are considered "fair game" (Welsh, 1999). Thus, the form and depth of degradation may vary from one part-time job to another, but as your own work experience may show, degradation seems to be a nearly universal feature of this type of deskilled work.

Web Research Project: Sexual Harassment at Work

A Critique of the Deskilling Thesis

The deskilling thesis undoubtedly captures one important tendency in the development of work. However, it does not paint a complete picture. That is because analyses like Braverman's and Zuboff's are too narrowly focused. They analyze specific job categories near the bottom of the occupational hierarchy. As a result, they do not allow us to form an impression of what is happening to the occupational structure as a whole. For instance, Zuboff analyzed lower-level service workers such as data-entry personnel and claims processors in a health insurance company. It is unclear from her research what is happening at the top of the service sector—whether, for example, good professional and managerial jobs are becoming more numerous relative to clerical jobs.

Subsequent analyses comparing the *entire* manufacturing and service sectors cast doubt on whether Zuboff's argument can be generalized. For example, one report prepared for the U.S. Bureau of Labor Statistics showed a wide range in the distribution of earnings in the service sector. Thus, although there are many low-paying jobs in services, there are also many high-paying jobs. Moreover, the distribution of income in services is practically the same as that in manufacturing. On some measures of job quality, such as job security, service jobs are on average better than manufacturing jobs. Thus, the decline of the manufacturing sector and the rise of the service sector do not imply a downward slide of the entire labor force. To be sure, there are many bad jobs in the service sector. Yet there are also many good jobs, so one should avoid generalizing about the entire sector from case studies of just a few job categories (Meisenheimer, 1998).

Even if we examine the least skilled service workers—fast-food servers, video store clerks, parking lot attendants, and the like—we find reason to question one aspect of the deskilling thesis, for although these jobs are dull and pay poorly, they are not as "dead-end" as they are frequently made out to be. One analysis of U.S. census and survey data showed that the least skilled service workers tend to be under the age of 25 and hold their jobs only briefly. After working in these entry-level jobs for a short time, most men move on to blue-collar jobs and most women move on to clerical jobs. These are not great leaps up the socioeconomic hierarchy, but the fact that they are common suggests that work is not all bleak and hopeless even at the bottom of the service sector (Jacobs, 1993; Myles and Turegun, 1994).

Braverman and Zuboff exaggerated the downward slide of the U.S. labor force because they underestimated the continuing importance of skilled labor in the economy. Assembly lines and computers may deskill many factory and office jobs. But if deskilling is to take place, then some members of the labor force must invent, design, advertise, market, install, repair, and maintain complex machines, including computerized and robotic systems.

Spencer Grant/PhotoEdit

The Toshiba circuit board assembly line in Irvine, California.

Most of these people have better jobs than the factory and office workers analyzed by Braverman and Zuboff. Moreover, although technological innovations kill off entire job categories, they also create entire new industries with many good jobs. Thirty years ago, Santa Clara County in California was best known for its excellent prunes. Today, it has been transformed into Silicon Valley, home to many tens of thousands of electronics engineers, computer programmers, graphics designers, venture capitalists, and so forth. Seattle, Austin, the Research Triangle in North Carolina, and Route 128 outside Boston are similar success stories. The introduction of new production techniques may even increase a country's competitive position in the world market. This can lead to employment gains at both the low and the high ends of the job hierarchy as consumers abroad rush to purchase relatively low cost goods.

Rather than involving a downward shift in the entire labor force, it seems more accurate to think of recent changes in work as involving a declining middle or (to say the same thing differently) a polarization between good and bad jobs. Many good jobs are opening up at the top of the socioeconomic hierarchy. Even more mediocre and bad jobs are opening up at the bottom. There are fewer new jobs in the middle (Myles, 1988).

Job polarization is nowhere more evident than in Silicon Valley. Top executives in Silicon Valley earned up to $121 million each in 1999, up 70 percent from the year before. Even at less lofty levels, there are many thousands of high-paying, creative jobs in the valley, more than half of them dependent on the high-tech sector (Bjorhus, 2000). Amid all this wealth, however, the electronics assembly factories in Silicon Valley are little better than high-tech sweatshops. Most workers in the electronics factories earn less than 60 percent of the Valley's average wage. They are mainly immigrant women of Hispanic and Asian origin. In the factories they work long hours and are frequently exposed to toxic solvents, acids, and gases. Semiconductor workers therefore suffer industrial illnesses at three times the average rate for other manufacturing jobs. Three separate studies have found significantly higher miscarriage rates among women working in chemical handling jobs than in other manufacturing jobs. There are no unions in Silicon Valley's electronics assembly factories, and the industry has fought efforts to introduce union-scale wages and working conditions (Corporate Watch, 2000). Although the opulent lifestyles of Silicon Valley's millionaires are often featured in the mass media, one must remember that a more accurate picture of the Valley—indeed, of work in the United States as a whole—is that of an increasingly polarized hierarchy.

Table 13.1 illustrates the trend toward job polarization in the United States. Based on projections by the U.S. Bureau of Labor Statistics, ▶Table 13.1 estimates growth of the 20 fastest-growing occupations for the period 1998–2008. Thirteen of the 20 jobs that are expected to grow fastest (65 percent of the total) require no higher education. Most of them demand only short-term on-the-job training. Twelve of them are low-level service jobs, including sales clerks, cashiers, home health workers, nursing aides, orderlies, janitors, waiters, and security guards. Five of the 20 fastest-growing jobs (25 percent) require higher education and are at the high end in terms of earning power. They include several computer-related occupations (systems analysts, computer engineers, and computer support specialists), managers, and administrators. Two of the 20 jobs (10 percent) require higher education but are in the middle in terms of earning power: registered nurses and high school teachers. Thus, Table 13.1 suggests that although the top of the service sector is growing, the bottom is growing more quickly and the middle is growing more slowly. This pattern is consistent with the pattern of growing income inequality discussed in Chapter 8 ("Stratification: United States and Global Perspectives"). There, you will recall,

Table 13.1
Expected Job Growth, Top 20 Occupations, United States, 1998–2008

Occupation	Estimated New Jobs, 1998–2008	Education Required
1. Systems analyst	577,000	Bachelor's degree
2. Retail salespersons	563,000	Short-term on-the-job training
3. All other sales and related	558,000	Moderate on-the-job training
4. Cashiers	556,000	Short-term on-the-job training
5. General managers, top executives	551,000	Work experience plus college degree
6. Truck drivers	493,000	Short-term on-the-job training
7. Office clerks	463,000	Short-term on-the-job training
8. Registered nurses	451,000	Associate degree
9. Computer support specialists	439,000	Associate degree
10. Personal care and home health aides	433,000	Short-term on-the-job training
11. Teacher assistants	375,000	Short-term on-the-job training
12. Janitors, cleaners, maids, housekeepers	365,000	Short-term on-the-job training
13. Nursing aides, orderlies, attendants	325,000	Short-term on-the-job training
14. Computer engineers	323,000	Bachelor's degree
15. Teacher, secondary school	322,000	Bachelor's degree
16. Office and administrative support supervisors and managers	313,000	Work experience in a related occupation
17. All other managers and administrators	305,000	Work experience plus college degree
18. Receptionists and information clerks	305,000	Short-term on-the-job training
19. Waiters and waitresses	303,000	Short-term on-the-job training
20. Security guards	294,000	Short-term on-the-job training

Source: U.S. Department of Labor (2000b).

we noted growing inequality between the top 20 percent of income earners and the remaining 80 percent.

Labor Market Segmentation

The polarization of jobs noted previously is taking place in the last of three stages of labor market development identified by David Gordon and his colleagues (Gordon, Edwards, and Reich, 1982). The period from about 1820 to 1890 was the period of *initial proletarianization* in the United States. During this period, craft workers in small workshops were replaced by a large industrial working class. Then, from the end of the 19th century until the start of World War II, the labor market entered the phase of *labor homogenization.* Extensive mechanization and deskilling took place during this stage. Finally, the third phase of labor market development is that of **labor market segmentation.** During this stage, which began after World War II and continues up to the present, the labor market has been divided into two distinct parts called the *primary* and *secondary* labor markets. In these different settings, workers have different characteristics.

The **primary labor market** is composed disproportionately of highly skilled or well-educated white males. They are employed in large corporations that enjoy high levels of capital investment. In the primary labor market, employment is relatively secure, earnings are high, and fringe benefits are generous. The **secondary labor market** contains a disproportionately large number of women and members of racial minorities, particularly African and Hispanic Americans. Employees in the secondary labor market tend to be unskilled and lack higher education. They work in small firms with low levels of capital investment. Employment is insecure, earnings are low, and fringe benefits are meager.

Labor market segmentation is the division of the market for labor into distinct settings. In these settings, work is found in different ways and workers have different characteristics. There is only a slim chance of moving from one setting to another.

The **primary labor market** is composed mainly of highly skilled or well-educated white males. They are employed in large corporations that enjoy high levels of capital investment. In the primary labor market, employment is secure, earnings are high, and fringe benefits are generous.

The secondary labor market contains a disproportionately large number of women and members of racial minorities, particularly African and Hispanic Americans. Employees in the secondary labor market tend to be unskilled and lack higher education. They work in small firms with low levels of capital investment. Employment is insecure, earnings are low, and fringe benefits are meager.

This characterization may seem to advance us only a little beyond our earlier distinction between good and bad jobs. However, proponents of labor market segmentation theory offer fresh insights into two important issues. First, they argue that work is found in different ways in the two labor markets. Second, they point out that social barriers make it difficult for individuals to move from one labor market to the other. To appreciate the significance of these points, it is vital to note that workers do more than just work. They also seek to control their work and prevent outsiders from gaining access to it. Some workers are more successful in this regard than others. Understanding the social roots of their success or failure permits us to see why the primary and secondary labor markets remain distinct. Therefore, we now turn to a discussion of forms of worker control.

Worker Resistance and Management Response

One criticism lodged against Braverman's analysis of factory work is that he inaccurately portrays workers as passive victims of management control. In reality, workers often resist the imposition of task specialization and mechanization by managers. They go on strike, change jobs, fail to show up for work, sabotage production lines, and so forth (Burawoy, 1979; Clawson, 1980; Gouldner, 1954).

Worker resistance has often caused management to modify its organizational plans. For example, Henry Ford was forced to double wages to induce his workers to accept the monotony, stress, and lack of autonomy associated with assembly-line production. Even so, gaining the cooperation of workers proved difficult. Therefore, beginning in the 1920s, some employers started to treat their employees more like human beings than cogs in a giant machine. They hoped to improve the work environment and thus make their employees more loyal and productive.

In the 1930s, the **human relations school of management** emerged as a challenge to Frederick W. Taylor's scientific management approach. Originating in studies conducted at the Hawthorne plant of the Western Electric Company near Chicago, the human relations school of management advocated less authoritarian leadership on the shop floor, careful selection and training of personnel, and greater attention to human needs and employee job satisfaction.

Over the next 70 years, owners and managers of big companies in all the rich industrialized countries realized they had to make more concessions to labor if they wanted a loyal and productive workforce. These concessions included not just higher wages, but more decision-making authority about product quality, promotion policies, job design, product innovation, company investments, and so forth. The biggest concessions to labor were made in countries with the most powerful trade union movements, such as Sweden (see Chapter 14, "Politics," and Chapter 21, "Collective Action and Social Movements"). In these countries, a large proportion of eligible workers are members of unions (about 80 percent in Sweden). Moreover, unions are organized in nationwide umbrella organizations that negotiate directly with centralized business organizations and governments over wages and labor policy in general. At the other extreme among highly industrialized countries is the United States. Here, fewer than 13 percent of eligible workers are members of unions and there is no centralized, nationwide bargaining among unions, businesses, and governments. Two indicators of the relative inability of American workers to wrest concessions from their employers are given in ▸Figures 13.1 and 13.2. On average, Americans work more hours per week than people in most of the other rich industrialized countries. Americans also have many fewer paid vacation days per year than workers in the other rich industrialized countries.

In the realm of industry-level decision making, too, American workers lag behind workers in western Europe and Japan. We can see this if we briefly consider the two main types of decision-making innovations that have been introduced into the factories of the rich industrialized countries since the early 1970s:

The **human relations school of management** emerged as a challenge to Taylor's scientific management approach in the 1930s. It advocated less authoritarian leadership on the shop floor, careful selection and training of personnel, and greater attention to human needs and employee job satisfaction.

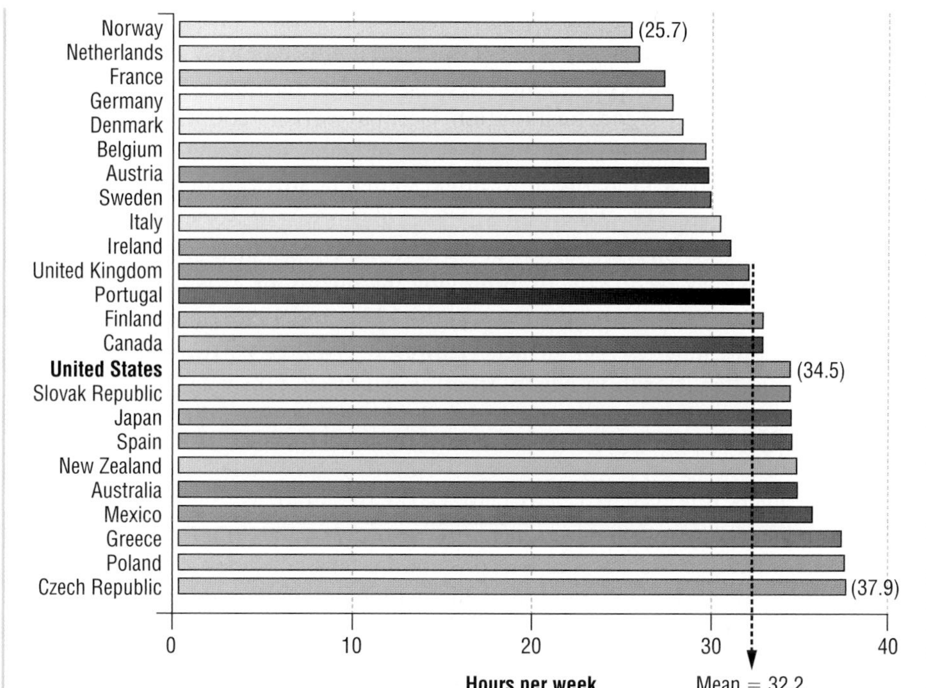

Figure 13.1
Average Hours Worked per Week, Selected Countries, 2003

Source: OECD (2005).

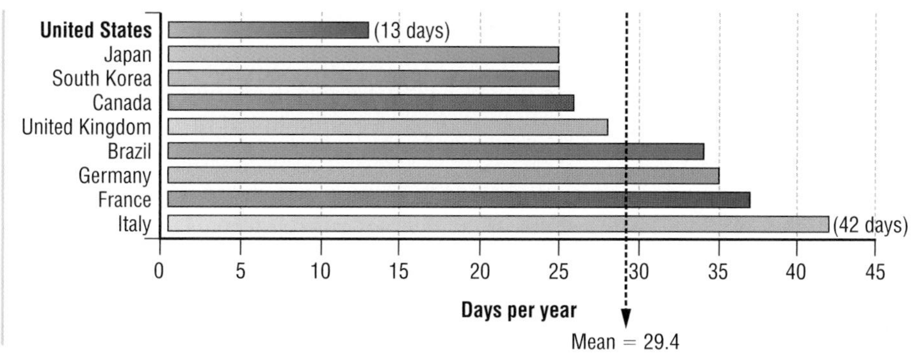

Figure 13.2
Average Paid Vacation Days per Year for Employees with One Year of Service, Selected Countries, 2004

Source: InfoPlease (2004).

1. *Reforms that give workers more authority on the shop floor* include those advanced by the **quality of work life** movement. "Quality circles" originated in Sweden and Japan. They involve small groups of a dozen or so workers and managers collaborating to improve both the quality of goods produced and communication between workers and managers. In some cases, this approach has evolved into a system that results in high productivity gains and worker satisfaction. For example, at Saab's main auto plant in Trolhattan, Sweden, the assembly line was eliminated (Krahn and Lowe, 1998: 239). Robots took over the arduous job of welding. In the welding area, groups of 12 workers program the computers, maintain the robots, ensure quality control, perform administrative tasks, and clean up. Elsewhere in the plant, autonomous teams of workers devote about 45 minutes to completing similar sets of integrated tasks. Workers build up inventory in "buffer zones" and decide themselves how to use it, thus introducing considerable flexibility into their schedule. All workers are encouraged to take advantage of many in-plant opportunities to upgrade their skills. Quality circles have been introduced in some American industries, including automotive and aerospace. However, they are less widespread in the United States than in western Europe and Japan.

The **quality of work life** movement originated in Sweden and Japan. It involves small groups of a dozen or so workers and managers collaborating to improve both the quality of goods produced and communication between workers and managers.

2. *Reforms that allow workers to help formulate overall business strategy* give workers more authority than quality circles. For example, in much of western Europe, workers are consulted not just on the shop floor but also in the boardroom. In Germany, this system is known as **codetermination.** German workers' councils review and influence management policies on a wide range of issues, including when and where new plants should be built and how capital should be invested in technological innovation. There are a few American examples of this sort of worker involvement in high-level decision making, mostly in the auto industry. Worker participation programs were widely credited with improving the quality of American cars and increasing the auto sector's productivity in the 1980s and 1990s, making it competitive again with Japanese car makers. However, "Worker participation programs have had a difficult birth in North America. No broad policy agenda guides their development, and no systematic social theory lights their way" (Hodson and Sullivan, 1995 [1990]: 449).

Internationally, unions have clearly played a key role in increasing worker participation in industrial decision making since the 1920s and especially since the 1970s. To varying degrees, owners and managers of big corporations have conceded authority to workers to create a more stable, loyal, and productive workforce. Understandably, workers who enjoy more authority in the workplace, whether unionized or not, have tried to protect the gains they have won. As we will now see, they have thereby contributed to the separation of primary from secondary labor markets.

Unions and Professional Organizations

Unions are organizations of workers that seek to defend and promote their members' interests. By bargaining with employers, unions have succeeded in winning improved working conditions, higher wages, and more worker participation in industrial decision making for their members.

With employers, unions have also helped to develop systems of labor recruitment, training, and promotion. These systems are sometimes called **internal labor markets** because they control pay rates, hiring, and promotions within corporations. At the same time, they reduce competition between a firm's workers and external labor supplies.

In an internal labor market, advancement through the ranks is governed by training programs that specify the credentials required for promotion. Seniority rules specify the length of time one must serve in a given position before being allowed to move up. These rules also protect senior personnel from layoffs according to the principle of "last hired, first fired." Finally, in internal labor markets, recruitment of new workers is usually limited to entry-level positions. In this way, the intake of new workers is controlled. Senior personnel are assured of promotions and protection from outside competition. For this reason, internal labor markets are sometimes called "labor market shelters."

Labor market shelters operate not only among unionized factory workers. Professionals, such as doctors, lawyers, and engineers, have also created highly effective labor market shelters. **Professionals** are people with specialized knowledge acquired through extensive higher education. They enjoy a high degree of work autonomy and usually regulate themselves and enforce standards through professional associations. (The American Medical Association is probably the best-known professional association in the United States.) Professionals exercise authority over clients and subordinates. They operate according to a code of ethics that emphasizes the altruistic nature of their work. Finally, they specify the credentials needed to enter their professions and thus maintain a cap on the supply of new professionals. This reduces competition, ensures high demand for their services, and keeps their earnings high. In this way, the professions act as labor market shelters, much like unions. (For further discussion of professionalization, see Chapter 17, "Education.")

Codetermination is a German system of worker participation that allows workers to help formulate overall business strategy. German workers' councils review and influence management policies on a wide range of issues, including when and where new plants should be built and how capital should be invested in technological innovation.

Unions are organizations of workers that seek to defend and promote their members' interests.

Internal labor markets are social mechanisms for controlling pay rates, hiring, and promotions within corporations while reducing competition between a firm's workers and external labor supplies.

Professionals are people with specialized knowledge acquired through extensive higher education. They enjoy a high degree of work autonomy and usually regulate themselves and enforce standards through professional associations.

Barriers Between the Primary and Secondary Labor Markets

We saw earlier that workers in the secondary labor market do not enjoy the high pay, job security, and benefit packages shared by workers in the primary labor market, many of whom are members of unions and professional associations. We may now add that workers find it difficult to exit the "job ghettos" of the secondary labor market. That is because three social barriers make the primary labor market difficult to penetrate: Often, there are (1) few entry-level positions in the primary labor market; (2) lack of informal networks linking the secondary and primary labor markets; and (3) lack of training and certification.

Few Entry-Level Openings in the Primary Labor Market

One set of circumstances that contributes to the lack of entry-level positions in the primary labor market is corporate "downsizing" and plant shutdowns. These took place on a wide scale in the United States throughout the 1980s and early 1990s (Box 13.1). The lack of entry-level positions is especially acute during periods of economic recession. A recession is usually defined as a period of six months or more during which the economy shrinks and unemployment rises. The United States was in the grip of recession in 27.9 percent of the 1200 months of the 20th century (38.2 percent of the months in the first half of the century, 17.7 percent of the months in the second half) (▸Figure 13.3). Big economic forces like plant shutdowns and recessions prevent upward mobility and often result in downward mobility. Consider the case of Jervis, a 19-year-old African American interviewed by Katherine Newman in her study of fast-food workers in Harlem. Like nearly all people in the secondary labor market, he wants steady work and upward mobility. Until 1991, he worked in a local fast-food restaurant and dreamed about moving up to a management position: "A decent job that gives you experience . . . that has a chain or something like that, that you could try to make some type of career where you don't always have to be locked into a position. It's a starting place. The bottom line was the overall benefit [of the job]: I moved ahead in life. That was good enough" (quoted in Newman, 1999: 252). Unfortunately, the recession of 1990–1991 cut short Jervis's dream. He lost his job.

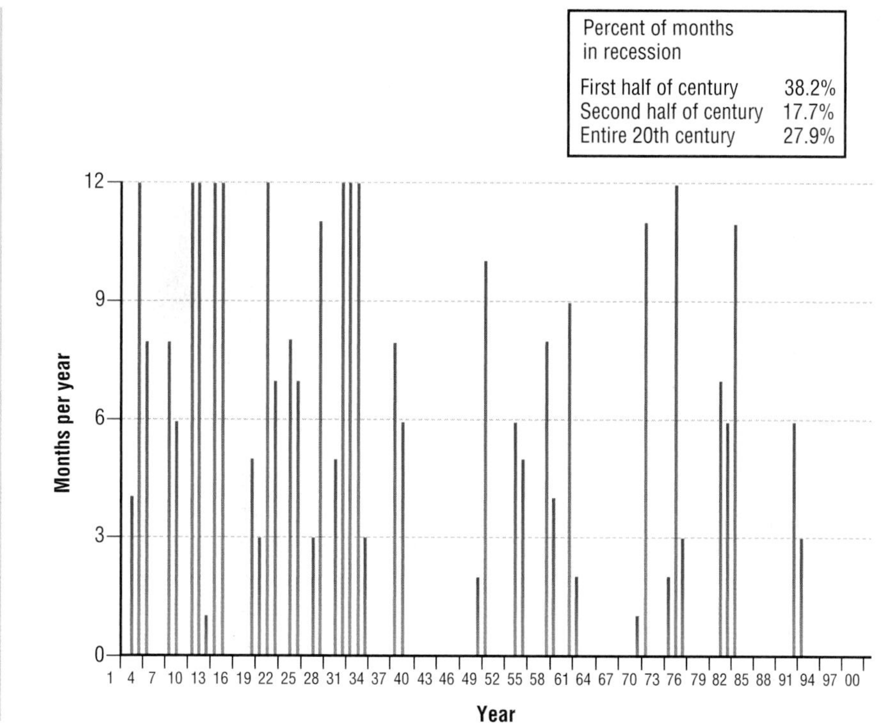

▸Figure 13.3
Months of Recession per Year, 1901–2000, United States

Source: Rex (2003).

BOX 13.1
Sociology at the Movies

Roger and Me (1989)

Scene 1: An auto factory in Flint, Michigan. Scene 2: The chief executive officer of General Motors (GM), Roger Smith, announces he's closing the factory. Scene 3: Newspaper headlines proclaim that GM is opening plants in Mexico. Scene 4: Michael Moore, the director of *Roger and Me* (1989), tries to interview Roger Smith so he can get him to face up to the consequences of his corporate decisions for the ordinary citizens of Flint. Moore is repeatedly rebuffed. Scene 5: During a gala Christmas party, Smith talks about generosity and "the total Christmas experience." The scene is interlaced with shots of the families of fired autoworkers being evicted from their homes. In one shot, a decorated Christmas tree is thrown on top of a family's belongings.

Roger and Me is an infuriating yet funny movie about deindustrialization and its impact on former GM workers in Flint. Beginning in the early 1980s, GM laid off tens of thousands of workers in the city and moved their jobs to Mexico. In 1980, Flint had 80,000 autoworkers. In 2000, it had 30,000 (Steinhart, 2000). Flint, once prosperous, was dubbed by *Money* magazine the worst place to live in America.

In the movie, Michael Moore connects corporate decision making to everyday life in the United States and industrial policy abroad. We see how multinational corporations can close factories in the United States, thereby exporting jobs to low-wage countries like Mexico and overturning the lives of ordinary American workers. Moore's attempts to interview Smith and discuss these issues are consistently irreverent and hilarious. Typically, when he tries to see Smith at his office, he offers the security guards his Chuck E. Cheese discount card for identification.

Roger and Me was released when many scholars and politicians were expressing fears that the U.S. labor force was on a downward slide due to deindustrialization. In cities like Flint, some laid-off workers moved away. Others stayed but were unable to find work and so contributed to a rising poverty rate. Flint has never recovered from deindustrialization. In 2000, the city's unemployment rate was 7.7 percent, more than two and a half times Michigan's 2.9 percent rate. *Roger and Me* remains a testament to the suffering of laid-off employees when neither corporations nor governments assume any responsibility for compensating, retraining, and relocating them.

The Everett Collection

Michael Moore directs *Roger and Me.*

Lack of Informal Networks

A second barrier preventing people in the secondary labor market from penetrating the primary labor market is their lack of informal networks linking them to good job openings. People often find out about job availability through informal networks of friends and acquaintances (Granovetter, 1995). These networks typically consist of people of the same ethnic and racial backgrounds. African and Hispanic Americans, who compose a disproportionately large share of workers in the secondary labor market, are less likely than whites to find out about job openings in the primary labor market, where the labor force is disproportionately white and non-Hispanic. Even within racial

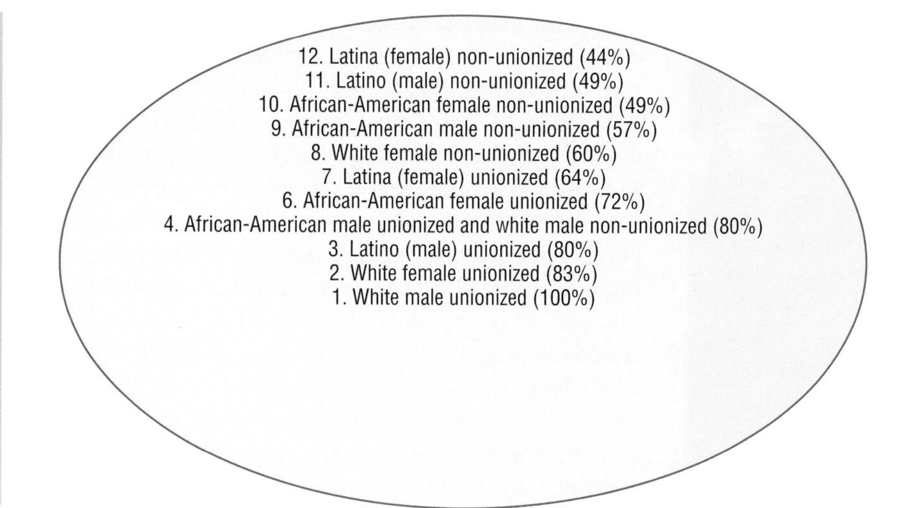

Figure 13.4
From the Primary to the Secondary Labor Market
Median weekly earnings in 1995 of full-time wage and salary workers in the United States by union status, sex, and race expressed as a percentage of the earning of unionized white male workers (=100%). Earnings decline when moving from the core of the primary labor market (category 1) to the periphery of the secondary labor market (category 12). What happens to union status, sex, and race as one moves from core to periphery?

Source: From *Working Women in America: Split Dreams* by Sharlene Hesse-Biber and Greg Lee Carter. Copyright © 2000 Oxford University Press. Used with permission.

groups, the difference between getting good work and having none is sometimes a question of having the right connections. Jervis, for instance, certainly knew about the pockets of middle-class African Americans living on the tree-lined streets near his apartment. He saw them dressing in stylish suits and driving to and from work in nice cars. But he had no contact with these people and no sense of how he might form ties with them that could lead him to good, steady work. When Newman asked him, "How do people meet job contacts in the first place?" Jervis replied: "I'd say luck and destiny"—hardly a realistic sense of the kind of networking often required to get good work (Newman, 1999: 252).

Lack of Training and Certification

Finally, mobility out of the secondary labor market is difficult because workers usually lack the required training and certification for jobs in the primary labor market. What is more, due to their low wages and scarce leisure time, they cannot usually afford to upgrade either their skills or their credentials. When Newman interviewed Jervis, his self-confidence and his hope of rising to a management position were fading fast. He was even unable to find steady minimum-wage work in retail shops and fast-food restaurants, partly because older and more experienced workers were taking many of the jobs that had formerly been open to teenagers. With only a tenth-grade education, his prospects were dim. He could not afford to go back to school. Due to impersonal economic forces, lack of network ties, and insufficient education, he was stuck, probably permanently, on the margins of the secondary labor market.

The dollar effects of these three barriers to entering the primary labor market are vividly illustrated by Figure 13.4. It shows that the median weekly earnings of full-time wage and salary workers in the United States decline sharply as one moves from the core of the primary labor market (category 1) to the periphery of the secondary labor market (category 12). Notice also that men, non-Hispanic whites, and unionized workers tend to be concentrated in the primary labor market. Women, African and Hispanic Americans, and nonunion members tend to be concentrated in the secondary labor market. We conclude that the barriers to entering the primary labor market are neither color-blind nor gender-neutral.

The Time Crunch and Its Effects

Although the quality of working life is higher for people who work in the primary than for those who work in the secondary labor market, one must be careful not to exaggerate the differences. Overwork and lack of leisure have become central features of our culture, and this is true in both labor markets. We are experiencing a growing time crunch. All the adults in most American households work full-time in the paid labor

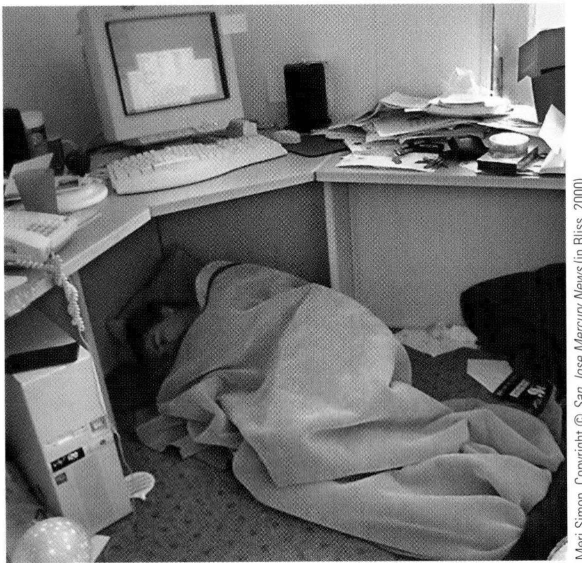
Meri Simon, Copyright © *San Jose Mercury News* (in Bliss, 2000)

Bedtime at Yahoo, Inc. In high-tech industries, working on very little sleep is common. David Filo, cofounder of Yahoo, naps under his desk.

force, and many adolescents work part-time. Some people work two jobs to make ends meet. Many office workers, managers, professionals, and farmers work 10, 12, or more hours a day due to tight deadlines, demands for high productivity, and a trimmed-down workforce. In 2001 nearly 11 percent of American workers (including part-time workers) put in 41 to 48 hours of paid work per week, nearly 11 percent put in 49 to 59 hours, and nearly 8 percent put in 60 hours or more (U.S. Department of Labor, 2003b). Add to this the heavy demands of family life—household chores, child care, and senior care—and one can readily understand why stress, depression, aggression, and substance abuse are on the rise, in both the secondary and the primary labor market. Three-quarters of American workers believe there is greater stress on the job now than there was two decades ago (Solomon, 1999: 51). According to one human resources expert:

> People are on the edge, and acting it out in the workplace. They're yelling obscenities at each other, coming into work chronically late, throwing food in the cafeteria, and crying in the hallways. It's not just a matter of acting a little inappropriately any more. No, it's gotten far worse than that. Nice people are having trouble with alcohol, drugs, depression and acting aggressively at work.
>
> And these aren't isolated instances—instead, they're a composite picture from chronic work distress, as well as difficulties trying to deal with personal life overload from marital problems, single parenthood, financial worries and the like. The stress is so great that people are snapping. And no one has to tell you that it's getting worse (Solomon, 1999: 48–9).

Stress is often defined as the feeling that one is unable to cope with life's demands given one's resources. Work is the leading source of stress throughout the world. In one study of office workers in 16 countries, including the United States, 54 percent of the respondents cited work as a current cause of stress in their lives and 29 percent cited money problems—which are also work related because their job is the main source of income for most people ("Work-related Stress," 1995). The results of a nationwide survey released in 2001 show that nearly half of U.S. employees feel they are overworked, overwhelmed by how much work they have to do, or unable to find the time to step back and process or reflect on the work they are doing. Those who often or very often use cell phones, beepers, pagers, computers, e-mail, and fax machines are 9 percent more likely than other employees to experience high levels of feeling overworked. Women, baby boomers (people born between 1946 and 1964), and managers and professionals also have significantly above average feelings of overwork (Galinsky, Kim, and Bond, 2001). The rate of severe depression is also on the rise. A survey of nine countries, including the United States, found that severe depression has increased in each succeeding generation since 1915. In some countries, people born after 1955 are three times more likely to experience serious depression than their grandparents ("The Changing Rate," 1992). According to a recent World Health Organization study, severe depression is the second leading contributor to "disease burden" (years lived with a disability) in the rich postindustrial countries. It is expected to rise to the number-one position by 2020 (Vernarec, 2000).

Three main reasons account for the decline of leisure and the more frantic pace of work for those employed in the paid labor force (Schor, 1992). First, big corporations are in a position to invest enormous and increasing resources in advertising. As we saw in Chapter 3 ("Culture"), advertising pushes Americans to consume goods and services at higher and higher levels all the time. Shopping has become an end in itself for many people, a form of entertainment, and, in some cases, a deeply felt "need." But consumerism requires money, so people have to work harder to shop more. Second, most corporate executives apparently think it is more profitable to have employees work more hours rather than hire more workers and pay expensive benefits for new employees.[1] Third, as we have just discussed, American workers are not in a position to demand reduced working hours and more vacation time because few of them are unionized. They lack clout and suffer the consequences in terms of stress, depression, and other work-related ailments.

The Problem of Markets

One conclusion we can draw from the preceding discussion is that the secondary labor market is a relatively **free market.** That is, the supply of, and demand for, labor regulates wage levels and other benefits. If labor supply is high and demand is low, wages fall. If labor demand is high and supply is low, wages rise. People who work in the secondary labor market lack much power to interfere in the operation of the forces of supply and demand.

In contrast, the primary labor market is a more **regulated market.** Wage levels and other benefits are established not just by the forces of supply and demand but also by the power of workers and professionals. As we have seen, they are in a position to influence the operation of the primary labor market to their own advantage.

This suggests that the freer the labor market, the higher the resulting level of social inequality. In fact, in the freest markets, many of the least powerful people are unable to earn enough to subsist, whereas a few of the most powerful people can amass unimaginably large fortunes. That is why the secondary labor market cannot be entirely free. The federal government had to establish a legal minimum wage to prevent the price of unskilled labor from dropping below the point at which people are literally able to make a living (Box 13.2). Similarly, in the historical period that most closely approximates a completely free market for labor—late-18th-century England—starvation became so widespread and the threat of social instability so great the government was forced to establish a system of state-run "poor houses" that provided minimal food and shelter for people without means (Polanyi, 1957 [1944]).

The question of whether free or regulated markets are better for society lies at the center of much debate in economics and politics (Kuttner, 1997). For many economic sociologists, however, that question is too abstract. In the first place, regulation is not an either-or issue but a matter of degree. A market may be more or less regulated. Second, markets may be regulated by different groups of people with varying degrees of power and different norms and values. Therefore, the costs and benefits of regulation may be socially distributed in many different ways. Only sociological analysis can sort out the costs and benefits of different degrees of market regulation for various categories of the population. These, then, are the main insights of economic sociology as applied to the study of markets: (a) The structure of markets varies widely across cultures and historical periods, and (b) the degree and type of regulation depend on how power, norms, and values are distributed among various social groups (Lie, 1992).

The economic sociologist's approach to the study of markets is different from that of the dominant trend in contemporary economics, known as the neoclassical school (Becker, 1976; Mankiw, 1998). Instead of focusing on how power, norms, and values shape markets, neoclassical economists argue that free markets maximize economic growth. We may use the minimum wage to illustrate their point. According to neoclassical economists, if the minimum wage is eliminated entirely, everyone will be better off. For example, in a situation where labor is in low demand and high supply, wages will fall. This will increase profits. Higher profits will in turn allow employers to invest more in expanding their businesses. The new investment will create new jobs, and the rising demand for labor will drive

In a **free market**, prices are determined only by supply and demand.

In a **regulated market**, various social forces limit the capacity of supply and demand to determine prices.

[1]However, medical economists have recently shown that work-related illnesses are actually costing businesses far more than they realize in terms of absenteeism and low productivity (Vernarec, 2000). It is too early to tell whether this research will have an impact on hiring policies.

BOX 13.2

SOCIAL POLICY: WHAT DO YOU THINK?

The Minimum Wage

"Flipping burgers at Mickey D's is no way to make a living," a young man once told John Lie. Having tried his hand at several minimum-wage jobs as a teenager, John knew the young man was right. At a little more than $5 an hour, a minimum-wage job may be fine for teenagers, many of whom are supported by their parents. However, it is difficult to live on one's own, much less support a family, on a minimum-wage job, even if you work full-time. This is the problem with the minimum wage. It does not amount to a living wage for many people.

In 1999, nearly 12 million American workers held minimum-wage jobs. Of this group, 58 percent were women. One million of them were single mothers (Bernstein, Hartmann, and Schmitt, 1999). In 1998 dollars, the minimum wage rose from about $3 an hour to about $7 an hour between 1938 and 1968. It fell to an inflation-adjusted $4.49 an hour by 2004 (▶Figure 13.6). Today, a single mother working full-time at the minimum wage does not make enough to lift a family of three (herself and two children) above the poverty level. In 1998, her earnings would have been 18 percent below the poverty level (Bernstein, Hartmann, and Schmitt, 1999). Because the minimum wage has fallen since the late 1960s, the percentage of workers earning poverty-level wages has increased. It now stands at about 29 percent. The percentage of workers earning poverty-level wages is much higher for women than men, and much higher for African Americans than others (▶Table 13.2).

Given a million single mothers who cannot lift themselves and their children out of poverty even if they work full-time, many scholars and policymakers suggest raising the minimum wage. Others disagree. They fear that raising the minimum wage would decrease the number of available jobs. Others disagree in principle with government interference in the economy. Some scholars and policymakers even advocate the abolition of the minimum wage.

What do you think? Should the minimum wage be raised? Should someone working full-time be entitled to live above the poverty level? Or should businesses be entitled to hire workers at whatever price the market will bear? In thinking about this question, you should bear in mind what happened between 1996 and 1999. In late 1996 and 1997, the minimum wage increased. In the next couple of years, the employment rate of low-wage workers, and particularly single mothers, followed suit. That is, due to the booming economy, the percentage of low-wage workers rose dramatically (Bernstein, Hartmann, and Schmitt, 1999). What does this say about the relationship between the minimum wage and the employment rate of low-wage workers?

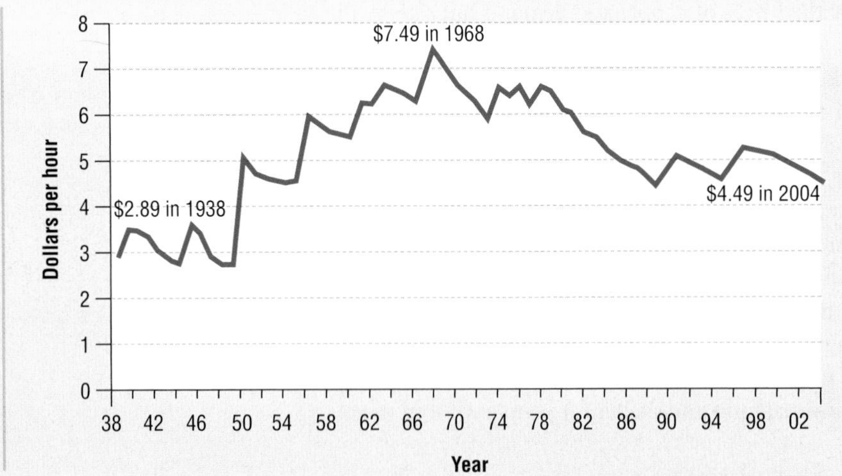

▶Figure 13.6
Value of the Federal Minimum Wage, United States, 1938–2004 (in 1998 dollars)

Source: Economic Policy Institute (2005); Infoplease.com (2003); U.S. Department of Labor (1999f; 2005a).

▶**Table 13.2**
Share of Workers Earning Poverty-Level Wages, by Gender and Race, United States, 1993–1997 (in percent)

		ALL AMERICANS		AFRICAN AMERICANS	
Year	Total	Men	Women	Men	Women
1973	23.5	12.8	39.1	28.4	50.07
1979	23.7	13.4	37.0	23.4	42.8
1989	28.5	21.2	36.8	33.2	43.6
1997	28.6	22.5	35.3	33.2	42.6

Source: Mishel, Bernstein, and Schmitt (1999: 136, 141).

ACCOUNT OF THE

SALE of a WIFE, by J. NASH,

IN THOMAS-STREET MARKET,

On the 29th of May, 1823.

This day another of those disgraceful scenes which of late have so frequently annoyed the public markets in this country took place in St. Thomas's Market, in this city; a man (if he deserves the name) of the name of John Nash, a drover, residing in Rosemary-street, appeared there leading his wife in a halter, followed by a great concourse of spectators; when arrived opposite the Bell-yard, he publicly announced his intention of disposing of his better half by Public Auction, and stated that the biddings were then open; it was a long while before any one ventured to speak, at length a young man who thought it a pity to let her remain in the hands of her present owner, generously bid 6d.! In vain did the anxious seller look around for another bidding, no one could be found to advance one penny, and after extolling her qualities, and warranting her sound, and free from vice, he was obliged, rather than keep her, to let her go at that price. The lady appeared quite satisfied, but not so the purchaser, he soon repented of his bargain, and again offered her to sale, when being bid nine-pence, he readily accepted it, and handed the lady to her new purchaser, who, not liking the transfer, made off with her mother, but was soon taken by her purchaser, and claimed as his property, to this she would not consent but by order of a magistrate, who dismissed the case. Nash, the husband, was obliged to make a precipitate retreat from the enraged populace.

Copy of Verses written on the Occasion:

COME all you kind husbands who have scolding wives,
Who tho' living together are tired of your lives,
If you cannot persuade her nor good natur'd make her
Place a rope round her neck & to market pray take her

Should any one bid, when she's offer'd for sale,
Let her go for a trifle lest she should get stale,
If six-pence be offer'd, & that's all can be had,
Let her go for the same rather than keep a lot bad.

Come all jolly neighbours, come dance sing & play,
Away to the wedding where we intend to drink tea;
All the world assembles, the young and the old,
For to see this fair beauty, as we have been told.

Here's success to this couple to keep up the fun,
May bumpers go round at the birth of a son;
Long life to them both, and in peace & content
May their days and their nights for ever be spent.

Shepherd, Printer, No. 6, on the Broad Weir, Bristol.

An unregulated market creates gross inequalities. Here, an account of a wife sold by her husband for sixpence in Britain in 1823.

wages up. Social inequality may increase, but eventually *everyone* will be better off thanks to the operation of the free market.[2]

One difficulty with the neoclassical theory is that, in the real world, resistance to the operation of free markets increases as you move down the social hierarchy from the wealthiest and most powerful members of society to the poorest and least powerful. People at the low end of the social hierarchy usually fight against falling wages. The resulting social instability can disrupt production and investment. A social environment full of strikes, riots, and industrial sabotage is unfavorable to high productivity and new capital investment. That is why the economic system of a society at a given time is a more or less stable set of compromises between advocates and opponents of free markets. Markets are only as free

[2]There is no necessary contradiction between one aspect of the sociological and the neoclassical approach to markets. As markets become less regulated, average wealth may grow and the gap between rich and poor may increase at the same time. As we saw in Chapter 8 ("Stratification: United States and Global Perspectives"), this is exactly what has been happening in the United States for nearly 30 years. Note also that a less influential school of economic thought, known as *institutionalism,* is much closer to the sociological perspective on markets than the neoclassical school.

as people are prepared to tolerate, and their degree of tolerance varies historically and between cultures (Berger and Dore, 1996; Doremus, Keller, Pauly, and Reich, 1998).

In the rest of this chapter, we offer several illustrations of how sociologists analyze markets. First, we compare *capitalism, communism,* and *democratic socialism,* the three main types of economic system in the 20th century. Second, we examine the ability of big corporations to shape markets in the United States today. Finally, we extend our analysis of corporate and free market growth to the global level. We identify advocates and opponents of these developments and sketch the main work-related decisions that face us in the early 21st century.

Economic Systems

Capitalism

The world's dominant economic system today is **capitalism.** Capitalist economies have two distinctive features:

1. *Private ownership of property.* In capitalist economies, individuals and corporations own almost all the means of producing goods and services. Individuals and corporations are therefore free to buy and sell land, natural resources, buildings, manufactured goods, medical services, and just about everything else. Like individuals, **corporations** are legal entities. They can enter contracts and own property. However, corporate ownership has two advantages over individual ownership. First, corporations are taxed at a lower rate than individuals. Second, the corporation's owners are not normally liable if the corporation harms consumers or goes bankrupt. Instead, the corporation itself is legally responsible for damage and debt.
2. *Competition in the pursuit of profit.* The second hallmark of capitalism is that producers, motivated by the prospect of profits, compete to offer consumers desired goods and services at the lowest possible price.

A purely capitalist economy is often called a laissez-faire system. *Laissez-faire* is French for "allow to do." In a laissez-faire system, the government does not interfere in the operation of the economy at all; it allows producers and consumers to do what they want. Adam Smith was an 18th-century Scottish economist who first outlined the operation of the ideal capitalist economy. According to Smith, everyone benefits from laissez-faire. The most efficient producers make profits while consumers can buy at low prices. If everyone pursues their narrow self-interest, unimpeded by government, the economy will achieve "the greatest good for the greatest number," said Smith (1981 [1776]).

In reality, no economy is purely laissez-faire. The state had to intervene heavily to create markets in the first place. For example, 500 years ago the idea that land is a commodity that can be bought, sold, and rented on the free market was utterly foreign to the native peoples who lived in the territory that is now North America. To turn the land into a marketable commodity, European armies had to force native peoples off the land and eventually onto reservations. Governments had to pass laws regulating the ownership, sale, and rent of land. Without the military and legal intervention of government, no market for land would exist.

Today, governments must also intervene in the economy to keep the market working effectively. For instance, governments create and maintain an economic infrastructure (roads, ports, etc.) to make commerce possible. They pass laws governing the minimum wage, occupational health and safety, child labor, and industrial pollution to protect workers and consumers from the excesses of corporations. If very large corporations get into financial trouble, they can expect the government to bail them out on the grounds that their bankruptcy would be devastating to the economy. For example, when Chrysler was facing financial ruin in the 1970s, the federal government stepped in with low-interest loans and other help to keep the corporation afloat. Similarly, in the 1980s, the government doled out $500 million of taxpayers' money to protect the nation's savings and loans institutions from collapse (Sherrill, 1990).

Capitalism is the dominant economic system in the world today. Capitalist economies are characterized by private ownership of property and competition in the pursuit of profit.

Corporations are legal entities that can enter into contracts and own property. They are taxed at a lower rate than individuals and their owners are normally not liable for the corporation's debt or any harm it may cause the public.

Table 13.3
Economic Competitiveness of Countries, 2004 ($N = 49$)

Country	Competitiveness Score	Country	Competitiveness Score
1. United States	100.0	26. Spain	67.414
2. Singapore	89.008	27. Israel	63.458
3. Canada	86.626	28. India	62.971
4. Australia	86.046	29. South Korea	62.201
5. Denmark	84.378	30. Portugal	58.485
6. Finland	83.636	31. Slovak Republic	57.462
7. Luxembourg	83.083	32. Colombia	57.370
8. Ireland	80.303	33. Hungary	57.209
9. Sweden	79.578	34. Czech Republic	56.440
10. Taiwan	79.543	35. Greece	56.346
11. Austria	78.933	36. Slovenia	55.498
12. Switzerland	78.809	37. Jordan	54.131
13. Netherlands	78.613	38. South Africa	53.786
14. Malaysia	75.919	39. Russia	52.140
15. Norway	75.468	40. Italy	50.307
16. New Zealand	75.394	41. Philippines	49.666
17. Germany	73.435	42. Brazil	48.130
18. United Kingdom	72.186	43. Romania	47.997
19. Japan	71.915	44. Turkey	43.459
20. China	70.725	45. Mexico	43.239
21. Belgium	70.324	46. Poland	41.593
22. Chile	69.901	47. Indonesia	38.095
23. Estonia	68.426	48. Argentina	36.937
24. Thailand	68.235	49. Venezuela	24.748
25. France	67.673		

Source: IMD International (2004).

Governments also play an influential role in establishing and promoting many leading industries, especially those that require large outlays on research and development. Until the 1960s, the U.S. government covered two-thirds of all research and development costs in the country, and it still covers nearly a third (Rosenberg, 1982; see Chapter 22, "Technology and the Global Environment," Figure 22.2). It developed the foundations of the Internet in the late 1960s. In the late 1980s, it gave $100 million a year to corporations working to improve the quality of the microchip (Reich, 1991). Occasionally, the government even resorts to armed force to ensure the smooth operation of the market, as, for example, when oil supplies are threatened by hostile foreign regimes (Lazonick, 1991).

As these examples of government intervention suggest, we should see Smith's ideal of a laissez-faire economy as just that: an ideal. In the real world, markets are free to varying degrees, but none is or can be entirely free. Which capitalist economies are the most free and which are the least free? The International Institute for Management Development in Switzerland publishes a widely respected annual index of competitiveness for capitalist countries. The index is based on the amount of state ownership of industry and many other indicators of market freedom. Table 13.3 gives the 2004 scores for the 49 countries on which data are available. The United States tops the list. It is the most competitive economy in the world, with a score of 100. Venezuela ranks 49th with a score of under 25.

Sociology Now™

Learn more about **Capitalism** by going through the Capitalism versus Socialism Learning Module.

Communism

Like laissez-faire capitalism, communism is an ideal. **Communism** is the name Karl Marx gave to the classless society that, he said, is bound to develop out of capitalism. Socialism is the name he gave to the transitional phase between capitalism and commu-

Communism is an economic system characterized by public ownership of property and government planning of the economy.

Sociology Now™

Learn more about **Communism** by going through the Capitalism versus Socialism Learning Module.

nism. No country in the world is communist in the pure sense of the term. About two dozen countries in Asia, Latin America, and Africa consider themselves socialist. They include China, North Korea, Vietnam, and Cuba. As an ideal, communism is an economic system with two distinct features:

1. *Public ownership of property.* Under communism, the state owns almost all the means of producing goods and services. Private corporations do not exist. Individuals are not free to buy and sell goods and services. The aim of public ownership is to ensure that all individuals have equal wealth and equal access to goods and services.
2. *Government planning.* Five-year state plans establish production quotas, prices, and most other aspects of economic activity. Political officials—not forces of supply and demand—design these state plans and determine what is produced, in what quantities, and at what prices. A high level of control of the population is required to implement these rigid state plans. As a result, democratic politics is not allowed to interfere with state activities. Only one political party exists—the Communist Party. Elections are held regularly, but only the Communist Party is allowed to run for office (Zaslavsky and Brym, 1978).

Until recently, the countries of Central and Eastern Europe were single-party, socialist societies. The most powerful of these countries was the Soviet Union, which comprised Russia and 14 other socialist republics. In perhaps the most surprising and sudden change in modern history, the countries of the region started introducing capitalism and holding multiparty elections in the late 1980s and early 1990s.

The collapse of socialism in Central and Eastern Europe was due to several factors. For one thing, the citizens of the region enjoyed few civil rights. For another, their standard of living was only about half as high as that of people in the rich industrialized countries of the West. The gap between East and West grew as the arms race between the Soviet Union and the United States intensified in the 1980s. The standard of living fell as the Soviet Union mobilized its economic resources to try to match the quantity and quality of military goods produced by the United States. Dissatisfaction was widespread and expressed itself in many ways, including strikes and political demonstrations. It grew as television and radio signals beamed from the West made the gap between socialism and capitalism more apparent to the citizenry. Eventually, the communist parties of the region felt they could no longer govern effectively and so began to introduce reforms.

In the 1990s, some Central and Eastern European countries were more successful than others in introducing elements of capitalism and raising their citizens' standard of living. The Czech Republic was most successful. Russia and most of the rest of the former Soviet Union were least successful. Many factors account for the different success rates, but perhaps the most important is the way different countries introduced reforms. The Czechs introduced both of the key elements of capitalism—private property and competition in the pursuit of profit. The Russians, however, introduced private property without much competition. Specifically, the Russian government first allowed prices to rise to market levels. This made many basic goods too expensive for a large part of the population, which was quickly impoverished. Next, the government sold state-owned property to individuals and corporations. However, the only people who could afford to buy the factories, mines, oil refineries, airlines, and other economic enterprises were organized criminals and former officials of the Communist Party. They alone had access to sufficient capital and insider information about how to make the purchases (Handelman, 1995). The effect was to make Russia's level of socioeconomic inequality among the highest in the world. A crucial element lacking in the Russian reform was competition. A few giant corporations control nearly every part of the Russian economy. They tend not to compete against each other. Competition would drive prices down and efficiency up. However, these corporations are so big that they can agree among themselves to set prices at levels that are most profitable for them. They thus have little incentive to innovate. Moreover, they are so big and rich that they have enormous influence over government. When a few corporations

Table 13.4
Public Ownership of Six Basic Industries in Five Countries

Postal	Service	Telecommunications	Electricity	Railways	Airlines	Steel	Public Ownership Score
United States	100	0	25	25	0	0	150
Canada	100	25	100	75	0	0	300
Japan	100	100	0	75	25	0	300
France	100	100	75	100	75	75	525
Sweden	100	100	50	100	100	75	525

Note: The values shown for each industry are as follows:
100 = all or nearly all government owned
75 = mixed ownership, government predominating
50 = approximately equal public and private ownership
25 = mixed ownership, private predominating
0 = all or nearly all private ownership.

We calculated the Public Ownership Score for each country by adding the values for each industry in that country. All estimates are ours.

are so big that they can behave in this way, they are called **oligopolies.**[3] Russia is full of them. The lack of competition in Russia has prevented the country from experiencing much economic growth since reforms began more than a decade ago (Brym, 1996a; 1996b; 1996c).

Democratic Socialism

Several prosperous and highly industrialized countries in northwestern Europe, such as Sweden, Denmark, and Norway, are **democratic socialist** societies. So are France and Germany, albeit to a lesser degree. Such societies have two distinctive features (Olsen, 2002).

1. *Public ownership of certain basic industries.* In democratic socialist countries, the government owns certain basic industries entirely or in part. These industries include telecommunications, electricity, railways, airlines, and steel (Table 13.4). Still, as a proportion of the entire economy, the level of public ownership is not high—far lower than the level of public ownership in socialist societies. The great bulk of property is privately owned, and competition in the pursuit of profit is the main motive for business activity, just as in capitalist societies.
2. *Government intervention in the market.* As the term "democratic socialism" implies, these countries enjoy regular, free, multiparty elections, just like the United States. However, unlike the United States, political parties backed by a strong trade union movement have formed governments in democratic socialist countries for much of the post–World War II period. As many as 80 percent of nonagricultural workers, including white-collar workers, are unionized in democratic socialist countries. (The comparable figure for the United States is less than 13 percent.) The governments that these unions back intervene strongly in the operation of markets for the benefit of ordinary workers. Taxes are considerably higher than in capitalist countries. Consequently, social services are much more generous. Looking at government expenditure as a percent of gross domestic product (GDP), one finds that the Swedish government spends more than twice as much as the United States on welfare, unemployment insurance, medical care, child care, worker training, and other social services. Due to the government's vigorous role in redistributing income, the gap between rich and poor is much narrower than in the capitalist countries. Moreover, on average, workers earn more, work fewer hours, and enjoy more paid vacation days.

[3]A monopoly is a single producer that completely dominates a market.

Oligopolies are giant corporations that control part of an economy. They are few and tend not to compete against one another. Instead, they can set prices at levels that are most profitable for them.

In **democratic socialist** countries, democratically elected governments own certain basic industries entirely or in part and intervene vigorously in the market to redistribute income.

Since the 1980s, the democratic socialist countries have moved in a somewhat more capitalist direction. In particular, they have privatized some previously government-owned industries and services. Still, these countries retain their distinct approach to governments and markets, which is why democratic socialism is sometimes called a "Third Way" between capitalism and socialism.

The Corporation

Having reviewed the major types of economic system, we now want to investigate the role of the dominant economic organization in almost all countries today: the corporation.

Giant Corporations

We noted in our previous discussion of the Russian economy that oligopolies typically force consumers to pay higher prices and exercise excessive influence on government. In the United States and other Western countries, "antitrust" laws limit the growth of oligopolies. The 1890 Sherman Antitrust Act and the 1914 Clayton Act are the basic U.S. antitrust laws. They have prevented the largest corporations from gaining much more control of specific industries than they had in the 1930s. Thus, in 1935, the four largest firms in each manufacturing industry controlled, on average, 37 percent of sales in that industry. In 1992 (remarkably, the last year for which data are available as of this writing), the figure stood at 40 percent (Hodson and Sullivan, 1995 [1990]: 393; U.S. Bureau of the Census, 2000e).

However, the law has been only partly effective in stabilizing the growth of oligopolies. For instance, the government managed to break the stranglehold of American Telephone and Telegraph (AT&T) on the telecommunications market in the 1980s, but its efforts to break Microsoft into two smaller corporations failed. Moreover, when the four biggest corporations in an industry make 4 out of every 10 dollars in sales, it is hard to deny they are enormously powerful. The top 500 corporations in the United States control more than two-thirds of business resources and profit. This is a world apart from the early 19th century, when most business firms were family owned and served only local markets.

The Growth of Conglomerates

It is also important to note that an important effect of U.S. antitrust law is to encourage big companies to diversify. That is, rather than increasing their share of control in their own industry, corporations often move into new industries. Big companies that operate in several industries at the same time are called **conglomerates.** For example, in 2000, America Online (AOL), the world's biggest Internet service provider, took over Time-Warner, the entertainment giant, thereby forming a conglomerate. The takeover allowed AOL's business to grow, but because it grew outside the Internet industry, AOL avoided the charge of forming an oligopoly. Unlike oligopolies, conglomerates are growing rapidly in the United States. Big companies are swallowed up by still bigger ones in wave after wave of corporate mergers (Mizruchi, 1982; 1992).

Interlocking Directorates

Outright ownership of a company by a second company in another industry is only one way that corporations may be linked. **Interlocking directorates** are another. Interlocking directorates are formed when an individual sits on the board of directors of two or more noncompeting companies. (As we saw, antitrust laws prevent an individual from sitting on the board of directors of a competitor.) For instance, in 1997 the board of directors of IBM included, among others, the chair and chief executive officer (CEO) of Mobil Oil, the chair and CEO of Ford, and the president of Mitsubishi ("IBM," 1997). In 2000 Ann McLaughlin, secretary of labor in the Reagan administration, served on the boards of directors of Nordstrom, Kellogg, Marriott, Fannie Mae (the Federal National Mortgage Association), and Microsoft, among other major corporations ("Ann

Conglomerates are large corporations that operate in several industries at the same time.

Interlocking directorates are formed when an individual sits on the board of directors of two or more noncompeting companies.

McLaughlin," 2000). Such interlocks enable corporations to exchange valuable information and form alliances for their mutual benefit. They also create useful channels of communication to, and influence over, government (Mintz and Schwartz, 1985; Mintz, 1989; Useem, 1984).

Small Firms and Big Corporations

Of course, small businesses continue to exist (Granovetter, 1984). In the United States, 85 percent of businesses have fewer than 20 employees. Fully 40 percent of the labor force works in firms with fewer than 100 employees. Small firms are particularly important in the service sector. However, compared with large firms, profits in small firms are typically low. Bankruptcies are common. Small firms usually use outdated production and marketing techniques. Jobs in small firms often offer low wages and meager benefits.

Most of the U.S. labor force now works in large corporations. About a third of the labor force is employed in the 1500 largest industrial, financial, and service firms. In the service sector, the biggest employer is Wal-Mart, with 1,341,500 employees in the United States in 2003. In manufacturing, the largest employer is General Motors, with 355,500 employees in the United States in 2003 ("Forbes 500," 2004). As we will now see, however, even these figures underestimate the global reach and influence of the biggest corporations.

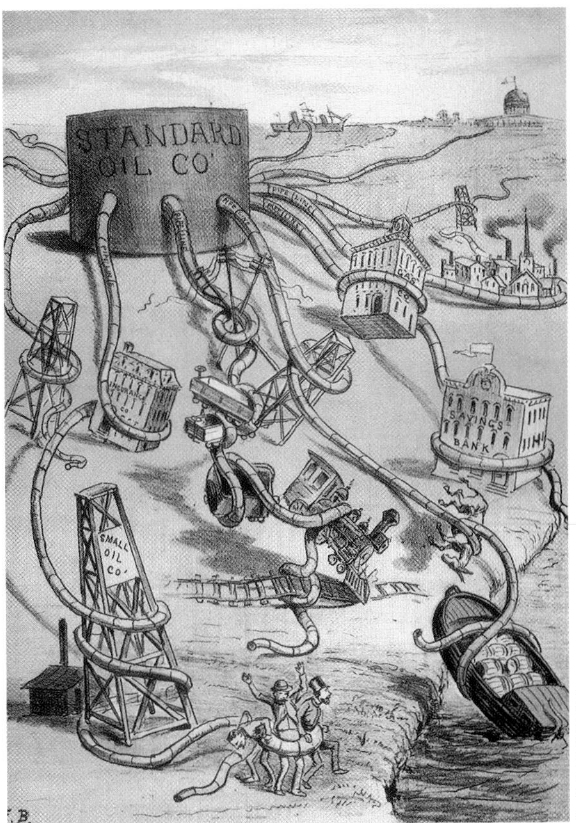

The Granger Collection

"The Monster Monopoly," an 1884 cartoon attacking John D. Rockefeller's Standard Oil Company. One of the most famous antitrust cases ever to reach the U.S. Supreme Court resulted in the breakup of Standard Oil in 1911. The Court broke new ground in deciding to dissolve the company into separate geographical units.

Globalization

In the 1980s and early 1990s, the United States was hit by a wave of corporate "downsizing" (Dudley, 1994; Gordon, 1996; Smith, 1990). Especially in the older manufacturing industries of the Northeast and Midwest—an area sometimes called the "rust belt"—hundreds of thousands of blue-collar workers and middle managers were fired. In places like Flint, Michigan, and Racine, Wisconsin, the consequences were devastating. Unemployment soared. Social problems such as alcoholism and wife abuse became acute.

Some people blamed government for the plant shutdowns. They said taxes were so high, big corporations could no longer make decent profits. Others blamed the unemployed workers themselves. They said powerful unions drove up the hourly wage to the point where companies like General Motors and Ford were losing money. Katherine Dudley studied the closure of an auto factory in Racine. She noticed that unemployed workers "who were once able to fulfill their obligations to family, community, and nation . . . have become culturally 'deviant.' . . . [They] are no longer perceived as . . . hardworking [and] self-sacrificing . . . [but are] the target of national—and now even international—ridicule, censure, and shame" (Dudley, 1994: 161). Still others blamed the corporations. As soon as big corporations closed plants in places like Racine and Flint, they opened new ones in places like northern Mexico. Mexican workers were happy to earn only one-sixth or one-tenth as much as their American counterparts. The Mexican government was delighted to make tax concessions to attract the new jobs.

In the 1980s, workers, governments, and corporations got involved as unequal players in the globalization of the world economy. Japan and Germany had fully recovered from the devastation of World War II. With these large and robust industrial economies now firing on all cylinders, American-based multinationals were forced to cut costs and become more efficient to remain competitive. On a scale far larger than ever before, they began to build branch plants in many countries to take advantage of inexpensive labor

AP/Wide World Photos

Some people in the rich, industrialized countries oppose the globalization of commerce. For example, the World Trade Organization (WTO) was set up by the governments of 134 countries in 1994 to encourage and referee global commerce. When the WTO met in Seattle in December 1999, 40,000 union activists, environmentalists, supporters of worker and peasant movements in developing countries, and other opponents of multinational corporations staged protests that caused property damage and threatened to disrupt the proceedings. Police and the National Guard replied with concussion grenades, tear gas, rubber bullets, and mass arrests. Similar protests have taken place at subsequent WTO meetings in other countries.

and low taxes. Multinational corporations based in Japan and other highly industrialized countries did the same.

However, although multinational corporations could easily move investment capital from one country to the next, workers were rooted in their communities, and governments were rooted in their nation-states. Multinationals thus had a big advantage over the other players in the globalization game. They could threaten to move plants unless governments and workers made concessions. They could play one government off against another in the bidding war for new plants. And they could pick up and leave when it became clear that relocation would do wonders for their bottom line.

Today, more than 25 years after the globalization game began in earnest, it is easier to identify the winners than the losers. The clear winners are the stockholders of the multinational corporations, whose profits have soared. The losers, at least initially, were American blue-collar workers. To cite just one example, between 1980 and 1993, General Motors cut its labor force by more than 30 percent. Between 1993 and 1999, the number of Americans employed by General Motors fell another 20 percent.

Even while these cuts were being made, however, some large American manufacturers were hiring. For instance, employment at Boeing grew more than 65 percent between 1993 and 1999. In the service sector, employment soared. For example, Wal-Mart employed twice as many people in 1999 as in 1993 ("Forbes 500," 1999; Hodson and Sullivan, 1995 [1990]: 393). In 2000, unemployment in the United States hit a 38-year low. As a result, many analysts believed that the 1980s was a period of extremely difficult economic restructuring rather than the beginning of the decline of the American economy, as some people warned at the time.

New worries surfaced by 2003, however, some linked to the rise of China as an economic powerhouse. Since 1979, Chinese economic growth has averaged about 10 percent a year, transforming China into the world's seventh largest economy and one of the world's major exporters. Americans now eagerly buy a quarter of China's exports. Why? Chinese wages are low, so Chinese manufactured goods are inexpensive, and American consumers like a bargain. The downside is that inexpensive Chinese goods have driven many American manufacturers out of business and left many American workers without jobs. Other American manufacturers have reduced operations in the United States and established Chinese branch plants, effectively exporting American jobs in the process. Manufacturing employment fell to 13 percent of total employment in the United States in 2000, half of what it was in 1970.

Defenders of free trade argue that most of the decline in manufacturing employment is not due to Chinese competition but to the increased productivity of American workers (Mankiw, 2003). From their point of view, an American worker can produce a lot more today than 30 years ago because of improved technology, so fewer workers are need to manufacture more goods. This argument is accurate but it ignores the fact that increased productivity itself is in part a response to competition from abroad; we invest more in technology to help overcome wage competition. Free trade supporters also note that most displaced workers eventually find other jobs. Again true. Yet, more often than not, the new jobs are inferior to the jobs that are lost. Recent research shows that the overall quality of American jobs (as measured by job stability, wages, and part-time vs. full-time employment) is declining (Tal, 2004). An American worker in a manufacturing plant may lose her job because of cheap Chinese imports and then find a new job at a Wal-Mart checkout, but the new job is more likely to be part-time, pay less, and offer fewer benefits. (Ironically, the checkout clerk will wind up scanning Chinese manufactured goods because Wal-Mart accounts for more than 10 percent of all sales of Chinese imports in the United States.)

Another issue ignored by supporters of free trade is that it is not just manufacturing jobs that are being lost to low-wage countries like China. In 2004, the *International Herald-Tribune* reported that in the next three years a major New York securities firm plans to replace its team of American software engineers (annual wage: $150,000) with equally competent engineers in India (annual wage: $20,000). Between 2004 and 2009, the number of radiologists in the United States is expected to decline "significantly" because magnetic resonance imaging (MRI) data can be sent to Asia over the Internet and diagnoses can be delivered at a fraction of the cost (Schumer and Roberts, 2004). These and many other examples of "outsourcing" point to a growing trend for high-wage, technical jobs to be lost to highly educated workers in India, Poland, and elsewhere. As Nobel Prize–winning economist Paul A. Samuelson recently noted, many mainstream economists prefer to ignore the negative effects of free trade on average income and class inequality in the United States (Samuelson, 2004; Public Citizen, 2004).

Globalization and the Less Developed Countries

In the globalization game, there are both winners and losers among the governments and citizens of the less developed countries too. On the one hand, it is hard to argue with the assessment of the rural Indonesian woman interviewed by Diane Wolf. She prefers the regime of the factory to the tedium of village life. In the village, the woman worked from dawn till dusk doing household chores, taking care of siblings, and feeding the family goat. In the factory, she earns less than a dollar a day sewing pockets on men's shirts. Yet because work in the factory is less arduous, pays something, and holds out the hope of even better work for future generations, the woman views it as nothing less than liberating (Wolf, 1992). Many workers in other regions of the world where branch plants of multinationals have sprung up in recent decades feel much the same way. A wage of $2 an hour is good pay in Mexico, and workers rush to fill jobs along Mexico's northern border with the United States.

Yet the picture is not all bright. The governments of developing countries attract branch plants by imposing few if any pollution controls on their operations. This has dangerous effects on the environment. Typically, fewer jobs are available than the number of workers who are drawn from the countryside to find work in the branch plants. This results in the growth of urban slums suffering from high unemployment and unsanitary conditions. High-value components are often imported. Therefore, the branch plants create few good jobs involving design and technical expertise. Finally, some branch plants—particularly clothing and shoe factories in Asia—exploit children and women, requiring them to put in long workdays at paltry wages and in unsafe conditions.

Companies such as Nike and The GAP have been widely criticized for conditions in their overseas sweatshops. Nike is the market leader in sports footwear. It has been in the forefront of moving production jobs overseas to places like Vietnam and Indonesia. In Indonesia, Nike factory workers make about 10 cents an hour. That is why labor costs account for only about 4 percent of the price of a pair of Nike shoes. Workdays in the factories stretch as long as 16 hours. Substandard air quality and excessive exposure to toxic chemicals like toluene are normal. An international campaign aimed at curbing Nike's labor practices has had only a modest impact. For example, in 1999 wages in the Indonesian factories were raised about a penny an hour ("The Nike Campaign," 2000). In that same year, a Nike vice president blasted human rights groups working to improve labor conditions in Nike's overseas factories. In a leaked letter to Vietnam's highest ranking labor official, he wrote that "United States human rights groups . . . are not friends of Vietnam." The letter also said their ultimate goal is to turn Vietnam into "a so-called democracy, modeled after the United States" (Press, 1999). In Vietnam, this amounts to a charge of subversion.

Meanwhile, The Gap has invested heavily in the Northern Mariana Islands near Guam. Strictly speaking, the Marianas are not a poor foreign country, because they are a

U.S. commonwealth territory with a status similar to that of Puerto Rico. But they might as well be. Garment manufacturing is the biggest source of income on the islands, and The Gap (which also owns Banana Republic and Old Navy) is the biggest employer. What attracts The Gap to the Marianas are below-minimum U.S. wage rates, duty-free access to U.S. markets, and the right to sew "Made in U.S.A." labels on clothes manufactured there (Bank of Hawaii, 1999). However, work conditions are horrific. Many workers live in guarded dormitories surrounded by barbed wire preventing their escape. They work 12 to 18 hours a day without overtime. Cases like The Gap in the Marianas and Nike in Indonesia and Vietnam suggest that the benefits of foreign investment are unlikely to be uniformly beneficial for the residents of developing countries in the short term.

The Future of Work and the Economy

Web Interactive Exercise: Social Capital

Although work and the economy have changed enormously over the years, one thing has remained constant for centuries. Businesses have always looked for ways to cut costs and boost profits. Three of the most effective means they have adopted to accomplish these goals involve introducing new technologies, organizing the workplace more efficiently, and exporting capital to take advantage of inexpensive labor abroad. Much is uncertain about the future of work and the economy. However, it is a pretty good bet that businesses will continue to follow these established practices.

Just how these practices will be implemented is less predictable. It is possible, for example, to use technology and improved work organization to increase productivity by complementing the abilities of skilled workers. Worldwide, the automotive, aerospace, and computer industries have tended to adopt this approach. They have introduced automation and robots on a wide scale. They constantly upgrade the skills of their workers. And they have proven the benefits of small autonomous work groups for product quality, worker satisfaction, and therefore the bottom line. Let's call this approach "competing with the high end of the wage scale."

On the other hand, new technology and more efficient work organization can be used to replace workers, deskill jobs, and employ low-cost labor—mainly women and minority group members—on a large scale. Women are entering the labor force at a faster rate than men. Hispanic American, Asian American, and African American workers are entering the labor force at a much faster rate than whites (Figure 13.5). Competition from low-wage industries abroad remains intense. Therefore, the second option is especially tempting in some industries. Let's call this approach "competing with the low end of the wage scale."

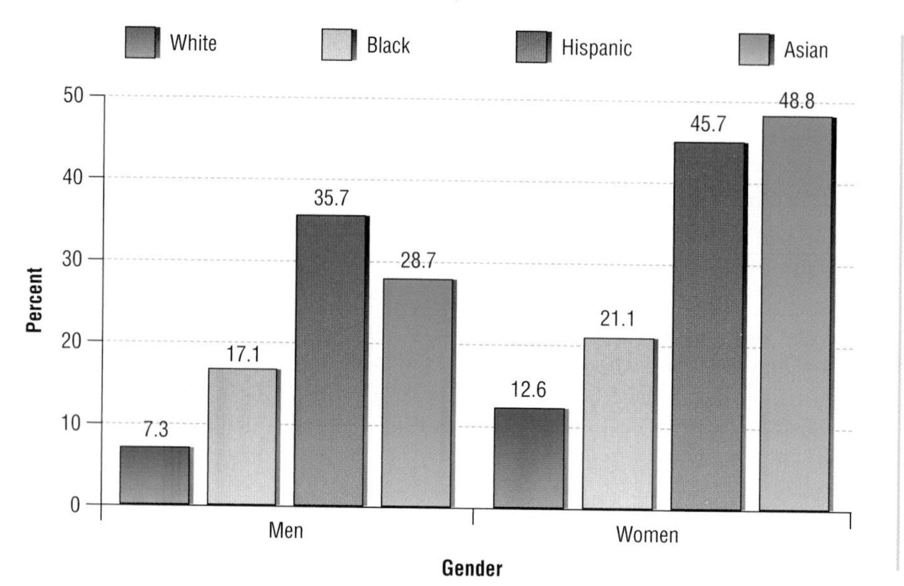

Figure 13.5
Percent Increase in U.S. Labor Force by Race and Hispanic Origin, 1998–2008 (projected; in percent)

Source: U.S. Department of Labor (1999d).

The future of work in the United States will be closely tied to which end of the wage scale we decide to compete with—whether we try to beat the Chinese and the Mexicans at their own game or, like the Germans and the Japanese, focus on using technology to continuously upgrade the skills of our labor force. Our analysis suggests that workplace and political struggles have no small bearing on how technologies are implemented, how work is organized, and what economic policies governments adopt. To a degree, therefore, the future of work and the economy is up for grabs.

Summary

Sociology Now™

Reviewing is as easy as ❶, ❷, ❸.

❶ Before you do your final review, take the SociologyNow diagnostic quiz to help you identify the areas on which you should concentrate. You will find information on how to access SociologyNow on the foldout at the front of the textbook.

❷ As you review, take advantage of SociologyNow's study aids to help you master the topics in this chapter.

❸ When you are finished with your review, take SociologyNow's post-test to confirm you are ready to move on to the next chapter.

1. What are the major work-related revolutions in human history and what are their consequences?

The first work-related revolution began about 10,000 years ago when people established permanent settlements and started herding and farming. The second work-related revolution began 225 years ago when various mechanical devices such as the steam engine greatly increased the ability of producers to supply markets. The third revolution in work was marked by growth in the provision of various services. It accelerated in the last decades of the 20th century with the widespread use of the computer. Each revolution in work increased productivity and the division of labor, caused a sectoral shift in employment, and made work relations more hierarchical. However, for the past 30 years, the degree of hierarchy has decreased in some industries, resulting in productivity gains and increased worker satisfaction.

2. What are "good" and "bad" jobs and which type of job is becoming more plentiful?

"Bad jobs" pay little and require the performance of routine tasks under close supervision. Working conditions are unpleasant and sometimes dangerous. Bad jobs require little formal education. In contrast, good jobs often require higher education. They pay well. They are not closely supervised, and they encourage the worker to be creative in pleasant surroundings. "Good jobs" offer secure employment, opportunities for promotion, health insurance, and other fringe benefits. In a bad job, you can easily be fired, you receive few if any fringe benefits, and the prospects for promotion are few. Deskilling and the growth of part-time jobs are two of the main trends in the workplace in the 20th century. However, skilled labor has remained very important in the economy. Good jobs are becoming more plentiful, but the number of bad jobs is growing even more rapidly. The result is polarization or segmentation of the labor force into primary and secondary labor markets. Various social barriers limit mobility from the secondary to the primary labor market.

3. What are the main types of markets?

Markets are free or regulated to varying degrees. No market that is purely free or completely regulated could function for long. A purely free market would create unbearable inequalities, and a completely regulated market would stagnate.

4. How have workers and professionals reacted to market conditions and economic change?

Workers have resisted attempts to deskill and control jobs. Consequently, business has had to make concessions by giving workers more authority on the shop floor and in formulating overall business strategy. Such concessions have been biggest in countries where workers are more organized and powerful. Unions and professional organizations have established internal labor markets to control pay rates, hiring, and promotions in organizations and reduce competition with external labor supplies.

5. What are corporations?

Corporations are legal entities that can enter into contracts and own property. They are taxed at a lower rate than individuals and their owners are normally not liable for the corporation's debt or any harm it may cause the public. Corporations are the dominant economic players in the world today. They exercise disproportionate economic and political influence by forming oligopolies, conglomerates, and interlocking directorates.

6. How does the growth of multinational corporations affect various social strata?

Growing competition between multinational corporations has led big corporations to cut costs by building more branch plants in low-wage, low-tax countries. Stockholders have profited from this strategy. However, the benefits for workers in both the industrialized and the less developed countries have been mixed.

Questions to Consider

1. Women are entering the labor force at a faster rate than men. Hispanic Americans, Asian Americans, and African Americans are entering the labor force at a much faster rate than whites. What policies must companies adopt if they hope to see women and members of ethnic and racial minorities achieve workplace equality with white men?
2. The computer is widely regarded as a laborsaving device and has been adopted on a wide scale. Yet, on average, Americans work more hours per week now than they did 20 or 30 years ago. How do you explain this paradox?
3. Most of the less developed countries have been eager to see multinational corporations establish branch plants on their soil. What sorts of policies must less developed countries adopt to ensure maximum benefits for their populations from these branch plants? Would it be beneficial if the less developed countries worked out a common approach to this problem rather than competing against each other for branch plants?

Web Resources
Companion Website for This Book

http://sociology.wadsworth.com

Begin by clicking on the Student Resources section of the website. Choose "Introduction to Sociology" and then the Brym and Lie book cover. Next, select the chapter you are currently studying from the pull-down menu. From the Student Resources page you will have easy access to InfoTrac® College Edition, MicroCase Online exercises, additional web links, and many other resources to aid you in your study of sociology, including practice tests for each chapter.

Recommended Websites

Provocative analyses of the future of work can be found at http://www.leftbusinessobserver.com/Work.html and http://www.leftbusinessobserver.com/Jobless_future.html.

The Bureau of Labor Statistics of the U.S. Department of Labor publishes estimates on the economy and labor market 10 years into the future, including projections of employment by industry and occupation. For these projections, go to http://www.bls.gov/emp/empbib05.htm. For the job outlook for college graduates to 2008, visit http://www.bls.gov/opub/ooq/2000/fall/art01.pdf.

Robert Kuttner, founder and coeditor of *The American Prospect* magazine, presents a thought-provoking analysis of "The Limits of Markets" at http://www.prospect.org/web/page.ww?section=root&name=ViewPrint&articleId=4845.

CHAPTER 14 Politics

Brooks Kraft/Corbis

In this chapter, you will learn that:

- Political sociologists analyze the distribution of power in society and its consequences for political behavior and public policy.
- Sociological disputes about the distribution of power often focus on how social structures, and especially class structures, influence political life.
- Some political sociologists analyze how state institutions and laws affect political behavior and public policy.
- Three waves of democratization have swept the world in the last 175 years.
- Societies become highly democratic only when their citizens win legal protections of their rights and freedoms. This typically occurs when their middle and working classes become large, organized, and prosperous.
- Enduring social inequalities limit democracy even in the richest countries.
- War and terrorism are means of conducting politics by other means.

Introduction

The Tobacco War

In the spring of 1998, the tobacco war reached a decisive stage. Congress was ready to pass a bill that would cost the tobacco companies $516 billion in damages. The bill would also raise tobacco taxes by $1.10 a pack, limit cigarette advertising, and give Washington broad new powers to regulate the tobacco industry.

The public seemed eager to support the legislation. After all, 75 percent of the people would never have to pay the new tax because only a quarter of American adults smoked. And there was widespread alarm in the land. More than three decades of educational work by governments, schools, and health professionals made it common knowledge that one out of three smokers would die prematurely and probably wretchedly due to illnesses caused by smoking. Well-informed citizens knew that more than 400,000 Americans die *annually* from tobacco-related illnesses, more than total American casualties in all 20th-century wars combined. They knew that about 90 percent of smokers started the habit by the age of 20. They knew that the percentage of grade 12 students who smoked rose from about 17 percent to nearly 25 percent between 1992 and 1997, mainly because of tobacco companies' marketing efforts.

Then, in 1998, the last straw: Documents released in a series of lawsuits against the tobacco industry revealed that tobacco companies were targeting teenagers in their ads, manipulating ammonia levels in tobacco to maximize nicotine addiction, and misrepresenting it all in public. (Some of these events were portrayed in the 1999 Oscar-winning film, *The Insider,* starring Russell Crowe.) Little wonder that polls showed strong public support for the antitobacco bill. The United States finally seemed ready to join the other rich industrialized countries in helping to stub out one of the world's leading health hazards.

Sociology Now™

Reviewing is as easy as ❶, ❷, ❸.

Use SociologyNow to help you make the grade on your next exam. When you are finished reading this chapter, go to the Chapter Summary for instructions on how to make SociologyNow work for you.

Chief executive officers of the major U.S. tobacco companies declare under oath at a 1994 congressional hearing that smoking is not addictive and does not cause any disease. This was a turning point in the battle against the tobacco industry.

John Duricka/AP/Wide World Photos

Representatives of the tobacco industry did not sit idly in the bleachers, however. They mobilized their allies, including retailers and smokers, to phone and write their members of Congress expressing outrage at the antitobacco bill. They tripled the budget for tobacco industry lobbyists. Legions of professional arm twisters wined, dined, and cajoled members of Congress to vote against the bill. And then the industry bankrolled a last-minute $40 million national advertising blitz. The ad campaign gnawed away at traditional American sore points. According to the ads, the antitobacco bill was really a government tax grab. It would increase government regulation at the expense of individual freedom. It would allow antitobacco industry lawyers to earn exorbitant fees. And, just as Prohibition had encouraged liquor smuggling and the production of moonshine whiskey in the 1920s and early 1930s, the new law would encourage the import of contraband cigarettes. These arguments worked. The bill was defeated in June 1998. Just before the final vote, a *Wall Street Journal*–NBC poll found that 70 percent of Americans thought the bill's real aim was to raise new revenue. Only 20 percent said its purpose was to curb teen smoking (Centers for Disease Control, 2000; "European Tobacco Ban," 1998; Kluger, 1996; Leman, 1998; McKenna, 1998; "Monitoring the Future Study," 1998).

In separate deals, the 50 states eventually agreed to sign agreements with the tobacco companies worth $246 billion, less than half the amount demanded in the federal bill. The money was intended to recover the cost of treating Medicaid-eligible smokers. The defeat of the federal bill raises important political questions, however. Does the outcome of the tobacco war illustrate the operation of "government of the people, by the people, for the people," as Abraham Lincoln defined democracy in his famous speech at Gettysburg? The tobacco war certainly allowed a diverse range of Americans to express conflicting views. It permitted them to influence their elected representatives. And, in the end, members of Congress did vote in line with the wishes of most American adults as expressed in public opinion polls. This suggests that Lincoln's characterization of American politics applied as well in 1998 as it did in 1863.

However, big business's access to a bulging war chest might lead one to doubt that Lincoln's definition applies. Few groups can put together virtually overnight $19 million for lobbyists, $3 million for political party contributions, and $40 million for public relations and advertising experts to sway the hearts and minds of the American people and their lawmakers. Should we therefore conclude that some people, especially big businessmen,[1] are more equal than others?

The tobacco war raises the question that lies at the heart of political sociology. What accounts for the degree to which a political system responds to the demands of all its citizens? As you will see, political sociologists have often answered this question by examining the effects of social structures, especially class structures, on politics. Although this approach contributes much to our understanding of political life, it is insufficient by itself. A fully adequate theory of democracy requires that we also examine how state institutions and laws affect political processes. By way of illustration, we show how voter registration laws bias American politics in favor of some groups at the expense of others.

From the mid-1970s till the early 1990s, a wave of competitive elections swept across many formerly nondemocratic countries. Most dramatically, elections were held in the former Soviet Union at the end of this period. Many Western analysts were ecstatic. Yet by the mid-1990s, it became clear that their optimism was naive. Often, the new regimes turned out to be feeble and limited democracies. Political sociologists therefore began to reconsider the social preconditions of democracy. We review some of their work in this chapter's third section. We conclude that genuine democracy is not based just on elections. In addition, large classes of people must win legal protection of their rights and freedoms for democracy to take root and grow. This has not yet happened in most of the world.

Some analysts believe that politics in the rich industrialized countries is less likely to be shaped by social inequality in the future. Others hold that the marriage of home computers and elections will allow citizens to get more involved in politics by voting in more elections and directly on the Internet. Our reading of the evidence is different. In concluding this chapter, we argue that persistent social inequality is the major barrier to the progress of democracy in countries like the United States.

Before developing these themes, however, we define some key terms.

Power and Authority

Politics is a machine that determines "who gets what, when, and how" (Lasswell, 1936). **Power** fuels the machine. Power is the ability to control others, even against their will (Weber, 1947: 152). Having more power than others gives you the ability to get more valued things sooner. Having less power than others means you get fewer valued things later. Political sociology's key task is figuring out how power drives different types of political machines.

The use of power sometimes involves force. For example, one way of operating a system for distributing jobs, money, education, and other valued things is by imprisoning people who don't agree with the system. In this case, people obey political rules because they are afraid to disobey. More often, people agree with the distribution system or at least accept it grudgingly. For instance, most people pay their taxes without much pressure from the IRS and their parking tickets without serving jail time. They recognize the right of their rulers to control the political machine. When most people basically agree with how the political machine is run, raw power becomes **authority.** Authority is legitimate, institutionalized power. Power is **legitimate** when people regard its use as valid or justified. Power is *institutionalized* when the norms and statuses of social organizations govern its use. These norms and statuses define how authority should be used, how individuals can achieve authority, and how much authority is attached to each status in the organization.

Sociology Now™

Learn more about **Power and Authority** by going through the Power and Authority Learning Module.

Power is the ability to control others, even against their will.

Authority is legitimate, institutionalized power.

Legitimate governments are those that enjoy a perceived right to rule.

[1]We say business*men* advisedly. In 2000, only 46 women were on the list of America's 400 richest people. Of these, a mere 6 were self-made women. This suggests where the real power lies (DiCarlo, 2000).

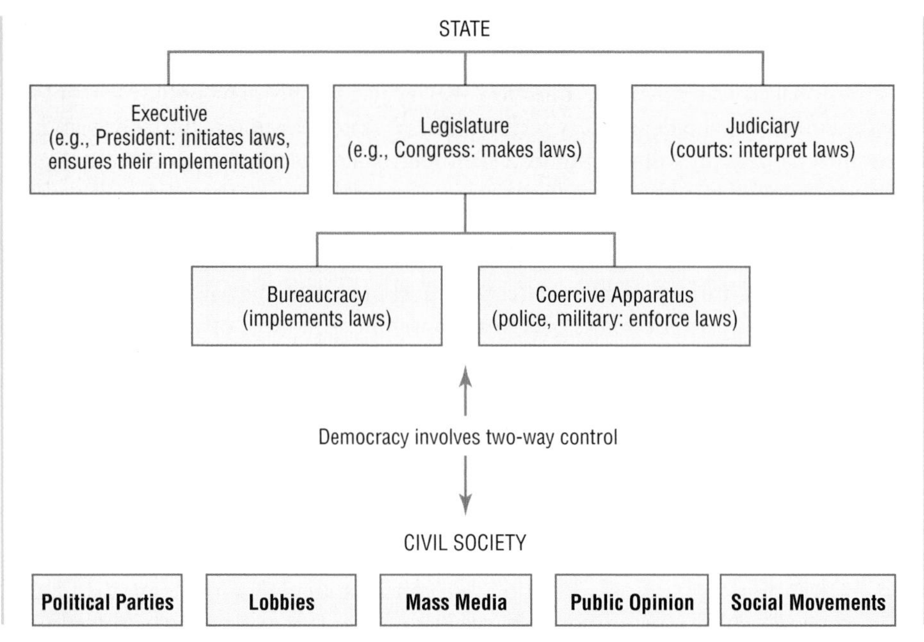

Figure 14.1
The Institutions of State and Civil Society

Sociology Now™

Learn more about **Revolutions** by going through the Revolutions Animation.

Traditional authority, the norm in tribal and feudal societies, involves rulers inheriting authority through family or clan ties. The right of a family or clan to monopolize leadership is widely believed to be derived from the will of a god.

Legal-rational authority is typical of modern societies. It derives from respect for the law. Laws specify how one can achieve office. People generally believe these laws are rational. If someone achieves office by following these laws, people respect their authority.

Charismatic authority is based on belief in the claims of extraordinary individuals to be inspired by God or some higher principle.

A **political revolution** is the overthrow of political institutions by an opposition movement and replacement by new institutions.

The **state** consists of the institutions responsible for formulating and carrying out a country's laws and public policies.

Civil society is the private sphere of social life.

In an **autocracy**, absolute power resides in the hands of a single person or party.

Authoritarian states sharply restrict citizen control of the state.

In a **democracy**, citizens exercise a high degree of control over the state. They do this mainly by choosing representatives in regular, competitive elections.

Types of Authority

Max Weber (1947) wrote that authority can have one of three bases:

1. **Traditional authority.** Particularly in tribal and feudal societies, rulers inherit authority through family or clan ties. The right of a family or clan to monopolize leadership is widely believed to originate from the will of a god.
2. **Legal-rational authority.** In modern societies, authority is derived from respect for the law. Laws specify how one can achieve office. People generally believe these laws are rational. If someone achieves office by following these laws, their authority is respected.
3. **Charismatic authority.** Sometimes extraordinary, charismatic individuals challenge traditional or legal-rational authority. They claim to be inspired by a god or some higher principle that transcends other forms of authority. Most people believe this claim. One such principle is the idea that all people are created equal. Charismatic figures sometimes emerge during a **political revolution,** an attempt by many people to overthrow existing political institutions and establish new ones. Political revolutions take place when widespread and successful movements of opposition clash with crumbling traditional or legal-rational authority.

Types of Political System

Politics takes place in all social settings, including intimate face-to-face relationships, families, and colleges. However, political sociology is mainly concerned with institutions that *specialize* in the exercise of power and authority. Taken together, these institutions form the **state.** The state consists of institutions that formulate and carry out a country's laws and public policies. In performing these functions, the state regulates citizens in **civil society,** the private sphere of social life (Figure 14.1).

Autocracies and Authoritarian States

Citizens in civil society control the state to varying degrees. In an **autocracy,** absolute power resides in the hands of a single person or party. In an **authoritarian** state, power is somewhat more widely shared but citizen control is still sharply restricted. In a **democracy,**

Archive Iconografico, S.A./Corbis

The three faces of authority according to Weber: traditional authority (King Louis XIV of France, circa 1670), charismatic authority (Vladimir Lenin, Bolshevik leader of the Russian Revolution of 1917), and legal-rational authority (Ronald Reagan, campaigning for the presidency of the United States).

SuperStock

Corbis

Table 14.1
Type of State, World, 1973 and 2004

Type of State	Percent of All States, 1973	Percent of All States, 2004	Percent of World Population, 2004
Autocratic	46	25	35
Authoritarian	25	29	21
Democratic	29	46	44
Total	100	100	100

Source: Freedom House (2004).

citizens exert a relatively high degree of control over the state. They do this partly by choosing representatives in regular, competitive elections (Table 14.1).

Sociology Now™

Learn more about **Social Movements** by going through the % of people who have taken part in a lawful demonstration Map Exercise.

Political parties are organizations that compete for control of government in regular elections. In the process, they give voice to policy alternatives and rally adult citizens to vote.

Lobbies are organizations formed by special interest groups to advise and influence politicians.

The **mass media** in a democracy help to keep the public informed about the quality of government.

Public opinion is composed of the values and attitudes of the adult population as a whole. It is expressed mainly in polls and letters to lawmakers and gives politicians a reading of citizen preferences.

Social movements are collective attempts to change all or part of the political or social order by stepping outside the rules of normal politics.

Pluralist theory holds that power is widely dispersed. As a result, no group enjoys disproportionate influence, and decisions are usually reached through negotiation and compromise.

Democracies

In modern democracies, citizens do not control the state directly. They do so through several organizations. **Political parties** compete for control of government in regular elections. They put forward policy alternatives and rally adult citizens to vote. Special interest groups such as trade unions and business associations form **lobbies.** They advise politicians about their members' desires. They also remind politicians how much their members' votes, organizing skills, and campaign contributions matter. The **mass media** keep a watchful and critical eye on the state. They keep the public informed about the quality of government. **Public opinion** refers to the values and attitudes of the adult population as a whole. It is expressed mainly in polls and letters to lawmakers. Public opinion gives politicians a reading of citizen preferences. Finally, when dissatisfaction with normal politics is widespread, protest sometimes takes the form of **social movements.** A social movement is a collective attempt to change all or part of the political or social order by stepping outside the rules of normal politics. As Thomas Jefferson wrote in a letter to James Madison in 1787, "a little rebellion now and then is a good thing" for democracy. It helps to keep government responsive to the wishes of the citizenry.

Bearing these definitions in mind, we now consider the merits and limitations of sociological theories of democracy.

Theories of Democracy

Pluralist Theory

In the early 1950s, New Haven, Connecticut, was a city of about 150,000 people. It had seen better times. As in many other American cities, post–World War II prosperity and new roads had allowed much of the white middle class to resettle in the suburbs. This eroded the city's tax base. It also left much of the downtown to poor and minority-group residents. Some parts of New Haven became slums.

Beginning in 1954, Mayor Richard Lee decided to do something about the city's decline. He planned to attract new investment, eliminate downtown slums, and stem the outflow of the white middle class. Urban renewal was a potentially divisive issue. But according to research conducted at the time, key decisions were made in a highly democratic manner. The city government listened closely to all major groups. It adopted policies that reflected the diverse wishes and interests of city residents.

The social scientists who studied New Haven politics in the 1950s are known as **pluralists** (Polsby, 1959; Dahl, 1961). They argued that the city was highly democratic because power was widely dispersed. They showed that few of the prestigious families in New Haven's *Social Register*[2] were economic leaders in the community. Moreover, neither economic leaders nor the social elite monopolized political decision making. Different groups of people decided various political issues. Some of these people had low status in the community. Moreover, power was more widely distributed than in earlier decades. The pluralists concluded that no single group exercised disproportionate power in New Haven.

The pluralists believed that politics worked much the same way in the United States as a whole. America, they said, is a heterogeneous society with many competing interests and centers of power. None of these power centers can consistently dominate. The owners of United States Steel, for instance, may want tariffs on steel imports to protect the company's U.S. market. The owners of General Motors may oppose tariffs on steel because they want to keep their company's production costs down. The idea that "industry" speaks with one voice is thus a myth. Competing interests exist even within one group. For instance, the automobile company with the lead in developing electric cars may favor clean-air legislation now. An auto company lagging in its research effort may favor a go-slow approach to such laws. Because there is so much heterogeneity between and within groups, no single group can control political life. Sometimes one category of voters or one set of interest groups wins a political battle, sometimes another. Most often, however, politics involves negotiation and compromise between competing groups. Because no one group of people is always able to control the political agenda or the outcome of political conflicts, democracy is guaranteed.

Elite Theory

Elite theorists, C. Wright Mills (1956) chief among them, sharply disagreed. According to Mills, **elites** are small groups that occupy the command posts of America's most influential institutions. These institutions include the country's two or three hundred biggest corporations, the executive branch of government, and the military. Mills wrote that the men who control these institutions make important decisions that profoundly affect all members of society. And they do so without much regard for elections or public opinion.

Mills showed how the corporate, state, and military elites are connected. People move from one elite to another during their careers. Their children intermarry. They maintain close social contacts. They tend to be recruited from upper-middle and upper classes. Yet Mills denied these connections turn the three elites into a **ruling class.** A ruling class is a self-conscious and cohesive group of people, led by owners of big business, who act to advance their common interests. The three elites are relatively independent of one another, Mills insisted. They may see eye-to-eye on many issues, but each has its own sphere of influence. Conflict between elite groups is frequent (Mills, 1956: 277; Alford and Friedland, 1985: 199).

Sociology Now™

Learn more about **Elite Theory** by going through the Power Elite Model Animation.

A Critique of Pluralism

Most political sociologists today question the pluralist account of American politics. That is because research has established the existence of large, wealth-based inequalities in political participation and political influence. As we will see, most political sociologists today are skeptical about some of C. Wright Mills's claims too. On the whole, however, they are more sympathetic to the elitist view.

Political Participation

Consider, for example, some results of the "Citizen Participation Study." In the early 1990s, a team of researchers surveyed a representative sample of more than 15,000 American adults. They asked respondents if they voted in the 1988 presidential campaign,

[2] The *Social Register* is a listing of America's highest-status families. First published in 1887, it now has about 40,000 entries.

Elite theory holds that small groups occupying the command posts of America's most influential institutions make the important decisions that profoundly affect all members of society. Moreover, they do so without much regard for elections or public opinion.

Elites are exclusive groups that control the command posts of an institution.

A **ruling class** is a self-conscious, cohesive group of people in elite positions. They act to advance their common interests and are led by corporate executives.

Pluralist theory portrays politics as a neatly ordered game of negotiation and compromise in which all players are equal. Brian Jones, "The Centre of the Universe," (1992).

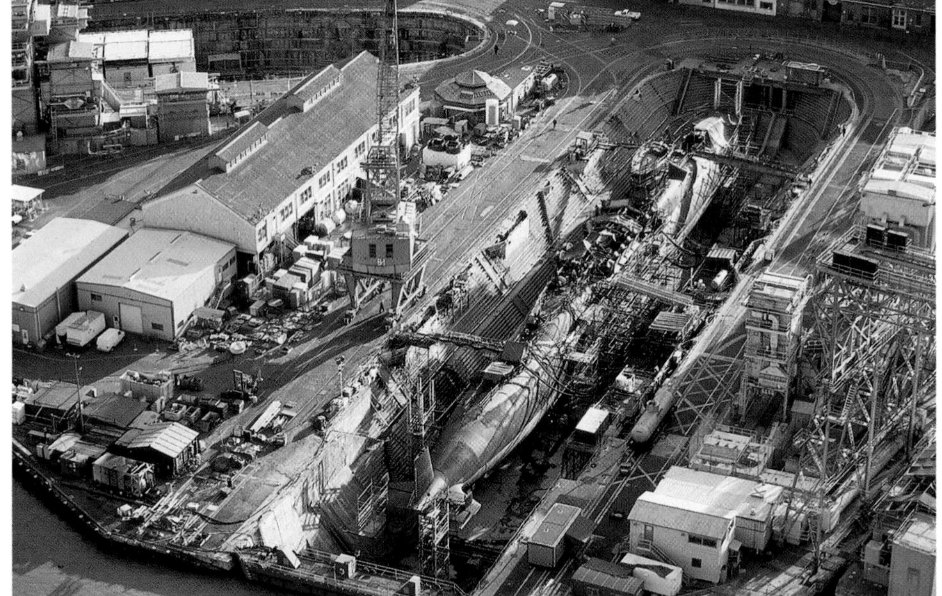

In a nationally televised address on January 17, 1961, President Eisenhower sounded much like C. Wright Mills and other elite theorists when he warned of the "undue influence" of the "military-industrial complex" in American society. Maintaining a large, permanent military establishment is "new in the American experience," he said. An "engaged citizenry" offers the only effective defense against the "misplaced power" of the military-industrial lobby, according to Eisenhower.

how many contacts they had with public officials, how many hours they worked in the election campaign, and how many dollars they contributed to it. Then they calculated the percentage of each political activity that was undertaken by people in each income group. They found that people with higher incomes are more politically active, especially in those forms of political activity that re most influential.

Figure 14.2 compares the political participation of rich and poor Americans. The rich were defined as those who had family incomes of $125,000 per year or more. The poor were defined as those who had family incomes of less than $15,000 per year. So defined, the rich composed 3 percent of American citizens, the poor 18 percent. The ratio of rich to poor was 0.17:1 (3/18 = 0.17). Note how the ratio of rich to poor is higher for more influential political activities. For voting, the ratio of rich to poor is 0.29:1. For contacts with public officials, the ratio is 0.50:1. For hours spent campaigning, the ratio is 0.62:1. And for dollars contributed to campaigns, the ratio is 17.5:1. In other words, the

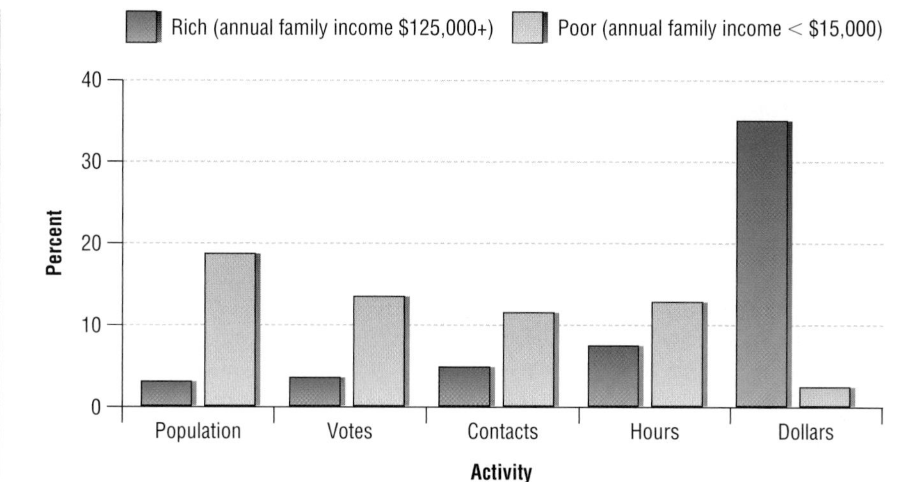

Figure 14.2
Percent of Political Activities Undertaken by Rich and Poor Americans

Source: From Sidney Verba, et al., "The Big Tilt," *The American Prospect*, Vol. 8, No. 32: May 1–June 1, 1997, p. 74–80. Copyright © 1997 American Prospect, 5 Broad Street, Boston, MA 02109. All rights reserved. Used with permission.

rich contribute 17.5 times more money to election campaigns than the poor, although the poor are 6 times more numerous than the rich.

Political Influence and PACs

If money talks, does it speak with a single voice? In principle, well-to-do Americans may contribute the same amount of money to all candidates in an election campaign. If so, we would be wrong to assume that rich people share political interests and act in concert. A study of **political action committees** (PACs) conducted by sociologist Dan Clawson and his associates shines light on this issue (Clawson, Neustadtl, and Scott, 1992).

In 1988, winning members of the House of Representatives spent an average of $388,000 on their election campaigns. Winning senators spent nearly 10 times as much—$3,745,000. Thus, *every week* of his or her term of office, a member of the House had to raise $3,700, and a senator $12,000, to finance an average winning campaign. PACs help them do that by collecting money from many contributors, pooling it, and then making donations to candidates (Box 14.1). What do contributors to PACs expect in return for their money? Republican senator and former presidential candidate Bob Dole answered the question delicately when he said, "They expect something in return other than good government." One business donor put it more bluntly: "One question . . . raised in recent weeks had to do with whether my financial support in any way influenced several political figures to take up my cause. I want to say in the most forceful way I can, I certainly hope so" (quoted in Clawson, Neustadtl, and Scott, 1992: 9).

Figure 14.3 shows the contributions of the PACs that gave more than $100,000 in 1984 to 455 political races. It divides the races into three groups. In 1 case out of 15, large corporations were politically divided, giving only one to two times more money to one candidate than the other. In 1 case out of 5, large corporations mainly supported one candidate, giving two to nine times more money to one candidate than the other. And in 3 cases out of 4, large corporations were politically unified, giving more than nine times more money to one candidate than the other. These data suggest that big business is for the most part unified in its political views. It tends to favor one candidate over another.

Which candidates do corporate PACs tend to favor? The Republicans. When Clawson and his associates analyzed contributions by *all* large PACS, they found a sharp split between a unified business-Republican group on one side and a labor-women-environmentalist-Democratic group on the other.

This raises an interesting question. According to elite theorists, the distribution of power in America is heavily skewed toward the wealthy. The wealthy tend to support the Republicans. Why then do Democrats often become presidents and get elected to Congress? As we will see, this question points to an important limitation of elite theory.

Political action committees (PACs) are organizations that raise funds for politicians who support particular issues.

BOX 14.1

SOCIAL POLICY: WHAT DO YOU THINK?

Financing Political Campaigns

"No matter what parliamentary tactics are used to prevent reform . . . no matter how fierce the opposition, no matter how personal, no matter how cynical this debate remains . . . I will persevere," proclaimed Senator John McCain, leader of the reform movement. Senator Mitch McConnell, who led the opposition, replied: "I'd call [the reform movement] no progress whatsoever. . . . I'd call it . . . pretty dead" (quoted in Mitchell, 1999).

What was the issue that generated such heated rhetoric in the fall of 1999? It was the effort, led by Senators McCain and Feingold, to reform political campaign financing. They were especially upset about the role of "soft money." At the time, campaign finance law capped each contribution to an individual candidate at $1000. The reasoning behind the law was that politics should be about ideas, policies, and personalities, not money. After all, we wouldn't have much of a democracy if individuals could "buy" elections. Paradoxically, however, the law also allowed soft money contributions. These are contributions to political parties, not individual candidates. Soft money is big money that comes from corporations, unions, and wealthy individuals. Soft money donations reached almost $500 million in the 2000 presidential campaign.

Why should we be concerned about the role of money in politics? The average winner in the 1998 Senate elections raised $5.2 million. The average loser raised $2.8 million. This suggests that money helps to win campaigns. Moreover, about 90 percent of the incumbents won in the 1996 and 1998 Senate elections. Thus, not only do candidates who raise more money tend to win, candidates with more money tend to be incumbents (Center for Responsive Politics, 2000).

In 2002, after seven years of intense debate in Congress, President Bush reluctantly signed into law a bill that restricts soft money contributions. Individual candidates can now receive campaign contributions of up to $2000, and state political parties can receive contributions of up to $10,000 for get-out-the-vote drives. The law does not allow soft money contributions. Nonetheless, many experts immediately said that clever campaign lawyers and donors would be able to subvert it (Seelye and Mitchell, 2002).

The experts were right. Early in the 2004 presidential election campaign, the Democrats successfully argued that section 527 of the tax code allowed soft money to finance "independent" advocacy groups. They proceeded to build a series of such groups to attack the president. The Republicans then adopted the same strategy. They managed to raise and spend only a third as much as the Democrats on so-called "527 groups," but theirs were more effective, especially the group that financed the notorious and devastating "swift boat" ads attacking John Kerry's military record and anti–Vietnam War activities. All told, soft money contributions in the 2004 presidential race were up more than 10 percent over 2000, hardly an indicator of success for Senators McCain and Feingold ("The Soft Money Boomerang," 2004).

In your opinion, should campaign financing be controlled to make it more difficult for soft money to influence politics? If so, then how might it be possible to overcome the addiction of both parties to soft money? Alternatively, is the right to free speech such an important principle that it should override any restrictions on campaign financing? Or does the right to free speech allow the voices of the rich to drown out the voices of the less well-to-do?

Figure 14.3
The Political Unity of Big Business, United States

Source: From *Money Talks: Corporate Pacs and Political Influence* by Dan Clawson, et al. Copyright © 1992 Basic Books. Reprinted by permission of the author.

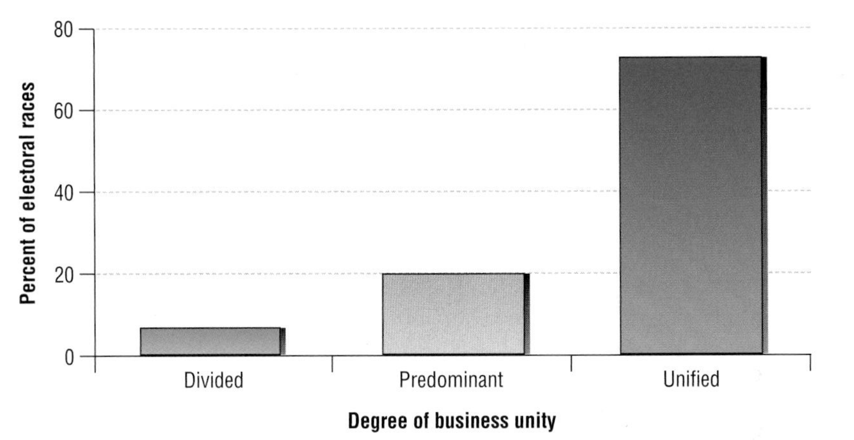

Power Resource Theory

In general, elite theorists believe it makes little difference whether Republicans or Democrats are in power. For them, elites always control society. Elections are little more than sideshows. Therefore, they believe, the victory of one party over another doesn't deserve much sociological attention.

We disagree. So do most political sociologists today. It matters a great deal to most citizens whether the party in office supports or opposes antitobacco laws, more military spending, weaker environmental standards, more publicly funded medical care, bigger government subsidies for child care, abortion on demand, gun control, and so forth. Elite theorists are correct to claim that most power is concentrated in the hands of the wealthy. But we still need a theory that accounts for the successes and failures of different parties and policies in different times and places. This is where **power resource theory** is helpful. It focuses on how *variations* in the distribution of power affect the fortunes of parties and policies.

To understand power resource theory we must first distinguish between liberal and conservative voters. *Liberal* (or left-wing) voters promote extensive government involvement in the economy. Among other things, this means they favor a strong "social safety net" of health and welfare benefits to help the less fortunate members of society. As a result, liberal policies often lead to less economic inequality. In contrast, *conservative* (or right-wing) voters favor a reduced role for government in the economy. They favor a smaller welfare state and emphasize the importance of individual initiative in promoting economic growth. Economic issues aside, liberals and conservatives also tend to differ on social or moral issues. Liberals tend to support equal rights for women and racial and sexual minorities. Conservatives tend to support more traditional social and moral values.[3] The top of ▶Table 14.2 shows how liberal and conservative sentiments were translated into support for Al Gore, a Democrat, and George W. Bush, a Republican, in 2000 according to the 2002 General Social Survey (GSS). Some 78 percent of liberals were Gore supporters, whereas 76 percent of conservatives were Bush supporters.

▶Table 14.2

Support for Democratic and Republican Presidential Candidates by Selected Social Characteristics and Attitudes, 2000 (in percent; all totals = 100)

	CANDIDATE	
	Bush	**Gore**
Political views (*N* = 835)		
Liberal	22	78
Moderate	48	52
Conservative	76	24
Total annual family income (*N* = 1505)		
<$25,000	42	58
$25,0001	57	43
Sex (*N* = 1666)		
Male	60	40
Female	48	52
Race/ethnicity (*N* = 1678)		
White	61	49
Black	10	90
Hispanic	43	57
Abortion acceptable if woman wants one for any reason (*N* = 550)		
Yes	44	56
No	62	38
Homosexual relations . . . (*N* = 545)		
Always or almost always wrong	64	36
Sometimes wrong or not wrong at all	39	61

Note: For simplicity, Table 14.2 does not include supporters of candidates other than Bush and Gore.

Source: National Opinion Research Center (2004).

Web Interactive Exercises: How Electoral Laws Affect Political Participation

Political Parties and Class Support

The policies favored by different parties have different effects on different groups of people. Therefore, different parties tend to be supported by different classes, religious groups, races, and other groups. In most Western democracies, the main factor that distinguishes parties is differences in *class* support (Korpi, 1983: 35; Lipset and Rokkan, 1967; Manza, Hout, and Brooks, 1995). For example, as Table 14.2 shows, in the 2000 presidential race, 58 percent of people with total family income below $25,000 a year supported Gore, compared with only 43 percent for Bush. For people with total family income of $25,000 a year or more, the pattern was reversed. As these figures show, low-income earners in the United States tend to support liberal, Democratic candidates. High-income earners tend to support conservative, Republican candidates. This stands to reason because most Democrats favor policies that promote less inequality in society, such as universal health care.

Power resource theory holds that the distribution of power between major classes partly accounts for the successes and failures of different political parties.

[3]Some people are liberal on economic issues and conservative on social issues or vice versa. Many such people call themselves "moderates" rather than liberals or conservatives.

Kevin Lamarque/Corbis

Bob Daemmrich/Corbis

In the 2004 presidential election, Democrat John Kerry and Republican George W. Bush appealed to different segments of the American population. Low-income earners, African Americans, Hispanic Americans, women, liberals, supporters of homosexuals, and supporters of reproductive choice tended to support Gore. High-income earners, whites, men, conservatives, opponents of homosexuals and opponents of reproductive choice tended to support Bush (Table 14.2). Compared to the 2000 election, Bush increased his support among blacks, Hispanics, voters in exurban and rural areas, women, and voters in the Deep South (Judis and Teixeira, 2005).

The tendency for people in different classes to vote for different parties varies from one country to the next. The strength of this tendency depends on many factors. One of the most important is how socially organized or cohesive classes are (Brym with Fox, 1989: 57–91; Brym, Gillespie, and Lenton, 1989). For example, an upper class that can create PACs to support Republican candidates and lobbies to support conservative laws is more powerful than an upper class that cannot take such action. If an upper class makes such efforts while a working class fails to organize itself, right-wing candidates have a better chance of winning office. Conservative policies are more likely to become law. Similarly, a working class that can unionize many workers is more powerful than one with few unionized workers. That is because unions often collect money for the party that is more sympathetic to union interests. They also lobby on behalf of their members and try to convince members to vote for the pro-union party. If workers become more unionized while an upper class fails to organize itself, then left-wing candidates have an improved chance of winning office. Liberal policies are more likely to become law.

Organization and Power

This, then, is the main insight of power resource theory: Organization is a source of power. Change in the distribution of power between major classes partly accounts for the fortunes of different political parties and different laws and policies (Korpi, 1983; Esping-Andersen, 1990; O'Connor and Olsen, 1998; Shalev, 1983).

Table 14.3
Some Consequences of Working-Class Power in 18 Rich Industrialized Countries, 1946–1976

	Percent of Nonagricultural Workforce Unionized	Socialist Share of Government	Percent of Total National Income to Top 10% Earners	Percent Poor
Mainly socialist countries (Sweden, Norway)	68.5	High	21.8	4.3
Partly socialist countries (Austria, Australia, Denmark, Belgium, UK, New Zealand, Finland)	46.6	Medium	23.6	7.8
Mainly nonsocialist countries (Ireland, West Germany, Netherlands, USA, Japan, Canada, France, Italy, Switzerland)	28.0	Low	28.3	10.8

Note: "Socialist share of government" is the proportion of seats in each cabinet held by socialist parties weighted by the socialist share of seats in parliament and the duration of the cabinet. "Percent poor" is the average percentage of the population living in relative poverty according to Organisation for Economic Co-operation and Development (OECD) standards, with poverty line standardized according to household size.

Source: Korpi (1983: 40, 196).

We can see how power resource theory works by looking at Table 14.3, which compares 18 industrialized democracies in the three decades after World War II. We divide the countries into three groups. Under "mainly socialist countries" are those (like Sweden) where socialist parties usually control governments. (Socialist parties are more left wing than the Democrats in the United States.) Under "partly socialist countries" are those (like Australia) where socialist parties *sometimes* control, or share in the control of, governments. And under "mainly nonsocialist countries" are those (like the United States) where socialist parties rarely or never share control of governments. The group averages in the first column show that socialist parties are generally more successful where workers are more unionized. The group averages in the third and fourth columns show that there is more economic inequality in countries that are weakly unionized and have no socialist governments. In other words, by means of taxes and social policies, socialist governments ensure that the rich earn a smaller percentage of national income and the poor form a smaller percentage of the population. Studies of pensions, medical care, and other state benefits in the rich industrialized democracies reach similar conclusions. In general, where working classes are more organized and powerful, disadvantaged people are economically better off (Myles, 1989 [1984]; O'Connor and Brym, 1988; Olsen and Brym, 1996; Korpi and Palme, 2003).

Other Party Differences: Religion, Race, and Gender

Class is not the only factor that distinguishes parties. Historically, *religion* has also been an important basis of party differences. For example, in Western European countries with large Catholic populations, such as Switzerland and Belgium, parties are distinguished partly by the religious affiliation of their supporters. In recent decades, *race* has become a cleavage factor of major and growing importance in some countries. In the United States in particular, African Americans have overwhelmingly supported the Democratic Party since the 1960s (Brooks and Manza, 1997b). According to the 2002 GSS, Al Gore won 90 percent of the black vote and 57 percent of the Hispanic vote (see Table 14.2). Race is an increasingly important division in French politics too. This is due to heavy Arab immigration from Algeria, Morocco, and Tunisia since the 1950s and growing anti-immigration sentiment among a substantial minority of whites (Veugelers, 1997). Finally, a political *gender* gap has emerged in some countries, such as the United States. The 2002 GSS indicates that 60 percent of male voters but only 48 percent of female voters chose Bush; 40 percent of male voters and 52 percent of female voters preferred Gore. Power resource theory focuses mainly on how the shifting distribution of power between working and upper classes affects electoral success. However, one can also use the theory to analyze the electoral fortunes of parties that attract different religious groups, races, gender groups, and so forth.

State-Centered Theory

Web Research Projects: Party Identification

Democratic politics is a contest among various classes, religious groups, races, and other collectivities to control the state for their own advantage. When power is substantially redistributed due to such factors as change in the cohesiveness of social groups, old ruling parties usually fall and new ones take office.

Note, however, that a winner-take-all strategy would be nothing short of foolish. If winning parties passed laws that benefit only their supporters, they might cause mass outrage and even violent opposition. Yet it would be bad politics to allow opponents to become angry, organized, and resolute. After all, winners want more than just a moment of glory. They want to be able to enjoy the spoils of office over the long haul. To achieve stability, they must give people who lose elections a say in government. That way, even determined opponents are likely to recognize the government's legitimacy. Pluralists thus make a good point when they say that democratic politics is about accommodation and compromise. They only lose sight of how accommodation and compromise typically give more advantages to some than to others, as both elite theorists and power resource theorists stress.

There is, however, more to the story of politics than conflict between classes, religious groups, races, and so forth. Theda Skocpol and other **state-centered theorists** show how the state itself can structure political life, no matter how power is distributed at a given moment (Block, 1979; Skocpol, 1979; Evans, Rueschemeyer, and Skocpol, 1985). Their argument is a valuable supplement to power resource theory.

To illustrate how state structures shape politics, consider a common American political practice: nonvoting. Just over half of eligible American adults have voted in recent presidential elections (although the highly charged 2004 election, with about 59 percent voter turnout, was an exception; Box 14.2). Apart from Switzerland, the United States has the lowest voter turnout of any rich democracy in the world (Piven and Cloward, 1989 [1988]: 5). How can we explain this troubling fact?

Voter Registration Laws

The high rate of nonvoting is partly a result of voter registration law, a feature of the American political structure, not of the current distribution of power. In every democracy, laws specify voter registration procedures. In some countries, citizens are registered to vote automatically when they receive state-issued identity cards at the age of 18. In other countries, state-employed canvassers go door to door before each election to register voters. Only in the United States do individual citizens have to take the initiative to go out and register themselves in voter registration centers. However, many American citizens are unable or unwilling to register. As a result, the United States has a proportionately smaller pool of eligible voters than the other democracies. Only about 70 percent of American citizens are registered to vote. True, since the National Registration Act was passed in 1993, it has been possible to register by mail, when renewing a driver's license, and when applying for welfare and disability services. However, the percentage of American adults registered to vote increased only about 2 percent between 1995 and 1998. The new "motor voter" law has had little impact on actual voter turnout (Ganz, 1996; Quinn, 2000).

State-centered theory holds that the state itself can structure political life to some degree independently of the way power is distributed between classes and other groups at a given time.

Apart from shrinking the pool of eligible voters, American voter registration law has a second important consequence. Because some *types* of people are less able and less inclined to register than others, a strong bias is introduced into the political system. Specifically, the poor are less likely to register than the better off. People without much

Box 14.2

YOU AND THE SOCIAL WORLD

Felon Disenfranchisement

In addition to the roughly half of eligible Americans who do not vote in presidential elections, some American adult citizens are not allowed to vote. The largest category of such people consists of convicted felons—criminals who have committed serious crimes. A patchwork of state laws prevents nearly all state and federal prisoners (currently about 2.1 million people) and many others who have finished serving their time (another 2.0 to 2.6 million people) from voting. That amounts to more than 2 percent of eligible voters. About 30 percent of disenfranchised felons are African Americans. The United States is the only country in the world that disenfranchises many nonimprisoned felons (The Sentencing Project, 2004). Figure 14.4 shows the distribution of types of felony disenfranchisement in the United States.

A unique poll on the subject of felon disenfranchisement was recently conducted. It found that 31 percent of Americans believe that currently imprisoned felons should not be allowed to vote. It also found that 60 percent of Americans believe that probationers and parolees should be allowed to vote (Manza, Brooks, and Uggen, 2004: 280).

In about 500 words, answer the following questions: Do you think that felons who are currently imprisoned should be allowed to vote? How about felons who are on probation or on parole? Justify your viewpoint by thinking about what consequences felon disenfranchisement has for American democracy and for felons themselves.

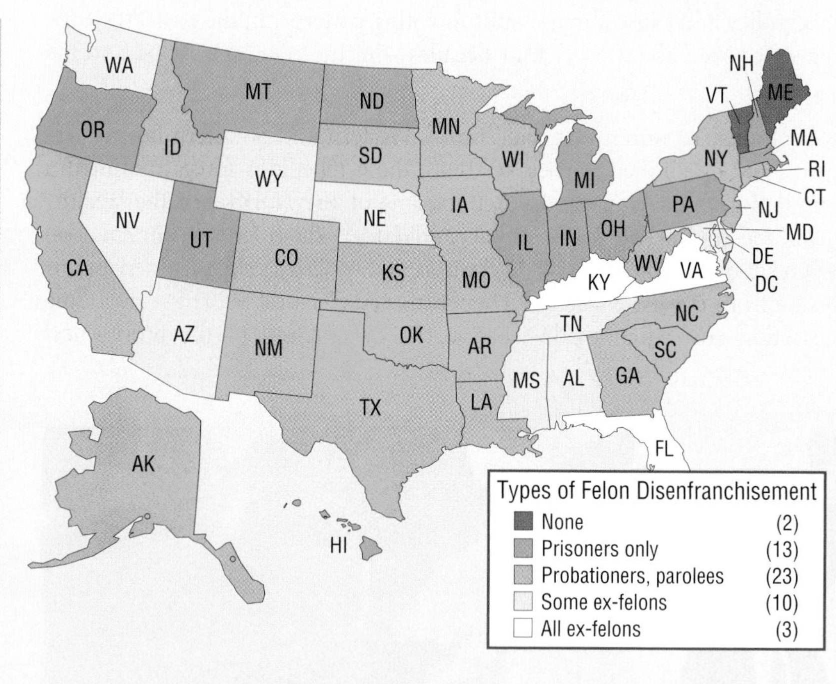

Figure 14.4
Types of Felony Disenfranchisement, United States, 2005

Note: Three states (Florida, Kentucky, and Virginia) disenfranchise all ex-felons for life. Ten states (Alabama, Arizona, Delaware, Maryland, Mississippi, Nebraska, Nevada, Tennessee, Washington, and Wyoming) disenfranchise some ex-felons after they leave prison and complete parole. Twenty-three states (Alaska, Arkansas, California, Colorado, Connecticut, Georgia, Idaho, Iowa, Kansas, Louisiana, Minnesota, Missouri, New Jersey, New Mexico, New York, North Carolina, Oklahoma, Rhode Island, South Carolina, South Dakota, Texas, West Virginia, and Wisconsin) disenfranchise imprisoned felons and felons who are on probation or parole. Twelve states (Hawaii, Illinois, Indiana, Massachusetts, Michigan, Montana, New Hampshire, North Dakota, Ohio, Oregon, Pennsylvania, and Utah) and the District of Columbia disenfranchise felons only when they are imprisoned. Vermont and Maine do not disenfranchise felons while in prison.

Source: The Sentencing Project (2005).

formal education are less likely to register than the better educated. Members of disadvantaged racial minority groups, especially African Americans, are less likely to register than whites. Thus, American voter registration law is a pathway to democracy for some, a barrier to democracy for others. Even the new motor voter law appears to have benefited mainly middle-class Americans rather than the disadvantaged (Brains, 1999; Grofman, 2000). Here we have "democracy's unresolved dilemma" (Lijphart, 1997). As Seymour Martin Lipset (1981 [1960]), America's leading political sociologist, explains:

> [W]hen the vote is low, this almost always means that the socially and economically disadvantaged groups are underrepresented in government. The combination of a low vote and a relative lack of organization among the lower-status groups means that they will suffer

> from neglect by the politicians who will be receptive to the wishes of the more privileged, participating, and organized strata. (226–7)

In short, the American political system is less responsive than other rich democracies to the needs of the disadvantaged for two main reasons. First, as we saw in our discussion of power resource theory, the working class is comparatively nonunionized and therefore weak. Second, as state-centered theory suggests, the law requires citizen-initiated voter registration, one result of which is that the vote is in effect taken away from many disadvantaged people.

Changing State Structures

In general, state structures resist change. *Constitutions* anchor their foundations. Only a large majority of federally elected representatives and state legislatures can change the constitution. *Laws* gird the upper stories of state structures. Some laws help to keep potentially disruptive social forces at bay. Voter registration law is a case in point. *Ideology* reinforces the whole edifice. All states create anthems, flags, ceremonies, celebrations, sporting events, and school curricula that stimulate patriotism and serve in part to justify existing political arrangements (Box 14.3).

Despite these anchors, girders, and reinforcements, big shocks do sometimes reorient American public policy and cause a major shift in voting patterns. In the past 110 years, these shocks have occurred about every four decades—in the 1890s, the 1930s, and the 1970s:

1. In the 1890s, industrial unrest was widespread. Western and Southern farmers revolted against the established parties. In 1896, these rebellious forces mounted a Democratic-Populist challenge to the Republicans of the North and the wealthy Democrats of the South. Their presidential candidate, William Jennings Bryan, won nearly 48 percent of the vote in the 1896 election. But America's elites learned an important lesson from Bryan's challenge. They instituted electoral reforms—including voter registration laws—that made possible the domination of the probusiness

AP/Wide World Photos

Lewis W. Hines/SuperStock

During the Great Depression (1929–1939), widespread poverty, unemployment, bankruptcy, and strike violence led to the election of Democrat Franklin Delano Roosevelt as president. Today's unemployment insurance, old-age pension, and public assistance programs all originated in Roosevelt's New Deal. The photo on the left shows FDR signing the Social Security Bill, August 14, 1935.

BOX 14.3
Sociology at the Movies

Gangs of New York (2002)

Most states may be imposing structures, but there was a time when little was solid, when every state lacked legitimacy in the eyes of many of its citizens, contending groups vied for dominance, violence was widely used to secure power, all was in the balance, and it was uncertain how things would turn out. Martin Scorsese's *Gangs of New York* recounts such a time in our history: New York between the 1840s and the Civil War.

"The forge of hell" is the way one character describes the city. Elections are rigged, city officials sell their services to the highest bidder, firefighters loot the buildings they "save," rival police forces brawl in the streets, the poor riot against the draft, Union soldiers force immigrants straight off the boat into uniform, and an audience greets an actor playing Abraham Lincoln with volleys of rotten fruit. At the center of it all are the mobs of Irish immigrants who battle second- and third-generation "nativists" for political control—the Irish, led by Priest Vallon (Liam Neeson), the nativists led by the vicious William Cutting, widely known as Bill the Butcher (played by Daniel Day-Lewis, who won an Oscar for Best Actor for his performance). With ferocity unrivaled in the history of cinema, the gangs attack each other on the Lower East Side of Manhattan. When the dust settles, Priest Vallon's young son, Amsterdam (Leonardo DiCaprio), swears to avenge his father's death.

At the level of its individual characters, the film is motivated by Amsterdam's quest. But because Scorsese is blessed with a deep sociological understanding of his subject matter, the individual characters become vehicles for a larger story, the chaotic origins of the American state. In the 1860s, New York was practically destroyed when the poor refused to be drafted to fight in the Civil War. They rioted and looted wealthy neighborhoods until government ships in the harbor fired their cannons on them and troops marched in to silence them once and for all. The American state was in fact weak well into the 1870s, when it was still common for independent militias funded by wealthy local capitalists to counter labor unrest and riots by the poor (Isaacs, 2002).

In the final scene of *Gangs of New York*, an adult Amsterdam stands in a graveyard with his girlfriend, Jenny Everdeane (Cameron Diaz), remembering the victims of the Draft Riots. The scene fast-forwards, and as it does, Amsterdam and Jenny disappear while the tombstones fade and vegetation grows over them. The camera pans up to focus on the familiar skyline of New York, solid and seemingly eternal. Thanks to *Gangs of New York*, however, we remember the frailty of all states and their conflict-ridden origins.

Miramax/Dimension Films/The Kobal Collection/Tursi, Mario

Leonardo DiCaprio leads his men into battle in *Gangs of New York*

Republican party in the North and the pro–plantation-owner Democratic party in the South. Low voter turnout and the effective disenfranchisement of many poor and black voters date from this era (Piven and Cloward, 1989 [1988]: 26–95).

2. In the 1930s during the Great Depression, widespread unemployment, bankruptcy, and strike violence led to the election of Democrat Franklin Delano Roosevelt as president. This time, America's business elite was too devastated and divided by the Depression to stave off the Democratic threat. Today's unemployment insurance, old-

AP/Wide World Photos

The Selma-to-Montgomery march for black voting rights was a pivotal moment in the Civil Rights movement. It led to Congress's enactment of the Voting Rights Act of 1965.

age pension, and public assistance programs all originated in Roosevelt's "New Deal" (Leman, 1977). The reform wave unleashed by the New Deal did not end until the 1960s. During that decade, the Civil Rights movement inspired sweeping constitutional change that extended many of the benefits of American citizenship to African Americans.

3. In the 1970s, America faced another economic crisis. Following their post–World War II reconstruction, Japan and West Germany emerged as major competitive threats to American manufacturers. Inflation became a serious issue after Middle East oil producers demanded much higher prices for oil. Industrial workers struck on a large scale for higher wages to compensate for inflation. Reacting to these threats, members of the American business elite became more politically organized and unified than ever before. Some sociologists argue that they formed a truly cohesive ruling class. These business leaders funded PACs, lobbies, and research institutes to promote conservative policies. Lowering taxes, cutting state funding of social programs, and creating a more favorable regulatory environment for business topped their list of priorities (Akard, 1992; Clawson, Neustadtl, and Scott, 1992; Domhoff, 1983; Schwartz, 1987; Useem, 1984; Vogel, 1996). The 1980 presidential victory of Republican Ronald Reagan capped the resurgence of post–World War II conservatism in American politics.

Summing Up

In sum, political sociology has made good progress since the 1950s. Each of the field's major schools has made a useful contribution to our appreciation of political life (▶Concept Summary 14.1). Pluralists teach us that democratic politics is about compromise and the accommodation of all group interests. Elite theorists teach us that despite compromise and accommodation, power is concentrated in the hands of high-status groups, whose interests the political system serves best. Power resource theorists teach us that despite the concentration of power in society, substantial shifts in the distribution of power do occur, with big effects on voting patterns and public policies. And state-centered theorists teach us that despite the influence of the distribution of power on political life, state structures exert an important effect on politics too.

We now turn to an examination of the historical development of democracy, its sociological underpinnings, and its future.

Concept Summary 14.1
Four Sociological Theories of Democracy Compared

	Pluralist	Elitist	Power Resource	State-Centered
How is power distributed?	Dispersed	Concentrated	Concentrated	Concentrated
Who are the main power holders?	Various groups	Elites	Upper class	State officials
On what is their power based?	Holding political office	Controlling major institutions	Owning substantial capital	Holding political office
What is the main basis of public policy?	The will of all citizens	The interests of major elites	The balance of power between classes, etc.	Influence of state structures
Do lower classes have much influence on politics?	Yes	No	Sometimes	Sometimes

The Future of Democracy

Two Cheers for Russian Democracy

Personal Anecdote

In 1989, the Institute of Sociology of the Russian Academy of Science invited Robert Brym and nine other sociologists to attend one of a series of seminars in Moscow. The seminars were designed to acquaint some leading sociologists in the Soviet Union with Western sociology. The country was in the midst of a great thaw. Totalitarianism was melting, leaving democracy in its place. Soviet sociologists had never been free to read and research what they wanted. Now they were eager to learn from North American and European scholars (Brym, 1990).

Or at least so it seemed. One evening about a dozen of the sociologists were sitting around comparing the merits of Canadian whiskey and Russian vodka. Soon, conversation turned from Crown Royal versus Moskovskaya to Russian politics. "You must be so excited about what's happening here," Robert said to his Russian hosts. "How long do you think it will be before Russia will have multiparty elections? Do you think Russia will become a liberal democracy like the United States or a socialist democracy like Sweden?"

One white-haired Russian sociologist slowly rose to his feet. His colleagues privately called him "the dinosaur." It soon became clear why. "*Nikogda*," he said calmly and deliberately—"never." "*Nikogda*," he repeated, his voice rising sharply in pitch, volume, and emphasis. Then, for a full minute he explained that capitalism and democracy were never part of Russia's history. Nor could they be expected to take root in Russian soil. "The Russian people," he proclaimed, "do not want a free capitalist society. We know 'freedom' means the powerful are free to compete unfairly against the powerless, exploit them, and create social inequality."

Everyone else in the room disagreed with the dinosaur's speech, in whole or in part. But not wanting to cause any more upset, we turned the conversation back to lighter topics. After 15 minutes, someone reminded the others that we had to rise early for tomorrow's seminars. The evening ended, its great questions unanswered.

Today, more than a decade later, the great questions of Russian politics remain unanswered. And it now seems that there was some truth in the dinosaur's speech after all. Russia first held multiparty elections in 1991. Surveys found that most Russians favored democracy over other types of rule. However, support for democracy soon fell because the economy collapsed.

While Versace does brisk business in Moscow, the streets are filled with homeless people. That is because the richest 10 percent of Russians earned 15 times more than the poorest 10 percent, making Russia one of the most inegalitarian countries in the world.

The government had formerly fixed prices. It now allowed prices to rise to levels set by the market. Consumer goods soon cost 10 or 12 times more than just a year earlier. Many enterprises shut down because they were too inefficient to stay in business under market conditions. This led to an unemployment rate of about 20 percent. Even when the state kept businesses alive by subsidizing them, they paid many workers irregularly. Sometimes workers went months without a paycheck. Many Russians were barely able to make ends meet. According to official estimates, 39 percent of the population lived below the poverty line in 1999.

At the other extreme, profitable businesses and valuable real estate formerly owned by the government were sold to private individuals and companies. The lion's share went to senior members of the Communist Party and organized crime syndicates. These were the only two groups with enough money and inside knowledge to take advantage of the sell-off. They became fantastically wealthy. Moscow is said to have more Mercedes-Benz automobiles per capita than any other city in the world. By 1994 the richest 10 percent of Russians earned 15 times more than the poorest 10 percent. The level of income inequality in Russia is one of the highest in the world (Brym, 1996a; 1996b; 1996c; Gerber and Hout, 1998; Handelman, 1995; Remnick, 1998).

Democratic sentiment weakened as economic conditions worsened (Whitefield and Evans, 1994). In elections held in 1995 and 1996, support for democratic parties plunged as support for communist and extreme right-wing nationalist parties surged (Brym, 1995; 1996d). Nationwide surveys conducted in 38 countries between 1995 and 1997 found that as many as 97 percent of the citizens of some countries viewed democracy as the ideal form of government. Russia ranked last, at a mere 51 percent (Klingemann, 1999). Democracy allowed a few people to enrich themselves at the expense of most Russians. Therefore, many citizens equated democracy not with freedom but with distress.

Russia's political institutions reflect the weakness of Russian democracy. Power is concentrated in the presidency to a much greater degree than in the United States. The parliament and the judiciary do not act as checks on executive power. Only a small number of Russians belong to political parties. Voting levels are low. Much of the mass media is state controlled. Minority ethnic groups are sometimes treated arbitrarily and cruelly. Clearly, Russian democracy has a long way to go before it can be considered on a par with democracy in the West.

The limited success of Russian democracy raises an important question. What social conditions must exist for a country to become fully democratic? That is the question to which we now turn. To gain some perspective, we first consider the three waves of democratization that have swept the world in the past 175 years (Huntington, 1991: 13–26) (Figure 14.5).

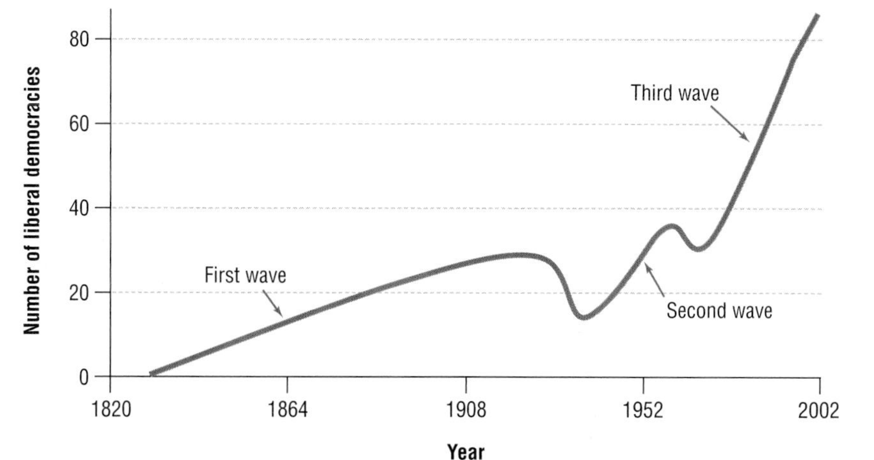

Figure 14.5
The Three Waves of Democratization, 1828–2002

Source: Diamond (1996: 28); Huntington (1991: 26); Marshall and Gurr (2003: 17).

The Three Waves of Democracy

The first wave of democratization began when more than half the white adult males in the United States became eligible to vote in the 1828 presidential election. By 1926, 33 countries enjoyed at least minimally democratic institutions. These countries included most of Western Europe, the British Dominions (Australia, Canada, and New Zealand), Japan, and four Latin American countries (Argentina, Colombia, Chile, and Uruguay). However, just as an undertow begins when an ocean wave recedes, a democratic reversal occurred between 1922 and 1942. During that period, fascist, communist, and militaristic movements caused two-thirds of the world's democracies to fall under authoritarian or totalitarian rule.

The second wave of democratization took place between 1943 and 1962. Allied victory in World War II returned democracy to many of the defeated powers, including West Germany and Japan. The beginning of the end of colonial rule brought democracy to some states in Africa and elsewhere. Some Latin American countries formed limited and unstable democracies. However, even by the late 1950s, the second wave was beginning to exhaust itself. Soon, the world was in the midst of a second democratic reversal. Military dictatorships replaced many democracies in Latin America, Asia, and Africa. A third of the democracies in 1958 were authoritarian regimes by the mid-1970s.

The third and biggest wave of democratization began in 1974 with the overthrow of military dictatorships in Portugal and Greece. It crested in the early 1990s. In Southern and Eastern Europe, Latin America, Asia, and Africa, a whole series of authoritarian regimes fell. In 1991, Soviet communism collapsed. By 1995, 117 of the world's 191 countries were democratic in the sense that their citizens could choose representatives in regular, competitive elections. That amounts to 61 percent of the world's countries, containing nearly 55 percent of the world's population (Diamond, 1996: 26).

The third wave seems less dramatic, however, if we bear in mind that these figures refer to **formal democracies**—countries that hold regular, competitive elections. Many of these countries are not **liberal democracies.** That is, like Russia, they lack the freedoms and constitutional protections that make political participation and competition meaningful. In formal but nonliberal democracies, substantial political power may reside with a military that is largely unaffected by the party in office. Certain cultural, ethnic, religious, or regional groups may not be allowed to take part in elections. The legislative and judicial branches of government may not constrain the power of the executive branch. Citizens may not enjoy freedom of expression, assembly, and organization. Instead, they may suffer from unjustified detention, exile, terror, and torture. At the end of 1995, 40 percent of the world's countries were liberal democracies, 21 percent were nonliberal democracies, and 39 percent were nondemocracies (Diamond, 1996: 28). The number of liberal

Formal democracy involves regular, competitive elections.

A **liberal democracy** is a country whose citizens enjoy regular, competitive elections *and* the freedoms and constitutional protections that make political participation and competition meaningful.

Nigeria celebrated independence from Britain in 1960 *(left)*. In 1993, General Sani Abacha *(right)* annulled the presidential election, became head of state, and began a reign of brutal civil rights violations. The world's third wave of democratization was ebbing.

democracies in the world fell nearly 2 percent between 1991 and 1995. Some new democracies, including large and regionally influential countries like Russia, Nigeria, Turkey, and Pakistan, experienced a decline in freedoms and protections. The third wave, it seems, was subsiding (U.S. Information Agency, 1998–1999).

The Social Preconditions of Democracy

Liberal democracies emerge and endure when countries enjoy considerable economic growth, industrialization, urbanization, the spread of literacy, and a gradual decrease in economic inequality (Huntington, 1991: 39–108; Lipset, 1981 [1960]: 27–63, 469–76; 1995; Moore, 1967; Rueschemeyer, Stephens, and Stephens, 1992; Zakaria, 1997). Economic development creates middle and working classes that are large, well organized, literate, and well-off. When these classes become sufficiently powerful, their demands for civil liberties and the right to vote and run for office have to be recognized. If powerful middle and working classes are not guaranteed political rights, they sweep away kings, queens, landed aristocracies, generals, and authoritarian politicians in revolutionary upsurges. In contrast, democracies do not emerge where middle and working classes are too weak to wrest big political concessions from predemocratic authorities. In intermediate cases—where, say, a country's military is about as powerful a political force as its middle and working classes—democracy is precarious and often merely formal. The history of unstable democracies is largely a history of internal military takeovers (Germani and Silvert, 1961).

Apart from the socioeconomic conditions noted previously, favorable external political and military circumstances help liberal democracy endure. Liberal democracies, even strong ones like France, collapse when they are defeated by fascist, communist, and military regimes and empires. They revive when democratic alliances win world wars and authoritarian empires break up. Less coercive forms of outside political intervention are sometimes effective too. For example, in the 1970s and 1980s, the European Union helped liberal democracy in Spain, Portugal, and Greece by integrating these countries into the Western European economy and giving them massive economic aid.

In sum, powerful, prodemocratic foreign states and strong, prosperous middle and working classes are liberal democracy's best guarantees. It follows that liberal democracy will spread in the less economically developed countries only if they prosper and enjoy support from a confident United States and European Union, the world centers of liberal democracy.

Recognizing the importance of the United States and the European Union in promoting democracy in many parts of the world should not obscure two important facts, however. First, the United States is not always a friend of democracy. For example, between the end of World War II and the collapse of the Soviet Union in 1991, democratic regimes that were sympathetic to the Soviet Union were often destabilized by the United States and replaced by antidemocratic governments. American leaders were willing to export arms and offer other forms of support to antidemocratic forces in Iran, Chile, Nicaragua, Guatemala, and other countries because they believed it was in the United States' political and economic interest to do so (Chapter 9, "Globalization, Inequality, and Development").

Second, just because the United States promotes democracy in many parts of the world, we should not assume that liberal democracy has reached its full potential in this country. We saw otherwise in our discussion of the limited participation and influence of disadvantaged groups in American politics. It seems fitting, therefore, to conclude this chapter by briefly assessing the future of liberal democracy in America.

Some analysts think that the home computer will soon increase Americans' political involvement. They think it will help solve the problem of unequal political participation by bringing more disadvantaged Americans into the political process. Others think that growing affluence means that there are fewer disadvantaged Americans to begin with. This makes economic or material issues less relevant than they used to be. In the concluding section, we raise questions about both contentions. We argue that political participation is likely to remain unequal in the foreseeable future. Meanwhile, issues concerning economic inequality are likely to remain important for most people. Liberal democracy can realize its full potential only if both problems—political and economic inequality—are adequately addressed.

Electronic Democracy

On October 20, 1935, the *Washington Post* ran a full-page story featuring the results of the first nationwide poll. The story also explained how the new method of measuring public opinion worked. George Gallup was the man behind the poll. In the article, he said that polls allow the people to reclaim their voice: "After one hundred and fifty years we return to the town meeting. This time the whole nation is within the doors" (quoted in London, 1994: 1). Gallup was referring to the lively New England assemblies that used to give citizens a direct say in political affairs. He viewed the poll as a technology that could bring the town hall to the entire adult population of the country.

Gallup's idea seems naive today. Social scientists have shown that polls often allow politicians to mold public opinion, not just reflect it. For example, they can word questions to increase the chance of eliciting preferred responses. They can then publicize the results to serve their own ends (Ginsberg, 1986; also Chapter 2, "How Sociologists Do Research"). From this point of view, polls are little different from other media events that are orchestrated by politicians to sway public opinion.

Recently, however, some people have greeted one new technology with the same enthusiasm that Gallup lavished on polls. Computers linked to the Internet could allow citizens to debate issues and vote on them directly. This could give politicians a clear signal of how public policy should be conducted. Some people think that in an era of low and declining political participation, computers can revive American democracy. Public opinion would then become the law of the land (Westen, 1998).

It is a grand vision, but flawed. Social scientists have conducted more than a dozen experiments with electronic public meetings. They show that even if the technology needed for such meetings were available to everyone, interest is so limited that no more than a third of the population would participate (Arterton, 1987).

Subsequent experience supports this conclusion. The people most likely to take advantage of electronic democracy are those who have access to personal computers and the

Table 14.4
The Digital Divide: Social Characteristics of Internet Users, United States, September 2001 (in percent; $N = 137{,}000$)

	Internet Use
Family income ($)	
Less than $15,000	25.0
15,000–24,999	33.4
25,000–34,999	44.1
35,000–49,999	57.1
50,000–74,999	67.3
75,000+	78.9
Educational attainment	
Less than high school	12.8
High school diploma/GED	39.8
Some college	62.4
Bachelor's degree	80.8
Beyond bachelor's degree	83.7
Race	
Asian American & Pacific Islander	60.4
White	59.9
Hispanic	31.6
Black	39.8
Age group (years)	
3–8	27.9
9–17	68.6
18–24	65.0
25–49	63.9
50+	37.1

Source: U.S. Department of Commerce (2002: 28).

Internet. They form a privileged and politically involved group. They are not representative of the American adult population. This is apparent from Table 14.4. The table contains data from a government-sponsored survey of World Wide Web users. Compared with the general population, American Web users are younger, better educated and wealthier and comprise a higher proportion of men, whites, and people in occupations requiring substantial computer use. It is also significant that nearly 83 percent of American Web users are registered to vote. That compares with about 70 percent in the voting-age population as a whole. We conclude that if electronic democracy becomes widespread, it will probably reinforce the same inequalities in political participation that plague American democracy today.

Postmaterialism

Believers in electronic democracy think that the computer will solve the problem of unequal political participation. **Postmaterialists** believe that economic or material issues are becoming less important in American politics. They argue as follows: Liberal democracies are less stratified than both nonliberal democracies and nondemocracies. That is, in liberal democracies, the gap between rich and poor is less extreme and society as a whole is more prosperous. In fact, say the postmaterialists, prosperity and the moderation of stratification have reached a point where they have fundamentally changed political life in America. They claim that as recently as 50 years ago, most people were politically motivated mainly by their economic or material concerns. As a result, parties were distinguished from one another chiefly by the way they attracted voters from different classes. Now, however, many if not most Americans have supposedly had their basic material wants satisfied. Particularly young people who grew up in prosperous times are less concerned with material issues, such as whether their next paycheck can feed and house their family. They are more concerned with postmaterialist issues, such as women's rights, civil rights, and the environment. The postmaterialists conclude that the old left–right political division, based on class differences and material issues, is being replaced. The new left–right political division, they say, is based on age differences and postmaterialist issues (Clark and Lipset, 1991; Clark, Lipset, and Rempel, 1993; Inglehart, 1997; Clark and Lipset, 2001).

Although America is certainly more prosperous and less stratified than the less developed countries of the world, the postmaterialists are wrong to think that affluence is universal in America. Nor is inequality decreasing. We saw in Table 14.3 that the United States has one of the highest poverty rates of the 18 rich industrialized countries (also Chapter 8, "Stratification: United States and Global Perspectives"). Here we may add that poverty is particularly widespread among youth—just the people who, in the postmaterialist view, are the most affluent and least concerned with material issues. In May 2003, the unemployment rate for people between the ages of 16 and 19 was more than three times the rate for the whole labor force (U.S. Department of Labor, 2003). Moreover, between the late 1960s and early 1990s, the percentage of children under the age of 18 living in poverty *after welfare payments* doubled, rising to about 22 percent. This makes the United States number one in child poverty among the 18 rich industrialized countries (Rainwater and Smeeding, 1995). These figures suggest that, today, more new voters are poor than at any time in the past 40 years.

In addition, inequality is not decreasing. People with a college degree have seen their real incomes rise substantially since the early 1970s. However, people without a college degree, who account for 70 percent of the American workforce, have seen their real incomes rise only slightly or decline (Bluestone and Rose, 1997).

Postmaterialism is a theory that claims that growing equality and prosperity in the rich industrialized countries have resulted in a shift from class-based to value-based politics.

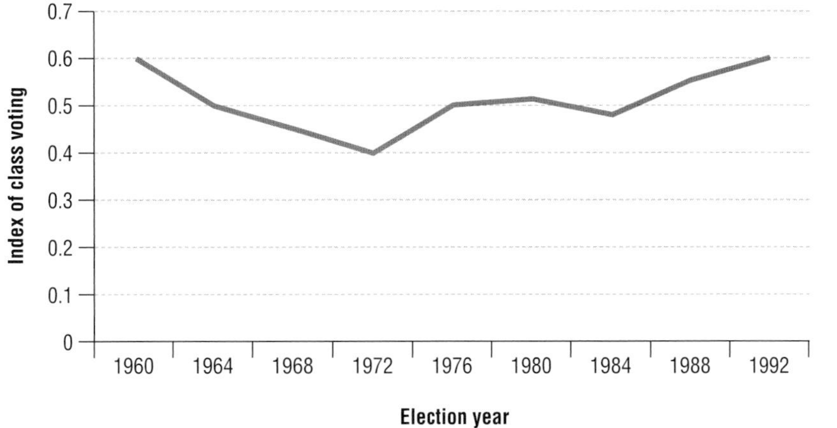

Figure 14.6
The Class Cleavage in U.S. Presidential Elections, 1960–1992

Source: Brooks and Manza (1997b).

Under these circumstances, we should not be surprised that bread-and-butter issues are still important for most voters. In fact, research shows that class is just as important in influencing voting as it was 40 years ago. Class voting fluctuates from election to election. It varies from one rich industrialized country to the next. But disadvantaged people still tend to vote for parties on the left and advantaged people for parties on the right. To varying degrees, Americans, and voters in other liberal democracies, still tend to vote according to their material interests. There is no denying that people are inserting postmaterialist issues into today's political debates. This seems especially true after the 2004 presidential election, in which moral issues featured prominently. It seems, however, that most people are layering these issues on top of old ones, not replacing them (Brooks and Manza, 1994; 1997a; 1997b; Hout, Brooks, and Manza, 1993; Manza, Hout, and Brooks, 1995; Weakliem, 1991). Figure 14.6 illustrates the American case. It shows change over time in the influence of class on voting. Although class voting dropped slightly from 1960 to 1972, it started to rise afterward. In 1992, class voting stood at the same level as in 1960.

So we arrive at the big dilemma of American politics. Problems of economic inequality continue to loom large, but it is doubtful that they will be addressed in a serious way unless disadvantaged Americans get more politically involved. Yet unequal political participation shows no sign of evaporating (Valelly, 1999). No mystery surrounds the identification of reforms needed to bring more disadvantaged people into the political process. For example, comparative research has determined that removing burdensome voter registration laws would increase participation rates in the United States by 8 percent to 15 percent (Lijphart, 1997). However, there is little political will to undertake such reforms now.

A solution may have to come from outside normal electoral politics, that is, from social movements (Chapter 21, "Collective Action and Social Movements"). We saw how big shocks reoriented American public policy in the 1890s, 1930s, and 1970s. Perhaps we need something similar to decrease political and economic inequality in the future. The full realization of Lincoln's "government of the people, by the people, for the people" may require one of those "little rebellions" called for by Thomas Jefferson. (As Figure 14.7 shows, some evidence suggests that Americans are heeding Jefferson's call.) As you think about this possibility, ask yourself some related questions: Who is most likely to lead such rebellions? Under what circumstances might such rebellions come about? To what extent would you support or oppose them? Why?

Sociology Now™

Learn more about **American Politics** by going through the Integrity of the Electoral Process Video Exercise.

Politics by Other Means

Participating in social movements is not the only way people step outside the rules of normal electoral politics to change society. We conclude our discussion by considering two other types of "politics by other means": war and terrorism.

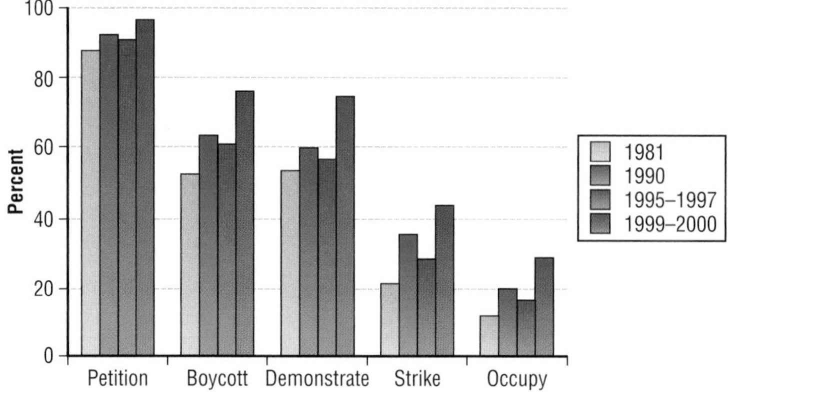

Figure 14.7
Nonconventional Political Action, United States, 1981–2000 (percent "actually done" and "might do"; $N = 6906$)

Note: The World Values Survey regularly asks the following question: "I'm going to read out some different forms of political action that people can take, and I'd like you to tell me, for each one, whether you have actually done any of these things, whether you might do it or would never, under any circumstances, do it: signing a petition, joining in boycotts, attending lawful demonstrations, joining unofficial strikes, occupying buildings or factories." Note the tendency for Americans to be increasingly willing to engage in all forms of nonconventional political behavior. Incidentally, this tendency is *not* evident in most other countries covered by the World Values Survey.

Source: World Values Survey (2003).

War

War Deaths

A **war** is a violent armed conflict between politically distinct groups who fight to protect or increase their control of territory. Humanity has spent much of its history preparing for war, fighting it, and recovering from it. Thus, recorded history shows that war has broken out some 14,000 times between 3600 B.C.E. and the present. It has killed roughly a billion soldiers and two billion civilians, approximately 3 percent of the people born in the last 5600 years. (This is an underestimate because it necessarily ignores armed conflict among people without a recorded history.) Wars have become more destructive over time with "improvements" in the technology of human destruction. Thus, the 20th century represents 1.7 percent of the time since the beginning of recorded war history but accounts for roughly 3.3 percent of the world's war deaths, military and civilian. About 100 million people died in 20th-century wars (Beer, 1974; Brzezinski, 1993; Haub, 2000).

War is a violent armed conflict between politically distinct groups who fight to protect or increase their control of territory.

The Business of War

War is an expensive business, and the United States spends far more than any other country financing it. With 4.5 percent of the world's population, the United States accounts for about a third of total military expenditures in the world. In addition, the United States is by far the largest exporter of arms, accounting for nearly 60 percent of world arms exports. The countries ranking distant second, third, and fourth are the United Kingdom (12 percent of world arms exports), France (11 percent), and Russia (4 percent). Most of the big arms importers are U.S. allies. In 2002, our best customers were France and Egypt. Next in order were Saudi Arabia, Israel, and Japan (U.S. Bureau of the Census, 2001: 327; 2002: 860; 2004–05: 329).

Types of War

Wars may take place between countries (interstate wars) and within countries (civil or societal wars). A special type of interstate war is the colonial war, which involves a colony engaging in armed conflict with an imperial power to gain independence. Figure 14.8 shows the magnitude of armed conflict in the world for each of these types of war from 1946 to 2002. (See the note accompanying Figure 14.8 for the definition of "magnitude of armed conflict.") You will immediately notice two striking features of the graph. First, after reaching a peak between the mid-1980s and early 1990s, the magnitude of armed conflict in the world dropped sharply. Second, since the mid-1950s, most armed conflict in the world has been societal rather than interstate. Today, countries rarely go to war against each other. They often go to war with themselves as contending political groups fight for state control or seek to break away and form independent states. Don't let the mass media distort your perception of global war. Wars like the 2003 U.S.–Iraq war account for little of the total magnitude of armed conflict, although they loom large in the mass media. Wars like the ongoing conflict in the Democratic Republic of the Congo ac-

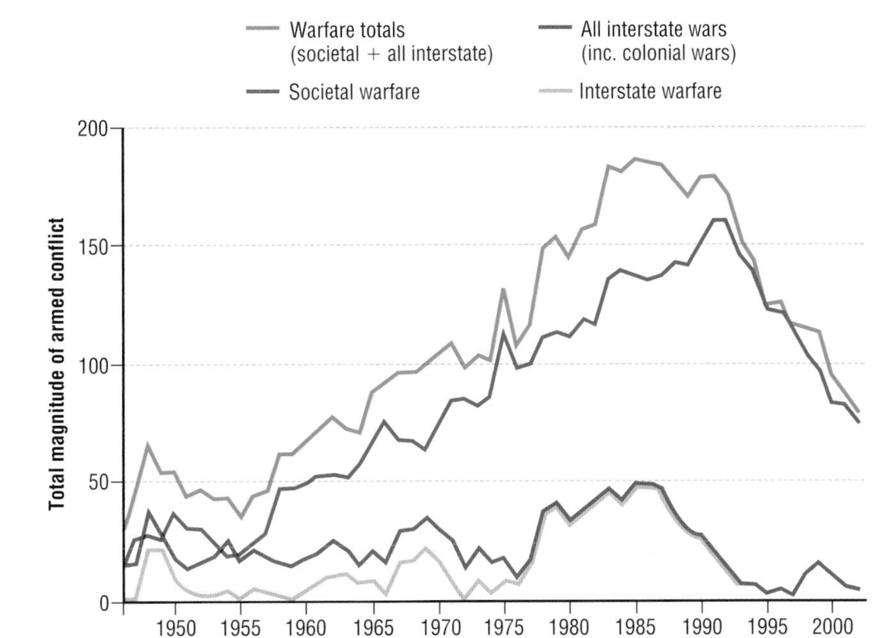

Figure 14.8
Global Trends in Violent Conflict, 1946–2002

Note: Data are for 158 countries with populations of at least 500,000 with 500 or more deaths directly related to war between 1946 and 2002. The magnitude of armed conflict is determined by the total number of combatants and casualties, the size of the area and dislocated populations, and the extent of infrastructure damage for each year the war is active.

Source: Monty G. Marshall and Ted Robert Gurr. *Peace and Conflict 2003.* College Park, MD: CIDCM, University of Maryland. www.cidcm.umedu/inser. Used by permission.

count for most of the total magnitude of armed conflict, yet are rarely mentioned in the mass media. The 2003 U.S.–Iraq war lasted six weeks, until the end of "major hostilities," and registered about 7000 military and civilian deaths. The civil war in the Democratic Republic of the Congo had dragged on for five years by August 2003 and registered about 3.5 million deaths ("Iraq Body Count," 2003; "Three Million Dead," 2003).

The Risk of War

War risk varies from one country to the next (Figure 14.9), but what factors determine the risk of war on the territory of a given country? Figure 14.10 helps us answer that question. Figure 14.10 classifies the countries of the world by type of government and level of prosperity. Government types include democracy, autocracy (absolute rule by a single person or party), and "intermediate" forms. Intermediate types of government include some elements of democracy (e.g., regular elections) and some elements of autocracy (e.g., no institutional checks on presidential power). In this graph, a country's gross domestic product per capita (GDPpc) indicates its level of prosperity. The graph divides the world's countries into quarters by GDPpc.

Given our earlier discussion of the social preconditions of democracy, it should come as no surprise that democracy is more common and autocracy less common in prosperous countries. What is particularly interesting about Figure 14.10 is the distribution of countries with *intermediate* types of government. Countries with intermediate types of government are at highest risk of war, especially societal or civil war. A democratic government tends to be stable because it enjoys legitimacy in the eyes of its citizens. An autocratic government tends to be stable because an iron hand rules it. In contrast, an intermediate type of government is characterized neither by high legitimacy nor iron rule. It is therefore most prone to collapsing into societal war, with armed political groups fighting each other for state control.

By August 2003, the civil war in the Democratic Republic of the Congo had dragged on for five years and registered about 3.5 million deaths.

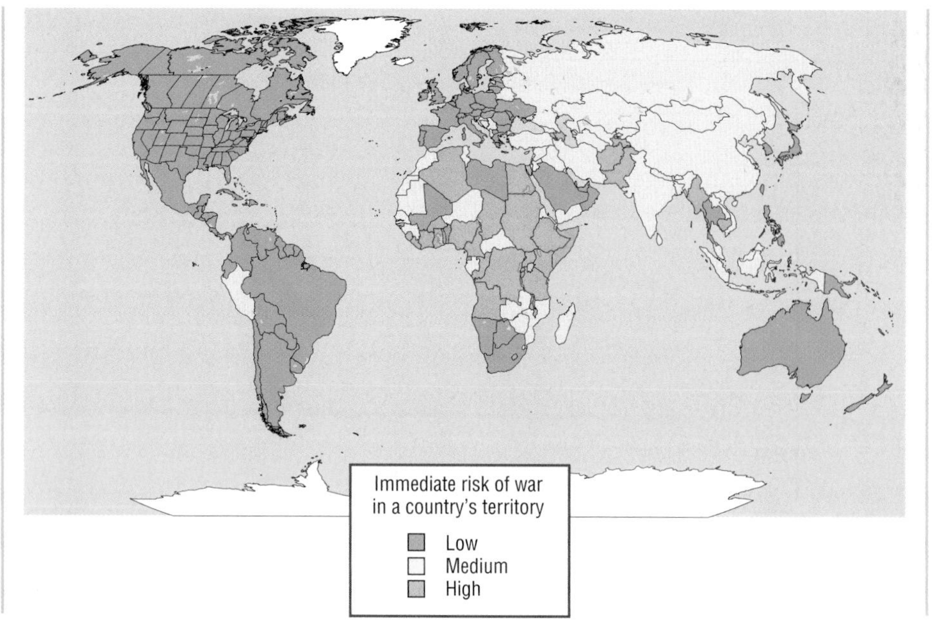

Figure 14.9
The Risk of War, 2002

Note: Data in this map are for 158 countries with populations of at least 500,000 with 500 or more deaths directly related to war between 1946 and 2002. A country's chance of avoiding war on its own territory is high if a country has no recent history of armed conflict, lacks or manages internal groups that wish to break away and form their own country, maintains stable and equitable democratic institutions, is well-to-do, and is free of serious external threats. Assigning a number to each of these factors and adding the numbers for each country yields a figure suggesting the risk of war breaking out in the territory of each country. Dividing the country scores into three categories (low-, medium-, and high-risk) and assigning different colors to each category allows us to create this world map. The 34 red-flagged countries, most in Africa, are at serious risk of war. The 50 yellow-flagged countries are at moderate risk of war. The 74 green-flagged countries are at low risk of war. Of course, countries that avoid war on their own territory may nonetheless engage in war elsewhere, the United States being the prime example. Since 1850, the United States has intervened militarily in other countries more than once a year on average (Kohn, 1988).

Source: Monty G. Marshall and Ted Robert Gurr. *Peace and Conflict 2003*. College Park, MD: CIDCM, University of Maryland. www.cidcm.umedu/inser. Used by permission.

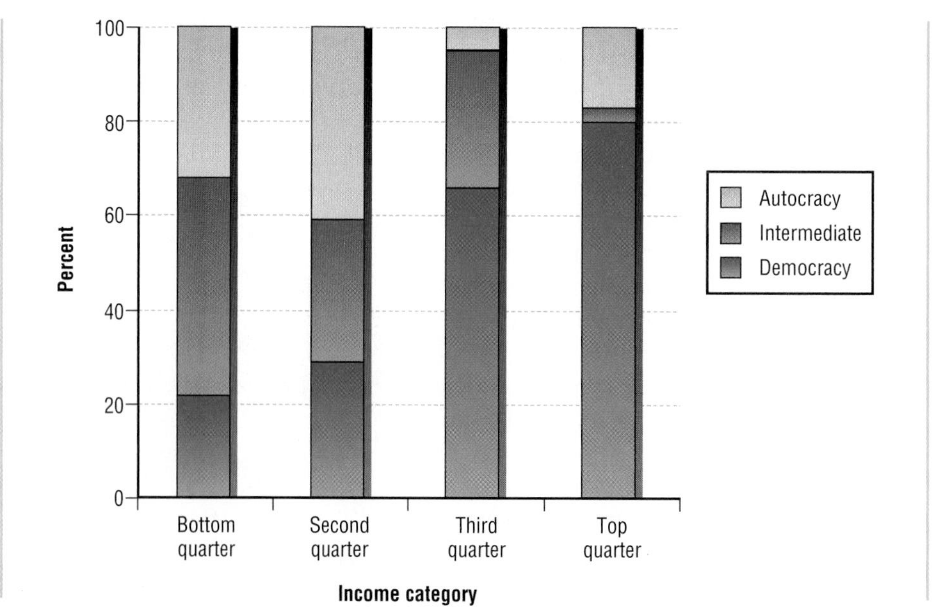

Figure 14.10
Type of Government by Income Category, 1990s

Source: Marshall and Gurr (2003: 21).

We conclude that economic development and democratization are the two main factors leading to less war in the world today. In a way, this is a deeply disturbing finding. In 2004, the United States allocated about 19 percent of its budget to war (or "defense") spending and only 0.8 percent to international development and humanitarian assistance (U.S. Bureau of the Census, 2004: 309). Public opinion polls show that the American public and its leaders regard the promotion of democracy and the improvement of living standards in other countries as the least important American goals, although, as we learned in Chapter 9 (see Table 9.2), the cost of achieving them would be relatively low (Bardes and Oldendick, 2003: 205). It thus seems that we prefer to spend very little money promoting higher living standards and democracy in other nations. This permits much armed conflict in the world. We then spend a great deal of money dealing with the armed conflict. Said differently, we have chosen to spend a lot to increase the sum total of human misery on the planet rather than spending a little to increase the welfare of humanity.

Terrorism and Related Forms of Political Violence

We can learn much about the sorry predicament of the world today by lingering a moment on the question of why societal warfare has largely replaced interstate warfare since World War II. As usual, historical perspective is useful (Tilly, 2002).

Historical Change

From the rise of the modern state in the 17th century until World War II, states increasingly monopolized the means of coercion in society. This had three important consequences. First, as various regional, ethnic, and religious groups came under the control of powerful central states, regional, ethnic, and religious wars declined and interstate warfare became the norm. Second, because states were powerful and monopolized the means of coercion, conflict became more deadly. Third, civilian life was pacified because the job of killing for political reasons was largely restricted to state-controlled armed forces. Thus, even as the death toll due to war rose, civilians were largely segregated from large-scale killing. As late as World War I (1914–1918), civilians composed only 5 percent of war deaths.

All this changed after World War II (1939–1945). Since then, there have been fewer interstate wars and more civil wars, guerrilla wars, massacres, terrorist attacks, and instances of attempted ethnic cleansing and genocide perpetrated by militias, mercenaries, paramilitaries, suicide bombers, and the like. Moreover, large-scale violence has increasingly been visited on civilian rather than military populations. By the 1990s, civilians composed fully 90 percent of war deaths. The mounting toll of civilian casualties is evident from U.S. Department of State data on casualties due to international **terrorist** attacks, among other sources of information (Figure 14.11).

Terrorism is defined by American law as premeditated, politically motivated violence against noncombatant targets including unarmed or off-duty military personnel by subnational groups or clandestine agents, usually intended to influence an audience. However, a broader definition would include the use by states of indiscriminate violence against civilians in order to achieve military and political goals. Moreover, what one side in a conflict would call terror might be regarded by the other side as legitimate resistance to occupation, as well as ethnic, religious, or national oppression.

Reasons for Change

Change in the form of collective violence came about for three main reasons (Tilly, 2002). First, decolonization and separatist movements roughly doubled the number of independent states in the world, and many of these new states, especially in Africa and Asia, were too weak to control their territories effectively. Second, especially during the Cold War (1946–1991), the United States, the Soviet Union, Cuba, and China often subsidized and sent arms to domestic opponents of regimes that were aligned against them. Third, the expansion of international trade in contraband provided rebels with new means of support. They took advantage of inexpensive international communication and travel to establish immigrant support communities abroad and export heroin, cocaine, diamonds, dirty money, and so forth. In sum, the structure of opportunities for engaging in collective violence shifted radically after World War II. As a result, the dominant form of collective violence changed from interstate to societal warfare.

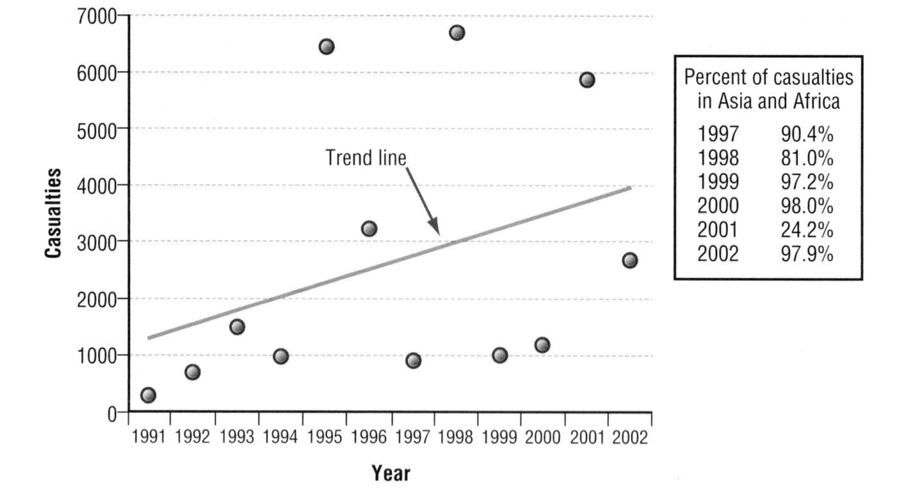

Figure 14.11
Casualties Due to International Terrorist Attacks, 1991–2002

Source: U.S. Department of State (1997; 2003: 163).

Al Qaeda and Contemporary Warfare

From this perspective, and whatever its individual peculiarities, al Qaeda is a typical creature of contemporary warfare. Al Qaeda originated in Afghanistan, a notoriously weak and dependent state. The United States supported al Qaeda's founders militarily in their struggle against the Soviet occupation of Afghanistan in the 1980s. Al Qaeda organized international heroin, diamond, and money laundering operations. It established a network of operatives around the world. All of this was made possible by changes in the structure of opportunities for collective violence after World War II.

International terrorists of the type the U.S. Department of State collects data on typically demand autonomy or independence for some country, population, or region (Pape, 2003). For example, among al Qaeda's chief demands are Palestinian statehood and the end of U.S. support for the wealthy regimes in Saudi Arabia, Kuwait, and the Gulf States. Al Qaeda has turned to terror as a means of achieving these goals because other ways of achieving them are largely closed off. The United States considers support for the oil-rich Arab countries to be in its national interest. It has so far done little to further the cause of Palestinian statehood. Staunch opponents of American policy cannot engage in interstate warfare with the United States, because they lack states of their own. At most, they are supported by states that lack the resources to engage in sustained warfare with the United States, including Iran, Syria, and the former regime of Iraq. Because the existing structure of world power closes off other possibilities for achieving political goals, terror emerges as a viable alternative for some people.

Summary

Sociology Now™

Reviewing is as easy as ❶, ❷, ❸.

❶ Before you do your final review, take the SociologyNow diagnostic quiz to help you identify the areas on which you should concentrate. You will find information on how to access SociologyNow on the foldout at the front of the textbook.

❷ As you review, take advantage of SociologyNow's study aids to help you master the topics in this chapter.

❸ When you are finished with your review, take SociologyNow's post-test to confirm you are ready to move on to the next chapter.

1. **What accounts for the level of democracy in a society?**

   The level of democracy in a society depends on the distribution of power. When power is concentrated in the hands of few people, society is less democratic.

2. **What are the major sociological theories of democratic politics?**

   Pluralists correctly note that democratic politics is about negotiation and compromise. However, they fail to appreciate that economically advantaged groups have more power than disadvantaged groups. *Elite theorists* correctly note that power is concentrated in the hands of advantaged groups. However, they fail to appreciate how variations in the distribution of power influence political behavior and public policy. *Power resource theorists* usefully focus on changes in the distribution of power in society and their effects. However, they fail to appreciate what *state-centered theorists* emphasize—that state institutions and laws also independently affect political behavior and public policy.

3. **What are the social preconditions of democracy?**

   Citizens win legal protection of rights and freedoms when their middle and working classes become large, organized, and prosperous and when powerful, friendly, prodemocratic foreign states support them.

4. **What does it mean to say that democracy has developed in waves?**

   During three periods, democracy rapidly spread in the world. Then the spurt of democratization slowed or reversed. We may regard each of these periods as a "wave" of democratization. The first wave began when more than half the white adult males in the United States became eligible to vote in the 1828 presidential election. By 1926, 33 countries enjoyed at least minimally democratic institutions. However, between 1922 and 1942, fascist, communist, and militaristic movements caused two-thirds of the world's democracies to fall under authoritarian or totalitarian rule. The second wave of democracy began after World War II, when Allied victory returned democracy to many fascist states, and wars of colonial independence led to the formation of a series of new democracies. Military dictatorships then replaced many of the new democracies; a third of the democracies in 1958 were authoritarian regimes by the mid-1970s. The third wave of democracy began in Portugal in 1974. It had slowed and in some cases reversed by the end of the 20th century.

5. **What is the difference between formal and liberal democracy and why is the distinction important for understanding the third wave of democratization?**

   Many new democracies that emerged from the most recent wave of democratization are formal, not liberal, democracies. Their citizens enjoy regular, competitive elections (the formal side of democracy) but lack legal protection of rights and freedoms (the liberal side).

6. **Do theories of electronic democracy and postmaterialism require a major rethinking of the way American democracy works?**

   Believers in electronic democracy and postmaterialism think that the United States has reached a new and higher stage of democratic development. However, enduring social inequalities prevent even the most advanced democracies from being fully democratic.

7. **What are the main causes of war?**

   The risk of war declines with a country's level of prosperity and its level of democratization.

8. **How has the nature of the state affected patterns of warfare?**

   The rise of the modern state in the 17th century led to the monopolization of the means of coercion in society. Once centralized state armies became the major military force in society, interstate warfare became the norm, warfare became more deadly, and proportionately few civilians died in war. However, the emergence of many weak states after World War II encouraged the outbreak of societal or civil wars, which are now the norm. Civilian deaths now account for most war deaths. Societal wars gain impetus when hostile outside powers get involved in them and rebels take advantage of increased opportunities to engage in illegal trade and establish support communities abroad. International terrorism has benefited greatly from the combination of weak states, outside support, and new ways of mobilizing resources.

Questions to Consider

1. **Analyze any recent election. What issues distinguish the competing parties or candidates? What categories of the voting population are attracted by each party or candidate? Why?**
2. **Younger people are less likely to vote than older people. How would power resource theory explain this?**
3. **Do you think the United States will become a more democratic country in the next 25 years? Will a larger percentage of the population vote? Will class and racial inequalities in political participation decline? Will public policy more accurately reflect the interests of the entire population? Why or why not?**

Web Resources

Companion Website for This Book

http://sociology.wadsworth.com

Begin by clicking on the Student Resources section of the website. Choose "Introduction to Sociology" and then the Brym and Lie book cover. Next, select the chapter you are currently studying from the pull-down menu. From the Student Resources page you will have easy access to InfoTrac® College Edition, MicroCase Online exercises, additional web links, and many other resources to aid you in your study of sociology, including practice tests for each chapter.

Recommended Websites

For data and analysis on the role that money plays in American elections, go to http://www.opensecrets.org.

For a useful listing of Web resources on American politics, visit http://sobek.colorado.edu/POLSCI/RES/amer.html.

The most important organization of international governance is the United Nations. For the UN website, go to http://www.un.org.

CHAPTER 15 Families

Digital Vision/Getty Images

In this chapter, you will learn that:

- The traditional nuclear family is less common than it used to be. Several new family forms are becoming more popular. The frequency of one family form or another varies by class, race and ethnicity, sexual orientation, and culture.
- One of the most important forces underlying change from the traditional nuclear family is the entry of most women into the paid labor force. Doing paid work increases women's ability to leave unhappy marriages and control whether and when they will have children.
- Marital satisfaction increases as one moves up the class structure, where divorce laws are liberal, when teenage children leave the home, in families where housework is shared equally, and among spouses who enjoy satisfying sexual relations.
- The worst effects of divorce on children can be eliminated if there is no parental conflict and the children's standard of living does not fall after divorce.
- The decline of the traditional nuclear family is sometimes associated with a host of social problems, such as poverty, welfare dependency, and crime. However, policies have been adopted in some countries that reduce these problems.

Introduction

Personal Anecdote

One Saturday morning, the married couple who lived next door to Robert Brym and his family asked Robert for advice on new speakers they wanted to buy for their sound system. Robert volunteered to go shopping with them at a nearby mall. They told him they also wanted to buy two outdoor garbage cans at a hardware store. Robert told them he didn't mind waiting.

"After they made the purchases, we returned to their minivan in the mall's parking lot," says Robert. "The wife opened the trunk, cleared some space, and said to her husband, 'Let's put the garbage cans back here.'

"Meanwhile, the husband had opened the side door. He had already put the speakers on the back seat and was struggling to do the same with the second garbage can. 'It's okay,' he said, 'I've already got one of them part way in here.'

"'Oh,' laughed the wife, 'I can judge space better than you and you'll never get that in there. Bring it back here.'

"'You know,' answered the husband, 'we don't always have to do things your way. I'm a perfectly intelligent person. I think there's room up here and that's where I'm going to put this thing. You can put yours back there or stick it anywhere else you like.'

"'Why are you yelling at me?' snapped the wife.

"'I'm not yelling,' shouted the husband. 'I'm just saying that I know as well as you what fits where. There's more than one way—your way—to do things.'

Sociology Now™

Reviewing is as easy as ❶, ❷, ❸.

Use SociologyNow to help you make the grade on your next exam. When you are finished reading this chapter, go to the Chapter Summary for instructions on how to make SociologyNow work for you.

"So, the wife put one garbage can in the trunk, the husband put one in the back seat (it was, by the way, a very tight squeeze) and we piled into the car for the drive home. The husband and the wife did not say a word to each other. When we got back to our neighborhood, I said I was feeling tired and asked whether I could perhaps hook up their speakers on Sunday. Actually, I wasn't tired. I just had no desire to referee round two. I went home, full of wonder at the occasional inability of presumably mature adults to talk rationally about something as simple as how to pack garbage cans into a minivan.

"However, trivializing the couple's argument in this way prevented me from thinking about it sociologically. If I had been thinking like a sociologist, I would have at least recognized that, for better or for worse, our most intense emotional experiences are bound up with our families. We love, hate, protect, hurt, express generosity toward, and envy nobody as much as our parents, siblings, children, and mates. Little wonder, then, that most people are passionately concerned with the rights and wrongs, the dos and don'ts, of family life. Little wonder that family issues lie close to the center of political debate in this country. Little wonder that words, gestures, and actions that seem trivial to an outsider can hold deep meaning and significance for family members."

Is the Family in Decline?

Because families are emotional minefields, few subjects of sociological inquiry generate as much controversy. Much of the debate centers on a single question: Is the family in decline, and if so, what should be done about it? The question is hardly new. A contributor to the *Boston Quarterly Review* of October 1859 wrote: "The family, in its old sense, is disappearing from our land, and not only our free institutions are threatened but the very existence of our society is endangered" (quoted in Lantz, Schultz, and O'Hara, 1977: 413). The same sentiment was expressed in September 1998. While criticizing President Clinton's affair with Monica Lewinsky, Democratic senator (later vice presidential hopeful) Joseph Lieberman of Connecticut stressed that "the decline of the family is one of the most pressing problems we are facing" (quoted in McKenna, 1998). This alarm, or one much like it, is sounded whenever the family undergoes rapid change, and particularly when the divorce rate increases.

Today, when some people speak about the decline of the family, they are referring to the **nuclear family.** The nuclear family is composed of a cohabiting man and woman who maintain a socially approved sexual relationship and have at least one child. Others are referring more narrowly to what might be called the **traditional nuclear family.** The traditional nuclear family is a nuclear family in which the wife works in the home without pay while the husband works outside the home for money. This makes him the "primary provider and ultimate authority" (Popenoe, 1988: 1).

Sociology Now™

Learn more about **Families** by going through the Family Structures Video Exercise.

In the 1940s and 1950s, many sociologists and much of the American public considered the traditional nuclear family the most widespread and ideal family form. However, for reasons we will examine later, the percentage of married-couple families with children living at home fell from 44 percent to just 24 percent of all households between 1960 and 2000 (Figure 15.1). Over the same period, the percentage of women over the age of 16 in the paid labor force increased from 38 percent to 60 percent. Consequently, only a minority of American adults live in traditional nuclear families today. Many new family forms have become popular in recent decades (Table 15.1).

A **nuclear family** consists of a cohabiting man and woman who maintain a socially approved sexual relationship and have at least one child.

A **traditional nuclear family** is a nuclear family in which the husband works outside the home for money and the wife works for free in the home.

Some sociologists, many of them functionalists, view the decreasing prevalence of the married-couple family and the rise of the "working mother" as an unmitigated disaster (e.g., Popenoe, 1998; 1996). In their view, rising rates of crime, illegal drug use, poverty, and welfare dependency (among other social ills) can be traced to the fact that so many American children are not living in two-parent households with stay-at-home mothers. They call for various legal and cultural reforms to shore up the traditional nuclear family.

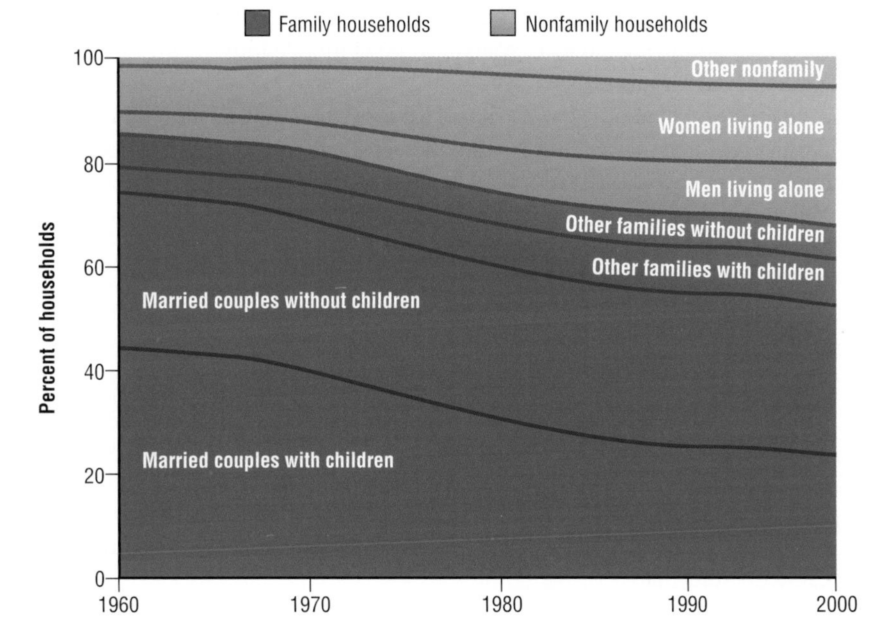

Figure 15.1
Household Types, United States, 1960–2000 (in percent)

Source: Suzanne M. Bianchi and Lynne M. Casper, "American Families," *Population Bulletin 55*, 4. Used with permission.

Table 15.1
The Traditional Nuclear Family and New Alternatives

Traditional Nuclear Family	New Alternatives
Legally married	Never-married singlehood, nonmarital cohabitation
With children	Voluntary childlessness
Two-parent	Single-parent (never married or previously married)
Permanent	Divorce, remarriage (including binuclear family involving joint custody, stepfamily or "blended" family)
Male primary provider, ultimate authority	Egalitarian marriage (including dual-career and commuter marriage)
Sexually exclusive	Extramarital relationships (including sexually open marriage, swinging, and intimate friendships)
Heterosexual	Same-sex intimate relationships or households
Two-adult household	Multi-adult households (including multiple spouses, communal living, affiliated families, and multigenerational families)

Source: Adapted from Macklin (1980: 906).

For instance, they want to make it harder to get a divorce, and they want people to place less emphasis on individual happiness at the expense of family responsibility.

Other sociologists, influenced by conflict and feminist theories, disagree with the functionalist assessment (e.g., Baca Zinn and Eitzen, 1993; Collins and Coltrane, 1991; Coontz, 1992; Skolnick, 1991). In the first place, they argue that it is inaccurate to talk about *the* family, as if this important social institution assumed or should assume only a single form. They emphasize that families have been structured in many ways and that the diversity of family forms is increasing as people accommodate the demands of new social pressures. Second, they argue that changing family forms do not necessarily represent deterioration in the quality of people's lives. In fact, such changes often represent *improvement* in the way people live. They believe that the decreasing prevalence of the traditional nuclear family and the proliferation of diverse family forms has benefited many men, women, and children and has not harmed other children as much as the functionalists think. They also believe that various economic and political reforms, such as the creation of an affordable nationwide day-care system, could eliminate most of the negative effects of single-parent households.

This chapter touches on divorce, reproductive choice, single-parent families, day care, and other topics in the sociology of families. However, we have structured this chapter around the debate about the so-called decline of the American family. We first outline the functional theory of the family because the issues raised by functionalism are still a focus of sociological controversy (Mann, Grimes, Kemp, and Jenkins, 1997). Borrowing from the work of conflict theorists and feminists, we next present a critique of functionalism. In particular, we show that the nuclear family became the dominant and ideal family form only under specific social and historical conditions. Once these conditions changed, the nuclear family became less prevalent, and a variety of new family forms proliferated. You will learn how these new family forms are structured and how their frequency varies by class, race, and sexual orientation. You will also learn that while postindustrial families solve some problems, they are hardly an unqualified blessing. The chapter's concluding section therefore considers the kinds of policies that might help alleviate some of the most serious concerns faced by families today. Let us, then, first review the functionalist theory of the family.

Functionalism and the Nuclear Ideal

Functional Theory

For any society to survive, its members must cooperate economically. They must have babies. And they must raise offspring in an emotionally supportive environment so the offspring can learn the ways of the group and eventually operate as productive adults. Since the 1940s, functionalists have argued that the nuclear family is ideally suited to meet these challenges. In their view, the nuclear family performs five main functions. It provides a basis for regulated sexual activity, economic cooperation, reproduction, socialization, and emotional support (Murdock, 1949: 1–22; Parsons, 1955).

Functionalists cite the pervasiveness of the nuclear family as evidence of its ability to perform the functions just listed. To be sure, other family forms exist. **Polygamy** expands the nuclear unit "horizontally" by adding one or more spouses (almost always wives) to the household. Polygamy is still legally permitted in many less industrialized countries of Africa and Asia. However, the overwhelming majority of families are monogamous, because they cannot afford to support several wives and many children. The **extended family** expands the nuclear family "vertically" by adding another generation—one or more of the spouse's parents—to the household. Extended families used to be common throughout the world. They still are in some places. However, according to the functionalists, the basic building block of the extended family (and of the polygamous family) is the nuclear unit.

George Murdock was a functionalist who conducted a famous study of 250 mainly preliterate societies in the 1940s. Murdock wrote, "Either as the sole prevailing form of the family or as the basic unit from which more complex familial forms are compounded, [the nuclear family] exists as a distinct and strongly functional group in every known society" (Murdock, 1949: 2). Moreover, the nuclear family, Murdock continued, is everywhere based on **marriage.** He defined marriage as a socially approved, presumably long-term sexual and economic union between a man and a woman. It involves rights and obligations between spouses and between spouses and their children.

Polygamy expands the nuclear family "horizontally" by adding one or more spouses (usually women) to the household.

The **extended family** expands the nuclear family "vertically" by adding another generation—one or more of the spouses' parents—to the household.

Marriage is a socially approved, presumably long-term sexual and economic union between a man and a woman. It involves reciprocal rights and obligations between spouses and between parents and children.

Functions of the Nuclear Family

Let us consider the five main functions of marriage and the nuclear family in more detail:

1. Sexual *regulation.* Imagine a world without an institution that defines the boundaries within which legitimate sexual activity is permitted. Such a world would be disrupted

by many people having sex wherever, whenever, and with whomever they pleased. An orderly social life would be difficult. Because marriage provides a legitimate forum for expressing the intense human need for sexual activity, says Murdock, it makes social order possible. Sex is not, however, the primary motive for marrying, he continues. After all, sex is readily available outside of marriage. Only 54 of Murdock's 250 societies forbade or disapproved of premarital sex between nonrelatives. In most of the 250 societies, a married man could legitimately have an extramarital affair with one or more female relatives (Murdock, 1949: 5–6). It is hardly news that premarital and extramarital sex is common in contemporary America and other postindustrial societies. As a president of the University of California once said in a *Time* magazine interview: "I find that the three major administrative problems on a campus are sex for the students, athletics for the alumni, and parking for the faculty" (quoted in Ember and Ember, 1973: 317).

2. *Economic cooperation.* Why then, apart from sex, do people marry? Murdock's answer is this: "By virtue of their primary sex difference, a man and a woman make an exceptionally efficient cooperating unit" (Murdock, 1949: 7). On average, women are physically weaker than men. Historically, pregnancy and nursing have restricted women in their activities. Therefore, writes Murdock, they can best perform lighter tasks close to home. These tasks include gathering and planting food, carrying water, cooking, making and repairing clothing, making pottery, and caring for children. Most men possess superior strength. They can therefore specialize in lumbering, mining, quarrying, land clearing, and house building. They can also range farther afield to hunt, fish, herd, and trade (Murdock, 1937). According to Murdock, this division of labor enables more goods and services to be produced than would otherwise be possible. People marry partly due to this economic fact. In Murdock's words, "Marriage exists only when the economic and the sexual are united into one relationship, and this combination occurs only in marriage" (Murdock, 1949: 8).
3. *Reproduction.* Before the invention of modern contraception, sex often resulted in the birth of a baby. According to Murdock, children are an investment in the future. Already by the age of 6 or 7, children in most societies do some chores. Their economic value to the family increases as they mature. When children become adults, they often help support their elderly parents. Thus, in most societies, there is a big economic incentive to having children.
4. *Socialization.* The investment in children can be realized only if adults rear the young to maturity. This involves not only caring for them physically but, as you saw in Chapter 4 ("Socialization"), teaching them language, values, beliefs, skills, religion, and much else. Talcott Parsons (1955: 16) regarded socialization as the "basic and irreducible" function of the family.
5. *Emotional support.* Parsons also noted that the nuclear family universally gives its members love, affection, and companionship. He stressed that in the nuclear family, it is mainly the mother who is responsible for ensuring the family's emotional well-being. She develops what Parsons calls the primary "expressive" role because she is the one who bears children and nurses them. It falls on the husband to take on the more "instrumental" role of earning a living outside the family (Parsons, 1955: 23). The fact that he is the "primary provider" makes him the ultimate authority.

Foraging Societies

Does functionalism provide an accurate picture of family relations at any point in human history? To assess the adequacy of the theory, let us briefly consider family patterns in the two settings that were apparently foremost in the minds of the functionalists. We first discuss families in preliterate, foraging societies. In such societies, people subsist by hunting animals and gathering wild edible plants. Most of the cases in Murdock's sample are foraging societies. We then discuss families in urban and suburban middle-class America in the 1950s. The functionalists whose work we are reviewing lived in such families themselves.

Peter Johnson/Corbis

There is rough gender equality among the !Kung-San, a foraging society in the Kalahari Desert in Botswana. That is partly because women play such a key economic role in providing food.

Foraging societies are nomadic groups of 100 or fewer people. As we would expect from Murdock's and Parsons' analyses, a gender division of labor exists among foragers. Most men hunt and most women gather. Women also do most of the child care. However, research on foragers conducted since the 1950s, on which we base our analysis, shows that men often tend babies and children in such societies (Leacock, 1981; Lee, 1979; Turnbull, 1961). They often gather food after an unsuccessful hunt. In some foraging societies, women hunt. In short, the gender division of labor is less strict than Murdock and Parsons thought. What is more, the gender division of labor is not associated with large differences in power and authority. Overall, men have few if any privileges that women don't also enjoy. Relative gender equality is based on the fact that women produce up to 80 percent of the food.

Foragers travel in small camps or bands. The band decides by consensus when to send out groups of hunters. When they return from the hunt, they distribute game to all band members based on need. Each hunter does not decide to go hunting based on his or her own nuclear family's needs. Each hunter does not distribute game just to his or her own nuclear family. Contrary to what Murdock wrote, it is the band, not the nuclear family, that is the most efficient social organization for providing everyone with his or her most valuable source of protein.

In foraging societies, children are considered an investment in the future. However, it is not true that people always want more children for purposes of economic security. In fact, too many children are considered a liability. Subsistence is uncertain in foraging societies, and when band members deplete an area of game and edible plants, they move elsewhere. As a result, band members try to keep the ratio of children to productive adults low. In a few cases, such as the pre-20th-century Inuit ("Eskimos"), newborns were occasionally allowed to die if the tribe felt its viability was threatened by having too many mouths to feed.

Life in foraging societies is highly cooperative. For example, women and men care for—and women even breast-feed—each other's children. Despite Parsons's claim that socialization is the "basic and irreducible" function of the nuclear family, it is the band, not the nuclear family, that assumes responsibility for child socialization in foraging societies. Socialization is more a public than a private matter. As a 17th-century Innu man from northern Quebec said to a French Jesuit priest who was trying to convince him to adopt European ways of raising children: "Thou hast no sense. You French people love only your own children; but we all love all the children of our tribe" (quoted in Leacock, 1981: 50).

In sum, recent research on foraging societies calls into question many of the functionalists' generalizations. In foraging societies, relations between the sexes are quite egalitarian. Children are not viewed just as an investment in the future. Each nuclear unit does not execute the important economic and socialization functions in private. On the contrary, cooperative band members execute most economic and socialization functions in public.

Let us now assess the functionalist theory of the family in the light of evidence concerning American middle-class families in the years just after World War II.

The American Middle Class in the 1950s

Functionalists recognized that the productive function of the family was less important after World War II than it had been in earlier times. In their view, the socialization and emotional functions of the family were now most important (Parsons, 1955). Thus, on the 19th-century family-owned farm or ranch, the wife had played an indispensable productive role while the husband was out in the field or on the range. She took responsibility for the garden, the dairy, the poultry, and the management of the household. The children also did crucial chores with considerable economic value. But in the typical urban or suburban nuclear family of the late 1940s and 1950s, noted the functionalists, only one person played the role of breadwinner. That was usually the husband. Children enjoyed more time to engage in the play and leisure-time activities that were now considered necessary for healthy development. For their part, most women got married, had babies, and stayed home to raise them. Strong normative pressures helped to keep women at home. Thus, in the 1950s, sociologist David Riesman called a woman's failure to obey the strict gender division of labor a "quasi-perversion." *Esquire* magazine called women's employment in the paid labor force a "menace." *Life* magazine called it a "disease" (quoted in Coontz, 1992: 32).

Web Interactive Exercise: Is the Woman's Place in the Home?

As a description of family patterns in the 15 years after World War II, functionalism has its merits. During the Great Depression (1929–1939) and the war (1939–1945), millions of Americans were forced to postpone marriage due to widespread poverty, government-imposed austerity, and physical separation. After this long and dreadful ordeal, many Americans just wanted to settle down, have children, and enjoy the peace, pleasure, and security that family life seemed to offer. Conditions could not have been better for doing just that. The immediate postwar era was one of unparalleled optimism and prosperity. Real per capita income rose 35 percent between 1945 and 1960. The percentage of Americans who owned their own homes jumped from 43 percent in 1940 to 62 percent in 1960. Government assistance in the form of the GI Bill and other laws helped to make the late 1940s and 1950s the heyday of the traditional nuclear family. This assistance took the form of guaranteed, tax-deductible mortgages, subsidized college education and health care for veterans, big income-tax deductions for dependents, and massive road-building projects that opened the suburbs for commuters. People got married younger. They had more babies. They got divorced less. Increasingly, they lived in married-couple families (Table 15.2). Middle-class women in the postwar years engaged in what has been called an "orgy of domesticity," devoting increasing attention to childrearing and housework. They also became increasingly concerned with the emotional quality of family life as love and companionship became firmly established as the main motivation for marriage (Coontz, 1992: 23–41; Skolnick, 1991: 49–74).

However, as sociologist Andrew J. Cherlin meticulously shows, the immediate postwar period was in many respects a historical aberration (Cherlin, 1992 [1981]: 6–30). Trends in divorce, marriage, and childbearing show a gradual *weakening* of the nuclear family from the second half of the 19th century until the mid-1940s, and continued weakening after the 1950s. Specifically, throughout the 19th century, the **divorce rate** rose. The divorce rate is the number of divorces that occur in a year for every 1000 people in the population.

The **divorce rate** is the number of divorces that occur in a year for every 1000 people in the population.

Table 15.2
The Family in Numbers: The 1940s and 1950s Compared

	1940s	1950s
Percent of women age 20–24 never married	48.0	20.0
Divorce rate (per 1000 population)	4.3	2.1
Total fertility rate for white women age 20	2.6	3.1
Total fertility rate for nonwhite women age 20	3.2	3.9
Married couples as percent of all families	84.4	87.8

Note: Most figures were read from graphs and are therefore approximate.
Sources: Adapted from Cherlin (1992 [1981]: 9, 19, 21); U.S. Census Bureau (1999a).

Meanwhile, the **marriage rate** fell. The marriage rate is the number of marriages that occur in a year for every 1000 people in the population. The **total fertility rate** also fell. The total fertility rate is the average number of children that would be born to a woman over her lifetime if she had the same number of children as women in each age cohort in a given year. In contrast, the divorce rate fell only between 1946 and 1958. The marriage rate took a big jump only in the two years following World War II. And the fertility rate rose only for women who reached childbearing age between 1930 and the mid-1950s. By the late 1950s or early 1960s, the earlier trends reasserted themselves. Only the peculiar historical circumstances of the postwar years, noted previously, temporarily reversed them (Figure 15.2).

The functionalists, we may conclude, generalized too hastily from the era they knew best—the period of their own adulthood. Contrary to what they thought, the big picture from the 19th century until the present is that of a gradually weakening nuclear family. Let us now consider the conditions that made other family forms more prevalent.

Conflict and Feminist Theories

> I hadn't really wanted to marry at all. I wanted to make something of myself, not just give it away. But I knew if I didn't marry I would be sorry. Only freaks didn't. I knew I had to do it quickly, too, while there was still a decent selection of men to choose from. . . . I was twenty. . . .
>
> Though I wanted to be a good wife, from the beginning I found it impossible to subdue my desires. I was in fierce competition with my husband, though Frank, completely absorbed in his own studies, was probably unaware of it. He believed he had married an impulsive girl, even a supergirl, but not a separate, feeling woman. . . . Though we had agreed to study like fury till our money ran out and then take turns getting jobs, at bottom we knew it would be he who would get the degrees and I who would get the jobs (Shulman, 1997 [1969]: 163, 173).

This passage is from *Memoirs of an Ex-Prom Queen* (Shulman, 1972), an exposé of the plight of the "all-American girl" in the 1950s. It shows a side of family life entirely obscured

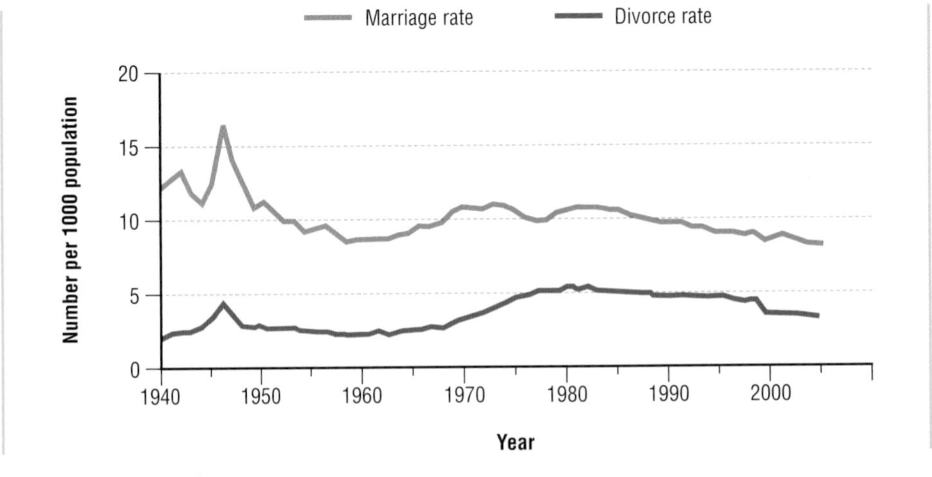

Figure 15.2
Marriages and Divorces, United States, 1940–2004 (per 1000 population)

Sources: Centers for Disease Control and Prevention (1995a, 1995b, 1998, 1999b, 2001, 2003, 2005a).

The **marriage rate** is the number of marriages that occur in a year for every 1000 people in the population.

The **total fertility rate** is the average number of children that would be born to a woman over her lifetime if she had the same number of children as women in each age cohort in a given year.

The Everett Collection

1950s TV classics such as *Father Knows Best* portrayed smoothly functioning, happy, white, middle-class, mother-householder, father-breadwinner families.

by the functionalists. As the passage suggests, and as the novel establishes in biting and sometimes depressing detail, postwar families did not always operate like the smoothly functioning, happy, white, middle-class, mother-householder, father-breadwinner household portrayed every week in 1950s TV classics such as *Leave It to Beaver, Father Knows Best,* and *The Adventures of Ozzie and Harriet.*

Many men and women felt coerced into getting married, trapped in their families, unable to achieve the harmony, security, and emotional satisfaction they had been promised. As a result, the nuclear family was often a site of frustration and conflict. Surveys show that only about a third of working-class couples and two-thirds of middle-class couples were happily married (Barnard, 1972). Wives were less satisfied with marriage than were husbands. They reported higher rates of depression, distress, and feelings of inadequacy. Dissatisfaction seems to have been especially high among the millions of women who, during World War II, had operated cranes in steel mills, greased locomotives, riveted the hulls of ships, worked the assembly lines in munitions factories, planted and harvested crops, and felled giant redwoods. They were universally praised for their dedication and industry during the war. Many of them did not want to leave these well-paying jobs that gave them gratification and independence. Management fired most of them anyway and downgraded others to lower-paying "women's" jobs to make room for returning soldiers. The tedium of domestic labor must have been especially difficult for many of these women to accept (Coontz, 1992: 23–41; Skolnick, 1991: 49–74).

Also, many families were simply too poor to participate in the functionalists' celebration of the traditional nuclear unit. For example, to support their families, some 40 percent of African American women with small children had to work outside their homes in the 1950s, usually as domestics in upper-middle-class and upper-class white households. A quarter of these black women headed their own households. Thus, to a degree not recognized by the functionalists, the existence of the traditional nuclear family among well-to-do whites depended in part on many black families *not* assuming the traditional nuclear form.

Unlike the functionalists, Marxists had long seen the traditional nuclear family as a site of gender conflict and a basis for the perpetuation of social inequality. In the 19th century, Marx's close friend and co-author, Friedrich Engels, argued that the traditional nuclear family emerged along with inequalities of wealth. Once wealth was concentrated in the hands of a man, wrote Engels, he became concerned about how to transmit it to his

In the 1950s, married women often hid their frustrations with family life.
Mimi Matte *Family Outing*

children, particularly his sons. How could a man safely pass on an inheritance, asked Engels? Only by controlling his wife sexually and economically. Economic control ensured that the man's property would not be squandered and would remain his and his alone. Sexual control, in the form of enforced female monogamy, ensured that his property would be transmitted only to his offspring. It follows from Engels' analysis that only the elimination of private property and the creation of economic equality—in a word, communism—can bring an end to the traditional nuclear family and the arrival of gender equality (Engels, 1970 [1884]: 138–9).

Engels was right to note the long history of male economic and sexual domination in the traditional nuclear family. After all, 100 years ago in the United States, any money a wife might earn typically belonged to her husband. As recently as 40 years ago, an American wife could not rent a car, take a loan, or sign a contract without her husband's permission. It was only about 25 years ago that it became illegal for a husband to rape his wife.

However, Engels was wrong to think that communism would eliminate gender inequality in the family. Gender inequality is as common in societies that call themselves communist as in those that call themselves capitalist. For example, as one American researcher concluded, the Soviet Union left "intact the fundamental family structures, authority relations, and socialization patterns crucial to personality formation and sex-role differentiation. Only a genuine sexual revolution [or, as we prefer to put it, a gender revolution] could have shattered these patterns and made possible the real emancipation of women" (Lapidus, 1978: 7).

Because gender inequality exists in noncapitalist (including precapitalist) societies, most feminists believe that something other than, or in addition to, capitalism accounts for gender inequality. In their view, *patriarchy*—male dominance and norms justifying that dominance—is more deeply rooted in the economic, military, and cultural history of humankind than the classical Marxist account allows (Chapter 11, "Sexuality and Gender"). For them, only a "genuine gender revolution" can alter this state of affairs.

Just such a revolution in family structures, authority relations, and socialization patterns picked up steam in the United States and other Western countries about 40 years ago, although its roots extend back to the 18th century. As you will see, the revolution is evident in the rise of romantic love and happiness as bases for marriage, women's increasing control over reproduction due to their use of contraceptives, and women's increasing participation in the system of higher education and the paid labor force, among other factors. We next consider some consequences of the gender revolution for the selection of mates, marital satisfaction, divorce, reproductive choice, housework, and child care. We begin by considering the sociology of mate selection.

Power and Families

Love and Mate Selection

Most Americans take for granted that marriage ought to be based on love (Figures 15.3 and 15.4). Our assumption is evident, for example, in the way that most popular songs in the United States celebrate love as the sole basis of long-term intimacy and marriage. In contrast, most of us view marriage devoid of love as tragic.

Yet in most societies throughout human history, love has had little to do with marriage. Some languages, such as the Chinese dialect spoken in Shanghai, even lack a word for love. Historically and across cultures, marriages were typically arranged by third parties, not by brides and grooms. The selection of marriage partners was based mainly on

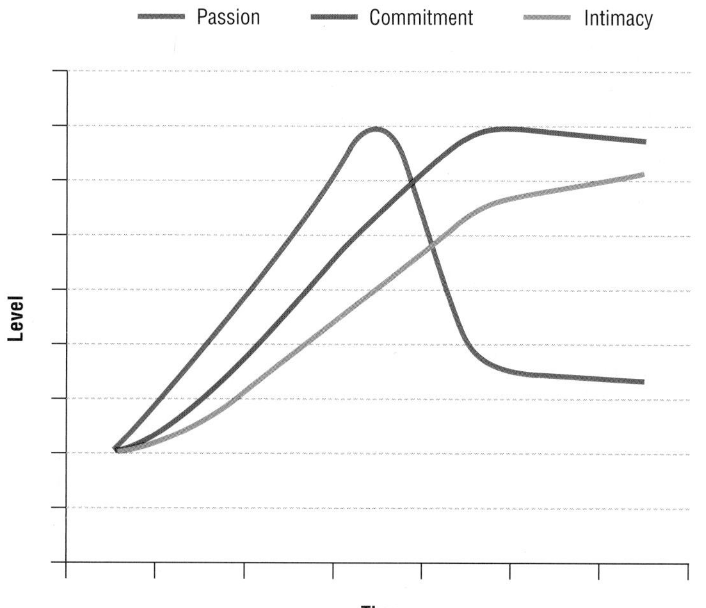

Figure 15.3
The Components of Love

According to psychologist Robert Sternberg, love can be built from three components: passion (erotic attraction), intimacy (confiding in others and shared feelings), and commitment (intention to remain in the relationship). In actual relationships, these components may be combined in various ways to produce different kinds of love. The fullest love requires all three components. Research shows that in long-term relationships, passion peaks fairly quickly and then tapers off. Intimacy rises more gradually but remains at a higher plateau. Commitment develops most gradually but also plateaus at a high level.

Source: Sternberg (1986).

calculations intended to maximize the prestige, economic benefits, and political advantages accruing to the families from which the bride and groom came. For a family of modest means, a small dowry might be the chief gain from allowing their son to marry a certain woman. For upper-class families, the benefits were typically bigger but no less strategic. In the early years of industrialization, for example, more than one old aristocratic family in economic decline scrambled to have its offspring marry into a family of the upstart bourgeoisie. Giuseppe di Lampedusa's *The Leopard,* probably the greatest Italian novel of the 20th century, deals in an especially moving way with just such a strategic marriage (di Lampedusa, 1991 [1958]).

The idea that love should be important in the choice of a marriage partner first gained currency in 18th-century England with the rise of liberalism and individualism, philosophies that stressed freedom of the individual over community welfare (Stone, 1977). However, the intimate linkage between love and marriage that we know today emerged only in the early 20th century, when Hollywood and the advertising industry began to promote self-gratification on a grand scale. For these new spinners of fantasy and desire, an important aspect of self-gratification was heterosexual romance leading to marriage (Rapp and Ross, 1986).

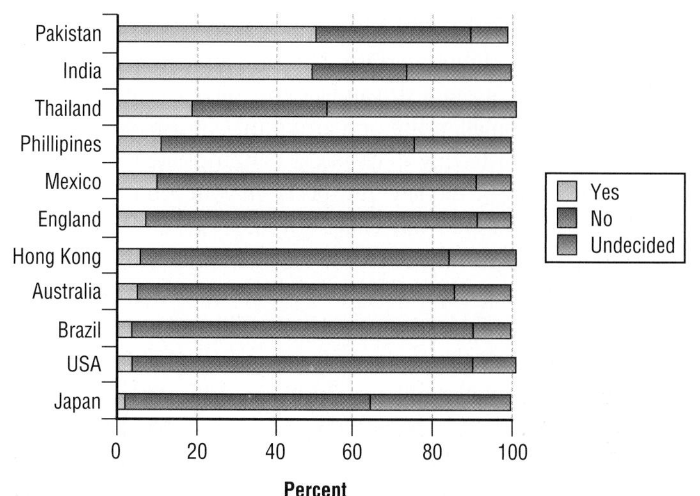

Figure 15.4
"If a man (woman) had all the other qualities you desired, would you marry this person if you were not in love with him (her)?"

Willingness to marry without love is more common in less Westernized and less modernized societies. In the United States, only 4 percent of respondents said they would definitely be willing to marry without love. (Some percentages do not equal 100 because of rounding.)

Levine et al (1995: 561).

Hollywood glamorized heterosexual, romantic love and solidified the intimate linkage between love and marriage that we know today.
Clark Gable and Vivien Leigh in *Gone with the Wind* (1939).

The Everett Collection

Social Influences on Mate Selection

Still, it would be a big mistake to think that love alone determines mate selection in our society—far from it. Three sets of social forces influence whom you are likely to fall in love with and marry (Kalmijn, 1998: 398–404).

1. *Marriage resources.* Potential spouses bring certain resources with them to the "marriage market." They use these resources to attract mates and compete against rivals. These resources include financial assets, status, values, tastes, and knowledge. Most people want to maximize the financial assets and status they gain from marriage, and they want a mate who has similar values, tastes, and knowledge. As a result, whom you fall in love with and choose to marry is determined partly by the assets you bring to the marriage market.
2. *Third parties.* A marriage between people from two different groups may threaten the internal cohesion of one or both groups. Therefore, to varying degrees, families, neighborhoods, communities, and religious institutions raise young people to identify with the groups they are members of and think of themselves as different from members of other groups. They may also apply sanctions to young people who threaten to marry outside the group. As a result, whom you fall in love with and choose to marry is determined partly by the influence of these third parties.
3. *Demographic and compositional factors.* The probability of marrying inside one's group increases with the group's size and geographical concentration. Conversely, if you are a member of a small group or a group that is dispersed geographically, you stand a greater chance of having to choose an appropriate mate from outside your group. There may simply be too few "prospects" in your group from which to choose (Brym, 1984; Brym, Gillespie, and Gillis, 1985). In addition, the ratio of men to women in a group influences the degree to which members of each sex marry inside or outside the group. For instance, war and incarceration may eliminate many male group members as potential marriage partners. This may encourage female group members to marry outside the group or forgo marriage altogether. Finally, because people usually meet potential spouses in "local marriage markets"—schools, colleges, places of work, neighborhoods, bars, and clubs—the degree to which these settings are socially segregated also influences mate selection. You are more likely to marry outside your group if local marriage markets are socially heterogeneous. As a result,

whom you fall in love with and choose to marry is determined partly by the size, geographical dispersion, and sex ratio of the groups you belong to and the social composition of the local marriage markets you frequent.

As a result of the operation of these three sets of social forces, the process of falling in love and choosing a mate is far from random. The percentage of people who marry inside their group is more than 90 percent for African Americans, 75 percent for Asian Americans, 65 percent for Hispanic Americans, and 25 percent for European Americans. About 80 percent of Protestants and Jews and 60 percent of Catholics marry within their group. There is also a fairly strong correlation (approximately $r = 0.55$) between the educational attainment of husbands and wives (Kalmijn, 1998: 406–8). We are freer than ever before to fall in love with and marry anyone we want. As in all things, however, social forces still constrain our choices to varying degrees (Box 15.1).

Marital Satisfaction

Just as mate selection came to depend more on romantic love over the years, so marital stability came to depend more on having a happy rather than a merely useful marriage. This change occurred because women in the United States and many other societies have become more autonomous, especially over the past 40 years or so. That is, one aspect of the gender revolution is that women are freer than ever to leave marriages in which they are unhappy.

One factor that contributed to women's autonomy was the introduction of the birth control pill in the 1960s. The birth control pill made it easier for women to delay childbirth and have fewer children. A second factor that contributed to their autonomy was the entry of millions of women into the system of higher education and the paid labor force (Cherlin, 1992 [1981]: 51–2, 56; Collins and Coltrane, 1991: xxv). Once women enjoyed a source of income independently of their husbands, they gained the means to decide the course of their own lives to a greater extent than ever before. A married woman with a job outside the home is less tied to her marriage by economic necessity than a woman who works only at home. If she is deeply dissatisfied with her marriage, she can more easily leave. Reflecting this new reality, laws were changed in the 1960s to make divorce easier and to divide property between divorcing spouses more equitably. The divorce rate rose 57 percent from 1960 to 1981, then declined 30 percent from 1981 to 2004 (see Figure 15.2). Women initiate most divorces.

The Social Roots of Marital Satisfaction

If marital stability now depends largely on marital satisfaction, what are the main factors underlying marital satisfaction? The sociological literature emphasizes five sets of forces (Collins and Coltrane, 1991: 394–406; 454–64):

1. *Economic forces.* Money issues are the most frequent subjects of family quarrels, and money issues loom larger when there isn't enough money to satisfy a family's needs and desires. Accordingly, marital satisfaction tends to fall and the divorce rate to rise as you move down the socioeconomic hierarchy. The lower the social class and the lower the educational level of the spouses, the more likely it is that financial pressures will make them unhappy and the marriage unstable. Marital dissatisfaction and divorce are also more common among groups with high poverty rates. Such groups include spouses who marry in their teens and African Americans. In contrast, the marital satisfaction of both husbands and wives generally *increases* when wives enter the paid labor force. This is mainly because of the beneficial financial effects. However, if *either* spouse spends so much time on the job that he or she neglects the family, marital satisfaction falls.
2. *Divorce laws.* Many surveys show that, on average, married people are happier than unmarried people. Moreover, when people are free to end unhappy marriages and remarry, the average level of happiness increases among married people. Thus, the level

BOX 15.1
Sociology at the Movies

My Big Fat Greek Wedding (2002)

"Nice Greek girls are expected to do three things: marry Greek boys, make Greek babies, and feed everyone until the day we die." So says Toula Portokalos (played by Nia Vardalos, who also wrote the film script). Toula is a nice Greek girl who is devoted to her big, animated family despite its quirks. Her father (Michael Constantine) sprays Windex on just about everything and believes that nearly every word has a Greek root, including the Japanese *kimono.* Her mother (Lainie Kazan) complains, "When I was your age, we didn't *have* food!" Both parents are deeply worried because Toula is unmarried and she is already 30 years old. She is seemingly content to work in the family restaurant and spend her time with her family, smelling faintly of garlic bread.

Toula's life changes one day when a handsome WASP high school teacher by the name of Ian Miller (John Corbett) comes into the restaurant for a meal. Miller dazzles her. He notices her too. She vows to change her life. She invests in new clothes, contact lenses, and a trip to the beauty salon. She starts taking college courses and gets a job at her aunt's travel agency. Miller sees her there and asks her out for a date. Things progress and they soon decide to get married.

A clash between cultures then ensues because, as Toula tells Ian, "No one in our family has ever gone out with a non-Greek." Ian is Protestant, Toula Greek Orthodox, and because Toula's parents feel a lot more strongly about their religion than Ian's parents feel about theirs, it is decided that Ian must undergo baptism and convert. Ian's parents are reserved, Toula's parents effusive. Ian's mother brings a Bundt cake to the wedding party but none of the Greeks knows what it is or what to do with it. The Greeks hold their *ouzo* but Ian's mother gets smashed. The groom's family fills a few rows on one side of the church, the bride's family overflows the other side. (Toula has 27 first cousins, most of them named Nick.) In the end, notwithstanding the misunderstandings and the obstacles, love conquers all and the happy couple wed.

The key to the movie comes right at the end, when Toula's father toasts the newlyweds. He explains that in Greek, Toula's family name (Portokalos), comes from the word for oranges (*portokali*), whereas Ian's surname (Miller) derives from the Greek word for apple (*milo*). "In the end," he concludes, "we're all fruit." It's a funny way of saying that despite seemingly insurmountable obstacles, mixing apples and oranges is indeed possible in the American melting pot.

My Big Fat Greek Wedding celebrates ethnic intermarriage but ignores two important issues. First, institutional racism, segregation, and other structural obstacles prevent some "apples" and "oranges" from intermarrying in American society (Chapter 10, "Race and Ethnicity"). The movie would be decidedly less plausible if the newlyweds were a Jamaican American from New York and a Swedish American from Minnesota. Second, some "apples" and "oranges" are happy to remain apples and oranges. They recognize that ethnic and racial intermarriage can lead to the loss of the minority group's cultural heritage and that much of the richness of American society and culture—our foods, our dance, our music, our languages—derives from the preservation of ethnic and racial ties. *My Big Fat Greek Wedding* reflects the opportunities and aspirations of only one segment of American society.

Nia Vardalos and John Corbett in *My Big Fat Greek Wedding* (2002).

of marital happiness has increased in the United States over the past few decades, especially for wives, partly because it has become easier to get a divorce. For the same reason, in countries where getting a divorce is more difficult (e.g., Italy, Spain), husbands and wives tend to be less happy than in countries where getting a divorce is easier (e.g., the United States, Canada) (Stack and Eshleman, 1998).

3. *The family life cycle.* About a quarter of divorces take place in the first 3 years of a first marriage, and half of all divorces take place by the end of the seventh year. However, for marriages that last longer, marital satisfaction reaches a low point after about 15 to 20 years. Marital satisfaction generally starts high, falls when children are born, reaches a low point when children are in their teenage years, and rises again when children reach adulthood (Rollins and Cannon, 1974). ▸Figure 15.5 illustrates the effect of the family life cycle on marital satisfaction using survey data. Nonparents and parents whose children have left home (so-called empty nesters) enjoy the highest level of marital satisfaction. Parents who are just starting families or who have adult children living at home enjoy intermediate levels of marital satisfaction. Marital satisfaction is lowest during the "establishment" years, when children are attending school. Although most people get married at least partly to have children, it turns out that children, and especially teenagers, usually put big emotional and financial strains on families. This results in relatively low marital satisfaction.
4. *Housework and child care.* Marital happiness is higher among couples who share housework and child care. The farther couples are from an equitable sharing of domestic responsibilities, the more tension there is among all family members (Hochschild with Machung, 1989).
5. *Sex.* Having a good sex life is associated with marital satisfaction. Contrary to popular belief, surveys show that sex generally improves during a marriage. Sexual intercourse is also more enjoyable and frequent among happier couples. From these findings, some experts conclude that general marital happiness leads to sexual compatibility (Collins and Coltrane, 1991: 344). However, the reverse may also be true. Good sex may lead to a good marriage. After all, sexual preferences are deeply rooted in our psyches and our earliest experiences. They cannot easily be altered to suit the wishes of our partners. If spouses are sexually incompatible, they may find it hard to change, even if they communicate well, argue little, and are generally happy on other grounds. On the other hand, if a husband and wife are sexually compatible, they may work harder to resolve other problems in the marriage for the sake of preserving their good sex life. Thus, the relationship between marital satisfaction and sexual compatibility is probably reciprocal. Each factor influences the other.

Religion, we note, has little effect on level of marital satisfaction. But religion does influence the divorce rate. Thus, states with a high percentage of regular churchgoers and a high percentage of fundamentalists have lower divorce rates than other states (Sweezy and

Sociology Now™

Learn more about **Divorce** by going through the Divorce and Remarriage Learning Module.

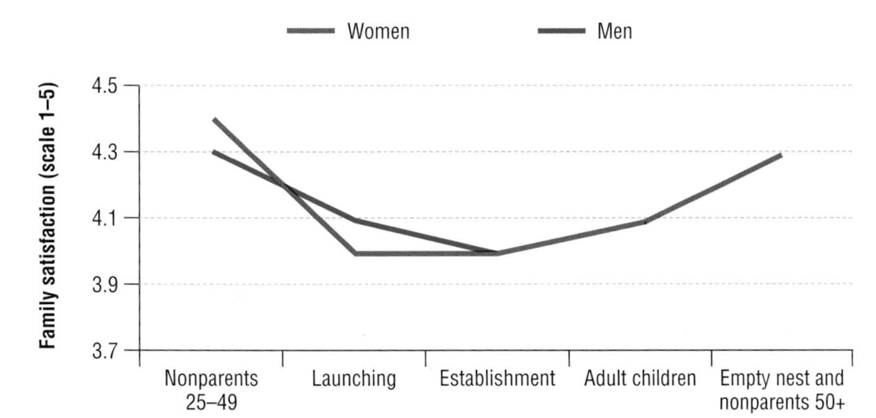

▸Figure 15.5
Family Satisfaction and the Family Life Cycle, United States

Note: Data are for 1998.
Source: Keller (2000).

The United States divorce rate reached a historic high in 1979 and has declined since then. Andrew Benyei *Pink Couch* (1993).

Tiefenthaler, 1996). Let us now see what happens when low marital satisfaction leads to divorce.

Divorce

Economic Effects

After divorce, the most common pattern is a rise in the husband's income and a decline in the wife's. That is because husbands tend to earn more, children typically live with their mother, and support payments are often inadequate. Characteristically, one study found that in the three months after separation, 38 percent of custodial mothers but just 10.5 percent of noncustodial fathers fell below the poverty line (Bartfeld, 2000: 209). Between 1996 and 1998, the Clinton administration created a new felony offense for people who flee across state lines to avoid paying child support. It also developed a new computerized collection system to track parents across state lines, and new penalties and incentives to get states to cooperate in tracking so-called deadbeat parents (Clinton, 1998). These policies forced many delinquent parents to pay child support. Consequently, whereas child support payments declined in the 1970s and 1980s, they rose in the 1990s (Case, Lin, and McLanahan, 2003).

Emotional Effects

While divorce enables spouses to leave unhappy marriages, serious questions have been raised about the emotional consequences of divorce for children, particularly in the long term. Some scholars claim that divorcing parents are simply trading the well-being of their children for their own happiness. What does research say about this issue?

Research shows that children of divorced parents tend to develop behavioral problems and do less well in school than children in intact families. They are more likely to engage in delinquent acts and to abuse drugs and alcohol. They often experience an emotional crisis, particularly in the first two years after divorce. What is more, when children of divorced parents become adults, they are less likely than children of nondivorced parents to be happy. They are more likely to suffer health problems, depend on welfare, earn low incomes, and experience divorce themselves. In one California study, almost half the children of divorced parents entered adulthood as worried, underachieving, self-deprecating, and sometimes angry young men and women (Wallerstein and Blakeslee, 1989; Wallerstein, Lewis, and Blakeslee, 2000). Clearly, divorce can have serious, long-term, negative consequences for children.

However, much of the research that seems to establish a link between divorce and long-term negative consequences for children is based on families who seek psychological counseling. Such families are a small and unrepresentative minority of the population. By definition, they have more serious emotional problems than the large majority, which does not need psychological counseling after divorce. One must be careful not to generalize from such studies. Another problem with much of this research is that some analysts fail to ask whether factors other than divorce might be responsible for the long-term distress experienced by many children of divorced parents.

SuperStock

A high level of parental conflict creates long-term distress among children. Divorce without parental conflict does children much less harm. Children in divorced families have a higher level of well-being on average than children in high-conflict intact families.

Factors Affecting the Well-Being of Children

Researchers who rely on representative samples and examine the separate effects of many factors on children's well-being provide the best evidence on the consequences of divorce for children. For example, Amato and Keith (1991) reanalyzed the results of 92 relevant studies. They showed that on average the overall effect of divorce on children's well-being is not strong and declines over time. They also found three factors that account for much of the distress among children of divorce:

1. *A high level of parental conflict* creates long-term distress among children. Divorce without parental conflict does children much less harm. In fact, children in divorced families have a higher level of well-being on average than children in high-conflict *intact* families. The effect of parental conflict on the long-term well-being of children is substantially greater than the effect of the next two factors discussed by Amato and Keith.
2. *A decline in living standards.* By itself, the economic disadvantage experienced by most children in divorced families exerts a small impact on their well-being. Nonetheless, it is clear that children of divorce who do not experience a decline in living standards suffer less harm.
3. *The absence of a parent.* Children of divorce usually lose a parent as a role model, source of emotional support, practical help, and supervision. By itself, this factor also has a small effect on children's well-being, even if the child has continued contact with the noncustodial parent.

Subsequent studies confirm these generalizations and add an important observation. Many of the behavioral and adjustment problems experienced by children of divorce existed before the divorce took place. We cannot therefore attribute them to the divorce itself (Cherlin et al., 1991; Entwisle and Alexander, 1995; Furstenberg and Cherlin, 1991; Stewart et al., 1997).

In sum, claiming that divorcing parents trade the well-being of their children for their own happiness is an exaggeration. High levels of parental conflict have serious negative consequences for children, even when they enter adulthood. In such high-conflict situations, divorce can benefit children. Increased state intervention, such as the initiatives taken by the Clinton administration, can ensure that children of divorce do not experience the decline in living standards that often has long-term negative consequences for them. By itself, the absence of a parent has a small negative effect on children's well-being. But this effect is getting smaller over time, perhaps in part because divorce is so common that it is no longer a stigma.

Reproductive Choice

We have seen that the power women gained from working in the paid labor force put them in a position to leave a marriage if it made them deeply unhappy. Another aspect of the gender revolution that women are experiencing is that they are increasingly able to de-

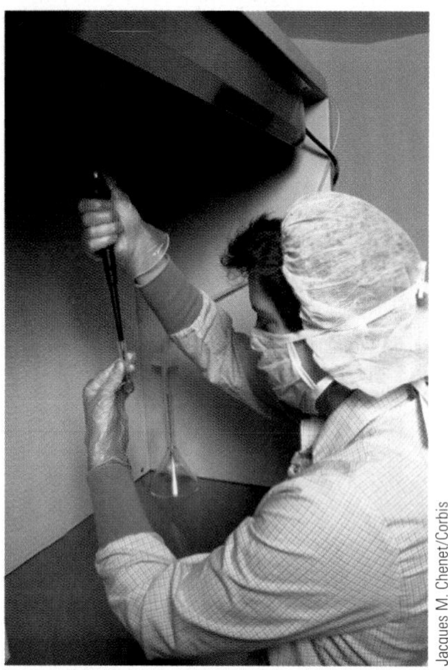

Fertilizing an egg *in vitro*.

cide what happens in the marriage if they stay. For example, women now have more say over whether they will have children and, if so, when they will have them and how many they will have.

Children are increasingly expensive to raise. They no longer give the family economic benefits as they did, say, on the family farm. Most women want to work in the paid labor force, many of them to pursue a career. As a result, most women decide to have fewer children, to have them farther apart, and to have them at an older age. Indeed, 1 out of 20 couples does not have children at all, and among college graduates the figure is 3 out of 20.

Women's reproductive decisions not to have children are carried out by means of contraception and abortion. The United States Supreme Court struck down laws prohibiting birth control in 1965. Abortion first became legal in various states around 1970. Today, public opinion polls show that most Americans think women should be free to make their own reproductive choices. A substantial minority, however, is opposed to abortion.

Because Americans are sharply divided on the abortion issue, "right-to-life" versus "pro-choice" advocates have been clashing since the 1970s. Right-to-life advocates want to repeal laws legalizing abortion. Pro-choice advocates want these laws preserved. Both groups have tried to influence public opinion and lawmakers to achieve their aims. For example, as a result of pressure from the right-to-life lobby, RU-486, the so-called morning-after pill, was introduced in the United States years after it was available in Western Europe. A few extreme right-to-life advocates (almost all men) have resorted to violence (Box 15.2).

Sociologists Randall Collins and Scott Coltrane believe that a repeal of abortion laws would return us to the situation that existed in the 1960s. Many abortions took place then, but because they were illegal, they were expensive and hard to obtain and posed more dangers to women's health. Clearly, if abortion laws were repealed, poor women and their unwanted children would suffer most. Taxpayers would wind up paying bigger bills for welfare and medical care.

Reproductive Technologies

Sociology Now™

Learn more about **Reproductive Technologies** by going through the % of population using contraceptives Map Exercise.

For most women, exercising reproductive choice means being able to prevent pregnancy and birth by means of contraception and abortion. For some women, however, it means *facilitating* pregnancy and birth by means of reproductive technologies. As many as 15 percent of couples are infertile. With a declining number of desirable children available for adoption, and a persistent and strong desire by most people to have children, demand is strong for techniques to help infertile couples, some lesbian couples, and some single women have babies.

There are four main reproductive technologies. In *artificial insemination,* a donor's sperm is inserted in a woman's vaginal canal or uterus during ovulation. In *surrogate motherhood,* a donor's sperm is used to artificially inseminate a woman who has signed a contract to surrender the child at birth in exchange for a fee. In *in vitro fertilization,* eggs are surgically removed from a woman and joined with sperm in a culture dish, and an embryo is then transferred back to the woman's uterus. Finally, various *screening techniques* are used on sperm and fetuses to increase the chance of giving birth to a baby of the desired sex and end pregnancies deemed medically problematic.

Social, Ethical, and Legal Issues

These procedures raise several sociological and ethical issues. We may mention two here (Achilles, 1993). The first problem is discrimination. Most reproductive technologies are expensive. Surrogate mothers charge $10,000 or more to carry a child. *In vitro* fertilization can cost $100,000 or more. Obviously, poor and middle-income earners who happen to be infertile cannot afford these procedures. In addition, there is a strong tendency for members of the medical profession to deny single women and lesbian couples access

Box 15.2

YOU AND THE SOCIAL WORLD

The Abortion Issue

There are many shades of opinion and ambiguities in people's attitudes toward the abortion issue (Chapter 18, "The Mass Media"). At the extremes, however, we may distinguish between right-to-life and pro-choice advocates. Right-to-life advocates argue that life begins at conception. Therefore, they say, abortion destroys human life and is morally indefensible. They advocate adoption instead of abortion. In their opinion, the pro-choice option is selfish, expressing greater concern for career advancement and sexual pleasure than moral responsibility. In contrast, pro-choice advocates argue that every woman has the right to choose what happens to her own body and that bearing an unwanted child can harm not only a woman's career but the child too. For example, unwanted children are more likely to be neglected or abused. They are more likely to get in trouble with the law due to inadequate adult supervision and discipline. Furthermore, according to pro-choice advocates, religious doctrines claiming that life begins at conception are arbitrary. In any case, they point out, such ideas have no place in law because they violate the constitutionally guaranteed separation of church and state.

What are your views on abortion? To what degree are your views influenced by your social characteristics (family income, education, religiosity, etc.)? How do your views compare with those of other Americans with social characteristics similar to yours (Table 15.3, computed from the 2002 General Social Survey)? Why do certain social characteristics influence public opinion on the abortion issue in more or less predictable ways? What variables other than those listed in Table 15.3 might influence public opinion on the abortion issue? Answer these questions in about 500 words.

Table 15.3

"Please tell me whether or not you think it should be possible for a pregnant woman to obtain a legal abortion if the woman wants it for any reason," United States, 2002 (in percent)

	Yes	No	%	*N*
Gender				
Male	44	56	100	484
Female	41	59	100	416
Highest year of schooling completed				
0–11	31	69	100	154
12	39	61	100	258
13+	49	51	100	485
Age				
18–29	42	58	100	168
30–39	46	54	100	184
40–49	52	48	100	170
50–59	44	56	100	143
60–69	35	65	100	114
70+	35	65	100	125
Region				
New England	67	33	100	54
Middle Atlantic	52	48	100	150
South Atlantic	44	56	100	156
East North Central	39	61	100	146
West North Central*	37	63	100	72
East South Central	17	83	100	71
West South Central	28	72	100	81
Mountain	51	49	100	51
Pacific	50	50	100	119
Total annual family income				
$0–49,999	37	63	100	488
$50,000+	53	47	100	317
Vote in 2000 presidential election				
Bush	37	63	100	297
Gore	56	44	100	255
Attendance at religious services				
Less than once a month	55	45	100	495
Once a month or more	28	72	100	400
Religiously fundamentalist/moderate/liberal				
Fundamentalist	25	75	100	234
Moderate	39	61	100	317
Liberal	61	39	100	261

Source: National Opinion Research Center (2004).

to reproductive technologies. In other words, the medical community discriminates not just against those of modest means but against nonnuclear families.

A second problem introduced by reproductive technologies is that they render the terms "mother" and "father" obsolete, or at least vague. Is the mother the person who donates the egg, carries the child in her uterus, or raises the child? Is the father the person who donates the sperm or raises the child? As these questions suggest, a child conceived through a combination of reproductive technologies and raised by a heterosexual couple could have as many as three mothers and two fathers! This is not just a terminological problem. If it were, we could just introduce new distinctions like "egg mother," "uterine mother," and "social mother" to reflect the new reality. The real problem is social and legal. The question of who has what rights and obligations to the child, and what rights and obligations the child has vis-à-vis each parent, is unclear. This lack of clarity has already caused anguished court battles over child custody (Franklin and Ragone, 1999).

Public debate on a wide scale is needed to decide who will control reproductive technologies and to what ends. On the one hand, reproductive technologies may bring the greatest joy to infertile people. They may also prevent the birth of children with diseases such as muscular dystrophy and multiple sclerosis. On the other hand, reproductive technologies may continue to benefit mainly the well-to-do, reinforce traditional family forms that are no longer appropriate for many people, and cause endless legal wrangling and heartache.

Housework and Child Care

As we have seen, women's increased participation in the paid labor force and the system of higher education and their increased control over reproduction have transformed several areas of family life. Despite this far-ranging gender revolution, however, the domains of housework and child and senior care remain largely resistant to change. This fact was first documented in detail by sociologist Arlie Hochschild. She showed that even women who work full-time in the paid labor force usually begin a "second shift" when they return home. There, they prepare meals, help with children's homework, do laundry, clean the toilets, and so forth (Hochschild with Machung, 1989).

To be sure, there has been *some* change as men take a more active role in the day-to-day running of the household. But the change is modest. For example, one study conducted in the late 1980s compared full-time female homemakers in first marriages with wives in first marriages who worked 30 hours or more per week outside the home. The wives working full-time in the paid labor force did only 1 hour and 10 minutes less housework per day than the full-time homemakers. Husbands of women working full-time in the paid labor force did a mere 37 minutes more housework per day than husbands of full-time homemakers (calculated from Demo and Acock, 1993) (◗Figure 15.6). Studies estimate that on average, American men now do 20–35 percent of the housework and child care (Shelton and John, 1996: 299).

Even these figures do not reveal the whole picture, however. Men tend to do low-stress chores that can often wait a day or a week. These jobs include mowing the lawn, repairing the car, and preparing income tax forms. They also play with their children more than they used to. In contrast, women tend to do higher-stress chores that cannot wait. These jobs include getting kids dressed and out the door to school every day, preparing dinner by 6:00 p.m., washing clothes twice a week, and the like. In short, the picture is hardly that of a revolution (Harvey, Marshall, and Frederick, 1991).

Two main factors shrink the gender gap in housework, child care, and senior care. First, the smaller the difference between the husband's and the wife's earnings, the more equal the division of household labor. Apparently, women are routinely able to translate earning power into domestic influence. Put bluntly, their increased status enables them to get their husbands to do more around the house. In addition, women who earn relatively high incomes are able to use some of their money to pay outsiders to do domestic work.

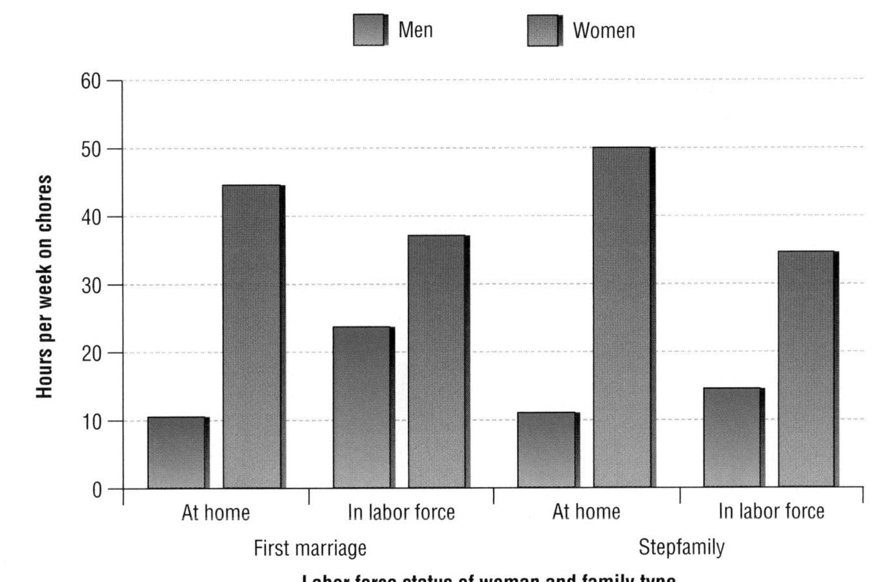

Figure 15.6
The Division of Domestic Labor by Woman's Work Status, United States

Note: Data are for 1987–1988.
Source: From "Family Division and the Division of Domestic Labor: How Have Things Really Changed" by David H. Demo and Alan C. Acock, *Family Relations*, Issue 9307, Vol. 42:3, 1993. Copyright © 1993 by the National Council on Family Relations, 3839 Central Ave., NE, Suite 550, Minneapolis, MN 55421.

Attitude is the second factor that shrinks the gender gap in domestic labor. The more husband and wife agree that there *should* be equality in the household division of labor, the more equality there is. Seeing eye to eye on this issue is often linked to both spouses having a college education (Greenstein, 1996). Thus, if there is going to be greater equality between men and women in doing household chores, two things have to happen. There must be greater equality between men and women in the paid labor force and broader cultural acceptance of the need for gender equality.

Michael Newman/PhotoEdit

The double day.

Domestic Violence

A 1997 Gallup poll found that 22 percent of women, compared with 8 percent of men, reported physical abuse by a spouse or companion at least once in the past (Figure 15.7). Some of that abuse is severe. For example, in about 2 percent of American families every year, husbands kick their wives, bite them, hit them with a fist, threaten to use a knife or gun, or use a knife or gun (Straus, 1995). Although about as many wives commit such acts of violence against their husbands, the husbands are about seven times more likely to injure their wives physically than vice versa.

There are three main types of domestic violence (Johnson and Ferraro, 2000):

1. *Common couple violence* occurs when partners have a specific argument and one partner lashes out physically at the other. For a couple that engages in this type of violence, violent acts are unlikely to occur often, to escalate over time, or to be severe. Both partners are equally likely to engage in common couple violence, regardless of their gender.
2. *Intimate terrorism* is part of a general desire of one partner to control the other. Where one partner engages in intimate terrorism, violent acts are likely to occur often, to escalate over time, and to be severe. Among heterosexual couples the aggressor is usually the man.

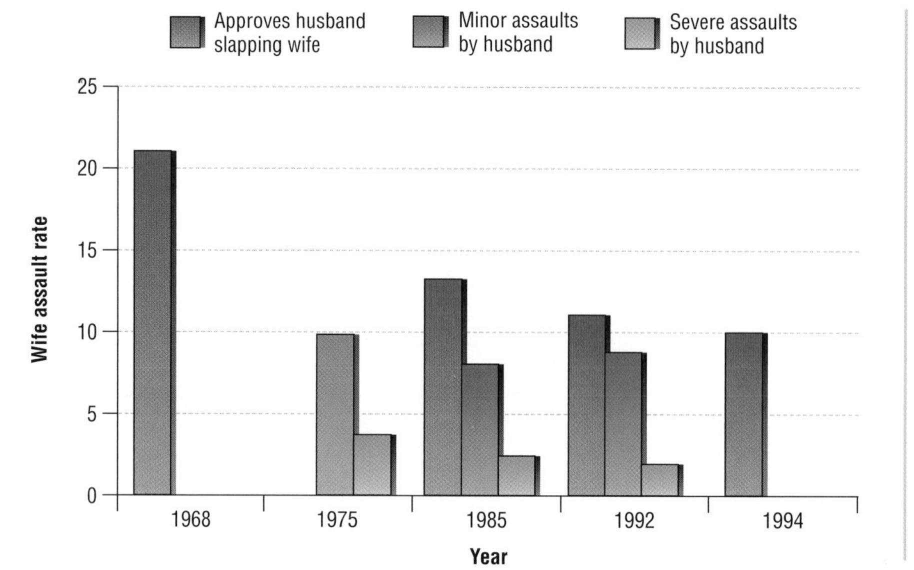

Figure 15.7
Spousal Violence Against Wives, United States, 1968–1994 (in percent)

Note: Figures include cohabiting but unmarried couples. Data on all indicators are not available for some years shown.
Source: Straus (1995).

3. *Violent resistance* is the third main type of domestic violence. Among heterosexual couples, it typically involves a woman violently defending herself against a man who has engaged in intimate terrorism.

Gender Inequality and Domestic Violence

For heterosexual couples, domestic violence seems to be associated with the level of gender equality in the family and in the larger society. The higher the level of gender inequality, the greater the frequency of domestic violence. Thus, severe wife assault is more common in lower-class, less highly educated families, where gender inequality tends to be high and men are more likely to believe that male domination is justified. Severe wife abuse is also more common among couples who as children witnessed their mothers being abused and were themselves abused, although research suggests that these socialization factors are considerably less influential than was once believed (Gelles, 1997 [1985]; Simons et al., 1995; Smith, 1990). Still, male domination in both childhood socialization and current family organization increases the likelihood of severe wife assault.

In addition, Straus (1994) has shown that wife assault is associated with gender inequality in the larger society. He first constructed a measure of wife assault for each U.S. state using data from a national survey. The measure shows the percentage of couples in each state in which the wife was physically assaulted by her partner during the 12 months preceding the survey. He then used government data to measure gender inequality in each state. His measure of gender inequality taps the economic, educational, political, and legal status of women. He found that as gender equality increases—as women and men become more equal in the larger society—wife assault declines. We conclude that for heterosexual couples, the incidence of domestic violence is highest where a big power imbalance between men and women exists, where norms justify the male domination of women, and, to a lesser extent, where early socialization experiences predispose men to behave aggressively toward women.

Summing up, we can say that conflict theorists and feminists have performed a valuable sociological service by emphasizing the importance of power relations in structuring family life. A substantial body of research shows that the gender revolution that has been taking place for nearly half a century has influenced the way we select mates, our reasons for being satisfied or dissatisfied with marriage, our propensity to divorce, the reproductive choices women make, the distribution of housework and child care, variations in the rate of severe domestic violence—in short, all aspects of family life. As you will now learn, the gender revolution has also created a much greater diversity of family forms.

Table 15.4
Unmarried Couples by Selected Characteristics, United States, 1970–1999

	1970	1980	1990	1999
Percent of households	0.8	2.0	3.1	4.3
Have children under 15 years old	37.5	27.1	31.2	33.5
Partners are under 25 years old	10.5	25.9	20.9	18.4
Partners are over 45 years old	69.6	21.3	17.0	24.7

Sources: U.S. Census Bureau (2001, 2002b).

Family Diversity

Cohabitation

In the last three decades, the number of American couples who are unmarried and cohabiting (or "living together") increased more than 500 percent (Table 15.4). By 2004, unmarried cohabiting couples probably comprised about 5 percent of all households. Five percent may not seem like a lot. However, if we examine the number of marriages and remarriages that begin as cohabiting relationships, the numbers grow much larger. About 10 percent of people who married between 1965 and 1974 cohabited before marrying. For people marrying between 1990 and 1994, the figure was more than 50 percent (Smock, 2000: 4). About half of cohabiting couples have children living with them. Once considered a disgrace, cohabitation has gone mainstream.

People who disapprove of cohabitation often do so because they oppose premarital sex. Often, they cite religious grounds for their opposition. In recent decades, however, the force of religious sanction has weakened. The sexual revolution and growing individualism have allowed people to pursue intimate relationships outside marriage if they so choose. Meanwhile, because women have pursued higher education and entered the paid labor force in increasing numbers, their gender roles are not so closely tied to marriage as they once were. These cultural and economic factors have all increased the rate of cohabitation.

Cohabitation is a relatively unstable relationship. Within 5 years of moving in together, about 55 percent of cohabiting couples marry and 40 percent split up. Moreover, marriages that begin with cohabitation are associated with a higher divorce rate than marriages that begin without cohabitation. This is true even when researchers compare couples at the same level of education and age at marriage.

Cohabitation and Marital Stability

The most often cited and best supported explanation for the association between cohabitation and marital instability is that people who cohabit before marriage differ from those who do not, and these differences increase the likelihood of divorce. Thus, people who cohabit before marriage tend to be less religious than those who do not, and religious people are less likely to divorce because they tend to believe that divorce is an unjustifiable solution to marital problems. Similarly, compared with people who do not cohabit, those who cohabit are more likely to be African American, occupy a lower class position, hold more liberal political and sexual views, and have parents who divorced. These factors are also associated with higher divorce rates (Starbuck, 2002: 239).

Sociological Significance of Cohabitation

Sociologists have debated the meaning and significance of cohabitation for two decades. Some view it as a prelude to marriage or a new form of marriage. Others think that cohabitation is more like being single. They regard it as a threat to family life, a temporary relationship without commitment or responsibility. Table 15.5 sheds light on this debate. Sociologists asked cohabiting couples how they think of their relationship—

Table 15.5
Perception and Outcome of Cohabiting Relationships

Type of Relationship, 1987–1988	Percent of Couples	OUTCOME OF RELATIONSHIP, 1992–1994, IN PERCENT		
		Still Live Together	Married	Separated
*Substitute for marriage**	10	39	25	35
Precursor to marriage	46	17	52	31
Coresidential dating	49	21	33	46
Trial marriage	15	21	28	51
Total	100			

*Percent does not equal 100 due to rounding.
Source: Bianchi and Casper (2000).

as a substitute for marriage, a precursor to marriage, a trial marriage, or merely as a form of serious dating. Interestingly, only 29 percent of respondents saw it as a form of serious dating and 15 percent as a marital experiment, suggesting that a minority of cohabiting individuals view cohabitation as a temporary relationship. The researchers also determined the outcome of the relationships after 5–7 years. Significantly, about two-thirds of the couples who thought of cohabitation as a precursor to marriage or a substitute for marriage were most likely to enjoy enduring relationships after 5–7 years; they either married or were still living together. On the other hand, only about half of the couples who thought of cohabitation as a more fleeting kind of relationship were still together after 5–7 years. We find this a surprisingly high number because roughly half of *all* recent marriages in the United States are likely to end in divorce. We conclude that although a substantial minority of people who cohabit do not see cohabitation as an enduring relationship, it nonetheless results in an enduring relationship for many of them. Moreover, most people see cohabitation as a prelude to marriage or a substitute for it. People who enter into a cohabiting relationship thinking of it in these ways are likely to enjoy an enduring relationship.

One of the lesbian couples who showed up at the Hawaii Department of Health in Honolulu in 1993 to apply for a marriage license. The couple was turned away on the grounds of improper sexual orientation. Their case ultimately made its way to the Supreme Court of Hawaii, which ruled in December 1999 that same-sex marriages are illegal.

AP/Wide World Photos

Same-Sex Unions and Partnerships

In February 2004, the mayor of San Francisco, Gavin Newsom, ordered his county clerk to begin issuing marriage licenses to gay and lesbian couples. Although the laws of the State of California do not allow such marriages, Newsom argued that these laws are discriminatory and contradict protections laid out in the state constitution. Thousands of marriages between same-sex couples took place in San Francisco, many on the steps of City Hall, in a massive showing of civil disobedience that soon spread across the United States. The events in California followed closely on the heels of a Massachusetts Supreme Court decision declaring the ban on gays and lesbians marrying in that state to be unconstitutional.

Opponents of same-sex marriage quickly began legal efforts to block the issuance of marriage licenses in California and to prevent same-sex marriages from taking place in their own states. Amidst heavy lobbying from religious conservatives, President Bush declared his support for an amendment to the federal constitution that would prevent same-sex marriages.

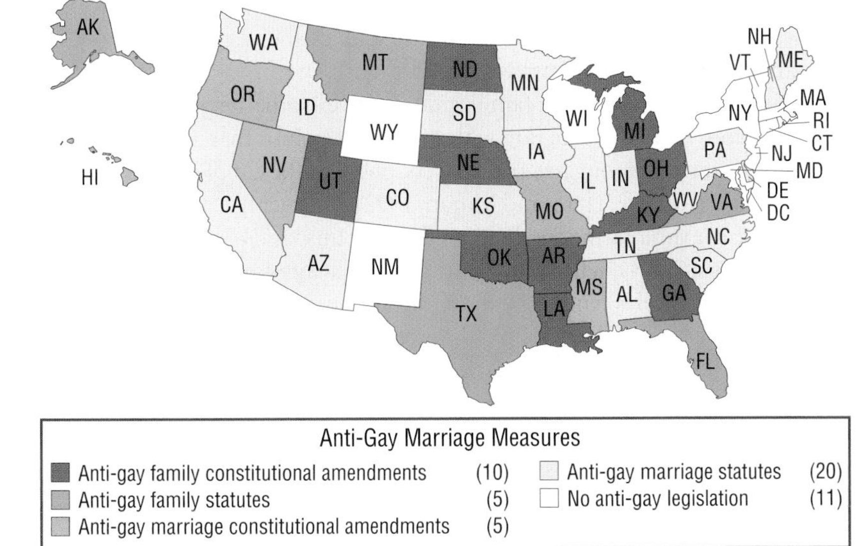

Figure 15.8
States with Laws Banning Same-Sex Marriages, 2005
In 1996 the federal "Defense of Marriage Act" became law. It requires no state to honor same-sex marriages performed in another state. It also defines marriage as "a legal union between one man and one woman." Since 1995, many states have passed laws or state constitutional amendments that either ban gay marriages or stipulate broader anti-gay family measures (e.g., they ban civil unions). As of February 2005, 40 states had passed anti-gay laws or constitutional amendments.

Source: National Gay and Lesbian Task Force (2005).

In some ways, those opposing same-sex marriage in the United States were swimming against the stream of cultural change (Figure 15.8). In 2001 the Netherlands became the first country in the world to legalize same-sex marriage. Belgium followed suit in 2003 and Spain and Canada legalized same-sex marriage in 2005. Eight other countries allow homosexuals to register their partnerships under the law in so-called *civil unions.* Civil unions recognize the partnerships as having some or all of the legal rights of marriage. These countries include Denmark (along with its dependency, Greenland), Hungary, Norway, Sweden, France, Iceland, Spain, and Germany. In the United States, there is more opposition to registered partnerships and same-sex marriages than in these countries. By 2005, 40 states had passed laws opposing gay marriage or banning civil unions. A nationwide Harris poll taken in January 2004 showed that 53 percent of Americans oppose same-sex marriages. Some 41 percent oppose civil unions for same-sex partners (Gallup Organization, 2004). Yet, despite continuing opposition to same-sex marriage, the direction of change in many parts of the world is clear. Amid sharp controversy, the legal and social definition of "family" is being broadened to include cohabiting same-sex partners in long-term relationships (ReligiousTolerance.org, 2005).

Research shows that most homosexuals, like most heterosexuals, want a long-term, intimate relationship with one other adult (Baca Zinn and Eitzen, 1993: 423). In fact, in Denmark, where homosexual couples can register partnerships under the law, the divorce rate for registered homosexual couples is lower than for heterosexual married couples (ReligiousTolerance.org, 2005). According to the U.S. census, just over 600,000 gay men and just under 600,000 gay women were living together in 2000 (U.S. Census Bureau, 2002). Educated guesses suggest that about half of these people were raising children. Most of these children are offspring of previous, heterosexual marriages. Some are adopted. Others result from artificial insemination.

Sociology Now™

Learn more about **Family Diversity** by going through the Diversity among Families Animation.

Raising Children in Homosexual Families

Many people believe that children brought up in homosexual families will develop a confused sexual identity, exhibit a tendency to become homosexuals themselves, and suffer discrimination from children and adults in the "straight" community. Unfortunately, there is little research in this area. Much of the research is based on small, unrepresentative samples. Nevertheless, the research findings are consistent. They suggest that children who grow up in homosexual families are much like children who grow up in heterosexual families. For example, a 14-year study assessed 25 young adults who were the offspring of lesbian families and 21 young adults who were the offspring of heterosexual families (Tasker and Golombok, 1997). The researchers found that the two groups were equally well adjusted and displayed little difference in sexual orientation. Two respondents from the lesbian families considered themselves lesbians, while all of the respondents from the heterosexual families considered themselves heterosexual.

Homosexual and heterosexual families do differ in some respects. Lesbian couples with children record higher satisfaction with their partnerships than lesbian couples without children. In contrast, among heterosexual couples, it is the childless who record higher marital satisfaction (Koepke, Hare, and Moran, 1992). On average, the partners of lesbian mothers spend more time caring for children than the husbands of heterosexual mothers. Because children usually benefit from adult attention, this must be considered a plus. Finally, homosexual couples tend to be more egalitarian than heterosexual couples, sharing most decision making and household duties equally. That is because they tend to consciously reject traditional marriage patterns. The fact that they have the same gender socialization and earn about the same income also encourages equality (Baca Zinn and Eitzen, 1993: 424). In sum, available research suggests that raising children in lesbian families has no apparent negative consequences for the children. Indeed, there may be some benefits for all family members.

Single-Parent Families: Racial and Ethnic Differences

We have seen how families differ from one another due to variations in the sexual orientation of adult family heads. Now let us examine how they vary across racial and ethnic groups in terms of the number of adults who head the family (Baca Zinn and Eitzen, 1993: 109–27; Cherlin 1981 [1992]: 91–123; Collins and Coltrane, 1991: 233–69). ▶Figure 15.9 focuses on the country's two most common family types (two-parent and single-mother) and three largest racial and ethnic categories (white, African American, and Hispanic American). It shows that whites have the lowest incidence of single-mother families. African Americans have by far the highest. In all racial and ethnic groups, the proportion of single-mother families has been increasing in recent decades, but the increase has been most dramatic among African Americans. Thus, among African Americans in 1970, there were 1.9 two-parent families for every single-mother family. By 2002 there were more than 1.2 single-mother families for every two-parent family, although that number had declined slightly since the late 1990s (Harden, 2001) (see also "Social Policy: What Do You Think?" in Chapter 8).

Some single-parent families result from separation, divorce, or death. Others result from people not getting married in the first place. Marriage is an increasingly unpopular

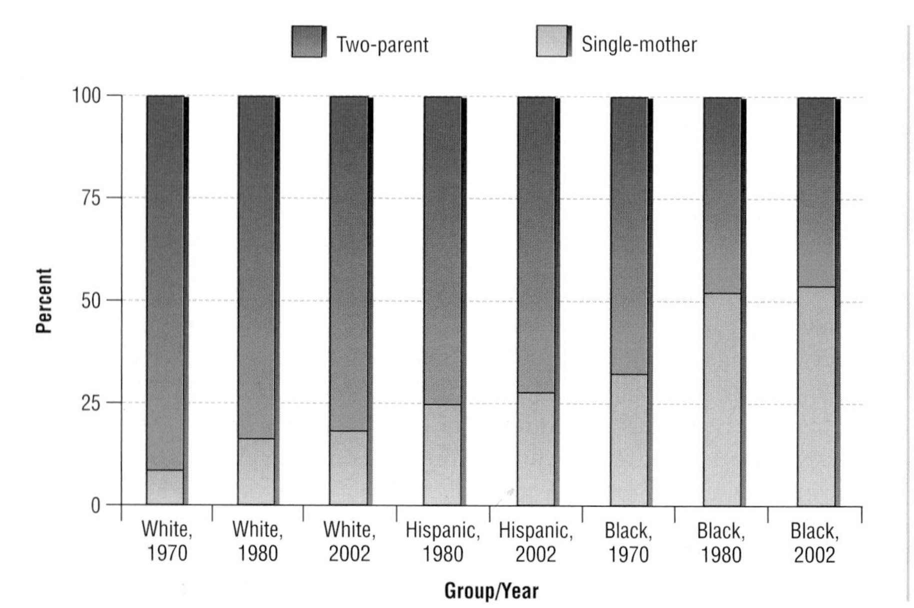

▶Figure 15.9
Families with Own Children under 18 by Race and Hispanic Origin, United States, 1970, 1980, and 2002

Note: 1970 data on Hispanics are not available.
Sources: Baca Zinn and Eitzen (1993: inside back cover); InfoPlease (2005).

institution, particularly among African Americans. This is clear from statistics on births to unmarried mothers. In 2001 among non-Hispanic whites, 22.5 percent of births were to unmarried mothers. Among Hispanics, the figure was nearly 42.5 percent. Among African Americans, it was just under 68.6 percent (Centers for Disease Control, 2002). Note, however, that the rate of unmarried cohabitation is about the same for African Americans and whites, but official statistics count the many births to unmarried cohabiting couples as births to unmarried women. Thus, the figures just cited exaggerate the differences between racial groups.

Laura Dwight/Corbis

By 2002, there were 1.2 single-mother families for every two-parent family in the African American community.

The Decline of the Two-Parent Family among African Americans

What accounts for the decline of the two-parent family among African Americans (Cherlin, 1992 [1981])? Although some scholars trace the decline of the African American two-parent family back to slavery (Jones, 1986), rapid decline began around 1925. By then, the mechanization of the cotton economy in the South had displaced many black agricultural laborers and sharecroppers. They were forced to migrate northward. In the North, they competed fiercely for industrial jobs. Due to discrimination, however, they suffered higher rates of unemployment than any other group in America. Thus, ever since about 1925, proportionately fewer black men have been able to help support a family. As a result, proportionately fewer stable two-person families have formed. Similarly, the decline of manufacturing industries in the Northeast and the movement of many blue-collar jobs to the suburbs in the 1970s and 1980s eliminated many secure, well-paying jobs for blacks and caused their unemployment rate to rise. It is precisely in this period that the rate of increase in African American single-mother families skyrocketed.

Apart from unemployment, a second factor explaining the decline of the two-parent family among African Americans is the declining ratio of eligible black men to black women. This has three sources. First, largely because of the disadvantaged economic and social position of the African American community, a disproportionately large number of black men are imprisoned, have been murdered, and suffer from drug addiction (Chapter 7, "Deviance and Crime"). Second, because the armed forces represent one of the best avenues of upward mobility for African American men, a disproportionately large number of them have enlisted and been killed in action. Third, a black man is nearly twice as likely as a black woman to marry a nonblack, and intermarriage has increased to nearly 10 percent of all marriages involving at least one black person. For all these reasons, there are relatively fewer black men available for black women to marry (Anderson, 1999; Cherlin, 1992 [1981]; Wilson, 1987).

The third main factor explaining the decline of the two-parent family in the black community concerns the relative earnings of women and men. In recent decades, the average income of African American women has increased. Meanwhile, the earning power of African American men has fallen. As a result, African American women are more economically independent than ever. On average, they have less to gain in purely economic terms from marrying a black man. Economically speaking, marriage has thus become a less attractive alternative for them (Cherlin, 1992 [1981]).

Adaptations to Poverty

Poor African American women have adapted to harsh economic realities in creative ways that testify to their resilience. In particular, they have developed strong kinship and friendship networks that enable them to survive with few resources. Members of the network help each other with child care. They share money when they have it. They lend each other household items. They give each other hand-me-downs. These adaptations to poverty have a downside, however. If an individual is lucky enough to come into a windfall that could help remove her and her children from poverty—a job or a modest inheritance, for example—the money is quickly used up by the network and the individual remains poor. In other words, the network is a source of survival *and* a shackle that helps keep poor black women impoverished (Stack, 1974).

Just as poor black women can keep their families functioning thanks to the assistance of grandmothers, aunts, sisters, and friends, so Hispanics have developed a family system in which the godfather (*padrino*) and godmother (*madrina*) often act as coparents, providing child care and emotional and financial support as needed. Indeed, Hispanic American families tend to rely on entire extended kin networks for social support.[1]

Note, however, that the tendency to rely on the extended kin network for social support declines with migration status and upward mobility. The best evidence for this comes from a study of extended kin among Mexican Americans (Glick, 1999). Using data from the census and a survey of income dynamics, the study found that reliance on extended kin is stronger among immigrant Hispanics than among those born in the United States. Moreover, reliance on extended kin declines as socioeconomic status increases. American-born, middle- and upper-class Hispanic Americans are less in need of social support from extended kin networks than are immigrant, working-class, and poor Hispanic Americans. These findings suggest that the prominence of extended kin networks among Hispanic Americans is at least in part a function of class. Just as the incidence of single-mother families and reliance on extended kin networks drops off sharply among middle-class and upper-middle-class black families, so in the Hispanic community, class position, and not just culture, shapes family structure.

Family Policy

Having discussed several aspects of the decline of the traditional nuclear family and the proliferation of diverse family forms, we can now return to the big question posed at the beginning of this chapter: Is the decline of the nuclear family a bad thing for society? Said differently, do two-parent families—particularly those with stay-at-home moms—provide the kind of discipline, role models, help, and middle-class lifestyle that children need to stay out of trouble with the law and grow up to become well-adjusted, productive members of society? Conversely, are family forms other than the traditional nuclear family the main source of teenage crime, poverty, welfare dependency, and other social ills (Box 15.3)?

Web Research Project: Family Values

The answer suggested by research is clear: yes and no (Houseknecht and Sastry, 1996; Popenoe, 1988; 1991; 1992, 1993, 1996; Sandqvist and Andersson, 1992). Yes, the decline of the traditional nuclear family can be a source of many social problems. No, it doesn't have to be that way.

Crossnational Differences: The United States and Sweden

The United States is a good example of how social problems can emerge from nuclear family decline. Sweden is a good example of how such problems can be averted. Table 15.6 illustrates this. The top panel of Table 15.6 shows that *on most indicators of nuclear family decline, Sweden leads the United States.* In Sweden, a smaller percentage of people get married. People usually get married at a later age than in the United States. The proportion of births outside marriage is twice as high as in the United States. A much larger proportion of Swedish than American women with children under the age of 3 work in the paid labor force.

The bottom panel of Table 15.6 shows that *on most measures of children's well-being, Sweden also leads the United States.* Thus, in Sweden, children enjoy higher average read-

[1]Strong social support from the extended family is, incidentally, one of the main reasons (along with a diet high in vegetable content) why Hispanic Americans have a lower death rate from cancer and heart disease than non-Hispanic whites (Braus, 1994).

BOX 15.3

SOCIAL POLICY: WHAT DO YOU THINK?

The Pro-Fatherhood Campaign

If you were watching David Letterman in September 1999, you might have seen a commercial that starts like this: "When young bull elephants from a national park in South Africa were moved to different locations without the presence of an adult male, they began to wantonly kill other animals. When an adult male was relocated with them, the delinquent behavior stopped." From a panoramic view of elephants, the commercial switched to a basketball court where an African American man is hugging an African American boy. The voice-over said: "Without the influence of their dads, kids are more likely to get into trouble, too. Just a reminder how important it is for fathers to spend time with their children" (quoted in Davidoff, 1999: 28).

The National Fatherhood Initiative sponsored the commercial. It is part of a nationwide campaign to emphasize the importance of fatherhood to family life in particular and society in general. The 1999 Responsible Fatherhood Act pledged more than $150 million to "allow states to implement programs that promote stable and married families and support responsible fatherhood" (quoted in Davidoff, 1999: 29).

Nobody can disagree with supporting fatherhood and stable family life. However, one problem with pro-fatherhood policies is that they are often intended to replace welfare programs, which are simultaneously being cut by both the federal and state governments. Moreover, critics of pro-fatherhood policies argue that what is essential for the healthy development of children is not just a father, or for that matter even a mother. As psychologists have shown, what is essential is a lasting and loving relationship with at least one adult (Silverstein and Auerbach, 1999). In other words, it is not necessarily the presence of a nuclear family that ensures healthy family life. Insofar as the Fatherhood Initiative supports only one kind of family, it devalues other forms of family life, including single-parent households, homosexual couples, and so on.

What do you think? Should we support the Fatherhood Initiative? Or does it lead us to ignore other social problems, such as poverty? Does the focus on fatherhood devalue other forms of family?

ing test scores than in this country. The poverty rate in two-parent families is only one-tenth the United States rate, while the poverty rate in single-parent families is only one-twelfth as high. The rate of infant abuse is one-eleventh the United States rate. The rate of juvenile drug offenses is less than half as high. Sweden does have a higher rate of juvenile delinquency than the United States. However, the lead is slight and concerns only minor offenses. Overall, then, the decline of the traditional nuclear family has gone farther in Sweden than in the United States, but children are much better off on average. How is this possible?

One possible explanation is that Sweden has something the United States lacks: a substantial family support policy. When a child is born in Sweden, a parent is entitled to 360 days of parental leave at 80 percent of his or her salary and an additional 90 days at a flat rate. Fathers can take an additional 10 days of leave with pay when the baby is born. Parents are entitled to free consultations at well-baby clinics. Like all citizens of Sweden, they receive free health care from the state-run

Table 15.6

Decline of the Nuclear Family and the Well-Being of Children: The United States (USA) and Sweden Compared

Indicators of Nuclear Family Decline	USA	Sweden	#1 Decline
Median age at first marriage			
Men	26.5	29.4	Sweden
Women	24.4	27.1	Sweden
Percentage of 45–49 population never married			
Men	5.7	15.4	Sweden
Women	5.1	9.1	Sweden
Nonmarital birthrate	25.7	50.9	Sweden
One-parent households with children <15 as a percent of all households with children <15	25.0	18.0	USA
Percent of mothers in labor force with children <3	51.0	84.0	Sweden
Total fertility rate	2.0	2.0	Tie
Average household size	2.7	2.2	Sweden
Indicators of Child Well-Being	**USA**	**Sweden**	**#1 Well-Being**
Mean reading performance score at 14	5.14	5.29	Sweden
Percent of children in poverty			
Single-mother households	59.5	2.2	Sweden
Two-parent households	11.1	2.2	Sweden
Death rate of infants from abuse (per 100,000)	9.8	0.9	Sweden
Suicide rate for children 15–19 (per 100,000)	11.1	6.2	Sweden
Juvenile delinquency rate (per 100,000)	11.6	12.0	USA
Juvenile drug offense rate (per 100,000)	558	241	Sweden

Source: Adapted from Houseknecht and Sastry (1996).

Jonathan Blair/Corbis

Painting class in a state-subsidized day care facility in Stockholm, Sweden.

system. Temporary parental benefits are available for parents with a sick child under the age of 12. One parent can take up to 60 days off per sick child per year at 80 percent of salary. All parents can send their children to heavily government subsidized, high-quality day care. Finally, Sweden offers its citizens generous direct cash payments based on the number of children in each family.[2]

Family Support Policies in the United States

Among industrialized countries, the United States stands at the other extreme. Since the Family and Medical Leave Act was passed in 1993, a parent is entitled to 12 weeks of *unpaid* parental leave. About 44 million citizens have no health-care coverage. Health care is at a low standard for many millions more. There is no system of state day care and no direct cash payments to families based on the number of children they have. The value of the dependent deduction on income tax has fallen by nearly 50 percent in current dollars since the 1940s. Thus, when an unwed Swedish woman has a baby, she knows she can rely on state institutions to maintain her standard of living and help give her child an enriching social and educational environment. When an unwed American woman has a baby, she is pretty much on her own. She stands a good chance of sinking into poverty, with all the negative consequences it has for her and her child.

In the United States, three criticisms are commonly raised against generous family support policies. First, some people say they encourage long-term dependence on welfare, illegitimate births, and the breakup of two-parent families. However, research shows that the divorce rate and the rate of births to unmarried mothers are not higher when welfare payments are more generous (Ruggles, 1997; Sweezy and Tiefenthaler, 1996).

Nor is welfare dependency widespread in America. African American teen mothers are often thought to be the group most susceptible to chronic welfare dependence. Kathleen Mullan Harris (1997) studied 288 such women in Baltimore. She found that 29 percent were never on welfare. Twenty percent were on welfare only once and for a very brief time. Twenty-three percent cycled on and off welfare—off when they could find work, on when they couldn't. The remaining 28 percent were long-term welfare users. However, most of these teen mothers said they wanted a decent job that would allow them

[2]We are grateful to Gregg Olsen for this information.

to escape life on welfare. That is why half of those on welfare in any given year were concurrently working.

A second criticism of generous family support policies focuses on child care. Some critics say nonfamily child care is bad for children under the age of 3. In their view, only parents can provide the love, interaction, and intellectual stimulation infants and toddlers need for proper social, cognitive, and moral development. The trouble with this argument is that, explicitly or implicitly, it compares the quality of child care in upper-middle-class families with the quality of child care in most existing day-care facilities in the United States. Yet, as we saw in Chapter 11 ("Sexuality and Gender"), existing child-care facilities in the United States are often of poor quality. They are characterized by high turnover of poorly paid, poorly trained staff and a high ratio of caregivers to children. When studies compare family care with day care involving a strong curriculum, a stimulating environment, plenty of caregiver warmth, low turnover of well-trained staff, and a low ratio of caregivers to children, they find that day care has no negative consequences for children older than 1 year (Clarke-Stewart, Gruber, and Fitzgerald, 1994). A recent study of more than 6000 American children found that a mother's employment outside the home does have a very small negative effect on the child's self-esteem, later academic achievement, language development, and compliance. However, this effect was apparent only if the mother returned to work within a few weeks or months of giving birth. Moreover, the negative effects usually disappeared by the time the child reached the age of 5 (Harvey, 1999). Research also shows that day care has some benefits for children, notably an enhanced ability to make friends. The benefits of high-quality day care are even more evident in low-income families, which often cannot provide the kind of stimulating environment offered by high-quality day care.

The third criticism lodged against generous family support policies is that they are expensive and have to be paid for by high taxes. This is true. Swedes, for example, are more highly taxed than the citizens of any other country. They have made the political decision to pay high taxes partly to avoid the social problems and associated costs that sometimes emerge when the traditional nuclear family is replaced with other family forms and no institutions are available to help family members in need. The Swedish experience teaches us, then, that there is a clear trade-off between expensive family support policies and low taxes. It is impossible to have both, and the degree to which any country favors one or the other is a political choice.

Summary

Sociology Now™

Reviewing is as easy as ❶, ❷, ❸.

❶ Before you do your final review, take the SociologyNow diagnostic quiz to help you identify the areas on which you should concentrate. You will find information on how to access SociologyNow on the foldout at the front of the textbook.

❷ As you review, take advantage of SociologyNow's study aids to help you master the topics in this chapter.

❸ When you are finished with your review, take SociologyNow's post-test to confirm you are ready to move on to the next chapter.

1. **What is the traditional nuclear family, and how prevalent is it compared with other family forms?**

The traditional nuclear family consists of a father-provider, mother-homemaker, and at least one child. Today, well under a quarter of American households are traditional nuclear families. Many different family forms have proliferated in recent decades, including cohabiting couples (with or without children), same-sex couples (with or without children), and single-parent families. The frequency of these forms varies by class, race, and sexual orientation.

2. **What is the functionalist theory of the family, and how accurate is it?**

The functionalist theory holds that the nuclear family is a distinct and universal family form because it performs five important functions in society: sexual regulation, economic cooperation, reproduction, socialization, and emotional support. The theory is most accurate in depicting families in the United States and other Western societies in the two decades after World War II. Families today and in other historical periods depart from the functional model in important respects.

3. **What are the emphases of Marxist and feminist theories of families?**

   Marxists stress how families are tied to the system of capitalist ownership. They argue that only the elimination of capitalism can end gender inequality in families. Feminists note that gender inequality existed before capitalism and in communist societies. They stress how the patriarchal division of power and patriarchal norms reproduce gender inequality.

4. **What consequences does the entry of women into the paid labor force have?**

   The entry of women into the paid labor force increases their power to leave unhappy marriages and control whether and when to have children. However, it does not have a big effect on the sexual division of labor in families.

5. **What accounts for variations in marital satisfaction?**

   Marital satisfaction is lower at the bottom of the class structure, where divorce laws are strict, when children reach their teenage years, in families where housework is not shared equally, and among couples who do not have a good sexual relationship.

6. **Under what circumstances are the effects of divorce on children worst?**

   The effects of divorce on children are worst if there is a high level of parental conflict and the children's standard of living drops.

7. **Does growing up in a lesbian household have any known negative effects on children?**

   Growing up in a lesbian household has no known negative effects on children.

8. **Are various social problems a result of the decline of the traditional nuclear family?**

   People sometimes blame the decline of the traditional nuclear family for increasing poverty, welfare dependence, and crime. However, some countries have adopted policies that largely prevent these problems. Therefore, the social problems are in a sense a political choice.

Questions to Consider

1. Do you agree with the functionalist view that the traditional nuclear family is the ideal family form for the United States today? Why or why not?
2. Ask your grandparents and parents how many people lived in their household when they were your age. Ask them to identify the role of each household member (mother, brother, sister, grandfather, boarder, etc.) and to describe the work done by each member inside and outside the household. Compare the size, composition, and division of labor of your household with that of your grandparents and parents. How have the size, composition, and division of labor of your household changed over three generations? Why have these changes occurred?

Web Resources

Companion Website for This Book

http://sociology.wadsworth.com

Begin by clicking on the Student Resources section of the website. Choose "Introduction to Sociology" and then the Brym and Lie book cover. Next, select the chapter you are currently studying from the pull-down menu. From the Student Resources page you will have easy access to InfoTrac® College Edition, MicroCase Online exercises, additional web links, and many other resources to aid you in your study of sociology, including practice tests for each chapter.

Recommended Websites

"Marriage and Family Processes" at http://www.trinity.edu/mkearl/family.html contains a wide range of valuable resources on family sociology.

You can find online tests and quizzes concerning love and relationships at http://d4.dir.dcn.yahoo.com/society_and_culture/relationships/quizzes_and_tests.

"Kinship and Social Organization" at http://www.umanitoba.ca/anthropology/kintitle.html is an online interactive tutorial that teaches you about variations in patterns of descent, marriage, and residence using five case studies.

Visit these sites for statistics on interracial families: http://www.census.gov/population/www/socdemo/interrace.html, divorce: http://www.cdc.gov/nchs/fastats/divorce.htm, and same-sex marriage and civil union: http://www.religioustolerance.org/hom_marr.htm.

CHAPTER 16 Religion

Charles O'Rear/Corbis

In this chapter, you will learn that:

- The structure of society and one's place in it influence one's religious beliefs and practices.
- Under some circumstances, religion creates societal cohesion, whereas under other circumstances it promotes social conflict. When religion creates societal cohesion, it also reinforces social inequality.
- Religion governs fewer aspects of many people's lives than in the past. However, a religious revival has taken place in the United States and other parts of the world in recent decades, and many people still adhere to religious beliefs and practices.
- Historical information suggests that the major world religions were movements of moral and social improvement that arose in times of great adversity and were led by charismatic figures. As they consolidated, they became more conservative.
- Adults who were brought up in religious families attend religious services more frequently than adults who were brought up in nonreligious families. Attendance also increases with age and varies by race.

Introduction

Personal Anecdote

Robert Brym started writing the first draft of this chapter just after coming home from a funeral. "Roy was a fitness nut," says Robert, "and cycling was his sport. One perfect summer day, he was out training with his team. I wouldn't be surprised if the sunshine and vigorous exercise turned his thoughts to his good fortune. At 41, he was a senior executive in a medium-size mutual fund firm. His boss, who treated him like a son, was grooming him for the presidency of the company. Roy had three vivacious children, ranging in age from 1 to 10, and a beautiful, generous, and highly intelligent wife. He was active in community volunteer work and everyone who knew him admired him. But on this particular summer day, he suddenly didn't feel well. He dropped back from the pack. He then suffered a massive heart attack. Within minutes, he was dead.

"During *shiva,* the ritual week of mourning following the death of a Jew, hundreds of people gathered in the family's home and on the front lawn. I never felt such anguish before. When we heard the steady, slow, clear voice of Roy's 10-year-old son solemnly intoning the mourner's prayer, we all wept. And we asked ourselves and each other the inevitable question: Why?"

In 1902, the great American psychologist William James observed that this question lies at the root of all religious belief. Religion is the common human response to the fact that we all stand at the edge of an abyss. Religion helps us cope with the terrifying

Sociology Now™

Reviewing is as easy as ❶, ❷, ❸.

Use SociologyNow to help you make the grade on your next exam. When you are finished reading this chapter, go to the Chapter Summary for instructions on how to make SociologyNow work for you.

fact that we must die (James, 1976 [1902]: 123, 139). It offers us immortality, the promise of better times to come, and the security of benevolent spirits who look over us. It provides meaning and purpose in a world that might otherwise seem cruel and senseless.

The motivation for religion may be psychological, as James argued. However, the content and intensity of our religious beliefs, and the form and frequency of our religious practices, are influenced by the structure of society and our place in it. In other words, the religious impulse takes literally thousands of forms. It is the task of the sociologist of religion to account for this variation. Why does one religion predominate here, another there? Why is religious belief more fervent at one time than another? Under what circumstances does religion act as a source of social stability and under what circumstances does it act as a force for social change? Are we becoming more or less religious? These are all questions that have occupied the sociologist of religion, and we will touch on all of them here.

Note that we will have nothing to say about the truth of religion in general or the value of any religious belief or practice in particular. These are questions of faith, not science; and faith, as the New Testament reminds us, "is the substance of things hoped for, the evidence of things not seen" (Hebrews 11:1). Sociology is a social science, and sociological truth is therefore based on theoretical interpretations of things that *are* seen. Questions of faith lie outside the province of sociology.

The cover of *Time* magazine once proclaimed "God is dead." As a sociological observation, the assertion is preposterous. In 2000, 66 percent of respondents who answered a General Social Survey (GSS) question on the subject had no doubt that God exists. Another 27 percent said they believe in God or some higher power at least some of the time. Only 4 percent said they don't know whether God exists. And a mere 3 percent said they don't believe in God (National Opinion Research Center, 2004). By this measure (and by other measures we will examine later), God is still very much alive in America. Nonetheless, as we will show, the scope of religious authority has declined in the United States and many other parts of the world. That is, religion governs fewer aspects of life than it used to. Some Americans still look to religion to deal with all of life's problems. But more and more Americans expect that religion can help them deal with only a restricted range of spiritual issues. Other institutions—medicine, psychiatry, criminal justice, education, and so forth—have grown in importance as the scope of religious authority has declined.

Sociology Now™

Learn more about **Religion and Society** by going through the Role of Religion Video Exercise.

Classical Approaches in the Sociology of Religion

Durkheim: A Functionalist Approach

Somebody once said that Super Bowl Sunday is second only to Christmas as a religious holiday in the United States. Do you agree with that opinion? Before making up your mind, consider the following facts. The largest TV audience in history was recorded in 1996, when 138.5 million Americans watched Super Bowl XXX, and the second largest was recorded in 2003, when 137.7 million watched Super Bowl XXXVII. It is clear that few events attract the attention and enthusiasm of Americans as much as the annual football classic ("Super Bowl Ratings," 2003).

Apart from drawing a huge audience, the Super Bowl generates a sense of what Durkheim would have called "collective effervescence." That is, the Super Bowl excites us by making us feel part of something larger than us: the New England Patriots, the Philadelphia Eagles, the institution of American football, the competitive spirit of the United States itself. For several hours each year, Super Bowl enthusiasts transcend their everyday lives and experience intense enjoyment by sharing the sentiments and values of a larger collectivity. In their fervor, they banish thoughts of their own mortality. They gain

Michael Newman/PhotoEdit

From a Durkheimian point of view, Super Bowl Sunday can be considered a religious holiday.

a glimpse of eternity as they immerse themselves in institutions that will outlast them and athletic feats that people will remember for generations to come.

So, do you think the Super Bowl is a religious event? There is no god of the Super Bowl (although some people wanted to elevate St. Louis's Cinderella quarterback Kurt Warner to that position in 2000 after his game-winning touchdown pass late in the game). Nonetheless, the Super Bowl meets Durkheim's definition of a religious experience. Durkheim said that when people live together, they come to share common sentiments and values. These common sentiments and values form a **collective conscience** that is larger than any individual. On occasion, we experience the collective conscience directly. This causes us to distinguish the secular, everyday world of the **profane** from the religious, transcendent world of the **sacred.** We designate certain objects as symbolizing the sacred. Durkheim called these objects **totems.** We invent certain public practices to connect us with the sacred. Durkheim referred to these practices as **rituals.** The effect (or function) of rituals and of religion as a whole is to reinforce social solidarity, said Durkheim. Durkheim would have found support for his theory in research showing that the suicide rate dips during the two days preceding Super Bowl Sunday and on Super Bowl Sunday itself, just as it does for the last day of the World Series, the Fourth of July, Thanksgiving Day, and other collective celebrations (Curtis, Loy, and Karnilowicz, 1986). This pattern is consistent with Durkheim's theory of suicide, which predicts a lower suicide rate when social solidarity increases (Chapter 1, "A Sociological Compass").

Durkheim would consider the Super Bowl trophy and the team logos to be totems. The insignias represent groups we identify with. The trophy signifies the qualities that professional football stands for: competitiveness, sportsmanship, excellence, and the value of teamwork. The football game itself is a public ritual that is enacted according to strict rules and conventions. We suspend our everyday lives as we watch the ritual being enacted. The ritual heightens our experience of belonging to certain groups, increases our respect for certain institutions, and strengthens our belief in certain ideas. These groups, institutions, and ideas all transcend us. Thus, the game is a sacred event in Durkheim's terms. It cements society in the way Durkheim said all religions do (Durkheim, 1976 [1915]). Do you agree with this Durkheimian interpretation of the Super Bowl? Why or why not? Do you see any parallels between the Durkheimian analysis of the Super Bowl and sports in your community or college?

- The **collective conscience** is composed of the common sentiments and values that people share as a result of living together.
- The **profane** refers to the secular, everyday world.
- The **sacred** refers to the religious, transcendent world.
- **Totems** are objects that symbolize the sacred.
- **Rituals** are public practices designed to connect people to the sacred.

Religion, Conflict Theory, and Feminist Theory

Durkheim's theory of religion is a functionalist account. It clearly offers some useful insights into the role of religion in society. However, critics lodge two main criticisms against it. First, it overemphasizes religion's role in maintaining social cohesion. In reality, religion often incites social conflict. Second, when religion does increase social cohesion, it often reinforces social inequality. Durkheim ignores this issue too.

Religion and Social Inequality

Consider first the role of religion in maintaining inequality. It was Marx who first stressed how religion often tranquilizes the underprivileged into accepting their lot in life. He called religion "the opium of the people" (Marx, 1970 [1843]: 131).

We can draw evidence for Marx's interpretation from many times, places, and institutions. For example, all the major world religions have traditionally placed women in a subordinate position. Consider these scriptural examples:

- Corinthians in the New Testament emphasizes that "women should keep silence in the churches. For they are not permitted to speak, but should be subordinate, as even the law says. If there is anything they desire to know, let them ask their husbands at home. For it is shameful for a woman to speak in church."
- The Sidur, the Jewish prayer book, includes this morning prayer: "Blessed are you, Lord our God, King of the Universe, who did not make me a woman."
- The Qur'an, the holy book of Islam, contains a Book of Women in which it is written that "righteous women are devoutly obedient. . . . As to those women on whose part you fear disloyalty and ill-conduct, admonish them, refuse to share their beds, beat them."

Catholic priests and Muslim mullahs must be men, as must Jewish rabbis in the Conservative and Orthodox denominations. Women have been allowed to serve as Protestant ministers only since the mid-19th century and as rabbis in the more liberal branches of Judaism since the 1970s. God is usually viewed as a man. In the 1980s, the GSS asked Americans: "When you think about God, how likely are each of these images to come to your mind?" Among the options offered were "father" and "mother." Respondents were almost twice as likely to think of God as a father than as a mother (National Opinion Research Center, 2004).

Religion and Class Inequality

If religion has traditionally supported gender inequality, it has also traditionally supported class inequality. In medieval and early modern Europe, Christianity promoted the view that the Almighty ordains class inequality. In the words of an Anglican verse:

The rich man at his castle,
The poor man at his gate.
God made them high or lowly
And ordered their estate.

Nor were Western Christians alone in justifying class hierarchy on religious grounds. In Russian and other Slavic languages, the words for rich (*bogati*) and God (*bog*) have the same root. This suggests that wealth is God-given and perhaps even that it makes the wealthy godlike. The Hindu scriptures say that the highest caste sprang from the lips of the supreme creator, the next highest caste from his shoulders, the next highest from his thighs, and the lowest, "polluted" caste from his feet. And the Qur'an, the holy book of Islam, says that social inequality is due to the will of Allah (Ossowski, 1963: 19–20).

In the United States today, most people do not think of social hierarchy in such rigid terms—quite the opposite. Most people celebrate the alleged *absence* of social hierarchy. This is part of what sociologist Robert Bellah calls our **civil religion,** a set of quasi-religious beliefs and practices that bind the population and justifies our way of life

A **civil religion** is a set of quasi-religious beliefs and practices that bind a population together and justify its way of life.

(Bellah, 1975). When we think of America as a land of golden opportunity, a country in which everyone can realize the American Dream (regardless of race, creed, or color), a place in which individualism and free enterprise ensure the maximum good for the maximum number, we are giving voice to America's civil religion. The National Anthem, the Stars and Stripes, and great public events like the Super Bowl help to make us feel at ease with our way of life, just as the Anglican verse helped the British feel comfortable with their stratification system hundreds of years ago. Paradoxically, however, our civil religion may also help to divert attention from the many inequalities that persist in American society. Strong belief in the existence of equal opportunity, for instance, may lead people to overlook the lack of opportunity that remains in our society (Chapter 8, "Stratification: United States and Global Perspectives"). In this manner, America's civil religion functions much like the old Anglican verse previously cited, although its content is markedly different.

Flip Schulke/Corbis

Black churches formed the breeding ground of the Civil Rights movement in the 1950s and 1960s.

Religion and Social Conflict

We can also find plenty of examples to illustrate religion's role in facilitating and promoting conflict. One case that springs immediately to mind is that of the African American community. In the South in the 1940s, whites sometimes allowed African Americans to sit at the back of their churches. More often, African Americans had to worship in separate churches of their own. These separate black churches formed the breeding ground of the Civil Rights movement in the 1950s and 1960s (Morris, 1984). Their impact was both organizational and inspirational. Organizationally, black churches supplied the ministers who formed the leadership of the Civil Rights movement. They also supplied the congregations within which marches, boycotts, sit-ins, and other forms of protest were coordinated. In addition, ideas from Christian doctrine inspired the protesters. Among the most powerful of these was the notion that African Americans, like the Jews in Egypt, were slaves who would be freed. It was, after all, Michael—regarded by Christians as the patron saint of the Jews—who rowed the boat ashore. Some white segregationists reacted strongly against efforts at integration, often meeting the peaceful protesters with deadly violence. But the South was never the same again. Religion had helped to promote the conflict needed to make the South a more egalitarian and racially integrated place.

In sum, religion can maintain social order under some circumstances, as Durkheim said. When it does, however, it often reinforces social inequality. Moreover, under other circumstances religion can promote social conflict.

Weber and the Problem of Social Change: A Symbolic Interactionist Interpretation

If Durkheim highlighted the way religion contributes to social order, Max Weber stressed the way religion can contribute to social change. Weber captured the core of his argument in a memorable image: If history is like a train, pushed along its tracks by economic and political interests, then religious ideas are like railroad switches, determining exactly which tracks the train will follow (Weber, 1946: 280).

Weber's most famous illustration of his thesis is his short book *The Protestant Ethic and Spirit of Capitalism.* Like Marx, Weber was interested in explaining the rise of modern capitalism. And, again like Marx, he was prepared to recognize the "fundamental importance of the economic factor" in his explanation (Weber, 1958 [1904–1905]: 26). But Weber was also bent on proving the one-sidedness of any *exclusively* economic interpretation.

Wendy Chan/The Image Bank/Getty Images

The port of Singapore. Some scholars argue that Confucianism in East Asia acted much like Protestantism in 19th-century Europe, invigorating rapid economic growth by virtue of its strong work ethic. This not only ignores the fact that Weber himself regarded Confucianism as a brake on economic growth in Asia, but also plays down the economic and political forces that stimulated economic development in the region. This is the kind of one-sided explanation that Weber warned against.

Weber made his case by first noting that the economic conditions Marx said were necessary for capitalist development existed in Catholic France during the reign of Louis XIV. Yet the wealth generated in France by international trade and commerce tended to be consumed by war and the luxurious lifestyle of the aristocracy rather than invested in the growth of capitalist enterprise. For Weber, what prompted vigorous capitalist development in non-Catholic Europe and North America was a combination of (a) favorable economic conditions such as those discussed by Marx and (b) the spread of certain moral values by the Protestant reformers of the 16th century and their followers.

For specifically religious reasons, wrote Weber, followers of the Protestant theologian John Calvin stressed the need to engage in intense worldly activity, to display industry, punctuality, and frugality in their everyday life. In the view of men like John Wesley and Benjamin Franklin, people could reduce their religious doubts and assure a state of grace by working diligently and living simply. Many Protestants took up this idea. Weber called it the Protestant ethic (Weber, 1958 [1904–1905]: 183).

According to Weber, the Protestant ethic had wholly unexpected economic consequences. Where it took root, and where economic conditions were favorable, early capitalist enterprise grew most robustly. Weber made his case even more persuasive by comparing Protestant Western Europe and North America with India and China. In Weber's view, Protestantism was constructed on the foundation of two relatively rational religions: Judaism and Catholicism. These religions were rational in two senses. First, their followers abstained from magic. Second, they engaged in legalistic interpretation of the holy writ. In contrast, said Weber, Buddhism in India and Confucianism in China had strong magical and otherworldly components. This hindered worldly success in competition and capital accumulation. As a result, capitalism developed slowly in Asia, said Weber (Weber, 1963).

In application, two problems have confronted Weber's argument. First, the correlation between the Protestant work ethic and the strength of capitalist development is weaker than Weber thought. In some places, Catholicism has coexisted with vigorous capitalist growth and Protestantism with relative economic stagnation (Samuelsson, 1961 [1957]).

Second, Weber's followers have not always applied the Protestant ethic thesis as carefully as Weber did. For example, since the 1960s, the economies of Taiwan, South Korea, Hong Kong, and Singapore have grown quickly. Some scholars argue that Confucianism in East Asia acted much like Protestantism in 19th-century Europe, invigorating rapid economic growth by virtue of its strong work ethic (Lie, 1998). This not only ignores the fact that Weber himself regarded Confucianism as a brake on economic growth in Asia, it also plays down the economic and political forces that stimulated economic development in the region. This is just the kind of one-sided explanation that Weber warned against (Chapter 9, "Globalization, Inequality, and Development").

Despite these problems, Weber's treatment of the religious factor underlying social change is a useful corrective to Durkheim's emphasis on religion as a source of social stability. Along with Durkheim's work, Weber's contribution stands as one of the most influential insights into the influence of religion on society.

The Rise, Decline, and Partial Revival of Religion

Secularization

In 1651, the British political philosopher Thomas Hobbes described life as "poore, nasty, brutish, and short" (Hobbes, 1968 [1651]: 150). His description fit the recent past. The standard of living in medieval and early modern Europe was abysmally low. On average, a person lived only about 35 years. The forces of nature and human affairs seemed entirely unpredictable. In this context, magic was popular. It offered easy answers to mysterious, painful, and capricious events.

As material conditions improved, popular belief in magic, astrology, and witchcraft gradually lost ground (Thomas, 1971). Christianity substantially replaced them. The better and more predictable times made Europeans more open to the teachings of organized religion. In addition, the Church campaigned vigorously to stamp out opposing belief systems and practices. The persecution of witches in this era was partly an effort to eliminate competition and establish a Christian monopoly over spiritual life.

The Church succeeded in its efforts. In medieval and early modern Europe, Christianity became a powerful presence in religious affairs, music, art, architecture, literature, and philosophy. Popes and saints were the rock musicians and movie stars of their day. The Church was the center of life in both its spiritual and worldly dimensions.

Bettmann/Corbis

The persecution of witches in the early modern era was partly an effort to eliminate competition and establish a Christian monopoly over spiritual life.
Burning of Witches by Inquisition in a German Marketplace. After a drawing by H. Grobert.

Church authority was supreme in marriage, education, morality, economic affairs, politics, and so forth. European countries proclaimed official state religions. They persecuted members of religious minorities.

In contrast, a few hundred years later, Max Weber remarked on how the world had become thoroughly "disenchanted." By the turn of the 20th century, he said, scientific and other forms of rationalism were replacing religious authority. His observations formed the basis of what came to be known as the **secularization thesis,** undoubtedly the most widely accepted argument in the sociology of religion until the 1990s. According to the secularization thesis, religious institutions, actions, and consciousness are unlikely to disappear, but they are certainly on the decline worldwide (Tschannen, 1991).

The **secularization thesis** says that religious institutions, actions, and consciousness are on the decline worldwide.

Religious Revival

Despite the consensus about secularization that was still evident in the 1980s, many sociologists modified their judgments in the 1990s. There were two reasons for this. In the first place, accumulated survey evidence showed that religion was not in an advanced state of decay. Actually, in many places, such as the United States, it was in robust health (Greeley, 1989).

Consider in this connection Figures 16.1 and 16.2. Figure 16.1 uses data from a large 2002 survey to compare the United States with 43 other countries. It divides the countries into three groups: (1) those listed by the United Nations as enjoying "high human development," (2) former Communist countries, and (3) all other countries (i.e., less developed countries that were never under Communist rule). The graph shows that less developed countries that were never under Communist rule have a relatively high percentage of citizens who consider religion important in their lives (between 57 percent and 97 percent). For the highly developed countries, the percentages are considerably lower (between 11 percent and 39 percent). They are lower still for the former Communist countries (between 11 percent and 36 percent). This suggests that religiosity is negatively correlated with level of economic development. It also suggests that Communist governments, which promoted atheism as state policy, did much to lower the level of religiosity in their

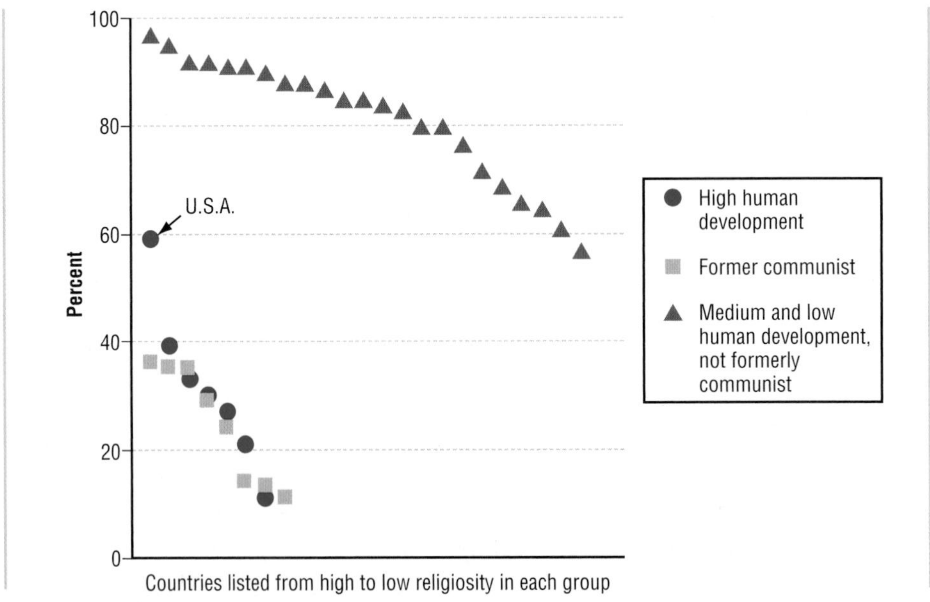

Figure 16.1
Percentage of People Who Think Religion Is Very Important, 44 Countries, 2002 (*N* = 38,000)

Note: Poland is a former Communist country and the UN ranks it 37th in its list of 53 countries in the "high human development" group. It is classified here as a former Communist country.
Sources: Pew Research Center (2002); United Nations (2002).

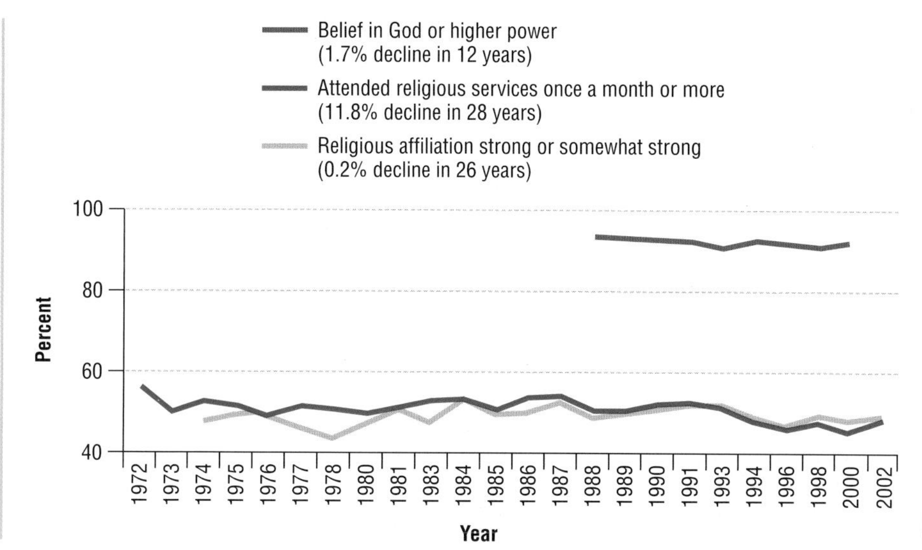

Figure 16.2
The Social Condition of Religion, United States, 1972–2002 (in percent; *N* = 43,698)

Source: National Opinion Research Center (2004).

countries. There is just one dramatic anomaly among the 44 countries: the United States. The United States has an unusually high level of religiosity for a rich country. Fully 59 percent of American adults considered religion important in their lives in 2002, compared with 33 percent for the United Kingdom and 30 percent for Canada, the next two countries in the "highly developed" group.

Figure 16.2 uses GSS data to show the resilience of religion in America over time. Whether we focus on the percentage of Americans who believe in God or some higher power or the percentage who claim that their religious affiliation is "strong" or "somewhat strong," we see little change over the past three decades. In 2000, 93 percent of Americans said they believed in God or some higher power. In 2002, 47 percent attended religious services once a month or more. Forty-eight percent felt at least somewhat strongly affiliated to their religion. True, the percentage of people who frequently attended religious services fell nearly 10 percent between 1972 and 2002. However, the percentage of people who believed in God or some higher power and the percentage who felt at least somewhat strongly affiliated to their religion barely changed. We conclude that although frequently attending religious services is less popular than it once was, religion is alive and well in the United States. This conclusion is reinforced when we compare the United States with other countries.

Religious Fundamentalism in the United States

The second reason many sociologists have modified their views about secularization is that an intensification of religious belief and practice has taken place among some people in recent decades. For example, since the 1960s, fundamentalist religious organizations have increased their membership in the United States, especially among Protestants (Finke and Starke, 1992).

Fundamentalists interpret their scriptures literally, seek to establish a direct, personal relationship with the higher being(s) they worship, are relatively intolerant of nonfundamentalists, and often support conservative social issues (Hunter, 1991) (Box 16.1). Attitudes toward abortion provide an example of American fundamentalist support for a conservative social issue. In 2000 only 23 percent of Americans identifying themselves as fundamentalists agreed with the statement that abortion is acceptable if "the woman wants it for any reason." In contrast, 43 percent of Americans who identify themselves as belonging to moderate or liberal denominations agreed with that statement (National Opinion Research Center, 2004). Similar divisions are evident with respect to other moral issues such as homosexuality and pornography.

Fundamentalists interpret their scriptures literally, seek to establish a direct, personal relationship with the higher being(s) they worship, and are relatively intolerant of nonfundamentalists.

BOX 16.1 Sociology at the Movies

Harry Potter and The Sorcerer's Stone (2001)

Harry Potter and the Sorcerer's Stone introduces us to Harry Potter (played by Daniel Radcliffe) who, at 1 year old, was orphaned when the most evil wizard of all, Lord Voldemort, murdered Harry's parents and tried to kill Harry. The infant Harry, who bears a thunderbolt-shaped scar on his forehead as the result of Lord Voldemort's attack, is deposited by the gentle giant Hagrid on the doorstep of his unwelcoming relatives, the Dursleys. He lives a miserable, lonely existence, ignored by his aunt and uncle and bullied by their fat and pampered son.

Shortly before his 11th birthday, Harry's life is turned upside down. He is summoned by a blizzard of letters to Hogwarts School of Witchcraft and Wizardry. The school, housed in a 1000-year-old castle and headed by the renowned Professor Dumbledore, provides select students with a seven-year program of instruction. At Hogwarts, Harry learns that he is a wizard. He makes friends but danger lurks, for Voldemort stalks Harry, determined that he will yet accomplish what he earlier failed to do—kill the young wizard. Harry battles Lord Voldemort and destroys him. Or does he?

When *Harry Potter and the Sorcerer's Stone* opened in North American theaters in November 2001, it enjoyed enormous box-office success. Children were prominent in the lineups, many of them outfitted as if for Halloween. Some of the children were there on organized school field trips after studying the book at school. Although most people viewed *Harry Potter and the Sorcerer's Stone* as harmless, others saw things differently. They denounced both the film and the book on which it was based as "demonic." Many of them were conservative Protestants who claimed that the book glorifies witchcraft, makes "evil look innocent," and subtly draws "children into an unhealthy interest in a darker world that is occultic and dangerous to physical, psychological and spiritual well-being" (Shaw, 2001). According to one Christian fundamentalist ministry, "the effect of [the movie] is undoubtedly to raise curiosity about magic and wizardry. And any curiosity raised on this front presents a danger that the world will satisfy it with falsehood before the church or the family can satisfy it with truth" (Ontario Consultants on Religious Tolerance, 2002). Scenes from *Harry Potter and the Sorcerer's Stone* were scrutinized for possible demonic messages. A similarity was proclaimed between the lightning bolt that appears on Harry Potter's forehead and the symbol adopted by Hitler's SS. The American Library Association reported that the *Harry Potter* series was attacked in 13 states, making the books the most challenged novels of 1999.

The Everett Collection

Daniel Radcliffe in *Harry Potter and The Sorcerer's Stone* (2001).

Do you agree with the decision to ban the *Harry Potter* books and condemn the movie? In general, do you think that religious organizations should be able to get schools to censor books and movies? If so, do you draw the line at some types of influence? Would it be acceptable if a white religious organization got a predominantly white school to ban the works of Toni Morrison and Maya Angelou because they say derogatory things about whites? Would it be acceptable if a Jewish religious organization got a predominantly Jewish school to ban Shakespeare's *The Merchant of Venice* because it portrays Jews in an unflattering way? Would it be acceptable if a religious organization strongly influenced by feminism convinced students in an all-girls school to ban the works of Ernest Hemingway ("too sexist") or if an antifeminist religious organization convinced students in an all-boys school to ban the writings of Margaret Atwood ("too anti-male")? In general, should religious organizations be allowed to influence schools to censor, or should censoring by religious organizations be banned?

Such social issues often spill over into political struggles, which is why religion in American politics is resurgent (Bruce, 1988) (Box 16.2). In 1980 and 1984, Jerry Falwell's conservative Moral Majority supported President Reagan's successful bids for the presidency. In 1988, conservative Christian Pat Robertson ran for the Republican presidential nomination, as did conservative Christian Pat Buchanan in 1992. The con-

Box 16.2

YOU AND THE SOCIAL WORLD

Religion, Politics, and You

Significant shifts in public opinion about the relationship between religion and politics have occurred in recent years. To learn about them, please read the brief summaries of nationwide polls conducted in 2000 and 2003 at http://people-press.org/reports/display.php3?ReportID532 and http://pewforum.org/docs/index.php?DocID526. Then write a 500-word essay summarizing the polls, outlining the degree to which your own attitudes mirror nationwide trends, and offering a sociological explanation as to why your attitudes mirror or differ from those of the public at large.

servative Christian Coalition continues to lobby hard in Washington today, and Christian conservatives are a major force in the Republican party. They accounted for about 40 percent of the votes that President Bush received in the 2000 presidential election and an even larger percentage of the votes he received in 2004 (Smith, 2000; Bumiller, 2003; Cooperman and Edsall, 2004). Many leading political figures, including the President himself, have expressed strong commitment to fundamentalist Christian principles.

Religious Fundamentalism Worldwide

The American experience is by no means unique. Fundamentalism has spread throughout the world since the 1970s. It is typically driven by politics. Hindu nationalists now form the government in India. Jewish fundamentalists were always important players in Israeli political life, often holding the balance of power in Israeli governments, but they have become even more influential in recent years (Kimmerling, 2001: 173–207). A revival of Muslim fundamentalism began in Iran in the 1970s. Led by the Muslim cleric the Ayatollah Khomeini, it was a movement of opposition to the repressive, American-backed regime of Shah Reza Pahlavi, which fell in 1979. Muslim fundamentalism then swept much of the Middle East, Africa, and parts of Asia. In Iran, Afghanistan, and Sudan, Muslim fundamentalists took power. Other predominantly Muslim countries' governments have begun to introduce elements of Islamic religious law (*shari'a*), either from conviction or as a precaution against restive populations (Lewis, 2002: 106). Religious fundamentalism has thus become a worldwide political phenomenon. In not a few cases it has taken extreme forms and involved violence as a means of establishing fundamentalist ideas and institutions (Juergensmeyer, 2000).

Fundamentalism and Extremist Politics in the Arab World

Personal Anecdote

In 1972, Robert Brym was doing his B.A. at the Hebrew University of Jerusalem. One morning at the end of May, he switched on the radio to discover that a massacre had taken place at Lod (now Ben-Gurion) International Airport outside Tel Aviv, just 26 miles from his apartment. Three Japanese men dressed in business suits had arrived on Air France flight 132 from Paris. They were members of the Japanese Red Army, a small, shadowy terrorist group with links to the General Command of the Popular Front for the Liberation of Palestine. Both groups wanted to help "liberate" Israel from Jewish rule.

After they picked up their bags, the three men pulled out automatic rifles and started firing indiscriminately. Before pausing to slip fresh clips into their rifles, they lobbed hand grenades into the crowd at the ticket counters. One man ran onto the tarmac, shot some disembarking passengers, and then blew himself up. This was the

Since the 1960s, fundamentalist religious organizations have rapidly increased their membership, especially among Protestants.

A. Ramey/PhotoEdit

first suicide attack in modern Middle East history.[1] Security guards shot a second terrorist and arrested the third, Kozo Okamoto. When the firing stopped, 26 people lay dead. Half were non-Jews. In addition to the two terrorists, 11 Catholics were murdered. They were Puerto Rican tourists who had just arrived on a pilgrimage to the Holy Land.

Both the Japanese Red Army and the General Command of the Popular Front for the Liberation of Palestine were strictly nonreligious organizations. Their members were atheists who quoted Marx and Lenin, not Jesus or Muhammad. Yet something unexpected happened to Kozo Okamoto, the sole surviving terrorist of the Lod massacre. Israel sentenced him to life in prison but freed him in 1985 in a prisoner exchange with Palestinian forces. Okamoto wound up living in Lebanon's Bekaa Valley, the main base of the Iranian-backed Hizbollah fundamentalist terrorist organization. At some point in the late 1980s or early 1990s, he was swept up in the Middle East's Islamic revival. Okamoto the militant atheist converted to Islamic fundamentalism.

Kozo Okamoto's life tells us something important and not at all obvious about religious fundamentalism and politics in the Middle East and elsewhere. Okamoto was involved in extremist politics first and became a religious fundamentalist later. Religious fundamentalism became a useful way for him to articulate and implement his political views. This is quite typical. Religious fundamentalism often provides a convenient vehicle for framing political extremism, enhancing its appeal, legitimizing it, and providing a foundation for the solidarity of political groups (Pape, 2003; Sherkat and Ellison, 1999: 370).

Many people fail to see the political underpinnings of religious fundamentalism. They regard religious fundamentalism as an independent variable and extremist politics as a dependent variable. In this view, some people happen to become religious fanatics and then their fanaticism commands them to go out and kill their opponents. For example, here is what President George W. Bush said when he addressed Congress on September 20, 2001, just nine days after al Qaeda's attacks on the World Trade Center and the Pentagon. Al Qaeda's goal, said President Bush, is "imposing its radical beliefs on people

[1]Notice the cultural continuity. Among the first suicide attackers in the twentieth century were the Japanese kamikaze pilots of World War II.

port. John Walker Lindh is a white American citizen who fought with the Taliban. There are many Filipino supporters of al Qaeda in the southern Philippines. Knowing that religious profiling is used to select people for scrutiny may encourage al Qaeda to elude detection by using more non-Muslims as terrorists in the United States.

- Religious profiling may make members of the Arab and Muslim communities reluctant to come forward with information, and it may create resentment and bitterness on the part of an overwhelmingly loyal community, thus hampering the investigation of terrorist activity. Anecdotal evidence suggests that this is precisely what is happening (Elgrably, 2002).

Where do you stand on the issue of religious profiling? Do you think that Muslim appearance should be used as a criterion to select people for special scrutiny? If so, how do you reconcile your opinion with the argument offered earlier, which led to the conclusion that religious profiling may not achieve what is intended and may in fact achieve the opposite? If you disagree with religious profiling, then what other criteria should be employed to select people for special scrutiny as potential threats to national security?

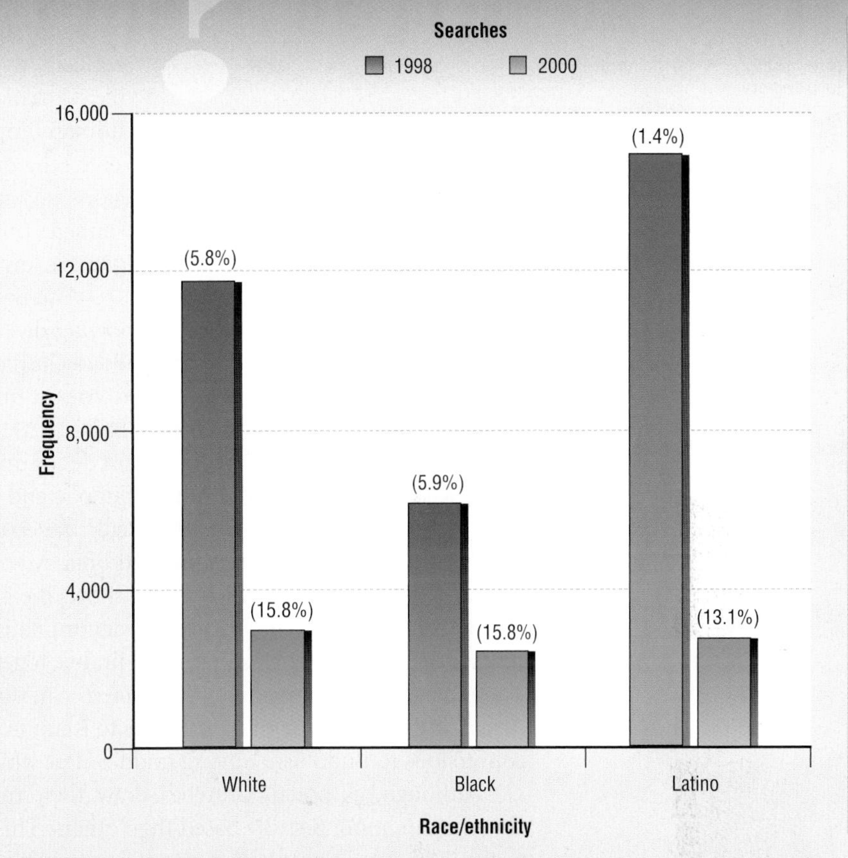

Figure 16.3
U.S. Customs Service Searches by Race and Ethnicity, 1998 and 2000 (hit rates in parentheses)

Source: Lamberth Consulting (2002).

The Structure of Religion in the United States and the World

Types of Religious Organization

The latest edition of the *Encyclopedia of American Religions* lists more than 2100 religious groups that are active in the United States (Melton, 1996 [1978]). Although each of these organizations is undoubtedly unique in some respects, sociologists generally divide religious groups into just three types: churches, sects, and cults (Troeltsch, 1931 [1923]; Stark and Bainbridge, 1979) (Concept Summary 16.1).

Sociology Now™

Learn more about **Types of Religious Organization** by going through the Types of Religious Organization Learning Module.

Church

In the sociological sense of the term, a **church** is any bureaucratic religious organization that has accommodated itself to mainstream society and culture. Because of this accommodation, it may endure for many hundreds if not thousands of years. The bureaucratic nature of a church is evident in the formal training of its leaders, its strict hierarchy of roles, and its clearly drawn rules and regulations. Its integration into mainstream society is evident in its teachings, which are generally abstract and do not challenge worldly

A **church** is a bureaucratic religious organization that has accommodated itself to mainstream society and culture.

Concept Summary 16.1
Church, Sect, and Cult Compared

	Church	Sect	Cult
Integration into society	High	Medium	Low
Bureaucratization	High	Low	Low
Longevity	High	Low	Low
Leaders	Formally trained	Charismatic	Charismatic
Class base	Mixed	Low	Various but segregated

authority. In addition, churches integrate themselves into the mainstream by recruiting members from all classes of society.

Churches take two main forms. First are **ecclesia,** or state-supported churches. For example, Christianity became the state religion in the Roman Empire in the 4th century, and Islam is the state religion in Iran and Sudan today. State religions impose advantages on members and disadvantages on nonmembers. Tolerance of other religions is low in societies with ecclesia.

Alternatively, churches may be pluralistic, allowing diversity within the church and expressing tolerance of nonmembers. Pluralism allows churches to increase their appeal by allowing various streams of belief and practice to coexist under their overarching authority. These subgroups are called **denominations.** Baptists, Methodists, Lutherans, Presbyterians, and Episcopalians form the major Protestant denominations in the United States. The major Catholic denominations are Roman Catholic and Orthodox. The major Jewish denominations are Orthodox, Conservative, Reform, Reconstructionist, and Chasidic. The major Muslim denominations are Sunni and Shia. Many of these denominations are divided into even smaller groups. Figure 16.4 shows the percentage of Americans who belonged to the major religions and Protestant denominations in 2002. (Note also the small percentages of Buddhists, Muslims, and Hindus, together totaling about 4.5 million people. These religions are growing in the United States because of immigration from Asia and Africa, and also because of conversions to Islam in the African American community and conversions to Buddhism among middle-class whites, especially on the West Coast.)

Although, as noted, churches draw their members from all social classes, some churches are more broadly based than others. This is clear from Figure 16.5. According to the 2002 GSS, about half the American population defines itself as lower or working class and half as middle or upper class (Chapter 8, "Stratification: United States and Global

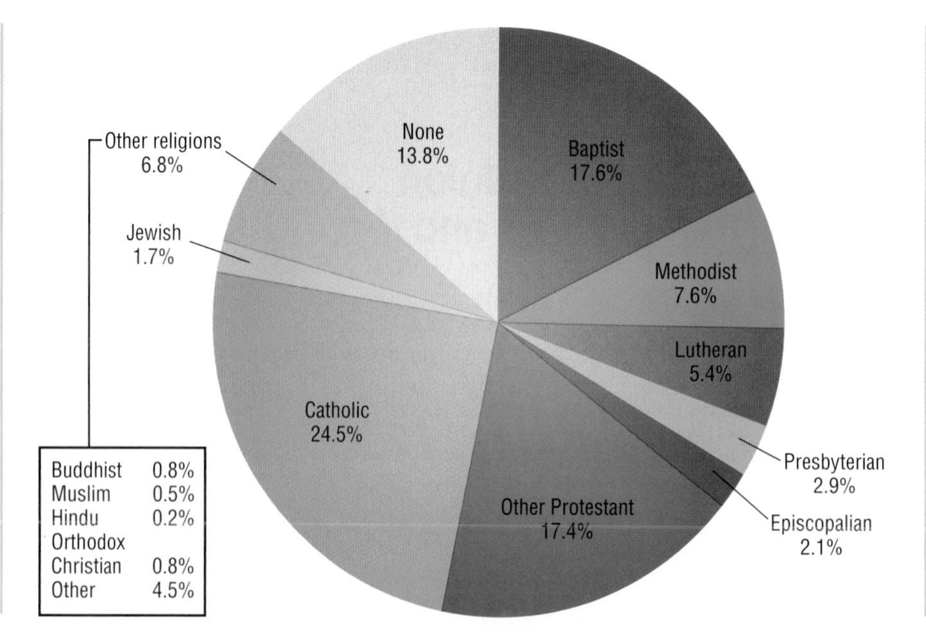

Figure 16.4
Religious Preference, United States, 2002 (in percent; $N = 2765$)

Does not add up to 100% due to rounding.
Source: National Opinion Research Center (2004).

Ecclesia are state-supported churches.

Denominations are the various streams of belief and practice that some churches allow to coexist under their overarching authority.

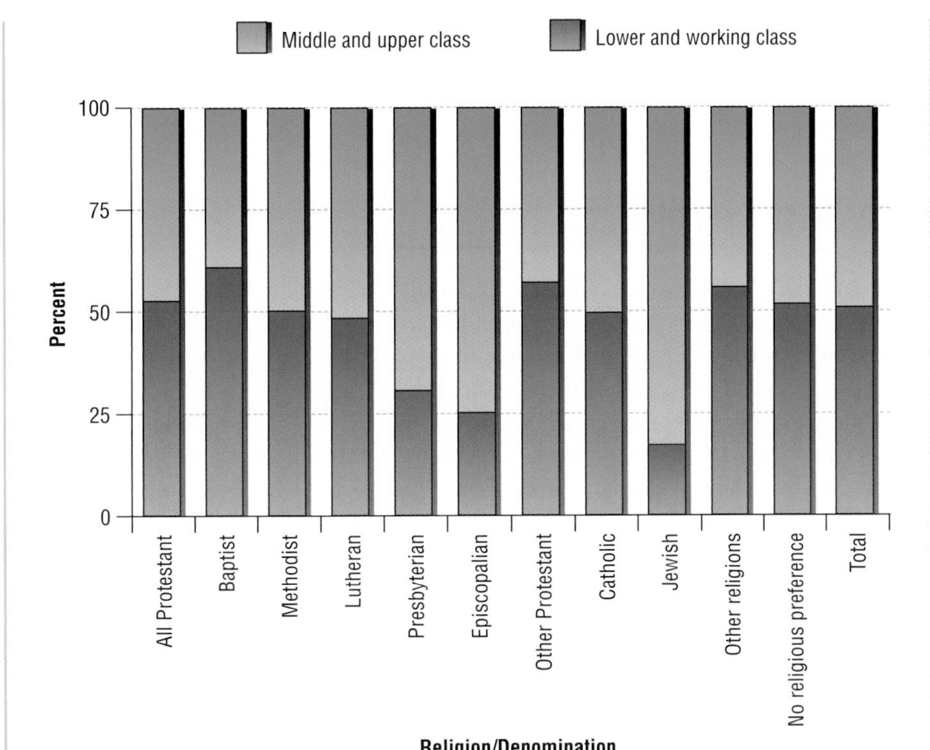

Figure 16.5
Religious Preference by Class, United States, 2002 (in percent; *N* = 2731)

Note: Class is defined subjectively here. Respondents were asked whether they considered themselves lower class, working class, middle class, or upper class.
Source: National Opinion Research Center (2004).

Perspectives"). However, the Baptist denomination is 61 percent lower or working class. At the other extreme, 75 percent of Episcopalians and 83 percent of Jews are middle and upper class.

Sect

Sects often form by breaking away from churches as a result of disagreement about church doctrine. Sometimes, sect members choose to separate themselves geographically, as the Amish do in their small farming communities in Pennsylvania, Ohio, and Indiana. However, even in urban settings, strictly enforced rules concerning dress, diet, prayer, and intimate contact with outsiders can separate sect members from the larger society. Chasidic Jews in New York and other large American cities prove the viability of this isolation strategy. Sects are less integrated into society and less bureaucratized than churches. They are often led by charismatic leaders, men and women who claim to be inspired by supernatural powers and whose followers believe them to be so inspired. These leaders tend to be relatively intolerant of religious opinions other than their own. They tend to recruit like-minded members mainly from lower classes and marginal groups. Worship in sects tends to be highly emotional and based less on abstract principles than on immediate personal experience (Stark, 1985: 314). Many sects are short-lived, but those that do persist tend to bureaucratize and turn into churches. If religious organizations are to enjoy a long life, they require rules, regulations, and a clearly defined hierarchy of roles.

Web Research Project: "Cults" and the Brainwashing Controversy

Cult

Cults are small groups of people deeply committed to a religious vision that rejects mainstream culture and society. Cults are generally led by charismatic individuals. They tend to be class-segregated groups. That is, a cult tends to recruit members from only one segment of the stratification system, high, middle, or low. For example, many American cults today recruit nearly all their members from among the college educated. Some of these cults seek converts almost exclusively on college campuses (Kosmin, 1991). Because they propose a radically new way of life, cults tend to recruit few members and soon disappear. There are, however, exceptions—and some extremely important ones at that. Jesus

- **Sects** usually form by breaking away from churches due to disagreement about church doctrine. Sects are less integrated into society and less bureaucratized than churches. They are often led by charismatic leaders, who tend to be relatively intolerant of religious opinions other than their own.
- **Cults** are small groups of people deeply committed to a religious vision that rejects mainstream culture and society.

and Muhammad were both charismatic leaders of sects. They were so compelling that they and their teachings were able to inspire a large number of followers, including rulers of states. Their cults were thus transformed into churches. We now examine how this came about in the cases of the world's major religions.

Sociology Now™

Learn more about **World Religions** by going through the Religion Map Exercise.

World Religions

There are five major world religions: Judaism, Christianity, Islam, Hinduism, and Buddhism (Figure 16.6). They are *major* in the sense that they have had a big impact on world history and, aside from Judaism, continue to have hundreds of millions of adherents. They are *world* religions in the sense that their adherents live in many countries.

Other religions have many adherents too. For example, some scholars consider Confucianism a major world religion. Certainly, Confucianism has had a tremendous impact on East Asian societies, and we still find East Asians who claim to be Confucian, especially in Singapore. Furthermore, some Confucian rituals such as ancestor worship

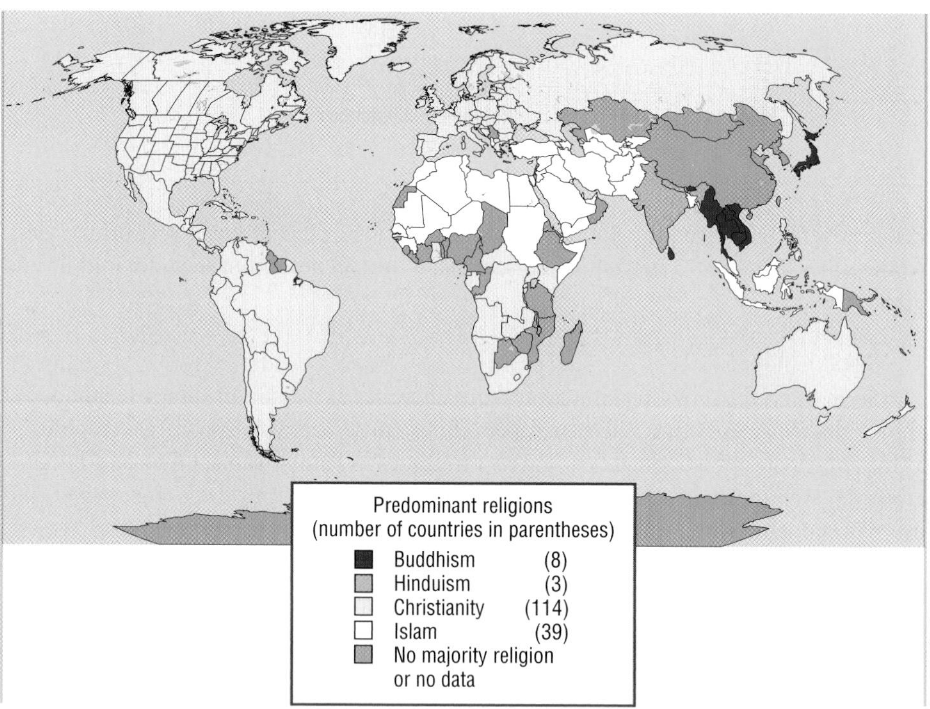

Figure 16.6
The World's Predominant Religions

This map shows the predominant religion in each of the world's countries, defined as the religion to which more than 50 percent of a country's population adhere. Nearly a third of the world's population (1.9 billion people) are Christian, and Christianity is the predominant religion in 114 countries. Christianity is most widespread in North and South America, Europe, Oceania (Australia, New Zealand, etc.), and southern Africa. The second most widespread religion in the world is Islam. Muslims comprise nearly a fifth of the world's population (1.2 billion people). Islam is the predominant religion in 39 countries in the Middle East, central and northern Africa, and parts of Asia. Hindus comprise the world's third largest religion. Although only 3 countries are predominantly Hindu, one of them is India, so nearly 13 percent of the world's population adheres to Hinduism (762 million people). The fourth major world religion is Buddhism, the predominant religion in 8 Asian countries. Six percent of the world's population are Buddhists (354 million people). Sikhism is often called the world's fifth major organized religion. It has 20 million adherents in the state of Punjab in India. Although numerically small on a world scale, Judaism must also be included in any list of major world religions because Christianity and Islam have their roots in Judaism. There are 14 million Jews in the world and Judaism is the predominant religion only in Israel.

Source: Adherents.com (2001).

resemble religious practice. However, we believe that Confucianism is best seen as a worldview or a philosophy of life (Ivanhoe, 2000 [1993]). Few East Asians regard Confucianism as a religion.

The five major world religions we survey are similar in three ways. First, with the exception of Hinduism, charismatic leaders helped to turn them into world religions. Max Weber defined charismatic leaders as men and women who claim to be inspired by supernatural or divine powers and whose followers believe them to be so inspired. Second, and again with the exception of Hinduism, all five world religions had egalitarian and emancipatory messages at their origins. That is, they claimed to stand for equality and freedom. Third, over time, the charismatic leadership of the world religions became routinized. The **routinization of charisma** is Weber's term for the transformation of divine enlightenment into a permanent feature of everyday life. It involves turning religious inspiration into a stable social institution with defined roles, such as interpreters of the divine message, teachers, dues-paying laypeople, and so forth. The routinization of charisma often involves the weakening of the ideals of freedom and equality. Under some conditions it gives way to their opposite: repression and inequality.

Judaism

According to the Bible, the first Jew was Abraham, who lived nearly 4000 years ago in ancient Mesopotamia, now Iraq (Roth, 1961; Gottwald, 1979). Abraham's unique contribution was to assert that there is only one God; before him, people believed in many gods. The Bible says that God promised Abraham abundant offspring and, for his descendants, a land of their own.

Abraham's great-grandson, Joseph, was sold by his brothers into slavery in Egypt. There, Joseph became a trusted adviser of the Pharaoh. However, a later Pharaoh enslaved the Jews, and about 400 years after Joseph's arrival in Egypt the prophet Moses led them out of bondage. The emancipation of the Jews from slavery still stands today as a defining moment in the history of Judaism. It has inspired generations of Jews and non-Jews to believe that God sanctions freedom and equality. Practicing Jews celebrate the emancipation annually during Passover, which has become one of the most important Jewish festivals.

After 40 years of wandering in the desert, during which time Moses received the Ten Commandments, the Jews arrived in Canaan, the land that God had promised Abraham. (Modern-day Israel and ancient Canaan occupy roughly the same territory.) Thereafter, Judaism spread throughout the Middle East and established itself as a major religion during Roman rule. Increased persecution in the Roman period led the religious leaders (rabbis) to settle in Galilee. There, between the second and the sixth centuries C.E., Judaism assumed its contemporary form, characterized by the centrality of the rabbis and the Torah (the five books of Moses, known to Christians as the Old Testament).[2]

The central teachings of Judaism rest above all on belief in one God, Yahweh. However, many commentators argue that the core of Judaism lies less in belief than in the performance of the 613 divine commandments, or *mitzvot,* mentioned in the Torah. The mitzvot (singular, *mitzvah*) include prescriptions for justice, righteousness, and observance: Rest and pray on the Sabbath, honor the old and the wise, do not wrong a stranger in buying or selling, do not seek revenge or hold a grudge, do not eat the meat of animals with hooves, and so forth. Temples or synagogues are common places of worship for Jews, but they are not essential. Worship can take place wherever there is an assembly of ten adult Jewish males (or females in the more liberal branches of Judaism).

Like some other Middle Eastern peoples, the Jews were involved in international trade long before the birth of Christ. They therefore settled in many places outside the area that is now Israel. However, after the Romans destroyed the Jewish temple in Jerusalem in 70 C.E., the Jews dispersed even more widely to the far reaches of Europe, Asia, and North

The **routinization of charisma** is Weber's term for the transformation of the unique gift of divine enlightenment into a permanent feature of everyday life. It involves turning religious inspiration into a stable social institution with defined roles (interpreters of the divine message, teachers, dues-paying laypeople, etc.).

[2]"C.E." stands for "common era" and is now preferred over the ethnocentric "A.D.," which stands for *anno Domini* (Latin for "the year of our Lord"). "B.C.E." stands for "before the common era."

Nathan Benn/Corbis

Even in urban settings, strictly enforced rules concerning dress, diet, prayer, and intimate contact with outsiders can separate sect members from the larger society.

Africa, forming what became known as the Diaspora ("dispersion"). Jews retained their identity in the Diaspora because in premodern times they specialized in mercantile activities that separated them from peasants and landowners, because they tended to be strictly observant, and because they were periodically persecuted, mainly by Christians. (In general, Muslims were more tolerant of Jews until recently.) In 1948, Israel was returned to Jewish sovereignty.

More than 5 million Jews live in the United States—about the same number as in Israel. The considerable diversity among Jews in the United States derives from disagreements that began nearly 400 years ago. In 17th-century Eastern Europe, ecstatic and mystical sects of Chasidim broke away from the staid and bookish Judaism of the time. Today in New York and other large American cities, one can still see Chasidic Jews dressed in their characteristic black robes and broad-brimmed hats (ironically, the garb of 17th-century Polish noblemen). In 19th-century Germany, the Reform movement also broke with traditional Jewish practice. Influenced by Lutheranism, Reform Judaism was a liberal movement that involved a loosening of strict rules of religious observance, prayer in the German language, the integration of men and women in worship, services that followed the kind of decorum associated with Protestant worship, and the introduction of choirs and organs to enhance prayer. The Reform movement transplanted itself to the United States with the immigration of German Jews and is now the most popular Jewish denomination in this country. Orthodox Judaism emerged as a reaction against the liberalizing tendencies of the Reform movement. It involved a return to traditional observance, including strict adherence to dietary rules, the segregation of men and women in prayer, and so forth. Conservative Judaism crystallized in Britain and the United States in the 19th century as an attempt to reconcile what its practitioners regard as the positive elements in Orthodoxy with the dynamism of Reform. Finally, Reconstructionist Judaism is a smaller, 20th-century American offshoot of Conservatism, known chiefly for its liberalism, social activism, and gender-egalitarianism.

Christianity

Observant Jews believe that God promised them a Messiah, a redeemer whose arrival would signal the beginning of an era of eternal peace, prosperity, and righteousness. Jews believe that the Messiah has not yet arrived. Christians believe he has. In the Christian view, the Messiah is Jesus (McManners, 1990; Brown, 1996).

Jesus was a poor Jew, and his early followers were all Jewish. Yet he criticized the Judaism of his time for its external conformity to tradition and ritual at the expense of developing a true relationship to God as demanded by the prophets. Believe in God and love him; love your neighbor—these are the two main lessons of Jesus. What made his teaching novel was his demand that people match outward performance with inner conviction. Thus, it was not enough not to murder. One could not even hate. Nor was it enough not to commit adultery. One could not even lust after a neighbor's wife (Matt. 5: 21–8).

These lessons made Jesus anti-authoritarian and even revolutionary. Criticizing ritual by rote put Jesus at odds with the established Judaism of his time. Admonishing people to love their neighbors impressed upon them the need to emancipate slaves and women. It also challenged people to recognize the essential equality of the beggar and the wealthy merchant in the eyes of God. Roman authorities could hardly be happy with such teachings, which attracted the poor and the dispossessed and were a direct challenge to the legitimacy of the Roman Empire, built as it was on slavery and privilege. Little wonder that the Romans persecuted Jesus and his followers, eventually executing him by crucifixion, a cruel and painful death reserved for slaves and the worst criminals. Christians interpret the death of Jesus as atonement for the sins of humanity.

For at least a century after Jesus' death, the Christians formed a minor Jewish sect. Aided by twelve of Jesus' main followers, the apostles, Christianity gained adherents largely from within the Jewish community. However, because they encountered opposition from Judaism, the Christians soon began to preach their message to non-Jews. This required that they redefine Jesus' message as a correction and fulfillment of Greek and Roman philosophy. This helped Christianity spread, but even as it did, the Roman Empire continued to persecute Christians. Many heresies and schisms also beset the Christians in these first centuries after the death of Jesus. There thus arose the need for stable, recognized leadership and set, recognized holy texts. The clerical rank of bishop and the canon date from this period. They were the routinization of Jesus' charisma in practice.

In 312 C.E., the Roman emperor Constantine I ("the Great") converted to Christianity. He soon turned Christianity into a state religion. It then spread rapidly throughout Europe. By allying itself first with the Roman Empire and then with other earthly powers—European royalty and the landowning class—the Church became the dominant institution, religious or secular, in Europe until the 16th century. It also contributed to gender inequality insofar as women played a marginal role in its affairs. Just as Judaism had been transformed from an emancipatory religion into one that could be criticized by Jesus for ritualistic staleness, so Christianity was transformed from a revolutionary force into a pillar of the existing order.

In the 16th century, a German priest by the name of Martin Luther challenged the Christian establishment by seeking to establish a more personal relationship between the faithful and God. At the time, ordinary people were illiterate, and they had to rely on priests to hear the holy word and have it interpreted for them. However, by insisting that Christians come to know God themselves, as Jesus demanded, Luther called into question the whole Church hierarchy. His protests and his ideas quickly captured the imagination of half of Europe and led to the split of Christianity into the two major branches that persist to this day, Catholicism and Protestantism.

This is by no means the only division within Christianity. In the Middle Ages, Christianity split into Western and Eastern halves, the former centered in Rome, the latter in Constantinople (now Istanbul, Turkey). Various Orthodox churches today derive from the Eastern tradition. Protestantism has been especially prone to splintering because it emphasizes the individual's relationship to God rather than a central authority. Today, there are hundreds of different Protestant churches.

Christians retained the Jewish Bible as the Old Testament, adding the Gospels and letters of the apostles as the New Testament. The Bible is the most important text for Christians. Especially for Protestants, reading the New Testament is an important part of what it means to be a Christian. Traditionally, the most important

Martin Luther (1483–1546). On October 31, 1517, Luther nailed his "Ninety-Five Theses" to a church door. They struck at the root of Church authority and led to Luther's excommunication.

SuperStock

holiday for Christians was Easter, but increasingly Christians have celebrated Christmas, originally a pagan holiday that became especially popular in the 19th century.

Christianity remains the dominant religion in the West. Because of its successful missionary efforts, it can be found virtually everywhere in the world. Yet Christianity remains a truly heterogeneous religion. Some Christians are fundamentalist and conservative, others are mainstream and more liberal, while still others are socialist and even revolutionary. Some support feminists and homosexuals, while others regard them as abominations. Its success is due in part to its ability to encompass diverse and even contradictory currents.

Islam

Christianity emerged in a society dominated by Roman conquerors, in which ordinary people were burdened by heavy taxes and temple levies. Several military revolts against the Romans took place in this period, and some scholars view Jesus' message as a religious response to the oppressive social conditions of his time.

Islam originated more than 600 years later in the city of Mecca, in what is now Saudi Arabia. It too may be seen as a religious response to a society in crisis. Mecca was a rich center of trade, and the powerful merchants of that city had become greedy, overbearing, and corrupt. The merchants ignored the traditional moral code that originated in the surrounding nomadic tribes (the Bedouin). The Bedouin themselves were heavily indebted to the merchants and became so poor that some of them were sold into slavery. On a larger canvas, many people in Arabia thought that the Persian and Roman Empires, which dominated the Middle East, might soon fall, heralding the end of the world (Rodinson, 1996).

Into this crisis stepped Muhammad, who claimed to have visions from God (Hodgson, 1974; Lapidus, 2002 [1988]). His teachings were later written down in the Qur'an. Certain episodes and personalities that appear in the holy books of Judaism and Christianity also appear in the Qur'an; Muslims acknowledge Abraham and recognize both Moses and Jesus as prophets, for example. The central belief of Islam is that there is one true God, Allah, and that the words of his prophet, Muhammad, must be followed. Islam also emphasizes important teachings of Christianity, such as egalitarianism and universal love. Even more than Judaism or Christianity, however, the Qur'an is important for devout Muslims because they believe it is the direct word of God.

Like Judaism with its Talmudic commentaries on the Bible, Islam stresses the significance of *hadith* (traditions), a corpus of anecdotes and commentaries about Muhammad. Like rabbis in Judaism and priests in Catholicism, *ulamas* (scholars) maintain religious authority in Islam. Unlike Christianity, however, Islam has never had a central Church. Although any place can become a site of prayer and learning, mosques are the traditional places of worship.

People who profess Islam have five duties. At least once in their life they must recite the Muslim creed aloud, correctly, with full understanding, and with heartfelt belief. (The creed is: "There is no god but Allah and Muhammad is his prophet.") Five times a day they must worship in a religious service. They must fast from sunrise to sunset every day during the ninth month of the lunar calendar (Ramadan). They must give to the poor. And at least once in their life they must make a pilgrimage to the holy city of Mecca.

By the time of his death, Muhammad had founded an empire, and a dispute broke out over how his followers could identify his successor (the *khalifa,* or caliph in English). One group claimed that the caliphate should be an elected office occupied by a member of a certain Meccan tribe. These were the Sunni Muslims. A second group claimed that the caliph should be the direct descendant of Muhammad. These were the Shia Muslims. Today, the great majority of Muslims are Sunni, while the Shia are concentrated in Iran and southern Iraq (▶Figure 16.7). The Shia are generally more conservative and fundamentalist than the Sunni.

Islam spread rapidly after Muhammad's death, replacing Christianity in much of the Middle East, Africa, and parts of southern Europe. It ushered in a great cultural flowering and an era of considerable religious tolerance by the standards of the time. Significantly,

Figure 16.7
Distribution of Sunni and Shia Muslims

Source: University of Texas (2003).

the Jews flourished in Muslim Spain and North Africa at the very time they were being persecuted and expelled from Christian Europe.

Only a few Islamic sects developed in the modern era. The most noteworthy was Wahhabism, an extreme fundamentalist movement that originated in the 18th century in what is now Saudi Arabia. Its founder, Muhammad ibn Abd al-Wahhab, upheld ritual over intentions, opposed reverence of the dead, and demanded that prayer and honors be extended only to Allah and not to Muhammad or the saints. He opposed all music and all books other than the Qur'an, and favored the extermination of anyone who disagreed with him, especially if they happened to be Shia. Characteristically, in 1801 Wahhabis stormed the Iraqi city of Karbala, wrecked and looted the sacred tomb of Muhammad's grandson, Hussein, and slaughtered thousands of the city's Shia residents.

Wahhabism might be ignored as a minor fanatical sect if it were not for one important historical fact. In 1747, al-Wahhab made a pact with Muhammad ibn Sa'ud whereby ibn Sa'ud became the political ruler of the Arabian peninsula and al-Wahhab its religious authority. In effect, Wahabbism became the state religion of what is now Saudi Arabia. This extreme form of fundamentalism continues to flourish in Saudi Arabia today, where its tenets are taught in the Saudi school system. It is not coincidental that nearly all of the suicide bombers who attacked the United States on September 11, 2001, were Saudi citizens, as is Osama bin Laden himself, and that Saudi money finances schools throughout the Muslim world that propagate anti-Western ideas (Schwartz, 2003).

Hinduism

Hinduism is the dominant religion of India, but it has no single founder and no books that are thought to be inspired by God. Its major texts are epic poems such as the *Bhagavad Gita,* the *Mahabharata,* and the *Ramayana.* Like Judaism, Hinduism originated nearly 4000 years ago, so we have little sense of the social context in which it first emerged (Flood, 1996). What is clear is that its otherworldliness and mystical tendencies make it very different from the three main Western religions.

Hindus believe in reincarnation, a cycle of birth, death, and rebirth. Only the body dies, according to Hindu belief. The soul returns in a new form after death. The form in which it returns depends on how one lived one's previous life. Hindus believe that people who live in a way that is appropriate to their position in society will live better future lives. In rare cases, one reaches a stage of spiritual perfection (*nirvana*) that allows the soul to escape the cycle of birth and rebirth and reunite with God. In contrast, people who do not live in a way that is appropriate to their position in society will supposedly live an inferior life when they are reincarnated. In the worst case, evildoers are expected to be reincarnated as nonhumans. This way of thinking helped to create a caste system, a rigid, religiously sanctioned class hierarchy. Vertical social mobility was nearly impossible because, according to Hindu belief, striving to move out of one's station in life is inappropriate and ensures reincarnation in a lower form.

There are many gods in the Hindu tradition, although all of them are thought to be aspects of the one true God. This too makes Hinduism different from the Western religions. A final difference between Hinduism and the Western religions is its propensity to assimilate rather than exclude other religious beliefs and practices. Traditionally, Jews, Christians, and Muslims tended to reject nonbelievers unless they converted. God tells Moses on Mount Sinai: "You shall have no other gods before me." In contrast, in the Bhagavad Gita, Krishna says that "whatever god a man worships, it is I who answer the prayer." This attitude of acceptance helped Hinduism absorb many of the ancient religions of the peoples on the Indian subcontinent. It also explains why there are such wide regional and class variations in Hindu beliefs and practices; Hinduism as it is practiced bears the stamp of many other religions.

Buddhism

By about 600 B.C.E., Hinduism had developed into a system of rituals and sacrifices that was widely considered a burden. For example, one could escape the consequences of committing inappropriate or evil acts but only by having priests perform a series of mechanical rituals. In a sense, Gautama Buddha was to Hinduism what Jesus was to Judaism. Like Jesus in Palestine 600 years later, Buddha objected to the stale ritualism of the established religion and sought to achieve a direct relationship with God (Gombrich, 1996; Lopez, 2001; Robinson and Johnson, 1997 [1982]). He rejected Hindu ideas of caste and reincarnation and offered a new way for everyone to achieve spiritual enlightenment. Rather than justifying inequality, he promised the possibility of salvation to people of low status and women, who were traditionally marginalized by Hinduism.

Buddha based his method of salvation on what he called the Four Noble Truths: (1) Life is suffering. There are moments of joy, but poverty, violence, and other sources of sorrow overshadow them. (2) All suffering derives from desire. We suffer when we fail to achieve what we want. (3) Suffering ceases by eliminating desire. If we can train ourselves not to lust, not to be greedy, not to crave pleasure, not even to desire material comforts, we will not suffer. (4) We can eliminate desire by behaving morally, focusing intently on our feelings and thoughts, meditating, and achieving wisdom. Nirvana can be achieved by "blowing out" the burning torch of futile passions of existence.

Gautama Buddha (about 563–480 B.C.E.).

The Lowe Art Museum, The University of Miami/SuperStock

Buddhism does not presume the existence of one true God. Rather, it holds out the possibility of everyone becoming a god of sorts. Similarly, it does not have a central church or text, such as the Bible. Not surprisingly, then, Buddhism is notable for its diversity of beliefs and practices. Numerous schools and scriptures make up the Buddhist tradition. Many Westerners are familiar with Zen Buddhism, especially influential in East Asia, which emphasizes the possibility of enlightenment through meditation.

Buddhism spread rapidly across Asia after India's ruler adopted it as his own religion in the third century B.C.E. He sent missionaries to convert people in Tibet, Cambodia (Kampuchea), Nepal, Sri Lanka (formerly Ceylon), Myanmar (formerly Burma), China, Korea, and Japan. Ironically, the influence of Buddhism in the land of its birth started to die out after the fifth century C.E. and is negligible in India today. In India, as we have seen, Hinduism predominates. One of the reasons for the popularity of Buddhism in East and Southeast Asia is that Buddhism is able to coexist with local religious practices. Unlike Western religions, Buddhism does not insist on holding a monopoly on religious truth.

Bases of Formation of World Religions

Little historical evidence helps us understand the social conditions that gave rise to the first world religions, Judaism and Hinduism. We are on safer ground when it comes to understanding the rise of Buddhism, Christianity, and Islam. Insofar as we dare to make sociological generalizations about thousands of years of complex religious development spanning many cultures and virtually the entire globe, we can venture four conclusions. First, new world religions are founded by charismatic personalities in times of great trouble. The impulse to find a better world is encouraged by adversity in this one. Second, the founding of new religions is typically animated by the desire for freedom and equality, always in the afterlife, and often in this one. Third, the routinization of charisma typically makes religion less responsive to the needs of ordinary people, and it often supports injustices. For this reason, and also because there is no lack of adversity in the world, movements of religious reform and revival are always evident, and they often spill over into politics. For example, the Catholic Church played a critically important role in undermining communism in Poland in the 1970s and the 1980s, and Catholic "liberation theology" animated the successful fight against right-wing governments in Latin America in the same period (Kepel, 1994 [1991]; Segundo, 1976 [1975]; Smith, 1991). It is for this reason, too, that we venture a fourth, speculative conclusion, namely that new world religions could well emerge in the future.

Religiosity

We have reviewed the major classical theories of religion and society, the modern debate about secularization, the major types of religious organizations, and the development of the world religions. It is now time to consider some social factors that determine how important religion is to people, that is, their **religiosity.**

We can measure religiosity in various ways. Strength of belief, emotional attachment to a religion, knowledge about a religion, frequency of performing rituals, and frequency of applying religious principles in daily life all indicate how religious a person is (Glock, 1962). Ideally, one ought to examine many measures to get a fully rounded and reliable picture of the social distribution of religiosity. For simplicity's sake, however, we focus on just one measure here: how often people attend religious services. Again, we turn to the GSS for insights.

Table 16.2 divides GSS respondents into two groups: those who said they attend religious services less than once a month and those who said they attend religious services once a month or more. It then subdivides respondents by their age, race, and whether their mother and their father attended religious services frequently when they were children.

Some fascinating patterns emerge from the data. First, older people attend religious services more frequently than younger people. There are two reasons for this. First, older people have more time and more need for religion. Because they are not usually in school, employed in the paid labor force, or busy raising a family, they have more opportunity than younger people to go to church, synagogue, mosque, or temple. And because elderly people are generally closer than younger people to illness and death, they are more likely to require the solace of religion. To a degree, then, attending religious services is a life-cycle phenomenon. That is, we can expect younger people to attend religious services more frequently as they age. But there is another issue at stake here too. Different age groups live through different times, and elderly people reached maturity when religion was a more authoritative force in society than it is today. A person's current religiosity depends partly on whether he or she grew up in more religious times. Thus, although young people are likely to attend services more often as they age, they are unlikely ever to attend services as frequently as elderly people do today.

Second, Table 16.2 shows that frequent church attendance is more common among African Americans than whites. This is undoubtedly because of the central political and

Religiosity refers to how important religion is to people.

Table 16.2
Social Factors Influencing How Often Americans Attend Religious Services (in percent)

ATTENDS RELIGIOUS SERVICES . . .	Less Than Once a Month	Once a Month or More	%
Age, yr (N = 2743)			
18–29	61	39	100
30–39	54	46	100
40–49	53	47	100
50–59	53	47	100
60–69	44	56	100
70+	45	55	100
Race (N = 2742)			
White	56	44	100
Black	37	63	100
Mother's attendance at religious services during respondent's youth (N = 8151)			
Less than once a month	65	35	100
Once a month or more	41	59	100
Father's attendance at religious services during respondent's youth (N = 7302)			
Less than once a month	59	41	100
Once a month or more	38	62	100

Note: Race and age are for 2002. Mother's attendance and father's attendance are for 1983–1989, the only years this variable was measured in the General Social Survey.
Source: National Opinion Research Center (2004).Source: U.S. Department of Health and Human Services (2004: 197).

cultural role played by the church historically in helping African Americans cope with, and combat, slavery, segregation, discrimination, and prejudice.

Third, respondents whose mothers and fathers attended religious services frequently are more likely to do so themselves. Religiosity is partly a *learned* behavior. Whether parents give a child a religious upbringing is likely to have a lasting impact on the child. Table 16.2 shows that children of frequent churchgoers are much more likely than children of infrequent churchgoers to become frequent churchgoers themselves.

This is by no means an exhaustive list of the factors that determine frequency of attending religious services. However, this brief overview suggests that religiosity depends on opportunity, need, and learning. The people who attend religious services most often are those who were taught to be religious as children, who need organized religion for political reasons or due to their advanced age, and who have the most time to go to services.

Religiosity is partly a learned behavior. Whether parents give a child a religious upbringing is likely to have a lasting impact on the child.

Myrleen Ferguson/PhotoEdit

The Future of Religion

Secularization is one of the two dominant trends influencing religion throughout the world. We can detect secularization in survey data that track religious attitudes and practices over time. For example, between 1972 and 2002, the percentage of Americans expressing no religious preference increased from 5 percent to 14 percent, whereas the percentage of people attending religious services once a month or more fell from 57 percent to 47 percent (National Opinion Research Center, 2004). The percentage of Americans who reported that religion is important in their lives fell from 75 percent in the middle of the 20th century to around 50 percent at century's end. We also know that various secular institutions are taking over some of the functions formerly performed by religion, thus robbing it of its once pervasive authority over all aspects of life. It is probably an exaggeration to claim, as Max Weber did, that the whole world is gradually becoming "disenchanted." But certainly part of it is.

We also know from survey data and other sources reviewed earlier that even as secularization grips many people, many others in the United States and throughout the world have been caught up by a religious revival of vast proportions. Religious belief and practice are intensifying for these people, in part because religion serves as a useful vehicle for political expression. The fact that this revival was quite unexpected just a few decades ago should warn us not to be overly bold in our forecasts. It seems to us, however, that the two contradictory social processes of secularization and revival are likely to persist for some time to come, resulting in a world that is neither more religious nor more secular, but one that is certainly more polarized.

Sociology Now™

Learn more about **the Future of Religion** by going through the Religion in the United States and the World Data Experiment.

Summary

Sociology Now™

Reviewing is as easy as ❶, ❷, ❸.

❶ Before you do your final review, take the SociologyNow diagnostic quiz to help you identify the areas on which you should concentrate. You will find information on how to access SociologyNow on the foldout at the front of the textbook.

❷ As you review, take advantage of SociologyNow's study aids to help you master the topics in this chapter.

❸ When you are finished with your review, take SociologyNow's post-test to confirm you are ready to move on to the next chapter.

1. **What was Durkheim's theory of religion and what are the main criticisms that have been lodged against it?**

   Durkheim argued that the main function of religion is to increase social cohesion by providing ritualized opportunities for people to experience the collective conscience. Critics note that Durkheim ignored the ways religion can incite social conflict and reinforce social inequality.

2. **What was Weber's theory of religion and what are the main criticisms that have been lodged against it?**

   Weber argued that religion acts like a railroad switch, determining the tracks along which history will be pushed by the force of political and economic interest. Protestantism, for example, invigorated capitalist development. Critics note that the correlation between economic development and the predominance of Protestantism is not as strong as Weber thought. They also note that some of Weber's followers offer one-sided explanations of the role of religion in economic development, which Weber warned against.

3. **What is the secularization thesis and what are the main criticisms that have been lodged against it?**

   The secularization thesis holds that religious institutions, actions, and consciousness are on the decline worldwide. Critics of the secularization thesis point out that there has been a religious revival in the United States and elsewhere over the past 30 years or so. They also note that survey evidence shows that religion in the United States is resilient.

4. **Does radical religious fundamentalism incite terrorist politics?**

   Religious fundamentalism often provides a convenient vehicle for framing political extremism, enhancing its appeal, legitimizing it, and providing a foundation for the solidarity of political groups. From this point of view, radical religious fundamentalism is more an effect of extremist politics than vice versa.

5. **What is the revised secularization thesis?**

   The revised secularization thesis recognizes the religious revival and the resilience of religion but still maintains that the scope of religious authority has declined over time. The revisionists say that religion is increasingly restricted to the realm of the spiritual; it governs fewer aspects of people's lives and is more a matter of personal choice than it used to be.

6. **What do the main world religions have in common?**

   The main world religions were founded by charismatic personalities in times of great trouble. The founding of new religions is typically animated by the desire for freedom and equality, always in the afterlife, and often in this one. The routinization of charisma typically makes religion less responsive to the needs of ordinary people and it often supports injustices. Therefore, the need for religious reform and revival—and the need for new religions—is likely to persist.

7. **What determines the frequency with which people attend religious services?**

   Among other factors, the frequency of attending religious services is determined by opportunity (how much time people have available for attending), need (whether people are in a social position that increases their desire for spiritual answers to life's problems), and learning (whether people were brought up in a religious household).

Questions to Consider

1. Does the sociological study of religion undermine one's religious faith, make one's religious faith stronger, or have no necessary implications for one's religious faith? On what do you base your opinion? What does your opinion imply about the connection between religion and science in general?
2. What influence did religion have on politics in the 2004 U.S. presidential election? Specifically, how influential was the religious right in influencing the platform of the Republican Party and attracting voters to that party? For information on this subject, use the CNN and *Time* search engines at http://www.cnn.com and http://www.time.com/time, respectively.

Web Resources

Companion Website for This Book

http://sociology.wadsworth.com

Begin by clicking on the Student Resources section of the website. Choose "Introduction to Sociology" and then the Brym and Lie book cover. Next, select the chapter you are currently studying from the pull-down menu. From the Student Resources page you will have easy access to InfoTrac® College Edition, MicroCase Online exercises, additional web links, and many other resources to aid you in your study of sociology, including practice tests for each chapter.

Recommended Websites

Search for "religion" and "education" using the search engine of the General Social Survey at http://www.icpsr.umich.edu/GSS99/search.htm. The search engine will retrieve dozens of research reports and tables based on this respected, ongoing survey of the American public.

Ontario Consultants on Religious Tolerance is an excellent website that provides basic, unbiased information on dozens of religions, religious tolerance and intolerance, religion and science, abortion and religion, and so forth. Visit the site at http://www.religioustolerance.org.

CHAPTER 17 Education

Bob Daemmrich/The Image Works

In this chapter, you will learn that:

- Schools perform important functions in society, including training and socializing the young, fostering social cohesion, transmitting culture from generation to generation, and sorting students, presumably by talent, for further training and employment.
- Schools do a far from perfect job of sorting students by talent. To a degree, they simply funnel poor and minority students into low-ability classes and more affluent students into high-ability classes. Eventually this results in children occupying positions in the occupational structure similar to those occupied by their parents.
- Standardized tests (IQ, SAT, and ACT) help to sort students by talent *and* reproduce the existing class structure.
- Student success in the education system is influenced by students' cognitive ability, the quality of the schools they attend, the material and emotional support offered by their families, the degree to which they learn elements of high-status culture in school, and the operation of self-fulfilling prophecies about which students are likely to succeed and which are not.
- Mass, compulsory education has its roots in the Protestant Reformation, democratic revolutions, the rise of the modern state, and globalization. These forces have spread mass, compulsory education throughout the world.
- Educational standards are very low in the bottom third of American schools. Proposed solutions to this problem include local initiatives aimed at improving schools, the use of vouchers that would allow students in inferior schools to attend private schools, redistributing existing resources and increasing education budgets, and substantially improving the social environment of young, disadvantaged children before and outside school.

Affirmative Action and Class Privilege

In December 1996, Barbara Grutter applied to the University of Michigan Law School. Although she had a 3.8 undergraduate grade point average and an LSAT score that placed her in the 86th percentile, the Law School rejected her application for admission. Around the same time, Patrick Hamacher and Jennifer Gratz applied to the University of Michigan's College of Literature, Science and the Arts. Hamacher had a 3.0 grade point average and an ACT score of 28, Gratz a 3.8 grade point average and an ACT score of 25. Both students were likewise denied admission.

Grutter, Hamacher, and Gratz are white. Following the rejection of their applications for admission, they brought suit against the University of Michigan. They charged that the university gave unlawful preference to Black, Hispanic, and Native American applicants. They also charged that as white students, they were denied equal protection under the law as guaranteed by the Constitution. Outside an immediate government interest, they argued, the Constitution prohibits the state from using race as a criterion for access to government programs and services. Their cases were heard by the U.S. Supreme Court (Gratz and Hamacher v. Bollinger et al., 1997; Grutter v. Bollinger et al., 1997).

At the time, a student needed 150 admission points to gain acceptance to the University of Michigan. Most points were awarded for academic achievement as signified by one's grade point average and score on a standardized test such as the Scholastic Aptitude Test (SAT). However, Black, Hispanic, and Native American applicants received an automatic 20 admission points because of their underprivileged status. That is the practice to which Grutter, Hamacher, and Gratz objected.

Middlebury College, Vermont

Peter Finger/Corbis

The practice of granting underprivileged students special privileges in college admissions has a long and distinguished history in the United States. Before World War II, hardly any American colleges were models of openness and generosity. Most colleges were the preserve of the children of a wealthy elite. In contrast, following the war many educators argued that the country would be stronger if colleges admitted capable students regardless of their ethnic or racial background and their ability to afford a higher education. Most colleges broadened recruitment efforts, found new ways to distribute information, advice, and encouragement to potential students who lacked such resources, and started offering financial support to needy students. Beginning with the GI Bill, federal and state governments eagerly assisted these efforts (Duffy and Goldberg, 1997).

A reaction against such openness and generosity has been growing since the 1970s. Increasingly, white students like Grutter, Hamacher, and Gratz have argued that they are discriminated against in a country that is supposed to oppose discrimination. Such claims have led others to make the counterclaim that privileged white students also benefit from the admissions process. For example, at the University of Michigan at the time of the Grutter, Hamacher, and Gratz trials, a review committee could award up to 20 admission points to children of donors and other key supporters.

Academic researchers, admissions officers at elite colleges, and investigative journalists at respected newspapers such as the *New York Times* have recently detailed the privileges of class in college admissions (Avery, Fairbanks, and Zeckhauser, 2003; Steinberg, 2003; Toor, 2001). Students routinely receive admission points if they have a parent who graduated from the college to which they are applying. Such students benefit from the so-called legacy factor. For example, President George W. Bush was accepted at Yale with an SAT score of 1206 (566 verbal, 640 math), which was below average at Yale for his admission year. He got a leg up because he was a legacy student. Both his father and his grandfather graduated from Yale (Kinsley, 2003). The class that entered highly selective Middlebury College in Vermont in 2002 is fairly typical in this regard. While 27 percent of applicants were accepted at Middlebury, the figure for legacy students was 45 percent (Steinberg, 2003). (For African American and Hispanic American applicants, the admission rate was nearly 59 percent.)

A second mechanism that bestows advantages on privileged students involves parents contributing money to the colleges that their children want to attend. This is the "development" factor, so called because such gifts aid the development of the colleges that receive them. As Mike Schoenfeld, dean of enrollment planning at Middlebury College, said

Box 17.1

YOU AND THE SOCIAL WORLD

Class Privilege vs. Affirmative Action

Congratulations. You have just been appointed a justice of the U.S. Supreme Court. Your first case involves the complaint of Barbara Grutter, Patrick Hamacher, and Jennifer Gratz that they were the objects of discrimination when they applied to the University of Michigan (discussed in text). You must now write a 500-word opinion on the merits of their complaint that deals in general with the following questions: Should colleges give any special treatment to specific individuals or groups in admission decisions? If so, who should receive special treatment and why? If not, why not? After writing your opinion, add a brief addendum reflecting on whether and how your race, ethnicity, and class influence your judgment.

in 2003, "I'm sure every admissions office in the country is paying attention to families' ability to make a major donation. . . . That's likely to be even more important in [a] down economy" (quoted in Steinberg, 2003).

What is your opinion about this highly politically charged subject? Before making up your mind, please consider the arguments for and against special treatment:

- *Advocates of affirmative action* say it compensates for historical injustices such as slavery and expulsion. From this point of view, affirmative action helps to create a level playing field for people of all races and ethnic groups. Advocates also argue that affirmative action enriches college campuses by encouraging racial and ethnic diversity. Finally, they say affirmative action creates a middle-class leadership group in minority communities. These leaders show by example, instruction, and advocacy how minority groups can raise their status in society. One study of black students who benefited from affirmative action shows that they are significantly more likely to "give back" to their communities through various forms of service than are white students who attended the same colleges (Bowen and Bok, 1998).
- *Opponents of affirmative action* contest each of these points. First, while they don't deny injustice, they emphasize its historical character. They argue they should not have to pay for wrongs committed as long as 300 years ago. Second, although they do not necessarily deny the benefits of cultural diversity, opponents of affirmative action uphold the importance of individual rights over decision making based on group characteristics. They note that colleges apply affirmative action criteria to all members of selected minority groups, distributing admission points to rich and poor alike. This approach ignores the circumstances and the rights of individuals. Finally, opponents of affirmative action note its costs. For example, Stephen Carter, an African American law professor at Yale, describes himself as an "affirmative action baby" (Carter, 1991). He is skeptical of affirmative action because he believes it demeans the individual achievements of African Americans. In his view, affirmative action lets people dismiss African American achievements because they are presumed to derive from preferential treatment rather than talent.
- *Advocates of special treatment for the well-to-do* make just one argument. They note that without the generosity of the alumni, college tuition could jump by as much as two-thirds (Steinberg, 2003). In other words, everyone who goes to college—especially *less* privileged students—benefits from the money brought in through legacy and development admissions.
- *Advocates of meritocracy* also make a single argument. A **meritocracy** is a stratification system in which equality of opportunity allows people to rise or fall to a position that matches their talent and effort. Advocates of meritocracy oppose special treatment for any group. They believe on principle that the only fair system is one in which talent alone determines college admission. Only a meritocracy allows the most talented individuals to contribute most to society (Box 17.1).

A **meritocracy** is a stratification system in which equality of opportunity allows people to rise or fall to a position that matches their talent and effort.

Ted Horowitz/Corbis

Schools encourage the development of a separate youth culture that often conflicts with parents' values.

Sociology Now™

Learn more about **the Functions of Education** by going through the Functions of Education Learning Module.

Macrosociological Processes

The Functions of Education

Awarding admission points based on cash gifts, family ties, and minority status is controversial because it strikes at the heart of a widespread belief about the American educational system. Many Americans believe that we enjoy equal access to basic schooling. They think that schools identify and sort students based on merit and effort. They regard the educational system as an avenue of upward mobility. From their point of view, the best and the brightest are bound to succeed whatever their economic, ethnic, racial, or religious background. The school system is the American Dream in action. In their view, **educational attainment** is largely an outcome of individual talent and hard work. (Educational attainment refers to number of years of school completed, and should not be confused with **educational achievement,** which refers to how much students actually learn.)

The view that the American educational system is responsible for *sorting* students based on talent and effort is a central component of the functional theory of education. The functional theory also stresses the *training* role of schools. That is, in schools, most people learn how to read, write, count, calculate, and perform other tasks essential to the workings of a modern industrial society. A third function of the educational system involves the *socialization* of the young (Durkheim, 1956; 1961). Schools teach the young to view their nation with pride, respect the law, think of democracy as the best form of government, and value capitalism. Finally, schools *transmit culture* from generation to generation, fostering a common identity and social cohesion in the process. Schools have played a particularly important role in assimilating the disadvantaged, minorities, and immigrants into American society (Fass, 1989), although in recent decades, our common identity has been based on respect for the cultural diversity of American society.

Sorting, training, socializing, and transmitting culture are *manifest* functions or positive goals that schools accomplish intentionally. Schools perform certain *latent,* or unintended, functions too. For example, schools encourage the development of a separate youth culture that often conflicts with parents' values (Coleman, 1966). Especially at the college level, educational institutions bring potential mates together, thus serving as a "marriage market." Schools perform a useful custodial service by keeping children under surveillance for much of the day and freeing parents to work in the paid labor force. By keeping millions of young people temporarily out of the full-time paid labor force, colleges restrict job competition and support wage levels (Bowles and Gintis, 1976). Finally, because they can encourage critical, independent thinking, educational institutions sometimes become "schools of dissent" that challenge authoritarian regimes and promote social change (Brower, 1975; Freire, 1972).

Web Interactive Exercise: The Effect of Economic Inequality from the Conflict Perspective

The Effect of Economic Inequality from the Conflict Perspective

Educational attainment refers to number of years of school students complete.

Educational achievement refers to how much students actually learn.

From the conflict perspective, the chief problem with the functionalist view is that it exaggerates the degree to which schools sort students by ability and thereby ensure that the most talented students eventually get the most rewarding jobs. Conflict theorists argue that, in fact, schools distribute the benefits of education unequally, allocating most of the benefits to children from upper classes and higher-status racial and ethnic groups. This

Mike Derer/AP/Wide World Photos

Many of the nation's schools are old and in a state of disrepair. In New York City, for example, more than half the schools are half a century old or more. At least a fifth of New York schools are in need of immediate major repairs. Leaky ceilings, broken lights, and crumbling walls are common. A girl was killed in 1997 when debris fell from a dilapidated school in Brooklyn (Tornquist, 1998). Here, Angelo Castucci, vice principal of Hawthorne Avenue School in Newark, New Jersey, stands in a closed classroom, next to a wooden brace used to keep the ceiling from falling down. Even though a wing of the 104-year-old school had to be closed and exterior scaffolding installed, the Hawthorne Avenue School is not on the top-10 list of school construction projects in Newark.

means that rather than functioning as a meritocracy, schools tend to reproduce the stratification system generation after generation (Jencks et al., 1972; Lucas, 1999; Oakes, 1985).

Schools reproduce the stratification system partly because they vary so much in quality. This was not always so. In the 1950s and 1960s, schools in the United States were surprisingly homogeneous in resources and educational outcomes. Because differences in school resources were not big, they accounted for little of the observed difference in educational achievement among students (Coleman et al., 1966; Swell and Hauser, 1993). Since the 1970s, however, big gaps have opened up in the quality of American schools. Consequently, school quality matters more in determining the quality of education (Fischer et al., 1996).

For example, Jonathan Kozol (1991) compared average spending in Chicago city schools with spending in an upper-middle-class, suburban Chicago school. He found that spending per pupil was 78 percent higher in the suburban school. The suburban school offered a wide range of college-level courses and boasted the latest audiovisual, computer, photographic, and sporting equipment. Meanwhile, many schools in inner-city Chicago neighborhoods lacked adequate furniture and books. On an average day in the city, even teachers were unavailable for 190 classrooms. Kozol described schools in East St. Louis, Illinois, where raw sewage backed up into classrooms and teachers were short of paper and chalk. In one high school, there wasn't a single VCR. This sort of disparity repeats itself

Diane Bondareff/AP/Wide World Photos

When a 2002 nationwide poll asked Americans to indicate the areas in which we are spending too little money, the nation's education system placed second by an insignificant amount to health care. Seventy-four percent of Americans with an opinion on the subject said that we are not spending enough on education, compared with 75 percent who said that we are not spending enough on health care. Here, Sumaya Jackson, 12, center, blows a whistle as she walks in front of her father, Robert, with schoolmates from Manhattan's P.S. 187 as they demonstrate at the New York State Supreme Court building in New York City, Tuesday, October 12, 1999. A coalition of public schools advocates argued in court for a change in funding formulas, saying that New York City's public school children were being cheated out of money for education. Robert Jackson helped spearhead the effort.

throughout the country. Why? Because school funding is almost always based mainly on local property taxes. In wealthy communities, where property is worth a lot, people can be taxed at a lower rate than in poor communities and still generate more school funding per pupil. (Do the arithmetic: Taxing $100,000 of property at 3 percent yields $3000, whereas taxing $50,000 of property at 5 percent yields only $2500.)

Thus, wide variations in the wealth of communities and a system of school funding based mainly on local property taxes ensure that most children from poor families learn inadequately in ill-equipped schools and most children from well-to-do families learn well in better-equipped schools. Disparities in the quality of schools go far beyond resources, however. Some schools, mainly those in poor neighborhoods, have a lot of students from disadvantaged homes, dropouts, and disciplinary problems. They pay less well than schools in richer districts and therefore tend to have weaker teachers. They are therefore less conducive to learning than schools with few disadvantaged students, dropouts, and disciplinary problems, and stronger teachers. Thus, apart from the distribution of educational resources, the kind of students who attend schools and the quality of its teachers (or the "social composition" of schools) influence the quality of education.

Standardized Tests

A second feature of schools that helps to reproduce the existing system of social stratification is the standardized test. In a society based on feudal privilege, there wasn't much point in testing people. You were born a king, a lord, or a peasant and there was no chance of becoming anything else, regardless of how bright or how dull you were. In contrast, in a society that is supposed to be based on equality of opportunity, you need testing to figure out who is good at what. You are supposed to become an engineer or a hairdresser because of your ability, and your ability must be judged by objective, standardized testing.

Tracking involves sorting students into high-ability, middle-ability, and low-ability classes based on the results of IQ and other tests.

Schools sort students into high-ability, middle-ability, and low-ability classes based on the results of intelligence quotient (IQ) and other tests. This is called **tracking.** IQ tests are supposed to measure only innate ability, although, as we will see, whether they do is doubtful. After high school, students are sorted into colleges of varying quality based on the results of SAT and ACT (American College Tests) exams. The SAT originally derived from IQ tests. However, its sponsors now claim it measures "developed verbal and mathematical reasoning abilities related to performance in college" (quoted in Zwick, 2002: 8). In other words, the SAT focuses more on what students have achieved in terms of their verbal and math reasoning rather than their innate abilities. In the 1950s, a statistician in Iowa considered the SAT to be geared too closely to the elite colleges of the Northeast and insufficiently tied to subjects students actually learn in school. He established the ACT as an alternative to the SAT. While the IQ test is supposed to focus on innate abilities, the ACT is supposed to be the most achievement-oriented of the three tests.

Sociology Now™

Learn more about **Tracking** by going through the Tracking in Education Animation.

How Do IQ and Social Status Influence Academic and Economic Success?

The best research on the subject shows that by itself, IQ contributes to academic success and to economic success later in life. This sorting by merit is what the functionalists would predict. However, a number of background factors also influence success (in ▶Figure 17.1, focus for the moment on the solid black lines). This is in accordance with what the conflict theorists would predict. One set of background factors derives from the home environment. Your success in school and your economic success later in life depend directly on how much money your parents earn, how many years of education they have, how much they encourage your creativity and studying, how many siblings you have, and so forth. In general, having encouraging parents with more education and higher income, and having fewer siblings, gives a person the greatest chance of success. A second set of background factors influencing success derives from the community environment. As we have already suggested, your academic and economic success depend directly on the percentage of disadvantaged students in your school, the dropout rate in your school, which part of the country you come from, and so forth. These home and community background factors bestow privileges and disadvantages on people independent of IQ. As the solid orange lines in Figure 17.1 suggest, they also influence how many years you spend in school, whether you are placed in high-ability, medium-ability, or low-ability classes, and whether you complete college (▶Figure 17.2 and ▶Table 17.1). These schooling variables influence your academic and economic success too.

Significantly, home environment, community environment, and schooling experience affect IQ test results (see the dashed black lines in Figure 17.1). Said differently, if you change a person's home environment, community environment, and schooling experience, you will probably change that person's IQ. A few scholars have argued that IQ is ge-

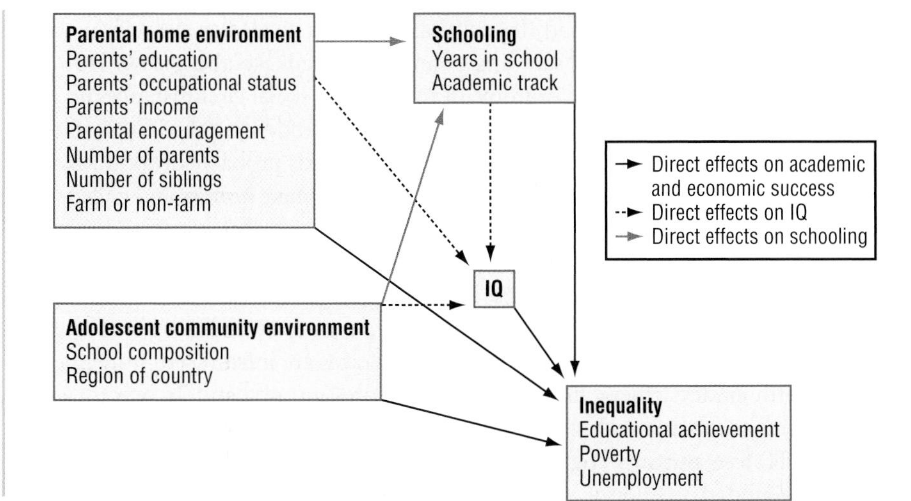

▶Figure 17.1
How Social Background and IQ Influence Inequality

Source: Adapted from Fischer et al. (1996: 74); Sewell and Hauser (1993).

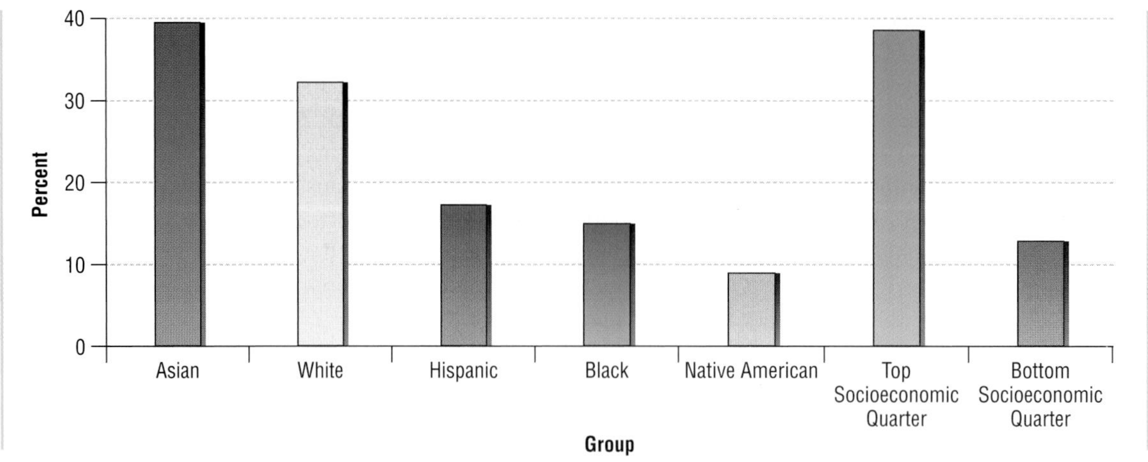

Figure 17.2
Percent of Eighth-Grade U.S. Public School Students in High-Ability Classes, by Ethnicity, Race, and Socioeconomic Status (N = 14,000)

Source: Kornblum (1997 [1988]: 542).

Table 17.1
Father's Occupation and Child's Chances of Completing College, United States

	CHILD'S CHANCES OF COMPLETING COLLEGE	
Father's Occupation	Students with Median Scores (in percent)	Students with Scores in Top 14% (in percent)
Professional	38	81
Manager	26	82
Clerical/sales	18	65
Skilled blue collar	15	60
Unskilled blue collar	12	54

Source: Hout, Raftery, and Bell (1993: 46).

netic in origin, that it cannot be changed, and that improving the economic circumstances and the quality of schooling of the underprivileged is therefore a waste of money (Herrnstein and Murray, 1994). However, the fact that changes in home environment, community environment, and schooling experience produce changes in IQ demonstrates that this argument must be qualified. IQ is partly genetic and partly social in origin.

To illustrate, consider two cases where changed social circumstances resulted in improved IQ scores. In the first decades of the 20th century, most Jewish immigrants in America tested well below average on IQ tests. This was sometimes used as an argument against Jewish immigration (Gould, 1996 [1981]). Today, most American Jews test above average in IQ. The genetic makeup of Jews hasn't changed in the past century. Why the change in IQ scores? Sociologists point to upward mobility. During the 20th century, most American Jews moved up the stratification system. As their fortunes improved, they made sure their children had the skills and the cultural resources needed to do well in school. Average IQ scores rose as the social standing of Jews improved (Steinberg, 1989 [1981]).

Here is an even more dramatic example of the effect of social circumstances on intelligence and school performance. The 300 black and Latino students at the Hostos-Lincoln Academy of Science in the South Bronx were all written off as probable dropouts by their eighth-grade counselors. Yet most seniors in the school now take honors and college-level classes. Eighty percent of them go to college, well above the national average. The reason? The City of New York designated Hostos-Lincoln a special school in 1987. It is small, well equipped, attentive to individual students, and demanding. It stresses team teaching and a safe, family-like environment. Is Hostos-Lincoln an exception? No. A study of 820 American high schools shows that where similar programs are introduced, students from eighth to twelfth grades achieve 30 percent higher scores in math and 24 percent higher scores in reading compared with students in traditional schools (Hancock, 1994).

In short, IQ tests measure cognitive ability *and* social status. They reflect genetic endowment *and* underlying social stratification (Schiff and Lewontin, 1986). Therefore, when schools use IQ tests to determine who should enter a high-ability track, they are not increasing the opportunities of only the most talented students. They are also increasing the opportunities of some less talented students who happen to come from advantaged backgrounds. Similarly, when schools use IQ tests to determine who should enter a low-ability track, they are not decreasing the opportunities of only the less talented students. They are also decreasing the opportunities of some talented students who happen to come from disadvantaged backgrounds. This is not a pure meritocratic sorting system.

Are SAT and ACT Tests Biased?

Just as IQ tests sort students in schools, SAT and ACT tests sort students for college entrance, helping to determining who gets into the better colleges and who does not. These tests (along with high school grades) usefully increase the ability of colleges to predict who will do well and who will graduate. However, some scholars charge the SAT and ACT tests with bias against disadvantaged minority groups. They base their argument on test scores like those in ▶Table 17.2, which shows that in 2001, white and Asian American students scored substantially higher than Native Americans, African Americans, and Hispanic Americans. Critics say that SAT and ACT tests presume background knowledge that minority students lack because they do not share intimate knowledge of the majority culture (Zwick, 2002: 109–42).

Can culturally biased items be found in the SAT and ACT tests? Yes, they can. A test item is defined as biased if equally skilled members of different groups have significantly different rates of correct response to the item. For example, the following item, taken from an actual SAT test, asks students to choose the opposite of "turbulent":

TURBULENT

- aerial
- compact
- pacific
- chilled
- sanitary

This item is biased in favor of Spanish-speaking students, probably because the correct answer, "pacific," has the same root as the Spanish *pacifico,* which means peaceful. Knowing the Spanish word gives Spanish-speaking students an advantage in answering

▶Table 17.2

Average SAT Scores By Race or Ethnicity for College-Bound Seniors, 2001 (N = 1,035,336)

	RACE OR ETHNICITY						
	Native American	Asian American	African American	Mexican American	Puerto Rican American	Other Hispanic American	Other White American
Test section							
Verbal	481	501	433	451	457	460	529
Mathematical	479	566	426	458	451	465	531
Total	960	1067	859	909	908	925	1060
Percentage of all test-takers (a)	0.7	9.9	11.6	4.5	1.4	3.9	67.8
Ethnic or racial group as percentage of population (b)	1.5	4.2	12.9	7.3	1.2	4.0	69.1
Ratio of (a) to (b)	0.47	2.36	0.90	0.61	1.17	0.98	0.981

Source: U.S. Census Bureau (2003); Zwick (2002: 15).

the item. They are significantly more likely to answer this question correctly than equally skilled members of other ethnic and racial groups. On the other hand, Spanish-speaking students are disadvantaged when English words look like a Spanish word but mean something different. In this case, their knowledge of Spanish apparently confuses some of them. Researchers have identified items that are biased in this manner for all racial and ethnic groups. They discovered two things. First, there are few such items. Second, removing them has little effect on test results. That is, the gap in test scores between different ethnic and racial groups is practically the same before and after biased items are removed (Zwick, 2002: 128–30).

A more serious criticism of the SAT is that more affluent students can afford coaching that boosts their scores. Large commercial coaching firms include Kaplan Inc., with 1200 locations worldwide, and the Princeton Review, with 70 offices across the United States. Coaching courses cost around $800, and private coaches can charge as much as $500 an hour. Research shows that students who pay for SAT coaching are more likely than uncoached students to be wealthy, highly motivated, and Asian American. Although coaching can boost the SAT scores of some students by 100 points or more, the average is about 25 points. According to some analysts, 25 points equals the gain one could normally expect over a full year of maturation and school instruction. Thus, coaching clearly discriminates against less affluent students. Recognizing this, more than half of secondary schools offer SAT coaching programs, many of which are free. Georgia funds free SAT prep software. This helps decrease coaching discrimination, but it is a far cry from the universally accessible and free test preparation that would be required to eliminate it (Zwick, 2002: 159–78).

Coaching discrimination aside, it seems to us that the biggest problem with the SAT and ACT tests is not the tests themselves. The biggest problem lies in the background factors (home environment, community environment, and schooling) that determine who gets to take the test in the first place and how well prepared different groups of test-takers are (Massey et al., 2002). The company that administers the SAT identifies various ethnic and racial groups among the test-takers. Members of three groups make it into the test room in far smaller numbers than one would expect. Compared with their representation in the population as a whole, Native American test-takers are only 47 percent as numerous as one would expect, Chicanos are just 61 percent as numerous, and African American are 90 percent as numerous (see Table 17.2). These figures strikingly reveal the toll of background factors on who gets to take the SAT in the first place. In addition, we must bear in mind that widespread poverty, inadequate schools, tracking, and other background factors make Native Americans, Chicanos, and African Americans less prepared for the SATs on average than non-Hispanic white and Asian American students. This accounts in large measure for differences between groups in test scores. Background disparities are later reflected in the proportion of people in various racial and ethnic groups who complete college, and this eventually translates into earnings disparities in the paid labor force (▸Figures 17.3 and 17.4).

Case Study: Functionalist versus Conflict Theories of the American Community College

We can more fully illustrate how sociologists use functionalist and conflict theories by applying them to the case of the American community college system.

Between the beginning and the end of the 20th century, the number of community colleges in the United States grew from zero to more than 1400. About 5.5 million students are enrolled in American community colleges today (Cohen and Brawer, 2003 [1981]: 1, 15, 37). Aside from the general population increase, two social forces contributed most heavily to the rise of the community college system. First, the country needed skilled workers in industry and services. Second, the belief grew that higher education would contribute to upward mobility and greater equality in American society. The accuracy of that belief has become a point of contention among sociologists.

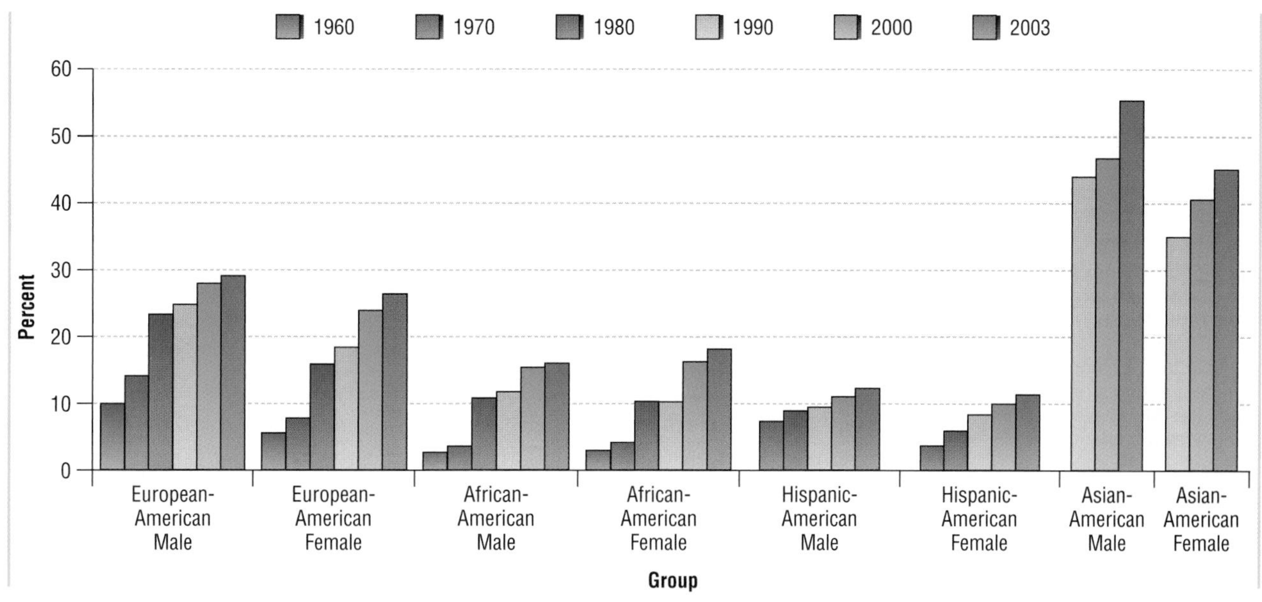

Figure 17.3
Percent with Four or More Years of College, United States, by Ethnicity, Race, and Gender, 1960-2003

Source: U.S. Census Bureau (2004-5: 141).

Functionalists examine the social composition of the student body in community colleges and find a somewhat disproportionate number of students from lower socioeconomic strata and minority ethnic groups. In 1997, for example, 46 percent of ethnic minority students in American higher education were enrolled in community colleges, whereas community colleges accounted for 38 percent of the total enrollment in American higher education (Cohen and Brawer, 2003 [1981]: 46). Although many community colleges are located in affluent or middle-class areas, others are located close to the neighborhoods of disadvantaged students, allowing them to live at home while studying. Tuition fees are generally lower than in four-year colleges. Graduates of community colleges are usually able to find relatively good jobs and steady employment. These facts seem to confirm the functionalist view that the community college system creates new opportunities for disadvantaged youth who might otherwise have less rewarding jobs.

Conflict theorists deny that the growth of community colleges increases upward mobility and equality in American society. In the long run, they argue, it is the entire stratifi-

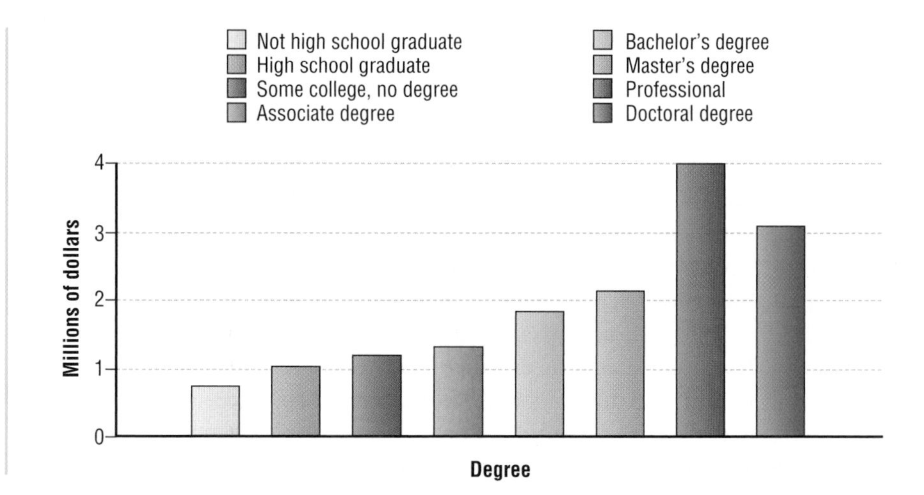

Figure 17.4
Estimated Lifetime Earnings by Educational Attainment, United States (in $ millions)

Source: Day and Newburger (2002).

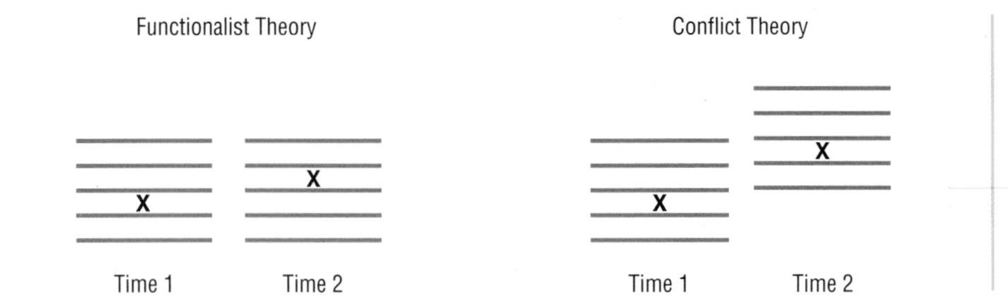

▸Figure 17.5
Functionalist vs. Conflict Theories of the Effect of Community Colleges on Social Mobility

According to functionalists, community colleges encourage upward mobility of students over time.

According to conflict theorists, in the long term, the entire stratification system produces higher-paying jobs, but the position of community college graduates relative to graduates of four-year institutions does not change.

cation system that is upwardly mobile. That is, the quality of nearly *all* jobs improves but the *relative* position of community college graduates versus graduates of four-year institutions remains the same (▸Figure 17.5). In fact, conflict theorists go so far as to argue that community colleges *reinforce* prevailing patterns of social and class inequality by directing students from disadvantaged backgrounds away from four-year institutions and thus decreasing the probability that they will earn a four-year degree and a high-status position in society (Karabel, 1986: 18). Only about 25 to 30 percent of community college students eventually transfer to universities, and they tend to be students from the most advantaged backgrounds.

Functionalists and conflict theorists both have a point. Community colleges *do* create opportunities for individual upward mobility that some students would otherwise not have. Community colleges *do not,* however, change the overall pattern of inequality in American society. In fact, expecting community colleges or, for that matter, any part of the institution of education to change the stratification system as a whole is probably naive. Decreasing the level of inequality in society requires passing laws that change people's entitlements and the rewards they receive for doing different kinds of work, not just increasing educational opportunities.

We conclude that functionalists paint a somewhat idealized picture of the education system. Although usefully identifying the manifest and latent functions of education, they fail to emphasize sufficiently the far-reaching effects of stratified home, community, and school environments on student achievement and placement. A similar conclusion is warranted if we examine the effects of gender on education.

Gender and Education: The Feminist Contribution

In some respects, women are doing better than men in the American education system. More American women than men have graduated from high school since about 1870. Overall, women in college have higher grade point averages than men and they complete their degrees faster. The number of women enrolled as college undergraduates has exceeded the number of men since 1978. By 1984, more women than men were enrolled in graduate and professional schools. The enrollment gap between women and men is growing—not just in the United States, but also in Britain, Canada, France, Germany, and Australia (Berliner, 2004; Jacobs, 1996; National Center for Educational Statistics, 2004a: 103, 115).

The facts just listed represent considerable improvement over time in the position of women in the education system. Yet feminists who have looked closely at the situation have established that women are still at a disadvantage (Spade, 2001). Consider field of study (▸Table 17.3). A disproportionately large number of men earn Ph.D.s and professional degrees in the physical sciences, engineering, computer science, dentistry, medicine, and law—all relatively high paying fields, most requiring a strong math and science background. A disproportionately large number of women earn Ph.D.s and professional degrees in home economics; area, ethnic, and cultural studies; education; English; foreign languages; and other relatively low paying fields requiring little background in math and science. Parents and teachers are partly responsible for these choices because they tend to direct boys and

Table 17.3
Overrepresentation of Women and Men among Ph.D.s and Selected Professional Degrees Awarded, United States, 2001

Female Overrepresentation	Male Overrepresentation
70%+ female overrepresentation	**70%+ male overrepresentation**
Home economics	Computer and information sciences
	Engineering and engineering technology
	Mathematics
	Philosophy, religion, and theology
	Physical science and science technologies
60–69.9%+ female overrepresentation	**60–69.9%+ male overrepresentation**
Area, ethnic, and cultural studies	Architecture and related programs
Education	Dentistry
English language and literature	Protective services
Foreign languages	
Health sciences	
Library science	
50–59.9%+ female overrepresentation	**50–59.9%+ male overrepresentation**
Liberal arts and sciences, general studies and humanities	Agricultural and natural resources
Psychology	Business management and administrative services
Public administration and services	Communications and technologies
Visual and performing arts	Law and legal studies
	Medicine
	Multi/interdisciplinary studies
	Parks and recreation
	Social sciences
	Unclassified (includes precision production trades and transportation and materials moving)

Note: "Overrepresentation" means that men or women compose more than 50 percent of a category.
Source: U.S. Census Bureau (2004: 193).

girls toward what they regard as masculine and feminine fields of study. Sex segregation in the labor market also influences choice of field of study. College students know that women are more likely to get jobs in certain fields than others and they make career choices accordingly (Spade, 2001) (Chapter 11, "Sexuality and Gender"). We conclude that like class and race, gender structures the educational experience and its consequences.

Microsociological Processes

The Stereotype Threat: A Symbolic Interactionist Perspective

Macrosociological issues such as the functions of education and the influence of class, race, and gender on educational achievement do not exhaust the interests of sociologists of education. They have also contributed much to our understanding of the face-to-face interaction processes that influence the educational process. Consider, for example, the self-fulfilling prophecy, which we first encountered in Chapter 4 ("Socialization"). There we discussed a famous experiment in which, at the beginning of a school year, researchers randomly identified students as high or low achievers to their teachers. At the end of the school year, they found that students arbitrarily singled out as high achievers scored higher on an IQ test than those arbitrarily singled out as low achievers. The researchers concluded that teachers' expectations influenced students' performance (Rosenthal and Jacobson, 1968; also Rist, 1970; Weinstein, 2002; Willis, 1984 [1977]).

Personal Anecdote

Or consider this personal example. John Lie's mother took him to school just a few days after the Lie family immigrated to Hawaii from Japan. "I was just ten years old," John recalls. "I knew only a few words of English. I knew almost nothing about American schools. So naturally, I was very quiet. The other students looked at me with a mixture of pity and curiosity. Months later I learned they assumed I was either mute or mentally retarded.

"I had to enroll in a special education class. It was the most insulting experience of my life. A few months earlier I had been doing algebra. Now the teacher was telling me that two pens plus three pens make five pens. I was so turned off I began to rebel. I would come to class late. I refused to make eye contact with the teacher. Finally, the teacher called in my parents. She told them I was mentally and emotionally deficient. I was at once furious and frustrated. I wanted to get back at her. So the next time she started her inane arithmetic exercise, asking me 'What would you like three of? What would you like four of?' I said I wanted three beers and four cigarettes. She hit the roof. She screamed, stormed out, and I never saw her again. I had flunked my special education class.

"I forgot this episode for years. Then, in an undergraduate sociology class where the prof was discussing self-fulfilling prophecies, I suddenly realized I had been the victim of one. Self-fulfilling prophecies are expectations that help cause what they predict. In my case, the special education teacher, like my classmates, had treated me like I was emotionally deficient and I started acting that way: showing up late, avoiding eye contact, remaining sullen, giving idiotic answers. My behavior confirmed the teacher's assessment of me. Treat someone like an idiot and often the person starts acting like one. Conversely, treat someone with respect, and useful learning can begin. This proved to be a very useful lesson for me as a teacher."

In general, low teacher expectations often encourage low student achievement. Thus, when black and white children begin school, their achievement test scores are similar. However, the longer they stay in school, the more black students fall behind. By the sixth grade, blacks in many school districts are two full grades behind whites in achievement. Why? Analysts suggest that teachers at all levels often expect African Americans, Latinos, and Native Americans to do poorly in school. Rather than being treated as a person with good prospects, a minority student is often under suspicion of intellectual inferiority and often feels rejected by teachers, white classmates, and the curriculum. This **stereotype threat** has a negative impact on the school performance of disadvantaged groups (Massey et al., 2002; Steele, 1997).

Sociology Now™

Learn more about **Norms and Values in the Education System** by going through the Roles in Education Video Exercise.

Too often, such alienation turns into resentment and defiance of authority. Many minority students even reject academic achievement because they see it as a goal of the dominant culture. Discipline problems, ranging from apathy to disruptive and illegal behavior, can result. The corollary of identifying one's race or ethnicity with poor academic performance is thinking of good academic performance as "selling out" to the dominant culture. That is precisely what is going on when an African American with poor grades calls an African American with good grades an "Oreo" (black on the outside, white on the inside) or when a Native American with poor grades calls a Native American with good grades an "apple" (red on the outside, white on the inside) (Ogbu, 2003).

In contrast, challenging minority students, giving them emotional support and encouragement, giving greater recognition in the curriculum to the accomplishments of their group, creating an environment in which they can relax and achieve—all these strategies explode the self-fulfilling prophecy and improve academic performance (Steele, 1992). Anecdotal evidence supporting this argument may be found in the brilliant 1988

Stereotype threat refers to the harmful impact of negative stereotypes on the school performance of disadvantaged groups.

BOX 17.2
Sociology at the Movies

Stand and Deliver (1988)

Garfield High School in East Los Angeles was on the verge of losing its accreditation in the early 1980s because so many of its students were failing. Because they were mostly Chicano and poor, the looming loss of accreditation was widely considered regrettable but hardly surprising.

Enter Jaime Escalante, a tough and engaging idealist who quit his promising job in the computer industry to work twice the hours and earn half the money teaching at Garfield. *Stand and Deliver* is the story of how Escalante, convincingly played by Edward James Olmos, inspired 18 students to study math in school, after school, on Saturdays, and during the summer so they could take the Advanced Placement Calculus Exam. Only 2 percent of high school students nationwide even attempt the exam. All 18 of the Garfield students passed, many with high grades.

Edward James Olmos in *Stand and Deliver.*

The Everett Collection

In two years, students with poor and failing grades—gang members, students with after-school jobs, students with onerous responsibilities taking care of their younger siblings and elder family members—registered the best performance in the Advanced Placement Calculus Exam in the southern California school system.

Stand and Deliver shows how hard their struggle was. Parents discouraged them. "Boys don't like you if you're too smart," one mother told her daughter. Other students scorned them. One gang member requested three copies of his math book—one for the classroom, one for his locker, one for home—so he could avoid being seen walking around with a book in his hand. Teachers doubted them. "Our kids can't handle calculus," one teacher told Escalante at a staff meeting. The testing service even distrusted the students' test results because they were so good, forcing the students to take the test a second time just to prove there was no cheating and that a group of poor Chicano kids living in a *barrio* really could excel in advanced math. Throughout, Escalante persisted. "Students will rise to the level of expectations," he told his fellow teachers. The principal asked: "What do you need, Mr. Escalante?" "*Ganas,*" Escalante replied. "That's all we need, is *ganas.*" *Ganas* is Spanish for desire. There must have been plenty of it around Garfield. The number of Garfield students who passed the Advanced Placement Calculus Exam rose every year from 18 in 1982 to 87 in 1987, the year before the movie was made.

Stand and Deliver is an inspiring story of how students' school performances can be improved if they are encouraged to think highly of themselves, if much is expected of them, and if they are inspired by a dedicated teacher. It also raises the question of how more extraordinary teachers like Jaime Escalante can be attracted to the teaching profession. We hold teachers to high standards. We expect them to get a college education. We entrust them with the intellectual and moral development of impressionable children. We expect them to work with dedication and inspiration. We expect their behavior to be morally impeccable and a model to their students. We maintain these high standards because we think education is so important and we love our children. Yet we pay teachers relatively poorly. How can this be explained? Why is there such a big gap between the high standards to which we hold teachers and the amount we're willing to pay them? What are the consequences for students in the public school system?

movie *Stand and Deliver,* based on the true-life story of high school math teacher Jaime Escalante. Escalante refused to write off his failing East Los Angeles Chicano pupils as "losers" and inspired them to remarkable achievements in calculus (Box 17.2).

Cultural Capital

Some students do better or worse in school than one would expect given their IQ and home, school, and community environments. That is because another variable, cultural capital, accounts for some differences in school performance. **Cultural capital** refers to

Cultural capital refers to widely shared, high-status cultural signals (attitudes, preferences, formal knowledge, behaviors, goals, and credentials) used for social and cultural exclusion.

"widely shared, high status cultural signals (attitudes, preferences, formal knowledge, behaviors, goals, and credentials) used for social and cultural exclusion" (Lamont and Lareau, 1988: 156). If you own a lot of cultural capital you have "highbrow" tastes in literature, music, art, dance, and even sports (rowing and fencing, for example). You behave according to established rules of etiquette. You value and pursue the goals of your school. You eventually earn a degree from one of the "right" colleges. People may earn cultural capital through socialization in high-status households or they may acquire it in school. Owning cultural capital increases one's chance of success in school and in the paid labor force.

The independent effect of cultural capital is sometimes exaggerated (Kingston, 2001). However, the most often cited American study on the subject suggests that cultural capital may be as important as measured ability (e.g., IQ) in determining grades (DiMaggio, 1982). Significantly, the original research on cultural capital was conducted in France, and it purported to show that possession of cultural capital is linked to being born in high-status families (Bourdieu and Passeron, 1990 [1977]). American research shows a much weaker link between the status of one's family and the acquisition of cultural capital. In this country, things are more fluid. Having parents with a higher education is no guarantee of acquiring cultural capital. Having parents who lack higher education is no guarantee you won't acquire it. In the United States, people acquire cultural capital mostly in school, where students can learn and display particular tastes, styles, and understandings that make communication easier with high-status individuals. Consequently, "[a]ctive participation in prestigious status cultures may be a practical and useful strategy for low status students who aspire towards upward mobility" (DiMaggio, 1982: 190). As an influence on success in school and the paid labor force that operates independently of the effect of ability and background factors, cultural capital has attracted the attention of many scholars in recent years.

Historical and Comparative Perspectives

The Rise of Mass Schooling

In Europe 300 years ago, the nobility and the wealthy usually hired personal tutors to teach their children to read and write, learn basic history, geography, and foreign languages, and study how to dress properly, conduct themselves in public, greet status superiors, and so on. Few people went to college. Only a few professions, such as theology and law, required extensive schooling. The great majority of Europeans were illiterate. As late as the 1860s, more than 80 percent of Spaniards and more than 30 percent of the French could not read (Vincent, 2000). Even as recently as a century ago, most people in the world never attended even a day of school. As late as 1950 only about 10 percent of the world's countries boasted systems of compulsory mass education (Meyer, Ramirez, and Soysal, 1992).

Sociology Now™

Learn more about **High School Graduates** by going through the % of person's who are high school graduates Map Exercise.

Today the situation is different. Compulsory mass education became a universal feature of European life by the early 20th century and nearly universal literacy was achieved by the middle of the 20th century (Vincent, 2000). Today, every country in the world has a system of mass schooling. In 2001, 88 percent of Americans in the 25–29 age group had completed high school (up from 78 percent in 1971) and nearly 29 percent had completed college (up from 17 percent in 1971) (National Center for Education Statistics, 2002a: 172, 174). By these measures, the United States is one of the most highly educated societies in the world.

What accounts for the spread of mass schooling? Sociologists usually highlight four factors: the Protestant Reformation, the democratic revolution, the rise of the modern state, and globalization.

The Protestant Reformation

The Catholic Church dominated preindustrial Europe. It did not want mass literacy because one of the sources of its power was the monopoly of priests over the ability to read and interpret the Bible. The Church even arrested and executed people who tried to translate the Latin Bible into national languages such as German, French, and English. That was because translations made the Bible more accessible. Thus, nearly all English-language translations of the Bible are based on the King James version, published in 1611. The King James version is based largely on William Tyndale's translation. Tyndale was burned at the stake in 1536 for translating and publishing the Bible in English.

In the early 16th century, Martin Luther, a German monk, began to criticize the Catholic Church. Protestantism grew out of his criticisms. The Protestants believed that the Bible alone, and not Church doctrines, should guide Christians. They expected Christians to have more direct contact with the word of God than was allowed by the Catholic Church. Accordingly, Protestants needed to be able to read the scriptures for themselves. The rise of Protestantism was thus a spur to popular literacy and the rise of mass education.

The Democratic Revolution

By the late 18th century, the populations of France, the United States, and other new democracies demanded access to centers of learning, which had previously been restricted to the wealthy. They also struggled for the right to public education.

In the United States, educational opportunities expanded gradually in the 19th and 20th centuries (Katznelson and Weir, 1985; Fass, 1989). Women began to enter colleges in the late 19th century and surged into the higher education system in the 1960s. Similarly, African Americans successfully fought for the right to enter schools previously restricted to white students (Kirp, 1982). The 1954 Supreme Court decision in *Brown v. Board of Education* was a watershed in the movement toward educational equality for African Americans. It struck down an 1896 Supreme Court decision that established the doctrine of "separate but equal," or segregation. According to the *Brown v. Board of Education* ruling, separate is unequal by definition (Orfield and Eaton, 1996). Significantly, between 1960 and 2000, the percentage of white Americans with at least four years of college tripled, but the percentage of African Americans increased more than 4.5 times (see Figure 17.3).

The Modern State

The modern state also encouraged compulsory mass education. It did so partly because education promoted loyalty and social order in an era when the grip of the Church was weakening. That is, state officials correctly viewed education as an instrument for instilling loyalty and teaching young people a common national language, identity, and values. Another important force that encouraged the state to promote compulsory mass education was industrialization. In preindustrial societies, people did not require formal education for many kinds of work (e.g., farming) and they could learn other occupations on the job (e.g., carpentry). In contrast, factories and offices called for literate and numerate workers. Hence, government leaders interested in economic development promoted mass education.

Globalization

Many of the conditions that contributed to mass education in the West now exist in the world's less developed countries. Religious authority is growing weaker, democracy is growing stronger, and new governments require the loyalty of their citizens and see education as necessary for economic development (McMahon, 1999). In addition, transnational corporations require a more literate and highly educated world population to do business, and transnational organizations such as the United Nations promote literacy and schooling. Still, this high moral principle remains a far-off goal. There are more than 860 million illiterate adults in the world, two-thirds of them women. In sub-Saharan Africa, 40 percent of primary school–age children do not attend school (UNESCO, 2002) (◗Figure 17.6).

Figure 17.6 Male and Female Illiteracy, Less Developed Countries, 2000.

Outside the richest countries of the world, female illiteracy is higher than male illiteracy. But is male illiteracy distributed across countries in much the same way as is female illiteracy? In other words, do high male and female illiteracy go hand in hand? Do low male and female illiteracy? Compare these two maps to find the answer. Do you think illiteracy is always a function of economic development? For example, compare illiteracy rates in Russia (with a gross domestic product per capita [GDPpc] in 2000 of $7700), Argentina (with a GDPpc in 2000 of $12,900), and Spain (with a GDPpc in 2000 of $18,000). Do illiteracy rates differ much in these countries? If so, why? If not, why not?

Source: UNESCO (2001).

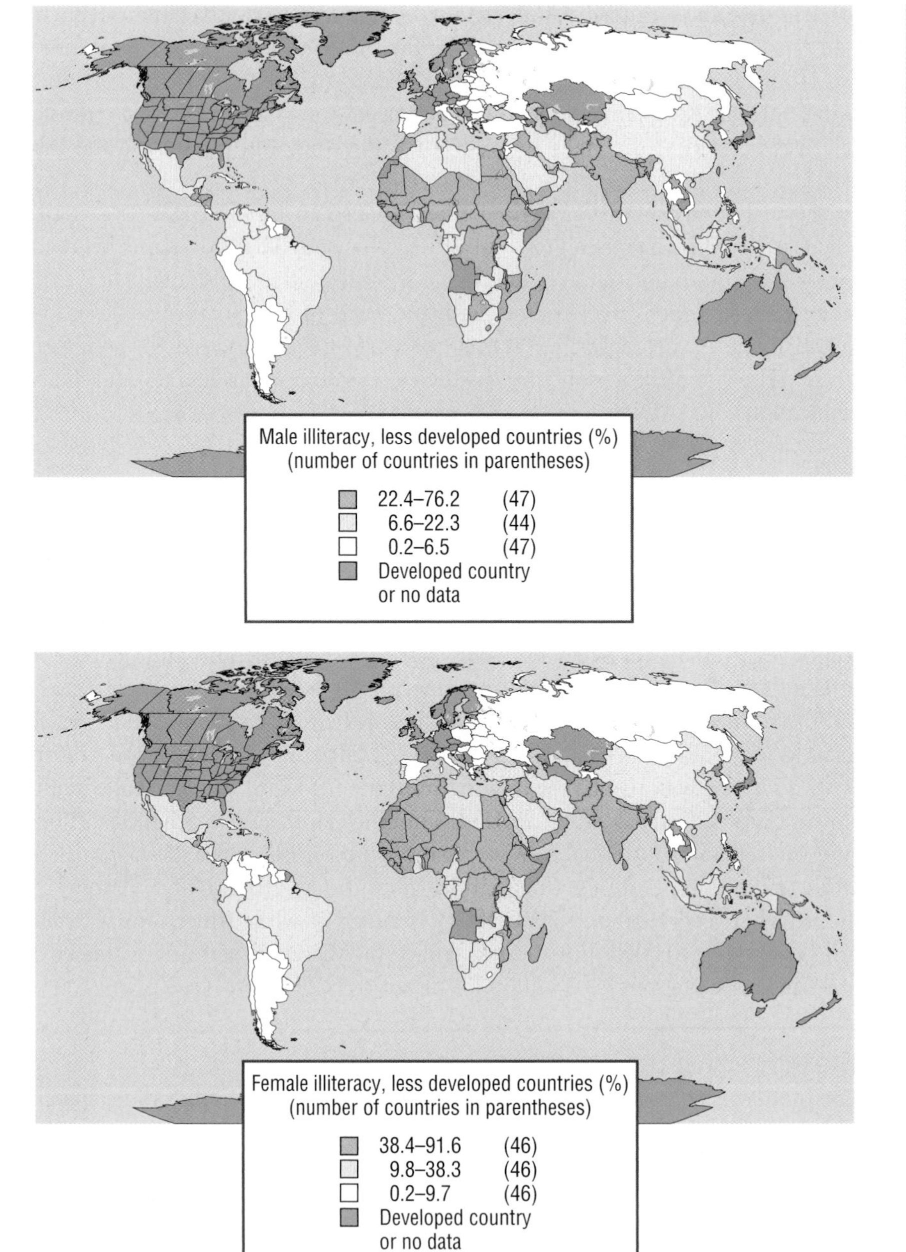

Credential Inflation and Professionalization

One of the most noteworthy recent developments in the field of education worldwide is what sociologist Randall Collins calls **credential inflation** (Collins, 1979). In Collins's usage, credential inflation refers to the fact that it takes ever more certificates and diplomas to qualify for a given job. For example, a century ago, many professors, even at the most prestigious universities, did not hold a doctoral degree. Today, most professors are Ph.D.s.

Credential inflation refers to the fact that it takes ever more certificates and diplomas to qualify for a given job.

Part of the reason for credential inflation is the increasing technical requirements of many jobs. For example, because aircraft engines and avionics systems are more complex than they were, say, 75 years ago, working as an airplane mechanic today requires more technical expertise. Certification ensures that the airplane mechanic can meet the higher technical demands of the job. As Collins points out, however, in many jobs there is a poor fit between one's credentials and one's specific responsibilities. On-the-job training, not a diploma or a degree, often gives people the skills they need to get the job done. Yet, according to Collins, credential inflation takes place partly because employers find it a convenient sorting mechanism. For example, an employer may assume that an Ivy League graduate has certain manners, attitudes, and tastes (cultural capital) that will be useful in a high-profile managerial position. Just as family background used to serve as a way of restricting high-status occupations to certain people, credentials serve that purpose today.

Corbis

Because professionalization promotes high standards and high earnings, it has spread widely. Even some clowns now consider themselves professionals. Thus, there is a World Clown Association (WCA) that has turned rubber noses and big shoes into a serious business.

Credential inflation is also fueled by **professionalization.** Professionalization occurs when members of an occupation insist that people earn certain credentials to enter the occupation. Professionalization ensures that standards are maintained. It also keeps earnings high. After all, if "too many" people enter a given profession, the cost of services offered by that profession is bound to fall. Professionalization thus serves as a way of safeguarding monopolies over knowledge and expertise (Derber, Schwartz, and Magrass, 1990). Professionalization helps explain why Ph.D.s typically earn less than professionals such as doctors and lawyers, even though Ph.D.s typically have more years of formal education than professionals (see Figure 17.4). The American Medical Association and the American Bar Association are powerful organizations that regulate and effectively limit entry into the medical and legal professions. American professors have never been in a position to form such powerful organizations.

Because professionalization promotes high standards and high earnings, it has spread widely. Even some clowns now consider themselves professionals. Thus, there is a World Clown Association (WCA) that has turned rubber noses and big shoes into a serious business. At its 18th annual conference in 2000, the WCA held seminars on a wide variety of subjects, including "character development," "on-target marketing," "incredible bubbles," and "simple but impressive balloons" (Prittie, 2000). Sociologist David K. Brown tells the story of a friend who had been a successful plumber for more than 20 years but tired of the routine and decided to become a clown (Brown, 1995: xvii). His friend quickly found that becoming a clown is not just a matter of buying a costume and acting silly. First, he had to enter the "Intensive Summer Clown Training Institute" at a local college. The institute awarded him a certificate signifying his competence as a clown. Based on his performance at the institute, the prestigious Ringling Brothers Clown School in Florida invited him to enroll. Even a clown, it seems, needs credentials these days, and a really good clown can dream of going on to clown graduate school.

Contested Terrain: Crisis and Reform in U.S. Schools

School Standards

Globalization of education has increased interest in comparing education in the United States with education in other countries. Such comparisons often show American students to be performing relatively poorly against their foreign counterparts. Consequently, many Americans believe that the U.S. school system has turned soft if not rotten. This has been true at least since the publication of *A Nation at Risk* (National Commission on Excellence in Education, 1983). Critics argue that the youth of Japan and South Korea spend long hours concentrating on the basics of math, science, and language. Meanwhile, American students spend fewer hours in school and study subjects that are supposedly of little practical value. From this point of view, art classes, drama programs, athletics, and sensitivity to cultural diversity in the American school curriculum are harmful distractions. If students don't spend more school time on subjects that "really" matter, the United States can expect to suffer declining economic competitiveness in the 21st century, the critics charge.

Professionalization takes place when members of an occupation insist that people earn certain credentials to enter the occupation. Professionalization ensures standards and keeps professional earnings high.

Many Americans believe that the public school system has turned soft if not rotten. They argue that the youth of Japan and South Korea spend long hours concentrating on the basics of math, science, and language. Meanwhile, American students spend fewer hours in school and study more nonbasic subjects that are supposedly of little practical value. As the text shows, the reality is more complex than this simple characterization.

The Purcell Team/Corbis

Evidence apparently showing the inferiority of the American school system emerges from the ongoing Trends in International Math and Science Study (TIMSS). This massive crossnational research effort regularly tests hundreds of thousands of students on their math and science skills in grades 4 and 8 and the final year of school. The 1999 study found that, compared with grade 4 students in 25 other countries, American fourth graders did respectably well. They placed eighth in their combined math and science scores. However, the relative standing of American students dropped as they progressed through the school system. By the time they reached the last year of school, Americans placed 18th out of 21 countries. Table 17.4 gives the 2003 results for eighth-grade students in 45 countries and the 1999 results for eighth-grade students in 6 other countries that did not participate in the 2003 survey. It shows that the United States ranked 18th in math and 13th in science. This places the United States in about the top third but significantly below a number of countries in Asia and Europe.

Despite the alarms sounded by the Department of Education and various business and political organizations in the United States, some analysts judge the TIMSS test results misleading (Bracey, 1998; Schrag, 1997). In the first place, some of the countries that participate in the study do not follow sampling guidelines. For political reasons, some of them exclude groups of students who educational administrators thought would do poorly on the exam. These countries artificially inflate their TIMSS scores. Second, different countries have different kinds of secondary school systems. For instance, some keep students in school for 14 years, while others, like the United States, have 12-year systems. Moreover, some countries have higher dropout rates than the United States and siphon off poor academic performers to trade schools and job-training programs before they graduate high school. This leaves only the top academic performers in the last year of high school. In contrast, the United States tries to make sure that as many students as possible graduate high school, because this enhances the quality of democracy, increases social cohesion in a culturally diverse society, and improves economic performance. If one compares twelfth-grade students from the inclusive United States system with fourteenth-grade students from an elite school system, the United States students are bound to look inferior. Thus, international comparisons, such as those made possible by the TIMSS study, are sometimes misleading.

Recognizing the limitations of international comparisons is not an excuse for ignoring the academic flabbiness of many U.S. schools. It is an opportunity to recognize where

Table 17.4
Math and Science Achievement Scores in 51 Countries, 2003 (* = 1999)

Math		Science		Math		Science	
Average	466	Average	473	Italy	484	Israel	488
Singapore	605	Singapore	578	Armenia	478	Thailand*	482
South Korea	589	South Korea	558	Serbia	477	Bulgaria	479
Hong Kong	586	Hong Kong	556	Bulgaria	476	Jordan	475
Taiwan	585	Estonia	552	Romania	475	Moldova	472
Japan	570	Japan	552	Thailand*	467	Romania	470
Belgium (Flemish)	537	Hungary	543	Norway	461	Serbia	468
Netherlands	536	Czech Republic*	539	Moldova	460	Armenia	461
Canada*	531	England*	538	Cyprus	459	Iran	453
Estonia	531	Netherlands	536	Macedonia	435	Macedonia	449
Hungary	529	Finland*	535	Lebanon	433	Cyprus	441
Finland*	520	**Canada***	**533**	Turkey*	429	Bahrain	438
Czech Republic*	520	**Australia**	**527**	Jordan	424	Turkey*	433
Latvia	**508**	**United States**	**527**	Indonesia	411	Palestine	435
Malaysia	**508**	**Sweden**	**524**	Iran	411	Egypt	421
Russia	**508**	**New Zealand**	**520**	Tunisia	410	Indonesia	420
Slovak Republic	**508**	**Slovenia**	**520**	Egypt	406	Chile	413
Australia	**504**	Lithuania	519	Bahrain	401	Tunisia	404
United States	**504**	Slovak Republic	517	Palestine	390	Saudi Arabia	398
Lithuania	**502**	Belgium (Flemish)	517	Chile	387	Lebanon	393
Sweden	**499**	Russia	514	Morocco	387	Morocco	396
Scotland	**498**	Latvia	512	Philippines	378	Philippines	397
England*	**496**	Scotland	512	Botswana	366	Botswana	365
Israel	**496**	Malaysia	510	Saudi Arabia	332	Ghana	255
New Zealand	**494**	Norway	494	Ghana	276	South Africa	244
Slovenia	493	Italy	491	South Africa	264		

Countries significantly higher than the United States.
Countries significantly lower than the United States.
Source: National Center for Education Statistics (2000; 2004b).

the real problem lies. As one educator notes, "The top third of American schools are world-class, . . . the next third are okay, and the bottom third are in terrible shape" (Bracey, 1998). It would probably do students a lot of good if expectations and standards were raised in the entire school system. However, the real crisis in American education can be found not in upper-middle-class suburban schools but in the schools that contain many disadvantaged minority students, most of them in the inner cities. We need to keep this in mind when discussing the sensitive issue of school reform.

Solutions to the School Crisis

At the clubhouse, I work with Lakesha. She is a mentor, which means she knows a lot about computers. When she is not at the clubhouse, she is an engineer. She shows me how to do lots of fun things with computers like controlling LEGO robots. I want to learn about engineering in college.

—Latoya Perry, age 13 (quoted in Resnick and Rusk, 1996)

Much research underscores the crisis in the American school system. Here we restrict ourselves to mentioning two especially discouraging sets of indicators. Four U.S. Department of Education surveys conducted between 1983 and 2000 contain data on twelfth-grade students' attitudes toward school (Figure 17.7). The results show that the percentage of

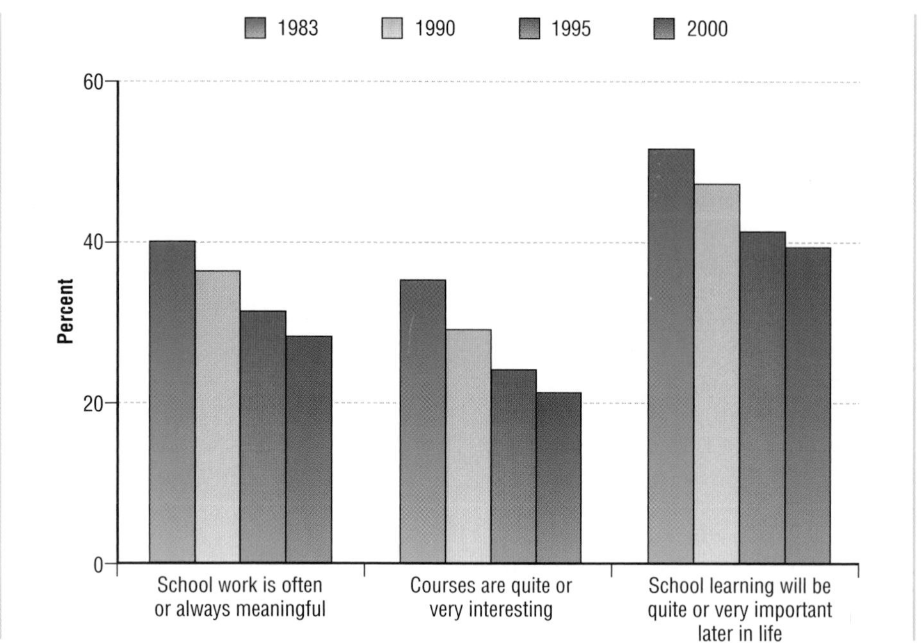

Figure 17.7
Interest in School, 12th Graders, United States, 1983-2000

Source: National Center for Education Statistics (2002b: 72).

twelfth graders who think school is always or often meaningful declined from 40 percent to 28 percent in that 17-year period. The percentage claiming that courses are quite or very interesting fell from 35 percent to 21 percent. The percentage who believe that what they learn in school will be quite or very important later in life dropped from 51 percent to 29 percent. Thus, an increasing number of students graduating from American high schools think that their education is meaningless, boring, and irrelevant.

Not surprisingly, discipline problems blossom in such an environment. A 1999 survey of principals in five countries shows that American principals were more likely to report disruptive behavior problems than principals from Canada, Russia, Italy, and Japan. Specifically, the United States ranked first on five of seven reported behavior problems, including classroom disturbance, theft, physical injury to other students, intimidation or verbal abuse of other students, and intimidation or verbal abuse of teachers or other staff. Italy took the dubious honor of beating the United States on cheating, and Canada edged the United States on vandalism (Figure 17.8). Interestingly, principals in the country with the fewest behavior problems, Japan, tended to regard the problems as most serious. Conversely, principals in the country with the most behavior problems, the United States, tended to regard them as least serious. You can exercise your sociological imagination to explain this negative correlation.

Statistics on completion and dropout rates suggest that the crisis of the American school system is strongly related to minority status, which in turn is strongly related to class position. For example, in 2002, more than 21 percent of non-Hispanic whites older than 24 had a B.A. In contrast, only 17 percent of African Americans and 11 percent of Hispanic Americans older than 24 had a B.A. Nearly 89 percent of non-Hispanic whites older than 25 had a high school diploma. This compares with just 79 percent of African Americans and 57 percent of Hispanic Americans (U.S. Census Bureau, 2002). The high school dropout rate for 16 to 24 year olds also varies sharply by racial and ethnic group. In 2002, the dropout rate for white non-Hispanics was less than 7 percent. For black students it was a little more than 13 percent, and for Latinos it was nearly 28 percent (National Center for Education Statistics, 2002a: 164).

Because educational attainment is the single most important factor that determines income, Americans have been trying for decades to figure out how to improve the educational attainment of disadvantaged minorities. For 40 years, the main strategy has been school desegregation by busing. By trying to make schools more racially and ethnically in-

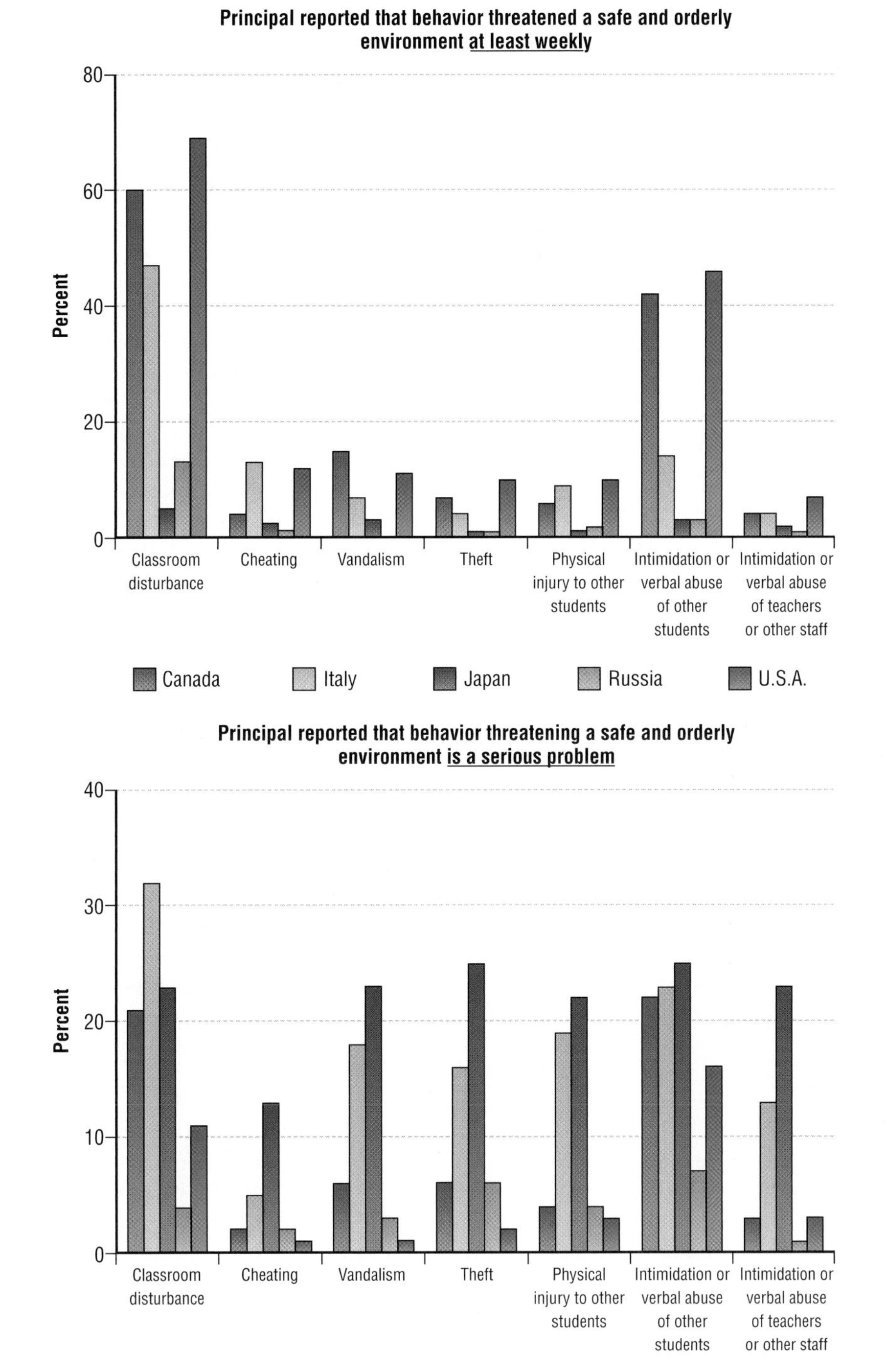

Figure 17.8
Eighth-Grade Student Behavior Problems, Five Countries

Note: Data are for 1999.
Source: National Center for Education Statistics (2003).

tegrated, many people hoped that the educational attainment of disadvantaged students would rise to the level of the more advantaged students.

Things did not work out as hoped. Instead of accepting busing and integration, many white families moved to all-white suburbs or enrolled their children in private schools, which are more racially and ethnically homogeneous than public schools. Meanwhile, in integrated public schools, gains on the part of minority students were limited by racial tensions, competition among academically mismatched students from widely divergent family backgrounds, and related factors (Parillo, Stimson, and Stimson, 1999). Research shows that desegregation closes only 10 percent to 20 percent of the academic gap between black and white students (Jencks et al., 1972).

BOX 17.3
SOCIAL POLICY: WHAT DO YOU THINK?

The No Child Left Behind Act

Despair over the quality of public education in the United States is widespread. Not surprisingly, therefore, people have suggested many ways of improving the schools. In the late 1990s, Caroline Minter Hoxby proposed an idea that is intriguing but controversial. She believes that many problems in the education system stem from the fact that bad schools are never punished. For instance, a school's budget is unaffected if it has a high dropout rate and many of its students do poorly on their SATs. Many people feel that this is as it should be. They think that slashing the budgets of bad schools would only serve to punish their students. Hoxby has no interest in punishing students for the poor quality of their schools. Instead, she argues that public funding of schools should be cut and parents should be given school vouchers valued at $2500 to $5000 a year per child. Parents would be free to enroll their children in any school they wanted by giving the vouchers to the schools of their choice. In this way, children would be moved out of bad schools and into good ones. Schools' budgets would depend mainly on how many students enroll in them. This system, which effectively promises to privatize public education, would help children go to better schools and would give the bad schools an incentive to improve, says Hoxby. Eventually, if administrators don't improve the bad schools, they will be forced to close them for lack of funding. Hoxby notes: "It would be like restaurants. . . . New, successful schools would take over from old failing schools, and the old ones would disappear" (quoted in Cassidy, 1999: 147).

Although President Bush likes the voucher idea, the desire to build a broad consensus led him to modify it and propose the No Child Left Behind Act, which was passed in 2002. The law seeks to make schools accountable for their performance. It establishes a national system of tests, graduation rates, and other indicators that allow administrators to track and compare the performance of schools. If a school fails to reach a state-defined level of proficiency for two years running, it is subjected to escalating sanctions and interventions. Children in low-performing schools must be provided extra help if they need it and they are allowed to transfer to better schools, to which the school district must provide transportation. The act also promises to provide the resources needed to place highly qualified teachers in every classroom by 2005–6.

A majority of Americans and their political representatives from both parties supported the No Child Left Behind Act when it was passed. Yet is has failed to achieved its aims. Some of the main reasons for its failure include the following:

1. The administration has not honored its funding commitments. States have been saddled with noble and expensive policy objectives, but the federal government has provided little money to meet them.
2. States are allowed to define their own proficiency standards (including the precise definition of a "highly qualified teacher"). As a consequence, some states have *lowered* their standards to make it *seem* as if they are meeting national goals. So, for example, while Michigan reported that 40 percent of its schools failed to reach their annual progress goals one year, Arkansas and Wyoming reported that *none* of its schools failed to meet those goals—yet school standards in Arkansas and Wyoming are actually lower than those in Michigan.
3. Many schools with a high proportion of minority students, especially students who speak little English, find it extremely difficult to reach mandated progress goals. Yet few minority students transfer to other schools because parents want their children to stay in the neighborhood, other schools are already overcrowded, or there are no alternatives in the area. In August 2003, 250,000 Chicago students were eligible for transfer but only 19,000 applied and a mere 1100 went to a new school (Mathis, 2003; Schrag, 2004).

Do you think the No School Left Behind Act is a good idea in principle? Why or why not? Could its implementation be modified to make it work better? If so, how? If not, why not?

The limited success of desegregation has convinced many people that instead of pouring money into busing, a wiser course of action would be to improve the quality of traditionally underfinanced minority schools. Many educators fear that ignoring integration will deny American students from different races and ethnic groups the opportunity to learn to work and live together. Nevertheless, the movement to focus on improving school quality is gaining momentum.

Proposals to improve school quality fall into four main categories (see Box 17.3 for the fourth of these). First are *local initiatives* that can improve schooling without changing the distribution or current cost of educational resources. Second are *government initiatives* that can improve schooling by redistributing existing resources and increasing school budgets. Third are solutions that look outside the school system and stress the need for *comprehensive preschools for children and job training and job creation for disadvantaged parents* as the main ways to encourage upward mobility and end poverty. Let us consider each of these types of reform in turn.

Local Initiatives

We already noted one local initiative in our discussion of the self-fulfilling prophecy that disadvantaged students are bound to do poorly in school. By challenging minority students, giving them emotional support and encouragement, preparing a curriculum that gives more recognition to the accomplishments of their group, and creating an environment in which they can relax and achieve, they do better in school.

A second local initiative is the mentoring movement. It involves community members volunteering to work with disadvantaged students in schools, church basements, community centers, and housing projects. There, the mentors tutor students, socialize with them, act as role models, and impart practical skills. An outstanding example of the mentoring movement is the series of drop-in "computer clubhouses" that have been established since 1993 for underprivileged youth between the ages of 10 and 16 (Resnick and Rusk, 1996). These computer clubhouses are not places of traditional classroom instruction with an emphasis on computers. Nor are they places where computer resources are simply made available to be used in whatever way young people want. Instead of surfing the Web, students make waves. That is, the mentors build on students' existing interests in cartoons, dance, music, and so forth to help them create specific design projects that use computer technology. Emilio sees a laser-light show and wants to create one himself. To do this, he has to glue mirrors to robotic motors and then learn how to write code to precisely control the motors' movements. That way, when he reflects laser light off the mirrors, he achieves the effects he wants. He has to learn much mathematical thinking along the way. Paul arrives from Trinidad without ever having used a computer, but he likes to draw cartoon characters. Two years later he has learned enough skills at the computer clubhouse to land a part-time job designing webpages for a local company. He hopes to pursue a career in computer animation and graphic design. Sandi, a Native American girl, wants to learn more about her heritage, so she creates a multimedia show using text, graphics, photographs, and sound. The project impresses her mentor, who hopes that her example will stimulate other pupils to create similar projects. Latoya Perry, the 13-year-old African American girl whose eloquent words opened this section, models herself after her mentor, Latesha, who teaches Latoya how to program robots. Latoya now wants to become an engineer herself. All these success stories come from the first year of operation of the computer clubhouse in Boston. They show how much creating an environment of trust and respect can accomplish. In many cases it is possible to partly erase the effects of material disadvantage by providing adequate resources for education, building on students' existing interests, and developing concrete design projects rather than expecting students either to passively absorb information or learn on their own.

Redistributing and Increasing School Budgets

Although we could tell more success stories, we need to stress that local initiatives can go only so far to improve the quality of education in the United States. Computer clubhouses require expensive equipment, software, and Internet connections. The teaching profession needs higher salaries to attract more inspiring teachers and reduce classroom size, because smaller classes have a positive, long-term impact on student achievement (Hacsi, 2002: 206). And no amount of teaching cultural diversity is going to buy desks, repair a leaky roof, or get rid of rodents scouring the school grounds for scraps of food. According to a report by the U.S. General Accounting Office, the investigative arm of Congress, a third of the country's schools need major repairs or outright replacement (Tornquist, 1998). This requires money. The second set of proposed educational reforms speaks to the need for increased investment in education and redistributing existing resources (Reich, 1991).

Where could the money come from and how could existing funds be redistributed? One possibility is that federal, state, or local governments collect school taxes and tie them at least in part to people's ability to pay. As in some other postindustrial countries, well-to-do people could be obliged to pay a higher *rate* of school tax than the less well-to-do, and funds for schools could then be distributed more equitably to communities, whatever their wealth. As things currently stand, the neediest schools in the country receive about $1000 less annually per student than schools with the fewest poor children. The greatest disparity is in New York State, where the gap was $2152 in 2002 (Schemo, 2002).

Economic Reform and Comprehensive Preschools

The problem with the first two proposed solutions for the school crisis is that attempts to implement them have met with limited success. Sociologists began to understand how little schools could do on their own to encourage upward mobility and end poverty in the 1960s, when sociologist James Coleman headed a monumental survey of American schools (Coleman et al., 1966). Coleman began his research convinced that the educational achievement of black children was due to the underfunding of their schools. What he found was that differences in the quality of schools—measured by assessments of such factors as school facilities and curriculum—accounted at most for about a third of the variation in students' academic performance. At least two-thirds of the variation in academic performance was due to inequalities imposed on children by their homes, neighborhoods, and peers.

Four decades later, little research contradicts Coleman's finding. Consider the results of Project Head Start, a federal government program that provides modest educational, medical, and social services to economically disadvantaged preschool children and their families. Head Start was often hailed as a major accomplishment of President Johnson's War on Poverty. Yet follow-up studies found that gains in cognitive and socioemotional functioning registered by Head Start graduates disappeared within about two years of their leaving the program.

Similarly, between 1964 and 2000, Congress dispensed more than $100 billion to schools under Title I of the Elementary and Secondary Education Act. Until 1994, most of the money was used for additional instruction for students who were falling behind, especially in reading. Yet a 1997 study conducted for the Department of Education found no academic differences between students who received assistance and those who did not. Money alone, it seems, does not buy educational equality (Traub, 2000: 55).

On the basis of the research just reviewed, we conclude that programs aimed at increasing school budgets and encouraging local school-reform initiatives need to be augmented by policies that improve the social environment of young, disadvantaged children *before and outside* school (Hertzman, 2000). Most children do not go to school until they are 5 years old. When they enroll, school takes up less than half their waking day.

Paul Conklin/PhotoEdit

Research suggests that programs aimed at increasing school budgets and encouraging local school reform initiatives need to be augmented by policies that improve the social environment of young, disadvantaged children *before and outside* school. Head Start is insufficient in this regard.

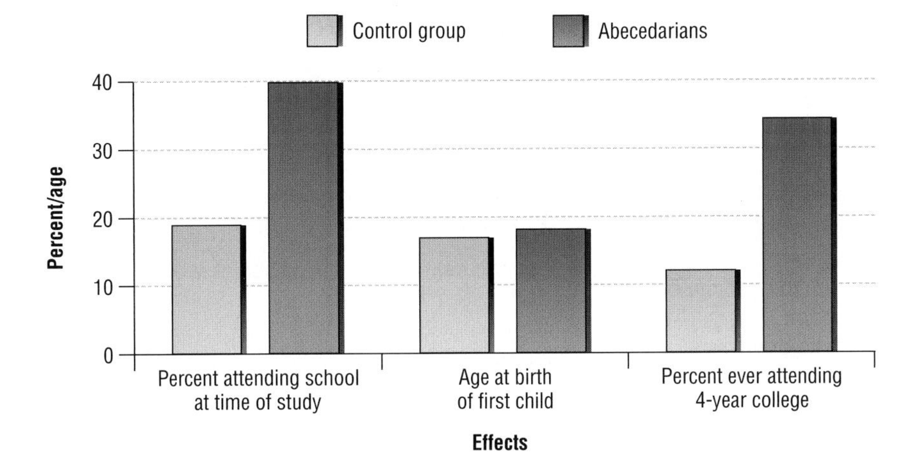

Figure 17.9
Some Effects of Comprehensive Preschool on 21-Year-Olds (98 percent African American, N = 111)

Source: F.A. Campbell, C.T. Ramey, E.P. Pungello, S. Miller-Johnson, and J. Sparling (1992), *Early Childhood Education: Young Adult Outcomes from the Abecedarian Project*, Applied Developmental Science. Used by permission.

Academic work takes up only half the school day. What happens the rest of the time, at home and with peers, has a major impact on academic and, eventually, occupational success. Children from disadvantaged homes do better in school if their parents create a healthy, supportive, and academically enriching environment at home and if peers do not lead children to a life of drugs, crime, and disdain for academic achievement. Policies aimed at helping to create these conditions—job training and job creation for parents, comprehensive child and family assistance programs that start when a child is born—would go a long way toward improving the success rate of programs that increase school budgets and encourage local reform initiatives (Hacsi, 2002; Meier, 2002; Wagner, 2002).

A few model comprehensive child and family assistance programs exist. One such program is the Abecedarian Project in North Carolina, funded by the federal government. Another is a state-funded preschool program in Vineland, New Jersey. Children are enrolled in these programs at birth. The programs last all day. The teachers are trained professionals. The programs offer individualized child care, primary health care, nutritional supplements, adult literacy courses, and education and counseling for pregnant teenagers. Classes are small and richly stocked with books, art supplies, and computers. In effect, these programs form alternative communities, replacements for the troubled neighborhoods in which the children and their mothers would otherwise spend their days. They yield impressive results. A 1999 study of the Abecedarian Project analyzed 21-year-old graduates of the program who had been enrolled as infants between 1972 and 1977. Ninety-eight percent of them were African American. Two decades after they first entered the preschool program, they scored higher than a control group on math and reading achievement tests. Those graduates who had their own child were nearly two years older when the child was born than comparable members of the control group. They were far more likely than members of the control group to be attending an educational institution at the time of the study and far more likely to have ever attended a four-year college (Campbell and Ramey, 1994; Campbell et al., 2002) (Figure 17.9).

Offering such programs to all poverty-level children in the United States would cost three times more than current Head Start programs, which are far inferior in quality and comprehensiveness. Effective job training and job creation programs for poor adults would also be very costly. Realistically speaking, many people are likely to oppose such reforms at this time. After all, this is an era when governments are cutting school budgets in the inner city. It is an era when many parents prefer to send their children to private schools or move to neighborhoods in the suburbs with excellent public schools if they can afford it (Orfield and Eaton, 1996). On the other hand, one must remember that 72 percent of Americans in the 2002 General Social Survey said too little money is being spent on the nation's schools. Education was the country's number two priority, just 1 percent behind "improving and protecting the nation's health" (Chapter 9, "Globalization, Inequality, and Development," Table 9.5). This suggests that most Americans still want the education system to live up to its ideals and serve as a path to upward mobility.

Summary

Sociology Now™

Reviewing is as easy as ❶, ❷, ❸.

❶ Before you do your final review, take the SociologyNow diagnostic quiz to help you identify the areas on which you should concentrate. You will find information on how to access SociologyNow on the foldout at the front of the textbook.

❷ As you review, take advantage of SociologyNow's study aids to help you master the topics in this chapter.

❸ When you are finished with your review, take SociologyNow's post-test to confirm you are ready to move on to the next chapter.

1. **What are the arguments for and against affirmative action?**

   Advocates of affirmative action say it compensates for historical injustices, encourages ethnic and racial diversity, and creates middle-class leaders in minority groups. Opponents argue that people should not have to compensate for injustices committed up to 300 years ago, that affirmative action ignores individual rights and differences, and that it diminishes the achievements of minority students.

2. **What are the manifest and latent functions of schools?**

   The manifest functions of schools include the training and socialization of students, the creation of social cohesion, the transmission of culture from generation to generation, and the sorting of students, presumably by merit. Latent functions include the creation of a youth culture, a marriage market, a custodial and surveillance system for children, a means of maintaining wage levels by keeping college students temporarily out of the job market, and occasionally becoming a "school of dissent" that opposes authorities.

3. **What are the sources and effects of economic inequality on education?**

   Economic inequality creates schools of widely differing quality, communities that are able to support or tend to discourage high-quality education, and families with varying degrees of access to material and emotional resources for the support of children. As a result of economic inequality, children enter school with widely differing levels of preparation and eagerness to learn. These differences increase as students work their way through school. Consequently, although schools do offer some opportunities for upward mobility, they also help to reproduce existing social inequalities.

4. **What do standardized tests measure and what are their effects?**

   Standardized tests are supposed to measure innate ability (IQ tests) or mathematical and reasoning abilities related to performance in college (the SAT and ACT tests). To some extent, they do. Thus, to a degree, they help to sort students by ability and aid in the creation of a meritocracy. However, they also measure students' preparedness to learn and thrive in school, and preparedness is strongly related to background factors such as family's race and class position and the social composition of schools. Therefore, standardized tests also help to reproduce existing social inequalities.

5. **How do self-fulfilling prophecies work in the education system?**

   Teachers' expectations that certain students will do poorly in school often result in poor student performance. Teachers' expectations that certain students will do well in school often result in good student performance. These expectations reinforce the effects of background factors and, like background factors, help to reproduce existing patterns of inequality.

6. **What is the role of cultural capital in academic achievement?**

   In the United States, cultural capital—elements of high-status culture—tends to be learned at school more than at home. Thus, acquiring cultural capital in schools is a path to upward mobility for low-status students.

7. **What factors account for the rise of mass, compulsory education?**

   The spread of mass, compulsory education has been encouraged by the Protestant Reformation, democratic revolutions, the modern state, industrialization, and globalization. Today, all countries have systems of mass, compulsory education, although illiteracy is still widespread in poor countries, especially in sub-Saharan Africa.

8. **What is credential inflation and how is it related to professionalization?**

   Credential inflation (the need for more certification and diplomas to qualify for a given job) has been fueled by the increasing technical requirements of many jobs. It has also been encouraged by the ability of people in certain occupations to exercise control over their occupations (professionalization). Credential inflation is thus a means of excluding people from the professions to maintain high standards and income levels.

9. **How do American schools compare with schools in other countries?**

   Although crossnational surveys show that the American school system is an average to poor performer, the surveys are often misleading because weaker students in some countries do not participate in the surveys. Only the bottom third of American schools have very low standards.

10. **What are the major reforms that have been proposed to deal with the crisis of American schools?**

    The major school reforms that have been proposed in recent years include local initiatives such as mentoring, giving students in poor schools vouchers that would allow them to attend private schools, redistributing and increasing school budgets, and substantially improving the social environment of young, disadvantaged children before and outside school.

Questions to Consider

1. In your opinion, how meritocratic were the schools you attended? Did the most talented students tend to perform best? Did material advantages and parental support help the best students? Did material disadvantages and lack of parental support hinder the achievements of weaker students?
2. How would you try to solve the problem of unequal access to education? Do you favor any of the older approaches, such as busing children from poor districts to wealthier districts or greater federal control over education budgets? What do you think about the solutions to the school crisis discussed at the end of this chapter? Do you have some suggestions of your own?

Web Resources

Companion Website for This Book

http://sociology.wadsworth.com

Begin by clicking on the Student Resources section of the website. Choose "Introduction to Sociology" and then the Brym and Lie book cover. Next, select the chapter you are currently studying from the pull-down menu. From the Student Resources page you will have easy access to InfoTrac® College Edition, MicroCase Online exercises, additional web links, and many other resources to aid you in your study of sociology, including practice tests for each chapter.

Recommended Websites

Search for "education" using the search engine of the General Social Survey at http://www.icpsr.umich.edu:8080/GSS/homepage.htm. The search engine will retrieve dozens of research reports and tables based on this respected, ongoing survey of the American public.

The United States Department of Education website at http://www.ed.gov/index.jsp contains a wealth of policy-related material and statistics on education, as does the website of the National Center for Education Statistics at http://www.nces.ed.gov.

UNESCO maintains an extremely informative website at http://www.unseco.org. It is full of revealing documents and startling statistics on international and comparative education and literacy.

CHAPTER 18 The Mass Media

John Henley/Corbis

In this chapter, you will learn that:

- Movies, television, and other mass media sometimes blur the distinction between reality and fantasy.
- The mass media are products of the 19th and especially the 20th centuries.
- Historically, the growth of the mass media is rooted in the rise of Protestantism, democracy, and capitalism.
- The mass media make society more cohesive.
- The mass media also foster social inequality.
- Although the mass media are influential, audiences filter, interpret, resist, and even reject media messages if they are inconsistent with their beliefs and experiences.
- The mass media misrepresent women and racial minorities in important ways.
- The interaction between producers and consumers of media messages is most evident on the Internet.

The Significance of the Mass Media

Illusion Becomes Reality

The turn of the 21st century was thick with movies about the blurred line separating reality from fantasy. *The Truman Show* (1998) gave us Jim Carrey as an insurance sales agent who discovers that everyone in his life is an actor. He is the unwitting subject of a television program that airs 24 hours a day. In *The Matrix* (1999), Keanu Reaves finds that his identity and his life are illusions. Like everyone else in the world, Reaves is hardwired to a giant computer that uses humans as an energy source. The computer supplies people with nutrients to keep them alive and simulated realities to keep them happy. Similar blurring between reality and media-generated illusion was evident in *Pleasantville* (1998), *EdTV* (1999), and *Nurse Betty* (2000).

The most disturbing movie in this genre is *American Psycho* (2000). Based on a novel banned in some parts of North America when it was first published in 1991, the movie is the story of Patrick Bateman, Wall Street yuppie by day, cold and meticulous serial killer by night. Unfortunately, the public outcry over the horrifying murder scenes virtually drowned out the book's important sociological point. *American Psycho* is really about how people become victims of the mass media and consumerism. Bateman, the serial murderer, says he is "used to imagining everything happening the way it occurs in movies." When he kisses his lover, he experiences "the 70-mm image of her lips parting and the subsequent murmur of 'I want you' in Dolby sound" (Ellis, 1991: 265). In Bateman's mind, his 14 murder victims are mere props in a movie in which he is the star. He feels no more empathy for them than an actor would for any other stage object. The mass media have so completely emptied him of genuine emotion, he even has trouble remembering his victims' names. At the same time, however, the mass media have so successfully infused him with consumer values he can describe his victims' apparel in great detail—styles, brand names, stores where they bought their clothes, and even prices. Thus, in *American Psycho,* killer and killed are both victims of consumerism and the mass media.

Sociology Now™

Reviewing is as easy as ❶, ❷, ❸.

Use SociologyNow to help you make the grade on your next exam. When you are finished reading this chapter, go to the Chapter Summary for instructions on how to make SociologyNow work for you.

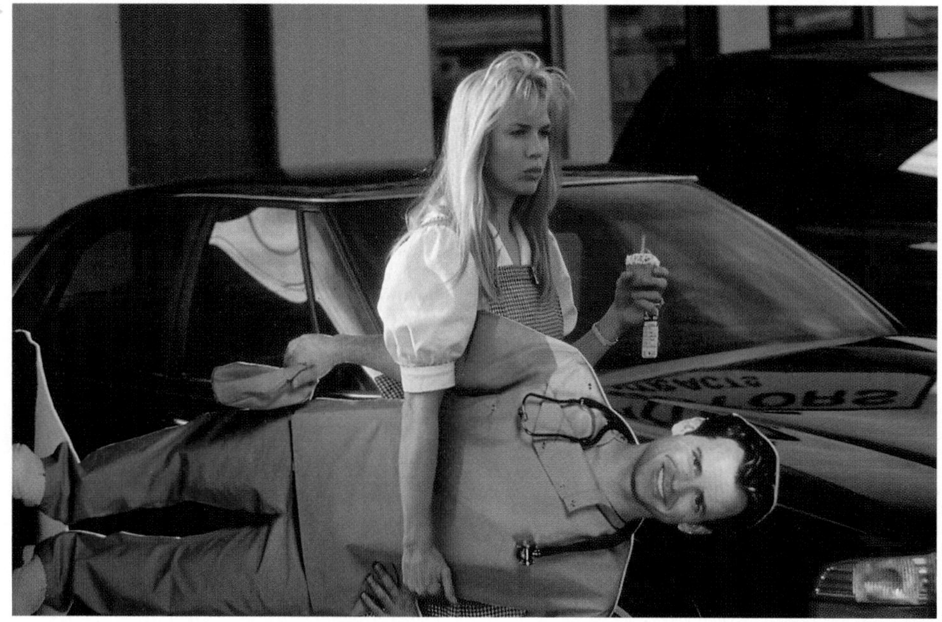

Many recent films—including *Nurse Betty,* starring Renée Zellweger and Morgan Freeman—suggest that the line between reality and media-generated illusion is becoming blurred.

The Everett Collection

In different ways, these movies suggest that the fantasy worlds created by the mass media are increasingly the only realities we know, every bit as pervasive and influential as religion was 500 or 600 years ago. Do you think this is an exaggeration dreamed up by filmmakers and novelists? If so, consider that of the 8760 hours in a year, the average American spends 3649 of them interacting with the mass media (Figure 18.1). That is 63 percent of our waking time, assuming we sleep an average of 8 hours a day. We spend more time watching TV, listening to the radio, going to the movies, reading newspapers and magazines, playing CDs, using the Internet, and so forth, than we do sleeping, work-

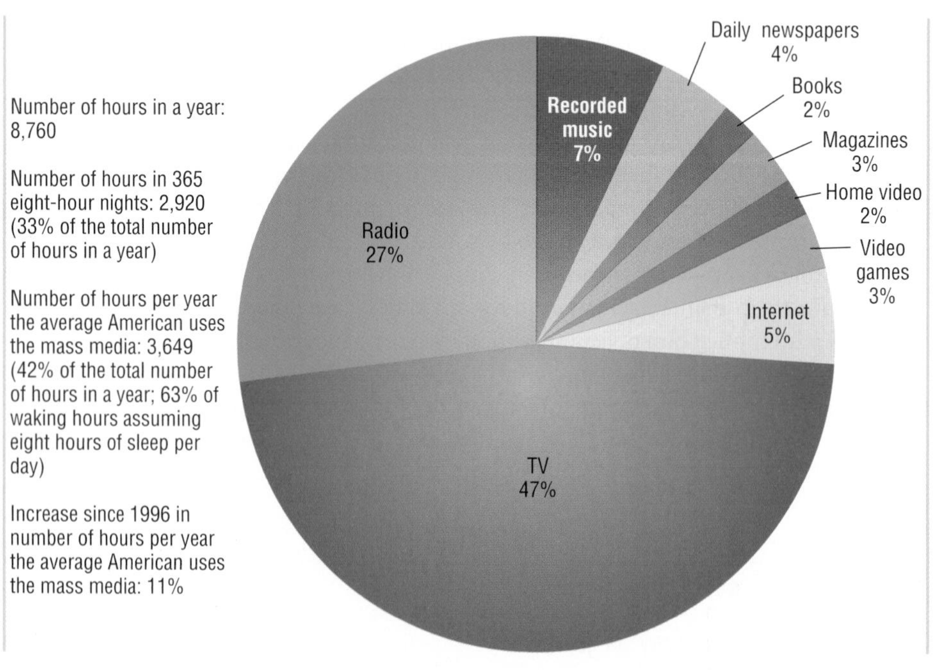

Figure 18.1
Media Usage, United States, 2005 (hours per capita, projected)

Source: U.S. Census Bureau (2002: 689).

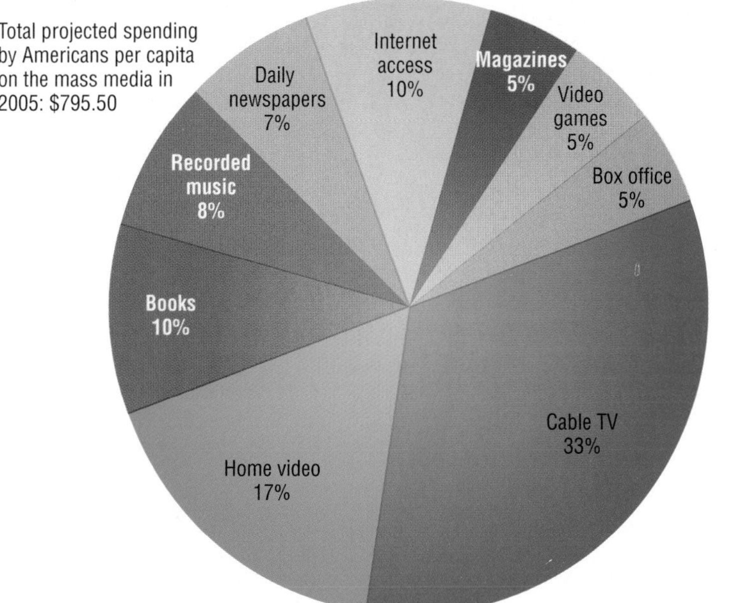

Figure 18.2
Media Usage, United States, 2005 (dollars per capita, projected)

Source: U.S. Census Bureau (2002: 689).

ing, or going to school. Figure 18.1 shows how we are projected to divide our time among the various mass media in 2005. Figure 18.2 shows how we are projected to spend our money. These graphs suggest that much of our reality is media generated. Media guru Marshall McLuhan, who coined the term "global village" in the early 1960s, said the media are extensions of the human body and mind (McLuhan, 1964). Forty years later, it is perhaps equally valid to claim that the human body and mind are extensions of the mass media (Baudrillard, 1983; 1988; Bourdieu, 1988) (Box 18.1).

What Are the Mass Media?

The term **mass media** refers to print, radio, television, and other communication technologies. Often, "mass media" and "mass communication" are used interchangeably to refer to the transmission of information from one person or group to another. The word "mass" implies that the media reach many people. The word "media" signifies that communication does not take place directly through face-to-face interaction. Instead, technology intervenes or mediates in transmitting messages from senders to receivers. Furthermore, communication via the mass media is usually one way, or at least one-sided. There are few senders (or producers) and many receivers (or audience members). Thus, most newspapers print a few readers' letters in each edition, but journalists and advertisers write virtually everything else. Ordinary people may appear on the Oprah Winfrey or David Letterman shows, enjoy play-along features on *Who Wants to Be a Millionaire?* and the *Today* show, and even delight in a slice of fame on *Survivor* or *The Amazing Race.* However, producers choose the guests and create the content for these programs. Similarly, a handful of people may visit your personal website, but an average of 87.5 million people visited the Yahoo search engine every month in 2002 ("Search Engines," 2002).

The **mass media** are print, radio, television, and other communication technologies. The word "mass" implies that the media reach many people. The word "media" signifies that communication does not take place directly through face-to-face interaction. Instead, technology intervenes or mediates in transmitting messages from senders to receivers. Furthermore, communication via the mass media is usually one way, or at least one-sided. There are few senders (or producers) and many receivers (or audience members).

Usually, then, members of the audience cannot exert much influence on the mass media. They can only choose to tune in or tune out. And even tuning out is difficult because it excludes one from the styles, news, gossip, and entertainment most people depend on to grease the wheels of social interaction. Few people want to be cultural misfits. This does not mean that people are always passive consumers of the mass media. As noted later, we filter, interpret, and resist what we see and hear if it contradicts our experience and beliefs. Even so, in the interaction between audiences and media sources, the media sources usually dominate.

Box 18.1

YOU AND THE SOCIAL WORLD

How the Mass Media Affect You

On a few sheets of paper, divide a 24-hour day into 288 segments, each segment representing five minutes. Keep a tally of your activities for a couple of days, allowing for the fact that you may engage in several activities simultaneously. (For example, you may listen to the radio while reading.) How much time do you devote to interacting with the mass media? How does the amount of time you spend interacting with the mass media compare with the amount of time you spend sleeping, studying, working for money, and participating in sports or physical exercise? Briefly describe how your use of the mass media compares with that of the average American as described in the text and in Figures 18.1 and 18.2.

Now consider the more difficult question of the nature of your interaction with the mass media. Do you think it is accurate to say that you get your ideas about how to dress, how to style your hair, and what music to listen to from the mass media? Or do you think it is more accurate to say that the mass media is like a smorgasbord from which you choose items that suit your taste? To what degree are your hopes, aspirations, and dreams truly your own? Answer these questions in about 500 words.

Spencer Grant/PhotoEdit

The newspaper was the dominant mass medium even as late as 1950.

To appreciate fully the impact of the mass media on life today, we need to trace their historical development. That is the first task we set ourselves in the following discussion. We then critically review theories of the mass media's effects on social life. As you will see, each of these theories contributes to our appreciation of media effects. Finally, we assess developments on the media frontier formed by the Internet, television, and other mass media. We show that to a degree, the new media frontier blurs the distinction between producer and consumer and has the potential to make the mass media somewhat more democratic for those who can afford access.

The Rise of the Mass Media

It may be difficult for you to imagine a world without the mass media. Yet, as Table 18.1 shows, most of the mass media are recent inventions. The first developed systems of writing appeared only about 5500 years ago in Egypt and Mesopotamia (now southern Iraq). The print media became truly mass phenomena only in the 19th century. The inexpensive daily newspaper, costing a penny, first appeared in the United States in the 1830s. At that time, long-distance communication required physical transportation. To spread the news, you needed a horse or a railroad. In 1794, it took 44 days for news from New York to reach Cincinnati. In 1841, it took a week. The slow speed of communication was costly. For instance, the last military engagement between Britain and the United States in the War of 1812 was the Battle of New Orleans. It took place 15 days *after* the combatants signed a peace treaty. The good news did not reach the troops near the mouth of the Mississippi until they had suffered 2100 casualties, including 320 dead.

The newspaper was the dominant mass medium even as late as 1950 (Smith, 1980; Schudson, 1991). However, change was in the air as early as 1876, when a nationwide system of telegraphic communication was established (Pred, 1973). From that time on, long-distance communication no longer required physical transportation. The transformative power of the new medium was soon evident. For example, until 1883, hundreds of local time zones existed in the United States—Michigan alone had 27. The correct time was determined by local solar time and was typically maintained by a clock in a church steeple or a respected jeweler's shop window. But virtually instant communication by telegraph made it possible to establish just four time zones in the United States (Carey, 1989). Railroad companies spearheaded the move to standardize time. A Canadian civil and railway engineer, Sir Sanford Fleming, was the main driving force behind the adoption of standard time in North America and worldwide (Blaise, 2001).

Table 18.1
The Development of the Mass Media

Year (C.E.)	Media Development
Circa 100	Papermaking developed in China
Circa 1000	Movable clay type used in China
Circa 1400	Movable metal type developed in Korea
1450	Movable metal type used in Germany, leading to the Gutenberg Bible
1702	First daily newspaper, London's *Daily Courant*
1833	First mass-circulation newspaper, *New York Sun*
1837	Louis Daguerre invents a practical method of photography in France
1844	Samuel Morse sends the first telegraph message between Washington and Baltimore
1875	Alexander Graham Bell sends the first telephone message
1877	Thomas Edison develops the first phonograph
1895	Motion pictures are invented
1901	Italian inventor Guglielmo Marconi transmits the first transatlantic wireless message from England to Newfoundland
1906	First radio voice transmission
1920	First regularly scheduled radio broadcast, Pittsburgh
1925	78 rpm records chosen as a standard
1928	First commercial TV broadcast
1949	Network TV begins in the United States
1952	VCR invented
1961	First cable television, San Diego
1969	First four nodes of the United States Defense Department's ARPANET (precursor of the Internet) set up at Stanford University, UCLA, UC Santa Barbara, and the University of Utah
1975	First microcomputer marketed
1983	Cell phone invented
1989	World Wide Web conceived by Tim Berners-Lee at the European Laboratory for Particle Physics in Switzerland

Source: Berners-Lee (1999); Croteau and Hoynes (1997:9–10); "The Silent Boom" (1998).

The telegraph and the railroad thus gave new meaning to the old expression "times change."

Most of the electronic media are creations of the 20th century. The first television signal was transmitted in 1925. Twenty-four years later, network TV began in the United States. The U.S. Department of Defense established ARPANET in 1969. It was designed as a system of communication between computers that would automatically find alternate transmission routes if one or more nodes in the network broke down due to, say, nuclear attack. The Internet grew out of ARPANET, which in turn begot the hyperlinked system of texts, images, and sounds known as the World Wide Web around 1991. By 2005, some 817 million people worldwide used the Web ("Top 20," 2005). It was a quick trip. A mere 140 years separate the Pony Express from the home videoconference.

Causes of Media Growth

The rise of the mass media can be explained by three main factors: one religious, one political, and one economic.

The Protestant Reformation

In the 16th-century Catholic Church, people relied on priests to tell them what was in the Bible. In 1517, however, Martin Luther protested certain practices of the Church. Among other things, he wanted people to develop a more personal relationship with the Bible. Within 40 years, Luther's new form of Christianity, known as Protestantism, was es-

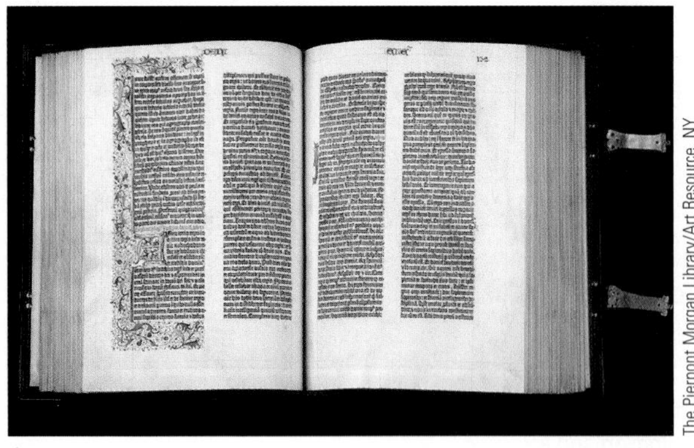
The Pierpont Morgan Library/Art Resource, NY

The Gutenberg Bible.

tablished in half of Europe. Suddenly, millions of people were being encouraged to read. The Bible became the first mass media product in the West and by far the best-selling book.

The diffusion of the Bible and other books was made possible by technological improvements in papermaking and printing (Febvre and Martin, 1976 [1958]). The most significant landmark was Johannes Gutenberg's invention of the printing press. In the 50 years after Gutenberg produced his monumental Bible in 1455, more books were produced than in the previous 1000 years. The printed book enabled the widespread diffusion and exchange of ideas. It contributed to the Renaissance (a scholarly and artistic revival that began in Italy around 1300 and spread to all of Europe by 1600) and the rise of modern science (Johns, 1998).

A remarkable feature of the book is its durability. Many electronic storage media became obsolete just a few years after being introduced. For instance, eight-track tapes are relics of the 1970s and $5\frac{1}{4}$-inch floppy disks are icons of the early 1980s. They are barely remembered today. In contrast, books are still being published today, nearly 550 years after Gutenberg published his Bible. In fact, 96,080 new books and editions were published in the United States in 2000 (U.S. Census Bureau, 2002: 703).

Democratic Movements

A second force that promoted the growth of the mass media was political democracy. From the 18th century on, the ordinary citizens of France, the United States, and other countries demanded and achieved representation in government. At the same time, they wanted to become literate and gain access to previously restricted centers of learning. Democratic governments, in turn, depended on an informed citizenry and therefore encouraged popular literacy and the growth of a free press (Habermas, 1989).

Today the mass media, and especially TV, mold our entire outlook on politics. Television was a source of presidential campaign news for only 51 percent of Americans in 1952, but by 1960 it was a source of campaign news for 87 percent of the public (Nie, Verba, and Petrocik, 1979 [1976]: 274). From a mass media point of view, the 1960 presidential election was significant in a second respect as well. It was the year of the first televised presidential debate—between John F. Kennedy and Richard Nixon. Sander Vanocur, one of the four reporters who asked questions during the debate, later recalled: "The people who watched the debate on their television sets apparently thought Kennedy came off better than Nixon. Those who heard the debate on radio thought Nixon was superior to Kennedy" (quoted in "The Candidates Debate," 1998). Kennedy smiled. Nixon perspired. Kennedy relaxed. Nixon fidgeted. The election was close, and most analysts believe that Kennedy got the edge simply because 70 million viewers thought he looked better on TV. Television was thus beginning to redefine the very nature of American politics. The 2000 presidential debates on TV were in some respects a rerun of 1960. Al Gore was widely perceived as wooden, nervous, pompous, and aggressive, George W. Bush as relaxed and friendly. Gore, the more experienced debater, was widely expected to win the debates, but many observers believe he lost them—and therefore the election—because of his problematic TV presence. Little wonder, then, that some analysts complain that television has oversimplified politics, reducing it to a series of catchy slogans and ever-shorter uninterrupted comments or "sound bites." (The average sound bite in nightly network

Americans now spend about 63 percent of their waking time interacting with the mass media.

Myrleen Ferguson/PhotoEdit

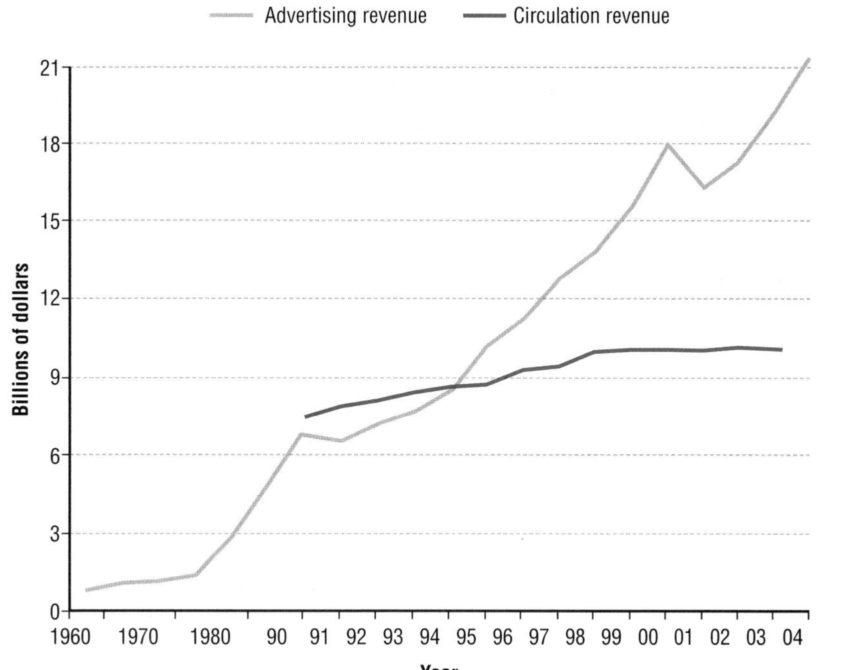

Figure 18.3
Magazine Advertising and Circulation Revenue, United States, 1960–2004

Source: Magazine Publishers of America (2005a, 2005b).

news shrank from 42.3 seconds in 1968 to 8.4 seconds in 1992; Thelen, 1996.) From this point of view, candidates are marketed for high office like Kellogg's sells breakfast cereal, and a politician's stage presence is more important than his or her policies in determining success at the polls.

Capitalist Industrialization

The third major force that stimulated the growth of the mass media was capitalist industrialization. Modern industries required a literate and numerate workforce. They also needed rapid means of communication to do business efficiently. Moreover, the mass media turned out to be a major source of profit in their own right. In 2002 box office receipts of the American movie industry reached nearly $10 billion ("2002 Marks Record Year," 2003). The global CD industry netted $32 billion in revenue (Waters, 2003). In 2003, American magazines generated revenues of more than $29 billion, two-thirds of which came from advertising (Figure 18.3). As these examples illustrate, the mass media form a big business.

We conclude that the sources of the mass media lie deeply embedded in the religious, political, and economic needs of our society. Moreover, the mass media are among the most important institutions in our society today. How, then, do sociologists explain the effects of the mass media on society? To answer this question, we now summarize the relevant sociological theories.

Theories of Media Effects

Functionalism

As societies develop, they become larger and more complex. The number of institutions and roles proliferate. Due to the sheer scale of the society, face-to-face interaction becomes less viable as a means of communication. As a result, the need increases for new means of coordinating the operation of the various parts of society. For example, people in Maine must have at least a general sense of what is happening in California and they need to share certain basic values with Californians if they are going to feel that they are citizens of the same country. The mass media do an important job in this regard. The 19th-century German philosopher Georg Hegel once said that the daily ritual of reading the newspaper unites the secular world, just as the ritual of daily prayer once united the Christian world. Stated more generally, his point is valid. The nationwide distribution of newspapers, magazines, movies, and television programming cements the large, socially diverse, and geographically far-flung population of the United States. In a fundamental sense, the nation is an imagined community, and the mass media make it possible for us to imagine it (Anderson, 1990).

Thus, the mass media perform an important function by *coordinating* the operation of industrial and postindustrial societies. But, according to functionalist theorists, their significance does not stop there (Wright, 1975). In addition, the mass media are also important agents of *socialization.* Families have relinquished their formerly nearly exclusive right to transmit norms, values, and culture. The mass media have stepped into the breach. They reinforce shared ideals of democracy, competition, justice, and so forth (see Chapter 4, "Socialization").

A third function of the mass media involves *social control.* That is, the mass media help to ensure conformity. For example, news broadcasts, TV dramas, and "docutainment" programs such as *Cops* pay much attention to crime, and they regularly sing the praises of heroes who apprehend and convict criminals. By exposing deviants and showcasing law enforcement officials and model citizens, the mass media reinforce ideas about what kinds of people deserve punishment and what kinds of people deserve rewards. In this way, they reproduce the moral order. Some people think *Jerry Springer, Jenny Jones,* and other similar shows are outlandish, and in a way they are. From a sociological point of view, however, they are also deeply conservative programs, for when television audiences get upset about marital infidelities and other outrages, they are reinforcing some of the most traditional norms of American society and thus serving as agents of social control. As Saul Bellow wrote in *Herzog,* "A scandal [is] after all a sort of service to the community" (Bellow, 1964: 18).

The mass media's fourth and final function is to provide *entertainment.* Television, movies, magazines, and so forth give us pleasure, relaxation, and momentary escape from the tension and tedium of everyday life. How often have you come home after a long and frustrating day at college or work, picked up the remote, channel-surfed, concluded there's nothing really worth watching, but settled for a soap opera or some other form of easily digestible entertainment? It is precisely because some products of the mass media require little effort on the part of the audience that they are important. They relieve stress. Moreover, they do so in a way that doesn't threaten the social order. Without such escapes, who knows how our daily tensions and frustrations might express themselves?

Conflict Theory

Clearly, functionalism offers valuable insights into the operation of the mass media. However, the functional approach has been criticized by conflict theorists for paying insufficient attention to the social inequality fostered by the mass media. Specifically, conflict theorists say that functionalism exaggerates the degree to which the mass media serve the interests of the entire society. They contend that some people benefit from the mass media more than others do. In particular, the mass media favor the interests of dominant classes and political groups (Gitlin, 1983; Herman and Chomsky, 1988; Iyengar, 1991; Horkheimer and Adorno, 1986 [1944]).

Conflict theorists maintain that there are two ways in which dominant classes and political groups benefit disproportionately from the mass media. First, the mass media broadcast beliefs, values, and ideas that create widespread acceptance of the basic structure of society, including its injustices and inequalities. Second, ownership of the mass media is highly concentrated in the hands of a small number of people and is highly profitable for them. Thus, the mass media are a source of economic inequality.

Media Ownership

Ownership of the mass media has certainly become more highly concentrated over time. Around 50 corporations controlled half of all media organizations in the United States in 1984. By 1993, about 20 corporations exercised this degree of media control (Bagdikian, 1997 [1983]). Between 1992 and 1996, the proportion of U.S. television stations owned by the 10 biggest owners nearly doubled; the proportion of radio stations owned by the 10 biggest owners more than quadrupled ("Media Mergers," 2000). U.S. book production, film production, newspaper publishing, and cable TV are each dominated by only six firms. A mere five firms dominate the U.S. music industry (McChesney, 1999).

It is not just the *degree* of media concentration that has changed. In addition, the *form* of media concentration began to shift in the 1990s (McChesney, 1999). Until the 1990s, media concentration involved mainly "horizontal integration." That is, a small number of firms tried to control as much production as possible in their particular fields (film production, newspapers, radio, television, etc.). In the 1990s, however, "vertical integration" became much more widespread. That is, media firms sought to control production and distribution in many fields. They became media conglomerates. Today, a media conglomerate may own any combination of television networks, stations, and production facilities; magazines, newspapers, and book publishers; cable channels and cable systems; movie studios, theaters, and video store chains; sports teams and stadiums; Web portals and software companies. A media conglomerate can create content and deliver it in a variety of forms. For instance, it can make a movie, promote it on its TV and radio networks, and then spin off a TV series, CD, book, and merchandise—all delivered to the consumer in outlets owned by the conglomerate itself. You can appreciate the scale of control we're talking about if we simply list some of the better-known companies owned by AOL/Time Warner, the biggest media conglomerate. These companies include America Online, Netscape, Amazon.com, Time-Life Books, Book-of-the-Month Club, Warner Books, Little, Brown and Company, HBO, CNN, Warner Bros., Castle Rock Entertainment, *Time, Fortune, Life, Sports Illustrated, Money, People, Entertainment Weekly,* Warner Brothers Records, Elektra Records, the Atlanta Braves, the Atlanta Hawks, the Atlanta Thrashers, and the Goodwill Games.

The Federal Communications Commission (FCC) is the federal watchdog over interstate and international communications by radio, television, wire, satellite, and cable. In the interest of maintaining diversity and competing viewpoints in a vibrant democracy, the FCC imposed ownership restrictions on radio, television, newspaper, and cable companies from its founding in 1934 until 1996. In 1996, it removed the restrictions for radio

Erik Freeland/Corbis

Steve Case, Gerald Levin, Ted Turner, and Richard Parson announce the AOL-Time Warner merger in 2000.

BOX 18.2
SOCIAL POLICY: WHAT DO YOU THINK?

Media Conglomerates

Who owns NBC? General Electric. Who owns ABC? Disney. Who owns CBS? Westinghouse. Who owns CNN? AOL/Time Warner, which also owns HBO, America Online, Warner Brothers, *Time, Sports Illustrated,* and so forth.

Hardly anyone doubts that the mass media, especially television, have a big impact on American society. However, people disagree whether concentrated ownership affects the content of broadcast news. Critics of concentrated ownership cite examples of corporations influencing the flow of information. For instance, Brian Ross, the leading investigative reporter for the TV newsmagazine *20/20,* prepared a segment about Disney World in 1998. Ross claimed that Disney was so lax in doing background checks on employees that it had hired pedophiles. ABC killed the story before airtime. ABC, you will recall, is owned by Disney (McChesney, 1999). Similarly, ABC News formally apologized to Philip Morris (a major TV advertiser through its subsidiary, Kraft Foods) for airing a report about the cigarette company's manipulation of ammonia levels in tobacco. Although the report was true, ABC apparently buckled to corporate pressure. One media critic has even gone so far as to suggest that there are few stories critical of the military on the news because two of the owners of the four major TV networks (General Electric and Westinghouse) are large defense contractors (Miller, 1996).

Other analysts suggest that corporate ownership does not have a major impact on the televised flow of information. They argue that despite some abuses, freedom of speech is protected. We hear plenty of stories critical of corporations and the government, and if we don't hear more it's mainly because the public isn't interested. Furthermore, they say, even if there were a problem with concentrated ownership, government intervention would likely create more problems than it solves.

What do you think? Does the high concentration of mass media ownership have an impact on the information we receive from the major television networks? Are different *types* of stories influenced to varying degrees and in different ways by the conglomeration of the mass media? If it could be determined that the problem is serious, what could you do about it?

stations. The result was unprecedented concentration of ownership in the radio industry. For example, between 1996 and 2003, one company, Clear Channel, increased its holdings from 40 radio stations to 1225. Its closest rival owned about 150 stations in 2003. In addition, Clear Channel syndicates more than 100 programs to 7800 stations in the United States, reaching 180 million listeners a week (Boehlert, 2003; Hogan, 2003). In 2003 the FCC decided to go even farther and relax the ownership cap on television, newspaper, and cable companies. It proposed that competing TV networks be allowed to merge and that just one company be allowed to control newspapers, radio and television stations, and cable TV outlets in the same market (e.g., in one city). In July 2003 the House of Representatives blocked the proposal, but because of strong support in the Senate and the White House, the stage seemed set for high political drama, with the precise outcome unclear. It seems reasonable to expect that if the FCC proposal is eventually accepted in some form, we will witness even more rapid consolidation in the mass media (Labaton, 2003).

Aside from AOL/Time Warner, the biggest media players in the United States include Disney, Viacom, and News Corporation. Other giant media conglomerates include Bertelsmann (Germany) and Sony (Japan).

Media Bias

Does the concentration of the mass media in fewer and fewer hands deprive the public of independent sources of information, limit the diversity of opinion, and encourage the public to accept their society as it is? Conflict theorists think so (Boxes 18.2 and 18.3). They believe that when a few conglomerates dominate the production of news in particular they squeeze out alternative points of view. Moreover, as Edward Hermann and Noam Chomsky (1988) argue, several mechanisms help to bias the news in a way that supports powerful corporate interests and political groups. These biasing mechanisms include:

- *Advertising.* Most of the revenue earned by television stations, radio stations, newspapers, and magazines comes from advertising by large corporations. According to Hermann and Chomsky, these corporations routinely seek to influence the news so it

BOX 18.3
Sociology at the Movies

The Fog of War (2003)

The Fog of War surveys the life of Robert McNamara, secretary of defense during the Kennedy and Johnson administrations and an architect of the Vietnam War. The film's format—a single talking head plus period film clips, photos, and audiotapes—might seem a recipe for tedium. Instead, master filmmaker Errol Morris presents us with a compelling work that won the 2003 Oscar for best documentary, a film that, if taken seriously by enough people, could change the world.

McNamara does what many people want but too few accomplish. He learns from the mistakes that he and others made or nearly made so he can clear away some of the "fog of war," the muddled thinking that accompanies all armed conflict. For example, a lesson McNamara derived from his involvement in the 1962 Cuban missile crisis is that you win most and lose least if you empathize with your enemy. He tells the story of how Tommy Thompson, former U.S. ambassador to Moscow, locked horns with President Kennedy. Kennedy believed that negotiating with the Russians to remove their missiles from Cuba was futile. He was prepared to start a nuclear war. Thompson, however, knew the Russian leader, Khrushchev, personally and understood that he would back down from his belligerent position if presented with an option that would allow him to remove the missiles from Cuba and still say to his hardline generals that he had won the confrontation with the United States. "The important thing for [him]," Thompson argued, "is to be able to say, 'I saved Cuba; I stopped the invasion.'" Thompson convinced Kennedy. Negotiations began and nuclear war was averted. "That's what I call empathy," McNamara observes. "We must try to put ourselves inside [the enemy's] skin and look at [ourselves] through their eyes."

Empathy, McNamara acknowledges, was absent a few years later when the United States began carpet-bombing and deploying hundreds of thousands of troops in Vietnam. The U.S. administration thought the war would prevent Vietnam from falling under Chinese Communist rule. It didn't appreciate that Vietnam had been a colony of China for a thousand years and of France for nearly a century and was now engaged in a bitter struggle for independence that the United States seemed eager to prevent. It was a costly misunderstanding. The American administration failed to appreciate the nationalist motives of the Vietnamese and underestimated their resolve. Fifty-eight thousand Americans and two million Vietnamese died in the war.

Some analysts argue that the mass media contribute to the fog of war today by obstructing the development of empathy with the enemies of the United States. (Remember, empathy does not mean having warm and fuzzy feelings about an enemy but understanding things from the enemy's perspective so we can design policies that enable us to win most and lose least.) Many representatives of the mass media are aware of their failings in this regard. Thus, once it became evident in 2004 that Iraq lacked weapons of mass destruction and had few or no ties with Osama bin Laden and that the Iraqi insurgency against the United States was determined and widespread, representatives of several major American newspapers and television networks virtually apologized to the public for not subjecting the administration's case for war with Iraq to sufficient scrutiny ("The *Times* and Iraq," 2004). Even after these admissions, media self-censorship was widespread according to some analysts. Some media outlets, notably Fox News, failed to present balanced coverage as a matter of editorial policy, while some journalists spoke in muted tones for fear of losing their assignments (Massing, 2004). Arguably, therefore, Americans have an imprecise picture of the situation on the ground in Iraq. We know little about the true impact of the war on civilians, the extent of popular support for the insurgency, the population's attitudes toward the American military presence, and other crucial matters. In *The Fog of War,* Robert McNamara argues that willingness to reexamine one's reasoning is an essential element of successful warfare. Arguably, however, media-imposed limits on our knowledge make it difficult for the American public to develop informed opinion about what to do in order to maximize our gains and minimize our losses in Iraq.

Secretary of Defense Robert McNamara and President Lyndon Johnson.

will reflect well on them. Thus, in one survey, 93 percent of newspaper editors said advertisers have tried to influence their news reports. Thirty-seven percent of newspaper editors admitted to actually being influenced by advertisers (Bagdikian, 1997 [1983]).

- *Sourcing.* Studies of news gathering show that most news agencies rely heavily for information on press releases, news conferences, and interviews organized by large corporations and government agencies. These sources routinely slant information to reflect favorably on government and corporate policies and preferences. Indeed, as has been well documented, industry often manipulates science, pays for objective-sounding third-party endorsements, and uses public relations experts to put a favorable "spin" on unpleasant facts—all for purposes of self-promotion (Rampton and Stauber, 2001). Unofficial news sources often lack the resources to engage in such practices. In any case, they are consulted less often and tend to be used only to provide reactions and minority viewpoints that are secondary to the official story.
- *Flak.* Governments and big corporations routinely attack journalists who depart from official and corporate points of view. For example, tobacco companies have systematically tried to discredit media reports that cigarettes cause cancer. In a notorious case, the respected public affairs show *60 Minutes* refused to broadcast a damaging interview with a former Philip Morris executive because CBS was threatened with legal action by the tobacco company. (This incident is the subject of the 2000 Oscar-nominated movie *The Insider.*)

On the whole, the arguments of the conflict theorists are compelling. We do not, however, find them completely convincing (Gans, 1979a). After all, if 37 percent of newspaper editors have been influenced by advertisers, 63 percent have not. News agencies may rely heavily on government and corporate sources, but this doesn't stop them from routinely biting the hand that offers to feed them and evading flak shot their way. Examples of mainstream journalistic opposition to official viewpoints are plentiful. In the early 1970s, investigative reporters from the *Washington Post* unleashed the Watergate scandal. It eventually resulted in the resignation of President Richard Nixon. In the late 1990s, *Time* published a cover story on "corporate welfare" in America. It showed that in the midst of the nation's biggest economic boom ever, and after nearly two decades of reducing welfare budgets to the poor, the federal government was giving $125 billion a year in subsidies to American businesses (Barlett and Steele, 1998). As these examples show, even mainstream news sources, although owned by media conglomerates, do not always act like the lap dogs of the powerful (Hall, 1980).

Still, conflict theorists make a valid point if they restrict their argument to how the mass media support core American values. In their defense of core values, the mass media *are* virtually unanimous. For example, the mass media enthusiastically support democracy and capitalism. We cannot think of a single instance of a major American news outlet advocating a fascist government or a socialist economy in the United States.

Similarly, the mass media virtually unanimously endorse consumerism as a way of life. As discussed in Chapter 3 ("Culture"), consumerism is the tendency to define oneself in terms of the goods and services one purchases. Endorsement of consumerism is evident in the fact that advertising fills the mass media and is its lifeblood. In the United States, expenditure on advertising is nearly a quarter the expenditure on all levels of education (United Nations, 1998). The average American is exposed to a staggering number of ads each day, some estimates placing the number in the thousands. Companies pay filmmakers to use their brand-name products conspicuously in their movies. The Orange Bowl is turned into the Federal Express Orange Bowl, and the Cotton Bowl becomes the Mobil Cotton Bowl. In some magazines, such as *Wired, Vogue,* and *Vanity Fair,* ads figure so prominently that one must search for the articles. The American public responds to the ongoing advertising blitz by participating in the biggest buying spree in recorded history, even at the cost of ballooning credit card debt. Specifically, credit card debt is growing at the rate of about 10 percent per household per year. The growth rate is nearly twice as high

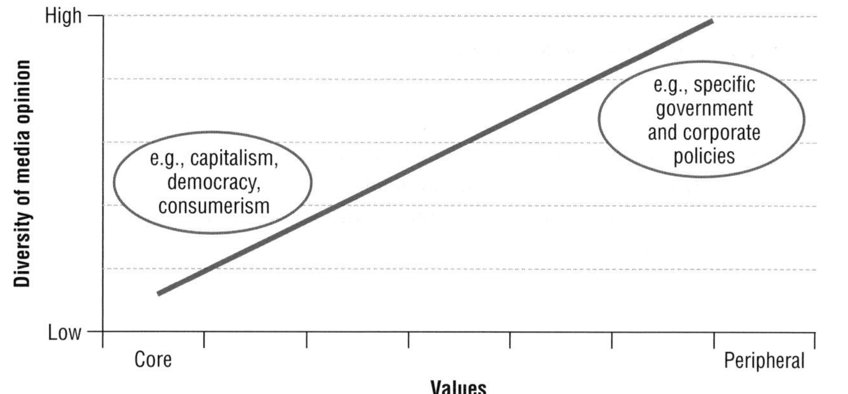

Figure 18.4
The Relationship Between Centrality of Values and Diversity of Media Opinion

among the bottom half of income earners as among the top half (Ewen, 1976; 1988; Twitchell, 1999; Yoo, 1998).

It is only when the mass media deal with news stories that touch on less central values that one may witness a diversity of media opinion. For example, as our Watergate and corporate welfare examples show, specific government and corporate policies are often the subject of heated debate in the mass media.[1] Thus, despite the indisputable concentration of media ownership, the mass media are diverse and often contentious on specific issues that do not touch on core values (Figure 18.4).

Interpretive Approaches

The final episode of *Buffy the Vampire Slayer* for the 1999 season included a scene of a shooting spree at a high school. However, the Fox TV network canceled the episode because it was scheduled to air shortly after two teenagers massacred 13 of their classmates at Columbine High School in Littleton, Colorado. In canceling the show, network officials were reacting to members of the public who shared the widespread belief that airing it might incite more school violence.

The view that the mass media powerfully influence a passive public is common. Many people believe that violence on TV causes violence in real life, pornography on the magazine stands leads to immoral sexual behavior, and adolescents are more likely to start smoking cigarettes when they see popular movie stars lighting up. In a 1995 CBS/*New York Times* poll, for example, respondents said that they thought television was the number one cause of teenage violence (Kolbert, 1995).

Functionalists and conflict theorists share this top-down, deterministic view; members of both schools of thought stress how the mass media bridge social differences and reinforce society's core values. True, the two schools of thought differ in that functionalists regard core values as serving everyone's interests, whereas conflict theorists regard them as favoring the interests of the rich and powerful. However, by focusing so tightly on core values, both approaches understate the degree to which audience members interpret media messages in different ways. The signal contribution of symbolic interactionist and related approaches is that they highlight the importance of such interpretive acts.

Just how much influence do the mass media actually exert over their audiences? The question is mired in controversy, but there is good reason to view the top-down, deterministic view as one-sided. For example, you may recall our discussion of media violence in Chapter 2 ("How Sociologists Do Research"). There we found that most research on the subject is plagued by a big validity problem. Simply stated, experiments on media violence may not be measuring what they say they are. Consequently, the degree to which TV vio-

[1]Conservatives often criticize the mass media for its allegedly liberal bias. We do not wish to enter this highly charged political debate. We note, however, that the conservative critique of the mass media focuses only on the media's analysis of specific government and corporate policies, not their analysis of stories relating to core values.

lence encourages violent behavior is unclear. The sociological consensus seems to be that TV violence has a weak effect on a small percentage of viewers (Felson, 1996: 123; Freedman, 2002).

There are other reasons for questioning the strength of media effects. For instance, half a century ago researchers showed that people do not change their attitudes and behaviors just because the media tell them to do so. That is because the link between persuasive media messages and actual behavior is indirect. A **two-step flow of communication** takes place (Katz, 1957; Schiller, 1989; Schudson, 1995). In step 1, respected people of high status evaluate media messages. They are the opinion leaders of a neighborhood or a community, people who are usually more highly educated, well-to-do, or politically powerful than others in their circle. Because of their high status, they exercise considerable independence of judgment. In step 2, opinion leaders *may* influence the attitudes and behaviors of others. In this way, opinion leaders filter media messages. The two-step flow of communication limits media effects. If people are influenced to vote for certain candidates, buy certain products, or smoke cigarettes, it is less because the media tell them to and more because opinion leaders suggest they should.

Yet another persuasive argument that leads one to question the effects of the mass media comes from interpretive sociologists such as symbolic interactionists and interdisciplinary **cultural studies** experts. They use in-depth interviewing and participant observation to study how people actually interpret media messages.

British sociologist Stuart Hall, one of the foremost proponents of this approach, emphasizes that people are not empty vessels into which the mass media pour a defined assortment of beliefs, values, and ideas. Rather, audience members take an active role in consuming the products of the mass media. They filter and interpret mass media messages in the context of their own interests, experiences, and values. Thus, in Hall's view, any adequate analysis of the mass media needs to take into account both the production and the consumption of media products. First, he says, we need to study the meanings intended by the producers. Then we need to study how audiences consume or evaluate media products. Intended and received meanings may diverge; audience members often interpret media messages in ways other than those intended by the producers (Hall, 1980; Seiter, 1999). For example, when Dale Earnhardt, Jr., won a NASCAR race at the track where his famous father had died in a crash a year earlier, some people cried "fix," just as they did when retiring baseball legend Cal Ripken, Jr., hit a home run in his last All-Star Game. This suggests that people are often skeptical if not downright cynical about what they see on TV (Brady, 2001).

The **two-step flow of communication** between mass media and audience members involves (1) respected people of high status and independent judgment evaluating media messages and (2) other members of the community being influenced to varying degrees by these opinion leaders. Due to the two-step flow of communication, opinion leaders filter media messages.

Cultural studies is an increasingly popular interdisciplinary area of media research. It focuses not just on the cultural meanings producers try to transmit but also on the way audiences filter and interpret mass media messages in the context of their own interests, experiences, and values.

Personal Anecdote

A personal example: When John Lie's parents were preparing to emigrate to the United States in the late 1960s, his mother watched many American movies and television shows. One of her favorite TV programs was *My Three Sons,* a sitcom about three boys living with their father and grandfather. From the show she learned that boys wash dishes and vacuum the house in the United States. When the Lie family immigrated to Hawaii, John and his brother—but not his sister—had to wash dishes every night. When John complained, his mother reassured him that "in America, only boys wash dishes."

Even children's television viewing turns out to be complex when viewed through an interpretive lens. Research shows that young children clearly distinguish "make-believe" media violence from real-life violence (Hodge and Tripp, 1986). That is one reason why watching episode after episode of *South Park* has not produced a nation of *South Park* clones. Similarly, research shows differences in the way working-class and middle-class women relate to TV. Working-class women tend more than middle-class women to evaluate TV programs in terms of how realistic they are. This critical attitude reduces their ability to identify strongly with many characters, personalities, and story lines. For instance, working-class women know from their own experience that families often don't work the way they are shown on TV. They view the idealized, middle-class nuclear family depicted in many television shows with a mixture of nostalgia and skepticism (Press, 1991). Age also affects how one relates to television. Elderly viewers tend to be selective and focused in their television viewing. In contrast, people who grew up with cable TV and remote control often engage in channel surfing, conversation, eating, and housework, zoning in and out of programs in anything but an absorbed fashion (Press, 1991). The idea that such viewers are sponges, passively soaking up the values embedded in TV programs and then mechanically acting on them, is inaccurate.

Feminist Approaches

Finally, let us consider feminist approaches to the study of mass media effects. In the 1970s, feminist researchers focused on the representation—more accurately, the misrepresentation—of women in the mass media. They found that in TV dramas women tended to be cast as housewives and secretaries and in other subordinate roles, whereas men tended to be cast as professionals and authority figures. Women usually appeared in domestic settings, men in public settings. Advertising targeted women only as purchasers of household products and appliances. Furthermore, researchers discovered that the news rarely mentioned issues of importance for many women, such as wage discrimination in the paid labor force, sexual harassment and abuse, child-care problems, and so forth. News reports sometimes trivialized or denounced the women's movement. Newsworthy issues (the economy, party politics, international affairs, and crime) were associated with men, and men were much more likely than women to be used as news sources and to deliver the news (Watkins and Emerson, 2000: 152–3).

Most of this early feminist research assumed that audiences are passive. Analysts argued that the mass media portray women in stereotypical fashion; audience members recognize and accept the stereotypes as normal and even natural; and the mass media thereby reinforce existing gender inequalities. However, in the 1980s and 1990s, feminist researchers criticized this simple formula. Much influenced by cultural studies, they realized that audience members selectively interpret media messages and sometimes even contest them.

A good example of this subtler and less deterministic approach is a study by Andrea Press and Elizabeth Cole (1999) of audience reaction to abortion as portrayed on TV shows. Over a four-year period, Press and Cole conducted 34 discussion groups involving 108 women. The women watched three TV programs focusing on abortion and then discussed their own attitudes and their reactions to the shows. The programs were "pro-choice" and dealt with women who chose abortion to avoid poverty.

Press and Cole found complex, ambivalent, and sometimes contradictory attitudes toward abortion among audience members. However, four distinct clusters of opinion emerged:

1. "Pro-life" women from all social classes form the most homogeneous group. They think that abortion is never justified. On principle, they reject the mass media's justifications for abortion.
2. Pro-choice working-class women who think of themselves as members of the working class adopt a pro-choice stand as a survival strategy, not on principle. They do not condone abortion, but they fear that laws restricting abortion would be applied prejudicially against women of their class. Therefore, they oppose any such restrictions. At the same time, they reject the TV message that financial hardship justifies abortion.
3. Pro-choice working-class women who aspire to middle-class status distance themselves from the "reckless" members of their own class who sought abortions on the TV shows. They tolerate abortion for such people but reject it for themselves and for other "responsible" women.

Missy "Misdemeanor" Elliott writes and produces her own music. She often directs her own videos. Her work is a running critical commentary on real-world issues confronting young black women. Thus, in terms of both production and content, her work breaks down the established roles and images of black women in America.

4. Pro-choice middle-class women believe that only an individual woman's feelings can determine whether abortion is right or wrong in her own case. Many pro-choice middle-class women have deep reservations about abortion, and many of them reject it as an option for themselves. However, they staunchly defend the right of all women, especially the kind of women portrayed in the TV shows they watched, to choose abortion.

One of the most striking aspects of Press and Cole's findings is that for different reasons, three of the four categories of audience members (categories 1, 2, and 3) are highly skeptical of TV portrayals of the abortion issue. Their class position and attitudes act as filters influencing how they react to TV shows and how they view the abortion issue. Moreover, three of the four categories of audience members (categories 2, 3, and 4) reject the simple pro-choice versus pro-life dichotomy often portrayed by the mass media. Many pro-choice advocates express ambivalence about abortion and even reject it as an option for themselves. One must conclude that real women are typically more complicated than the stereotypes promoted in the mass media, and that women in the audience typically know that.

In recent years, some feminists have focused on the capacity of the mass media to reproduce and change the system of racial inequality in American society. In the work of these scholars, the twin issues of female misrepresentation and active audience interpretation reappear, this time with a racial twist. On the one hand, they find that certain stereotypical images of women of color recur in the mass media. African American women, for example, often appear in the role of the welfare mother, the highly sexualized Jezebel, and the mammy. On the other hand, they recognize that some mass media, especially independent filmmaking and popular music, have enabled women of color to challenge these stereotypes. The music and videos of Erykah Badu, Missy "Misdemeanor" Elliott, and Lauryn Hill are especially noteworthy in this regard. These artists write and produce their own music. They often direct their own videos. Their work is a running critical commentary on real-world issues confronting young black women. Thus, in terms of both production and content, their work breaks down the established roles and images of black women in America (Watkins and Emerson, 2000).

Two studies systematically tracked some of the issues raised by feminist researchers and so may usefully conclude this part of our discussion. The studies analyzed the characterization of women and racial minorities (and also disabled and poor people) as fictional television characters in prime-time and daytime series, films, and animated cartoons in the periods 1991–1992, 1994–1997, and 2001–2002 (*Fall Colors,* 2002; Gerbner, 1998). (The 2001–2002 data are for prime-time shows only.) Among other things, they compare the percentage size of minority groups in the American population with their percentage representation in fictional TV roles.

In Figure 18.5, which summarizes part of the studies, a score of 100 indicates that the percentage of a group in the American population is the same as its percentage in fictional TV roles. A score of more than 100 indicates the degree of overrepresentation, a score of less than 100 the degree of underrepresentation. Figure 18.5 shows that in 1994–1997 there were 29 percent more white men in fictional TV roles than white men in the U.S. population. At the same time, there were 28 percent fewer women in fictional TV roles than in the population, 46 percent fewer Native Americans, 61 percent fewer Asian/Pacific Island Americans, 66 percent fewer people 60 years of age or older, 76 percent fewer Hispanics, 88 percent fewer people with disabilities, and 89 percent fewer poor people. These figures demonstrate that TV falls far short of reflecting the diversity of American society in many crucial aspects.

In some ways, the situation is improving. For example, comparing 1991–1992 with 1994–1997 and 2001–2002, we see that the representation of African Americans is now

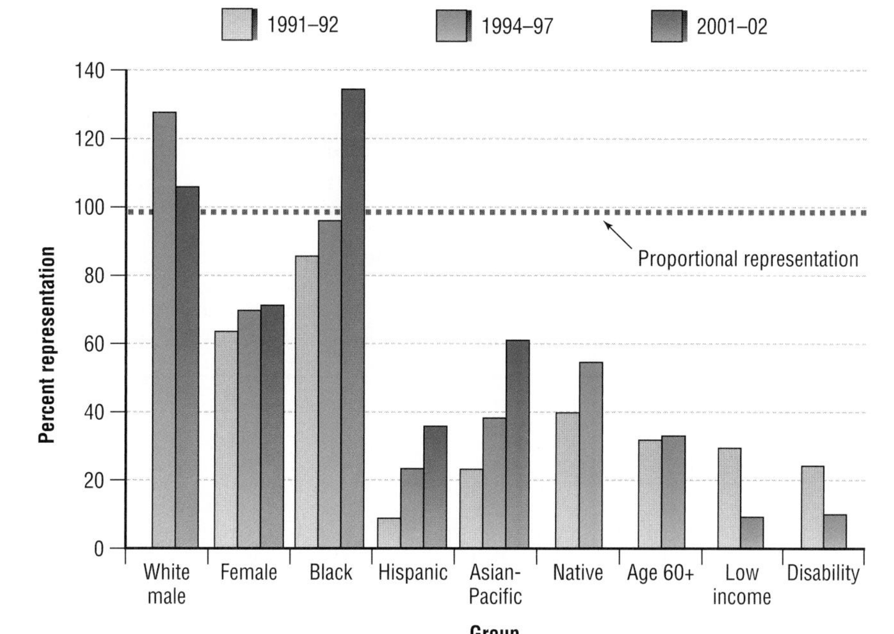

Figure 18.5
Representation of Minority Groups in Prime-Time and Daytime Television, United States, 1991–1992 and 1994–1997

Note: 2001–2002 data are for prime time only. A 2001–2002 figure for Native Americans was reported but it is so small that it is not visible on the graph. For other groups, missing bars indicate missing data.
Sources: *Fall Colors* (2002: 4, 17); Gerbner (1998).

substantially greater than their representation in the U.S. population. The representation of Asian Americans more than tripled between 1991–1992 and 2001–2002. On the other hand, the underrepresentation of Hispanic Americans and Native Americans remains startling, with Hispanic Americans underrepresented by 64 percent and Native Americans practically disappearing from prime-time TV in 2001–2002. Moreover, the portrayal of women, racial minorities, the poor, and people with disabilities still tends to reinforce traditional, mainstream, negative stereotypes. For instance, fully 60 percent of fictional TV characters suffering from mental illness in the period 1994–1997 were involved in crime or violence. Nonwhites tended to play comical or criminal characters rather than serious, heroic types.[2] Characters of different races often interacted professionally, sometimes interacted socially, but were rarely romantically involved. Asian Americans were typically portrayed as nerdy students, inscrutable martial arts masters, seductive Dragon Ladies, or clueless immigrants. Women were valued chiefly for their youth, sex appeal, and beauty, whereas older women were often associated with evil. Thus, in the 1990s, women playing fictional TV roles were on average younger than men, and they became still younger relative to men in the course of the decade. Young women tended to play mainly romantic roles, but the proportion of villains among women increased with age. The tendency for villainy to increase among older men was much weaker (*Fall Colors,* 2002; Gerbner, 1998).

The research findings reviewed here point to positive change in the way the mass media treat women and various minority groups. We have come a long way since the 1950s, when virtually the only blacks on TV were men who played butlers and buffoons (the prototype being Rochester, Jack Benny's butler). Research suggests that the mass media still have a long way to go before they cease reinforcing traditional stereotypes in American society (Dines and Humez, 1995). However, research also suggests that audiences and artists are hardly passive vehicles of these stereotypes, instead struggling to diversify the way the mass media characterize them.

[2]Similarly, in a study of four years of stories in *Time, Newsweek,* and *U.S. News & World Report,* one researcher found that 53–66 percent of poor people were portrayed as African Americans, whereas the actual rate was about 29 percent. This media bias reinforces the false idea that most poor people are African American (Fitzgerald, 1997).

Concept Summary 18.1
Theories of Mass Media Effects

Theory	Mass Media Effects
Functionalism	Coordination, socialization, social control, entertainment—all of which make social order possible
Conflict theory	Reinforcement of economic inequality and the core values of a stratified society
Interpretive approaches	Ambiguous because audiences filter, interpret, resist, and sometimes reject media messages
Feminist approaches	Maintenance of gender and racial inequalities through perpetuation of stereotypes in the mass media, although resistance and innovation result in change

Summing Up

We conclude that each of the theoretical approaches reviewed earlier contributes to our understanding of how the mass media influence us (Concept Summary 18.1):

- *Functionalism* usefully identifies the main social effects of the mass media: coordination, socialization, social control, and entertainment. By performing these functions, the mass media help make social order possible.
- *Conflict theory* offers an important qualification. As vast moneymaking machines controlled by a small group of increasingly wealthy people, the mass media contribute to economic inequality and to maintaining the core values of a stratified social order.
- *Interpretive approaches* offer a second qualification. They remind us that audience members are people, not programmable robots. We filter, interpret, resist, and sometimes reject media messages according to our own interests and values. A full sociological appreciation of the mass media is obliged to recognize the interaction between producers and consumers of media messages.
- *Feminist approaches* offer yet a third qualification. They highlight the misrepresentation of women and racial minorities in the mass media. They also emphasize the ways in which women and racial minorities have successfully challenged these misrepresentations and sought to diversify the characterization of race and gender by the mass media.

Domination and Resistance on the Internet

We have emphasized the interaction that normally takes place between the mass media and its audiences. To drive the point home, we now offer an in-depth analysis of domination and resistance on the Internet. As you will see, the Internet provides fresh opportunities for media conglomerates to restrict access to paying customers and accumulate vast wealth. Simultaneously, however, it gives consumers new creative capabilities, partially blurring the distinction between producer and consumer. The Internet, we conclude, has the potential to make the mass media more democratic—at least for those who can afford access.

Sociology Now™

Learn more about **Technology and Change** by going through the Technology and Change Learning Module.

To develop this idea, let us first consider the forces that restrict Internet access and augment the power of media conglomerates. We then discuss some countertrends.

Access

The Internet requires an expensive infrastructure of personal computers, servers, and routers; an elaborate network of fiber-optic, copper twist, and coaxial cables; and many other components. People must pay for this infrastructure. Consequently, access is not open to everyone. In the United States, for example, college-educated whites with

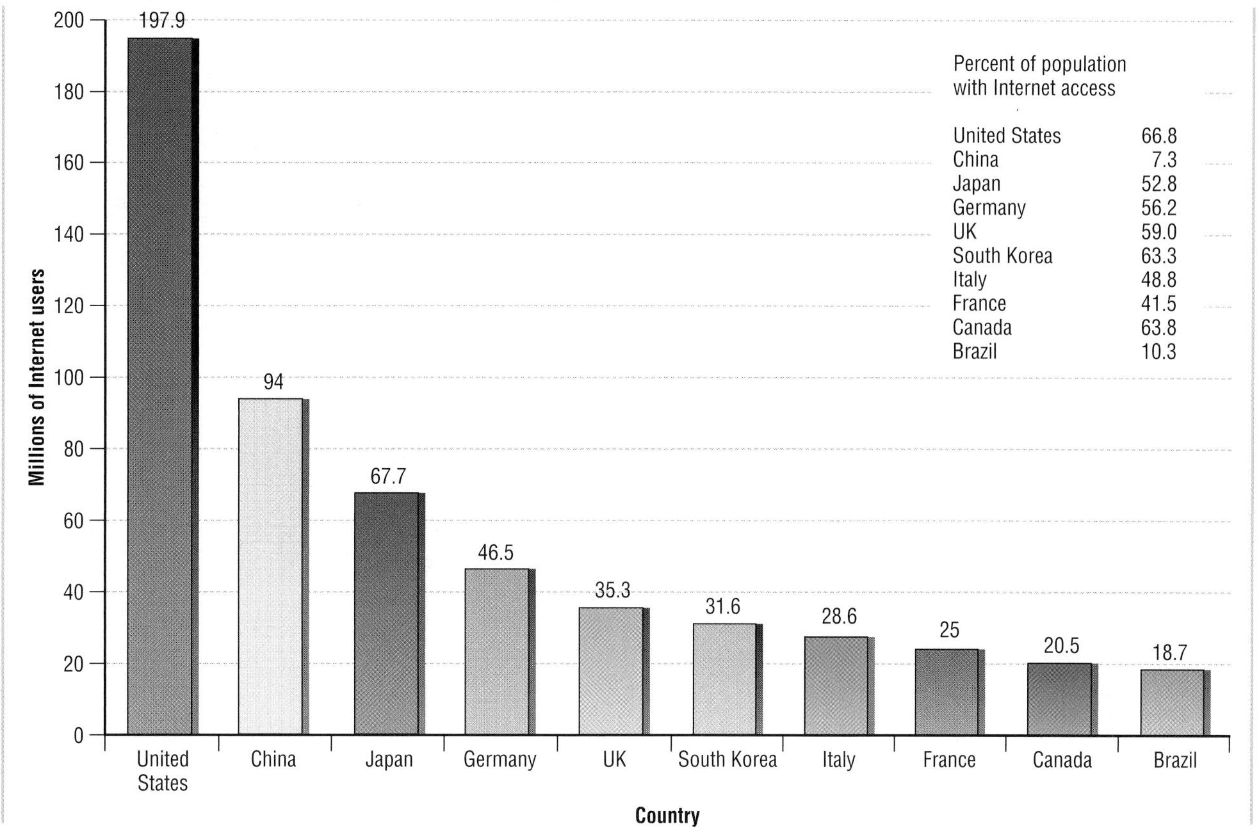

Figure 18.6
Top 10 Countries, Millions of Internet Users, 2005

Source: "Top 20" (2005).

above-average incomes are most likely to enjoy Internet access (see Chapter 14, "Politics," Table 14.5).

Nor is Internet access evenly distributed globally. In 2005 the United States was the overwhelming leader in getting its population connected (Figure 8.6). Globally, the rate of Internet connectivity is much higher in rich countries than in poor countries (Figure 18.7; Chapter 8, "Stratification: United States and Global Perspectives").

Web Interactive Exercises icon: Is the Digital Divide Growing?

Content

U.S. domination is equally striking when it comes to Internet content. In rough terms, perhaps 60 percent of the servers that provide content on the Internet are in the United States, as are 15 of the 20 most popular search engines (Internet Software Consortium, 2003; "Search Engines," 2002). Some analysts say that American domination of the Web is an example of **media imperialism,** or control of a mass medium by a single national culture and the undermining of other national cultures. France and Canada are perhaps the countries that have spoken out most strongly against the perceived American threat to their national culture and identity. Some media analysts in those countries deeply resent the fact that the United States is the world's biggest exporter and smallest importer of mass media products, including Web content.[3]

Web Research Project: The Mass Media, Globalization, and Localization

Media imperialism is the domination of a mass medium by a single national culture and the undermining of other national cultures.

[3]The problem of media imperialism is felt particularly acutely in Canadian broadcasting because the country has a small population (slightly smaller than California's), is close to the United States, and is about 75 percent English speaking. Private broadcasters dominate the Canadian market and rely mainly on American entertainment programming. Moreover, the widespread use of cable and satellite dishes permits most Canadians to receive American programming directly from source (Knight, 1998 [1995]: 109–10).

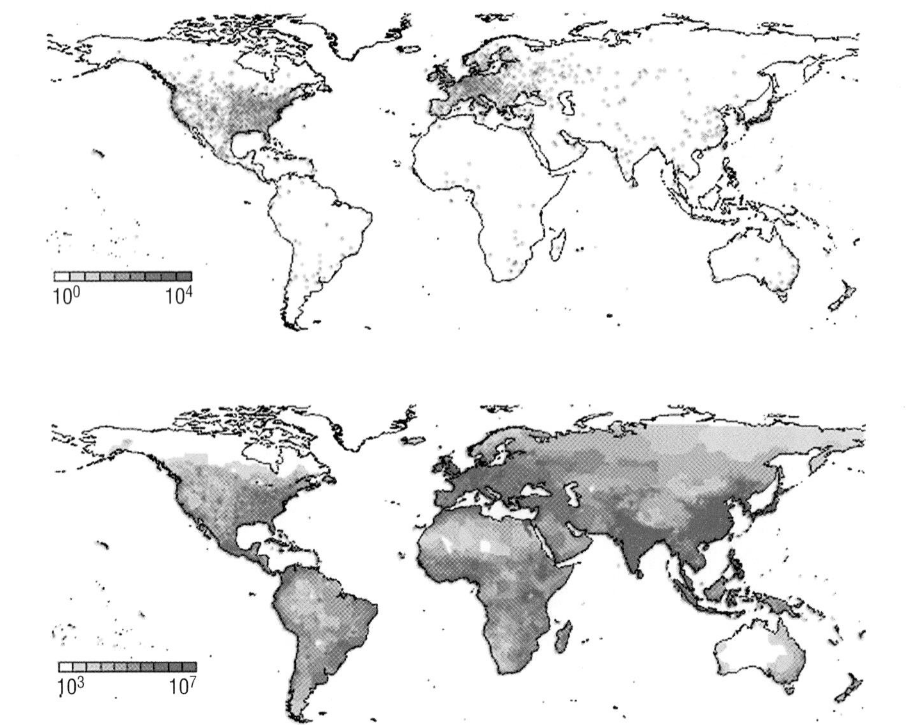

Figure 18.7
Internet Connectivity and Population Density

Nearly a quarter of a million routers in the world connected local computer networks (such as the computer network of a company or a university) to the Internet in 2001. The top map shows the number of routers per geographical area (1° of latitude by 1° of longitude) in the world. The bottom map shows the number of people per geographical area (1° of latitude by 1° of longitude) in the world. Darker areas indicate higher density. Based on these maps, what can you conclude about the relationship between Internet usage and population density in the highly developed countries and in the world as a whole?

Source: "Modeling the Internet's Large-Scale Topology," Department of Physics. University of Notre Dame. Reprinted with permission.

According to some media analysts, the Internet not only restricts access and promotes American content, it also increases the power of media conglomerates. That is evident, for example, in the realm of **media convergence,** or the blending of the World Wide Web, television, telephone, and other communications media into new, hybrid media forms. Many consumers find a personal computer (PC) too complex to operate and find TV limited in its functionality and entertainment value. Media convergence is intended to appeal mainly to such people (Dowling, Lechner, and Thielmann, 1998). The most visible form of media convergence today is interactive TV.

Interactive TV involves the reception of digital signals via cable, satellite dish, or fiber-optic telephone line. An interactive TV is connected to the Web through a built-in computer and contains a hard drive big enough to record 12 to 30 hours of TV programs. If you have interactive TV in your household, you can instruct it to record the exact blend of programs you want to watch from a long menu of specialty channels. Existing or upcoming features include ordering feature-length movies, using e-mail, holding videoconferences with people in remote locations, visiting websites that offer information on the programs or movies you're watching, shopping for a wide variety of goods, doing your banking, and paying your utility bills—all from the convenience of your couch. Audience research shows that customers are eager to adopt interactive TV. Therefore, all the media giants are scrambling for market share. After all, the media conglomerates stand to earn many billions of dollars laying fiber-optic cable, building new media appliances, writing software, creating TV programs, and selling goods and services online (Cable and Telecommunications Association for Marketing, 2001; Davis, 2000; Jupiter Media Metrix, 2001).

The control of interactive TV by huge media conglomerates may seem like an old story. In some respects it is. Ownership of every mass medium has become more concentrated over time. Interactive TV seems poised to repeat the pattern more quickly than any other mass medium. Because entry costs are so high, only media giants can get involved.

Media convergence is the blending of the World Wide Web, television, and other communications media as new, hybrid media forms.

However, this is a media story with a twist. The big media conglomerates may be able to carve out a new and lucrative niche for themselves by merging the Internet and television. However, they can never fully dominate the Internet. That is because it is the first mass medium that makes it relatively easy for consumers to become producers.

Melbourne Age/AP/Wide World Photos

Media Imperialism on Enaotai Island, West Papua New Guinea.

Millions of people are not just passive users of the Internet. Instead, they help to create it. For example, the Web boasts millions of personal websites. As of May 2003, it contained about 1600 role-playing communities, or multiple user dimensions (MUDs), with nearly 67,000 registered users ("The MUD Connector," 2005). WebcamSearch.com lists more than 42,000 public-access cameras connected to the Web. They open a window to the goings-on in people's offices, college dorms, apartments, and frat houses ("WebcamSearch.com," 2005). There are thousands of formal discussion groups on the Internet (e.g., "Google Groups," 2005). Each discussion group is composed of people who discuss defined subjects by e-mail or in real time. Some discussion groups focus on particle physics. Others are devoted to banjoes, lawyer jokes, Russian politics, sadomasochism, and just about every other human activity imaginable. The groups are self-governing bodies with their own rules and norms of "netiquette" (McLaughlin, Osborne, and Smith, 1995; Sudweeks, McLaughlin, and Rafaeli, 1999). Entry costs are relatively low. All you need to join or create a website, a MUD, or a discussion group is your own PC, some free or inexpensive software, and an Internet connection. AOL/Time Warner and all the other media conglomerates have no control over the many thousands of communities that are proliferating online.

So we see that we must temper the image of the Internet as a medium that is subject to increasing domination by large conglomerates. There is also a contrary trend. Individual users are making independent, creative contributions to Internet growth. Similarly, the view that the growth of the Web is an example of American media imperialism has not gone unchallenged. Where some media analysts see American media imperialism, others see globalization and postmodernization, social processes we introduced in Chapter 3 ("Culture"). From the latter point of view, all cultures, including that of the United States, are becoming less homogeneous and more fragmented as they borrow elements from each other. Inexpensive international travel and telecommunications make this cultural blending possible. Thus, if you look carefully at the Web, you will see that even U.S. sites adopt content liberally from Latin America, Asia, and elsewhere. Just as international influences can be seen in today's hairstyles, clothing fashions, foods, and popular music, so can the Internet be seen as a site of globalization (Hall, 1992).

Of course, nobody knows exactly how the social forces we have outlined will play themselves out. A phenomenon like Napster emerges, enabling millions of people to

Reuters NewMedia, Inc./Corbis

Shawn Fanning, creator of Napster. Napster allowed people to share recorded music on the Web freely, but media conglomerates took Napster to court, forcing it to stop the giveaway on the grounds that it was effectively stealing royalties from musicians and profits from music companies. Since then, alternative music-sharing programs, such as Kazaa, have become popular, suggesting that there is no end in sight to the tug-of-war between the forces for and against media conglomerates on the Web.

share recorded music freely on the Web using a central server. Some analysts point to Napster as evidence of Internet democratization. Then the media conglomerates take Napster to court, forcing it to stop the giveaway on the grounds that it is effectively stealing royalties from musicians and profits from music companies. Some analysts see the court case as evidence of inevitable conglomeration on the Web. Then new Napster-like programs, such as Kazaa, Limewire, and Gnutella emerge (the latter is named after the popular chocolate and hazelnut paste because both spread easily). These programs allow people to share recorded music on the Web *without* a central server, making them virtually impossible to shut down. A group of record companies tried to close Kazaa by taking it to an American court for copyright infringement late in 2002. However, they discovered that the distributor of its software is incorporated in the tiny South Pacific island of Vanuatu. Kazaa is managed from Australia, its servers are in Denmark, its source code is stored in Estonia, its developers live in the Netherlands, and it has 60 million users in 150 countries. It is therefore highly doubtful that a decision on behalf of the record companies could be enforced ("Digital Dilemmas," 2003). And so the tug-of-war between conglomeration and democratization continues, with no end in sight. As one observer noted: "The recording industry's long-running battle against online music piracy has come to resemble one of those whack-a-mole arcade games, where the player hammers one rubber rodent's head with a mallet only to see another pop up nearby" (Lohr, 2003). One thing is clear, however. The speed of technological innovation and the many possibilities for individual creativity on the Internet make this an exciting era to be involved in the mass media and to study it sociologically.

Summary

Sociology Now™

Reviewing is as easy as ❶, ❷, ❸.

❶ Before you do your final review, take the SociologyNow diagnostic quiz to help you identify the areas on which you should concentrate. You will find information on how to access SociologyNow on the foldout at the front of the textbook.

❷ As you review, take advantage of SociologyNow's study aids to help you master the topics in this chapter.

❸ When you are finished with your review, take SociologyNow's post-test to confirm you are ready to move on to the next chapter.

1. What are the mass media?

The mass media are means of transmitting information and entertainment from one person or group to another. The communication is typically from a few senders to many receivers. The mass media sometimes blur the line between reality and fantasy.

2. Which historical forces stimulated the growth of the mass media?

Three main historical forces stimulated the growth of the mass media. The Protestant Reformation of the 16th century encouraged people to read the Bible themselves. The democratic movements that began in the late 18th century encouraged people to demand literacy. Beginning in the late 19th century, capitalist industrialization required rapid means of communication and fostered the mass media as important sources of profit. However, the mass media became truly large-scale only when penny newspapers were published in the first half of the 19th century. The electronic media are products of the 20th century.

3. What are the main theories of mass media effects?

Functionalism stresses that the mass media act to coordinate society, exercise social control, and socialize and entertain people. *Conflict theory* stresses that the mass media reinforce social inequality. They do this both by acting as sources of profit for the few people who control media conglomerates and by promoting core values that help legitimize the existing social order. *Interpretive* approaches to studying the mass media stress that audiences actively filter, interpret, and sometimes even resist and reject media messages according to their interests and values. *Feminist* approaches concur and also emphasize the degree to which the mass media perpetuate gender and racial stereotypes.

Questions to Consider

1. Locate webcams in your region or community by searching http://www.webcamsearch.com. Who has set up these webcams? For what purposes? How might they be used to strengthen ties among family members, friendship networks, special-interest groups, and so forth?
2. By phoning TV and radio stations in your community, find out which are locally controlled and which are controlled by large media companies. If there are no locally controlled stations, does this have implications for the kind of news coverage and public affairs programming you may be watching? If there are locally controlled stations in your community, do they differ in terms of programming content, audience size, and audience type from stations owned by large media companies? If you can observe such differences, why do they exist?
3. The 1999 movie *Enemy of the State*, starring Will Smith, depicts one drawback of new media technologies: the possibility of intense government surveillance and violation of privacy rights. Do you think that new media technologies are unqualified blessings, or could they limit our freedom and privacy? If so, how? How could the capacity of new media technologies to limit freedom and privacy be constrained?

Web Resources

Companion Website for This Book

http://sociology.wadsworth.com

Begin by clicking on the Student Resources section of the website. Choose "Introduction to Sociology" and then the Brym and Lie book cover. Next, select the chapter you are currently studying from the pull-down menu. From the Student Resources page you will have easy access to InfoTrac® College Edition, MicroCase Online exercises, additional web links, and many other resources to aid you in your study of sociology, including practice tests for each chapter.

Recommended Websites

"The Media and Communications Studies Site" at the University of Wales is one of the best sites on the web devoted to the mass media. Among other interesting sections, it includes useful theoretical materials and resources on class, gender, and race in the mass media, mainly from a British perspective. Visit it at http://www.aber.ac.uk/media/Functions/mcs.html.

Although the University of Wales website contains some American materials, you might want to supplement it by visiting the University of Iowa's "Gender, Ethnicity and Race in Media" site at http://www.uiowa.edu/%7Ecommstud/resources/GenderMedia.

For useful resources on corporate ownership of the mass media and the social and political problems that result from increasingly concentrated media ownership, go to http://www.fair.org/media-woes/corporate.html. See also http://www.thenation.com/special/bigten.html for listings of the companies owned by the big-10 media conglomerates in the United States.

"Project Censored" is run out of Sonoma State University in California. It defines its main goal as "to explore and publicize the extent of censorship in our society by locating stories about significant issues of which the public should be aware, but is not." Visit the project site at http://www.projectcensored.org.

CHAPTER 19 Health and Medicine

Corbis Sygma

In this chapter, you will learn that:

- Health risks are unevenly distributed in human populations. Men and women, upper and lower classes, rich and poor countries, and privileged and disadvantaged racial and ethnic groups are exposed to health risks to varying degrees.
- The average health status of Americans is lower than the average health status of people in other rich postindustrial countries. That is partly because the level of social inequality is higher in the United States and partly because the health-care system makes it difficult for many people to receive adequate care in this country.
- The dominance of medical science is due to its successful treatments and the way doctors excluded competitors and established control over their profession and their clients.
- In some respects, modern medicine is a victim of its own success. For example, the wide availability of antibiotics has led to lax hygiene in hospitals, an epidemic of hospital infections leading to death, and the spread of drug-resistant germs.
- Patient activism, alternative medicine, and holistic medicine promise to improve the quality of health care in the United States and worldwide.

Health

The Black Death

In 1346, rumors reached Europe of a plague sweeping the East. Originating in Asia, the epidemic spread along trade routes to China and Russia. A year later, 12 galleys sailed from southern Russia to Italy. Diseased sailors were aboard. Their lymph nodes were terribly swollen and eventually burst, causing painful death. Anyone who came into contact with the sailors was soon infected. As a result, their ships were driven out of several Italian and French ports in succession. Still, the disease spread relentlessly, again moving along trade routes to Spain, Portugal, and England. Within two years, the Black Death, as it came to be known, killed a third of Europe's population. Six hundred and fifty years later, the plague still ranks as the most devastating catastrophe in human history (Herlihy, 1998; McNeill, 1976; Zinsser, 1935).

Today we know the cause of the plague was a bacillus that spread from lice to rats to people. It spread so efficiently because many people lived close together in unsanitary conditions. In the middle of the 14th century, however, nobody knew anything about germs. Therefore, Pope Clement VI sent a delegation to Europe's leading medical school in Paris to find out the cause of the plague. The learned professors studied the problem. They reported that a particularly unfortunate conjunction of Saturn, Jupiter, and Mars in the sign of Aquarius had occurred in 1345. The resulting hot, humid conditions caused the earth to emit poisonous vapors. To prevent the plague, they said, people should refrain from eating poultry, waterfowl, pork, beef, fish, and olive oil. They should not sleep during the daytime or engage in excessive exercise. Nothing should be cooked in rainwater. Bathing should be avoided at all costs.

Museo del Prado, Madrid, Spain/Giraudon, Paris/SuperStock

The Black Death.

We do not know if the pope followed the professors' advice. We do know he made a practice of sitting between two large fires to breathe pure air. Because the plague bacillus is destroyed by heat, the practice may have saved his life. Other people were less fortunate. Some rang church bells and fired cannons to drive the plague away. Others burned incense, wore charms, and cast spells. But, apart from the pope, the only people to have much luck in avoiding the plague were the well-to-do (who could afford to flee the densely populated cities for remote areas in the countryside) and the Jews (whose religion required that they wash their hands before meals, bathe once a week, and conduct burials soon after death).

Sociological Issues of Health and Medicine

Some of the main themes of the sociology of health and medicine are embedded in the story of the Black Death. First, recall that some groups were more likely to die of the plague than others were. This is a common pattern. Health risks are always unevenly distributed. Women and men, upper and lower classes, rich and poor countries, and privileged and disadvantaged racial and ethnic groups are exposed to health risks to varying degrees. This suggests that health is not just a medical question but also a sociological issue. The first task we set ourselves in the following is to examine the sociological factors that account for the uneven distribution of health in society.

The story of the Black Death also suggests that health problems change over time. Epidemics of various types still break out, but there can be no Black Death where sanitation and hygiene prevent the spread of disease.[1] Today we are also able to treat many infectious diseases such as tuberculosis and pneumonia with antibiotics. Twentieth-century medical science developed these wonder drugs and many other lifesaving therapies. **Life expectancy** is the average age at death of the members of a population. Life expectancy in the United States was 47 years in 1900. In 2004, it was 77 years. As a result of increased life expectancy, degenerative conditions such as cancer and heart disease have an opportunity to develop in a way that was not possible a century ago (Table 19.1, column 1).

Sociology Now™

Learn more about **Life Expectancy** by going through the Average Life Expectancy Map Exercise.

In contrast, we cannot help being struck by the superstition and ignorance surrounding the treatment of the ill in medieval times. Remedies were often herbal but also included earthworms, urine, and animal excrement. People believed it was possible to maintain good health by keeping body fluids in balance. Therefore, cures that released body fluids were common. These included hot baths, laxatives, and diuretics, which increase the flow of urine. If these treatments didn't work, bloodletting was often prescribed. No special qualifications were required to administer medical treatment. Barbers doubled as doctors.

However, the backwardness of medieval medical practice, and the advantages of modern scientific medicine, can easily be exaggerated. For example, medieval doctors stressed the importance of prevention, exercise, a balanced diet, and a congenial environment in maintaining good health. We now know that this is sound advice. On the other hand, one of the great shortcomings of modern medicine is its emphasis on high-tech cures rather than preventive and environmental measures. Therefore, we investigate in the

Life expectancy is the average age at death of the members of a population.

[1]One case that may approximate the Black Death in the 21st century is the spread of AIDS in sub-Saharan Africa. At the end of 2003, there were 10 African countries in which more than 10 percent of the adult population aged 15–49 was infected with HIV. In 5 of those countries, the figure exceeded 20 percent. In Botswana and Swaziland, the figure was 38 percent, the highest in the world (Population Reference Bureau, 2004).

following pages not just modern scientific medicine's many wonderful cures and treatments, but also its weaknesses. We also examine how the medical professions gained substantial control over health issues and promoted their own approach to well-being.

Health and Inequality

Defining and Measuring Health

According to the World Health Organization (WHO), **health** is

> the ability of an individual to achieve [his or her] potential and to respond positively to the challenges of the environment. . . . The basic resources for health are income, shelter and food. Improvement in health requires a secure foundation in these basics, but also information and life skills; a supportive environment, providing opportunities for making health choices among goods, services and facilities; and conditions in the economic, social and physical environments . . . that enhance health (World Health Organization, 2000).

The WHO definition lists in broad terms the main factors that promote good health. However, when it comes to *measuring* the health of a population, sociologists typically examine the negative: rates of illness and death. They reason that healthy populations experience less illness and longer life than unhealthy populations. This is the approach we follow here.

Assuming ideal conditions, how long can a person live? So far, the record is held by Jeanne Louise Calment, a French woman who died in 1997 at the age of 122. (Other people claim to be older, but they lack authenticated birth certificates or other proof.) Calment was an extraordinary individual. She took up fencing at 85, rode a bicycle until she was 100, gave up smoking at 120, and released a rap CD at 121 (Matalon, 1997). In contrast, only one in a hundred people in the world's rich countries now lives to be 100. Medical scientists tell us that the **maximum average human life span**—the average age of death for an entire population under *ideal* conditions—is likely to increase in this century. Now it is about 85 years (Olshansky, Carnes, and Desesquelles, 2001).

Table 19.1
Leading Causes of Death, United States, 1900 and 2001

	Deaths per 100,000 Population	Percentage of Deaths
1900		
1. Pneumonia/influenza	202.2	11.8
2. Tuberculosis	194.4	11.3
3. Diarrhea/other intestinal	142.7	8.3
4. Heart disease	137.4	8.0
5. Stroke	106.9	6.2
6. Kidney disease	88.6	5.2
7. Accidents	72.3	4.2
8. Cancer	64.0	3.7
9. Senility	50.2	2.9
10. Bronchitis	40.3	2.3
All other causes	620.1	36.1
Total	1719.1	100.0
2001		
1. Heart disease	245.7	28.9
2. Cancer	194.3	22.9
3. Stroke	57.4	6.8
4. Chronic lung disease	43.5	5.1
5. Accidents	34.3	4.0
6. Diabetes	25.0	2.9
7. Pneumonia/influenza	21.8	2.6
8. Alzheimer's disease	18.8	2.2
9. Kidney disease	13.9	1.6
10. Blood poisoning	11.3	1.3
11. Suicide	10.3	1.2
12. Liver diseases	9.4	1.1
13. Homicide	6.9	0.8
14. High blood pressure	6.7	0.8
15. Lung inflammation	6.1	0.7
All other causes	143.3	16.9
Total	849.0	100.0

Sources: Arias and Smith (2003: 4); National Office of Vital Statistics (1947).

Unfortunately, conditions are nowhere ideal. Throughout the world, life expectancy—the average number of years a person can *actually* expect to live—is less than 85 years. Figure 19.1 shows life expectancy in selected countries. Leading the list is Japan, where life expectancy was 81 years in 2004. Among the world's 20 or so rich countries, the United States had the lowest life expectancy at 77 years. In India, life expectancy was only 62 years. The poor African country of Sierra Leone suffered the world's lowest life expectancy at just 35 years, the same as Europe in 1600 (Population Reference Bureau, 2004).

Accounting for the difference between the maximum average human life span and life expectancy is one of the main tasks of the sociologist of health. For example, while the maximum average human life span is 85 years, life expectancy in the United States is 77 years. This implies that on average, Americans are being deprived of 8 years of life due to

Health, according to the World Health Organization, is "the ability of an individual to achieve his [or her] potential and to respond positively to the challenges of the environment."

The **maximum average human life span** is the average age of death for a population under ideal conditions. It is currently about 85 years.

avoidable *social* causes. Avoidable social causes deprive the average citizen of Sierra Leone of 50 years of life (85 – 35 = 50). Clearly, social causes have a big—and variable—impact on illness and death. We must therefore discuss them in detail.

Launette/AP/Wide World Photo

In the 21st century, the maximum human life span might increase because of medical advances. So far, however, the record is held by Jeanne Louise Calment, a French woman who died in 1997 at the age of 122.

The Social Causes of Illness and Death

People get sick and die partly due to natural causes. One person may have a genetic predisposition to cancer. Another may come in contact with the deadly Ebola virus in the environment. However, over and above such natural causes of illness and death, we can single out three types of *social* causes.

Lifestyle Factors

Cigarette smoking, excessive use of alcohol and drugs, poor diet, lack of exercise, and social isolation are among the chief lifestyle factors associated with poor health and premature death.

Official statistics such as those in Table 19.1 list immediate rather than background causes of death. For example, the immediate cause of death for some people is pneumonia, but that condition is sometimes brought on by congestive heart failure, which in turn may be precipitated by obesity. The best available estimates for the United States suggest that the three leading background causes of death in 2000 were tobacco use (18.1 percent of the total), poor diet and physical inactivity (16.6 percent), and alcohol consumption (3.5 percent) (Mokdad et al., 2004).

A third of the people who smoke are likely to die prematurely from smoking-related illnesses. This amounts to about 435,000 Americans *annually* (Mokdad et al., 2004). To put that number in perspective, consider that *total* American combat deaths since the Civil War stand at about 428,000 ("Statistical Summary," 2001; "Iraq Coalition," 2005). About 30 percent of all cancer deaths in the United States result from tobacco use, and smoking is also a leading cause of heart disease (Remennick, 1998: 17).

Many experts believe that poor diet and physical inactivity will soon overtake tobacco as the leading background cause of death. Today, poor diet and physical inactivity are responsible for about 400,000 American deaths every year (Mokdad et al., 2004). They are leading causes of heart disease and are implicated in about 35 percent of all cancer deaths (Remennick, 1998: 17). Poor diet and physical inactivity are especially big problems in poor neighborhoods. Typically, near the corner of East 149th Street and Southern

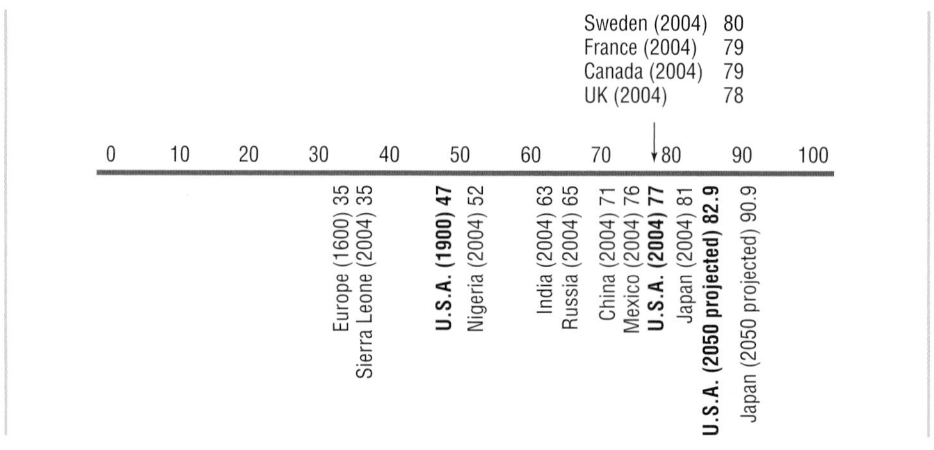

Figure 19.1
Life Expectancy, Selected Countries and Years

Sources: Population Reference Bureau (2004); Tuljapurkar, Li, and Boe (2000).

Boulevard in the South Bronx, you'll find a Dunkin' Donuts, a Popeyes Chicken and Biscuits, a White Castle, a pizzeria, a bakery offering sugary cakes, a Latin restaurant selling curled slabs of fried pork skin—and epidemic rates of heart disease, diabetes, and asthma, even among young people (Gonzalez, 2005).

Social isolation, too, affects one's chance of becoming ill and dying prematurely. Thus, unmarried people have a greater chance of dying prematurely than older people. At any age, the death of a spouse increases one's chance of dying, while remarrying decreases one's chance of dying (Helsing, Szklo, and Comstock, 1981). Social isolation is a particularly big problem among elderly people who retire, lose a spouse and friends, and cannot rely on family members or state institutions for social support. Such people are prone to fall into a state of depression that contributes to ill health. Some recent research suggests, however, that marriage has a protective benefit only for men (Statistics Canada, 2003).

Human-Environmental Factors

The environment constructed by humans poses major health risks. For example, more than 100 oil refineries and chemical plants are concentrated in a 75-mile strip between New Orleans and Baton Rouge. The area is commonly known as "Cancer Alley." That is because the petrochemical plants spew cancer-causing pollutants into the air and water. Local residents, overwhelmingly African American, are more likely than other Americans to get cancer because they breathe and drink high concentrations of these pollutants (Bullard, 1994 [1990]). Cancer Alley is a striking illustration of how human-environmental conditions can cause illness and death (Chapter 22, "Technology and the Global Environment").

The Public Health and Health-Care Systems

The state of a nation's health depends partly on public and private efforts to improve people's well-being and treat their illnesses. The **public health system** is composed of government-run programs that ensure access to clean drinking water, basic sewage and sanitation services, and inoculation against infectious diseases. The absence of a public health system is associated with high rates of disease and low life expectancy. The **health-care system** is composed of a nation's clinics, hospitals, and other facilities for ensuring health and treating illness. The absence of a system that ensures access by citizens to a minimum standard of health care is also associated with high rates of disease and low life expectancy.

Exposure to all three sets of social causes of illness and death is strongly related to country of residence, class, race, and gender. We now consider the impact of these factors, beginning with country of residence.

Global Health Inequalities

Web Research Project: The Changing Social Distribution of HIV/AIDS

Acquired immune deficiency syndrome (AIDS) is the leading cause of death in urban Haiti. Extreme poverty has forced many Haitians to become prostitutes. They cater mainly to tourists from North America and Europe. Some of those tourists carry the human immunodeficiency virus (HIV), which leads to AIDS, and they introduced it into Haiti. The absence of adequate health care and medical facilities makes the epidemic's impact all the more devastating (Farmer, 1992).

AIDS is also the leading cause of death in the poverty-stricken part of Africa south of the Sahara desert. ▶Figure 19.2 shows that in December 2004, 7.4 percent of sub-Saharan Africans—25.4 million people—were living with HIV/AIDS. In contrast, 0.6 percent of North Americans and 0.3 percent of Western Europeans were living with HIV/AIDS. This means that HIV/AIDS is more than 12 times more common in sub-Saharan Africa than in North America and nearly 25 times more common than in Western Europe. Despite the much greater prevalence of HIV/AIDS in sub-Saharan Africa, however, spending on research and treatment is concentrated overwhelmingly in the rich countries of North America and Western Europe. As the case of HIV/AIDS illustrates, global inequality influences the exposure of people to different health risks.

The **public health system** is composed of government-run programs that ensure access to clean drinking water, basic sewage and sanitation services, and inoculation against infectious diseases.

The **health-care system** is composed of a nation's clinics, hospitals, and other facilities for ensuring health and treating illness.

Mike Hutchings/Reuters/Corbis

A health worker at Nazareth House in Cape Town, South Africa, lavishes care and attention on some of the 41 infected children in her care. Nearly a fifth of South Africa's adult population is infected with HIV/AIDS.

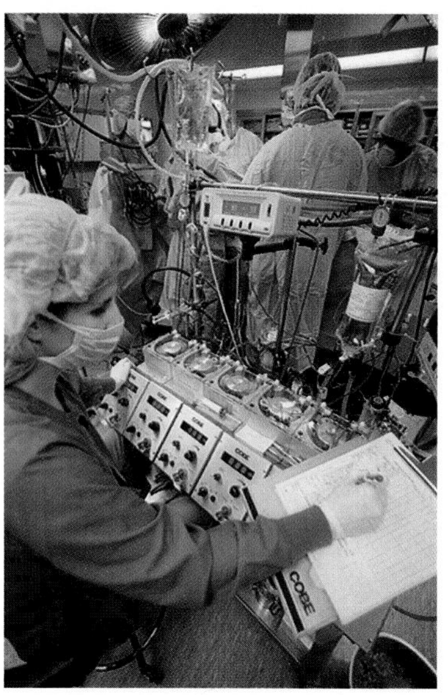

Mark Richards/PhotoEdit

Biomedical advances increase life expectancy, but the creation of a sound public health system has even more dramatic effects.

You might think that prosperity increases health due to biomedical advances, such as new medicines and diagnostic tools. If so, you are only partly correct. Biomedical advances do increase life expectancy. In particular, vaccines against infectious diseases have done much to improve health and ensure long life. However, the creation of a sound public health system is even more important in this regard. If a country can provide its citizens with clean water and a sewage system, epidemics decline in frequency and severity and life expectancy soars.

The industrialized countries started to develop their public health systems in the mid-19th century. Social reformers, concerned citizens, scientists, and doctors joined industrialists and politicians in urging governments to develop health policies that would help create a healthier labor force and citizenry (Goubert, 1989 [1986]; McNeill, 1976; Rosenberg, 1962). But what was possible in North America and Western Europe 150 years ago is not possible in many of the developing countries today. Most of us take clean water for granted. In contrast, more than 1 billion of the world's 6 billion people do not have access to a sanitary water supply (de Villiers, 1999).

We show other indicators of health inequality for selected countries in Table 19.2. We immediately see the positive correlation between national wealth and good health. The United States, Japan, and Canada are rich countries. They spend a substantial part of their wealth on health care. Many physicians and nurses service their populations. As a result,

Figure 19.2
Number of People with HIV/AIDS, December 31, 2004 (adult prevalence in parentheses)

Source: UNAIDS (2002). Reprinted with permission.

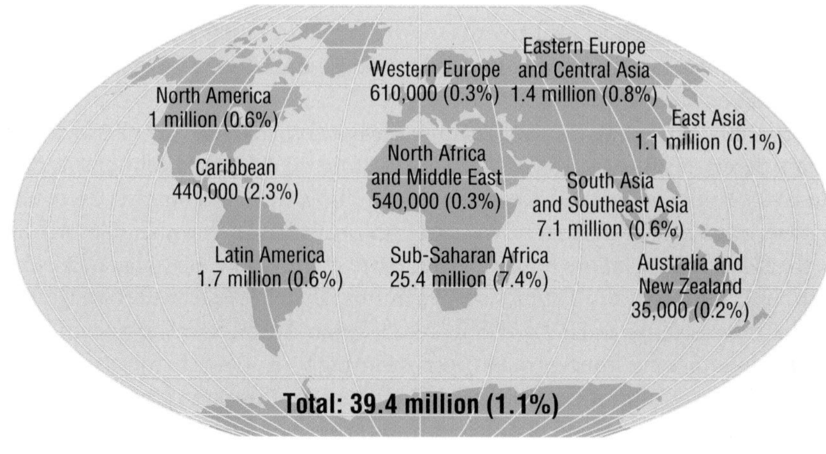

Table 19.2
Health Indicators, Selected Countries, 1999–2002

	Health Expenditure per Capita ($US)	Physicians/ 100,000 Population	Nurses/ 100,000 Population	Infant Mortality/ 1000 Live Births	Children Immunized against Measles (%)
United States	4499	279	972	6.0	92
Japan	2009	193.2	744.9	3.2	94
Canada	2534	229.1	897.1	5.3	96
Mexico	483	186.4	86.5	25	95
Zambia	49	6.9	113.1	95	90

Sources: Population Reference Bureau (2002); United Nations (2003b); World Health Organization (2002).

infant mortality (the annual number of deaths before the age of 1 year for every 1000 live births) is low. As noted previously, rich countries also enjoy high life expectancy. Mexico, however, is poorer than the United States, Japan, and Canada and spends much less per capita on health care. Accordingly, its population is less healthy in several respects. Sub-Saharan Zambia is one of the poorest countries in the world. It spends little on health care, has few medical personnel, and suffers a very high infant mortality rate.

Closer inspection of Table 19.2 reveals an anomaly, however. The United States spends more than twice as much per person on health care as Japan and 78 percent more than Canada. On average, Americans work nearly two months a year just to pay their medical bills. The United States has 22 percent more doctors per 100,000 people than Canada and 44 percent more than Japan. Yet the United States has a lower life expectancy than Canada or Japan, it immunizes a smaller percentage of its children against measles, and it has a higher rate of infant mortality. On one measure—immunization of children against measles—the United States falls behind Mexico. If the United States had an infant mortality rate as low as Cuba's, we would save more than 2200 babies a year (Kristof, 2005).

An efficient health-care system achieves relatively high outputs (e.g., high life expectancy) given the resources invested in it. An inefficient health-care system achieves relatively low outputs (e.g., low life expectancy) given the resources invested in it. In the first attempt by researchers to measure the efficiency of health-care systems in these terms, the most efficient health-care systems turned out to include those of France, Italy, Spain, Japan, and Saudi Arabia. The American health-care system was only a little above average among 191 countries, in the same league as health-care systems in most of South America, Eastern Europe, and India (Evans et al., 2001: 309).

Class Inequalities in Health Care

What accounts for the American anomaly? Why do we spend far more on health care than any other country in the world yet wind up with a population that on average is less healthy than the population of other rich countries? Part of the answer is that the gap between rich and poor is greater in the United States than in Japan, Sweden, Canada, France, and other rich countries. In general, the higher the level of inequality in a country, the more unhealthy its population (Wilkinson, 1996). Because the United States contains a higher percentage of poor people than do other rich countries, its average level of health is lower. Moreover, because income inequality has widened in the United States since the early 1970s, health disparities between income groups have grown (Williams and Collins, 1995).

Health inequality manifests itself in many ways. For instance, the infant mortality rate in Harlem, New York's main African American ghetto, is higher than in Bangladesh (Shapiro, 1992). Male life expectancy in poor African American areas of Washington, D.C., is 15 years lower than in the nearby rich suburb of Fairfax, Virginia (Epstein, 1998).

Infant mortality is the number of deaths before the age of 1 year for every 1000 live births in a population in 1 year.

Table 19.3
Leading Causes of Death: Ratios for Sex, Race, and Ethnicity, 2000

	RATIO				
Cause	Female:Male	African American: White	Hispanic American: White	Asian American/ Pacific Islander: White	Native American: White
Heart disease	1.02	0.90	0.80	0.87	0.71
Cancer	0.90	0.94	0.84	1.14	0.72
Stroke	1.53	0.96	0.82	1.34	0.71
Lung disease	1.00	0.50	0.45	0.59	0.70
Accidents	0.52	1.08	2.16	1.20	2.98
Influenza/pneumonia	1.25	0.75	0.86	1.14	0.89
Diabetes	1.15	1.56	1.92	1.19	2.00
Alzheimer's disease	2.42	0.45	0.48	0.36	0.36
Kidney disease	1.07	1.71	1.07	1.07	1.36
Suicide	0.25	0.54	1.30	1.38	2.00
Liver disease	0.53	0.91	3.00	0.82	4.27
Blood poisoning	1.27	n.a.	n.a.	n.a.	n.a.
Infant diseases	n.a.	4.50	6.67	2.75	2.75
Homicide	n.a.	7.00	9.00	2.50	4.50
HIV/AIDS	n.a.	9.00	9.00	0.67	1.67

n.a. = not available, because of the way the Centers for Disease Control and Prevention constructs its tables.
We calculated the ratios by dividing the percent of total deaths resulting from each cause (heart disease, etc.) for population categories (female versus male, etc.).
Source: Anderson (2002: 8, 9).

One reason for this disparity is that the poor are more likely than others to be exposed to violence, high-risk behavior, and environmental hazards, and they are more likely to do physical labor in which accidents are common. As you know, poverty is more common among African Americans, Hispanic Americans, and Native Americans than white non-Hispanic Americans. One would therefore expect these minority racial and ethnic groups to have higher mortality rates than non-Hispanic whites for some causes of death and lower mortality rates for other causes of death. Table 19.3 shows just such a pattern. In general, non-Hispanic whites have a relatively high mortality rate for degenerative diseases associated with old age (heart disease, cancer, stroke, Alzheimer's disease). African Americans, Hispanic Americans, and Native Americans have a relatively high mortality rate reflecting their lower class standing. They are much more likely than non-Hispanic whites to die from accidents, infant diseases, homicide, and HIV/AIDS.

As noted previously, poor diet is an especially big problem among the less well-to-do, and it contributes heavily to poor health. Typically, poor people do not have access to healthy food, and in any case they cannot afford it. The most dangerous consequence of poor diet is obesity. Nearly a third of Americans are obese, making them the most overweight people in the world (see Chapter 12, "Sociology of the Body: Disability, Aging, and Death," Figure 12.2). Recent research suggests that the high rate of obesity in this country is a substantial part of the reason why Americans live on average about two years less than people in other rich countries (Torrey and Haub, 2004).

A second reason why the poor are less healthy than the well-to-do is that they cannot afford adequate, and in some cases even minimal, health care. Thus, in 1996, 24.3 percent of households with incomes less than $25,000 had no health insurance, compared with only 7.6 percent of households with incomes of $75,000 or more (U.S. Department of Commerce, 1997). In spite of Medicaid, most poor people are inadequately served. Only about half the poor receive Medicaid assistance. Furthermore, poor people typically live in areas where medical treatment facilities are inadequate. This is especially true in recent decades, when many public hospitals that served the poor were closed due to government budget cuts (Albelda and Folbre, 1996).

If poor people have less access to doctors and hospitals than the well-to-do, they also tend to have less knowledge about healthy lifestyles. For example, they are less likely to know what constitutes a nutritious diet. This, too, contributes to their propensity to illness. Illness, in turn, makes it more difficult for poor people to escape poverty (Abraham, 1993).

Racial Inequalities in Health Care

Racial disparities in health status are largely, though not entirely, due to economic differences between racial groups. Thus, most studies show that blacks and whites *at the same income level* have similar health statuses. Nonetheless, the health status of African Americans is somewhat lower than the health status of European Americans even within the same income group. This suggests that racism affects health. It does so in three ways. First, income and other rewards do not have the same value across racial groups. For instance, due to discrimination, each year of education completed by an African American results in smaller income gains than it does for white Americans. Because, as we have seen, income is associated with good health, blacks tend to be worse off than whites at the same income level. Second, racism affects access to health services. That is because African Americans at all income levels tend to live in racially segregated neighborhoods with fewer health-related facilities. Third, the experience of racism induces psychological distress that has a negative effect on health status. For example, racism increases the likelihood of drug addiction and engaging in violence (Williams and Collins, 1995).

Increases in income have a bigger positive health impact on below-median income earners than on above-median income earners (▶Figure 19.3). But inequality is not simply a matter of differential access to resources such as medical care and knowledge. Even among people who have the *same* access to medical resources, people of higher rank tend to live healthier and longer lives. Why? Researchers in the United States, Britain, and Canada have argued that people of high rank experience less stress because they are more in control of their lives. If you can decide when to work, how to work, and what to work on, if you can exercise autonomy and creativity at work, you are likely to be healthier than someone who lacks these freedoms. You not only have the resources to deal with stress, you also have the ability to turn it off. In contrast, subordinates in a hierarchy have little control over their work environment. They experience a continuous sense of vulnerability that results in low-level stress. Continuous low-level stress, in turn, results in reduced immune function, increased hardening of the arteries, increased chance of heart attack, and other ailments. In short, if access to medical resources is associated with improved health, so is lower stress—and both are associated with higher positions in the socioeconomic hierarchy (Epstein, 1998; Evans, 1999).

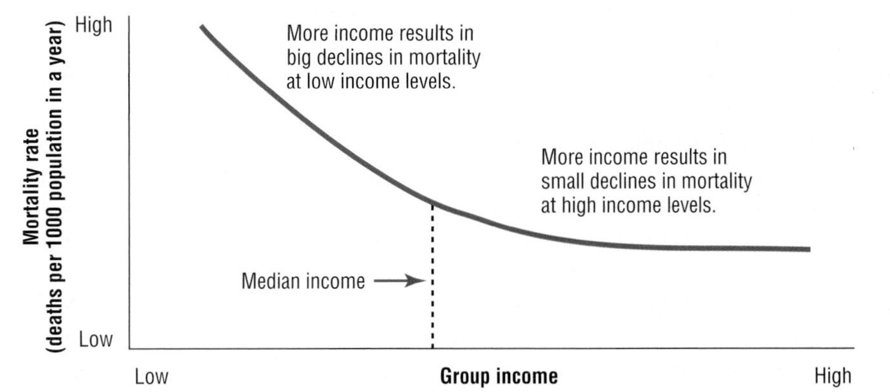

▶Figure 19.3
Mortality Rate by Group Income

Gender Inequalities in Health Care: The Feminist Contribution

Feminist scholars have brought health inequalities based on gender to the attention of the sociological community in recent decades. In a review of the relevant literature in the *New England Journal of Medicine,* one researcher concluded that such gender inequalities are substantial (Haas, 1998). Specifically:

- Gender bias exists in medical research. Thus, more research has focused on "men's diseases" (such as cardiac arrest) than on "women's diseases" (such as breast cancer). Similarly, medical research is only beginning to explore the fact that women may react differently than men to some illnesses and may require different treatment regimes.
- Gender bias also exists in medical treatment. For example, women undergo fewer kidney transplants, various cardiac procedures, and other treatments than men.
- Because women live longer than men, they experience greater lifetime risk of functional disability and chronic illness and greater need for long-term care. Yet more is spent on men's than women's health care in this country. (In contrast, Canadian health-care spending for women and men, excluding expenditures related to childbirth, is about equal. This is probably due to the fact that Canada, unlike the United States, has a system of universal health insurance for a comprehensive range of health-care services [Mustard et al., 1998]. We analyze the American health-care system later.)
- There are 40 percent more poor women than poor men in the United States (Casper, McLanahan, and Garfinkel, 1994: 597). Because, as we have seen, poverty contributes to ill health, we could expect improvements in women's economic standing to be reflected in improved health status for women.

In sum, although women live longer than men, gender inequalities have a negative impact on women's health. Women's health is negatively affected by differences between women and men in access to gender-appropriate medical research and treatment as well as the economic resources needed to secure adequate health care (see Table 19.3, column 1).

Health and Politics: The United States from Conflict and Functional Perspectives

Earlier we noted the existence of an "American anomaly." We spend more on health care than any other country, yet all the other rich postindustrial societies have healthier populations. One reason for this anomaly, as we have seen, is the relatively high level of social inequality in the United States. A second reason, which we will now examine, is the nature of the American health-care system.

Sociology Now™

Learn more about **the American Health-Care System** by going through the Health Care in the United States Learning Module.

You will recall from our discussion in Chapter 1 ("A Sociological Compass") and elsewhere that conflict theory is concerned mainly with the question of how privileged groups seek to maintain their advantages and subordinate groups seek to increase theirs. As such, conflict theory is an illuminating approach to analyzing the American health-care system. We can usefully see health care in the United States as a system of privilege for some and disadvantage for others. It therefore contributes to the poor health of less well-to-do Americans.

Consider, for example, that the United States lacks a system of health insurance that covers the entire population. Only the elderly, the poor, and veterans receive medical benefits from the government under the Medicare, Medicaid, and military health-care programs. All told, the American government pays about 45 percent of all medical costs out of taxes. In the United Kingdom, Sweden, and Denmark the comparable figure is about 85 percent; in Japan and Germany it is around 80 percent; and in France, Canada, Italy, and Australia it is around 70 percent. The governments of Germany, Italy, Belgium, Denmark, Finland, Greece, Iceland, Luxembourg, Norway, and Spain cover almost all health-care costs, including drugs, eyeglasses, dental care, and prostheses. About 40 million Americans lack health insurance. Another 40 million are inadequately covered ("Health Care Systems," 2001; Starr, 1994 [1992]).

Sociology Now™

Learn more about **Health Insurance** by going through the % of persons not covered by health insurance Map Exercise.

Problems with Private Health Insurance and Health Maintenance Organizations

Private insurance programs run by employers and unions cover most Americans, although some people buy their own private coverage. About 85 percent of employees receive their health coverage through health maintenance organizations (HMOs) (Gorman, 1998). HMOs are private corporations. They collect regular payments from employers and employees. When an employee needs medical treatment, an HMO administers it.

Like all corporations, HMOs pursue profit. They employ four main strategies to keep their shareholders happy. Unfortunately, all four strategies lower the average quality of health care in the United States (Kuttner, 1998a; 1998b):

1. Some HMOs avoid covering sick people and people who are likely to get sick. This keeps their costs down. For example, if an HMO can show that you had a medical condition before you came under its care, the HMO won't cover you for that condition.
2. HMOs try to minimize the cost of treating sick people they can't avoid covering. Thus, HMOs have doctor-compensation formulas that reward doctors for withholding treatments that are unprofitable.
3. There have been allegations that some HMOs routinely inflate diagnoses to maximize reimbursements. In 2000, Columbia/HCA, the largest for-profit hospital chain in the nation, agreed to pay the federal government $745 million to settle a federal billing fraud investigation (Galewitz, 2000).
4. HMOs keep overhead charges high. Administrative costs are higher in the private sector of the health-care system than in the public sector in almost every country for which data are available. Administrative costs are highest in the private sector of the American health-care system. In 1996, administrative costs in Medicare and Medicaid were only 37 percent as high as administrative costs in the private sector of American health care (▶Figure 19.4).

Advantages of Private and For-Profit Health-Care Institutions

Despite these drawbacks, running HMOs and other health-care institutions such as hospitals as for-profit organizations has one big advantage. It is an advantage that functionalists would undoubtedly highlight because they tend to emphasize the con-

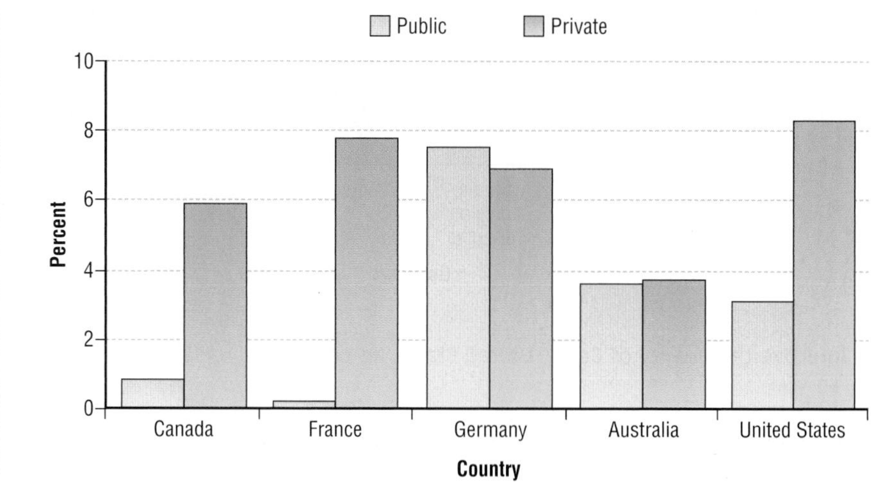

▶Figure 19.4
Administrative Costs as a Percentage of Health-Care Spending, Selected Countries

Note: Data are for mid-1990s.

Source: "Health Care Systems" (2001: 8)

tribution of social institutions to the smooth operation of society: Health organizations are so profitable that they can invest enormous sums in research and development, the latest diagnostic equipment, and high salaries to attract many of the best medical researchers and practitioners on the planet. Significantly, for people with adequate coverage, waiting times to see doctors and receive treatment are very short by international standards. Compare the United States with Canada in these respects. The United States has twice as many magnetic resonance imaging (MRI) machines per million people as Canada. The extended waiting time for nonemergency surgery in Canada has become a hot political issue, and it is not unusual for well-to-do Canadians needing elective surgery to travel to the United States and pay for it here. All this suggests that the United States enjoys the best health-care system in the world—for those who can afford it (Box 19.1).

The main supporters of the current United States health-care system are the stockholders of the 1500 private health-insurance companies and the physicians and other health professionals who get to work with the latest medical equipment, conduct cutting-edge research, and earn high salaries. Thus, HMOs and the American Medical Association (AMA) have been at the forefront of attempts to convince Americans that the largely private system of health care serves the public better than any state-run system could. Their efforts have been only partly successful. The 1998 General Social Survey asked a nationwide sample of Americans whether HMOs improve the quality of medical care. As ▶Figure 19.5 shows, only 22 percent of Americans agreed or strongly agreed that they do. In contrast 41 percent disagreed or strongly disagreed. Nearly twice as many Americans disapprove of HMOs as approve of them (National Opinion Research Center, 2004).

A National Health-Care System for the United States?

Despite this overall negative evaluation of the private health-care system, attempts to create a national system of health care in which everyone is covered regardless of his or her employment status or income level have failed (Hacker, 1997; Marmor, 1994; Quadagno, 1988). Most recently, Congress rejected President Clinton's 1993–1994 Health Security

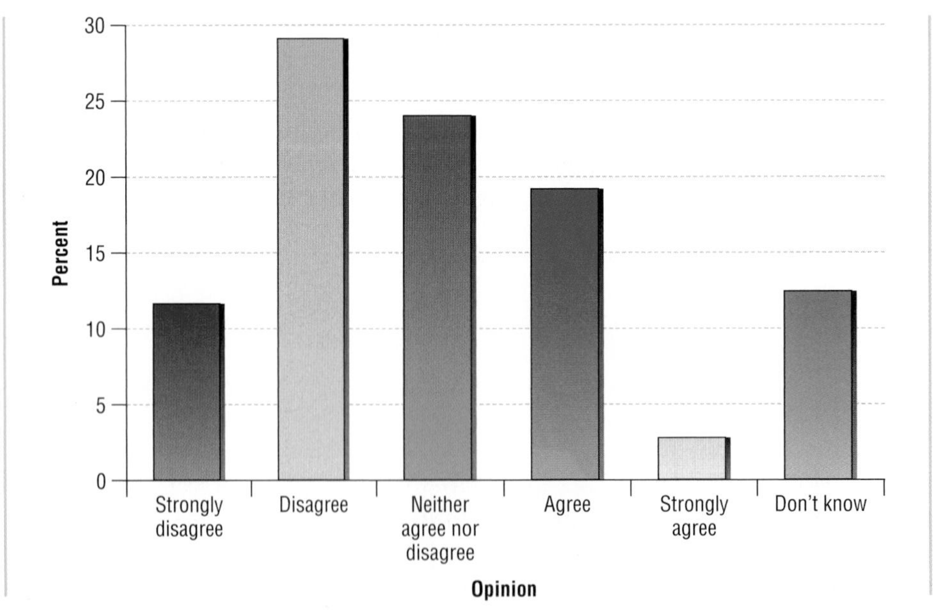

▶Figure 19.5
"HMOs Improve the Quality of Care," United States (in percent; *N* = 1382)
Data are for 1998.

Source: National Opinion Research Center (1999).

Box 19.1

YOU AND THE SOCIAL WORLD

What Kind of Health-Care System Do You Prefer?

A recent survey asked 8688 American and Canadian adults to evaluate their respective health-care systems. Among other things, they were asked if, over the past year, they did not receive a health-care service when they felt they needed it. If they did not, they were counted as having an "unmet health-care need."

Table 19.4 shows the percentage of four categories of respondents with an unmet health-care need: all Americans, Americans with health insurance, Americans without health insurance, and all Canadians (all of whom have state health insurance). Please inspect the table. Is there a meaningful difference between Canadians and Americans? If so, why? If not, why not? State your answers in a paragraph.

Equipped with the conclusions you draw from Table 19.4 and from the discussion of health-care systems in the text, explain in about 500 words why you favor a largely private health-care system like that of the United States or a state-managed system like that of Canada. In framing your answer, you may wish to consult the brief comparison of American, Canadian, Japanese, and German health-care systems at http://www.context.org/ICLIB/IC39/CoopTalr.htm.

Table 19.4

Individuals Reporting an Unmet Health-Care Need, United States and Canada, 2002–2003 (in percent; $N = 8688$)

	Percent with Unmet Health-Care Need	Main Reason for Unmet Health-Care Need
Canada	10.7	Waiting time
United States	13.1*	Cost
Insured	11.3**	Cost
Uninsured	40.0***	Cost

*Statistically significant difference between Canadians and Americans. This means that the sample difference between Canadians and Americans is probably meaningful; the chance that the difference does not exist in the population is less than 5 percent.
**No statistically significant difference between Canadians and insured Americans. This means that the sample difference between Canadians and insured Americans is probably not meaningful; the chance that the difference exists in the population is 5 percent or *greater.*
***Statistically significant difference between Canadians and uninsured Americans. This means that the sample difference between Canadians and uninsured Americans is probably meaningful; the chance that the difference does not exist in the population is less than 5 percent.
Source: Sanmartin et al. (2004: 18, 30).

proposal. Clinton was unable to unify political and public support for his proposal, partly because of the massive media campaign bankrolled by health-insurance companies and Clinton's political opponents (Skocpol, 1996).

Summing up, we may say that the apparently *natural* processes of health and illness are in fact deeply *social* processes. Social circumstances account for variations in life expectancy and rates of mortality due to various causes. These social circumstances include a country's standard of living, level of inequality, and type of health-care system and a person's gender, class, race, and ethnicity.

In the next section of this chapter, we make a similar argument about medicine. **Medicine** is a social institution devoted to prolonging life by fighting disease and promoting health. It may seem to lie squarely in the realm of pure science. However, as you will now learn, society shapes medical practice every bit as much as it influences health processes. We can see this clearly by examining how the medical and psychiatric professions have increased their control over people in the past 150 years or so. In the following discussion, we begin this task by showing how forms of deviance that used to be considered the province of morality and the law have come increasingly under the sway of psychiatry. This, we argue, is only partly due to the scientifically proven benefits of psychiatric care. We then examine how medicine drove other competing professions out of the health-care market. Again, this demonstrates that the type of health care we receive is a product not just of scientific considerations but of social forces too.

Medicine is an institution devoted to fighting disease and promoting health.

Medicine

The Medicalization of Deviance: A Symbolic Interactionist Approach

You may recall from our discussion of deviance that one of the preoccupations of symbolic interactionism is the labeling process (Chapter 7, "Deviance and Crime"). According to symbolic interactionists, deviance results not just from the actions of the deviant but also from the responses of others, who define some actions as deviant and other actions as normal.

Here we may add that the *type* of label applied to a deviant act may vary widely over time and from one society to another, depending on how that act is interpreted. Consider, for instance, the **medicalization of deviance.** The medicalization of deviance refers to the fact that over time, "medical definitions of deviant behavior are becoming more prevalent in . . . societies like our own" (Conrad and Schneider, 1992 [1980]: 28–9). In an earlier era, much deviant behavior was labeled evil. Deviants tended to be chastised, punished, and otherwise socially controlled by members of the clergy, neighbors, family members, and the criminal justice system. Today, however, a person prone to drinking sprees is more likely to be declared an alcoholic and treated in a detoxification center. A person predisposed to violent rages is more likely to be medicated. A person inclined to overeating is more likely to seek therapy and, in extreme cases, surgery. A heroin addict is more likely to seek the help of a methadone program. As these examples illustrate, what used to be regarded as willful deviance is now often regarded as involuntary deviance. Increasingly, what used to be defined as "badness" is defined as "sickness." As our definitions of deviance change, deviance is increasingly coming under the sway of the medical and psychiatric establishments (Figure 19.6).

How did the medicalization of deviance come about? What are the major social forces responsible for the growing capacity of medical and psychiatric establishments to control our lives? To answer this question, we first examine the fascinating case of mental illness. As you will see, our changing definitions of mental illness show perhaps more clearly than any other aspect of medicine how thin a line separates science from politics in the field of health care.

"Now here's a young woman in her twenties, let's call her Betty Smith . . . she has never had a job, and she doesn't seem to want to go out and look for one. She is a very quiet girl, she doesn't talk much to anyone—even her own family, and she acts like she is afraid of people, especially young men her own age. She won't go out with anyone, and whenever someone comes to visit her family, she stays in her own room until they leave. She just stays by herself and daydreams all the time and shows no interest in anything or anybody."

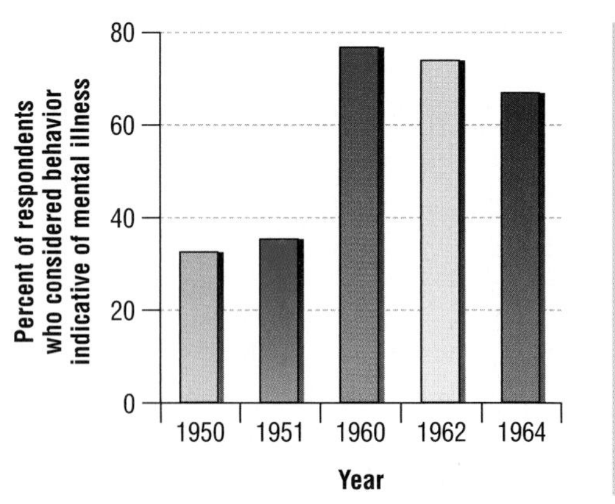

Figure 19.6
An Example of the Medicalization of Deviance

Five North American surveys conducted in the 1950s and 1960s presented respondents with the anecdote to the left. The graph shows the percentage of respondents who considered the behavior described in the anecdote evidence of mental illness. Notice the difference between the 1950s and the 1960s. (Nearly 100 percent of psychiatrists who evaluated the anecdote thought it illustrated "simple schizophrenia.")

Source: Conrad and Schneider (1992 [1980]: 59).

The **medicalization of deviance** is the tendency for medical definitions of deviant behavior to become more prevalent over time.

The Political Sociology of Mental Illness

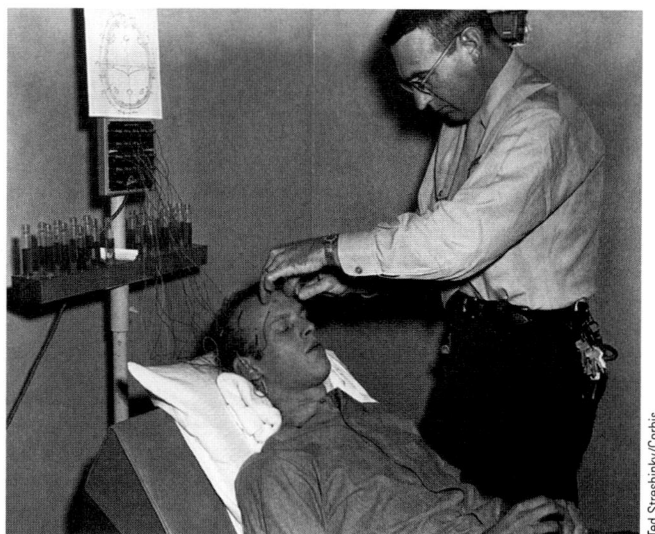

Ted Streshinky/Corbis

An example of the medicalization of deviance. A lobotomy is performed in Vacaville State Prison in California in 1961 to "cure" the inmate of criminality.

In 1974, a condition that had been considered a psychiatric disorder for more than a century ceased to be labeled as such by the American Psychiatric Association (APA). Did the condition disappear because it had become rare to the point of extinction? No. Did the discovery of a new wonder drug eradicate the condition virtually overnight? Again, no. In fact, in 1974 the condition was perhaps more widespread and certainly more public than ever before. But, paradoxically, just as the extent of the condition was becoming more widely appreciated, the "bible" of the APA, the *Diagnostic and Statistical Manual of Mental Disorders* (*DSM*), ceased to define it as a psychiatric disorder.

The "condition" we are referring to is homosexuality. In preparing the third edition of the *DSM* for publication, a squabble broke out among America's psychiatrists over whether homosexuality is in fact a psychiatric disorder. Gay and lesbian activists, who sought to destigmatize homosexuality, were partly responsible for a shift in the views of many psychiatrists on this subject. In the end, the APA decided that homosexuality is not a psychiatric disorder. They deleted the entry. The APA membership confirmed the decision in 1974.

The controversy over homosexuality was only one of several *political* debates that erupted among American psychiatrists in the 1970s and 1980s (Shorter, 1997: 288–327). Among others:

- The *DSM* task force initially decided to eliminate the term "neurosis" on the grounds that its role as a cause of mental disorder had never been proven experimentally. The decision outraged the psychoanalytic community. "Neurosis" was a keystone of their Freudian theories. So the psychoanalysts threatened to block publication of the third edition of the *DSM* unless they were appeased. In 1979, the APA board backed down, placing "neurosis" in parentheses after "disorder." This compromise had nothing to do with science.
- When veterans of the war in Vietnam started returning to the United States, they faced great difficulty reentering American society. The war was unpopular, so they were not universally greeted as heroes. The economy went into a tailspin in 1973, making jobs difficult to find. And the veterans had suffered high levels of stress during the war itself. Many of them believed that their troubles were psychiatric in nature, and soon a nationwide campaign was under way, urging the APA to recognize post-traumatic stress disorder (PTSD) in their manual. Many psychiatrists were reluctant to do so. Nonetheless, PTSD was listed in the third edition of the *DSM*. To be sure, the campaign succeeded partly on the strength of evidence that extreme trauma has psychological (and at times physiological) effects. But in addition, as one activist later explained, the PTSD campaign succeeded because "[we] were better organized, more politically active, and enjoyed more lucky breaks than [our] opposition" (Chaim Shatan, quoted in Scott, 1990: 308). Again, politics and not just science helped to shape the definition of a mental disorder.
- Feminists were unhappy that the 1987 edition of the *DSM* contained listings such as "self-defeating personality disorder." The *DSM* said that this disorder is twice as common among women as men. Feminists countered that the definition is an example of victim blaming. Under pressure from the feminists, the 1994 edition of the *DSM* dropped the concept.

Some mental disorders have obvious organic causes, such as chemical imbalances in the brain. These organic causes can often be precisely identified. Often they can be treated with drugs or other therapies. Moreover, experiments can be conducted to verify their existence and establish the effectiveness of one treatment or another. However, the examples listed previously show that the definition of a host of other mental disorders depends not just on scientific evidence but also on social values and political compromise.

In the mid-19th century there was just one mental disorder recognized by the federal government: idiocy/insanity. By 1975, the *DSM* recognized 106 mental disorders. The 1994 *DSM* lists 297 mental disorders. As the number of mental disorders has grown, so has the proportion of Americans presumably affected by them. In the mid-19th century, few people were defined as suffering from mental disorders. However, one respected survey conducted in 1992 found that fully 48 percent of Americans would suffer from a mental disorder—very broadly defined, of course—during their lifetimes. The most common mental disorder, supposedly affecting 17 percent of the population at some point in their lives, is severe depression (Blazer, Kessler, McGonagle and Swartz, 1994; Shorter, 1997: 294).

Edward Shorter, one of the world's leading historians of psychiatry, notes that in psychiatric practice, definitions of mental disorders are often expanded to include ailments with dubious or unknown biological foundations (e.g., "minor depression," "borderline schizophrenia") (Shorter, 1997: 228). In addition, as we have seen, the sheer number of conditions labeled "mental disorder" increased rapidly during the 20th century. We suggest four main reasons for expansion in the number and scope of such labels.

1. As we saw in Chapter 13 ("Work and the Economy"), Americans are now experiencing more stress and depression than ever before, due mainly to the increased demands of work and the growing time crunch. Mental health problems are thus more widespread than they used to be. At the same time, traditional institutions for dealing with mental health problems are less able to cope with them. The weakening authority of the church and the weakening grip of the family over the individual leave the treatment of mental health problems more open to the medical and psychiatric establishments.
2. The number of mental disorders has increased because powerful organizations demand it. The U.S. Census Bureau first asked the American Medico-Psychological Association to classify mental disorders in 1908, and HMOs today demand precise diagnostic codes before paying for psychiatric care. Because public and private organizations find the classification of mental disorders useful, they have proliferated.
3. The cultural context stimulates inflation in the number and scope of mental disorders. Probably more than any other people, Americans are inclined to turn their problems into medical and psychological issues, sometimes without inquiring deeply into the disadvantages of doing so. For example, in 1980 the term "attention deficit disorder" (ADD) was coined to label hyperactive and inattentive schoolchildren, mainly boys.[2] By the mid-1990s, doctors were writing 6 million prescriptions a year for Ritalin, an amphetamine-like compound that controls ADD. Evidence shows that some children diagnosed with ADD have problems absorbing glucose in the brain or suffer from imbalances in chemicals that help the brain regulate behavior (Optometrists Network, 2000). Yet the diagnosis of ADD is typically conducted clinically, i.e., by interviewing and observing children to see if they exhibit signs of serious "inattention," "hyperactivity," and "impulsivity." This means that many children diagnosed with ADD may have no organic disorder at all. Some cases of ADD may be due to the school system failing to capture children's imagination. Some may involve children acting out because they are deprived of attention at home. Some may involve plain, old-fashioned youthful enthusiasm. A plausible case could be made that Tom Sawyer or Winnie the Pooh suffered from ADD (Shea et al., 2000). However, once hyperactivity and inattentiveness in school are defined as a medical and psychiatric condition, officials routinely prescribe drugs to control the problem and tend to ignore possible social causes.

[2]Psychiatrists now recognize several types of ADD, but the distinguishing features of each type are not relevant to our discussion.

4. The fourth main reason for inflation in the number and scope of mental disorders is that various professional organizations have promoted it. Consider PTSD. There is no doubt that PTSD is a real condition and that many veterans suffer from it. However, once the disorder was officially recognized in the 1970s, some therapists trivialized the term. By the mid-1990s some therapists were talking about PTSD "in children exposed to movies like *Batman*" (Shorter, 1997: 290). Some psychiatric social workers, psychologists, and psychiatrists may magnify the incidence of such mental disorders because doing so increases their stature and their patient load. Others may do so simply because the condition becomes trendy. Whatever the motive, overdiagnosis is the result.

The Professionalization of Medicine

The preceding discussion shows that the diagnosis and treatment of some mental disorders is not a completely scientific enterprise. Social processes are at least as important as scientific principles in determining how we treat some mental disorders. Various mental health professions compete for patients, as do different schools of thought within professions. Practitioners offer a wide and sometimes confusing array of treatments and therapies. The American public spends billions of dollars a year on supposed cures, but in some cases their effectiveness is debatable, and some people remain skeptical about their ultimate worth.

In 1850, the practice of medicine was in an even more chaotic state. Herbalists, faith healers, midwives, druggists, and medical doctors vied to meet the health needs of the American public. A century later, the dust had settled. Medical science was victorious. Its first series of breakthroughs involved identifying the bacteria and viruses responsible for various diseases and then developing effective procedures and vaccines to combat them. These and subsequent triumphs in diagnosis and treatment convinced most people of the superiority of medical science over other approaches to health. Medical science worked, or at least it seemed to work more effectively and more often than other therapies.

It would be wrong, however, to think that scientific medicine came to dominate health care only because it produced results. A second, sociological reason for the rise and dominance of scientific medicine is that doctors were able to professionalize. As noted in Chapter 13 ("Work and the Economy"), a profession is an occupation requiring extensive formal education. Professionals regulate their own training and practice. They restrict competition within the profession, mainly by limiting the recruitment of practitioners. They maximize competition with some other professions, partly by laying exclusive claim to a field of expertise. Professionals are usually self-employed. They exercise considerable authority over their clients. And they profess to be motivated mainly by the desire to serve their community, although they earn a lot of money in the process. Professionalization, then, is the process by which people gain control and authority over their occupation and their clients. It results in professionals enjoying high occupational prestige and income and considerable social and political power (Johnson, 1972; Friedson, 1986; Starr, 1982).

The American Medical Association

The professional organization of American doctors is the AMA, founded in 1847. It quickly set about broadcasting the successes of medical science and criticizing alternative approaches to health as quackery and charlatanism. By the early years of the 20th century, the AMA had convinced state licensing boards to certify only doctors who had been trained in programs recognized by the AMA. Soon, schools teaching other approaches to health care were closing down across the country. Doctors had never earned much. In the 18th century it was commonly said that "few lawyers die well, few physicians live well" (Illich, 1976: 58). But once it was possible to lay virtually exclusive claim to health care, big financial rewards followed. Today, American doctors working full-time in private practice earn on average about $200,000 a year, although income varies substantially by specialty.

Sociology Now™

Learn more about **Physicians** by going through the # of physicians per 100,000 Map Exercise.

The Rise of Modern Hospitals

The modern hospital is the institutional manifestation of the medical doctor's professional dominance. Until the 20th century, most doctors operated small clinics and visited patients in their homes. Medicine's scientific turn in the mid-19th century guaranteed the rise of the modern hospital. Expensive equipment for diagnosis and treatment had to be shared by many physicians. This required the centralization of medical facilities in large, bureaucratically run institutions that strongly resist deviations from professional conduct (Box 19.2). Practically nonexistent until the Civil War, hospitals are now widespread. Yet despite their undoubted benefits, economic as well as health related, hospitals and the medicine practiced in them are not an unqualified blessing, as you are about to learn.

The Social Limits of Modern Medicine

Around February 5, 2003, a 64-year-old professor of medicine from Guangzhou, the capital of Guangdong Province in South China, came down with an unidentified respiratory ailment. It did not bother him enough to cancel a planned trip to Hong Kong, so on February 12 he checked into that city's Metropole Hotel. Ironically, as it turned out, the desk clerk assigned him room 911. Other ninth-floor guests included an elderly couple from Toronto and three young women from Singapore. All of these people, along with a local resident who visited the hotel during this period, fell ill between February 15 and 27 with the same respiratory ailment as the professor. The professor died on March 4. The Canadian woman returned to Toronto on February 23 and died at her home on March 5. The eventual diagnosis: severe acute respiratory syndrome, or SARS, a new (and in 9 percent of cases, deadly) pneumonia-like illness for which there is no vaccine and no cure.

SARS originated in Guangdong Province. By June 12, 8445 cases of SARS had been identified in 29 countries, and 790 people had died of the disease. Quickly and efficiently, global travel spread HIV/AIDS, West Nile virus, and now SARS from remote and isolated locales to the world's capitals. The United Nations has labeled Toronto the world's most multicultural city. It has a large Chinese population, mainly from Hong Kong. It is therefore not surprising that outside of China, Hong Kong, and Taiwan, Toronto became the world's number one SARS hot spot (Abraham, 2003; World Health Organization, 2003).

Once identified as a potential SARS case, a person is quarantined at home for 10 days. However, if people exhibit symptoms of the disease, they go to a poorly ventilated institution where the air is maintained at a constant warm temperature that is ideal for the multiplication of germs. In this institution, many young and elderly people with weakened immune systems congregate. A steady stream of germs pours in round the clock. Staff members too often fail to follow elementary principles of good hygiene. That institution is a hospital. There, germs spread. Most of the 238 people in Toronto who caught SARS as of June 12, 2003, did so in the hospital before stringent isolation and disinfection procedures were imposed.

Our characterization of hospitals as ideal environments for the spread of germs may seem harsh. It is not. In fact, among the world's rich countries, the hospital system in the United States is perhaps the most dangerous in this respect. The *Chicago Tribune* published a major investigative report on the problem in 2002 (Berens, 2002a; 2002b; 2002c). Adopting the same methods used by epidemiologists, the *Tribune* analyzed records from 75 federal and state agencies. It also examined hospital files, patient databases, and court cases to produce the most comprehensive analysis of preventable patient deaths linked to infections in 5810 hospitals nationwide. It found that of the 35 million Americans admitted to a hospital each year, about 6 percent contract a hospital-acquired infection such as pneumonia, influenza, or staphylococcus. In 2000, an estimated 103,000 people died due to hospital infections. (The Centers for Disease Control estimated 90,000 such deaths in 2000, but its research extrapolated from just 315 hospitals.) If the government classified death due to hospital infection, it would be the fourth leading cause of death in the nation, behind heart disease, cancer, and stroke. The *Tribune* identified about a quarter of hospital infection deaths as nonpreventable because they resulted from problems that had

BOX 19.2
Sociology at the Movies

Patch Adams (1998)

Patch Adams, played by Robin Williams, is suicidal. Checking into a mental hospital, he finds that the doctors, who are supposed to be helping him, are indifferent. In contrast, other patients help him overcome his suicidal urges. He resolves to become a doctor to help other patients.

Patch Adams, based on a real person of the same name, breathes humor and life into the dreary world of the modern hospital. As an intern, Adams finds that patients are identified by their ID number and disease. Doctors and nurses seem more concerned about medical charts than about their patients. Finally, Adams startles a nurse by asking her about a patient: "What's her name?" The very idea that a patient may be something more than his or her medical records reveals the impersonal and bureaucratic nature of the modern doctor–patient relationship.

Patch Adams is intent on infusing personal care, humor, and humanity into the doctor–patient relationship. In dealing with children whose hair had fallen out because of chemotherapy, Patch plays a clown in order to bring smiles to their faces. He believes humor and laughter can be a great cure.

Not surprisingly, Adams faces resistance from medical school administrators. After all, he deviates significantly from the norm of impersonal professionalism. They attempt to expel him from the medical school. With support from his friends and patients, however, he manages to win a court battle to remain in medical school. In real life, Patch Adams goes on to become a medical doctor, who not only maintains a sense of humor but also continues to live up to his ideals, including helping poor patients around the world.

Patch Adams is a sentimental movie, pitting the humorous individual against the grim organization. However, the critic Roger Ebert (1998) wrote, "To himself . . . [Patch Adams is] an irrepressible bundle of joy, a zany live wire who brings laughter into the lives of the sick and dying. To me, he's a pain in the wazoo. If this guy broke into my hospital room and started tap-dancing with bedpans on his feet, I'd call the cops." Here Ebert is saying that the norm of professionalism—grim and impersonal though it may be—may be preferable to the antics of Patch Adams. Do you agree with Ebert? Would you prefer your doctor to be a "human being" or simply to play his professional role efficiently and effectively? What are the health advantages and disadvantages of each approach to doctoring?

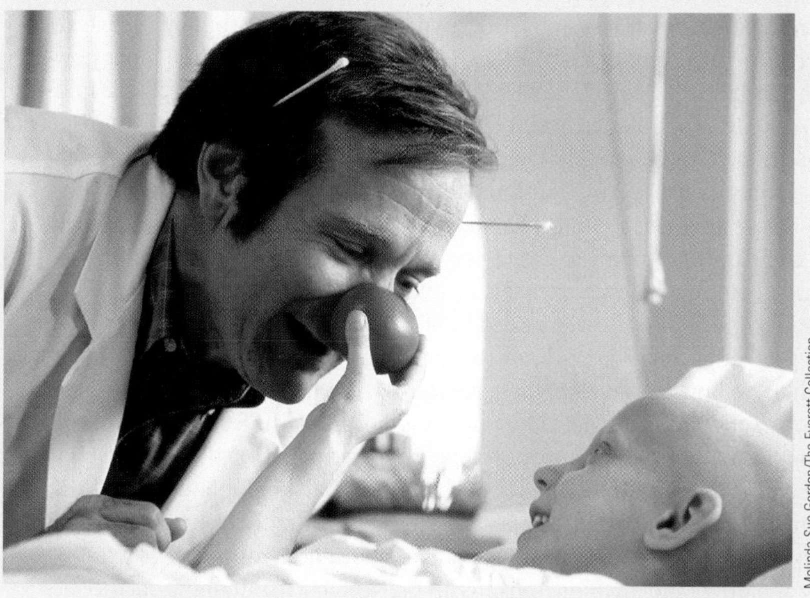

Melinda Sue Gordon/The Everett Collection

Patch Adams, starring Robin Williams.

not been detected beforehand by state, federal, or health-care investigators. That leaves 75,000 preventable deaths, caused by problems investigators had already identified. These deaths could have been avoided if we kept hospital rooms and operating theaters cleaner, sterilized all instruments after use, fixed ventilation problems, routinely flushed water pipes, ensured that all doctors and nurses disinfect their hands after examining each patient, and so forth.

Cutting costs and catering to paying customers are among the chief means of keeping shareholders happy in a health-care system driven chiefly by profitability. This means investing disproportionately in expensive, high-tech, cutting-edge diagnostic equipment and treatment for those who can afford it. It also means scrimping on simple, labor-intensive, time-consuming hygiene for those who cannot. About a third of all hospitals in America are operating at a loss and a third are on the edge of bankruptcy, according to the American Hospital Association. Particularly in hospitals in financial distress, cleaning staffs are too small and insufficiently trained. Nurses are too few. According to research by the Harvard School of Public Health, these are the kinds of factors correlated with hospital-acquired infections. As San Francisco registered nurse Trande Phillips says: "When you have less time to save lives, do you take 30 seconds to wash your hands? When you're speeding up you have to cut corners. We don't always wash our hands. I'm not saying it's right, but you've got to deal with reality" (quoted in Berens, 2002a).

This was not always the reality. From the 1860s to the 1940s, American hospital staffs were obsessed with cleanliness. They had to be. In the era before the widespread use of penicillin and antibiotics, infection often meant death. In the 1950s, however, the prevention of infections in hospitals became less of a priority because penicillin and antibiotics became widely available. It was less expensive to wait until a patient got sick and then respond to symptoms by prescribing drugs than preventing the sickness in the first place. Doctors and nurses have grown lax about hygiene over the past half-century. For example, the *Chicago Tribune* investigators found a dozen recent health-care studies showing that about half of doctors and nurses do not disinfect their hands between patients (Berens, 2002a).

Using penicillin and antibiotics indiscriminately has its own costs. When living organisms encounter a deadly threat, only the few mutations that are strong enough to resist the threat survive and go on to reproduce. Accordingly, if you use a lot of penicillin and antibiotics, "super germs" that are resistant to these drugs multiply. This is just what has happened.[3] Penicillin could kill nearly all staphylococcus germs in the 1940s, but by 1982 it was effective in fewer than 10 percent of cases. In the 1970s, doctors turned to the more powerful methicillin, which in 1974 could kill 98 percent of staphylococcus germs. By the mid-1990s, it could kill only about 50 percent. It has thus come about that various strains of drug-resistant germs now cause pneumonia, blood poisoning, tuberculosis, and other infectious diseases. Drug-resistant germs that could formerly survive only in the friendly hospital environment have now adapted to the harsher environment outside the hospital walls. Pharmaceutical companies are racing to create new antibiotics to fight drug-resistant bugs, but germs mutate so quickly that our arsenal is shrinking (Berens, 2002c).

The epidemic of infectious diseases caused by slack hospital hygiene and the overuse of antibiotics suggests that social circumstances constrain the success of modern medicine. It is difficult to see how we can solve these problems without setting up a government watchdog to oversee and enforce strict rules regarding hospital disinfection. That would require considerable tax money and it would impose a principle other than profitability on the health-care system. It would therefore be a political hot potato (Box 19.3). Meanwhile, many people are growing skeptical of the claims of modern medicine. They are beginning to challenge traditional medicine and explore alternatives that rely less on high technology and drugs and are more sensitive to the need for maintaining balance between humans and their environment in the pursuit of good health. In concluding this chapter, we explore some of these challenges and alternatives.

Recent Challenges to Traditional Medical Science

Patient Activism

By the mid-20th century, the dominance of medical science in the United States was virtually complete. Any departure from the dictates of scientific medicine was considered deviant. Thus, when sociologist Talcott Parsons defined the **sick role** in 1951, he first pointed out that illness suspends routine responsibilities and is not deliberate. Then he

Playing the **sick role**, according to Talcott Parsons, involves the nondeliberate suspension of routine responsibilities, wanting to be well, seeking competent help, and cooperating with health-care practitioners at all times.

[3]It surely has not helped that antibiotics are routinely added to cattle and chicken feed to prevent disease and thereby lower production costs. This only builds up resistance to antibiotics in humans.

BOX 19.3

SOCIAL POLICY: WHAT DO YOU THINK?

The High Cost of Prescription Drugs

Americans pay more for prescription drugs than anyone else in the world. In 2002, prescription drug prices in other rich countries were 37 percent to 53 percent below American prices (Figure 19.7). Between 1998 and 2002, the price of prescription drugs in the United States increased three times faster than the rate of inflation and faster than any other item in the nation's health-care budget. Elderly people and people with chronic medical conditions such as diabetes feel the burden most acutely because they are the biggest prescription drug users.

Other rich countries keep prescription drug prices down through some form of government regulation. For example, since 1987, Canadian drug companies have not been able to increase prices of brand-name drugs above the inflation rate. New brand-name drugs cannot exceed the highest Canadian price of comparable drugs used to treat the same disease. For new brand-name drugs that are unique and have no competitors, the price must be no higher than the median price for that drug in the United Kingdom, France, Italy, Germany, Sweden, Switzerland, and the United States. If a company breaks the rules, the government requires a price adjustment. If the government deems that a company has deliberately flouted the law, it imposes a punitive fine (Patent Medicine Prices Review Board, 2002). Not surprisingly, more than a million Americans now regularly buy their brand-name prescription drugs directly from Canadian pharmacies.

American drug manufacturers justify their high prices by claiming they need the money for research and development (R&D). The American public benefits from R&D, they say, while lower drug prices impair R&D in other countries.

Their argument would be more convincing if evidence showed that price curbs actually hurt R&D. In Britain, however, where the government regulates prescription drug costs, drug companies spend 20 percent of their sales revenue on R&D. In the United States, the figure is just 12.5 percent. In Canada, expenditure on R&D has increased 1500 percent since the beginning of government regulation (1987–2002). This hardly suggests that price regulation hurts R&D (Barry, 2002c; Patent Medicine Prices Review Board, 2002: 49).

What we can say with confidence is that the American pharmaceutical industry is by far the most profitable industry in the country. In 2001, profit as a percentage of revenue was 18.5 percent—four times higher than that of all other industries combined. We also know that drug companies spend about half as much on advertising and promotions as they do on R&D. This drives up drug prices. Finally, we know that the pharmaceutical industry spends more on lobbying and political campaign contributions than any other industry. Spending $197 million in 1999–2000, it hired 625 Washington lobbyists, more than one for each member of Congress. Most of the lobbying effort is aimed at influencing members of Congress to maintain a free market in drug prices (Barry, 2002a; 2002b; 2002c).

The drug companies' lobbying efforts have proved only partly successful (Gearon, 2002; Saunders 2003). In the fall of 2001, Maine, New Hampshire, and Vermont formed a drug purchasing pool. By buying drugs in bulk, the three states saved 10–15 percent on Medicaid drugs. These states belong to a larger coalition including Connecticut, Massachusetts, New York, Pennsylvania, and Rhode Island. The group wants to negotiate directly with drug companies for deep discounts. Arkansas, Idaho, and Texas are also exploring drug-purchasing alliances. Some states, such as Michigan, California, and Florida, are trimming costs by specifying the drugs available through Medicaid and discouraging doctors from prescribing expensive brand-name drugs over generics. The governors of Minnesota, Illinois, Iowa, and Wisconsin, and city officials from New York, Boston, and Springfield, Massachusetts, said they want to import less expensive medicines from Canada to save their state budgets and their citizens millions of dollars.

The drug companies, backed by the Food and Drug Administration (FDA), are fighting these maneuvers. Maine has already defeated one drug-company challenge in the courts. The Pharmaceutical Research and Manufacturers of America has gone to court to stop the use of Medicaid drug lists in Florida, Michigan, and Maine. Meanwhile, major drug manufacturers are limiting production and sales in Canada to cover only domestic needs, thus choking off exports to the United States. The FDA won an injunction in November 2003 to close down Tulsa-based R_x Depot, which had grown to 88 branches in 27 states by giving customers access to online pharmacies in Manitoba. The struggle over prescription drug prices is becoming a major policy debate focused squarely on the advisability of allowing a free market to operate unchallenged in the health field.

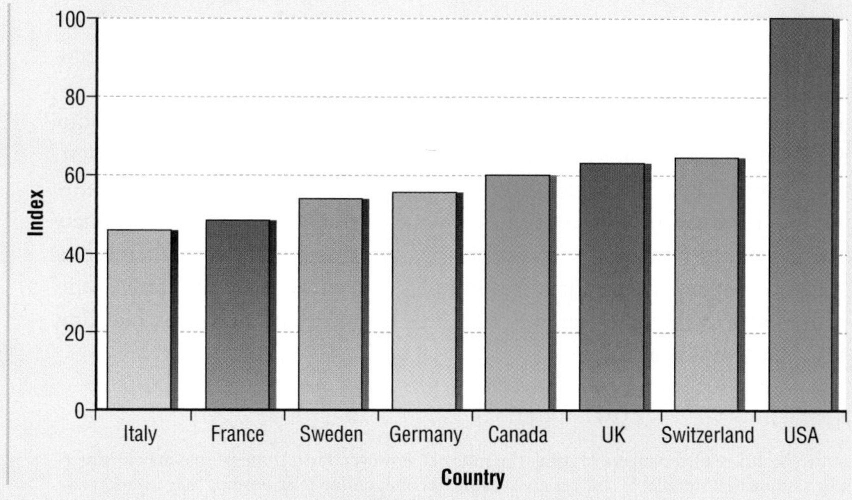

Figure 19.7
Prescription Drug Costs in Eight Rich Countries, 2002 (Index, United States = 100)

Source: Patent Medicine Prices Review Board (2002: 23).

stressed that people playing the sick role must want to be well and must seek competent help, cooperating with health-care practitioners at all times (Parsons, 1951: 428ff.). Must they? By Parsons' definition, a competent person suffering from a terminal illness cannot reasonably demand that doctors refrain from using heroic measures to prolong his or her life. And by his definition, a patient cannot reasonably question doctors' orders, no matter how well educated the patient and how debatable the effect of the prescribed treatment. Although Parsons' definition of the sick role may sound plausible to many people born before World War II, it probably sounds authoritarian and foreign to most younger people. Today, it corresponds best to elderly people and to patients in intensive care units who are too weak and disoriented to take a more active role in their own care (Rier, 2000).

That is because things have changed. The American public is more highly educated now than it was 50 years ago. Many people now possess the knowledge, the vocabulary, the self-confidence, and the political organization to participate in their own health care rather than passively accept whatever experts tell them. Increasingly, patients are taught to perform simple, routine medical procedures themselves. Many people now use the Internet to seek information about various illnesses and treatments.[4] Increasingly, they are uncomfortable with doctors acting like patriarchal fathers and patients like dutiful children. Doctors now routinely seek patients' informed consent for some procedures rather than deciding what to do on their own. Similarly, most hospitals have established ethics committees, which were unheard of only 25 years ago (Rothman, 1991). These are responses to patients wanting a more active role in their own care.

Some recent challenges to the authority of medical science are organized and political. For example, when AIDS activists challenge the stereotype of AIDS as a "gay disease" and demand more research funding to help find a cure, they change research and treatment priorities in a way that could never have happened in, say, the 1950s or 1960s (Epstein, 1996). Similarly, when feminists support the reintroduction of midwifery and argue against medical intervention in routine childbirth, they are challenging the wisdom of established medical practice. The previously male-dominated profession of medicine considered the male body the norm and paid relatively little attention to women's diseases, such as breast cancer, and women's issues, such as reproduction. This, too, is now changing thanks to feminist intervention (Boston Women's Health Collective, 1998; Rothman, 1982; 1989; Schiebinger, 1993). And although doctors and the larger society traditionally treated people with disabilities like incompetent children, various movements now seek to empower them (Charlton, 1998; Zola, 1982). As a result, attitudes toward the disabled are changing (Chapter 12, "Sociology of the Body: Disability, Aging, and Death").

Recent challenges to the pharmaceutical industry may also be mentioned here. Major drug companies fund most of the research that government panels use to decide whether new drugs should be allowed to go to market. The drug companies also hire many panel members as consultants and researchers. Understandably, some observers worry about bias in the drug approval process. They ask whether drug companies sometimes rush new drugs to market and heavily promote them without proper safeguards.

For example, some years ago a whole class of painkillers was designed for people who suffer ulcers and bleeding when taking traditional painkillers like aspirin and ibuprofen. The drugs were approved with little difficulty. The pharmaceutical companies then unleashed major advertising campaigns to promote them. People soon started asking their doctors for the new drugs by name: Vioxx, Celebrex, and Bextra. Doctors, themselves bombarded by drug company sales pitches, free samples, and educational programs, were only too happy to oblige. Millions of patients with little or no risk of ulcers and bleeding started taking the new painkillers.

---

[4]There are at least two health-related dangers in using the Internet, however. First, some people may misinterpret information or assume that unreliable sources are reliable. Second, online relationships may lead to real-world meetings and so contribute to the spread of HIV/AIDS and other sexually transmitted diseases. Epidemiologists at the U.S. Centers for Disease Control and Prevention are now conducting a study on this subject (Roberts, 2000).

Then, in 2004, researchers disclosed that taking these drugs raises the risk of heart attack. The drugs were promptly whisked off the market. A few months later, a government advisory panel ruled that they could be marketed with appropriate warnings. But data compiled by the Center for Science in the Public Interest show that at least 10 of the 32 panel members had consulted or received support from the companies that make Vioxx, Celebrex, and Bextra or from a company that is seeking approval for a similar drug. Given such close ties between industry and government, many observers are calling for a more independent approval process that will consider only the public interest in reaching its decisions ("Experts," 2005; "Prescription," 2004).

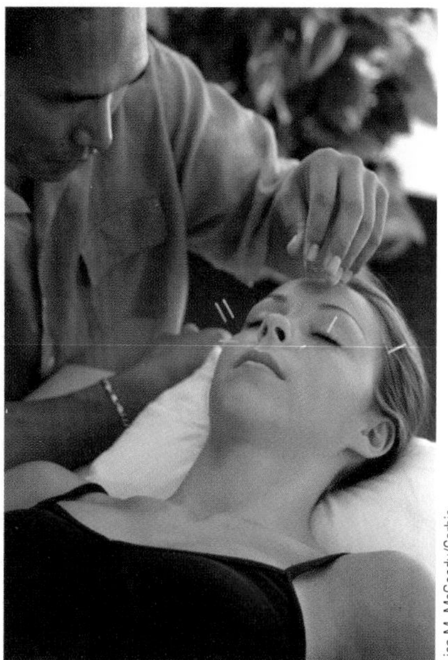

Lisa M. McGeody/Corbis

Acupuncture is one of the most widely accepted forms of alternative medicine.

Alternative Medicine

Other challenges to the authority of medical science are less organized and less political than those just mentioned. Consider, for example, alternative medicine. The most frequently used types of alternative medicine are chiropractic, acupuncture, massage therapy, and various relaxation techniques. Alternative medicine is used mostly to treat back problems, chronic headache, arthritis, chronic pain, insomnia, depression, and anxiety. Especially popular in the western states, alternative medicine is most often used by highly educated, upper-income white Americans in the 25-to-49 age group. A nationwide poll conducted in 1990 showed that 34 percent of Americans had tried alternative medicine in the year before the survey. Most of them had *not* lost all faith in traditional medical science. Thus, 83 percent of them tried alternative medicine in conjunction with treatment from a medical doctor (Eisenberg, Kessler, Foster, Norlock, Calkins, and Delbanco, 1993). Two surveys conducted in 1998 show that demand for alternative care is rising (American Chiropractic Association, 1999).

Sociology Now™

Learn more about **Alternative Medicine** by going through the Alternative Health Care Methods Video Exercise.

Despite its growing popularity, many medical doctors were hostile to alternative medicine until recently. They lumped all alternative therapies together and dismissed them as unscientific (Campion, 1993). By the late 1990s, however, a more tolerant attitude was evident in many quarters. For some kinds of ailments, physicians began to recognize the benefits of at least the most popular forms of alternative medicine. For example, a 1998 editorial in the respected *New England Journal of Medicine* admitted that the beneficial effect of chiropractic on low back pain is "no longer in dispute" (Shekelle, 1998). This change in attitude was due in part to new scientific evidence showing that spinal manipulation is a relatively effective and inexpensive treatment for low back pain (Manga, Angus, and Swan, 1993).

The medical profession's grudging acceptance of chiropractic in the treatment of low back pain indicates what we can expect in the uneasy relationship between scientific and alternative medicine in coming decades. Doctors will for the most part remain skeptical about alternative therapies unless properly conducted experiments demonstrate their beneficial effects. Most Americans probably agree with this cautious approach.

Holistic Medicine

Medical doctors understand that a positive frame of mind often helps in the treatment of disease. For example, research shows that strong belief in the effectiveness of a cure can by itself improve the condition of about a third of people suffering from chronic pain or fatigue (Campion, 1993). This is known as the **placebo effect.** Doctors also understand that conditions in the human environment affect people's health. There is no dispute, for example, about why so many people in Cancer Alley develop malignancies. However, despite their appreciation of the effect of mind and environment on the human body, traditional scientific medicine tends to respond to illness by treating disease symptoms as a largely physical and individual problem. Moreover, scientific medicine keeps subdividing into more specialized areas of practice that rely more and more heavily on drugs and high-tech machinery. Most doctors are less concerned with maintaining and

The **placebo effect** is the positive influence on healing of strong belief in the effectiveness of a cure.

improving health by understanding the larger mental and social context within which people become ill.

Traditional East Indian and Chinese medical practices take a different approach. India's Ayurvedic medical tradition sees individuals in terms of the flow of vital fluids, or "humors," and their health in the context of their environment. In this view, maintaining good health requires not only balancing fluids in the individual but also balancing the relationship between individuals and the world around them (Zimmermann, 1987 [1982]). In spite of significant differences, the fundamental outlook is similar in traditional Chinese medicine. Chinese medicine and its remedies, ranging from acupuncture to herbs, seek to restore individuals' internal balance, as well as their relationship to the outside world (Unschuld, 1985). Contemporary **holistic medicine,** the third and final challenge to traditional scientific medicine we will consider, takes a similar approach to these "ethnomedical" traditions. Practitioners of holistic medicine argue that good health requires maintaining a balance between mind and body, and between the individual and the environment.

Most holistic practitioners do not reject scientific medicine. However, they emphasize disease prevention. When they treat patients, they take into account the relationship between mind and body and between the individual and his or her social and physical environment. Holistic practitioners thus seek to establish close ties with their patients and treat them in their homes or other relaxed settings. Rather than expecting patients to react to illness by passively allowing a doctor to treat them, they expect patients to take an active role in maintaining their good health. And, recognizing that industrial pollution, work-related stress, poverty, racial and gender inequality, and other social factors contribute heavily to disease, holistic practitioners often become political activists (Hastings, Fadiman, and Gordon, 1980).

Holistic medicine emphasizes disease prevention. Holistic practitioners treat disease by taking into account the relationship between mind and body and between the individual and his or her social and physical environment.

In sum, patient activism, alternative medicine, and holistic medicine represent the three biggest challenges to traditional scientific medicine today. Few people think of these challenges as potential replacements for scientific medicine. However, many people believe that together with traditional scientific approaches, these challenges will help to improve the health status of people in the United States and throughout the world in the 21st century.

Summary

Sociology Now™

Reviewing is as easy as ❶, ❷, ❸.

❶ Before you do your final review, take the SociologyNow diagnostic quiz to help you identify the areas on which you should concentrate. You will find information on how to access SociologyNow on the foldout at the front of the textbook.

❷ As you review, take advantage of SociologyNow's study aids to help you master the topics in this chapter.

❸ When you are finished with your review, take SociologyNow's post-test to confirm you are ready to move on to the next chapter.

1. Aren't all causes of illness and death biological?

Ultimately, yes. However, *variations* in illness and death rates are often due to social causes. The social causes of illness and death include human-environmental factors, lifestyle factors, and factors related to the public health and health-care systems. All three factors are related to country of residence, class, race, and gender. Specifically, health risks are lower among upper classes, rich countries, and privileged racial and ethnic groups than among lower classes, poor countries, and disadvantaged racial and ethnic groups. In some respects related to health, men are in a more advantageous position than women.

2. Doesn't the United States have the world's best health-care system?

In some ways the United States does have the world's most advanced health-care system. Cutting-edge research, abundant high-tech diagnostic equipment, and exceptionally well trained medical practitioners help make it so. However, the average health status of Americans is lower than the average health status of people in other rich postindustrial countries. That is partly because the level of social inequality is higher in the United States and partly because the private health-care system in this country makes it difficult for many people to receive adequate care.

3. **In what sense does a thin line separate science from politics in health care?**

   Power struggles have taken place throughout the history of medicine. For example, deviant acts once regarded as evil have come under the increasing sway of medical professionals. The recent history of psychiatry shows that social values and political compromise are at least as important as science in determining the classification of some mental disorders. Medical science came to dominate health care partly because it proved to be so successful in treating the ill. In addition, dominance was ensured by doctors excluding competitors and establishing control over their profession and their clients.

4. **What are the "social limits of medicine"?**

   A health-care system based on profitability and a free market creates large inequalities in treatment. In this sense, the market acts to limit the benefits of medicine. For example, cost cutting has encouraged the overuse of antibiotics and the neglect of basic hygiene in hospitals, resulting in more hospital-caused infections and the spread of drug-resistant germs. Similarly, a free market in prescription drugs has made medication prohibitively expensive for many Americans.

5. **What are the main challenges and alternatives to traditional medicine?**

   Several challenges to traditional scientific medicine promise to improve the quality of health care in the United States and worldwide. These include patient activism, alternative medicine, and holistic medicine.

Questions to Consider

1. Because health resources are scarce, tough decisions must be made about how they are allocated. For example, drug companies, physicians, hospitals, government research agencies, and other components of the health-care system have to decide how much to invest in trying to prolong the life of the elderly versus how much to invest in improving the health of the poor. What do you think are the main factors that help different components of the health-care system decide how to allocate resources between these two goals? Specifically, how important is the profit motive? Political pressure? Which components of the health-care system does the profit motive influence most? Which are most influenced by political pressure? If you were in charge of a major hospital or government funding for health research, how would you divide your scarce resources between trying to prolong the life of the elderly and improving the health of the poor? Why? What pressures might be placed on you to act differently than you want?
2. Do you believe that patient activism and alternative medicine improve health care or detract from the efforts of scientifically trained physicians and researchers to do the best possible research and administer the best possible treatments? Because patient activists may not be scientifically trained and because alternative therapies may not be experimentally proven, are there dangers inherent in these challenges to traditional medicine? On the other hand, do biases in traditional medicine detract from health care by ignoring the needs of patient activists and the possible benefits of alternative therapies?

Web Resources

Companion Website for This Book

http://sociology.wadsworth.com

Begin by clicking on the Student Resources section of the website. Choose "Introduction to Sociology" and then the Brym and Lie book cover. Next, select the chapter you are currently studying from the pull-down menu. From the Student Resources page you will have easy access to InfoTrac® College Edition, MicroCase Online exercises, additional web links, and many other resources to aid you in your study of sociology, including practice tests for each chapter.

Recommended Websites

How fast are you aging? What can you do to slow the aging process? Find out by answering questions on your health risks at www.realage.com.

The U.S. Centers for Disease Control and Prevention maintains a rich website at http://www.cdc.gov. It is full of up-to-date health information and statistics.

Physicians for a National Health Program advocates a universal, comprehensive, national health-care program. It has more than 10,000 members and chapters across the United States. Visit the website of these medical activists at http://www.pnhp.org.

The American Iatrogenic Association devotes itself to studying and reporting illness caused by the health-care system. Its website challenges the view that the health-care system always promotes good health. Go to http://www.iatrogenic.org/index.html.

CHAPTER 20 Population and Urbanization

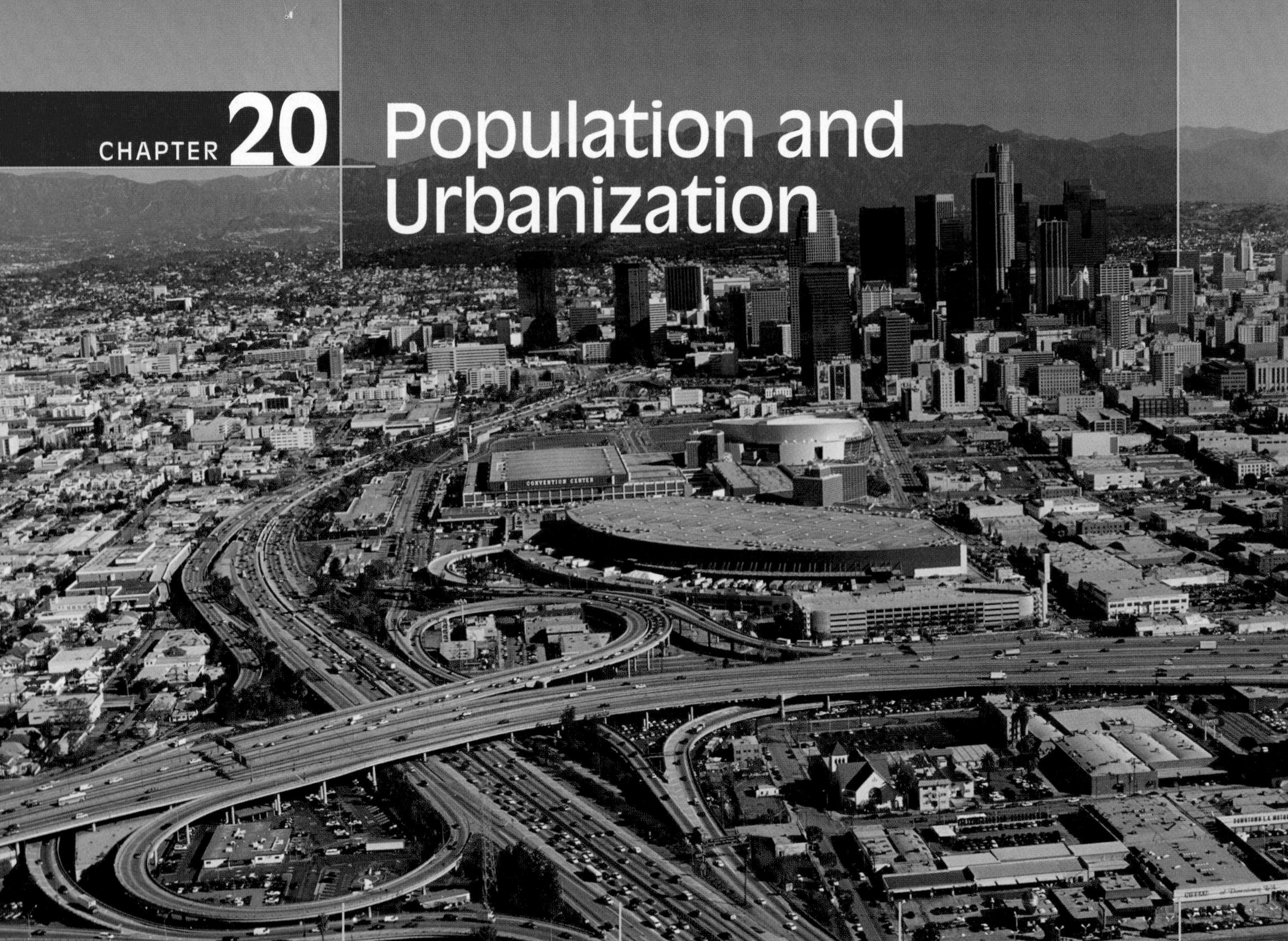

Douglas Slone/Corbis

In this chapter, you will learn that:

- Many people think that only natural conditions influence human population growth. However, social forces are important influences too.
- In particular, sociologists have focused on two major social determinants of population growth: industrialization and social inequality.
- Industrialization also plays a major role in causing the movement of people from countryside to city.
- Cities are not as anonymous and alienating as many sociologists once believed them to be.
- The spatial and cultural forms of cities depend largely on the level of development of the societies in which they are found.

Population

The City of God

Rio de Janeiro, Brazil, is one of the world's most beautiful cities. Along the warm, blue waters of its bays lie flawless beaches, guarded by four- and five-star hotels and pricey shops. Rising abruptly behind them is a mountain range, partly populated, partly covered by luxuriant tropical forest. The climate seems perpetually balanced between spring and summer. The inner city of Rio is a place of great wealth and beauty, devoted to commerce and the pursuit of leisure.

Rio is also a large city. With a metropolitan population of more than 12 million people in 2002, it is the eighteenth biggest metropolitan area in the world, larger than Chicago, Paris, and London (Brinkhoff, 2002). Not all of its 12 million inhabitants are well-off, however. Brazil is characterized by more inequality of wealth than any other country in the world. Slums started climbing up the hillsides of Rio about a century ago. Fed by a high birth rate and people migrating from the surrounding countryside in search of a better life, slums are now home to about 20 percent of the city's inhabitants (Jones, 2003).

Sociology Now™

Learn more about **Population** by going through the Population per square mile of land Map Exercise.

Some of Rio's slums began as government housing projects designed to segregate the poor from the rich. One such slum, as famous in its own way as the beaches of Copacabana and Ipanema, is Cidade de Deus (the "City of God"). Founded in the 1960s, it became one of the most lawless and dangerous parts of Rio by the 1980s. It is a place where some families of four live on $50 a month in houses made of cardboard and discarded scraps of tin, a place where roofs leak and rats run freely. For many inhabitants, crime is survival. Drug traffickers wage a daily battle for control of territory, and children as young as 6 perch in key locations with walkie-talkies to feed information to their bosses on the comings and goings of passersby.

Sociology Now™

Reviewing is as easy as ❶, ❷, ❸.

Use SociologyNow to help you make the grade on your next exam. When you are finished reading this chapter, go to the Chapter Summary for instructions on how to make SociologyNow work for you.

Donald Klein/SuperStock

One of Rio de Janeiro's biggest slums.

Cidade de Deus is also the name of a widely acclaimed movie released in 2002. Based in part on the true-life story of Paulo Lins, who grew up in a slum and became a novelist, *Cidade de Deus* chronicles the gang wars of the 1970s and 80s. It leaves us with the nearly hopeless message that in a war without end, each generation of drug traffickers is younger and more ruthless than its predecessor.

Paulo Lins escaped Brazil's grinding poverty. So did Luiz Inácio Lula da Silva, Brazil's current president. They are inspiring models of what is possible. They are also depressing reminders that the closely related problems of population growth and urbanization are more serious now than ever. Brazil's 41 million people in 1940 multiplied to about 180 million in 2004. The country is now more urbanized than the United States, with more than three-quarters of its population living in urban areas (estimated from Lahmeyer, 2003; Ministério de Ciência e Tecnologia Brasil, 2002).

This chapter tackles the closely connected problems of population growth and urbanization. We first show that population growth is a process governed less by natural laws than by social forces. We argue that these social forces are not related exclusively to industrialization, as social scientists commonly believed just a few decades ago. Instead, social inequality also plays a major role in shaping population growth. We next turn to the problem of urbanization. Today, population growth is typically accompanied by the increasing concentration of the world's people in urban centers. As recently as 40 years ago, sociologists typically believed that cities were alienating and anomic (or normless). We argue that this view is an oversimplification. We also outline the social roots of the city's physical and cultural evolution from preindustrial to postindustrial times.

Sociology Now™

Learn more about **Demographics** by going through the Three Basic Demographic Processes Animation.

The Population "Explosion"

Web Interactive Exercises: World Population

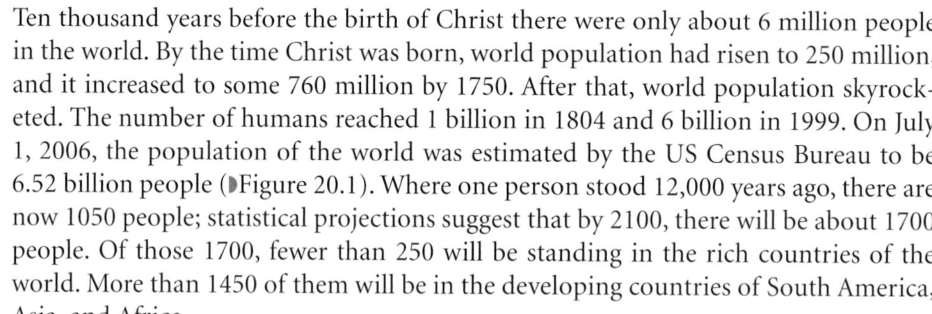

Ten thousand years before the birth of Christ there were only about 6 million people in the world. By the time Christ was born, world population had risen to 250 million, and it increased to some 760 million by 1750. After that, world population skyrocketed. The number of humans reached 1 billion in 1804 and 6 billion in 1999. On July 1, 2006, the population of the world was estimated by the US Census Bureau to be 6.52 billion people (Figure 20.1). Where one person stood 12,000 years ago, there are now 1050 people; statistical projections suggest that by 2100, there will be about 1700 people. Of those 1700, fewer than 250 will be standing in the rich countries of the world. More than 1450 of them will be in the developing countries of South America, Asia, and Africa.

Many analysts project that after passing the 10 billion mark around 2100, world population will level off. But given the numbers cited previously, is it any wonder that some population analysts say we're now in the midst of a population "explosion"? Explosions are horrifying events. They cause widespread and severe damage. They are fast and unstoppable. And that is exactly the imagery some population analysts, or **demographers,** wish to convey (e.g., Ehrlich, 1968; Ehrlich and Ehrlich, 1990) (Figure 20.2). They have written many books, articles, and television programs dealing with the population explosion. You may have encountered some of these in your school or church. Images of an overflowing multitude in, say, Bangladesh, Nigeria, or Brazil remain fixed in our minds. Some people are frightened enough to refer to overpopulation as catastrophic. They link it to recurrent famine, brutal ethnic warfare, and other massive and seemingly intractable problems.

Demographers are social-scientific analysts of human population.

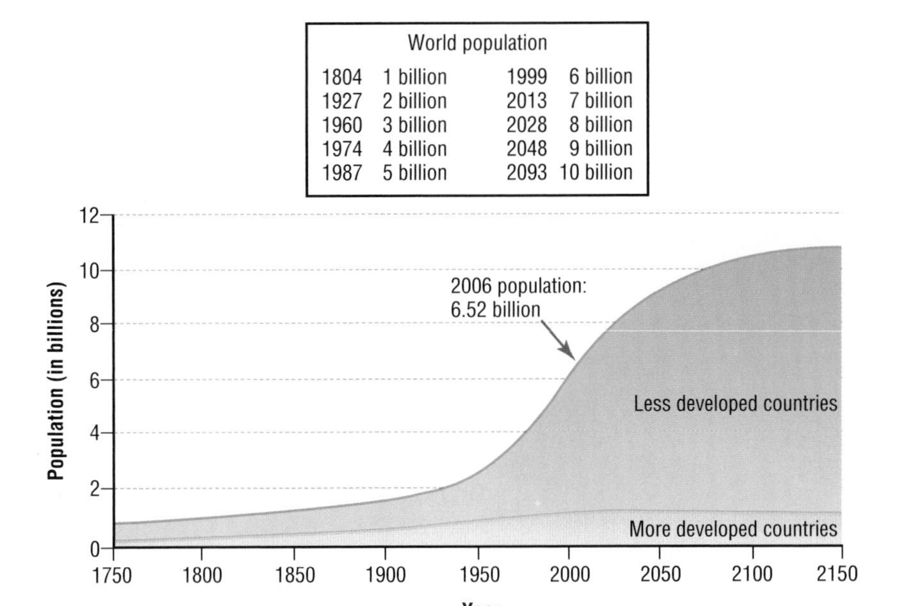

Figure 20.1
World Population, 1750–2150 (in billions, projected)

Sources: Livi-Bacci (1992: 31); Population Reference Bureau (2003b).

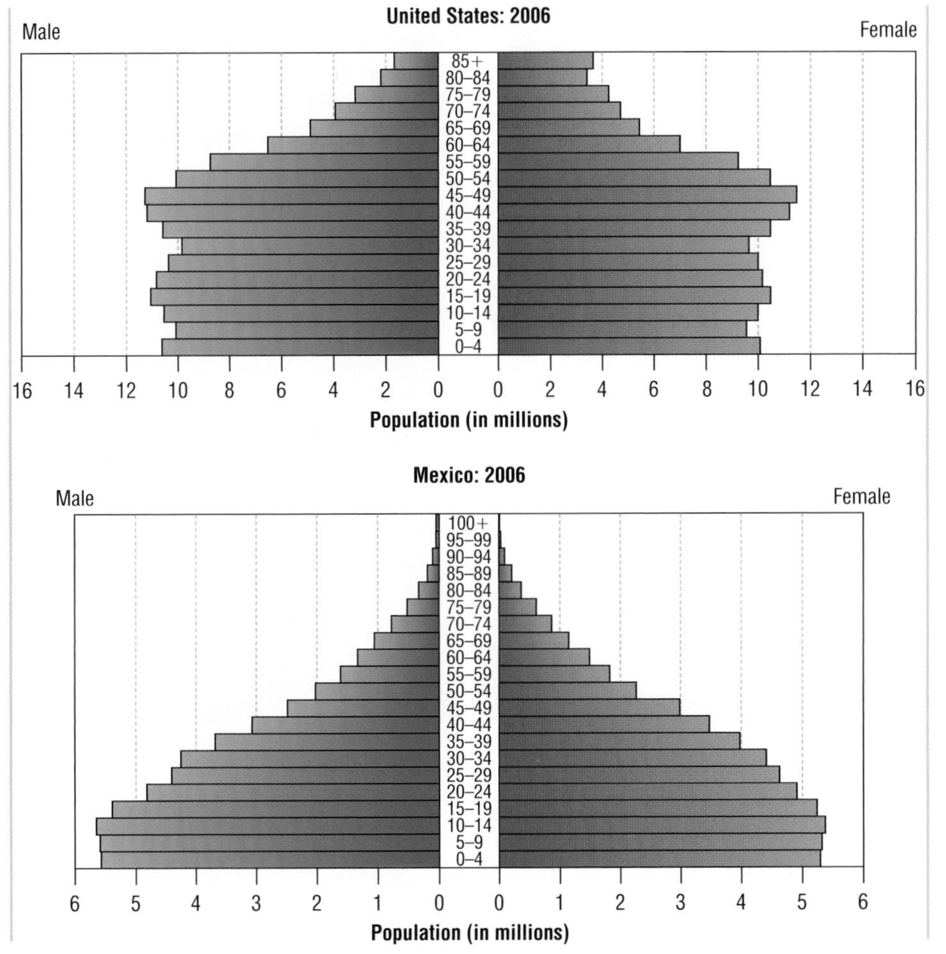

Figure 20.2
How Demographers Analyze Population Changes and Composition

The main purpose of demography is to figure out why the size, geographical distribution, and social composition of human populations change over time. The basic equation of population change is $P_2 = P_1 + B - D + I - E$, where P_2 is population size at a given time, P_1 is population size at an earlier time, B is the number of births in the interval, D is the number of deaths in the interval, I is the number of immigrants arriving in the interval, and E is the number of emigrants leaving in the interval. One basic tool for analyzing the composition of a population is the age–sex pyramid, which shows the number of males and females in each age cohort of the population at a given point in time. Age–sex pyramids for the United States and Mexico are shown here, projected by the U.S. Census Bureau for 2006. Why do you think they look so different? Compare your answer with that of the theory of the demographic transition, discussed in this section.

Source: U.S. Census Bureau (2000c).

A "population explosion"? Hong Kong is one of the most densely populated places on earth.

The Purcell Group/Corbis

If this imagery makes you feel that the world's rich countries must do something about overpopulation, you're not alone. In fact, concern about the population "bomb" is as old as the social sciences. In 1798, Thomas Robert Malthus, a British clergyman of the Anglican faith, proposed a highly influential theory of human population (Malthus, 1966 [1798]). As you will soon see, contemporary sociologists have criticized, qualified, and in part rejected his theory. But because much of the sociological study of population is, in effect, a debate with Malthus's ghost, we must confront the man's ideas squarely.

Theories of Population Growth

The Malthusian Trap

Scala/Art Resource, NY

Albrecht Dürer, *The Four Horsemen of the Apocalypse* (woodcut, 1498). According to Malthus, only war, pestilence, and famine could keep population growth in check.

Malthus's theory rests on two undeniable facts and a questionable assumption. The facts: People must eat and they are driven by a strong sexual urge. The assumption: While food supply increases slowly and arithmetically (1, 2, 3, 4, etc.), population size grows quickly and geometrically (1, 2, 4, 8, etc.). Based on these ideas, Malthus concluded that "the superior power of population cannot be checked without producing misery or vice" (Malthus, 1966 [1798]: 217–18). Specifically, only two forces can hold population growth in check. First are "preventive" measures, such as abortion, infanticide, and prostitution. Malthus called these "vices" because he morally opposed them and thought that everyone else ought to also. Second are "positive checks," such as war, pestilence, and famine. Malthus recognized that positive checks create much suffering. Yet he felt that they were the only forces that could be allowed to control population growth. Here, then, is the so-called **Malthusian trap:** A cycle of population growth followed by an outbreak of war, pestilence, or famine that keeps population growth in check. Population size might fluctuate, said Malthus, but it has a natural upper limit that Western Europe has reached.

Although many people supported Malthus's theory, others reviled him as a misguided prophet of doom and gloom (Winch, 1987). For example, people who wished to help the poor disagreed with Malthus. He felt that such aid was counterproductive. Welfare, he said, would enable the poor to buy more food. With more food, they would have more children. And having more children would only make them poorer than they already were. Better leave them alone, said Malthus. That will reduce the sum of human suffering in the world.

Sociology Now™

Learn more about **the Malthusian Trap** by going through the Malthusian Perspective Learning Module.

A Critique of Malthus

Although Malthus's ideas are in some respects compelling, events have cast doubt on several of them. Specifically:

- Ever since Malthus proposed his theory, technological advances have allowed rapid growth in how much food is produced for each person on the planet. This is the opposite of the slow growth Malthus predicted. For instance, in the period 1991–1993, India produced 23 percent more food per person than it did in 1979–1981, and China produced 39 percent more. Moreover, except for Africa south of the Sahara, the largest increases in the food supply are taking place in the developing countries (Sen, 1994).
- If, as Malthus claimed, there is a natural upper limit to population growth, it is unclear what that limit is. Malthus thought the population couldn't grow much larger in late-18th-century Western Europe without "positive checks" coming into play. Yet the Western European population increased from 187 million people in 1801 to 321 million in 1900. It has now stabilized at about half a billion (McNeill, 1990). The Western European case suggests that population growth has an upper limit far higher than that envisaged by Malthus.
- Population growth does not always produce misery. For example, despite its rapid population increase over the past 200 years, Western Europe is one of the most prosperous regions in the world.
- Helping the poor does not generally result in the poor having more children. For example, in Western Europe, social welfare policies (unemployment insurance, state-funded medical care, paid maternity leave, pensions, etc.) are the most generous on the planet. Yet the size of the population is quite stable. In fact, as you will learn in the following, some forms of social welfare produce rapid and large *decreases* in population growth, especially in the poor, developing countries.

The **Malthusian trap** refers to a cycle of population growth followed by an outbreak of war, pestilence, or famine that keeps population growth in check.

- Although the human sexual urge is as strong as Malthus thought, people have developed contraceptive devices and techniques to control the consequences of their sexual activity (Szreter, 1996). There is no necessary connection between sexual activity and childbirth.

The developments listed here all point to one conclusion. Malthus's pessimism was overstated. Human ingenuity seems to have enabled us to wriggle free of the Malthusian trap, at least for the time being.

We are not, however, home free. Today there are renewed fears that industrialization and population growth are putting severe strains on the planet's resources. As Chapter 22 ("Technology and the Global Environment") establishes, we must take these fears seriously. It is encouraging to learn that the limits to growth are as much social as natural, and therefore avoidable rather than inevitable. However, we will see that our ability to avoid the Malthusian trap in the 21st century will require all the ingenuity and self-sacrifice we can muster. For the time being, however, let us consider the second main theory of population growth, the theory of the demographic transition.

Demographic Transition Theory

According to **demographic transition theory,** the main factors underlying population dynamics are industrialization and the growth of modern cultural values (Notestein, 1945; Coale, 1974; Chesnais, 1992) (Figure 20.3). The theory is based on the observation that the European population developed in four distinct stages.

The Preindustrial Period

In the first, preindustrial stage of growth, a large proportion of the population died every year due to inadequate nutrition, poor hygiene, and uncontrollable disease. In other words, the **crude death rate** was high. The crude death rate is the annual number of deaths (or "mortality") per 1000 people in a population. During this period, the **crude birth rate** was high too. The crude birth rate is the annual number of live births per 1000 people in a population. In the preindustrial era, most people wanted to have as many children as possible. That was partly because relatively few children survived till adulthood. In addition, children were considered a valuable source of agricultural labor and a form of old age security in a society consisting largely of peasants and lacking anything resembling a modern welfare state.

Demographic transition theory explains how changes in fertility and mortality affected population growth from preindustrial to postindustrial times.

The **crude death rate** is the annual number of deaths per 1000 people in a population.

The **crude birth rate** is the annual number of live births per 1000 women in a population.

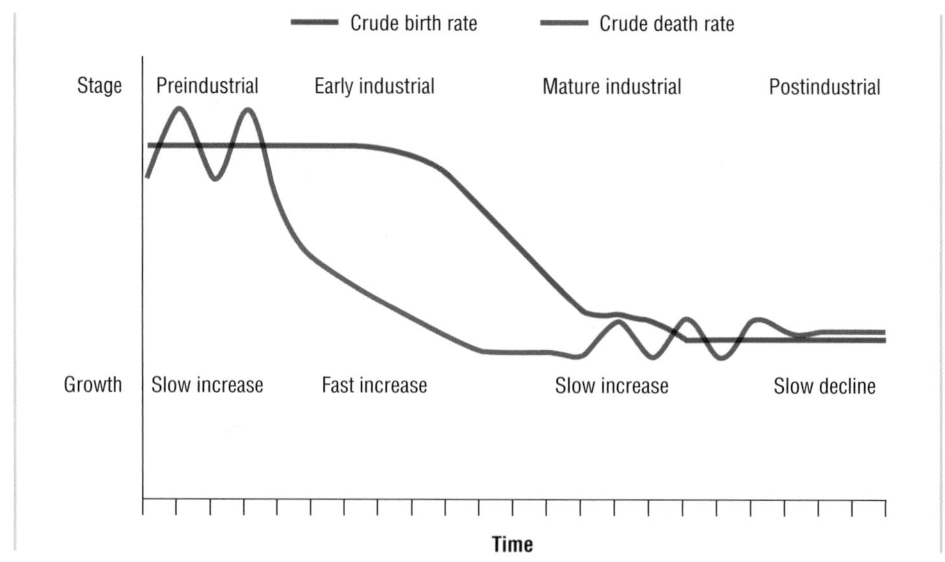

Figure 20.3
Demographic Transition Theory

The Early Industrial Period

The second stage of European population growth was the early industrial, or transition, period. At this stage, the crude death rate dropped. People's life expectancy, or average life span, increased because economic growth led to improved nutrition and hygiene. However, the crude birth rate remained high. With people living longer and women having nearly as many babies as in the preindustrial era, the population grew rapidly. Malthus lived during this period of rapid population growth, and that accounts in part for his alarm.

The Mature Industrial Period

The third stage of European population growth was the mature industrial period. At this stage, the crude death rate continued to fall. The crude birth rate fell even more dramatically. The crude birth rate fell because economic growth eventually changed people's traditional beliefs about the value of having many children. Having lots of children made sense in an agricultural society, where, as we have seen, children were a valuable economic resource. In contrast, children were more of an economic burden in an industrial society. That is because breadwinners worked outside the home for a wage or a salary and children contributed little if anything to the economic welfare of the family. Note, however, that the crude birth rate took longer to decline than the crude death rate did. That is because people's values often change more slowly than their technologies. People can put in a sewer system or a water purification plant to lower the crude death rate faster than they can change their minds about something as fundamental as how many children to have. Eventually, however, the technologies and outlooks that accompany modernity led people to postpone getting married and to use contraceptives and other birth-control methods. As a result, population stabilized during the mature industrial period. This demonstrates the validity of one of the demographer's favorite sayings: "Economic development is the best contraceptive."

The Postindustrial Period

In the last decades of the 20th century, the total fertility rate continued to fall. (We defined the total fertility rate in Chapter 15 as the average number of children that would be born to a woman over her lifetime if she had the same number of children as women in each age cohort in a given year.) In fact, it fell below the **replacement level** in some countries. The replacement level is the number of children each woman must have on average for population size to remain stable. Ignoring any inflow of settlers from other countries (**immigration,** or **in-migration**) and any outflow to other countries (**emigration,** or **out-migration**), the replacement level is 2.1. This means that on average, each woman must give birth to slightly more than the two children needed to replace her and her mate. Slightly more than two children are required because some children die before they mature and reach reproductive age.

By the 1990s, some Europeans were worrying about declining fertility and its possible effects on population size. As you can see in ▶Table 20.1, 28 countries, 23 of them in Europe, now have fertility rates below 1.4. The number of countries or areas of the world with a fertility level below the replacement level rose from 51 in 1995 to 68 in 2004 (United Nations, 1998a; Population Reference Bureau, 2004). The United States is a member of this group. In 2004, the United States had a fertility rate of 2.0. Due to the proliferation of low-fertility societies, some scholars suggest that we have now entered a fourth, postindustrial stage of population development. In this fourth stage of the demographic transition, the number of deaths per year exceeds the number of births (Van de Kaa, 1987).

A Critique of Demographic Transition Theory

As outlined earlier, the demographic transition theory provides a rough picture of how industrialization affects population growth. However, research has revealed a number of inconsistencies in the theory. Most of them are due to the theory placing too much emphasis on industrialization as the main force underlying population growth (Coale and

The **replacement level** is the number of children that each woman must have on average for population size to remain stable. Ignoring any inflow of population from other countries and any outflow to other countries, the replacement level is 2.1.

Immigration, or in-migration, is the inflow of people into one country from one or more other countries and their settlement in the destination country.

In-migration (*see* immigration).

Emigration, or out-migration, is the outflow of people from one country and their settlement in one or more other countries.

Out-migration (*see* emigration).

Table 20.1
Countries with the Lowest and Highest Total Fertility Rates, 2003

TOTAL FERTILITY RATE <1.4		Total Fertility Rate >5.9	
Country	**Fertility Rate**	**Country**	**Fertility Rate**
Armenia	1.2	Niger	8.0
Belarus	1.2	Guinea-Bissau	7.1
Bosnia-Herzegovina	1.2	Somalia	7.1
Bulgaria	1.2	Mali	7.0
Czech Republic	1.2	Yemen	7.0
Latvia	1.2	Uganda	6.9
Moldova	1.2	Afghanistan	6.8
Poland	1.2	Angola	6.8
Romania	1.2	Comoros	6.8
San Marino	1.2	Congo, Democratic Republic of	6.8
Slovakia	1.2		
Slovenia	1.2	Liberia	6.8
South Korea	1.2	Chad	6.6
Taiwan	1.2	Malawi	6.6
Ukraine	1.2	Sierra Leone	6.5
Andorra	1.3	Congo	6.3
Croatia	1.3	Burkina Faso	6.2
Greece	1.3	Burundi	6.2
Hungary	1.3	Guinea	6.0
Italy	1.3		
Estonia	1.3	**World's Three Biggest Countries**	
Germany	1.3		
Hungary	1.3	China	1.7
Japan	1.3	United States	2.0
Latvia	1.3	India	3.1
Singapore	1.3		
Spain	1.3		

Source: Population Reference Bureau (2004).

Watkins, 1986). For example, demographers have found that reductions in fertility sometimes occur when standards of living stagnate or decline, not just when they improve due to industrialization. Thus, in Russia and some developing countries today, declining living standards have led to a deterioration in general health and a subsequent decline in fertility. Because of such findings, many scholars have concluded that an adequate theory of population growth must pay more attention to social factors other than industrialization and in particular to the role of social inequality.

Population and Social Inequality

Karl Marx

One of Malthus's staunchest intellectual opponents was Karl Marx. Marx argued that the problem of overpopulation is specific to capitalism (Meek, 1971). In his view, overpopulation is not a problem of too many people. Instead, it is a problem of too much poverty. Do away with the exploitation of workers by their employers, said Marx, and poverty will disappear. If a society is rich enough to eliminate poverty, then by definition its population is not too large. By eliminating poverty, one also solves the problem of overpopulation in Marx's view.

Marx's analysis makes it seem that capitalism can never generate enough prosperity to solve the overpopulation problem. He was evidently wrong. Overpopulation is not a serious problem in the United States or Japan or Germany today.[1] It *is* a problem in most of Africa, where capitalism is weakly developed and the level of social inequality is much higher than in the postindustrial societies. Still, a core idea in Marx's analysis of the overpopulation problem rings true. As some contemporary demographers argue, social inequality is a main cause of overpopulation. In the following, we illustrate this argument by first considering how gender inequality influences population growth. Then we discuss the effects of class inequality on population growth.

Gender Inequality and Overpopulation

The effect of gender inequality on population growth is well illustrated by the case of Kerala, a state in India with more than 30 million people. Kerala had a total fertility rate of 1.8 in 1991, half of India's national rate and far less than the replacement level of 2.1. How did Kerala achieve this remarkable feat? Is it a highly industrialized oasis in the midst of a semi-industrialized country, as one might expect given the arguments of demographic transition theory? To the contrary, Kerala is not highly industrialized. In fact, it is

[1]However, because Americans in particular consume so much energy and other resources, we have a substantial negative impact on the global environment (Chapter 22, "Technology and the Global Environment").

among the poorer Indian states, with a per capita income less than the national average. Then has the government of Kerala strictly enforced a state childbirth policy similar to China's? The Chinese government strongly penalizes families that have more than one child, and it allows abortion at 8½ months. As a result, China had a total fertility rate of just 2.0 in 1992 (Wordsworth, 2000). In Kerala, however, the government keeps out of its citizens' bedrooms. The decision to have children remains a strictly private affair.

The women of Kerala achieved a low total fertility rate because their government purposely and systematically raised their status over a period of decades (Franke and Chasin, 1992; Sen, 1994). The government helped to create a realistic alternative to a life of continuous childbearing and childrearing. It helped women understand that they could achieve that alternative if they wanted to. In particular, the government organized successful campaigns and programs to educate women, increase their participation in the paid labor force, and make family planning widely available. These government campaigns and programs resulted in Keralan women enjoying the highest literacy rate, the highest labor force participation rate, and the highest rate of political participation in India. Given their desire for education, work, and political involvement, most Keralan women want small families, so they use contraception to prevent unwanted births. Thus, by lowering the level of gender inequality, the government of Kerala solved its overpopulation problem. In general, where

> women tend to have more power [the society has] low rather than high mortality and fertility. Education and employment, for example, often accord women wider power and influence, which enhance their status. But attending school and working often compete with childbearing and child rearing. Women may choose to have fewer children in order to hold a job or increase their education (Riley, 1997) (Box 20.1).

Class Inequality and Overpopulation

Unraveling the Keralan mystery is an instructive exercise. It establishes that population growth depends not just on a society's level of industrialization but also on its level of gender inequality. *Class* inequality influences population growth too. We turn to the South Korean case to illustrate this point.

In 1960 South Korea had a total fertility rate of 6.0. This prompted one American official to remark that "if these Koreans don't stop overbreeding, we may have the choice of supporting them forever, watching them starve to death, or washing our hands of the problem" (quoted in Lie, 1998: 21). Yet by 1989, South Korea's total fertility rate had dropped to a mere 1.6. By 2003 it fell to 1.3. Why? The first chapter in this story involves land reform, not industrialization. The government took land from big landowners and gave it to small farmers. Consequently, the standard of living of small farmers improved. This eliminated a major reason for high fertility. Once economic uncertainty decreased, so did the need for child labor and support of elderly parents by adult offspring. Soon, the total fertility rate began to fall. Subsequent declines in the South Korean total fertility rate were due to industrialization, urbanization, and the higher educational attainment of the population. But a decline in class inequality in the countryside first set the process in motion.

The reverse is also true. Increasing social inequality can lead to overpopulation, war, and famine. For example, in the 1960s the governments of El Salvador and Honduras encouraged the expansion of commercial agriculture and the acquisition of large farms by wealthy landowners. The landowners drove peasants off the land. The peasants migrated to the cities. There they hoped to find employment and a better life. Instead, they often found squalor, unemployment, and disease. Suddenly, two countries with a combined population of less than 5 million people had a big "overpopulation" problem. Competition for land increased and contributed to rising tensions. This eventually led to the outbreak of war between El Salvador and Honduras in 1969 (Durham, 1979).

Similarly, economic inequality helps to create famines. As Nobel Prize–winning economist Amartya Sen notes, "Famine is the characteristic of some people not *having* enough

BOX 20.1

SOCIAL POLICY: WHAT DO YOU THINK?

How Can We Find 100 Million Missing Women?

The U.S. Census Bureau expects the 6.4 billion inhabitants of the planet in 2004 to multiply to 9.1 billion by 2050. Yet demographers are fairly confident that world population will level off sometime between 2070 and 2100, reaching its peak at about 10.2 billion people.

Two main factors are causing the rate of world population growth to fall: economic development and the emancipation of women. Agricultural societies need many children to help with farming, but industrial societies require fewer children. Because many countries in the so-called Third World are industrializing, the rate of world population growth is falling apace. The second main factor responsible for the declining growth rate is the improving economic status and education of women. Once women become literate and enter the nonagricultural paid labor force, they quickly recognize the advantages of having few children. The birth rate plummets. In many Third World countries, that is just what is happening.

In other Third World countries, the position of women is less satisfactory. We can see this by examining the ratio of women to men, or the **sex ratio** (United Nations, 2000). In the United States in 2000, the sex ratio was about 1.03. That is, there were 103 women for every 100 men. This is about average for a highly developed country. (The sex ratio for Germany was also 1.03 and for Japan it was 1.04.) The surplus of women reflects the fact that men are more likely than women to be employed in health-threatening occupations, consume a lot of cigarettes and alcohol, and engage in riskier and more violent behavior, whereas women are the hardier sex, biologically speaking.

In the world as a whole, the picture is reversed. There were just 98 women for every 100 men in 2000. In India and China, there were only 94 women for every 100 men. Apart from Asia, North Africa is the region that suffers most from a deficit of women.

What accounts for variation in the sex ratio? According to Amartya Sen (1990; 2001), the sex ratio is low where women have less access to health services, medicine, and adequate nutrition than do men. These factors are associated with high female mortality. Another factor is significant in China and India. In those countries, some parents so strongly prefer sons over daughters that sex-selective abortion contributes to the low sex ratio. Parents who strongly prefer sons over daughters are inclined to abort female fetuses. In contrast, in highly developed countries, women and men have approximately equal access to health services, medicine, and adequate nutrition, and sex-selective abortion is very rare. Therefore, there are more women than men. By this standard, the world as a whole is "missing" about 5 women for every 100 men (because $103 - 98 = 5$). This works out to about 100 million women missing in 2000 due to sex-selective abortion and unequal access to resources of the most basic sort.

We can "find" many of the missing 100 million women partly by eliminating gender inequalities in access to health services, medicine, and adequate nutrition. Increased female literacy and employment in the paid labor force are the most effective paths to eliminating such gender inequalities. That is because literate women who work in the paid labor force are in a stronger position to demand equal rights and are more likely to be married to men with similar sympathies.

The question of how to eliminate sex-selective abortion is more difficult. Economic factors do not account for variations in sex-selective abortion. In some affluent parts of Asia with high levels of female education and economic participation, sex-selective abortion is common; India's lowest sex ratio—at 0.79, it is in fact the lowest sex ratio in the world—can be found in the wealthy northern states of Punjab and Haryana (Rahman, 2004).[3] In some other parts of Asia with low levels of female education and economic participation, sex-selective abortion is relatively rare. The best explanation for variations in sex-selective abortion seems to be that preference for sons is a strong *cultural* tradition in some parts of Asia. In India, for example, it may not be coincidental that sex-selective abortion is most widespread in the north and the west, where the ruling nationalist and fundamentalist Hindu party, BJP (the Bharatiya Janata Party), is most popular. Hindu nationalism and religious fundamentalism may feed into a strong preference for sons over daughters.

This leaves open the question of how, if at all, reformers inside and outside the region might rectify the situation. Cultural and religious traditions do not easily give way to economic forces. Meanwhile, a strong preference for sons over daughters adds millions to the number of missing women every year. What do you think should be done?

[3]This is true not just in some of the more affluent states of India. Sex-selective abortion is also relatively common in well-to-do Singapore, Taiwan, and South Korea.

food to eat. It is not the characteristic of there not *being* enough food to eat" (Sen, 1981: 1; our emphasis). Sen's distinction is crucial, as his analysis of several famines shows. Sen found that in some cases, although food supplies did decline, enough food was available to keep the stricken population fed. However, suppliers and speculators took advantage of the short supply. They hoarded grain and increased prices beyond the means of most people. In other cases, there was no decline in food supply at all. Food was simply withheld for political reasons, that is, to bring a population to its knees, or because many people

The **sex ratio** is the ratio of women to men in a geographical area.

were not considered entitled to receive it by the authorities. For example, in 1932–33 Stalin instigated a famine in Ukraine that killed between 7 million and 10 million people. His purpose: to bring the Ukrainian population to its knees and force them to give up their privately owned farms in order to join state-owned, collective farms. The source of famine, Sen concludes, is not underproduction or overpopulation but inequality of access to food (Drèze and Sen, 1989). In fact, even when starving people gain access to food, other forms of inequality may kill them. For instance, when food relief agencies delivered 3 million sacks of grain to prevent famine in western Sudan in the mid-1980s, tens of thousands of people died anyway because they lacked clean water, decent sanitation, and vaccinations against various diseases (de Waal, 1989). Western aid workers failed to listen carefully to the Sudanese people about their basic medical and sanitary needs, with disastrous results.

Summing Up

A new generation of demographers has begun to explore how class inequality and gender inequality affect population growth (Levine, 1987; Seccombe, 1992; Szreter, 1996). Their studies drive home the point that population growth and its negative consequences do not stem from natural causes (as Malthus held). Nor are they only responses to industrialization and modernization (as demographic transition theory suggests). Instead, population growth is influenced by a variety of social causes, social inequality chief among them.

Some undoubtedly well-intentioned Western analysts continue to insist that people in the developing countries should be forced to stop multiplying at all costs. Some observers even suggest diverting scarce resources from education, health, and industrialization into various forms of birth control, including, if necessary, forced sterilization (Riedmann, 1993). They regard the presumed alternatives—poverty, famine, war, ethnic violence, and the growth of huge, filthy cities—as too horrible to contemplate. However, they fail to see how measures that lower social inequality help to control overpopulation and its consequences. Along with industrialization, lower levels of social inequality cause total fertility rates to fall.

Urbanization

We have seen that overpopulation remains a troubling problem due to lack of industrialization and too much gender and class inequality in much of the world. We may now add that overpopulation is in substantial measure an *urban* problem. Driven by lack of economic opportunity in the countryside, political unrest, and other factors, many millions of people flock to big cities in the world's poor countries every year. Thus, most of the fastest-growing cities in the world today are in semi-industrialized countries where the factory system is not highly developed. As Table 20.2 shows, in 1900, 9 of the 10 biggest cities in the world were in industrialized Europe and the United States. By 2015, in contrast, 6 of the world's 10 biggest cities will be in Asia, 2 will be in Africa, and 2 will be in Latin America. Only 1 of the 10 biggest cities—Tokyo—will be in a highly industrialized country. Urbanization is, of course, taking place in the world's rich countries too. For instance, in North America, the urban population is expected to increase from 76 percent to 84 percent of the total population between 1996 and 2030 (Figure 20.4). In Africa and Asia, however, the urban population is expected to increase much faster—from about

Table 20.2
World's 10 Largest Metropolitan Areas, 1900 and 2015, Projected (in millions)

1900		2015	
London, England	6.5	Tokyo, Japan	28.7
New York, USA	4.2	Mumbai (Bombay), India	27.4
Paris, France	3.3	Lagos, Nigeria	24.4
Berlin, Germany	2.4	Shanghai, China	23.4
Chicago, USA	1.7	Jakarta, Indonesia	21.2
Vienna, Austria	1.6	São Paulo, Brazil	20.8
Tokyo, Japan	1.5	Karachi, Pakistan	20.6
Saint Petersburg, Russia	1.4	Beijing, China	19.4
Philadelphia, USA	1.4	Dhaka, Bangladesh	19.0
Manchester, England	1.3	Mexico City, Mexico	18.8

Sources: Department of Geography, Slippery Rock University (1997, 2003).

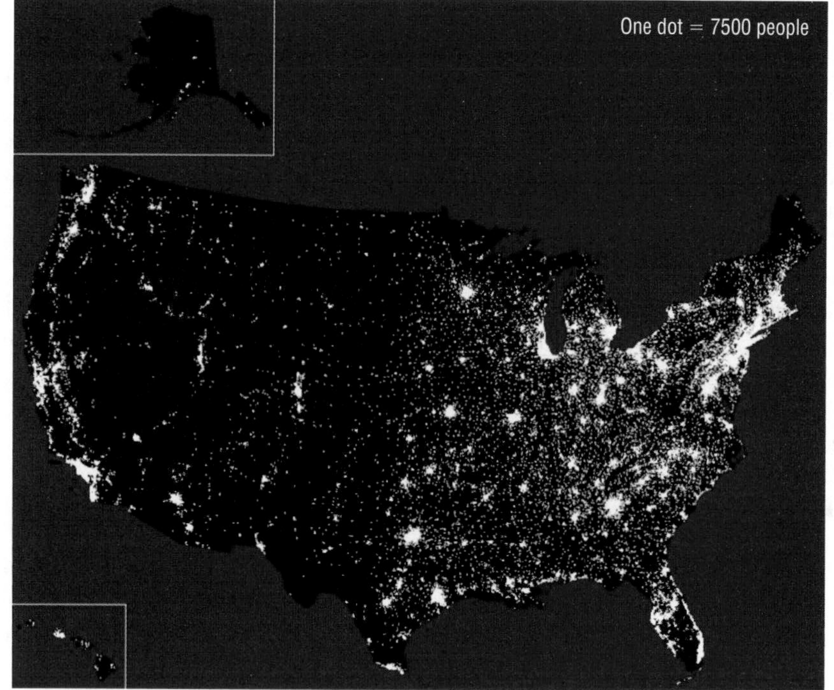

Figure 20.4
2000 Population Distribution in the United States

If you were circling the Earth in a satellite on a clear night, the United States would look like this map, in which each dot of light represents 7500 people. The dense clusters of light are the most urbanized areas.

Source: U.S. Census Bureau (2003a).

Christopher Morris/Black Star Publishing/Picture Quest

Mexico City during one of its frequent smog alerts. Of the world's 10 biggest cities in 2015, only one—Tokyo—will be in a highly industrialized country. All the others, including Mexico City, will be in developing countries.

35 percent to 55 percent of the total population in the same time period (United Nations, 1997b).

From the Preindustrial to the Industrial City

To a degree, urbanization results from industrialization. As you will learn in the following discussion, many great cities of the world grew up along with the modern factory, which drew hundreds of millions of people out of the countryside and transformed them into urban, industrial workers. Industrialization is not, however, the whole story behind the growth of cities. As we have just seen, the connection between industrialization and urbanization is weak in the world's less developed countries today. Moreover, cities first emerged in Syria, Mesopotamia, and Egypt 5000 or 6000 years ago, long before the growth of the modern factory. These early cities served as centers of religious worship and political administration. Similarly, it was not industry but international trade in spices, gold, cloth, and other precious goods that stimulated the growth of cities in preindustrial Europe and the Middle East. Thus, the correlation between urbanization and industrialization is far from perfect (Bairoch, 1988 [1985]; Jacobs, 1969; Mumford, 1961; Sjöberg, 1960).

The Granger Collection

Carcassonne, France, a medieval walled city.

Preindustrial cities differed from those that developed in the industrial era in several ways. Preindustrial cities were typically smaller, less densely populated, built within protective walls, and organized around a central square and places of worship. The industrial cities that began to emerge at the end of the 18th century were more dynamic and complex social systems. A host of social problems, including poverty, pollution, and crime, accompanied their growth. The complexity, dynamism, and social problems of the industrial city were all evident in Chicago at the turn of the 20th century. Not surprisingly, therefore, it was at the University of Chicago that American urban sociology was born.

The Chicago School and the Industrial City

From the 1910s to the 1930s, the members of the **Chicago school** of sociology distinguished themselves by their vividly detailed descriptions and analyses of urban life, backed up by careful in-depth interviews, surveys, and maps showing the distribution of various features of the social landscape, all expressed in plain yet evocative language (Lindner, 1996 [1990]). Three of its leading members, Robert Park, Ernest Burgess, and Roderick McKenzie, proposed a theory of **human ecology** to illuminate the process of urbanization (Park, Burgess, and McKenzie, 1967 [1925]). Borrowing from biology and ecology, the theory highlights the links between the physical and social dimensions of cities and identifies the dynamics and patterns of urban growth.

The Concentric Zone Model

The theory of human ecology, as applied to urban settings, holds that cities grow in ever-expanding concentric circles. It is sometimes called the "concentric zone model" of the city. Three social processes animate this growth (Hawley, 1950). **Differentiation** refers to the process by which urban populations and their activities become more complex and heterogeneous over time. For instance, a small town may have a diner, a pizza parlor, and a Chinese restaurant. But if that small town grows into a city, it will likely boast a variety of ethnic restaurants reflecting its more heterogeneous population. Moreover, in a city, members of different ethnic and racial groups and socioeconomic classes may vie with one another for dominance in particular areas. Businesses may also try to push residents out of certain areas to establish commercial zones. When this happens, people are engaging in **competition,** an ongoing struggle by different groups to inhabit optimal locations. Finally, **ecological succession** takes place when a distinct group of people moves from one area to another and another group moves into the old area to replace the first group. For example, a recurrent pattern of ecological succession involves members of the middle class moving to the suburbs, with working-class and poor immigrants moving into the inner city.

In Chicago in the 1920s, differentiation, competition, and ecological succession resulted in the zonal pattern illustrated by Figure 20.5:

The **Chicago school** founded urban sociology in the United States in the first decades of the 20th century. Its members distinguished themselves by their vivid and detailed descriptions and analyses of urban life and their development of the theory of human ecology.

Human ecology is a theoretical approach to urban sociology that borrows ideas from biology and ecology to highlight the links between the physical and social dimensions of cities and identify the dynamics and patterns of urban growth.

Differentiation in the theory of human ecology refers to the process by which urban populations and their activities become more complex and heterogeneous over time.

Competition in the theory of human ecology refers to the struggle by different groups for optimal locations in which to reside and set up their businesses.

Ecological succession in the theory of human ecology refers to the process by which a distinct urban group moves from one area to another and a second group comes in to replace the group that has moved out.

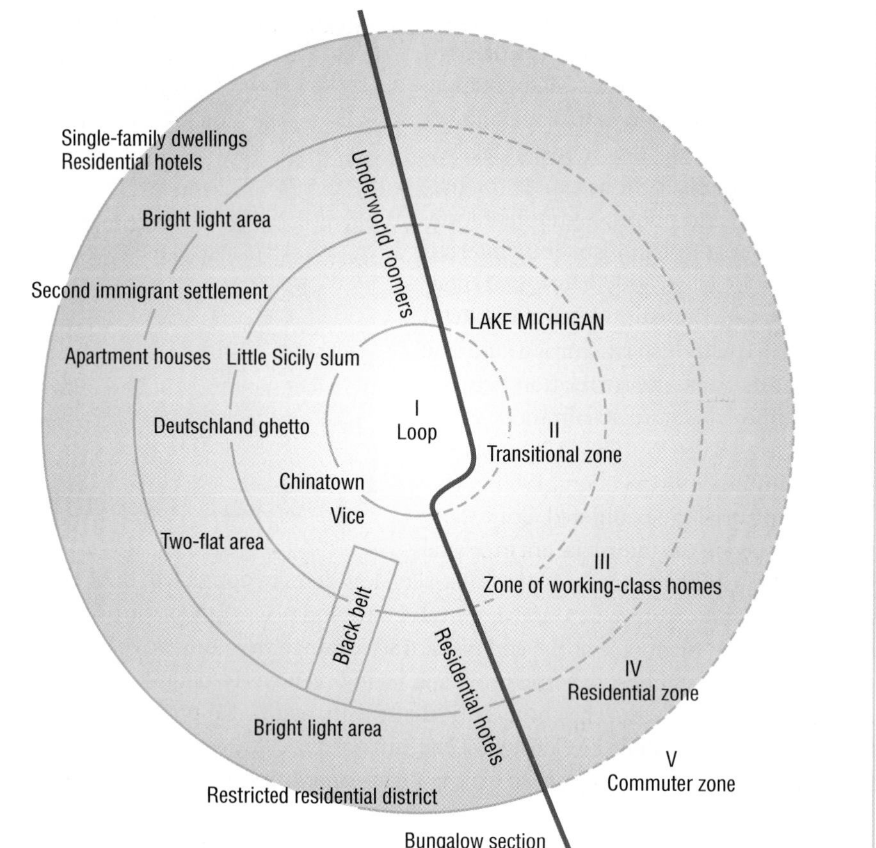

Figure 20.5
The Concentric Zone Model of Chicago, about 1920

Source: From "The Growth of the City: An Introduction to a Research Project," Ernest W. Burgess, p. 47–62, in *The City* by Robert E. Park, et al. Copyright © 1967 University of Chicago Press. Used with permission.

1. Zone 1 was the central business district (known in Chicago as "the Loop"). It contained retail shopping areas, office buildings, and entertainment centers. The land in this zone was the most valuable in the city.
2. Zone 2, the "zone of transition," was the area of most intense competition between residential and commercial interests. The businesses usually succeeded in driving out middle-class residents by bidding up the price of land and getting city governments to rezone the area for commercial use. Homes then declined in value, and cheap rental housing came to predominate. Eventually, the housing deteriorated into slums because speculators spent little on maintenance. They merely held on to the buildings until they could sell them for the commercial value of the land on which they stood. Where commercial development failed to materialize, the slums endured, attracting new immigrants, the poor, people with physical and mental disabilities, the unemployed, and criminals. The Chicago sociologists viewed this zone as "socially disorganized" because of its high level of deviance and crime.
3. When members of ethnic groups in Zone 2 could afford better housing, they moved to Zone 3, the "zone of working-class homes." These were mostly inexpensive, semi-detached buildings.
4. The upwardly mobile offspring of families in Zone 3 usually moved to Zone 4, the "residential zone," containing small, middle-class, detached homes.
5. Zone 5, the "commuter zone," was where middle-, upper-middle-, and upper-class families lived in more expensive detached homes. These people also owned cars and commuted to work in the city.

Urbanism is a way of life that, according to Louis Wirth, involves increased tolerance but also emotional withdrawal and specialized, impersonal, and self-interested interaction.

Urbanism: A Way of Life

For members of the Chicago school, the city was more than just a collection of socially segregated buildings, places, and people. It also involved a way of life they called **urbanism.** They defined urbanism as "a state of mind, a body of customs[,] . . .

traditions, . . . attitudes and sentiments" specifically linked to city dwelling (Park, Burgess, and McKenzie, 1967 [1925]: 1). Louis Wirth (1938) developed this theme. According to Wirth, rural life involves frequent face-to-face interaction among a few people. Most of these people are familiar with each other, share common values and a collective identity, and strongly respect traditional ways of doing things. Urban life, in contrast, involves the absence of community and of close personal relationships. Extensive exposure to many socially different people leads city dwellers to become more tolerant than rural folk, said Wirth. However, urban dwellers also withdraw emotionally and reduce the intensity of their social interaction with others. In Wirth's view, interaction in cities is therefore superficial, impersonal, and focused on specific goals. People become more individualistic. Weak social control leads to a high incidence of deviance and crime.

After Chicago: A Critique

The Chicago school dominated American urban sociology for decades. It still inspires much interesting research (e.g., Anderson, 1990). However, three major criticisms of this approach to understanding city growth have gained credibility over the years.

One criticism focuses on Wirth's characterization of the "urban way of life." Research shows that social isolation, emotional withdrawal, stress, and other problems may be just as common in rural as in urban areas (Webb and Collette, 1977; 1979; Crothers, 1979). After all, in a small community a person may not be able to find anyone with whom to share a particular interest or passion. Moreover, farmwork can be every bit as stressful as work on an assembly line. Research also shows that urban life is less impersonal, anomic, and devoid of community than the Chicago sociologists made it appear. True, newcomers (of whom there were admittedly many in Chicago in the 1920s) may find city life bewildering if not frightening. Neighborliness and friendliness to strangers are less common in cities than in small communities (Fischer, 1981). However, even in the largest cities, most residents create social networks and subcultures that serve functions similar to those performed by the small community. Friendship, kinship, ethnic, and racial ties, as well as work and leisure relations, form the bases of these urban networks and subcultures (Fischer, 1984 [1976]; Jacobs, 1961; Wellman, 1979). Cities, it turns out, are clusters of many different communities. Sociologist Herbert Gans found such a rich assortment of close social ties in his research on Italian Americans that he was prompted to call them "urban villagers" (Gans, 1962).

A second major criticism of the Chicago school's approach to urban sociology focuses on the concentric zone model. True, specific activities and groups are concentrated in distinct areas of American cities. For example, except for a handful of integrated communities, such as Shaker Heights, Ohio, racial segregation in housing remains a prominent feature of American urban life (Massey and Denton, 1993). However, the specific patterns discovered by the Chicago sociologists are most applicable to American industrial cities in the first quarter of the 20th century. Thus, in preindustrial cities, slums are more likely to be found on the outskirts and wealthy districts in the city core, whereas commercial and residential buildings are often not segregated (Sjöberg, 1960) In many parts of the world, city cores are reserved for the well-to-do while the poor live in the suburbs (for example, see the description of Rio de Janeiro with which we began this chapter). After the automobile became a major means of transportation, some American cities expanded not in concentric circles but in wedge-shaped sectors along natural boundaries and transportation routes (Hoyt, 1939). Others grew up around not one but many nuclei, each attracting similar kinds of activities and groups (Harris and Ullman, 1945). Recent models of urban growth emphasize the expansion of services from the city core to the city periphery, aided by the construction of radial highways (Harris, 1997) (▶Figure 20.6).

The third main criticism of the human ecology approach is that it presents urban growth as an almost natural process, slighting its historical, political, and economic foundations in capitalist industrialization. The Chicago sociologists' analysis of competition in the transitional zone came closest to avoiding this problem. However, their discussions of

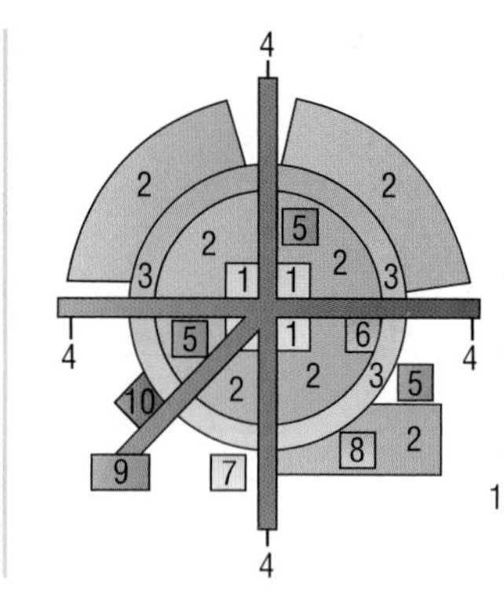

District

1 Central city
2 Suburban residential areas
3 Circumferential highway
4 Radial highway
5 Shopping mall
6 Industrial district
7 Office park
8 Service center
9 Airport complex
10 Combined employment and shopping center

Figure 20.6
The Peripheral Model of Cities

Source: Harris (1997).

differentiation and ecological succession made the growth of cities seem almost like a force of nature rather than a process rooted in power relations and the urge to profit.

The Conflict View and the New Urban Sociology

The so-called **new urban sociology,** heavily influenced by conflict theory, sought to correct this problem (Gottdiener and Hutchison, 2000 [1994]; Zukin, 1980). For new urban sociologists, urban space is not just an arena for the unfolding of social processes like differentiation, competition, and ecological succession. Instead, they see urban space as a set of *commodified* social relations. That is, urban space, like all commodities, can be bought and sold for profit. As a result, political interests and conflicts shape the growth pattern of cities. John Logan and Harvey Molotch (1987), for example, portray cities as machines fueled by a "growth coalition." This growth coalition is composed of investors, politicians, businesses, property owners, real estate developers, urban planners, the mass media, professional sports teams, cultural institutions, labor unions, and universities. All these partners try to get government subsidies and tax breaks to attract investment dollars. Reversing the pattern identified by the Chicago sociologists, this investment has been used to redevelop decaying downtown areas in many American cities since the 1950s. The Faneuil Hall Marketplace in Boston, Harborplace in Baltimore, South Street Seaport in New York, Grand Avenue in Milwaukee, Union Station in St. Louis, Bayside in Miami, and the Aloha Tower in Honolulu are all examples of such redevelopment projects from the 1970s.

According to Logan and Molotch, members of the growth coalition present redevelopment as a public good that benefits everyone. This tends to silence critics, prevent discussions of alternative ideas and plans, and veil the question of who benefits and who does not. In reality, the benefits of redevelopment are often unevenly distributed. Most redevelopments are "pockets of revitalization surrounded by areas of extreme poverty" (Hannigan, 1998a: 53). That is, local residents often enjoy few if any direct benefits from redevelopment. Indirectly, they may suffer when budgets for public schooling, public transportation, and other amenities are cut to help pay for development subsidies and tax breaks.

We can see the workings of the growth coalition in the histories of Los Angeles (Abelmann and Lie, 1995), Chicago (Cronon, 1991), Miami (Portes and Stepick, 1993), and other great American cities. This does not mean that the growth coalition is all-powerful. Community activism often targets local governments and corporations that seek unrestricted growth. Sometimes activists meet with success (Castells, 1983). Yet for the past 50 years, the growth coalition has managed to reshape the face of American cities, more or less in its own image.

The **new urban sociology** emerged in the 1970s and stressed that city growth is a process rooted in power relations and the urge to profit.

The **corporate city** refers to the growing post–World War II perception and organization of the North American city as a vehicle for capital accumulation.

Suburbanism is a way of life outside city centers that is organized mainly around the needs of children and involves higher levels of conformity and sociability than life in the central city.

The Corporate City

Due to the efforts of the growth coalition, the North American industrial city, typified by Chicago in the 1920s, gave way after World War II to the **corporate city.** John Hannigan defines the corporate city as "a vehicle for capital accumulation—that is, . . . a money-making machine" (Hannigan, 1998b [1995]: 345).

The Growth of Suburbs

In the suburbs—urbanized areas outside the political boundaries of cities—developers built millions of single-family detached homes for the corporate middle class. These homes boasted large backyards and a car or two in every garage. A new way of life developed, which sociologists, appropriately enough, dubbed **suburbanism.** Every bit as distinctive as urbanism, suburbanism organized life mainly around the needs of children. It also involved higher levels of conformity and sociability than life in the central city (Fava, 1956). Suburbanism became fully entrenched as developers built shopping malls to serve the needs of the suburbanites. This reduced the need to travel to the central city for consumer goods.

The suburbs were at first restricted to the well-to-do. However, following World War II, brisk economic growth and government assistance to veterans put the suburban lifestyle within the reach of middle-class Americans. The lack of housing in city cores, extensive road-building programs, the falling price of automobiles, and the baby boom that began in 1946 also stimulated mushroom-like suburban growth. By 1970, more Americans lived in suburbs than in urban core areas. This remains the case today.

The corporate city: New York.

Gated Communities, Exurbs, and Edge Cities

Due to the expansion of the suburbs, urban sociologists today often focus their attention not on cities but on entire **metropolitan areas.** Metropolitan areas include downtown city cores and their surrounding suburbs. They also include three recent developments that indicate the continued decentralization of urban America: the growth of **gated communities,** in which upper-middle-class residents pay high taxes to keep the community patrolled by security guards and walled off from the outside world; the spread of **exurbs,** or rural residential areas, within commuting distance of the city; and the formation of **edge cities,** or exurban clusters of malls, offices, and entertainment complexes that arise at the convergence point of major highways (Garrau, 1991). The spread of gated communities is motivated above all by fear of urban crime. The growth of exurban residential areas and edge cities since the 1970s is motivated mainly by the mounting costs of operating businesses in city cores and the growth of new telecommunication technologies that allow businesses to operate in the exurbs. Home offices, mobile employees, and decentralized business locations are all made possible by these technologies. Some sociologists, urban and regional planners, and others lump all these developments together as indicators of **urban sprawl,** the spread of cities into ever larger expanses of the surrounding countryside.

City cores continued to decline as the middle class fled, pulled by the promise of suburban and exurban lifestyles and pushed by racial animosity and crime. Many middle-class people went farther afield, abandoning the Snowbelt cities in America's traditional industrial heartland and migrating to the burgeoning cities of the American Sunbelt in the South and the West (Table 20.3 and Figure 20.7). As a result, particularly in northeastern and midwestern cities, tax revenues in the city core fell, even as more money was needed to sustain social welfare programs for the poor.

Table 20.3
The 20 Largest Cities in the United States, 2003

Rank	City	Population
1	New York, NY	8,085,742
2	Los Angeles, CA	3,819,951
3	Chicago, IL	2,869.121
4	Houston, TX	2,009,690
5	Philadelphia, PA	1,479,339
6	Phoenix, AZ	1,388,416
7	San Diego, CA	1,266,753
8	San Antonio, TX	1,214,725
9	Dallas, TX	1,208,318
10	Detroit, MI	911,402
11	San Jose, CA	898,349
12	Indianapolis, IN	783,438
13	Jacksonville, FL	773,781
14	San Francisco, CA	751,682
15	Columbus, OH	728,432
16	Austin, TX	672,011
17	Memphis, TN	645,978
18	Baltimore, MD	628,670
19	Milwaukee, WI	586,941
20	Fort Worth, TX	585,122

Note: Figures are for incorporated cities, not metropolitan areas.
Source: U.S. Census Bureau (2004).

- **Metropolitan areas** include downtown city cores and their surrounding suburbs.
- In **gated communities,** upper-middle-class residents pay high taxes to keep the community patrolled by security guards and walled off from the outside world.
- **Exurbs** are rural residential areas within commuting distance of the city.
- **Edge cities** are exurban clusters of malls, offices, and entertainment complexes that arise at the convergence point of major highways.
- **Urban sprawl** is the spread of cities into ever larger expanses of the surrounding countryside

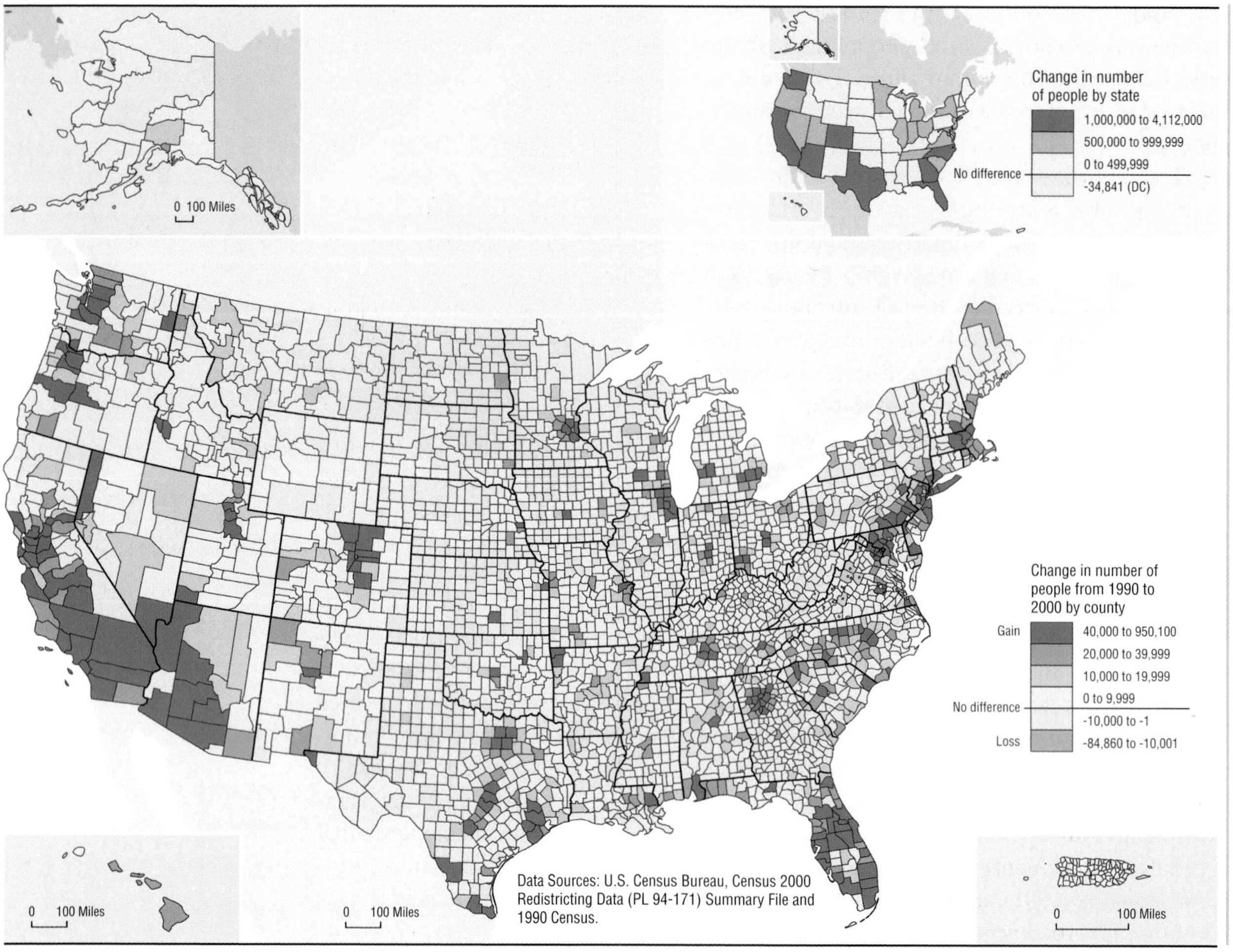

Change in number of people by state
1,000,000 to 4,112,000
500,000 to 999,999
0 to 499,999
No difference
-34,841 (DC)

Change in number of people from 1990 to 2000 by county
Gain
40,000 to 950,100
20,000 to 39,999
10,000 to 19,999
0 to 9,999
No difference
-10,000 to -1
Loss
-84,860 to -10,001

0 100 Miles

Data Sources: U.S. Census Bureau, Census 2000 Redistricting Data (PL 94-171) Summary File and 1990 Census.

Figure 20.7
Change in Number of People, United States, 1990 to 2000

U.S. Census Bureau (2001).

Urban Renewal

Web Research Projects: Welfare Reform

In a spate of urban renewal in the 1950s and 1960s, many homes in low-income and minority-group neighborhoods were torn down and replaced by high-rise apartment buildings and office towers in the city core. In the 1970s and 1980s, some middle-class people moved into rundown areas and restored them in a process called **gentrification.** Still, large residential sections of downtown Detroit, Baltimore, Cleveland, and other cities remained in a state of decay. The number of Americans living in high-poverty neighborhoods doubled as recessions and economic restructuring closed factories in inner cities. The situation of the inner cities improved during the economic boom of the 1990s but it is unclear whether the improvement will last (Box 20.2).

The Urbanization of Rural America

Napa Valley used to be a rural agricultural backwater on the northern edge of San Francisco Bay. Then, in the 1960s, two things happened to change the landscape forever. First, San Franciscans started moving to Napa, in some cases to establish second residences. Second, good-quality wine became an increasingly popular beverage and Napa attracted substantial outside investment because it was already a center for premium wines. Housing developments, hotels, vineyards, wineries, storage facilities, and bottling operations popped up everywhere. Wineries started selling wine, wine paraphernalia, T-shirts,

Gentrification is the process of middle-class people moving into rundown areas of the inner city and restoring them.

food, and cookbooks to the tourists who flocked to the valley. To attract customers, they also hosted banquets, opened restaurants, held art exhibits and concerts, and even sponsored a wine train that allowed passengers to stop at wineries and sample the wares. By the mid-1980s, the county's main highway was a traffic nightmare. Not without a fight from old-time residents, Napa Valley was transformed into an exurban offshoot of San Francisco (Friedland, 2002: 354–8).

Napa Valley is an extreme case of what has happened throughout rural America since World War II. To varying degrees, agriculture has been industrialized and rural areas have been partly urbanized. How industrialized is agriculture? One indicator of the extent of agricultural industrialization is that between 1935 and 1997, the average American farm grew from about 100 to 500 acres while the number of farms fell from about 7 million to 2 million, just 800,000 of which were responsible for 90 percent of the value of agricultural production.[2] Because expensive buildings and mechanized and computerized equipment are necessary for most commercial farming today, two-thirds of all agricultural production now comes from nonfamily farms and large family farms with sales of $250,000 a year or more. Many farms have thousands of acres and millions of dollars of capitalization (Hoppe, 2001: iv, 6). How urbanized are farmers? One indicator of the extent of urbanization is that more than half of all U.S. farm operators work off-farm, nearly 80 percent of them at full-time jobs, often in towns and cities. Remarkably, almost 90 percent of farm household income originates from off-farm sources, including jobs, businesses, pensions, and investments (Mishra et al., 2002). Satellite TV and Internet access are common in rural America, adding further to its urban flavor.

To varying degrees, many small towns, especially those within an hour's drive of big cities, have also been partly urbanized. They have lost much of their small-town character and have been turned into exurbs. Small towns hold many attractions for the city dweller. The cost of housing is lower than in the city. So is the crime rate. Public schools are often better. The air is cleaner. Nature is close to hand. In the last decades of the 20th century, many small towns became even more desirable as places of residence due to improvements in rural road systems and the centralization of shopping, entertainment, services, and work in regional centers. In the 1990s, America's nonurban population increased by 3 million people, reversing a 50-year trend (Salamon, 2003: 10).

Newcomers to small-town America tend to be younger, more highly educated, more ethnically diverse, and more urbane than old-time residents. They are also less committed to their adopted communities. Old-timers tend to live downtown, attend church regularly, volunteer for community activities, and take a strong interest in local politics, education, youth employment, and youth leisure activities. Newcomers tend to reside in segregated housing developments outside the downtown. They are less likely to do volunteer work and are more likely to move if they can profit from selling their homes. They are also less likely to take responsibility for the well-being of other people's children. Their own children therefore enjoy few organized activities and few of them have part-time jobs. The absence of community, of trust and mutual responsibility, is a breeding ground for violent behavior. It may be more than coincidental that the string of high school killings that shook the nation from 1997 to 1999 took place in the recently suburbanized or exurbanized towns of Pearl, Mississippi; West Paducah, Kentucky; Jonesboro, Arkansas; Edinboro, Pennsylvania; Springfield, Oregon; and Littleton, Colorado (Salamon, 2003: 18).

The Postmodern City

Many of the conditions that plagued the industrial city—poverty, inadequate housing, structural unemployment—are still evident in cities today. However, since about 1970, a new urban phenomenon has emerged alongside the legacy of old urban forms. This is the **postmodern city** (Hannigan, 1995a). The postmodern city has three main features:

The **postmodern city** is a new urban form that is more privatized and socially and culturally fragmented and globalized than the corporate city.

[2]About 62 percent of the farms in the country are basically residences that produce little for market.

BOX 20.2
Sociology at the Movies

8 Mile (2002)

8 Mile Road is a depressing stretch of rundown buildings, gas stations, fast-food outlets, and strip malls that separates the rich and poor areas of Detroit, the most racially segregated city in the country. South of 8 Mile Road, Detroit is overwhelmingly African American. The suburbs north of 8 Mile Road are overwhelmingly European American.

Jimmy "Rabbit" Smith (Eminem) is a member of the white minority in the poor area. He has left his pregnant girlfriend to move in with his unemployed mother (Kim Basinger), his little sister, and his mother's unemployed boyfriend, with whom he went to high school and whom he hates. They live in a trailer park near 8 Mile Road. Rabbit operates a punch press by day. He is thin and pale. He slouches and keeps a beanie pulled low over his head, as if hiding from the world or keeping it at bay. His close friends think he is a gifted rap artist, but his more numerous enemies think a white boy rapping is a travesty. They ridicule Rabbit and call him Elvis. He meets Alex (Brittany Murphy), a beautiful young woman who discerns his talent, but she betrays him. No wonder Rabbit is sullen and angry, reserving his rare smiles for his little sister.

Two rap competitions bracket the movie. In the first session, Rabbit has 45 seconds to out-insult his opponent. Instead, he freezes and is laughed and booed off the stage. In the second session, near the end of the movie, he has 90 seconds to prove his mettle. This time, he succeeds brilliantly.

The lyrics in the second rap session are worth heeding for their sociological implications. Rabbit first anticipates the attack of his African American opponent, Papa Doc, by listing his own deficiencies: I'm a white rapper, I'm a bum, I live in a trailer with my mom, my friend is so dumb he shot himself, my girlfriend betrayed me, and so forth. "Tell these people something they *don't* know about me," Rabbit taunts. After taking the wind out of his opponent's sails, he dissects Papa Doc with the following words:

But I know something about you:
You went to Cranbrook, that's a private school.
What's the matter dawg, you embarrassed?
This guy's a gangster?
His real name's Clarence.
And Clarence lives at home with both parents.
And Clarence's parents have a real good marriage.

The Everett Collection

Eminem (Marshall Mathers) and Brittany Murphy in *8 Mile*.

Cranbrook is a prep school in Bloomfield Hills, a Detroit suburb north of 8 Mile Road. Clarence, it turns out, is not quite the streetwise gang member from a single-parent family he makes himself out to be.

Clarence, however, represents big news. Between 1970 and 1990, the number of Americans living in high-poverty neighborhoods (where the poverty rate is 40 percent or higher) doubled. During the economic boom of the 1990s, the process of poverty concentration went into reverse. Two and a half million Americans moved out of high-poverty neighborhoods and mostly into less impoverished, older, inner-ring suburbs around major metropolitan areas. Nowhere was the reversal more dramatic than in Detroit (Jargowsky, 2003) (Figures 20.8 and 20.9).

The phenomenon represented by Clarence is encouraging yet fraught with danger. On the one hand, poor people are better off if they are widely dispersed rather than spatially concentrated. Spatially concentrated poor people have to deal not only with their own poverty but with a violent environment, low-performing schools, and a lack of positive role models. Thus, the exodus from the inner city is a positive development because it puts poor people in neighborhoods where poverty is less concentrated. On the other hand, the exodus increases the percentage of poor people in many of the older, inner-ring suburbs around major metropolitan areas. Particularly now that the economic boom of the 1990s has ended, this may result in the migration of many of the social ills of the inner city into the inner suburbs. If Clarence's children grow up where he did, they may be less susceptible to the ridicule of an Eminem.

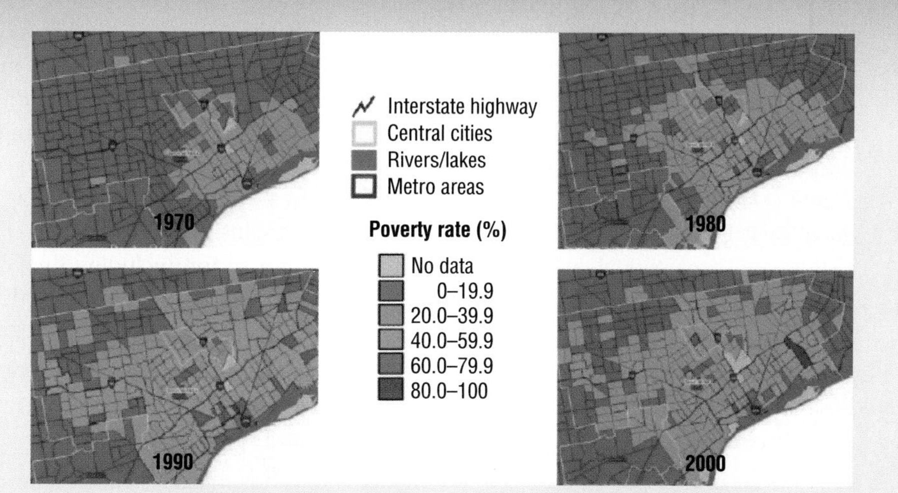

Figure 20.8

The Distribution of Poverty in Detroit, 1970–2000

Source: From Paul A. Jargowski, *Stunning Process, Hidden Problems*, p. 7. © 2003 The Brookings Institution. Reprinted with permission.

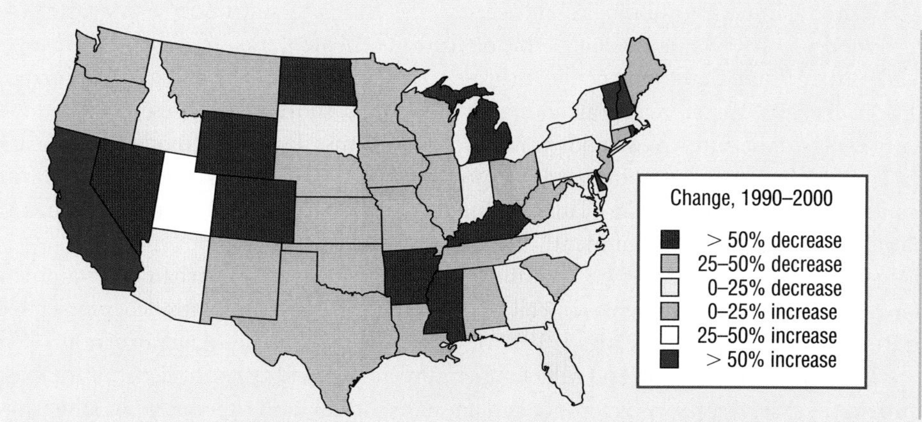

Figure 20.9

Percent Change in Population of High-Poverty Neighborhoods, United States, 1990–2000

Most states experienced a substantial decrease in the percent of people living in high-poverty neighborhoods in the 1990s. Decreases were biggest in the Midwest and the South. In a few states in the Northeast and the West, the percent of people living in high-poverty neighborhoods *increased*. California, for example, experienced an increase of 87 percent, largely due to immigration of low-income people from Latin America and the growth of *barrios*, predominantly Hispanic high-poverty neighborhoods.

Source: From Paul A. Jargowski, *Stunning Process, Hidden Problems*, p. 7. © 2003 The Brookings Institution. Reprinted with permission.

1. The postmodern city is more *privatized* than the corporate city because access to formerly public spaces is increasingly limited to those who can afford to pay. Privatization is evident in the construction of closed-off gated communities in the suburbs. They have round-the-clock security guards stationed at controlled-access front gates and foot patrols on the lookout for intruders. Privatization is also apparent in downtown cores. There, gleaming office towers and shopping areas are built beside slums. Yet the two areas are separated by the organization of space and access. For instance, a series of billion-dollar, block-square structures have been built around Bunker Hill in Los Angeles. Nearly all pedestrian linkages to the surrounding poor immigrant neighborhoods have been removed. Barrel-shaped, "bum-proof" bus benches prevent homeless people from sleeping on them. Trash cans are designed to be "bag lady–proof." Overhead sprinklers in Skid Row Park discourage overnight sleeping. Public toilets and washrooms have been removed in areas frequented by vagrants (Davis, 1990). Like the Renaissance Center in Detroit, the private areas of downtown Los Angeles are increasingly intended for exclusive use by middle-class visitors and professionals who work in the information sector, including financial services, the computer industry, telecommunications, entertainment, and so forth.
2. The postmodern city is also more *fragmented* than the corporate city. That is, it lacks a single way of life, such as urbanism or suburbanism. Instead, a great variety of lifestyles and subcultures proliferate in the postmodern city. They are based on race, ethnicity, immigrant status, class, sexual orientation, and so forth.
3. The third characteristic of the postmodern city is that it is more *globalized* than the corporate city. According to Saskia Sassen (1991), New York, London, and Tokyo epitomize the global city. They are centers of economic and financial decision making. They are also sites of innovation, where new products and fashions originate. In short, they have become the command posts of the globalized economy and its culture.

The processes of privatization, fragmentation, and globalization are evident in the way the postmodern city has come to reflect the priorities of the global entertainment industry. Especially in the 1990s, the city and its outlying districts came to resemble so many Disneyfied "Magic Kingdoms." These Magic Kingdoms are based on capital and technologies from the United States, Japan, Canada, Britain, and elsewhere. Here we find the latest entertainment technologies and spectacular thrills to suit nearly every taste. The postmodern city gets its distinctive flavor from its theme parks, restaurants and night clubs, waterfront developments, refurbished casinos, giant malls, megaplex cinemas, IMAX theaters, virtual reality arcades, ride simulators, sports complexes, book and CD megastores, aquariums, and hands-on science "museums." In the postmodern city, nearly everything becomes entertainment or, more accurately, combines entertainment with standard consumer activities. This produces hybrid activities like "shoppertainment," "eatertainment," and "edutainment."

John Hannigan has shown how the new venues of high-tech urban entertainment manage to provide excitement—but all within a thoroughly clean, controlled, predictable, and safe environment (Hannigan, 1998a). Not only are the new Magic Kingdoms kept spotless, in excellent repair, and fully temperature and humidity controlled, but they also provide a sense of security by touting familiar name brands and rigorously excluding anything and anybody that might disrupt the fun. For example, entertainment developments often enforce dress codes, teenager curfews, and rules banning striking workers and groups espousing social or political causes from their premises. The most effective barriers to potentially disruptive elements, however, are affordability and access. User surveys show that the new forms of urban entertainment tend to attract middle- and upper-middle-class patrons, especially whites. That is because they are pricey and many of them are in places that lack public transit and are too expensive for most people to reach by taxi.

Referring to the major role played by the Disney corporation in developing the new urban entertainment complexes, an architect once said that our downtowns would be "saved by a mouse" (quoted in Hannigan, 1998a: 193). But do the new forms of entertainment that dot the urban landscape increase the economic well-being of the communities in which they are established? Not much, beyond creating some low-level, dead-end jobs (security guard, waiter, janitor). Do they provide ways of meeting new people, seeing old friends and neighbors, and in general improving urban sociability? Not really. You visit a theme park with family or friends, but you generally stick close to your group and rarely have chance encounters with other patrons or bump into acquaintances. Does the high-tech world of globalized urban entertainment enable cities and neighborhoods to retain and enhance their distinct traditions, architectural styles, and ambience? It would be hard to destroy the distinctiveness of New York, San Francisco, or Vancouver, but many large North American cities are becoming homogenized as they provide the same entertainment services—and the same global brands—as Tokyo, Paris, and Sydney. If the mouse is saving our cities, perhaps he is also gnawing away at something valuable in the process.

Summary

Sociology Now™

Reviewing is as easy as ❶, ❷, ❸.

❶ Before you do your final review, take the SociologyNow diagnostic quiz to help you identify the areas on which you should concentrate. You will find information on how to access SociologyNow on the foldout at the front of the textbook.

❷ As you review, take advantage of SociologyNow's study aids to help you master the topics in this chapter.

❸ When you are finished with your review, take SociologyNow's post-test to confirm you are ready to move on to the next chapter.

1. **What is the Malthusian theory of population growth? Does it apply today?**

   Thomas Malthus argued that while food supplies increase slowly, populations grow quickly. Because of these presumed natural laws, only war, pestilence, and famine can keep human population growth in check. Several developments have cast doubt on Malthus's theory. Food production has increased rapidly. The limits to population size are higher than Malthus expected. Some populations are large yet prosperous. Some countries provide generous social welfare and still maintain low population growth rates. The use of contraception is widespread.

2. **What is demographic transition theory?**

   Demographic transition theory holds that the main factors underlying population dynamics are industrialization and the growth of modern cultural values. In the preindustrial era, both crude birth rates and crude death rates were high, and population growth was therefore slow. In the first stages of industrialization, crude death rates fell, so population growth was rapid. As industrialization progressed and people's values about having children changed, the crude birth rate fell, resulting in slow growth again. Finally, in the postindustrial era, the crude death rate has risen above the crude birth rate in many societies. As a result, their populations slowly shrink unless in-migration augments their numbers.

3. **What factors aside from the level of industrialization affect population growth?**

   The level of social inequality between women and men, and between classes, affects population dynamics, with lower levels of social inequality typically resulting in lower crude birth rates and therefore lower population growth rates.

4. **Is urbanization a function of industrialization?**

   Much urbanization is associated with the growth of factories. However, religious, political, and commercial need gave rise to cities in the preindustrial era. Moreover, the fastest-growing cities in the world today are in semi-industrialized countries.

5. **What did members of the Chicago school contribute to our understanding of the growth of cities?**

   The members of the Chicago school famously described and explained the spatial and social dimensions of the industrial city. They developed a theory of human ecology that explained urban growth as the outcome of differentiation, competition, and ecological succession. They described the spatial arrangement of the industrial city as a series of expanding concentric circles. The main business, entertainment, and shopping area stood in the center, with the class position of residents increasing as one moved from inner to outer rings.

6. **What are the main weaknesses of the Chicago school's analysis of cities?**

   Subsequent research showed that the city is not as anomic as the Chicago sociologists made it appear. Moreover, the concentric zone pattern applies best to the American industrial city in the first quarter of the 20th century. Preindustrial cities, non-American cities, and contemporary cities do not fit the concentric zone pattern well. Moreover, the new urban sociology criticized the Chicago school for making city growth seem like an almost natural process, playing down the power conflicts and profit motives that prompted the evolution of cities.

7. **What are corporate and postmodern cities?**

   The corporate city that emerged after World War II was a vehicle for capital accumulation that stimulated the growth of the suburbs and resulted in the decline of inner cities. The postmodern city that took shape in the last decades of the 20th century is characterized by the increased globalization of culture, fragmentation of lifestyles, and privatization of space.

8. **What are some of the other major changes that have taken place in city life in recent decades?**

   Cities have become suburbanized and exurbanized as they sprawl into the surrounding countryside. In the process, rural areas and small towns have been partly urbanized.

Questions to Consider

1. Do you think that rapid global population growth is cause for alarm? If not, why not? If so, what aspects of global population growth are especially worrisome? What should be done about them?
2. Do you think of the city mainly as a place of innovation and tolerance or as a site of crime, prejudice, and anomie? Where does your image of the city come from?—your own experience? the mass media? your sociological reading?

Web Resources

Companion Website for This Book

http://sociology.wadsworth.com

Begin by clicking on the Student Resources section of the website. Choose "Introduction to Sociology" and then the Brym and Lie book cover. Next, select the chapter you are currently studying from the pull-down menu. From the Student Resources page you will have easy access to InfoTrac® College Edition, MicroCase Online exercises, additional web links, and many other resources to aid you in your study of sociology, including practice tests for each chapter.

Recommended Websites

The U.S. Census Bureau supports perhaps the richest demographic site on the World Wide Web. It contains mainly United States but also some international data, all easily accessible. Visit the site at http://www.census.gov.

Another useful, data-rich site is run by the Population Reference Bureau at http://www.prb.org.

The World Wide Web Library: Demography and Population Studies at http://demography.anu.edu.au/VirtualLibrary offers a comprehensive list of international websites devoted to the social-scientific study of population.

The Urban Institute is a nonpartisan economic and social policy research organization located in Washington, D.C. Visit its website at http://www.urban.org.

For information on all 409 urban areas in the world with a population of more than 1 million, go to http://www.citypopulation.de/Country.html?E+World.

http://www.urbanpoverty.net is an excellent interactive website that lets you create your own maps showing the distribution of poverty and other variables for any metropolitan area in the United States from 1970 to 2000.

CHAPTER 21 Collective Action and Social Movements

Mark Peterson/Corbis

In this chapter, you will learn that:

- People sometimes lynch, riot, and engage in other forms of nonroutine group action to correct perceived injustices. Such events are rare, short-lived, spontaneous, and often violent. They subvert established institutions and practices. Nevertheless, most nonroutine collective action requires social organization, and people who take part in collective action often act in a calculated way.
- Collective action can result in the creation of one or more formal organizations or bureaucracies to direct and further the aims of its members. The institutionalization of protest signifies the establishment of a social movement.
- People are more inclined to rebel against existing conditions when strong social ties bind them to many other people who feel similarly wronged, when they have the time, money, and other resources needed to protest, and when political structures and processes give them opportunities to express discontent.
- For social movements to grow, members must make the activities, goals, and ideology of the movement consistent with the interests, beliefs, and values of potential recruits.
- The history of social movements is a struggle for the acquisition of constantly broadening citizenship rights—and opposition to those struggles.

How to Spark a Riot

Personal Anecdote

Robert Brym almost sparked a small riot once. "It happened in grade eleven," says Robert, "shortly after I learned that water combined with sulfur dioxide produces sulfurous acid. The news shocked me. To understand why, you have to know that I lived sixty miles east of the state of Maine and about a hundred yards downwind of one of the largest pulp and paper mills in Canada. Waves of sulfur dioxide billowed day and night from the mill's smokestacks. The town's pervasive rotten-egg smell was a long-standing complaint in the area. But, for me, disgust turned to upset when I realized the fumes were toxic. Suddenly it was clear why many people I knew—especially people living near the mill—woke up in the morning with a kind of 'smoker's cough.' By the simple act of breathing we were causing the gas to mix with the moisture in our bodies and form an acid that our lungs tried to expunge, with only partial success.

"Twenty years later, I read the results of a medical research report showing that area residents suffered from rates of lung disease, including emphysema and lung cancer, significantly higher than the North American average. But even in 1968 it was evident my hometown had a serious problem. I therefore hatched a plan. Our high school was about to hold its annual model parliament. The event was notoriously boring, partly because, year in, year out, virtually everyone voted for the same party, the Conservatives. But here was an issue, I thought, that could turn things around. A local man, K. C. Irving, owned the pulp and paper mill. *Forbes* business magazine ranked him as one of the richest men in the world. I figured that when I

Sociology Now™

Reviewing is as easy as ❶, ❷, ❸.

Use SociologyNow to help you make the grade on your next exam. When you are finished reading this chapter, go to the Chapter Summary for instructions on how to make SociologyNow work for you.

---

told my fellow students what I had discovered, they would quickly demand the closure of the mill until Irving guaranteed a clean operation.

"Was *I* naive. As head of the tiny Liberal party, I had to address the entire student body during assembly on election day to outline the party platform and rally votes. When I got to the part of my speech explaining why Irving was our enemy, the murmuring in the audience, which had been growing like the sound of a hungry animal about to pounce on its prey, erupted into loud boos. A couple of students rushed the stage. The principal suddenly appeared from the wings and commanded the student body to settle down. He then took me by the arm and informed me that for my own safety, my speech was finished. So, I discovered on election day, was our high school's Liberal party. And so, it emerged, was my high school political career.

"This incident troubled me for many years, partly due to the embarrassment it caused, partly due to the puzzles it presented. Why did I almost spark a small riot? Why didn't my fellow students rebel in the way I thought they would? Why did they continue to support an arrangement that was enriching one man at the cost of a community's health? Couldn't they see the injustice? Other people did. Nineteen sixty-eight was not just the year of my political failure in high school. It was also the year that student riots in France nearly toppled that country's government. In Mexico, the suppression of student strikes by the government left dozens of students dead. In the United States, students at Berkeley, Michigan, and other colleges demonstrated and staged sit-ins with unprecedented vigor. They supported free speech on their campuses, an end to American involvement in the war in Vietnam, increased civil rights for American blacks, and an expanded role for women in public affairs. It was, after all, the Sixties."

The Study of Collective Action and Social Movements

Robert didn't know it at the time, but by asking why students in Paris, Mexico City, and Berkeley rebelled while his fellow high school students did not, he was raising the main question that animates the study of collective action and social movements. Under what social conditions do people act in unison to change, or resist change to, society? That is the main issue we address in this chapter. We have divided the chapter into three sections:

1. We first discuss the social conditions leading to the formation of lynch mobs, riots, and other types of nonroutine **collective action.** When people engage in collective action, they act in unison to bring about or resist social, political, and economic change (Schweingruber and McPhail, 1999: 453). Some collective actions are "routine" and others are "nonroutine" (Useem, 1998: 219). Routine collective actions are usually nonviolent and follow established patterns of behavior in bureaucratic social structures. For instance, when Mothers Against Drunk Driving (MADD) lobbies for tougher laws against driving under the influence of alcohol, when members of a community organize a campaign against abortion or for freedom of reproductive choice, or when workers decide to form a union, they are engaging in routine collective action. Sometimes, however, "usual conventions cease to guide social action and people transcend, bypass, or subvert established institutional patterns and structures" (Turner and Killian, 1987 [1957]: 3). On such occasions, people engage in nonroutine collective action, which is often short-lived and sometimes violent. They may, for example, form lynch mobs and engage in riots. Until the early 1970s, it was widely believed that people who engage in nonroutine collective action lose their individuality and capacity for reason. Lynch mobs and riots were often seen as wild and uncoordinated affairs, more like stampedes of frightened cattle than structured social processes. As you will see, however, sociologists later showed that this portrayal is an exaggeration. It deflects at-

Collective action occurs when people act in unison to bring about or resist social, political, and economic change. Some collective actions are routine. Others are nonroutine. Routine collective actions are typically nonviolent and follow established patterns of behavior in existing types of social structures. Nonroutine collective actions take place when usual conventions cease to guide social action and people transcend, bypass, or subvert established institutional patterns and structures.

tention from the social organization and inner logic of extraordinary sociological events.[1]

2. We next outline the conditions underlying the formation of **social movements.** Social movements are enduring and usually bureaucratically organized collective attempts to change (or resist change to) part or all of the social order. This is achieved by petitioning, striking, demonstrating, and establishing lobbies, unions, and political parties. We will see that an adequate explanation of institutionalized protest also requires the introduction of a set of distinctively sociological issues. These concern the distribution of power in society and the framing of political issues in ways that appeal to many people.
3. Finally, we make some observations about the changing character of social movements. We argue that the history of social movements is the history of attempts by underprivileged groups to broaden their members' citizenship rights and increase the scope of protest from the local to the national to the global level.

We begin by considering the lynch mob, a well-studied form of nonroutine collective action.

Nonroutine Collective Action: The Lynch Mob

The Lynching of Claude Neal

On October 27, 1934, a black man was lynched near Greenwood, a town in Jackson County, Florida. Claude Neal, 23, was accused of raping and murdering 19-year-old Lola Cannidy, a pretty white woman and the daughter of his employer. The evidence against Neal was not totally convincing. Some people thought he confessed under duress. But Neal's reputation in the white community as a "mean n_____," "uppity," "insolent," and "overbearing" helped to seal his fate (McGovern, 1982: 51). He was apprehended and jailed. Then, for his own safety, he was removed to the jailhouse in Brewton, Alabama, about 120 miles northwest of the crime scene.

When the white residents in and around Greenwood found that Neal had been taken from the local jail, they quickly formed a lynch mob to find him. Once word mysteriously leaked out that Neal was in Brewton, 15 men in three cars immediately headed west. They figured out a way to get the sheriff out of the Brewton jail by sending him on a wild goose chase. Then, entering the jail holding guns and dynamite, they threatened to blow up the place if the lone jailer did not hand over Neal. He complied. They then tied Neal's hands with a rope. Finally, they dumped him in the back seat of a car for the ride back to Jackson County. There, a mob of two or three thousand people soon gathered near the Cannidy house.

Web Interactive Exercises: Is the Klan History?

The mob was in a state of violent agitation. Drinking moonshine whiskey and shouting "We want the n_____," many of them "wanted to get their hands on [Neal] so bad they could hardly stand it," according to one bystander. However, the jail raiders feared the mob was uncontrollable and that its members might injure each other in the frenzy to get at Neal. So they led their prisoner into the woods. An investigator interviewed one member of the company 10 days later. He described what next happened in these horrifying words:

> After taking the n_____ to the woods about four miles from Greenwood, they cut off his penis. He was made to eat it. Then they cut off his testicles and made him eat them and say he liked it. Then they sliced his sides and stomach with knives and every now and then somebody would cut off a finger or a toe. Red hot irons were used on the n_____ to burn him from top to bottom.

Social movements are enduring collective attempts to change part or all of the social order by means of rioting, petitioning, striking, demonstrating, and establishing lobbies, unions, and political parties.

[1]Reflecting new research and theoretical perspectives, the older term, *collective behavior,* fell into disfavor in the 1990s. That is because *behavior* suggests a relatively low level of consciousness of self and therefore conduct that is not entirely rational. Following Weber (1947), *action* denotes greater consciousness of self and therefore more rationality.

The National Association for the Advancement of Colored People (NAACP) took out this full-page ad in the *New York Times* on November 23, 1922, to encourage people to support passage of the Dyer anti-lynching bill in Congress. The bill was passed in the House of Representatives but defeated in the Senate. Despite the NAACP's vigorous efforts throughout the 1920s and 1930s, Congress never outlawed lynching.

THE SHAME OF AMERICA

Do you know that the United States is the Only Land on Earth where human beings are BURNED AT THE STAKE?

In Four Years, 1918-1921, Twenty-Eight People Were Publicly BURNED BY AMERICAN MOBS

3436 People Lynched 1889 to 1922

For What Crimes Have Mobs Nullified Government and Inflicted the Death Penalty?

The Alleged Crimes	The Victims
Murder	1288
Rape	571
Crimes against the Person	615
Crimes against Property	333
Miscellaneous Crimes	453
Absence of Crime	176
	3436

Why Some Mob Victims Died:
Not turning out of road for white boy in auto
Being a relative of a person who was lynched
Jumping a labor contract
Being a member of the Non-Partisan League
"Talking back" to a white man
"Insulting" white man.

Is Rape the "Cause" of Lynching?

Of 3,436 people murdered by mobs in our country, only 571, or less than 17 per cent., were even accused of the crime of rape.

83 WOMEN HAVE BEEN LYNCHED IN THE UNITED STATES

Do lynchers maintain that they were lynched for "the usual crime"?

AND THE LYNCHERS GO UNPUNISHED

THE REMEDY

The Dyer Anti-Lynching Bill Is Now Before the United States Senate

The Dyer Anti-Lynching Bill was passed on January 26, 1922, by a vote of 230 to 119 in the House of Representatives

The Dyer Anti-Lynching Bill Provides:
That culpable State officers and mobbists shall be tried in Federal Courts on failure of State courts to act, and that a county in which a lynching occurs shall be fined $10,000, recoverable in a Federal Court.

The Principal Question Raised Against the Bill is upon the Ground of Constitutionality.

The *Constitutionality* of the Dyer Bill Has Been Affirmed by—
The Judiciary Committee of the House of Representatives
The Judiciary Committee of the Senate
The United States Attorney General, legal adviser of Congress
Judge Guy D. Goff, of the Department of Justice

The Senate has been petitioned to pass the Dyer Bill by—
29 Lawyers and Jurists, including two former Attorneys General of the United States
19 State Supreme Court Justices
24 State Governors
3 Archbishops, 85 bishops and prominent churchmen
29 Mayors of large cities, north and south.

The American Bar Association at its meeting in San Francisco, August 9, 1922, adopted a resolution asking for further legislation by Congress to punish and prevent lynching and mob violence.

Fifteen State Conventions of 1922 (3 of them Democratic) have inserted in their party platforms a demand for national action to stamp out lynchings.

The Dyer Anti-Lynching Bill is not intended to protect the guilty, but to assure to every person accused of crime trial by due process of law.

THE DYER ANTI-LYNCHING BILL IS NOW BEFORE THE SENATE
TELEGRAPH YOUR SENATORS TODAY YOU WANT IT ENACTED

If you want to help the organization which has brought to light the facts about lynching, the organization which is fighting for 100 per cent. Americanism, not for some of the people some of the time, but for all of the people, white or black, all of the time

Send your check to J. E. SPINGARN, Treasurer of the

NATIONAL ASSOCIATION FOR THE ADVANCEMENT OF COLORED PEOPLE
70 FIFTH AVENUE, NEW YORK CITY

THIS ADVERTISEMENT IS PAID FOR IN PART BY THE ANTI-LYNCHING CRUSADERS.

National Association for the Advancement of Colored People, 1922

"From time to time during the torture," continued the investigator, "a rope would be tied around Neal's neck and he was pulled up over a limb and held there until he almost choked to death, then he would be let down and the torture [would] begin all over again" (McGovern, 1982: 80).

Having thus disposed of Neal, the jail raiders tied a rope around his body. They attached the rope to a car and dragged the body several miles to the mob in front of the Cannidy house. There, several people drove knives into the corpse, "tearing the body almost to shreds" according to one report (McGovern, 1982: 81). Lola Cannidy's grandfather took his .45 and pumped three bullets into the corpse's forehead. Some people started kicking the body. Others drove cars over it. Children were encouraged to take sharpened sticks and drive them deep into the flesh of the dead man. Then some members of the crowd rushed to a row of nearby shacks inhabited by blacks and burned the dwellings to the ground. Others took the nude and mutilated body of Claude Neal to the lawn of the Jackson County courthouse, where they strung it up on a tree. Justice, they apparently felt, had now been served. Later, they sold a photograph of Neal's hanging body as a postcard.

Breakdown Theory and Functionalism

Until about 1970, most sociologists believed that at least one of three conditions must be met for nonroutine collective action, such as Claude Neal's lynching, to emerge. First, a group of people must be economically deprived or socially rootless. Second, their norms must be strained or disrupted. Third, they must lose their capacity to act rationally by getting caught up in the supposedly inherent madness of crowds. Following Charles Tilly and his associates, we may group these three factors together as the **breakdown theory** of collective action. That is because all three factors assume that collective action results from the disruption or breakdown of traditional norms, expectations, and patterns of behavior (Tilly, Tilly, and Tilly, 1975: 4–6). At a more abstract level, breakdown theory may be seen as a variant of functionalism, for it regards collective action as a form of social imbalance that results from various institutions functioning improperly (Chapter 1, "A Sociological Compass").

Achmad Ibrahim/AP/Wide World Photo

A demonstration becomes heated in Jakarta, Indonesia, as a crowd burns the flag of an Islamic opposition party.

Deprivation, Crowds, and the Breakdown of Norms

Most pre-1970 sociologists would have said that Neal's lynching was caused by one or more of the following factors:

1. *A background of economic deprivation experienced by impoverished and marginal members of the community.* The very year of Neal's lynching signals deprivation: 1934, the midpoint of the Great Depression of 1929–1939. As fast as farm income fell in the collapsed cotton economy of Jackson County, bankruptcies and rural unemployment rose. Severe economic deprivation may have resulted in a general rise in tensions in the area, requiring only a spark for ignition. Moreover, blacks may have become the collective target of white frustration because fully a quarter of all black farmers in Jackson County owned their own land and received government aid from the Farm Credit Administration. In contrast, many whites were landless migrants from other states or dispossessed sharecroppers who may have resented blacks receiving federal funds (McGovern, 1982: 39–41). Often, say proponents of breakdown theory, it is not grinding poverty, or **absolute deprivation,** that generates collective action so much as **relative deprivation.** Relative deprivation refers to the growth of an intolerable gap between the social rewards people expect to receive and those they actually receive. Social rewards are widely valued goods, including money, education, security, prestige, and so forth. Accordingly, people are most likely to rebel when rising expectations (brought on by, say, rapid economic growth and migration) are met by a sudden decline in received social rewards (due to, say, economic recession or war) (Davies, 1969; Gurr, 1970). From this point of view, the rapid economic growth of the Roaring Twenties, followed by the economic collapse of 1929, would likely have caused widespread relative deprivation in Jackson County.
2. *The inherent irrationality of crowd behavior* is a second factor likely to be stressed in any pre-1970 explanation of the Neal lynching. Gustave Le Bon, an early interpreter of crowd behavior, wrote that an isolated person may be a cultivated individual. But in a crowd, the individual is transformed into a "barbarian," a "creature acting by instinct" possessing the "spontaneity, violence," and "ferocity" of "primitive beings" (Le Bon, 1969 [1895]: 28). Le Bon argued that this transformation occurs because people lose their individuality and will power when they join a crowd. Simultaneously, they gain a sense of invincible group power that derives from the crowd's sheer size. Their feeling of invincibility allows them to yield to instincts they would normally hold in check. Moreover, if people remain in a crowd long enough, they enter something like a hypnotic state. This makes them particularly open to the suggestions of manipulative leaders and ensures that extreme passions spread through the crowd like a contagious disease. (Sociologists call Le Bon's argument the **contagion** theory of crowd behavior.) For all these reasons, Le Bon held, people in crowds are often able to perform

Sociology Now™

Learn more about **Demonstrations** by going through the % of people who have taken part in a lawful demonstration Map Exercise.

Breakdown theory suggests that social movements emerge when traditional norms and patterns of social organization are disrupted.

Absolute deprivation is a condition of extreme poverty.

Relative deprivation is an intolerable gap between the social rewards people feel they deserve and the social rewards they actually receive.

Contagion is the process by which extreme passions supposedly spread rapidly through a crowd like a contagious disease.

Sociology Now™

Learn more about **Crowd Behavior** by going through the Crowd Behavior Learning Module.

extraordinary and sometimes outrageous acts. "Extraordinary" and "outrageous" are certainly appropriate terms for describing the actions of the citizens of Jackson County in 1934.

3. *The serious violation of norms* is the third factor that pre-1970s sociologists would likely have stressed in trying to account for the Neal lynching. In the 1930s, intimate contact between blacks and whites in the South was strictly forbidden. In that context, black-on-white rape and murder were not just the most serious of crimes but the deepest possible violation of the region's norms. Neal's alleged crimes were therefore bound to evoke a strong reaction on the part of the dominant race. In general, before about 1970, sociologists highlighted the breakdowns in traditional norms that preceded group unrest, sometimes referring to them as indicators of **strain** (Smelser, 1963: 47–8, 75).

Assessing Breakdown Theory

Can deprivation, contagion, and strain really explain what happened in the backwoods of Jackson County in the early hours of October 27, 1934? Can breakdown theory adequately account for collective action in general? The short answer is no. Increasingly since 1970, sociologists have uncovered flaws in all three elements of breakdown theory. They have proposed alternative frameworks for understanding collective action. To help you appreciate the need for these alternative frameworks, let us reconsider the three elements of breakdown theory in the context of the Neal lynching.

Deprivation

Research shows no clear association between fluctuations in economic well-being (as measured by, say, the price of cotton in the South) and the number of lynchings that took place each year between the 1880s and the 1930s (Mintz, 1946). Moreover, in the case of the Neal lynching, the main instigators were not especially economically deprived. The men who seized Neal from the Brewton jail were middle- to lower-middle-class farmers, merchants, salesmen, and the like. They were economically solvent, even at the height of the Great Depression, with enough money, cars, and free time to take two days off work to organize the lynching. Nor were they socially marginal "outside agitators" or rootless, recent migrants to the region. They enjoyed good reputations in their communities as solid citizens, churchgoing men with a well-developed sense of civic responsibility. Truly socially marginal individuals did not take part in the lynching at all (McGovern, 1982: 67–8, 85). This fits a general pattern. In most cases of collective action, leaders and early joiners are well-integrated members of their communities, not outsiders (Brym, 1980; Brym and Economakis, 1994; Economakis and Brym, 1995; Lipset, 1971 [1951]). Levels of deprivation, whether absolute or relative, are not commonly associated with the frequency or intensity of outbursts of collective action (McPhail, 1994).

Contagion

Despite its barbarity, the Neal lynching was not a spontaneous and unorganized affair. Sophisticated planning went into the Brewton jail raid. For example, decoying the sheriff took much cunning. Even the horrific events in front of the Cannidy home did not just erupt suddenly because of the crowd's "madness." For example, hours before Neal's body was brought to the Cannidy home, some adults got the idea of sharpening some long sticks, stacking them, and instructing children to use them to pierce the body. As this example shows, and as research on riots, crowds, and demonstrations has consistently confirmed, nonroutine collective action may be wild but it is usually structured. In the first place, nonroutine collective action is structured by ideas and norms that emerge in the crowd itself, such as the idea of preparing sharp sticks in the Neal lynching (Turner and Killian, 1987 [1957]). Second, nonroutine collective action is structured by the predispositions that unite crowd members and predate their collective action. The participants in the Neal lynching, for instance, were all predisposed to take part in it by racist attitudes. If

Strain refers to breakdowns in traditional norms that precede collective action.

they had not been similarly predisposed, they would never have assembled for the lynching in the first place (Berk, 1974; Couch, 1968; McPhail, 1991). Third, nonroutine collective action is structured by the *degree* to which different types of participants adhere to emergent and preexisting norms. Leaders, rank-and-file participants, and bystanders adhere to such norms to varying degrees (Zurcher and Snow, 1981). Fourth, preexisting social relationships among participants structure nonroutine collective action. For instance, relatives, friends, and acquaintances are more likely than strangers to cluster together and interact in crowds, riots, demonstrations, and lynchings (McPhail, 1991; McPhail and Wohlstein, 1983; Weller and Quarantelli, 1973).

Strain

The alleged rape and murder of Lola Cannidy by Claude Neal did violate the deepest norms of the Old South in a pattern that was often repeated. Thus, data exist on 4752 lynchings that took place in the United States between 1882 and 1964, when the last lynching was recorded. Three-quarters of them were white lynchings of blacks. Nearly two-thirds were motivated by alleged rapes or murders (calculated from Williams, 1970: 12–15).

However, lynching had deeper roots than the mere violation of norms governing black–white relations. Significantly, it was a means by which black farmworkers were disciplined and kept tied to the southern cotton industry after the abolition of slavery threatened to disrupt the industry's traditional, captive labor supply.

▶Figure 21.1, which contains data on the annual frequency and geographical distribution of lynching, supports this interpretation. Note first that there were more white than black lynching victims in the first half of the 1880s. That is because, originally, lynching was not just an expression of a racist system of labor control. It was also a means by which people sought quick and brutal justice in areas with little government or police, whatever the alleged criminal's race. Only in 1886 did the number of black victims exceed the number of white victims for the first time. After that, as state control over criminal justice became more widespread, the number of white lynchings continued to decline. The practice was soon used almost exclusively to control blacks.

Second, notice that nearly two-thirds of all lynchings took place in just eight contiguous southern states that formed the center of the cotton industry. Other data show that the great majority of lynchings took place in rural areas, where, of course, cotton is farmed (McAdam, 1982: 89–90). Where cotton was king, lynching was its handmaiden.

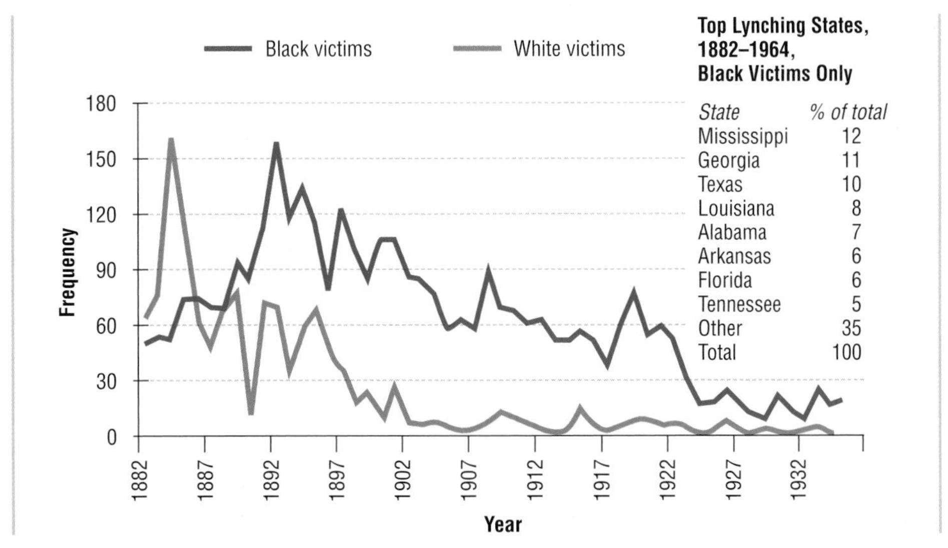

▶Figure 21.1
Frequency of Lynching, United States, 1882–1935

Source: Williams (1970: 8–11).

Third, observe how the annual number of lynchings rose when the cotton industry's labor supply was most threatened. The peak in black lynchings occurred between 1891 and 1901: an annual average of 112. These were the years when the Populist Party, a coalition of black and white farmers, threatened to radically restructure southern agriculture and eliminate many of the white plantation owners' privileges. More lynching was one reaction to this danger to the traditional organization of agricultural labor. Finally, note that lynching disappeared as a form of collective action when the cotton industry lost its economic significance and its utter dependence on dirt-cheap black labor. Specifically, the organization of the southern cotton industry began to change after 1915 due to mechanization, the mass migration of black workers to jobs in northern industry, and other factors. By 1935, when the cotton industry's economic significance had substantially declined, "only" 18 lynchings of blacks took place. After that, the figure never again reached double digits, finally dropping to zero in 1965.

We conclude that lynching was a two-sided phenomenon. Breakdown theory alerts us to one side. Lynching was partly a reaction to the violation of norms that threatened to *disorganize* traditional social life in the South. But breakdown theory deflects attention from the other side of the phenomenon. Lynching was also a form of collective action that grew out of, and was intended to maintain, the traditional *organization* of the South's cotton industry. Without that organization, there was no lynching (Tolnay and Beck, 1995; Soule, 1992).

Rumors and Riots

The study of rumor reinforces the idea that social organization underlies all collective action, even its apparently most fleeting and unstructured forms. **Rumors** are claims about the world that are not supported by authenticated information. They are a form of communication that takes place when people try to construct a meaningful interpretation of an ambiguous situation. They are often short-lived, although they may recur (Shibutani, 1966: 17).

While rumor transmission is a form of collective action in its own right, it typically intensifies just before and during riots. That is because tension and uncertainty about the near future mount at such times. In turn, increased rumor transmission often incites more rioting. Thus, rumors significantly aggravated tensions in about two-thirds of the roughly 300 race riots that rocked American cities between 1964 and 1968. Similarly, rumors inflamed the three-day riot that broke out in Los Angeles in April 1992, when four white members of the Los Angeles Police Department were acquitted of the videotaped beating of Rodney King, an unarmed black motorist. The LA riot involved about 45,000 active participants and 100,000 onlookers. Forty-five people were killed and 2400 were injured. Thirteen thousand police officers arrested 10,000 blacks and Latinos. Insured damage alone totaled $1 billion. Public fears were stoked and rioting intensified when the extent of the violence was initially exaggerated, when word spread that the authorities were using the riot as an excuse for cracking down on illegal immigrants, when firefighters were said to be saving only non-black businesses, when poor people heard rumors of looting and decided they didn't want to miss the opportunity, and when they heard the police were not responding to the looting (Fine and Turner, 2001: 29–39, 55, 58–9).

"Rumor," wrote Shakespeare in *Henry IV*, "is a pipe blown by surmises, jealousies, conjectures." More recent analyses link rumors to the primary emotions of hope, fear, and anger. *Hope* gives rise to "pipe dreams," a sort of public wish fulfillment. Rumors of the death of Black Muslim leader Louis Farrakhan might serve as a pipe dream for some white Americans. *Fear* gives rise to "bogey rumors." Claims of Farrakhan's death that might circulate in the African American community would qualify as bogey rumors. *Anger* gives rise to "wedge-driving rumors," which divide populations into antagonistic groups. The assertion that Farrakhan's death was due to his being poisoned by white doctors under government orders would constitute a wedge-driving rumor (Fine and Turner, 2001: 64; Knapp, 1944).

Rumors are claims about the world that are not supported by authenticated information. They are a form of communication that takes place when people try to construct a meaningful interpretation of an ambiguous situation.

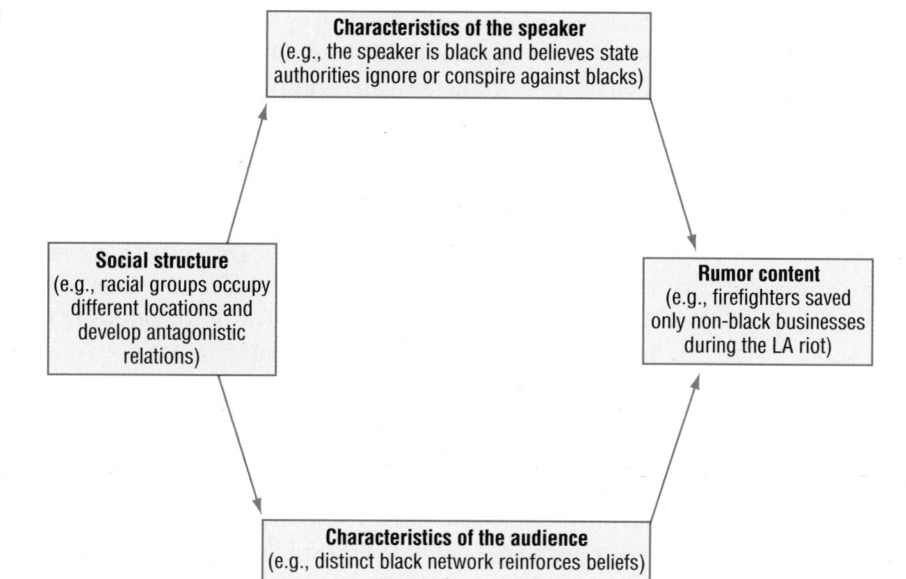

Figure 21.2
The Social Determinants of Rumors

Source: Adapted from Fine and Turner (2001: 79).

Rumors are often false or unverifiable. They seem credible to insiders but preposterous to outsiders. Yet nobody should dismiss them as frivolous. For just as x-rays can reveal flaws in the structure of a bridge, so can a sociological understanding of rumors reveal the distribution of hope, fear, and anger in society and the structural flaws that lie beneath these emotions. For example, there is no evidence to corroborate the rumor that firefighters saved only non-black businesses during the LA riot. Yet the rumor is a faithful expression of three underlying social facts (Figure 21.2). First, blacks and whites occupy different structural locations in American society and, to varying degrees, have developed antagonistic relations (Chapter 10, "Race and Ethnicity"). Second, that antagonism has given rise to the belief among some members of the African American community that white-dominated corporations and state institutions ignore or conspire against them. A third reality reinforces that view: There is little overlap between black and white networks of communication. The absence of much candid talk between the two communities reinforces the beliefs of each community about the other. In the context of these three undeniable social facts, the rumor that firefighters saved only non-black businesses during the LA riot gained credibility in the African American community, even if it was factually inaccurate.

Our discussion of lynching led us to conclude that collective action is rarely a short-term reaction to disorganization and deprivation. Instead, it is usually part of a long-term attempt to correct perceived injustice that requires a sound social-organizational basis. We now see that a second form of collective action—rumor transmission—is also rooted in the hard facts of social organization. We thus arrive at the starting point of post-1970 theories of collective action and social movements. For more than 30 years, most students of the subject have recognized that we can best understand collective action by focusing on its social-organizational roots.

Social Movements

According to breakdown theory, people usually rebel soon after social breakdown occurs. In this view, rapid urbanization, industrialization, mass migration, unemployment, and war often lead to the buildup of deprivations or the violation of important norms. Under these conditions, people soon take to the streets.

In reality, however, people often find it difficult to turn their discontent into an enduring social movement. Social movements emerge from collective action only when the

Table 21.1
Correlates of Collective Violence, France, 1830–1960

Variable	Correlation with Frequency of Collective Violence
Breakdown variables	
Number of suicides	0.00
Number of major crimes	−0.16
Deprivation variables	
Manufactured goods prices	0.05
Food prices	0.08
Value of industrial production	0.10
Real wages	0.03
Organizational variable	
Number of union members	0.40
Political process variable	
National elections	0.17
State response variable	
Days in jail	−0.22

Notes: (1) Correlations can range from −1.0 (indicating a perfect inversely proportional relationship) to 1.0 (indicating a perfect directly proportional relationship). A correlation of 0 indicates no relationship. (2) The correlation between major crimes and the rate of collective violence is negative, but it should be positive according to breakdown theory. (3) The exact years covered by each correlation vary.
Source: Adapted from Tilly, Tilly, and Tilly (1975: 81–2).

discontented succeed in building up a more or less stable membership and organizational base. Once this is accomplished, they typically move from an exclusive focus on short-lived actions such as demonstrations to more enduring and routine activities. Such activities include establishing a publicity bureau, founding a newspaper, and running for public office. These and similar endeavors require hiring personnel to work full-time on various movement activities. Thus, the creation of a movement bureaucracy takes time, energy, and money. On these grounds alone, one should not expect social breakdown quickly to result in the formation of a social movement.

Solidarity Theory: A Variant of Conflict Theory

Research conducted since 1970 shows that, in fact, social breakdown often does not have the expected short-term effect. That is because several social-structural factors modify the effects of social breakdown on collective action. For example, Charles Tilly and his associates studied collective action in France, Italy, and Germany in the 19th and 20th centuries (Lodhi and Tilly, 1973; Snyder and Tilly, 1972; Tilly, 1979a; Tilly, Tilly, and Tilly, 1975). They systematically read newspapers, government reports, and other sources so they could analyze a representative sample of strikes, demonstrations, and acts of collective violence. (They defined acts of collective violence as events in which groups of people seized or damaged persons or property.) They measured social breakdown by collecting data on rates of urban growth, suicide, major crime, prices, wages, and the value of industrial production. Breakdown theory would be supported if they found that levels of social breakdown rose and fell with rates of collective action. But they did not. For example, as the top panel of Table 21.1 shows for France, nearly all the correlations between collective violence and indicators of breakdown are close to zero. This means that acts of collective violence did not increase in the wake of mounting social breakdown, nor did they decrease in periods marked by less breakdown.

Significantly, however, Tilly and his associates found stronger correlations between collective violence and some other variables. You will find them in the bottom panel of Table 21.1. These correlations hint at the three fundamental lessons of the **solidarity theory** of social movements, a variant of conflict theory (Chapter 1, "A Sociological Compass") and the most influential approach to the subject since the 1970s.

Sociology Now™

Learn more about **Unions** by going through the % of workers that are unionized Map Exercise.

ry suggests
nents are
that
ial mem-
mov rces,
increasin iti-
terial, and oth
movement memb

Resource Mobilization

Inspecting Table 21.1, we first observe that collective violence in France increased when the number of union members rose, and such violence decreased when the number of union members fell. Why? Because union organization gave workers more power, which increased their capacity to pursue their aims—if necessary, by demonstrating, striking, and engaging in collective violence. We can generalize from the French case as follows. Most collective action is part of a power struggle. The struggle usually intensifies as disadvantaged groups become more powerful relative to privileged groups. How do disadvantaged groups become more powerful? By gaining new members, getting better organized, and increasing their access to scarce resources, such as money, jobs, and means of communication (Bierstedt, 1974). French unionization is thus only one example of **source mobilization,** a process by which groups engage in more collective action as power increases due to their growing size and increasing organizational, material, er resources (Gamson, 1975; Jenkins, 1983; McCarthy and Zald, 1977; Oberschall, , 1978; Zald and McCarthy, 1979).

Political Opportunities

Table 21.1 also shows that there was somewhat more collective violence in France when national elections were held. Again, why? Because elections gave people new political opportunities to protest. In fact, election campaigns often serve as invitations to engage in collective action by providing a focus for discontent and a chance to put new representatives with new policies in positions of authority. When else do new political opportunities open up for the discontented? Chances for protest also emerge when influential allies offer support, when ruling political alignments become unstable, and when elite groups get divided and come into conflict with one another (Tarrow, 1994: 86–9; Useem, 1998). Said differently, collective action takes place and social movements crystallize not just when disadvantaged groups become more powerful but when privileged groups and the institutions they control get divided and therefore become weaker. As Harvard economist John Kenneth Galbraith once said about the weakness of the Russian ruling class at the time of the 1917 revolution, if someone manages to kick in a rotting door, some credit has to be given to the door. In short, this second important insight of solidarity theory links the timing of collective action and social movement formation to the emergence of new **political opportunities** (McAdam, 1982; Piven and Cloward, 1977; Tarrow, 1994).

Social Control

The third main lesson of solidarity theory is that government reactions to protest influence subsequent protest (Box 21.1). Specifically, governments can try to lower the frequency and intensity of protest by taking various **social-control** measures (Oberschall, 1973: 242–83). These measures include making concessions to protesters, co-opting the most troublesome leaders (for example, by appointing them advisers), and violently repressing collective action. The last point explains the modest correlation in Table 21.1 between frequency of collective violence and governments throwing more people into jail for longer periods of time—for in France, more violent protest often resulted in more state repression. However, the correlation is modest because social-control measures do not always have the desired effect. For instance, if grievances are very deeply felt, and yielding to protesters' demands greatly increases their hopes, resources, and political opportunities, government concessions may encourage protesters to press their claims further. And although the firm and decisive use of force usually stops protest, using force moderately or inconsistently often backfires. That is because unrest typically intensifies when protesters are led to believe that the government is weak or indecisive (Piven and Cloward, 1977: 27–36; Tilly, Tilly, and Tilly, 1975: 244).

Discussions of strain, deprivation, and contagion dominated analyses of collective action and social movements before 1970. Afterward, analyses of resource mobilization, political opportunities, and social control dominated the field. Let us now make the new ideas more concrete. We do so by analyzing the ups and downs of one of the most important social movements in the 20th-century United States, the union movement, and its major weapon, the strike.

Case Study: Strikes and the Union Movement in the United States

Workers have traditionally drawn three weapons from the arsenal of collective action to advance their interests: unions, political parties, and strikes. Unions enable groups of workers to speak with one voice and thus to bargain more effectively with their employers for better wages, working conditions, and benefits. The union movement brought us many things we take for granted today, such as the eight-hour day, two-day weekends, health insurance, and pensions. In most of the advanced industrial democracies (although, as we saw in Chapter 14, not in the United States), workers have also created and supported labor or socialist parties. Their hope has been that by gaining political influence, they can get laws passed that favor their interests. Finally, when negotiation and po-

Political opportunities for collective action and social-movement growth occur during election campaigns, when influential allies offer insurgents support, when ruling political alignments become unstable, and when elite groups become divided and conflict with one another.

Social control refers to the means by which authorities seek to contain collective action, including co-optation, concessions, and coercion.

BOX 21.1

SOCIAL POLICY: WHAT DO YOU THINK?

Government Surveillance of Social Movements

A public policy debate has emerged over whether the U.S. government should have a free hand to spy on the activities of social movements. The debate was provoked by Osama bin Laden's terrorist network, al Qaeda. Al Qaeda originated in Afghanistan in the 1980s, where it helped radical anti-Western Muslim fundamentalists wage a successful war against the Soviet Union ("Hunting bin Laden," 1999). By 2000, bin Laden had operatives in 60 countries. He often used a satellite phone to communicate with them—until U.S. law enforcement officials revealed they were tapping his calls. Once he learned of these taps, bin Laden increased his use of another, more effective means of communication: sending messages that are easily encrypted but difficult to decode via the Internet (Kelley, 2001; McCullagh, 2000a). Such messages may have been used to help plan and coordinate the complex, almost simultaneous jet hijackings that resulted in the crash of an airliner in Pennsylvania and the destruction of the World Trade Center and part of the Pentagon on September 11, 2001.

Hiding messages in innocent-looking packages is an old practice made easier by computers (Johnson and Jajodia, 1998). For example, using steganography programs freely available on the Web, one can hide messages inside photographs or MP3 files and then place the files on a publicly accessible website, where they can be downloaded by operatives ("Welcome," 2001; "MP3stego," 2001).* A digitized photograph or an MP3 file is made up of many zeros and ones. Steganography programs insert a small number of additional zeros and ones into such files. The added zeros and ones are invisible when the photograph is seen and inaudible when the MP3 file is heard. When decoded, however, they form a message. Such messages have been found in files posted in sports chat rooms and pornography sites, for example.

Neville Elder/Corbis

The remains of the twin towers of the Word Trade Center, destroyed by al Qaeda on September 11, 2001.

From the point of view of law enforcement officials, the problem is to figure out how to decode the messages. Supercomputers can be used, but they can take months to decode a single message. That is why some people think we need more government regulation, such as laws requiring that all steganography and other encryption programs be built with a "backdoor," or an encryption key that would allow officials to read coded messages quickly and easily.

Right-wing free-market organizations oppose this idea. They are wary of giving the government more power to invade people's privacy (McCullagh, 2000b). Opposition may be found on the left, too, because many liberals know that the government has a history of surveillance of popular social movements. For example, the Civil Rights movement, led by Dr. Martin Luther King, Jr., and others, was under constant surveillance by the FBI and other law enforcement agencies. Many people now decry the government's attempt to control and even suppress the Civil Rights movement. They worry that government access to backdoors and encryption keys will only enhance its ability to spy on popular American movements and suppress them.

Here, then, we face one of democracy's great dilemmas: Democracy empowers both its citizens and its enemies. So should the government be allowed to increase its surveillance of social movements? Increasing government surveillance in general is likely to harm both the enemies of democracy and its champions. But failing to increase government surveillance is likely to help democracy's enemies. It may be that the only way out of this dilemma is to allow increased surveillance only of those movements that most Americans consider to be dangerous to society. Identifying those movements will require vigorous political debate and tough political decisions.

**Steganography* comes from the Greek for "hidden writing."

litical influence get them nowhere, workers have tried to extract concessions from employers by withholding their labor. That is, they have gone on strike:

May 22, Minneapolis

A crowd of 20,000 striking workers clashes with police, who kill 2 workers and wound 50. The workers virtually control the city. They are supporting truck drivers in the coal yards, whose employers have rejected their attempts to form a union. In their next confrontation, the police kill 2 workers and wound 67. The governor declares martial law. The employers finally accept a federal plan that leads to collective bargaining agreements with 500 Minneapolis employers.

May 23, Toledo

Ten thousand militant workers assemble outside the Electric Auto-Lite factory, imprisoning 1500 strikebreakers inside the plant. The sheriff orders his deputies to attack. The crowd fights back and several people are seriously injured. The authorities call in the Ohio National Guard, armed with machine guns and bayoneted rifles. The Guard fires into the crowd. They kill two people and wound many more, but the crowd refuses to disperse. Four more companies of Guardsmen are called up. With workers now threatening to shut down the entire city, Auto-Lite finally agrees to get rid of the strikebreakers and engage in federal mediation. The strikers win a 22 percent wage increase and limited union recognition.

July 2, San Francisco

Seven hundred police officers storm dockworkers who have been on strike for 45 days. Twenty-five people are hospitalized. Two days later, the police charge again, hospitalizing 155 people and killing 2. The National Guard is called in to restore order. An eyewitness describes the funeral procession for the slain strikers as follows:

> *In solid ranks, eight to ten abreast, thousands of strike sympathizers. . . . Tramp-tramp-tramp. No noise except that. The band with its muffled drums and somber music. . . . On*

AP/Wide World Photos

North Carolina mill workers on strike, 1934.

the marchers came—hour after hour—ten, twenty, thirty thousand of them. . . . A solid river of men and women who believed they had a grievance and who were expressing their resentments in this gigantic demonstration (quoted in Piven and Cloward, 1977: 125).

September 1, nationwide

More than 375,000 textile workers are on strike. Employers hire armed guards who, with the National Guard, keep the mills open in Alabama, Mississippi, Georgia, and the Carolinas. The governor of Georgia declares martial law and sets up a detention camp for 2000 strikers. Six strikers are killed in clashes with police in South Carolina. Another 9 are killed elsewhere in the country. This brings the annual total of slain strikers to at least 40. Riots break out in Rhode Island, Connecticut, and Massachusetts, and National Guardsmen are on duty throughout New England.

It was 1934, one the bloodiest years of collective violence in American history. What spurred the mass insurgency? As we might suspect from our knowledge of resource mobilization theory, an important underlying cause was the rapid growth of the industrial working class over the preceding half century. By 1920, industrial workers made up 40 percent of the American labor force and were central to the operation of the economy. As a result, strikes were never more threatening to political and industrial leaders. On the other hand, workers were not well organized. Until 1933, they did not have the right to bargain collectively with their employers, so fewer than 12 percent of America's nonfarm workers were union members. And they were anything but well-to-do. The economic collapse that began in 1929 brought unemployment to a full third of the workforce and severely depressed the wages of those lucky enough to have jobs.

More than their ability to mobilize organizational and material resources, it was a new law that galvanized industrial workers by opening vast political and economic opportunities for them. In 1932, a nation in despair swept Franklin Delano Roosevelt into the White House. To end the Great Depression, Roosevelt forged his New Deal legislation, one of the earliest pieces of which was the 1933 National Industrial Recovery Act (NIRA). Section 7(a) of the NIRA specified workers' minimum wages and maximum hours of work. It also gave them the right to form unions and bargain collectively with their employers. Not surprisingly, industrial workers hailed the NIRA as a historic breakthrough. But employers challenged the law in the courts. And so the seesaw was set in motion. First the Supreme Court invalidated the NIRA. Then Congress reinstated the basic terms of the NIRA by passing the Wagner Act in 1935. Then the Wagner Act was largely ignored in practice. Finally, in 1937, the Supreme Court ruled the Wagner Act constitutional. In the interim, from 1933 to 1937, the promise of the new pro-union laws gave industrial workers new hope and determination. With the workers thus armed, open class war rocked America.

From 1933 until the end of World War II, many millions of American workers joined unions. In 1945, unionization reached its historic peak (Figure 21.3). In that year, 35.5 percent of nonfarm employees were union members. Then the figure began to drop. **Union density** (union members as a percentage of nonfarm workers) remained above 30 percent until the early 1960s. By 2004, it stood at a mere 12.5 percent. Today, the United States has the lowest union density of any rich industrialized country. What accounts for the post-1945 drop? Focusing on resource mobilization and political opportunities takes us a long way toward answering this question.

Strikes and Resource Mobilization

Union density is the number of union members in a given location and time as a percentage of nonfarm workers. It measures the organizational power of unions.

The post-1945 drop in union density is partly a result of changes in America's occupational structure. The industrial working class has shrunk and therefore become weaker (Troy, 1986). In 1900, there was roughly one blue-collar (nonagricultural, manual) job in America for every white-collar (service) job. By 1970, the ratio of blue-collar to white-collar jobs dropped to about 1:2. By 2000, it fell to about 1:3. These figures show that blue-

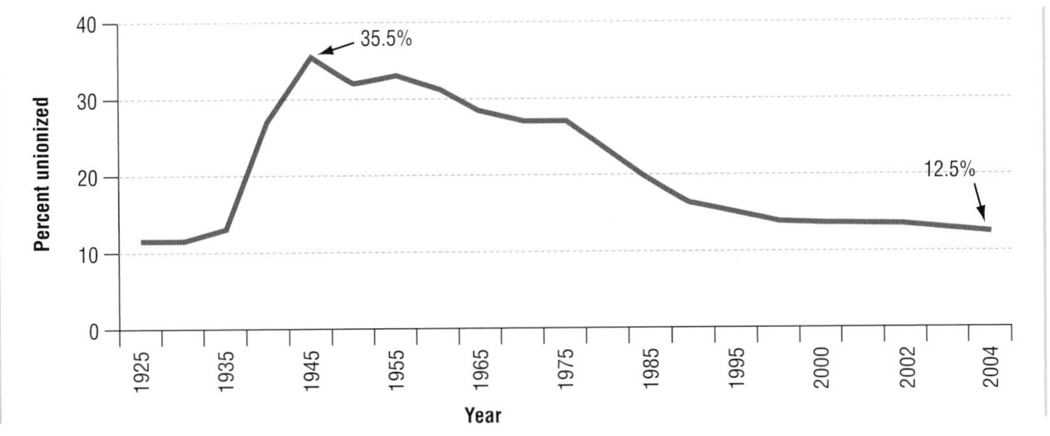

Figure 21.3
Unionization as Percent of Nonfarm Workers, United States, 1925–2004

Source: U.S. Department of Labor (2000b, 2000c, 2001, 2003g, 2005c).

collar workers are an increasingly rare species. Yet it is precisely among blue-collar workers that unionism is strongest. True, unionization has increased among government workers. Since the early 1960s, they have enjoyed limited union rights. But this gain has not offset losses due to the decline in the size of the industrial working class. Meanwhile, unions have scarcely penetrated the rapidly growing ranks of white-collar workers in the private sector. Usually better educated and higher paid and with more prestige attached to their occupations compared with blue-collar workers, American private-sector white-collar workers have traditionally resisted unionization.

The industrial working class has also been weakened by globalization and employer hostility to unions. As we saw in Chapter 13 ("Work and the Economy"), the globalization of production that began in the 1970s put American blue-collar workers in direct competition for jobs with overseas workers. Employers could now close American factories and relocate them in Mexico, China, and other countries unless American workers were willing to work for lower wages, fewer benefits, and less job security. American plant closings became increasingly common in the 1970s and 1980s, and unions were often forced to make concessions on wages and benefits. Growing ineffectiveness weakened unions and made them less popular among some workers. In addition, beginning in the 1970s, many American employers began to contest unionization elections legally. They also hired consulting firms in anti-union "information" campaigns aimed at keeping their workplaces union free. In some cases, they used outright intimidation to prevent workers from unionizing. Thus, a decline in organizational resources available to industrial workers was matched by an increase in anti-union resources mobilized by employers (Clawson and Clawson, 1999: 97–103).

An abandoned factory in East St. Louis. The globalization of production that began in the 1970s put American blue-collar workers in direct competition for jobs with overseas workers. Employers could now close American factories and relocate them in Mexico, China, and other countries unless American workers were willing to work for lower wages, fewer benefits, and less job security. American plant closings became increasingly common in the 1970s and 1980s, and unions were often forced to make concessions on wages and benefits.

Joseph Sohm, ChromoSohm, Inc./Corbis

Strikes and Political Opportunities

In addition to the erosion of the union movement's mass base, government action has erected barriers to union growth since the end of World War II. This was evident as early as 1947, when Congress passed the Taft-Hartley Act in reaction to a massive post–World War II strike wave. Unions were no longer allowed to force employees to become members or to require union membership as a condition

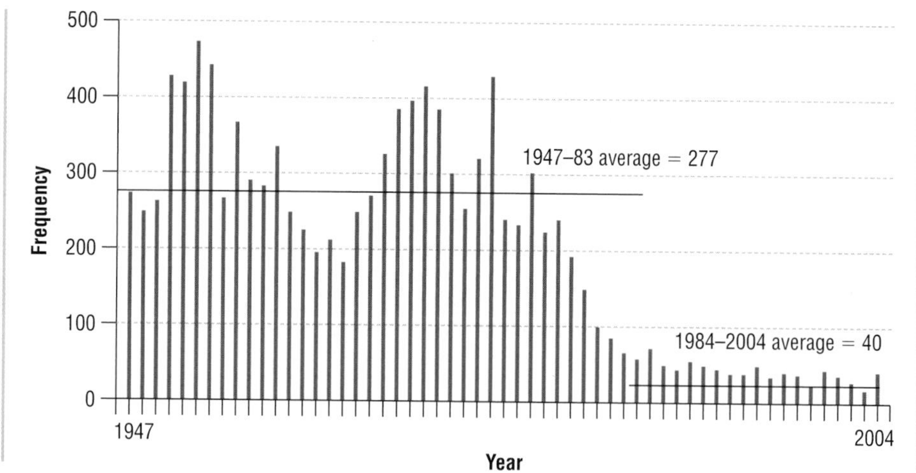

Figure 21.4
Frequency of Strikes with 1000+ Workers, United States, 1947–2004

Source: U.S. Department of Labor (2005b).

of being hired. The Taft-Hartley Act also allowed employers to replace striking workers. Unions thus became less effective—and therefore less popular—as vehicles for achieving workers' aims. Taft-Hartley remains the basic framework for industrial relations in America.

Resource mobilization theory teaches us that social organization usually facilitates collective action. The opposite also holds. Less social organization typically means less protest. We can see this by examining the frequency of strikes over time. Comparing historical periods, we see that unusually low union density has helped to virtually extinguish the strike as a form of collective action in the United States.[2] This is apparent from Figure 21.4, which shows the annual number of strikes involving 1000 or more workers from 1947 to 2004. Between 1947 and 1983, an annual average of 277 big strikes took place. In contrast, between 1984 and 2004, there was an annual average of only 40 big strikes. This indicates a major historical shift.

Over the short term, strikes have usually been more frequent during economic booms and less frequent during economic busts (Kaufman, 1982). That is the main reason we see year-to-year fluctuations in strike frequency in Figure 21.4. With more money, more job opportunities, and bigger strike funds in good times, workers can better afford to go out on strike to press their claims than during periods of high unemployment.

However, Figure 21.5 shows that the relationship between unemployment and strike frequency changed after 1983. Thus, the dots representing the years 1948–1983 slope downward. This means that whenever unemployment increased, big strikes were less common. In contrast, for the period 1984–2002, a flat line replaces the downward slope. This means that even in good times, workers avoided striking. It seems that, due partly to weak unions, many workers are now unable to use the strike weapon as a means of improving their wages and benefits (Brym, 2003; Cramton and Tracy, 1998; Schor and Bowles, 1987).[3]

Recent Developments in the Union Movement

Since the mid-1990s, the American Federation of Labor–Congress of Industrial Organizations (AFL-CIO), the largest union umbrella organization in the United States, has sought to reverse the trends in unionization described earlier. Specifically, it has tried

[2]Note, however, that strike activity also tends to be low in countries with *high* levels of unionization. Sweden, for example, has the world's highest union density. Its strike rate is low, however, because workers and their representatives have been involved in government policymaking since World War II. Decisions about wages and benefits tend to be made in negotiations among unions, employer associations, and governments rather than on the picket line. We conclude that strike activity is highest in countries with intermediate levels of unionization.

[3]Another important factor making strikes less sensitive to the business cycle since the early 1980s is that they have become riskier from the worker's viewpoint because of (1) the willingness of employers to fire strikers and replace them with other workers (though outlawed in much of Western Europe and some Canadian provinces, using replacement workers was legalized in 1938 by the Supreme Court and became widespread after President Reagan fired the nation's striking air traffic controllers in 1980) and (2) cuts in income-replacing social welfare benefits.

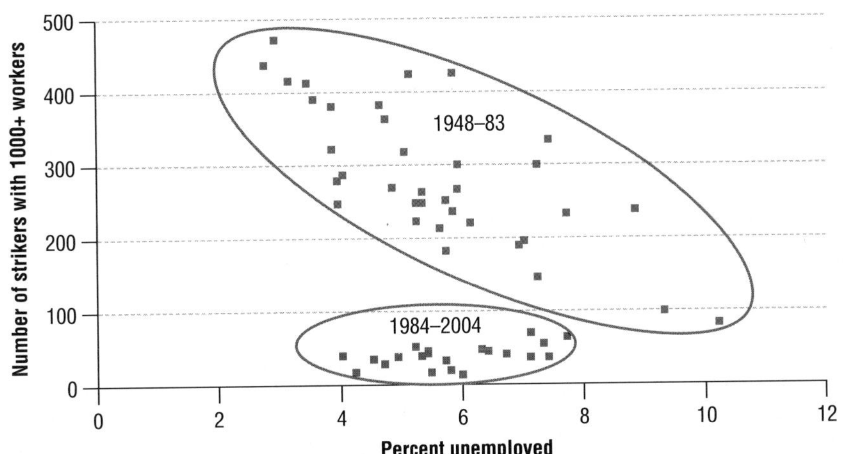

Figure 21.5
Unemployment and the Frequency of Big Strikes, United States, 1948–2004

Source: United States Department of Labor (2004, 2005a, 2005c).

to organize immigrants, introduce more feminist issues into its program in a bid to attract more women, and develop new forms of employee organization and representation that are more appropriate to a postindustrial society. The latter include coalitions with other social movements, and councils that bring together all the unions in a city or other geographical area (Clawson and Clawson, 1999: 112–15). Whether these strategies will succeed in revitalizing the American union movement is unclear. Figures on union density as of 2004 suggest that they have not yet reversed the downward slide of the union movement (see Figure 21.3).

Framing Discontent: The Contribution of Symbolic Interactionism

As we have seen, solidarity theory helps to explain the emergence of many social movements. Still, the rise of a social movement sometimes takes strict solidarity theorists by surprise. So does the failure of an aggrieved group to press its claims by means of collective action. It seems, therefore, that something lies between (a) the capacity of disadvantaged people to mobilize resources for collective action and (b) the recruitment of a substantial number of movement members. That something is **frame alignment** (Benford, 1997; Goffman, 1974; Snow, Rochford Jr., Worden, and Benford, 1986; Valocchi, 1996). Frame alignment is the process by which individual interests, beliefs, and values either become congruent with the activities, ideas, and goals of the movement. Thanks to the efforts of scholars operating mainly in the symbolic interactionist tradition (Chapter 1, "A Sociological Compass"), frame alignment has recently become the subject of sustained sociological investigation.

Examples of Frame Alignment

Frame alignment can be encouraged in several ways. For example, social-movement leaders can reach out to other organizations that, they believe, contain people who may be sympathetic to their movement's cause. Thus, leaders of an antinuclear movement may use the mass media, telephone campaigns, and direct mail to appeal to feminist, antiracist, and environmental organizations. In doing so, they assume that these organizations are likely to have members who would agree at least in general terms with the antinuclear platform.

Frame alignment is the process by which individual interests, beliefs, and values become congruent and complementary with the activities, goals, and ideology of a social movement.

Movement activists can also stress popular values that have so far not featured prominently in the thinking of potential recruits. They can also elevate the importance of positive beliefs about the movement and what it stands for. For instance, in trying to win new recruits, movement members might emphasize the seriousness of the social movement's

Hintz Ditz/Corbis

The "Human Rights Now" tour, Los Angeles, 1988. *Left to right:* Peter Gabriel, Tracy Chapman, Youssou N'Dour, Sting, Joan Baez, and Bruce Springsteen. When bands play at protest rallies or festivals, it is not just for entertainment and not just because the music is relevant to a social movement's goals. The bands also attract nonmembers to the movement. This is one way of framing a social movement's goals to make them appealing to nonmembers.

purpose. They might analyze the causes of the problem the movement is trying solve in a clear and convincing way. Or they might stress the likelihood of the movement's success. By doing so, they can increase the movement's appeal to potential recruits and perhaps win them over to the cause.

Finally, social movements can stretch their objectives and activities to win recruits who are not initially sympathetic to the movement's original aims. This may involve a "watering down" of the movement's ideals. Alternatively, movement leaders may decide to take action calculated to appeal to nonsympathizers on grounds that have little or nothing to do with the movement's purpose. When rock, punk, or reggae bands play at nuclear disarmament rallies or gay liberation festivals, it is not necessarily because the music is relevant to the movement's goals. Nor do bands play just because movement members want to be entertained. The purpose is also to attract nonmembers. Once attracted by the music, nonmembers may make friends and acquaintances in the movement and then be encouraged to attend a more serious-minded meeting.

In short, there are many ways that social movements can make their ideas more appealing to more people. All of them involve increasing the alignment between the way movement members and potential recruits frame issues (Box 21.2).

An Application of Frame Alignment Theory: Back to 1968

Frame alignment theory stresses the face-to-face interaction strategies employed by movement members to recruit nonmembers who are like-minded, apathetic, or even initially opposed to the movement's goals. Resource mobilization theory focuses on the broad social-structural conditions that facilitate the emergence of social movements. One theory usefully supplements the other.

The two theories certainly help clarify the 1968 high school incident described at the beginning of this chapter. In light of our discussion, it seems evident that two main factors prevented Robert Brym from influencing his classmates when he spoke to them about the dangers of industrial pollution from the local pulp and paper mill.

First, he lived in a poor and relatively unindustrialized region of Canada where people had few resources they could mobilize on their own behalf. Per capita income and the level of unionization were among the lowest of any state or province in North America.

BOX 21.2
Sociology at the Movies

The Day after Tomorrow (2004)

Most summers, Hollywood releases a disaster movie in which a highly implausible catastrophe serves as the backdrop for heroism and hope. Audiences return home momentarily frightened but ultimately safe in the knowledge that the chance of any such cataclysm is vanishingly remote.

The Day after Tomorrow follows the usual script. The movie opens with a sequence of bizarre meteorological events. A section of ice the size of Rhode Island breaks off the Antarctic ice cap. Snow falls in New Delhi. Hail the size of grapefruits pounds Tokyo. Enter Jack Hall (Dennis Quaid), a scientist whose research suggests an explanation: Sudden climate change is a very real possibility. The idea becomes a political football when it is ridiculed by the vice president of the United States, but once torrential rains and a tidal wave flood New York City, Hall's theories are vindicated. In a matter of days, temperatures plummet—at one point, they fall 10 degrees a minute to 150 below the freezing point. The entire Northern Hemisphere is plunged into a new ice age. Almost everyone freezes to death in the northern United States, while millions of desperate southerners flee into Mexico.

Although is seems like standard fare, *The Day after Tomorrow* is a Hollywood disaster movie with a difference, for it is based on a three-part idea with considerable scientific support. Part 1 is agreed upon by everyone: Since the Industrial Revolution, humans have released increasing quantities of carbon dioxide into the atmosphere as we burn more and more fossil fuels to operate our cars, furnaces, and factories. Part 2 is agreed upon by the overwhelming majority of—but not all—scientists: The accumulation of carbon dioxide allows more solar radiation to enter the atmosphere and less heat to escape. This contributes to global warming. As temperatures rise, more water evaporates and the polar ice caps begin to melt. This causes more rainfall, bigger storms, and more flooding. Part 3 is the most recent and controversial part of the argument: The melting of the polar ice caps may be adding enough freshwater to the oceans to disrupt the flow of the Gulf Stream, the ocean current that carries warm water up the east coast of North America and the west coast of Europe. Computer simulations suggest that decreased salinity could push the Gulf Stream southward, causing average winter temperatures to drop by 10 degrees Fahrenheit in the northeastern United States and other parts of the Northern Hemisphere. A recent Pentagon study suggests that such a climate change could cause droughts, storms, flooding, border raids, large-scale illegal migration from poor regions, and even war between nuclear powers over scarce food, drinking water, and energy (Joyce and Keigwin, 2004; Stipp, 2004). *The Day after Tomorrow* greatly exaggerates the suddenness and magnitude of what scientists mean by abrupt climate change. "Abrupt" can mean centuries to climatologists, and temperature drops of 10 degrees a minute are pure fantasy. Still, at the movie's core lies an ominous and real possibility.

The Day after Tomorrow also teaches us an important sociological lesson about the framing of issues by social movements and their opponents. Environmental problems do not become social issues spontaneously. They are socially constructed in what might be called a "framing war." Just as Jack Hall and the vice president spar over the credibility of Hall's prediction of sudden climate change, so do groups with different interests dispute all environmental problems, framing them in different ways so as to win over public opinion.

The Day after Tomorrow became involved in the framing war because in the month leading up to the release of the movie, environmentalists started piggybacking their message on it. They bombarded journalists with e-mails explaining global warming and offering interviews with leading scientists on the subject. Newspapers and magazines around the world subsequently carried stories on the issue. Environmentalists then distributed flyers to moviegoers leaving theaters (Houpt, 2004). In this way, *The Day after Tomorrow* became not just another disaster movie but part of the framing war around one of the major environmental issues of the day.

20th Century Fox/The Kobal Collection

The Day after Tomorrow (2004).

The unemployment rate was among the highest. In contrast, K. C. Irving, who owned the pulp and paper mill, was so powerful that most people in the region could not even conceive the need to rebel against the conditions of life that he created for them. He owned most of the industrial establishments in the province. Every daily newspaper, most of the weeklies, all of the TV stations, and most of the radio stations were his too. Little wonder one rarely heard a critical word about his operations. Many people believed that Irving could make or break local governments single-handed. Should one therefore be surprised that mere high school students refused to take him on? In their reluctance, Robert's fellow students were only mimicking their parents, who, on the whole, were as powerless as Irving was mighty (Brym, 1979).

Second, many of Robert's classmates did not share his sense of injustice. Most of them regarded Irving as the great provider. They thought his pulp and paper mill, as well as his myriad other industrial establishments, gave many people jobs. They regarded that fact as more important for their lives and the lives of their families than the pollution problem Robert raised. Frame alignment theory suggests that Robert needed to figure out ways of building bridges between their understanding and his. He did not. Therefore, he received an unsympathetic hearing.

Where Do You Fit In?

Try applying solidarity and frame alignment theories to times when *you* felt a deep sense of injustice against an institution such as a school, an organization, a company, or a government. Did you do anything about your upset? If not, why not? If so, what did you do? Why were you able to act in the way you did? Did you try to get other people to join you in your action? If not, why not? If so, how did you manage to recruit them? Did you reach the goal you set out to achieve? If not, why not? If so, what enabled you to succeed? If you've never been involved in a collective action to correct a perceived injustice, try analyzing a movie about collective action using insights gleaned from solidarity and frame alignment theories. A classic is *Norma Rae* (1979), starring Sally Field. Field won the Best Actress Oscar for her performance in this film as a southern textile worker who joins with a labor organizer to unionize her mill.

Social Movements from the 18th to the 21st Century

Attempts to synthesize theories of collective action and social movement formation are in their infancy (Diani, 1996; McAdam, McCarthy, and Zald, 1996; Tarrow, 1994). Still, we can briefly summarize what we have learned about the causes of collective action and social movement formation with the aid of Figure 21.6. Breakdown theory partly answers the question of *why* discontent is sometimes expressed collectively and in nonroutine ways. Industrialization, urbanization, mass migration, economic slowdown, and other social changes often cause dislocations that engender feelings of strain, deprivation, and injustice. Solidarity theory focuses on *how* these social changes may eventually facilitate the emergence of social movements. They cause a reorganization of social relations, shifting the balance of power between disadvantaged and privileged groups. Solidarity theory also speaks to the question of *when* collective action erupts and social movements emerge. The opening and closing of political opportunities, as well as the exercise of social control by authorities, helps to shape the timing of collective action. Finally, by analyzing the day-to-day strategies employed to recruit nonmembers, frame alignment theory directs our attention to the question of *who* is recruited to social movements. Altogether, then, the theories we have considered provide a comprehensive picture of the why, how, when, and who of collective action and social movements.

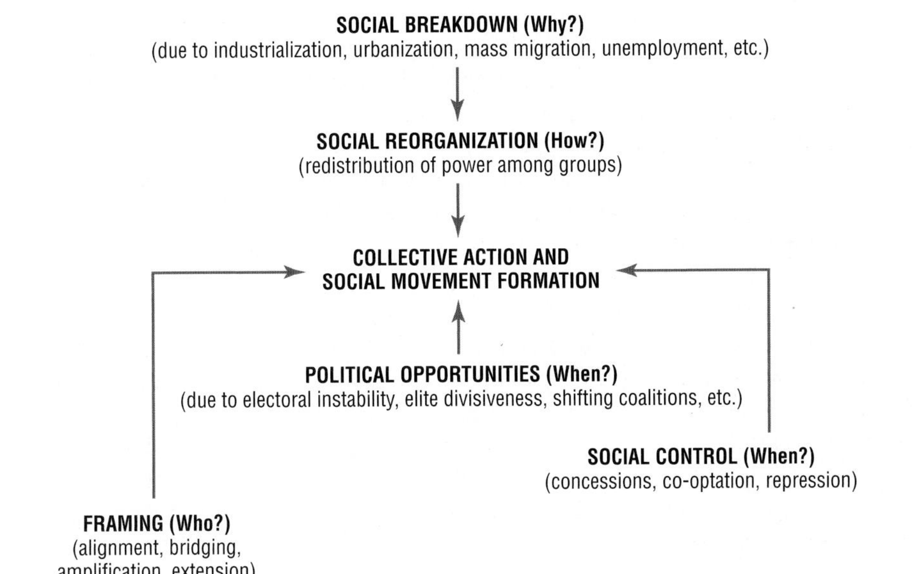

Figure 21.6
Determinants of Collective Action and Social-Movement Formation

Bearing this summary in mind, we can now turn to this chapter's final goal: sketching the historical development and future prospects of social movements in broad, rapid strokes. We begin three centuries ago.

The History of Social Movements

In 1700, social movements were typically small, localized, and violent. In Europe, poor residents of a city might riot against public officials in reaction to a rise in bread prices or taxes. Peasants on an estate might burn their landowner's barns (or their landowner) in response to his demand for a larger share of the crop. However, as the state grew, the form of protest changed. The state started taxing nearly all its citizens at higher and higher rates as government services expanded. It imposed a uniform language and often a common curriculum in a compulsory education system. It drafted most young men for army service. It instilled in its citizens all the ideological trappings of modern nationalism, from anthems to flags to historical myths.

As the state came to encompass most aspects of life, social movements changed in three ways. First, they became national in scope. That is, they typically directed themselves against central governments rather than local targets. Second, their membership grew. That was partly because potential recruits were now literate and could communicate using the printed word. In addition, big new social settings—factories, offices, densely populated urban neighborhoods—could serve as recruitment bases. Third, social movements became less violent. That is, their size and organization often allowed them to bureaucratize, stabilize, and become sufficiently powerful to get their way without frequent resort to extreme measures (Tilly, 1978; 1979a; 1979b; Tilly, Tilly, and Tilly, 1975).

Social movements often used their power to expand the rights of citizens. We may identify four stages in this process, focusing on Britain and the United States. In Britain, rich property owners fought against the king in the 18th century for **civil citizenship.** Civil citizenship is the right to free speech, freedom of religion, and justice before the law. The male middle class and the more prosperous stratum of the working class fought against rich property owners in the 19th century for **political citizenship.** Political citizenship is the right to vote and run for office. In early-20th-century Britain, women and poorer workers succeeded in achieving these same rights despite the opposition of well-to-do men in particular. During the remainder of the century, blue- and white-collar workers fought against the well-to-do for **social citizenship.** Social citizenship is the right

Civil citizenship recognizes the right to free speech, freedom of religion, and justice before the law.

Political citizenship recognizes the right to run for office and vote.

Social citizenship recognizes the right to a certain level of economic welfare security and full participation in the social life of the country.

Bibliotheque Nationale de France

In medieval Europe, social movements were small, localized, and violent. For example, a medieval French historian reported that in 1358, "there were very strange and terrible happenings in several parts of the kingdom. . . . They began when some of the men from the country towns came together in the Beauvais region. They had no leaders and at first they numbered scarcely a hundred. One of them got up and said that the nobility of France . . . were disgracing and betraying the realm, and that it would be a good thing if they were all destroyed. At this they all shouted: 'He's right! He's right! Shame on any man who saves the gentry from being wiped out!' They banded together and went off, without further deliberation and unarmed except for pikes and knives, to the house of a knight who lived near by. They broke in and killed the knight, with his lady and children, big and small, and set fire to the house" (Froissart, 1968 [c. 1365]: 151).

AP/Wide World Photos

The 15th Amendment to the Constitution gave African Americans the right to vote in 1870. However, most of them were unable to exercise that right, at least in the South, from the late 19th century until the 1960s because of poll taxes and literacy tests. The Civil Rights movement of the 1960s was in part a struggle over that issue. Some African Americans claim that they were effectively denied the right to vote in the 2000 presidential election.

Web Research Projects: Ballots or Bricks? The Role of Violence in African American Politics

to a certain level of economic security and full participation in social life with the help of the modern welfare state (Marshall, 1965).

The timing of the struggle for citizenship rights was different in the United States. In particular, universal suffrage for white males was won earlier in the 19th century than in Europe. This accounts in part for the greater radicalism of the European working class. It had to engage in a long and bitter struggle for the right to vote while its United States counterpart was already incorporated into the political system (Lipset, 1977). Another important distinguishing feature of the United States concerns African Americans. The 15th Amendment to the Constitution gave them the right to vote in 1870. However, most of them were unable to exercise that right, at least in the South, from the late 19th century until the 1960s. That was because of various restrictions on voter registration, including poll taxes and literacy tests. The Civil Rights movement of the 1960s was in part a struggle over this issue. It helped to create a community that is more politically radical than its white counterpart.

The Future of Social Movements

Partly because the success of the Civil Rights movement inspired them, so-called **new social movements** emerged in the 1970s (Melucci, 1980; 1995). What is new about new social movements is the breadth of their goals, the kinds of people they attract, and their potential for globalization. Let us consider each of these issues in turn.

Goals

Some new social movements promote the rights not of specific groups but of humanity as a whole to peace, security, and a clean environment. Such movements include the peace movement, the human rights movement, and the environmental movement. Other new social movements, such as the women's movement and the gay rights movement, promote the rights of particular groups that have been excluded from full social participation. Accordingly, the more than 1200 gay rights groups in the United States have fought for laws that eliminate all forms of discrimination based on sexual orientation. They have also fought for the repeal of laws that discriminate on the basis of sexual orientation, such as antisodomy laws and laws that negatively affect parental custody of children. They have succeeded mainly at the county and local government levels. The women's movement has been more successful in getting laws passed.[4] In 1982, it even came close to having the Equal Rights Amendment (ERA) recognized as the 27th Amendment to the Constitution. The ERA is intended to eliminate discrimination based on sex. Approved by the House of Representatives in 1971 and by the Senate in 1972, the ERA fell just 3 states short of the 38 needed for ratification in 1982. For the past 30 years, the women's movement has been most successful in getting admission practices altered in professional schools, winning more freedom of reproductive choice for women, and opening up opportunities for women in the political, religious, military, educational, medical, and business systems (Whittier, 1995). The emergence of the peace, environmental, human rights, gay rights, and women's movements marked the beginning of a fourth stage in the history of social movements. This fourth stage involves the promotion of **universal citizenship,** or the extension of citizenship rights to all adult members of society and to society as a whole (Roche, 1995; Turner, 1986: 85–105).

Membership

New social movements are also novel in that they attract a disproportionately large number of highly educated, relatively well-to-do people from the social, educational, and cultural fields. Such people include teachers and college professors, journalists, social workers, artists, actors, and writers (as well as student apprentices to these occupations), who, for several reasons, are more likely to participate in new social movements than are people in other occupations. Their higher education exposes them to radical ideas and makes those ideas appealing. They tend to hold jobs outside the business community, which often opposes their values. And they often get personally involved in the problems of their clients and audiences, sometimes even becoming their advocates (Brint, 1984; Rootes, 1995).

New social movements became prominent in the 1970s. They attract a disproportionately large number of highly educated people in the social, educational, and cultural fields and universalize the struggle for citizenship.

Universal citizenship recognizes the right of marginal groups to full citizenship and the rights of humanity as a whole.

[4] The first wave of the women's movement dates back to the 19th-century "suffragettes," who struggled to win the vote for women. In the text, we refer to the second wave that emerged in the 1960s and whose goals were much broader.

Globalization Potential

Finally, new social movements are new in that they have more potential for globalization than did old social movements.

Up until the 1960s, social movements were typically *national* in scope. That is why, for example, the intensity and frequency of urban race riots in the United States in the 1960s did not depend on such local conditions as the degree of black–white inequality in a given city (Spilerman, 1970; 1976). Instead, African Americans came to believe that racial problems were nationwide and capable of solution only by the federal government. Congressional and presidential action (and lack of action) on civil rights issues, national TV coverage of race issues, and growing black consciousness and solidarity helped to create this belief.[5]

Many new social movements that gained force in the 1970s increased the scope of protest beyond the national level. For example, members of the peace movement viewed federal laws banning nuclear weapons as necessary. Environmentalists felt the same way about federal laws protecting the environment. However, environmentalists also recognized that the condition of the Brazilian rain forest affects climatic conditions worldwide. Similarly, peace activists understood that the spread of weapons of mass destruction can destroy all of humanity. Therefore, members of the peace and environmental movements pressed for *international* agreements binding all countries to protect the environment and stop the spread of nuclear weapons. Social movements went global.

The globalization of social movements was facilitated by inexpensive international travel and communication. New technologies made it easier for people in various national movements to work with like-minded activists in other countries. In the age of CNN, inexpensive jet transportation, fax machines, websites, and e-mail, it was possible not only to see the connection between apparently local problems and their global sources. It was also possible to act both locally and globally.

An Environmental Social Movement

Consider the case of Greenpeace. Greenpeace is a highly successful environmental movement that originated in Vancouver in the mid-1970s. It now has offices in 38 countries, with its international office in Amsterdam (Greenpeace, 2000). Among many other initiatives, it has mounted a campaign to eliminate the international transportation and dumping of toxic wastes. Its representatives visited local environmental groups in African and other developing countries. They supplied the Africans with organizing kits to help them tie their local concerns to global political efforts. They also published a newsletter to keep activists up-to-date about legal issues. Thus, Greenpeace coordinated a global campaign that enabled weak environmental organizations in developing countries to act more effectively. Their campaign also raised the costs of continuing the international trade in toxic waste.

Greenpeace is a highly successful global environmental movement that originated in Vancouver in the mid-1970s and now has offices in 38 countries. Here, Greenpeace activists try to stop a whaling ship.

Canadian Press Picture Archives/Sean White

Greenpeace is hardly alone in its efforts to go global. In 1953, 110 international social-movement organizations spanned the globe. By 1993, there were 631. About a quarter were human rights organizations, and about a seventh were environmental organizations. The latter are by far the fastest-growing organizational type (Smith, 1998: 97).

Even "old" social movements can go global due to changes in the technology of mobilizing supporters. In 1994, for example, the peasants of Chiapas, a southern Mexican province, started an uprising against the Mexican government. Oppressed by Europeans and their descendants for nearly 500 years, the poor, indigenous people of southern Mexico were now facing a government

[5] In the 1990s, improved data and statistical techniques confirmed that local levels of strain and deprivation did not influence the incidence of rioting. However, later analyses did find two local effects. First, race riots were more frequent where job competition between blacks and whites was more intense. Second, riots tended to occur in geographical proximity, suggesting that favorable outcomes of collective action in one city provided a model that was learned by people in neighboring cities (Myers, 1997; Olzak and Shanahan, 1996; Olzak, Shanahan, and McEneaney, 1996).

edict preventing them from gaining access to formerly communal farmland. They wanted the land for subsistence agriculture. But the government wanted to make sure the land stayed in the hands of large, Hispanic ranchers and farmers, who could earn foreign revenue by exporting goods to the United States and Canada under the terms of the new North American Free Trade Agreement. The peasants seized a large number of ranches and farms. A mysterious masked man known simply as "Subcomandante Marcos" was their leader. Effectively using the Internet and the international mass media as his secret weapon against the Mexican government, Marcos led what the *New York Times* called "the first postmodern revolution," combining a peasant uprising with the World Wide Web, short-wave radio, and photo spreads in *Marie Claire.* By ingeniously keeping the movement in the international public eye using modern technologies of communication, Marcos mobilized support abroad and limited the retaliatory actions of the Mexican government (*A Place*. . . , 1998; Jones, 1999).

Reuters News Media, Inc./Corbis

Effectively using the Internet and the international mass media as his secret weapon against the Mexican government, Marcos led what the *New York Times* called "the first postmodern revolution," combining a peasant uprising with the World Wide Web, short-wave radio, and photo spreads in *Marie Claire*. By keeping the movement in the international public eye using modern technologies of communication, Marcos mobilized support abroad and limited the retaliatory actions of the Mexican government.

The globalization of social movements can be further illustrated by coming full circle and returning to the anecdote with which we began this chapter. In 1991, Robert Brym visited his hometown. He hadn't been back in years. As he entered the city he vaguely sensed that something was different. "I wasn't able to identify the change until I reached the pulp and paper mill," says Robert. "Suddenly, it was obvious. The rotten-egg smell was virtually gone. I discovered that in the 1970s a local woman whose son developed a serious case of asthma took legal action against the mill and eventually won. The mill owner was forced by law to install a 'scrubber' in the main smokestack to remove most of the sulfur dioxide emissions. Soon, the federal government was putting pressure on the mill owner to purify the polluted water that poured out of the plant and into the local river system." Apparently, local citizens and the environmental movement had caused a deep change in the climate of opinion. This influenced the government to force the mill owner to spend millions of dollars to clean up his operation. It took decades, but what was political heresy in 1968 became established practice by 1991. That is because environmental concerns had been amplified by the voice of a movement that had grown to global proportions. In general, as this case illustrates, globalization helps to ensure that many new social movements transcend local and national boundaries and promote universalistic goals.

Summary

Sociology Now™

Reviewing is as easy as ❶, ❷, ❸.

❶ Before you do your final review, take the SociologyNow diagnostic quiz to help you identify the areas on which you should concentrate. You will find information on how to access SociologyNow on the foldout at the front of the textbook.

❷ As you review, take advantage of SociologyNow's study aids to help you master the topics in this chapter.

❸ When you are finished with your review, take SociologyNow's post-test to confirm you are ready to move on to the next chapter.

1. **Common sense and some sociological theories suggest that lynch mobs, riots, and other forms of collective action are irrational and unstructured and take place when people are angry and deprived. Is this view accurate?**

   In the short term, deprivation and strain due to rapid social change are generally *not* associated with increased collective action and social movement formation. Mobs, riots, and other forms of collective action may be wild and violent but social organization and rationality underlie much crowd behavior.

2. **Which aspects of social organization facilitate rebellion against the status quo?**

   People are more inclined to rebel against the status quo when social ties bind them to many other people who feel similarly wronged and when they have the time, money, organization, and other resources needed to protest. In addition, collective action and social movement formation are more likely to occur when political opportunities allow them. Political opportunities emerge due to elections, increased support by influential allies, the instability of ruling political alignments, and divisions among elite groups.

3. How do the attempts of authorities to control unrest affect collective action?

Authorities' attempts to control unrest influence mainly the timing of collective action. They may offer concessions to insurgents, co-opt leaders, and employ coercion.

4. What is "framing"?

For social movements to grow, members must make the activities, goals, and ideology of the movement congruent with the interests, beliefs, and values of potential new recruits. Doing so is known as "framing."

5 How have social movements changed in the past three centuries?

In 1700, social movements were typically small, localized, and violent. By the mid-20th century, social movements were typically large, national, and less violent. In the late 20th century, new social movements developed broader goals, recruited more highly educated people, and developed global potential for growth.

6. How is the history of social movements tied to the struggle for the acquisition of citizenship rights?

The history of social movements is a struggle for the acquisition of constantly broadening citizenship rights. These rights include (a) the right to free speech, religion, and justice before the law (civil citizenship), (b) the right to vote and run for office (political citizenship), (c) the right to a certain level of economic security and full participation in the life of society (social citizenship), and (d) the right of marginal groups to full citizenship and the right of humanity as a whole to peace and security (universal citizenship).

Questions to Consider

1. **How would you achieve a political goal? Map out a detailed strategy for reaching a clearly defined aim, such as a reduction in income tax or an increase in government funding of colleges. Who would you try to recruit to help you achieve your goal? Why? What collective actions do you think would be most successful? Why? To whose attention would these actions be directed? Why? Write a manifesto that frames your argument in a way that is culturally appealing to potential recruits.**
2. **Do you think that social movements will be more or less widespread in the 21st century than they were in the 20th century? Why or why not? What kinds of social movements are likely to predominate?**

Web Resources

Companion Website for This Book

http://sociology.wadsworth.com

Begin by clicking on the Student Resources section of the website. Choose "Introduction to Sociology" and then the Brym and Lie book cover. Next, select the chapter you are currently studying from the pull-down menu. From the Student Resources page you will have easy access to InfoTrac® College Edition, MicroCase Online exercises, additional web links, and many other resources to aid you in your study of sociology, including practice tests for each chapter.

Recommended Websites

The use of the Internet to mobilize social-movement support worldwide is well demonstrated by Mexico's Zapatista National Liberation Army in the southern province of Chiapas. Visit their website (mostly in Spanish, but with sections in English, French, and Portuguese) at http://www.ezln.org and read about their information warfare in *Wired* magazine at http://www.wired.com/news/politics/0,1283,10769,00.html.

The environmental movement Greenpeace is one of the most successful cases of globalized protest. Greenpeace now has offices in 38 countries. Its website is at http://adam.greenpeace.org/information.shtml.

The National Organization for Women (NOW) is the largest organization of feminist activists in the United States, with half a million members and 550 chapters in all 50 states and the District of Columbia. Founded in 1966, NOW's goal has been to take action to bring about equality for all women. NOW's website is at http://www.now.org.

Formed in 1955, the American Federation of Labor–Congress of Industrial Organizations (AFL-CIO) is the largest union umbrella organization in the United States, with 13 million members in 68 unions. Its stated goal is "to bring social and economic justice to our nation by enabling working people to have a voice on the job, in government, in a changing global economy and in their communities." To learn more about the AFL-CIO, visit http://www.aflcio.org.

CHAPTER 22

Technology and the Global Environment

NASA

In this chapter, you will learn that:

- Some people think of technology as a useful magic that drives history forward.
- Others think of technology as a monster that has escaped human control and causes more harm than good.
- In reality, technology does transform society and history, but it is under human control because human need shapes technological growth.
- Increasingly, technological development has come under the sway of large multinational corporations and the military establishments of the major world powers.
- Widespread environmental degradation is the main negative consequence of technological development.
- Policy-oriented scientists, the environmental movement, the mass media, and respected organizations have to discover and promote environmental issues if they are to be framed as social problems. In addition, the public must connect the information learned from these groups to real-life events.
- Economically disadvantaged groups experience more environmental risks than economically advantaged groups.
- Most Americans are not prepared to pay the price of creating a safe environment, but repeated environmental catastrophes could easily change their minds.
- By helping to make the public aware of the environmental and other choices we face in the 21st century, sociology can play an important role in the evolution of human affairs.

Technology: Savior or Frankenstein?

On August 6, 1945, the United States Air Force dropped an atomic bomb on Hiroshima, Japan. The bomb killed about 200,000 Japanese, almost all civilians. It hastened the end of World War II, thus making it unnecessary for American troops to suffer heavy losses in a land invasion of Japan.

Scholars interested in the relationship between technology and society also recognize that Hiroshima divided the 20th century into two distinct periods. We may call the period before Hiroshima the era of naive optimism. During that time, technology could do no wrong, or so at least it seemed to nearly all observers. **Technology** was widely defined as the application of scientific principles to the *improvement* of human life. It seemed to be driving humanity down a one-way street named progress, picking up speed with every passing year thanks to successively more powerful engines: steam, turbine, internal combustion, electric, jet, rocket, and nuclear. Technology produced tangible benefits. Its detailed workings rested on scientific principles that were mysterious to all but those with advanced science degrees. Therefore, most people regarded technologists with reverence and awe. They were viewed as a sort of priesthood whose objectivity allowed them to stand outside the everyday world and perform near-magical acts.

Web Interactive Exercises: The Surveillance Society

With Hiroshima, the blush was off the rose. Growing pessimism was in fact evident three weeks earlier when the world's first nuclear bomb exploded at the Alamogordo Bombing Range in New Mexico. The bomb was the child of J. Robert Oppenheimer, who had been appointed head of the top-secret Manhattan Project just 28 months earlier. After recruiting what General Leslie Groves called "the greatest collection of eggheads ever," including three past and seven future Nobel Prize winners, Oppenheimer organized the largest and most sophisticated technological project in human history up to that time. As an undergraduate at Harvard, Oppenheimer had studied Indian philosophy, among other subjects. On the morning of July 16, 1945, as the flash of intense white light faded and the purplish fireball rose, sucking desert sand and debris into a mushroom cloud more than seven and a half miles high, Oppenheimer

Technology is the practical application of scientific principles.

J. Robert Oppenheimer, the "father" of the atom bomb.

Sociology Now™

Learn more about **Technology and Change** by going through the Technology and Change Learning Module.

quoted from Hindu scripture, "I am become Death, the shatterer of worlds" (quoted in Parshall, 1998).

Oppenheimer's misgivings continued after the war. Having witnessed the destructive power he helped unleash, Oppenheimer wanted the United States to set an example to the only other nuclear power at the time, the Soviet Union. He wanted both countries to halt thermonuclear research and refuse to develop the hydrogen bomb. But the governments of the United States and the Soviet Union had other plans. When Secretary of State Dean Acheson brought Oppenheimer to meet President Truman in 1946, Oppenheimer said, "Mr. President, I have blood on my hands." Truman later told Acheson, "Don't bring that fellow around again" (quoted in Parshall, 1998).

Overall, Americans value science and technology highly. By a wide margin, the United States is the world leader in scientific research, publications, and elite achievements; and its lead is growing (Figure 22.1). In 2000, more than 70 percent of Americans agreed that science does more good than harm or felt neutral about the benefits of science. Nearly 30 percent thought that science does more harm than good (World Values Survey, 2003) (Figure 22.2). In the postwar years, however, a growing number of people, including Nobel Prize winners who worked on the bomb, have come to share Oppenheimer's doubts (Feynman, 1999: 9–10). Indeed, they have extended those doubts not just to the peaceful use of nuclear energy but also to technology in general. Increasingly, ordinary citizens—and a growing chorus of leading scientists—are beginning to think of technology as a monster run amok, a Frankenstein rather than a savior (Box 22.1).

The Environmental Awakening

It was only in the 1970s that a series of horrific disasters awoke many people (including some sociologists) to the fact that technological advance is not always beneficial, not even always benign. The most infamous technological disasters of the 1970s and 1980s include the following:

- An outbreak of Legionnaires Disease in a Philadelphia hotel in 1976 killed 34 people. It alerted the public to the possibility that the very buildings they live and work in can harbor toxic chemicals, lethal molds, and dangerous germs.
- In 1977, dangerously high levels of toxic chemicals were discovered leaking into the basements and drinking water of the residents of Love Canal, near Niagara Falls, New

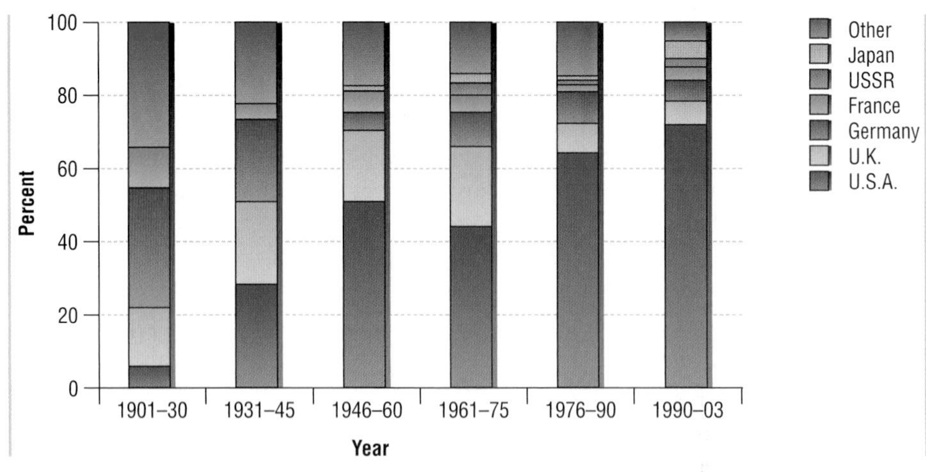

Figure 22.1
Nobel Prizes in Physics, Chemistry, and Physiology/Medicine by Country, 1901–2003 (in percent)

Source: U.S. Census Bureau, 2004–5: 520.

Figure 22.2
"In the long run, do you think the scientific advances we are making will help or harm mankind?"1981–2000 (percent "harm")

This graph shows that (1) Americans are among the most optimistic people in the world concerning the effects of science on humanity; (2) Americans and the citizens of other rich countries were more pessimistic about the effects of science on humanity at the end of the 20th century than they were in 1981; (3) the citizens of countries that are not rich were more *optimistic* about the effects of science on humanity at the end of the 20th century than they were in 1981; and (4) whatever the overall tendency, attitudes toward science fluctuate over time.

World Values Survey (2003).

York. This led to the immediate shutdown of an elementary school and the evacuation of residents from their homes.

- The partial meltdown of the reactor core at the Three Mile Island nuclear facility in Pennsylvania in 1979 caused lethal radioactive water and gas to pour into the environment. (A 1974 report by the Atomic Energy Commission said that such an accident would likely occur only once in 17,000 years.)
- A gas leak at a poorly maintained Union Carbide pesticide plant in Bhopal, India, killed about 4000 people in 1984 and injured 30,000, a third of whom died excruciating deaths in the following years.
- In 1986, the No. 4 reactor at Chernobyl, Ukraine, exploded, releasing 30 to 40 times the radioactivity of the blast at Hiroshima. It resulted in mass evacuations, more than 10,000 deaths, countless human and animal mutations, and hundreds of square miles of unusable cropland.
- In 1989, the Exxon Valdez ran aground in Prince William Sound, Alaska, spilling 11 million gallons of crude oil, producing a dangerous slick more than 1000 miles long, causing billions of dollars of damage, and killing hundreds of thousands of animals.

Robert Llewellyn/SuperStock

The Three Mile Island nuclear facility.

Normal Accidents and the Risk Society

By the mid-1980s, sociologist Charles Perrow was referring to events such as those listed previously as **normal accidents.** The term "normal accident" recognizes that the very complexity of modern technologies ensures that they will *inevitably* fail, although in unpredictable ways (Perrow, 1984). For example, a large computer program contains many thousands of conditional statements. They take the form "If $x = y$, do z; if $a = b$, do c." When in use, the program activates many billions of *combinations* of conditional statements. Consequently, complex programs cannot be tested for all possible eventualities. Therefore, when rare combinations of conditions occur, they have unforeseen consequences that are usually minor, occasionally amusing, sometimes expensive, and too often dangerous. You experience normal accidents when your home computer crashes or

Normal accidents are those that occur inevitably though unpredictably due to the very complexity of modern technologies.

BOX 22.1
Sociology at the Movies

The Matrix (1999), *The Matrix Reloaded* (2003), *The Matrix Revolutions* (2003), *The Terminator* (1984), *Terminator 2: Judgment Day* (1991), *Terminator 3: Rise of the Machines* (2003)

"Have you ever felt that there's something not right in the world?" With these words, *The Matrix* introduces Thomas Anderson, played by Keanu Reeves. Respectable software programmer by day, notorious hacker by night, Anderson, who goes by the handle "Neo," is plagued by the thought that there is something wrong with the world. "You don't know what it is," he says, "but it's there, like a splinter in your mind."

Neo discovers what's wrong when he encounters two legendary hackers, Morpheus (Laurence Fishburne) and Trinity (Carrie-Anne Moss). They introduce him to the secret of his day world and the reality of the Matrix. It turns out that the "reality" lived by Neo and others is a form of collective imagination made possible by a gigantic computer, the Matrix. Neo and other people are nothing more than power supplies housed in liquid-filled containers. They supply the Matrix with energy. The Matrix, in turn, supplies these "batteries" with images—making them feel that they are living, not merely dreaming.

Only about 250,000 humans have managed to escape the Matrix. Morpheus, Neo, Trinity, and the others live in Zion, near the Earth's core. In *The Matrix Revolutions,* machines are rapidly drilling toward Zion, the destruction of which—and therefore the end of humanity—is just 36 hours away.

The imminent destruction of humanity by machines is also the theme of the *Terminator* series. The first *Terminator* movie begins in Los Angeles in 2029, where a battle rages between superintelligent machines that rule the world and the few surviving humans. A killing machine known as a Terminator (played by Arnold Schwarzenegger) is sent back to 1984 to eliminate John Connor, the future leader of the human resistance movement, thus ensuring the victory of the machines in 2029. The Terminator does not accomplish his goal, and in *Terminator 2: Judgment Day,* a kinder, gentler Terminator is sent back to 1995, this time to protect John Connor from a more advanced Terminator machine, the T-1000 (Robert Patrick). About 10 years later, in *Terminator 3: Rise of the Machines,* Schwarzenegger is once again transported to the past to protect John Connor, this time against a still more advanced Terminator, the T-X. He saves O'Connor and O'Connor's future wife, but the deadly machine complex known as Skynet manages to destroy almost every human on Earth.

hangs. A few years ago, the avionics software for the F-16 jet fighter caused the jet to flip upside down whenever it crossed the equator. In January 1990, the entire long-distance network of American Telephone & Telegraph was crippled for nine hours due to a bug in the software for its routing switches. In Perrow's sense of the term, these are all normal accidents, although not as dangerous as the chemical and nuclear mishaps mentioned previously.

German sociologist Ulrich Beck also coined a term that stuck when he said we live in a **risk society.** A risk society is one in which technology distributes danger among all categories of the population. Some categories, however, are more exposed to technological danger than others. Moreover, in a risk society, danger does not result from technological accidents alone. In addition, increased risk is due to mounting *environmental* threats. Environmental threats are more widespread, chronic, and ambiguous than technological accidents. They are therefore more stressful (Beck, 1992; 1986; Freudenburg, 1997). New and frightening terms—"greenhouse effect," "global warming," "acid rain," "ozone depletion," "endangered species"—have entered our vocabulary. To many people, technology seems to be spinning out of control. From their point of view, it enables the production of ever more goods and services but at the cost of

A **risk society** is a society defined by the way risk is distributed as a side effect of technology.

The idea of people inventing machines that try to destroy their creators is not new. It first surfaced in 1580, when, according to legend, a rabbi in Prague built a creature known as a Golem to protect the Jews from their enemies. The Golem ran amok and had to be destroyed. Mary Shelley revived this theme in her 1818 novel, *Frankenstein,* and it has been with us ever since. *The Matrix* and *Terminator* series increase the scale of the problem and add extraordinary special effects, but beyond the glitz is the same anxiety. We continue to suspect that technology is not always a means of improving human life. Increasingly, we think it is antagonistic to human values and that, unless we are careful, it can destroy us.

The Matrix Reloaded, starring Keanu Reaves.

The Everett Collection

These anxieties are not just the stuff of fiction. Bill Joy, chief scientist at Sun Microsystems and codeveloper of the Java programming language, predicts that within decades new genetic entities and miscroscopic robots ("nanobots") will be routinely programmed to make copies of themselves. For example, nanobots designed to make water from hydrogen and oxygen, or viruses engineered to kill crop-damaging insects, will be programmed to self-replicate so the job can be accomplished faster. However, a programming error or a genetic mutation could result in the copying process getting out of control. An out-of-control virus mutation could kill crops rather than harmful insects. An out-of-control water-manufacturing nanobot could flood the world and leave it without land. Another danger, Joy asserts, is that the new technologies democratize the ability of people to do evil. Unlike the construction of a nuclear bomb, the creation of a deadly virus requires relatively inexpensive equipment that is commercially available. Therefore, a single crazed or politically motivated scientist can do much damage to the world. Joy's remarks have gained credibility since the terrorist attacks of September 11, 2001, and the subsequent anthrax attacks in the United States (Joy, 2000; Kurzweil, 1999: 137–42). *The Matrix* and *Terminator* series give voice to these growing concerns.

breathable air, drinkable water, safe sunlight, plant and animal diversity, and normal weather patterns. In the same vein, Neil Postman (1992) refers to the United States as a **technopoly.** He argues that the United States is the first country in which technology has taken control of culture. Technology, he says, compels people to try to solve all problems using technical rather than moral criteria, although technology is often the source of the problems.

The latest concern of technological skeptics is biotechnology. Molecular biologists have mapped the entire human gene structure and are also mapping the gene structure of selected animals and plants. They can splice genes together, creating plants and animals with entirely new characteristics. As we will see, the ability to create new forms of life holds out incredible potential for advances in medicine, food production, and other fields. That is why the many advocates of this technology speak breathlessly of a "second genesis" and "the perfection of the human species." Detractors claim that without moral and political decisions based on a firm sociological understanding of who benefits and who suffers from these new techniques, the application of biotechnology may be a greater threat to our well-being than any other technology ever developed.

Technopoly is a form of social organization in which technology compels people to try to solve all problems using technical rather than moral criteria, even though technology is often the source of the problems.

These considerations suggest five tough questions (all of which we tackle later):

A sea otter covered in oil spilled by the Exxon Valdez in 1989.

Gary Braasch/Corbis

1. Is technology *the* great driving force of historical and social change? This is the opinion of cheerleaders and naysayers, those who view technology as our savior and those who fear it as a Frankenstein. In contrast, we argue that technology is able to transform society only when it is coupled with a powerful social need. People control technology as much as technology transforms people.
2. If some people do control technology, then exactly who are they? We argue against the view that scientific and engineering wizards are in control. The military and big corporations now decide the direction of most technological research and its application.
3. What are the most dangerous spin-offs of technology and how is risk distributed among various social groups? We focus on global warming, industrial pollution, the decline of biodiversity, and genetic pollution. We show that although these dangers put all of humanity at risk, the degree of danger varies by class, race, and country. In brief, the socially and economically disadvantaged are most at risk.
4. How can we overcome the dangers of environmental degradation? We argue that market and technological solutions are insufficient by themselves. In addition, much self-sacrifice and cooperation will be required.
5. The fifth and final question underlies all the others. Of what use is sociology in helping us solve the world's technological and environmental problems?

Technology *and* People Make History

Russian economist Nikolai Kondratiev was the first social scientist to notice that technologies are invented in clusters. As Table 22.1 shows, a new group of major inventions has cropped up every 40 to 60 years since the Industrial Revolution. Kondratiev argued that these flurries of creativity cause major economic growth spurts beginning 10 to 20 years later and lasting 25 to 35 years each. Thus, Kondratiev subscribed to a form of **technological determinism,** the belief that technology is the major force shaping human society and history (see also Ellul, 1964 [1954]).

Technological determinism is the belief that technology is the main factor shaping human society and history.

Is it true that technology helps shape society and history? Of course it is. James Watt developed the steam engine in Britain in the 1760s. It was the main driving force in the mines, mills, factories, and railways of the Industrial Revolution. Gottlieb Daimler invented the internal combustion engine in Germany in 1883. It was the foundation stone

Table 22.1
"Kondratiev Waves" of Modern Technological Innovation and Economic Growth

Wave	Invention Dates	New Technologies	Base	Economic Growth Spurt
1	1760s–70s	Steam engine, textile manufacturing, chemistry, civil engineering	Britain	1780–1815
2	1820s	Railways, mechanical engineering	Britain, Continental Western Europe	1840–70
3	1870s–80s	Chemistry, electricity, internal combustion engine	Germany, United States	1890–1914
4	1930s–40s	Electronics, aerospace, chemistry	United States	1945–70
5	1970s	Microelectronics, biotechnology	United States, Japan	1985–?

Source: Modified from Pacey (1983, 32).

of two of the world's biggest industries, automobiles and petroleum. John Atanasoff was among the first people to invent the computer in 1939 at Iowa State College (now University). It utterly transformed the way we work, study, and entertain ourselves. It also put the spurs to one of the most sustained economic booms ever. We could easily cite many more examples of how technology shapes history and transforms society.

However, if we probe a little deeper into the development of any of the technologies mentioned previously, we notice a pattern: They did not become engines of economic growth until *social* conditions allowed them to do so. The original steam engine, for instance, was invented by Hero of Alexandria in the first century C.E. He used it as an amusing way of opening a door. People then promptly forgot the steam engine. Some 1700 years later, when the Industrial Revolution began, factories were first set up near rivers and streams, where waterpower was available. That was several years before Watt patented his steam engine. Watt's invention was all the rage once its potential became evident. But it did not cause the Industrial Revolution, and it was adopted on a wide scale only after the social need for it emerged (Pool, 1997: 126–7).

Similarly, Daimler's internal combustion engine became the basis of the automobile and petroleum industries thanks to changes in the social organization of work introduced by Henry Ford; the self-defeating business practice of Ford's main competitors, the Stanley brothers; and, oddly enough, an epidemic of hoof-and-mouth disease. When Ford incorporated his company in 1903, a steam-driven automobile, the Stanley Steamer, was his main competition. Many engineers then believed that the Stanley Steamer was the superior vehicle on purely technical grounds. Many engineers still think so today. (For one thing, the Stanley Steamer didn't require a transmission system.) But while the Stanley brothers built a finely tooled automobile for the well-to-do, Ford tried to figure out a way of producing a cheap car for the masses. His inspiration was the meat-packing plants of Cincinnati and Chicago. In 1913, he modeled the first car assembly line after those plants. Only then did he open a decisive lead in sales over the Stanleys. The Stanleys were finally done in a few years later. An outbreak of hoof-and-mouth disease led officials to close down the public watering troughs for horses that were widely used in American cities. Owners of the Stanley Steamer used the troughs to replenish its water supply. So we see it would be wrong to say, along with strict technological determinists, that Daimler's internal combustion engine *caused* the growth of the car industry and then the petroleum industry. The car and petroleum industries grew out of the internal combustion engine only because an ingenious entrepreneur efficiently organized work in a new way and because a chance event undermined access to a key element required by his competitor's product (Pool, 1997: 153–5).

U.S. Army Photo

ORDVAC, an early computer developed at the University of Illinois, was delivered to the Ballistic Research Laboratory at the Aberdeen Proving Ground of the United States Army. Technology typically advances when it is coupled to an urgent social need.

Regarding the computer, Atanasoff stopped work on it soon after the outbreak of World War II. However, once the military potential of the computer became evident, its development resumed. The British computer Colossus helped to decipher secret German codes in the last two years of the war and played an important role in the Allied victory. The University of Illinois delivered one of the earliest computers, the ORDVAC, to the Ballistic Research Laboratory at the Aberdeen Proving Ground of the United States Army. Again we see how a new technology becomes a major force in society and history only after it is coupled with an urgent social need. We conclude that technology and society influence each other. Scientific discoveries, once adopted on a wide scale, often transform societies. But scientific discoveries are turned into useful technologies only when social need demands it.

How High Tech Became Big Tech

The 19th Century

Enjoying a technological advantage usually translates into big profits for businesses and military superiority for countries. In the 19th century, gaining technological advantage was still inexpensive. It took only modest capital investment, a little knowledge about the best way to organize work, and a handful of highly trained workers to build a shop to manufacture stirrups or even steam engines. In contrast, mass-producing cars, sending a man to the moon, and other feats of 20th- and 21st-century technology require enormous capital investment, detailed attention to the way work is organized, and legions of technical experts. Add to this the intensely competitive business and geopolitical environment of the 20th and 21st centuries, and one can readily understand why ever larger sums have been invested in research and development over the past 100 years.

It was in fact already clear in the last quarter of the 19th century that turning scientific principles into technological innovations was going to require not just genius but substantial resources, especially money and organization. Thus, Thomas Edison established the first "invention factory" at Menlo Park, New Jersey, in the late 1870s. Historian of science Robert Pool notes that

> the most important factor in Edison's success—outside of his genius for invention—was the organization he had set up to assist him. By 1878, Edison had assembled at Menlo Park a staff of thirty scientists, metalworkers, glassblowers, draftsmen, and others working under his close direction and supervision. With such support, Edison boasted that he could turn out "a minor invention every ten days and a big thing every six months or so." (Pool, 1997: 22)

The phonograph and the electric lightbulb were two such "big things." Edison inspired both. Both, however, were also expensive team efforts, motivated by vast commercial possibilities. (Edison founded General Electric, the most profitable company in the world and the second most valuable based on market capitalization; see "Global 1000," 1999.)

The 20th and 21st Centuries

By the beginning of the 20th century, the scientific or engineering genius operating in isolation was only rarely able to contribute much to technological innovation. By mid-century, most technological innovation was organized along industrial lines. Entire armies of experts and vast sums of capital were required to run the new invention factories. The prototype of today's invention factory was the Manhattan Project, which built the nuclear bomb in the last years of World War II. By the time of Hiroshima, the U.S. nuclear industry was one of the country's biggest. The era of big science and big technology had arrived. Only governments and, increasingly, giant multinational corporations could afford to sustain the research effort of the second half of the 20th century.

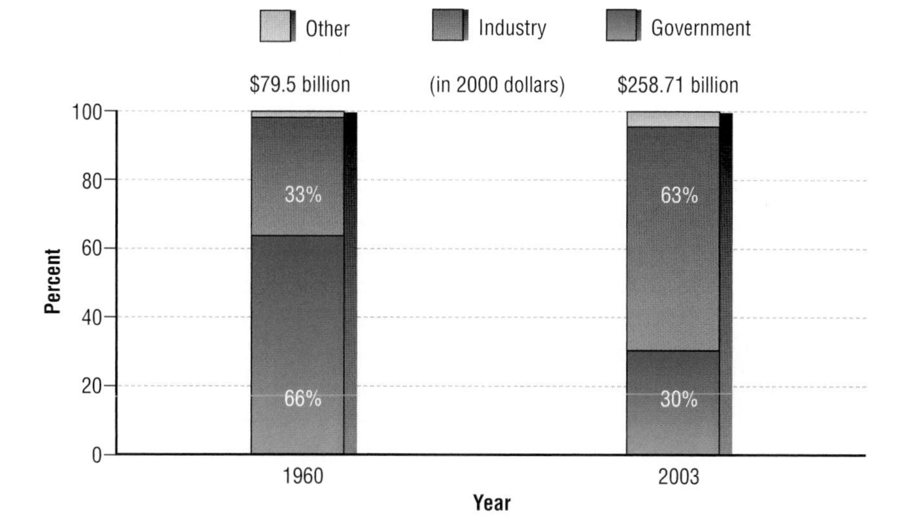

Figure 22.3
Research and Development, United States, 1960 and 2003, by Source (in percent)

Source: U.S. Census Bureau, 2004–5: 510; U.S. Department of Labor, 2005.

At the beginning of the 21st century, there seemed to be no upper limit to the amount that could be spent on research and development. The United States had fewer than 10,000 research scientists before World War I. Today, it has well over a million (Hobsbawm, 1994: 523). In 2003, American research and development spending reached $259 billion, up from $80 billion in 1960 (calculated in 2002 dollars to take account of inflation). During that same period, industry's share of spending rose from 33 percent to 63 percent of the total, whereas government's dropped from 66 percent to 30 percent (Figure 22.3).

Because large multinational corporations now routinely invest astronomical sums in research and development to increase their chance of being the first to bring innovations to market, the time lag between new scientific discoveries and their technological application is continuously shrinking. That is clear from Figure 22.4, which shows how long it took five of the most popular new consumer products of the 1980s and 1990s to penetrate the U.S. market. It was fully 38 years after the videocassette recorder (VCR) was invented in 1952 that the device achieved 25 percent market penetration. It took 18 years before the

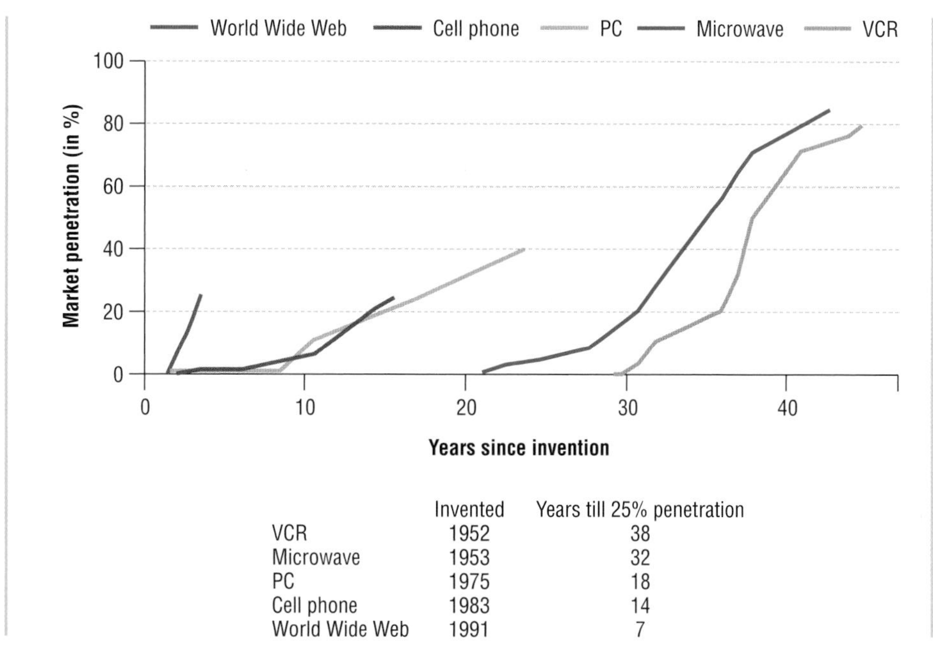

	Invented	Years till 25% penetration
VCR	1952	38
Microwave	1953	32
PC	1975	18
Cell phone	1983	14
World Wide Web	1991	7

Figure 22.4
Market Penetration by Years since Invention, United States

Source: From "The Silent Boom" *Forbes*, July 7, 1997. Reprinted by permission of Forbes Magazine © 1997 Forbes, Inc.

SuperStock

Research in biotechnology is big business. Even in the late 1980s, nearly 40 percent of the biotechnology scientists who belonged to the prestigious National Academy of Sciences had industry affiliations.

Kurt Scholz/SuperStock

Due to global warming, glaciers are melting, the sea level is rising, and extreme weather events are becoming more frequent.

personal computer, invented in 1975, was owned by 25 percent of Americans. The World Wide Web, invented in 1991, took only 7 years to reach that level of market penetration.

Because of these developments, it should come as no surprise that military and profit-making considerations now govern the direction of most research and development. A reporter once asked a bank robber why he robbed banks. "Because that's where the money is," the robber answered. This is hardly the only motivation prompting scientists and engineers to research particular topics. Personal interests, individual creativity, and the state of a field's intellectual development still influence the direction of inquiry. This is especially true for theoretical work done in colleges, as opposed to applied research funded by governments and private industry. It would, however, be naive to think that practicality doesn't also enter the scientist's calculation of what he or she ought to study. Even in a more innocent era, Sir Isaac Newton studied astronomy partly because the explorers and mariners of his day needed better navigational cues. Michael Faraday was motivated to discover the relationship between electricity and magnetism partly by his society's search for new forms of power (Bronowski, 1965 [1956]: 7–8). The connection between practicality and research is even more evident today. Many researchers—even many of those who do theoretically driven research in colleges—are pulled in particular directions by large research grants, well-paying jobs, access to expensive state-of-the-art equipment, and the possibility of winning patents and achieving commercial success. For example, many leading molecular biologists in the United States today have established genetic engineering companies, serve on their boards of directors, or receive research funding from them. In not a few cases, major pharmaceutical and agrochemical corporations have bought out these companies because they see their vast profit potential (Rural Advancement Foundation International, 1999). Even in the late 1980s, nearly 40 percent of the biotechnology scientists who belonged to the prestigious National Academy of Sciences had industry affiliations (Rifkin, 1998: 56).

Economic lures, increasingly provided by the military and big corporations, have generated moral and political qualms among some researchers. Some scientists and engineers wonder whether work on particular topics achieves optimum benefits for humanity. Certain researchers are troubled by the possibility that some scientific inquiries may be harmful to humankind. However, a growing number of scientists and engineers recognize that to do cutting-edge research they must still any residual misgivings, hop on the bandwagon, and adhere to military and industrial requirements and priorities. That, after all, is where the money is.

Environmental Degradation

The side effect of technology that has given people the most serious cause for concern is environmental degradation. It has four main aspects: global warming, industrial pollution, the decline of biodiversity, and genetic pollution. Let us briefly consider each of these problems.

Global Warming

Ever since the Industrial Revolution, humans have been burning increasing quantities of fossil fuels (coal, oil, gasoline, natural gas, etc.) to drive their cars, furnaces, and factories. Burning these fuels releases carbon dioxide into the atmosphere. The accumulation

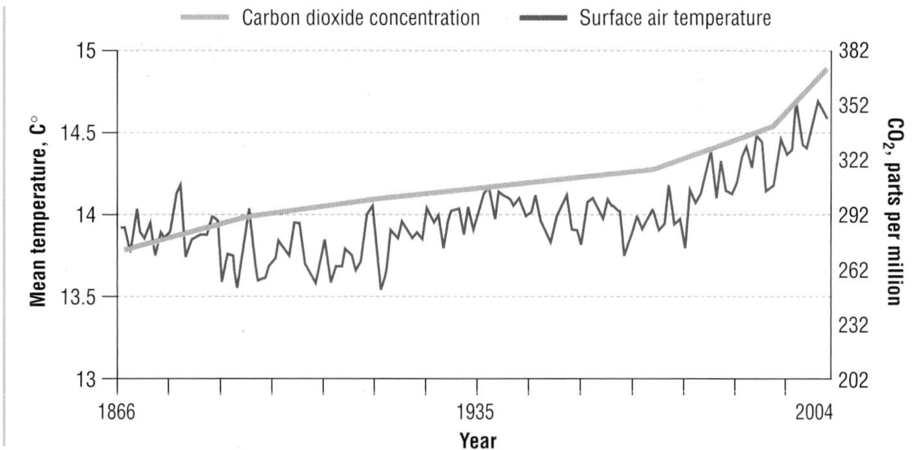

▸Figure 22.5
Annual Mean Global Surface Air Temperature and Carbon Dioxide Concentration, 1866–2004

Source: Goddard Institute for Space Studies, 2003; 2005; Karl and Trenberth, 1999: 102; Quaschning, 2003.

of carbon dioxide allows more solar radiation to enter the atmosphere and less heat to escape. This is the so-called **greenhouse effect.** Most scientists believe that the greenhouse effect contributes to **global warming,** a gradual increase in the world's average surface temperature. Using data from the National Aeronautics and Space Administration's (NASA) Goddard Institute for Space Studies, ▸Figure 22.5 graphs the world's annual average surface air temperature from 1866 to 2004 and the concentration of carbon dioxide in the atmosphere from 1866 to 2000. The graph shows a warming trend that mirrors the increased concentration of carbon dioxide in the atmosphere. It also shows that the warming trend began in the last third of the 20th century. In the century between 1866 and 1965, average surface air temperature fluctuated from year to year but was almost the same at the end of the 100-year period as at the beginning. From 1966 to 2004, average surface air temperature rose at a rate of 1.76 degrees Celsius per century.

Sociology Now™

Learn more about the **Greenhouse Effect** by going through the Share of Global Emissions Map Exercise.

Many scientists believe that global warming is already producing serious climatic change, for as temperatures rise, more water evaporates. This causes more rainfall and bigger storms, which leads to more flooding and soil erosion, which in turn leads to less cultivable land. People suffer and die all along the causal chain. This was tragically evident in 2005, when hurricanes Katrina and Rita delivered knockout punches to coastal Louisiana, Alabama, Mississippi, and Texas, killing more than a thousand and causing hundreds of billions of dollars of damage. Hurricane intensity and duration have been increasing with global warming for the past 30 years (Emanuel, 2005).

▸Figure 22.6 graphs the worldwide dollar cost of damage due to so-called natural disasters from 1980 to 2004. (As we have just seen, an increasingly large number of meteorological events deemed "natural" are rendered extreme by human action.) Clearly, the damage caused by extreme meteorological events is on the upswing. This, however, may be only the beginning. It seems that global warming is causing the oceans to rise. That is partly because warmer water expands and partly because the partial melting of the polar ice caps puts more water in the oceans. In the 21st century, this may result in the flooding of some heavily populated coastal regions throughout the world. Just a one-yard rise in the sea level would flood about 12 percent of the surface area of Egypt and Bangladesh and 0.5 percent of the surface area of the United States (Kennedy, 1993: 110).

Industrial Pollution

Industrial pollution is the emission of various impurities into the air, water, and soil due to industrial processes. It is a second major form of environmental degradation. Every day, we release a witch's brew into the environment, the more common ingredients of which include household trash, scrap automobiles, residue from processed ores, agricultural runoff containing dangerous chemicals, lead, carbon monoxide, carbon dioxide, sulfur dioxide, ozone, nitrogen oxide, various volatile organic compounds, chlorofluorocarbons (CFCs), and various solids mixed with liquid droplets floating in the air. Most

The **greenhouse effect** is the accumulation of carbon dioxide in the atmosphere that allows more solar radiation to enter the atmosphere and less solar radiation to escape.

Global warming is the gradual worldwide increase in average surface temperature.

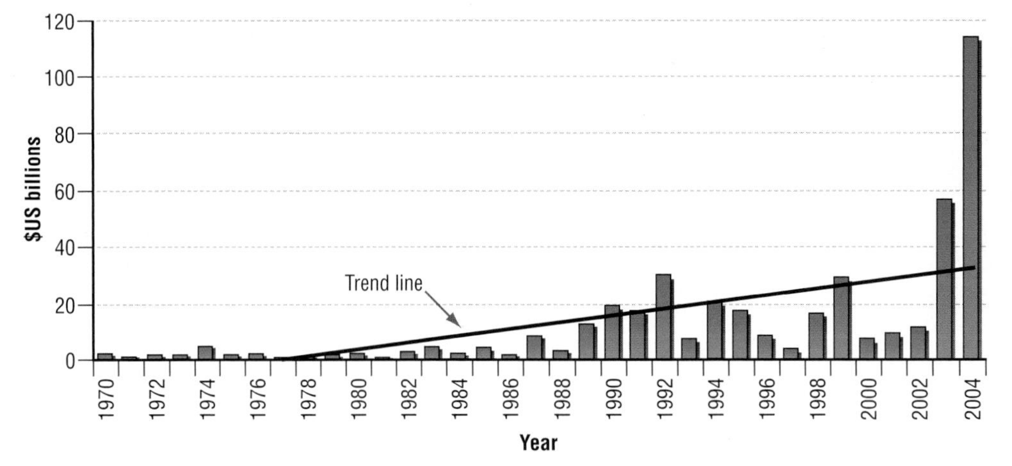

Figure 22.6

Worldwide Damage Due to "Natural" Disasters, 1970–2004 (in 2002 U.S. dollars)

Note: One might think that the 2004 figure is wildly inflated due to the Asian tsunami, but because the disaster hit poor regions, just $13.3 billion of insured damage (in 2002 dollars) was attributed to it. Much of the 2004 figure is due to a series of hurricanes in Florida and typhoons in Japan.
Source: Swiss Re, 2003: 8; 2004: 7; 2005: 5; U.S. Department of Labor, 2005.

pollutants are especially highly concentrated in the U.S. Northeast and around the Great Lakes. These densely populated areas are identified with the proliferation of old, heavy, dirty industries (United States Environmental Protection Agency, 2000).

Pollutants may affect us directly. For example, they seep into our drinking water and the air we breathe, causing a variety of ailments, particularly among the young, the elderly, and the ill. A dramatic natural experiment demonstrating the direct effect of air pollution on health occurred during the 1996 Atlanta Olympics. For the 17 days of the Olympics, asthma attacks among children in the Atlanta area plummeted 42 percent. When the athletes went home, the rate of asthma attacks among children immediately bounced back to normal levels. Epidemiologists soon figured out why. During the Olympics, Atlanta closed the downtown to cars and operated public transit around the clock. Vehicle exhaust fell, with an immediate benefit to children's health. Children's health deteriorated as soon as normal traffic resumed (Mittelstaedt, 2001).

Pollutants may also affect us indirectly. For instance, sulfur dioxide and other gases are emitted by coal-burning power plants, pulp and paper mills, and motor-vehicle exhaust. They form **acid rain.** This is a form of precipitation whose acidity eats away at, and eventually destroys, forests and the ecosystems of lakes. Another example: CFCs are used in industry and by consumers, notably in refrigeration equipment. They contain chlorine, which is responsible for the depletion of the **ozone layer** 5 to 25 miles above the Earth's surface. Ozone is a form of oxygen that blocks ultraviolet radiation from the sun. Let more ultraviolet radiation reach ground level and, as we are now witnessing, rates of skin cancer and crop damage increase.

Acid rain is precipitation whose acidity destroys forests and the ecosystems of lakes. It is formed by sulfur dioxide and other gases emitted by coal-burning power plants, pulp and paper mills, and motor-vehicle exhaust.

The **ozone layer** lies 5 to 25 miles above the Earth's surface. (Ozone is a gas composed of three oxygen atoms.) It is depleted by CFCs. The depletion of the ozone layer allows more ultraviolet light to enter the Earth's atmosphere. This increases the rate of skin cancer and crop damage.

Radioactive waste deserves special attention. Some 103 nuclear reactors are now generating commercial electricity in the United States. They run on enriched uranium or plutonium fuel rods. Once these fuel rods decay beyond the point where they are useful in the reactor, they become waste material. This waste is highly radioactive. It must decay about 10,000 years before humans can be safely exposed to it without special protective equipment. The spent fuel rods need to be placed in sturdy, watertight copper canisters and buried deep in granite bedrock where the chance of seismic disturbance and water seepage is small. The trouble is, most Americans are petrified at the prospect of having a nuclear waste facility anywhere near their families. As a result, spent fuel rods have been accumulating since the 1950s in "temporary" facilities. These are mainly pools of water near nuclear reactors. These facilities are a safety threat. The first long-term nuclear waste

Dave Martin/AP/Wide World Photo

A makeshift grave near New Orleans following Hurricane Katrina in 2005. Scientists predicted the effects of a major hurricane on low-lying New Orleans years before Katrina hit. They also knew *why* thousands might die: (1) The levee system along the Mississippi River eliminated the city's first line of defense against storm surge. Without levees along the Mississippi, silt from the river's floodwaters would stabilize land along the riverside and stop or at least slow down the sinking of coastal wetlands into the Gulf of Mexico. With the levees, silt is diverted into the Gulf, so the wetlands, which protect New Orleans from storm surge, are disappearing at an alarming rate. (2) Levees along Lake Pontchartrain, to the north of the city, were last reinforced with higher walls by the Army Corps of Engineers in 1965, when they were built to withstand a category 3 storm. It was therefore only a matter of time before a more severe storm would cause water to break through the levees. (3) Climate change, due in part to the burning of fossil fuels, has resulted in more intense hurricanes of longer duration. Hurricane Katrina was not just a natural disaster. It was also a *social* disaster caused by poor planning, neglect, and careless disregard for the human impact on nature.

repository in the United States, deep inside Nevada's Yucca Mountain, is scheduled to open in 2010, but delays are expected due to numerous regulatory and legal battles (Mitchell, 2002).

The Decline of Biodiversity

The third main form of environmental degradation is the decline in **biodiversity,** the enormous variety of plant and animal species inhabiting the Earth. Biodiversity changes as new species emerge and old species die off because they cannot adapt to their environment. This is all part of the normal evolutionary process. However, in recent decades the environment has become so inhospitable to so many species that the rate of extinction has greatly accelerated. Examination of fossil records suggests that for millions of years, an average of 1–3 species became extinct annually. Today, about 1000 species are becoming extinct annually (Tuxill and Bright, 1998: 41). In 11 countries, 10 percent or more of bird species are threatened with extinction. In 29 countries, 10 percent or more of mammal species are similarly threatened (Kidron and Segal, 1995: 14–15).

The extinction of species is impoverishing in itself, but it also has practical consequences for humans. For example, each species of animal and plant has unique properties. When scientists discover that a certain property has a medically useful effect, they get busy trying to synthesize the property in the laboratory. Treatments for everything from headaches to cancer have been found in this way. Indeed, about a quarter of all drugs prescribed in the United States today (including 9 of the top 10 in sales) include compounds first found in wild organisms. The single richest source of genetic material with pharmaceutical value is found in the world's rain forests, particularly in Brazil, where more than 30 million species of life exist. However, the rain forests are being rapidly destroyed by strip mining, the construction of huge pulp and paper mills and hydroelectric projects, and the deforestation of land by farmers and cattle grazers.

Similarly, fleets of trawlers belonging to the highly industrialized countries are now equipped with sonar to help them find large concentrations of fish. Some of these ships use fine-mesh nets to increase their catch. They have been enormously "successful." Trawlers have depleted fish stocks in some areas of the world. In North America, for example, the depletion of cod, salmon, bluefin tuna, and shark stocks has devastated fishing communities and endangered one of the world's most important sources of protein. All told, 11 of the world's 15 main fishing grounds and 69 percent of the world's main fish species are in decline (McGinn, 1998: 60; Ransom and Worm, 2003).

Biodiversity refers to the enormous variety of plant and animal species inhabiting the Earth.

Artificially splicing genes together may yield benefits as well as dangers. Woody Allen in *Sleeper* (1973).

The Everett Collection

Genetic Pollution

Genetic pollution is the fourth main form of environmental degradation. It refers to the health and ecological dangers that may result from artificially splicing genes together (Rifkin, 1998).

The genetic information of all living things is coded in a chemical called deoxyribonucleic acid (DNA). When members of a species reproduce, the characteristics of the mates are naturally transmitted to their offspring through DNA. **Recombinant DNA,** in contrast, is a technique developed by molecular biologists in the last few decades. It involves artificially joining bits of DNA from a donor to the DNA of a host. Donor and host may be of the same or different species. The donor DNA grows along with the host DNA, in effect creating a new form of life. For example, scientists inserted the gene that makes fireflies sparkle at night into a tobacco plant. The offspring of the plant had leaves that glowed in the dark. Researchers inserted human growth hormone into a mouse embryo. This created mice that grew twice as big and twice as fast as ordinary mice. Biologists combined embryo cells from a sheep and a goat and placed them in the womb of a surrogate animal. The surrogate animal then gave birth to an entirely new species, half sheep, half goat.

These wonders of molecular biology were performed in the mid-1980s and helped to dramatize and publicize the potential of recombinant DNA. Since 1990, governments and corporations have been engaged in a multibillion-dollar international effort to create a complete genetic map of humans and various plants, microorganisms, and animal species. With human and other genetic maps in hand, and using recombinant DNA and related techniques, it is possible to design what some people regard as more useful animals and plants and superior humans. By 2000, scientists had identified the location and chemical structure of every one of the approximately 40,000 human genes. This will presumably enable them to understand the function of each gene. They can then detect and eliminate hereditary propensities to a wide range of diseases. Recombinant DNA will also enable farmers to grow disease- and frost-resistant crops with higher yields. It will allow miners to pour ore-eating microbes into mines, pump the microbes above ground after they have had their fill, and then separate out the ore. This will greatly reduce the cost and danger of mining. Recombinant DNA will allow companies to grow plants that produce cheap biodegradable plastic and microorganisms that consume oil spills and absorb radioactivity. The potential health and economic benefits to humankind of these and many other applications of recombinant DNA are truly startling.

Genetic pollution refers to the potential dangers of mixing the genes of one species with those of another.

Recombinant DNA involves removing a segment of DNA from a gene or splicing together segments of DNA from different living things, thus effectively creating a new life form.

So are the dangers genetic pollution poses to human health and the stability of ecosystems (Rifkin, 1998: 67–115; Tokar, 2001). For example, when a nonnative organism enters a new environment, it usually adapts without problem. Sometimes, however, it unexpectedly wreaks havoc. Kudzu vine, Dutch elm disease, the gypsy moth, chestnut blight, starlings, Mediterranean fruit flies, zebra mussels, rabbits, and mongooses have all done just that. Now, however, the potential for ecological catastrophe has multiplied. That is because scientists are regularly testing genetically altered plants (effectively, nonnative organisms) in the field. Some have gone commercial, and many more will soon be grown on a wide scale. These plants are resistant to insects, disease, and frost. However, once their pollen and seeds escape into the environment, weeds, insects, and microorganisms will eventually build up resistance to the genes that resist herbicides, pests, and viruses. Thus, superbugs, superweeds, and superviruses will be born. We cannot predict the exact environmental consequences of these developments. However, the insurance industry refuses to insure genetically engineered crops against the possibility of their causing catastrophic ecological damage.

Global warming, industrial pollution, the decline of biodiversity, and genetic pollution threaten everyone. However, as you will now see, the degree to which they are perceived as threatening depends on certain social conditions being met. Moreover, the threats are not evenly distributed in society.

The Social Construction of Environmental Problems

Environmental problems do not become social issues spontaneously. Before they can enter the public consciousness, policy-oriented scientists, the environmental movement, the mass media, and respected organizations must discover and promote them. People have to connect real-life events to the information learned from these groups. Because some scientists, industrial interests, and politicians dispute the existence of environmental threats, the public can begin to question whether environmental issues are in fact social problems that require human intervention. We must not, then, think of environmental issues as inherently problematic. Rather, they are contested phenomena. They can be socially constructed by proponents. They can be socially demolished by opponents. This is the key insight of the school of thought known as *social constructionism* (Hannigan, 1995b).

The Case of Global Warming

The controversy over global warming is a good example of how people create and contest definitions of environmental problems (Gelbspan, 1999; 1997; Hart and Victor, 1993; Mazur, 1998; Ungar, 1999; 1998; 1995; 1992). The theory of global warming was first proposed about a century ago. However, an elite group of scientists began serious research on the subject only in the late 1950s. They attracted no public attention until the 1970s. That is when the environmental movement emerged. The environmental movement gave new legitimacy and momentum to the scientific research and helped to secure public funds for it. Respected and influential scientists now began to promote the issue of global warming. The mass media, always thirsting for sensational stories, were highly receptive to these efforts. Newspaper and television reports about the problem began to appear in the late 1970s. They proliferated in the mid- to late-1980s. Between 1988 and 1991, the public's interest in global warming reached an all-time high. That was because frightening events helped to make the media reports more believable. For example, the summer of 1988 brought the worst drought in half a century. As crops failed, New York sweltered, and huge fires burned in Yellowstone National Park, *Time* magazine ran a cover story entitled "The Big Dry." It drew the connection between global warming and extreme weather. Many people got worried. Soon, respected organizations outside the scientific community, the mass media, and environmental movement—such as the insurance industry and the United Nations—expressed concern about the effects of global warming. By 1994, 59 percent of Americans with an opinion on the subject thought that using coal, oil, and gas contributes to the greenhouse effect (calculated from National Opinion Research Center, 2004).

By 1994, however, public concern with global warming had already passed its peak. The eruption of Mount Pinatubo in the Philippines pumped so much volcanic ash into the atmosphere, clouding the sunshine, that global surface air temperatures fell in 1992–93. The media thought the story had grown stale, and reports about global warming sharply declined. Some scientists, industrialists, and politicians began to question whether global warming was in fact taking place. They cited satellite data showing that the Earth's lower atmosphere had cooled in recent decades. They published articles and took out ads to express their opinion, thus increasing public skepticism.

With surface temperatures showing warming and lower atmospheric temperatures showing cooling, different groups lined up on different sides of the global warming debate. Those who had most to lose from carbon emission cuts emphasized the lower atmospheric data. This group included Western coal and oil companies, the member states of the Organization of Petroleum Exporting Countries (OPEC), and other coal- and oil-exporting nations. Those who had most to lose from the consequences of global warming or least to lose from carbon emission cuts emphasized the surface data. This group included insurance companies, an alliance of small island states, the European Union, and the United Nations. In the United States, the division was sufficient to prevent the government from acting. The Clinton administration pushed for a modest 7 percent cut in carbon emissions between 1990 and 2012. But the Republican-controlled Congress blocked the proposal. As a result, the United States is now the only industrialized country that has failed to legislate cuts in carbon emissions. The Bush administration is relying on a voluntary strategy that allows industry to decide how much it will cut carbon emissions.

In 1998 the global warming skeptics were dealt a blow when their satellite data were shown to be misleading. Until then, no one had taken into account that the satellites were gradually slipping from their orbits due to atmospheric friction, thus causing imprecise temperature readings. Allowing for the slippage, scientists from NASA and private industry now calculate that temperatures in the lower atmosphere are rising, just like temperatures on the Earth's surface (Wentz and Schabel 1998; Hansen et al. 1998). But these new findings did not lay to rest the claims of the global warming skeptics. They now claim that even if global warming is occurring, it may be due to natural rather than human causes. Thus, as social constructionists suggest, the power of competing interests to get their definition of reality accepted as the truth continues to influence public perceptions of global warming.

In addition to being socially defined, environmental problems are socially distributed. That is, environmental risks are greater for some groups than others. Let us now examine this issue.

The Social Distribution of Environmental Risk

You may have noticed that after a minor twister touches down on some unlucky community in Texas or Kansas, TV reporters often rush to interview the surviving residents of trailer parks. The survivors stand amid the rubble that was their lives. They heroically remark on the generosity of their neighbors, their good fortune in still having their family intact, and our inability to fight nature's destructive forces. Why trailer parks? Small twisters aren't particularly attracted to them, but reporters are. That is because trailers are pretty flimsy in the face of a small tornado. They often suffer a lot of damage from twisters. They therefore make a more sensational story than the minor damage typically inflicted on upper-middle-class homes with firmly shingled roofs and solid foundations. This is a general pattern. Whenever disaster strikes—from the sinking of the *Titanic* to the fury of

Hurricane Katrina—economically and politically disadvantaged people almost always suffer most. That is because their circumstances render them most vulnerable.

Philip Gould/Corbis

Petrochemical plants between New Orleans and Baton Rouge form what local residents call "Cancer Alley." Here, the Union Carbide plant in Taft, Louisiana.

Environmental Racism

In fact, the advantaged often consciously put the disadvantaged in harm's way to avoid risk themselves. For example, oil refineries, chemical plants, toxic dumps, garbage incinerators, and other environmentally dangerous installations are more likely to be built in poor communities with a high percentage of African Americans or Hispanic Americans than in more affluent, mainly white communities. That is because disadvantaged people are often too politically weak to oppose such facilities, and some may even value the jobs they create. Thus, in a study conducted in the mid-1980s, the number and size of hazardous waste facilities were recorded for every ZIP code area in the United States. At a time when about 20 percent of Americans were of African or Hispanic origin, ZIP code areas lacking any such facilities had, on average, a 12 percent minority population. ZIP code areas with one such facility had about a 24 percent minority population on average. And ZIP code areas with more than one such facility or with one of the five largest landfills in the United States had on average a 38 percent minority population. The study concluded that three out of five African Americans and Hispanic Americans lived in communities with uncontrolled toxic waste sites (Szasz and Meuser, 1997: 100; Stretesky and Hogan, 1998). Similarly, the 75-mile strip along the lower Mississippi River between New Orleans and Baton Rouge has been nicknamed "Cancer Alley" because the largely black population of the region suffers from unusually high rates of lung, stomach, pancreatic, and other cancers (Chapter 19, "Health and Medicine"). The main reason? This small area is the source of fully one-quarter of the petrochemicals produced in the country, containing more than 100 oil refineries and chemical plants (Bullard, 1994 [1990]). A final example: Some poor Native American reservations have been targeted as possible interim nuclear waste sites. That is partly because states have little jurisdiction over reservations, so the usual state protests against such projects are less likely to prove effective. In addition, the Goshute tribe in Utah and the Mescalero Apaches in New Mexico have expressed interest in the project because of the money it promises to bring into their reservations (Pool, 1997: 247–8). Here again we see the recurrent pattern of what some analysts call **environmental racism** (Bullard, 1994 [1990]). This is the tendency to heap environmental dangers on the disadvantaged, and especially on disadvantaged racial minorities.

Environmental Risk and the Less Developed Countries

What is true for disadvantaged classes and racial groups in the United States also holds for the world's less developed countries. The underprivileged face more environmental dangers than the privileged (Kennedy, 1993: 95–121). In North America, Western Europe, and Japan, population growth is low and falling. Industry and government are eliminating some of the worst excesses of industrialization. In contrast, world population will grow from about 6 to 7 billion between 2000 and 2010, and nearly all of that growth will be in the less developed. Moreover, Mexico, Brazil, China, India, and many other countries are industrializing rapidly. This is putting tremendous strain on their natural resources. Rising demand for water, electricity, fossil fuels, and consumer products is creating more polluted rivers, dead lakes, and industrial waste sites. At a quickening pace, rain forests, grazing land, cropland, and wetlands are giving way to factories, roads, airports, and housing complexes. Smog-blanketed megacities continue to sprawl. Eighteen of the world's 21 biggest cities are in less developed countries.

Environmental racism is the tendency to heap environmental dangers on the disadvantaged, especially on disadvantaged racial minorities.

Given the picture sketched previously, it should come as no surprise that on average, people in less developed countries are more concerned about the environment than people in rich countries (Brechin and Kempton, 1994). However, the developing countries cannot afford much in the way of pollution control, so antipollution regulations are lax by North American, Western European, and Japanese standards. This is an incentive for some multinational corporations to site some of their most environmentally unfriendly operations in the less developed countries (Clapp, 1998). It is also the reason the industrialization of the less developed countries is proving so punishing to the environment. When car ownership grows from less than 1 percent to 10 percent of the population in China, and when 50–75 million Indians with motor scooters upgrade to cars, environmental damage may well be catastrophic. That is because the Chinese and the Indians simply cannot afford catalytic converters and electric cars. They have no regulations phasing in the use of these and other devices that save energy and pollute less.

For the time being, however, the rich countries do most of the world's environmental damage. That is because their inhabitants earn and consume more than the inhabitants of less developed countries. How much more? The richest one-fifth of humanity earns about 80 times more than the poorest one-fifth (up from 30 times more in 1950). In the past half century, the richest one-fifth doubled its per capita consumption of energy, meat, timber, steel, and copper and quadrupled its car ownership. In that same period, the per capita consumption of the poorest one-fifth hardly changed. The United States has only 4.5 percent of the world's population, but it uses about 25 percent of the Earth's resources. It also produces more than 20 percent of global emissions of carbon dioxide, the pollutant responsible for about one-half of global warming (Ehrlich et al., 1997). Thus, the inhabitants of the developed countries cause a disproportionately large share of the world's environmental problems, enjoy a disproportionate share of the benefits of technology, and live with fewer environmental risks than do people in the less developed countries.

Social inequalities are also apparent in the field of biotechnology. For instance, the large multinational companies that dominate the pharmaceutical, seed, and agrochemical industries now routinely send anthropologists, biologists, and agronomists to all corners of the world. There they take samples of wild plants, the crops people grow, and human blood. They hope to find genetic material with commercial value in agriculture and medicine. If they discover genes with commercial value, the company they work for patents the discovery. This gives the company the exclusive legal right to manufacture and sell the genetic material without compensating the donors. For example, Indian farmers and then scientists worked for a hundred generations discovering, skillfully selecting, cultivating, and developing techniques for processing the neem tree, which has powerful antibacterial and pesticidal properties. However, a giant corporation based in a rich country is now the sole commercial beneficiary of their labor. Monsanto (United States), Novartis (Switzerland), Glaxo Wellcome (United Kingdom), and other prominent companies in the life sciences call this "protection of intellectual property." Indigenous peoples and their advocates call it "biopiracy" (Rifkin, 1998: 37–66).

Genetic Engineering and Social Class

Finally, consider the possible consequences for the class structure of people having their babies genetically engineered. This should be possible on a wide scale in 10 or 20 years. Free of inherited diseases and physical abnormalities, and perhaps genetically programmed to enjoy superior intellectual and athletic potential, these children would, in effect, speed up and improve the slow and imperfect process of natural evolution. That, at least, is the rosy picture sketched by proponents of the technology. In practice, because only the well-to-do are likely to be able to afford fully genetically engineered babies, the new technology could introduce an era of increased social inequality and low social mobility. Only the economically underprivileged would bear a substantial risk of genetic inferiority. This future was foreseen in the 1997 movie *Gattica.* The plot revolves around the tension between a society that genetically engineers all space pilots to perfection and a young man played by Ethan Hawke, who was born without the benefit of genetic engi-

BOX 22.2

SOCIAL POLICY: WHAT DO YOU THINK?

Web-Based Learning and Higher Education

"I love it," says Carol Thibeault, a student at Central Connecticut State University. "With online classes, there's no set time that I have to show up. Sometimes I lug a heavy laptop onto the commuter bus and work on course files I've downloaded while I ride along. I even take my computer to the beach" (quoted in Maloney 1999, 19). Carol is not alone in expressing her enthusiasm for the new information technology in higher education. E-mail and the World Wide Web are now about as exotic as the telephone in the United States and other rich countries. Some scholars see "online education" as the future of higher education.

The advantages of Web-based education are many. Parents with children, or students with jobs, can learn at their own speed, on their own schedule, and in their own style. This will make learning easier and more enjoyable. Potentially, many students can be taught efficiently and effectively. This will lower the cost of higher education.

Although few would argue for its elimination, many people think that too much dependence on the "virtual classroom" has drawbacks. Some professors argue that it is difficult to control the quality of online educational materials and instruction. That is why dropout rates for distance education courses tend to be significantly higher than rates in conventional classrooms (Merisotis, 1999). Others suggest that distance education will spread primarily among low-cost, low-status institutions. At elite institutions, they say, classroom contact and discussion will become even more important. So while one group of students will enjoy a great deal of personal attention from faculty members, another group will receive only cursory and impersonal attention.

What do you think the role of Web-based learning should be in higher education? Do you think that distance learning is superior or inferior to traditional classroom learning? Is it better to learn from a professor and other students in a "real" classroom as opposed to a "virtual" classroom? Do you think that online education will lead to an increase in social inequality?

neering yet aspires to become a space pilot. Hawke's character manages to overcome his genetic handicap. It is clear from the movie, however, that his success is both illegal and extremely rare. The norm is rigid genetic stratification, and it is strongly sanctioned by state and society. (For other examples of how new technologies can contribute to social inequality, see the discussion of job polarization in Chapter 13, of electronic democracy in Chapter 14, and Box 22.2.)

What Is to Be Done?

The Market and High-Tech Solutions

Some people believe we already have two weapons that will work together to end the environmental crisis: the market and high technology. The case of oil illustrates how these weapons can combine forces. If oil reserves drop or producers withhold oil from the market for political reasons, oil prices go up. This makes it worthwhile for oil exploration companies to develop new technologies to recover more oil. When they discover more oil and bring it to market, prices fall. This is what happened following the oil crises of 1973 (when prices tripled) and 1978–9 (when prices tripled again) due to turmoil in the Middle East. Reserves eventually rose and prices fell. Similarly, if too little rice and wheat are grown to meet world demand, the price of these grains goes up. This prompts agrochemical companies to invent higher-yield grains. Farmers use the new grain seed to grow more wheat and rice, and prices eventually fall. This is what happened during the so-called "green revolution" of the 1960s. Projecting these experiences into the future, optimists believe that we will deal similarly with global warming, industrial pollution, and other forms of environmental degradation. In their view, human inventiveness and the profit motive will combine to create the new technologies we need to survive and prosper in the 21st century.

Some evidence supports this optimistic scenario. In recent years, we have adopted new technologies to combat some of the worst excesses of environmental degradation. For example, we have replaced brain-damaging leaded gas with unleaded gas. We have devel-

Figure 22.7
Air Pollutant Emission Projections, United States, 1990–2010 (in million short tons, projected)

Source: Office of Air Quality Planning and Standards, 1998: 5.4–5.8.

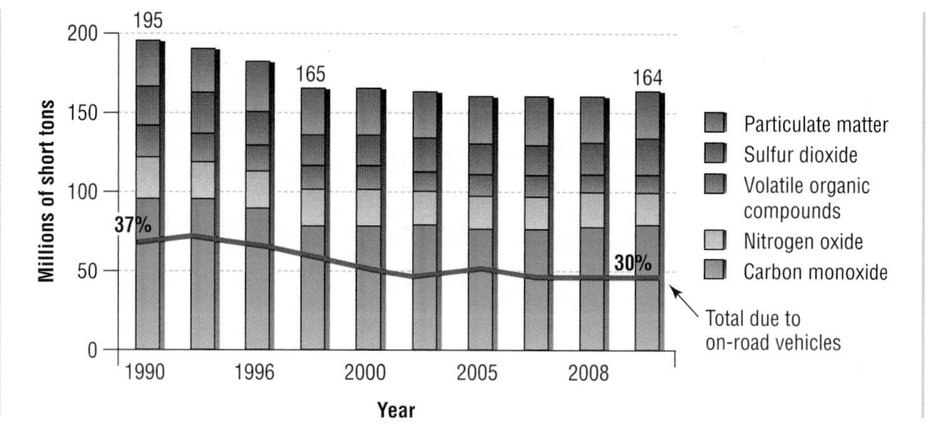

oped environmentally friendly refrigerants, allowing the production of ozone-destroying CFCs to plummet. In a model of international cooperation, rich countries have subsidized the cost of replacing CFCs in the developing countries. Efficient windmills and solar panels are now common. More factories are equipped with high-tech pollution control devices, preventing dangerous chemicals from seeping into the air and water. We have introduced cost-effective ways of recycling metal, plastic, paper, and glass. We are developing new methods for decreasing carbon dioxide emissions from the burning of fossil fuels (Parson and Keith, 1998). Figure 22.7 uses data from the U.S. Environmental Protection Agency to illustrate one consequence of such efforts. It shows production of five of the most common air pollutants in the United States between 1990 and 1998 and expected pollutant production between 1998 and 2010. Production of the five pollutants fell more than 15 percent between 1990 and 1999.

Problems with the Market and High-Tech Solutions

Clearly, market forces are helping to bring environmentally friendly technologies online. However, three factors suggest that market forces cannot solve environmental problems on their own. First, price signals often operate imperfectly. Second, political pressure is often required to stimulate policy innovation. Third, markets and new technologies are not working quickly enough to deal adequately with the environmental crisis. Let us consider each of these issues in turn.

Web Research Projects: Who Are the Environmentalists?

- *Imperfect price signals.* The price of many commodities does not reflect their actual cost to society. Gasoline in the United States costs more than $2 a gallon on average at the time of this writing, but the *social* cost, including the cost of repairing the environmental damage caused by burning the gas, is another $2.50. To avoid popular unrest, the government of Mexico City charges consumers only about 10 cents per cubic meter for water. The actual cost is about 10 times higher (Ehrlich et al., 1997). Due to these and other price distortions, markets sometimes fail to send signals that might result in the speedy adoption of technological and policy fixes.
- *Political pressure.* Political pressure exerted by environmental social-movement activists, community groups, and public opinion is often necessary to motivate corporate and government action on environmental issues. For instance, organizations like Greenpeace have successfully challenged the practices of logging companies, whalers, the nuclear industry, and other groups engaged in environmentally dangerous practices. Many less famous community associations have also played an important role in this regard (Brown, 1997). Antinuclear activism is an outstanding example of a movement that forced a substantial turnaround in government and corporate policy. For instance, in Germany, which obtains one-third of its electricity from nuclear power, the antinuclear movement has had a major effect on public opinion. In June 2000, the government decided to phase out all of the country's nuclear power plants within

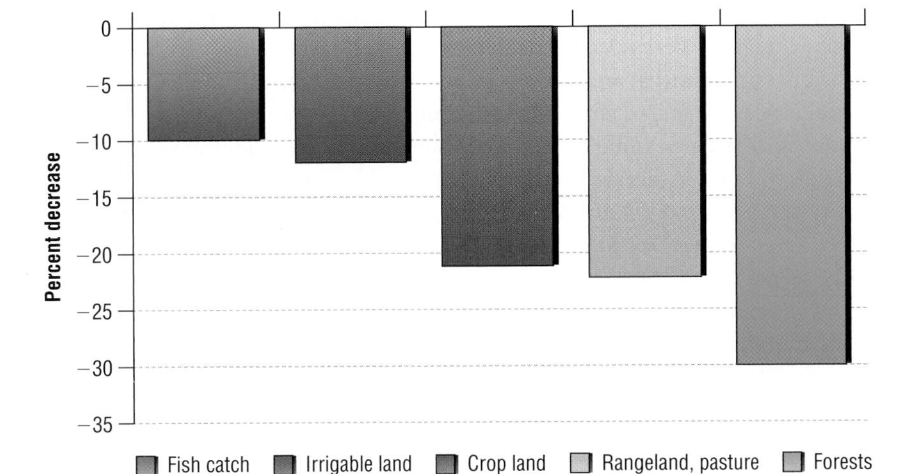

Figure 22.8
Renewable Resources, World, Percent Change, 1990–2010 (projected)

Source: Postel 1994, 11.

about 20 years. In the United States, no new nuclear power plants are planned. Again, the antinuclear movement must be credited with helping to change the public mood and bring about a halt in construction of new nuclear facilities.[1] All told, about 8 percent of Americans belong to groups committed to protecting the environment. About 10 percent have contributed money to such organizations (National Opinion Research Center, 2004). Without the political efforts of pro-environment individuals, organizations, and social movements, it is doubtful that many environmental issues would be defined as social problems by corporations and governments.

- *The slow pace of change.* We saw previously how price signals and new technologies have created pockets of environmental improvement, especially in the rich countries. However, it is unclear whether they can deal with the moral and political issues raised by biotechnology. Moreover, our efforts so far to clean up the planet are just not good enough. Returning to Figure 22.7, we observe that after improving somewhat in the 1990s, U.S. air pollution is not expected to get any better between 1999 and 2010. Glancing back at Figure 22.5, we note that global warming continues to accelerate. Examining Figure 22.8, we see that we can expect a substantial decrease in all of the world's renewable resources until at least the end of the decade. In 1993, 1680 of the world's leading scientists, including 104 Nobel Prize winners, signed the "World Scientists' Warning to Humanity." It stated: "A great change in our stewardship of the earth and the life on it is required, if vast human misery is to be avoided and our global home on this planet is not to be irretrievably mutilated. . . . Human beings and the natural world are on a collision course" (Union of Concerned Scientists, 1993). Evidence suggests we still are.

These considerations imply that market and high-tech solutions are inadequate by themselves when it comes to dealing with the environmental problems we face.

The Cooperative Alternative

The alternative to the pure market and high-tech approach involves a high degree of cooperation among citizens, governments, and corporations. This cooperative strategy includes investing heavily in energy-saving technologies, environmental cleanup, and subsidized, environmentally friendly industrialization in the developing countries. It would require renewed commitment to voluntary efforts, new laws and enforcement bodies to ensure compliance, increased environmentally related research and development by industry and government, more environmentally directed foreign aid, and new taxes to help pay for some of

[1]Statements by the Bush administration suggest this could change, however.

it (Livernash and Rodenburg, 1998). In addition, a cooperative strategy would entail careful assessment of the risks associated with biotechnology projects and consultation with the public before such projects are allowed to go forward. Profits from genetic engineering would also have to be shared equitably with donors of genetic material.

Before such policies could be adopted, four conditions would have to be met. The broad public in North America, Western Europe, and Japan would have to be (1) aware of the gravity of environmental problems, (2) confident in the capacity of citizens, their governments, and corporations to solve them, (3) willing in some cases to make substantial economic sacrifices to get the job done, and (4) able to overcome resistance to change on the part of interest groups with a deep stake in things as they are.

Data from the General Social Survey allow us to see whether the first three conditions are being met in the United States. They paint a good news/bad news scenario. Nearly all

Table 22.2
Public Opinion on Environmental Issues, United States (in percent)

Proportion of Americans who think the following environmental problems are extremely/very/somewhat dangerous:	
1. Air pollution caused by cars	91
2. Air pollution caused by industry	94
3. Nuclear power stations	84
4. A rise in the world's temperature caused by the "greenhouse effect"	82
5. Pollution of America's rivers, lakes, and streams	95
6. Pesticides and chemicals used in farming	84
"It is just too difficult for someone like me to do much about the environment."	
Strongly agree/agree	27
Neither agree nor disagree	17
Disagree/strongly disagree	56
"Government should let businesses decide for themselves how to protect the environment, even if it means they don't always do the right thing, or government should pass laws to make businesses protect the environment, even if it interferes with businesses' right to make their own decisions."	
Government should let businesses decide	11
Government should pass laws	89
"We are faced with many problems in this country, none of which can be solved easily or inexpensively. I'm going to name some of these problems, and for each one I'd like you to tell me whether you think we're spending too much money on it, too little money, or about the right amount. Are we spending too much money, too little money, or about the right amount on improving and protecting the environment?"	
Too little	61
About right	30
Too much	9
"How willing would you be to pay much higher prices in order to protect the environment?"	
Very/fairly willing	47
Neither willing nor unwilling	25
Not very/not at all willing	28
"How willing would you be to accept cuts in your standard of living in order to protect the environment?"	
Very/fairly willing	32
Neither willing nor unwilling	23
Not very/not at all willing	45
"How willing would you be to pay much higher taxes in order to protect the environment?"	
Very/fairly willing	34
Neither willing nor unwilling	21
Not very/not at all willing	45
"How often do you make a special effort to sort glass or cans or plastic or papers and so on for recycling?"	
Always/often/sometimes	87
Never	13
"How often do you cut back on driving a car for environmental reasons?"	
Always/often/sometimes	36
Never	67

Source: National Opinion Research Center (2002).
Note: Data are for 1994.

Americans are aware of the environmental problem. As Table 22.2 shows, between 82 percent and 95 percent of Americans consider pollution and other environmental problems to be dangerous. Moreover, most people (56 percent) think they can do something about environmental issues themselves, while a huge majority (89 percent) believes the government should pass more laws to protect the environment. Most Americans (61 percent) even say we are spending too little on environmental cleanup. All this is encouraging.

However, expressing environmental awareness and agreeing on the need for action is one thing. Biting the bullet is another. Fewer than half of Americans (47 percent) are willing to pay much higher prices to protect the environment. Fewer than a third (32 percent) are willing to accept cuts in their standard of living. Barely a third (34 percent) are willing to pay much higher taxes. Most Americans are prepared to protect the environment if it does not inconvenience them too much. Thus, 87 percent say they sort glass, cans, plastic, or paper for recycling. Only 36 percent say they have ever cut back on driving for environmental reasons.

Other surveys conducted in the United States and elsewhere reveal much the same pattern. Most people know about the environmental crisis. They want it dealt with. Yet they are unwilling to pay much of the cost themselves. In 2005, researchers at Yale and Columbia Universities calculated an "environmental sustainability index" for each of the world's countries and found that the United States ranked only 47th—behind Russia (33rd) and the Congo (39th)—in its effort to protect the environment (Yale Center, 2005). The situation reminds one of American attitudes toward involvement in World War II. In 1939, when Britain and France went to war with Germany, most Americans considered the Nazi threat remote and abstract. They refused to go to war until the Japanese attacked Pearl Harbor in 1941, destroying U.S. naval power in the Pacific and making it clear that America had to fight to survive. This episode teaches us that people are not usually prepared to make big personal sacrifices for seemingly remote and abstract goals. However, they are prepared to sacrifice a great deal if the goals become less remote and abstract. Analogously, more and bigger environmental catastrophes may have to occur before more people are willing to take remedial action. It may take one or more environmental Pearl Harbors to get most Americans to make the necessary commitment to help save the planet.

The Resistance of Powerful Interest Groups

The fourth and final obstacle to implementing the cooperative alternative is the resistance of interest groups who benefit from things the way they are. Two of the most powerful such interest groups are the oil and automobile industries in the United States. We focus on them to illustrate the problem.

The United States imports about one-half the petroleum it uses. About a third of its imports comes from the Middle East, especially Saudi Arabia, Iraq, and Kuwait. American allies in Western Europe depend somewhat more heavily on Middle East oil, whereas Japan buys more than three-quarters of its oil from the region. We pay a heavy price economically, environmentally, and politically for this dependence. We even went to war with Iraq in 1991 and 2003 partly (some would say mainly) to protect our oil interests in the region. Meanwhile, the United States is the world's second biggest producer of automobiles. Cars use much of our petroleum and are the single biggest source of carbon emissions. As has often been said, Americans have a love affair with the automobile and we consider it our right to enjoy among the lowest gas prices in the world. The oil and automobile industries employ millions of Americans, and its owners exert enormous influence on the White House. The Bush and Cheney families earned their personal wealth in the oil industry and they have many friends and business associates in it.

Given the extent and depth of these interests in the status quo, proposals to adopt cleaner and less expensive alternatives to gas-burning cars have met with little response from government or industry. (The movement to improve fuel efficiency gained force after the 1973 oil crisis, but even it petered out when the government allowed sport utility vehicles [SUVs] to be classified as small trucks and thereby avoid stringent fuel-economy regulation. SUVs are now the country's most popular type of vehicle.) For example, it has

Courtesy of Shell Hydrogen

Western Europe and Japan are well ahead of the United States in the development of hydrogen-powered cars. Shown here is the first hydrogen filling station in Japan, opened in 2003.

been proposed that we use ethanol (made from corn) and the abundant North American supply of natural gas to produce hydrogen for fuel cells that could power electric cars. Experts estimate that such cars would cost 40 percent less to run than cars fueled by gasoline and would produce only one-third the carbon emissions. Start-up costs for production facilities, pipelines, and retrofitting filling stations are no more than what major oil companies now spend on oil exploration and production over a decade. If half the cars in the United States ran on hydrogen fuel cells, we would no longer need to import any oil. Some analysts say we could reach that goal by 2015 if government and industry really got behind the project.[2] However, neither has expressed much interest (King, 2003).

Contrasting Cases: Japan and Iceland

In contrast, a company owned partly by Shell Oil opened Japan's first hydrogen filling station in 2003 as part of that country's Hydrogen and Fuel Cell Demonstration Project. Financially backed by the Japanese Ministry of Economy, Trade, and Industry, the project will build five filling stations in metro Tokyo to provide hydrogen for a fleet of prototype fuel cell vehicles made by several automotive companies (Shell Hydrogen, 2003). Iceland also opened its first commercial hydrogen fueling station in 2003, when three hydrogen-fueled buses started running regular routes on the streets of its capital city, Reykjavik. Four years earlier, the citizens and government of Iceland formed and financially backed a consortium called New Icelandic Energy. Its members include DaimlerChrysler in Germany, Shell Oil in the Netherlands and the United Kingdom, and Norsk Hydro in Norway. The consortium plans to eliminate the use of all fossil fuels in the country by 2030 and turn the country into a major exporter of hydrogen, the "Kuwait of the North." In Western Europe, similar cooperative efforts are expected to put millions of hydrogen-powered cars on the roads in the next decade (Árnason and Sigfússon, 2000; Canadian Broadcasting Corporation, 2003). Here then are stunning examples of how cooperation among citizens, governments, and corporations can make progress in the fight against environmental degradation. They stand in sharp contrast to the market and high-technology approach favored by American interest groups with a deep stake in things as they are. Short of a change in the context of U.S. politics, perhaps motivated by repeated environmental catastrophes, it is difficult to imagine what could make the cooperative strategy more popular here.

Evolution and Sociology

For many thousands of years, humans have done well on this planet. That is because we have created cultural practices, including technologies, that allowed us to adapt to and thrive in our environment. Nonetheless, there have been some failures along the way. Many tribes and civilizations are extinct. And our success to date as a species is no warrant for the future. If we persist in using technologies that create an inhospitable environment, nature will deal with us in the same way it always deals with species that cannot adapt.

Broadly speaking, we have two survival strategies to cope with the challenges that lie ahead: competition and cooperation. Charles Darwin wrote famously about competition in *On the Origin of Species* (1859). He observed that members of each species struggle

[2]There is an additional benefit. The fuel cells would produce more electricity than needed to run the cars. If just 12 million hydrogen-powered cars released this energy into the power grid at night, they would generate as much electricity as all of today's power plants.

against each other and against other species in their struggle to survive. Most of the quickest, the strongest, the best camouflaged, and the smartest live long enough to bear offspring. Most of the rest are killed off. Thus, the traits passed on to offspring are those most valuable for survival. Ruthless competition, it turns out, is a key survival strategy of all species, including humans.

In *The Descent of Man,* Darwin mentioned our second important survival strategy: cooperation. In some species mutual assistance is common. The species members that flourish are those that best learn to help each other (Darwin, 1871: 163). The Russian geographer and naturalist Peter Kropotkin (1908 [1902]) elaborated this idea. After spending five years studying animal life in Siberia, he concluded that "mutual aid" is at least as important a survival strategy as competition. Competition takes place when members of the same species compete for limited resources, said Kropotkin. Cooperation occurs when species members struggle against adverse environmental circumstances. According to Kropotkin, survival in the face of environmental threat is best assured if species members help each other. Kropotkin also showed that the most advanced species in any group—ants among insects, mammals among vertebrates, humans among mammals—are the most cooperative. Many evolutionary biologists now accept Kropotkin's ideas (Gould, 1988). Recently, based on computer simulations involving competitive and cooperative strategies, mathematicians concluded that "cooperation [is] as essential for evolution as . . . competition" (Nowak, May, and Sigmund, 1995: 81).

As we have seen, a strictly competitive approach to dealing with the environmental crisis—relying on the market alone to solve our problems—now seems inadequate. Instead, it appears we require more cooperation and self-sacrifice. This involves substantially reducing consumption, paying higher taxes for environmental cleanup and energy-efficient industrial processes, subsidizing the developing countries to industrialize in an environmentally friendly way, and so forth. Previously, we outlined some grave consequences of relying too little on a cooperative survival strategy at this historical juncture. But which strategy you emphasize in your own life is, of course, your choice.

Similarly, throughout this book—when we discussed families, gender inequality, crime, race, population, and many other topics—we raised social issues lying at the intersection point of history and biography—yours and ours. We set out alternative courses of action and outlined their consequences. We thus followed our disciplinary mandate: helping people make informed choices based on sound sociological knowledge (Wilensky, 1997) (Figure 22.9). In the context of the present chapter, however, we can make an even bolder claim for the discipline. Conceived at its broadest, sociology promises to help in the rational and equitable evolution of humankind.

Figure 22.9
The Advantage of Sociological Knowledge

Summary

Sociology Now™

Reviewing is as easy as ❶, ❷, ❸.

❶ Before you do your final review, take the SociologyNow diagnostic quiz to help you identify the areas on which you should concentrate. You will find information on how to access SociologyNow on the foldout at the front of the textbook.
❷ As you review, take advantage of SociologyNow's study aids to help you master the topics in this chapter.
❸ When you are finished with your review, take SociologyNow's post-test to confirm you are ready to move on to the next chapter.

1. Does technology shape society?

   New technologies routinely transform societies, but this does not mean that technology is beyond human control. Human control is evident in the fact that new technologies are adopted only when there is a widespread social need for them. It is also evident in the fact that since the last third of the 19th century, technological development has increasingly come under the control of multinational corporations and the military establishments of the major world powers. Research scientists and engineers who work for these organizations must normally adhere to their research priorities.

2. Why are a growing number of people skeptical about the benefits of technology?

   Skepticism about the benefits of technology derive from the fact that new technologies have unintended, negative consequences. These negative consequences include "normal accidents" (the inevitable but unpredictable failure of complex technologies) and growing environmental risks.

3. What are the major forms of environmental degradation and who is most exposed to the risks associated with them?

   The major forms of environmental degradation include global warming, industrial pollution, the destruction of biodiversity, and genetic pollution. The people most exposed to the risks associated with the various forms of environmental degradation include members of racial minorities, lower classes, and less developed societies.

4. How is social constructionism applied to the study of environmental problems?

   Social constructionism emphasizes that social problems do not emerge spontaneously. Instead, they are contested phenomena whose prominence depends on the ability of supporters and detractors to make the public aware of them.

5. Are market and high-tech solutions capable of dealing with the problem of environmental degradation?

   Market and high-tech solutions can help solve many environmental problems. However, three issues suggest they are insufficient by themselves. First, price signals do not always mirror market conditions. Second, political pressure is often needed to motivate governments and corporations to take action on environmental issues. Third, the pace of change is too slow.

6. What needs to be done to solve the problem of environmental degradation?

   Increased cooperation among citizens, governments, and corporations is required to solve the environmental crisis. This cooperative strategy involves renewed commitment to voluntary efforts, new laws and enforcement bodies to ensure compliance, increased investment in energy-saving research and development by industry and government, more environmentally directed foreign aid, and new taxes to help pay for some if it. A cooperative strategy is more evident in Western Europe and Japan than in the United States, where market solutions are more popular. Many Americans are unwilling to undergo the personal sacrifices and changes in lifestyle required to deal with the problem of environmental degradation, and some interest groups with a deep stake in things as they are resist change. However, repeated environmental catastrophes could change the context of U.S. politics in a way that would make the cooperative strategy more popular in this country.

7. What role can sociology play in helping people deal with environmental and other important social problems?

   Sociology can play an important role in sensitizing the public to the social issues and choices humanity faces in the 21st century. For example, sociology poses the choice between more competition and more cooperation as ways of solving the environmental crisis.

Questions to Consider

1. What are the main environmental problems in your community? How are they connected to global environmental issues? (See "Recommended Websites," following, for useful leads.)
2. Take an inventory of your environmentally friendly and environmentally dangerous habits. In what ways can you act in a more environmentally friendly way?

http://sociology.wadsworth.com

Begin by clicking on the Student Resources section of the website. Choose "Introduction to Sociology" and finally the Brym and Lie book cover. Next, select the chapter you are currently studying from the pull-down menu. From the Student Resources page you will have easy access to InfoTrac®

College Edition, MicroCase Online exercises, additional web links, and many other resources to aid you in your study of sociology, including practice tests for each chapter.

Recommended Websites

The website of the U.S. Environmental Protection Agency at http://www.epa.gov is a valuable educational tool. Of particular interest is the search engine at http://www.epa.gov/epahome/commsearch.htm. It allows you to discover the environmental issues in your community.

For links to environmental movements, go to http://www.wsu.edu/amerstu/ce/env_jus.html.

Against All Reason is a provocative electronic journal devoted to "the radical nature of science as a route to knowledge and the radical critique of the social, political, and economic roles of science and technology." Go to http://www.human-nature.com/reason/index.html.

Good lists of web links pertaining to the sociology of science and technology can be found at http://WWW.Trinity.Edu/mkearl/science.html.

Glossary

A

Ablism is prejudice and discrimination against disabled people.

Absolute deprivation is a condition of extreme poverty.

Abstraction is the human capacity to create general ideas, or ways of thinking that are not linked to particular instances. For example, languages, mathematical notations, and signs allow us to classify experience and generalize from it.

An **achieved status** is a status that depends on the capabilities and efforts of the individual.

An **achievement-based** stratification system is one in which the allocation of rank depends on a person's accomplishments.

Acid rain is precipitation whose acidity destroys forests and the ecosystems of lakes. It is formed by sulfur dioxide and other gases emitted by coal-burning power plants, pulp and paper mills, and motor-vehicle exhaust.

Affirmative action is a policy that gives preference to women or minority group members if equally qualified people are available for a position, thus compensating for past discrimination.

An **age cohort** is a category of people born in the same range of years.

Age roles are norms and expectations about the behavior of people of different age cohorts.

Age stratification is social inequality between age cohorts.

Ageism is prejudice and discrimination against elderly people.

Agricultural societies are those in which plows and animal power are used to substantially increase food supply and dependability as compared with horticultural and pastoral societies. Agricultural societies first emerged about 5000 years ago. Average settlement size and permanence, the division of labor, productivity, and inequality are higher in agricultural societies than in horticultural and pastoral societies.

Altruistic suicide is Durkheim's term for suicide that occurs in high-solidarity settings, where norms tightly govern behavior. Altruism means devotion to the interests of others. Altruistic suicide is suicide in the group interest.

Anomic suicide is Durkheim's term for suicide that occurs in low-solidarity settings, where norms governing behavior are vaguely defined. Anomie means "without order."

Anticipatory socialization involves taking on the norms and behaviors of a role to which one aspires but does not yet occupy.

Apartheid was a caste system based on race that existed in South Africa from 1948 until 1992. It consigned the large black majority to menial jobs, prevented marriage between blacks and whites, and erected separate public facilities for members of the two races. Asians and people of "mixed race" enjoyed privileges between these two extremes.

An **ascribed status** is a status that does not depend on the capabilities and efforts of the individual.

An **ascription-based** stratification system is one in which the allocation of rank depends on the characteristics a person is born with.

Assimilation is the process by which a minority group blends into the majority population and eventually disappears as a distinct group.

An **association** exists between two variables if the value of one variable changes with the value of the other.

Authoritarian states sharply restrict citizen control of the state.

Authoritarian leadership demands strict compliance from subordinates. Authoritarian leaders are most effective in a crisis such as a war or the emergency room of a hospital.

Authority is legitimate, institutionalized power.

In an **autocracy,** absolute power resides in the hands of a single person or party.

B

Breakdown theory suggests that social movements emerge when traditional norms and patterns of social organization are disrupted.

Biodiversity refers to the enormous variety of plant and animal species inhabiting the Earth.

Bisexuals are people who prefer sexual partners of both sexes.

The **bourgeoisie** are owners of the means of production, including factories, tools, and land. They do not do any physical labor. Their income derives from profits.

A **bureaucracy** is a large, impersonal organization composed of many clearly defined positions arranged in a hierarchy. A bureaucracy has a permanent, salaried staff of qualified experts and written goals, rules, and procedures. Staff members always try to find ways of running the bureaucracy more efficiently.

Bureaucratic inertia refers to the tendency of large, rigid bureaucracies to continue their policies even when their clients' needs change.

Bureaucratic ritualism involves bureaucrats getting so preoccupied with rules and regulations that they make it difficult for the organization to fulfill its goals.

C

Capitalism is the dominant economic system in the world today. Capitalist economies are characterized by private ownership of property and competition in the pursuit of profit.

A **caste** system is an almost pure ascription-based stratification system in which occupation and marriage partners are assigned on the basis of caste membership.

Causality refers to the analysis of causes and their effects. Specifically, causality means that a change in the independent variable (x) produces a change in the dependent variable (y). In analyzing survey data, we establish causality by demonstrating that

- an association exists between x and y;
- x precedes y; and
- the introduction of a causally prior control variable does not result in the original association disappearing.

Charismatic authority is based on belief in the claims of extraordinary individuals to be inspired by God or some higher principle.

The **Chicago school** founded urban sociology in the United States in the first decades of the 20th century. Its members distinguished themselves by their vivid and detailed descriptions and analyses of urban life and their development of the theory of human ecology.

A **church** is a bureaucratic religious organization that has accommodated itself to mainstream society and culture.

Civil citizenship recognizes the right to free speech, freedom of religion, and justice before the law.

A **civil religion** is a set of quasi-religious beliefs and practices that bind a population together and justify its way of life.

Civil society is the private sphere of social life.

Class in Marx's sense of the term is determined by one's relationship to the means of production. In Weber's usage, class is determined by one's "market situation." Wright distinguishes classes on the basis of relationship to the means of production, amount of property owned, organizational assets, and skill. For Goldthorpe, classes are determined mainly by one's "employment relations."

Class conflict is the struggle between classes to resist and overcome the opposition of other classes.

Class consciousness refers to being aware of membership in a class.

Codetermination is a German system of worker participation that allows workers to help formulate overall business strategy. German workers' councils review and influence management policies on a wide range of issues, including when and where new plants should be built and how capital should be invested in technological innovation.

Collective action occurs when people act in unison to bring about or resist social, political, and economic change. Some collective actions are routine. Others are nonroutine. Routine collective actions are typically nonviolent and follow established patterns of behavior in existing types of social structures. Nonroutine collective actions take place when usual conventions cease to guide social action and people transcend, bypass, or subvert established institutional patterns and structures.

The **collective conscience** is composed of the common sentiments and values that people share as a result of living together.

Colonialism refers to the control of developing societies by more developed, powerful societies.

Communism is an economic system characterized by public ownership of property and government planning of the economy.

Comparable worth refers to the equal dollar value of different jobs. It is established in gender-neutral terms by comparing jobs in terms of the education and experience needed to do them and the stress, responsibility, and working conditions associated with them.

Competition is a mode of interaction in which power is unequally distributed, but the degree of inequality is less than in systems of domination. Envy is an important emotion in competitive interactions. Competition in the theory of human ecology refers to the struggle by different groups for optimal locations in which to reside and set up their businesses.

Contagion is the process by which extreme passions supposedly spread rapidly through a crowd like a contagious disease.

Conflict crimes are illegal acts that many people consider harmful to society. However, many people think they are not very harmful. They are punishable by the state.

Conflict theories of deviance and crime hold that deviance and crime arise out of the conflict between the powerful and the powerless.

Conflict theories of social interaction emphasize that when people interact, their statuses are often arranged in a hierarchy. Those on top enjoy more power than those on the bottom. The degree of inequality strongly affects the character of social interaction between the interacting parties.

Conflict theory generally focuses on large, macro-level structures, such as the relations between classes. It shows how major patterns of inequality in society produce social stability in some circumstances and social change in others. It stresses how members of privileged groups try to maintain their advantages while subordinate groups struggle to increase theirs. It typically leads to the suggestion that eliminating privilege will lower the level of conflict and increase the sum total of human welfare.

Conglomerates are large corporations that operate in several industries at the same time.

Consensus crimes are illegal acts that nearly all people agree are bad and harm society greatly. The state inflicts severe punishment for consensus crimes.

Constraint theories identify the social factors that impose deviance and crime (or conventional behavior) on people.

A **contingency table** is a cross-classification of cases by at least two variables that allows you to see how, if at all, the variables are associated.

A **control group** in an experiment is the group that is not exposed to the independent variable.

Control theory holds that the rewards of deviance and crime are ample. Therefore, nearly everyone would engage in deviance and crime if they could get away with it. The degree to which people are prevented from violating norms and laws accounts for variations in the level of deviance and crime.

A **control variable** is a variable whose influence is removed from the association between an independent and a dependent variable.

Consumerism is the tendency to define ourselves in terms of the goods we purchase.

Cooperation is a basis for social interaction in which power is more or less equally distributed between people of different status. The dominant emotion in cooperative interaction is trust. Cooperation may also be defined more generally as the human capacity to create a complex social life.

The **core** capitalist countries are rich countries, such as the United States, Japan, and Germany, that are the major sources of capital and technology in the world.

The **corporate city** refers to the growing post–World War II perception and organization of the North American city as a vehicle for capital accumulation.

Corporations are legal entities that can enter into contracts and own property. They are taxed at a lower rate than individuals and their owners are normally not liable for the corporation's debt or any harm it may cause the public.

A **counterculture** is a subversive subculture that opposes dominant values and seeks to replace them.

Credential inflation refers to the fact that it takes ever more certificates and diplomas to qualify for a given job.

Crossnational variations in internal stratification are differences between countries in their stratification systems.

The **crude birth rate** is the annual number of live births per 1000 women in a population.

The **crude death rate** is the annual number of deaths per 1000 people in a population.

Cults are small groups of people deeply committed to a religious vision that rejects mainstream culture and society.

Cultural capital refers to widely shared, high-status cultural signals (attitudes, preferences, formal knowledge, behaviors, goals, and credentials) used for social and cultural exclusion.

Cultural relativism is the belief that all cultures have equal value.

Cultural studies is an increasingly popular interdisciplinary area of media research. It focuses not just on the cultural meanings producers try to transmit but also on the way audiences filter and interpret mass media messages in the context of their own interests, experiences, and values.

Culture is the sum of socially transmitted practices, languages, symbols, beliefs, values, ideologies, and material objects that people create to deal with real-life problems. Cultures enable people to adapt to, and thrive in, their environments.

Cultural lag is the tendency of symbolic culture to change more slowly than material culture.

D

Dehumanization occurs when bureaucracies treat clients as standard cases and personnel as cogs in a giant machine. This treatment frustrates clients and lowers worker morale.

In a **democracy,** citizens exercise a high degree of control over the state. They do this mainly by choosing representatives in regular, competitive elections.

Democratic leadership offers more guidance than the *laissez-faire* variety but less control than the authoritarian type. Democratic leaders try to include all group members in the decision-making process, taking the best ideas from the group and molding them into a strategy with which all can identify. Outside of crisis situations, democratic leadership is usually the most effective leadership style.

The **Democratic Revolution** began about 1750, during which time the citizens of the United States, France, and other countries broadened their participation in government. This revolution also suggested that people organize society and that human intervention can therefore resolve social problems.

In **democratic socialist** countries, democratically elected governments own certain basic industries entirely or in part and intervene vigorously in the market to redistribute income.

Demographers are social-scientific analysts of human population.

Demographic transition theory explains how changes in fertility and mortality affected population growth from preindustrial to postindustrial times.

Denominations are the various streams of belief and practice that some churches allow to coexist under their overarching authority.

Dependency theory views economic underdevelopment as the result of exploitative relations between rich and poor countries.

A **dependent variable** is the presumed effect in a cause-and-effect relationship.

Deskilling refers to the process by which work tasks are broken into simple routines requiring little training to perform. Deskilling is usually accompanied by the use of machinery to replace labor wherever possible and increased management control over workers.

Deviance occurs when someone departs from a norm.

Differential association theory holds that people learn to value deviant or nondeviant lifestyles depending on whether their social environment leads them to associate more with deviants or nondeviants.

Differentiation in the theory of human ecology refers to the process by which urban populations and their activities become more complex and heterogeneous over time.

Disabled people who are incapable of performing within the range of "normal" human activity.

Discrimination is unfair treatment of people because of their group membership.

The **division of labor** refers to the specialization of work tasks. The more specialized the work tasks in a society, the greater the division of labor.

The **divorce rate** is the number of divorces that occur in a year for every 1000 people in the population.

Domination is a mode of interaction in which nearly all power is concentrated in the hands of people of similar status. Fear is the dominant emotion in systems of interaction based on domination.

Dramaturgical analysis views social interaction as a sort of play in which people present themselves so that they appear in the best possible light.

A **dyad** is a social relationship between two "nodes," or social units (e.g., people, firms, organizations, countries, etc.).

Dysfunctions are effects of social structures that create social instability.

E

Ecclesia are state-supported churches.

Ecological succession in the theory of human ecology refers to the process by which a distinct urban group moves from one area to another and a second group comes in to replace the group that has moved out.

The **ecological theory** of ethnic succession argues that ethnic groups pass through five stages in their struggle for territory: invasion, resistance, competition, accommodation and cooperation, and assimilation.

The **economy** is the institution that organizes the production, distribution, and exchange of goods and services.

Edge cities are exurban clusters of malls, offices, and entertainment complexes that arise at the convergence point of major highways.

Educational achievement refers to how much students actually learn.

Educational attainment refers to the number of years of school students complete.

The **ego,** according to Freud, is a psychological mechanism that balances the conflicting needs of the pleasure-seeking id and the restraining superego.

Egoistic suicide results from a lack of integration of the individual into society because of weak social ties to others.

Elite theory holds that small groups occupying the command posts of America's most influential institutions make the important decisions that profoundly affect all members of society. Moreover, they do so without much regard for elections or public opinion.

Elites are exclusive groups that control the command posts of institutions.

Emigration, or out-migration, is the outflow of people from one country and their settlement in one or more other countries.

Emotion management involves people obeying "feeling rules" and responding appropriately to the situations in which they find themselves.

Emotion labor is emotion management that many people do as part of their job and for which they are paid.

Environmental racism is the tendency to heap environmental dangers on the disadvantaged, especially on disadvantaged racial minorities.

Essentialism is a school of thought that sees gender differences as a reflection of biological differences between women and men.

An **ethnic enclave** is a spatial concentration of ethnic group members who establish businesses that serve and employ mainly members of the ethnic group and reinvest profits in community businesses and organizations.

An **ethnic group** is composed of people whose perceived cultural markers are deemed socially significant. Ethnic groups differ from one another in terms of language, religion, customs, values, ancestors, and the like.

Ethnocentrism is the tendency to judge other cultures exclusively by the standards of one's own.

An **ethnographic** researcher spends months or even years living with people to learn their language, values, and mannerisms—their entire culture—and develop an intimate understanding of their behavior.

Ethnomethodology is the study of how people make sense of what others do and say by adhering to preexisting norms.

Euthanasia (also known as mercy killing and assisted suicide) involves a doctor prescribing or administering medication or treatment that is intended to end a terminally ill patient's life.

Exchange theory holds that social interaction involves trade in valued resources.

An **experiment** is a carefully controlled artificial situation that allows researchers to isolate hypothesized causes and measure their effects precisely.

An **experimental group** in an experiment is the group that is exposed to the independent variable.

Exploratory research is an attempt to describe, understand, and develop theory about a social phenomenon in the absence of much previous research on the subject.

Expulsion is the forcible removal of a population from a territory claimed by another population.

The **extended family** expands the nuclear family "vertically" by adding another generation—one or more of the spouses' parents—to the household.

Exurbs are rural residential areas within commuting distance of the city.

F

The **female–male earnings ratio** is women's earnings expressed as a percentage of men's earnings.

Feminist theory claims that patriarchy is at least as important as class inequality in determining a person's opportunities in life. It holds that male domination and female subordination are determined not by biological necessity but by structures of power and social convention. It examines the operation of patriarchy in both micro- and macro-level settings, and it contends that existing patterns of gender inequality can and should be changed for the benefit of all members of society.

Feudalism was a legal arrangement in preindustrial Europe that bound peasants to the land and obliged them to give their landlords a set part of the harvest. In exchange, landlords were required to protect peasants from marauders and open their storehouses and feed the peasants if crops failed.

Field research is research based on the observation of people in their natural settings.

Folkways are the least important norms and violating them evokes the least severe punishment.

Foraging societies are those in which people live by searching for wild plants and hunting wild animals. Such societies predominated until about 10,000 years ago. Inequality, the division of labor, productivity, and settlement size are very low in such societies.

Formal democracy involves regular, competitive elections.

Formal organizations are secondary groups designed to achieve explicit objectives.

Formal punishment takes place when the judicial system penalizes someone for breaking a law.

Frame alignment is the process by which individual interests, beliefs, and values become congruent with and complementary to the activities, goals, and ideology of a social movement.

In a **free market,** prices are determined only by supply and demand.

The **functional theory of stratification** argues that (a) some jobs are more important than others, (b) people have to make sacrifices to train for important jobs, and (c) inequality is required to motivate people to undergo these sacrifices.

Functionalist theory stresses that human behavior is governed by relatively stable social structures. It underlines how social structures maintain or undermine social stability. It emphasizes that social structures are based mainly on shared values or preferences and suggests that reestablishing equilibrium can best solve most social problems.

Fundamentalists interpret their scriptures literally, seek to establish a direct, personal relationship with the higher being(s) they worship, and are relatively intolerant of nonfundamentalists.

G

In **gated communities,** upper-middle-class residents pay high taxes to keep the community patrolled by security guards and walled off from the outside world.

Your **gender** is your sense of being male or female and your playing masculine and feminine roles in ways defined as appropriate by your culture and society.

Gender discrimination involves rewarding men and women differently for the same work.

Gender identity is one's identification with, or sense of belonging to, a particular sex—biologically, psychologically, and socially.

A **gender ideology** is a set of ideas about what constitutes appropriate masculine and feminine roles and behavior.

A **gender role** is the set of behaviors associated with widely shared expectations about how males or females are supposed to act.

Generalizability exists when research findings apply beyond the specific case examined.

The **generalized other,** according to Mead, is a person's image of cultural standards and how they apply to him or her.

A **generation** is an age group that has unique and formative historical experiences.

Genetic pollution refers to the potential dangers of mixing the genes of one species with those of another.

Genocide is the intentional extermination of an entire population defined as a "race" or a "people."

Gentrification is the process of middle-class people moving into rundown areas of the inner city and restoring them.

A **gerontocracy** is a society ruled by elderly people.

The **Gini index** is a measure of income inequality. Its value ranges from 0 (which means that every household earns exactly the same amount of money) to 1 (which means that all income is earned by a single household).

The **glass ceiling** is a social barrier that makes it difficult for women to rise to the top level of management.

A **global commodity chain** is a worldwide network of labor and production processes whose end result is a finished commodity.

Global inequality refers to differences in the economic ranking of countries.

Global structures are patterns of social relations that lie outside and above the national level. They include international organizations, patterns of worldwide travel and communication, and the economic relations between countries.

Global warming is the gradual worldwide increase in average surface temperature.

Globalization is the process by which formerly separate economies, states, and cultures are tied together and people become increasingly aware of their growing interdependence.

Glocalization is the simultaneous homogenization of some aspects of life and the strengthening of some local differences under the impact of globalization.

The **greenhouse effect** is the accumulation of carbon dioxide in the atmosphere that allows more solar radiation to enter the atmosphere and less solar radiation to escape.

Groupthink is group pressure to conform despite individual misgivings.

H

Hate crimes are criminal acts motivated by a person's race, religion, or ethnicity.

Health, according to the World Health Organization, is "the ability of an individual to achieve his [or her] potential and to respond positively to the challenges of the environment."

The **health-care system** is composed of a nation's clinics, hospitals, and other facilities for ensuring health and treating illness.

Heterosexuals are people who prefer members of the opposite sex as sexual partners.

A **hidden curriculum** teaches students what will be expected of them as conventionally good citizens once they leave school.

High culture is culture consumed mainly by upper classes.

Holistic medicine emphasizes disease prevention. Holistic practitioners treat disease by taking into account the relationship between mind and body and between the individual and his or her social and physical environment.

Homophobic people are afraid of homosexuals.

Homosexuals are people who prefer sexual partners of the same sex. People usually call homosexual men gay and homosexual women lesbians.

Horticultural societies are those in which people domesticate plants and use simple hand tools to garden. Such societies first emerged about 10,000 years ago. Horticulture increases the food supply and makes it more dependable. This increases average settlement size and permanence, the division of labor, productivity, and inequality above the levels typical of foraging societies.

Hostile-environment sexual harassment involves sexual jokes, comments, and touching that interfere with work or create an unfriendly work setting.

Human ecology is a theoretical approach to urban sociology that borrows ideas from biology and ecology to highlight the links between the physical and social dimensions of cities and identify the dynamics and patterns of urban growth.

The **human relations school of management** emerged as a challenge to Taylor's scientific management approach in the 1930s. It advocated less authoritarian leadership on the shop floor, careful selection and training of personnel, and greater attention to human needs and employee job satisfaction.

A **hypothesis** is an unverified but testable statement about the relationship between two or more variables.

I

The **I,** according to Mead, is the subjective and impulsive aspect of the self that is present from birth.

The **id,** according to Freud, is the part of the self that demands immediate gratification.

Immigration, or in-migration, is the inflow of people into one country from one or more other countries and their settlement in the destination country.

Impaired people are considered deficient in physical or mental capacity.

Imperialism is the economic domination of one country by another.

An **independent variable** is the presumed cause in a cause-and-effect relationship.

The **Industrial Revolution** refers to the rapid economic transformation that began in Britain in the 1780s. It involved the large-scale application of science and technology to industrial processes, the creation of factories, and the formation of a working class. It created a host of new and serious social problems that attracted the attention of many social thinkers.

Industrial societies use machines and fuel to greatly increase the supply and dependability of food and finished goods. The first such societies emerged about 225 years ago in Great Britain. Productivity, the division of labor, and average settlement size increased substantially in industrial societies compared with agricultural societies. While social inequality was substantial during early industrialism, it declined as the industrial system matured.

Infant mortality is the number of deaths before the age of 1 year for every 1000 live births in a population in 1 year.

Informal punishment involves a mild sanction that is imposed during face-to-face interaction, not by the judicial system.

In-group members are people who belong to a group.

An **initiation rite** is a ritual that signifies the transition of the individual from one group to another and ensures his or her loyalty to the new group.

In-migration (*see* immigration).

Institutional racism is bias that is inherent in social institutions and is often not noticed by members of the majority group.

Intergenerational mobility is social mobility that occurs between generations.

Interlocking directorates are formed when an individual sits on the board of directors of two or more noncompeting companies.

Internal colonialism involves one race or ethnic group subjugating another in the same country. It prevents assimilation by segregating the subordinate group in terms of jobs, housing, and social contacts.

Internal labor markets are social mechanisms for controlling pay rates, hiring, and promotions within corporations while reducing competition between a firm's workers and external labor supplies.

Intersexed people are born with ambiguous genitals because of a hormone imbalance in their mother's womb or some other cause.

Intragenerational mobility is social mobility that occurs within a single generation.

L

Labeling theory holds that deviance results not so much from the actions of the deviant as from the response of others, who label the rule breaker a deviant.

Labor market segmentation is the division of the market for labor into distinct settings. In these settings, work is found in different ways and workers have different characteristics. There is only a slim chance of moving from one setting to another.

A **language** is a system of symbols strung together to communicate thought.

Latent functions are invisible and unintended effects of social structures.

***Laissez-faire* leadership** allows subordinates to work things out largely on their own, with almost no direction from above. It is the least effective type of leadership.

Legal-rational authority is typical of modern societies. It derives from respect for the law. Laws specify how one can achieve office. People generally believe these laws are rational. If someone achieves office by following these laws, people respect their authority.

Legitimate governments are those that enjoy a perceived right to rule.

A **liberal democracy** is a country whose citizens enjoy regular, competitive elections and the freedoms and constitutional protections that make political participation and competition meaningful.

The **life course** refers to the distinct phases of life through which people pass. These stages vary from one society and historical period to another.
Life expectancy is the average age at death of the members of a population.
Lobbies are organizations formed by special interest groups to advise and influence politicians.
The **looking-glass self** is Cooley's description of the way our feelings about who we are depend largely on how we see ourselves evaluated by others.

M

Macrostructures are overarching patterns of social relations that lie outside and above one's circle of intimates and acquaintances. Macrostructures include classes, bureaucracies, and power systems such as patriarchy.
The **Malthusian trap** refers to a cycle of population growth followed by an outbreak of war, pestilence, or famine that keeps population growth in check.
Manifest functions are visible and intended effects of social structures.
Markets are social relations that regulate the exchange of goods and services. In a market, the prices of goods and services are established by how plentiful they are (supply) and how much they are wanted (demand).
Marriage is a socially approved, presumably long-term sexual and economic union between a man and a woman. It involves reciprocal rights and obligations between spouses and between parents and children.
The **marriage rate** is the number of marriages that occur in a year for every 1000 people in the population.
Mass culture (*see* Popular culture).
The **mass media** are print, radio, television, and other communication technologies. The word "mass" implies that the media reach many people. The word "media" signifies that communication does not take place directly through face-to-face interaction. Instead, technology intervenes or mediates in transmitting messages from senders to receivers. Furthermore, communication via the mass media is usually one way, or at least one-sided. There are few senders (or producers) and many receivers (or audience members). In a democracy, mass media help to keep the public informed about the quality of government.
One's **master status** is the status that is most influential in shaping one's life at a given time and hence one's overriding public identity.
Material culture is composed of the tools and techniques that enable people to get tasks accomplished.
The **maximum average human life span** is the average age of death for a population under ideal conditions. It is currently about 85 years.
McDonaldization is a form of rationalization. Specifically, it refers to the spread of the principles of fast-food restaurants, such as efficiency, predictability, and calculability, to all spheres of life.
The **me,** according to Mead, is the objective component of the self that emerges as people communicate symbolically and learn to take the role of the other.
Media convergence is the blending of the World Wide Web, television, and other communications media as new, hybrid media forms.
Media imperialism is the domination of a mass medium by a single national culture and the undermining of other national cultures.
The **medicalization of deviance** is the tendency for medical definitions of deviant behavior to become more prevalent over time.
Medicine is an institution devoted to fighting disease and promoting health.
A **meritocracy** is a stratification system in which equality of opportunity allows people to rise or fall to a position that matches their talent and effort.
Metropolitan areas include downtown city cores and their surrounding suburbs.
Microstructures are the patterns of relatively intimate social relations formed during face-to-face interaction. Families, friendship circles, and work associations are all examples of microstructures.
A **minority group** is a group of people who are socially disadvantaged, even if they are in the numerical majority (e.g., the blacks of South Africa).
Modernization theory holds that economic underdevelopment results from poor countries lacking Western attributes. These attributes include Western values, business practices, levels of investment capital, and stable governments.
A **moral panic** occurs when many people fervently believe that some form of deviance or crime poses a profound threat to society's well-being.
Mores are core norms that most people believe are essential for the survival of their group or society.
Motivational theories identify the social factors that drive people to commit deviant and criminal acts.
Supporters of **multiculturalism** argue that the curricula of America's public schools and colleges should reflect the country's ethnic and racial diversity and recognize the equality of all cultures.

N

Neoliberal globalization is a policy that promotes private control of industry, minimal government interference in the running of the economy, the removal of taxes, tariffs, and restrictive regulations that discourage the international buying and selling of goods and services, and the encouragement of foreign investment.
New social movements became prominent in the 1970s. They attract a disproportionately large number of highly educated people in the social, educational, and cultural fields and universalize the struggle for citizenship.
The **new urban sociology** emerged in the 1970s and stressed that city growth is a process rooted in power relations and the urge to profit.
Nonmaterial culture is composed of symbols, norms, and other nontangible elements of culture.
Normal accidents are those that occur inevitably though unpredictably due to the very complexity of modern technologies.
Norms are generally accepted ways of doing things.
A **nuclear family** consists of a cohabiting man and woman who maintain a socially approved sexual relationship and have at least one child.

O

Oligarchy means "rule of the few." All bureaucracies have a supposed tendency for power to become increasingly concentrated in the hands of a few people at the top of the organizational pyramid.
Oligopolies are giant corporations that control part of an economy. They are few and tend not to compete against one another. Instead, they can set prices at levels that are most profitable for them.
Operationalization is the procedure by which researchers establish criteria for assigning observations to variables.
An **organizational environment** is composed of a host of economic, political, cultural, and other factors that lie outside an organization and affect the way it works.
Out-group members are people who are excluded from the in-group.
Out-migration (*see* emigration).
The **ozone layer** lies 5 to 25 miles above the Earth's surface. (Ozone is a gas composed of three oxygen atoms.) It is depleted by CFCs. The depletion of the ozone layer allows more ultravio-

let light to enter the Earth's atmosphere. This increases the rate of skin cancer and crop damage.

P

Participant observation involves carefully observing people's face-to-face interactions and actually participating in their lives over a long period, thus achieving a deep and sympathetic understanding of what motivates them to act in the way they do.

Parties, in Weber's usage, are organizations that seek to impose their will on others.

Pastoral societies are those in which people domesticate cattle, camels, pigs, goats, sheep, horses, and reindeer. Such societies first emerged about 10,000 years ago. Domesticating animals increases the food supply and makes it more dependable. This increases average settlement size and permanence, the division of labor, productivity, and inequality above the levels typical of foraging societies.

Patriarchy is the traditional system of economic and political inequality between women and men.

One's **peer group** is composed of people who are about the same age and of similar status as the individual. The peer group acts as an agent of socialization.

The **peripheral** countries are former colonies that are poor and are major sources of raw materials and cheap labor.

The **petty bourgeoisie,** in Marx's usage, is the class of small-scale capitalists who own means of production but employ only a few workers or none at all, forcing them to do physical work themselves.

The **placebo effect** is the positive influence on healing of strong belief in the effectiveness of a cure.

Pluralism is the retention of racial and ethnic culture combined with equal access to basic social resources.

Pluralist theory holds that power is widely dispersed. As a result, no group enjoys disproportionate influence, and decisions are usually reached through negotiation and compromise.

Political action committees (PACs) are organizations that raise funds for politicians who support particular issues.

Political citizenship recognizes the right to run for office and vote.

Political opportunities for collective action and social-movement growth occur during election campaigns, when influential allies offer insurgents support, when ruling political alignments become unstable, and when elite groups become divided and conflict with one another.

Political parties are organizations that compete for control of government in regular elections. In the process, they give voice to policy alternatives and rally adult citizens to vote.

A **political revolution** is the overthrow of political institutions by an opposition movement and replacement by new institutions.

Polygamy expands the nuclear family "horizontally" by adding one or more spouses (usually women) to the household.

Popular culture (or **mass culture**) is culture consumed by all classes.

A **population** is the entire group about which a researcher wishes to generalize.

The **Postindustrial Revolution** refers to the technology-driven shift from manufacturing to service industries and the consequences of that shift for virtually all human activities.

Postindustrial societies are those in which most workers are employed in the service sector and computers spur substantial increases in the division of labor and productivity. Shortly after World War II, the United States became the first postindustrial society. Gender inequality is reduced in postindustrial societies, partly because so many women are brought into the system of higher education and into the paid labor force. Class inequality increases in some postindustrial societies.

Postmaterialism is a theory that claims that growing equality and prosperity in the rich industrialized countries have resulted in a shift from class-based to value-based politics.

Postmodernism is characterized by an eclectic mixing of cultural elements and the erosion of consensus.

The **postmodern city** is a new urban form that is more privatized and socially and culturally fragmented and globalized than the corporate city.

Postnatural societies are those in which genetic engineering enables people to create new life forms. While genetic engineering holds out much promise for improving productivity, feeding the poor, ridding the world of disease, and so forth, social inequality could increase in postnatural societies unless people democratically decide on the acceptable risks of genetic engineering and the distribution of its benefits.

The **poverty rate** is the percentage of people living below the poverty threshold, which is three times the minimum food budget established by the U.S. Department of Agriculture.

Power is the probability that one actor within a social relationship will be in a position to carry out his or her own will despite resistance.

Power resource theory holds that the distribution of power between major classes partly accounts for the successes and failures of different political parties.

Prejudice is an attitude that judges a person according to his or her group's real or imagined characteristics.

Primary groups are groups whose members agree upon norms, roles, and statuses without putting them in writing. Social interaction leads to strong emotional ties, extends over a long period, involves a wide range of activities, and results in group members knowing one another well.

The **primary labor market** is composed mainly of highly skilled or well-educated white males. They are employed in large corporations that enjoy high levels of capital investment. In the primary labor market, employment is secure, earnings are high, and fringe benefits are generous.

Primary socialization is the process of acquiring the basic skills needed to function in society during childhood. Primary socialization usually takes place in a family.

In a **probability sample,** the units have a known and nonzero chance of being selected.

Production is the human capacity to make and use tools. It improves our ability to take what we want from nature.

Productivity refers to the amount of goods or services produced for every hour worked.

The **profane** refers to the secular, everyday world.

Professionalization takes place when members of an occupation insist that people earn certain credentials to enter the occupation. Professionalization ensures standards and keeps professional earnings high.

Professionals are people with specialized knowledge acquired through extensive higher education. They enjoy a high degree of work autonomy and usually regulate themselves and enforce standards through professional associations.

The **proletariat,** in Marx's usage, is the working class. Members of the proletariat perform physical labor but do not own means of production. They are thus in a position to earn wages.

The **Protestant ethic** is the 16th- and 17th-century Protestant belief that religious doubts can be reduced, and a state of grace assured, if people work diligently and live ascetically.

According to Weber, the Protestant ethic had the unintended effect of increasing savings and investment and thus stimulating capitalist growth.

The **public health system** is composed of government-run programs that ensure access to clean drinking water, basic sewage and sanitation services, and inoculation against infectious diseases.

Public opinion is composed of the values and attitudes of the adult population as a whole. It is expressed mainly in polls and letters to lawmakers and gives politicians a reading of citizen preferences.

Public policy involves the creation of laws and regulations by organizations and governments.

Q

The **quality of work life** movement originated in Sweden and Japan. It involves small groups of a dozen or so workers and managers collaborating to improve both the quality of goods produced and communication between workers and managers.

Quid pro quo sexual harassment takes place when sexual threats or bribery are made a condition of employment decisions.

R

Race is a social construct used to distinguish people in terms of one or more physical markers, usually with profound effects on their lives.

Racism is the belief that a visible characteristic of a group, such as skin color, indicates group inferiority and justifies discrimination.

Random means "by chance"—for example, having an equal and nonzero probability of being sampled.

Randomization involves assigning individuals to groups by chance processes.

Rational choice theory focuses on the way interacting people weigh the benefits and costs of interaction. According to rational choice theory, interacting people always try to maximize benefits and minimize costs.

Rationalization is the application of the most efficient means to achieve given goals and the unintended, negative consequences of doing so.

Recombinant DNA involves removing a segment of DNA from a gene or splicing together segments of DNA from different living things, thus effectively creating a new life form.

A **reference group** is composed of people against whom an individual evaluates his or her situation or conduct.

Regionalization is the division of the world into different and often competing economic, political, and cultural areas.

In a **regulated market,** various social forces limit the capacity of supply and demand to determine prices.

Rehabilitation involves curing disabilities to the extent possible through medical and technological intervention; trying to improve the lives of the disabled by means of care, training, and education; and integrating the disabled into "normal" society.

Relative deprivation is an intolerable gap between the social rewards people feel they deserve and the social rewards they actually receive.

Reliability is the degree to which a measurement procedure yields consistent results.

Religiosity refers to how important religion is to people.

The **replacement level** is the number of children that each woman must have on average for population size to remain stable. Ignoring any inflow of population from other countries and any outflow to other countries, the replacement level is 2.1.

Research is the process of systematically observing reality to assess the validity of a theory.

Resocialization occurs when powerful socializing agents deliberately cause rapid change in one's values, roles, and self-conception, sometimes against one's will.

Resource mobilization refers to the process by which social movements crystallize due to the increasing organizational, material, and other resources of movement members.

The **revised secularization thesis** holds that worldly institutions break off from the institution of religion over time. As a result, religion governs an ever smaller part of most people's lives and becomes largely a matter of personal choice.

The **rights revolution** is the process by which socially excluded groups have struggled to win equal rights under the law and in practice since the 1960s.

A **risk society** is a society defined by the way risk is distributed as a side effect of technology.

Rites of passage are cultural ceremonies that mark the transition from one stage of life to another (e.g., baptisms, confirmations, weddings) or from life to death (e.g., funerals).

Rituals are public practices designed to connect people to the sacred.

A **role** is the behavior expected of a person occupying a particular position in society.

Role conflict occurs when two or more statuses held at the same time place contradictory role demands on a person.

Role distancing involves giving the impression that one is just "going through the motions" and actually lacks serious commitment to a role.

A **role set** is a cluster of roles attached to a single status.

Role strain occurs when incompatible role demands are placed on a person in a single status.

The **routinization of charisma** is Weber's term for the transformation of the unique gift of divine enlightenment into a permanent feature of everyday life. It involves turning religious inspiration into a stable social institution with defined roles (interpreters of the divine message, teachers, dues-paying laypeople, etc.).

A **ruling class** is a self-conscious, cohesive group of people in elite positions. They act to advance their common interests and are led by corporate executives.

Rumors are claims about the world that are not supported by authenticated information. They are a form of communication that takes place when people try to construct a meaningful interpretation of an ambiguous situation.

S

The **sacred** refers to the religious, transcendent world.

A **sample** is the part of the population of research interest that is selected for analysis.

Sanctions are rewards and punishments intended to ensure conformity to cultural guidelines.

The **Sapir-Whorf thesis** holds that we experience certain things in our environment and form concepts about them. We then develop language to express our concepts. Finally, language itself influences how we see the world.

A **scapegoat** is a disadvantaged person or category of people whom others blame for their own problems.

Scientific management is a system of improving productivity developed in the 1910s by Frederick W. Taylor. After analyzing the movements of workers as they did their jobs, Taylor trained them to eliminate unnecessary actions and greatly improve their efficiency.

The **Scientific Revolution** began in Europe about 1550. It encouraged the view that sound conclusions about the workings of society must be based on solid evidence, not just speculation.

Secondary groups are larger and more impersonal than primary groups. Compared with primary groups, social interaction in secondary groups creates weaker emotional ties, extends over a shorter period, involves a narrow range of activities, and results in most group members having at most a passing acquaintance with one another.

The secondary labor market contains a disproportionately large number of women and members of racial minorities, particularly African and Hispanic Americans. Employees in the secondary labor market tend to be unskilled and lack higher education. They work in small firms with low levels of capital investment. Employment is insecure, earnings are low, and fringe benefits are meager.

Secondary socialization is socialization outside the family after childhood.

Sects usually form by breaking away from churches due to disagreement about church doctrine. Sects are less integrated into society and less bureaucratized than churches. They are often led by charismatic leaders, who tend to be relatively intolerant of religious opinions other than their own.

The **secularization thesis** says that religious institutions, actions, and consciousness are on the decline worldwide.

Segregation involves the spatial and institutional separation of racial or ethnic groups.

The **self** consists of one's ideas and attitudes about who one is.

A **self-fulfilling prophecy** is an expectation that helps bring about what it predicts.

In **self-report surveys,** respondents are asked to report their involvement in criminal activities, either as perpetrators or victims.

Self-socialization involves choosing socialization influences from the wide variety of mass media offerings.

The **semiperipheral** countries, such as South Korea, Taiwan, and Israel, consist of former colonies that are making considerable headway in their attempts to industrialize.

Your **sex** depends on whether you were born with distinct male or female genitals and a genetic program that released either male or female hormones to stimulate the development of your reproductive system.

The **sex ratio** is the ratio of women to men in a geographical area.

Playing the **sick role,** according to Talcott Parsons, involves the nondeliberate suspension of routine responsibilities, wanting to be well, seeking competent help, and cooperating with health-care practitioners at all times.

Significant others are people who play important roles in the early socialization experiences of children.

Slavery is the ownership and control of people.

A **social category** is composed of people who share a similar status but do not identify with one another.

Social citizenship recognizes the right to a certain level of economic welfare security and full participation in the social life of the country.

Social constructionism is a school of thought that sees gender differences as a reflection of the different social positions occupied by women and men.

Social constructionists argue that apparently natural or innate features of life are often sustained by social processes that vary historically and culturally.

Social control refers to the social sanctions by means of which conformity to cultural guidelines is ensured. It refers to the means by which authorities seek to contain collective action, including co-optation, concessions, and coercion.

Social deviations are noncriminal departures from norms that are nonetheless subject to official control. Some members of the public regard them as somewhat harmful, whereas other members of the public do not.

A **social diversion** is a minor act of deviance that is generally perceived as relatively harmless and that evokes, at most, a mild societal reaction such as amusement or disdain.

A **social group** is composed of one or more networks of people who identify with one another and adhere to defined norms, roles, and statuses.

Social interaction involves people communicating face-to-face and acting and reacting in relation to other people. It is structured around norms, role, and statuses.

Social movements are enduring collective attempts to change part or all of the social order by means of rioting, petitioning, striking, demonstrating, and establishing lobbies, unions, and political parties.

A **social network** is a bounded set of individuals who are linked by the exchange of material or emotional resources. The patterns of exchange determine the boundaries of the network. Members exchange resources more frequently with each other than with nonmembers. They also think of themselves as network members. Social networks may be formal (defined in writing), but they are more often informal (defined only in practice).

Social solidarity refers to (1) the degree to which group members share beliefs and values and (2) the intensity and frequency of their interaction.

Social stratification refers to the way society is organized in layers, or strata.

Social structures are stable patterns of social relations.

Socialization is the process by which people learn their culture. They do so by entering and disengaging from a succession of roles and becoming aware of themselves as they interact with others.

A **society** is composed of people who interact, usually in a defined territory, and share a culture.

Blau and Duncan's **socioeconomic index of occupational status (SEI)** combines, for each occupation, average earnings and years of education of men employed full time in the occupation.

Socioeconomic status (SES) combines income, education, and occupational prestige data in a single index of one's position in the socioeconomic hierarchy.

The **sociological imagination** is the quality of mind that enables one to see the connection between personal troubles and social structures.

Sociology is the systematic study of human behavior in social context.

Solidarity theory suggests that social movements are social organizations that emerge when potential members can mobilize resources, take advantage of new political opportunities, and avoid high levels of social control by authorities.

In **split labor markets,** low-wage workers of one race and high-wage workers of another race compete for the same jobs. High-wage workers are likely to resent the presence of low-wage competitors, and conflict is bound to result. Consequently, racist attitudes develop or get reinforced.

A **spurious association** exists between an independent and a dependent variable when the introduction of a causally prior control variable makes the initial association disappear.

The **state** consists of the institutions responsible for formulating and carrying out a country's laws and public policies.

State-centered theory holds that the state itself can structure political life to some degree independently of the way power is distributed between classes and other groups at a given time.

Statistical significance exists when a finding is unlikely to occur by chance, usually in 19 out of every 20 samples of the same size.

Status refers to a recognized social position that an individual can occupy.

Status cues are visual indicators of other people's social position.

Status groups differ from one another in terms of the prestige or social honor they enjoy and in terms of their lifestyle.

A **status set** is the entire ensemble of statuses occupied by an individual.

Stereotype threat refers to the harmful impact of negative stereotypes on the school performance of disadvantaged groups.

Stereotypes are rigid views of how members of various groups act, regardless of whether individual group members really behave that way.

People who are **stigmatized** are negatively evaluated because of a marker that distinguishes them from others.

Strain refers to breakdowns in traditional norms that precede collective action.

Strain theory holds that people may turn to deviance when they experience strain. Strain results when a culture teaches people the value of material success and society fails to provide enough legitimate opportunities for everyone to succeed.

Street crimes include arson, burglary, assault, and other illegal acts disproportionately committed by people from lower classes.

Structural mobility refers to the social mobility that results from changes in the distribution of occupations.

Subcultural theory argues that gangs are a collective adaptation to social conditions. Distinct norms and values that reject the legitimate world crystallize in gangs.

A **subculture** is a set of distinctive values, norms, and practices within a larger culture.

Suburbanism is a way of life outside city centers that is organized mainly around the needs of children and involves higher levels of conformity and sociability than life in the central city.

The **superego,** according to Freud, is a part of the self that acts as a repository of cultural standards.

A **survey** asks people questions about their knowledge, attitudes, or behavior, in either a face-to-face interview, telephone interview, or paper-and-pencil format.

A **symbol** is anything that carries a particular meaning, including the components of language, mathematical notations, and signs. Symbols allow us to classify experience and generalize from it.

Symbolic ethnicity is a nostalgic allegiance to the culture of the immigrant generation, or that of the old country, that is not usually incorporated into everyday behavior.

Symbolic interactionist theory focuses on interpersonal communication in micro-level social settings. It emphasizes that an adequate explanation of social behavior requires understanding the subjective meanings people attach to their social circumstances. It stresses that people help to create their social circumstances and do not merely react to them. And, by underscoring the subjective meanings people create in small social settings, it validates unpopular and nonofficial viewpoints. This increases our understanding and tolerance of people who may be different from us.

T

Taboos are among the strongest norms. When someone violates a taboo, it causes revulsion in the community, and punishment is severe.

Techniques of neutralization are the rationalizations that deviants and criminals use to justify their activities. Techniques of neutralization make deviance and crime seem normal, at least to the deviants and criminals themselves.

Technological determinism is the belief that technology is the main factor shaping human society and history.

Technology is the practical application of scientific principles.

Technopoly is a form of social organization in which technology compels people to try to solve all problems using technical rather than moral criteria, even though technology is often the source of the problems.

Terrorism is defined by American law as premeditated, politically motivated violence against noncombatant targets including unarmed or off-duty military personnel by subnational groups or clandestine agents, usually intended to influence an audience. However, a broader definition would include the use by states of indiscriminate violence against civilians in order to achieve military and political goals. Moreover, what one side in a conflict would call terror might be regarded by the other side as legitimate resistance to occupation, as well as ethnic, religious, or national oppression.

A **theory** is a tentative explanation of some aspect of social life that states how and why certain facts are related.

The **Thomas theorem** states: "Situations we define as real become real in their consequences."

The **total fertility rate** is the average number of children that would be born to a woman over her lifetime if she had the same number of children as women in each age cohort in a given year.

Total institutions are settings where people are isolated from the larger society and under the strict control and constant supervision of a specialized staff.

Totems are objects that symbolize the sacred.

Tracking involves sorting students into high-ability, middle-ability, and low-ability classes based on the results of IQ and other tests.

Traditional authority, the norm in tribal and feudal societies, involves rulers inheriting authority through family or clan ties. The right of a family or clan to monopolize leadership is widely believed to be derived from the will of a god.

A **traditional nuclear family** is a nuclear family in which the husband works outside the home for money and the wife works for free in the home.

Transgendered people are individuals who break society's gender norms by defying the rigid distinction between male and female.

Transnational communities are communities whose boundaries extend between countries.

Transnational corporations are large businesses that rely increasingly on foreign labor and foreign production; skills and advances in design, technology, and management; world markets; and massive advertising campaigns. They are increasingly autonomous of national governments.

Transsexuals believe they were born with the "wrong" body. They identify with, and want to live fully as, members of the "opposite" sex, and to do so they often change their appearance or resort to medical intervention.

A **triad** is a social relationship among three "nodes," or social units (e.g., people, firms, organizations, countries, etc.).

The **two-step flow of communication** between mass media and audience members involves (1) respected people of high status and independent judgment evaluating media messages and (2) other members of the community being influenced to varying degrees by these opinion leaders. Due to the two-step flow of communication, opinion leaders filter media messages.

U

The **unconscious,** according to Freud, is the part of the self that contains repressed memories that we are not normally aware of.

Union density is the number of union members in a given location and time as a percentage of nonfarm workers. It measures the organizational power of unions.

Unions are organizations of workers that seek to defend and promote their members' interests.

Universal citizenship recognizes the right of marginal groups to full citizenship and the rights of humanity as a whole.

Urban sprawl is the spread of cities into ever larger expanses of the surrounding countryside.

Urbanism is a way of life that, according to Louis Wirth, involves increased tolerance but also emotional withdrawal and specialized, impersonal, and self-interested interaction.

V

Values are ideas about what is right and wrong.

Validity is the degree to which a measure actually measures what it is intended to measure.

A **variable** is a concept that can take on more than one value.

Vertical social mobility refers to movement up or down the stratification system.

Victimless crimes involve violations of the law in which no victim has stepped forward and been identified.

A **virtual community** is an association of people, scattered across the country, continent, or planet, who communicate via computer and modem about a subject of common interest.

W

War is a violent armed conflict between politically distinct groups who fight to protect or increase their control of territory.

White-collar crime refers to an illegal act committed by a respectable, high-status person in the course of work.

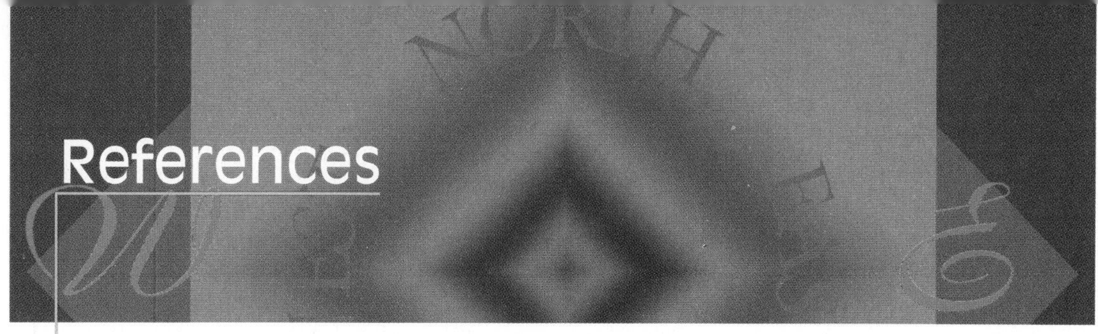

References

Abelmann, Nancy, and John Lie. 1995. *Blue Dreams: Korean Americans and the Los Angeles Riots.* Cambridge, MA: Harvard University Press.

Abraham, Carolyn. 2003. "Hong Kong Hotel is Focus of Pneumonia Investigation." *globeandmail.com* 20 March. On the World Wide Web at http://globeandmail.workopolis.com/servlet/Content/fasttrack/20030320/UBUGGN?section=Healthcare (16 June 2003).

Abraham, Laurie Kaye. 1993. *Mama Might Be Better Off Dead: The Failure of Health Care in Urban America.* Chicago: University of Chicago Press.

Abramsky, Sasha. 1999. "When They Get Out." *The Atlantic Monthly* June. On the World Wide Web http://www.theatlantic.com/issues/99jun/9906prisoners.htm (29 April 2000).

Achilles, Rhona. 1993. "Desperately Seeking Babies: New Technologies of Hope and Despair." Pp. 214–229 in Bonnie J. Fox, ed. *Family Patterns, Gender Relations.* Toronto: Oxford University Press.

Adams, Henry E., Lester W. Wright, Jr., and Bethany A. Lohr. 1998. "Is Homophobia Associated with Homosexual Arousal?" *Journal of Abnormal Psychology* 105: 440–445.

AdCritic.com. 2000. "Ad Critic: All Ads, All the Time." On the World Wide Web at http://www.adcritic.com (16 May 2000).

Adherents.com. 2001. "Religion Statistics: Predominant Religions." On the World Wide Web at http://www.adherents.com/adh_predom.html (30 November 2001).

Adler, Patricia A., and Peter Adler. 1998. *Peer Power: Preadolescent Culture and Identity.* New Brunswick, NJ: Rutgers University Press.

Ahmad, Imad-ad-Dean. 2000. "Female Genital Mutilation: An Islamic Perspective." On the World Wide Web at http://www.minaret.org/fgm-pamphlet.htm (20 January 2003).

Akard, Patrick J. 1992. "Corporate Mobilization and Political Power: The Transformation of US Economic Policy in the 1970s." *American Sociological Review* 57: 587–615.

Akin, David. 2002 "Kevin Warwick is a Borg." *Globe and Mail* June 1: F7.

Albas, Daniel, and Cheryl Albas. 1989. "Modern Magic: The Case of Examinations." *The Sociological Quarterly* 30: 603–613.

Albelda, Randy, and Nancy Folbre. 1996. *The War on the Poor: A Defense Manual.* New York: New Press.

Albrow, Martin. 1997. *The Global Age: State and Society Beyond Modernity.* Stanford CA: Stanford University Press.

Aldrich, Howard E. 1979. *Organizations and Environments.* Englewood Cliffs, NJ: Prentice-Hall.

Alford, Robert R., and Roger Friedland. 1985. *Powers of Theory: Capitalism, the State, and Democracy.* Cambridge: Cambridge University Press.

Amato, Paul R., and Bruce Keith. 1991. "Parental Divorce and the Well-Being of Children: A Meta-Analysis." *Psychological Bulletin* 110: 26–46.

American Association of Retired People. 1999. *A Profile of Older Americans.* Washington, DC. On the World Wide Web at http://research.aarp.org/general/profile99.pdf (13 August 2000).

__________. 2001. *A Profile of Older Americans 2001.* On the World Wide Web at http://research.aarp.org/general/profile_2001.pdf (23 April 2003).

American Chiropractic Association. 1999. "Two More Surveys Show Demand for Alternative Care is Rising." On the World Wide Web at http://www.amerchiro.org/research/new_research.html (2 May 2000).

American Psychological Association. "Answers to Your Questions About Sexual Orientation and Homosexuality." 1998. On the World Wide Web at http://www.apa.org/pubinfo/orient.html (14 June 2000).

American Society of Plastic Surgeons. 2003. "Plastic Surgery Information Service." On the World Wide Web at http://www.plasticsurgery.org/mediactr/expanded2001.cfm (1 March 2003).

__________. 2004. "National Clearinghouse of Plastic Surgery Statistics." On the World Wide Web at http://www.plasticsurgery.org/public_\education/Statistical-Trends.cfm (27 August 2004).

American Sociological Association. 2005. *Careers in Sociology.* On the World Wide Web at http://www.asanet.org/student/career/homepage.html (23 January 2005).

Anderson, Ben. 1999. "GOP Combats Census Sampling with Money, Logistics." *Conservative News Service.* On the World Wide Web at http://www.conservativenews.net/InDepth/archive/199903/IND19990316b.html (6 May 2000).

Anderson, Benedict O. 1991. *Imagined Communities: Reflections on the Origin and Spread of Nationalism.* London: Verso.

Anderson, Elijah. 1990. *Streetwise: Race, Class, and Change in an Urban Community.* Chicago: University of Chicago Press.

Anderson, Margo, and Stephen E. Feinberg. 2000. "Race and Ethnicity and the Controversy over the US Census." *Current Sociology* 48, 3: 87–110.

Anderson, Robert N. 2002. "Deaths: Leading Causes for 2000." *National Vital Statistics Reports* 50, 16. On the World Wide Web at http://www.cdc.gov/nchs/data/nvsr/nvsr50/nvsr50_16.pdf (13 June 2003).

Angier, Natalie. 2000. "Do Races Differ? Not Really, DNA Shows." *The New York Times on the Web* 22 August. On the World Wide Web at http://www.nytimes.com/library/national/science/082200sci-genetics-race.html (24 August 2000).

"Ann McLaughlin Named to Microsoft Board of Directors." 2000. *Microsoft PressPass* 27 January. On the World Wide Web at http://www.microsoft.com/PressPass/press/2000/Jan00/annMcLaughlinPR.asp (30 April 2000).

Annie E. Casey Foundation. 1998. *Child Care You Can Count On: Model Programs and Policies.* Baltimore. On the World Wide Web at http://www.kidscount.org/publications/child/afford.htm (30 April 2000).

Antonius, George. 1939. *The Arab Awakening: The Story of the Arab National Movement.* Philadelphia: J. B. Lipincott.

"Aqua Singer Defends Breast Implants." 2000. *National Post* 6 April: F11.

Arendt, Hannah. 1977 [1963]. *Eichmann in Jerusalem: A Report on the Banality of Evil,* rev. ed. Harmondsworth, UK: Penguin.

Arias, Elizabeth, and Betty L. Smith. 2003. "Deaths: Preliminary Data for 2001." *National Vital Statistics Reports* 51, 5. On the World Wide Web at http://www.cdc.gov/nchs/data/nvsr/nvsr51/nvsr51_05.pdf (13 June 2003).

Ariès, Phillipe. 1962 [1960]. *Centuries of Childhood: A Social History of Family Life,* Robert Baldick, trans. New York: Knopf.

__________. 1982. *The Hour of Our Death.* New York: Knopf.

Armour, Stephanie. 2005. "Pregnant workers report growing discrimination." *USA Today* 17 February: 1B–2B.

Árnason, Bragi, and Thorsteinn I. Sigfússon. 2000. "Iceland—A Future Hydrogen Economy." *International Journal of Hydrogen Energy* 25:389–394.

Arnett, Jeffrey Jensen. 1995. "Adolescents' Uses of Media for Self-Socialization." *Journal of Youth and Adolescence* 24: 519–533.

Arterton, F. Christopher. 1987. *Teledemocracy: Can Technology Protect Democracy?* Newbury Park, CA: Sage Publications.

Asch, Solomon. 1955. "Opinion and Social Pressure." *Scientific American* July: 31–35.

Associated Press. 2002. "More People on Welfare after Years of Decline." *New York Times* December 31. On the World Wide Web at http://www.nytimes.com (31 December 2002).

Averett, Susan, and Sanders Korenman. 1996. "The Economic Reality of *The Beauty Myth.*" *Journal of Human Resources* 31: 304–330.

Avery, Christopher, Andrew Fairbanks, and Richard Zeckhauser. 2003. *The Early Admission Game: Joining the Elite.* Cambridge, MA: Harvard University Press.

Babbie, Earl. 2000 [1973]. *The Practice of Social Research,* rev. ed. of 9th ed. Belmont CA: Wadsworth.

Baca Zinn, M., and D. Stanley Eitzen. 1993 [1988]. *Diversity in American Families,* 3rd ed. New York: HarperCollins.

Bagdikian, Ben H. 1997 [1983]. *The Media Monopoly,* 5th ed. Boston: Beacon.

Bairoch, Paul. 1988 [1985]. *Cities and Economic Development: From the Dawn of History to the Present.* Christopher Braider, trans. Chicago: University of Chicago Press.

Bales, Kevin. 1999. *Disposable People: New Slavery in the Global Economy.* Berkeley: University of California Press.

__________. 2002. "The Social Psychology of Modern Slavery." *Scientific American* 286, 4: 80–88.

Ball, Howard, S. D. Berkowitz, and Mbulelo Mzamane, eds. 1998. *Multicultural Education in Colleges and Universities: A Transdisciplinary Approach.* Mahwah, NJ: Lawrence Erlbaum Associates.

Baltzell, E. Digby. 1964. *The Protestant Establishment: Aristocracy and Caste in America.* New York: Vintage.

Bank of Hawaii. 1999. *Commonwealth of the Northern Mariana Islands: Economic Report, October 1999.* On the World Wide Web at http://www.boh.com/econ/pacific/cnmi/1999/cnmi1999.pdf (23 June 2000).

Banner, Lois W. 1992. *In Full Flower: Aging Women, Power, and Sexuality.* New York: Knopf.

Bannon, Lisa. 2000. "Why Girls and Boys Get Different Toys." *The Wall Street Journal* 14 February: B1, B4.

Baran, Paul A. 1957. *The Political Economy of Growth.* New York: Monthly Review Press.

Barash, David. 1981. *The Whispering Within.* New York: Penguin.

Barber, Benjamin. 1992. "Jihad vs. McWorld," *The Atlantic Monthly* March. On the World Wide Web at http://www.theatlantic.com/politics/foreign/barberf.htm (28 April 2000).

__________. 1996. *Jihad vs. McWorld: How Globalism and Tribalism are Reshaping the World.* New York: Ballantine Books.

Bardes, Barbara A., and Robert W. Oldendick. 2003. *Public Opinion: Measuring the American Mind.* Belmont, CA: Wadsworth.

Barlett, Donald L., and James B. Steele. 1998. "Corporate Welfare." *Time* 152, 19: 9 November. On the World Wide Web at http://www.time.com/time/magazine/1998/dom/981109/cover.html (11 January 2000).

__________, and __________. 2002. "Playing the Political Slots." *Time* December 23: 38–47.

Barnard, Chester I. 1938. *The Functions of the Executive.* Cambridge, MA: Harvard University Press.

Barnard, Jessie. 1972. *The Future of Marriage.* New York: World.

Barnes, Jessica S., and Chandler E. Bennett. 2002. "The Asian Population: 2000." Washington, DC: U.S. Census Bureau. On the World Wide Web at http://www.census.gov/prod/2002pubs/c2kbr01-16.pdf (22 June 2003).

Barnett, Cynthia. n.d. "The Measurement of White-Collar Crime Using Uniform Crime Reporting (UCR) Data." U.S. Department of Justice, Federal Bureau of Investigation, Criminal Justice Information Services (CJIS) Division. On the World Wide Web at http://www.fbi.gov/ucr/whitecollarforweb.pdf (21 February 2003).

Bar-On, Dan. 1999. *The Indescribable and the Undiscussable: Reconstructing Human Discourse After Trauma.* Ithaca, NY: Cornell University Press.

Barry, Patricia. 2002a. "Ads, Promotions Drive up Drug Costs." *AARP.* On the World Wide Web at http://www.aarp.org/bulletin/departments/2002/medicare/0310_medicare_1.html (17 June 2003).

__________. 2002b. "Drug Industry Spends Huge Sums Guarding Prices." *AARP.* On the World Wide Web at http://www.aarp.org/bulletin/departments/2002/medicare/0510_medicare_1.html (17 June 2003).

__________. 2002c. "Drug Profits vs. Research." *AARP.* On the World Wide Web at http://www.aarp.org/bulletin/departments/ 2002/medicare/0605_medicare_1.html (17 June 2003).

Bartfeld, Judi. 2000. "Child Support and the Postdivorce Economic Well-Being of Mothers, Fathers, and Children." *Demography* 37: 203–213.

Barth, Fredrik, ed. 1969. *Ethnic Groups and Boundaries: The Social Organization of Cultural Difference.* Boston: Little, Brown.

Batalova, Jeanne A., and Philip N. Cohen. 2002. "Premarital Cohabitation and Housework: Couples in Cross-National Perspective" *Journal of Marriage and the Family* 64: 743–755.

Baudrillard, Jean. 1983. *Simulations.* New York: Semiotext(e).

__________. 1988 [1986]. *America.* Chris Turner, trans. London: Verso.

__________. 1988. *Selected Writings,* Mark Poster, ed. Stanford, CA: Stanford University Press.

Bauman, Zygmunt. 1991 [1989]. *Modernity and the Holocaust.* Ithaca, NY: Cornell University Press.

__________. 1998. *Globalization: The Human Consequences.* New York: Columbia University Press.

Bayer, Ada-Helen, and Leon Harper. 2000. *Fixing to Stay: A National Survey of Housing and Home Modification Issues.* Washington, DC: AARP. On the World Wide Web at http://research.aarp.org/il/home_mod.pdf (13 August 2000).

Bean, Frank D., and Marta Tienda. 1987. *The Hispanic Population of the United States.* New York: Russell Sage Foundation.

Beauvoir, Simone de. 1972 [1970]. *The Coming of Age,* Patrick O'Brian, trans. New York: G.P. Putnam's Sons.

Beck, Tobin. 2002. "Quayle 10 years after Murphy Brown." *The Washington Times.* May 9. On the World Wide Web at http://www.washtimes.com/upi-breaking/09052002-060042-8923r.htm (25 January 2003).

Beck, Ulrich. 1992 [1986]. *Risk Society: Towards a New Modernity.* Mark Ritter, trans. London: Sage.

Becker, Elizabeth. 2003. "U.S. Ready to End Tariffs on Textiles in Hemisphere." *New York Times* February 11. On the World Wide Web at http://www.nytimes.com (13 February 2003).

__________. 2005. "U.S. Nearly Triples Tsunami Aid Pledge, to $950 Million." *New York Times* 10 February. On the World Wide Web at www.nytimes.com (10 February 2005).

Becker, Ernest. 1973. *The Denial of Death.* New York: Free Press.

Becker, Gary. 1976. *The Economic Approach to Human Behavior.* Chicago: University of Chicago Press.

Becker, Gaylene. 1980. *Growing Old in Silence.* Berkeley: University of California Press.

Becker, Howard. 1963. *Outsiders: Studies in the Sociology of Deviance.* New York: Free Press of Glencoe.

Beer, Frances A. 1974. *How Much War in History: Definitions, Estimates, Extrapolations and Trends.* Beverly Hills, CA: Sage.

Bell, Daniel. 1960. *The End of Ideology.* New York: Collier.

__________. 1973. *The Coming of Post-Industrial Society: A Venture in Social Forecasting.* New York: Basic Books.

Bell, Wendell, and Robert V. Robinson. 1980. "Cognitive Maps of Class and Racial Inequalities in England and the United States." *American Journal of Sociology* 86: 320–349.

Bellah, Robert A. 1975. *The Broken Covenant: American Civil Religion in a Time of Trial.* New York: Seabury Press.

Bellow, Saul. 1964. *Herzog.* New York: Fawcett World Library.

Belluck, Pam. 2002. "New Wave of the Homeless Floods Cities' Shelters." *New York Times* December 18. On the World Wide Web at http://www.nytimes.com (18 December 2002).

Benford, Robert D. 1997. "An Insider's Critique of the Social Movement Framing Perspective." *Sociological Inquiry* 67: 409–439.

Benson, April Lane, ed. 2000. *I Shop, Therefore I Am: Compulsive Buying and the Search for Self.* Northvale, NJ: Jason Aronson Inc.

Benson, John M. 1999. "End-of-Life Issues." *Public Opinion Quarterly* 63: 263–277.

Berens, Michael J. 2002a. "Infection Epidemic Carves Deadly Path." *Chicago Tribune* July 21. On the World Wide Web at http://www.chicagotribune.com/news/specials/chi-0207210272jul21.story (16 June 2003).

__________. 2002b. "Lax Procedures Put Infants at High Risk." *Chicago Tribune* July 22. On the World Wide Web at http://www .chicagotribune.com/news/specials/chi-0207220180jul22.story (16 June 2003).

__________. 2002c. "Drug-resistant Germs Adapt, Thrive Beyond Hospital Walls." *Chicago Tribune* June 23. On the World Wide Web at http://www.chicagotribune.com/news/specials/chi-0207230231jul23.story (16 June 2003).

Berger, Peter L. 1967. *The Sacred Canopy: Elements of a Sociological Theory of Religion.* Garden City, NY: Doubleday.

__________. 1986. *The Capitalist Revolution: Fifty Propositions About Prosperity, Equality, and Liberty.* New York: Basic Books.

__________, and Thomas Luckmann. 1966. *The Social Construction of Reality: A Treatise in the Sociology of Knowledge.* Garden City, NY: Doubleday.

Berger, Suzanne, and Ronald Dore, eds. 1996. *National Diversity and Global Capitalism.* Ithaca, NY: Cornell University Press.

Berk, Richard A. 1974. *Collective Behavior.* Dubuque, IA: Wm. C. Brown.

Berkowitz, S. D. 1982. *An Introduction to Structural Analysis: The Network Approach to Social Research.* Toronto: Butterworths.

Berliner, Wendy. 2004. "Where Have all the Young Men Gone? *Manchester Guardian* 8 May: 8.

Berners-Lee, Tim. "Tim Berners-Lee." 1999. On the World Wide Web at http://www.w3.org/People/Berners-Lee/Overview.html (2 May 2000).

Bernstein, Aaron. 2000. "Down and Out in Silicon Valley." *Business Week* 27 March: 76–92.

Bernstein, Jared, Heidi Hartmann, and John Schmitt. 1999. "The Minimum Wage Increase: A Working Woman's Issue." On the World Wide Web at http://www.epinet.org/Issuebriefs/Ib133.html (30 April 2000).

Bianchi, Suzanne M., and Lynne M. Casper. 2000. "American Families." *Population Bulletin* 55, 4. On the World Wide Web at http://www.ameristat.org/Template.cfm?Section=Population_Bulletin1&template=/ContentManagement/ContentDisplay.cfm&ContentID=5885 (9 June 2003).

Bianchi, Suzanne M., and Daphne Spain. 1996. "Women, Work, and Family in America." *Population Bulletin* 51, 3: 2–48.

Biegler, Rebecca S. 1999. "Psychological Interventions Designed to Counter Sexism in Children: Empirical Limitations and Theoretical Foundations." Pp. 129-152 in W. B. Swann, Jr., J.H. Langlois, and L.A. Gilbert, eds. *Sexism and Stereotypes in Modern Society: The Gender Science of Janet Taylor Spence.* Washington, DC: American Psychological Association.

Bierstedt, Robert. 1963. *The Social Order.* New York: McGraw-Hill.

———. 1974. "An Analysis of Social Power." Pp. 220–241 in *Power and Progress: Essays in Sociological Theory.* New York: McGraw-Hill.

Birdwhistell, Ray L. 1970. *Kinesics and Context.* Philadelphia: Pennsylvania State University.

Bjorhus, Jennifer. 2000. "Gap Between Execs, Rank-and-file Grows Wider." *San Jose Mercury News* 18 June. On the World Wide Web at http://www.mercurycenter.com/premium/business/docs/disparity18.htm (20 June 2000).

Björklund, Anders, and Markus Jäntti. 2000. "Intergenerational Mobility of Socio-Economic Status in Comparative Perspective." *Nordic Journal of Political Economy* 26, 1: 3–33.

Black, Donald. 1989. *Sociological Justice.* New York: Oxford University Press.

Blaise, Clark. 2000. *Time Lord: The Remarkable Canadian Who Missed His Train and Changed the World.* Toronto: A.A. Knopf Canada.

Blake, Michael. 2000. "Rights for People, Not Cultures." *National Post* 18 August: A16.

Blau, Peter M. 1963 [1955]. *The Dynamics of Bureaucracy: A Study of Interpersonal Relationships in Two Government Agencies,* rev. ed. Chicago: University of Chicago Press.

———. 1964. *Exchange and Power in Social Life.* New York: Wiley.

———, and Otis Dudley Duncan. 1967. *The American Occupational Structure.* New York: Wiley.

Blauner, Robert. 1972. *Racial Oppression in America.* New York: Harper & Row.

Blazer, Dan G., Ronald C. Kessler, Katherine A. McGonagle, and Marvin S. Swartz. 1994. "The Prevalence and Distribution of Major Depression in a National Community Sample: The National Comorbidity Survey." *American Journal of Psychiatry* 151: 979–986.

Bliss, Jeff. 2000. "Getting a Life Offline." *Financial Post* 29 June: C3.

Block, Fred. 1979. "The Ruling Class Does Not Rule." Pp. 128–140 in R. Quinney, ed. *Capitalist Society.* Homewood, IL: Dorsey Press.

Bloom, Richard. 2003. "Urban Legend's Accuracy Rate is 81 Per Cent." *Globe and Mail* January 25: S5.

Bluestone, Barry, and Bennett Harrison. 1982. *The Deindustrialization of America.* New York: Basic Books.

———, and Stephen Rose. 1997. "Overworked and Underemployed: Unraveling an Economic Enigma." *The American Prospect* 31: 58–69. On the World Wide Web at http://www .prospect.org/archives/31/31bluefs.html (1 May 2000).

Blum, Deborah. 1997. *Sex on the Brain: The Biological Differences Between Men and Women.* New York: Penguin.

Blumberg, Paul. 1989. *The Predatory Society: Deception in the American Marketplace.* New York: Oxford University Press.

Blumer, Herbert. 1969. *Symbolic Interactionism: Perspective and Method.* Englewood Cliffs, NJ: Prentice-Hall.

Boal, Mark. 1998. "Spycam City." *The Village Voice.* 30 September–6 October. On the World Wide Web at http://www.villagevoice.com/issues/9840/boal.shtml (29 April 2000).

Bobo, Lawrence, and James R. Kluegel. 1993. "Opposition to Race Targeting: Self-Interest, Stratification Ideology, or Racial Attitudes." *American Sociological Review* 58: 443–464.

Boehlert, Eric. 2003. "Clear Channel's Big, Stinking Deregulation Mess." *AlterNet.org* February 28. On the World Wide Web at http://www.alternet.org/story.html?StoryID=15281 (2 June 2003).

Bogus, Carl T. 1992. "The Strong Case for Gun Control." *The American Prospect* 10: 19–28.

Bonacich, Edna. 1972. "A Theory of Ethnic Antagonism: The Split Labor Market." *American Sociological Review* 37: 547–559.

———. 1973. "A Theory of Middleman Minorities." *American Sociological Review* 38: 583–594.

Bornschier, Volker, and Christopher Chase-Dunn. 1985. *Transnational Corporations and Underdevelopment.* New York: Praeger.

Boston Women's Health Book Collective, ed. 1998. *Our Bodies, Our Selves for the New Century: A Book by and for Women.* New York: Simon & Schuster.

Boswell, A. Ayres, and Joan Z. Spade. 1996. "Fraternities and Collegiate Rape Culture: Why Are Some Fraternities More Dangerous Places for Women." *Gender and Society* 10: 133–147.

Bouchard, Thomas J., Jr., David T. Lykken, Matthew McGue, Nancy L. Segal, and Auke Tellegen. 1990. "Sources of Human Psychological Differences: The Minnesota Study of Twins Reared Apart." *Science* 250, 4978: 223–226.

Boulding, Elise. 1976. *The Underside of History.* Boulder, CO: Westview.

Bourdieu, Pierre. 1977 [1972]. *Outline of a Theory of Practice,* Richard Nice, trans. Cambridge: Cambridge University Press.

———. 1984 [1979]. *Distinction: A Social Critique of the Judgment of Taste,* Richard Nice, trans. Cambridge, MA: Harvard University Press.

———. 1998 [1996]. *On Television.* New York: New Press.

———. 1998. *Acts of Resistance: Against the Tyranny of the Market,* Richard Nice, trans. New York: New Press.

Bourdieu, Pierre, and Jean-Claude Passeron. 1990 [1977]. *Reproduction in Education, Society and Culture,* 2nd ed., Richard Nice, trans. London: Sage.

Bowen, William G., and Derek Bok. 1998. *The Shape of the River: Long-Term Consequences of Considering Race in College and University Admissions.* Princeton, NJ: Princeton University Press.

Bowles, Samuel, and Herbert Gintis. 1976. *Schooling in Capitalist America: Educational Reform and the Contradictions of Economic Life.* New York: Basic Books.

Bowles, Samuel, and Herbert Gintis. 2002. "The Inheritance of Inequality." *Journal of Economic Perspectives* 16, 3: 3–30.

Boyd, Monica. 1997. "Feminizing Paid Work." *Current Sociology* 45: 49–73.

———, John Goyder, Frank Jones, Hugh A. McRoberts, Peter Pineo, and John Porter. 1985. *Ascription and Achievement: Studies in Mobility and Status Attainment in Canada.* Ottawa: Carleton University Press.

Bracey, Gerald W. 1998. "Are U.S. Students Behind?" *The American Prospect* 37, March–April: 54–70. On the World Wide Web at http://www.prospect.org/archives/37/37bracfs.html (1 May 2000).

Bradburn, Norman M., and Seymour Sudman, 1979. *Improving Interview Method and Questionnaire Design.* San Francisco: Jossey-Bass.

Bradbury, Bruce, and Markus Jäntti. 2001. "Child Poverty across Twenty-five Countries." Pp. 62–91 in Bruce Bradbury, Stephen P. Jenkins, and John Micklewright, eds. *The Dynamics of Child Poverty in Industrialised Countries.* Cambridge: Cambridge University Press.

Brady, Erik. 2001. "Too Good To Be True?" *USA Today* (12 July 2001). On the World Wide Web at http://www.usatoday.com/sports/stories/2001-07-12-cover.htm (21 July 2001).

Brains, Craig Leonard. 1999. "When Registration Barriers Fall, Who Votes?" *Public Choice* 35: 161–176.

Braithwaite, John. 1981. "The Myth of Social Class and Criminality Revisited." *American Sociological Review* 46: 36–57.

———. 1989. *Crime, Shame and Reintegration.* New York: Cambridge University Press.

Braus, Patricia. 1994. "Why Do Hispanics Have Lower Death Rates?" *American Demographics* 16, 5: 18–19.

Brave, Ralph. 2003. "James Watson Wants to Build a Better Human." *AlterNet* 29 May. On the World Wide Web at http://www.alternet.org/story/16026/ (19 January 2005).

Braverman, Harry. 1974. *Labor and Monopoly Capital: The Degradation of Work in the Twentieth Century.* New York: Monthly Review Press.

Brazzini, D. G., W. D. McIntosh, S. M. Smith, S. Cook, and C. Harris. 1997. "The Aging Woman in Popular Film: Underrepresented, Unattractive, Unfriendly, and Unintelligent." *Sex Roles* 36: 531–543.

Breault, K. D. 1986. "Suicide in America: A Test of Durkheim's Theory of Religious and Family Integration, 1933–1980." *American Journal of Sociology* 92: 628–656.

Brechin, Steven R., and Willett Kempton. 1994. "Global Environmentalism: A Challenge to the Postmaterialism Thesis." *Social Science Quarterly* 75: 245–269.

Breen, Richard, and David B. Rottman. 1995. *Class Stratification: A Comparative Perspective.* New York: Harvester Wheatsheaf.

Brennan, Teresa. 2003. *Globalization and Its Terrors: Daily Life in the West.* London: Routledge.

Brinkhoff, T. 2002. "The Principal Agglomerations of the World." On the World Wide Web at http://www.citypopulation.de (2 August 2003).

Brint, Stephen. 1984. "New Class and Cumulative Trend Explanations of the Liberal Political Attitudes of Professionals." *American Journal of Sociology* 90: 30–71.

———, and Jerome Karabel. 1989. *The Diverted Dream: Community Colleges and the Promise of Educational Opportunity in America, 1900–1985.* New York: Oxford University Press.

"British Magazines Agree to Ban Ultra-thin Models." 2000. *National Post* 23 June: A2.

Brokaw, Tom. 2001. "Into an Unknowable Future." *The New York Times on the Web.* On the World Wide Web at http://www.nytimes.com/2001/09/28/opinion/28BROK.html?todaysheadlines (28 September 2001).

Bromley, Julian V. 1982 [1977]. *Present-Day Ethnic Processes in the USSR.* Moscow: Progress Publishers.

Bronowski, J. 1965 [1956]. *Science and Human Values,* revised ed. New York: Harper & Row.

Brooks, Clem, and Jeff Manza. 1994. "Do Changing Values Explain the New Politics? A Critical Assessment of the Postmaterialist Thesis." *Sociological Quarterly* 35: 541–570.

———, and ———. 1997a. "Class Politics and Political Change in the United States, 1952–1992." *Social Forces* 76: 379–408.

———, and ———. 1997b. "Social Cleavages and Political Alignments: U.S. Presidential Elections, 1960 to 1992." *American Sociological Review* 62: 937–946.

Brouwer, Steve. 1998. *Sharing the Pie: A Citizen's Guide to Wealth and Power in America.* New York: Holt.

Brower, David. 1975. *Training the Nihilists: Education and Radicalism in Tsarist Russia.* Ithaca, NY: Cornell University Press.

Brown, David K. 1995. *Degrees of Control: A Sociology of Educational Expansion and Occupational Credentialism.* New York: Teachers College Press.

Brown, Dee A. 1970. *Bury My Heart at Wounded Knee: An Indian History of the American West.* New York: Henry Holt.

Brown, Lester R., Christopher Flavin, Hilary French, et al. 2000. *State of the World 2000.* New York: W. W. Norton.

Brown, Lyn Mikel, and Carol Gilligan. 1992. *Meeting at the Crossroads: Women's Psychology and Girls' Development.* Cambridge, MA: Harvard University Press.

Brown, Peter. 1996. *The Rise of Western Christendom: Triumph and Diversity,* A.D. *200–1000.* Oxford: Blackwell.

Brown, Phil. 1997. "Popular Epidemiology Revisited." *Current Sociology* 45, 3: 137–156.

Browne, Malcolm W. 1998. "From Science Fiction to Science: The Whole Body Transplant." *New York Times* 5 May: B, 16.

Browning, Christopher R. 1992. *Ordinary Men: Reserve Police Battalion 101 and the Final Solution in Poland.* New York: HarperCollins.

Bruce, Steve. 1988. *The Rise and Fall of the New Christian Right: Conservative Protestant Politics in America 1978–1988.* Oxford: Clarendon Press.

———. 2002. *God Is Dead: Secularization in the West.* Oxford: Blackwell.

Brumberg, Joan Jacobs. 1997. *The Body Project: An Intimate History of American Girls.* New York: Random House.

Brym, Robert J. 1979. "Political Conservatism in Atlantic Canada." Pp. 59–79 in Robert J. Brym and R. James Sacouman, eds. *Underdevelopment and Social Movements in Atlantic Canada.* Toronto: New Hogtown Press.

———. 1980. *Intellectuals and Politics.* London: George Allen & Unwin.

———. 1984. "Cultural versus Structural Explanations of Ethnic Intermarriage in the USSR: A Statistical Reanalysis." *Soviet Studies* 36: 594–601.

———. 1990. "Sociology, *Perestroika,* and Soviet Society." *Canadian Journal of Sociology* 15: 207–215.

———. 1995. "Voters Quietly Reveal Greater Communist Leanings." *Transition: Events and Issues in the Former Soviet Union and East-Central and Southeastern Europe* 1, 16: 32–35.

———. 1996a. "The Ethic of Self-reliance and the Spirit of Capitalism in Russia." *International Sociology* 11: 409–426.

———. 1996b. "Reevaluating Mass Support for Political and Economic Change in Russia." *Europe–Asia Studies* 48: 751–766.

———. 1996c. "'The Third Rome' and 'The End of History': Notes on Russia's Second Communist Revolution." *Canadian Review of Sociology and Anthropology* 33: 391–406.

———. 1996d. "The Turning Point in the Presidential Campaign." Pp. 44–49 in *The 1996 Presidential Election and Public Opinion.* Moscow: VTsIOM. [In Russian.]

———. 2001. "Hip-Hop from Dissent to Commodity: A Note on Consumer Culture." Pp. 78–81 in R. Brym, ed. *New Society: Sociology for the 21st Century,* 3rd ed. Toronto: Nelson. On the World Wide Web at http://www.nelson.com/nelson/harcourt/sociology/newsociety3e/article.htm (25 January 2003).

———. 2003. "Affluence, Power, and Strikes in Canada, 1973–2000." Pp. 55–67 in James Curtis, , Edward Grabb, and Neil Guppy, eds. *Social Inequality in Canada: Patterns, Problems, and Policies.* Toronto: Pearson Prentice-Hall.

———, and Evel Economakis. 1994. "Peasant or proletarian? Blacklisted Pskov Workers in St. Petersburg, 1913." *Slavic Review* 53: 120–139.

——— et al. 2005. "In faint praise of the World Bank's gender development policy." *Canadian Journal of Sociology* 30: 95-111.

——— with Bonnie J. Fox. 1989. *From Culture to Power: The Sociology of English Canada.* Toronto: Oxford University Press.

———, Michael Gillespie, and A. Ron Gillis. 1985. "Anomie, Opportunity, and the Density of Ethnic Ties: Another View of Jewish Outmarriage in Canada." *Canadian Review of Sociology and Anthropology* 22: 102–112.

———, Michael Gillespie, and Rhonda L. Lenton. 1989. "Class Power, Class Mobilization, and Class Voting: The Canadian Case." *Canadian Journal of Sociology* 14: 25–44.

———, and Rhonda Lenton. 2001. *Love Online: A Report on Digital Dating in Canada.* Toronto: MSN.CA. On the World Wide Web at http://www.nelson.com/nelson/harcourt/sociology/newsociety3e/socplus.htm (30 December 2003).

——— with the assistance of Rozalina Ryvkina. 1994. *The Jews of Moscow, Kiev and Minsk: Identity, Antisemitism, Emigration.* New York: New York University Press.

Brzezinski, Zbigniew. 1993. *Out of Control: Global Turmoil on the Eve of the Twenty-first Century.* New York: Scribner.

———. 2002. "Confronting Anti-American Grievances." *New York Times* 1 September. On the World Wide Web at http://www.nytimes.com (1 September 2002).

Buckley, Neil. 2003. "Unhealthy Food is Everywhere, 24 Hours a Day, and Inexpensive." *Financial Times* 18 February, p. 13.

Bullard, Robert D. 1994 [1990]. *Dumping in Dixie: Race, Class and Environmental Quality,* 2nd ed. Boulder, CO: Westview Press.

Bumiller, Elisabeth. 2003. "Evangelicals Sway White House on Human Rights Issue Abroad." New York Times October 26. On the World Wide Web at http://www.nytimes.com (26 October 2003).

Burawoy, Michael. 1979. *Manufacturing Consent: Changes in the Labor Process under Monopoly Capitalism.* Chicago: University of Chicago Press.

——— et al. 2000. *Global Ethnography: Forces, Connections and Imaginations in a Postmodern World.* Berkeley: University of California Press.

Bureau of Justice Statistics. 2002. *Sourcebook of Criminal Justice Statistics, 2001.* On the World Wide Web at http://www.albany.edu/sourcebook/1995 (15 February 2003).

Burgess, Ernest. W. 1967 [1925]. "The Growth of the City: An Introduction to a Research Project." Pp. 47–62 in Robert E. Park, Ernest W. Burgess, and Roderick D. McKenzie. *The City.* Chicago: University of Chicago Press.

Burleigh, Michael. 2000. *The Third Reich: A New History.* New York: Hill & Wang.

Burns, Tom, and G. M. Stalker. 1961. *The Management of Innovation.* London: Tavistock.

Bush, George W. 2001. "Address to a Joint Session of Congress and the American People." On the World Wide Web at http://www.whitehouse.gov/news/releases/2001/09/20010920-8.html (22 December 2002).

Buss, D. M. 1994. *The Evolution of Desire.* New York: Basic Books.

———. 1998. "The Psychology of Human Mate Selection: Exploring the Complexity of the Strategic Repertoire." Pp. 405–29 in C. Crawford and D. L. Krebs, eds. *Handbook of Evolutionary Psychology: Ideas, Issues, and Applications.* Mahwah, NJ: Erlbaum.

Butterfield, Fox. 2001. "Killings Increase in Many Big Cities." *New York Times on the Web.* On the World Wide Web at http://www.nytimes.com/2001/12/21/national/21CRIM.html?todaysheadlines (21 December 2001).

Buxton, L. H.Dudley. 1963. "Races of Mankind," *Encyclopaedia Britannica* 18: 864–866. Chicago: Encyclopaedia Britannica, Inc.

Cable and Telecommunications Association for Marketing. 2001. "Cable Television Customers Say 'Yes' to Interactive Television." On the World Wide Web at http://www.ctam.com/ctam/about/pressreleases/images/010110.pdf (28 May 2003).

Cahill, Spencer. E. 1999. "Emotional Capital and Professional Socialization: The Case of Mortuary Science Students (and Me)." *Social Psychology Quarterly* 62: 101–16.

Callahan, Raymond E. 1962. *Education and the Cult of Efficiency: A Study of the Social Forces That Have Shaped the Administration of the Public Schools.* Chicago: University of Chicago Press.

Camarillo, Albert. 1979. *Chicanos in a Changing Society: From Mexican Pueblos to American Barrios in Santa Barbara and Southern California, 1848–1930.* Cambridge, MA: Harvard University Press.

Campbell, Donald, and Julian Stanley. 1963. *Experimental and Quasi-Experimental Designs for Research.* Chicago: Rand McNally.

Campbell, Frances A., and Craig T. Ramey. 1994. "Effects of Early Intervention on Intellectual and Academic Achievement: A Follow-up Study of Children from Low-income Families." *Child Development* 65: 684–698.

Campbell, Jane, and Mike Oliver. 1996. *Disability Politics: Understanding Our Past, Changing Our Future.* London: Routledge.

———, E. P. Pungello, J. Sparling, and S. Miller-Johnson. 2002. "Early Childhood Education: Young Adult Outcomes from the Abecedarian Project." *Applied Developmental Science* 6, 1: 42–57.

Campion, Edward W. 1993. "Why Unconventional Medicine?" *New England Journal of Medicine* 328: 282.

Canadian Broadcasting Corporation. 2003. "H2 Powering the Future." On the World Wide Web at http://www.cbc.ca/venture/fuelcell/hopes_drhydrogen.html (30 July 2003).

Cancio, A. Silvia, T. David Evans, and Daivd J. Maume, Jr. 1996. "Reconsidering the Declining Significance of Race: Racial Differences in Early Career Wages." *American Sociological Review* 61: 541–556.

"The Candidates Debate." 1998. *MSNBC News.* On the World Wide Web at http://msnbc.com/onair/msnbc/TimeAndAgain/archive/ken-nix/Default.asp?cp1=1 (2 May 2000).

Cardoso, Fernando Henrique, and Enzo Faletto. 1979. *Dependency and Development in Latin America,* Marjory Mattingly Urquidi, trans. Berkeley: University of California Press.

Carey, James. 1989. *Culture as Communication.* Boston: Unwin Hyman.

Carpenter, Dave. 2003. "McDonald's High-Tech With Kitchen, Kiosks." *KioskCom.* On the World Wide Web at http://www.kioskcom.com/articles_detail.php?ident=1856 (23 October 2003).

Carter, Stephen L. 1991. *Reflections of an Affirmative Action Baby.* New York: Basic Books.

Case, Anne C., I-Fen Lin, and Sara S. McLanahan. 2003. "Explaining Trends in Child Support: Economic, Demographic, and Policy Effects." *Demography* 40: 171–189.

Casper, Lynne M., Sara S. McLanahan, and Irwin Garfinkel. 1994. "The Gender-Poverty Gap: What Can We Learn From Other Countries?" *American Sociological Review* 59: 594–605.

Cassidy, John. 1999. "Schools Are Her Business." *New Yorker* 18–25 October: 144–160.

Castells, Manuel. 1983. *The City and the Grassroots: A Cross-Cultural Theory of Urban Social Movements.* Berkeley: University of California Press.

__________. 1996. *The Information Age: Economy, Society and Culture: The Rise of the Network Society,* Vol. 1. Oxford: Blackwell.

"Category: Disasters." 2005. On the World Wide Web at http://www.nationmaster.com/cat/disasters (1 February 2005).

Cavalli-Sforza, L. Luca, Paolo Menozzi, and Alberto Piazza. 1994. *The History and Geography of Human Genes.* Princeton, NJ: Princeton University Press.

Cavanagh, John, and Richard J. Barnet. 1994. *Global Dreams: Imperial Corporations and the New World Order.* New York: Simon & Schuster.

Center for Responsive Politics. 2000. "Election Statistics at a Glance." On the World Wide Web at http://www.opensecrets.org/pubs/bigpicture2000/overview/stats.ihtml (1 May 2000).

Centers for Disease Control and Prevention, National Center for Health Statistics. 1995a. *Monthly Vital Statistics Report* 43, 9(S): 22 March.

__________. 1995b. *Monthly Vital Statistics Report* 43, 12(S): 14 July.

__________. 1997. *Monthly Vital Statistics Report* 46, 1(S)2: 11 September.

__________. 1998. *Monthly Vital Statistics Report* 46, 12: 28 July.

__________. 1999a. *National Vital Statistics Reports* 47, 19: 30 June.

__________. 1999b. *National Vital Statistics Reports* 47, 25: 5 October. On the World Wide Web at http://www.cdc.gov/nchs/data/ nvs47_25.pdf (27 April 2000).

__________. 2000. "Cumulative Age of Initiation of Cigarette Smoking—United States, 1991." On the World Wide Web at http://www.cdc.gov/tobacco/init.htm (1 May 2000).

__________. 2001. 49, 6. On the World Wide Web at http://www .cdc.gov/nchs/data/nvsr/nvsr49/nvsr49_06.pdf (8 June 2003).

__________. 2002. *Monthly Vital Statistics Report* 51, 2: 18 December. On the World Wide Web at http://www.cdc.gov/nchs/data/nvsr/nvsr51/nvsr51_02.pdf (11 December 2003).

__________. 2003. *National Vital Statistics Report.* 51, 6. On the World Wide Web at http://www.cdc.gov/nchs/data/nvsr/ nvsr51/nvsr51_06.pdf (8 June 2003).

__________. 2005a. *National Vital Statistics Reports* 53, 16. On the World Wide at http://www.cdc.gov/nchs/data/nvsr/nvsr53/nvsr53_16.pdf (7 March 2005).

__________. 2005b. "Self-inflicted Injury/Suicide." On the World Wide Web at http://www.cdc.gov/nchs/fastats/suicide.htm (24 January 2005).

Central Intelligence Agency. 2002. *The World Factbook 2002.* On the World Wide Web at http://www.cia.gov/cia/publications/factbook (6 February 2003).

Centre for Economic Policy Research. 2002. *Making Sense of Globalization: A Guide to the Economic Issues.* London.

Chambliss, Daniel F. 1996. *Beyond Caring: Hospitals, Nurses, and the Social Organization of Ethics.* Chicago: University of Chicago Press.

Chambliss, William J. 1989. "State-Organized Crime." *Criminology* 27: 183–208.

Chang, Ha-Joon. 2002. *Kicking Away the Ladder: Development Strategy in Historical Perspective.* London: Anthem Press.

"The Changing Rate of Major Depression." 1992. *Journal of the American Medical Association* 268, 21: 3098–4005.

Charlton, James I. 1998. *Nothing About Us Without Us: Disability Oppression and Empowerment.* Berkeley: University of California Press.

Chase, Steven. 2003. "Free-trade Talks in Jeopardy." *Globe and Mail* 8 September: B1, B6.

Chaves, Mark. 1994. "Secularization as Declining Religious Authority." *Social Forces* 72: 749–774.

Cherlin, Andrew J. 1992 [1981]. *Marriage, Divorce, Remarriage,* rev. ed. Cambridge, MA: Harvard University Press.

__________, Frank F. Furstenberg, Jr., P. Lindsay Chase-Lansdale, Kathleen E. Kiernan, Philip K. Robins, Donna Ruane Morrison, and Julien O. Teitler. 1991. "Longitudinal Studies of Effects of Divorce on Children in Great Britain and the United States." *Science* 252: 1386–1389.

Chesnais, Jean-Claude. 1992 [1986]. *The Demographic Transition: Stages, Patterns, and Economic Implications.* Elizabeth Kreager and Philip Kreager, trans. Oxford: Clarendon Press.

Choldin, Harvey M. 1994. *Looking for the Last Percent: The Controversy over Census Undercounts.* New Brunswick, NJ: Rutgers University Press.

Chomsky, Noam. 1991. *Deterring Democracy.* London: Verso.

Cicourel, Aaron V. 1968. *The Social Organization of Juvenile Justice.* New York: Wiley.

Clapp, Jennifer. 1998. "Foreign Direct Investment in Hazardous Industries in Developing Countries: Rethinking the Debate." *Environmental Politics* 7, 4: 92–113.

Clark, Terry Nichols, and Seymour Martin Lipset. 1991. "Are Social Classes Dying?" *International Sociology* 6: 397–410.

__________. 2001. *The Breakdown of Class Politics: A Debate on Post-Industrial Stratification.* Washington, DC: Woodrow Wilson Center Press and Baltimore: Johns Hopkins University Press.

__________, __________, and Michael Rempel. 1993. "The Declining Political Significance of Class." *International Sociology* 8: 293–316.

Clarke-Stewart, K. Alison, Christian P. Gruber, and Linda May Fitzgerald. 1994. *Children at Home and in Day Care.* Hillsdale, NJ: Lawrence Erlbaum.

Clawson, Dan. 1980. *Bureaucracy and the Labor Process: The Transformation of U.S. Industry, 1860–1920.* New York: Monthly Review Press.

__________, and Mary Ann Clawson. 1999. "What Has Happened to the US Labor Movement? Union Decline and Renewal" *Annual Review of Sociology* 25: 95–119.

__________, Alan Neustadtl, and Denise Scott. 1992. *Money Talks: Corporate PACS and Political Influence.* New York: Basic Books.

Clemens, Michael, Steve Radelet, and David Roodman. 2004. "U.S. aid, global poverty, and the earthquake/tsunami death toll." On the World Wide Web at http://www.cgdev.org/Publications/index.cfm?PubID=187 (2 February 2005).

Clement, Wallace, and John Myles. 1994. *Relations of Ruling: Class and Gender in Postindustrial Societies.* Montreal: McGill-Queen's University Press.

Clemetson, Lynette. 2003. "More Americans in Poverty in 2002, Census Study Says." *New York Times.* On the World Wide Web at http://www.nytimes.com (27 September 2003).

Clinard, Marshall B., and Peter C. Yeager. 1980. *Corporate Crime.* New York: Free Press.

Clinton, William J. 1998. "Statement on Signing the Child Support Performance and Incentive Act of 1998." *Weekly Compilation of Presidential Documents* 34, 29: 1396.

Cloward, Richard A., and Lloyd E. Ohlin. 1960. *Delinquency and Opportunity: A Theory of Delinquent Gangs.* New York: Free Press.

CNN.COM. 1998. "Prosecutor: Attackers Planned to Rob Gay Student." On the World Wide Web at http://www.htt.com/cnn.com/US/9811/19/shepard.01/index.html (1 November 1999).

__________. 1999. "From Little League to Madness: Portraits of the Littleton Shooters." 30 April. On the World Wide Web at http://www.cnn.com/SPECIALS/1998/schools/they.hid.it.well/index.html (29 April 2000).

Coale, Ansley J. 1974. "The History of Human Population." *Scientific American* 23, 3: 41–51.

__________, and Susan C. Watkins, eds. 1986. *The Decline of Fertility in Europe.* Princeton, NJ: Princeton University Press.

Cohen, Adam. 2000. "The Big Easy on the Brink." *Time* (Canadian edition) 10 July: 43.

Cohen, Albert. 1955. *Delinquent Boys: The Subculture of a Gang.* New York: Free Press.

Cohen, Arthur M., and Florence B. Brawer. 2003 [1981]. *The American Community College*, 4th ed. San Francisco: Jossey-Bass.

Cohen, Stanley. 1972. *Folk Devils and Moral Panics: The Creation of the Mods and Rockers.* London: MacGibbon & Kee.

Colapinto, John. 1997. "The True Story of John/Joan." *Rolling Stone* 11 December: 54–73, 92–97.

__________. 2001. *As Nature Made Him: The Boy Who was Raised as a Girl.* New York: Harper Collins.

Cole, Michael. 1995. *Cultural Psychology.* Cambridge, MA: Harvard University Press.

__________ et al. 1966. *Equality of Educational Opportunity.* Washington, DC: United States Department of Health, Education, and Welfare, Office of Education.

Coleman, James S. 1961. *The Adolescent Society.* New York: Free Press.

__________. 1990. *Foundations of Social Theory.* Cambridge, MA: Harvard University Press.

__________ et al. 1966. *Equality of Educational Opportunity.* Washington, DC: U.S. Department of Health, Education, and Welfare, Office of Education.

Collins, Randall. 1975. *Conflict Sociology: Toward an Explanatory Science.* New York: Academic Press.

__________. 1979. *The Credential Society: An Historical Sociology of Education and Stratification.* New York: Academic Press.

__________. 1982. *Sociological Insight: An Introduction to Nonobvious Sociology.* New York: Oxford University Press.

__________, and Scott Coltrane. 1991 [1985]. *Sociology of Marriage and the Family: Gender, Love, and Property,* 3rd ed. Chicago: Nelson-Hall.

Colson, Elizabeth. 1982. *Planned Change: The Creation of a New Community.* Berkeley: Institute of International Studies, University of California.

Columbia Broadcasting System. "Super Bowl Ratings Up From Last Year." 2000. On the World Wide Web at http://cbs.sportsline.com/u/ce/multi/0,1329,1959564_59,00.html (9 August 2000).

Commins, Patricia. 1997. "Foreign Sales Prop Up McDonald's." *Globe and Mail* 26 August: B8.

"Comparable Worth." 1990. *Issues in Ethics* 3, 2. On the World Wide Web at http://www.scu.edu/SCU/Centers/Ethics/publications/iie/v3n2/comparable.shtml (30 April 2000).

Comte, Auguste. 1975. *Auguste Comte: The Foundation of Sociology,* Kenneth Thompson, ed. New York: Wiley.

Condry, J., and S. Condry. 1976. "Sex Differences: The Eye of the Beholder." *Child Development* 47: 812–819.

Conley, Dalton. 1999. *Being Black, Living in the Red: Race, Wealth, and Social Policy in America.* Berkeley: University of California Press.

Conrad, Peter, and Joseph W. Schneider. 1992 [1980]. *Deviance and Medicalization: From Badness to Sickness,* expanded ed. Philadelphia: Temple University Press.

Converse, Jean M., and Stanley Presser. 1986. *Survey Questions: Handrcrafting the Standardized Questionnaire.* Newbury Park, CA: Sage.

Conwell, Chic. 1937. *The Professional Thief: By a Professional Thief,* annotated and interpreted by Edwin H. Sutherland. Chicago: University of Chicago Press.

Cooley, Charles Horton. 1902. *Human Nature and the Social Order.* New York: Scribner's.

Coontz, Stephanie. 1992. *The Way We Never Were: American Families and the Nostalgia Trip.* New York: Basic Books.

__________. 1997. *The Way We Really Are: Coming to Terms with America's Changing Families.* New York: Basic Books.

__________, and Peta Henderson, eds. 1986. *Women's Work, Men's Property: The Origins of Gender and Class.* London: Verso.

Cooperman, Alan, and Thomas B. Edsall. 2004. "Evangelicals Say They Led Charge for the GOP." *Washington Post.com* 8 November. On the World Wide Web at http://www.washingtonpost.com/wp-dyn/articles/A32793-2004Nov7.html (7 March 2005).

Cornell, Stephen. 1988. *The Return of the Native: American Indian Political Resurgence.* New York: Oxford University Press.

Corporate Watch. 2000. "High Tech Sweatshops." On the World Wide Web at http://www.corpwatch.org/trac/gallery/sweat/27a.html (20 June 2000).

Coser, Rose Laub. 1960. "Laughter among Colleagues: A Study of the Functions of Humor among the Staff of a Mental Hospital." *Psychiatry* 23: 81–95.

Costello, Carrie Yang. 2004. "Changing Clothes: Gender Inequality and Professional Socialization." *NWSA Journal* 16: 138–155.

Couch, Carl J. 1968. "Collective Behavior: An Examination of Some Stereotypes." *Social Problems* 15: 310–322.

Council of American-Islamic Relations. 2002. *The Status of Muslim Civil Rights in the United States 2002.* Washington, DC. On the World Wide Web at http://www.cair-net.org/civilrights2002/civilrights2002.pdf (21 December 2002).

Coupland, Douglas. 1991. *Generation X: Tales for An Accelerated Culture.* New York: St. Martin's Press.

Cramton, Peter, and Joseph Tracy. 1998. "The Use of Replacement Workers in Union Contract Negotiations: The U.S. Experience, 1980–1989." *Journal of Labor Economics* 16: 667–701.

Creedon, Jeremiah. 1998. "God With a Million Faces." *Utne Reader* July–August: 42–48.

Creighton, Sarah, and Catherine Mihto. 2001. "Managing Intersex." *British Medical Journal* 323, 7324: 1264-5.

Critser, Greg. 2003. *Fat Land: How Americans Became the Fattest People in the World.* Boston: Houghton Mifflin.

Crompton, Rosemary. 1993. *Class and Stratification: An Introduction to Current Debates.* Cambridge: Polity Press.

__________, and Michael Mann, eds. 1986. *Gender and Stratification.* Cambridge: Polity Press.

Cronon, William. 1991. *Nature's Metropolis: Chicago and the Great West.* New York: W. W. Norton.

Croteau, David, and William Hoynes. 1997. *Media/Society: Industries, Images, and Audiences.* Thousand Oaks, CA: Pine Forge Press.

Crothers, Charles. 1979. "On the Myth of Rural Tranquility: Comment on Webb and Collette." *American Journal of Sociology* 84: 429–437.

Crozier, Michel. 1964 [1963]. *The Bureaucratic Phenomenon.* Chicago: University of Chicago Press.

Crucefix, Linda. 2003. "Do the Inuit really have 200 words for 'snow'?" *Ask Us @ U of T.* On the World Wide Web at http://www.newsandevents.utoronto.ca/bios/askus3.htm (22 January 2003).

Curtis, James, John Loy, and Wally Karnilowicz. 1986. "A Comparison of Suicide-Dip Effects of Major Sport Events and Civil Holidays." *Sociology of Sport Journal* 3: 1–14.

Dahl, Robert A. 1961. *Who Governs?* New Haven, CT: Yale University Press.

Darder, Antonia, and Rodolfo D. Torres, eds. 1998. *The Latino Studies Reader: Culture, Economy and Society.* Malden, MA: Blackwell.

Darwin, Charles. 1859. *On the Origin of Species by Means of Natural Selection.* London: John Murray.

__________. 1871. *The Descent of Man.* London: John Murray.

Davidoff, Judith. 1999. "The Fatherhood Industry: Welfare Reformers Set Their Sights on Wayward Dads." *The Progressive.* November: 28–31.

Davies, Christine. 1998. *Jokes and Their Relation to Society.* Berlin: Mouton de Gruyter.

Davies, James C. 1969. "Toward a Theory of Revolution." Pp. 85–108 in Barry McLaughlin, ed. *Studies in Social Movements: A Social Psychological Perspective.* New York: Free Press.

Davies, Mark, and Denise B. Kandel. 1981. "Parental and Peer Influences on Adolescents' Educational Plans: Some Further Evidence." *American Journal of Sociology* 87: 363–387.

da Vinci, Leonardo. 1970. The Notebooks of Leonardo da Vinci, *Volume 1,* Jean Paul Richter, ed. New York: Dover.

Davis, Fred. 1992. *Fashion, Culture, and Identity.* Chicago: University of Chicago Press.

Davis, Jim. 2000. "AOL Previews TV Plans." *CNET News* 6 January. On the World Wide Web at http://news.cnet.com/category/0-1006-200-1516271.html (2 May 2000).

Davis, Kingsley, and Wilbert E. Moore. 1945. "Some Principles of Stratification." *American Sociological Review* 10: 242–249.

Davis, Lennard J. 1995. *Enforcing Normalcy: Disability, Deafness, and the Body.* London: Verso.

Davis, Mike. 1990. *City of Quartz: Excavating the Future in Los Angeles.* New York: Verso.

Dawidowicz, Lucy S. 1975. *The War Against the Jews, 1933–1945.* New York: Holt, Rinehart & Winston.

Day, Jennifer Cheeseman, and Eric Newburger. 2002. "The Big Payoff: Educational Attainment and Synthetic Estimates of Work-Life Earnings." On the World Wide Web at http://www.census.gov/prod/2002pubs/p23-210.pdf (27 April 2003).

Dean, John. 2000. "Why Americans Don't Vote—And How that Might Change." On the World Wide Web at http://www.cnn.com/2000/LAW/11/columns/fl.dean.voters.02.11.07 (8 November 2000).

DeFine, Michael Sullivan. 1997. "A History of Governmentally Coerced Sterilization: The Plight of the Native American Woman." On the World Wide Web at http://www.geocities.com/CapitolHill/9118/mike2.html (24 April 2003).

DeFrancis, Marc. 2002. "U.S. Elder Care is in a Fragile State." *Population Today* 30, 1: 1–3.

de la Garza, Rodolpho O., Luis DiSipio, F. Chris Garcia, John Garcia, and Angelo Falcon. 1992. *Latino Voices: Mexican, Puerto Rican, and Cuban Perspectives on American Politics.* Boulder, CO: Westview.

Delmos, Monika. 2002. "Mangled words divide generations in Japan." *Globe and Mail* 24 August: A 14.

De Long, J. Bradford. 1998. *Global Trends: 1980–2015 and Beyond.* Ottawa: Industry Canada.

Demo, David H., and Alan C. Acock. 1993. "Family Diversity and the Division of Domestic Labor: How Much Have Things Really Changed? *Family Relations* 42: 323–331.

Democratic National Committee. 1998. "Democrats Fight to Make Sure that Every American is Counted in the 2000 Census." On the World Wide Web at http://www.democrats.org/archive/news/rel1998/rel060298.html (6 May 2000).

Denzin, Norman K. 1992. *Symbolic Interactionism and Cultural Studies: The Politics of Interpretation.* Oxford: Blackwell.

Department of Geography, Slippery Rock University. 1997. "World's Largest Cities, 1900." On the World Wide Web at http://www.sru.edu/depts/artsci/ges/discover/d-6-8.htm (2 May 2000).

———. 2003. "World's Largest Urban Agglomerations, 2015." On the World Wide Web at http://www1.sru.edu/gge/faculty/hughes/100/100-6/d-6-9b.htm (2 August 2003).

Department of Justice, Canada. 1995. "A Review of Firearm Statistics and Regulations in Selected Countries." On the World Wide Web at http://www.cfc-ccaf.gc.ca/research/publications/reports/1990%2D95/reports/siter_rpt_en.html (29 April 2000).

Derber, Charles. 1979. *The Pursuit of Attention: Power and Individualism in Everyday Life.* New York: Oxford University Press.

———, William A. Schwartz, and Yale Magrass. 1990. *Power in the Highest Degree: Professionals and the Rise of a New Mandarin Order.* New York: Oxford University Press.

DeSoya, Indra, and John Oneal. 1999. "Boon or Bane? Reassessing the Productivity of Foreign Direct Investment with New Data." *American Sociological Review* 64: 766–782.

de Villiers, Marq. 1999. *Water.* Toronto: Stoddart Publishing.

de Waal, Alexander. 1989. *Famine That Kills: Darfur, Sudan, 1984–1985.* Oxford: Clarendon Press.

Diamond, Larry. 1996. "Is the Third Wave Over?" *Journal of Democracy* 7, 3: 20–37. On the World Wide Web at http://muse.jhu.edu/demo/jod/7.3diamond.html (1 May 2000).

Diani, Mario. 1996. "Linking Mobilization Frames and Political Opportunities: Insights from Regional Populism in Italy." *American Sociological Review* 61: 1053–1069.

Dibbell, Julian. 1993. "A Rape in Cyberspace." *The Village Voice* 21 December: 36–42. On the World Wide Web at http://www.levity.com/julian/bungle.html (29 April 2000).

DiCarlo, Lisa. 2000. "32 Debut on the List." *Forbes.com.* On the World Wide Web at http://www.forbes.com/tool/toolbox/rich400 (22 September 2000).

Dickerson, Marla. 1998. "A New Army of Child-Care Workers." *Los Angeles Times* (12 August) 1A.

Dietz, Tracy L. 1998. "An Examination of Violence and Gender Role Portrayals in Video Games: Implications for Gender Socialization and Aggressive Behavior." *Sex Roles* 38: 425–442.

"Digital Dilemmas." 2003. *The Economist* 25 January: 3–26. On the World Wide Web at http://www.economist.com/displayStory.cfm?Story_id=1534303 (29 May 2003).

di Lampedusa, Giuseppe Tomasi. 1991 [1958]. *The Leopard.* New York: Pantheon.

DiMaggio, Paul. 1982. "Cultural Capital and School Success: The Impact of Status Culture Participation on the Grades of U.S. High School Students." *American Sociological Review* 47: 189–201.

———, and Walter W. Powell. 1983. "The Iron Cage Revisited: Institutional Isomorphism and Collective Rationality in Organizational Fields." *American Sociological Review* 48: 147–160.

Dines, Gail, and Jean McMahon Humez. 1995. *Gender, Race, and Class in Media: A Text-Reader.* Thousand Oaks, CA: Sage.

Doberman, John. 1997. *Darwin's Athletes: How Sport Has Damaged Black America and Preserved the Myth of Race.* Boston: Houghton Mifflin.

Domhoff, G. William. 1983. *Who Rules America Now? A View for the 1980s.* New York: Touchstone.

Donahue III, John J., and Steven D. Levitt. 2001. "The Impact of Legalized Abortion on Crime." *Quarterly Journal of Economics* 116: 379–420.

Dolnick, Edward. 1993. "Deafness as culture." *The Atlantic Monthly* 272, 3: 37–48.

Dore, Ronald. 1983. "Goodwill and the Spirit of Market Capitalism." *British Journal of Sociology* 34: 459–482.

Doremus, Paul N., William W. Keller, Louis W. Pauly, and Simon Reich. 1998. *The Myth of the Global Corporation.* Princeton, NJ: Princeton University Press.

Douglas, Jack D. 1967. *The Social Meanings of Suicide.* Princeton, NJ: Princeton University Press.

Douglas, Susan J. 1994. *Where the Girls Are: Growing Up Female with the Mass Media.* New York: Random House.

Dowling, Michael, Christian Lechner, and Bodo Thielmann. 1998. "Convergence—Innovation and Change of Market Structures Between Television and Online Services." *Electronic Marketing* 8, 4. On the World Wide Web at http://www.electronicmarkets.com/netacademy/publications.nsf/all_pk/1124 (2 May 2000).

Dreyfus, Robert. 1999. "Money 2000." *The Nation* 18 October: 11–17.

Drèze, Jean, and Amartya Sen. 1989. *Hunger and Public Action.* Oxford: Clarendon Press.

DuBois, W. E. B. 1967 [1899]. *The Philadelphia Negro: A Social Study.* New York: Schocken.

Dudley, Kathryn Marie. 1994. *The End of the Line: Lost Jobs, New Lives in Postindustrial America.* Chicago: University of Chicago Press.

Duffy, Elizabeth A., and Idana Goldberg. 1997. *Crafting a Class: Admissions and Financial Aid, 1955–1994.* Princeton, NJ: Princeton University Press.

Duffy, Jim, Georg Gunther, and Lloyd Walters. 1997. "Gender and Mathematical Problem Solving." *Sex Roles* 37: 477–494.

Durham, William H. 1979. *Scarcity and Survival in Central America: Ecological Origins of the Soccer War.* Stanford, CA: Stanford University Press.

Durkheim, Émile. 1951 [1897]. *Suicide: A Study in Sociology,* G. Simpson, ed., J. Spaulding and G. Simpson, trans. New York: Free Press.

———. 1956. *Education and Sociology,* Sherwood D. Fox, trans. New York: Free Press.

———. 1961 [1925]. *Moral Education: A Study in the Theory and Application of the Sociology of Education,* Everett K. Wilson and Herman Schnurer, trans. New York: Free Press.

———. 1973 [1899–1900]. "Two Laws of Penal Evolution." *Economy and Society* 2: 285–308.

———. 1976 [1915]. *The Elementary Forms of the Religious Life,* Joseph Ward Swain, trans. New York: Free Press.

Dutton, Judy. 2000. "Detect His Lies Every Time." *Cosmopolitan* April: 126.

Dyson, Freeman. 1999. *The Sun, the Genome, and the Internet.* New York: Oxford University Press.

Eagley, Alice H., and Wendy Wood.1999. "The Origins of Sex Differences in Human Behavior: Evolved Dispositions Versus Social Roles." *American Psychologist* 54: 408–423.

Ebert, Roger. 1998. "Patch Adams." *Chicago Sun-Times.* On the World Wide Web at http://www.suntimes.com/ebert/ebert_reviews/1998/12/122504.html (2 May 2000).

Eccles, J. S., J. E. Jacobs, and R. D. Harold. 1990. "Gender Role Stereotypes, Expectancy Effects and Parents' Socialization of Gender Differences." *Journal of Social Issues* 46: 183–201.

Economakis, Evel, and Robert J. Brym. 1995. "Marriage and Militance in a Working Class District of St. Petersburg, 1896–1913." *Journal of Family History* 20: 23–43.

Economic Policy Institute. 2005. "Minimum Wage." On the World Wide Web at http://www.epinet.org/content.cfm/issueguides_minwage_minwage (30 March 2005).

Edel, Abraham. 1965. "Social Science and Value: A Study in Interrelations." Pp. 218–38 in Irving Louis Horowitz, ed. *The New Sociology: Essays in Social Science and Social Theory in Honor of C. Wright Mills.* New York: Oxford University Press.

Edelman, Peter. 2002. "The True Purpose of Welfare Reform." *New York Times* May 29. On the World Wide Web at http://www.nytimes.com (29 May 2002).

e.Harlequin.com. 2000. "About eHarlequin.com." On the World Wide Web at http://eharlequin.women.com/harl/globals/about/00bkrd11.htm (17 May 2000).

Ehrenreich, Barbara. 2001. *Nickel and Dimed: On (Not) Getting by in America.* New York: Henry Holt.

Ehrlich, Paul R. 1968. *The Population Bomb.* New York: Ballantine.

———, and Anne H. Ehrlich. 1990. *The Population Explosion.* New York: Simon & Schuster.

———, Gretchen C. Daily, Scott C. Daily, Norman Myers, and James Salzman. 1997. "No Middle Way on the Environment." *Atlantic Monthly* 280, 6: 98–104. On the World Wide Web at http://www.theatlantic.com/issues/97dec/enviro.htm (2 May 2000).

Eichler, Margrit. 1988. *Nonsexist Research Methods: A Practical Guide.* Boston: Unwin Hyman.

Einstein, Albert. 1954. *Ideas and Opinions,* Carl Seelig, ed., Sonja Bargmann, trans. New York: Crown.

Eisenberg, David M., Ronald C. Kessler, Cindy Foster, Frances E. Norlock, David R. Calkins, and Thomas L. Delbanco. 1993. "Unconventional Medicine in the United States—Prevalence, Costs, and Patterns of Use." *New England Journal of Medicine* 328: 246.

Eisenstadt, S. N. 1956. *From Generation to Generation.* New York: Free Press.

Eisler, Riane. 1995 [1987]. *The Chalice and the Blade: Our History, Our Future.* New York: HarperCollins.

"Ejército Zapatista de Liberación Nacional." 2000. On the World Wide Web at http://www.ezln.org (30 July 2000).

Ekman, Paul. 1978. *Facial Action Coding System.* New York: Consulting Psychologists Press.

Elgrably, Jordan. 2002. "Facts of Racial Profiling Prove Eye-Opening Even to Defense Attorneys." *Criminal Defense Weekly* 1, 9. On the World Wide Web at http://www.criminaldefense.com/magazine1.9/elgrably1.html (December 23 2002).

Elias, Norbert. 1985 [1982]. *The Loneliness of the Dying,* Edmund Jephcott, trans. Oxford: Blackwell.

———. 1994 [1939]. *The Civilizing Process,* Edmund Jephcott, trans. Oxford: Blackwell.

Elliott, H. L. 1995. "Living Vicariously Through Barbie." On the World Wide Web at http://ziris.syr.edu/path/public_html/barbie/main.html (19 November 1998).

Ellis, Brett Easton. 1991. *American Psycho.* New York: Vintage.

Ellul, Jacques. 1964 [1954]. *The Technological Society,* trans. John Wilkinson. New York: Vintage.

Emanuel, Kerry. 2005. "Increasing destructiveness of tropical cyclones over the past 30 years." *Nature,* 436, 4:686–688.

Ember, Carol, and Melvin Ember. 1973. *Anthropology.* New York: Appleton-Century-Crofts.

Engels, Frederick. 1970 [1884]. *The Origins of the Family, Private Property and the State,* Eleanor Burke Leacock, ed., Alec West, trans. New York: International Publishers.

England, Charles R. n.d. "A Look at the Indian Health Service Policy of Sterilization, 1972–1976." On the World Wide Web at http://www.dickshovel.com/IHSSterPol.html (23 April 2003).

England, Paula. 1992a. *Comparable Worth: Theories and Evidence.* Hawthorne, NY: Aldine de Gruyter.

———. 1992b. "From Status Attainment to Segregation and Devaluation." *Contemporary Sociology* 21: 643–647.

Entine, John. 2000. *Taboo: Why Black Athletes Dominate Sports and Why We Are Afraid to Talk About It.* New York: Public Affairs.

Entwisle, Doris R., and Karl L. Alexander. 1995. "A Parent's Economic Shadow: Family Structure versus Family Resources as Influences on Early School Achievement." *Journal of Marriage and the Family* 57: 399–410.

Epstein, Helen. 1998. "Life and Death on the Social Ladder." *New York Review of Books* 45, 12: 16 July: 26–30.

Epstein, Steven. 1996. *Impure Science: AIDS, Activism, and the Politics of Knowledge.* Berkeley: University of California Press.

Ericksen, Julia. 1998. "With Enough Cases, Why Do You Need Statistics? Revisiting Kinsey's Methodology." *Journal of Sex Research* 35: 132-40.

Erikson, Robert, and John H. Goldthorpe. 1992. *The Constant Flux: A Study of Class Mobility in Industrial Societies.* Oxford: Clarendon Press.

Esping-Andersen, Gøsta. 1990. *The Three Worlds of Welfare Capitalism.* Princeton, NJ: Princeton University Press.

Estrich, Susan. 1987. *Real Rape.* Cambridge, MA: Harvard University Press.

"Ethnic Groups in the World." 2001. *Scientific American.* On the World Wide Web at http://www.sciam.com/1998/0998issue/0998numbers.html (4 December 2001).

Etzioni, Amitai. 1975. *A Comparative Analysis of Complex Organizations,* 2nd ed. New York: Free Press.

"European Tobacco Ban Receives Fresh Approval." 1998. *The Lancet* 351: 1568.

Evans D. B., A. Tandon, C. J. Murray, and J. A. Lauer. 2001. "Comparative Efficiency of National Health Systems: Cross National Econometric Analysis." *BMJ* 323, 7308:307–310. On the World Wide Web at http://bmj.com/cgi/content/full/323/7308/307?maxtoshow=&HITS=10&hits=10&RESULTFORMAT=&searchid=1008146506125_662&stored_search=&FIRSTINDEX=0&volume=323&firstpage=307 (13 June 2003).

Evans, Peter B., Dietrich Rueschemeyer, and Theda Skocpol. 1985. *Bringing the State Back In.* Cambridge: Cambridge University Press.

Evans, Robert G. 1999. "Social Inequalities in Health." *Horizons* (Policy Research Secretariat, Government of Canada) 2, 3: 6–7.

Ewen, Stuart. 1976. *Captains of Consciousness: Advertising and the Social Roots of Consumer Culture.* New York: McGraw-Hill.

——————. 1988. *All Consuming Images: The Politics of Style in Contemporary Culture.* New York: Basic Books.

——————. 1997. *PR! A Social History of Spin.* New York: Basic Books.

"Excerpts from Justices' Opinions on Michigan Affirmative Action Cases." 2003. *New York Times* June 24. On the World Wide Web at http://www.nytimes.com (24 June 2003).

Executive Office of the President of the United States. 2000. "A Citizen's Guide to the Federal Budget." On the World Wide Web at http://usgovinfo.about.com/newsissues/usgovinfo/gi/dynamic/offsite.htm?site=http://w3.access.gpo.gov/usbudget(8 June 2000).

"Experts and the Drug Industry." 2005. *New York Times* 4 March. On the World Wide Web at www.nytimes.com (4 March 2005).

Fagot, Beverly I., Caire S. Rodgers, and Mary D. Leinbach. 2000. "Theories of Gender Socialization." Pp. 65–89 in Thomas Eckes, ed. *The Developmental Social Psychology of Gender.* Mahwah NJ: Lawrence Erlbaum Associates.

Fall Colors II: Exploring the Quality of Diverse Portrayals on Prime Time Television. 2000. Oakland, CA: Children Now. On the World Wide Web at http://www.childrennow.org/media/fall-colors-2k/fc2-2k.pdf (5 August 2000).

Fall Colors 2001–02: Prime Time Diversity Report. 2002. Oakland and Los Angeles: Children Now. On the World Wide Web at http://www.childrennow.org/media/fc2002/fc-2002-report.pdf (29 May 2002).

Farley, Christopher John. 1998. "Rock Star." *Time* July 20. On the World Wide Web at http://www.time.com/time/sampler/article/0,8599,166239,00.html (9 April 2003).

Farmer, Paul. 1992. *AIDS and Accusation: Haiti and the Geography of Blame.* Berkeley: University of California Press.

Fass, Paula S. 1989. *Outside In: Minorities and the Transformation of American Education.* New York: Oxford University Press.

Fava, Sylvia Fleis. 1956. "Suburbanism as a Way of Life." *American Sociological Review* 21: 34–37.

Feagin, Joe R., and Melvin P. Sikes. 1994. *Living with Racism: The Black Middle-Class Experience.* Boston: Beacon Press.

Fearon, E. R. 1997. "Human Cancer Syndrome: Clues to the Origin and Nature of Cancer." *Science* 278: 1043–1050.

Featherman, David L., and Robert M. Hauser. 1976. "Sexual Inequalities and Socioeconomic Achievement in the U.S., 1962–1973." *American Sociological Review* 41: 462–483.

——————, and ——————. 1978. *Opportunity and Change.* New York: Academic Press.

——————, F. Lancaster Jones, and Robert M. Hauser. 1975. "Assumptions of Mobility Research in the United States: The Case of Occupational Status." *Social Science Research* 4: 329–360.

Febvre, Lucien, and Henri-Jean Martin. 1976 [1958]. *The Coming of the Book: The Impact of Printing 1450–1800,* David Gerard, trans. London: NLB.

Federal Reserve Bank of San Francisco. 2000. "Tech Stocks and House Prices in California." September 15. On the World Wide Web at http://www.frbsf.org/econrsrch/wklyltr/2000/el2000-27.html (11 March 2003).

Feeley, Malcolm M., and Jonathan Simon. 1992. "The New Penology: Notes on the Emerging Strategy of Corrections and Its Implications." *Criminology* 30: 449–474.

Fein, Helen. 1979. *Accounting for Genocide: National Responses and Jewish Victimization During the Holocaust.* New York: Free Press.

Felson, Richard B. 1996. "Mass Media Effects on Violent Behavior." *Annual Review of Sociology* 22: 103–128.

Feminist.com. "The Wage Gap." 1999. On the World Wide Web at http://www.feminist.com/wgot.htm (30 April 2000).

Fernandez-Dols, Jose-Miguel, Flor Sanchez, Pilar Carrera, and Maria-Angeles Ruiz-Belda. 1997. "Are Spontaneous Expressions and Emotions Linked? An Experimental Test of Coherence." *Journal of Nonverbal Behavior* 21: 163–177.

Ferrante, Joan, and Pierre Brown Jr., eds. 2001 [1998]. *The Social Construction of Race and Ethnicity in the United States,* 2nd ed. Upper Saddle River, NJ: Prentice Hall.

Feynman, Richard P. 1999. *The Pleasure of Finding Things Out.* Jeffrey Robbins, ed. Cambridge, MA: Perseus Books.

Figart, Deborah M., and June Lapidus. 1996. "The Impact of Comparable Worth on Earnings Inequality." *Work and Occupations* 23: 297–318.

Fine, Gary Alan, and Patricia A. Turner. 2001. *Whispers on the Color Line: Rumor and Race in America.* Berkeley: University of California Press.

Finke, Roger, and Rodney Starke. 1992. *The Churching of America, 1776–1990: Winners and Losers in Our Religious Economy.* New Brunswick, NJ: Rutgers University Press.

Finsterbusch, Kurt. 2001. *Clashing Views on Controversial Social Issues.* Guilfod, CT: Dushkin.

Firebaugh, Glenn, and Frank D. Beck. 1994. "Does Economic Growth Benefit the Masses? Growth, Dependence and Welfare in the Third World." *American Journal of Sociology* 59: 631–653.

Fischer, Claude S. 1981. "The Public and Private Worlds of City Life." *American Sociological Review* 46: 306–316.

——————. 1984 [1976]. *The Urban Experience,* 2nd ed. New York: Harcourt Brace Jovanovich.

——————, Michael Hout, Martín Sánchez Jankowski, Samuel R. Lucas, Ann Swidler, and Kim Voss. 1996. *Inequality by Design: Cracking the Bell Curve Myth.* Princeton, NJ: Princeton University Press.

Fisher, Lawrence M. 2003. "Job-Rich Silicon Valley Has Turned Fallow, Survey Finds." *New York Times* January 20. On the World Wide Web at http://www.nytimes.com (10 March 2003).

Fitzgerald, Mark. 1997. "Media Perpetuate a Myth." *Editor and Publisher* 130, 33: 13.

Flexner, Eleanor. 1975. *Century of Struggle: The Woman's Rights Movement in the United States,* rev. ed. Cambridge, MA: Harvard University Press.

Flood, Gavid D. 1996. *An Introduction to Hinduism.* Cambridge: Cambridge University Press.

Floud, Roderick, Kenneth Wachter, and Annabel Gregory. 1990. *Height, Health and History: Nutritional Status in the United Kingdom, 1750–1980.* Cambridge: Cambridge University Press.

Fludd, Robert. 1617–1619. *Utriusque Cosmi Maioris Scilicet et Minoris Metaphysica, Physica Atqve Technica Historia.* Oppenheim, Germany: Johan-Theodori de Bry.

Fong, Eric W., and William T. Markham. 2002. "Anti-Chinese Politics in California in the 1870s: An Intercounty Analysis." *Sociological Perspectives* 45: 183–210.

Forbes.com. 1999. "Forbes: The World's Richest People: USA 98." On the World Wide Web at http://www.forbes.com/tool/toolbox/billnew/country98.asp?country=United%20States,98 (1 May 2000).

__________. 2000. "Forbes 500 Annual Directory." On the World Wide Web at http://www.forbes.com/tool/toolbox/forbes500s/asp/rankindex.asp (30 April 2000).

__________. 2003a. "Forbes 500." On the World Wide Web at http://www .forbes.com/2003/03/26/500sland.html (22 May 2003).

__________. 2003b. "General Electric." On the World Wide Web at http://www.forbes.com/finance/lists/38/2003/LIR.jhtml?passListId=38&passYear=2003&passListType=Company&uniqueId=UINP&datatype=Company (11 December 2003).

__________. 2004. "The 400 Richest Americans." On the World Wide Web at http://www.forbes.com/lists/forbes400/2004/09/22/rl04land.html (30 January 2005).

Foucault, Michel. 1977 [1975]. *Discipline and Punish: The Birth of the Prison.* Alan Sheridan, trans. New York: Pantheon.

__________. 1990 [1978]. *The History of Sexuality: An Introduction,* Vol. 1. Robert Hurley, trans. New York: Vintage.

Frank, André Gundar. 1967. *Capitalism and Underdevelopment in Latin America: Historical Studies of Chile and Brazil.* New York: Monthly Review Press.

Frank Porter Graham Child Development Center. 1999. "Early Learning, Later Success: The Abecedarian Study." On the World Wide Web at http://www.fpg.unc.edu/~abc/abcedarianWeb/index.htm (10 August 2000).

Frank, Robert H. 1988. *Passions Within Reason: The Strategic Role of the Emotions.* New York: W. W. Norton.

Frank, Thomas. 1997. *The Conquest of Cool.* Chicago: University of Chicago Press.

__________, and Matt Weiland, eds. 1997. *Commodify Your Dissent: Salvos from the Baffler.* New York: W. W. Norton.

Franke, Richard H., and James D. Kaul. 1978. "The Hawthorne Experiments: First Statistical Interpretation." *American Sociological Review* 43: 623–643.

Franke, Richard W., and Barbara H. Chasin. 1992. *Kerala: Development Through Radical Reform.* San Francisco: Institute for Food and Development Policy.

Frankel, Glenn. 1996. "U.S. Aided Cigarette Firms in Conquests Across Asia." *Washington Post* November 17: A01. On the World Wide Web at http://www.washington-post.com/wp-srv/national/longterm/tobacco/stories/asia.htm (8 February 2003).

Franklin, Karen. 1998. "Psychosocial Motivations of Hate Crime Perpetrators." Paper presented at the annual meetings of the American Psychological Association (San Francisco: 16 August).

Franklin, Sara, and Helen Ragone, eds. 1999. *Reproducing Reproduction.* Philadelphia: University of Pennsylvania Press.

Freedom House. 2004. *Freedom in the World 2004.* On the World Wide Web at http://www.freedomhouse.org/research/survey2004.htm (2 May 2004).

Freedman, Jonathan L. 2002. *Media Violence and Its Effect on Aggression: Assessing the Scientific Evidence.* Toronto: University of Toronto Press.

Freidson, Eliot. 1986. *Professional Powers: A Study of the Institutionalization of Formal Knowledge.* Chicago: University of Chicago Press.

Freire, Paulo. 1972. *The Pedagogy of the Oppressed.* New York: Herder and Herder.

French, Howard W. 2001. "The Japanese, it seems, are out-growing Japan." *New York Times* 1 February: 4A.

Freud, Sigmund. 1962 [1930]. *Civilization and Its Discontents.* James Strachey, trans. New York: W. W. Norton.

__________. 1973 [1915–17]. *Introductory Lectures on Psychoanalysis.* James Strachey, trans., James Strachey and Angela Richards, eds. Harmondsworth, UK: Penguin.

Freudenburg, William R. 1997. "Contamination, Corrosion and the Social Order: An Overview." *Current Sociology* 45, 3: 19–39.

Fried, Martha Nemes, and Morton H. Fried. 1980. *Transitions: Four Rituals in Eight Cultures.* New York: W. W. Norton.

Friedenberg, Edgar Z. 1959. *The Vanishing Adolescent.* Boston: Beacon Press.

Friedkin, Noah E. 1998. *A Structural Theory of Social Influence.* Cambridge: Cambridge University Press.

Friedland, William. H. 2002. "Agriculture and Rurality: Beginning the 'Final Separation'?" *Rural Sociology* 67: 350–371.

Fröbel, Folker, Jürgen Heinrichs, and Otto Kreyre. 1980. *The New International Division of Labour: Structural Unemployment in Industrialised Countries and Industrialisation in Developing Countries.* Pete Burgess, trans. Cambridge: Cambridge University Press.

Froissart, Jean. 1968 [c. 1365]. *Chronicles,* selected and translated by Geoffrey Brereton. Harmondsworth, UK: Penguin.

Furstenberg, Frank F., Jr., and Andrew Cherlin. 1991. *Divided Families: What Happens to Children When Parents Part.* Cambridge, MA: Harvard University Press.

Gado, Mark. 2003. "A Cry in the Night: The Kitty Genovese Murder." *Court TV's Crime Library.* On the World Wide Web at http://www.crimelibrary.com/serial_killers/predators/kitty_genovese/1.html (23 July 2003).

Gaines, Donna. 1990. *Teenage Wasteland: Suburbia's Dead End Kids.* New York: Pantheon.

Galewitz, Phil. 2000. "Firm Settles Fraud Case: Hospital Chain Columbia/HCA to pay $745 Million." *ABC-NEWS.com* May 18. On the World Wide Web at http://abcnews.go.com/sections/business/DailyNews/columbiahca_990518.html (12 June 2003).

Galinsky, Ellen, Stacy S. Kim, and James T. Bond. 2001. *Feeling Overworked: When Work Becomes Too Much.* New York: Families and Work Institute.

"A Gallery of Social Structures: Structures of World Trade." 2003. *Max-Planck-Institut für Gesselschaftsforschung.* On the World Wide Web at http://www.mpi-fg-koeln.mpg.de/~lk/netvis/trade (9 February 2003).

Gallup Organization. 2004. "American Public Opinion about Gay and Lesbian Marriages." On the World Wide Web at http://www.gallup.com/poll/focus/sr040127.asp (5 March, 2004).

Galper, Joseph. 1998. "Schooling for Society." *American Demographics* 20, 3: 33–34.

Galt, Virginia. 1999. "Jack Falling Behind Jill in School, Especially in Reading." *Globe and Mail* 30 October: A10.

Gambetta, Diego, ed. 1988. *Trust: Making and Breaking Cooperative Relations.* Oxford: Blackwell.

Gamson, William A. 1975. *The Strategy of Social Protest.* Homewood, IL: Dorsey Press.

__________, Bruce Fireman, and Steven Rytina. 1982. *Encounters with Unjust Authority.* Homewood, IL: Dorsey Press.

"Gangs: Going Global." 2005. *The Economist* 25 February – 4 March: 29.

Gans, Herbert. 1962. *The Urban Villagers: Group and Class in the Life of Italian-Americans.* New York: Free Press.

__________. 1979a. *Deciding What's News: A Study of CBS Evening News, NBC Nightly News, Newsweek and Time.* New York: Pantheon.

__________. 1979b, "Symbolic Ethnicity: The Future of Ethnic Groups and Cultures in America." Pp. 193–220 in Herbert Gans et al., eds. *On the Making of Americans: Essays in Honor of David Reisman.* Philadelphia: University of Pennsylvania Press.

__________. 1995. *The War Against Poverty: The Underclass and Antipoverty Policy.* New York: Basic Books.

Ganz, Marshall. 1996. "Motor Voter or Motivated Voter?" *The American Prospect* 28. On the World Wide Web at http://www.prospect.org/archives/28/28ganz.html (21 November 2000).

Gap.com. "Gap." 1999. On the World Wide Web at http://www.gap.com/onlinestore/gap/advertising/khakitv.asp (28 April 2000).

"Garciaparra Explains His Superstitions." 2000. On the World Wide Web at http://www.geocities.com/Colosseum/Track/4242/nomar3.wav (28 April 2000).

Garfinkel, Harold. 1967. *Studies in Ethnomethodology.* Englewood Cliffs, NJ: Prentice-Hall.

Garfinkel, Simson. 2000. *Database Nation: The Death of Privacy in the 21st Century.* Cambridge, MA: O'Reilly & Associates.

Garkawe, Sam. 1995. "The Impact of the Doctrine of Cultural Relativism on the Australian Legal System." *E Law* 2, 1. On the World Wide Web at http://www.murdoch.edu.au/elaw/issues/v2n1/garkawe.txt (10 May 2000).

Garland, David. 1990. *Punishment and Modern Society: A Study in Social Theory.* Chicago: University of Chicago Press.

Garner, David M. 1997. "The 1997 Body Image Survey Results." *Psychology Today* 30, 1: 30–44.

Garrau, Joel. 1991. *Edge City: Life on the New Frontier.* New York: Doubleday.

Garson, Barbara. 2001. *Money Makes the World Go Around.* New York: Viking.

Gates, Bill, with Nathan Myhrvold and Peter Rinearson. 1996. *The Road Ahead.* New York: Penguin.

Gaubatz, Kathlyn Taylor. 1995. *Crime in the Public Mind.* Ann Arbor: University of Michigan Press.

Gauvain, Mary, Beverly I. Fagot, Craige Leve, and Kate Kavanagh. 2002. "Instruction by Mothers and Fathers During Problem Solving with Their Young Children." *Journal of Family Psychology* 6: 81–90.

Gearon, Christopher J. 2002. "States Forming Alliances to Deal with Drugmakers." *AARP.* On the World Wide Web at http://www.aarp.org/bulletin/departments/2002/medicare/0410_medicare_1.html (17 June 2003).

Gebhard, Paul H., and Alan B. Johnson, 1979. *The Kinsey Data: Marginal Tabulations of the 1938-1963 Interviews Conducted by the Institute for Sex Research.* Philadelphia: W. B. Saunders.

Geertz, Clifford. 1973. *The Interpretation of Cultures: Selected Essays.* New York: Basic Books.

Gelbspan, Ross. 1997. *The Heat Is On: The High Stakes Battle over Earth's Threatened Climate.* Reading, MA: Addison-Wesley.

__________. 1999. "Trading Away Our Chances to End Global Warming." *Boston Globe* 16 May: E2.

Gelles, Richard J. 1997 [1985]. *Intimate Violence in Families,* 3rd ed. Thousand Oaks, CA: Sage.

Gellner, Ernest. 1988. *Plough, Sword and Book: The Structure of Human History.* Chicago: University of Chicago Press.

General Motors. 2000. "GM Energy and Environment Vehicle Strategy: Creating Products and Options." On the World Wide Web at http://www.gm.com/environment/products/chart/index.html (2 August 2000).

Gerber, Theodore P., and Michael Hout. 1998. "More Shock than Therapy: Market Transition, Employment, and Income in Russia, 1991–1995." *American Journal of Sociology* 104: 1–50.

Gerbner, George. 1998. "Casting the American Scene: A Look at the Characters on Prime Time and Daytime Television from 1994–1997." *The 1998 Screen Actors Guild Report.* On the World Wide Web at http://www.media-awareness.ca/eng/issues/minrep/resource/reports/gerbner.htm (5 August 2000).

Germani, Gino, and Kalman Silvert. 1961. "Politics, Social Structure and Military Intervention in Latin America." *European Journal of Sociology* 11: 62–81.

Gerschenkron, Alexander. 1962. *Economic Backwardness in Historical Perspective: A Book of Essays.* Cambridge, MA: Harvard University Press.

Giddens, Anthony. 1987. *Sociology: A Brief but Critical Introduction,* 2nd ed. New York: Harcourt Brace Jovanovich.

__________. 1990. *The Consequences of Modernity.* Stanford, CA: Stanford University Press.

Gille, Zsuzsa, and Seán Ó Riain. 2002. "Global Ethnography." *Annual Review of Sociology* 28: 271–295.

Gilligan, Carol. 1982. *In a Different Voice: Psychological Theory and Women's Development.* Cambridge, MA: Harvard University Press.

__________, Nona P. Lyons, and Trudy J. Hanmer, eds. 1990. *Making Connections: The Relational Worlds of Adolescent Girls at Emma Willard School.* Cambridge, MA: Harvard University Press.

Gillis, John R. 1981. *Youth and History: Tradition and Change in European Age Relations, 1770–Present,* expanded student ed. New York: Academic Press.

Gilman, Sander L. 1991. *The Jew's Body.* New York: Routledge.

Gilpin, Robert. 2001. *Global Political Economy: Understanding the International Economic Order.* Princeton, NJ: Princeton University Press.

Gimbutas, Marija. 1982. *Goddesses and Gods of Old Europe.* Berkeley: University of California Press.

Ginsberg, Benjamin. 1986. *The Captive Public: How Mass Opinion Promotes States Power.* New York: Basic Books.

Gitlin, Todd. 1983. *Inside Prime Time.* New York: Pantheon.

Gladwell, Malcolm. 2002. "The Politics of Politesse." *New Yorker* December 23–30: 57–58.

Glaser, Barney, and Anselm Straus. 1967. *The Discovery of Grounded Theory.* Chicago: Aldine.

Glazer, Nathan. 1997. *We Are All Multiculturalists Now.* Cambridge, MA: Harvard University Press.

__________, and Daniel Patrick Moynihan. 1963. *Beyond the Melting Pot.* Cambridge, MA: MIT Press.

Glick, Jennifer E. 1999. "Economic Support from and to Extended Kin: A Comparison of Mexican Americans and Mexican Immigrants." *The International Migration Review* 33: 745–765.

Gleick, James. 2000 [1999]. *Faster: The Acceleration of Just About Everything.* New York: Vintage.

"Global 1000." 1999. *Business Week Online.* On the World Wide Web at http://www.businessweek.com (12 July 1999).

Glock, Charles Y. 1962. "On the Study of Religious Commitment." *Religious Education* 62, 4: 98–110.

"Gnutella." 2000. On the World Wide Web at http://gnutella.wego.com (7 August 2000).

Goddard Institute for Space Studies. 2003. "Annual Mean Temperature Anomalies in .01 C: Selected Zonal Means." http://www.giss.nasa.gov/data/update/gistemp/ZonAnn.Ts.txt (29 July 2003).

__________. 2005. "Global Surface Air Temperature Anomaly (C) (Base: 1951-1980)." http://www.giss.nasa.gov/data/update/gistemp/graphs/Fig.A.txt (2 March 2005).

Goffman, Erving. 1959 [1956]. *The Presentation of Self in Everyday Life.* Garden City, NY: Anchor.

__________. 1961. *Asylums: Essays on the Social Situation of Mental Patients and Other Inmates.* Garden City, NY: Anchor Books.

__________. 1963a. *Behavior in Public Places: Notes on the Social Organization of Gatherings.* New York: Free Press.

__________. 1963b. *Stigma: Notes on the Management of Spoiled Identity.* Englewood Cliffs, NJ: Prentice-Hall.

__________. 1971. *Relations in Public: Microstudies of the Public Order.* New York: Basic Books.

__________. 1974. *Frame Analysis.* Cambridge, MA: Harvard University Press.

__________. 1981. *Forms of Talk.* Philadelphia: University of Pennsylvania Press.

"Going Native: Dinner Parties." 2003. On the World Wide Web at http://www.geocities.com/robainsley2/eikoku/1.html (3 February 2003).

Goldenberg, Suzanne. 2002. "Arabs and Muslims to be Fingerprinted at US Airports." *Guardian Unlimited* 2 October. On the World Wide Web at http://www.guardian.co.uk/september11/story/0,11209,802771,00.html (24 December 2002).

Goldhagen, Daniel Jonah. 1996. *Hitler's Willing Executioners: Ordinary Germans and the Holocaust.* New York: Knopf.

Goldthorpe, John H. in collaboration with Catriona Llewellyn and Clive Payne. 1987 [1980]. *Social Mobility and Class Structure in Modern Britain,* 2nd ed. Oxford: Clarendon Press.

Goll, David. 2002. "True Priority of Office Ethics Clouded by Scandals." *East Bay Business Times* August 19. On the World Wide Web at http://eastbay.bizjournals.com/eastbay/stories/2002/08/19/smallb3.html (13 January 2003).

Gombrich, Richard Francis. 1996. *How Buddhism Began: The Conditioned Genesis of the Early Teachings.* London: Athlone.

Gonzalez, David. 2005. "Paying a Price for Doughnuts, Burgers and Pizza." *New York Times* 25 January. On the World Wide Web at http://www.nytimes.com (25 January 2005).

Goode, Erich, and Nachman Ben-Yehuda. 1994. *Moral Panics: The Social Construction of Deviance.* Cambridge, MA: Blackwell.

Goodell, Jeff. 1999. "Down and Out in Silicon Valley." *Rolling Stone* 9 December: 64–71.

"Google Groups." 2005. On the World Wide Web at http://groups.google.ca/groups?group=*&hl=en (14 February 2005).

Gordon, David M. 1996. *Fat and Mean: The Corporate Squeeze of Working Americans and the Myth of Managerial "Downsizing."* New York: Free Press.

__________, Richard Edwards, and Michael Reich. 1982. *Segmented Work, Divided Workers: The Historical Transformation of Labor in the United States.* New York: Cambridge University Press.

Gordon, Linda. 1990 [1976]. *Woman's Body, Woman's Right: Birth Control in America.* Harmondsworth, UK: Penguin.

Gordon, Sarah. 1984. *Hitler, Germans, and the Jewish Question.* Princeton, NJ: Princeton University Press.

Gorden-Larsen, Penny. 2003. "Obesity: Research Shows Less Income, Education Not Always Top Factors in Child Obesity." *Obesity, Fitness and Wellness Weekly* 1 February: 21–22.

Gorman, Christine. 1998. "Playing the HMO Game." *Time* 152, 2: 13 July. On the World Wide Web at http://www.time.com/time/magazine/1998/dom/980713/cover1.html (2 May 2000).

Gormley, Jr., William T. 1995. *Everybody's Children: Child Care as a Public Problem.* Washington, DC: Brookings Institution.

Gortmaker, S. L., A. Must, J. M. Perrin, A. M. Sobol, and W. H. Dietz. 1993. "Social and Economic Consequence of Overweight in Adolescence and Young Adulthood." *New England Journal of Medicine* 329, 14: September 30: 1008–1012.

Gottdiener, Mark, and Ray Hutchison. 2000 [1994]. *The New Urban Sociology,* 2nd ed. Boston: McGraw-Hill.

Gottfredson, Michael, and Travis Hirschi. 1990. *A General Theory of Crime.* Stanford, CA: Stanford University Press.

Gottwald, Norman K. 1979. *The Tribes of Yahweh: A Sociology of the Religion of Liberated Israel, 1250–1050 B.C.E.* Maryknoll, NY: Orbis.

Goubert, Jean-Pierre. 1989 [1986]. *The Conquest of Water,* Andrew Wilson, trans. Princeton, NJ: Princeton University Press.

Gould, S. J., and R. C. Lewontin. 1979. "The Spandrels of San Marco and the Panglossian Paradigm: A Critique of the Adaptationist Programme." *Proceedings of the Royal Society of London,* Series B, Biological Sciences 205, 1161, September 21: 581–598.

__________. 1988. "Kropotkin Was No Crackpot." *Natural History* 97, 7: 12–18.

__________. 1996 [1981]. *The Mismeasure of Man,* rev. ed. New York: W. W. Norton.

Gouldner, Alvin W. 1954. *Patterns of Industrial Bureaucracy: A Case Study of Modern Factory Administration.* New York: Free Press.

Government of Canada. 2002. "Study Released on Firearms in Canada." On the World Wide Web at http://www.cfc-ccaf.gc.ca/en/general_public/news_releases/survey-08202002.asp (22 February 2003).

Graff, Harvey J. 1995. *Conflicting Paths: Growing Up in America.* Cambridge, MA: Harvard University Press.

Granovetter, Mark. 1973. "The Strength of Weak Ties." *American Sociological Review* 78: 1360–1380.

__________. 1995 [1974]. *Getting a Job: A Study of Contacts and Careers.* Cambridge, MA: Harvard University Press.

__________. 1984. "Small is Bountiful." *American Sociological Review* 49: 323–334.

Gratz and Hamacher v. Bollinger et al. 1997. Supreme Court of the United States. On the World Wide Web at http://www.moraldefense.com/Campaigns/Equality/gratz_v_bollinger_SupCt_brief.pdf (30 December 2003).

Greeley, Andrew. 1974. *Ethnicity in the United States.* New York: John Wiley.

__________. 1989. *Religious Change in America.* Cambridge, MA: Harvard University Press.

Greenberg, David F. 1988. *The Construction of Homosexuality.* Chicago: University of Chicago Press.

Greenpeace. 2000. "Greenpeace Contacts Worldwide." 2000. On the World Wide Web at http://adam.greenpeace.org/information.shtml (2 May 2000).

Greenstein, Theodore N. 1996. "Husbands' Participation in Domestic Labor: Interactive Effects of Wives' and Husbands' Gender Ideologies." *Journal of Marriage and the Family* 58: 585–595.

Grescoe, P. 1996. *The Merchants of Venus: Inside Harlequin and the Empire of Romance.* Vancouver: Raincoast.

Griffin, John Howard. 1961. *Black Like Me.* Boston: Houghton Mifflin.

Grindstaff, Laura. 1997. "Producing Trash, Class, and the Money Shot: A Behind-the-Scenes Account of Daytime TV Talk Shows." Pp. 164–202 in James Lull and Stephen Hinerman, eds. *Media Scandals: Morality and Desire in the Popular Culture Marketplace.* Cambridge: Polity Press.

Griswold, Wendy. 1994. *Cultures and Societies in a Changing World.* Thousand Oaks, CA: Pine Forge Press.

Groce, Nora Ellen. 1985. *Everyone Here Spoke Sign Language: Hereditary Deafness on Martha's Vineyard.* Cambridge, MA: Harvard University Press.

Grofman, Bernard. 2000. "Questions and Answers About Motor Voter." On the World Wide Web at http://www.fairvote.org/reports/1995/chp6/grofman.html (21 November 2000).

Grusky, David B., and Robert M. Hauser. 1984. "Comparative Social Mobility Revisited: Models of Convergence and Divergence in 16 Countries." *American Sociological Review* 49: 19–38.

Grutter v. Bollinger et al., 1997. Supreme Court of the United States. On the World Wide Web at http://www.moraldefense.com/Campaigns/Equality/grutter_v_bollinger_SupCt_brief.pdf (30 December 2003).

Guillén, Mauro F. 2001. "Is Globalization Civilizing, Destructive or Feeble? A Critique of Five Key Debates in the Social Science Literature." *Annual Review of Sociology* 27. On the World Wide Web at http://knowledge.wharton.upenn.edu/PDFs/938.pdf (6 February 2003).

Gunderson, Edna, Bill Keveney, and Ann Oldenburg. 2002. "'The Osbournes' Find a Home in America's Living Rooms." *USA Today* April 19: 1A, 2A.

Gurr, Ted Robert. 1970. *Why Men Rebel.* Princeton, NJ: Princeton University Press.

Gusfield, Joseph R. 1963. *Symbolic Crusade: Status Politics and the American Temperance Movement.* Urbana, IL: University of Illinois Press.

Gutiérrez, David G. 1995. *Walls and Mirrors: Mexican Americans, Mexican Immigrants, and the Politics of Ethnicity.* Berkeley: University of California Press.

"GVU's WWW User Survey." 1999. On the World Wide Web at http://www.cc.gatech.edu/gvu/user_surveys/survey-1998-10/graphs/graphs.html#general (11 February 2001).

Haas, Jack, and William Shaffir. 1987. *Becoming Doctors: The Adoption of a Cloak of Competence.* Greenwich, CT: JAI Press.

Haas, Jennifer. 1998. "The Cost of Being a Woman." *New England Journal of Medicine* 338: 1694–1695.

Habermas, Jürgen. 1989. *The Structural Transformation of the Public Sphere,* Thomas Burger, trans. Cambridge, MA: MIT Press.

Hacker, Andrew. 1992. *Two Nations: Black and White, Separate, Hostile, Unequal.* New York: Ballantine Books.

__________. 1997. *Money: Who Has How Much and Why.* New York: Scribner.

Hacker, Jacob S. 1997. *The Road to Nowhere: The Genesis of President Clinton's Plan for Health Security.* Princeton, NJ: Princeton University Press.

Hacsi, Timothy A. 2002. *Children as Pawns: The Politics of Educational Reform.* Cambridge, MA: Harvard University Press.

Hagan, John. 1989. *Structuralist Criminology.* New Brunswick, NJ: Rutgers University Press.

__________. 1994. *Crime and Disrepute.* Thousand Oaks, CA: Pine Forge Press.

__________, John Simpson, and A.R. Gillis. 1987. "Class in the Household: A Power-Control Theory of Gender and Delinquency." *American Journal of Sociology* 92: 788–816.

Haiken, Elizabeth. 1997. *Venus Envy: A History of Cosmetic Surgery.* Baltimore: Johns Hopkins University Press.

Haines, Herbert H. 1996. *Against Capital Punishment: The Anti-Death Penalty Movement in America, 1972–1994.* New York: Oxford University Press.

Hall, Edward. 1959. *The Silent Language.* New York: Doubleday.

__________. 1966. *The Hidden Dimension.* New York: Doubleday.

Hall, Stuart. 1980. "Encoding/Decoding." Pp. 128–138 in Stuart Hall, Dorothy Hobson, Andrew Lowe, and Paul Willis, eds. *Culture, Media, Language: Working Papers in Cultural Studies, 1972–79.* London: Hutchinson.

__________. 1992. "The Question of Cultural Identity." Pp. 274–313 in Stuart Hall, David Held, and Tim McGrew, eds. *Modernity and its Futures.* Cambridge: Polity and Open University Press.

Hamachek, D. 1995. "Self-concept and School Achievement: Interaction Dynamics and a Tool for Assessing the Self-concept Component." *Journal of Counseling and Development* 73: 419–425.

Hammer, Michael. 1999. "Is Work Bad for You?" *The Atlantic Monthly* August: 87–93.

Hampton, Janie, ed. 1998. *Internally Displaced People: A Global Survey.* London: Earthscan.

Hancock, Graham. 1989. *Lords of Poverty: The Power, Prestige, and Corruption of the International Aid Business.* New York: Atlantic Monthly Press.

Hancock, Lynnell. 1994. "In Defiance of Darwin: How a Public School in the Bronx Turns Dropouts into Scholars." *Newsweek* 24 October: 61.

Handelman, Stephen. 1995. *Comrade Criminal: Russia's New Mafiya.* New Haven, CT: Yale University Press.

Handy, Bruce. 2002. "Glamour with Altitude." *Vanity Fair* October: 214–228.

Haney, Craig, W. Curtis Banks, and Philip G. Zimbardo. 1973. "Interpersonal Dynamics in a Simulated Prison." *International Journal of Criminology and Penology* 1: 69–97.

Hanke, Robert. 1998. "'Yo Quiero Mi MTV!' Making Music Television for Latin America." Pp. 219–245 in Thomas Swiss, Andrew Herman, and John M. Sloop, eds. *Mapping the Beat: Popular Music and Contemporary Theory.* Oxford: Blackwell.

Hannigan, John. 1995a. "The Postmodern City: A New Urbanization?" *Current Sociology* 43, 1: 151–217.

__________. 1995b. *Environmental Sociology: A Social Constructionist Perspective.* London: Routledge.

__________. 1998a. *Fantasy City: Pleasure and Profit in the Postmodern Metropolis.* New York: Routledge.

__________. 1998b [1995]. "Urbanization." Pp. 337–359 in Robert J. Brym, ed. *New Society: Sociology for the 21st Century,* 2nd ed. Toronto: Harcourt Brace Canada.

Hannon, Roseann, David S. Hall, Todd Kuntz, Van Laar, and Jennifer Williams. 1995. "Dating Characteristics Leading to Unwanted vs. Wanted Sexual Behavior." *Sex Roles* 33: 767–783.

Hansen, James E., Makiko Sato, Reto Ruedy, Andrew Lacis, and Jay Glascoe. 1998. "Global Climate Data and Models: A Reconciliation." *Science* 281: 930–932.

Hao, Xiaoming. 1994. "Television Viewing Among American Adults in the 1990s." *Journal of Broadcasting and Electronic Media* 38: 353–360.

Harden, Blaine. 2001. "Two-Parent Families Rise After Changes in Welfare." *The New York Times Online.* On the World Wide Web at http://www.nytimes.com/2001/08/12/national/12FAMI.html?todaysheadlines=&pagewanted=print (August 12 2001).

Harrigan, Jinni A., and Kristy T. Tiang. 1997. "Fooled by a Smile: Detecting Anxiety in Others." *Journal of Nonverbal Behavior* 21: 203–221.

Harris, Chauncy D. 1997. "The Nature of Cities and Urban Geography in the Last Half Century." *Urban Geography* 18: 15–35.

__________, and Edward L. Ullman. 1945. "The Nature of Cities." *Annals of the American Academy of Political and Social Science* 242: 7–17.

Harris, David. A. 2001. "'Flying While Arab,' Immigration Issues, and Lessons from the Racial Profiling Controversy." Briefing for the United States Commission on Civil Rights, 12 October. On the World Wide Web at http://www.profilesininjustice.com/images/02_flyingwhilearab.pdf (22 December 2002).

Harris, Kathleen Mullan. 1997. *Teen Mothers and the Revolving Welfare Door.* Philadelphia: Temple University Press.

Harris, Marvin. 1974. *Cows, Pigs, Wars and Witches: The Riddles of Culture.* New York, Random House.

Harrison, Bennett. 1994. *Lean and Mean: The Changing Landscape of Corporate Power in the Age of Flexibility.* New York: Basic Books.

Hart, David M., and David G. Victor. 1993. "Scientific Elites and the Making of US Policy for Climate Change Research, 1957–74." *Social Studies of Science* 23:643–80.

Harvey, Andrew S., Katherine Marshall, and Judith A. Frederick. 1991. *Where Does the Time Go?* Ottawa: Statistics Canada.

Harvey, Elizabeth. 1999. "Short-term and Long-term Effects of Early Parental Employment on Children of the National Longitudinal Survey of Youth." *Developmental Psychology* 35: 445–449.

Hastings, Arthur C., James Fadiman, and James C. Gordon, eds. 1980. *Health for the Whole Person: The Complete Guide to Holistic Medicine.* Boulder, CO: Westview Press.

Haub, Carl. 2000. "How Many People Have Ever Lived on Earth?" On the World Wide Web at http://www.discover.com/ask/main57.html (5 June 2003).

Hauser, Robert M., John Robert Warren, Min-Hsiung Huang, and Wendy Y. Carter. 2000. "Occupational Status, Education, and Social Mobility in the Meritocracy." Pp. 179–229 in Kenneth Arrow, Samuel Bowles, and Steven Durlauf, eds. *Meritocracy and Economic Inequality.* Princeton, NJ: Princeton University Press.

Hawley, Amos. 1950. *Human Ecology: A Theory of Community Structure.* New York: Ronald Press.

Häyry, Matti, and Tuija Lehto. 1998. "Genetic Engineering and the Risk of Harm." Paper presented at the 20th World Congress of Philosophy. Boston. On the World Wide Web at http://www.bu.edu/wcp/Papers/Bioe/BioeHay2.htm (17 January 2005).

Haythornwaite, Caroline, and Barry Wellman. 2002. "The Internet in Everyday Life: An Introduction." Pp. 3–41 in *The Internet in Everyday Life.* Oxford: Blackwell.

"Health Care Systems: An International Comparison." 2001. Ottawa: Strategic Policy and Research, Intergovernmental Affairs. On the World Wide Web at http://www.pnrec.org/2001papers/DaigneaultLajoie.pdf (13 June 2003).

Hechter, Michael. 1974. *Internal Colonialism: The Celtic Fringe in British National Development, 1536–1966.* Berkeley: University of California Press.

__________. 1987. *Principles of Group Solidarity.* Berkeley: University of California Press.

Hein, Simeon. 1992. "Trade Strategy and the Dependency Hypothesis: A Comparison of Policy, Foreign Investment, and Economic Growth in Latin America and East Asia." *Economic Development and Cultural Change* 40: 495–521.

Helsing, Knud J., Moyses Szklo, and George W. Comstock. 1981. "Factors Associated with Mortality After Widowhood." *American Journal of Public Health* 71: 802–809.

Henson, Kevin Daniel. 1996. *Just a Temp.* Philadelphia: Temple University Press.

Henwood, Doug. 1999. "The Material Rewards for Labor Over Time: Earnings." *Left Business Observer* 27 November. On the World Wide Web at http://www.panix.com/~dhenwood/Stats_earns.html (29 April 2000).

Herlihy, David. 1998. *The Black Death and the Transformation of the West.* Cambridge, MA: Harvard University Press.

Herman, Edward S., and Noam Chomsky. 1988. *Manufacturing Consent: The Political Economy of the Mass Media.* New York: Pantheon.

__________, and Gerry O'Sullivan. 1989. *The "Terrorism" Industry: The Experts and Institutions That Shape Our View of Terror.* New York: Pantheon.

Herrnstein, Richard J., and Charles Murray. 1994. *The Bell Curve: Intelligence and Class Structure in American Life.* New York: Free Press.

Hersch, Patricia. 1998. *A Tribe Apart: A Journey into the Heart of American Adolescence.* New York: Ballantine Books.

Hertzman, Clyde, 2000. "The Case for Early Childhood Development Strategy." *Isuma: Canadian Journal of Policy Research* 1, 2: 11–18.

Hesse-Biber, Sharlene. 1996. *Am I Thin Enough Yet? The Cult of Thinness and the Commercialization of Identity.* New York: Oxford University Press.

__________, and Gregg Lee Carter. 2000. *Working Women in America: Split Dreams.* New York: Oxford University Press.

Hilberg, Raoul. 1961. *The Destruction of the European Jews.* Chicago: Quadrangle Books.

Hirschi, Travis. 1969. *Causes of Delinquency.* Berkeley: University of California Press.

__________, and Hanan C. Selvin. 1972. "Principles of Causal Analysis." Pp. 126–47 in Paul F. Lazarsfeld, Ann K. Pasanella, and Morris Rosenberg, eds. *Continuities in the Language of Social Research.* New York: Free Press.

Hirschman, Albert O. 1970. *Exit, Voice, and Loyalty: Responses to Decline in Firms, Organizations, and States.* Cambridge, MA: Harvard University Press.

Hirst, Paul, and Grahame Thompson. 1999. *Globalization in Question: The International Economy and the Possibilities of Governance,* 2nd ed. London: Polity.

Hobbes, Thomas. 1968 [1651]). *Leviathan.* Middlesex, UK: Penguin.

Hobsbawm, Eric. 1994. *Age of Extremes: The Short Twentieth Century, 1914–1991.* London: Abacus.

Hochberg, Fred P. 2002. "American Capitalism's Other Side." *New York Times* July 25. On the World Wide Web at http://www.nytimes.com (25 July 2002).

Hochschild, Arlie Russell. 1973. *The Unexpected Community: Portrait of an Old Age Subculture.* Berkeley: University of California Press.

__________. 1979. "Emotion Work, Feeling Rules, and Social Structure." *American Journal of Sociology* 85: 551–575.

__________. 1983. *The Managed Heart: Commercialization of Human Feeling.* Berkeley: University of California Press.

__________ with Anne Machung. 1989. *The Second Shift: Working Parents and the Revolution at Home.* New York: Viking.

Hochstetler, Andrew L., and Neal Shover. 1998. "Street Crime, Labor Surplus, and Criminal Punishment, 1980–1990." *Social Problems* 44: 358–368.

Hodge, Robert, and David Tripp. 1986. *Children and Television: A Semiotic Approach.* Cambridge: Polity.

Hodgson, Marshall G. S. 1974. *The Venture of Islam: Conscience and History in a World Civilization,* 3 vols. Chicago: University of Chicago Press.

Hodson, Randy, and Teresa Sullivan. 1995 [1990]. *The Social Organization of Work,* 2nd ed. Belmont, CA: Wadsworth.

Hoffman, Donna L., and Thomas P. Novak. 1998. "Bridging the Racial Divide on the Internet." *Science* 280: 390–391.

Hogan, John. 2003. "Radio." *Clear Channel.* On the World Wide Web at http://www.clearchannel.com/radio (2 June 2003).

Hoggart, Richard. 1958. *The Uses of Literacy.* Harmondsworth, UK: Penguin.

Holloway, Marguerite. 1999. "The Aborted Crime Wave?" *Scientific American* 281, 6: 23–24.

"Hollywood Lights Up." *The Washington Post* 30 August 1997: A26.

Homans, George Caspar. 1950. *The Human Group.* New York: Harcourt, Brace.

__________. 1961. *Social Behavior: Its Elementary Forms.* New York: Harcourt, Brace and World.

hooks, bell. 1984. *Feminist Theory: From Margin to Center.* Boston: South End Press.

Hoover, Robert N. 2000. "Cancer—Nature, Nurture, or Both." *New England Journal of Medicine* 343, 2. On the World Wide Web at http://www.nejm.org/content/2000/0343/0002/0135.asp (16 July 2000).

Hopkins, Terence K., and Immanuel Wallerstein. 1986. "Commodity Chains in the World Economy Prior to 1800." *Review* 10: 157–170.

Hoppe, Robert A., ed. 2001. *Structural and Financial Characteristics of U.S. Farms: 2001 Family Farm Report.* Washington, DC: U.S. Department of Agriculture. On the World Wide Web at http://www.ers.usda.gov/publications/aib768/aib768.pdf (2 August 2003).

Horan, Patrick M. 1978. "Is Status Attainment Research Atheoretical?" *American Sociological Review* 43: 534–541.

Horkheimer, Max, and Theodor W. Adorno. 1986 [1944]. *Dialectic of Enlightenment,* John Cumming, trans. London: Verso.

Houpt, Simon. 2004. "Pass the popcorn, save the world." *Globe and Mail* 29 May: R1, R13.

Houseknecht, Sharon K., and Jaya Sastry. 1996. "Family 'Decline' and Child Well-Being: A Comparative Assessment." *Journal of Marriage and the Family* 58: 726–739.

Hout, Michael. 1988. "More Universalism, Less Structural Mobility: The American Occupational Structure in the 1980s." *American Journal of Sociology* 93: 1358–1400.

__________, and William R. Morgan. 1975. "Race and Sex Variations in the Causes of the Expected Attainments of High School Seniors." *American Journal of Sociology* 81: 364–394.

__________, Clem Brooks, and Jeff Manza. 1993. "The Persistence of Classes in Post-Industrial Societies." *International Sociology* 8: 259–277.

__________, Adrian E. Raftery, and Eleanor O. Bell. 1993. "Making the Grade: Educational Stratification in the United States, 1925–1989." Pp.25–49 in *Persistent Inequality: Changing Inequality in 13 Countries,* Yossi Shavit and Hans-Peter Blossfeld, eds. Boulder, CO: Westview.

"How to Tell Your Friends from the Japs." 1941. *Time* 22 December: 33.

Hoyt, Homer. 1939. *The Structure and Growth of Residential Neighborhoods in American Cities.* Washington, DC: Federal Housing Authority.

Hughes, Fergus P. 1995 [1991]. *Children, Play and Development,* 2nd ed. Boston: Allyn and Bacon.

Hughes, H. Stuart. 1967. *Consciousness and Society: The Reorientation of European Social Thought, 1890–1930.* London: Macgibbon and Kee.

Human Rights Campaign. 1999. "The Hate Crime Prevention Act of 1999." On the World Wide web at http://www.hrc.org/issues/leg/hcpa/index.html (30 April 2000).

Human Rights Watch. 1995. *The Human Rights Watch Global Report on Women's Human Rights.* New York: Human Rights Watch.

Hunter, James Davison. 1991. *Culture Wars: The Struggle to Define America.* New York: Basic Books.

Hunter, Shireen T. 1998. *The Future of Islam and the West: Clash of Civilizations or Peaceful Coexistence?* Westport, CT: Praeger.

"Hunting bin Laden." 1999. On the World Wide Web at http://www.pbs.org/wgbh/pages/frontline/shows/binladen (13 September 2001).

Huntington, Samuel. 1968. *Political Order in Changing Societies.* New Haven, CT: Yale University Press.

__________. 1991. *The Third Wave: Democratization in the Late Twentieth Century.* Norman: University of Oklahoma Press.

IBM. 1997. *IBM Annual Report 1997.* On the World Wide Web at http://www.ibm.com/annualreport/1997/arbm.html (30 April 2000).

Ignatieff, Michael. 2000. *The Rights Revolution.* Toronto: Anansi.

Ignatiev, Noel. 1995. *How the Irish Became White.* New York: Routledge.

Illich, Ivan. 1976. *Limits to Medicine: Medical Nemesis: The Expropriation of Health.* New York: Penguin.

IMD International. 2004. "The World Competitiveness Scoreboard 2004." On the World Wide Web at http://www02 imd .ch/documents/wcc/content/ranking.pdf (31 March 2005).

Infocom. 2003. "Bureaucracy." On the World Wide Web at http://infocom.elsewhere.org/gallery/bureaucracy/bureaucracy.html (14 March 2003).

Infoplease. 2003. "Federal Minimum Wage Rates, 1955–2002." On the World Wide Web at http://www.infoplease.com/ipa/A0774473.html (21 May 2003).

__________. 2005a. "The Death Penalty Worldwide." On the World Wide Web at http://www.infoplease.com/ipa/A0777460.html (11 March 2005).

__________. 2005b. "Families by Type, Race, and Hispanic Origin, 2002." On the World Wide Web at http://www.infoplease .com/ipa/A0880691.html (8 March 2005).

Inglehart, Ronald. 1997. *Modernization and Postmodernization: Cultural, Economic, and Political Change in 43 Societies.* Princeton, NJ: Princeton University Press.

__________, and Wayne E. Baker. 2000. "Modernization, Cultural Change, and the Persistence of Traditional Values." *American Sociological Review* 65: 19–51.

Inkeles, Alex, and David H. Smith. 1976. *Becoming Modern: Individual Change in Six Developing Countries.* Cambridge, MA: Harvard University Press.

International Civil Aviation Organization. 2002. "World Scheduled Airlines: System Scheduled Traffic and Operations, 1929–2001E." http://www.air-transport.org/public/industry/bin/ICAOTraffic.pdf (6 February 2003).

"International Organizations by Year and Type (Table 2)." 2001. *Yearbook of International Organizations.* On the World Wide Web at http://www.uia.org/uiastats/ytb299.htm (6 February 2003).

"Internet Growth." 2000. On the World Wide Web at http://citywideguide.com/InternetGrowth.html (29 April 2000).

Internet Movie Database. 2003. On the World Wide Web at http://us.imdb.com (13 March 2003).

Internet Software Consortium 2003. "Internet Domain Survey Host Count." On the World Wide Web at http://www.isc.org/ds/hosts.html (6 February 2003).

Inter-University Consortium for Political and Social Research. 1992. "Description-Study No. 9593." On the World Wide Web at http://www.icpsr.umich.edu:8080/ICPSR-STUDY/09593.xml (9 March 2003).

"Internet Usage Statistics—The Big Picture." 2005. On the World Wide Web at http://www.internetworldstats.com/stats.htm (25 February 2005).

"Iraq Body Count." 2003. On the World Wide Web at http://www.iraqbodycount.net/bodycount.htm (7 June 2003).

"Iraq Coalition Casualty Count." 2005. On the World Wide Web at http://icasualties.org/oif/ (14 March 2005).

Isaacs, Larry. 2002. "To Counter 'The Very Devil' and More: The Making of Independent Capitalist Militias in the Gilded Age." *American Journal of Sociology* 108: 353-405.

Isajiw, W. Wsevolod. 1978. "Olga in Wonderland: Ethnicity in a Technological Society." Pp. 29–39 in Leo Driedger, ed. *The Canadian Ethnic Mosaic: A Quest for Identity.* Toronto: McClelland & Stewart.

Ivanhoe, P. J. 2000 [1993]. *Confucian Moral Self-Cultivation.* Indianapolis, IN: Hackett.

Iyengar, Shanto. 1991. *Is Anyone Responsible? How Television Frames Political Issues.* Chicago: University of Chicago Press.

Jackman, Mary R., and Robert W. Jackman. 1983. *Class Awareness in the United States.* Berkeley: University of California Press.

Jacobs, Jane. 1961. *The Death and Life of Great American Cities.* New York: Random House.

__________. 1969. *The Economy of Cities.* New York: Random House.

Jacobs, Jerry A. 1993. "Careers in the U.S. Service Economy." Pp. 195–224 in Gøsta Esping-Andersen, ed. *Changing Classes: Stratification and Mobility in Post–Industrial Societies.* London: Sage.

__________. 1996. "Gender Inequality and Higher Education." *Annual Review of Sociology* 22: 153-85.

James, Harold. 2001. *The End of Globalization: Lessons from the Great Depression.* Cambridge, MA: Harvard University Press.

James, William. 1976 [1902]. *The Varieties of Religious Experience: A Study in Human Nature.* New York: Collier Books.

Janis, Irving. 1972. *Victims of Groupthink.* Boston: Houghton Mifflin.

Jargowsky, Paul A. 2003. *Stunning Progress, Hidden Problems: The Dramatic Decline of Concentrated Poverty in the 1990s.* Washington, DC: Center on Urban and Metropolitan Policy, The Brookings Institution. On the World Wide Web at http://www.brookings.edu/dybdocroot/es/urban/publications/jargowskypoverty.pdf (5 August 2003).

Jayson, Sharon. 2005. "Yep, life'll burst that self-esteem bubble." *USA Today* 16 February: 1D-2D.

Jencks, Christopher. 1994. *The Homeless.* Cambridge, MA: Harvard University Press.

———, Marshall Smith, Henry Acland, Mary Jo Bane, David Cohen, Herbert Gintis, Barbara Heyns, and Stephan Michelson. 1972. *Inequality: A Reassessment of the Effect of Family and Schooling in America.* New York: Basic Books.

Jeness, Valerie. 1995. "Social Movement Growth, Domain Expansion, and Framing Processes: The Gay/Lesbian Movement and Violence Against Gays and Lesbians as a Social Problem." *Social Problems* 42: 145–170.

Jenkins, J. Craig. 1983. "Resource Mobilization Theory and the Study of Social Movements." *Annual Review of Sociology* 9: 527–553.

Jensen, Margaret Ann. 1984. *Love's Sweet Return. The Harlequin Story.* Toronto: Women's Press.

Johansen, Bruce E. 1998. "Sterilization of Native American Women." On the World Wide Web at http://www.ratical.org/ratville/sterilize.html (24 April 2003).

Johns, Adrian. 1998. *The Nature of the Book: Print and Knowledge in the Making.* Chicago: University of Chicago Press.

Johnson, Chalmers. 2000. *Blowback: The Costs and Consequences of American Empire.* New York: Metropolitan Books.

Johnson, Michael P., and Kathleeen J. Ferraro. 2000. "Research on Domestic Violence in the 1990s: Making Distinctions." *Journal of Marriage and the Family* 62: 948-63.

Johnson, Neil F., and Sushil Jajodia. 1998. "Exploring Steganography: Seeing the Unseen." *IEEE Computer* February. On the World Wide Web at http://www.jjtc.com/pub/r2026a.htm (13 September 2001).

Johnson, Terence J. 1972. *Professions and Power.* London: Macmillan.

Jones, Brian J., Bernard J. Gallagher III, and Joseph A. McFalls, Jr. 1995. *Sociology: Micro, Macro, and Mega Structures.* Fort Worth, TX: Harcourt Brace College Publishers.

Jones, Christopher. 1999. "Chiapas' Well-Connected Rebels." *Wired News* 1 February. On the World Wide Web at http://www.wired.com/news/print/0,1294,17633,00.html (30 July 2000).

Jones, Jacqueline. 1986. *Labor of Love, Love of Sorrow: Black Women, Work and Slavery from Slavery to the Present.* New York: Random House.

Jones, Patrice M. 2003. "Drug Lords Do What Officials Don't—Control Brazil's Slums." *Chicago Tribune* 2 February. On the World Wide Web at http://www.il-rs.com.br/ilingles/informative/marco_2003/informative_drugs.htm (5 August 2003).

Jordan, Penny. 1999. *A Treacherous Seduction.* Toronto: Harlequin.

Joy, Bill. 2000. "Why the Future Doesn't Need Us." *Wired* 8, 4. On the World Wide Web at http://wired.com/wired/archive/8.04/joy_pr.html (4 November 2000).

Joyce, Terrence, and Lloyd Keigwin. 2004. "Abrupt Climate Change: Are We on the Brink of a New Little Ice Age?" Ocean and Climate Change Institute, Woods Hole Oceanographic Institution. On the World Wide Web at http://www.whoi.edu/institutes/occi/currenttopics/abrupt-climate_joyce_keigwin.html (29 May 2004).

Judis, John B., and Ruy Teixeira. 2005. "Movement Interruptus." *American Prospect Online* January 4. On the World Wide Web at http://www.prospect.org/web/printfriendly-view.ww?id=8955 (3 April 2005).

Juergensmeyer, Mark. 2000. *Terror in the Mind of God: The Global Rise of Religious Violence.* Berkeley: University of California Press.

Jupiter Media Metrix. 2001. "Waiting for Critical Mass: Timing Investments in the Fragmented iTV Market." *iTV Marketer.* On the World Wide Web at http://www.itvmarketer.com/mktres/northamerica/mrna03.htm (28 May 2003).

Kalmijn, Matthijs. 1996. "The Socioeconomic Assimilation of Caribbean American Blacks." *Social Forces* 74: 911–930.

———. 1998. "Intermarriage and Homogamy: Causes, Patterns, Trends." *Annual Review of Sociology* 24: 395–421.

Kanter, Rosabeth Moss. 1977. *Men and Women of the Corporation.* New York: Basic Books.

———. 1983. *The Change Masters: Innovation and Entrepreneurship in the American Corporation.* New York: Simon & Schuster.

———. 1989. *When Giants Learn to Dance: Mastering the Challenges of Strategy, Management, and Careers in the 1990s.* New York: Simon & Schuster.

Karabel, Jerome. 1986. "Community Colleges and Social Stratification in the 1980s." In L. S. Zwerling, ed. *The Community College and its Critics.* San Francisco: Jossey-Bass.

Karklins, Rasma. 1986. *Ethnic Relations in the USSR: The Perspective from Below.* London: Unwin Hyman.

Karl, Thomas R., and Kevin E. Trenberth. 1999. "The Human Impact on Climate." *Scientific American* 281, 6: September: 100–105.

Karp, David A., and William C. Yoels. 1976. "The College Classroom: Some Observations on the Meaning of Student Participation." *Sociology and Social Research* 60: 421–439.

Karraker, Katherine Hildebrandt, Dena Ann Vogel, and Margaret Ann Lake. 1995. "Parents' Gender-Stereotyped Perceptions of Newborns: The Eye of the Beholder Revisited." *Sex Roles* 33: 687–701.

Kasindorf, Martin, Stephanie Armour, and Andrea Stone. 1998. "In Work World, Affairs Can Drag Down People at the Top." *USA Today* 24 August: 8A.

Katz, Elihu. 1957. "The Two-Step Flow of Communication: An Up-to-Date Report on an Hypothesis." *Public Opinion Quarterly* 21: 61–78.

Katznelson, Ira, and Margaret Weir. 1985. *Schooling for All: Class, Race, and the Decline of the Democratic Ideal.* New York: Basic Books.

Kaufman, Bruce E. 1982. "The Determinants of Strikes in the United States, 1900–1977." *Industrial and Labor Relations Review* 35: 473–490.

Keister, Lisa A. 2000. *Wealth in America: Trends in Wealth Inequality.* Cambridge: Cambridge University Press.

———, and Stephanie Moller. 2000. "Wealth Inequality in the United States." *Annual Review of Sociology* 26: 63–81.

Keller, Larry. 2000. "Dual Earners: Double Trouble." On the World Wide Web at http://www.cnn.com/2000/CAREER/trends/11/13/dual.earners (13 November 2000).

Kelley, Jack. 2001. "Terror Groups Hide Behind Web Encryption." *USA Today* (June 19). On the World Wide Web at http://www.usatoday.com/life/cyber/tech/2001-02-05-binladen.htm (13 September 2001).

Kellner, Douglas. 1995. *Media Culture: Cultural Studies, Identity and Politics Between the Modern and the Postmodern.* New York: Routledge.

Kemper, T. D. 1978. *A Social Interactional Theory of Emotion.* New York: Wiley.

__________. 1987. "How Many Emotions Are There? Wedding the Social and Autonomic Components." *American Journal of Sociology* 93: 263–289.

Kennedy, Paul. 1993. *Preparing for the Twenty-First Century.* New York: HarperCollins.

Kepel, Gilles. 1994 [1991]. *The Revenge of God: The Resurgence of Islam, Christianity and Judaism in the Modern World,* Alan Braley, trans. University Park: Pennsylvania State University Press.

Kerig, Patricia K., Philip A. Cowan, and Carolyn Pape Cowan. 1993. "Marital Quality and Gender Differences in Parent-Child Interaction." *Developmental Psychology* 29: 931–939.

Kett, Joseph F. 1977. *Rites of Passage: Adolescence in America, 1790 to the Present.* New York: Basic Books.

Kevles, Daniel J. 1999. "Cancer: What Do They Know?" *New York Review of Books* 46, 14: 14–21.

Kidron, Michael, and Ronald Segal. 1995. *The State of the World Atlas,* 5th ed. London: Penguin.

Kilborn, Peter T. 2001. "Recession is Stretching the Limit on Welfare Benefits." *New York Times* December 9. On the World Wide Web at http://www.nytimes.com (9 December 2001).

Kimmerling, Baruch, ed. 1996. "Political Sociology at the Crossroads." *Current Sociology* 44, 3: 1–176.

__________. 2001. *The Invention and Decline of Israeliness: State, Society, and the Military.* Berkeley: University of California Press.

__________. 2003. *Politicide: Ariel Sharon's War against the Palestinians.* London: Verso.

Kinder, Donald R., and Lynn M. Sanders. 1996. *Divided by Color: Racial Politics and Democratic Ideals.* Chicago: University of Chicago Press.

King, Ralph. 2003. Mary Tolan's Modest Proposal. *Business* June: 116–122.

Kingston, Peter. 1997. "Top of the Class." *The Guardian* 21 October: 2.

Kingston, Paul W. 2001. "The Unfulfilled Promise of Cultural Capital Theory." *Sociology of Education* Supplement: 88–91.

Kinsey, Alfred C., Wardell B. Pomeroy, Clyde E. Martin, and Paul H. Gebhard. 1953. *Sexual Behavior in the Human Female.* Philadelphia: W. B. Saunders.

__________, Wardell B. Pomeroy, and Clyde E. Martin. 1948. *Sexual Behavior in the Human Male.* Philadelphia: W. B. Saunders.

Kinsley, Michael. 2003. "How Affirmative Action Helped George W." On the World Wide Web at http://www.cnn.com/2003/ALLPOLITICS/01/20/timep.affirm.action.tm (1 February 2003).

Kirp, David L. 1982. *Just Schools: The Idea of Racial Equality in American Education.* Berkeley: University of California Press.

Klee, Kenneth. 1999. "The Siege of Seattle." *Newsweek* 13 December. On the World Wide Web at http://server5.ezboard.com/fdrugpolicytalkwtoseattleriots1999.showMessage?topicID=12.topic (3 August 2000).

Kleege, Georgina. 1999. *Sight Unseen.* New Haven, CT: Yale University Press.

Klein, Naomi. 2000. *No Logo: Taking Aim at the Brand Bullies.* New York: HarperCollins.

Kling, Kristen C., Janet Shibley Hyde, Carolin J. Showers, and Brenda N. Buswell. 1999. "Gender Differences in Self-Esteem: A Meta-Analysis." *Psychological Bulletin* 125, 4: 470–500.

Klingemann, Hans-Dieter. 1999. "Mapping Political Support in the 1990s: A Global Analysis." In Pippa Norris, ed. *Critical Citizens: Global Support for Democratic Governance.* Oxford: Oxford University Press. On the World Wide Web at http://ksgwww.harvard.edu/people/pnorris/Chapter_2.htm (20 October 1999).

Klockars, Carl B. 1974. *The Professional Fence.* New York: Free Press.

Kluegel, James R., and Eliot R. Smith. 1986. *Beliefs about Inequality: Americans' Views of What Is and What Ought to Be.* New York: Aldine de Gruyter.

Kluger, Richard. 1996. *Ashes to Ashes: America's Hundred-Year Cigarette War, the Public Health, and the Unabashed Triumph of Philip Morris.* New York: Knopf.

Knapp, Robert H. 1944. "A Psychology of Rumor." *Public Opinion Quarterly* 8: 23–37.

Knight, Graham. 1998 [1995]. "The Mass Media." Pp. 103–27 in Robert J. Brym, ed. *New Society: Sociology for the 21st Century,* 2nd ed. Toronto: Harcourt Brace Canada.

Koepke, Leslie, Jan Hare, and Patricia B. Moran. 1992. "Relationship Quality in a Sample of Lesbian Couples with Children and Child-Free Lesbian Couples." *Family Relations* 41: 224–229.

Kohlberg, Lawrence. 1981. *The Psychology of Moral Development: The Nature and Validity of Moral Stages.* New York: Harper & Row.

Kohn, Alfie. 1988. "Make Love, Not War." *Psychology Today* 22, 6: 35–38.

Kolbert, E. 1995. "Americans Despair of Popular Culture." *New York Times* 20 August: Section 2: 1, 23.

Kolko, Gabriel. 2002. *Another Century of War?* New York: New Press.

Kornblum, William. 1997 [1988]. *Sociology in a Changing World,* 4th ed. Fort Worth, TX: Harcourt Brace College Publishers.

Korpi, Walter. 1983. *The Democratic Class Struggle.* London: Routledge & Kegan Paul.

__________, and Joakim Palme. 2003. "New Politics and Class Politics in the Context of Austerity and Globalization: Welfare State Regress in 18 Countries, 1975-95." *American Political Sceince Review* 97: 425-446.

Kosmin, Barry A. 1991. *Research Report of the National Survey of Religious Identification.* New York: CUNY Graduate Center.

Koss, Mary P., Christine A. Gidycz, and Nadine Wisniewski. 1987. "The Scope of Rape: Incidence and Prevalence of Sexual Aggression and Victimization in a National Sample of Higher Education Students." *Journal of Consulting and Clinical Psychology* 55: 162–170.

Kozol, Jonathan. 1991. *Savage Inequalities: Children in America's Schools.* New York: Crown.

Krahn, Harvey, and Graham Lowe. 1998. *Work, Industry, and Canadian Society,* 3rd ed. Toronto: ITP Nelson.

Kreisi, Hanspeter, et al. 1995. *New Social Movements in Western Europe.* Minneapolis: University of Minnesota Press.

Kristof, Nicholas D. 1997. "With Stateside Lingo, Valley Girl Goes Japanese," *New York Times* 19 October: Section 1, 3.

__________. 2005a. "Health Care? Ask Cuba." *New York Times* 12 January. On the World Wide Web at www.nytimes.com (12 January 2005).

__________. 2005b. "Land of Penny Pinchers." *New York Times* 5 January. On the World Wide Web at www.nytimes.com (5 January 2005).

Kropotkin, Petr. 1908 [1902]. *Mutual Aid: A Factor of Evolution,* revised ed. London: W. Heinemann.

Kübler-Ross, Elisabeth. 1969. *On Death and Dying.* New York: Macmillan.

Kuhn, Thomas. 1970 [1962]. *The Structure of Scientific Revolutions,* 2nd ed. Chicago: University of Chicago Press.

Kurzweil, Ray. 1999. *The Age of Spiritual Machines: When Computers Exceed Human Intelligence.* New York: Viking Penguin.

Kuttner, Robert. 1997. "The Limits of Markets." *The American Prospect* 31: 28–41. On the World Wide Web at http://www.prospect.org/archives/31/31kuttfs.html (30 April 2000).

__________. 1998a. "In This For-Profit Age, Preventive Medicine Means Avoiding Audits." *Boston Globe* 22 March: E7.

__________. 1998b. "Toward Universal Coverage." *The Washington Post* 14 July: A15.

Labaton, Stephen. 2003. "F.C.C. Media Rule Blocked in House in a 400-to-21 Vote." *New York Times* July 24. On the World Wide Web at http://www.nytimes.com (24 July 2003).

LaFeber, Walter. 1993. *Inevitable Revolutions: The United States in Central America,* 2nd ed. New York: W. W. Norton.

__________. 1999. *Michael Jordan and the New Global Capitalism.* New York: W. W. Norton.

LaFree, Gary D. 1980. "The Effect of Sexual Stratification by Race on Official Reactions to Rape." *American Sociological Review* 45: 842–854.

__________. 1998. "Social Institutions and the Crime 'Bust' of the 1990s." *Journal of Criminal Law and Criminology* 88: 1325–1368.

Lahmeyer, Jan. 2003. "Brazil: Historical Demographical Data of the Whole Country." On the World Wide Web at http://www.library.uu.nl/wesp/populstat/Americas/brazilc.htm (5 August 2003).

Lamberth Consulting. 2002. "Racial Profiling Doesn't Work." On the World Wide Web at http://www.lamberthconsulting.com/research_work.asp (December 22 2002).

Lamont, Michele. 1992. *Money, Morals, and Manners.* Chicago: University of Chicago Press.

__________, and Annette Lareau. 1988. "Cultural Capital: Allusions, Gaps, and Glissandos in Recent Theoretical Developments." *Sociological Theory* 6: 153–168.

Lane, Harlan. 1992. *The Mask of Benevolence: Disabling the Deaf Community.* New York: Alfred A. Knopf.

Lantz, Herman, Martin Schultz, and Mary O'Hara. 1977. "The Changing American Family from the Preindustrial to the Industrial Period: A Final Report." *American Sociological Review* 42: 406–421.

Lapidus, Gail Warshofsky. 1978. *Women in Soviet Society: Equality, Development, and Social Change.* Berkeley: University of California Press.

Lapidus, Ira M. 2002 [1998]. *A History of Islamic Societies,* 2nd ed. Cambridge: Cambridge University Press.

Laslett, Peter. 1991 [1989]. *A Fresh Map of Life: The Emergence of the Third Age.* Cambridge, MA: Harvard University Press.

Lasswell, Harold. 1936. *Politics: Who Gets What, When and How.* New York: McGraw-Hill.

Laumann, Edward O., John H. Gagnon, Robert T. Michael, and Stuart Michaels. 1994. *The Social Organization of Sexuality: Sexual Practices in the United States.* Chicago: University of Chicago Press.

Laxer, Gordon. 1989. *Open for Business: The Roots of Foreign Ownership in Canada.* Toronto: Oxford University Press.

Lazare, Daniel. 1999. "Your Constitution Is Killing You: A Reconsideration of the Right to Bear Arms." *Harper's* 299, 1793, October: 57–65.

Lazonick, William. 1991. *Business Organization and the Myth of the Market Economy.* Cambridge: Cambridge University Press.

Leacock, Eleanor Burke. 1981. *Myths of Male Dominance: Collected Articles on Women Cross-Culturally.* New York: Monthly Review Press.

Lears, Jackson. 1994. *Fables of Abundance: A Cultural History of Advertising in America.* New York: Basic Books.

Le Bon, Gustave. 1969 [1895]. *The Crowd: A Study of the Popular Mind.* New York: Ballantine Books.

Lee, Richard B. 1979. *The !Kung San: Men, Women and Work in a Foraging Society.* Cambridge: Cambridge University Press.

Lefkowitz, Bernard. 1997a. "Boys Town: Did Glen Ridge Raise Its Sons to Be Rapists?" *Salon* August 13. On the World Wide Web at http://www.salon.com/aug97/mothers/guys970813.html (20 March 2003).

__________. 1997b. *Our Guys: The Glen Ridge Rape and the Secret Life of the Perfect Suburb.* Berkeley: University of California Press.

Lehoczky, Etelka. 2003. "Stewardess Chic." *Chicago Tribune Online Edition.* April 2. On the World Wide Web at http://www.chicagotribune.com/shopping/chi-0304020338apr02,0,3507596.story?coll=chi-shopping-hed (6 April 2003).

Leidner, Robin. 1993. *Fast Food, Fast Talk: Service Work and the Routinization of Everyday Life.* Berkeley: University of California Press.

Leman, Christopher. 1977. "Patterns of Policy Development: Social Security in the United States and Canada." *Public Policy* 25: 26–291.

Leman, Nicholas. 1998. "I'd Walk a Mile for a Fee." *New York Review of Books* 45, 11: 33–35.

Lemann, Nicholas. 1999. *The Big Test: The Secret History of the American Meritocracy.* New York: Farrar, Straus & Giroux.

Lemon, Alaina. 2000. *Between Two Fires: Gypsy Performance and Romani Memory from Pushkin to Postsocialism.* Durham, NC: Duke University Press.

Lenski, Gerhard. 1966. *Power and Privilege: A Theory of Social Stratification.* New York: McGraw-Hill.

__________, Patrick Nolan, and Jean Lenski. 1995. *Human Societies: An Introduction to Macrosociology,* 7th ed. New York: McGraw-Hill.

Lerner, Gerda. 1986. *The Creation of Patriarchy.* New York: Oxford University Press.

"The Lesson Nobody Learns." 1999. *The Economist* 24 April: 25–26.

Levine, David. 1987. *Reproducing Families: The Political Economy of English Population History.* Cambridge: Cambridge University Press.

Levine, David I., and Bhashkar Mazumder. 2002. "Choosing the Right Parents: Changes in the Intergenerational Transmission of Inequality Between 1980 and the Early 1990s." *Working Paper Series, Federal Reserve Bank of Chicago.* On the World Wide Web at http://www.chicagofed.org/publications/workingpapers/papers/wp2002-08.pdf (26 March 2003).

Levine, R. A., and D. T. Campbell. 1972. *Ethnocentrism: Theories of Conflict, Ethnic Attitudes, and Group Behavior.* New York: Wiley.

Levy, Frank. 1998. *The New Dollars and Dreams: American Incomes and Economic Change.* New York: Russell Sage Foundation.

Lewis, Bernard. 2002. *What Went Wrong? Western Impact and Middle Eastern Response.* New York: Oxford University Press.

Lewontin, Richard C. 1991. *Biology as Ideology: The Doctrine of DNA.* New York: HarperCollins.

Lichtenstein, Paul, Niels V. Holm, Pia K. Verkasalo, Anastasia Iliadou, Jaakko Kaprio, Markku Koskenvuo, Eero Pukkala, Axel Skytthe, and Kari Hemminki. 2000. "Environment and Heritable Factors in the Causation of Cancer—Analyses of Cohorts of Twins from Sweden, Denmark, and Finland." *New England Journal of Medicine* 343, 2. On the World Wide Web at http://content.nejm.org/cgi/content/short/343/2/78 (12 July 2001).

Lie, John. 1992. "The Concept of Mode of Exchange." *American Sociological Review* 57: 508–523.

__________. 1998. *Han Unbound: The Political Economy of South Korea.* Stanford, CA: Stanford University Press.

__________. 2001. *Multiethnic Japan.* Cambridge, MA: Harvard University Press.

Lieberson, Stanley. 1980. *A Piece of the Pie: Blacks and White Immigrants Since 1880.* Berkeley: University of California Press.

__________.1991. "A New Ethnic Group in the United States." Pp. 444–57 in Norman R. Yetman, ed. *Majority and Minority: The Dynamics of Race and Ethnicity in American Life,* 5th ed. Boston: Allyn & Bacon.

__________, and Mary Waters. 1986. "Ethnic Groups in Flux: The Changing Ethnic Responses of American Whites." *Annals of the American Academy of Social and Political Science* 487: 79–91.

Liebow, Elliot. 1967. *Tally's Corner: A Study of Negro Street-Corner Men.* Boston: Little, Brown.

__________. 1993. *Tell Them Who I Am: The Lives of Homeless Women.* New York: Free Press.

Light, Ivan. 1991. "Immigrant and Ethnic Enterprise in North America." Pp. 307–318 in Norman R. Yetman, ed. *Majority and Minority: The Dynamics of Race and Ethnicity in American Life,* 5th ed. Boston: Allyn & Bacon.

Lightfoot-Klein, Hanny, Cheryl Chase, Tim Hammond, and Ronald Goldman. 2000. "Genital Surgery on Children below the Age of Consent." Pp. 440–479 in Lenore T. Szuchman and Frank Muscarella, eds. *Psychological Perspectives on Human Sexuality.* New York: John Wiley & Sons.

Lijphart, Arend. 1997. "Unequal Participation: Democracy's Unresolved Dilemma." *American Political Science Review* 91: 1–14.

Lindner, Rolf. 1996 [1990]. *The Reportage of Urban Culture: Robert Park and the Chicago School.* Adrian Morris, trans. Cambridge: Cambridge University Press.

Linton, Ralph. 1936. *The Study of Man.* New York: Appleton-Century-Croft.

Lipset, Seymour Martin. 1971 [1951]. *Agrarian Socialism: The Cooperative Commonwealth Federation in Saskatchewan,* rev. ed. Berkeley: University of California Press.

__________. 1977. "Why No Socialism in the United States?" Pp. 31–363 in Seweryn Bialer and Sophia Sluzar, eds. *Sources of Contemporary Radicalism.* Boulder, CO: Westview Press.

__________. 1981 [1960]. *Political Man: The Social Bases of Politics,* 2nd ed. Baltimore: Johns Hopkins University Press.

__________. 1994. "The Social Requisites of Democracy Revisited." *American Sociological Review* 59: 1–22.

__________, and Reinhard Bendix. 1963. *Social Mobility in Industrial Society.* Berkeley: University of California Press.

__________, and Stein Rokkan. 1967. "Cleavage Structures, Party Systems, and Voter Alignments: An Introduction." Pp. 1–64 in Seymour Martin Lipset and Stein Rokkan, eds. *Party Systems and Voter Alignments: Cross-National Perspectives.* New York: Free Press.

__________, Martin A. Trow, and James S. Coleman. 1956. *Union Democracy: The Internal Politics of the International Typographical Union.* Glencoe, IL: Free Press.

Liptak, Adam. 2003. "Death Row Numbers Decline as Challenges to System Rise." *New York Times* January 11. On the World Wide Web at http://www.nytimes.com (11 January 2003).

__________. 2005. "Court Takes Another Step in Reshaping Capital Punishment." *New York Times* 2 March. On the World Wide Web at www.nytimes.com (2 March 2005).

Lisak, David. 1992. "Sexual Aggression, Masculinity, and Fathers." *Signs* 16: 238–262.

Livernash, Robert, and Eric Rodenburg. 1998. "Population Change, Resources, and the Environment." *Population Bulletin* 53, 1. On the World Wide Web at http://www.prb.org/pubs/population_bulletin/bu53-1.htm (25 August 2000).

Livi-Bacci, Massimo. 1992. *A Concise History of World Population.* Cambridge, MA: Blackwell.

Lock, Margaret. 1993. *Encounters with Aging: Mythologies of Menopause in Japan and North America.* Berkeley: University of California Press.

__________. 2002. *Twice Dead: Organ Transplants and the Reinvention of Death.* Berkeley: University of California Press.

Lodhi, Abdul Qaiyum, and Charles Tilly. 1973. "Urbanization, Crime, and Collective Violence in 19th Century France." *American Journal of Sociology* 79: 296–318.

Lofland, John, and Lyn H. Lofland. 1995 [1971]. *Analyzing Social Settings: A Guide to Qualitative Observation and Analysis,* 3rd ed. Belmont, CA: Wadsworth.

Lofland, L. H. 1985. "The Social Shaping of Emotion: Grief in Historical Perspective." *Symbolic Interaction* 8: 171–190.

Logan, John R., and Harvey L. Molotch. 1987. *Urban Fortunes: The Political Economy of Place.* Berkeley: University of California Press.

London, Scott. 1994. "Electronic Democracy—A Literature Survey." On the World Wide Web at http:///www.west.net/~insight/london/ed.htm (15 August 1998).

Long, Elizabeth. 1997. *From Sociology to Cultural Studies.* Malden, MA: Blackwell.

Longwood College Library. 2000. "Doctor-Assisted Suicide—Chronology of the Issue." On the World Wide Web at http://web.lwc.edu/administrative/library/death.htm (2 May 2000).

Lopez, Donald S. 2001. *The Story of Buddhism: A Concise Guide to Its History and Teachings.* San Francisco: Harper.

Lowe, Graham. 2000. *The Quality of Work: A People-Centred Agenda.* Toronto: Oxford University Press.

Lucas, Samuel Roundfield. 1999. *Tracking Inequality: Stratification and Mobility in American High Schools.* New York: Teachers College Press.

Lurie, Alison. 1981. *The Language of Clothes.* New York: Random House.

Lynch, Michael, and David Bogen. 1997. "Sociology's Asociological 'Core': An Examination of Textbook Sociology in Light of the Sociology of Scientific Knowledge." *American Sociological Review* 62: 481–493.

Lyon, David, and Elia Zureik, eds. 1996. *Computers, Surveillance, and Privacy.* Minneapolis: University of Minnesota Press.

MacCarthy, Fiona. 1999. "Skin Deep." *New York Review of Books* 46, 15: 19–21.

MacDonald, K., and R. D. Parke. 1986. "Parent-Child Physical Play: The Effects of Sex and Age on Children and Parents." *Sex Roles* 15: 367–378.

Machung, Anne. 1989. "Talking Career, Thinking Jobs: Gender Differences in Career and Family Expectations of Berkeley Seniors." *Feminist Studies* 15: 35–58.

Macionis, John J. 1997 [1987]. *Sociology,* 6th ed. Upper Saddle River, NJ: Prentice-Hall.

MacKinnon, Catharine A. 1979. *Sexual Harassment of Working Women.* New Haven, CT: Yale University Press.

Macklin, Eleanor D. 1980. "Nontraditional Family Forms: A Decade of Research." *Journal of Marriage and the Family* 42: 905–922.

Magazine Publishers of America. 2005a. "Combined Circulation Revenue for All ABC Magazines, 1988-2003." On the World Wide Web at http://www.magazine.org/Circulation/circulation_trends_and_magazine_handbook/1412.cfm (13 February 2005).

__________. 2005b. "Magazine Advertising Revenue and Pages for PIB Measured Magazines, 1960-2004." On the World Wide Web at http://www.magazine.org/Advertising_and_ PIB/ad_trends_and_magazine_handbook/1238.cfm (13 February 2005).

Maguire, Kathleen, and Ann L. Pastore, eds. 1998. *Sourcebook of Criminal Justice Statistics 1997.* On the World Wide Web at http://www.albany.edu/sourcebook/1995/pdf/t256.pdf (29 April 2000).

Maguire, Mike, Rod Morgan, and Robert Reiner, eds. 1994. *The Oxford Handbook of Criminology.* Oxford: Clarendon Press.

Mahony, Rhona. 1995. *Kidding Ourselves: Breadwinning, Babies, and Bargaining Power.* New York: Basic Books.

Maloney, Wendi A. 1999. "Brick and Mortar." *Academe* 85, 5: 19–24.

Malthus, Thomas Robert. 1966 [1798]. *An Essay on the Principle of Population.* J.R. Bodnar, ed. London: Macmillan.

Manga, Pran, Douglas E. Angus, and William R. Swan. 1993. "Effective Management of Low Back Pain: It's Time to Accept the Evidence." *Journal of the Canadian Chiropractic Association* 37: 221–229.

Mankiw, N. Gregory. 1998. *Principles of Macroeconomics.* Fort Worth, TX: The Dryden Press.

__________. 2003. "China's Trade and U.S. Manufacturing Jobs." Testimony Before the House Committee on Ways and Means. Washington D.C. 30 October. On the World Wide Web at http://www.whitehouse.gov/cea/mankiw_testimony_house_ways_and_means_oct_30.pdf (1 April 2005).

Mann, Susan A., Michael D. Grimes, Alice Abel Kemp, and Pamela J. Jenkins. 1997. "Paradigm Shifts in Family Sociology? Evidence From Three Decades of Family Textbooks." *Journal of Family Issues* 18: 315–349.

Mannheim, Karl. 1952. "The Problem of Generations." Pp. 276–320 in *Essays on the Sociology of Knowledge,* Paul Kecskemeti, ed. New York: Oxford University Press.

Manza, Jeff, Michael Hout, and Clem Brooks. 1995. "Class Voting in Capitalist Democracies Since World War II: Dealignment, Realignment, or Trendless Fluctuation?" *Annual Review of Sociology* 21: 137–162.

__________, Clem Brooks, and Christopher Uggen. 2004. "Public Attitudes Toward Felon Disenfranchisement in the United States." *Public Opinion Quarterly* 68: 275-86.

Marger, Martin M. 2003. *Race and Ethnic Relations: American and Global Perspectives,* 6th ed. Belmont, CA: Wadsworth.

Marklein, Mary Beth. 2002. "Students Say College Studies Take a Back Seat to Longer Work Hours." *USA Today* April 17: 8D.

Markoff, John, and Matt Richtel. 2002. "Silicon Valley without Trimmings." *New York Times* July 15. On the World Wide Web at http://www.nytimes.com (15 July 2002).

Markowitz, Fran. 1993. *A Community in Spite of Itself: Soviet Jewish Émigrés in New York.* Washington, DC: Smithsonian Institute Press.

Marmor, Theodore R. 1994. *Understanding Health Care Reform.* New Haven, CT: Yale University Press.

Marrus, Michael Robert. 1987. *The Holocaust in History.* Hanover, NH: University Press of New England for Brandeis University Press.

Marshall, Gordon, Howard Newby, David Rose, and Carolyn Vogler. 1988. *Social Class in Modern Britain.* London: Hutchinson.

__________, Adam Swift, and Stephen Roberts. 1997. *Against the Odds? Social Class and Social Justice in Industrial Societies.* Oxford: Clarendon Press.

Marshall, Monty G., and Ted Robert Gurr. 2003. *Peace and Conflict 2003.* College Park: Department of Government and Politics, University of Maryland. On the World Wide Web at http://www.cidcm.umd.edu/inscr/PC03print.pdf (3 June 2003).

Marshall, S. L. A. 1947. *Men Against Fire: The Problem of Battle Command in Future War.* New York: Morrow.

Marshall, T. H. 1965. "Citizenship and Social Class." Pp. 71–134 in T. H. Marshall, ed. *Class, Citizenship, and Social Development: Essays by T. H. Marshall.* Garden City, NY: Anchor.

Martin, Joanne. 1992. *Cultures in Organizations.* New York: Oxford University Press.

Martin, Patricia Yancey, and Robert A. Hummer. 1989. "Fraternities and Rape on Campus." *Gender and Society* 3: 457–473.

Martineau, Harriet. 1985. *Harriet Martineau on Women,* Gayle Graham Yates, ed. New Brunswick, NJ: Rutgers University Press.

Marx, Gary T. 1997. "Of Methods and Manners for Aspiring Sociologists: 36 Moral Imperatives." *The American Sociologist* 28: 102–125. On the World Wide Web at http://web.mit.edu/gtmarx/www/37moral.html (27 April 2000).

Marx, Karl. 1904 [1859]. *A Contribution to the Critique of Political Economy,* N. Stone, trans. Chicago: Charles H. Kerr.

__________. 1970 [1843]. *Critique of Hegel's "Philosophy of Right,"* Annette Jolin and Joseph O'Malley, trans. Cambridge: Cambridge University Press.

__________, and Friedrich Engels. 1972 [1848]. "Manifesto of the Communist Party." Pp. 331–62 in R. Tucker, ed. *The Marx-Engels Reader.* New York: W. W. Norton.

Massey, Douglas S., and Nancy A. Denton. 1987. "Trends in the Residential Segregation of Blacks, Hispanics, and Asians: 1970–1980." *American Sociological Review* 52: 802–825.

__________, and __________. 1993. *American Apartheid: Segregation and the Making of the Underclass.* Cambridge, MA: Harvard University Press.

__________, Camille Z. Charles, Garvey F. Lundy, and Mary J. Fischer. 2003. *The Source of the River: The Social Origins of Freshmen at America's Selective Colleges and Universities.* Princeton, NJ: Princeton University Press.

Massing, Michael. 1999. "The End of Welfare?" *New York Review of Books* 46, 15: 22–26.

__________ et al. 1999. "Beyond Legalization: New Ideas for Ending the War on Drugs." *The Nation* 20 September: 11–48.

__________. 2004. "Iraq, the Press and the Election." *New York Review of Books* 51, 20. On the World Wide Web at http://www.nybooks.com/articles/17633 (31 December 2004).

Masters, W. H., and V. E. Johnson. 1966. *Human Sexual Response.* Boston: Little, Brown.

Matalon, Jean-Marc. 1997. "Jeanne Calment, World's Oldest Person, Dead at 122." *The Shawnee News-Star* 5 August. On the World Wide Web at http://www.news-star.com/stories/080597/life1.html (2 May 2000).

Mathis, William J. 2003. "No Child Left Behind: Costs and Benefits." *Phi Delta Kappan* May. On the World Wide Web at http://www.pdkintl.org/kappan/k0305mat.htm (29 March 2005).

Matsueda, Ross L. 1988. "The Current State of Differential Association Theory." *Crime and Delinquency* 34: 277–306.

__________. 1992. "Reflected Appraisals, Parental Labeling, and Delinquency: Specifying a Symbolic Interactionist Theory." *American Journal of Sociology* 97: 1577–1611.

Mauer, Marc. 1994. "Americans Behind Bars: The International Use of Incarceration, 1992–1993." On the World Wide Web at http://www.druglibrary.org/schaffer/Other/sp/abb.htm (29 April 2000).

Maume, David J., Jr., A. Silvia Cancio, and T. David Evans. 1996. "Cognitive Skills and Racial Wage Inequlity: Reply to Farkas and Vicknair." *American Sociological Review* 61: 561–564.

Mayer, J. P. 1944. *Max Weber and German Politics.* London: Faber and Faber.

Mazur, Allan. 1998. "Global Environmental Change in the News: 1987–90 vs. 1992–96." *International Sociology* 13: 457–472.

McAdam, Doug. 1982. *Political Process and the Development of Black Insurgency, 1930–1970.* Chicago: University of Chicago Press.

__________, John D. McCarthy, and Mayer N. Zald. 1996. "Introduction: Opportunities, Mobilizing Structures, and Framing Processes—Toward a Synthetic, Comparative Perspective on Social Movements." Pp. 1–20 in Doug McAdam, John D. McCarthy, and Mayer N. Zald, eds. *Comparative Perspectives on Social Movements: Political Opportunities, Mobilizing Structures, and Cultural Framing.* New York: Cambridge University Press.

__________, and Douglas A. Snow. 1997. *Social Movements: Readings on Their Emergence, Mobilization and Dynamics.* Los Angeles: Roxbury.

McAllister, Sue. 2003. "Home Sales Up 22 Percent from Year Ago." *Mercury News.* January 25. On the World Wide Web at http://www.bayarea.com/mld/mercurynews/business/5029311.htm (11 March 2003).

__________. 2005. "Median house price rises 15%." *The Mercury News* 20 January. On the World Wide Web at http://www.mercurynews.com/mld/mercurynews/news/local/10688627.htm?1c (30 January 2005).

McCarthy, John D., and Mayer N. Zald. 1977. "Resource Mobilization and Social Movements: A Partial Theory." *American Journal of Sociology* 82: 1212–1241.

McCarthy, Shawn. 1999. "Entrepreneurs Find a Place in the Sun." *Globe and Mail* 31 July: B1, B5.

McChesney, Robert W. 1999. "Oligopoly: The Big Media Game Has Fewer and Fewer Players." *The Progressive* November: 20–24 On the World Wide Web at http://www.progressive.org/mcc1199.htm (7 August 2000).

McClendon, McKee J. 1976. "The Occupational Status Attainment Processes of Males and Females." *American Sociological Review* 41: 52–64.

McConaghy, Nathaniel. 1999. "Unresolved Issues in Scientific Sexology." *Archives of Sexual Behavior* 28, 4: 285–318.

McCrum, Robert, William Cran, and Robert MacNeil. 1992. *The Story of English,* new and rev. ed. London: Faber and Faber.

McCullagh, Declan. 2000a. "Bin Laden: Steganography Master?" *Wired* February 7. On the World Wide Web at http://www.wired.com/news/print/0.1294.41658.00.html (13 September 2001).

__________. 2000b. "Regulating Privacy: At What Cost?" *Wired* September 19. On the World Wide Web at http://www.wired.com/news/print/0.1294.38878.00.html (13 September 2001).

McCulloch, Robert. 2003. "Height and Income." On the World Wide Web at http://gsbwww.uchicago.edu/fac/robert.mcculloch/research/BusinessStatistics/robHw/SimpleLinearRegression/q11.html (22 April 2003).

McGinn, Anne Platt. 1998. "Promoting Sustainable Fisheries." Pp. 59–78 in Lester R. Brown, Christopher Flavin, Hilary French, et al. *State of the World 1998.* New York: W. W. Norton.

McGovern, James R. 1982. *Anatomy of a Lynching: The Killing of Claude Neal.* Baton Rouge: Louisiana State University Press.

McGuire, Meredith B. 2002. *Religion: The Social Context,* 5th ed. Belmont, CA: Wadsworth.

McKelvey, Bill. 1982. *Organizational Systematics: Taxonomy, Evolution, Classification.* Berkeley: University of California Press.

McKenna, Barrie. 1998. "Clinton Says, 'I'm Sorry' as Old Allies Desert Him." *Globe and Mail* 5 September: A1, A12.

__________. 1998. "How a US Tobacco Settlement Was Stubbed Out." *Globe and Mail* 20 June: A15.

McLaughlin, Margaret L., Kerry K. Osborne, and Christine B. Smith. 1995. "Standards of Conduct on Usenet." Pp. 90–111 in Steven G. Jones, ed., *CyberSociety.* Thousand Oaks, CA: Sage.

McLuhan, Marshall. 1964. *Understanding Media: The Extensions of Man.* New York: McGraw-Hill.

McMahon, Walter W. 1999. *Education and Development: Measuring the Social Benefits.* Oxford: Oxford University Press.

McManners, John, ed. 1990. *Oxford Illustrated History of Christianity.* Oxford: Oxford University Press.

McNeill, William H. 1976. *Plagues and Peoples.* Garden City, NY: Anchor Press.

__________. 1990. *Population and Politics since 1750.* Charlottesville, WV: University Press of Virginia.

McPhail, Clark. 1991. *The Myth of the Madding Crowd.* New York: Aldine de Gruyter.

__________. 1994. "The Dark Side of Purpose: Individual and Collective Violence in Riots." *The Sociological Quarterly* 35: 1–32.

__________, and Ronald T. Wohlstein. 1983. "Individual and Collective Behaviors Within Gatherings, Demonstrations, and Riots." *Annual Review of Sociology* 9: 579–600.

Mead, G. H. 1934. *Mind, Self and Society.* Chicago: University of Chicago Press.

Medawar, Peter. 1996. *The Strange Case of the Spotted Mice and Other Classic Essays on Science.* New York: Oxford University Press.

"Media Mergers, Consolidation, and Conglomeration." 2000. On the World Wide Web at http://fargo.itp.tsoa.nyu.edu/~walter/foi/index.htm (2 May 2000).

Meek, Ronald L., ed. 1971. *Marx and Engels on the Population Bomb: Selections from the Writings of Marx and Engels Dealing with the Theories of Thomas Robert Malthus.* Dorothea L. Meek and Ronald L. Meek, trans. Berkeley, CA: Ramparts Press.

Meier, Deborah. 2002. *In Schools We Trust.* Boston: Beacon Press.

Meier, Diane E., Carol-Ann Emmons, Sylvan Wallenstein, Timothy Quill, R. Sean Morrison, and Christine K. Cassell. 1998, "A National Survey of Physician Assisted Suicide and Euthanasia in the United States." *New England Journal of Medicine* 338: 1193–1201.

Meisenheimer II, Joseph R. 1998. "The Service Industry in the 'Good' Versus 'Bad' Jobs Debate." *Monthly Labor Review.* 121, 2: 22–47.

Melton, J. Gordon. 1996 [1978]. *Encyclopedia of American Religions,* 5th ed. Detroit: Gale.

Melucci, Alberto. 1980. "The New Social Movements: A Theoretical Approach." *Social Science Information* 19: 199–226.

__________. 1995. "The New Social Movements Revisited: Reflections on a Sociological Misunderstanding." Pp. 107–19 in Louis Maheu, ed. *Social Classes and Social Movements: The Future of Collective Action.* London: Sage.

Merisotis, Jamie P. 1999. "The 'What's-The-Difference?' Debate." *Academe* 85, 5: 47–51.

"Meritocracy in America." 2005. *The Economist* 1 January: 22–24.

Merton, Robert K. 1938. "Social Structure and Anomie." *American Sociological Review* 3: 672–682.

__________. 1968 [1949]. *Social Theory and Social Structure.* New York: Free Press.

Messner, Michael. 1995 [1989]. "Boyhood, Organized Sports, and the Construction of Masculinities." Pp. 102–14 in Michael S. Kimmel and Michael A. Messner *Men's Lives,* 3rd ed. Boston: Allyn & Bacon.

Metropolitan Museum of Art. 2000. "Mrs. Charles Dana Gibson (1873–1956)" On the World Wide Web at http://costumeinstitute.org/gibson.htm (13 June 2000).

Meyer, John W., Francisco O. Ramirez, and Yasemin Nuhoglu Soysal. 1992. "World Expansion of Mass Education, 1870–1980." *Sociology of Education* 65: 128–149.

__________, and W. Richard Scott. 1983. *Organizational Environments: Ritual and Rationality.* Beverly Hills, CA: Sage.

Michael, Robert T., John H. Gagnon, Edward O. Laumann, and Gina Kolata. 1994. *Sex in America: A Definitive Survey.* Boston: Little, Brown.

Michels, Robert. 1949 [1911]. *Political Parties: A Sociological Study of the Oligarchical Tendencies of Modern Democracy,* E. and C. Paul, trans. New York: Free Press.

Milem, Jeffrey F. 1998. "Attitude Change in College Students: Examining the Effect of College Peer Groups and Faculty Normative Groups." *The Journal of Higher Education* 69: 117–140.

Miles, Robert. 1989. *Racism.* London: Routledge.

Milgram, Stanley. 1974. *Obedience to Authority: An Experimental View.* New York: Harper.

Miller, Jerome G. 1996. *Search and Destroy: African-American Males in the Criminal Justice System.* New York: Cambridge University Press.

Miller, Mark Crispin. 1996. "Free the Media." *The Nation* 3 June 3: 9–15.

Miller, Robert L. 1998. "The Limited Concerns of Social Mobility Research." *Current Sociology* 46, 4: 145–170.

Mills, C. Wright. 1956. *The Power Elite.* New York: Oxford University Press.

__________. 1959. *The Sociological Imagination.* New York: Oxford University Press.

Mills, Janet Lee. 1985. "Body Language Speaks Louder Than Words." *Horizons* (February) 8–12.

Ministério de Ciência e Tecnologia Brasil. 2002. "Brazil Urban Population." On the World Wide Web at http://www.mct.gov.br/clima/ingles/comunic_old/res7_1_1.htm (5 August 2003).

Minkel, J. R. 2002. "A Way with Words." *Scientific American* 25 March. On the World Wide Web at http://www.mit.edu/~lera/sciam (21 January 2003).

Minois, George. 1989 [1987]. *History of Old Age: From Antiquity to the Renaissance,* Sarah Hanbury Tenison, trans. Chicago: University of Chicago Press.

Mintz, Alexander. 1946. "A Re-Examination of Correlations Between Lynchings and Economic Indices." *Journal of Abnormal and Social Psychology* 41: 154–160.

Mintz, Beth. 1989. "United States of America." Pp. 207–236 in Tom Bottomore and Robert J. Brym, eds. *The Capitalist Class: An International Study.* New York: New York University Press.

__________, and Michael Schwartz. 1985. *The Power Structure of American Business.* Chicago: University of Chicago Press.

Mishel, Lawrence, Jared Bernstein, and John Schmitt. 1999. *The State of Working America, 1998–99.* Ithaca, NY: Cornell University Press.

Mishra, Ashok K., Hisham S. El-Osta, Mitchell J. Morehart, James, D. Johnson, and Jeffrey W. Hopkins. 2002. *Income, Wealth, and the Economic Well-Being of Farm Households.* Washington, DC: U.S. Department of Agriculture. On the World Wide Web at http://ers.usda.gov/publications/aer812/aer812.pdf (2 August 2003).

Mitchell, Alison. 1999. "Vote on Campaign Finances Is Blocked by Senate GOP." *New York Times on the Web* 20 October. On the World Wide Web at http://www.nytimes.com/library/politics/102099campaign-finance.html (1 May 2000).

__________. 2002. "Senate Approves Nuclear Waste Site in Nevada." *New York Times* 10 July. On the World Wide Web at http://www.nytimes.com (10 July 2002).

Mitford, Jessica. 1998 [1963]. *The American Way of Death Revisited.* New York: Vintage.

Mittelman, James H. 2000. *The Globalization Syndrome: Transformation and Resistance.* Princeton, NJ: Princeton University Press.

Mittelstaedt, Martin. 2001. "When a Car's Tailpipe Is More Lethal Than a Car Crash." *Globe and Mail* 29 September: F9.

Mizruchi, Mark S. 1982. *The American Corporate Network, 1904–1974.* Beverly Hills, CA: Sage.

__________. 1992. *The Structure of Corporate Political Action: Interfirm Relations and Their Consequences.* Cambridge, MA: Harvard University Press.

Moffett, Sebastian. 2003. "For Ailing Japan, Longevity Begins to Take Its Toll." *Wall Street Journal* 11 February: A1, A12.

Mokdad, Ali H., James S. Marks, Donna F. Stroup, and Julie L. Gerberding. 2004. "Actual Causes of Death in the United States, 2000." *JAMA* 291, 10: 1238-45.

Molloy, Beth L., and Sharon D. Herzberger. 1998. "Body Image and Self-Esteem: A Comparison of African-American and Caucasian Women." *Sex Roles* 38: 631–643.

Molm, Linda D. 1997. *Coercive Power in Social Exchange.* Cambridge: Cambridge University Press.

"Monitoring the Future Study." 1998. On the World Wide Web at http://monitoringthefuture.org (1 May 2000).

Moore, Barrington, Jr. 1967. *Social Origins of Dictatorship and Democracy: Lord and Peasant in the Making of the Modern World.* Boston: Beacon Press.

Moore, David S. 1995. *The Basic Practice of Statistics.* New York: W. H. Freeman.

Morris, Aldon D. 1984. *The Origins of the Civil Rights Movement: Black Communities Organizing for Change.* New York: Free Press.

Morris, Charles R. 1996. *The AARP: America's Most Powerful Lobby and the Clash of Generations.* New York: Times Books.

Morris, Norval, and David J. Rothman, eds. 1995. *The Oxford History of the Prison: The Practice of Punishment in Western Society.* New York: Oxford University Press.

Mortimer, Jeylan T., and Roberta G. Simmons. 1978. "Adult Socialization." *Annual Review of Sociology* 4: 421–454.

"MP3stego." 2001. On the World Wide Web at http://www.cl.cam.ac.uk/~fapp2/steganography/mp3stego (13 September 2001).

"The MUD Connector." 2005. On the World Wide Web at http://www.mudconnect.com/ (14 February 2005).

Mumford, Lewis. 1961. *The City in History: Its Origins, Its Transformations, and Its Prospects.* New York: Harcourt, Brace, & World.

Mundell, Helen. 1993. "How the Color Mafia Chooses Your Clothes." *American Demographics* November. On the World Wide Web at http://www.demographics.com/publications/ad/93_ad/9311_ad/ad281.htm (2 May 2000).

Murdoch, Guy. 1995. "Child Care Centers." *Consumers' Research Magazine* 78, 10: 2.

Murdock, George Peter. 1937. "Comparative Data on the Division of Labor by Sex." *Social Forces* 15: 551–553.

__________. 1949. *Social Structure.* New York: Macmillan.

Mustard, Cameron A., Patricia Kaufert, Anita Kozyrskyj, and Teresa Mayer. 1998. "Sex Differences in the Use of Health Care Services." *New England Journal of Medicine* 338: 1678–1683.

Myerhoff, Barbara. 1978. *Number Our Days.* New York: Dutton.

Myers, Daniel J. 1997. "Racial Rioting in the 1960s: An Event History Analysis of Local Conditions." *American Sociological Review* 62: 94–112.

Myers, Ransom A., and Boris Worm. 2003. "Rapid Worldwide Depletion of Predatory Fish Communities." *Nature* 423:280–3.

Myles, John. 1988. "The Expanding Middle: Some Canadian Evidence on the Deskilling Debate." *Canadian Review of Sociology and Anthropology* 25: 335–364.

__________. 1989 [1984]. *Old Age in the Welfare State: The Political Economy of Public Pensions,* 2nd ed. Lawrence: University Press of Kansas.

__________, and Adnan Turegun. 1994. "Comparative Studies in Class Structure." *Annual Review of Sociology* 20: 103–124.

Nagel, Joane. 1996. *American Indian Ethnic Renewal: Red Power and the Resurgence of Identity and Culture.* New York: Oxford University Press.

National Association for the Advancement of Colored People. 1934. *The Lynching of Claude Neal.* New York: National Association for the Advancement of Colored People.

National Basketball Association. 2000. "New York Knicks History." On the World Wide at http://nba.com/knicks/00400499.html#2 (29 May 2000).

National Center for Educational Statistics. "Trends in International Mathematics and Science Study." 2000. On the World Wide Web at http://nces.ed.gov/timss/results.asp (12 May 2003).

__________. 2002a. *The Condition of Education 2002.* Washington, DC. On the World Wide Web at http://nces.ed.gov/ pubs2002/2002025.pdf (17 May 2003).

———. 2002b. "Student Effort and Educational Progress." On the World Wide Web at http://www.nces.ed.gov/programs/coe/2002/section3/indicator18.asp (6 February 2003).

———. 2003. *Comparative Indicators of Education in the United States and Other G8 Countries: 2002.* Washington, DC. On the World Wide Web at http://nces.ed.gov/pubs2003/2003026.pdf (12 May 2003).

National Center for Education Statistics. 2004a. *The Condition of Education, 2004.* Washigngton, DC: Institute of Education Sciences, U.S. Department of Education. On the World Wide Web at http://nces.ed.gov/programs/coe (12 June 2004).

———. 2004b. *Highlights from the Trends in International Mathematics and Science Study: TIMSS 2003.* On the World Wide Web at http://nces.ed.gov/pubsearch/pubsinfo.asp?pubid=2005005 (29 March 2005).

National Center for Injury Prevention and Control. 2000. "Suicide in the United States." On the World Wide Web at http://www.cdc.gov/ncipc/factsheets/suifacts.htm (27 April 2000).

National Commission on Excellence in Education. 1983. *A Nation at Risk.* Washington, D.C.

"The National Council of La Raza." 2000. On the World Wide Web at http://www.nclr.org/about (29 April 2000).

National Energy Information Center. 2005. "International Gross Domestic Product Information." On the World Wide Web at http://www.eia.doe.gov/emeu/international/other.html#IntlGDP (26 January 2004).

National Gay and Lesbian Task Force. 2005. "Anti-Gay Marriage Measures in the U.S. as of February 2005." On the World Wide Web at http://www.thetaskforce.org/downloads/marriagemap.pdf (8 March 2005).

National Office of Vital Statistics. 1947. "Deaths and Death Rates for Leading Causes of Death: Death Registration States, 1900–1940." Special tabulation prepared for the authors.

National Opinion Research Center. 2004. *General Social Survey, 1972-2002.* Chicago: University of Chicago. Machine readable file.

"National Organization for Men Against Sexism." 2000. On the World Wide Web at http://nomas.idea-net.com (15 June 2000).

Nestle, Marion. 2002. *Food Politics.* Chicago: University of Chicago Press.

Neugarten, Bernice. 1974. "Age Groups in American Society and the Rise of the Young Old." *Annals of the American Academy of Political and Social Science* 415: 187–198.

Nevitte, Neil. 1996. *The Decline of Deference.* Peterborough, Canada: Broadview Press.

Newcomb, Theodore M. 1943. *Personality and Social Change: Attitude Formation in a Student Community.* New York: Holt, Rinehart & Winston.

Newman, Katherine S. 1988. *Falling From Grace: The Experience of Downward Mobility in the American Middle Class.* New York: Free Press.

———. 1999. *No Shame in My Game: The Working Poor in the Inner City.* New York: Knopf and the Russell Sage Foundation.

Newport, Frank. 2000. "Support for Death Penalty Drops to Lowest Level in 19 Years, Although Still High at 66%." The Gallup Organization. On the World Wide Web at http://www.gallup.com/poll/releases/pr000224.asp (8 August 2000).

New York City Department of Consumer Affairs. 2001. "The High Cost of Dying." On the World Wide Web at http://home.nyc.gov/html/dca/html/dcafuneralreport.html (27 April 2003).

Nie, Norman H., Sidney Verba, and John R. Petrocik. 1979 [1976]. *The Changing American Voter,* rev. ed. Cambridge, MA: Harvard University Press.

Nielsen Media Research. 2005. "Top 20." On the World Wide Web at http://tv.yahoo.com/nielsen/ (11 March 2005).

"The Nike Campaign." 2000. On the World Wide Web at http://www.web.net/~msn/3nike.htm (23 June 2000).

Nisbett, Richard E., Kaiping Peng, Incheol Choi, and Ara Norenzayan. 2001. "Culture and Systems of Thought: Holistic Versus Analytic Cognition." *Psychological Review* 108: 291–310.

———, and Lee Ross. 1980. *Human Inference: Strategies and Shortcomings of Social Judgment.* Englewood Cliffs, NJ: Prentice-Hall.

Nolen, Stephanie. 1999. "Gender: The Third Way." *Globe and Mail* September 25: D1, D4.

"Nomar Garciaparra Pictures." 1998. On the World Wide Web at http://www.geocities.com/Colosseum/Track/4242/photos.htm (15 June 1998).

Notestein, F. W. 1945. "Population—The Long View." Pp.36–57 in T. W. Schultz, ed. *Food for the World.* Chicago: University of Chicago Press.

Nowak, Martin A., Robert M. May, and Karl Sigmund. 1995. "The Arithmetics of Mutual Help." *Scientific American* 272, 6: 76–81.

Nowell, Amy, and Larry V. Hedges. 1998. "Trends in Gender Differences in Academic Achievement from 1960 to 1994: An Analysis of Differences in Mean, Variance, and Extreme Scores." *Sex Roles* 39: 21–43.

Nuland, Sherwin B. 1993. *How We Die: Reflections on Life's Final Chapter.* New York: Vintage.

Oakes, Jeannie. 1985. *Keeping Track: How Schools Structure Inequality.* New Haven, CT: Yale University Press.

Oates, Joyce Carol. 1999. "The Mystery of JonBenét Ramsey." *New York Review of Books* 24 June: 31–37.

Oberschall, Anthony. 1973. *Social Conflict and Social Movements.* Englewood Cliffs, NJ: Prentice-Hall.

O'Connor, Julia S., and Robert J. Brym. 1988. "Public Welfare Expenditure in OECD Countries: Towards a Reconciliation of Inconsistent Findings." *British Journal of Sociology* 39: 47–68.

———, and Gregg M. Olsen, eds. 1998. *Power Resources Theory and the Welfare State: A Critical Approach.* Toronto: University of Toronto Press.

OECD. 2005. "Labour Market Statistics – DATA." On the World Wide Web at http://www1.oecd.org/scripts/cde/members/lfsdataauthenticate.asp (30 March 2005).

Ogbu, John U. 2003. *Black American Students in an Affluent Suburb: A Study of Academic Disengagement.* Mahwah, NJ: L. Erlbaum Associates.

Ogburn, William F. 1966 [1922]. *Social Change with Respect to Culture and Original Nature.* New York: Dell.

Ogunwole, Stella U. 2002. "The American Indian and Alaska Native Population: 2000." Washington, DC: U.S. Census Bureau. On the World Wide Web at http://www.census.gov/prod/2002pubs/c2kbr01-15.pdf (21 June 2003).

O'Hare, William P. 1996. "A New Look at Poverty in America." *Population Bulletin* 51, 2: 2–46.

O'Neil, Dennis. 2004. "Patterns of Subsistence: Classification of Cultures Based on the Sources and Techniques of Acquiring Food and other Necessities." On the World Wide Web at http://anthro.palomar.edu/subsistence/ (12 January 2005).

Oliver, Melvin L., and Thomas M. Shapiro. 1995. *Black Wealth/White Wealth: A New Perspective on Racial Inequality.* New York: Routledge.

Oliver, Mike. 1996. *Understanding Disability: From Theory to Practice.* Basingstoke, UK: Macmillan.

Olsen, Gregg. 2002. *The Politics of the Welfare State: Canada, Sweden, and the United States.* Toronto: Oxford University Press.

Olsen, Gregg, and Robert J. Brym. 1996. "Between American Exceptionalism and Swedish Social Democracy: Public and Private Pensions in Canada." Pp. 261–279 in Michael Shalev, ed. *The Privatization of Social Policy? Occupational Welfare and the Welfare State in America, Scandinavia and Japan.* London: Macmillan.

Olshansky, S. Jay. 1997. "Infectious Diseases: New and Ancient Threats to World Health." *Population Bulletin* 52, 2.

__________, Bruce A. Carnes, and Christine Cassel. 1990. "In Search of Methusaleh: Estimating the Upper Limits of Human Longevity." *Science* 250: 634–640.

__________, __________, and Aline Desesquelles. 2001. "Prospects for Human Longevity." *Science* 291, 5508: 1491–1492.

Olzak, Susan, and Suzanne Shanahan. 1996. "Deprivation and Race Riots: An Extension of Spilerman's Analysis." *Social Forces* 74: 931–962.

__________, __________, and Elizabeth H. McEneaney. 1996. "Poverty, Segregation, and Race Riots: 1960 to 1993." *American Sociological Review* 61: 590–614.

Omega Foundation. 1998. "An Appraisal of the Technologies of Political Control: Summary and Options Report for the European Parliament." On the World Wide Web at http://home.icdc.com/~paulwolf/eu_stoa_2.htm (29 April 2000).

Omi, Michael, and Howard Winant. 1986. *Racial Formation in the United States.* New York: Routledge.

Ontario Consultants on Religious Tolerance. 2002. "The Harry Potter Books: Efforts to Ban Books." On the World Wide Web at http://www.religioustolerance.org/potter3.htm (17 May 2002).

Optometrists Network. 2000. "Attention Deficit Disorder." On the World Wide Web at http://www.add-adhd.org/ADHD_attention-deficit.html (14 August 2000).

Oregon Department of Human Services. 2004. *Sixth Annual Report on Oregon's Death with Dignity Act.* Portland OR. On the World Wide Web at http://www.dhs.state.or.us/publichealth/chs/pas/03pasrpt.pdf (5 February 2005).

Organization of African Unity. 2000. *Rwanda: The Preventable Genocide.* New York. On the World Wide Web at http://www.visiontv.ca/RememberRwanda/Report.pdf (15 January 2005).

Orfield, Gary, and Susan E. Eaton. 1996. *Dismantling Desegregation: The Quiet Reversal of Brown v. Board of Education.* New York: Free Press.

Ornstein, Michael. 1998. "Survey Research." *Current Sociology* 46, 4: 1–87.

Ossowski, Stanislaw. 1963. *Class Structure in the Social Consciousness,* S. Patterson, trans. London: Routledge & Kegan Paul.

Pacey, Arnold. 1983. *The Culture of Technology.* Cambridge, MA: MIT Press.

Pammett, Jon H. 1997. "Getting Ahead Around the World." Pp. 67–86 in Alan Frizzell and Jon H. Pammett, eds. *Social Inequality in Canada.* Ottawa: Carleton University Press.

Pape, Robert A. 2003 "The Strategic Logic of Suicide Terrorism." *American Political Science Review* 97: 343–361.

Parillo, Vincent N., John Stimson, and Ardyth Stimson. 1999. *Contemporary Social Problems,* 4th ed. Boston: Allyn & Bacon.

Park, Robert Ezra. 1914. "Racial Assimilation in Secondary Groups." *Publications of the American Sociological Society* 8: 66–72.

__________. 1950. *Race and Culture.* New York: Free Press.

__________, Ernest W. Burgess, and Roderick D. McKenzie. 1967 [1925]. *The City.* Chicago: University of Chicago Press.

Parshall, Gerald. 1998. "Brotherhood of the Bomb." *US News & World Report* 125, 7 (17–24 August) 64–68.

Parson, E. A., and D. W. Keith. 1998. "Fossil Fuels Without CO_2 Emissions." *Science* 282: 1053–1054.

Parsons, Talcott. 1942. "Age and Sex in the Social Structure of the United States." *American Sociological Review* 7: 604–616.

__________. 1951. *The Social System.* New York: Free Press.

__________. 1955. "The American Family: Its Relation to Personality and to the Social Structure." Pp. 3–33 in Talcott Parsons and Robert F. Bales, eds. *Family, Socialization and Interaction Process.* New York: Free Press.

Pascual, Brian. 2002. "Avril Lavigne Hates Britney Spears." *ChartAttack* April 19. On the World Wide Web at http://teenmusic.about.com/gi/dynamic/offsite.htm?site=http%3A%2F%2Fwww.chartattack.com%2Fdamn%2F2002%2F04%2F1901.cfm (15 January 2003).

Patent Medicine Prices Review Board. 2002. *Annual Report.* Ottawa. On the World Wide Web at http://www.pmprb-cepmb.gc.ca/CMFiles/ar-2002e21IRA-6162003-196.pdf (17 June 2003).

Patterson, Orlando. 1982. *Slavery and Social Death.* Cambridge, MA: Harvard University Press.

__________. 1997. *The Ordeal of Integration: Progress and Resentment in America's "Racial" Crisis.* Washington, DC: Civitas.

Payer, Cheryl. 1982. *The World Bank: A Critical Analysis.* New York: Monthly Review Press.

Peacock, Mary. 2000. "The Cult of Thinness." On the World Wide Web at http://www.womenswire.com/image/toothin.html (13 June 2000).

Pear, Robert. 2003. "House Endorses Stricter Work Rules for Poor." *New York Times* February 14. On the World Wide Web at http://www.nytimes.com (14 February 2003).

Perrow, Charles B. 1984. *Normal Accidents.* New York: Basic Books.

Perry-Castañeda Library Map Collection. 2000. "Comparative Soviet Nationalities by Republic." On the World Wide Web at http://www.lib.utexas.edu/Libs/PCL/Map_collection/commonwealth/USSR_NatRep_89.jpg (23 November 2000).

Pessen, Edward. 1984. *The Log Cabin Myth: The Social Backgrounds of the Presidents.* New Haven, CT: Yale University Press.

Peters, John F. 1994. "Gender Socialization of Adolescents in the Home: Research and Discussion." *Adolescence* 29: 913–934.

Pettersson, Jan. 2003. "Democracy, Regime Stability, and Growth." *Scandinavian Working Papers in Economics.* On the World Wide Web at http://swopec.hhs.se/sunrpe/abs/sunrpe2002_0016.htm (13 February 2003).

Pew Research Center for the People and the Press. 2001. "Overwhelming Support for Bush, Military Response but . . . American Psyche Reeling from Terror Attacks." On the World Wide Web at http://www.pewtrusts.com/pubs/pubs_item.cfm?image=img5&content_item_id=749&content_type_id=18&page=p1 (22 December 2002).

__________. 2002. "Among Wealthy Nations U.S. Stands Alone in its Embrace of Religion." On the World Wide Web at http://people-press.org/reports/display.php3?ReportID=167 (3 May 2003).

Phillips, Kevin. 1990. *The Politics of Rich and Poor: Wealth and the American Electorate in the Reagan Aftermath.* New York: Random House.

Piaget, Jean, and Bärbel Inhelder. 1969. *The Psychology of the Child,* Helen Weaver, trans. New York: Basic Books.

Pinker, Steven. 1994. "Apes—Lost for Words." *New Statesman and Society* 15 April: 30–31.

__________. 1994b. *The Language Instinct.* New York:Morrow.

__________. 2001. "Talk of Genetics and Vice Versa." *Nature* 413, 6855, 4 October: 465–466.

__________. 2002. *The Blank Slate: The Modern Denial of Human Nature.* New York: Viking.

Piven, Frances Fox, and Richard A. Cloward. 1977. *Poor People's Movements: Why They Succeed, How They Fail.* New York: Vintage.

__________, and __________. 1989 [1988]. *Why Americans Don't Vote.* New York: Pantheon.

__________, and __________. 1993 [1971]. *Regulating the Poor: The Functions of Public Welfare,* updated ed. New York: Vintage.

A Place Called Chiapas. 1998. Vancouver: Canada Wild Productions. (Movie).

Podolny, Joel M., and Karen L. Page. 1998. "Network Forms of Organization." *Annual Review of Sociology* 24: 57–76.

Polanyi, Karl. 1957 [1944]. *The Great Transformation: The Political and Economic Origins of Our Time.* Boston: Beacon Press.

PollingReport.com. 2005. "Crime." On the World Wide Web at http://www.pollingreport.com/crime.htm (11 March 2005).

Polsby, Nelson W. 1959. "Three Problems in the Analysis of Community Power." *American Sociological Review* 24: 796–803.

Pool, Robert. 1997. *Beyond Engineering: How Society Shapes Technology.* New York: Oxford University Press.

Popenoe, David. 1988. *Disturbing the Nest: Family Change and Decline in Modern Societies.* New York: Aldine de Gruyter.

__________. 1991. "Family Decline in the Swedish Welfare State." *Public Interest* 102: 65–78.

__________. 1992. "Family Decline: A Rejoinder." *Public Interest* 109: 116–118.

__________. 1993. "American Family Decline, 1960–1990: A Review and Appraisal." *Journal of Marriage and the Family* 55: 527–555.

__________. 1996. *Life Without Father: Compelling New Evidence that Fatherhood and Marriage are Indispensable for the Good of Children and Society.* New York: Martin Kessler Books.

Population Reference Bureau. 2003a. "2003 World Population Data Sheet." On the World Wide Web at http://www.prb.org/pdf/WorldPopulationDS03_Eng.pdf (2 August 2003).

__________. 2003b. "Human Population: Fundamentals of Growth." On the World Wide Web at http://www.prb.org/Content/NavigationMenu/PRB/Educators/Human_Population/Population_Growth/Population_Growth.htm (2 August 2003).

__________. 2004. *2004 World Population Data Sheet.* On the World Wide Web at http://www.prb.org/pdf04/04WorldDataSheet_Eng.pdf (4 March 2005).

Portes, Alejandro. 1996. "Global Villagers: The Rise of Transnational Communities." *The American Prospect* 25: 74–77. On the World Wide Web at http://www.prospect.org/archives/25/25port.html (29 April 2000).

__________, and Robert D. Manning. 1991. "The Immigrant Enclave: Theory and Empirical Examples." Pp. 319–32 in Norman R. Yetman, ed. *Majority and Minority: The Dynamics of Race and Ethnicity in American Life,* 5th ed. Boston: Allyn & Bacon.

__________, and Rubén G. Rumbaut. 1990. *Immigrant America: A Portrait.* Berkeley: University of California Press.

__________, and Alex Stepick. 1993. *City on the Edge: The Transformation of Miami.* Berkeley: University of California Press.

__________, and Cynthia G. Truelove. 1991. "Making Sense of Diversity: Recent Research on Hispanic Minorities in the United States." Pp. 402–19 in Norman R. Yetman, ed. *Majority and Minority: The Dynamics of Race and Ethnicity in American Life,* 5th ed. Boston: Allyn & Bacon.

Postel, Sandra. 1994. "Carrying Capacity: Earth's Bottom Line." Pp. 3–21 in Linda Starke, ed. *State of the World 1994.* New York: W. W. Norton.

Postman, Neil. 1982. *The Disappearance of Childhood.* New York: Delacorte.

__________. 1992. *Technopoly: The Surrender of Culture to Technology.* New York: Vintage.

Powers, Elizabeth T. 1994. "The Impact of AFDC on Birth Decisions and Program Participation." *Working Papers.* Cleveland: Federal Reserve Bank of Cleveland.

Pred, Allan R. 1973. *Urban Growth and the Circulation of Information.* Cambridge, MA: Harvard University Press.

Prejean, Sister Helen. 2005. "Death in Texas." *New York Review of Books* 52, 1: 13 January. On the World Wide Web at http://www.nybooks.com/articles/17670 (12 March 2005).

"Prescription for Confusion." 2004. *New York Times* 28 December. On the World Wide Web at http://www.nytimes.com (28 December 2004).

Press, Andrea. 1991. *Women Watching Television: Gender, Class and Generation in the American Television Experience.* Philadelphia: University of Pennsylvania Press.

__________, and Elizabeth R. Cole. 1999. *Speaking of Abortion: Television and Authority in the Lives of Women.* Chicago: University of Chicago Press.

Press, Eyal. 1999. "A Nike Sneak." *The Nation* 5 April. On the World Wide Web at http://www.thenation.com/issue/990405/0405press.shtml (8 August 2000).

Priestly, Mark. 2001. "Introduction: The Global Context of Disability." Pp. 3–25 in *Disability and the Life Course: Global Perspectives,* Mark Priestly, ed. Cambridge: Cambridge University Press.

Prigerson, Holly Gwen. 1992. "Socialization to Dying: Social Determinants of Death Acknowledgement and Treatment among Terminally Ill Geriatric Patients." *Journal of Health and Social Behavior* 33: 378–395.

Princeton Review. 2003 "Where Will Your SAT or ACT Scores Take You?" On the World Wide Web at http://www.princetonreview.com/college/testprep/testprep.asp?TPRPAGE=295&TYPE=SAT (13 May 2003).

Prittie, Jennifer. 2000. "The Serious Business of Rubber Noses and Big Shoes." *National Post* 29 March: A1–A2.

Proctor, Robert N. 1988. *Racial Hygiene: Medicine under the Nazis.* Cambridge, MA: Harvard University Press.

Provine, Robert R. 2000. *Laughter: A Scientific Investigation.* New York: Penguin.

Public Broadcasting System. 1997. "Double Talk?" On the World Wide Web at http://www.pbs.org/newshour/bb/education/july-dec97/bilingual_9-21.html (11 August 2000).

Public Citizen. 2004. "U.S. Workers' Jobs, Wages and Economic Security." On the World Wide Web at http://www.citizen.org/documents/NAFTA_10_jobs.pdf (31 March 2005).

Quadagno, Jill. 1988. *The Transformation of Old Age Security: Class and Politics in the American Welfare State.* Chicago: University of Chicago Press.

__________. 1994. *The Color of Welfare: How Racism Undermined the War on Poverty.* New York: Oxford University Press.

Quaschning, Volker. 2003. "Development of Global Carbon Dioxide Emissions and Concentration in Atmosphere." On the World Wide Web at http://www.volker-quaschning.de/datserv/CO2/index_e.html (2 March 2005).

Quinn, Tom. 2000. "The 'Motor Voter' Question." *Ford Foundation Report.* On the World Wide Web at http://www.fordfound.org/publications/ff_report/view_ff_report_detail.cfm?report_index=247 (21 November 2000).

Raag, Tarja, and Christine L. Rackliff. 1998. "Preschoolers' Awareness of Social Expectations of Gender: Relationships to Toy Choices." *Sex Roles* 38: 685–700.

Rahman, Shaikh Azizur. 2004. "Where the girls aren't." *Globe and Mail* 16 October: F2.

Rainwater, Lee, and Timothy M. Smeeding. 1995. "Safety Nets for Children are Weakest in US." United National Children's Emergency Fund. On the World Wide Web at http://www.unicef.org:80/pon96/indust4.htm (1 May 2000).

Ramirez, Roberto R., and G. Patricia de la Cruz. 2003. "The Hispanic Population in the United States: March 2002." Washington, DC: U.S. Census Bureau. On the World Wide Web at http://www.census.gov/prod/2002pubs/c2kbr01-16.pdf (24 June 2003).

Rampton, Sheldon, and John Stauber. 2001. *Trust Us, We're Experts: How Industry Manipulates Science and Gambles with Your Future.* New York: Jeremy P. Tarcher/Putnam.

Rank, Mark Robert. 1994. *Living on the Edge: The Politics of Welfare in America.* New York: Columbia University Press.

Rapp, R., and E. Ross. 1986. "The 1920s: Feminism, Consumerism and Political Backlash in the U.S." Pp. 52–62 in J. Friedlander, B. Cook, A. Kessler-Harris, and C. Smith-Rosenberg, eds. *Women in Culture and Politics.* Bloomington: Indiana University Press.

Reed, Dan. 2000. "Janitors Get Pay Increase; Strike Averted." *San Jose Mercury News* 4 June. On the World Wide Web at http://www.mercurycenter.com/premium/local/docs/janitors04.htm (6 June 2000).

Reeves, Terrance, and Chandler E. Bennett. 2003. "The Asian and Pacific Islander Population in the United States; March 2000." Washington, DC: U.S. Census Bureau. On the World Wide Web at http://www.census.gov/prod/2003pubs/p20-540.pdf (22 June 2003).

Reich, Robert B. 1991. *The Work of Nations: Preparing Ourselves for 21st-Century Capitalism.* New York: Knopf.

Reiman, Jeffrey H. 1995 [1979]. *The Rich Get Richer and the Poor Get Prison: Ideology, Class, and Criminal Justice,* 4th ed. Boston: Allyn & Bacon.

__________. 1996. *And the Poor Get Prison: Economic Bias in American Criminal Justice.* Boston: Allyn & Bacon.

Reinarman, Craig, and Harry G. Levine, eds. 1999. *Crack in America: Demon Drugs and Social Justice.* Berkeley: University of California Press.

Reiter, Ester. 1991. *Making Fast Food: From the Frying Pan into the Fryer.* Montreal: McGill-Queen's University Press.

Religious Tolerance.org. 2005. "Same-Sex Marriages (SSM) and Civil Unions." On the World Wide Web at http://www.religioustolerance.org/hom_marr.htm (9 March 2005).

Remennick, Larissa I. 1998. "The Cancer Problem in the Context of Modernity: Sociology, Demography, Politics." *Current Sociology* 46, 1: 1–150.

__________. 2002. "Transnational Community in the Making: Russian-Jewish Immigrants of the 1990s in Israel." *Journal of Ethnic and Migration Studies* 28: 515–530.

Remnick, David. 1998. "How Russia Is Ruled." *New York Review of Books* 45, 6: 10–15.

Renner, Michael. 2000. "Creating Jobs, Preserving the Environment." Pp. 162–83 in Lester R. Brown et al., eds. *State of the World 2000.* New York: Norton.

Rennison, Callie. 2002. "Criminal Victimization 2001: 2000–2001 Changes with Trends 1993–2001." U.S. Department of Justice. Office of Justice Programs. Bureau of Justice Statistics. On the World Wide Web at http://www.ojp.usdoj.gov/bjs/abstract/cv01.htm (16 February 2003).

Reskin, Barbara, and Irene Padavic. 2002 [1994]. *Women and Men at Work,* 2nd ed. Thousand Oaks, CA: Pine Forge.

Resnick, Mitchell, and Natalie Rusk. 1996. "Access Is Not Enough: Computer Clubhouses in the Inner City." *The American Prospect* 27, July–August: 60–68. On the World Wide Web at http://www.prospect.org/archives/27/27resn.html (2 May 2000).

Rex, T. 2003. "Economic Recessions in the United States." On the World Wide Web at http://www.quinnell.us/politics/knowledge/recessions.html (21 May 2003).

Richer, Stephen. 1990. *Boys and Girls Apart: Children's Play in Canada and Poland.* Ottawa: Carleton University Press.

Richtel, Matt. 2002. "Bay Area Real Estate Prices Too Hot for Some to Touch." *New York Times* May 29. On the World Wide Web at http://www.nytimes.com (11 March 2003).

Ridgeway, Cecilia L. 1983. *The Dynamics of Small Groups.* New York: St. Martin's Press.

Riedmann, Agnes. 1993. *Science That Colonizes: A Critique of Fertility Studies in Africa.* Philadelphia: Temple University Press.

Rier, David A. 2000. "The Missing Voice of the Critically Ill: A Medical Sociologist's First-Person Account." *Sociology of Health and Illness* 22: 68–93.

Rifkin, Jeremy. 1995. *The End of Work: The Decline of the Global Labor Force and the Dawn of the Post Market Era.* New York: G. P. Putnam's Sons.

__________. 1998. *The Biotech Century: Harnessing the Gene and Remaking the World.* New York: Jeremy P. Tarcher/Putnam.

Riley, Matilda White, Anne Foner, and Joan Waring. 1988. "Sociology of Age." Pp. 243–90 in Neil Smelser, ed. *Handbook of Sociology.* Newbury Park, CA: Sage.

Riley, Nancy. 1997. "Gender, Power, and Population Change." *Population Bulletin* 52, 1. On the World Wide Web at http://www.prb.org/pubs/population_bulletin/bu52-1.htm (25 August 2000).

Rist, Ray C. 1970. "Student Social Class and Teacher Expectations: The Self-fulfilling Prophecy in Ghetto Education." *Harvard Educational Review* 40: 411–451.

Ritzer, George. 1996a. *The McDonaldization of Society,* rev. ed. Thousand Oaks, CA: Pine Forge Press.

__________. 1996b. "The McDonaldization Thesis: Is Expansion Inevitable?" *International Sociology* 11: 291–307.

Robbins, Liz. 2005. "Nash Displays Polished Look: On the Court, of Course." *New York Times* 19 January. On the World Wide Web at http://ww.nytimes.com (19 January 2005.)

Roberts, D. F. 1995. "The Pervasiveness of Plasticity." Pp.1–17 in *Human Variability and Plasticity,* C. G. N. Mascie-Taylor, and Barry Bogin, eds. Cambridge: Cambridge University Press.

Roberts, Siobhan. 2000. "Web Could Be New STD Breeding Ground." *National Post* 17 June: A2.

Robertson, Roland. 1992. *Globalization: Social Theory and Global Culture.* Newbury Park, CA: Sage.

Robinson, John P., and Suzanne Bianchi. 1997. "The Children's Hours." *American Demographics* December: 20–24.

Robinson, Richard H., and Willard L. Johnson. 1997 [1982]. *The Buddhist Religion: A Historical Introduction,* 4th ed. Belmont, CA: Wadsworth.

Robinson, Robert V., and Wendell Bell. 1978. "Equality, Success, and Social Justice in England and the United States." *American Sociological Review* 43: 125–143.

Roche, Maurice. 1995. "Rethinking Citizenship and Social Movements: Themes in Contemporary Sociology and Neoconservative Ideology." Pp. 186–219 in Louis Maheu, ed. *Social Classes and Social Movements: The Future of Collective Action.* London: Sage.

Rodinson, Maxime. 1996. *Muhammad,* 2nd ed. Anne Carter, trans. London: Penguin.

Roediger, David R. 1991. *The Wages of Whiteness: Race and the Making of the American Working Class.* London: Verso.

Roethlisberger, Fritz J., and William J. Dickson. 1939. *Management and the Worker.* Cambridge, MA: Harvard University Press.

Rogers, Everett M. 1995. *Diffusion of Innovations,* 4th ed. New York: Free Press.

Rogers, Jackie Krasas, and Kevin D. Henson. 1997. "'Hey, Why Don't You Wear a Shorter Skirt?' Structural Vulnerability and the Organization of Sexual Harassment in Temporary Clerical Employment." *Gender and Society* 11: 215–237.

Rollins, Boyd C., and Kenneth L. Cannon. 1974. "Marital Satisfaction over the Family Life Cycle." *Journal of Marriage and the Family* 36: 271–284.

Rones, Philip L., Randy E. Ilg, and Jennifer M. Gardner. 1997. "Trends in Hours of Work Since the Mid-1970s." *Monthly Labor Review* April. On the World Wide Web at http://www.bls.gov/opub/mlr/1997/04/art1full.pdf (26 January 2003).

Rootes, Chris. 1995. "A New Class? The Higher Educated and the New Politics." Pp. 220–235 in Louis Maheu, ed. *Social Classes and Social Movements: The Future of Collective Action.* London: Sage.

Roscoe, Lori A., L. J. Dragovic, and Donna Cohen. 2000. "Dr. Jack Kevorkian and Cases of Euthanasia in Oakland County, Michigan, 1990–1998." On the World Wide Web at http://www.nejm.org/content/2000/0343/0023/1735.asp (12 July 2000).

Rose, Michael S. 2001. "The Facts Behind the Massacre." *Catholic World News* 17 October. On the World Wide Web at http:///www.cwnews.com/news/viewstory.cfm?recnum=20654 (15 January 2005).

Rosenberg, Charles E. 1962. *The Cholera Years: The United States in 1832, 1849, and 1866.* Chicago: University of Chicago Press.

__________. 1987. *The Care of Strangers: The Rise of America's Hospital System.* New York: Basic.

Rosenberg, Nathan. 1982. *Inside the Black Box: Technology and Economics.* Cambridge: Cambridge University Press.

Rosenthal, Robert, and Lenore Jacobson. 1968. *Pygmalion in the Classroom: Teacher Expectation and Pupils' Intellectual Development.* New York: Holt, Rinehart, & Winston.

Roslin, Alex. 2000. "Black & Blue." *Saturday Night* 23 September: 44–49.

Rossi, Peter H. 1989. *Down and Out in America: The Origins of Homelessness.* Chicago: University of Chicago Press.

Rostow, W. W. 1960. *The Stages of Economic Growth: A Non-Communist Manifesto.* New York: Cambridge University Press.

Roth, Cecil. 1961. *A History of the Jews.* New York: Schocken.

Rothman, Barbara Katz. 1982. *In Labor: Women and Power in the Birthplace.* New York: W. W. Norton.

__________. 1989. *Recreating Motherhood: Ideology and Technology in a Patriarchal Society.* New York: W. W. Norton.

Rothman, David J. 1991. *Strangers at the Bedside: A History of How Law and Bioethics Transformed Medical Decision Making.* New York: Basic Books.

__________. 1998. "The International Organ Traffic." *New York Review of Books* 45, 5: 14–17.

Rothman, Stanley, and Amy E. Black. 1998. "Who Rules Now? American Elites in the 1990s." *Society* 35, 6: 17–20.

Rubin, J. Z., F. J. Provenzano, and Z. Lurra. 1974. "The Eye of the Beholder." *American Journal of Orthopsychiatry* 44: 512–519.

Rubin, Lillian B. 1994. *Families on the Fault Line.* New York: HarperCollins.

Rueschemeyer, Dietrich, Evelyne Huber Stephens, and John Stephens. 1992. *Capitalist Development and Democracy.* Chicago: University of Chicago Press.

Ruggles, Steven. 1997. "The Effects of AFDC on American Family Structure, 1940–1990." *Journal of Family History* 22: 307–325.

Rural Advancement Foundation International. 1999. "The Gene Giants." On the World Wide Web at http://www.rafi.org/web/allpub-one.shtml?dfl=allpub.db&tfl=allpub-one-frag.ptml&operation=display&ro1=recNo&rf1=34&rt1=34&usebrs=true (2 May 2000).

Rushton, J. Phillipe. 1995. *Race, Evolution, and Behavior: A Life History Perspective.* New Brunswick, NJ: Transaction Publishers.

Russett, Cynthia Eagle. 1966. *The Concept of Equilibrium in American Social Thought.* New Haven, CT: Yale University Press.

Ryan, Kathryn M., and Jeanne Kanjorski. 1998. "The Enjoyment of Sexist Humor, Rape Attitudes, and Relationship Aggression in College Students." *Sex Roles* 38: 743–756.

Rytina, Steve. 1992. "Scaling the Intergenerational Continuity of Occupation: Is Occupational Inheritance Ascriptive After All?" *American Journal of Sociology* 97: 1658–1688.

Sahlins, Marshall D. 1972. *Stone Age Economics.* Chicago: Aldine.

Salamon, Sonya. 2003. "From Hometown to Nontown: Rural Community Effects of Suburbanization." *Rural Sociology* 68: 1–24.

Sampson, Robert. 1997. "The Embeddedness of Child and Adolescent Development: A Community-Level Perspective on Urban Violence." Pp. 31–77 in Joan McCord, ed. *Violence and Childhood in the Inner City.* Cambridge: Cambridge University Press.

__________, and John H. Laub. 1993. *Crime in the Making: Pathways and Turning Points through Life.* Cambridge, MA: Harvard University Press.

__________, and William J. Wilson. 1995. "Toward a Theory of Race, Crime and Urban Inequality." Pp. 37–54 in John Hagan and Ruth D. Peterson, eds., *Crime and Inequality.* Stanford, CA: Stanford University Press.

Samuelsson, Kurt. 1961 [1957]. *Religion and Economic Action,* E. French, trans. Stockholm: Scandinavian University Books.

Samuelson, Paul A. 2004. "Where Ricardo and Mill Rebut and Confirm Arguments of Mainstream Economists Supporting Globalization." *Journal of Economic Perspectives* 18: 135-46.

Sanday, Peggy Reeves. 1990. *Fraternity Gang Rape: Sex, Brotherhood, and Privilege on Campus.* New York: New York University Press.

Sanderson, Stephen K. 1995. *Macrosociology: An Introduction to Human Societies,* 3rd ed. New York: HarperCollins.

Sandqvist, Karin, and Bengt-Erik Andersson. 1992. "Thriving Families in the Swedish Welfare State." *Public Interest* 109: 114–116.

Sanmartin, Claudia, Edward Ng, Debra Blackwell, Jane Gentleman, Michael Martinez, Catherine Simile. 2004. *Joint Canada/United States Survey of Health.* Ottawa: Statistics Canada. On the World Wide Web at http://www.statcan.ca/english/freepub/82M0022XIE/2003001/pdf/82M0022XIE2003001.pdf (14 March 2005).

"Santa Clara County Less Crowded Since Demise of Dot-Coms." 2002. *Silicon Valley/San Jose Business Journal* 3 May. On the World Wide Web at http://sanjose.bizjournals.com/sanjose/stories/2002/05/06/weekinbiz.html (11 March 2003).

Sapiro, Virginia, Steven J. Rosenstone, and the National Election Studies. 2004. *American National Election Studies Cumulative Data File, 1948-2002* [Computer file]. 12th ICPSR version. Ann Arbor MI: University of Michigan, Center for Political Studies [producer] and Inter-university Consortium for Political and Social Research [distributor].

Sartre, Jean-Paul. 1965 [1948]. *Anti-Semite and Jew,* George J. Becker, trans. New York: Schocken.

Sassen, Saskia. 1991. *The Global City: New York, London, Tokyo.* Princeton, NJ: Princeton University Press.

Saunders, Doug. 2003. "U.S. FDA Blocks Mail-Order Drugs." *Globe and Mail* November 8: A23.

Savelsberg, Joachim, with contributions by Peter Brühl. 1994. *Constructing White-Collar Crime: Rationalities, Communication, Power* (Philadelphia: University of Pennsylvania Press).

Sax, Leonard. 2002. "How Common is Intersex? A Response to Anne Fausto-Sterling." *The Journal of Sex Research* 39, 3: 174–178.

Saxton, Alexander. 1900. *The Rise and Fall of the White Republic.* London: Verso.

Scarr, Sandra, and Richard A. Weinberg. 1978. "The Influence of 'Family Background' on Intellectual Attainment." *American Sociological Review* 43: 674–692.

Schemo, Diana Jean. 2002. "Neediest Schools Receive Less Money, Report Finds." *New York Times* August 9. On the World Wide Web at http://www.nytimes.com (9 August 2002).

Schiebinger, Londa L. 1993. *Nature's Body: Gender in the Making of Modern Science.* Boston : Beacon Press.

Schiff, Michel, and Richard Lewontin. 1986. *Education and Class: The Irrelevance of IQ Genetic Studies.* Oxford: Clarendon Press.

Schiller, Herbert I. 1969. *Mass Communications and American Empire.* New York: Kelley.

__________. 1989. *Culture Inc.: The Corporate Takeover of Public Expression.* New York: Oxford University Press.

__________. 1996. *Information Inequality: The Deepening Social Crisis in America.* New York: Routledge.

Schippers, Mimi. 2002. *Rockin' Out of the Box: Gender Maneuvering in Alternative Hard Rock.* New Brunswick, NJ: Rutgers University Press.

Schlesinger, Arthur. 1991. *The Disuniting of America: Reflections on a Multicultural Society.* New York: W. W. Norton.

Schlosser, Eric. 1998. "The Prison-Industrial Complex." *The Atlantic Monthly* December. On the World Wide Web at http://www.theatlantic.com/issues/98dec/prisons.htm (29 April 2000).

Schluchter, Wolfgang. 1981 [1980]. *The Rise of Western Rationalism: Max Weber's Developmental History,* Guenther Roth, trans. Berkeley: University of California Press.

Schneider, Barbara, and David Stevenson. 1999. *The Ambitious Generation: America's Teenagers: Motivated But Directionless.* New Haven, CT: Yale University Press.

Schor, Juliet B. 1992. *The Overworked American: The Unexpected Decline of Leisure.* New York: Basic Books.

__________. 1999. *The Overspent American: Why We Want What We Don't Need.* New York: Harper.

———, and Samuel Bowles. 1987. "Employment Rents and the Incidence of Strikes." *The Review of Economics and Statistics* 69: 584–592.

Schrag, Peter. 1997. "The Near-Myth of Our Failing Schools." *The Atlantic Monthly* October. On the World Wide Web at http://www.theatlantic.com/issues/97oct/fail.htm (2 May 2000).

———. 2004. "Bush's Education Fraud." *The American Prospect* 1 February. On the World Wide Web at http://www.prospect.org/web/page.ww?section=root&name=ViewPrint& articleId=6998 (29 March 2005).

Schudson, Michael. 1991. "National News Culture and the Rise of the Informational Citizen." Pp. 265–82 in Alan Wolfe, ed., *America at Century's End.* Berkeley: University of California Press.

———. 1995. *The Power of News.* Cambridge, MA: Harvard University Press.

Schuettler, Darren. 2002. "Earth Summit Bogs Down in Bitter Trade Debate." *Yahoo! Canada News* 28 August. On the World Wide Web at http://ca.news.yahoo.com/020828/5/olia.html (12 February 2002).

Schumer, Charles, and Paul Craig Roberts. 2004. "Exporting jobs is not free trade." *International Herald Tribune* 7 January. On the World Wide Web at http://www.iht.com/articles/123898.html (31 March 2005).

Schwartz, James. 1999. "Oh My Darwin! Who's the Fittest Evolutionary Thinker of Them All?" *Lingua Franca* 9, 8. On the World Wide Web at http://www.mit.edu/~pinker/darwin_wars.html (20 January 2003).

Schwartz, Michael, ed. 1987. *The Structure of Power in America: The Corporate Elite as a Ruling Class.* New York: Holmes & Meier.

Schwartz, Stephen. 2003. *The Two Faces of Islam: The House of Sa'ud from Tradition to Terror.* New York: Doubleday.

Schweingruber, David, and Clark McPhail. 1999. "A Method for Systematically Observing and Recording Collective Action." *Sociological Methods and Research* 27: 451–498.

Scoon-Rogers, Lydia, and Gordon H. Lester. 1995. *Child Support for Custodial Mothers and Fathers: 1991.* Washington, DC: U.S. Bureau of the Census, U.S. Department of Commerce, Economics and Statistics Administration. On the World Wide Web at http://www.census.gov/prod/2/pop/p60/p60-187.pdf (1 May 2000).

Scott, James C. 1998. *Seeing Like a State: How Certain Schemes to Improve the Human Condition Have Failed.* New Haven, CT: Yale University Press.

Scott, Janny. 1998. "Manners and Civil Society." *Journal* 2, 3. On the World Wide Web at http://www.civnet.org/journal/issue7/ftjscott.htm (12 April 2003).

———. 2001. "Study Puts Census Errors at $4 Billion." *The New York Times Online* August 8, 2001. On the World Wide Web at http://www.nytimes.com/2001/08/08/nyregion/08CENS.html?searchpv=day04&pagewanted=print (12 August 2001).

Scott, Peter Dale, and Jonathan Marshall. 1991. *Cocaine Politics: Drugs, Armies, and the CIA in Central America.* Berkeley: University of California Press.

Scott, Shirley Lynn. 2003. "The Death of James Bulger." *Court TV's Crime Library.* On the World Wide Web at http://www.crimelibrary.com/notorious_murders/young/bulger/1.html?sect=10 (23 July 2003).

Scott, Wilbur J. 1990. "PTSD in *DSM-III:* A Case in the Politics of Diagnosis and Disease." *Social Problems* 37: 294–310.

Scully, Diana. 1990. *Understanding Sexual Violence: A Study of Convicted Rapists.* Boston: Unwin Hyman.

"Search Engines of All Countries in All Languages as of August 27, 2002." 2002. *WebMasterAid.com.* On the World Wide Web at http://webmasteraid.com/cgi-bin/d.cgi (25 May 2003).

Seccombe, Wally. 1992. *A Millennium of Family Change: Feudalism to Capitalism in Northwestern Europe.* London: Verso.

Seelye, Katherine Q., and Alison Mitchell. 2002. "Pocketing Soft Money Before Pockets are Sewn Up." *New York Times.* On the Wolrld Wide Web at http://www.nytimes.com (4 March 2002).

Seeman, Neil. 2000. "Capital Questions." *The National Post* 6 August: B1, B6.

Segundo, Juan Luis, S. J. 1976 [1975]. *The Liberation of Theology,* John Drury, trans. Maryknoll, NY: Orbis.

Seiter, Ellen. 1999. *Television and New Media Audiences.* Oxford: Clarendon Press.

Selznick, Philip. 1957. *Leadership in Administration: A Sociological Interpretation.* New York: Harper & Row.

Sen, Amartya. 1981. *Poverty and Famines: An Essay on Entitlement and Deprivation.* Oxford: Clarendon Press.

———. 1990. "More than 100 Million Women are Missing." *New York Review of Books* 20 December: 61–66.

———. 1994. "Population: Delusion and Reality." *New York Review of Books* 41, 15: 62–71.

———. 2001. "Many Faces of Gender Inequality." *The Frontline* 9 November. On the Worldwide Web at http://www.ksg .harvard.edu/gei/Text/Sen-Pubs/Sen_many_faces_of_gender_inequality.pdf (5 August 2003).

The Sentencing Project. 1997. "Americans Behind Bars: U.S. and International Use of Incarceration, 1995." On the World Wide Web at http://www.sentencingproject.org/pubs/tsppubs/9030data.html (29 April 2000).

———. 2001. "U.S. Surpasses Russia as World Leader in Rate of Incarceration." On the World Wide Web at http://www .sentencingproject.org/brief/usvsrus.pdf (27 June 2001).

———. 2005. "Felony Disenfranchisement Laws in the United States." On the World Wide Web at http://www.sentencingproject.org/pdfs/1046.pdf (14 October 2005).

Sewell, William H. 1958. "Infant Training and the Personality of the Child." *American Journal of Sociology* 64: 150–159.

———, and Robert Hauser. 1993. "A Review of the Wisconsin Longitudinal Study of Social and Psychological Factors in Aspirations and Achievements 1963–1992." *CDE Working Paper No. 92–01.* Center for Demography and Ecology, University of Wisconsin-Madison. On the World Wide Web at http://www.ssc.wisc.edu/cde/ cdewp/92-01.pdf (15 May 2003).

"SexQuiz.org." 2000. On the World Wide Web at http://www.sexquiz.org (12 August 2000).

Shakur, Sanyika (a.k.a. Monster Kody Scott). 1993. *Monster: The Autobiography of an L.A. Gang Member.* New York: Penguin.

Shalev, Michael. 1983. "Class Politics and the Western Welfare State." Pp. 27–50 in S. E. Spiro and E. Yuchtman-Yaar, eds. *Evaluating the Welfare State: Social and Political Perspectives.* New York: Academic Press.

Shanker, Stuart. 2002. "The Generativist-Interactionist Debate over Specific Language Impairment: Psycholinguistics at a Crossroad." *American Journal of Psychology* 115: 415–450.

Shapiro, Andrew L. 1992. *We're Number One.* New York: Vintage.

Shapiro, Joseph P. 1993. *No Pity: People with Disabilities Forging a New Civil Rights Movement.* New York: Times Books.

Shattuck, Roger. 1980. *The Forbidden Experiment: The Story of the Wild Boy of Aveyron.* New York: Farrar, Straus, & Giroux.

Shaw, Karen. 2001. "Harry Potter Books: My Concerns." On the World Wide Web at http://www.shaw.ca/investors/Annual_Report/01/ShawAR.pdf (17 May 2002).

Shaw, Martin. 2000. *Theory of the Global State: Globality as Unfinished Revolution.* Cambridge: Cambridge University Press.

Shea, Christopher. 1994. "'Gender Gap' on Examinations Shrank Again This Year." *Chronicle of Higher Education* 41, 2: A54.

Shea, Sarah E., Kevin Gordon, Ann Hawkins, Janet Kawchuk, and Donna Smith. 2000. "Pathology in the Hundred Acre Wood: A Neurodevelopmental Perspective on A. A. Milne." *Canadian Medical Association Journal* 2000,163 (12):1557–9. On the World Wide Web at http://www.cma.ca/cmaj/vol-163/issue-12/1557.htm (12 December 2000).

Shekelle, Paul G. 1998. "What Role for Chiropractic in Health Care?" *New England Journal of Medicine* 339: 1074–1075.

Shell Hydrogen. 2003. "Showa Shell Opens Tokyo's First Liquid Hydrogen Station." http://www.shell.com/home/Framework?siteId=hydrogen-en&FC1=&FC2=&FC3=%2Fhydrogen-en%2Fhtml%2Fiwgen%2Fnews_and_library%2Fpressreleases%2F2003%2Fshowa_station_1806_1100.html&FC4=&FC (31 July 2003).

Shelton, Beth Anne, and Daphne John. 1996. "The Division of Household Labor." *Annual Review of Sociology* 22: 299–322.

Sherif, M., L. J. Harvey, B. J. White, W. R. Hood, and C. W. Sherif. 1988 [1961]. *The Robber's Cave Experiment: Intergroup Conflict and Cooperation.* Middletown, CT: Wesleyan University Press.

Sherkat, Darren E. 1998. "Counterculture or Continuity? Competing Influences on Baby Boomers' Religious Orientations and Participation." *Social Forces* 76: 1087–1115.

__________, and Christopher G. Ellison. 1999. "Recent Developments and Current Controversies in the Sociology of Religion." *Annual Review of Sociology* 25: 363–394.

Sherrill, Robert. 1990. "The Looting Decade: S&Ls, Big Banks and Other Triumphs of Capitalism." *The Nation* 251, 17: 19 November: 589–623.

__________. 1997. "A Year in Corporate Crime." *The Nation* 7 April: 11–20.

Shibutani, Tamotsu. 1966. *Improvised News: A Sociological Study of Rumor.* Indianapolis, IN: Bobbs-Merrill.

Shipler, David K. 1997. *A Country of Strangers: Blacks and Whites in America.* New York: Knopf.

"Short Guys Finish Last." 1995–96. *The Economist,* 23 December – 5 January: 19–22.

Short, James F., Jr., and Fred L. Strodtbeck. 1965. *Group Process and Gang Delinquency.* Chicago: University of Chicago Press.

Shorter, Edward. 1997. *A History of Psychiatry: From the Era of the Asylum to the Age of Prozac.* New York: Wiley.

Shulman, Alix Kates. 1997 [1969]. *Memoirs of an Ex-Prom Queen.* New York: Penguin.

Siegel, Jacob. 1996. "Aging into the 21st Century." Administration on Aging. On the World Wide Web at http://www.aoa.dhhs.gov/aoa/stats/aging21/default.htm (2 May 2000).

Silberman, Steve. 2000. "Talking to Strangers." *Wired* 8, 5: 225–233, 288–296. On the World Wide Web at http://www.wired.com/wired/archive/8.05/translation.html (23 May 2000).

"The Silent Boom." 1998. *Fortune* 7 July: 170–171.

Silverstein, Louise B., and Carl F. Auerbach. 1999. "Reconstructing the Essential Father." *American Psychologist* 54: 397–407.

Simmel, Georg. 1950. *The Sociology of Georg Simmel,* Kurt H. Wolff, trans and ed. New York: Free Press.

Simon, Jonathan. 1993. *Poor Discipline: Parole and the Social Control of the Underclass, 1890–1990.* Chicago: University of Chicago Press.

Simons, Ronald L., Chyi-In Wu, Christine Johnson, and Rand D. Conger. 1995. "A Test of Various Perspectives on the Intergenerational Transmission of Domestic Violence." *Criminology* 33: 141–160.

Sjöberg, Gideon. 1960. *The Preindustrial City: Past and Present.* New York: Free Press.

Skinner, B. F. 1953. *Science and Human Behavior.* New York: Macmillan.

Skocpol, Theda. 1979. *States and Revolutions: A Comparative Analysis of France, Russia, and China.* Cambridge: Cambridge University Press.

__________. 1996. *Boomerang: Clinton's Health Security Effort and the Turn Against Government in U.S. Politics.* New York: W. W. Norton.

Skolnick, Arlene. 1991. *Embattled Paradise: The American Family in an Age of Uncertainty.* New York: Basic Books.

Skolnick, Jerome K. 1997. "Tough Guys." *The American Prospect* 30: 86–91. On the World Wide Web at http://www.prospect.org/archives/30/fs30jsko.html (29 April 2000).

Smeeding, Timothy M. 2004. "Public Policy and Economic Inequality: The United States in Comparative Perspective." Working Paper No. 367. Syracuse NY: Maxwell School of Citizenship and Public Affairs, Syracuse University. On the World Wide Web at http://www.lisproject.org/publications/LISwps/367.pdf (30 January 2005).

Smelser, Neil. 1963. *Theory of Collective Behavior.* New York: Free Press.

Smith, Adam. 1981 [1776]. *An Inquiry into the Nature and Causes of the Wealth of Nations,* 2 vols. Indianapolis, IN: Liberty Press.

Smith, Anthony. 1980. *Goodbye Gutenberg: The Newspaper Revolution of the 1980s.* New York: Oxford University Press.

Smith, Christian. 1991. *The Emergence of Liberation Theology: Radical Religion and Social Movement Theory.* Chicago: University of Chicago Press.

__________. 2000. *Christian America? What Evangelicals Really Want.* Berkeley: University of California Press.

Smith, Jackie. 1998. "Global Civil Society? Transnational Social Movement Organizations and Social Capital." *American Behavioral Scientist* 42: 93–107.

Smith, Michael. 1990. "Patriarchal Ideology and Wife Beating: A Test of a Feminist Hypothesis." *Violence and Victims* 5: 257–273.

Smith, Tom W. 1992. "A Methodological Analysis of the Sexual Behavior Questions on the GSS." *Journal of Official Statistics* 8: 309–325.

__________. 1999. "1998 National Gun Policy Survey of the National Opinion Research Center: Research Findings." On the World Wide Web at http://www.norc.uchicago.edu/new/guns98.pdf (22 February 2003).

Smith, Vicki. 1990. *Managing in the Corporate Interest: Control and Resistance in an American Bank.* Berkeley: University of California Press.

Smock, Pamela J. 2000. "Cohabitation in the United States: An Appraisal of Research Themes, Findings, and Implications." *Annual Review of Sociology* 26: 1–20.

Snow, David A., E. Burke Rochford, Jr., Steven K. Worden, and Robert D. Benford. 1986. "Frame Alignment Processes, Micromobilization, and Movement Participation." *American Sociological Review* 51: 464–481.

Snyder, David. 1979. "Collective Violence Processes: Implications for Disaggregated Theory and Research." Pp. 35–61 in Louis Kriesberg, ed. *Research in Social Movements, Conflict and Change: A Research Annual.* Greenwich, CT: JAI Press.

__________, and Charles Tilly. 1972. "Hardship and Collective Violence in France, 1830–1960." *American Sociological Review* 37: 520–532.

Sofsky, Wolfgang. 1997 [1993]. *The Order of Terror: The Concentration Camp,* William Templer, trans. Princeton, NJ: Princeton University Press.

"The Soft Money Boomerang." 2004. *New York Times* 29 December. On the World Wide Web at http://www.nytimes.com (29 December 2004).

Solomon, Charlene Marmer. 1999. "Stressed to the Limit." *Workforce* 78, 9: 48–54.

Sørenson, Aage B. 1992. "Women, Family and Class." *Annual Review of Sociology* 18: 39–61.

Sorenson, Elaine. 1994. *Comparable Worth: Is It a Worthy Policy?* Princeton, NJ: Princeton University Press.

Soule, Sarah A. 1992. "Populism and Black Lynching in Georgia, 1890–1900." *Social Forces* 71: 431–449.

Southern Poverty Law Center. 2005. "Active U.S. Hate Groups in 2003." On the World Wide Web at http://www.splcenter.org/intel/map/hate.jsp (16 February 2005).

Spade, Joan Z. 2001. "Gender and Education in the United States." In Jeanne H. Ballantine and Joan Z. Spade, eds. *Schools and Society: A Sociological Approach to Education.* Belmont, CA: Wadsworth. Pp. 270–278.

Spencer, Herbert. 1975 [1897–1906]. *The Principles of Sociology,* 3rd ed. Westport, CT: Greenwood Press.

Spigel, Lynn. 1992. *Make Room for TV.* Chicago: University of Chicago Press.

Spilerman, Seymour. 1970. "The Causes of Racial Disturbances: A Comparison of Alternative Explanations." *American Sociological Review* 35: 627–649.

__________. 1976. "Structural Characteristics of Cities and the Severity of Racial Disorders." *American Sociological Review* 41: 771–793.

__________. 2000. "Wealth and Stratification Processes." *Annual Review of Sociology* 26: 497–524.

Spitz, René A. 1945. "Hospitalism: An Inquiry into the Genesis of Psychiatric Conditions in Early Childhood." Pp. 53–74 in *The Psychoanalytic Study of the Child,* Vol. 1. New York: International Universities Press.

__________. 1962. "Autoerotism Re-examined: The Role of Early Sexual Behavior Patterns in Personality Formation." Pp. 283–315 in *The Psychoanalytic Study of the Child,* Vol. 17. New York: International Universities Press.

Spitzer, Allan. 1973. "The Historical Problem of Generations." *American Historical Review* 78: 1353–1385.

Spitzer, Steven. 1980. "Toward a Marxian Theory of Deviance." Pp. 175–91 in Delos H. Kelly, ed. *Criminal Behavior: Readings in Criminology.* New York: St. Martin's Press.

Srinivas, M. N. 1952. *Religion and Society among the Coorgs of South India.* Oxford: Oxford University Press.

Stacey, Judith. 1991. *Brave New Families.* New York: Basic Books.

Stack, Carol. 1974. *All Our Kin: Strategies for Survival in a Black Community.* New York: Harper.

Stack, Stephen, and J. Ross Eshleman. 1998. "Marital Status and Happiness: A 17-Nation Study." *Journal of Marriage and the Family* 60: 527–536.

Stanfield, Rochelle L. 1997. "Blending of America." *National Journal* 13 September: 1780–1782.

Staples, Brent. 2004. "Why Some Politicians Need Their Prisons to Stay Full." *New York Times* 27 December. On the World Wide Web at http://www.nytimes.com (27 December 2005).

Starbuck, Gene H. 2002. *Families in Context.* Belmont, CA: Wadsworth.

Stark, Rodney. 1985. *Sociology.* Belmont, CA: Wadsworth.

__________, and William Sims Bainbridge. 1979. "Of Churches, Sects, and Cults: Preliminary Concepts for a Theory of Religious Movements." *Journal for the Scientific Study of Religion* 18: 117–131.

Starr, Paul. 1982. *The Social Transformation of American Medicine.* New York: Basic Books.

__________. 1994 [1992]. *The Logic of Health Care Reform: Why and How the President's Plan Will Work,* rev. ed. New York: Penguin.

"Statistical Summary: America's Major Wars." 2001. On the World Wide Web at http://www.cwc.lsu.edu/cwc/other/stats/warcost.htm (14 March 2005).

Statistics Canada. 2002. "Highlights for: Greater Napanee (Town), Ontario." On the World Wide Web at http://www12.statcan.ca/english/Profil01/Details/details1.cfm?SEARCH=BEGINS&ID=6690&PSGC=35&SGC=3511015&DataType=1&LANG=E&Province=35&PlaceName=napanee&CMA=&CSDNAME=Greater%20Napanee&A=&TypeNameE=Town&Prov= (15 January 2003).

__________. 2003. "Social Support and Mortality among Seniors." *The Daily* May 23. On the World Wide Web at http://www .statcan.ca/Daily/English/030523/d030523a.htm (17 June 2003).

Stearns, Carol Zisowitz, and Peter N. Stearns. 1985. "Emotionology: Clarifying the History of Emotions and Emotional Standards." *American Historical Review* 90: 813–836.

__________, and __________. 1986. *Anger: The Struggle for Emotional Control in America's History.* Chicago: University of Chicago Press.

Steckel, Richard H. 1995. "Stature and the Standard of Living." *Journal of Economic Literature* 33: 1903–1940.

Steele, Claude M. 1992. "Race and the Schooling of Black Americans." *The Atlantic Monthly* April. On the World Wide Web at http://www.theatlantic.com/unbound/flashbks/blacked/steele.htm (2 May 2000).

__________. 1997. "A Threat in Thin Air: How Stereotypes Shape Intellectual Identity and Performance." *American Psychologist* 52: 613–629.

Steinberg, Jacques. 2003. "Of Sheepskins and Greenbacks." *New York Times* February 13: A20.

Steinberg, Stephen. 1989 [1981]. *The Ethnic Myth: Race, Ethnicity, and Class in America,* updated ed. Boston: Beacon Press.

__________. 1995. *Turning Back: The Retreat from Racial Justice in American Thought and Policy.* Boston: Beacon Press.

Steinhart, David. 2000. "Flint's Future: Meaner than a Junkyard Bond." *Financial Post* 12 August: D1, D6.

Stenger, Richard. 2003. "NASA Chief Blasted over Shuttle Memos." On the World Wide Web at http://www.cnn.com/2003/TECH/space/02/27/sprj.colu.memo/index.htm (6 March 2003).

Stephens, W. Richard, Jr. 1999. *Careers in Sociology,* 2nd ed. Boston: Allyn and Bacon.

Sternberg, Robert J. 1986. "A Triangular Theory of Love." *Psychological Review* 93: 119–135.

__________. 1998 [1995]. *In Search of the Human Mind,* 2nd ed. Fort Worth, TX: Harcourt Brace.

Stewart, Abigail, Anne P. Copeland, Nia Lane Chester, Janet E. Malley, and Nicole B. Barenbaum. 1997. *Separating Together: How Divorce Transforms Families.* New York: The Guilford Press.

Stiglitz, Joseph E. 2002. *Globalization and Its Discontents.* New York: W. W. Norton.

Stiker, Henri-Jacques. 1999 [1982]. *A History of Disability,* William Sayers, trans. Ann Arbor: University of Michigan Press.

Stipp, David. 2003. "The Pentagon's Weather Nightmare." *Fortune* 26 January. On the World Wide Web at http://paxhumana.info/article.php3?id_article=400 (29 May 2004).

Stolzenberg, Ross. M. 1990. "Ethnicity, Geography, and Occupational Achievement of Hispanic Men in the United States." *American Sociological Review* 55: 143–154.

Stone, Lawrence. 1977. *The Family, Sex and Marriage in England, 1500–1800.* New York: Harper & Row.

Stopford, John M., and Susan Strange. 1991. *Rival States, Rival Firms: Competition for World Market Shares.* Cambridge: Cambridge University Press.

Stotsky, Sandra. 1999. *Losing Our Language: How Multicultural Classroom Instruction Is Undermining Our Children's Ability to Read, Write, and Reason.* New York: Free Press.

Stouffer, Samuel A., et al. 1949. *The American Soldier,* 4 vols. Princeton, NJ: Princeton University Press.

Straus, Murray A. 1994. "State-to-State Differences in Social Inequality and Social Bonds in Relation to Assaults on Wives in the United States." *Journal of Comparative Family Studies* 25: 7–24.

__________. 1995. "Trends in Cultural Norms and Rates of Partner Violence: An Update to 1992." Pp. 30–33 in Sandra M. Stith and Murray A. Straus, eds. *Understanding Partner Violence: Prevalence, Causes, Consequences, and Solutions.* Minneapolis, MN: National Council on Family Relations.

Strauss, Anselm L. 1993. *Continual Permutations of Action.* New York: Aldine de Gruyter.

Stretesky, Paul, and Michael J. Hogan. 1998. "Environmental Justice: An Analysis of Superfund Sites in Florida." *Social Problems* 45: 268–287.

Sudweeks, Fay, Margaret McLaughlin, and Sheizaf Rafaeli, eds. 1999. *Network and Netplay: Virtual Groups on the Internet.* Menlo Park, CA: AAAI Press.

Sullivan, Amy, Katrina Hedberg, and David W. Fleming. 2000. "Legalized Physician-Assisted Suicide in Oregon—The Second Year." *New England Journal of Medicine* 342, 8. On the World Wide Web at http://www.nejm.org/content/2000/0342/0008/0598.asp (14 August 2000).

Sumner, William Graham. 1940 [1907]. *Folkways.* Boston: Ginn.

"Super Bowl Ratings Up 1 Percent." 2003. *CNN Sports Illustrated* January 27. On the World Wide Web at http://sportsillustrated.cnn.com/football/2003/playoffs/news/2003/01/27/superbowl_ratings_ap (3 May 2003).

Supreme Court of the State of Hawai'i. 1999. "Baehr v. Miike." On the World Wide Web at http://www.state.hi.us/jud/20371.htm (19 August 2000).

"Supreme Court Rejects Census Sampling." 1999. *Catalog Age Weekly.* On the World Wide Web at http://www.catalogagemag.com/content/Weekly/1999/1999012803.HTM (6 May 2000).

"A Survey of Human Rights Law." 1998. *The Economist.* December 5.

Sutherland, Edwin H. 1939. *Principles of Criminology.* Philadelphia: Lippincott.

__________. 1949. *White Collar Crime.* New York: Dryden.

Suttles, Gerald D. 1968. *The Social Order of the Slum: Ethnicity and Territory in the Inner City.* Chicago: University of Chicago Press.

Sweezy, Kate, and Jill Tiefenthaler. 1996. "Do State-Level Variables Affect Divorce Rates?" *Review of Social Economy* 54: 47–65.

Swidler, Ann. 1980. "Love and Adulthood in American Culture." Pp. 120–147 in Neil J. Smelser and Erik H. Erikson, eds. *Themes of Work and Love in Adulthood.* Cambridge, MA: Harvard University Press.

Swiss Re. 2003. *Natural Catastrophes and Reinsurance.* Zurich. On the World Wide Web at http://www.swissre.com (28 July 2003).

__________. 2004. *Sigma: Natural Catastrophes and Man-Made Disasters in 2003.* On the World Wide Web at http://www.swissre.com/ (2 March 2005).

__________. 2005. *Sigma: Natural Catastrophes and Man-Made Disasters in 2004.* On the World Wide Web at http://www.swissre.com/ (2 March 2005).

Sykes, Gresham, and David Matza. 1957. "Techniques of Neutralization: A Theory of Delinquency." *American Sociological Review* 22: 664–670.

Sylwester, Kevin. 2002. "Democracy and Changes in Income Inequality." *International Journal of Business and Economics* 1: 167–178.

Szasz, Andrew, and Michael Meuser. 1997. "Environmental Inequalities: Literature Review and Proposals for New Directions in Research and Theory." *Current Sociology* 45, 3: 99–120.

Szreter, Simon. 1996. *Fertility, Class and Gender in Britain, 1860–1940.* Cambridge: Cambridge University Press.

Tajfel, Henri. 1981. *Human Groups and Social Categories: Studies in Social Psychology.* Cambridge: Cambridge University Press.

Takaki, Ronald. 1989. *Strangers from a Different Shore: A History of Asian Americans.* New York: Penguin.

Tal, Benjamin. 2004. "Assessing US Job Quality." *CIBC World Markets: Economics and Strategy* 21 June. On the World Wide Web at http://research.cibcwm.com/economic_public/download/eqi-us-062004.pdf (1 April 2005).

Tannen, Deborah. 1990. *You Just Don't Understand Me: Women and Men in Conversation.* New York: William Morrow.

__________. 1994a. *Talking from 9 to 5: How Women's and Men's Conversational Styles Affect Who Gets Heard, Who Gets Credit, and What Gets Done at Work.* New York: William Morrow.

__________. 1994b. *Gender and Discourse.* New York: Oxford University Press.

Tarmann, Allison. 2003. "Obesity in the U.S.: Reaching a Critical Mass." On the World Wide Web at http://www.prb.org/Template.cfm?Section=PRB&template=/Content/ContentGroups/Articles/03/Obesity_in_the_U_S_Reaching_a_Critical_Mass.htm (12 February 2003).

Tarrow, Sidney. 1994. *Power in Movement: Social Movements, Collective Action and Politics.* Cambridge: Cambridge University Press.

Tasker, Fiona L., and Susan Golombok. 1997. *Growing Up in a Lesbian Family: Effects on Child Development.* New York: The Guilford Press.

Tavris, Carol. 1992. *The Mismeasure of Woman.* New York: Simon & Schuster.

Taylor, Ronald, ed. 1998 [1994]. *Minority Families in the United States: A Multicultural Perspective.* Upper Saddle River, NJ: Prentice-Hall.

Tec, Nechama. 1986. *When Light Pierced the Darkness: Christian Rescue of Jews in Nazi-Occupied Poland.* New York: Oxford University Press.

Television Bureau of Advertising. 2005. "Gross Domestic Product, Total Ad Volume, and Television Ad Volume 1960-2003." On the World Wide Web at http://www.tvb.org/nav/build_frameset.asp?url=/rcentral/mediatrendstrack/gdpvolume/gdp.asp?c=gdp1 (27 January 2005).

Terry, Jennifer, and Jacqueline Urla, eds. 1995. *Deviant Bodies: Critical Perspectives on Difference in Science and Popular Culture.* Bloomington: Indiana University Press.

"That's AOL Folks . . ." 2000. *CNNfn.* On the World Wide Web at http://cnnfn.com/2000/01/10/deals/aol_warner (2 May 2000).

Thelen, David. 1996. *Becoming Citizens in the Age of Television: How Americans Challenged the Media and Seized Political Initiative during the Iran-Contra Debate.* Chicago: University of Chicago Press.

Theodore, Peter S. 1998. "Heterosexual Masculinity and Homophobia: A Reaction to the Self?" Paper presented at the annual meetings of the American Psychological Association (San Francisco: 17 August).

Thernstrom, Stephan, and Abigail Thernstrom. 1997. *America in Black and White: One Nation, Indivisible.* New York: Simon & Schuster.

Thoits, Peggy A. 1989. "The Sociology of Emotions." *Annual Review of Sociology* 15: 317–342.

Thomas, Keith. 1971. *Religion and the Decline of Magic.* London: Weidenfeld and Nicholson.

Thomas, William Isaac. 1966 [1931]. "The Relation of Research to the Social Process." Pp. 289–305 in Morris Janowitz, ed. *W.I. Thomas on Social Organization and Social Personality.* Chicago: University of Chicago Press.

__________, and Florian Znaniecki. 1958 [1918–20]. *The Polish Peasant in Europe and America: Monograph of an Immigrant Group,* 2 vols., 2nd ed. New York: Dover Publications.

Thompson, E. P. 1967. "Time, Work Discipline, and Industrial Capitalism." *Past and Present* 38: 59–67.

__________. 1968. *The Making of the English Working Class.* Harmondsworth, UK: Penguin.

Thorne, Barrie. 1993. *Gender Play: Girls and Boys in School.* New Brunswick, NJ: Rutgers University Press.

"Three Million Dead in Congo War: Relief Group." 2003. *CBC News* April 9. On the World Wide Web at http://www.cbc.ca/stories/2003/04/08/congo030407 (7 June 2003).

Thurow, Lester. 1999. "Building Wealth." *The Atlantic Monthly* June. On the World Wide Web at http://www.theatlantic.com/issues/99jun/9906thurow.htm (2 May 2000).

Tienda, Marta, and Ding-Tzann Lii. 1987. "Minority Concentration and Earnings Inequality: Blacks, Hispanics, and Asians Compared." *American Journal of Sociology* 93: 141–165.

Tilly, Charles. 1978. *From Mobilization to Revolution.* Reading, MA: Addison-Wesley.

__________. 1979a. "Collective Violence in European Perspective." Pp. 83–118 in H. Graham and T. Gurr, eds. *Violence in America: Historical and Comparative Perspective,* 2nd ed. Beverly Hills, CA: Sage.

__________. 1979b. "Repertoires of Contention in America and Britain, 1750–1830." Pp. 126–55 in Mayer N. Zald and John D. McCarthy, eds. *The Dynamics of Social Movements: Resource Mobilization, Social Control, and Tactics.* Cambridge, MA: Winthrop Publishers.

__________. 2002. "Violence, Terror, and Politics as Usual." Unpublished paper, Department of Sociology, Columbia University.

__________, Louise Tilly, and Richard Tilly. 1975. *The Rebellious Century, 1830–1930.* Cambridge, MA: Harvard University Press.

Tilly, Chris. 1996. *Half a Job: Bad and Good Part-Time Jobs in a Changing Labor Market.* Philadelphia: Temple University Press.

Tillyard, E. M. W. 1943. *The Elizabethan World Picture.* London: Chatto and Windus.

"The *Times* and Iraq." 2004. *New York Times* 26 May. On the World Wide Web at www.nytimes.com (3 January 2005).

Tkacik, Maureen. 2002. "The Return of Grunge." *Wall Street Journal* December 11: B1, B10.

Toffler, Alvin. 1990. *Powershift: Knowledge, Wealth, and Violence at the Edge of the 21st Century.* New York: Bantam.

Tokar, Brian, ed. 2001. *Redesigning Life: The Worldwide Challenge to Genetic Engineering.* Montreal: McGill-Queen's University Press.

Tolnay, Stewart E., and E. M. Beck. 1995. *A Festival of Violence: An Analysis of Southern Lynchings, 1882–1930.* Urbana: University of Illinois Press.

Tong, Rosemarie. 1989. *Feminist Thought: A Comprehensive Introduction.* Boulder, CO: Westview.

Tönnies, Ferdinand. 1957 [1887]. *Community and Society.* Charles P. Loomis, ed. and trans. East Lansing, MI: Michigan State University Press.

__________. 1988 [1887]. *Community and Society (Gemeinschaft und Gesselschaft).* New Brunswick, NJ: Transaction.

Tonry, Michael. 1995. *Malign Neglect: Race, Crime, and Punishment in America.* New York: Oxford University Press.

Tooby, John, and Leda Cosmides. 1992. "The Psychological Foundations of Culture." Pp. 19–136 in Jerome Barkow, Lea Cosmides, and John Tooby, eds. *The Adapted Mind: Evolutionary Psychology and the Generation of Culture.* New York: Oxford University Press.

Toor, Rachel. 2001. *Admissions Confidential: An Insider's Account of the Elite College Selection Process.* New York: St. Martin's Press.

"Top 20 Countries with the Highest Number of Internet Users." 2005. On the World Wide Web at http://www.internetworldstats.com/top20.htm (13 February 2005).

Tornquist, Cynthia. 1998. "Students Head Back to Decaying Classrooms." *Cnn.com* 30 August. On the World Wide Web http://www.cnn.co.uk/US/9808/30/hazardous.schools (11 August 2000).

Torret, Barbara Boyle, and Carl Haub. 2004. "Why Do Canadians Outlive Americans?" On the World Wide Web at http://www.prb.org/Template.cfm?Section=PRB&template=/ContentManagement/ContentDisplay.cfm&ContentID=11868 (14 March 2005).

Traub, James. 2000. "What No School Can Do." *New York Times Magazine* 16 January: 52–57, 68, 81, 90–91.

Travers, Jeffrey, and Stanley Milgram. 1969. "An Experimental Study of the Small World Problem." *Sociometry* 32: 425–443.

Troeltsch, Ernst. 1931 [1923]. *The Social Teaching of the Christian Churches,* Olive Wyon, trans. 2 vols. London: George Allen and Unwin.

Troy, Leo. 1986. "The Rise and Fall of American Trade Unions: The Labor Movement from FDR to RR." Pp. 75–109 in Seymour Martin Lipset, ed. *Unions in Transition: Entering the Second Century.* San Francisco: ICS Press.

Tschannen, Olivier. 1991. "The Secularization Paradigm: A Systematization." *Journal for the Scientific Study of Religion* 30: 395–415.

Tsutsui, William M. 1998. *Manufacturing Ideology: Scientific Management in Twentieth-Century Japan.* Princeton, NJ: Princeton University Press.

Tuljapurkar, Shripad, Nan Li, and Carl Boe. 2000. "A Universal Pattern of Mortality Decline in the G7 Countries." *Nature* 405: 789–792.

Tumin, Melvin. 1953. "Some Principles of Stratification: A Critical Analysis." *American Sociological Review* 18: 387–394.

Turkle, Sherry. 1995. *Life on the Screen: Identity in the Age of the Internet.* New York: Simon & Schuster.

Turnbull, Colin M. 1961. *The Forest People.* New York: Doubleday.

Turner, Bryan S. 1986. *Citizenship and Capitalism: The Debate over Reformism.* London: Allen & Unwin.

__________. 1996 [1984]. *The Body and Society: Explorations in Social Theory,* 2nd ed. London: Sage.

Turner, Ralph H., and Lewis M. Killian. 1987 [1957] *Collective Behavior,* 3rd ed. Englewood Cliffs, NJ: Prentice-Hall.

Tuxill, John, and Chris Bright. 1998. "Losing Strands in the Web of Life." Pp. 41–58 in Lester R. Brown, Christopher Flavin, Hilary French, et al. *State of the World 1998.* New York: W. W. Norton.

Twenge, Jean M. 1997. "Changes in Masculine and Feminine Traits Over Time: A Meta-analysis." *Sex Roles* 36: 305–325.

Twitchell, James B. 1999. *Lead Us into Temptation: The Triumph of American Materialism.* New York: Columbia University Press.

"2002 Marks Record Year at the Box Office." 2003. *Movies.com.* On the World Wide Web at http://movies.go.com/news/2003/1/endoftheyearroundup010203.html (25 May 2003).

Tyree, Andrea, Moshe Semyonov, and Robert W. Hodge. 1979. "Gaps and Glissandos: Inequality, Economic Development, and Social Mobility in 24 Countries." *American Sociological Review* 44: 410–424.

Ungar, Sheldon. 1992. "The Rise and (Relative) Decline of Global Warming as a Social Problem." *Sociological Quarterly* 33: 483–501.

__________. 1995. "Social Scares and Global Warming: Beyond the Rio Convention." *Society and Natural Resources* 8: 443–456.

__________. 1998. "Bringing the Issue Back In: Comparing the Marketability of the Ozone Hole and Global Warming." *Social Problems* 45: 510–527.

__________. 1999. "Is Strange Weather in the Air? A Study of U.S. National Network News Coverage of Extreme Weather Events." *Climatic Change* 41: 133–150.

Union of Concerned Scientists. 1993. "World Scientists' Warning to Humanity." On the World Wide Web at http://www.englib.cornell.edu/scitech/u95/warn.html (2 May 2000).

United Airlines. 2003. "Flight Attendant History." On the World Wide Web at http://www.ual.com/page/article/0,1360,3191,00.html (6 April 2003).

United Nations. 1994. *Human Development Report 1994.* New York: Oxford University Press.

__________. 1997. "Percentage of Population Living in Urban Areas in 1996 and 2030." On the World Wide Web at http://www .undp.org/popin/wdtrends/ura/uracht1.htm (2 May 2000).

__________. 1998a. "Below-Replacement Fertility." On the World Wide Web at http://www.popin.org/pop1998/7.htm (4 July 1999).

__________. 1998b. *Human Development Report 1998.* New York: Oxford University Press.

__________. 1998c. "Universal Declaration of Human Rights." On the World Wide Web at http://www.un.org/Overview/rights.html (25 January 2003).

__________. 2004. *Human Development Report 2004.* On the World Wide Web at http://hdr.undp.org/reports/global/2004/ (26 January 2005).

UNESCO (United Nations Educational, Scientific and Cultural Organization). 2001a. "Indicators on Illiteracy." On the World Wide Web at http://www.un.org/depts/unsd/social/literacy.htm (29 November 2001).

__________. 2001b. "World Culture Report 2000—Cultural Trade and Communications Trends: International Tourism." On the World Wide Web at http://www.unesco.org/culture/worldreport/html_eng/stat2/table18.pdf (6 February 2003).

__________. 2002. "Education Goals Remain Elusive in More Than 70 Countries." On the World Wide Web at http://portal.unesco.org/uis/ev.php?URL_ID=5175&URL_DO=DO_TOPIC&URL_SECTION=201&reload=1044561680 (6 February 2003).

"The US and Foreign Aid Assistance." 2005. On the World Wide Web at http://www.globalissues.org/TradeRelated/Debt/USAid.asp?so=p2003#oda (2 February 2005).

"The United States of the World." 2003. *Globe and Mail* March 8: F1.

U.S. Administration on Aging. 1999. "Older Population by Age: 1900 to 2050." On the World Wide Web at http://www.aoa.dhhs.gov/aoa/stats/AgePop2050.html (2 May 2000).

U.S. Census Bureau. 1993. "We the American . . . Foreign Born." On the World Wide Web at http://www.census.gov/apsd/wepeople/we-7.pdf (29 April 2000).

———. 1995. *Statistical Abstract of the United States 1995.* On the World Wide Web at http://www.census.gov/prod/1/gen/95statab/pop.pdf (9 June 2003).

———. 1997. "Country of Origin and Year of Entry into the US of the Foreign Born, by Citizenship Status: March 1997." On the World Wide Web at http://www.bls.census.gov/cps/pub/1997/for_born.htm (29 April 2000).

———. 1998a. "Population of the 100 Largest Urban Places: 1950." On the World Wide Web at http://www.census.gov/population/documentation/twps0027/tab18.txt (2 May 2000).

———. 1998b. "Total Midyear Population for the World: 1950–2050." On the World Wide Web at http://www.census.gov/ipc/www/worldpop.html (2 May 2000).

———. 1999a. "Households, by Type: 1940 to Present." On the World Wide Web at http://www.census.gov/population/socdemo/hh-fam/htabHH-1.txt (1 May 2000).

———. 1999b. "Money Income in the United States." *Current Population Reports, 1998.* On the World Wide Web at http://www.census.gov/prod/99pubs/p60-206.pdf (6 June 2000).

———. 1999c. *1997 Census of Agriculture: Agricultural Economics and Land Ownership Survey.* On the World Wide Web at http://www.nass.usda.gov/census/census97/aelos/aelos.htm (2 August 2003).

———. 1999d. "Population Estimates for Cities with Populations of 100,000 and Greater." On the World Wide Web at http://www.census.gov/population/estimates/metro-city/SC100K98-T1-DR.txt (2 May 2000).

———. 1999e. "Region and Country or Area of Birth of the Foreign-Born Population, With Geographic Detail Shown in Decennial Census Publications of 1930 or Earlier: 1850 to 1930 and 1960 to 1999." On the World Wide Web at http://www.census.gov/population/www/documentation/twps0029/tab04.html (29 April 2000).

———. 2000a. "Historical National Population Estimates: July 1, 1900 to July 1, 1999." On the World Wide Web at http://www.census.gov/population/estimates/nation/popclockest.txt (29 April 2000).

———. 2000b. "IDB Population Pyramids." On the World Wide Web at http://www.census.gov/ipc/www/idbpyr.html (4 August 2003).

———. 2000c. "National Population Projections." On the World Wide Web at http://www.census.gov/population/www/projections/natproj.html (28 April 2000).

———. 2000d. "Percentage of Industry Statistics Accounted for by Largest Companies: 1992." On the World Wide Web at http://www.census.gov:80/mcd/mancen/download/mc92cr.sum (30 April 2000).

———. 2000e. "Resident Population Estimates of the United States by Age and Sex: April 1, 1990 to July 1, 1999, with Short-Term Projection to March 1, 2000." On the World Wide Web at http://www.census.gov/population/estimates/nation/intfile2-1.txt (29 April 2000).

———. 2000f. "Resident Population Estimates of the United States by Sex, Race, and Hispanic Origin: April 1, 1990 to July 1, 1999, with Short-Term Projection to March 1, 2000." On the World Wide Web at http://www.census.gov/population/estimates/nation/intfile3-1.txt (25 May 2000).

———. 2000g. "Selected Characteristics of Households, by Total Money Income in 1998." On the World Wide Web at http://ferret.bls/census.gov/macro/031999/hhinc/new01_001.htm (7 June 2000).

———. 2000h. "State Population by Rank, Percent Change, and Population Density, 1980 to 1999." Statistical Abstract of the United States: 2000: 24. On the World Wide Web at http://www.census.gov/prod/2001pubs/statab/sec01.pdf (26 November 2001).

———. 2001a. *Mapping Census 2000: The Geography of U.S. Diversity.* On the World Wide Web at http://www.census.gov/population/cen2000/atlas/censr01-1.pdf (12 August 2003).

———. 2001b. "Profile of the Foreign-Born Population in the United States: 2000." *Current Population Reports.* On the World Wide Web at http://www.census.gov/prod/2002pubs/p23-206.pdf (19 June 2003).

———. 2001c. *Statistical Abstract of the United States 2001.* On the World Wide Web at http://www.census.gov/prod/ 2002pubs/01statab/pop.pdf (9 June 2003).

———. 2001d. "Table 1.1: Population by Sex, Age, and Citizenship Status: March 2000 (Numbers in Thousands)." On the World Wide Web at http://www.census.gov/population/socdemo/foreign/p20-534/ tab0101.txt (19 June 2003).

———. 2002a. "Selected Characteristics of Households, by Total Money Income in 2000." On the World Wide Web at http://ferret.bls.census.gov/macro/032002/hhinc/new01_001.htm (8 March 2003).

———. 2002b. *Statistical Abstract of the United States 2002.* On the World Wide Web at http://www.census.gov/prod/ www/statistical-abstract-02.html (19 June 2003).

———. 2002c. "Table 18.2: Population by Region, Sex, and Hispanic Origin Type, with Percent Distribution by Region: March 2002." On the World Wide Web at http://www .census.gov/population/socdemo/hispanic/ppl-165/tab18-2.txt (20 June 2003).

———. 2002d. "Table FINC-02. Age of Reference Person, by Total Money Income in 2001, Type of Family, Race and Hispanic Origin of Reference Person." On the World Wide Web at http://ferret.bls.census.gov/macro/032002/faminc/new02_000.htm (19 June 2003).

———. 2002e. "Table 1.1: Population by Sex, Age, and Citizenship Status: March 2002 (Numbers in Thousands)." On the World Wide Web at http://www.census.gov/population/ socdemo/foreign/ppl-162/tab01-01.txt (19 June 2003).

__________. 2002f. "Table 1. United States—Race and Hispanic Origin: 1790 to 1990." On the World Wide Web at http://www.census.gov/population/documentation/twps0056/tab01.pdf (20 June 2003).

__________. 2002g. "Total Midyear Population for the World: 1950–2050." On the World Wide Web at http://blue.census.gov/ipc/www/worldpop.html (1 December 2002).

__________. 2003. "DP-1. Profile of General Demographic Characteristics: 2000." On the World Wide Web at http://factfinder.census.gov/servlet/QTTable?ds_name=DEC_2000_SF1_U&geo_id=01000US&qr_name=DEC_2000_SF1_U_DP1 (15 February 2003).

__________. 2003a. "2000 Population Distribution in the United States." On the World Wide Web at http://www.census.gov/geo/www/mapGallery/images/2k_night.jpg (3 August 2003).

__________. 2004a. "Annual Estimates of the Population for Incorporated Places Over 100,000, Ranked by July 1, 2003 Population: April 1, 2000 to July 1, 2003." On the World Wide Web at http://www.census.gov/popest/cities/ SUB-EST2003.html (4 March 2005).

__________. 2004b. "Historical Poverty Tables—People." On the World Wide Web at http://www.census.gov/hhes/poverty/histpov/perindex.html (30 January 2005).

__________. 2004c. "Poverty Thresholds 2003." On the World Wide Web at http://www.census.gov/hhes/poverty/threshld/thresh03.html (31 January 2005).

__________. 2004d. *Statistical Abstract of the United States, 2004–2005*. Washington D.C. On the World Wide Web at http:// www.census.gov/prod/www/statistical-abstract-04.html (21 February 2005).

__________. 2004e. "Table 9. Poverty of People, by Region: 1959 to 2003 (Numbers in Thousands)." On the World Wide Web at http://www.census.gov/hhes/poverty/histpov/hstpov9.html (30 January 2005).

__________. 2005. "Countries Ranked by Population: 2006." On the World Wide Web at http://www.census.gov/cgi-bin/ipc/idbrank.pl (11 March 2005).

U.S. Code. 1998. "Title 18—Crimes and Criminal Procedure. Part I. Crimes. Chapter 7, Assault." On the World Wide Web at http://www.fgm.org/USCode.html (20 January 2003).

U.S. Department of Commerce. 1997. *News*. September 29. On the World Wide Web at http://www.census.gov/Press-Release/cb97-162.html (15 June 2003).

__________. 2002a. *A Nation Online: How Americans are Expanding Their Use of the Internet*. Washington DC. On the World Wide Web at http://www.ntia.doc.gov/ntiahome/dn (5 June 2003).

U.S. Department of Education. 1997. *Pursuing Excellence: A Study of U.S. Fourth Grade Mathematics and Science Achievement in International Context*. On the World Wide Web at http://nces.ed.gov/timss/report/97255-01.html (2 May 2000).

__________. 1998. *Pursuing Excellence: A Study of U.S. Twelfth Grade Mathematics and Science Achievement in International Context*. On the World Wide Web at http://nces.ed.gov/timss/twelfth/index.html (2 May 2000).

__________. 1999. *The Condition of Education 1999*. On the World Wide Web at http://nces.ed.gov/pubs99/condition99 (2 May 2000).

__________. 2000. *The Digest of Education Statistics 1999*. Washington, DC: National Center for Education Statistics. On the World Wide Web at http://www .nces.ed.gov/pubs2000/digest99 (15 January 2001).

U.S. Department of Health and Human Services. 1999. *Healthy People 2000: National Health Promotion and Disease Prevention Objectives*. Hyattsville, MD: Centers for Disease Control and Prevention, National Center for Health Statistics. On the World Wide Web at http://www.cdc.gov/nchswww/data/hp2k99.pdf (29 April 2000).

__________. 2002. "Child Health USA 2002." On the World Wide Web at http://www.mchb.hrsa.gov/chusa02/main_pages/page_12.htm (25 April 2003).

__________. 2004a. *Health, United States, 2004*. Hyattsville MD: Centers for Disease Control and Prevention, National Center for Health Statistics. On the World Wide Web at http://www.cdc.gov/nchs/data/hus/hus04trend.pdf#046 (24 January 2005).

__________. 2004b. "Youth Risk Behavior Surveillance—United States, 2003." *Morbidity and Mortality Weekly Report* 53, 21 May. On the World Wide Web at http://www.cdc.gov/ mmwr/PDF/SS/SS5302.pdf (24 February 2005).

U.S. Department of Labor, Bureau of Labor Statistics. 1998. "Table A-1. Employment Status of the Civilian Population by Sex and Age." On the World Wide Web at http://stats .bls.gov/webapps/legacy/cpsatab1.htm (30 April 2000).

__________. 1999a. "Comparative Civilian Labor Force Statistics: Ten Countries, 1959–1998." On the World Wide Web at ftp://ftp.bls.gov/pub/special.requests/ForeignLabor/flslforc.txt (30 April 2000).

__________. 1999b. "Employment Status of the Civilian Population by Sex and Age." On the World Wide Web at http://stats.bls .gov/news.release/empsit.t01.htm (30 April 2000).

__________. 1999c. *Report on the American Workforce*. Washington D.C. On the World Wide Web at http://www.bls.gov/opub/rtaw/pdf/rtaw1999.pdf (26 January 2003).

__________. 1999d. "Table 5. Civilian Labor Force by Sex, Age, Race, and Hispanic Origin, 1978, 1988, 1998, and projected 2008." On the World Wide Web at http://stats .bls.gov/emplt985 .htm (30 April 2000).

__________. 1999e. "Usual Weekly Earnings Summary." On the World Wide Web at http://stats.bls.gov/news.release/wkyeng.nws.htm (30 April 2000).

__________. 1999f. "Value of the Federal Minimum Wage." On the World Wide Web at http://www.dol.gov/dol/esa/public/minwage/chart2.htm (30 April 2000).

__________. 2000a. "Labor Force Statistics from the Current Population Survey." On the World Wide Web at http://146.142.4.24/cgi-bin/surveymost (31 December 2000).

__________. 2000b. "Union Members Summary." On the World Wide Web at http://stats.bls.gov/news.release/union2.nws.htm (30 April 2000).

__________. 2000c. "Union Membership Data from the National Directory Series." On the World Wide Web at ftp://146.142 .4.23/pub/special.requests/collbarg/unmem.txt (1 August 2000).

__________. 2000d. "Work Stoppages Involving 1,000 Workers or More, 1947–2000." On the World Wide Web at http://stats .bls.gov/news.release/wkstp.t01.htm (28 April 2000).

__________. 2001. "Union Members Summary." On the World Wide Web at http://stats.bls.gov/news.release/union2.nws.htm (22 March 2000).

———. 2002. "Median usual weekly earnings of full-time wage and salary workers by detailed occupation and sex." On the World Wide Web at ftp://ftp.bls.gov/pub/special.requests/lf/aat39.txt (16 April 2003).

———. 2003a. "Comparative Civilian Labor Force Statistics, Ten Countries, 1959–2002." Bureau of Labor Statistics, Office of Productivity and Technology. On the World Wide Web at ftp://ftp.bls.gov/pub/special.requests/ForeignLabor/flslforc.txt (21 May 2003).

———. 2003b. "Labor force data files." On the World Wide Web at http://www.bls.gov/emp/emplab1.htm (16 April 2003).

———. 2003c. "Persons at work in agriculture and nonagricultural industries by hours of work." Bureau of Labor Statistics. On the World Wide Web at ftp://ftp.bls.gov/ pub/special.requests/lf/aat19.txt (21 May 2003).

———. 2003d. "Table A-1. Employment Status of the Civilian Population by Sex and Age." On the World Wide Web at http://www.bls.gov/news.release/empsit.t01.htm (8 June 2003).

———. 2003e. "20 Leading Occupations of Employed Women, 2001 Annual Averages." On the World Wide Web at http://www .dol.gov/wb/wb_pubs/20lead2001.htm (17 April 2003).

———. 2003f. "Union Members Summary." On the World Wide Web at http://www.bls.gov/news.release/union2.nr0.htm (8 August 2003).

———. 2003g. "Work Stoppage Data." On the World Wide Web at http://data.bls.gov/labjava/outside.jsp?survey=ws (8 August 2003).

———. 2004a. "Median weekly earnings of full-time wage and salary workers by detailed occupation and sex." On the World Wide Web at ftp://ftp.bls.gov/pub/special.requests/lf/aat39.txt (23 February 2005).

———. 2004b. "Where Can I Find the Unemployment Rate for Previous Years?" On the World Wide Web at http://www.bls.gov/cps/prev_yrs.htm (3 March 2005).

———. 2004c. National Compensation Survey: Occupational Wages in the United States, July 2003 Supplementary Tables. Washington D.C.: Bureau of Labor Statistics. On the World Wide Web at http://www.bls.gov/ncs/ocs/sp/ncbl0636.pdf (30 January 2005).

———. 2005a. "Consumer Price Index." On the World Wide Web at ftp://ftp.bls.gov/pub/special.requests/cpi/cpiai.txt (30 March 2005).

———. 2005b. "Employed and unemployed full- and part-time workers by age, sex, race, and Hispanic or Latino ethnicity (in thousands)." On the World Wide Web at ftp://ftp.bls.gov/pub/special.requests/lf/aat8.txt (31 March 2005).

———. 2005c. "Labor Force Statistics from the Current Population Survey." On the World Wide Web at http://www.bls.gov/cps/home.htm (3 March 2005).

———. 2005d. "Union Members Summary." On the World Wide Web at http://www.bls.gov/news.release/union2 .nr0.htm (3 March 2005).

———. 2005e. "Work Stoppage Data." On the World Wide Web at http://data.bls.gov/PDQ/outside.jsp?survey=ws (3 March 2005).

U.S. Department of State. 1997. *1996 Patterns of Global Terrorism Report.* On the World Wide Web at http://www.state.gov/www/global/terrorism/1996Report/1996index.html#table (3 June 2003).

———. 2003. *Patterns of International Terrorism 2002.* On the World Wide Web at http://www.state.gov/s/ct/rls/pgtrpt/2002/pdf (3 June 2003).

———. 1998. *National Air Pollutant Emission Trends Report, 1900–1996.* On the World Wide Web at http://www.epa.gov/ttn/chief/trends96/chapter5.pdf (29 April 1999).

———. 2000. *National Air Pollutant Emission Trends, 1900–1998.* On the World Wide Web at http://www.epa.gov/ttn/chief/trends98/emtrnd.html (3 August 2000).

U.S. Federal Bureau of Investigation. 1997. *Hate Crime Statistics 1997.* On the World Wide Web at http://www.fbi.gov/ucr/hc97all.pdf (29 April 2000).

———. 1999. *Uniform Crime Reports for the United States 1998.* On the World Wide Web at http://www.fbi.gov/ ucr/98cius.htm (25 May 2000).

———. 2002a. *Crime in the United States, 2001.* On the World Wide Web at http://www.fbi.gov/ucr/cius_01/01crime.pdf (15 February 2003).

———. 2002b. "Hate Crime Fact Sheet." On the World Wide Web at http://www2.fbi.gov/pressrel/pressrel02/2001hc .htm (7 August 2003).

———. 2003. *Uniform Crime Reports 2003.*On the World Wide Web at http:// www.fbi.gov/ucr/03cius.htm (11 March 2005).

———. 2004a. *Hate Crime Statistics 2003.* Washington D.C. On the World Wide Web at http://www.fbi.gov/ucr/03hc.pdf (21 February 2005).

———. 2004b. *Uniform Crime Reports 2003.* On the World Wide Web at http://www.fbi.gov/ucr/03cius.htm (23 February 2005).

U.S. Information Agency. 1998–1999. *The People Have Spoken: Global Views of Democracy,* 2 vols. Washington, DC: Office of Research and Media Reaction.

U.S. Secretary of State. 1999. "Summary and Highlights: International Affairs (Function 150): Fiscal Year 2000 Budget Request." On the World Wide Web at http://www.state.gov/www/budget/2000_budget.html (2 May 2000).

United Nations. 2004. *Human Development Report 2004.* On the World Wide Web at http://hdr.undp.org/reports/global/2004/ (23 February 2004).

University of Texas. 2003. "Muslim Distribution." Perry-Castañeda Library Map Collection http://www.lib.utexas.edu/maps/world_maps/muslim_distribution.jpg (7 May 2003).

University of Virginia. 2003. "The Oracle of Bacon at Virginia." On the World Wide Web at: http://www.cs.virginia.edu/oracle (13 March 2003).

Unschuld, Paul. 1985. *Medicine in China.* Berkeley: University of California Press.

Urban Institute. 1998. "Policy Challenges Posed by the Aging of America." On the World Wide Web at http://www.urban.org/health/oldpol.html (2 May 2000).

Useem, Bert. 1998. "Breakdown Theories of Collective Action." *Annual Review of Sociology* 24: 215–238.

Useem, Michael. 1984. *The Inner Circle: Large Corporations and the Rise of Business Political Activity in the U.S. and U.K.* New York: Oxford University Press.

Valelly, Richard. 1999. "Voting Rights in Jeopardy." *The American Prospect* 46: 43–49 On the World Wide Web at http://www.prospect.org/archives/46/46valelly.html (14 January 2001).

Valocchi, Steve. 1996. "The Emergence of the Integrationist Ideology in the Civil Rights Movement." *Social Problems* 43: 116–130.

Van de Kaa, Dirk. 1987. "Europe's Second Demographic Transition." *Population Bulletin* 42, 1: 1–58.

van Kesteren, John, Pat Mayhew, Paul Nieuwbeerta. 2001. "Criminal Victimisation in Seventeen Industrialised Countries: Key Findings from the 2000 International Crime Victims Survey." On the World Wide Web at http://www.minjust.nl:8080/b_organ/wodc/reports/ob187i.htm (11 March 2005).

Vanneman, Reeve, and Lynn Weber Cannon. 1987. *The American Perception of Class.* Philadelphia: Temple University Press.

Veblen, Thorstein. 1899. *The Theory of the Leisure Class.* On the World Wide Web at http://socserv2.socsci.mcmaster.ca/~econ/ugcm/3ll3/veblen/leisure/index.html (29 April 2000).

Verba, Sidney, Kay Lehman Schlozman, and Henry E. Brady. 1997. "The Big Tilt: Participatory Inequality in America." *The American Prospect* 32: 74–80.

Vernarec, Emil. 2000. "Depression in the Work Force: Seeing the Cost in a Fuller Light." *Business and Health* 18, 4: 48–55.

Veugelers, John. 1997. "Social Cleavage and the Revival of Far Right Parties: The Case of France's National Front." *Acta Sociologica* 40: 31–49.

Vincent, David. 2000. *The Rise of Mass Literacy: Reading and Writing in Modern Europe.* Cambridge: Polity Press.

Vitale, Ami. 2000. "The War Next Door." *Saturday Night* 3 June: 48–54.

Vogel, David J. 1996. "The Study of Business and Politics." *California Management Review* 38, 3: 146–165.

Vogel, Ezra F., ed. 1975. *Modern Japanese Organization and Decision-Making.* Berkeley: University of California Press.

Vygotsky, Lev S. 1987. *The Collected Works of L. S. Vygotsky,* Vol. 1, N. Minick, trans. New York: Plenum.

Wade, Robert, and Martin Wolf. 2002. "Are Global Poverty and Inequality Getting Worse?" *Prospect* March: 16–21.

Wagner, Tony. 2002. *Making the Grade: Reinventing America's Schools.* New York: RoutledgeFalmer.

Wald, Matthew L., and John Schwartz. 2003. "Alerts Were Lacking, NASA Shuttle Manager Says." *New York Times* July 23. On the World Wide Web at http://www.nytimes.com (23 July 2003).

Waldfogel, Jane. 1997. "The Effect of Children on Women's Wages." *American Sociological Review* 62: 209–217.

Wallace, Anthony F. C. 1993. *The Long, Bitter Trail: Andrew Jackson and the Indians.* New York: Hill & Wang.

Wallace, James. 1997. *Overdrive: Bill Gates and the Race to Control Cyberspace.* New York: Wiley.

__________, and Jim Erickson. 1992. *Hard Drive: Bill Gates and the Making of the Microsoft Empire.* New York: Wiley.

Wallerstein, Immanuel. 1974–1989. *The Modern World-System,* 3 vols. New York: Academic Press.

Wallerstein, Judith S., and Sandra Blakeslee. 1989. *Second Chances: Men, Women, and Children a Decade After Divorce.* New York: Ticknor & Fields.

__________, Julia Lewis, and Sandra Blakeslee. 2000. *The Unexpected Legacy of Divorce: A 25 Year Landmark Study.* New York: Hyperion.

Wasserman, Stanley, and Katherine Faust. 1994. *Social Network Analysis: Methods and Applications.* Cambridge: Cambridge University Press.

Waters, Darren. 2003. "Music Industry Plots Sales Recovery." *BBC News.* April 9. On the World Wide Web at http://news.bbc.co.uk/1/low/entertainment/music/2931897.stm (25 May 2003).

Waters, Mary C. 1990. *Ethnic Options: Choosing Identities in America.* Berkeley: University of California Press.

__________. 2000. *Black Identities.* Cambridge, MA: Harvard University Press.

Watkins, S. Craig, and Rana A. Emerson. 2000. "Feminist Media Criticism and Feminist Media Practices." *Annals of the American Academy of Political and Social Science* 571: 151–166.

Watson, James D. 1968. *The Double Helix: A Personal Account of the Discovery of the Structure of DNA.* New York: Atheneum.

__________. 2000. *A Passion for DNA: Genes, Genomes, and Society.* Cold Spring Harbor NY: Cold Spring Harbor Laboratory Press.

Watson, James L., ed. 1997. *Golden Arches East: McDonald's in East Asia.* Stanford, CA: Stanford University Press.

Watts, Duncan J. 2003. *Six Degrees: The Science of a Connected Age.* New York: W. W. Norton.

Weakliem, David L. 1991. "The Two Lefts? Occupation and Party Choice in France, Italy, and the Netherlands." *American Journal of Sociology* 96: 1327–1361.

"WebcamSearch.com." 2005. On the World Wide Web at http://webcamsearch.com/ (14 February 2005).

Webb, Eugene J., Donald T. Campbell, Richard D. Schwartz, and Lee Sechrest. 1966. *Unobtrusive Measures: Nonreactive Research in the Social Sciences.* Chicago: Rand McNally.

Webb, Stephen D., and John Collette. 1977. "Rural–Urban Differences in the Use of Stress-Alleviating Drugs." *American Journal of Sociology* 83: 700–707.

__________. 1979. "Reply to Comment on Rural–Urban Differences in the Use of Stress-Alleviating Drugs." *American Journal of Sociology* 84: 1446–1452.

Weber, Max. 1946. *From Max Weber: Essays in Sociology,* Hans Gerth and C. Wright Mills, ed. and trans. New York: Oxford University Press.

__________. 1947. *The Theory of Social and Economic Organization,* T. Parsons, ed., A. M. Henderson and T. Parsons, trans. New York: Free Press.

__________. 1958 [1904–5]. *The Protestant Ethic and the Spirit of Capitalism.* New York: Charles Scribner's Sons.

__________. 1963. *The Sociology of Religion,* Ephraim Fischoff, trans. Boston: Beacon Press.

__________. 1964 [1949]. "'Objectivity' in Social Science and Social Policy." Pp. 49–112 in Edward A. Shils and Henry A. Finch, trans. and eds. *The Methodology of the Social Sciences.* New York: Free Press of Glencoe.

__________. 1978 [1968]. *Economy and Society,* Guenther Roth and Claus Wittich, eds. Berkeley: University of California Press.

Weeks, Jeffrey. 1986. *Sexuality.* London: Routledge.

Weinstein, Rhona S. 2002. *Reaching Higher: The Power of Expectations in Schooling.* Cambridge, MA: Harvard University Press.

Weis, Joseph G. 1987. "Class and Crime." Pp. 71–90 in Michael Gottfredson and Travis Hirschi, eds. *Positive Criminology.* Beverly Hills, CA: Sage.

Weisbrot, Mark, and Dean Baker. 2002. "The Relative Impact of Trade Liberalization on Developing Countries." Center for Economic and Policy Research, 11 June. Washington, DC. On the World Wide Web at http://www.cepr.net/relative_impact_of_trade_liberal.htm (10 February 2003).

__________, __________, Egor Kraev, and Judy Chen. 2001. "The Scorecard on Globalization, 1980–2000." Center for Economic and Policy Research. Washington, DC. On the World Wide Web at http://www.cepr.net/globalization/scorecard_on_globalization.htm (10 February 2003).

Welch, Michael. 1997. "Violence Against Women by Professional Football Players: A Gender Analysis of Hypermasculinity, Positional Status, Narcissism, and Entitlement." *Journal of Sport and Social Issues* 21: 392–411.

Weller, Jack M., and E. L. Quarantelli. 1973. "Neglected Characteristics of Collective Behavior." *American Journal of Sociology* 79: 665–685.

Wellman, Barry. 1979. "The Community Question: The Intimate Networks of East Yorkers." *American Journal of Sociology* 84: 201–231.

__________, and Stephen Berkowitz, eds. 1997 [1988]. *Social Structures: A Network Approach,* updated ed. Greenwich, CT: JAI Press.

__________ et al. 1996. "Computer Networks as Social Networks: Collaborative Work, Telework, and Virtual Community." *Annual Review of Sociology* 22: 213–238.

__________, Peter J. Carrington, and Alan Hall. 1997 [1988]. "Networks as Personal Communities." Pp. 130–184 in Barry Wellman and S. D. Berkowitz, eds., *Social Structures: A Network Approach,* updated ed., Greenwich, CT: JAI Press.

Wells, H. G. 1927. "The Country of the Blind." Pp. 123–146 in *Selected Short Stories.* Harmondsworth, UK: Penguin. On the World Wide Web at http://www.fantasticfiction.co.uk/etexts/y3800.htm (24 April 2003).

Welsh, Sandy. 1999. "Gender and Sexual Harassment." *Annual Review of Sociology* 25: 169–190.

Wente, Margaret. 2000. "How David Found His Manhood." *Globe and Mail* 29 January: A15–A16.

Wentz, Frank J., and Matthias Schabel. 1998. "Effects of Orbital Decay on Satellite-Derived Lower-Tropospheric Temperature Trends." *Nature* 394: 661–664.

West, Candace, and Don Zimmerman. 1987. "Doing Gender." *Gender and Society* 1: 125–151.

Westen, Tracy. 1998. "Can Technology Save Democracy?" *National Civic Review* 87: 47–56.

"Whatever Will Be Will Be Free on the Internet." 2003. *New York Times* 14 September. On the World Wide Web at http://www.nytimes.com (14 September 2003).

Wheeler, Stanton. 1961. "Socialization in Correctional Communities." *American Sociological Review* 26: 697–712.

Whitefield, S., and G. Evans. 1994. "The Russian Election of 1993: Public Opinion and the Transition Experience." *Post-Soviet Affairs* 10: 38–60.

Whitman, David. 2000. "When East Beats West Old Money Bests New." *Business Week* 8 May: 28.

Whittier, Nancy. 1995. *Feminist Generations: The Persistence of the Radical Women's Movement.* Philadelphia: Temple University Press.

Whorf, Benjamin Lee. 1956. *Language, Thought, and Reality,* John B. Carroll, ed. Cambridge, MA: MIT Press.

"Why Britney Spears Matters." 2001. *The Laughing Medusa: Online Literary Journal of Women's Studies.* On the World Wide Web at http://www.gwu.edu/~medusa/2001/britney.html (17 January 2003).

Whyte, William Foote. 1981 [1943]. *Street Corner Society: The Social Structure of an Italian Slum,* 3rd revised and expanded ed. Chicago: University of Chicago Press.

Wiegand, Steve, and Steve Gibson. 1999. "The Business of Death." On the World Wide Web at http://www.sacbee.com/static/archive/news/projects/cost_of_dying (16 February 2003).

Wilder, D. A. 1990. "Some Determinants of the Persuasive Power of Ingroups and Outgroups: Organization of Information and Attribution of Independence." *Journal of Personality and Social Psychology* 59:1202–13.

Wilensky, Harold L. 1967. *Organizational Intelligence: Knowledge and Policy in Government and Industry.* New York: Basic Books.

__________. 1997. "Social Science and the Public Agenda: Reflections on the Relation of Knowledge to Policy in the United States and Abroad." *Journal of Health Politics, Policy and Law* 22: 1241–1265.

Wiley, Norbert. 1994. *The Semiotic Self.* Chicago: University of Chicago Press.

Wilgoren, Jodi. 2003. "Governor Assails System's Errors as He Empties Illinois Death Row." *New York Times* 12 January. On the World Wide Web at http://www.nytimes.com (12 January 2003).

Wilkinson, Richard G. 1996. *Unhealthy Societies: The Afflictions of Inequality.* London: Routledge.

Willardt, Kenneth. 2000. "The Gaze He'll Go Gaga For." *Cosmopolitan* April: 232–237.

Williams, Daniel T. 1970. "The Lynching Records at Tuskegee Institute." *Eight Negro Bibliographies.* New York: Kraus Reprint Co.

Williams, David R., and Chiquita Collins. 1995. "U.S. Socioeconomic and Racial Differences in Health: Patterns and Explanations." *Annual Review of Sociology* 21: 349–386.

Williams, Jr., Robin M. 1951. *American Society: A Sociological Interpretation.* New York: Knopf.

Willis, Paul. 1984 [1977]. *Learning to Labour: How Working-Class Kids Get Working-Class Jobs.* New York: Columbia University Press.

Wilson, Edward O. 1975. *Sociobiology: The New Synthesis.* Cambridge, MA: Belknap Press of the Harvard University Press.

Wilson, William Julius. 1980 [1978]. *The Declining Significance of Race: Blacks and Changing American Institutions,* 2nd ed. Chicago: University of Chicago Press.

__________. 1987. *The Truly Disadvantaged: The Inner City, the Underclass, and Public Policy.* Chicago: University of Chicago Press.

__________. 1996. *When Work Disappears: The World of the New Urban Poor.* New York: Knopf.

Winch, Donald. 1987. *Malthus.* Oxford: Oxford University Press.

Wirth, Louis. 1938. "Urbanism as a Way of Life." *American Journal of Sociology* 44:1–24.

Witt, Susan D. 1997. "Parental Influence on Children's Socialization to Gender Roles." *Adolescence* 32: 253–259.

Wolf, Diane Lauren. 1992. *Factory Daughters: Gender, Household Dynamics, and Rural Industrialization in Java.* Berkeley: University of California Press.

Wolf, Eric R. 1982. *Europe and the People without History.* Berkeley: University of California Press.

Wolf, Martin. 2003. "An Unfinished Revolution." *Financial Times,* 23 January, special report on "World 2003": 3.

Wolf, Naomi. 1997. *Promiscuities: The Secret Struggle for Womanhood.* New York: Vintage.

Wolff, Edward N. 1996 [1995]. *Top Heavy: The Increasing Inequality of Wealth in America and What Can Be Done About It,* expanded ed. New York: New Press.

Wolff, Janet. 1999. "Cultural Studies and the Sociology of Culture." *Invisible Culture* 1. On the World Wide Web at http://www.rochester.edu/in_visible_culture/issue1/wolff/wolff.html (10 May 2000).

"Woman soldier in abuse spotlight." 2004. *BBC News World Edition* 7 May. On then World Wide Web at http://news.bbc.co.uk/2/hi/americas/3691753.stm (3 March 2005).

Wood, Julia. 1999 [1996]. *Everyday Encounters: An Introduction to Interpersonal Communication,* 2nd ed. Belmont, CA: Wadsworth.

Wood, W., F. Y. Wong, and J. G. Chachere. 1991. "Effects of Media Violence on Viewers' Aggression in Unconstrained Social Interaction." *Psychological Bulletin* 109: 371–383.

Woodbury, Anthony. 2003. "Endangered Languages." *Linguistic Society of America.* On the World Wide Web at http://www.lsadc.org/web2/endangeredlgs.htm (19 July 2003).

Wooton, Barbara H. 1997. "Gender Differences in Occupational Employment." *Monthly Labor Review* 120, 4: 15–24. On the World Wide Web at http://stats.bls.gov/opub/mlr/1997/04/art2full.pdf (30 April 2000).

Wordsworth, Araminta. 2000. "Family Planning Officials Drown Baby in Rice Paddy." *National Post* 25 August: A10.

"Work-related Stress: A Condition Felt 'Round the World.'" 1995. *HR Focus* 72, 4: 17.

World Bank. 1999. "GNP Per Capita 1997, Atlas Method and PPP." http://www.worldbank.org/data/databytopic/GNPPC97.pdf (10 July 1999).

__________. 2002. "6.1: Integration with the World Economy." On the World Wide Web at http://www.worldbank.org/data/wdi2002/tables/table6-1.pdf (6 February 2003).

World Health Organization. 1999a. "WHO Estimates of Health Personnel." On the World Wide Web at http://www.who.int/whosis/healthpersonnel/index.html (19 November 1999).

__________. 1999b. "World Health Report 1999: Basic Indicators for all Member States." On the World Wide Web at http://www .who.int/whr/1999/en/indicators.htm (2 May 2000).

__________. 2000. "WHO Terminology Information System." On the World Wide Web at http://www.who.int/terminology/ ter/wt001.html#health (2 May 2000).

__________. 2001. "Female Genital Mutilation." On the World Wide Web at http://www.who.int/frh-whd/FGM (20 January 2003).

__________. 2002. *World Health Report 2002.* On the World Wide Web at http://www.who.int/whr/2002/en (13 June 2003).

__________. 2003. "Cumulative Number of Reported Probable Cases." On the World Wide Web at http://www.who.int/csr/sars/country/2003_06_16/en (16 June 2003).

World Health Organization. 2004. *Aids Epidemic Update: December 2004.* Geneva. On the World Wide Web at http://www.unaids.org/wad2004/EPI_1204_pdf_en/EpiUpdate04_en.pdf (13 March 2005).

World Tourism Organization. 2002. "International Tourist Arrivals by (Sub)region." On the World Wide Web at http://www.world-tourism.org/market_research/facts&figures/latest_data/tita01_07-02.pdf (6 February 2003).

World Values Survey. 2003. Machine readable data set.

Worth, Robert. 1995. "A Model Prison." *The Atlantic Monthly* November. On the World Wide Web at http://www.theatlantic.com/issues/95nov/prisons/prisons.htm (28 May 2000).

__________. 2001. "The Deep Intellectual Roots of Islamic Terror." *The New York Times on the Web* 13 October 2001. On the World Wide Web at http://www.nytimes.com/2001/10/13/arts/13ROOT.html?searchpv=past7days (15 October 2001).

Wright, Charles Robert. 1975. *Mass Communication: A Sociological Perspective.* New York: Random House.

Wright, Erik Olin. 1985. *Classes.* London: Verso.

__________. 1997. *Class Counts: Comparative Studies in Class Analysis.* Cambridge: Cambridge University Press.

Wright, John W., ed. 1998. *The New York Times Almanac 1999.* New York: Penguin.

X, Malcolm. 1965. *The Autobiography of Malcolm X.* New York: Grove.

Yale Center for Environmental Law and Policy and Center for International Earth Science Information Network, Columbia University. 2005. *2005 Environmental Sustainability Index.* On the World Wide Web at http://www.yale.edu/esi/ (2 March 2005).

Yamane, David. 1997. "Secularization on Trial: In Defense of a Neosecularization Paradigm." *Journal for the Scientific Study of Religion* 36: 109–122.

Yancey, William L., Eugene P. Ericksen, and George H. Leon. 1976. "Emergent Ethnicity: A Review and Reformulation." *American Sociological Review* 41: 391–403.

Yoo, Peter S. 1998. "Still Charging: The Growth of Credit Card Debt Between 1992 and 1995." *Review* (Federal Reserve Bank of St. Louis). January–February.

Yook, Soon-Hyung, Hawoong Jeong, and Albert-László Barabási. 2001. "Modeling the Internet's Large-Scale Topology." Department of Physics, University of Notre Dame. On the World Wide Web at http://arxiv.org/PS_cache/cond-mat/pdf/0107/0107417.pdf (23 May 2003).

Zakaria, Fareed. 1997. "The Rise of Illiberal Democracy." *Foreign Affairs* 76, 6: 22–43.

Zald, Meyer N., and John D. McCarthy. 1979. *The Dynamics of Social Movements.* Cambridge, MA: Winthrop.

Zangwill, Israel. 1909. *The Melting Pot: Drama in Four Acts.* New York: Macmillan.

Zaslavsky, Victor, and Robert J. Brym. 1978. "The Functions of Elections in the USSR." *Soviet Studies* 30: 62–71.

———, and ———. 1983. *Soviet Jewish Emigration and Soviet Nationality Policy.* London: Macmillan.

Zerubavel, Eviatar. 1981. *Hidden Rhythms: Schedules and Calendars in Social Life.* Chicago: University of Chicago Press.

Zijderveld, Anton C. 1983. "The Sociology of Humour and Laughter." *Current Sociology* 31, 3: 1–103.

Zimbardo, Philip G. 1972. "Pathology of Imprisonment." *Society* 9, 6: 4–8.

Zimmerman, Rachel. 2002. "Botox Gives a Special Life to These Soirees." *Wall Street Journal* April 16: B1, B3.

Zimmermann, Francis. 1987 [1982]. *The Jungle and the Aroma of Meats: An Ecological Theme in Hindu Medicine,* Janet Lloyd, trans. Berkeley: University of California Press.

Zimring, Franklin E., and Gordon Hawkins. 1995. *Incapacitation: Penal Confinement and the Restraint of Crime.* New York: Oxford University Press.

Zinsser, Hans. 1935. *Rats, Lice and History.* Boston: Little, Brown.

Zogby International. 2001. "Arab American Institute Polls Results: Arab Americans are strong advocates of war against terrorism; Overwhelmingly endorse President Bush's actions; Significant numbers have experienced discrimination since Sept. 11." On the World Wide Web at http://www.zogby.com/news/ReadNews.dbm?ID=487 (21 December 2002).

Zogby, James J. 2002. *What Arabs Think: Values, Beliefs and Concerns.* Utica, NY: Zogby International / The Arab Thought Foundation. On the World Wide Web at http://www.zogby.com/about/index.cfm (21 December 2002).

Zola, Irving Kenneth. 1982. *Missing Pieces: A Chronicle of Living with a Disability.* Philadelphia: Temple University Press.

Zolberg, Aristide. 2001. "Guarding the Gates in a World on the Move." On the World Wide Web at http://www.ssrc.org/sept11/essays/zolberg_text_only.htm (21 December 2002).

Zuboff, Shoshana. 1988. *In the Age of the Smart Machine: The Future of Work and Power.* New York: Basic Books.

Zukin, Sharon. 1980. "A Decade of the New Urban Sociology." *Theory and Society* 9: 539–574.

Zurcher, Louis A., and David A. Snow. 1981. "Collective Behavior and Social Movements." Pp. 447–482 in Morris Rosenberg and Ralph Turner, eds. *Social Psychology: Sociological Perspectives.* New York: Basic Books.

Zussman, Robert. 1992. *Intensive Care: Medical Ethics and the Medical Profession.* Chicago: University of Chicago Press.

———. 1997. "Sociological Perspectives on Medical Ethics and Decision-Making." *Annual Review of Sociology* 23: 171–189.

Zwick, Rebecca. 2002. *Fair Game? The Use of Standardized Academic Tests in Higher Education.* New York: RoutledgeFalmer.

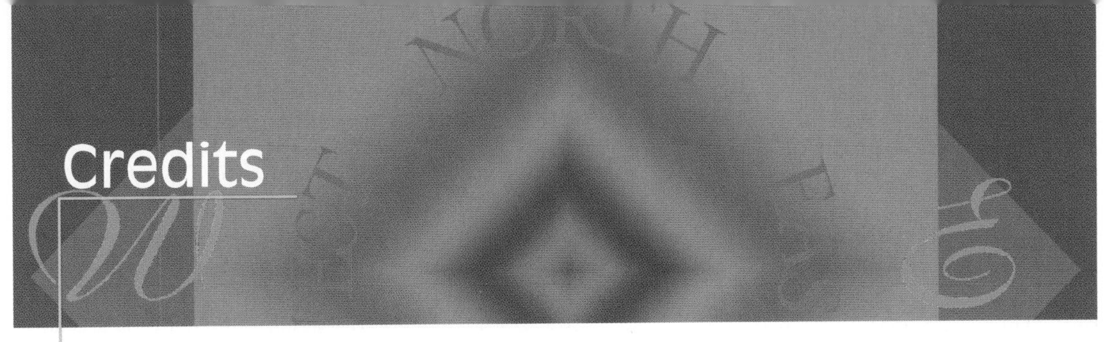

Credits

This page constitutes an extension of the copyright page. We have made every effort to trace the ownership of all copyrighted material and to secure permission from copyright holders. In the event of any question arising as to the use of any material, we will be pleased to make the necessary corrections in future printings. Thanks are due to the following authors, publishers, and agents for permission to use the material indicated.

Frontmatter

x: Canadian Press Picture Archives/Kevin Frayer **xiii:** The Everett Collection **xvi:** Luis Castaneda/Getty Images **xx:** Douglas Slone/Corbis

Chapter 1

xxxvi: Zigy Kaluzny/Stone/Getty Images **3:** Courtesy of A. C. Fine Art, Nova Scotia **4:** Bettmann/ Corbis **5:** SuperStock **7:** Brown Brothers **8:** 20th Century Fox/Dreamworks/ The Kobal Collection **9:** By permission of the Houghton Library, Harvard University **10, left:** Paul Almasy/Corbis **10, right:** Musee du Louvre, Paris/Giraudon, Paris/SuperStock **11:** Photograph copyright © 2001 The Detroit Institute of the Arts **14, top:** Courtesy of Columbia University **14, bottom:** Archivo Iconografico, S.A./Corbis **15, top:** Brown Brothers **15, bottom:** Bettmann/Corbis **17:** The Granger Collection **19:** Hulton-Deutsch Collection/Corbis **22, left:** Reuters NewMedia Inc./Corbis **22, right:** David Bergman/Corbis **24, 25:** Copyright © 2001 Time, Inc. Reprinted by permission.

Chapter 2

32: Richard Lord/PhotoEdit **35:** Archivo Iconografico, S.A./Corbis **36:** Bettmann/ Corbis **38:** Courtesy of Carol Wainio, London, Ontario, Canada **39:** Courtesy of Lillian B. Rubin **46:** David Tumely/Corbis **47:** Jeff Greenberg/PhotoEdit **48:** AP/Wide World Photos **55:** Charles Tilly by John Sheretz. © 1997 CASAS **57:** 20th Century Fox/ American Zoetrope/The Kobal Collection

Chapter 3

62: Canadian Press Picture Archives/Kevin Frayer **64:** Andrew Woolley **66:** Mark Richards/PhotoEdit **69:** The National Post, Toronto, Canada, 2000 **72:** Hulton-Deutsch Collection/Corbis **73:** The Everett Collection **75:** © 1992 Joel Gordon **79:** Courtesy of Kelloggs **81:** Owen Franklin/Corbis **87:** The Everett Collection **88:** Brown Brothers **89:** Bill Aron/PhotoEdit **91:** Reuters NewMedia Inc./Corbis

Chapter 4

94: Zena Holloway/Getty Images **96:** © Martin Rogers/Stock Boston **98:** Corbis **100:** MDP/ New Market/Page, Gene/The Kobal Collection **102:** Myrleen F. Cate/PhotoEdit **103:** Courtesy of Carol Gilligan. Photo by Jerry Bauer **104:** Spencer Grant/PhotoEdit **106:** SuperStock **107:** Paul Conklin/PhotoEdit **111:** Mills and Boon **113, top:** Jonathan Blair/Corbis **113, bottom:** Oleg Popov/Reuters/Corbis **120:** Spencer Grant/PhotoEdit

Chapter 5

124: Henry Diltz/Corbis **127, top:** Bettmann/ Corbis **127, bottom:** The Everett Collection **130:** Sipkin Corey/Sygma/Corbis **132:** D. Le Strat/Sygma/Corbis **134:** SuperStock **137:** Bettmann/Corbis **138:** Castle Rock/Fortis/ The Kobal Collection **142:** Robert J. Brym **144:** Appeared in *Leatherneck,* March, 1945 **147, left:** Erich Lessing/Art Resource, NY **147, right:** Charles & Josette Lenars/Corbis

Chapter 6

150: Yellow Dog Productions/Getty Images **152:** The Everett Collection **155:** Time Inc./ Time Life Pictures/Getty Images **157:** Reuters NewMedia Inc./Corbis **158:** The Everett Collection **162:** Brian Leng/Corbis **165:** Hulton-Deutsch Collection/Corbis **166:** JPL/NASA **168:** Brandom Films/The Everett Collection **170:** Rob Lewine/Corbis **175:** Dave Jacobs/ Index Stock Imagery **176:** Archivo Iconografico, S. A./Corbis **178:** Jacob Halaska/Index Stock Imagery

Chapter 7

182: The Everett Collection **184:** Bettmann/ Corbis **185:** National Library of Medicine, Washington, DC **186:** WeeGEE/ICP/Getty Images **187:** New York Public Library **193:** The Everett Collection **197:** The Everett Collection **202:** New York Public Library **205:** Alliance Atlantis/Dog Eat Dog/United Broadcasting/The Kobal Collection **208:** The Everett Collection

Chapter 8

212: Bruce Ayers/Getty Images **214:** The Granger Collection **218, left:** Reuters NewMedia, Inc./Corbis **218, right:** Reuters NewMedia, Inc./Corbis **220:** Touchstone/ The Kobal Collection/Ovino, Peter **222, left:** *Saturday Night,* June 3, 2000. "The War Next Door," p. 48 story and photographs by Ami Vitale **222, right:** *Saturday Night,* June 3, 2000. "The War Next Door," p. 48 story and photographs by Ami Vitale **226:** Art Resource, NY **231, left:** © 1995 Alex Webb/Magnum Photos, Inc. **231, right:** Richard T. Nowitz/ Corbis **238, left:** Robert Hepier/The Everett Collection **238, right:** Gerrit Greve/Corbis **240:** Copyright © the Dorothea Lange Collection, Oakland Museum of California

Chapter 9

248: Jeremy Horne/Getty Images **250, left:** Chuck Savage/Corbis **250, right:** Ariel Skelley/Corbis **252, left:** Brian A. Vikander/ Corbis **252, right:** Roy McMahon/Corbis **253:** Sandy Felsenthal/Corbis **255:** John Van Hasselt/ Corbis Sygma **261:** Photo by Miro Cernetig, *Globe & Mail,* Toronto, Canada **264:** From Eric Hobsbawm's *The Age of Empire* (Vintage Books). Copyright © 1987 by E. J. Hobsbawm. **267:** Brent Patterson/Corbis **272:** Warner Brothers/The Kobal Collection/Close, Murray

Chapter 10

276: Jack Moebes/Corbis **278:** From *The Mismeasurement of Man* by Stephen Jay Gould (Norton) **280, top:** Duomo/Corbis **280, bottom:** Annie Sachs/Archive Photo/Getty Images **281:** Time Inc./Time Life Pictures/ Getty Images **285:** Heather Titus/Photo Resource Hawaii **291:** Bettmann/Corbis **293:** Jacob A. Riis Collection, Museum of the City of New York **294:** Saskatchewan Archives Board, R-A8223 1&2 **297:** © 2004 Gwendolyn Knight Lawrence/Artists Rights Society, New York **298:** Museum of History and Industry, Seattle **299:** David Tumley/Corbis **301:** Lewis H. Hines/Hulton Archive/ Getty Images **309:** Lions Gate/The Kobal Collection

Chapter 11

312: Rob Lewine/Corbis **317:** Frans Lemmens/Getty Images **319:** Pascal Le Segretain/Corbis Sygma **320:** Reuters News Media, Inc./Corbis **321:** © David Young-Wolff/PhotoEdit **323, top left:** Courtesy of the White Rock Beverage Company **323, top right:** Courtesy of the White Rock Beverage Company **323, bottom:** Lou Chardonnay/Corbis **327:** Peter Dejong/AP/Wide World Photos **329:** Joseph Sohm, ChromoSohm, Inc./Corbis **330:** Fox Searchlight/The Kobal Collection/Matlock, Bill **332:** The Granger Collection **333:** Janette Beckman/Corbis **335:** Rachel Epstein/PhotoEdit **340:** SuperStock **341:** Joseph Sohm, ChromoSohm, Inc./Corbis

Chapter 12

346: Tim Waele/Isosport/Corbis **348:** Matthew Mendelsohn/Corbis **350:** Bridgeman Art Library, London/SuperStock **351:** Sinibaldi/Corbis **352:** Rob Lewine/Corbis **354:** Lightscapes Photography, Inc./Corbis **355:** 20th Century Fox/Conundrument/The Kobal Collection/Watson, Glenn **357:** Schwartzwald Lawrence/Corbis Sygma **358:** Erica Lansner/Stone/Getty Images **364:** Stone/Getty Images **366:** John Hillary/Reuters/Corbis

Chapter 13

370: Luis Castaneda/Getty Images **372, top:** From *In the Age of the Smart Machine: The Future of Work and Power* by Shoshana Zuboff, © 1988 by Basic Books, Inc. Reprinted by permission of Basic Books, a Member of Persens Books, L.L.C. **372, bottom:** Courtesy of *Wired* Magazine, Condé Nast Publications, Dec. 1999, James Porto photo **373:** © Sebastiao Salgado/Contact Press Images **375:** The Everett Collection **378:** Spencer Grant/PhotoEdit **384:** The Everett Collection **386:** Meri Simon Copyright © *San Jose Mercury News* (in Bliss, 2000) **395:** The Granger Collection **396:** AP/Wide World Photos

Chapter 14

402: Brooks Kraft/Corbis **404:** John Duricka/AP/Wide World Photos **407, top:** Archive Iconografico, S.A./Corbis **407, center:** SuperStock **407, bottom:** Corbis **410, top:** Courtesy of the Artist and the Nancy Poole Studios, Toronto, Ontario **410, bottom:** Mark Richards/PhotoEdit **414, top:** Kevin Lamarque/Corbis **414, bottom:** Bob Daemmrich/Corbis **418, left:** AP/Wide World Photos **418, right:** Lewis W. Hines/SuperStock **419:** Miramax/Dimension Films/The Kobal Collection/Tursi, Mario **420:** AP/Wide World Photos **422, left:** © 1995 Lise Sarfati/Magnum Photos, Inc. **422, right:** Bojan Brecelj/Corbis **424, left:** Marc Riboud/Magnum Photos, Inc. **424, right:** Jeremiah Kamau/Reuters/Corbis **429:** Patrick Robert/Corbis

Chapter 15

434: Digital Vision/Getty Images **440:** Peter Johnson/Corbis **443:** The Everett Collection **444:** Courtesy of the artist and Bau-Xi Gallery, Toronto, ON, Canada **446:** The Everett Collection **448:** IFC Films/The Kobal Collection/Giraud, Sophie **450:** Andrew Benyei, *Pink Couch,* 1993. Fiberglass, 24 × 15 × 19 inches Photo: Ron Giddings. Reproduced with permission of the artist. **451:** SuperStock **452:** Jacques M. Chenet/Corbis **455:** Michael Newman/PhotoEdit **458:** AP/Wide World Photos **461:** Laura Dwight/Corbis **464:** Jonathan Blair/Corbis

Chapter 16

468: Charles O'Rear/Corbis **471:** Michael Newman/PhotoEdit **473:** Flip Schulke/Corbis **474:** Wendy Chan/The Image Bank/Getty Images **475:** Bettmann/Corbis **478:** The Everett Collection **480:** A. Ramey/PhotoEdit **488:** Nathan Benn/Corbis **489:** SuperStock **492:** The Lowe Art Museum, The University of Miami/SuperStock **494:** Myrleen Ferguson/PhotoEdit

Chapter 17

498: Bob Daemmrich/The Image Works **500:** Peter Finger/Corbis **502:** Ted Horowitz/Corbis **503:** Mike Derer/AP/Wide World Photos **504:** Diane Bondareff/AP/Wide World Photos **513:** The Everett Collection **517:** Corbis **518:** The Purcell Team/Corbis **524:** Paul Conklin/PhotoEdit

Chapter 18

528: John Henley/Corbis **530:** The Everett Collection **532:** Spencer Grant/PhotoEdit **534, top:** The Pierpont Morgan Library/Art Resource, NY **534, bottom:** Myrleen Ferguson/PhotoEdit **537:** Erik Freeland/Corbis **539:** The Everett Collection **544:** The Everett Collection **549:** *Melbourne Age*/AP/Wide World Photos **550:** Reuters NewMedia, Inc./Corbis

Chapter 19

552: Corbis Sygma **554:** Museo del Prado, Madrid, Spain/Giraudon, Paris/SuperStock **556:** Launette/AP/Wide World Photo **558, left:** Mike Hutchings/Reuters/Corbis **558, right:** Mark Richards/PhotoEdit **567:** Ted Streshinsky/Corbis **571:** Melinda Sue Gordon/The Everett Collection **575:** Lisa M. McGeady/Corbis

Chapter 20

578: Douglas Slone/Corbis **580:** Donald Klein/SuperStock **582:** The Purcell Group/Corbis **583:** Scala/Art Resource, NY **590:** Christopher Morris/Black Star Publishing/Picture Quest **591:** The Granger Collection **595:** © 1996 Joel Gordon **598:** The Everett Collection

Chapter 21

604: Mark Peterson/Corbis **608:** National Association for the Advancement of Colored People, 1922 **609:** Achmad Ibrahim/AP/Wide World Photo **616:** Neville Elder/Corbis **617:** AP/Wide World Photos **619:** Joseph Sohm, ChromoSohm, Inc./Corbis **622:** Hintz Diltz/Corbis **623:** 20th Century Fox/The Kobal Collection **626, top:** Bibliotheque Nationale de France **626, bottom:** AP/Wide World Photos **628:** Canadian Press Picture Archives/Sean White **629:** Reuters NewsMedia, Inc./Corbis

Chapter 22

632: NASA **634:** Bettmann/Corbis **635:** Robert Llewellyn/SuperStock **637:** The Everett Collection **638:** Gary Braasch/Corbis **640:** U.S. Army Photo **642, top:** SuperStock **642, bottom:** Kurt Scholz/SuperStock **645:** Dave Martin/AP/Wide World Photo **646:** The Everett Collection **649:** Philip Gould/Corbis **656:** Courtesy of Shell Hydrogen **657:** Copyright © 1990 Watterson. Reprinted with permission of Universal Press Syndicate. All rights reserved.

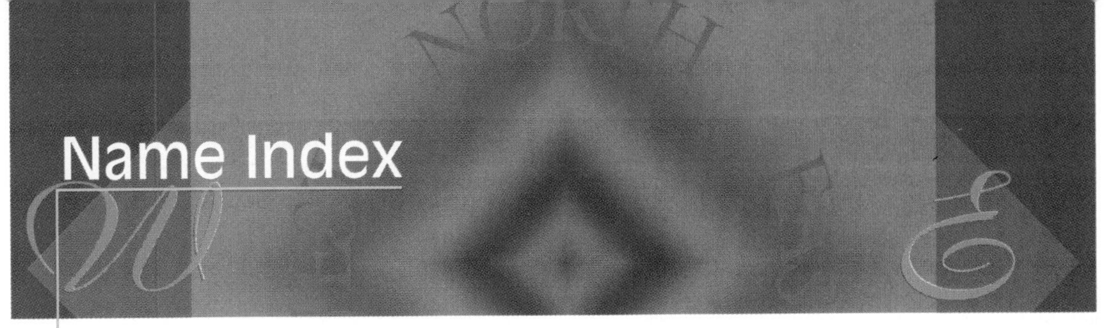

Name Index

Subject Index

A

D